HANDBOOK OF SEPARATION TECHNIQUES FOR CHEMICAL ENGINEERS

Philip A. Schweitzer,
Editor in Chief

1,120 pages, 660 illustrations

This timely work shows you the applications of all the industrially accepted techniques for separating any and all chemicals from each other *without the use of chemical reactions.*

Unlike other works, this important Handbook covers all types of methods and all types of mixtures—liquid-liquid, liquid-solid, solid-solid, gas-liquid, and gas-solid. It is packed with quickly found answers to the chemical engineer's everyday problems, including the removal of noxious materials from wastewater streams.

Written for everyone concerned with practical applications of separation techniques, the Handbook provides specific information to help you choose the techniques to use. By referring to specific types of mixtures, you will find discussions of available separation methods *for those types of mixtures,* plus guidance on choosing the best method to meet your requirements. The emphasis throughout is on start-up procedures, troubleshooting, and the proper selection of the accessory equipment. A wealth of illustrative problems shows how the basic theory is applied.

The Handbook is essential to the performance of many functions. For the **process engineer** who selects the separation technique and sizes the equipment, the book provides all the design data and techniques needed for all methods. For the **project manager and project engineer** who must convert the design of the process engineer into mechanical equipment and accessories that will do what they're supposed to do, all the requirements of design and how-to are supplied. The **environmental engineer,** who has to identify and eliminate pollution and who must determine the most economical and practical approach, will find here all of today's available techniques for removing undesirable ingredients from waste streams. And the **plant engineer** who installs, operates, and maintains separation equipment will find every procedure needed to forestall and solve problems.

Whatever your separation problem— removing organic solvents or solids from a liquid stream, evaluating the suitability of a complete distillation installation for its application to a completely different separation, specifying a unit, or any other problem— you'll find the procedure you need carefully detailed, the steps clearly outlined, and the optimum equipment fully described. All the methods given have been proven in practice.

Whenever you have a mixture to separate and are not familiar enough with the various separation techniques available to allow you to refer to it immediately, all you have to do is look up the type of mixture you're concerned with. There you'll find all the possible separation techniques for that type of mixture discussed and illustrated.

For example, the chapters on liquid-liquid mixtures describe *continuous distillation... batch distillation ... steam stripping ... solvent recovery...packed tower design...extraction ... decantation ... ion exchange ... carbon adsorption ... dialysis and electrodialysis...parametric pumping...*and more.

The chapters on liquids with dissolved solids include *membrane filtration...extraction ... evaporation ... crystallization* and *foam separation.* In the gas mixtures section, *carbon adsorption and absorption tray column design* are covered. Under solid-liquid mixtures, *centrifugation ... felt straining ... sedimentation...hydroclones* and *drying* are covered. *Leaching* and *flotation* are discussed in the chapter on solid mixtures. A discussion on *electrostatic precipitation* is included for gas-liquid and gas-solid mixtures.

This single source for separation techniques—offering the knowledge of 36 experts and providing you with a broad range of information—will save you work. You will discover new methods as well as some new applications for well-known techniques. By checking the advantages and disadvantages of each method described, you can make more accurate decisions. And you will avoid costly errors by using the guidelines for proper application of each piece of equipment.

Handbook of
Separation Techniques
for Chemical Engineers

OTHER McGRAW-HILL HANDBOOKS OF INTEREST

Handbook of
Separation Techniques
for Chemical Engineers

PHILIP A. SCHWEITZER *Editor-in-chief*
Contract Manager, Chem-Pro Equipment Corp.

A James Peter Book
James Peter Associates, Inc.

McGRAW-HILL BOOK COMPANY

New York St. Louis San Francisco Auckland Bogotá
Düsseldorf Johannesburg London Madrid
Mexico Montreal New Delhi Panama
Paris São Paulo Singapore
Sydney Tokyo Toronto

Library of Congress Cataloging in Publication Data

Main entry under title:

Handbook of separation techniques for chemical
engineers.

"A James Peter book."
Includes index.
1. Separation (Technology)—Handbooks, manuals,
etc. I. Schweitzer, Philip A.
TP156.S45H35   660.2'842   79-4096
ISBN 0-07-055790-X

1234567890   KPKP   7865432109

*The editors for this book were Harold B. Crawford and Joseph Williams,
the designer was Naomi Auerbach, and the production supervisor
was Sally Fliess. It was set in Caledonia
by University Graphics, Inc.*

Printed and bound by The Kingsport Press.

Contents

TP
156
S45
H35

PART THREE   Gas (Vapor) Mixtures

PART FOUR   Solid-Liquid Mixtures

PART FIVE   Solid Mixtures

PART SIX   Gas-Solid Mixtures

Index follows Section 6.2

Contributors

Charles M. Ambler *Registered Professional Engineer*

Robert E. Anderson *Senior Research Scientist, Functional Polymers Division, Diamond Shamrock Corporation* (SEC. 1.12)

Frank F. Aplan *Mineral Processing Section, Department of Mineral Engineering, The Pennsylvania State University* (SEC. 5.2)

Allan W. Ashbrook *Manager, Research and Development Division, Eldorado Nuclear, Ltd.*

R. C. Bennett *Manager, Swanson Division of Whiting Corporation* (SEC. 2.3)

David M. Boyd, Jr. *Manager, Instrument Design & Service Department, UOP Process Division, UOP Inc.* (SEC. 1.5)

Clifford W. Cain, Jr. *Senior Research Engineer, Filtration and Minerals Research, Johns-Manville* (SECS. 4.1 AND 4.2)

H. T. Chen *Professor of Chemical Engineering, New Jersey Institute of Technology* (SEC. 1.15)

Eng J. Chou *Sr. Development Engineer, Research and Development, Research Cottrel Inc.*

Roger W. Day *Marketing Director, Solids Control Operations, Geosource, Inc.* (SEC. 4.9)

John W. Drew *Director of Engineering, Chem-Pro Equipment Corp.* (SEC. 1.6)

John S. Eckert *Consulting Engineer* (SEC. 1.7)

R. W. Ellerbe *Project Manager, The Rust Engineering Company* (SECS. 1.3 AND 1.4)

Robert C. Emmett, Jr. *Program Manager, Industrial Processing Technology, Eimco-BSP Division of Envirotech Corporation* (SEC. 4.4)

Charles N. Grichar *Supervisor of Application Engineering, Solids Control Operations, Geosource, Inc.* (SEC. 4.9)

William B. Hooper *Engineering Fellow, Process Design, Corporate Engineering Department, Monsanto Company* (SEC. 1.11)

T. I. Horzella *Enviro Energy Corporation* (SEC. 6.2)

Roy A. Hutchins *Technical Services Superintendent, Agricultural Chemicals Division, Texas Gulf, Inc.* (SEC. 1.14)

Louis J. Jacobs, Jr. *Engineering Specialist, Corporate Engineering Department, Monsanto Company* (SEC. 1.11)

E. G. Kominek *Technical Director, Industrial Water and Waste Operations, Eimco-PMD Division, Envirotech Corporation* (SEC. 4.8)

J. Louis Kovach *President, Nuclear Consulting Services* (SEC. 3.1)

L. D. Lash *Systems Project Manager, Eimco-PMD Division, Envirotech Corporation* (SEC. 4.8)

Teh Cheng Lo *Technical Fellow, Development Department, Hoffmann-La Roche, Inc.* (SEC. 1.10)

Joseph H. Maas *Independent Chemical Engineer and Consultant* (SECS. 1.1 AND 6.1)

John E. McCormick *Associate Professor of Chemical Engineering, New Jersey Institute of Technology* (SEC. 1.2)

Nicholas Nickolaus *Vice-President Marketing, Pall Corporation* (SEC. 4.6)

Paul G. Nygren   *Senior Staff Engineer, Union Carbide Corporation* (SEC. 1.8)

Yoshiyuki Okamoto   *Professor of Chemistry, Polytechnic Institute of New York* (SEC. 2.5)

Dr. M. C. Porter   *Vice President of Research & Development, Nuclepore Corporation* (SEC. 2.1)

Rajaram K. Prabhudesai   *Senior Process Engineer, Stauffer Chemical Company* (SEC. 5.1)

Gordon M. Ritcey   *Head, Extractive Metallurgy Section, Mineral Sciences Laboratory, Canada Centre for Mineral and Energy Technology* (SEC. 2.2)

Lanny A. Robbins   *Associate Scientist, Dow Chemical USA* (SEC. 1.9)

Edward C. Roche, Jr.   *Professor of Chemical Engineering, New Jersey Institute of Technology* (SEC. 1.2)

G. G. Schneider   *Enviro Energy Corporation* (SEC. 6.2)

Gyanendra Singh   *Section Head, Food Products Research. The Procter & Gamble Company* (SEC. 2.4)

P. J. Striegl   *Amex Corporation* (SEC. 6.2)

John R. Thygeson, Jr.   *Associate Professor of Chemical Engineering, Drexel University* (SEC. 4.10)

Theodore H. Wentz   *Sales Manager, Proctor and Schwartz Division, SCM Corp.* (SEC. 4.10)

Arthur C. Wrotnowski   *Consultant* (SEC. 4.7)

F. A. Zenz   *Vice-President, Engineering, The Ducon Company* (SEC. 3.2)

James F. Zievers   *Vice-President, Industrial and Pump Mfg. Co.* (SEC. 4.3)

Preface

Many textbooks have been written on the subject of separation techniques. These books are usually devoted to the derivation of correlations, are highly theoretical, and provide relatively few concrete examples of the application of theory to practical everyday problems. Very seldom is more than one separation technique covered in a single volume, which makes it somewhat difficult to evaluate which technique is best suited for a particular application. This handbook, however, includes all the major separation techniques which are used industrially.

This handbook has been designed to provide the chemical engineer with sufficient information to evaluate which technique is best suited for his or her specific requirements and then, by means of illustrative problems, to show how the theory is applied. Since an understanding of the theory is necessary for proper application, the basic theory is presented and ample references are supplied for those interested in further theoretical study and in the derivation of the correlations used.

For the purpose of this book, *separation techniques* are defined as those operations which isolate specific ingredients of a mixture without a chemical reaction taking place. One deviation has been made from this principle by the inclusion of the section dealing with *ion exchange*. This was done because of the importance of ion exchange to the field of separation techniques.

The separation techniques covered are widely used in chemical manufacturing operations as well as in the design of pollution control equipment. The latter application usually involves the greatest degree of evaluation of one technique versus another.

This handbook should be helpful to chemical engineers, consultants, environmentalists, government officials, and others who are involved in the separation of mixtures of ingredients whether for manufacturing operations or for pollution control.

The editor-in-chief wishes to thank the many contributors who made their time available and were willing to share their expertise with other members of the engineering profession through the sections of the handbook which they contributed. Thank yous are also extended to the many organizations, companies, and individuals who graciously permitted use of their charts, data, photographs, and other pertinent information. An additional thank you is extended to the editor-in-chief's wife, for her understanding and many hours of typing throughout the course of the preparation of this handbook.

Philip A. Schweitzer

International System (SI) of Units and Conversion Factors

This coherent system of measurement, designated "SI" in all languages, has been accepted as the preferred system of units by 36 countries, including the United States.

Since most nations have or are in the process of converting from their individual national systems of measurement to SI units, it will only be a matter of time until the conversion is made in all nations. Many industries in the United States are already in the process of converting.

Throughout this handbook a dual system of measurement has been utilized—the English (U.S. Customary) and SI systems.

Tables giving SI base units and prefixes to be used in forming multiples and submultiples of SI units are given. And to assist the reader in making conversions, a table has been included which provides conversion factors for the more common English units to their equivalent SI units, and vice versa.

SI Base Units

Quantity	Base unit	Symbol
Length	meter	m
Mass	kilogram	kg
Time	second	s
Electric current	ampere	A
Thermodynamic temp.	kelvin	K
Amount of substance	mole	mol
Luminous intensity	candela	cd

SI Prefixes

Multiple	SI prefix	Symbol
10^{18}	exa	E
10^{15}	peta	P
10^{12}	tera	T
10^{9}	giga	G
10^{6}	mega	M
10^{3}	kilo	k
10^{2}	hecto	h
10	deka	da
10^{-1}	deci	d
10^{-2}	centi	c
10^{-3}	milli	m
10^{-6}	micro	μ
10^{-9}	nano	n
10^{-12}	pico	p
10^{-15}	femto	f
10^{-18}	atto	a

System of Consistent Units

Measurement	English	Metric
Absolute temperature	°R or °F abs.	K or °C abs.
Area	square inch	square centimeter
	square foot	square meter
Capacity	quart	liter
	gallon	liter
Density	pound per cubic foot	gram per cubic centimeter
Force of gravity conversion factor	$32.17 \left(\dfrac{\text{lb mass}}{\text{lb force}}\right)\left(\dfrac{\text{ft}}{\text{s}^2}\right)$	$980.6 \left(\dfrac{\text{g mass}}{\text{g force}}\right)\left(\dfrac{\text{cm}}{\text{s}^2}\right)$
Gas constant	$1546 \dfrac{\text{ft-lb force}}{\text{lb}\cdot\text{mol}\cdot{}^\circ\text{F}}$	$84{,}400 \dfrac{\text{cm}\cdot\text{g force}}{\text{g}\cdot\text{mol}\cdot{}^\circ\text{C}}$
Gas flow rate	cubic foot per second	cubic centimeter per second
Length	inch	millimeter
	foot	centimeter
	foot	meter
Molecular weight	lb·mol	g·mol
Number of gas molecules in a mole	$2.76 \times 10^{26}/\text{lb}\cdot\text{mol}$	$6.06 \times 10^{23}/\text{g}\cdot\text{mol}$
Pressure	pound per square inch	kilogram per square centimeter
	pound per square foot	kilogram per square meter
Specific heat	Btu/(°F)(lb)	cal/(°C)(g)
Thermal conductivity	Btu/(s)(ft²)(°F/ft)	cal/(s)(cm²)(°C/cm)
Velocity	foot per second	centimeter per second
Viscosity	pound per foot-second	poise
		pascal-second
Volume	cubic inch	cubic centimeter
	cubic foot	cubic meter
	gallon	liter
Weight	ounce	gram
	pound	kilogram

Conversion Factors

To convert from:	to	Multiply by
atmosphere (atm)	millimeter of mercury (mmHg) at 32°F	760
atmosphere (atm)	dyne per square centimeter (dyn/cm²)	1.1033×10^6
atmosphere (atm)	foot of water at 39.1°F (ftH$_2$O)	33.90
atmosphere (atm)	gram per square centimeter (g/cm²)	1033.3
atmosphere (atm)	inch of mercury at 32°F (inHg)	29.921
atmosphere (atm)	pound per square foot (lb/ft²)	2116.3
atmosphere (atm)	pound per square inch (lb/in²)	14.696
Btu (British thermal unit)	foot-pound (ft·lb)	777.9
Btu	horsepower-hour (hp·h)	3.929×10^{-4}
Btu	joule (J)	1055.1
Btu	kilowatthour (kWh)	2.93×10^{-4}
Btu/ft³	joule per cubic meter (J/m³)	37,260
Btu/h	watt (W)	0.29307
Btu/min	horsepower (hp)	0.02357
Btu/lb	joule per kilogram (J/kg)	2326
Btu/(lb)(°F)	calorie per gram degree Celsius [cal/(g)(°C)]	1
Btu/(lb)(°F)	joule per kilogram kelvin [J/(kg)(K)]	4186.8
Btu/s	watts (W)	1054.4
Btu/(ft²)(h)	joules per square meter per second [J/(m²)(s), or W/m²]	3.1546
Btu/(ft²)(min)	kilowatt per square foot (kW/ft²)	0.1758
Btu(60°F)/°F	calorie per degree Celsius (cal/°C)	543.6
calorie (gram)	Btu	3.968×10^{-3}
calorie (gram)	joule (J)	4.186
centigrade heat unit	Btu	1.8
centimeter (cm)	foot (ft)	0.03281
centimeter (cm)	inch (in)	0.3937
centimeter (cm)	meter (m)	0.01
centimeter (cm)	micron	10,000
cubic centimeter (cm³)	cubic foot (ft³)	3.532×10^{-5}
cubic centimeter (cm³)	gallon (gal)	2.6417×10^{-4}
cubic foot (ft³)	cubic centimeter (cm³)	28,317
cubic foot (ft³)	cubic meter (m³)	0.028317
cubic foot (ft³)	gallon (gal)	7.481
cubic foot (ft³)	liter (L)	28.316
cubic foot per minute (ft³/min)	cubic centimeters per second (cm³/s)	472
cubic inch (in³)	cubic meter (m³)	1.6387×10^{-5}
degree Celsius (°C)	kelvin (K)	K = °C + 273
degree Celsius (°C)	degree Fahrenheit (°F)	°F = 9/5(°C) + 32
degree Fahrenheit (°F)	degree Celsius (°C)	°C = (°F − 32)/1.8
degree Fahrenheit (°F)	kelvin (K)	K = (°F + 459)/1.8
degree Rankine (°R)	kelvin (K)	K = °R/1.8
dyne per square centimeter (dyn/cm²)	pascal (Pa)	0.1
foot (ft)	meter (m)	0.3048
foot per minute (ft/min)	centimeter per second (cm/s)	0.5080
foot per square second (ft/s²)	meter per square second (m/s²)	0.3048
gallon (U.S.) (gal)	cubic meter (m³)	0.003785
gallon (gal)	liter (L)	3.785
gallon per minute (gal/min)	cubic foot per hour (ft³/h)	8.021
gallon per minute (gal/min)	cubic meter per hour (m³/h)	0.227
gallon per minute per square foot (gal/min·ft²)	meter per hour (m/h)	2.44

Conversion Factors (Continued)

To convert from:	to	Multiply by
grain (gr)	gram (g)	0.06480
grain per cubic foot (gr/ft³)	gram per cubic meter (g/m³)	2.2884
grain per gallon (gr/gal)	parts per million (ppm)	17.118
gram (g)	kilogram (kg)	0.001
gram per cubic centimeter (g/cm³)	pound per cubic foot (lb/ft³)	62.43
gram per cubic centimeter (g/cm³)	pound per gallon (lb/gal)	8.345
gram per liter (g/L)	pound per cubic foot (lb/ft³)	0.0624
gram per square centimeter (g/cm²)	pound per square foot (lb/ft²)	2.0482
gram per square centimeter (g/cm²)	pound per square inch (lb/in²)	0.014223
inch (in)	meter (m)	0.0254
kilogram (kg)	pound (lb avoirdupois)	2.2046
kilogram per square centimeter (kg/cm²)	pounds per square inch (lb/in²)	14.223
liter (L)	cubic meter (m³)	0.001
micron	micrometer (μm)	1
millimeter (mm)	meter (m)	0.001
millimeter mercury at 0°C (mmHg)	foot of water at 39.1°F (ftH₂O)	0.446
millimeter mercury at 0°C (mmHg)	pound per square inch (lb/in²)	0.1934
pound (lb avoirdupois)	grain (gr)	7000
pound (lb avoirdupois)	kilogram (kg)	0.454
pound per cubic foot (lb/ft³)	gram per cubic centimeter (g/cm³)	0.016
pound per cubic foot (lb/ft³)	kilogram per cubic meter (kg/m³)	16.018
pound per cubic foot (lb/ft³)	gram per liter (g/L)	16
pound per gallon (lb/gal)	gram per liter (g/L)	120
pound per square foot (lb/ft²)	atmosphere (atm)	4.725×10^{-4}
pound per square foot (lb/ft²)	kilogram per square meter (kg/m²)	4.882
pound per square inch (lb/in²)	atmosphere (atm)	0.068
pound per square inch (lb/in²)	kilogram per square centimeter (kg/cm²)	0.07
pound per square inch per foot	kilogram per square centimeter per meter	0.23
square centimeter (cm²)	square foot (ft²)	1.08×10^{-3}
square foot (ft²)	square meter (m²)	0.0929
square foot per hour (ft²/h)	square meter per second	2.581×10^{-5}
square inch (in²)	square centimeter (cm²)	6.452
square inch (in²)	square meter (m²)	6.452×10^{-4}
tons (metric)	kilogram (kg)	1000
tons (metric)	pound (lb)	2204.6

Handbook of
Separation Techniques
for Chemical Engineers

Part 1

Liquid-Liquid Mixtures

Continuous Distillation: Separation of Binary Mixtures

JOSEPH M. MAAS, Ch.E., P.E. *Independent Chemical Engineer and Consultant.*

NOMENCLATURE

For McCabe-Thiele Diagram and Smoker Equation

x Mole fraction of lower-boiling (light) component in liquid phase

y Mole fraction of lower-boiling (light) component in vapor phase

x' Mole-fraction difference, on the x axis, of the transposed coordinate, when using the Smoker equation

y' Mole-fraction difference, on the y axis, of the transposed coordinate, when using the Smoker equation

α Relative volatility of light to heavy component

m Slope of operating lines, taken as straight lines, in the McCabe-Thiele construction or as expressions of such in the Smoker equation

b Intercept of straight operating lines with y axis, used in the McCabe-Thiele construction or as expressions of such in the Smoker equation

k Real root of Eq. (1) falling between 0 and 1 (see Fig. 1a and b)

n Number of theoretical stages, determined by the number of rectangular "steps" drawn or calculated, between any two points of the significant parts of the distillation diagram

R Reflux ratio or moles of reflux per unit of time, divided by moles of distillate per unit of time

SUBSCRIPTS

F Refers to feed

P Refers to product or distillate

W Refers to bottom product or residue

rect   Refers to expressions confined to rectifying-section stages

strp   Refers to expressions confined to stripping-section stages

r Refers (as n_r) to rectifying-section stages

s Refers (as n_s) to stripping-section stages

m Refers (as R_m) to minimum reflux ratio

NUMERICAL SUBSCRIPTS

See Fig. 1a.

The x,y coordinates indicate mole-fraction concentrations of liquids and vapors at points of flow from stage to stage, starting with the condenser and counting the rectifying stages downward, and the stripping stages from the reboiler, in an upward direction to the feed stage.

This nomenclature is best identified with the actions in progress within the tower.

x_0, y_0 where x_0 refers to concentration of liquid from total condensation of overhead vapors, part of which returns to the top stage of the tower as reflux; where y_0 refers to the concentration of vapor from the top stage of the tower

x_1, y_0 where x_1 refers to the concentration of the liquid leaving the top stage of the tower (after entering the stage at x_0 concentration) and at the concentration of x_1. The liquid leaving is in equilibrium with the vapors leaving (at concentration y_0)

x_1, y_1 where x_1 is the concentration of the liquid entering the stage below the top stage, and which changes in concentration from x_1 to x_2 in the course of passing through this stage. Where y_1 is the concentration of the vapor leaving this stage (second from top of tower) and at point it is not in equilibrium with the entering liquid x_1

x_2, y_1 where x_2 is the concentration of the liquid as it leaves (this second stage from the top of the tower) and is in equilibrium with the vapors leaving this stage, which is in accordance with the definition of the theoretical stage

x_2, y_2 continuing in this manner to the other stages, according to the number (or fraction thereof) present, and using similar procedures for the stripping section permits a visual concept of the methods to be aquired

NOTE: This type of subscript nomenclature from stage to stage finds a use in writing of computer programs or in the calculation of stage to stage when this is done by hand. It also serves in examination of computer print-out results when they come from a source where results are not carefully labeled.

Additional Nomenclature

FOR "CALCULATION FORM FOR STAGES, SMOKER EQUATION" (see Tables 3 and 4)

A, B, C Constants that apply in the polynomial-curve equation, used to calculate relative volatility from concentration

F A factor in computation steps 31 to 35 of Table 3

FOR THE FENSKE EQUATION

n_{LK_D} Moles of light key component in distillate

n_{HK_D} Moles of heavy key component in distillate

n_{HK_B} Moles of heavy key component in bottoms

n_{LK_B} Moles of light key component in bottoms

α Relative volatility of the key components

N_m Number of minimum stages of separation

FOR CORRELATION OF THEORETICAL STAGES WITH REFLUX RATIO (see Fig. 2).

O/D Moles of reflux per moles of distillate or reflux ratio required for a given number of stages S

$(O/D)_M$ Minimum reflux ratio corresponding to $S = \infty$

S_M Minimum number of stages corresponding to $O/D = \infty$

$\phi(S)$ Function value plotted on y axis corresponding to function value of $F(O/D)$

$F(O/D)$ Function value plotted on x axis corresponding to function value of S

S Number of stages required when reflux rate is a given value for O/D

FOR TWO-COMPONENT FLASH SEPARATION

W Vertical distance for tangent slope of vaporization operating line

D Horizontal distance for tangent slope of vaporization operating line, sometimes regarded as negative, in a directional sense

y_0 Distance y, to equilibrium line, corresponding to some given value of x

PRACTICAL DESIGN APPROACH

Distillation and Separation

Variables and definitions used in distillation are best considered in relation to the steady conditions of continuous distillation. An acceptable definition of distillation is: "The object of distillation is the separation of a volatile liquid from a nonvolatile substance or, more frequently, the separation of two or more liquids of different volatility."[80] Separation here refers to a division into parts, each of different composition as the result of relative differences in volatility of the components of the mixture being separated.

Application of Distillation

Continuous distillation may be used when the components involved have adequate volatility differences and these differences extend over the entire concentration range of the intended separation. Under this requirement, no condition must ever exist where vapor and liquid phases in contact have identical composition, or separation will stop at this point. This condition occurs where azeotropic or a constant-boiling mixture can form as well as when a pure-product condition is approached. An example is found in the distillation of the ethyl alcohol and water system, where the maximum purity of the alcoholic product is limited to the concentration of the azeotrope.

    Another limitation can exist if the distillation is set too close to the critical-state conditions of the mixtures involved, since the two-phase state merges into a single fluid condition with no separation possible at the critical state. Some distillations have been carried out as close as 25° F from the critical temperature.

Process Description

Separation by distillation may be thought of as depending upon two factors, the relative volatility of the materials to be separated and the compounding of this initial condition through the use of the distillation apparatus. The first factor is controllable only to the extent that the pressure and temperature of the distillation may be modified to give the most favorable relative-volatility differences; the second factor, the functions of the apparatus, is not particularly limited except for the costs involved.

    The influence of the inherent volatility properties of the components under separation and the functions of the apparatus work in combination. When a mixture is put into a state of partial vaporization, the vapor and liquid phases have compositional differences, as the first separation tendency. The apparatus will then be able to function to increase this initial state of compositional differences through the effects of the internal vapor and liquid stream flows established through the stages provided. This compounding of the single-stage effect must be done by the combination of the action of an extra flow that is induced within the apparatus and the establishment of a series of stages of contact (or their equivalent), where vapor and liquid streams alternately intermingle and then separate at each stage, while flowing in a generally countercurrent direction. In a continuous distillation the mixture being separated enters the system at some selected stage location where it will cause a minimum of concentration disturbance to the internal streams, and thus cause a maximum amount of separation to take place.

Single-Stage or Flash Separation

A flash separation takes place when a mixture under partially vaporized conditions has compositional differences between the vapor and liquid phases. Under complete mixing conditions and over a sufficient duration of time, a state of equilibrium is attained. The phase compositions at equilibrium are defined at the pressure and temperature of the system, and the separation is called an *equilibrium stage.*

A single-stage separation may be considered as a distillation in itself and may be called a flash separation. The performance of such a distillation compared with the separation obtained under equilibrium conditions is described as the efficiency of what is termed the actual stage, to a theoretical stage. Both are compared at the same conditions of temperature and pressure; the efficiency is referred to as stage, plate, or tray efficiency, alternately, and its measurement numerically as fractions of a stage is described in the method which follows.

In distillation, the state of equilibrium that exists between the vapor and liquid streams as they leave the stage from separate places is the basis of practical measurement in the attainment of stage-equilibrium condition, while the equilibrium stage requires that the phases be in a state of coexistence for its definition. An actual distillation tray may sometimes produce a separation between its vapor and liquid streams (as they leave the tray) that even exceeds the separation attainable by an equilibrium stage (or tray).

The theoretical stage or tray then is not the maximum separation attainable from an actual stage. This is possible because there is a concentration gradient across the tray that is set up from the liquid inlet to the outlet overflow. When the path of the liquid is long and the tray efficiency is normally high, the vapors entering the tray from below are mixed to a uniform composition (the entering gas is the same at both ends of the flow path) and as the gas passes upward through the flowing liquid, equilibrium is attained at each point, just over the surface of the flowing liquid. The gas leaving the tray near the liquid inlet, which contains the richest liquid, will be richer than the gas leaving the tray at the liquid outlet. It follows that if the tray is efficient enough for the production of equilibrium gas at the overflow and is still giving even richer gas upstream and at the inlet, the average gas composition will exceed the equilibrium composition based on equilibrium of gas with the outlet liquid. Some trays such as the dual-flow trays have liquid flows that are vertical with the tower as to towers packed with rings. These do not have the crossflow condition and should not exceed the equilibrium performance of 100% stage efficiency.

In order to rationalize the fact that an overall tray efficiency can exceed 100%, the point efficiency taken over the whole concentration gradient may be mathematically weighted, or averaged if sufficient data are available at the tray. Such an investigation would be reserved for research rather than design. It should be mentioned that if the relative volatility of the components under separation is underestimated, the calculated tray efficiency may exceed 100% through error.

The apparatus used in a flash separation is mainly a vessel used to hold the partially vaporized mixture charged into the system. The liquid level is usually on control. A heater may be located on the entering charge or within the vessel. A condenser with an appropriately located back-pressure valve is used to permit the conversion of the vapor separated into a liquid product. This is an approximate description for either the small laboratory still or a large-scale plant separator.

Where a vessel is used with no entrainment devices, the separation of the vapor and liquid may be accomplished by allowing sufficient residence time for the passage of both phases through the vessel so that satisfactory separation takes place. This type of design is based on the use of Stokes' law and the choice of liquid-droplet diameter for the settling of the liquid phase, with a similar calculation for the rise of bubbles of gas from the liquid layer.

In calculating the relative amounts of vapor and liquid produced from a phase separation when a mixture is partially vaporized, the McCabe-Thiele diagram is used for two-component mixtures. The flash separation at equilibrium is an example of a meeting point of three evaluations, since it could be considered as a vapor-liquid equilibrium stage, a theoretical stage, or a minimum stage.

In relation to design, the flash separation is usually considered as a single vapor-liquid equilibrium stage, particularly when residence time and mixing are provided and are followed by good separation. There is also reason to refer to this condition as a theoretical stage, since it would be based on the operating conditions. Finally it is also a minimum stage like the stages that are defined at an infinite reflux ratio, also called a condition of total reflux and more often described with multistage distillation.

Minimum-Stage and Flash Separation

The McCabe-Thiele diagram (see Fig. 3) used to calculate the vapor and liquid separation shows that the origin of the short operating line starts at the diagonal (at the feed composition point), and then the line is extended to the equilibrium curve. Since the diagonal is used to represent a total reflux condition as a reflux operating line, a flash separation is a minimum stage of separation when taken to equilibrium.

It can be demonstrated that any minimum stage may be described as or found equivalent to an equilibrium flash stage or vice versa. This can be seen by using the usual block diagram of a stage, with the four arrows for the liquid and vapor flows involved. The two streams entering the distillation stage may be taken as a partially vaporized mixture, such as would enter a flash, and the outlet streams of vapor and liquid would be the equilibrium separation obtained. This equivalence may be extended to apply in multistage distillation if the stages are visualized as a succession of flash separations. Due allowance should be made in this analogy that the vapor and liquid leaving any given stage (these being considered as stage products) will be routed to the proper stages to form stage charges, simulating flash separations as follows: Vapor from a given stage, passing to the stage above, is considered part of the charge to the stage it is entering. This vapor is joined by a liquid stream to make a total flash charge. This liquid stream must be the liquid stream that normally would enter this stage (from the stage above it) in order to complete the simulated flash separation.

Possible limits in the selection of stage and reflux combinations for a given separation in distillation are the extremes of operation at minimum reflux on the one hand and at the minimum stage requirement on the other. The first is predicated on minimum reflux with an infinite number of available stages, the second upon minimum number of stages with an infinite reflux ratio, also called a "total" reflux. Operation at total reflux demands that the product withdrawn has been reduced to zero rate, and no feed is entering the tower.

While these extreme conditions are in themselves outside the range of practical operation, they are used as design concepts and find their use in the calculation of reflux and stage requirements at finite conditions.

When operations are attempted at reflux conditions lower than the minimum, or with fewer stages than required for a desired separation, the degree of separation will be reduced below that which was expected.

In order to visualize flow conditions at the tower stages under total reflux conditions, when each stage is in effect a minimum stage, a number of conditions must be understood. First, the following statements apply to the distillation stages, as minimum stages under total reflux.

1. Reflux entering the top of any given tower stage is identical in concentration and flow rate to vapors leaving the top.

2. Vapors entering the bottom of the stage are identical in concentration and flow rate to the liquid leaving the stage.

3. Any minimum stage (under total reflux) may be considered making a separation equivalent to an equilibrium flash. This amounts to the net effect of regarding the vapor and liquid streams entering any given stage as a vaporized charge mixture entering a flash vessel and, after flow and mixing, passing separated products on to the next stage.

4. Where the tower boilup is high (that is, when the tower boilup is increased,

producing more reflux), any given tower stage becomes equivalent to a flash which is producing *relatively more* overhead product. Reflux entering and vapors leaving each stage will be relatively richer in the higher-boiling component (or components). Here the reflux has been increased, but not the reflux ratio; so there is no improvement in overhead composition or product, if such were visualized as being withdrawn in an infinitely small amount.

5. As a corollary to items 1 and 2 above, the concentration difference between vapors entering and those leaving a given stage must be equal to the concentration difference between the liquids entering and leaving the stage.

6. When minimum stages are used as a unit to measure separation, it is possible to use fractional parts of a stage. It is possible to express separation in fractional parts of a minimum stage through the use of the exponent of the relative volatility of the components (see Fenske equation).

In order to visualize conditions at the stages in a tower under finite-reflux conditions for comparison with the above-mentioned minimum-stage operation at total reflux, the following similar statements are expressed. This detailed description will be found useful in designs related to heat and material balances later.

1. Reflux entering the top of any given tower stage above the feed stage is *less* than the flow rate of the vapors leaving the top of the stage. The component difference between the vapor and the liquid mentioned must be equal to the net product taken overhead, when flow rates and compositions of these streams are taken into account.

2. Vapors entering and liquid leaving the bottom of the stage will also reflect differences in composition and flow rates equivalent to the net overhead product being taken. Any given stage below the feed stage will exhibit similar material-balance differences between the streams that will account for the passage of the bottoms product downward through the tower.

3. A theoretical stage (defined at the existing or intended reflux rate) may be considered as a unit stage that will produce a state of equilibrium resulting between the vapor leaving and the liquid leaving the given stage.

4. Increased tower boilup (that is, where reboil heat has been increased) will require additional reflux, and as a consequence a better separation is produced, based of course on the same amount of net product being taken.

5. As a corollary to items 1 and 2, the concentration differences between the vapors entering and leaving a given stage will not be equal to those between the liquid streams entering and leaving, but when differences in flow rates are taken into account, the net component differences will account for the passage of net overhead product upward, or net bottoms downward through the stages, according to whether the stage is above or below the feed position.

6. The theoretical stage unit, when used to measure separation (always identified with reflux ratio in use), may also be expressed as equivalent to a given number of minimum stages of separation. This aspect of analysis is used in the preliminary steps in design.

Multistage Separation by Continuous Distillation

A multistage separation by distillation requires three combined effects from the apparatus: extra flow of internal vapor and liquid streams, establishment of a series of contact steps (stages) between the vapor and liquid streams within, and that the general nature of the vapor and liquid streams is countercurrent.

The apparatus used would include a column or tower, both terms suggesting a vertical cylindrical vessel, within which the multistage effect is produced through the installation of what are called tower trays, these spaced to make possible a series of contacts between the vapor and liquid streams flowing within the tower. This does not exclude the use of tower packing or other devices, since the basic functions are similar for all.

The general nature of the internal flow of vapor and liquid within the tower is the following: The feed mixture in a partially vaporized state, usually between the bubble-point or dew-point conditions, enters the tower at some selected stage location and merges into vapor and liquid streams that are established within the tower. An extra flow of vapor and liquid in excess of the net amount from the feed is generated through the combined effects of the vaporization of some of the liquid at the base of the tower and the return of part of the vapors that are leaving at the top after they have been condensed to a liquid state. The resulting flows of liquid and vapor are countercurrent, the liquid flowing downward and across each tray, while the vapor is being forced upward through the caps or orifices in the trays, resulting in mixing of the vapor and liquid by the action of the bubbles formed. The liquid flowing downward from stage to stage passes through a seal at each stage which provides the hydraulic head required to force the upward-flowing gas through the caps or orifices located in the trays.

The vapor generation at the base of the tower is reboiling, and the return of condensed liquid to the top of the tower is refluxing. Reboiling is heat addition through vaporization, and refluxing is a heat-removal effect from condensed vapors being reevaporated, as evidenced by the reflux liquid entering at the top stage of the tower and being evaporated while flowing downward. These two effects, reboiling and refluxing, must equate to a heat balance when the heat input of the feed and the heat output of the products and separated streams, as well as other minor heat quantities, are taken into account.

An examination of the heat and mass transfer taking place during the counteraction of reboiling and refluxing going on from stage to stage will show that the liquid reflux is first depleted of lower-boiling components by their vaporization and thus passes the higher-boiling components downward, during a process of heat exchange with the rising vapors from the reboiler. The latter are progressively losing their higher-boiling components in the downflowing liquid and thus passing the lower-boiling components upward. This goes on during the succession of contacts and vapor-liquid separations from stage to stage during the countercurrent flowing condition that exists.

The mass transfer that takes place causes a composition gradient to be formed with the highest-boiling materials concentrating at the base of the tower; the tray liquids will show progressively decreasing boiling points, with decreasing stage temperatures. The pressure drop that normally exists from the base to the top of the tower will accentuate the boiling-point differences mentioned. The stage temperatures will also be slightly elevated by the stage pressure drop mentioned. During all proper conditions of distillation there must be a continuous temperature drop from the lower to the upper stages of the tower.

Variables in Continuous Distillation

A design requires that variables be selected, then quantified, and that the results be properly calculated. Variables required in formulating a design are related to flow, composition, enthalpy, energy, and stages. Selection of the variables comes from visualizing the expected conditions with a knowledge of the mechanism of distillation processes and with the aid of a flow diagram to establish a record of the points of importance set by construction.[87] During this procedure a principle is applied which may be called "problem description," which requires that a minimum number of independent variables be satisfied in order to describe the distillation. In a mathematical way, a choice always exists among the variables that may be selected, just as in the phase rule when it is applied in a more basic situation. The independent variables are those quantities which are fixed in the distillation and are analogous to the degrees of freedom, as used in phase-rule application. It is obvious that there are limitations on the values that may be assigned to the variables, such as, for example, a reflux rate set below the minimum, which will be found in the course of a computation.

This subject is covered in the texts of McCabe and Smith,[53] Hanson, Duffin, and Somerville,[32] King,[44] and Van Winkle.[76]

Stage and Reflux Computation

An important element of distillation design is the application of methods used in the computation of the various stage and reflux combinations that may be employed in obtaining some given separation. The alternate problem is the separation expected from the employment of some given number of stages under a fixed reflux rate.

The methods vary in complexity, time required in their application, and accuracy. The variation in methods is desirable, since a solution to the problem of the selection of a required stage-reflux combination is generally obtained only through a series of trial calculations that usually employ more than one method.

The nature of distillation is such that there is a multiplicity of solutions to the problem of finding some stage and reflux combination that will produce a given separation. The nature of any exact or rigorous calculation method, upon which the final answer depends, is that the method permits, through calculation, a demonstration that an answer previously obtained from successive trials is correct. Whether the calculation is by hand or by a computer-program method, it follows that the answer is obtained before the final calculation and not directly from it. It therefore becomes important to describe the procedure for approximations used, before a final computation can be made (which will confirm the solution).

The procedures for preliminary computation of suitable stage and reflux combinations are the following:

1. Fenske equation. For minimum stages.
2. McCabe-Thiele diagram. For minimum reflux.
3. Gilliland reflux correlation. For theoretical stages.
4. McCabe-Thiele method. For alternate method when small number of stages and reasonably high relative volatility are considered.

The procedures for the rigorous and the more exact methods, starting with the more accurate one, are:

5. Multicomponent stage-to-stage computer programs. For two-component distillations when applicable.
6. Stage count, based on McCabe-Thiele method. Given in Fortran IV programming for computer calculation.
7. Smoker equation. Calculation method for use of equation, by sections of tower stages, thus adjusting for variations in relative volatility in the tower.

The steps of the procedure for the preliminary computation begin with the calculation of minimum stages required to produce the separation desired. The Fenske equation is used, and some convenience is possible if the equation is used with actual moles of components rather than the mole-fraction compositions, particularly if adjustments in composition are going to be explored in relation to stage and reflux requirements and the final purity and recovery have not been decided. It is also possible that more than one particular separation may be desired during the operations.

The Fenske equation—used to calculate the number of minimum stages of separation between two components—is

$$\frac{n_{LK_D}}{n_{HK_D}} \frac{n_{HK_B}}{n_{LK_B}} = \alpha^{N_m}$$

where n_{LK_D} = moles of light key component in distillate
n_{HK_D} = moles of heavy key component in distillate
n_{HK_B} = moles of heavy key component in bottoms
n_{LK_B} = moles of light key component in bottoms
α = relative volatility of the key components, according to the conditions of the problem; usually an average value is used
N_m = number of minimum stages of separation

The second step is to calculate the minimum reflux. This is based upon the composition of the feed and its state of vaporization as it would enter the tower. Under reflux conditions that fall below this value, theoretically it becomes impossible to obtain the separation indicated by the product and feed compositions, even if an infinite number of stages were used for the separation.

The conventions of the McCabe-Thiele diagram are available in many references[53,54,61,62] for detailed description. The diagrams here (Fig. 1a and b) include both the McCabe-Thiele and the Smoker analytic relations; so they may be easily compared. The relations used for the calculation of minimum reflux are identified in this diagram, to be examined with the text.

In the preliminary calculation, the minimum reflux is determined at bubble-point feed (feed at 0% vaporized) and a straight operating line is assumed. The operating line is represented graphically on the y-x (rectangular coordinates) plot by drawing a straight line starting at the intersection of the diagonal and the x coordinate of the product composition x_P down to the intersecting point determined by the feed composition x_F coordinate and the equilibrium curve. The minimum reflux is found from the slope of the operating line, which is easily found if the y distance can be calculated at the intersection of the x_F coordinate and the equilibrium curve. This y distance is obtained from the formula for the equilibrium curve:

$$y = \frac{x_F}{1 + (\alpha - 1)x_F}$$

where y is now used to determine the slope m of the operating line from the tangent as follows:

$$m = \frac{x_P - y}{x_P - x_F}$$

The reflux is related to the slope of the operating line by the expression

$$m = \frac{R}{R + 1}$$

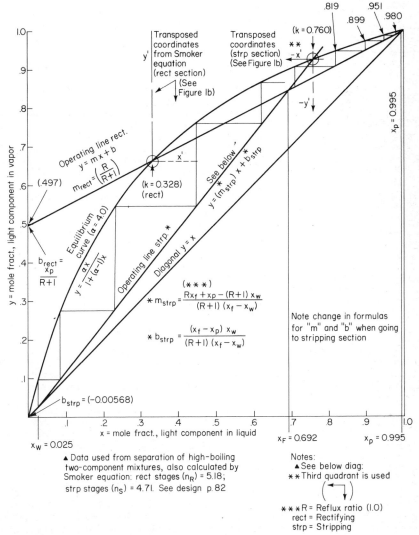

Fig. 1a McCabe-Thiele diagram of a higher alcohol separation.

or, at minimum reflux,

$$m = \frac{R_m}{R_m + 1}$$

or

$$R_m = \frac{m}{1 - m}$$

With the minimum stages and the minimum reflux values, it is now possible to obtain the theoretical stages at finite reflux rates, those between the minimum reflux and the infinite rate. This is done by the use of the Gilliland function curve; see Fig. 2. The method is evident from Fig. 2.

An alternate possibility of obtaining the preliminary stage and reflux requirements is to

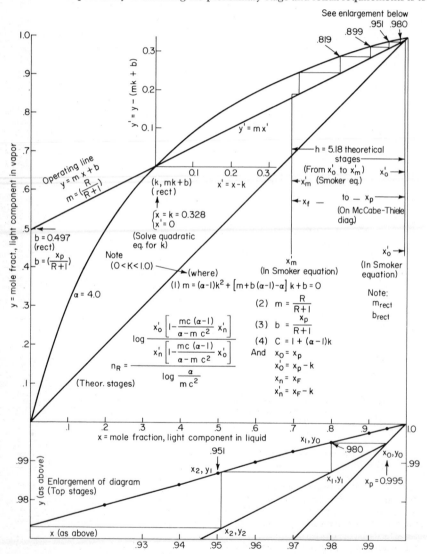

Fig. 1b Plot of Smoker equation results on McCabe-Thiele diagram for rectifying section of higher alcohol separation.

use the McCabe-Thiele construction to approximate the stages by the graphical method of stepping these off between the operating line and the equilibrium curve. This follows the conventions of the McCabe-Thiele constructions (Fig. 1) but is limited in the number of stages handled.

The preliminary stage-reflux estimates may be used as input in more exact computation in the form of either the so-called "shortcut" distillation programs on many computer

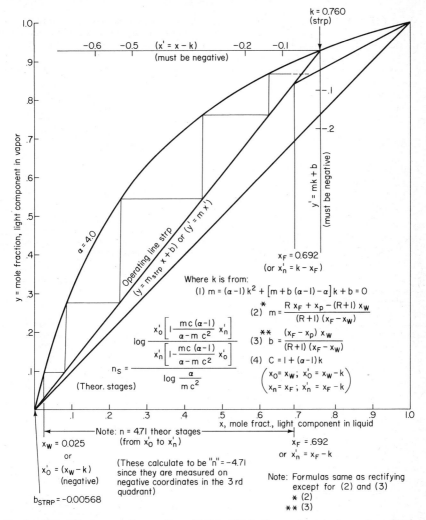

Fig. 1c Plot of Smoker equation results on McCabe-Thiele diagram for stripping section of higher alcohol separation.

facilities or the rigorous multicomponent "stage-to-stage" methods, equally available. These are generally applicable to the two-component separations when the volatilities may be expressed as the conventional K equilibrium constants (Table 1), dependent upon pressure and temperature conditions. It is also possible that the program will converge satisfactorily if the average stage temperature drop is above approximately $-0.111°C$ ($0.2°$ F). The advantage in using the computer programs is particularly great when the equilibrium and enthalpy data for the computation are available in the computer storage.

The stage-count program uses the usual McCabe-Thiele conventions, and the relative volatility is in a polynomial form or is used by deleting some of the constants involved. The formulas used are deducible from the Fortran sentences in the written program. This program is not rigorous but is exact. It has no heat balance programmed.

The Smoker equation (Fig. 1b) is used in a nonrigorous form because the variations in relative volatility must be adjusted for in steps (Figs. 5 and 6). The analytic expressions used in the Smoker method are described here without the mathematical derivation. The analytic expressions used in the formulation[71] of the Smoker equation, as well as the equation itself, with the coordinate system used, are drawn to scale (see Fig. 1a and b) with specific values placed on the diagrams to illustrate the relations used. It can be seen that there is close agreement, with only drafting inaccuracy, but when a partial stage is involved, the Smoker equation will calculate the fractional part of a stage, which can be approximated only by a graphical construction, since the stages are not a linear function of the mole-fraction change in concentration.

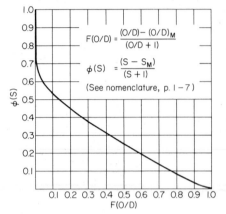

Fig. 2 Correlation of theoretical stages with reflux.

$$F(O/D) = \frac{(O/D)-(O/D)_M}{(O/D + I)}$$

$$\phi(S) = \frac{(S - S_M)}{(S + I)}$$

(See nomenclature, p. 1−7)

EXAMPLES OF DESIGN PROCEDURES

Two-Component Flash Separation

EXAMPLE

It is desired to remove part of a solvent from a heavier oil by means of a flash separation. A vessel with a controlled heat supply will be used, and the amount of feed that is vaporized can be regulated by control of the bottoms removed and maintenance of a constant liquid level in the vessel. Heat will be added or reduced, according to the liquid level.

A ratio-type flow controller could be used to compensate for changes in charge rate, for any set ratio of charge to bottoms.

Assume 0.40 is the mole fraction of solvent as the light component. The relative volatility of the solvent to the oil is taken as 2.0 and is constant over the operating range. It is desired to find the

TABLE 1   Propylene-Propane System, Relative Volatility of Propylene/Propane

$P = \text{lb/in}^2(\text{abs})$
$x_1 = \text{mole fraction propylene in liquid}$
$\alpha = \text{relative volatility}$

x_1	$P = 250$ α	$P = 300$ α
0.999	1.0848	1.0780
0.995	1.0852	1.0784
0.750	1.1076	1.0989
0.500	1.1283	1.1176
0.250	1.1475	1.1342
0.010	1.1624	1.1481

compositions and the amounts of vapor and liquid phases separated, between the limits of the bubble point and the dew point of the charge. The temperatures of the operation are not to be found, since the volatility is expressed as a ratio and not in absolute terms here.

The diagram shown in Fig. 3 gives an equilibrium curve and the diagonal, typical of the McCabe-Thiele diagram (see Fig. 1a and b). When the feed mixture is at its bubble point, no vapor is formed, but the composition of the first (hypothetical) bubble of gas formed may be determined from the coordinate of $x_F = 0.40$, the feed composition, and the point where it intersects the equilibrium line sets the vapor composition at the start of any evaporation. Similarly a horizontal line from the point of intersection of the coordinate $x = 0.40$ and the diagonal (out to the left) will give the dew-point composition of the first droplet of liquid condensed, from the intersection of the horizontal at the equilibrium curve.

The composition of the vapor is calculated by using the relation

$$y = \frac{\alpha x}{1 + (\alpha - 1)} = 0.5714 \qquad (x = 0.40, \alpha = 2.00)$$

The horizontal intersection which gives the dew-point composition of the first drop of condensed liquid is calculated by using

$$x = \frac{y}{\alpha - (\alpha - 1)\,y} = 0.25 \qquad (y = 0.40, \alpha = 2.00)$$

This would be the lowest value of the mole fraction of the light component in the liquid if the mixture was almost completely vaporized, making a minimum amount of liquid bottoms.

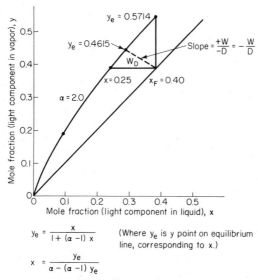

$$y_e = \frac{x}{1 + (\alpha - 1)\,x}$$ (Where y_e is y point on equilibrium line, corresponding to x.)

$$x = \frac{y_e}{\alpha - (\alpha - 1)\,y_e}$$

Fig. 3 Graphical method for two-component flash separation.

Flash separations between all-liquid and all-vapor limits of operation may be readily calculated as follows, taking into account that the liquid composition must vary from $x = 0.25$ to $x = 0.40$, depending upon the vaporization ratio set for the feed:

If the liquid composition is at the mole-fraction value of $x = 0.30$, what will be the relative amounts of vapor and liquid formed? What is the vapor composition? When $x = 0.30$,

$$y = \frac{\alpha x}{1 + (\alpha - 1)\,x} = 0.4615 \qquad (\alpha = 2.00)$$

$$\text{Ratio of moles liquid to moles vapor} = \frac{y_e(\text{at } x = 0.30) - x_F(\text{at } x = 0.40)}{x_F(\text{at } x = 0.40) - x(\text{at } x = 0.30)} = 0.615$$

The mole ratio is converted to mole percent:

$$\text{Percent liquid} = 0.615/1.615 = 38.1\%$$

The ratio of liquid to vapor is commonly described as being equal to $(-W/D)$ on McCabe-Thiele diagrams (see Fig. 3).

Calculation of Stages with Reflux

The calculation of theoretical stages (between concentration limits) for some given reflux rate must be done with analytical equations when the physical act of drafting a McCabe-Thiele diagram becomes impractical (see Fig. 1), as in the case of large numbers of stages involved, or extremely low values of relative volatility. The McCabe-Thiele method is the basis of the analytical equations used, and constant reference to such a diagram is a useful part of design (see Fig. 1a and b).

The analytical method of Smoker[71] may be used but sometimes requires calculation by sections for variations in relative volatility over the concentration range; it may be similarly used for variations in internal reflux rates (variable operating-line slope).

This method, demonstrated in the design of the propylene-propane separation, has the

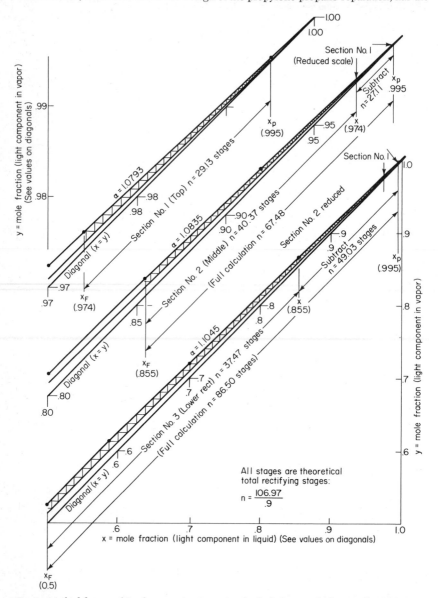

Fig. 4 Method for use of Smoker equation in sectional calculations applied to rectifying section.

advantage that a major computer facility is not needed for good results, just a good electronic desk-type calculator, preferably with about 10 storage spaces.

Figures 4 and 5 show how the sections are taken when using the Smoker equation. It is important after the first section is computed to calculate the next section as described under Separation of Propylene and Propane.

A simple method is also available using a computer facility, where it is possible to avoid the calculation by sections and adjust the relative volatility at each stage. This may be readily written into a stage-count type of program using the basic FORTRAN expressions in an abbreviated fashion and not trying to find some average alpha volatility for the midportion of the stage, which would not have much effect on the answer in any case.

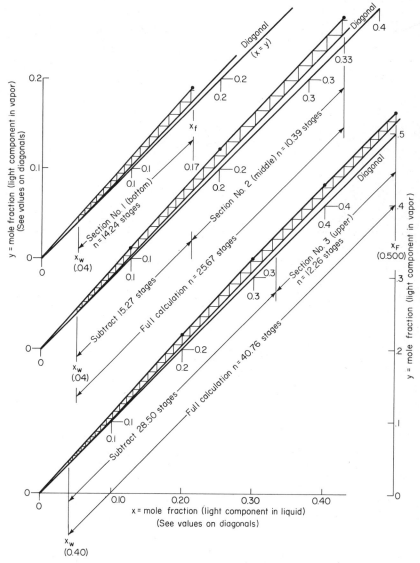

Fig. 5 Method for use of Smoker equation in sectional calculations applied to stripping section.

The following computer program can be used for the propylene-propane distillation when the polynomial constants A, B, and C (Table 2) are included as input. Other systems may be used by arranging to put similarly computed relative volatility values in as compositionally dependent equations, or for simpler systems, to make the A constant the volatility, and the B and C constants zero. The FORTRAN program is shown in Fig. 6.

Fig. 6 McCabe-Thiele stage count program in FORTRAN.

```
        DIMENSION   AAA(20)
C       PROGRAM FOR STAGE COUNT
      1 FORMAT (26X,28HA DISTILLATION METHOD/26X,32HFOR PROPYLENE-PROPANE
       1SEPARATION///)
      2 FORMAT (4X,32HTRAY COUNT TWO COMPONENT SYSTEM./)
      3 FORMAT (4X,14HRUN IDENTITY   ,20A4)
      4 FORMAT (4X,10HRUN DATE   ,I2,I3,I5//)
      5 FORMAT (2F20.10)
      6 FORMAT (F20.10)
      7 FORMAT (20A4)
      8 FORMAT (I2,I3,I5)
      9 FORMAT (4X7HACONST=F20.10,7HBCONST=F20.10)
     10 FORMAT (4X7HCCONST=F20.10)
     12 FORMAT (3F10.8,F12.8)
     13 FORMAT (4X5HXPROD,F10.8,1X3HXFD,F10.8,1X4HXBTM,F10.8,1X2HRR,F12.8/1/)
     14 FORMAT (4X31HMINIMUM REFLUX AT FEED POSITION,F12.8)
     15 FORMAT (4X31HY DISTANCE ON EQUIL.LINE AT FD. 4XF10.8/)
     16 FORMAT (4X30HY DISTANCE ON OPER.LINE AT FD. 4XF10.8)
     17 FORMAT (4X28HREFLUX RATE IS BELOW MINIMUM)
     18 FORMAT (7X8HTRAY NO., 4X5HX ONE,6X5HX TWO,6X5HY ONE,6X5HY TWO,3X11H
       1ALPHA AT X1)
     19 FORMAT (I12,F15.8,4F11.8)
     20 FORMAT (4X18HRECTIFYING SECTION)
     21 FORMAT (4X17HSTRIPPING SECTION)
     22 FORMAT (/)
     23 FORMAT (4X11HSLOPE RECT.,F10.8,2X12HSLOPE STRIP.,F15.8,4X10HMIN.RE
       1FL. ,F10.8)
     24 FORMAT (1H1)
     25 FORMAT (//T5,'MAXIMUM OF 250 STAGES REACHED—OPERATING TOO CLOSE
       1TO MINIMUM REFLUX')
C        TITLES DATES
         IN=1
         IOUT=3
    100 WRITE (3,1)
        WRITE (3,2)
     99 READ (1,7) (AAA(I),I=1,20)
        WRITE (3,3) (AAA(I),I=1,20)
        READ (1,8)IDAT1,IDAT2,IDAT3
        WRITE (3,4)IDAT1,IDAT2,IDAT3
C       READ AND WRITE VOLATILITY CONSTANTS
        READ (1,5)ACONST,BCONST
        READ (1,6)CCONST
        WRITE (3,9)ACONST,BCONST
        WRITE (3,10)CCONST
C       READ AND WRITE XPROD,XFD,XBTM,RR.
        READ(1,12)XPROD,XFD,XBTM,RR
        WRITE(3,13)XPROD,XFD,XBTM,RR
        SLOPE=RR/(RR+1.0)
    101 ALPHA=ACONST+BCONST*XFD+CCONST*XFD**2
    102 YALP=(ALPHA*XFD/(1.0+(ALPHA-1.0)*XFD))
    103 SLMN=((XPROD-YALP)/(XPROD-XFD))
    104 RMIN=(SLMN/(1.0-SLMN))
        WRITE(3,14)RMIN
    105 YFD=XPROD-(SLOPE*(XPROD-XFD))
        WRITE (3,15)YALP
        WRITE (3,16)YFD
        IF(YFD-YALP)107,106,106
    106 WRITE (3,17)
        CALL EXIT
C       RECTIFYING STAGE COUNT X,Y,ARE STEP OFF POINTS.
    107 WRITE (3,20)
```

```
        WRITE (3,18)
        N=0
        YONE=XPROD
        XONE=XPROD
108     N=N+1
        IF(N−250) 210,210,200
210     ALONE=ACONST+BCONST*XONE+CCONST*XONE**2
109     XTWO=(YONE/(ALONE−(ALONE−1.0)*YONE))
110     YTWO=(YONE−SLOPE*(XONE−XTWO))
        WRITE(3,19)N,XONE,XTWO,YONE,YTWO,ALONE
112     XONE=XTWO
        YONE=YTWO
        IF(XONE−XFD)113,108,108
C       STRIPPING STAGE COUNT
113     STRSL=(YFD−XBTM)/(XFD−XBTM)
        SYONE=XBTM
        SXONE=XBTM
        NS=0
        WRITE(3,21)
        WRITE (3,18)
114     NS=NS+1
        ALSONE=ACONST+BCONST*SXONE+CCONST*SXONE**2

115     SYTWO=(ALSONE*SXONE)/1.0+(ALSONE−1.0)*SXONE)
116     SXTWO=SXONE+((SYTWO−SYONE)/STRSL)
        WRITE(3,19) NS,SXONE,SXTWO,SYONE,SYTWO,ALSONE
        SXONE=SXTWO
        SYONE=SYTWO
        IF(SXONE−XFD)114,117,117
117     WRITE(3,23)
118     CALL EXIT
200     WRITE (3,25)
118     CALL EXIT
```

FORTRAN NOMENCLATURE
    Terms are in order of their occurrence in program write up.

ACONST	Constant in polynomial equation for relative volatility.
BCONST	
CCONST	
XPROD	Distillate product composition, x_p.
XFD	Feed composition, x_f.
XBTM	Bottoms composition, x_w.
RR	Reflux ratio, R.
SLOPE	Operating line slope, rectifying section.
ALPHA	Relative volatility, .
YALP	Coordinate "Y" , distance on equilibrium curve at x_f.
SLMN	Slope of operating line for minimum reflux conditions
RMIN	Minimum reflux ratio.
XONE	Step off position, x_1, in rectifying stage count.
YONE	" y_1 "
XTWO	" x_2 "
YTWO	" y_2 "
STRSL	Operating line slope in stripping section.
SYONE	Step off position y_1 in stripping stage count.
SXONE	" x_1 "
SYTWO	" y_2 "
SXTWO	" x_2 "
NS	Stage count in program loop calculation.
ALSONE	Alpha sub one refering to alpha at x_1.

EXAMPLES OF DESIGNS

Comparison of Batch and Continuous Distillation

The first comparison is made when the same number of stages and the same reflux ratio is used in both methods. Using the Smoker method (Fig. 1b) for the case of a two-component distillation, and the calculation form prepared for the c_3 splitter design, the following operation is compared:

$$x_P = 0.900 \qquad \alpha = 2.00 \text{ (constant)}$$
$$x_F = 0.500 \qquad \text{Reflux} = 20.0 \text{ (ratio)}$$
$$x_W = 0.100$$

With the above conditions as a continuous operation, using rectifying and stripping stages, it is calculated by the Smoker equation that 3.30 rectifying and 3.4 stripping stages

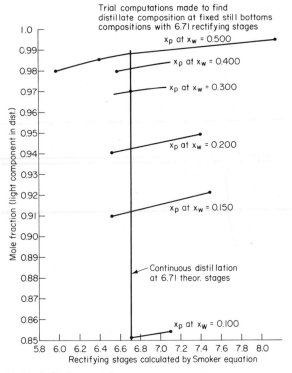

Fig. 7   Batch-distillation trial computations to match continuous operation.

are required when a bubble-point feed is used. (Note that the stages here will differ slightly.)

Since the batch process involves the withdrawal of portions of the total product at varying compositions, it is necessary to take the products as increments and sum up the amounts received until the remaining bottoms of the batch distillation is down to the 0.100 mole fraction expected in the continuous operation. The amount distilled off in the batch operation need not be the same as the relative amount of the overhead product taken in the continuous operation, for the same product purity.

In making the trial calculations for the increments representing the progress of the batch distillation, it is best to start with some fixed bottoms composition (x_F in this case because the batch distillation is all rectifying stages, unlike the continuous operation. The top composition may be plotted versus the number of theoretical stages calculated, for each mole fraction of x_F obtained progressively in the bottoms, taking the value of $x_F = 0.5$

as the start. The calculated number of stages are plotted to obtain the curve that intersects the 6.71 theoretical stages obtained for the continuous operation (see Fig. 7).

A curve for the change in overhead composition with the bottoms at the constant separation of 6.71 theoretical stages is obtained (see Fig. 8), and subsequent calculations are made to determine the material balances in a two-component separation, which shows the incremental separations and the amounts of each. A sufficient number are made, in this case four, from 0.500 down to 0.100 in increments of 0.100 (on the pot or bottoms composition).

This approximate evaluation shows that the 6.71 stages (all rectifying) will produce a product of 96% purity, compared with the 90% obtained in the continuous operation, under the same reflux and stage conditions. In order to get the same product in the continuous case, the continuous distillation would require approximately two more rectifying stages.

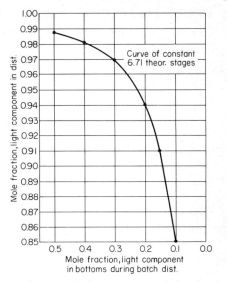

Fig. 8 Batch-distillation compositions based on zero holdup.

Assuming that there is no holdup to consider, the batch operation will start with an overhead of 0.988 mole fraction of light component when the bottoms are 0.500. The progress of the separation will be itemized:

	Mole fraction x_P	Mole fraction x_F
Start	0.988	0.500
0.5–0.4	0.980	0.400
0.4–0.3	0.970	0.300
0.3–0.2	0.940	0.200
0.2–0.15	0.911	0.150
0.10–0.10	0.850	0.100

From 100 mol of feed the material balances show the following product withdrawal:

Increment	Mol, light component	Mol, heavy component
0.5–0.4	16.83	0.26
0.4–0.3	11.97	0.29
0.3–0.2	8.94	0.41
0.2–0.1	6.90	0.80
Totals	44.64	1.76
Product % light	96.2%	
Recovery % light	89.3%	

Continuous distillation has 90% recovery and 90% purity.

Separation of Propylene and Propane

This is a two-component distillation where up to 160 stages may be used. This method makes use of a graphical solution such as the McCabe-Thiele seem inaccurate by contrast. The method here is applicable to hand calculation with a small electronic calculator, preferably one with about 10 storage spaces. When familiarity is obtained, a design may be completed in several hours, and the method is often easier to use in approximate computations than constructing a graphical McCabe-Thiele diagram.

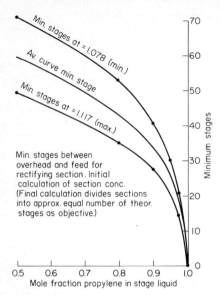

Min. stages between overhead and feed for rectifying section. Initial calculation of section conc. (Final calculation divides sections into approx. equal number of theor. stages as objective.)

Fig. 9 Preliminary estimate of rectifying-stage requirements.

The analytical method of Smoker is used over the column in sections so that changes in relative volatility[33] may be taken into account. An average value of relative volatility is taken over the range of concentration at each section, and using this value, the number of stages are computed.

Several sections are usually taken at stage liquid concentrations, first from the top product down to the feed. Similarly sections are taken for the stripper stages, from the bottoms up to the feed inlet. In this method, unlike the multicomponent methods, the feed stage is exactly established. In this way, approximately 100 stages may be handled with good accuracy in about 20 to 30 stages per section. Refer to Figs. 9 and 10.

It is necessary at the beginning to divide the sections into approximately the same number of stages rather than use a division based on equal differences in concentration. This is done by first making a division based on having the same number of minimum stages in each section and later making the proper adjustments for obtaining the same number of theoretical stages (approximately) for each section. With the relative volatility averaged in this way for each section, each having approximately the same number of theoretical stages, the most accurate effect is obtained.

The accuracy of dividing the column into sections will depend upon the number of sections taken and how evenly the sections are divided with respect to the number of stages included in each. The latter seems less important than the number of sections used, after a fair approximation is made to an equal division as theoretical stages. The accuracy

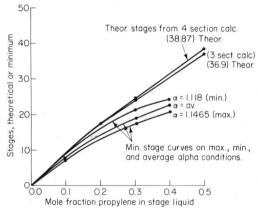

Fig. 10 Preliminary estimate of stripping-stage requirements.

of this technique has been tested by a second method that requires a computer program to be written, and here the relative-volatility values are corrected as the concentration changes at each stage, in the course of the computation.

The relative volatility of the propylene-propane system (Table 2) has been found to vary with concentration when only the two components are present, or when some smaller amounts of butanes, approximately less than 5%, are present. Very small amounts of lighter hydrocarbons are also not effective in preventing the use of this two-component calculation method. The small amounts of these outside materials may be lumped into the two main components or carried outside of the calculation and given an approximate separation, just to keep the bookkeeping aspects of the situation in order.

The nonideal aspect of the system was something of a secret for many years. It was found out that for safe designs a relative volatility of 1.08 must be used,[81] even though a volatility of as high as 1.14 could be found at distillation conditions in the equilibrium data for multicomponent hydrocarbon mixtures in existing K systems.

Relative-volatility values are given for this design in the form of polynomial expressions based upon the concentration of the stage liquid, in the mole fraction of propylene present

TABLE 2   Propylene-Propane Polynomial Constants

Operation, lb/in²(abs)	Constants		
	A	B	C
300	1.1486215	−0.0532519	−0.0174405
250	1.1630570	−0.0608571	−0.0171428
225	1.1730967	−0.0621885	−0.0217089
185	1.1930008	−0.0750549	−0.0205426

α = relative volatility
x = mole fraction of propylene in stage liquid
$\alpha = A + Bx + Cx^2$

(see Table 2). This amounts to having a concentration-directed volatility, different from the usual basis using pressure and temperature as arguments (see Table 1).

The internal reflux, taken at several sections throughout the column (see Fig. 11), was found to be almost constant, decreasing only slightly near the top stage. If the reflux induced through heat losses along the outside surface of a column of this type were examined, there could be an unbalancing of the reflux condition to some slight extent. The enthalpy values for the two-component system were taken from the article of Stupin,[72] but for design purposes the pure-component enthalpy of Project 44[2] would have been about the same, for heat balances.

The calculation may be carried out best by using the prepared calculating form (Table 3). It is based on the nomenclature used in the original article of Smoker, except for the use of capitalized (A, B, C) constants to simplify the binomial equation in the computation. These will hardly be confused with the constants used in the polynomial expression for relative volatility.

The same form is used for both the rectifying (Table 3) and stripping sections (Table 4) when taking into account the operating-line slopes and the b constant, and noting when to use x_P and x_W for the x_0 value as required. Note, for example, that when item 8 is multiplied by item 9, the item numbers are *not* used. When actual numerical values are required, these are underlined and decimal points are used.

It is necessary here and in many other preliminary situations to start by calculating the minimum stages of separation, in this case for the stages between the top product and the feed composition, or physically between the top stage and the feed stage. This is done by using the Fenske equation, which is an extension of the definition of relative volatility, in the sense that it is applied to more than one stage, which has been discussed earlier.

DESIGN PROBLEM

Find the stage and reflux requirements for a propylene-propane splitter with the following product and bottoms compositions. The rectifying and stripping stages are calculated separately. Preliminary reflux ratio and operating pressure are based on previous knowledge, but these could be determined by trials if necessary.

x_P = 0.995 mole fraction propylene product
x_F = 0.500 mole fraction propylene in feed
x_W = 0.040 mole fraction propylene in bottoms
Operating pressure = 300 lb/in²(abs)
Reflux ratio = 28.0
Rectifying stages = approximately 110 theoretical
Stripping stages = approximately 35 theoretical

(The above symbols are those used in Tables 3 and 4.) The relative volatility of the propylene-propane system was obtained from the data of Hill.[33] It was put into the form of a polynomial (Table 2) by means of a curve-fit computer program. Four pressures were used that could be of interest in a propylene-propane splitter operation. The example used is for a 300 lb/in²(abs) operation.

The relative volatility at the top stage, where the propylene content of the liquid is $x = 0.995$ as the mole fraction from the design example, is calculated as follows:

$$\alpha = 1.14862 - 0.053252(0.995) - 0.01744(0.995)^2$$
$$= 1.078 \text{ at } 300 \text{ lb/in}^2(\text{abs})$$

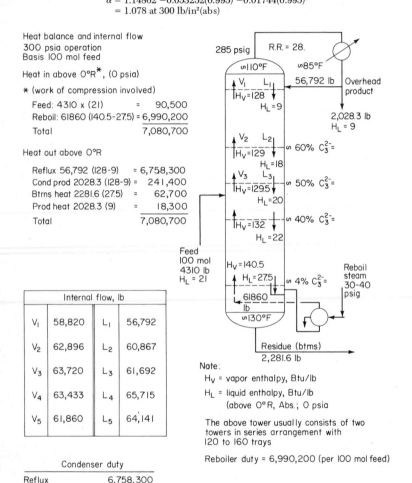

Heat balance and internal flow
300 psia operation
Basis 100 mol feed

Heat in above 0°R*, (0 psia)

* (work of compression involved)

Feed: 4310 x (21)	=	90,500
Reboil: 61860 (140.5-27.5)	= 6,990,200	
Total	7,080,700	

Heat out above 0°R

Reflux 56,792 (128-9)	= 6,758,300	
Cond prod 2028.3 (128-9)	= 241,400	
Btrns heat 2281.6 (27.5)	= 62,700	
Prod heat 2028.3 (9)	= 18,300	
Total	7,080,700	

Internal flow, lb			
V_1	58,820	L_1	56,792
V_2	62,896	L_2	60,867
V_3	63,720	L_3	61,692
V_4	63,433	L_4	65,715
V_5	61,860	L_5	64,141

Condenser duty

Reflux	6,758,300
Cond. prod.	241,400
(Per 100 mol feed)=	6,999,700

285 psig   R.R. = 28.

∽110°F   ∽85°F

V_1 L_1 56,792 lb   Overhead product
$H_V = 128$
$H_L = 9$

2,028.3 lb
$H_L = 9$

V_2 L_2
$H_V = 129$ ∽ 60% C_3^{2-} =
$H_L = 18$

V_3 L_3
$H_V = 129.5$ ∽ 50% C_3^{2-} =
$H_L = 20$

$H_V = 132$ ∽ 40% C_3^{2-} =
$H_L = 22$

Feed
100 mol
4310 lb $H_V = 140.5$
$H_L = 21$ $H_L = 27.5$ ∽ 4% C_3^{2-} =

61860
lb
∽130°F

Reboil steam
30-40 psig

Residue (btms)
2,281.6 lb

Note:

H_V = vapor enthalpy, Btu/lb
H_L = liquid enthalpy, Btu/lb
(above 0°R, Abs.; 0 psia

The above tower usually consists of two towers in series arrangement with 120 to 160 trays

Reboiler duty = 6,990,200 (per 100 mol feed)

Fig. 11   Flow diagram of propylene-propane fractionator.

If similarly calculated for the feed stage where $x = 0.500$, the value of α becomes 1.117 (where the number of significant figures are reduced).

This example divides the rectifying section into three subsections (see Figs. 4 and 5); the accuracy obtained with more divisions is described later. The division (Figs. 9 and 10) is based on the choice of concentrations that are approximately the same number of minimum stages apart, because the minimum stages are easy to calculate and are proportionately near enough to the division in terms of the theoretical stages obtained later. Final adjustments can be made if required by readjusting the concentration ranges first used (from the minimum stage division).

It is sufficiently accurate at the start to plot curves of minimum stages (Figs. 9 and 10) with

TABLE 3 Calculation Form for Stages, Smoker Equation

Item	C_3 splitter rectifying section	Section 1 full	Section 2 full	Section 2 subtract	Section 3 full	Section 3 subtract
1	x_P	0.995	0.995	0.995	0.995	0.995
2	x_W					
3	x_F	0.974	0.855	0.974	0.500	0.855
4	α	1.0793	1.0835	1.0835	1.1045	1.1045
5	R	28.0	28.0	28.0	28.0	28.0
6	$R + 1$	29.0	29.0	29.0	29.0	29.0
7	$b_{rect} = \frac{1}{6}$ $b_{strp} =$ $\dfrac{(3-1) \times 2}{6 \times (3-2)} =$	0.03431	0.03431	0.03431	0.03431	0.03431
8	$m_{rect} = \frac{5}{6}$ $m_{strp} = 2$ $\dfrac{(5 \times 3) + 1 - (6 \times 2)}{6 \times (3-2)} =$	0.9655	0.9655	0.9655	0.9655	0.9655
9	$(\alpha - 1) = 4 - 1.$	0.0793	0.0835	0.0835	0.1045	0.1045
10	$m(\alpha - 1) = 8 \times 9$	0.07627	0.08063	0.08063	0.10090	0.10090
11	$= 8 + (7 \times 9) - 4$	-0.11077	-0.11512	-0.11512	-0.13539	-0.13539
12	$A = 1. = \frac{10}{10}$	1.00	1.00	1.00	1.00	1.00
13	$B = \frac{11}{10}$	-1.4523	-1.4279	-1.4279	-1.3419	-1.3419
14	$C = \frac{7}{10}$	0.44982	0.42558	0.42558	0.34005	0.34005
15	$Ak^2 + Bk + C = 0$	0	0	0	0	0
16	$B^2 = 13 \times 13$	2.1090	2.0389	2.0389	1.8008	1.8008
17	$4. AC = 4 \times 12 \times 13$	1.7993	1.7023	1.7023	1.3602	1.3602
18	$B^2 - 4. AC = 16 - 17$	0.30978	0.33657	0.33657	0.34005	0.34005
19	$(B^2 - 4. AC)^{1/2} = 18^{1/2}$	0.55658	0.58015	0.58015	0.66374	0.66374
20	$2A = 2. = 2 \times 12$	2	2	2	2	2
21	$-B = -13$	1.4523	1.4279	1.4279	1.3419	1.3419
22	$-B \pm (B^2 - 4. AC)^{1/2} = 21$ $\pm 19 < 2.$	0.89572	0.84774	0.84774	0.66374	0.66374
23	$k = \frac{21}{20} = <1.0$	0.44786	0.42387	0.42387	0.33909	0.33909
24	$x'_{0rect} = 1 - 23$ $x_{0\,strp} = 2 - 23$	0.54716	0.57113	0.57113	0.65591	0.65591
25	$x'_n = 3 - 23$	0.52616	0.43113	0.55013	0.16091	0.51591
26	$c = 1. + (9 \times 23)$	1.03538	1.03539	1.03539	1.03543	1.03543
27	$mc = 8 \times 26$	0.99968	0.99969	0.99969	0.99973	0.99973
28	$mc^2 = 27 \times 26$	1.03504	1.03507	1.03507	1.03516	1.03516
29	$mc\,(\alpha - 1) = 27 \times 9$	0.078974	0.083474	0.083474	0.104472	0.104472
30	$\alpha - mc^2 = 4 - 28$	0.043955	0.048427	0.048427	0.069341	0.069341
31	$F = \frac{29}{30}$	1.7967	1.7237	1.7237	1.5066	1.5066
32	$F \times x'_n = 31 \times 25$	0.94535	0.74313	0.94825	0.24243	0.77728
33	$1. - (F \times x'_n) = (1. - 32)$	0.05465	0.25687	0.051749	0.75757	0.22271
34	$F \times x'_0 = 31 \times 24$	0.98308	0.98445	0.98445	0.98821	0.98822
35	$1. - (F \times x'_0) = (1. - 34)$	0.01692	0.01555	0.01555	0.01179	0.01179
36	$x'_0 \times 33 = 24 \times 33$	0.02990	0.14670	0.029555	0.49690	0.14608
37	$x'_n \times 35 = 25 \times 35$	0.0089037	0.0067045	0.008555	0.001896	0.006080
38	$= \frac{36}{37}$	3.3586	21.881	3.4547	262.08	24.026
39	$\log 38 = *$	1.2115	3.0857	1.2397	5.5686	3.1791
40	$\alpha/mc^2 = \frac{4}{28}$	1.0425	1.0468	1.0468	1.0670	1.0670
41	$\log 40 =$	0.041590	0.045725	0.045725	0.064838	0.064838
42	$n =$ theoretical stages $=$ $\frac{39}{41}$	29.13	67.48	27.11	86.50	49.03
	Net per section	29.13	(40.37)		(37.47)	
	$\alpha = A + Bx + Cx^2$	Constants at 300 lb/in²(abs) operation				
43	$A = 1.1486$					
44	$B = -0.05325$					
45	$C = -0.01744$					
46	$x_{av} =$	0.9845	0.9145	0.9145	0.6775	0.6775
47	$\alpha_{calc} =$	1.0793	1.0835	1.0835	1.1045	1.1045

*Natural or common, but consistent.

TABLE 4   Calculation Form for Stages, Smoker Equation

Item	C_3 splitter stripping section	Section 1 full	Section 2 full	Section 2 subtract	Section 3 full	Section 3 subtract
1	x_P	0.995	0.995	0.995	0.995	0.995
2	x_W	0.0400	0.0400	0.0400	0.0400	0.0400
3	x_F	0.170	0.330	0.170	0.500	0.330
4	α	1.1428	1.1342	1.1342	1.1235	1.1235
5	R	28.0	28.0	28.0	28.0	28.0
6	$R+1$	29.0	29.0	29.0	29.0	29.0
7	$b_{rect} = \frac{1}{6}$					
	$b_{strp} =$		Use true feed $x_F = 0.500$ for b_{strp}			
	$\dfrac{(3-1) \times 2}{6 \times (3-2)} =$	-0.001484	-0.001484	-0.001484	-0.001484	-0.001484
8	$m_{rect} = \frac{5}{6}$					
	$m_{strp} =$		Use true feed $x_F = 0.500$ for m_{strp}			
	$\dfrac{(5 \times 3) + 1 - (6 \times 2)}{6 \times (3-2)} =$	1.03711	1.03711	1.03711	1.03711	1.03711
9	$(\alpha - 1) = 4 - 1.$	0.1428	0.1342	0.1342	0.1235	0.1235
10	$m(\alpha - 1) = 8 \times 9$	0.14810	0.13918	0.13918	0.12808	0.12808
11	$= 8 + (7 \times 9) - 4$	-0.10590	-0.09729	-0.09729	-0.08658	-0.08658
12	$A = 1. = \frac{10}{10}$	1.0	1.0	1.0	1.0	1.0
13	$B = \frac{11}{10}$	-0.71510	-0.69904	-0.69904	-0.67595	-0.67595
14	$C = \frac{7}{10}$	-0.010022	-0.010664	-0.010664	-0.011588	-0.011588
15	$Ak^2 + Bk + C = 0$	0	0	0	0	0
16	$B^2 = 13 \times 13$	0.51137	0.48866	0.48866	0.45690	0.45690
17	$4.AC = 4 \times 12 \times 13$	-0.040088	-0.042657	-0.042657	-0.046353	-0.046353
18	$B^2 - 4.AC = 16 - 17$	0.55146	0.53132	0.53132	0.50325	0.50325
19	$(B^2 - 4.AC)^{1/2} = 18^{1/2}$	0.74260	0.72892	0.72892	0.70940	0.70940
20	$2A = 2. = 2 \times 12$	2	2	2	2	2
21	$-B = -13$	0.71510	0.69904	0.69904	0.67595	0.67595
22	$(-B) \pm (B^2 - 4AC)^{1/2} =$	1.4577	1.4280	1.4280	1.3853	1.3853
	$21 \pm 19 < 2.0$					
23	$k = \frac{21}{20} = <1.0$	0.72885	0.71398	0.71398	0.69267	0.69267
24	$x'_{0\ rect} = 1 - 23$					
	$x'_{0\ strp} = 2 - 23$	-0.68885	-0.67398	-0.67398	-0.65267	-0.65267
25	$x'_n = 3 - 23$	-0.55885	-0.38398	-0.54398	-0.19267	-0.36267
26	$c = 1. + (9 \times 23)$	1.10408	1.09582	1.09582	1.08555	1.08555
27	$mc = 8 \times 26$	1.14505	1.13648	1.13648	1.12583	1.12583
28	$mc^2 = 27 \times 26$	1.26422	1.24537	1.24537	1.22214	1.22214
29	$mc(\alpha - 1) = 27 \times 9$	0.16351	0.15252	0.15252	0.13904	0.13904
30	$\alpha - mc^2 = 4 - 28$	-0.12142	-0.11117	-0.11117	-0.098635	-0.098635
31	$F = \frac{29}{30}$	-1.3466	-1.3719	-1.3719	-1.4096	-1.4096
32	$F \times x'_n = 31 \times 25$	0.75250	0.52678	0.74629	0.27160	0.51124
33	$1. - (F \times x'_n) = (1. - 32) =$	0.24744	0.47322	0.25371	0.72840	0.48876
34	$F \times x'_0 = 31 \times 24$	0.92762	0.92463	0.92463	0.92003	0.92003
35	$1. - (F \times x'_0) = (1. - 34)$	0.072381	0.075367	0.075367	0.079968	0.079968
36	$x'_0 \times 33 = 24 \times 33$	0.17045	0.31894	0.17100	-0.47541	-0.31900
37	$x'_n \times 35 = 25 \times 35$	-0.04045	-0.02894	-0.04041	-0.01541	-0.02900
38	$= \frac{36}{37}$	4.2138	11.021	4.1709	30.855	10.999
39	$\log 38^*$	1.4384	2.3998	1.4281	3.4293	2.3978
40	$\alpha/mc^2 = \frac{4}{28}$	0.90395	0.91073	0.91073	0.91929	0.91929
41	$\log 40 \dagger$	-0.10093	-0.09351	-0.09351	-0.08415	-0.08415
42	$n = $ theoretical stages $=$	14.244	25.665	15.273	40.752	28.494
	$\frac{39}{41}$					
	Net per section	14.244	(10.392)		(12.258)	
	$\alpha = A + Bx + Cx^2$	0	0	0	0	0
43	$A = 1.1486$					
44	$B = -0.05325$					
45	$C = -0.01744$					
46	$x_{av} =$	0.105	0.250	0.250	0.415	0.415
47	α_{calc}	1.1428	1.1342	1.1342	1.1235	1.1235

*Natural or common, but consistent.

†Negative sign indicates direction of stage count; drop it.

concentration of the stage liquid at the extreme values of the relative volatilities, at several different concentrations between the top stage and the feed stage. Two curves are plotted, one for the lowest alpha value of 1.078 (at the top), and the highest value in the rectifying section, 1.117, which occurs at the feed-stage concentration; see Table 2 with its sample calculation.

The two curves are obtained by using the Fenske equation and selecting arbitrary concentrations between the top and feed stages to plot the curves. For example, the minimum stages occurring from the top-stage concentration, 0.995, down to the stage concentration of 0.800 is

$$\frac{0.995}{1.00 - 0.995} \times \frac{1.00 - 0.800}{0.800} = (1.078)^{Nm}$$

For the upper curve, $N_m = 52.01$.

$$\frac{0.995}{1.00 - 0.995} \times \frac{1.00 - 0.800}{0.800} = (1.117)^{Nm}$$

For the lower curve, $N_m = 35.31$.

With these two curves an average curve is plotted between (see Fig. 9), and this is accurate enough to divide the three subsections into parts of approximately 20 minimum stages each, from the total of 60 minimum stages calculated. The three subsections will be calculated according to the following concentration ranges given in mole fractions:

Top   0.995 to 0.974, average 0.9845
Middle   0.974 to 0.855, average 0.9145
Feed   0.855 to 0.500, average 0.6775

It is sufficiently accurate to use the mean of the concentration limits of each section as shown above, for the stages calculated in the following calculation form, based on the use of the Smoker equation.

Five calculations (see Figs. 4 and 5), are involved for the three sections. The first will be what can be called a full calculation, from 0.995 to the 0.974 range of x_P to x_F, using $\alpha = 1.079$.

The second will require a full calculation from 0.995 to 0.855 at the base of this section, but the alpha value used is based on the mean concentration of the *middle section*. This is called a full calculation to distinguish it from the subtracting calculation that follows. The subtracting calculation amounts to a recalculation of the top section at the alpha value used for the middle section, since the net stages of the middle section must be calculated by *difference* where the start is always from the top stage of the tower.

The lower section (section 3 on Fig. 4) is similarly calculated, the full calculation from the top ($x = 0.995$) and down to the feed ($x = 0.500$), but at alpha values from the average concentration of the third (feed) section. The subtraction calculation is then made from the top ($x = 0.995$) and down to ($x = 0.855$), which is the base of the middle section or the top of the feed section.

This may be somewhat involved, but the principle is to work downward through the tower and calculate each section, after the top one, by a subtraction of the stages above the one being computed, and to use the proper value of alpha in each case.

The following computation of theoretical stages is limited by the use of a constant internal reflux. In practical design, especially for this operation, this is justified (see Fig. 11).

The procedure for the sectional calculation of the rectifying and stripping stages is illustrated in Figs. 4 and 5, respectively. These diagrams approximate graphically the analytic method of solution of the problem and are particularly drawn to show how each sectional calculation is made to obtain the number of stages in that section at the average relative volatility that applies.

The general expression for the product composition x_P is x_0' in the Smoker equation, and similarly the x_F values taken for the lower concentration of light component for each section are regarded as x_n' in the formula shown in Fig. 1a and b. Note that the calculation steps follow this plan: section 1, where $n = 29.13$ theoretical stages, uses the average value 1.0793 for this top section from $x_P = 0.995$ to $x_F = 0.974$ and needs no subtraction correction.

Section 2 is calculated for the new volatility of 1.085, and it is necessary to make two stage calculations and to subtract the difference to obtain the stages between $x = 0.974$ and $x = 0.855$. This is done as follows:

Calculate the stages for $x_P = 0.995$ and $x_F = 0.855$, which is $n = 67.48$ stages, and this is called the "full" calculation.

Calculate the stages from $x_P = 0.995$ to $x_F = 0.974$, which is 27.11 stages (at 1.0835), and call this the "subtract" calculation. The difference in the number of stages is 40.37, and this is the stages required for section 2; note that the original 29.13 stages for the top section at 1.0793 (which stands as the top section stage requirement) is now 27.11 stages at the 1.0835 value as the proper correction factor for subtraction.

Section 3 similarly requires a full calculation from $x_P = 0.995$ to $x_F = 0.500$, giving $n = 86.50$ (at 1.1045) and requires the subtraction of $n = 49.04$ stages (at the value 1.1045) for the number of stages from $x = 0.855$ down to $x = 0.500$. This is equal to $n = 37.47$ stages for this section, as required. The total stages for the three sections as arbitrarily divided here is 106.97 stages, comparing very well with the tray-count program where volatility corrections are made through computer programming at each stage.

The following calculation (Table 4) involves the use of the calculation form for the stripping section. As shown, new constants b_{strp} and m_{strp} (items 7 and 8) are computed. In both cases the value of x_F used must take the true feed value, here $x = 0.500$, and not the individual values of the concentrations dividing the subsections. This is the only place where it becomes necessary to recall that the slope of the operating line, being constant for all sections, needs but one calculation and is easily performed using the true feed composition.

In this computation, for either rectifying or stripping, the calculated value in line 38 must never be a negative number. Aside from the impossibility of such a number having a logarithm, when this happens an impossible distillation situation will exist in the formulation of the problem as given.

In another instance, the stripping section is counted from a weak to a richer concentration, involving a few *negative differences;* these can be reconciled by the stages coming up *negatively,* indicating the *direction* of count. For this reason calculation line 40, which comes up negative, has the sign dropped or could be described as using the absolute value.

In order to limit the calculation to three parts and demonstrate the method, previously determined concentrations are used.

The preceding calculation of 107 rectifying stages using only three sections compares well with a four-section calculation which showed 105.2 stages based on a balanced division of 26 stages plus or minus 0.5 stages in each section. This also compared well with a computer-calculated program, where the alpha value was corrected at each stage and 106 theoretical stages were found. The stripping section also showed similar agreement.

If the tower loading is described in terms of the Badger F factor,* the stage (or tray) efficiency will vary, from 110% plus at an F of 0.6, to 107% at F = 0.8, to 101% at F = 1.0. This is an overall tower efficiency and can be attained only in large-diameter towers with long flow paths and at least 18-in tray spacing. The trays may be valve trays, but sieve trays would not give as high an efficiency in this service. Short flow paths in small-diameter towers may reduce the tray efficiency to as low as 70%, even though the materials distilled are of very low viscosity.

Taking the theoretical stages for the whole tower at 142 a stage efficiency at 101% would require 141 stages, by the count, but the richness of the feed will also influence the requirement; so some explorative calculations are necessary to cover the desired range of operations. The optimum economic choice will not be determined by the tower alone but will involve the condensing system and other parts as well.

Physical description of the splitter   A two-tower arrangement in series is usually required when more than approximately 125 stages are supplied. An 18-in tray spacing is usual, but as low as 12-in has been used. Valve trays, sometimes with perforations (small-diameter holes) between the valves, are used to reduce the pressure drop through the tower. Tower pressure drops usually are less than 3 in of water per stage, based on valve-tray use. The high pressure of operation [300 lb/in²(gage) here] does not make it very important to avoid reasonable pressure drop as in the case of low-pressure operations.

Tray design should take into account the usual concepts of hydraulic flow, and it is well to calculate the usual conditions related to flooding and downcomer areas; but in order to obtain a good tray efficiency, it is well to keep the vapor rate no higher than that which produces an F of 1.0.

The heating media used for reboiling should be of a sufficiently low temperature to avoid vapor-blanketing conditions at the surface of the tubes for heat exchange. If steam is used, a steam supply at 30 to 40 lb/in²(gage) is recommended to still obtain heat exchange near the maximum flux rate. It is possible to consider the use of hot water as a heating medium under some conditions, where the splitters may be a part of an ethylene cracking unit (cracking to produce ethylene and propylene). Here there may be a good use of hot water from the quenching of furnace effluents.

If the shell-and-tube type of condenser is used, the design method of Donahue[15] is particularly applicable. The high purity of the condensing propylene makes it possible to get excellent physical data for the computation of heat exchange.

Heat and material balances for this case are shown in Fig. 11 and Table 5. These are based upon 100 mol of feed.

The uniformity of the liquid-vapor ratio as an internal flow pattern, previously mentioned, is shown in units of pounds. Since the molecular weights of the stage vapor and the liquid are almost the same, the ratios in pounds or moles are almost the same. This will allow a check to be made on the validity of the use of a straight operating line.

The factor mentioned as the Badger F is the gross vapor velocity through the tower, in feet per second, divided by the square root of the specific volume of the flowing vapor, all at the temperature and pressure conditions existing at the point of flow. See design summary, Table 6.

Condensers and reboilers   The condensers may be water-cooled[15,24,30] bundles or in cooler areas even air-cooled.[24,70] The designs for either case are conventional and available.

The style of condensing system may vary. The usual elevated bundles are located closer to the ground in recent constructions, with flooded tubes. The accumulators are elevated enough to provide good pump suction conditions. In some plants the top product is removed several trays below the top tray, in order to reduce to some extent any light ends occurring in the overhead product. These go out with some gas vented at the accumulator, from which the reflux is circulated.

*Tower vapor, $[(ft/s) \div (sp. vol.)^{1/2}]$ at flowing conditions.

TABLE 5   Material Balance*
(Based on 100 mol of feed)

Composition	Mol wt	Mol %	mol/100 mol feed	lb 100 moles feed	lb/bbl	bbl/100 mol feed
			Feed			
C_3^{2-}	42.1	50.0	50.0	2105	182.7	11.52
C_3^+	44.1	50.0	50.0	2205	177.7	12.41
Tot./avg	43.1	100.0	100.0	4310	180.1	23.93
			Distillate Product			
C_3^{2-}	42.1	99.5	47.93	2018	182.7	11.044
C_3^+	44.1	0.5	0.241	10.6	177.7	0.060
Tot./avg	42.1	100.0	48.171	2028.6	182.7	11.104
			Bottoms Residue			
C_3^{2-}	42.1	4.0	2.073	87.3	182.7	0.48
C_3^+	44.1	96.0	49.76	2194	177.7	12.35
Tot./avg	44.0	100.0	51.833	2281.3	177.8	12.83

*See Fig. 11 for internal flow.
NOTE: The "fulcrum rule" is used to compute the amount of distillate and bottoms, by the inverse differences between the compositions of the feed and each product.

Thermosiphon reboilers are commonly used,[24,26,27,41,43] with the bottoms often removed separately from the base of the tower.

The flux rates experienced on the heat-exchange equipment are tabulated for large plants, that is, those producing 100 million pounds per year or more of propylene product.

The reboiler heat exchange may be expected to approach a maximum flux of approximately 14,000; some plants have exceeded 12,000 without loading the reboiler to find the maximum possible. This rate is based on bare surface with tubes having extended surface of 2.5 times the bare surface.

The condenser surface must take into account the seasonal changes in water temperature; so an overall coefficient of heat exchange (including a fouling factor of 0.001) of 140 + is possible. This would

TABLE 6   Summary of Distillation Data

Distillation: C_3 splitting, purity % = 99.5 C_3^{2-}, recovery % = 95.86

Stages	Actual	Min	Theoretical	Keys	Vol. (K)	Reflux (RR)
Rectifying	100	60	107	$C_3^=$	Variable	28.0
Stripping	35	23	37	C_3^+	Variable	$L/V = 1.0371$

Tower factor: top 0.82, Bottom 0.82 for 107% tray efficiency
Material balance, moles/100 moles feed (design)

Components	mol wt	Feed	Overhead	Bottom
Ethane and light		Trace		
Propylene	42.1	50.00	47.930	2.073
Propane	44.1	50.00	0.241	49.760
Butanes	72.1	(up to 4% tolerated without correction)		
Total		100.00	48.171	51.833

Operating pressure, lb/in²(abs): top 297, middle 304, bottom 312
Operating temp, °F: top 121, middle 132, bottom 143
Feed: temp approx 132°F, mol % vapor bubble point
Tray liquid, mol/h/stage: top approx 9000, bottom approx 9000
Trays (type): perforated valve
Tray space: 18 in throughout
Flow top: split
Flow bottom: split
Tower diam design: top 0.41 ft/s, base 0.47 ft/s
Condenser type: air-cooled or shell and tube water-cooled
Reboiler type: low-pressure steam or hot water

be for possibly four bundles, and two in series would be required, all based on a plant making over 100 million pounds per year of propylene. The flux that would be attainable for a 270 lb/in²(abs) operation with a 43°C (110°F) overhead, using 21°C (70°F) cooling water with the outlet at 43°C (110°F), would be about 3600 Btu/ft² of bare tube surface, with no extended surface used in this case. See Table 2.

Plant operation The operation is conventional, meaning that flow recorders and controllers are used on all the usual streams, including the reflux. Product purity is monitored, usually with chromatograph instrument methods, and the product is temporarily routed to high-pressure holding tanks before storage in the main areas. These may be underground caverns.

The product is usually given a wash and then a final drying by molecular-sieve methods, but the same effect could be obtained economically through a redistillation, removing the propylene from the base of the distillation tower, and the water would boil up overhead (because of the nonideal volatility of water in hydrocarbons).

Start-up operation This also is conventional in the sense that the equipment is pressure-tested after construction, cleaned by flushing with water, and finally purged, possibly using inert gas from a generator. The equipment is usually tested, using M. S. A. equipment to see if the oxygen has been properly displaced, before charging the equipment with feed mixture. In the course of a transfer of propane-propylene materials to fill equipment, care should be taken to avoid local freezing where the pressure drop is high in the release of the volatile material, and where possible the equipment should be dried as much as possible.

A complete operating directive is not possible here, but to complete this short description, it is necessary to add enough feed to prime the equipment so the flow of reflux can begin with the gradual addition of steam or the heat source at the reboiler. As the heat is added, attention should be given to the cooling water at the condenser, along with the pressure developing in the course of the start-up. There will be no evidence of a liquid layer being present, either in the base of the tower or in the reflux accumulator, until the pressure is near the 300 lb/in²(abs) operating condition, consistent with the temperature of the system.

Each plant may use its own method to bring the initial off-test products back into the system for recirculation. This is often accomplished through the installation of piping at the time of plant construction, so that recirculation is possible during the start-up periods.

The planned layout of distillation towers follows some general rules. Fired heaters are kept at least 50 ft away from the tower. Safety valves, now often installed in the vapor lines, may be discharged at some high elevation or even directly into the atmosphere in some isolated locations. It is more usual to discharge to a flare system some distance away.

Usually the pumps, towers, and exchangers are kept in separate lines, with room to withdraw and clean bundles during the shutdown periods. Switch gear is usually located in strategic positions in case of emergency shutdown. Similarly, when possible, important valves are also located where they may be used under emergency conditions.

Control rooms are located with regard to vision of the equipment and safety in case of emergency. A well-designed plant layout results when many conditions are taken into account.

Separation of Ethylbenzene from Styrene (or Vinylbenzene)

This separation is from a material balance that should be applicable to a two-component separation, but since small amounts of other components were present and a noticeable temperature change over the tower existed, a multicomponent distillation program was used for analysis. See Table 8.

Equilibrium data were prepared by using vapor pressures of the keys, ethylbenzene and styrene, in a plot of vapor pressure vs. the reciprocal temperature factor[85] which is 1/ (300 + °F). This is analogous to a reduced temperature, but the 300 value is more practical for correlation purposes. K values were then calculated for styrene using the vapor-pressure ratios with actual K values on ethylbenzene as a reference component, the latter from Lenoir.[5] The K values for xylene and toluene were available (Lenoir[5]); and the others, which are less important, were approximated from boiling points.

With these so approximated equilibrium K values a plausible stage efficiency may be obtained and at least a model of the operation is simulated, in an orderly fashion.

The liquid heats were approximated, since they do not greatly affect the heat balance. Latent heats were taken from Driesbach[16] and developed according to the method of Watson.[62,93]

The data were used as computer input. Three values each of volatility, liquid enthalpy, and vapor enthalpy are customarily curve-fitted for use within the programs normally available. In this case it would have been more convenient if the components present had been available in the systems included in most computer storage banks.

Runs show (see Fig. 12) that the best feed location in a 50-theoretical-stage tower would be very close to the thirty-fifth stage above the reboiler. It is expected from other

TABLE 7 Summary of Distillation Data

Distillation: C_3 splitting, purity % = 99.5 $C_3^=$, recovery % =82.3

Stages	Actual	Min	Theoretical	Keys	Vol. (K)	Reflux (RR)
Rectifying	125	$N = 110$	133	$C_3^=$		19.2
Stripping	17	$\alpha_{av} = 1.08$	18	C_3^+		$L/V = 1.0554$

Tower factor: top 0.82, bottom approx 0.82
Material balance, moles/100 moles feed (operation)

Components	mol wt	Feed	Overhead	Bottom
Ethane and light		Trace		
Propylene	42.1	57.00	46.90	10.10
Propane	44.1	43.00	0.24	42.76
Butanes	72.1	(Up to 4%)		
Total		100.00	47.14	52.86

Operating pressure, lb/in^2(abs): top 270, middle 277, bottom 285
Operating temp, °F: top 110°, middle 121°, bottom 132°
Feed: Temp approx 120°F, mol % vapor at bubble point
Tray Liquid, mol/h/stage: top 8460, bottom 8954
Trays (type): perforated valve trays
Tray space 18 in top, 24 in low
Flow top: split
Flow bottom: split
Tower diam: top 10 ft 0 in, base 10 ft 0 in
Condenser type: shell and tube, water (72°F in), flux 2380
Reboiler type: shell and tube, steam heated, low pressure, flux 11,900

TABLE 8 Summary of Distillation Data

Distillation (ethylbenzene/styrene): purity % 86.0/99.5; recovery % = 99.9/95.5

Stages	Actual	Min	Theoretical	Keys	Vol. (K)	Reflux (RR)
Rectifying	23	$N = 28.5$	15	ethylbenzene/styrene	1.014/0.665	8.22
Stripping	49	$\alpha_{av} = 1.478$	36	ethylbenzene/styrene	1.436/1.003	$L/V = 1.091$

Tower factor: top 2.12, bottom 1.19
Material balance, moles/100 moles feed (operation)

Components	mol wt	Feed	Overhead	Bottom
Toluene	92	0.57	0.57	0.00
270°F BP	106	1.71	1.71	0.00
Ethylbenzene	106	45.37	45.36	0.01425
Para and				
metaxylenes	106	2.89	2.87	0.0218
Styrene	104	49.33	2.21	47.12
Cumene	120	0.02	0.00	0.02
Methyl styrene	118	0.12	0.00	0.12
Total		100.00	52.72	47.28

Operating pressure, lb/in^2(abs): top 0.90, middle 2.02, bottom 3.10
Operating temp, °F: top, 126, middle 181, bottom 204
Feed temp 163°F, mol % vapor 1.0%
Tray liquid, mol/h/stage: top 1000, bottom 1263
Trays (type): valve trays
Tray space: 24 in
Flow top: split
Flow bottom: split
Tower diam: top 13 ft 0 in, base 13 ft 0 in
Condenser type: shell and tube water-cooled
Reboiler type: steam heated

distillations of similar aromatic materials that the stage efficiency will be near 70%. This is overall proportionate to using a 72-stage tower with a feed location at the forty-ninth stage above the reboiler, thus having 49 stripping and 23 rectifying trays.

It is evident from the curves mentioned above that, by arbitrarily assuming a feed location halfway, the twenty-fifth stage location would result in poorer separation. The optimum feed location may be determined for some given separation by observing the change in slope at the stages just above and just below the feed stage computed, as shown

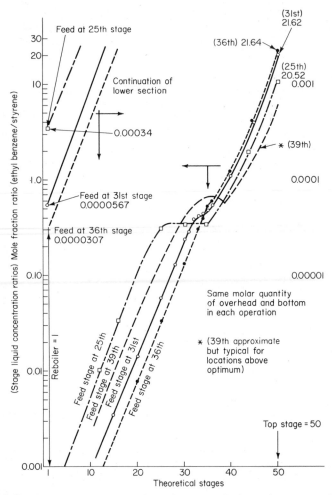

Fig. 12 Ethylbenzene-styrene separations obtained at various feed-stage locations as calculated at constant (8.22) reflux ratio.

in Fig. 12. Full details of the method used, which may be applied to either two-component or multicomponent cases, is described in Ref. 84.

It may be seen from the key-component ratios (of their mole fractions in the overhead and the bottoms) that the feed location at the thirty-sixth stage produces the highest ratio of ethylbenzene to xylene in the top product, and also the lowest similar ratio of these components in the bottoms, of any of the feed positions shown. The thirty-sixth-stage feed (position curve) also approximates a straight line which is indicative of the optimum feed location. Further runs at feed positions above the thirty-sixth would show less separation, forming curves of typical shape indicated in the diagram but not calculated for this particular problem.

The assumption of a feed position at the twenty-fifth stage may seem plausible at the start, since almost equal amounts of the key components are present in the feed. This does not take into account the objective of a very rich styrene desired from the bottoms, however, and a more tolerable amount of impurity acceptable in the overhead product; hence more lower stages are made available, 35 stripping stages. The problem is also influenced by having a lower relative volatility in the stripping section. The influence of the reflux ratio used could also modify the stage position. Thus no answer based on judgment is possible; only calculation remains to answer the problem.

Aromatic distillations pose many problems, particularly when the feed contains fair amounts of paraffinic components. Certain equilibrium conditions exist that are best explained in terms of design precautions by Lenoir.[49]

Separation of Benzene and Toluene

A nitration-grade benzene requires a high-purity separation of benzene from toluene, and commonly a relatively benzene-free toluene is also required. This two-component separation has a history of problems that stem from both mechanical and process occurrences that are of value to use in design examples.

In past designs it was found expedient to pressurize the reflux accumulator with a blanket of flue gas, to assure a subcooled liquid reflux, below about 104°C (220°F), in order to avoid pumping difficulties. This led to the removal of the overhead benzene as a

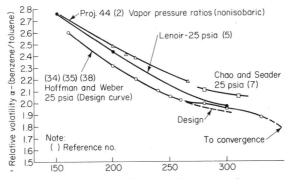

Fig. 13 Relative volatility of benzene-toluene system from various sources.

side stream, often taken from the tray just below the top tray, to permit the product to be freed from the inert gas, even though a tray of separation was sacrificed. (The mechanical implications will be covered later.)

A process difficulty caused by contamination of the feed by an outside component which was of a volatility almost identical to benzene and came from the same source (re-forming of petroleum fractions) also led to extensive analyses of tray liquid composition, over several intervals from the top to the bottom of the tower (during operation, of course).

Data of this sort were formerly computed by hand calculation, called "tray-by-tray" computations, using the principle of making heat and material balances around each stage to determine the internal flow rates of liquid and vapor. This was combined with the equilibrium calculations that verify the stage temperatures, and in all was a very laborious procedure, especially without modern computers.

Multicomponent distillation programs are not adaptable for computer calculation of problems of this nature (broken into parts of the tower operation) but will give overall computation balances, if they converge at all, on only two-component applications.

In this case the Smoker equation was used in five sections of computation to obtain the stages required, as is necessarily done by the procedures demonstrated (see Figs. 5 and 6).

Figure 13 gives the sources of the volatility values, which are the basis of the stage efficiencies calculated. The curve marked "design" was used; the other curves show variations in this type of data. The overall operation is summarized (see Table 10).

This distillation was investigated in five separate sections of the tower of 45 actual stages or 43 "trays" and separate stage efficiencies (Table 9). The two noticeable places where poor stage efficiencies exist are explainable. They are at the very top of the tower and also just above the feed inlet.

At the top of the tower a very cold reflux, 66°C (150°F), enters the top plate, which is being held at 96°C (204°F). This quenching effect sets up local recycle conditions that may be interpreted as overloading the top tray and resulting in poor stage efficiency in this area. Similarly, poor tray efficiency is caused just above the feed stage because of the vaporized feed entering and causing local recycling of higher-boiling material moving upward from the feed tray, which must eventually return to the base of the tower and similarly result in overload at the trays nearby.

Some of the practical reasons for this adverse operation stem from underdesign of the reflux pump capacity, justified by a cold reflux condition. Also mechanical difficulty may possibly be avoided if the reflux pump can operate in a temperature low enough to make the pump seals operate with less severe temperature conditions.

TABLE 9   Separation of Benzene and Toluene

Calculation of theoretical stages per section, in the 43-tray column. Reboiler is considered equal to 0.5, or half of an actual stage. Bottom tray is tray 1.

Tower section at trays	Actual trays per section	Theoretical stages/section	Tray efficiency, %
40–43, incl.	4	2.192 (calc.)	43.8
37–39, incl.	3	2.160 (calc.)	72.0
30–36, incl.	7	5.028 (calc.)	71.8
23–29, incl.	7	3.195 (calc.)	45.6
Feed above 22d:			
1–22, incl.	22	15.9 (calc.)	70.7
Reboiler = ½ tray	½		

NOTE: Total overhead condensation did not justify rectification credit as a stage or fraction thereof.

The theoretical stages for each section were computed by using the Smoker equation, as shown in detail for the propylene-propane separation. Here the variation of the operating-line slopes was taken into account by first computing each section separately, after making a heat balance to determine the internal reflux, and then the slope of the operating line from the relation $m = R/(R + 1)$.

Separation of Ethylbenzene and Metaxylene

This distillation is also a case where a multicomponent computer program may be used. There is a reasonably large temperature difference between the top and base of the tower, and thus K values such as temperature- and pressure-dependent volatility may be used. The predominance of ethylbenzene and metaxylene gives a two-phase condition, which is easily calculated because of the relatively constant alpha values, 1.088 (rectifying) and 1.081 (stripping).

The multicomponent calculation requires several trial calculations to determine the feed location and the number of total stages required, but it gives heat balances and all tower temperatures, including pressures from an assumed pressure drop over the tower. The Smoker method has the advantage of giving very exact feed positions, implied by the separate calculation of rectifying and stripping stages, but it requires separate calculation of heat balances and temperature calculations. These additional calculations are minor, and a saving of computer costs can be made by using the Smoker method to get the feed position.

In Fig. 14 the sixty-eighth and forty-fifth stage positions of the feed are shown in contrast to the twenty-first feed position, the latter determined by the Smoker method. It may be seen that the forty-eighth feed position, which was the feed stage arbitrarily chosen, uses an excess of stages.

This distillation has some aspects of interest as an example of some extreme situations not commonly encountered in both a physical operation and the calculation of its performance. The summary sheet employed for this separation (Table 9) shows that the plant in this case will use 300 actual stages, with a reflux ratio of 46.5, and with relative-volatility values ranging from 1.072 to 1.096 at the base and top, respectively. It is also somewhat extreme to require the high purity and low recovery of 99.99% pure and only 37.7% as acceptable recovery. The stage efficiency, based on using the optimum feed location at the twenty-first theoretical stage above the reboiler is calculated as 69%, which is near the usual overall efficiency regarded as about 70% for many aromatic separations.

Three columns are usually required to keep the height of the construction within reasonable limits, just as in the propylene-propane splitter described.

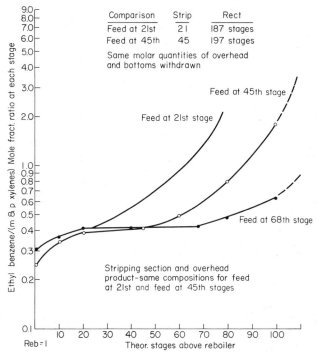

Fig. 14 Analysis of ethylbenzene-xylene separations at various feed-stage locations and at constant (45.8) reflux ratio.

The close separation results in a temperature change of only $-22.8°C$ ($41°F$) over the tower, or $-0.11°C$ ($0.19°F$) per theoretical stage. This temperature differential is quite low but is a little higher than that of the propylene-propane splitter, which has a differential of $-0.08°C$ ($0.14°F$) per theoretical stage. The splitter could not be run on the multicomponent programs, using K values based on temperature- and pressure-directed computation methods, and thus the Smoker equation is used. This failure in the use of the multicomponent type of program for the splitter led to the recommendation that a minimum of $-0.11°C$ ($0.20°F$) be observed for successful application of two-component separations to the usual computer method. In this separation, however, the presence of other components in more than trace amounts may have contributed to produce mixtures of which the stage-liquid bubble points could be calculated according to the procedure within the program, not necessarily with the relatively slight advantage of the temperature differential involved.

Plants built around 1960 did not have the benefit of fine computer techniques, and operations were often carried out at improper feed-stage locations, sometimes because of changes in feed composition after design. The result of this condition is exhibited in Fig. 14. With proper computer facilities a series of curves of optimum feed locations could be calculated for various feed compositions, if extra feed nozzles and valves were provided in design.

Tables 10 to 12 provide enough variety for most operations for qualitative reference when starting a design project.

TABLE 10   Summary of Distillation Data

Distillation: benzene/toluene for 5°C Freezing point, purity 99.923%

Stages	Actual	Min	Theoretical	Keys	Vol. (K)	Reflux (RR)
Rectifying	22	$N = 16.7$	12.2	Benzene/toluene	See curve	(hot) = 9.1
Stripping	23	$\alpha_{av} = 2.15$	15.9	Benzene/toluene	See curve	$L/V = 1.195$

Tower factor: top 1.23, bottom 1.37
Material balance, moles/100 moles feed (operation)

Components	mol wt	Feed	Overhead	Bottom
Saturates	78	0.030	0.022	0.008
Benzene	78.1	36.28	35.01	1.27
Toluene	92.1	63.69	0.0051	63.685
Totals		100.00	35.037	64.96

Operating pressure, lb/in²(abs): top 23.0, middle 25.0, bottom 27.2
Operating temp, °F: top 204, middle 230, bottom 285
Feed [100 lb/in²(gage)]: temp 265°F, mol % vapor 25
Tray liquid, mol/h/stage: top 1270, bottom 1800
Trays/packing (type): caps 4 in diam
Tray space: 24 in
Flow top: single
Flow bottom: single
Tower diam: top 8 ft 0 in, base 8 ft 0 in
Condenser type: shell and tube water-cooled
Reboiler type: shell and tube steam-heated

NOTE: Reflux is 150°F from a flue gas pressurized accumulator, with cold reflux ratio = 7.6; hot = 9.1 (ratio). Stage liquid rate is for feed at 457.3 moles/h.

Separation of High-Boiling Two-Component Mixtures

In many situations a practical distillation design is possible because the mixture may be regarded as consisting of two components. The relative volatility in this case may be determined from an Oldershaw distillation run and the component distribution from a true-boiling-point distillation. The results are fitted into a McCabe-Thiele diagram as shown in Fig. 1 in order to complete the design. In many processes the result of the reactions reduces the feed to a simple mixture of product and residue which may be dealt with as a two-component mixture.

In the refining industry, catalytic polymerization, naphtha re-forming, and demethylization to produce naphthalene may lead to this condition. Among chemical-plant separations the procedure may be of value in the separation of fatty acids, higher alcohols, and many other products.

The following example, typical of this procedure, illustrates how a polymer product is removed from a residue produced in the reaction and how these components, with properties and compositions as they occur in the mixture, are separated in a distillation plant.

EXAMPLE

No appreciable amount of middle-boiling substances was present between the polymer product and the residue, and although the residue was of indefinite composition, it could be regarded as a pseudo-component. A satisfactory separation of the feed was obtained from an Oldershaw test distillation[88] using five rectifying and five stripping stages and a reflux ratio of 1.0, the feed position being a first approximation on this trial. Other Oldershaw runs could be made if desired to confirm this as being acceptable.

The Oldershaw still has a glass column provided with a heated, silvered jacket to reduce heat losses, and operation is carried out as a continuous distillation. Batch distillation on such mixtures is also useful, as by means of a Peters true-boiling-point apparatus, described by Nelson.[86] This apparatus distills the material under inspection into close-boiling fractions, and then the physical properties for

TABLE 11   Summary of Distillation Data

Distillation: ethylbenzene-metaxylene, purity = 99.99/recovery = 37.8%

Stages	Actual	Min	Theoretical	Keys	Vol. (K)	Reflux (RR)
Rectifying	270	N = 144	187	ethylbenzene/metaxylene	Lenoir*	45.83
Stripping	30	α_{av} = 1.08	21	ethylbenzene/metaxylene	Lenoir*	L/V = 1.175

Tower factor: top 1.30, bottom 1.30
Material balance, moles/1000 moles feed (design)

Components	mol wt	Feed	Overhead	Bottom
Ethylbenzene	106.2	269.46	101.89	167.57
Paraxylene	106.2	15.0000	0.0013	14.9987
Metaxylene	106.2	659.64	0.00637	659.63363
Orthoxylene	106.2	55.90	0.00	55.90
Totals		1000.00	101.89767	898.10233

Operating pressure, lb/in²(abs): top 22.0, middle 34.55, bottom 47.0
Operating temp, °F: top 298, middle 339, bottom 374
Feed: temp 340°F, mol % vapor at bubble point
Tray liquid, mol/h/stage: top 4740, bottom 6080
Trays/packing (type): valve trays
Tray space: 24 in min
Flow top: split
Flow bottom: split
Tower diam: top 16 ft 0 in, base 14 ft 6 in
Condenser type: air-cooled is possible.
Reboiler type: shell and tube/heating oil or high-pressure steam

*Average relative volatilities for rectifying and stripping sections are 1.088 and 1.081, respectively, and are relatively constant.

TABLE 12   Summary of Distillation Data

Distillation: C_{10} alcohol (bp 460°F/residue 555°F bp)

Stages	Actual	Min	Theoretical	Keys	Vol. (K)	Reflux (RR)
Rectifying	12	N = 7.08	5.18	C_{10} alcohol	α = 4.0	1.0
Stripping	12	α_{av} = 3.9	4.97	residue	α = 3.8	L/V = 1.26

Tower factor: top 1.21, bottom 1.47
Material balance, moles/100 moles feed

Components	mol wt	Feed	Overhead	Bottom
Decyl alcohol	158	69.20	68.73	0.47
Residue (555°)	374	30.80	0.29	30.51
Total		100.00	69.02	30.98

Oldershaw distillation conditions:
    Overhead: 295°F at 48 mm, equivalent to 460°F atm bp
    Bottoms: 384°F at 56 mm, equivalent to 555°F atm bp
    Bottoms contained 2% overhead; overhead contained a trace of bottoms
Operating pressure lb/in²(abs): top 1.16, middle 1.54, bottom 1.93
Operating temp, °F: top 280, bottom 370
Feed: 19.38 moles/h, temp 300°F, mol % vapor 0%
Tray liquid, moles/h/stage: top 13.40, bottom 23.20
Trays/packing (type) perforated valve
Tray space: 20 in
Flow top: single
Flow bottom: single
Tower diam: top 3 ft 0 in, base 3 ft 0 in
Condenser type: shell and tube
Reboiler type: shell and tube on 300 lb/in² steam

each fraction, such as density and viscosity, are determined for the design uses. Viscosity values are especially important in vacuum-distillation design, as low temperatures used here may result in poor stage efficiencies if viscosity is too high in the operation.

The first need is to approximate the relative-volatility value that will fit into a McCabe-Thiele diagram first using a constant alpha curve, but with exact feed, product and bottoms, and also drawing the operating lines for rectifying and stripping sections. It is not difficult to do this at fairly high alpha values, but it may be necessary to enlarge the scale as shown (see Fig. 1a). This type of calculation assumes at the start that the relative volatility is a constant value, which is not generally true, but more Oldershaw tests, run at other cut points and different feed concentrations, will reveal this, and the equilibrium curve can then be modified accordingly.

The analysis of these data must include the consideration of the possibility of the efficiency of the Oldershaw column stages and whether their calibration by distillation with other materials has been properly tested. The five stages indicated for each section are taken as stages of 100% efficiency, which is theoretical stages at the reflux stated. If this was an Oldershaw test using separate plates in the column and five plates of lower efficiency were counted as theoretical stages, the design would still be safe but probably too conservative.

The summary of the data for this distillation is shown in Table 12 and Fig. 1a and b. This comparison is made so that the Smoker method may be examined directly with the McCabe-Thiele diagram in order to observe how the transfer of coordinates is made. Note that even on a five-stage drawing it is difficult to distinguish between relative volatilities of 4.0 and 3.8. With the use of the Smoker equation there is no particular limit to the accuracy obtained, and stages may be counted in fractional parts with no difficulty.

EQUILIBRIUM AND ENTHALPY DATA AND SOURCES

Equilibrium-Data Sources

Equilibrium-data requirements for distillation designs extend to a large number of possible compounds and mixtures, with the possibility of many ranges of temperature and pressure adding to the complexity of these demands. This often results in a search for some available data and the use of prediction methods to extend these data. It is seldom possible to make extensive researches on a system before design, but occasionally over the years an organization may aquire its own source of data in this respect.

Either the equilibrium data used must be kept within the scope of the correlation, or if used on an approximate basis, the resulting design should be carefully examined in terms of the technical and economic results of possible data inaccuracy.

Valuable sources of equilibrium data are quite scattered. A summary of the literature sources could start with methods applicable to hydrocarbons of many types in multicomponent mixtures.[3,5] Some of these include hydrogen, common gases, and water. They correlate vapor-liquid equilibrium ratios (K values) over various ranges of temperature and pressure. As mentioned in the paragraph above, the equilibrium data derived from these methods must not violate the restrictions of the method. The restrictions of a commonly accepted method, that of Chao and Seader,[7] have been given special treatment in an article of Lenoir.[49] In general all methods are restricted to conditions that do not approach critical-state pressures and temperatures too closely or violate certain conditions where the mixture contains a preponderance of one component such as hydrogen or other single-component entities. These methods (listed below) may be programmed into the computers, but two, Chao and Seader as well as Grayson and Streed, must have computer facilities. These, unlike the others listed below, have given equilibrium ratios that are "directed" by the composition of the particular mixture under consideration. The other methods use convergence pressure as a means of computation, which is practical for hand-computation procedures.

Other important sources of equilibrium data which include to a large extent the "nonideal" mixtures are not as easily calculated. These are limited to some degree in the number of components present in the mixture to be separated. These data seldom deal with mixtures containing more than three components, and the possibility of finding exact data in the range of composition or pressure-temperature conditions for a design is not good. Nevertheless it is necessary to obtain approximate answers in preliminary designs, and no particular source should ever be ignored.

Another source of data includes pure-component properties for cases where approximations are possible when ideal mixtures are formed (or may be assumed to form) or when the components of the mixture are similar. See Table 13.

TABLE 13   Equilibrium Constants (K) And Enthalpy

Units T,°F	K, equilibrium constant P, lb/in²(abs)				Enthalpy Btu/lb	
	14.7	100	200	300	H_{liq}†	H_{vap}
Hydrogen, H_2 in saturated C_8 to C_{12}'s*						
100	1300	190	95	63		1895
200	970	140	70	47		2235
300	730	105	52	35		2570
400		(Note declining K values with temp.)				
Methane, 5000 lb/in²(abs) conv.*						
100	180	27	14.5	9.5	275	355
200	215	33	17.0	11.5	345	415
300	245	34.5	17.8	12.0	420	475
400	270	37	18.8	12.9	500	543
Ethylene, 5000 lb/in²(abs) conv.*						
100	49	8.0	4.3	3.0	222	315
200	82	14	7.4	5.3	265	355
300	108	19.8	11.0	7.8	315	400
400	135	25.5	14.1	9.8	365	450
Ethane, 5000 lb/in²(abs) conv.*						
100	37	5.7	3.0	2.2	240	335
200	87	10.2	5.3	3.7	285	383
300	92	14.3	7.4	5.2	340	435
400	105	17.2	9.2	6.4	405	492
Propylene, 5000 lb/in²(abs) conv.*						
100	13.2	2.1	1.17	0.85	170	320
200	33.0	4.8	2.65	1.95	260	360
300	50.0	7.9	4.35	3.15	320	405
400	65.0	11.1	6.20	4.50	370	455
Propane, 5000 lb/in²(abs) conv.*						
100	10.6	1.75	0.97	0.69	170	320
200	24.5	4.05	2.25	1.62	255	363
300	41.0	7.10	3.90	2.85	330	413
400	73.0	9.10	5.00	3.65	390	470
Isobutane, 5000 lb/in²(abs) conv.*						
100	4.50	0.77	0.44	0.33	150	295
200	13.0	2.30	1.28	0.92	217	337
300	24.5	4.40	2.45	1.72	315	388
400	34.5	5.90	3.40	2.50	370	445
n-Butane, 5000 lb/in²(abs) conv.*						
100	3.5	0.56	0.32	0.24	160	312
200	9.9	1.70	0.96	0.70	225	355
300	20.5	3.55	2.00	1.50	305	407
400	30.0	5.30	3.00	2.22	385	465
Isopentane, 5000 lb/in²(abs) conv.*						
100	1.3	0.23	0.14	0.108	148	295
200	5.1	0.90	0.52	0.39	212	340
300	12.0	2.15	1.28	0.93	280	392
400	21.0	3.55	2.05	1.55	380	437
n-Pentane, 5000 lb/in²(abs) conv.*						
100	1.06	0.192	0.112	0.087	152	307
200	4.15	0.76	0.43	0.32	215	352
300	10.60	1.92	1.12	0.85	288	400
400	18.80	3.40	1.88	1.42	390	457

TABLE 13   Equilibrium Constants (K) and Enthalpy (Continued)

Units T,°F	K, equilibrium constant P, lb/in²(abs)				Enthalpy Btu/lb	
	14.7	100	200	300	H_{liq}†	H_{vap}
Hexane, 5000 lb/in²(abs) conv.*						
100	0.330	0.066	0.040	0.031	145	300
200	1.90	0.330	0.195	0.150	207	342
300	5.60	1.020	0.600	0.460	275	394
400	12.00	2.070	1.21	0.920	350	450
Heptane, 5000 lb/in²(abs) conv.*						
100	0.116	0.0215	0.0135	0.0110	140	295
200	0.820	0.1450	0.0860	0.0660	200	338
300	3.500	0.5600	0.3300	0.240	265	387
400	7.300	1.3400	0.8100	0.740	340	445
Octane, 5000 lb/in²(abs) conv.*						
100	0.040	0.0080	0.0051	0.0042	138	288
200	0.340	0.0620	0.0370	0.0290	195	333
300	1.800	0.2950	0.1650	0.1220	258	380
400	4.8500	0.8600	0.4900	0.3600	330	440
Nonane, 5000 lb/in²(abs) conv., 303°F bp*						
100	0.0142	0.0029	0.00195	0.0017	48	200
200	0.1600	0.0280	0.01650	0.0132	102	240
300	0.9200	0.1550	0.08900	0.0670	162	288
400	3.1000	0.5200	0.30000	0.2400	227	340
Decane, 5000 lb/in²(abs) conv., 345°F bp*						
100	0.0055	0.00115	0.00080	0.00072	46	196
200	0.0075	0.01260	0.00770	0.00610	99	238
300	0.4900	0.08400	0.04900	0.03600	157	284
400	1.9500	0.31000	0.17300	0.13000	221	335

	15	20	30	50	H_{liq}	H_{vap}
Benzene, 300 lb/in²(abs) conv.‡						
100	0.23	0.180	0.122	0.082	107	290
150	0.64	0.470	0.340	0.222	130	303
200	1.47	1.10	0.760	0.490	152	318
250	3.00	2.20	1.53	0.960	177	335
Toluene, 300 lb/in²(abs) conv.‡						
100	0.076	0.0580	0.0415	0.029	102	280
150	0.245	0.1820	0.1300	0.088	123	295
200	0.630	0.470	0.3250	0.212	147	312
250	1.400	1.040	0.7200	0.455	172	330
Ethylbenzene, 300 lb/in²(abs) conv.*						
100	0.0285	0.0211	0.0152	0.0108	37	206
200	0.2850	0.2180	0.1510	0.1020	83	239
300	1.4200	1.0600	0.7300	0.4750	135	277
400	4.4000	3.3000	2.2300	1.4300	194	320
Meta- or paraxylene, 300 lb/in²(abs) conv. (enthalpy mp)‡						
100	0.0240	0.0188	0.0132	0.0093	40	210
200	0.2580	0.1960	0.1380	0.0910	83	242
300	1.3200	0.9700	0.6800	0.4350	135	276
400	4.1000	3.1000	2.1000	1.3200	194	320
Orthoxylene, 300 lb/in²(abs) conv.‡						
100	0.0185	0.0146	0.0104	0.0073	40	214
200	0.2120	0.164	0.1150	0.0770	86	248
300	1.1300	0.860	0.6000	0.3830	136	286
400	3.6500	2.800	1.9000	1.1900	195	328

*NGAA.
†$H_{liq}^{(-200°F)}$, except for nonane and decane ($H_{liq}^{(rF)}$).
‡Lenoir.

Equilibrium Systems

(Vapor-Liquid Equilibrium Ratios, K Values) The following correlations with their characteristics noted apply to multicomponent hydrocarbon mixtures; some also include hydrogen, simple gases, and water as extra components.

Chao and Seader Method[7]   Requires computer program, contains 53 hydrocarbons, 10 nonhydrocarbons, which include hydrogen and water. For restrictions see Ref. 49.

Grayson and Streed Method[31]   Requires computer program. This method extends the temperature range of the Chao and Seader method, deals with temperatures that exceed 316°C (600°F); 44 hydrocarbon components, hydrogen, and 22 petroleum fractions are available as component entities.

DePriester Equilibrium K Data[89]   Easily available in Ref. 96. Contains 14 hydrocarbons, suitable for approximate use. Subcritical conditions required in application without expression of how close to critical is allowed.

Winn Vapor-Liquid Equilibria Charts[77]   This method gives K values from a series of nomographs at various convergence pressures. Fifteen hydrocarbons are included with 17 special systems that include hydrogen and nonhydrocarbon components as well as correlations on a molecular-weight basis.

K_{10} Charts by Cajander, Hipkin, and Lenoir[5]   This method includes 21 hydrocarbons (aliphatic), 10 petroleum fractions from 177 to 427°C (350 to 800°F) boiling ranges, 8 naphthene-type hydrocarbons, 7 aromatic-type hydrocarbons, 6 acetylenes, and 17 unsaturated hydrocarbons. For hydrogen-paraffin mixtures see Refs. 48 and 45. This method uses the convergence-pressure concept for determination of the K equilibrium ratios.

Natural Gasoline Supply Men's Association   "Engineering Data Book"[58] Hydrogen and two nonhydrocarbons; covers a wide range of convergence-pressure parameters. Special systems with hydrogen sulfide are included.

This summary does not cover all possible correlations for vapor-liquid equilibrium ratios that would be useful under the many applications required.

Distillation Equilibrium Data[8]   The first compilation; includes 176 systems, of which 147 are two-component mixtures. These data are reported as vapor and liquid compositions, measured at the reported temperature-pressure conditions.

Vapor-Liquid Equilibrium Data[9]   The second compilation; brings the number of systems reported up to 466, in completion of the distillation equilibrium data above. These data are reported as vapor and liquid compositions, measured at reported temperature-pressure conditions.

Azeotropic Data for Equilibrium Sources[40]   These data may be used to develop vapor-liquid equilibrium relations based on one analysis. They may be used to determine whether an azeotrope might occur in a mixture, and also report on nonazeotropic mixtures that may be found. The "Chemical Engineers' Handbook"[61,62,96] also includes most of the data from Ref. 40 mentioned above. Reference 40 brings the reported number of systems up to 14,900.

Pure-Component Data[2,29,16]   The first reference, commonly known as the American Petroleum Institute Research Project 44, contains thermodynamic data on 935 hydrocarbons and related compounds of interest in the petroleum industry. There are vapor-pressure compilations of immediate value to equilibrium needs, and heat values necessary for design; see below. The other references also are usable sources of data.

Enthalpy Data Sources,[2,56] **Enthalpy Data**   Usual design procedures in distillation permit the use of pure-component enthalpy values, except in rare cases where the mixing causes significant release or absorption of heat. It is customary to construct enthalpy curves with the zero-enthalpy condition at some sufficiently low temperature and in a liquid-state condition. Computation is more easily examined and heat values are less confusing in the course of practical design of equipment later.

The enthalpy data may be in the convenient curve form as in Ref. 56, or it may be necessary to make them up in the form suggested from the separate sources of heat values available, through the subtraction, say, of heats of vaporization from total heats of the vapor, as the case may be.

Section 13 of Ref. 96 is a good source of data that is easily available. Nonhydrocarbon enthalpy data, particularly if extreme nonideal conditions exist, will require special data for the system considered to take into account mixing effects. The effects of pressure on enthalpy for these systems would be very small at the pressures generally used in distillation.

Hydrocarbons may require some consideration for pressure effects on enthalpy[56,64] and possible mixture effects, although the latter are not significant in practical distillation design.

Equilibrium data are given in Table 13 in an abbreviated form suitable for preliminary investigations often made for economic evaluations. Enthalpy values have been placed directly in the table to provide convenient access if this system is placed in a computer where three or four points are sufficient.

Interpolation of these data is best done by plotting curves for each component using the relation of $1/(°F + 300)$ at constant pressures, and using the product of the K value times the pressure, at constant temperature. The latter is referred to as the $K\pi$ plot. The range chosen was based on the most likely temperatures and pressures used in the distillations anticipated.

Equilibrium data for nonideal binary mixtures are commonly expressed in the form of binary constants for use in equations to obtain the activity coefficients at various mixture concentrations. Table 14 is a common example of such a source. References for other sources are included later.

METHODS FOR ESTIMATING VOLATILITY IN BINARY SYSTEMS

Engineering Approach

The use of activity coefficients to express the volatility properties of binary systems that exhibit nonideal vapor-liquid behavior involves a number of equations, from which the most suitable may be selected according to the character of the nonideal system and the accuracy expected of the design.

Activity coefficients may be regarded as correction factors that may be applied to ideal conditions to obtain "real" system properties under proper temperature and pressure conditions. They are also of use, when applied, to visualize the nature of any physical variations that may occur in a system. A pressure change in a real system may be shown to have a tendency to *increase* above that produced when an ideal state of mixing takes place, as the activity coefficient becomes numerically greater than unity. This is true for either vapor- or liquid-phase activity coefficients.

The activity coefficients predicted from the following equations are for the effects of nonideal mixing in binary systems, these being the systems of mixtures of widely dissimilar components where volatility conditions are strongly affected by composition, especially by liquid-phase composition. The effect of the mixing of components of the vapor phase is considered negligible; so the vapor-phase activity coefficient may be considered unity. It is fortunate that near-atmospheric or generally low pressures are acceptable generally in the distillation of these systems, to permit an uncorrected or assumed ideal vapor state.

The following equations are among the most commonly used. All are based upon thermodynamic principles that an activity coefficient can be related to the Gibbs energy of mixing, and the Van Laar and Margules forms of these are derivable from the Gibbs equations, as shown by Wohl.[78,79] Many excellent publications[86,90,91] are available to justify the methods and equations shown.

These are:
1. Van Laar equations (Carlson and Colburn forms)
2. Margules (three- and four-suffix forms)
3. Wilson equation
4. Nonrandom two-liquid equation
5. Analytical solution of groups method

The first four methods are for use with data specific to the components of the system of interest, while the fifth method is an attempt to generalize all volatility effects from the chemical structure of the mixed components and their molecular "size" contribution to the activity coefficients for the mixture. The principles of this approach were developed by Derr and Deal,[95] from previous work on group contribution by Wilson, Pierotti, and Deal and Derr.[93,94] This is the most abstract method and would be used only for approximate work, although its usefulness can be acknowledged. This method would not be used if specific data for the system under consideration were available.

TABLE 14   Binary Constants for Equations for Activity Coefficients*

NOTE: These constants were obtained for the Van Laar equations but can be utilized in the Margules equations where the ratio of A_{1-2} to A_{2-1} is within the range of about 0.75 to 1.3.

Where the data cover a rather wide range of temperature, the equations can be considered only approximate. The temperature and pressure ranges are indicated with the first and last numbers referring to the pure components; if there is a constant-boiling mixture, its temperature is given between the other two.

The mixtures are given with the lower-boiling component stated first, and the temperatures and constants apply in the same order. Pressure is 760 mmHg except where noted.

Mixture	Temp, °C	A_{1-2}	A_{2-1}	Ref.
Acetaldehyde-ethanol	19.8–78.2	−0.10	−0.20	28
Acetaldehyde-water	19.8–100	0.69	0.78	6, 36, 41
Acetone-benzene	56.1–80.1	0.176	0.176	49, 51
Acetone-methanol	56.1–55.5–64.6	0.243	0.243	3, 18, 39
Acetone-water	56.1–100	0.89	0.65	3, 8, 18, 57
Acetone-water*	25	0.82	0.72	2, 52
Benzene-isopropanol	80.1–71.9–82.3	0.591	0.845	38
n-Butane-furfural†	37.8	1.10	1.26	34
	51.7	1.05	1.17	
	66.6	1.00	1.11	
	93.3	0.91	0.98	
Butanol-butyl acetate	117.7–116.6–126.1	0.22	0.24	5
Butene-l-furfural†	37.8	0.84	1.03	34
	51.7	0.80	0.99	
	66.6	0.76	0.95	
	93.3	0.70	0.90	
Carbon disulfide-acetone	46.3–39.5–56.1	0.556	0.778	19, 43, 59
Carbon disulfide-carbon tetrachloride	46.3–76.7	0.10	0.07	40, 42, 43
Carbon tetrachloride-benzene	76.4–80.2	0.052	0.046	45
Carbon tetrachloride-ethylene dichloride	76.4–74.5–83.5	0.334	0.258	25, 58
Ethanol-benzene	78.3–67.0–80.1	0.845	0.699	15, 29, 53
Ethanol-cyclohexane	78.3–66.3–80.8	0.913	0.751	37, 54
Ethanol-toluene	78.3–76.4–110.7	0.763	0.763	56
Ethanol-trichloroethylene	78.3–70.0–87.5	0.845	0.653	15
Ethanol-water‡	25	0.67	0.42	11, 48
Ethyl acetate-benzene	77.2–71.1–80.2	0.50	0.40	46
Ethyl acetate-ethanol	77.2–71.7–78.3	0.389	0.389	16, 24
Ethyl acetate-toluene	77.2–110.7	0.04	0.25	30
Ethyl ether-acetone	34.6–56.1	0.322	0.322	9, 18, 44
Ethyl ether-ethanol	34.6–78.3	0.42	0.55	10, 24, 31
n-Hexane-ethanol	68.9–59.3–78.3	0.68	1.12	22
Isobutane-furfural†	37.8	1.14	1.31	34
	51.7	1.09	1.23	
	66.6	1.04	1.16	
	93.3	0.96	1.03	
Isopropanol-water	82.3–100	1.042	0.492	26
Isopropyl ether-isopropanol	68.5–66.1–82.3	0.42	0.60	35
Methanol-benzene	56.1–55.5–64.6	0.243	0.243	14, 15, 27
Methanol-ethyl acetate	64.6–62.1–77.1	0.505	0.505	4
Methanol-trichloroethylene	64.6–59.8–87.5	0.845	0.845	15
Methanol-water	64.6–100	0.36	0.22	
Methanol-water§	25	0.25	0.20	7
Methyl acetate-methanol	57.2–53.7–64.6	0.462	0.462	3, 4
Methyl acetate-water	57.0–100	1.30	0.82	32, 33
Methyl ethyl ketone-water	79.6–73.6–100	1.50	0.75	33
Propanol-water	97.3–88.0–100	1.10	0.492	7, 17, 55
Water-Cellosolve	100–134.5	0.26	0.88	1, 12
Water-p-dioxane	100–87.7–101.3	0.66	0.87	20, 21, 50
Water-phenol	100–181	0.36	1.40	47, 49
Water-pyridine	100–115.5	0.38	0.62	13

*Pressure = 23.8 to 229.6 mmHg.
†Pressure = 5.9 to 13.260 mmHg.
‡Pressure = 23.8 to 59 mmHg.
§Pressure = 23.8 to 123.5 mmHg.

The first two equation-sets, those of Van Laar and Margules, are based on a thermodynamic expression of the relation between the Gibbs free energy and activity coefficient which follows a polynomial form of expansion from which both sets of equations may be derived and, by expansion, be used to account for multicomponent effects. The Wilson equations are based on a thermodynamic expression in the form of a logarithmic function, which has some advantage in the expansion into multicomponent applications.

Many types of comparisons are made concerning the relative accuracy and merits of these two general approach methods, those based on the Wohl expansion and those on the Wilson, but the design engineer is fortunate if any of these methods are applicable and contain data sources connected with whatever applicable method is available. For this reason a review of at least five methods is shown.

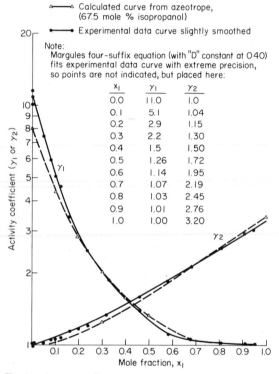

Fig. 15   Activity coefficients, isopropanol-water system.

Identification of the above equations by the use of the term "suffix" is not immediately evident, since it refers to the order of the interaction constants in the thermodynamic function from the Wohl type of expansion and not the final equation form. Here the simplest type of Van Laar equation comes from a two-suffix free-energy function; the simplest Margules is the three-suffix equation, and its higher-order form containing the D constant is the four-suffix equation.

The Van Laar, Margules, and a third not previously mentioned because it falls between the first two, the Scatchard-Hamer equations, have their first two constants equal to the logarithms of the activity coefficient taken at the points of infinite dilution, or in other words at the end of the curve condition, when the logarithm of the activity coefficients are plotted with the intermediate mixture compositions between the two components of the system. This is the commonly used plot of mole fraction of the light component in the liquid, or x_1. The constants of the Wilson equation do not have this simple relation, nor do the nonrandom two-liquid equations of Renon and Prausnitz.

In a pragmatic sense the equations above may be considered as activity-coefficient correlation methods, but when data are fitted, the thermodynamic consistency can be examined as a proof of the accuracy if enough data are at hand. The following examples illustrate how the initial vapor-liquid equilibrium data of limited amounts, and then where full amounts of data are available, are used to obtain activity coefficients for use in establishing a consistent set of values through correlation with the equations selected (see Fig. 15).

Calculation of Activity Coefficients from Data

A calculation based on data from an azeotrope is sufficient to perform a calculation of activity-coefficient values over the complete range of compositions, although confirming data are desirable. An azeotrope of isopropyl alcohol and water is considered. The atmospheric boiling point is 80.3°C (177°F); weight percent water is 12.6%. On a mole percent basis the light component, isopropyl alcohol, is 67.51%. This converted value will not be required until later, however, since at azeotropic compositions the vapor and liquid are of the same composition and the activity coefficient is simply expressed as a ratio of the total system pressure and the vapor pressure of the pure component.

Activity coefficients are defined for each component of a mixture as

$$Py_1 = \gamma_1 P_1 x_1 \quad \text{and} \quad Py_2 = \gamma_2 P_2 x_2 \tag{1}$$

where subscripts refer to the light and heavy components respectively.

P = total system pressure

P_1, P_2 = pure-component vapor pressures at system temperature

x_1, x_2, y_1, y_2 = mole fractions of the light and heavy components in the liquid and vapor phases

γ_1, γ_2 = activity coefficients for components as subscripted

It should be noted that Py_1 and Py_2 in Eq. (1) could be corrected for vapor-phase nonideality with an activity coefficient, but the vapor phase is considered ideal ($\gamma = 1.0$).

At azeotropic composition the values $x_1 = y_1$ and $x_2 = y_2$; so the activity coefficients become

$$\gamma_1 = \frac{P}{P_1} \quad \text{and} \quad \gamma_2 = \frac{P}{P_2}$$

At the azeotropic boiling point ($P = 760$ mm) of 80.3°C (177°F), the vapor pressures of the pure components, isopropyl alcohol and water, are 685 and 359 mm, respectively; so

$$\gamma_1 = 760/685 = 1.1095 \quad \text{with log} = 0.045127$$
$$\gamma_2 = 760/359 = 2.1170 \quad \text{with log} = 0.32572$$

These activity coefficients may be used with the azeotropic composition at which they have been determined to calculate the binary constants and then the complete range of activity coefficient values from $x = 0$ to $x = 1$.

The normal form of the Van Laar equations is

$$\log \gamma_1 = \frac{A_{12}}{[1 + (A_{12}x_1/A_{21}x_2)]^2} \tag{2}$$

$$\log \gamma_2 = \frac{A_{21}}{[1 + (A_{21}x_2/A_{12}x_1)]^2} \tag{2a}$$

where $A_{12} = \log \gamma_1$ (when $x_1 = 0$)

$A_{21} = \log \gamma_2$ (when $x_2 = 0$); note $x_2 = 1 - x_1$

The constants may be calculated in a direct fashion as

$$A_{12} = \log \gamma_1 \left(1 + \frac{x_2 \log \gamma_2}{x_1 \log \gamma_1}\right)^2 \tag{3}$$

$$A_{21} = \log \gamma_2 \left(1 + \frac{x_1 \log \gamma_1}{x_2 \log \gamma_2}\right)^2 \tag{3a}$$

The numerical solution is

$$A_{12} = 0.045127 \left(1 + \frac{0.3249 \times 0.3257}{0.6751 \times 0.045127}\right)^2$$
$$= 0.9032 \quad \text{or} \quad \gamma_1 = 8.00$$

$$A_{21} = 0.32570 \left(1 + \frac{0.6751 \times 0.045127}{0.3249 \times 0.32572} \right)^2$$
$$= 0.5402 \quad \text{or} \quad \gamma_2 = 3.47$$

With the A constants evaluated, it is possible to assume some fixed values of x_1 and calculate activity coefficients over the whole concentration range of the mixture (see Fig. 15) by going back to the original Van Laar equations (2) and (2a). It would be dangerous to use this single data point for a design; so more equilibrium data will be examined on this system to show possible variations in activity coefficients. The eventual use of the coefficients is to determine relative-volatility values for the distillation.

It can be shown by Eq. (1) and from the definition of relative volatility that the relative volatility of isopropanol to water is obtained as follows:

$$\alpha_{12} = \frac{y_1/x_1}{y_2/x_2} \tag{4}$$

From Eq. (1), it can be shown that the relative volatility becomes

$$\alpha_{12} = \frac{\gamma_1 P_1}{\gamma_2 P_2} \tag{5}$$

More complete data, used later, show a 30 mol % isopropanol mixture to have a boiling point of 81.3°C (178°F). Vapor pressures of the pure components at this temperature are P_1 = 720 mm, P_2 = 370 mm; from the Van Laar equation the activity coefficients are $\gamma_1 = 2.0$, $\gamma_2 = 1.3$; so the numerical value of the relative volatility α_{12} is

$$\alpha_{12} = \frac{2.0 \times 720}{1.3 \times 370} = 2.99$$

This procedure may be used to construct a relative-volatility curve, eventually showing the variation of alpha with composition (use the x_1 mole fraction as coordinate). At the azeotropic composition (67.51 mol % isopropanol) the relative volatility becomes equal to 1.

The simple form of the Margules equation (three-suffix) was not used. It was not expected to be as good, but later examples will show the *four*-suffix form to be adequate. The benefit of the Margules form, if the higher order is used, is that the expansion into multicomponent methods may be better facilitated, according to the method of Wohl.

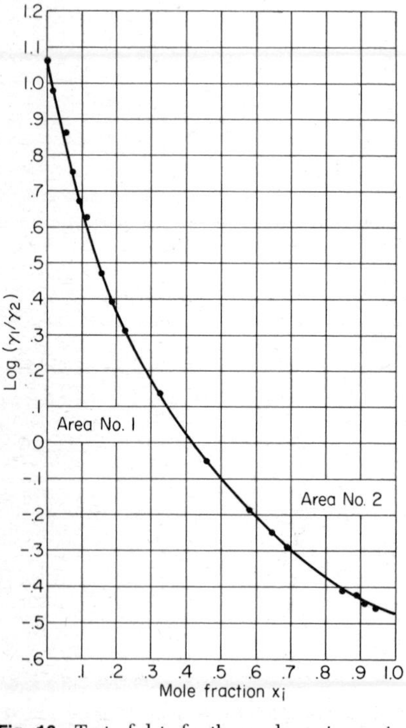

Testing of Data, Thermodynamic Consistency

Two sources of available equilibrium data are Ref. 8, page 501, and Ref. 9, page 502. These were examined separately, as they came from different sources. A plot of the logarithm of the ratio of (γ_1/γ_2) vs. x_1 is made (see Fig. 16) and the areas indicated are compared. If these are equal or nearly so, the data are good.

Both sets were satisfactory on this test, but there were points of irregularity near the very dilute region where the isopropyl alcohol content is near zero ($x_1 = 0$). Irregularities of this type may be revealed by plotting $(\log \gamma_1)/(x_2)^2$ vs. x_1 and noting where the curve of the data intercepts the axis at $x_1 = 0$ (see Fig. 17), and similarly for (γ_2/x^2_1).

Fig. 16 Test of data for thermodynamic consistency, isopropanol-water system.

The original data are given in the form of mole percent of the light component in the vapor and liquid at constant pressure (760 mm) with the equilibrium temperatures. For this system the temperature range is fairly small and no correction is considered. The temperatures are required essentially to obtain the vapor-pressure values in the calculation of the relative volatility later as well as for immediate calculation of the activity coefficients based on the data.

A simple procedure for calculation of the necessary data items is to divide the calculation into two parts, the γ_1 and γ_2 activity coefficients from the original composition and temperature values as follows (experimental data from Refs. 8 and 9):

P = system pressure = 760 mm

P_1 = vapor pressure isopropanol, at temperature T, °C

γ_1 = activity coefficient, isopropanol in water

The essential calculation (for γ_1) by columns (see Fig. 15) is as follows:

x_1	y_1	T	P_1	y_1P	x_1P_1	γ_1	$\log \gamma_1$
0.0083	0.1473	95.3	1270	112	10.55	10.6	1.026
0.0136	0.2244	93.2	1150	171	15.65	10.9	1.040
0.0254	0.3399	89.0	990	258	25.2	10.2	1.010
0.0570	0.4565	84.6	825	348	47.0	7.4	0.870
0.0843	0.5024	82.6	765	382	64.5	5.9	0.774
0.1000	0.5015	82.7	765	381	76.5	5.0	0.698
0.1232	0.5378	81.4	720	408	88.8	4.6	0.664
0.1629	0.5298	81.4	720	402	117.3	3.42	0.534
0.1986	0.5444	81.2	715	413	142.0	2.91	0.464
0.2387	0.5559	81.1	710	423	169.0	2.50	0.399
0.3314	0.5654	80.8	700	430	232.0	1.85	0.266
0.4597	0.5939	80.4	690	451	317.0	1.42	0.153
0.5838	0.6358	80.1	685	484	399.0	1.21	0.083
0.6813	0.6813	80.1	685	517	467.0	1.11	0.046
0.8100	0.7698	80.3	690	585	559.0	1.045	0.0191
0.9153	0.8801	81.0	705	669	645.0	1.035	0.0149
0.9535	0.9325	81.5	725	709	690.0	1.025	0.0107

Similarly (for γ_2) the corresponding curve (see Fig. 15):

P = system pressure = 760 mm

P_2 = vapor pressure water at temperature T, °C

γ_2 = activity coefficient, water in isopropanol

x_2	y_2	T	P_2	y_2P	x_2P	γ_2	$\log \gamma_2$
0.9917	0.8527	95.3	650	648	643	1.01	0.00432
0.9864	0.7756	93.2	630	590	621	0.95	−0.0223
0.9746	0.6601	89.0	490	502	477	1.05	0.0211
0.9430	0.5435	84.6	430	413	405	1.02	0.0086
0.9157	0.4976	82.6	390	378	357	1.06	0.0253
0.9000	0.4985	82.7	395	378	356	1.06	0.0253
0.8768	0.4622	81.4	370	352	324	1.085	0.0354
0.8371	0.4702	81.4	370	358	310	1.155	0.0626
0.8014	0.4556	81.9	370	346	296	1.17	0.0682
0.7613	0.4441	81.1	365	337	276	1.22	0.0863
0.6686	0.4346	80.8	365	330	244	1.35	0.131
0.5403	0.4061	80.4	360	309	195	1.585	0.201
0.4162	0.3642	80.1	355	277	148	1.87	0.272
0.3187	0.3887	80.2	355	242	113	2.14	0.331
0.1900	0.2302	80.3	360	175	68.4	2.56	0.409
0.0847	0.1199	80.0	365	91	30.9	2.94	0.469
0.0465	0.0675	81.5	370	51.3	17.2	2.99	0.475

The Margules form of the equations, derivable by the Wohl type of polynomial expansion of the free-energy or Gibbs energy equations, is shown below in the simple *three-suffix* form and also in the four-suffix form, which is preferable for data where another constant is required to obtain good curve fitting with the experimental data. The three-suffix form is given in two types of expressions, the second derivable by using $x_1 = 1 - x_2$ or its alternate for the binary formulas. Another set of extremely useful formulas has the A constants in a similarly direct form as given for the Van Laar equations.

$$\log \gamma_1 = x_2^2[A_{12} + 2x_1(A_{21} - A_{12})] \tag{6}$$
$$\log \gamma_2 = x_1^2[A_{21} + 2x_2(A_{12} - A_{21})] \tag{6a}$$

or in also derivable alternate form, where $x_1 = 1 - x_2$,

$$\log \gamma_1 = x_2^2[(2A_{21} - A_{12}) + 2x_2^3(A_{12} - A_{21})] \tag{7}$$
$$\log \gamma_2 = x_1^2[(2A_{12} - A_{21}) + 2x_1^3(A_{21} - A_{12})] \tag{7a}$$

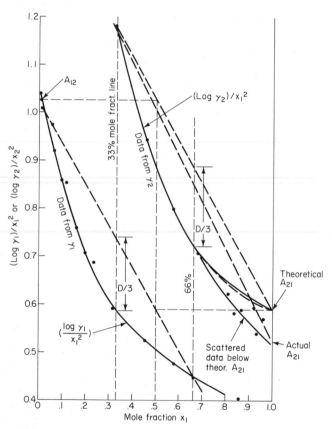

Fig. 17   Test of data for thermodynamic consistency, and to determine the value of the Margules D constant, isopropanol-water system.

The direct formula for constants, useful when only small amounts of data are available, such as azeotropic composition, is the following:

$$A_{12} = \left(\frac{x_2 - x_1}{x_2^2}\right) \log \gamma_1 + \frac{2 \log \gamma_2}{x_1} \tag{8}$$
$$A_{21} = \left(\frac{x_1 - x_2}{x_1^2}\right) \log \gamma_2 + \frac{2 \log \gamma_1}{x_2} \tag{8a}$$

The Margules form of the equations similarly derived by the Wohl type of expansion is shown in the four-suffix form, which was used to correlate the experimental data above. This four-suffix form is also used for expansion into multicomponent application (see Ref. 1, which also gives details for evaluation of the third constant D added to the expressions).

$$\log \gamma_1 = x_2^2[A_{12} + 2(A_{21} - A_{12} - D_{12})x_1 + 3D_{12}x_1^2] \tag{9}$$

$$\log \gamma_1 = x_1^2[A_{21} + 2(A_{12} - A_{21} - D_{12})x_2 + 3D_{12}x_2^2] \tag{9a}$$

These are the binary-system equations, and the D constant so subscripted would be used in the multicomponent forms, but in the binary form it is a single constant and could simply be called D. The normal procedure would be to calculate the binary A constants on the three-suffix equations and to use these, after smoothing the data by a plot of $\log \gamma_1/ x_1^2$, as on Fig. 17, to obtain the best value of the dilute (end-of-the-curve) conditions, where the greatest error is possible, as well as a plot of all data to check their consistency thermodynamically.

The plot in Fig. 17 was suggested by Gearhart[1] and accomplishes the purpose of also predicting the value of the D constant of the four-suffix Margules equation. The plot of the data from the activity coefficient A_{12} is continued until it passes the $2/3$ distance (0.667) on the x_1 mole fraction axis. A diagonal is drawn from A_{12} down to the intersection of the data curve and the $2/3$ distance. Assuming these data are correct, the following points are of interest:

The intersection of the lower diagonal and the 0.50 mole fraction vertical line should be the A_{21} value and should meet the data curve from the activity coefficient γ_2.

A second diagonal can be established from the point where this second data curve meets the $1/3$ distance (0.333 mole fraction on x_1) and from this diagonal and the 0.50 mole fraction vertical the consistency of A_{12} may be tested.

As shown in Fig. 17, the value of the D constant on the plot may be determined from its $1/3$ value, at the 0.333 mole fraction vertical and the 0.666 vertical.

Deviation from the typical curves with these relations between activity-coefficient values and so constructed diagonals is the result of inconsistant data values. The use of raw data in this case compares favorably (for design) with more sophisticated methods.[91] The use of smoothed data (least-squares values) assumes that the methods have been very accurate at the start in the experimental investigations, but this assumption cannot be made, as has often been shown.

Additional Equation Forms for Estimation of Volatility, Binary Applications

These forms are the more recent ones and are not derivable by the Wohl type of thermodynamic expansion. The Wilson equation, with its individual basis, requires molar volumes of the two pure components, at the temperatures of the system under the equilibrium conditions of interest. Equations are shown for the comparison of the Wilson parameters with the older forms (Van Laar, etc.) in order to check the calculations made during design [see Eqs. (12) and (12a)].

Note that the Wilson parameters and the activity coefficients are used with *natural* logarithmic values, whereas the older equation forms traditionally use the common logarithm base.

Here for the binary systems it is possible to use a hand-calculation procedure with the excellent small calculators available. Final computation may be as described for the Van Laar equation, the computation of relative-volatility values related to composition before the stage requirements are approached in design.

The Wilson equations express the natural logarithm of the activity coefficients in terms of the parameters lambda Λ_{12}, Λ_{21} and the concentrations x_1, x_2, for binary systems, as follows:

$$\ln \gamma_{12} = -\ln (x_1 + \Lambda_{12}x_2) + x_2 \left(\frac{\Lambda_{12}}{x_1 + \Lambda_{12}x_2} - \frac{\Lambda_{21}}{\Lambda_{21}x_1 + x_2} \right) \tag{10}$$

$$\ln \gamma_{21} = -\ln (x_2 + \Lambda_{21}x_1) - x_1 \left(\frac{\Lambda_{12}}{x_1 + \Lambda_{12}x_2} - \frac{\Lambda_{21}}{\Lambda_{21}x_1 + x_2} \right) \tag{10a}$$

where the lambda values are obtained from the molecular volumes of the pure components (at some average system temperature) and characteristic energy differences (denoted by lowercase lambda terms $\lambda_{12} - \lambda_{11}$ and $\lambda_{12} - \lambda_{22}$) as used in the following equations:

$$\Lambda_{12} = \frac{\nu_2}{\nu_1} \exp - \left(\frac{\lambda_{12} - \lambda_{11}}{R \times T} \right) \tag{11}$$

$$\Lambda_{21} = \frac{\nu_1}{\nu_2} \exp - \left(\frac{\lambda_{12} - \lambda_{22}}{R \times T} \right) \tag{11a}$$

where R is the gas constant (1.987) when used with the temperature T in kelvins. The molecular volumes nu (ν) are in units of cm^3/g mole, with subscripts indicating the pure components. Data made available would include the characteristic energy differences and molecular volumes over the required temperature range. The energy differences would have been adjusted to fit the experimental data of the binary system examined. It should be noted that the Wilson equations are not symmetrical in the sense that Eqs. (10) and (10a) have identical terms in brackets, but the last equation has a negative sign before the multiplier x_1 just before the parentheses. It should also be noted that the characteristic energy differences are subtractions of subscripted terms that are not symmetrical.

Equations (11) and (11a) use the symbol "exp," which denotes that the term following is to be taken as the exponent of the natural logarithm base e, in this case a negative exponent.

Equations (10) and (10a) may be reduced to a more simplified form when applied to infinite-dilution conditions, making the Wilson lambda constants comparable with the Van Laar A constants after noting that different logarithmic bases are used and that the ensuing Wilson expressions are not as simple as the Van Laar equations. These equations for infinite dilution are indicated by the infinity sign:

$$\ln \gamma_{12}^{\infty} = 1 - \ln \Lambda_{12} - \Lambda_{21} \tag{12}$$

$$\ln \gamma_{21}^{\infty} = 1 - \ln \Lambda_{21} - \Lambda_{12} \tag{12a}$$

which may be compared with the Van Laar A constants where $\log \gamma_{12}^{\infty} = A_{12}$ and $\log \gamma_{21}^{\infty} = A_{21}$.

The binary system isopropanol-benzene is taken for a numerical calculation of activity coefficients over the range of molar concentrations of isopropanol in benzene, by means of the Wilson equation, from data given in Ref. 39, and then this plot (Fig. 18) is compared with data from another source[90] giving Van Laar constants.

EXAMPLE

The system isopropanol-benzene has Van Laar constants determined as $A_{12} = 0.6723$, $A_{21} = 0.4638$, from data at a system pressure of 760 mm abs. It is also reported as the following Wilson data with characteristic energy differences $\lambda_{12} - \lambda_{11} = 1008$, $\lambda_{12} - \lambda_{22} = 160.5$; and molar volumes nu ($\nu$) are taken at 353 K as a close average system temperature; based on average boiling points of the components of 82 and $80°C$ for the isopropanol and benzene, and $v_1 = 83.3$, $v_2 = 95.5$ $cm^3/g \cdot$ mol, respectively.

The Wilson solution starts with the calculation of the lambda Λ_{12}, Λ_{21} values from Eqs. (11) and (11a) as follows:

$$\Lambda_{12} = \frac{95.5}{83.3} \exp - \frac{1008}{1.987 \times 353}$$
$$= 0.27609$$
$$\Lambda_{21} = \frac{83.3}{95.5} \exp - \frac{160.5}{1.987 \times 353}$$
$$= 0.69536$$

These values of lambda may be used with Eqs. (12) and (12a) for comparison with the Van Laar data, which are at infinite dilution, or they may be used with Eqs. (10) and (10a) to plot the activity-coefficient values over the whole concentration range of binary mixtures as in Fig. 18.

For a comparison with the Van Laar data Eqs. (12) and (12a) are used as follows:

$$\ln \gamma_{12} = 1 + 1.28703 - 0.69536$$
$$= 1.59167 (\gamma_{12} = 4.912) \tag{12}$$
$$\ln \gamma_{21} = 1 + 0.36333 - 0.27609$$
$$= 1.08724 (\gamma_{21} = 2.9661) \tag{12a}$$

Then comparing with the Van Laar values:

$$\log 4.912 = 0.6913 \text{ vs. } 0.6723 \text{ (Van Laar)}$$
$$\log 2.9661 = 0.4722 \text{ vs. } 0.4638 \text{ (Van Laar)}$$

The activity coefficients from the Van Laar constants are 4.70 vs. 4.91 (Wilson) and 2.91 vs. 2.97 (Wilson) at the infinite-dilution conditions and move closer at intermediate concentrations, which is within most design requirements for stage computations.

A typical calculation to obtain values for the plot in Fig. 18 follows for the value of x_1 (mole fraction) = 0.10, using Eqs. (10) and (10a). The terms with asterisks are the same for both equations. Note at $x_1 = 0.10$, $x_2 = 0.90$; then

$$\ln \gamma_{12} = -\ln (0.10 + 0.27609 \times 0.90) + 0.90 \times 0.07506^*$$
$$= 1.05417 \qquad\qquad + 0.067554$$
$$= 1.12172 \,(\text{or} \quad \gamma_{12} = 3.070)$$

$$0.07506^* = \frac{0.27609}{0.10 + 0.27609 \times 0.90} - \frac{0.69536}{0.69536 \times 0.10 + 0.9}$$

$$\ln \gamma_{21} = -\ln (0.90 + 0.69536 \times 0.10) - 0.1 \times 0.07506^*$$
$$= 0.030937 \qquad\qquad - 0.007506$$
$$= 0.023432 \,(\text{or} \quad \gamma_{21} = 1.024)$$

The nonrandom two-liquid equation of Renon[65,66] also offers a data source, but the parameters require considerable attention for their use in design. Although the use of the Wilson equation is excluded from the prediction of near immiscibility,[92] the Renon equation is effective in this area. Reference to this technique is given, but data sources are

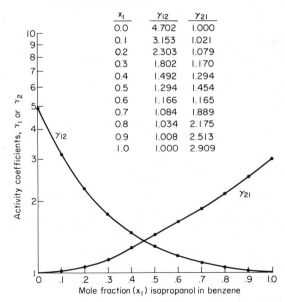

– – – Calculated curve from Van Laar constants
  $A_{12} = 0.6723$ $A_{21} = 0.4638$

●—— Calculated curve from Wilson characteristic
  energy differences: $(\lambda_{12} - \lambda_{11}) = 1008.0$
  $(\lambda_{12} - \lambda_{22}) = 160.5$
  and pure component molar volumes:
  $v_1 = 83.3 \text{ cm}^3/\text{g·mol}$ (isopropanol)
  $v_2 = 95.5 \text{ cm}^3/\text{g·mol}$ (benzene)

Note:
Plot of activity coefficients from Wilson equation is very close to those from the Van Laar equations. For this reason the Van Laar values are shown in tabulated form below:

x_1	γ_{12}	γ_{21}
0.0	4.702	1.000
0.1	3.153	1.021
0.2	2.303	1.079
0.3	1.802	1.170
0.4	1.492	1.294
0.5	1.294	1.454
0.6	1.166	1.165
0.7	1.084	1.889
0.8	1.034	2.175
0.9	1.008	2.513
1.0	1.000	2.909

Activity coefficients, γ_1 or γ_2

Mole fraction (x_1) isopropanol in benzene

Fig. 18 Comparison of activity coefficients calculated by Van Laar A constants and Wilson characteristic energy differences for the isopropanol-benzene system.

still limited, and it may be in the province of design not to carry out primary investigations in such areas but to take a more pragmatic course of experimental data from Oldershaw distillations.

The *analytical solution of groups* method, the most abstract, is still limited in data available and requires extensive computer facilities to develop the general relations necessary. The most readily available published data are those of Palmer.[12,60] where some structural group data are made available and the calculation technique is described in some detail.

PRACTICAL EQUIPMENT DESIGN METHODS

Categories of Design

This discussion covers what is conventionally called the process engineering design of the distillation plant. Even so, a multitude of practical considerations must be taken into account that are seemingly outside the science of distillation but will affect the choice of the distillation equipment used and the capacities finally selected.

Usually a set of job specifications are compiled to serve as a guide for all technologies involved. The specifications are often indexed according to the equipment items of interest using an alphabetical code to designate each piece of equipment with its number, such as T-1 for "tower 1." Besides desired equipment characteristics the specifications will include the utility conditions that prevail locally, available plant area, prevailing wind conditions, seasonal temperatures, code requirements for mechanical design, safety requirements, and many other conditions that may affect the final design in some fashion or other. The process design will be a starting point for many technical and commercial activities in the building of the plant.

A typical process design may include the following information categories in answer to the request of a plant for a given capacity that follows the designated specifications:

1. Material and heat balance
2. Process and instrument flow diagram
3. Plot plan
4. Utility requirements
5. Basic specifications for equipment
6. Auxiliary plant needs such as disposal, emergency pump-out systems, flares, and cooling-water circulation systems
7. Chemical usage for water treatment, stabilizers to maintain purity, etc.

The material balance will find general use and should include normal and maximum flow rates, in pounds per hour. Molar flow rates, volumetric flows such as gallons per hour, and such items as specific gravity, molecular weight, and even normal boiling points of the components, are not precluded. The recirculated streams such as reflux can also be found a convenience in the balance. Flowing temperatures and pressures may be found on the process flow diagram to complete the needs of the designers who follow. It is best to have the material balances based on normal and maximum plant design rates, not on some unit rate which can cause confusion if mutipliers must be used at every step.

The heat balance is best made from component enthalpy curves, all starting with the component in the liquid state at some low temperature, commonly − 0 or −200°C (0 or −200°F), the latter base temperature to be used if low plant temperatures are prevalent. Hydrogen may be an exception, but in computer computation it is sometimes expedient to invent some insignificantly small liquid-heat value for the hydrogen component to permit the program to function. If a number of unlike components are collected from various sources with enthalpy bases at mixed starting points, a heat balance from a computer may have a variety of unrecognizable values, including negative ones, making inspections difficult.

The process and instrument flow diagram should methodically include all the equipment and functional instrumentation that is necessary to indicate the plant operation and utility requirements. The degrees of freedom discussed relative to distillation may be extended here to see if the plant may be logically expected to operate during normal conditions and what the effects of plant failures will have on the equipment. The instrumentation should be arranged to give the safest conditions possible, for example, if there is a water or power failure, control valves would open (or close) to prevent extraordinary pressure or temperature conditions from developing. Sometimes it is possible to provide a master switch at some safe area which may be manipulated by hand to open and close a series of critical valves for emergency pump-out of the plant in the event of failures or fire. An emergency power generator may be required for this arrangement.

The plot plan should be considered for its functional arrangement relative to safety, start-up, and operating convenience. Other factors such as space allowance to remove equipment during maintenance operations and to provide for future plant expansion may also be considered. It is customary to arrange equipment in rows to line up all towers, heat exchangers, or pumps, so that they are less confusing to find and possibly have a more simplified piping arrangement. The spacing of equipment, particularly heaters and towers, is a point of safety; often a minimum distance is set for these. Much attention is also given to the placing of pipe racks, a function of the mechanical design.

Utility requirements usually are described as fuel, power, cooling water, steam, and instrument air. These are broken down into their respective divisions according to voltage source; circulated or makeup water and temperature rise; steam supply pressure, or exhaust steam make or supply; and instrument air supply pressure. Credit may be taken if the design includes heat recovery through steam generation or other means. It is also well to have a set of unit costs of utilities to establish possible pay-off times for equipment choices made during design. The length of these time periods is usually expressed according to the economy.

Basic specifications for equipment are a translation of the process-design abstractions into reality. Equipment specifications imply that the size has been determined, or the information presented will enable some specialist to do this. The specifications must also include all proper needs in terms of materials and other properties and construction arrangements to make a workable device for the purpose intended. One function of specifications is to aid in purchasing, by giving all vendors the same information for making bids. Later bid analysis should show how closely the equipment conforms with the original request. The process engineer need not be an equipment specialist but is required to anticipate the specialist's needs, and thus a rather complete set of traditional requirements exists for many types of equipment. Because standardization of equipment is also common, many choices in sizes and suppliers are available if a plant is to be built where others are already established.

Auxiliary plant needs such as disposal systems must be anticipated. No procedure can predict every contingency, but a figure is usually included in estimates for the possibility of extra costs. This should also include extra provision for start-up such as extra piping for recirculation and extra equipment that may be necessary; this can include items regularly required during operations but more like a utility item, such as the chemicals used in water treatment. It could also include product stabilizers, demulsifiers sometimes added in distillations, etc.

Procedures in Design

The material balance is made according to a service factor that takes into account the shutdown time necessary for routine plant maintenance. The plant "stream," or operating capacity, will be in excess of the average or calendar rate of production anticipated.

The feed and product quantities in the balance should show expected purity and recovery unless the preliminary anticipations of these lie beyond economic realization; then adjustments must be made.

The top-product composition, taken from the balance, may be used as a starting point to determine the top tower temperature and the reflux-accumulator condensing temperature. The general principle is to use the dew-point temperature of the overhead vapor for the top tower temperature, and the bubble-point temperature of the condensed liquid (in this case assumed to be liquid product) to set these respective temperatures. Similarly the bottom-product bubble-point temperature will be a means of obtaining the tower base temperature, but this must be done with the selection of a suitable tower pressure.

The tower pressure selected is based on the temperatures of the cooling and heating sources to be used in the distillation. These may be the commonly used circulated-cooling-water or air-cooled condensers, but for extremely low- or high-temperature practice other means of condensation cooling can be used. Similarly the heating source used in reboiling must correspond to the intended operating pressure of the tower, but also based on the anticipated condensing temperature of the overhead vapor. The reboiling heat source most commonly used is steam or hot oil (circulated) but it can also be derived from other sources for various reasons. In some cases the heating medium may be too hot for the service, for example, in the reboiling of relatively low temperature distillations such as propylene-propane splitters or deethanizers, where either exhaust steam or even hot water may be used as a heat supply.

A further balancing of the tower pressure is made by considering the effect of the operating pressure on the relative volatility of the components to be separated and the attendant equipment costs, which usually increase when producing a favorable volatility condition. For example, a pressure lowering may improve the relative volatility for the separation but will tend to increase the tower diameter required and may lower the heat-transfer possibilities at the condenser.

The best arrangement for the cooling of products and the possibility of utilizing the heat in exchange should also be considered. This will become evident in the arrangement of the process flow diagram and the calculation of the utility requirements as these points are considered.

With tower pressures and temperatures established, a suitable stage and reflux combination to be used in the distillation is calculated. Examples of designs employing the methods used have already been given. The heat balance is then continued to establish some feasible range that will combine the use of a tower of some reasonable size with a heat requirement that is also reasonable for the economy of the separation.

The heat balance may be expressed in terms of the enthalpy of the various streams, as "heat in" and "heat out" and use of the reflux to permit the calculation of the reboiler duty by difference (see Fig. 11).

The process flow diagram is indicative of the plant flowing streams but also includes the intended instrumentation, equipment items by number, heat-exchange duties, and flowing-stream temperatures and pressures. It becomes the "work sheet" of the job. It is a good plan to permit an instrument specialist to set the instrument designations at the start with the approved symbols used by the Instrument Society of America. Similarly item numbers and other symbols and designations should follow some pattern of usage.[97,98]

While the procedures in design given here cannot cover all conventional information available in the many sources mentioned below, some useful points will be considered. The principles of the sizing of equipment must be considered at several levels of accuracy. The approximate tower "size" may be calculated with the Badger F factor or more accurately determined by the various tray hydraulic methods described below. The heat exchange, including condensers and reboilers, also follows with conventional designs but the size may be approximated for several purposes such as the preparation of plot plans and estimation of costs or preliminary piping arrangements.

Methods of Distillation-Equipment Design

One part of the conventional design methods of chemical engineering is particularly adaptable to distillation. This ranges from the calculation of stage and reflux requirements and methods for the computation of volatility properties to some specific rules whereby useful equipment sizes may be determined. Among these the Badger (Company) F factor for evaluating tower vapor loading during operation or estimating the tower diameter required has a special use. This method should not be regarded as a final result, since a more rigorous method of hydraulic examination for the type of tray selected should follow it.

The value of this method or of similar ones is to set a diameter that can be reasonably expected to conform with normal liquid-flow patterns within the tower to give an operable distillation. This does not preclude some variation later according to the tray spacing selected or the size of the downcomer required. It follows that this initial criterion would differ if applied to absorbers or devices without vapor-liquid flow ratios similar to those of distillations.

The tower F factor is the actual tower vapor velocity based on the gross tower area, divided by the square root of the specific volume of the vapor, taken at the tower location of interest, which is usually stated as a point just below the top tray and at other locations such as the bottom tray and the feed tray to take into account differences in flow. High-pressure operations such as depropanizers or propylene-propane splitters will require an F value as low as 0.8 to obtain good stage efficiency, while atmospheric distillations such as crude-petroleum towers may be designed for an F of 1.8 or higher. In vacuum distillations the F should generally be taken back to the value of 1.0 for safety.

Most distillations carried out at the vapor velocities based on the F values stated above will have liquid flow rates across the trays that will not exceed a very rough maximum that can be crudely taken as gallons per hour per foot of the tower diameter. This rate seldom exceeds 4000 gal/(h)(ft) in distillation, but it may be as high as 8000 gal/(h)(ft) in absorbers. The flow rates are based on single flow paths, and if split into two streams, a favorable condition will exist for the flow but the tray efficiency would normally be reduced. This description is based upon towers with fair tray spacing of at least 18 up to 24 in, and with adequate downcomer capacity between the trays.

The downcomer capacity is based on some suitable conditions whereby the liquid leaving the tray may be reasonably depleted of entrained vapor as foam. It is common to assume 50% aeration and set about 10 s residence time for practical separation to take place before the liquid passes to the tray below.

The most common type of downcomer used, called an "apron downcomer," is made by cutting off a segmental part of the tray and containing the overflowing liquid between the sidewall of the tower and the so-called "apron" sheet extending from the overflow weir to the tray below. A seal is formed with the liquid on the tray below where the apron sheet ends just above that tray. The area of the segment of the downcomer is a function of the residence time required but also must be large enough to provide sufficient length for the overflow weir and enough segment distance (height) for the "throw distance" required for liquid discharge. Tray hydraulic computations and the degree of standardization of commercial trays usually permit an economic selection in this respect.

The computation methods and experimental data related to tray hydraulics, including downcomer flow, have been carried out as a private enterprise and a project financed by a number of companies in order to receive the results of periodic work done, but a degree of secrecy is obligatory. Design procedures for hand calculation or computer programs are often made available by the tray vendors in order to facilitate their services.

Tray hydraulic methods become a somewhat complicated design procedure as developed in the past for bubble-cap trays. Many such trays are still in use, but valve-type trays and more sophisticated types of sieve trays do not have the design complications from unbalanced flow conditions possible from bubble-cap shapes.

Tray hydraulic methods predict the limits for internal flow capacities, but except for special distillations the results cannot predict the tray efficiency resulting from the final choice of tray area (and downcomer) used. This is a natural result when so many unlike systems are separated. The volatility properties cannot be firmly evaluated for each case, and the tray efficiency will vary with the shape (factor) of the equipment. It is fortuitous that designs may include some extra reflux capacity as well as extra trays to hedge against the problem of only approximate expressions of stage efficiency. The relation between efficiency and the viscosity of the liquid under distillation conditions has been covered.[56]

A final design of a distillation unit will include instrumentation; piping and pumping requirements; heat exchange, including condensers and reboilers; considerations of residence time in holding the products in the reflux accumulator and the base of the tower; location of safety valves, drains, steam snuffing lines, and other safety provisions; and where valves should be placed to facilitate shutdown and operation.

The reflux and product condensation system is an important part of the design and involves choices of cooling methods, such as the use of either water-cooled tube bundles or air cooling on finned tubes, omitting the possibility of other methods. The method can be modified by placing the condenser bundles on the ground, particularly in the case of water-cooled bundles, and operating with the bundles holding condensed liquid at all times. When the bundles are placed above the reflux and product accumulator vessel, the liquid drops into it, giving less resistance to the flow and some difference in the heat-exchange rate, but the chief reason for this innovation is the savings in supporting the heavy tube bundles at the higher elevation. This procedure may require an insight into the two-phase flow conditions in the condenser piping (see the summary below).

Normally the condenser area may be approximated with some expected heat-exchange coefficient if the condensed product has some viscosity value near that of water or less, and an exchange coefficient no lower than 60 Btu/(h)(ft^2) is reasonable. It is convenient to prepare or obtain a set of tables to convert heat-exchange areas into bundle size, based on the tube outside diameter and length.

The reboiler heat-exchange surface may be approximated by noting that the maximum flux obtainable for induced-circulation types is approximately 14,000 Btu/(h)(ft^2) of tube surface, and some relatively lower amount can be used in the estimate. Similarly the bundle size should be examined because there are practical size limits for the handling and cleaning of heat-exchange bundles according to the local facilities.

Summary of Physical Properties Sources for Design   The following list of references includes data sources for properties usually required, in order of most probable usage.

Enthalpy and Thermal Properties. Refs. 2, 16, 29, 109, 61, 93, 62, 64, 56, 72, 20, 6.

Vapor Pressure. Refs. 2, 16, 29, 56, 57, 61, 93, 62, 64, 8, 9, 6, 104.

Critical-State Properties. Refs. 2, 16, 29, 64, 61, 93, 62, 56, 48, 20, 6.

Vapor-Liquid Equilibrium Systems and Data

NONIDEAL: Refs. 90, 91, 92, 8, 9, 1, 78, 79, 39, 65, 66, 94, 28, 42.

GROUP CONTRIBUTION IN MIXTURE VOLATILITY: Refs. 60, 12, 95, 93, 92, 91.

HYDROCARBON SYSTEMS WITH HYDROGEN AND OTHER SIMPLE COMPONENTS ADDED: Refs. 7, 31, 5, 77, 58, 95, 3, 10, 45, 46, 47, 48, 49, 50, 63, 81, 33, 34, 35, 36, 37, 38, 104.

EQUILIBRIUM IN CALCULATION: Refs. 82, 71, 61, 93, 62, 90, 67, 55, 59, 13.

Azeotropic Data. Refs. 40, 61, 93, 62, 80.

Density. Refs. 2, 16, 61, 93, 62, 64, 29, 56, 109.

Viscosity. Refs. 2, 16, 29, 56, 64, 61, 93, 62.

Surface Tension. Refs. 2, 16, 29, 56, 64, 61, 93, 92.

Summary of Design Methods   The following list of references emphasizes most probable usage.

General Equipment. Refs. 24, 85, 76, 53, 69, 73, 44, 86, 110, 75.

Fractionation Devices. Refs. 85, 76, 53, 69, 73, 44, 86, 32, 43, 73, 24.

Fluid Flow and Piping. Refs. 11, 53, 56.

Fluid Flow Two-Phase. Refs. 14, 53.

Flow-Diagram Construction. Refs. 97, 98, 43, 87.

Bubble-Cap Tray Hydraulics. Refs. 99, 100.

Valve-Cap Tray Hydraulics. Refs. 101, 102.

Sieve and Perforated Tray Hydraulics. Refs. 106, 107, 108, 85, 76.

Packed Tower (Hydraulics) and Design. Refs. 51, 17, 18, 19, 21, 22, 62.

Condensers (Tubular) Water-Cooled. Refs. 24, 43, 112, 15, 30, 52, 103.

Condensers, Exchangers, Air-Cooled. Refs. 70, 43, 52, 62, 112.

Reboilers, Reboil Systems. Refs. 43 (Chap. 15), 26, 27, 41, 52, 105, 24, 103.

Heat Exchangers, Coolers (Tubular). Refs. 112, 43, 24, 52, 62, 103.

Instrumentation. Refs. 97, 113, 98, 73, 18, 4, 87.

Pumps, Centrifugal and Positive Displacement (Steam Drive). Refs. 24, 61, 93, 62.

Steam Ejectors (Vacuum Distillation). Refs. 24, 61, 93, 62.

Designs by Unit Process Identification.

GENERAL DISTILLATION: Refs. 85, 76, 44, 53, 54, 73, 80, 56, 86, 69, 75, 84, 71, 82, 32, 87, 83, 110.

STEAM DISTILLATION: Refs. 23, 73, 26, 53.

HEAT EXCHANGE, SHOWER DECKS: Ref. 68.

GAS COOLING: Refs. 52, 25.

BATCH DISTILLATION: Refs. 4, 73, 74, 22.

Approximate Design Factors

TRAY EFFICIENCY IN DISTILLATION: Refs. 56, 111, 110, 19, 51.

HETP (height equivalent to a theoretical plate): Refs. 51, 17, 18, 19, 22, 62.

HEAT-EXCHANGE COEFFICIENTS: Refs. 112, 61, 93, 62.

REFERENCES

1. Adler, S. B., Friend, L., and Pigford, R. L., *Am. Inst. Chem. Eng. J.*, **12**, 629 (1966).
2. American Petroleum Institute, "Selected Values of Physical and Thermodynamical Properties of Hydrocarbons and Related Compounds," Project 44, Carnegie Press, Pittsburgh, Pa., 1953, and supplements.
3. Benedict, M., Webb, G. B., Rubin, L. C., and Friend, L., *Chem. Eng. Prog.*, **47**, 571, 609 (1951), 2 parts.
4. Block, B., *Chem. Eng.*, Jan. 16, 1967, p. 147.
5. Cajander, B. C., Hipkin, H. G., and Lenoir, J. M., *J. Chem. Eng. Data*, **5**, 251 (1960).
6. Canjar, L. M., and Manning, F. S., *Hydrocarbon Process. Pet. Refiner*, **44**(14), 121 (1965 and earlier).

7. Chao, K. C., and Seader, J. D., *Am. Inst. Chem. Eng. J.*, **7**, 598 (1961).
8. Chu, J. C., Getty, R. J., Brennecke, L. F., and Paul, R., "Distillation Equilibrium Data," Reinhold, New York, 1950.
9. Chu, J. C., Wang, S. L., Levy, S. L., and Paul, R., "Vapor-Liquid Equilibrium Data," Edwards, Ann Arbor, Mich., 1950.
10. Cooper, H. W., and Goldfrank, J. C., *Hydrocarbon Process.*, **46**, 141 (1967).
11. Crane Co., Flow of Fluids through Valves, Fittings and Pipe, Technical Paper 410, Crane Co., Chicago, Ill., 1969.
12. Deal, C. H., and Derr, E. L., *Ind. Eng. Chem.*, **60**, 28 1968.
13. Deam, J. R., and Maddox, R. N., *Hydrocarbon Process.*, **48**, 163 (1969).
14. Degance, A. E., and Atherton, R. W., "Chemical Engineering Aspects of Two-Phase Flow," parts 1–8, McGraw-Hill, New York, 1963.
15. Donahue, D. A., *Ind. Eng. Chem.*, **39**, 62 (1947).
16. Driesbach, R. R., "Physical Properties of Chemical Compounds," American Chemical Society, Washington, D.C., 1955.
17. Eckert, J. S., *Chem. Eng.*, Apr. 14, 1975, p. 70.
18. Eckert, J. S., and Walter, L. F., *Chem. Eng.*, Mar. 30, 1964, p. 79.
19. Eckert, J. S., and Walter, L. F., *Hydrocarbon Process. Pet. Refiner*, **45**, 107 (1964).
20. Edmister, W. C., "Applied Hydrocarbon Thermodynamics," vol. 1, 1961; vol. 2, 1974, Gulf Publishing Co., Houston, Tex.
21. Eichel, F. G., *Chem. Eng.*, Sept. 12, 1966, p. 197.
22. Eichel, F. G., *Chem. Eng.*, July 18, 1966, p. 159.
23. Ellerbe, R. W., *Chem. Eng.*, May 28, 1973, p. 110.
24. Evans, F. L., "Equipment Design Handbook (for Refineries and Chemical Plants)," vols. 1, 2, Gulf Publishing Co., Houston, Tex.
25. Fair, J. R., *Petro/Chem. Eng.*, August 1961, p. 57.
26. Fair, J. R., *Hydrocarbon Process. Pet. Refiner*, **42**, 159 (1963).
27. Fair, J. R., *Chem. Eng.*, (1), 119 (July 8, 1963); (2), 101 (Aug. 5, 1963).
28. Fleck, R. N., and Prausnitz, J. M., *Chem. Eng.*, May 20, 1968, p. 157.
29. Gallant, R. W., Physical Properties of Hydrocarbons, *Hydrocarbon Process.*, **44** (July 1965) to **42** (January 1970).
30. Gloyer, W., *Hydrocarbon Process.*, (1), 103 (June 1970); (2), 107 (July 1970).
31. Grayson, H. G., and Streed, C. W., Paper 20, Sixth World Petroleum Conference, Frankfurt, June 1963, or see Ref. 64.
32. Hanson, D. N., Duffin, J. H., and Somerville, G. F., "Computation of Multistage Separation Processes," Reinhold, New York, 1962.
33. Hill, A. B., Proposed paper for presentation at AIChE meeting, Atlantic City, N.J., Mar. 18, 1959. Hill data are given in Table 3 as polynomials.
34. Hoffman, D. W., and Weber, J. H., *Pet. Refiner*, **34**, 137 (1955).
35. Hoffman, D. W., and Weber, J. H., *Pet. Refiner*, **35**, 213 (1956).
36. Hoffman, D. W., and Weber, J. H., *Pet. Refiner*, **35**, 163 (1956).
37. Hoffman, D. W., and Weber, J. H., *Pet. Refiner*, **37**, 143 (1958).
38. Hoffman, D. W., and Weber, J. H., *Pet. Refiner*, **38**, 137 (1959).
39. Holmes, M. J., and Van Winkle, M., *Ind. Eng. Chem.*, **62**, 21 (1970).
40. Horsley, L. H., "Azeotropic Data," American Chemical Society, Washington, D.C., June 1952.
41. Jacobs, J. K., *Hydrocarbon Process. Pet. Refiner*, **40**, 189 (1961).
42. Kahre, L. C., and Hankinson, R. H., *Hydrocarbon Process.*, March 1972, p. 94.
43. Kern, D. Q., "Process Heat Transfer," McGraw-Hill, New York, 1950.
44. King, C. J., "Separation Processes," McGraw-Hill, New York, 1971.
45. Lenoir, J. M., *Hydrocarbon Process.*, **44**, 139 (1965).
46. Lenoir, J. M., *Hydrocarbon Process.*, **46**, 191 (1967).
47. Lenoir, J. M., *Hydrocarbon Process.*, (1), 167 (September 1969); (2), 121 (October 1969).
48. Lenoir, J. M., and Hipkin, H. G., *Am. Inst. Chem. Eng. J.*, **3**, 318 (1957).
49. Lenoir, J. M., and Koppany, C. R., *Hydrocarbon Process.*, **46**, 249 (1967).
50. Lenoir, J. M., and White, G. A., *Pet. Refiner*, **37**, 173 (1958).
51. Leva, M., "Tower Packings and Packed Tower Design," The United States Stoneware Co., 1951.
52. McAdams, W. H., "Heat Transmission," 3d ed., McGraw-Hill, New York, 1954.
53. McCabe, W. L., and Smith, J. C., "United Operations of Chemical Engineering," 3d ed., McGraw-Hill, New York, 1976.
54. McCabe, W. L., and Thiele, E. W., *Ind. Eng. Chem.*, **17**, 605 (1925).
55. Manley, D. B., *Hydrocarbon Process.*, January 1972, p. 113.
56. Maxwell, J. B., "Data Book on Hydrocarbons," Van Nostrand, Princeton, N.J., 1951.
57. Maxwell, J. B., and Bonnell, L. S., "Vapor Pressure Charts for Petroleum Hydrocarbons," Esso Research and Engineering Co., 1951.
58. Natural Gasoline Supply Men's Association, "Engineering Data Book," Natural Gasoline Association of America, Tulsa, Okla., 1957.
59. Osborne, A., *Chem. Eng.*, December 1964, p. 97.

60. Palmer, D. A., *Chem. Eng.*, June 9, 1975, p. 80.
61. Perry, J. H., "Chemical Engineers' Handbook," 3d ed., McGraw-Hill, New York, 1950.
62. Perry, R. H., "Chemical Engineers' Handbook," 4th ed., McGraw-Hill, New York, 1963.
63. Prausnitz, J. M., and Duffin, J. H., *Pet. Refiner*, **39**, 213 (May 1960).
64. Reid, R. C., and Sherwood, T. K., "The Properties of Gases and Liquids," 2d ed., McGraw-Hill, New York, 1960.
65. Renon, H., and Prausnitz, J. M., *Am. Inst. Chem. Eng. J.*, **14**, 135 (January 1968).
66. Renon, H., Doctorate Dissertation, University of California Berkeley, Appendix G has been deposited as document 9672, American Documentation Institute, Photoduplication Service, Library of Congress, Washington 25, D.C.
67. Schechter, R. S., and Van Winkle, M., *Pet. Refiner*, **36**, 301 (September 1957).
68. Scheiman, A. D., *Petro/Chem. Eng.*, (1), 28 (March 1965); (2), 75 (April 1965).
69. Skelland, A. H. P., "Diffusion and Mass Transfer," Wiley, New York, 1974.
70. Smith, E. C., *Chem. Eng.*, Nov. 17, 1958, p. 145.
71. Smoker, E. H., *Trans. Am. Inst. Chem. Eng.*, **34**, 165 (1938).
72. Stupin, W. J., *Hydrocarbon Process.* **45**, 222 (May 1956).
73. Treybal, R. E., "Mass Transfer Operations," 2d. ed., McGraw-Hill, New York, 1963.
74. Treybal, R. E., *Chem. Eng.*, Oct. 5, 1970, p. 95.
75. Van Winkle, M., *Pet. Refiner*, (1952).
76. Van Winkle, M., "Distillation," McGraw-Hill, New York, 1967.
77. Winn, F. W., *Refiner*, **33**, 482, 131 (1952).
78. Wohl, Kurt, *Chem. Eng. Prog.*, **49**, 218 (April 1953).
79. Wohl, Kurt, *Trans. Am. Inst. Chem. Eng.*, **42**, 216 (April 1946).
80. Young, S., "Distillation Principles and Processes," Macmillan, London, 1922.
81. Funk, E. W., and Prausnitz, J. M., *Am. Inst. Chem. Eng. J.*, **17**, 254 (1971).
82. Fenske, M. R., *Ind. Eng. Chem.*, **24**, 482 (1932).
83. Gilliland, E. R., *Ind. Eng. Chem.*, **32**, 1220 (1940).
84. Maas, J. H., *Chem. Eng.*, Apr. 16, 1973, p. 96.
85. Smith, B. D., "Design of Equilibrium Stage Processes," McGraw-Hill, New York, 1963.
86. Nelson, W. L., "Petroleum Refinery Engineering," 3d ed., McGraw-Hill, New York, 1949.
87. Kwauk, M., *Am. Inst. Chem. Eng. J.*, **2**, 240 (1956).
88. Berg, C., *Chem. Eng. Prog.*, **44**, 307 (April 1948).
89. DePriester, C. L., *Chem. Eng. Prog. Symp. Ser.* 7, **491** (1953).
90. Hala, E., Pick, J., Fried, V. and Vilim, O., "Vapor-Liquid Equilibrium," 2d ed., Pergamon, New York, 1967.
91. Null, H. R., "Phase Equilibrium in Process Design," Wiley, New York, 1970.
92. Prausnitz, J. M., "Molecular Thermodynamics of Fluid-Phase Equilibria," Prentice-Hall, Englewood Cliffs, N.J., 1969.
93. Pierotti, G. J., and Deal, C. H., *Ind. Eng. Chem.*, **51**, 95 (1959).
94. Wilson, G. M., and Deal, C. H., *Ind. Eng. Chem. Fundam.* **1**, 20 (1962).
95. Derr, E. L., and Deal, C. H., International Symposium on Distillation, Brighton England Proceedings, *Inst. Chem. Eng. Symp. Ser.*, **32**, 3:40 (1969).
96. Perry, R. H., "Chemical Engineers' Handbook," 5th ed., McGraw-Hill, New York, 1973.
97. Heitner, Irving, *Chem. Eng.*, **42**(10), 145 (October 1963).
98. Hill, R. G., *Chem. Eng.*, Jan. 1, 1968, p. 84.
99. Davies, James A., *Pet. Refiner*, parts I, II, August, September 1950.
100. Bolles, W. L., *Pet. Process.*, February 1956, p. 65; March 1956, p. 82; April 1956, p. 72; May 1956, p. 109.
101. "Flexitray Design Manual," Koch Engineering Co.
102. "Ballast Tray Design Manual," F. W. Glitch Co.
103. Standards of the Tubular Exchanger Manufacturers Association.
104. Wilson, O. G., *J. Ind. Eng. Chem.* 1928, p. 1363.
105. Palen, J. W., and Small, W. M., *Hydrocarbon Process.*, **43**, (11), 199 (November 1964).
106. Chase, J. D., July 31, 1967 p. 105; Aug. 28, 1967, p. 139.
107. Hughmark, G. A., and O'Connell, H. E., *Chem. Eng. Prog.*, **53**(3), (March 1957).
108. Weiler, D. W., Bonnet, F. W., and Leavitt, F. W., *Chem. Eng. Prog.*, **67**(9), (September 1971).
109. U.S. Dept. Commer. Circ. 500, parts I, II.
110. Gerster, J. A., *Ind. Eng. Chem.*, **52**(8), 645 (August 1960).
111. O'Connell, H. E., *Pet. Eng.*, **18**, (1947).
112. Sieder, E. N., Heat Transfer Tables, Alco Products Division, 1952.
113. Danatos, S., Hughson, R. V., Steyman, E. H., and Mamelian, A., *Chem. Eng.*, June 2, 1969.

Continuous Distillation: Separation of Multicomponent Mixtures

JOHN E. McCORMICK, B.Sc., Ph.D. *Associate Professor of Chemical Engineering, New Jersey Institute of Technology, Newark, N.J.; member, American Institute of Chemical Engineers; Licensed Professional Engineer; Consultant in Simulation, Process Design, and Mass Transfer.*

EDWARD C. ROCHE, Jr., M.E., M.S., Sc.D. *Professor of Chemical Engineering, New Jersey Institute of Technology, Newark, N.J.; member, American Institute of Chemical Engineers; Licensed Professional Engineer; Consultant in Computer Simulation, Distillation, and Process Design.*

GENERAL PRINCIPLES

In this section we are concerned with the fractional distillation of multicomponent mixtures, i.e., those with three or more components. The simple word "distillation" will be used in place of the more accurate phrase "fractional distillation." The illustrative examples will be from the petroleum and chemical process industries, both of which use distillation as a matter of course.

Multicomponent distillation is more difficult than binary distillation in that graphical techniques are not really useful, except in special cases. The computations require much more labor as well as a knowledge of the necessary approximation methods. Even before high-speed, large electronic computers can be used, one must do a great deal of hand calculation in order to develop necessary input data for an existing computer program. For multicomponent calculations we use the following:

- Material balances
- Energy balances
- Vapor-liquid equilibrium
- Estimation procedures

- Facilities limitations (cooling/heating restrictions)
- A well-organized approach

Multicomponent systems cannot be separated into the individual components through use of one single column or tower. Instead, the separation occurs between two of the components which are concentrated—one overhead in the distillate and one below in the bottoms. The basic ideas of binary distillation still apply but, necessarily, are adapted or extended to fit the new and more complex situation.

The two components upon which we focus our attention are called the "key components." The design aims at separating the "keys" to the extent desired in the product specifications. In other words, while we *can* specify the split of the keys, top and bottom, we *cannot* as well specify the distribution of the nonkey components. The latter distribution depends primarily on vapor-liquid equilibrium, the number of contact units, and to a lesser extent enthalpy considerations. The more volatile of the keys is called the "light key" (LK) and the less volatile is called the "heavy key" (HK). Selection of the keys will be discussed later.

The computational approach used is based on the ideas of binary distillation, with the "keys" being akin to the binary components. Even though we use shortcut or approximate methods for the hand calculations, the preliminary work has many steps. The overall approach is as follows:

1. Establish preliminary compositions for the top and bottoms products using available specifications for the keys and assuming a distribution for the nonkey components.

2. Determine column operating conditions (temperature and pressure) at the top and bottom, using appropriate vapor-liquid equilibrium data and limitations imposed by the facilities available.

3. Select and verify the key components.

4. Estimate the minimum number of equilibrium contact stages required for the desired separation.

5. Make sure there is consistency between the column operating conditions, product compositions (both top and bottom), and equilibrium data for the system.

6. Estimate the minimum reflux ratio for the required separation. Inherent in the estimate is the need to establish the thermal condition of the feed stream (or what is the degree of vaporization?).

7. Using available stage reflux correlations, obtain the number of equilibrium stages and the reflux ratio required to reach product specifications.

8. Estimate the feed-tray location within the column.

9. Estimate tray efficiency, column diameter and height, and overall column pressure drop. When column pressure drop is established, it should be compared with the assumed pressure drop.

Note that the procedure is not necessarily a single-pass calculation. In step 1 product compositions were assumed. In step 5 we require consistency, at which point we may have to revise the initial estimates used in step 1 and repeat the calculations. Similar comments apply to step 9.

This approach, while general, may not be the most appropriate. Sometimes the equilibrium relations are very complex. Sometimes the system is very nonideal and azeotropes are involved. Sometimes extraneous materials are deliberately added. Such systems and problems often arise in the manufacture of dyes, drugs, and chemical intermediates. These are covered later under Special Separations.

After the preliminary/approximate/shortcut calculations are finished, the work should be verified using a computer.

BASIC DATA

It is obvious that good design requires good data—especially so with multicomponent calculations. In this section we consider some basic relations and sources of data, and give references to textbooks and technical articles. The reason for such a limited treatment, quite simply, is that the estimation/prediction/extrapolation/interpolation of equilibrium data is an immense problem with no clear-cut unequivocal method of solution. Entire textbooks have been written on this subject alone—and doubtless more will be printed.

Vapor Pressure

Before we discuss vapor-liquid equilibrium, we must discuss vapor pressure. If the system is ideal, Raoult's and Dalton's laws apply and vapor-liquid equilibria can be calculated from vapor pressures alone. For nonideal systems the vapor pressure still is very important.

If possible we start with measured data. Most chemical reference books contain such data—often the very same data. The principal sources are listed in Refs. 13, 17, 25, 28, 40, 47, 50, 60, and 61.

Tabulated data may be useful for some hand calculations, but equations of representation are even better and more useful. Many correlations have been used, but only three or four are common.

1. Antoine equation (the most common)

$$\log P^0 = A - \frac{B}{T + C}$$

Antoine constants for many compounds are tabulated in Refs. 13, 17, and 61. Articles in the literature on vapor-liquid equilibrium often require vapor-pressure data, and frequently Antoine constants are listed. Hala et al.[27] tabulated a great many binary systems and included Antoine constants for all the compounds involved. If data are taken from a source such as Stull,[52] they may be reduced to the Antoine form using the procedures recommended by Thomson.[56]

2. An alternative (for computer use) is an equation of the form

$$\ln P^0 = a + bT + cT^2$$

These constants are not readily available and are usually fitted to the data by other computer programs (least-squares fit).

3. If data are very limited (at least one point is required, since vapor pressures cannot be calculated with any accuracy from structures, etc.), we can use the Clapeyron or Clausius-Clapeyron equations. The data point commonly available is the normal-boiling-point temperature, but we need the latent heat of vaporization as well. This approach usually is not very accurate:

$$\frac{dP^0}{dT} = \frac{\Delta H}{T \Delta V} \quad \text{or} \quad \frac{d \ln P^0}{d (1/T)} = - \frac{\Delta H}{R}$$

4. If data are available from several sources, a best set of data can be developed by a Cox plot[11] or using Stull's method.[52]

5. In case of desperation, use the methods evaluated in Reid[46] and Prausnitz.[43] The procedures outlined have been tested very reasonably, but there are exceptions in every approach; so caution is advised.

Relative Volatilities

If the multicomponent system is ideal (low pressures, not near critical temperature and pressure, and chemically similar components), we can use vapor-pressure data to get relative volatilities. From Raoult's and Dalton's laws:

$$x_i p_i^0 = p_i = P y_i \quad \text{or} \quad \frac{y_i}{x_i} = \frac{P_i^0}{P}$$

which holds for any component. Relative volatility is defined as

$$\alpha_{1-2} = \frac{y_1/x_1}{y_2/x_2} = \frac{y_1}{x_1} \frac{x_2}{y_2}$$

If $\alpha > 1$, substance 1 is the more volatile. The greater the value of α, the easier the separation. The relative volatility does not remain constant even in ideal systems because vapor pressures do not change proportionally between two components. Often the change in ratio is small and average relative volatilities are used.

We combine the Raoult-Dalton expression with the definition of relative volatility to get

$$\alpha_{1-2} = \frac{y_1}{x_1} \frac{x_2}{y_2} = \frac{P_1^0}{P} \frac{P}{P_2^0} = \frac{P_1^0}{P_2^0}$$

Relative volatilities can be calculated between any pair of components; however, as will be seen, one substance is chosen as a reference to which all other components are referred.

$$\alpha_{ir} = \frac{y_i}{x_i}\frac{x_r}{y_r} = \frac{P_i^0}{P_r^0}$$

Equilibrium Ratios

For many systems the equilibrium ratio K can be used. This is especially so for mixtures of hydrocarbons because the nonideality effects due to structural differences are minimized. By definition:

$$K_i = \frac{y_i}{x_i}$$

We observe that

$$\alpha_{ir} = \frac{K_i}{K_r}$$

Hence equilibrium ratios are easily converted into relative volatilities.

The real value of K lies in the fact that charts of $K = f(T,P)$ are available. Such charts are for systems of hydrocarbons only, which is somewhat restrictive. The De Priester charts[14] are reproduced here as Figs. 3 and 4.

Estimation Methods

The simple approach to equilibrium using K factors does not apply to a large number of systems. In particular, the nonhydrocarbon systems with components of different chemical classes are often not even remotely ideal. For such mixtures, the developments of thermodynamics are used. The superficially simple equation used is

$$K_i = \frac{y_i}{x_i} = \frac{\gamma_i P_i^0}{\nu_i P}$$

If the system is ideal, $\gamma_i = \nu_i = 1$. The real problem is estimation of γ_i, the activity coefficient, and ν_i, the fugacity coefficient. This task is complex because a large number of methods have been proposed over the years. Some of these methods are cited below, but it is beyond the scope of this text to treat the estimation of coefficients in even the sparsest manner. However, most texts on thermodynamics treat the subject in detail.[16,26,41,42,50,59] Besides this, Refs. 68 to 76 contain similar coverage with respect to distillation problems. Finally, there is a constant flow of new, "better," faster techniques. The common methods and the original references are:

Van Laar[59]
Margules[37]
Wohl[66]
Redlich and Kister[45]
Scatchard and Hamer[49]
Bonham[8]
Black[6,7]
Wilson[64]
NRTL[46]
ASOG[15]
UNIQUAC[1]
UNIFAC[21]

The last five of these methods are more difficult to use than earlier methods, but the results seem to be far superior. Articles illustrating the techniques are: Wilson,[29] NRTL,[33] ASOG,[39] and UNIFAC.[22] For a slightly less accurate approach, one can use a modified Wilson equation proposed by Tassios.[53] The Wilson equation does not apply to immiscible systems but has proved very good in predicting multicomponent behavior from binary vapor-liquid equilibrium data.

The amount of work implied and the data required seem most formidable. This is true.

But accurate calculations and successful designs need such data—otherwise the designer is faced with disastrous underdesign or uneconomic overdesign as real possibilities.

BASIC PROCEDURES

A few techniques and ideas are used time and time again. All-important are the ideas of equilibrium on a stage, mass balances, and energy (enthalpy) balances. Bubble and dew points are used frequently as well.

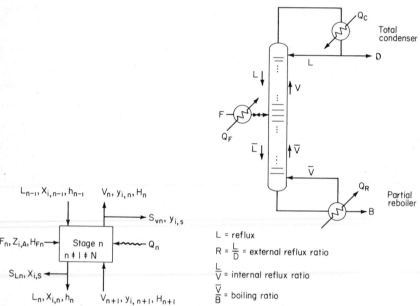

Fig. 1 The general equilibrium stage.

L = reflux

$R = \dfrac{L}{D}$ = external reflux ratio

$\dfrac{L}{V}$ = internal reflux ratio

$\dfrac{\overline{V}}{B}$ = boiling ratio

Fig. 2 Typical distillation tower.

The Equilibrium Stage

Figure 1 is a general equilibrium stage n, with components i. By convention, streams leaving a stage are identified with the stage number. Streams L_n and V_n are in equilibrium. P_n and T_n are constant. Streams F, L_{n-1}, and V_{n+1} enter the stage. Sidestreams S_{Ln} and S_{Vn} may be withdrawn. These are total streams in quantities such as moles per hour.

Mole fractions in the streams are identified as, say, $y_{i,n}$ in the V_n stream and $x_{i,n}$ in the L_n stream.

Componential flow rates are used sometimes. The notation is $V_n y_{i,n} = \gamma_{i,n}$, etc. In some cases the stage identifier is omitted but the meaning is made clear in context. The units are, say, moles of component i per unit time.

Sidestreams complicate the notation excessively as well as the calculations. The meaning of sidestream symbols will be made clear by the text. Q_n is enthalpy added or removed. The total vapor enthalpy is H_n and the liquid stream enthalpy is h_n. Sometimes componential enthalpies are used with the notation $H_{i,n}$ and $h_{i,n}$, but care must be taken to identify these clearly as enthalpies per unit mass of component i or enthalpy per total mass of component i in the stream.

The Mass Balance

Referring to Fig. 1, the total moles (or mass) entering the stage must equal the total moles leaving the stage. Naturally this applies to each component as well. Ignoring the sidestreams for simplicity, the overall balance is

$$L_n + V_n = F + L_{n-1} + V_{n+1}$$

and for each component:

$$L_n x_{i,n} + V_n y_{i,n} = F z_{i,F} + L_{n-1} x_{i,n-1} + V_{n+1} y_{i,n+1}$$

The Enthalpy Balance

Referring to Fig. 1, the corresponding enthalpy balances are

$$h_n + H_n = h_F + h_{n-1} + H_{n+1} + Q_n$$

or

$$L_n h_n + V_n H_n = F h_F + L_{n-1} h_{n-1} + V_{n+1} H_{n+1} + Q_n$$

depending on the definition used for h and H.

Extension of the Basic Equilibrium Stage

The use of multiple equilibrium stages to make the desired separation is an obvious extension. The arrangement of the equilibrium stage units can result in parallel, cross-, or countercurrent flow. For the distillation process the most efficient configuration is that of a countercurrent arrangement (vapor and liquid flowing in opposite directions). A schematic representation is given in Fig. 2.

Bubble Points

By definition a saturated-liquid stream is at the boiling (or bubble) point. The first and tiniest of bubbles formed has a composition different from the liquid, but the amount of material in the bubble is too small to change the composition of the liquid. Obviously, the liquid and the bubble are in equilibrium—a restriction that enables us to determine vapor composition and pressure (or composition and temperature if the liquid is at a prechosen pressure).

One of the components is chosen as a reference substance—often this is the heavy key. In general the reference component is the dominant component in the mixture or that component having the greatest contribution to the design equation. In arriving at the bubble-point relationship, the following definitions are used:

$$\frac{y_i}{x_i} = K_i$$

and

$$\alpha_{ir} = \frac{K_i}{K_r}$$

Mathematically, the design equation bubble-point calculation is

$$\Sigma y_i = \Sigma K_i x_i = 1.0$$

Using the relative volatility α_{ir}, the design equation becomes

$$\Sigma y_i = K_r \Sigma \alpha_{ir} x_i = 1.0$$

from which

$$K_r = \frac{1}{\Sigma \alpha_{ir} x_i}$$

These formulas can be used to ease and accelerate hand calculations of the bubble point even though these are trial-and-error operations. The format shown in Fig. 5 is useful for the bubble-point calculations. Part of the computation procedure is the comparison of K_r and $1/\Sigma \alpha_{ir} x_i$. If K_r and the reciprocal of the summation $\Sigma \alpha_{ir} x_i$ are equal (within the accuracy of the data), we can calculate y_i:

$$y_i = \frac{\alpha_{ir} x_i}{\Sigma \alpha_{ir} x_i}$$

If $K_r \neq 1/\Sigma \alpha_{ir} x_i$, use the reciprocal summation as the new estimate of K_r and thus find the new estimated temperature (if the pressure is fixed), since the reference component's vapor-liquid equilibrium constant is functionally dependent on temperature. If the temperature is preset, the revised/new pressure is obtained using

$$P_{new} = \frac{P_{assumed} K_{r,assumed}}{K_{r,calculated}}$$

where

$$K_{r,calculated} = \frac{1}{\Sigma \alpha_{ir} x_i}$$

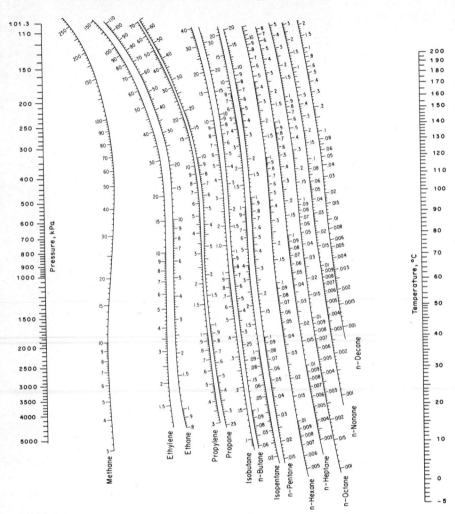

Fig. 3a DePriester chart—high-temperature range: metric units. (*From* Chem. Eng. Prog., *April 1978. Courtesy of American Institute of Chemical Engineers.*)

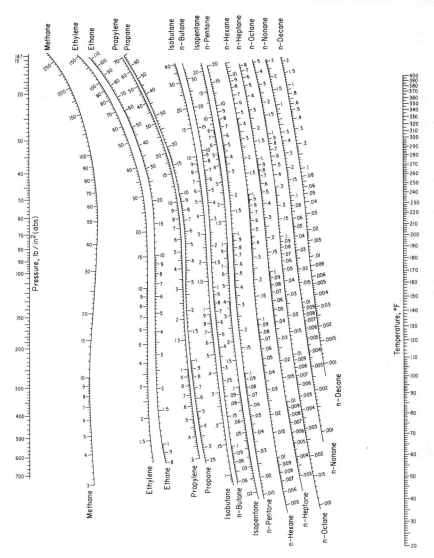

Fig. 3b De Priester chart—high-temperature range: English units. [*From* Chem. Eng. Prog., 49(7), 1 (1953). *Courtesy of American Institute of Chemical Engineers.*]

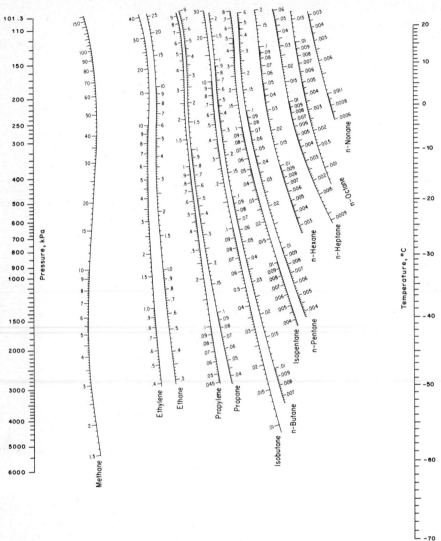

Fig. 4a DePriester chart—low-temperature range: Metric Units. (*Reprinted from* Chem. Eng. Prog., *April 1978. Courtesy of American Institute of Chemical Engineers.*)

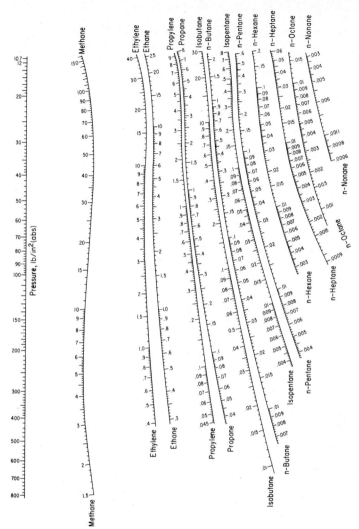

Fig. 4b DePriester chart—low-temperature: English units. [*Reprinted from* Chem. Eng. Prog., *Symp. Ser.* **49**, *1 (1953). Courtesy of American Institute of Chemical Engineers.*]

Project _____

Tower Designation _____

Stage Location _____

    Temperature (Set/Assumed) _____

    Pressure (Assumed/Set) _____

Component Name	X_i	K_i	α_{ir}	$\alpha_{ir} X_i$	y_i

---(REF)	$K_i = K_r$	1.00			

Total	1.000			$\Sigma\, \alpha_{ir}\, X_i = 1/K_r$	1.000

Fig. 5 Bubble-point format.

EXAMPLE 1

Determine the bubble-point temperature of the following mixture, the pressure being previously set at 100 lb/in²(abs).

Component	F, mol	mol wt
C_3	10	44.1
iC_4	30	58.1
C_4	50	58.1
iC_5	40	72.1
C_5	10	72.1
	140	

a. Determination of the reference component:

$$\text{lb} \cdot \text{mol} = 140$$
$$\text{lb} = \Sigma f_i \times (\text{mol wt})_i = 8694.0$$
$$\text{Avg mol wt} = \frac{8694.0}{140.0} = 62.1$$

The average molecular weight is between C_4 and iC_5. Since a bubble point is required, the more volatile of the pair is selected as the reference component, namely, C_4.

   b. Initial temperature estimate for the bubble point. Using vapor-pressure data, one obtains a temperature of 146°F (from Fig. 3 with $K_{C_4} = 1$).

$$P^0_{C_4}(T) = 100 \text{ lb/in}^2(\text{abs})$$

   c. Iterative calculation for the bubble-point temperature (K data from Fig 3).

Component	F, mol	$T = 146°F$		$T = 145°F$	
		K	α	K	α
C_3	10	2.80	2.80	2.77	
iC_4	30	1.35	1.35	1.34	
C_4 (R)	50	1.00	1.00	0.99	1.00
iC_5	40	0.48	0.48	0.475	
C_5	10	0.39	0.39	0.385	
	140				

For $T = 146°F$:

$$\Sigma \alpha_i x_i = \frac{141.600}{140} = 1.011$$

$$K_{R, \text{ calc}} = \frac{1}{1.011} = 0.989 \neq K_{C_4} = 1.0$$

$$T_{\text{calc}} = 145°F$$

For $T = 145°F$: Since the previous iteration had 1.011, which is nearly unity,

$$\Sigma K_i x_i = \frac{140.25}{140} = 1.002$$

The bubble-point temperature is 145°F.

d. Evaluation of vapor composition in equilibrium with the bubble-point liquid; equilibrium temperature is 145°F. The vapor composition can be obtained using the relation $y_i = K_i x_i$.

Component	F, mol	x_i	K_i	$\sim y_i$	y_i
C_3	10	0.071	2.77	0.197	0.197
iC_4	30	0.215	1.34	0.288	0.287
C_4	50	0.357	0.99	0.354	0.353
iC_5	40	0.286	0.475	0.136	0.136
C_5	10	0.071	0.385	0.027	0.027
	140	1.000		1.002	1.000

Dew Points

For dew-point calculations we have vapor compositions available and desire to obtain the pressure or temperature at which the first drop (dew) of liquid is formed. Using the basic equations presented in the development of the bubble-point design equation, an analogous design equation for dew-point calculations can be obtained.

$$\Sigma x_i = \Sigma \frac{y_i}{K_i} = 1.000$$

Again using relative-volatility data the dew-point design equation becomes

$$\Sigma x_i = \frac{1}{K_r} \Sigma \frac{y_i}{\alpha_{ir}} = 1.0$$

From which

$$K_r = \Sigma \frac{y_i}{\alpha_{ir}}$$

For dew-point calculations the format shown in Fig. 6 is helpful.

Again, we compare K_r and $\Sigma(y_i/\alpha_{ir})$ and correct the temperature or the pressure as in the bubble-point format. When the assumed and calculated values of K_r agree,

$$x_i = \frac{y_i/\alpha_{ir}}{\Sigma(y_i/\alpha_{ir})}$$

Even though the dew-point calculation, like the bubble-point calculation, is trial-and-error, the use of relative volatilities usually results in two to four trials.

EXAMPLE 2

Determine the dew-point temperature of Example 1 at the same system pressure of 100 lb/in²(abs).

Component	F, mol	mol wt
C_3	10	44.1
iC_4	30	58.1
C_4	50	58.1
iC_5	40	72.1
C_5	10	72.1
	140	

a. Determination of the reference component:

$$\text{lb·mol} = 140$$
$$\text{lb} = \Sigma f_i \, (\text{mol wt})_i = 8694.0$$
$$\text{Avg mol wt} = \frac{8694.0}{140.0} = 62.1$$

Since a dew-point calculation is required, the less volatile component of the pair, C_4/iC_5, is selected as the reference component, namely, iC_5.

Project _____

Tower Designation_____

Stage Location_____

Temperature (Set/Assumed)_____

Pressure (Assumed/Set)_____

Component Name	y_i	K_i	α_{ir}	$\dfrac{y_i}{\alpha_{ir}}$ ·	x_i
——					
——					
——					
—— (REF)		$K_i = K_r$	1.00		
Total	1.000			$\Sigma \dfrac{y_i}{\alpha_{ir}} = K_r$ 1.000	

Fig. 6 Dew-point format.

b. Initial temperature estimate for the dew point. Using vapor-pressure data, one obtains a temperature of 214°F (Fig. 3 with $K_{iC_5} = 1.0$).

$$P^0_{iC_5}(T) = 100 \ \text{lb/in}^2(\text{abs})$$

c. Iterative calculation for the dew-point temperature (K data from Fig. 3).

Component	F, mol	$T = 214°F$		$T = 173°F$		$T = 170°F$	
		K	α	K	α	K	α
C_3	10	4.60	4.60	3.50	5.38	3.40	
iC_4	30	2.50	2.50	1.78	2.74	1.75	
C_4	50	2.00	2.00	1.36	2.09	1.33	
iC_5 (R)	40	1.00	1.00	0.65	1.00	0.63	1.00
C_5	10	0.86	0.86	0.55	0.85	0.53	
	140						

For $T = 214°F$:

$$\Sigma \frac{y_i}{\alpha_i} = \frac{90.802}{140} = 0.649$$
$$K_{R,\text{calc}} = 0.649 \neq K_{iC_5} = 1.00$$
$$T_{\text{calc}} = 173°F$$

For $T = 173°F$:

$$\Sigma \frac{y_i}{\alpha_i} = \frac{88.496}{140} = 0.632$$
$$K_{R,\text{calc}} = 0.632 \neq K_{iC_4} = 0.63$$
$$T_{\text{calc}} = 170°F$$

For $T = 170°F$: Since the iteration $|\Delta T| = 3°F$, the dew-point design equation will be evaluated.

$$\Sigma \frac{y_i}{K_i} = \frac{140.038}{140} = 1.0$$

Dew-point temperature is 170°F.

d. Evaluation of liquid composition in equilibrium with the dew-point vapor; equilibrium temperature is 170°F. The liquid composition can be obtained using the relation

$$x_i = \frac{y_i}{K_i}$$

Component	F, mol	y_i	K_i	x_i
C_3	10	0.071	3.40	0.021
iC_4	30	0.215	1.75	0.123
C_4	50	0.357	1.33	0.268
iC_5	40	0.286	0.63	0.454
C_5	10	0.071	0.53	0.134
	140	1.000		1.000

Multicomponent Flash Calculations

In addition to the basic equilibrium calculations (bubble points and dew points) previously discussed, the equilibrium flash calculation must also be covered. This calculation concerns the determination of the equilibrium separation of a stream, with the result being a liquid-vapor equilibrium mixture. In distillation systems, this condition arises in determining the thermal condition of the feed, and occasionally in determining the phase condition of the effluent from the condenser and the reboiler.

The flash calculations encountered can be classified into the following four categories:
- Pressure and temperature specified
- Temperature and degree of vaporization specified
- Pressure and degree of vaporization specified
- Pressure and stream enthalpy specified

Pressure and Temperature Specified This specific flash calculation is generally encountered when analyzing the effluents from the distillation column's condenser and reboiler. For the case of pressure and temperature being specified the problem must be bounded. The bounding is necessary so as to avoid the nonconvergent trial-and-error calculation resulting from the assumption of a two-phase condition when in reality the result is a subcooled liquid or a superheated vapor. The bounding is accomplished by determining the bubble-point and/or dew-point temperature corresponding to the specified pressure. For a valid two-phase equilibrium calculation the following relationship must be satisfied:

$$T_{\text{bubble point}} < T_{\text{specified}} < T_{\text{dew point}}$$

The existence of a valid two-phase flash can also be verified using the design equations for the bubble point and the dew point, and equilibrium data evaluated at the specified pressure and temperature. The design equations to be used are

Bubble point: $\qquad f = \Sigma K_i x_i$

Dew point: $\qquad F = \Sigma \dfrac{y_i}{K_i}$

The appropriate compositional values to be used in the above relationships are the composite stream values z_i. Thus for the bubble-point relationship $x_i = z_i$, and for the dew point $y_i = z_i$. Using the vapor-liquid data associated with the specified pressure and temperature, the tabular summary shown in Table 1 yields the phase condition.

TABLE 1 Equilibrium Flash Criteria

	$f = \Sigma K_i z_i$	$F = \Sigma \dfrac{z_i}{K_i}$
Subcooled liquid	<1	>1
Bubble point	=1	>1
Two-phase condition	>1	>1
Dew point	>1	=1
Superheated vapor	>1	<1

Thus, as can be seen from Table 1, the condition for a valid two-phase equilibrium flash is

$$f = \Sigma K_i z_i > 1.0$$

$$F = \Sigma \frac{z_i}{K_i} > 1.0$$

For a two-phase vapor-liquid equilibrium calculation the vapor-liquid separation must be evaluated. Using the componential material balance:

$$Fz_i = Vy_i + Lx_i$$

and the definition of the vapor-liquid equilibrium value:

$$K_i = \frac{y_i}{x_i}$$

the resulting design equation is

$$V = \Sigma \frac{Fz_i K_i}{K_i + N}$$

where $N = L/V$. The solution methodology is to assume successive values of N and calculate the resulting value of N:

$$N_{calc} = \frac{V_{calc}}{F - V_{calc}}$$

Initial values of N that should be considered are $N = 0.01$ and $N = 100$. Successive values of N are obtained by interpolation/extrapolation using the plot shown in Fig. 7. Because of the wide range of values for N, the plot should be made on log-log paper (preferably 3 cycle × 3 cycle).

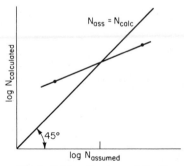

Fig. 7 Trial-and-error search for N.

EXAMPLE 3

Determine the equilibrium separation of the following mixture, the equilibrium conditions being 160°F and 100 lb/in²(abs) (K data from Fig. 3).

Component	F, mol	K [160°F, 100 lb/in² (abs)]
C_3	10	3.20
iC_4	30	1.60
C_4	50	1.20
iC_5	40	0.57
C_5	10	0.47
	140	

a. Verification of a two-phase flash. From Examples 1 and 2 the bubble-point and dew-point temperatures are available.

$$T = 160°F > 145°F = T_{\text{bubble point}}$$
$$T = 160°F < 170°F = T_{\text{dew point}}$$

Since $T_{\text{bubble point}} < T < T_{\text{dew point}}$, a two-phase condition exists. As an alternate the bubble-point and dew-point design equations can be evaluated, and based on these results the phase condition can be determined:

$$f = \Sigma K_i z_i = \frac{167.50}{140} = 1.196 > 1.0$$

$$F = \Sigma \frac{z_i}{K_i} = \frac{154.994}{140} = 1.107 > 1.0$$

Since both f and F are greater than 1.00, a two-phase condition exists.

    b. Determination of vapor/liquid separation. A value of $N = L/V$ is assumed. The design equation

$$V = \frac{F z_i K_i}{K_i + N}$$

is evaluated and N is calculated:

$$N_{\text{calc}} = \frac{V_{\text{calc}}}{F - V_{\text{calc}}}$$

As an initial estimate is made, $N_{\text{assumed}} = 1.0$

Component	Fz_i	K_i	Fz_iK_i	Tabulated values of $K_i + N$			
				$N = 1.0$	$N = 0.5$	$N = 0.6$	$N = 0.6_{i5}$
C_3	10	3.20	32.0	4.20	3.70	3.80	3.815
iC_4	30	1.60	48.0	2.60	2.10	2.20	2.215
C_4	50	1.20	60.0	2.20	1.70	1.80	1.815
iC_5	40	0.57	22.8	1.57	1.07	1.17	1.185
C_5	10	0.47	4.7	1.47	0.97	1.07	1.085
	140						

For $N = 1.0$:
$$V = \Sigma \frac{F z_i K_i}{K_i + N} = 71.073$$
$L = F - V = 68.927$
$N_{\text{calc}} = 0.9698$
Reduce N_{assumed} to 0.5

For $N = 0.5$:
$$V = \Sigma \frac{F z_i K_i}{K_i + N} = 92.954$$
$L = F - V = 47.046$
$N_{\text{calc}} = 0.5061$
Increase N_{assumed} to 0.60

For $N = 0.60$:
$$V = \Sigma \frac{F z_i K_i}{K_i + N} = 87.452$$
$L = F - V = 52.548$
$N_{\text{calc}} = 0.6009$
Increase N to 0.615

For $N = 0.615$:
$$V = \Sigma \frac{F z_i K_i}{K_i + N} = 86.689$$
$L = F - V = 53.311$
$N_{\text{calc}} = 0.6150$

Calculation completed: $L = 53.311$ mol and $V = 86.689$ mol.

    c. Determination of resulting vapor and liquid separation. The following design equations are necessary to evaluate v_i and l_i:

$$v_i = \frac{F z_i K_i}{K_i + N}$$
$$l_i = F z_i - v_i$$
$$y_i = \frac{v_i}{V}$$
$$x_i = \frac{l_i}{L}$$

Component	Feed		Vapor		Liquid	
	f_i	Z_i	v_i	y_i	l_i	x_i
C_3	10	0.071	8.388	0.097	1.612	0.030
iC_4	30	0.215	21.670	0.250	8.330	0.156
C_4	50	0.357	33.058	0.381	16.942	0.318
iC_5	40	0.286	19.241	0.222	20.759	0.390
C_5	10	0.071	4.332	0.050	5.668	0.106
	140	1.000	86.689	1.000	53.311	1.000

Temperature and Degree of Vaporization Specified   For this case the pressure as well as the component distribution between the equilibrium phases is to be determined. The basic relationships are the componential material balance and the definition of the vapor-liquid equilibrium constant. With N again defined as the ratio of L/V, the following design equation is obtained:

$$L_{\text{calc}} = \sum \frac{Fz_iN}{K_i + N}$$

The computational procedure is to assume a pressure P, obtain the necessary equilibrium data, and evaluate the above design relationship. Comparison of the calculated and known value of the liquid-phase rate makes available sufficient data to correct the assumed pressure via the following guidelines:

$$L_{\text{calc}} > L_{\text{given}} : \text{decrease } P$$
$$L_{\text{calc}} < L_{\text{given}} : \text{increase } P$$

EXAMPLE 4

Determine the equilibrium pressure for the following mixture, the temperature being 100°F and 10 mol of vapor.

Component	F, mol
C_3	10
iC_4	30
C_4	50
iC_5	40
C_5	10
	140

a. Estimation of equilibrium pressure to start the trial-and-error calculation. The following estimate of the resulting liquid is used in conjunction with vapor-pressure data to generate the initial pressure value. (Vapor-pressure data are obtained from Fig. 3 with $K = 1$.)

Component	Moles	$P^0(100°\text{F})$, lb/in²(abs)
C_3		
iC_4	30	76
C_4	50	50
iC_5	40	19.5
C_5	10	15
	130	

Using Dalton's and Raoult's laws,

$$P = \Sigma P_i^0 x_i = \frac{5710}{130} = 43.9 \text{ lb/in}^2(\text{abs})$$

Since C_3 will exist in the liquid, the iterative computation will be started with $P = 45$ lb/in²(abs).
b. Determination of equilibrium pressure (K data from Fig. 3). $N = 130/10 = 13$.

Component	Fz_i, mol	Fz_iN	$P =$ 45 lb/in²(abs) K	$K + N$	$P = 48$ lb/in²(abs) K	$K + N$	$P =$ 50 lb/in²(abs) K	$K + N$
C_3	10	130	3.60	16.60	3.40	16.40	3.30	16.30
iC_4	30	390	1.55	14.55	1.50	14.50	1.40	14.40
C_4	50	650	1.15	14.15	1.05	14.05	1.00	14.00
iC_5	40	520	0.47	13.47	0.425	13.425	0.42	13.42
C_5	10	130	0.37	13.37	0.35	13.35	0.33	13.33
	140							

For $P = 45$ lb/in²(abs):

$$L = \sum \frac{Fz_iN}{K_i \neq N} = 128.899$$

$L_{given} = 130.0$
Increase P to 48 lb/in²(abs).
For $P = 48$ lb/in²(abs):

$$L = \sum \frac{Fz_iN}{K_i + N} = 129.558$$

$L_{given} = 130.0$
Increase P to 50 lb/in²(abs).

For $P = 50$ lb/in²(abs):

$$L = \sum \frac{Fz_iN}{K_i + N} = 129.988$$

$$L \simeq L_{given}$$

Equilibrium pressure for 7.14% vaporization is 50 lb/in²(abs).

c. Determination of resulting vapor and liquid separation. The following design equations are necessary to evaluate v_i and l_i:

$$l_i = \frac{Fz_iN}{K_i + N}$$
$$v_i = Fz_i - l_i$$
$$x_i = \frac{l_i}{L}$$
$$y_i = \frac{v_i}{V}$$

Component	Feed		Liquid		Vapor	
	Fz_i	z_i	l_i	x_i	v_i	y_i
C_3	10	0.071	7.975	0.061	2.025	0.202
iC_4	30	0.215	27.083	0.208	2.917	0.291
C_4	50	0.357	46.430	0.358	3.570	0.357
iC_5	40	0.286	38.748	0.298	1.252	0.125
C_5	10	0.071	9.752	0.075	0.248	0.025
	140	1.000	129.988	1.000	10.012	1.000

Pressure and Degree of Vaporization Specified The solution method for this case or category is very similar to the previous case; in fact the same design relationship is applicable. The computational procedure is to assume a temperature T, obtain the necessary equilibrium data, and evaluate the design relationships $N = L/V$, or $L = NF/(N + 1)$.

$$L_{calc} = \sum \frac{Fz_iN}{K_i + N}$$

Again, comparison of the calculated and known value of the liquid-phase rate makes available sufficient data to correct the assumed temperature. The following guidelines can be used to improve the next trial temperature

$$L_{calc} > L_{given}: \text{increase } T$$
$$L_{calc} < L_{given}: \text{decrease } T$$

EXAMPLE 5

Determine the equilibrium temperature for the following mixture, the pressure being 75 lb/in²(abs) and 50 mol of liquid.

Component	F, mol
C_3	10
iC_4	30
C_4	50
iC_5	40
C_5	10
	140

a. Estimation of equilibrium temperature to start the trial-and-error calculation. Via the following liquid-phase estimate an initial value for the equilibrium temperature can be obtained, in conjunction with vapor-pressure data.

Component	L, mol
C_3	
iC_4	
C_4	
iC_5	40
C_5	10
	50

Reference component is iC_5, by inspection, and

$$P^0_{iC_5}(T) = 75 \text{ lb/in}^2 \text{ (abs)}$$
$$T = 190°F$$

The temperature was obtained using vapor-pressure data (from Fig. 3 with $K_{iC_5} = 1$).
b. Determination of equilibrium temperature (K data from Fig. 3). $N = 50/90 = 0.556$.

	Fz_i	Fz_iN	$T = 190°F$		$T = 150°F$		$T = 135°F$		$T = 140°F$	
			K	$K + N$	K	$K + N$	K	$K + N$	K	$K + N$
C_3	10	5.556	5.0	5.556	3.75	4.306	3.3	3.856	3.45	4.006
iC_4	30	16.667	2.625	3.181	1.825	2.381	1.55	2.106	1.65	2.206
C_4	50	27.778	2.05	2.606	1.475	2.031	1.15	1.706	1.233	1.789
iC_5	40	22.222	1.0	1.556	0.63	1.186	0.52	1.076	0.555	1.111
C_5	10	5.556	0.84	1.396	0.52	1.076	0.43	0.981	0.455	1.011
	140									

For $T = 190°F$:
$$L = \sum \frac{Fz_iN}{K_i + N} = 35.160$$
$L_{\text{given}} = 50.0$
Decrease T to $150°F$
For $T = 150°F$:
$$L = \sum \frac{Fz_iN}{K_i + N} = 45.868$$
$L_{\text{given}} = 50.0$
Decrease T to $135°F$

For $T = 135°F$:
$$L = \sum \frac{Fz_iN}{K_i + N} = 51.953$$
$L_{\text{given}} = 50.0$
Increase T to $140°F$
For $T = 140°F$:
$$L = \sum \frac{Fz_iN}{K_i + N} = 49.967$$
$L \simeq L_{\text{given}}$
Equilibrium temperature is $140°F$

c. Determination of resulting vapor and liquid separation. The necessary design equations are given as part of Example 4, part *c.*

Component	Feed		Liquid		Vapor	
	Fz_i	z_i	l_i	x_i	v_i	y_i
C_3	10	0.071	1.387	0.028	8.613	0.096
iC_4	30	0.215	7.555	0.151	22.445	0.249
C_4	50	0.357	15.527	0.311	34.473	0.383
iC_5	40	0.286	20.002	0.400	19.998	0.222
C_5	10	0.071	5.496	0.110	4.504	0.050
	140	1.000	49.967	1.000	90.033	1.000

Pressure and Stream Enthalpy Specified   This specific flash calculation is normally encountered in determining the feed condition as it enters the column. Of the four types of flash calculations this final case is the most difficult in that for a two-phase condition the calculations are a double trial-and-error procedure.

To bound the problem, the stream's bubble-point and dew-point temperatures are evaluated at the specified pressure. For the bubble-point and dew-point temperatures the corresponding enthalpy values are evaluated using standard mixing rules:

$$h_L = \Sigma h_i(T_{\text{bubble point}})z_i$$
$$H_v = \Sigma H_i(T_{\text{dew point}})z_i$$

From the given feed enthalpy the phase condition of the resulting flash can be determined. Table 2 summarizes the various feed conditions that can occur.

The two-phase feed condition requires the dual trial-and-error determination of the temperature and the liquid-vapor phases. The most expedient hand-oriented procedure is

TABLE 2 Feed-Phase Conditions and Enthalpy

Phase condition	Enthalpy value
Subcooled liquid	$h_{feed} < h_L$
Bubble-point feed	$h_{feed} = h_L$
Two-phase feed	$h_{feed} > h_L$
	$h_{feed} < H_V$
Dew-point feed	$H_{feed} = H_V$
Superhead vapor	$H_{feed} > H_V$

to estimate the equilibrium temperature and then utilize the computational procedure as described under Pressure and Temperature Specified. Use of the Case 1 procedure requires a minimal number of vapor-liquid equilibrium determinations.

The computational procedure to be used in conjunction with the "PT flash" requires an estimate of the equilibrium temperature. To obtain the required temperature, the following can be used:

1. Estimate the degree of vaporization using the material balance and enthalpy balance:

$$F = L + V$$
$$Fh_{feed} = Lh + VH$$
$$\frac{V}{F} = \frac{h_{feed} - h}{H - h}$$

2. Estimate the equilibrium temperature from a temperature-fraction vaporized curve. The bounds on the plot have been previously determined, and are

$$\frac{V}{F} = 0.0 \qquad T = T_{bubble\ point}$$
$$\frac{V}{F} = 1.0 \qquad T = T_{dew\ point}$$

Figure 8 is an auxiliary curve relating V/F and T. As the calculations proceed, the updated curve is developed and is used to more easily estimate the temperature required for a chosen value of V/F.

3. Using the computational procedure of Case 1 requires iterative values of $N = L/V$ to be assumed and verified by direct evaluation. As discussed previously, the initial pair of assumptions of $N = 0.01$ and $N = 100.0$ is arbitrary. For the "PH flash" the initial estimate

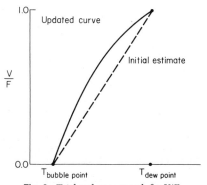

Fig. 8 Trial-and-error search for V/F.

of N can be obtained from the enthalpy data

$$N = \frac{L}{V} = \frac{H - h_{feed}}{h_{feed} - h}$$

4. The initial set of enthalpy data that should be used are

$$h = h_L(\text{bubble point})$$
$$H = H_V(\text{dew point})$$

For subsequent iterations the values of H and h can be evaluated directly from the PT flash

$$h = \Sigma h_i(T)x_i$$
$$H = \Sigma H_i(T)y_i$$

NOTE: The convention used in this section is that componential enthalpy data is temperature dependent. The enthalpy terms are functionally represented by $H_i(T)$ and $h_i(T)$.) Using these equations requires use of the following expressions to obtain the liquid and vapor compositions:

$$y_i = \frac{v_i}{V} = \frac{Fz_iK_i}{V(K_i + N)}$$
$$x_i = \frac{l_i}{L} = \frac{Fz_i - Vy_i}{F - V}$$

EXAMPLE 6

Determine the resulting equilibrium temperature and associated vapor-liquid separation (if any) for the following mixture, the pressure being 100 lb/in²(abs) and the enthalpy being 15,000 Btu/lb mol. The enthalpy datum is saturated liquid at −200°F, with the vapor enthalpy data modified via the partial-enthalpy method.

| Component | f, mol | mol wt | Enthalpy, Btu/lb | | | |
			h_L 145°F	h_L 170°F	H_V 145°F	H_V 170°F
C_3	10	44.1	195	210	328	340
iC_4	30	58.1	182	195	298	312
C_4	50	58.1	189	205	319	331
iC_5	40	72.1	175	192	303	314
C_5	10	72.1	180	195	309	321
	140					

a. Flash-calculation bounds. From Example 1, the bubble-point temperature for a pressure of 100 lb/in²(abs) has been determined to be 145°F. For this temperature the mixture per mole is obtained as follows:

$$h(\text{bubble point}) = \frac{\Sigma f_i h_i(145°F)(\text{mol wt})_i}{\Sigma f_i}$$
$$= \frac{1,586,746}{140} = 11,334 \text{ Btu/lb} \cdot \text{mol}$$

The feed enthalpy $h_F = 15,000$ Btu/lb mol; and, therefore, since $h_L(\text{bubble point}) < h_F$, the given stream is either partially or totally vaporized.

From Example 2, the dew-point temperature for a pressure of 100 lb/in²(abs) has been determined to be 170°F. For this temperature the enthalpy per mole of mixture is obtained as follows:

$$H_V(\text{dew point}) = \frac{\Sigma f_i H_i(170°F)(\text{mol wt})_i}{f}$$
$$= \frac{2,792,328}{140} = 19,945 \text{ Btu/lb} \cdot \text{mol}$$

Again comparing the dew-point enthalpy and the feed enthalpy $h_F < H_V$ (dew point), one obtains the result that the given stream is a liquid-vapor equilibrium mixture.

b. To minimize the computational effort, the following procedure will be used:

- Estimate the flash temperature:

$$T_{\text{bubble point}} < T < T_{\text{dew point}}$$

- Solve the "*PT*" flash for the liquid-vapor split. The initial value of $N = L/V$ can be obtained from an overall enthalpy balance

$$\frac{L}{V} = \frac{H_V - h_F}{h_F - h_L}$$

From the above flash calculation obtain the vapor and liquid enthalpies H_v and h_L. In addition, obtain the fraction vaporized V/F.

- From the "*PT*" flash update the "temperature-fraction vaporized" plot, whose bounds are already established.

$$\frac{V}{F} = 0.0 \qquad T = T_{\text{bubble point}} = 145°F$$

$$\frac{V}{F} = 1.0 \qquad T = T_{\text{dew point}} = 170°F$$

For the first trial the vapor phase will be approximated by the dew point and the liquid phase will be approximated by the bubble point.

  c. Trial 1.

$$\frac{V}{F} = \frac{h_F - h}{H - h} = \frac{15,000 - 11,334}{19,945 - 11,334} = 0.426$$

Assuming a linear relationship between temperature and fraction vaporized, the flash temperature can be obtained:

$$\frac{T - 145}{0.426 - 0.0} = \frac{170 - 145}{1.0 - 0.0}$$
$$T = 155.6°F$$

For the flash calculation, $T = 155°F$ will be assumed, and the assumed value of N will be $N = 1.35$, since

$$N = \frac{L}{V} = \frac{1 - V/F}{V/F} = \frac{1.0 - 0.426}{0.426} = 1.347$$

Component	f, mol	$K_i(155°F)$	$F \times K_i$	$K_i + N$
C_3	10	30.5	30.5	4.40
iC_4	30	1.50	45.0	2.85
C_4	50	1.15	57.5	2.50
iC_5	40	0.53	21.2	1.88
C_5	10	0.44	4.4	1.79

$$N = \frac{L}{V} = \frac{1 - V/F}{V/F} = \frac{1.0 - 0.426}{0.426} = 1.347$$

$$V_{\text{calc}} = \sum \frac{Fz_i K}{K + N} = 59.456 \text{ mol}$$

$$L_{\text{calc}} = F - V_{\text{calc}} = 140.0 - 59.456 = 80.544$$

$$N_{\text{calc}} = \frac{L_{\text{calc}}}{V_{\text{calc}}} = \frac{80.544}{59.456} = 1.355\ldots\text{good agreement}$$

Determination of flash breakdown and resulting stream enthalpy data for $T = 155°F$.

Component	f, mol	mol wt	Vapor		Liquid	
			v_i	H_i	l_i	h_i
C_3	10	44.1	6.932	332.8	3.068	201.0
iC_4	30	58.1	15.789	303.6	14.211	187.2
C_4	50	58.1	23.000	323.8	27.000	195.4
iC_5	40	72.1	11.277	307.4	28.723	181.8
C_5	10	72.1	2.458	313.8	7.542	186.0
	140		59.456		80.544	

Vapor enthalpy = $\Sigma H_i(\text{mol wt})v_i = 1,118,486$ Btu
Liquid enthalpy = $\Sigma h_i \times (\text{mol wt})_i \times l_i = 965,920$ Btu

$$h_{F,\text{calc}} = \frac{2,084,406}{140} = 14,889 \text{ Btu/mol}$$

$$H_V = \frac{1,118,486}{59.456} = 18,812 \text{ Btu/mol}$$

$$h_L = \frac{965,920}{80.544} = 11,992 \text{ Btu/mol}$$

By inspection the flash temperature must be increased.
  d. Trial 2.

$$\frac{V}{F} = \frac{h_F - h_L}{H_V - h_L} = \frac{15,000 - 11,992}{18,812 - 11,992} = 0.441$$

Again assuming a linear relationship between temperature and fraction vaporized, the next flash temperature is therefore,

$$\frac{T - 145}{0.441 - 0.0} = \frac{170 - 145}{1.0 - 0.0}$$

$T = 156°F$. The value of N is calculated:

$$N = \frac{L}{V} = \frac{1 - V/F}{V/F} = \frac{1 - 0.441}{0.441} = 1.268$$

Component	f, mol	$K_i(156°F)$	fz_iK
C_3	10	3.10	31.0
iC_4	30	1.53	45.9
C_4	50	1.16	58.0
iC_5	40	0.54	21.6
C_5	10	0.445	4.45

Component	$N = 1.268$	$N = 1.0$	$N = 1.15$	$N = 1.155$
		$K_i + N$		
C_3	4.368	4.100	4.250	4.255
iC_4	2.798	2.530	2.680	2.685
C_4	2.428	2.160	2.310	2.315
iC_5	1.808	1.540	1.690	1.695
C_5	1.713	1.445	1.595	1.600

For $N = 1.268$:
  $V_{\text{calc}} = 61.934$
  $L_{\text{calc}} = 140 - V_{\text{calc}} = 78.066$
  $N_{\text{calc}} = 1.260$ $N > N_{\text{calc}}$ Reduce N to 1.0
For $N = 1.0$:
  $V_{\text{calc}} = 69.661$
  $L_{\text{calc}} = 70.339$ $N_{\text{calc}} = 1.010$
  $N < N_{\text{calc}}$ Increase N to 1.15
For $N = 1.15$:
  $V_{\text{calc}} = 65.10$
  $L_{\text{calc}} = 74.900$ $N_{\text{calc}} = 1.155$
  $N < N_{\text{calc}}$ Assume $N = N_{\text{calc}} = 1.155$
For $N = 1.155$:
  $V_{\text{calc}} = 64.959$
  $L_{\text{calc}} = 75.041$ $N_{\text{calc}} = 1.155$

Determination of flash breakdown and resulting stream enthalpy data for $T = 156°F$.

Component	f, mol	mol wt	Vapor		Liquid	
			v_i	H_i	l_i	h_i
C_3	10	44.1	7.286	333.28	2.714	201.60
iC_4	30	58.1	17.095	304.16	12.905	187.72
C_4	50	58.1	25.054	324.28	24.946	196.04
iC_5	40	72.1	12.473	307.84	27.257	182.48
C_5	10	72.1	2.781	314.28	7.219	186.60
			64.959		75.041	

Vapor enthalpy $= \Sigma H_i \times (\text{mol wt})_i \times v_i = 1{,}227{,}069$ Btu
Liquid enthalpy $= \Sigma h_i \times (\text{mol wt})_i \times l_i = 904{,}749$ Btu

$$h_{F,\,\text{calc}} = \frac{2{,}131{,}818}{140} = 15{,}227 \text{ Btu/mol}$$

$$H_V = \frac{1{,}227{,}069}{64.959} = 18{,}890 \text{ Btu/mol}$$

$$h_L = \frac{904{,}749}{75.041} = 12{,}057 \text{ Btu/mol}$$

The solution to the problem lies between 155 and 156°F.
  e. Trial 3.

$$\frac{V}{F} = \frac{h_F - h_L}{H_V - h_L} = \frac{15{,}000 - 12{,}057}{18{,}890 - 12{,}057} = 0.431$$

We find the flash temperature as before:

$$\frac{T - 145}{0.431 - 0.0} = \frac{170 - 145}{1.0 - 0.0}$$

and so $T = 155.8°F$. The initial value of N is calculated as before:

$$N = \frac{L}{V} = \frac{1 - 0.431}{0.431} = 1.320$$

Component	f, mol	$K_i(155.8°F)$	Fz_iK
C_3	10	3.075	30.75
iC_4	30	1.525	45.75
C_4	50	1.160	58.00
iC_5	40	0.535	21.40
C_5	10	0.445	4.45

Component	$K_i + N$		
	$N = 1.30$	$N = 1.20$	$N = 1.22$
C_3	4.375	4.275	4.295
iC_4	2.825	2.725	2.745
C_4	2.460	2.360	2.380
iC_5	1.835	1.735	1.755
C_5	1.745	1.645	1.665

For $N = 1.30$:
  $V_{\text{calc}} = 61.013$
  $L_{\text{calc}} = 140 - V_{\text{calc}} = 78.987$
  $N_{\text{calc}} = 1.295$ $N > N_{\text{calc}}$ Reduce N to 1.20
For $N = 1.20$:
  $V_{\text{calc}} = 63.598$
  $L_{\text{calc}} = 76.402$ $N_{\text{calc}} = 1.201$
  $N < N_{\text{calc}}$ increase N to 1.22
For $N = 1.22$:
  $V_{\text{calc}} = 63.063$
  $L_{\text{calc}} = 76.937$ $N_{\text{calc}} = 1.22$

Determination of flash breakdown and resulting stream enthalpy data for $T = 155.8°F$.

Component	F, mol	mol wt	Vapor		Liquid	
			v_i	H_i	l_i	h_i
C_3	10	44.1	7.159	333.17	2.841	201.46
iC_4	30	58.1	16.667	304.03	13.333	187.60
C_4	50	58.1	24.370	324.17	25.630	195.89
iC_5	40	72.1	12.194	307.74	27.806	182.32
C_5	10	72.1	2.673	314.17	7.327	186.46
			63.063		76.937	

$$\text{Vapor enthalpy} = \Sigma H_i \times (\text{mol wt})_i \times v_i = 1{,}189{,}694 \text{ Btu}$$
$$\text{Liquid enthalpy} = \Sigma h_i \times (\text{mol wt})_i \times l_i = 926{,}285 \text{ Btu}$$
$$h_{F,\text{calc}} = \frac{2{,}115{,}979}{140} = 15{,}114 \text{ Btu/mol}$$
$$H_V = \frac{1{,}189{,}694}{63.063} = 18{,}865 \text{ Btu/mol}$$
$$h_L = \frac{926{,}285}{76.937} = 12{,}040 \text{ Btu/mol}$$

RESULTS: $T = 155.8°F$ and 45.05 mol % vaporization. The componential distribution between the liquid and vapor phases is shown above.

SHORTCUT DESIGN PROCEDURES (NEW COLUMNS)

Shortcut methods are used to develop preliminary column estimates before going to a computer or for screening studies. The following procedures are most useful with hydrocarbon systems and other systems for which K data are available or can be calculated easily (such as, say, chlorobenzene mixtures). If relative volatilities are less than 1.05 between the keys, the column may become too tall and special techniques may be needed. If the relative volatility varies widely (it may even go from over 1.0 to less than 1.0), the shortcut methods are not so useful. Other restrictions include no azeotropes, no reactions, no decomposition, and "reasonable" temperatures and pressures.

Preliminary Estimate of Products

The preliminary estimate of the products for a conventional fractionation column requires that the column feed be defined and also the column specifications. The feed can be defined in volumetric, mass, or molar units, but in general, fractionation calculations are carried out in molal units because of the definition of vapor-liquid equilibrium data. Therefore, as part of establishing the product estimates, componential molecular weight and reference liquid densities should be available for necessary unit conversions.

Column specifications usually are given in terms of specific components but can be in terms of groups of two or more components. When specifications are given in terms of component groupings, the process calculations become increasingly difficult. Column specifications can be expressed as:
1. Recovery of certain components
2. Product impurity limitations
3. Product purity limitations
Calculations are easier if one arranges the feed components by increasing molecular weight, decreasing vapor pressure, increasing normal boiling point, or preferentially by decreasing volatility. Example 7 illustrates the arranging as well as preliminary estimation of the column's products.

EXAMPLE 7

A butane-pentane splitter is to be designed to process 4200 lb/h of C_3-C_5 feed subject to specifications of:
- A maximum of 3 wt % iC_3 in the distillate product
- A maximum of 1 wt % C_4 in the bottoms product

The heat sink will be air (process design temperature to be 130°F), and 100 lb/in² (gage) steam will be used as the heat source in the reboiler. The feed to the column is tabulated below in terms of weight percent.

Component	wt%
Propane	5
Isobutane	15
Butane	25
Isopentane	20
Pentane	35

Using the appropriate physical properties, convert the column feed into both molal and liquid volume units:

Component	wt %	lb/h	mol wt	mol/h	mol %
C_3	5	210.0	44.1	4.762	7.25
iC_4	15	630.0	58.1	10.843	16.50
C_4	25	1050.0	58.1	18.072	27.50
iC_5	20	840.0	72.1	11.651	17.73
C_5	35	1470.0	72.1	20.388	31.02
	100	4200.0		65.716	100.00

Component	wt %	lb/h	lb/gal*	gal/h	vol %
C_3	5	210.0	4.22	49.76	5.92
iC_4	15	630.0	4.69	134.33	15.97
C_4	25	1050.0	4.87	215.61	25.63
iC_5	20	840.0	5.20	161.54	19.20
C_5	35	1470.0	5.25	280.00	33.28
	100	4200.0		841.24	100.00

*Reference liquid density is 60°F.

$$\text{Average molecular weight is } \frac{4200.0}{65.716} = 63.91$$

$$\text{Average liquid density is } \frac{4200.0}{841.24} = 4.993$$

Based on the given specifications the following variables are assigned:

$$x = \text{mass, lb } C_4 \text{ in the bottoms product (1 wt \%)}$$
$$y = \text{mass, lb } iC_5 \text{ in the distillate product (3 wt \%)}$$

Assuming (subject to later verification) that the C_3 and iC_4 have negligible concentration in the bottoms, and that C_5 has negligible concentration in the distillate, the following material balance can be prepared yielding the overall separation:

	Flow rates, lb/h		
Component	Feed	Distillate	Bottoms
C_3	210	210	0
iC_4	630	630	0
C_4	1050	$1050 - x$	x
iC_5	840	y	$840 - y$
C_5	1470	0	1470
	4200	$1890 - x + y$	$2310 + x - y$

Distillate specification: 3% iC_5 in distillate

$$\frac{3}{100} = \frac{y}{1890 - x + y}$$

Bottoms specification: 1% C_4 in bottoms

$$\frac{1}{100} = \frac{x}{2310 + x - y}$$

Solving these simultaneous linear equations yields

$$x = 22.75$$
$$y = 57.75$$

The estimated separation based on the given specifications and on the assumed separation of C_3, iC_4, and C_5 is:

	Flow rates, lb/h		
Component	Feed	Distillate	Bottoms
C_3	210	210.00	0.00
iC_4	630	630.00	0.00
C_4	1050	1027.25	22.75
iC_5	840	57.75	782.25
C_5	1470	0.00	1470.00
	4200	1925.00	2275.00

Since most fractional-distillation calculations require molar units, the material balance is also given in terms of moles per hour:

	Flow rates, lb·mol/h		
Component	Feed	Distillate	Bottoms
C_3	4.762	4.762	0.0
iC_4	10.843	10.843	0.0
C_4	18.072	17.680	0.0
iC_5	11.651	0.801	10.850
C_5	20.388	0.0	20.388
	65.716	34.086	31.630

Column Operating Conditions

The design is started with estimates of top and bottom compositions. The next step is to estimate the column operating conditions, i.e., find/set the temperatures and pressures at the top and bottom of the tower. Actually, the conditions that are determined are those of the reflux drum or accumulator, and these are primarily set by the cooling medium available for condensation of the reflux. The most common coolant is water, which has an upper limit of about 120°F whether or not a cooling tower is used (the limit is set by potential growth of algae and metallurgical problems). To allow for the necessary thermal driving force, the drum temperature is, say, 110 to 135°F (i.e., an approach Δt of 10°F).

Usually the distillate product is considered a saturated liquid. If it is saturated, this liquid at the estimated composition is at the temperature set above (110 to 135°F) and the pressure in the reflux drum is set by a bubble-point calculation using the format and procedure given earlier. If the distillate stream is to be taken as a vapor, the pressure is set via a dew-point calculation. The pressure and temperature developed so far are in the reflux drum. There is a pressure drop in the condenser and associated piping of approximately 1 to 10 lb/in². This pressure is added to the pressure in the reflux drum to give an estimate of pressure at the top of the tower. The overhead temperature is now determined from a dew-point calculation.

Assuming a nominal pressure drop through the column (2 to 15 lb/in²), a bubble-point calculation on the bottoms product yields the desired operating temperature. The assumed pressure drop through both the condenser and the column itself should be based on previous designs of similar systems. Verification of the condenser pressure drop cannot be completed until a detailed equipment design has been done, but past designs provide adequate data. For the column pressure drop the main variables are the number of theoretical stages and the associated stage efficiency. Revision of the overall column pressure drop can be done once the number of theoretical stages required is estimated along with a projected tray efficiency.

$$\text{Actual trays} = \frac{\text{theoretical stages}}{\text{tray efficiency}/100}$$

Using an estimate of the tray pressure drop obtained from previous similar designs, the column pressure drop can be revised. It should be noted the tray pressure drop will normally be in the range of 0.1 to 0.5 lb/in² per tray.

An alternate method of determining the column operating conditions is to start at the bottom of the column by assigning a temperature and obtaining the required pressure via a bubble-point calculation. The basis for assignment of the temperature might be thermal degradation of the bottoms product (thermally induced polymerization) or to avoid operating a reboiler in the vicinity of the critical temperature. The latter case yields problems from an operational point of view, since most reboilers operate on the basis of fixed heat input. Thus small variations in the bottoms-product composition can radically change the mixture's latent heat, resulting in large variations in the stripping vapor rate. To obtain the conditions at the top of the column and the reflux-distillate accumulator, the appropriate equilibrium calculations are performed to obtain the temperature.

In summary the operating conditions for the top and bottom equilibrium stages of a fractionation system can be obtained using the appropriate equilibrium bubble-point or dew-point calculation. Example 8 illustrates the computational aspects of evaluating these conditions.

EXAMPLE 8

For the problem as initially presented in Example 7 find the terminal operating conditions of the distillation system. The terminal conditions desired are the temperature of (1) distillate—reflux drum, (2) column overhead, and (3) column bottom/reboiler.

a. The distillate as well as the reflux will be a bubble-point liquid. The criteria for evaluation of the bubble-point conditions are:
- Use of an air-fin condenser to liquefy the distillate and reflux. The drum temperature will be assumed to be 130°F (design temperature for air is about 120°F).
- The bubble-point design equation with temperature specified is

$$\Sigma K_i x_i = 1.0$$

It is helpful to convert the distillate molar flow rates (from Example 7) into liquid-phase mole fractions.

Component	D, mol	x, mole fraction	mol wt
C_3	4.762	0.1397	44.1
iC_4	10.834	0.3181	58.1
C_4	17.680	0.5187	58.1
iC_5	0.801	0.0235	72.1
C_5	0	0	72.1
	34.086	1.0000	

For a fixed temperature of 130°F, the bubble-point calculation requires that the pressure be determined. From Example 7, the average molecular weight of the distillate is

$$\text{Avg mol wt} = \frac{1925.00}{34.086} = 56.46$$

The characteristic component is iC_4:

$$P_{\text{estimated}} = P^0_{iC_4}(130°F) = 120 \text{ lb/in}^2(\text{abs})$$

Component	x_i	$K[120 \text{ lb/in}^2(\text{abs})]$
C_3	0.1397	2.10
iC_4	0.3181	1.00
C_4	0.5187	0.73
iC_5	0.0235	0.33
C_5	0	0.27

$$\Sigma K_i x_i = 0.998 \text{ for } P_{\text{estimated}} \text{ of } 120 \text{ lb/in}^2(\text{abs})$$
$$P_{\text{calc}} = \frac{120 - 0.998}{1.000} = 119.7 \text{ lb/in}^2(\text{abs})$$

The result of the distillate bubble-point calculation is

$$T = 130°F$$
$$P = 120 \text{ lb/in}^2(\text{abs})$$

b. The conditions at the top of the column must be evaluated, since the fractionator has a total condenser and not an equilibrium stage. The criteria for evaluation of the dew-point conditions are:
- Assumption of the pressure drop through the total condenser of 2.5 lb/in². Thus the pressure at the top of the column is established as

$$P = 120 \text{ lb/in}^2(\text{abs}) + 2.5 \text{ lb/in}^2 = 122.5 \text{ lb/in}^2(\text{abs})$$

- The dew-point design equation with the pressure specified is

$$\sum \frac{y_i}{K_i} = 1.0$$

Since the fractionator has a total condenser, the column top vapor composition is identical to the distillate composition. Again using iC_4 as the characteristic component, vapor-pressure data can be used to estimate the column top temperature.

$$P^0_{iC_4}(T) = 122.5 \text{ lb/in}^2(\text{abs})$$
$$T \simeq 132°F$$

Component	y_i	$K_i(132°F)$	$K_i(145°F)$
C_3	0.1397	2.20	2.40
iC_4	0.3181	1.00	1.15
C_4	0.5187	0.73	0.85
iC_5	0.0235	0.34	0.42
C_5	0		0.33

For $T = 132°F$:
$$\sum \frac{y_i}{K_i} = 1.160$$
$$K_{iC_4} = 1.160 \quad \text{and} \quad T_{\text{calc}} = 145°F$$

For $T = 145°F$:
$$\sum \frac{y_i}{K_i} = 1.000$$

The result of the column top dew-point calculation is

$$T = 145°F$$
$$P = 122.5 \text{ lb/in}^2(\text{abs})$$

c. The column bottom conditions are evaluated at the reboiler, which is assumed to be an equilibrium stage. The criteria for establishing the conditions at the bottom of the column are: Assume a column pressure drop of 5 lb/in². Thus the column bottom pressure is

$$P = 122.5 + 5.0 = 127.5 \text{ lb/in}^2(\text{abs})$$

The bubble-point design equation with temperature specified is

$$\Sigma K_i x_i = 1.0$$

Convert the bottoms streams to a mole-fraction basis:

Component	B, mol	x, mole fraction	mol wt
C_3	0.0	0.0	44.1
iC_4	0.0	0.0	58.1
C_4	0.392	0.0124	58.1
iC_5	10.850	0.3430	72.1
C_5	20.388	0.6446	72.1
	31.630	1.0000	

The bubble-point calculations for a fixed pressure of 127.5 lb/in²(abs) require that the temperature be determined. As before, the average molecular weight can be determined from the product summary of Example 7.

$$\text{Avg mol wt} = \frac{2275.00}{31.630} = 71.93$$

The characteristic component is iC_5. Using vapor-pressure data, an estimated bottoms temperature is obtained:

$$P_{iC_5}(T) = 127.5 \text{ lb/in}^2(\text{abs})$$
$$T \simeq 238°F$$

Component	x_i	$K_i(238°F)$	$K_i(245°F)$
C_3	0.0		4.50
iC_4	0.0		2.55
C_4	0.0124	1.925	2.05
iC_5	0.3430	1.0	1.075
C_5	0.6446	0.87	0.94

For $T = 238°F$:

$$\Sigma K_i x_i = 0.927$$
$$K_{iC_5} = 1.08 \quad \text{and} \quad T_{\text{calc}} = 245°F$$

For $T = 245°F$:

$$\Sigma K_i x_i = 1.000$$

The result of the column bottom bubble-point calculation is

$$T = 245°F$$
$$P = 127.5 \text{ lb/in}^2(\text{abs})$$

Determination of the Key Components

The characteristic binary pair or key components is obvious in most cases. Normally the separation is required between a specific pair of adjacent components and these are the light and heavy keys. By implication, members of the pair to be split are major fractions of the feed—say a minimum of 0.25 mole fraction units each, and preferably higher.

Example 9 is based on Example 7, where the specifications of 3 wt % iC_5 in the distillate and 1 wt % C_4 in the bottoms coupled with the rather uniform distribution of components in the feed makes the key selection obvious.

The material balance calculations carried out in Example 7 assumed that the nonspecified components (C_3, iC_4, and C_5) underwent perfect separation, as is clearly seen in the last two tables in Example 7. Clearly, the key components are light key C_4 and heavy key iC_5.

It should be noted that the separation assumed for the nonkey components was arbitrary. It will be left to Examples 10 and 11 to illustrate how these assumptions will be verified as in the case of C_3, and revised as in the case of iC_4 and C_5. The overall result of the separation analysis will not affect the initial choice of C_4 and iC_5 as the keys.

EXAMPLE 9

For the separation described in Example 7, determine the key components and the average-relative-volatility data for the column using the heavy key as the column reference component.

a. From the molar material-balance summary, the separation clearly is between butane and isopentane.

Component	Flow rates, lb · mol/h		
	Feed	Distillate	Bottoms
C_3	4.762	4.762	0.0
iC_4	10.843	10.843	0.0
C_4	18.072	17.680	0.392
iC_5	11.651	0.801	10.850
C_5	20.388	0.0	20.388
	65.716	34.086	31.630

Reference component: isopentane (iC_5)
Key components:
   Light key (LK)—butane (C_4)
   Heavy key (HK)—isopentane (iC_5)

b. The average-relative-volatility data for the column will be generated using a three-point geometric average:

$$\alpha_{\text{av}} = \sqrt[3]{\alpha_1 \times \alpha_2 \times \alpha_3}$$

Since the condenser is not an equilibrium stage, the data for point 1 are at the top of the fractionating column (condenser inlet). Point 3 is taken at the reboiler. Data for point 2 are obtained using the arithmetic mean of the conditions at points 1 and 3.

$$T_2 = 0.5(T_1 + T_3) = 0.5(145 + 245) = 195°F$$
$$P_2 = 0.5(P_1 + P_3) = 0.5(122.5 + 127.5)$$
$$= 125 \text{ lb/in}^2(\text{abs})$$

Summary of operating conditions:

	Point 1	Point 2	Point 3
Temp, °F	145	195	245
Pressure, lb/in² (abs)	122.5	125	127.5

Summary of vapor-liquid equilibrium data and associated relative-volatility data:

	Point 1		Point 2		Point 3		Average
Component	K_i	α_i	K_i	α_i	K_i	α_i	$\alpha_{(av)i}$
C_3	2.40	5.71	3.35	4.93	4.50	4.19	4.90
iC_4	1.15	2.74	1.75	2.57	2.50	2.37	2.56
C_4	0.85	2.02	1.35	1.99	2.05	1.91	1.97
iC_5	0.42	1.00	0.68	1.00	1.075	1.00	1.00
C_5	0.33	0.79	0.58	0.85	0.94	0.87	0.84

Sometimes, one of the components to be split is a very small fraction of the feed. In such case the nearest large fraction component is chosen as the key. This is the case of split keys and leads to much more work—not so much in the preliminary estimates as in the final stages where the product specifications become all-important. If, in Example 9, the specification required separation of C_3 and C_4, the split-key situation is clear. The mole fraction of C_3 is reasonable, that of iC_4 is rather low, while that of C_4 is reasonable. In such case the choice of keys is light key C_3, heavy key C_4. See also under Split Keys below.

Minimum Stages at Total Reflux—The Fenske Equation[20]

The Fenske (or Fenske-Underwood) equation estimates the minimum number of theoretical stages at total or infinite reflux. This equation uses the desired separation between two components in a binary system and assumes that the relative volatility remains constant throughout the column.

If the equilibrium data have some interaction between components, it is desirable to determine a third set of equilibrium data. The third set of data can be obtained by using the arithmetic average of the conditions (temperature and pressure) obtained for the equilibrium stages at the top and bottom of the column. Then

$$\alpha_{LK,av} = \sqrt[3]{\left(\frac{K_{LK}}{K_{HK}}\right)_{top} \left(\frac{K_{LK}}{K_{HK}}\right)_{middle} \left(\frac{K_{LK}}{K_{HK}}\right)_{bottom}}$$

Otherwise, the average relative volatility can be obtained using a two-point geometric mean:

$$\alpha_{LK,av} = \sqrt{\left(\frac{K_{LK}}{K_{HK}}\right)_{top} \left(\frac{K_{LK}}{K_{HK}}\right)_{bottom}}$$

It should be noted that by definition the value of $\alpha_{HK,av}$ is 1.0.

The Fenske equation yields the minimum number of equilibrium stages via the equation

$$N_{min} = \frac{\ln[(\text{moles LK/moles HK})_{dist}(\text{moles HK/moles LK})_{bottom}]}{\ln(\alpha_{LK,av}/\alpha_{KH,av})}$$

Since the numerator of the above expression contains molar ratios, mass or volumetric (liquid) values or equivalent fractional values can be used directly, as all conversion factors cancel out. See Example 10.

Distribution of Nonkey Components

By combining the Fenske equation with a component material balance, the fractionation of the nonkey components can be predicted. The relationships to be used are the component material balance

$$f_i = d_i + b_i$$

and the original form of the Fenske equation written in terms of an arbitrary component i and a reference component r:

$$\left(\frac{d}{b}\right)_i = \left(\frac{\alpha_{i,\text{av}}}{\alpha_{r,\text{av}}}\right)^{N_\text{min}} \left(\frac{d}{b}\right)_r$$

In determining the product composition values using the combination of these two equations, one takes advantage of whether a component is very volatile or not very volatile. This decision as to the degree of volatility is aided by defining, arbitrarily, a mean α value

$$\alpha_{\text{mean,av}} = \frac{\alpha_{\text{LK,av}} + \alpha_{\text{HK,av}}}{2}$$

The following set of equations can be used to revise the estimate of the distillate and bottoms products:

Light components $(d_i > b_i)$:

$$\alpha_{i,\text{av}} > \alpha_{\text{mean,av}}$$

Reference component r is the heavy key (HK)

$$b_i = \frac{f_i}{1 + \left(\dfrac{d}{b}\right)_{\text{HK}} \left(\dfrac{\alpha_{i,\text{av}}}{\alpha_{\text{HK,av}}}\right)^{N_\text{min}}}$$

$$d_i = f_i - b_i$$

Heavy components $(b_i > d_i)$:

$$\alpha_{i,\text{av}} < \alpha_{\text{mean,av}}$$

Reference component r is the light key (LK)

$$d_i = \frac{f_i}{1 + \left(\dfrac{b}{d}\right)_{\text{LK}} \left(\dfrac{\alpha_{\text{LK,av}}}{\alpha_{i,\text{av}}}\right)^{N_\text{min}}}$$

$$b_i = f_i - d_i$$

(Note the inversion of the α ratio.) Using the revised product distributions, the appropriate specifications can be verified. If necessary the distillate and bottoms products are adjusted, and the equilibrium data/column operating conditions are verified. The resulting cyclic computational procedure is repeated until the minimum number of theoretical stages for two trials agrees to within $\sim 5\%$. Example 10 illustrates the steps in arriving at a consistent set of distillate and bottoms products.

EXAMPLE 10

For the problem started with Example 7, determine the minimum number of equilibrium stages using the Fenske equation. In addition, using the Fenske equation, estimate the degree of separation of the nonkey components. Comparison of the calculated separation of the nonkey components with that assumed indicates whether the relative-volatility data must be redetermined and the separation reassumed.

a. The minimum number of stages is obtained using the Fenske equation. The data required are tabulated in Example 9. Thus:

$$N_\text{min} = \frac{\ln\left[(17.680/0.801) \times (10.850/0.392)\right]}{\ln(1.97/1.00)} = 9.461$$

b. The degree of separation for the nonkey components is estimated using the Fenske equation coupled with the overall material balance. To aid in the computational effort, the mean relative volatility is calculated.

$$\alpha_{mean} = \frac{\alpha_{LK} + \alpha_{HK}}{2} = \frac{1.97 + 1.00}{2} = 1.485$$

The separation of the nonkey components uses the data of Example 9.
Propane (C_3): $\alpha_{C_3} > \alpha_{mean}$

$$\alpha_{C_3} = 4.90$$
$$f_{C_3} = 4.762 \text{ mol/h}$$
$$b_{C_3} = \frac{4.762}{1 + (0.801/10.850)(4.90/1.00)^{9.461}}$$
$$= 1.904 \times 10^{-5} \text{ mol/h}$$
$$\simeq 0.000 \text{ mol/h}$$
$$d_{C_3} = f_{C_3} - b_{C_3} = 4.762 \text{ mol/h}$$

The assumed and calculated separations of propane agree.
Isobutane (iC_4): $\alpha_{iC_4} > \alpha_{mean}$

$$\alpha_{iC_4} = 2.56$$
$$f_{iC_4} = 10.843$$
$$b_{iC_4} = \frac{10.843}{1 + (0.801/10.850)(2.56/1.00)^{9.461}}$$
$$= 0.020 \text{ mol/h}$$
$$d_{iC_4} = f_{iC_4} - b_{iC_4} = 10.823 \text{ mol/h}$$

The assumed separation had all of the iC_4 in the distillate. Since there is some iC_4 in the bottoms product, the column operating conditions and fractionation at infinite reflux must be recalculated.
Pentane (C_5): $\alpha_{C_5} < \alpha_{mean}$

$$\alpha_{C_5} = 0.84$$
$$f_{C_5} = 20.388 \text{ mol/h}$$
$$d_{C_5} = \frac{20.388}{1 + (0.392/17.680)\,(1.97/0.84)^{9.461}}$$
$$= 0.285 \text{ mol/h}$$
$$b_{C_5} = f_{C_5} - d_{C_5} = 20.103 \text{ mol/h}$$

As with the iC_4 separation, the C_5 assumed and calculated separations do not agree.
 c. The calculated separation is summarized below in both molar and mass units.
Molar units, lb mol/h:

Component	Feed	Distillate	Bottoms
C_3	4.762	4.762	0.0
iC_4	10.834	10.823	0.020
C_4	18.072	17.680	0.392
iC_5	11.651	0.801	10.850
C_5	20.388	0.285	20.103
	65.716	34.351	31.365

Mass units, lb/h:

Component	Feed	Distillate	Bottoms
C_3	210.0	210.00	0.0
iC_4	630.0	628.84	1.16
C_4	1050.0	1027.25	22.75
iC_5	840.0	57.75	782.25
C_5	1470.0	20.55	1449.45
	4200.0	1944.39	2255.61

 d. For the calculated separation the specifications are compared and evaluated.
 ▪ Maximum of 3 wt % iC_5 in distillate

$$\text{Calculated value} = \frac{57.75}{1944.39} \times 100$$
$$= 2.97 \text{ wt \%}$$

since $2.97 < 3.00\%$, the specification is satisfied.

- Maximum of 1 wt % C_4 in the bottoms

$$\text{Calculated value} = \frac{22.75}{2255.61} \times 100$$
$$= 1.01 \text{ wt \%}$$

since $1.01 > 1.00\%$, the specification is not satisfied.

When the specifications are not satisfied, the separation assumed is incorrect. A new separation must be assumed and the calculations of Example 10 must be repeated until the specifications are satisfied. Example 11 illustrates the procedure.

EXAMPLE 11

For the problem started in Example 7, the results of Example 10 show that the set of specifications were not met. Therefore, in this example the column parameters will be reevaluated so as to meet the desired specifications. As part of the reevaluation, the procedures of Examples 7, 8, 9, and 10 must be repeated.

a. Using the estimated separation obtained in Example 10, part *c*, revise the butane (light key) and isopentane (heavy key) separation to satisfy the required column specifications.

Component	Flow rates, lb/h		
	Feed	Distillate	Bottoms
C_3	210	210	0
iC_4	630	628.84	1.16
C_4	1050	$1050 - x$	x
iC_5	840	y	$840 - y$
C_5	1470	20.55	1449.45
	4200	$1909.39 - x + y$	$2290.61 + x - y$

Then

$$\frac{3}{100} = \frac{y}{1909.39 - x + y}$$

and

$$\frac{1}{100} = \frac{x}{2290.61 + x - y}$$

Solving these for x and y yields

$$x = 22.55$$
$$y = 58.36$$

The revised separation summaries are tabulated below:

Component	Flow rates, lb/h		
	Feed	Distillate	Bottoms
C_3	210	210.00	0.00
iC_4	630	628.84	1.16
C_4	1050	1027.45	22.55
iC_5	840	58.36	781.64
C_5	1470	20.55	1449.45
	4200	1945.20	2254.80

Component	Flow rates, lb·mol/h		
	Feed	Distillate	Bottoms
C_3	4.762	4.762	0.00
iC_4	10.843	10.823	0.020
C_4	18.072	17.684	0.388
iC_5	11.651	0.809	10.842
C_5	20.388	0.285	20.103
	65.716	34.363	31.353

b. Using these product separations, recalculate the column operating conditions as was done in Example 8.

Distillate bubble-point calculation with temperature specified at 130°F.

Component	Distillate, mol/h	$K[120 \text{ lb/in}^2(\text{abs})]$
C_3	4.762	2.10
iC_4	10.823	1.00
C_4	17.684	0.73
iC_5	0.809	0.33
C_5	0.285	0.27
	34.363	

Design equation:

$$\Sigma K_i x_i = \frac{\Sigma K_i d_i}{D} = 1.00$$

$$\Sigma K_i x_i = \frac{34.076}{34.363} = 0.992$$

Conditions of the distillate/reflux are as before: 130°F and 120 lb/in²(abs). Overhead or column top conditions are found by way of a dew-point calculation at the specified pressure of 122.5 lb/in²(abs) [$\Delta P_{\text{condenser}} = 2.5 \text{ lb/in}^2(\text{abs})$]

Component	Distillate, mol/h	$K(145°F)$	$K(148°F)$
C_3	4.762	2.40	2.425
iC_4	10.823	1.15	1.175
C_4	17.684	0.85	0.87
iC_5	0.809	0.42	0.425
C_5	0.285	0.33	0.34
	34.363		

Design equation:

$$\sum \frac{y_i}{K_i} = \frac{\Sigma(d_i/K)_i}{D} = 1.00$$

For $T = 145°F$:

$$\sum \frac{y_i}{K_i} = \frac{34.990}{34.363} = 1.018$$
$$K = 1.175 \qquad T_{\text{calc}} = 148°F$$

For $T = 148°F$:

$$\sum \frac{y_i}{K_i} = \frac{34.243}{34.363} = 0.997$$

Conditions of the column overhead are revised to 148°F and 122.5 lb/in²(abs).

Column bottom conditions are found by way of a bubble-point calculation on the bottoms product at a specified pressure of 127.5 lb/in²(abs) [$\Delta P_{\text{column}} = 5 \text{ lb/in}^2(\text{abs})$].

Component	Bottoms, mol/h	$K(245°F)$
C_3	0.0	4.50
iC_4	0.020	2.55
C_4	0.388	2.05
iC_5	10.842	1.075
C_5	20.103	0.94

Design equation:

$$\Sigma K_i x_i = \frac{\Sigma K_i b_i}{B} = 1.00$$

For $T = 245°F$ (assumed):

$$\Sigma Kx = \frac{31.398}{31.353} = 1.001$$

Conditions of the bottoms are as before, namely, 245°F and 127.5 lb/in²(abs).

c. As in Example 9, use the revised terminal operating conditions to determine the column average-relative-volatility data.

At the column midpoint conditions are

$$T = 0.5(148 + 245) = 196.5°F$$
$$P = 0.5(122.5 + 127.5) = 125 \text{ lb/in}^2(\text{abs})$$

Summary of operating conditions:

	Point 1	Point 2	Point 3
Temp, °F	148	196.5	245
Pressure, lb/in²(abs)	122.5	125	127.5

Equilibrium K data for the column midpoint conditions are to be found in the following summary of the vapor-liquid equilibrium data and the associated relative-volatility data.

	Point 1		Point 2		Point 3		
Component	K	α	K	α	K	α	Avg α
C_3	2.425	5.71	3.40	4.86	4.50	4.19	4.88
iC_4	1.175	2.76	1.80	2.57	2.55	2.37	2.56
$C_4(LK)$	0.87	2.05	1.38	1.97	2.05	1.91	1.98
$iC_5(HK)$	0.425	1.00	0.70	1.00	1.075	1.00	1.00
C_5	0.34	0.80	0.60	0.86	0.94	0.87	0.84

d. Using the established separation and relative-volatility data, find the minimum stages using the Fenske equation as in Example 10:

$$N_{\min} = \frac{\ln\left[(17.684/0.809) \times (10.842/0.388)\right]}{\ln(1.98/1.00)} = 9.391$$

e. As in Example 10, calculate the separation of the nonkey components using the equilibrium data and the minimum stages.

$$\alpha_{\text{mean}} = \frac{\alpha_{LK} + \alpha_{HK}}{2} = \frac{1.98 + 1.00}{2} = 1.49$$

Propane (C_3):

$$\alpha_{C_3} = 4.88$$
$$f_{C_3} = 4.762 \text{ mol/h}$$
$$b_{C_3} = \frac{4.762}{1 + (0.809/10.842)(4.88/1.00)^{9.391}}$$
$$= 2.188 \times 10^{-5} \text{ mol/h}$$
$$\approx 0.000 \text{ mol/h}$$
$$d_{C_3} = 4.762 \text{ mol/h}$$

The assumed and calculated separations agree.
Isobutane (iC_4):

$$\alpha_{iC_4} = 2.56$$
$$f_{iC_4} = 10.843$$
$$b_{iC_4} = 0.021$$
$$d_{iC_4} = 10.843 - 0.021 = 10.822 \text{ mol/h}$$

The assumed and calculated separations are essentially in agreement.
Pentane (C_5):

$$\alpha_{C_5} = 0.84$$
$$f_{C_5} = 20.388 \text{ mol/h}$$

$$d_{C_5} = 0.292 \text{ mol/h}$$
$$b_{C_5} = 20.388 - 0.292 = 20.096 \text{ mol/h}$$

The assumed and calculated separations are essentially in agreement. The revised separation is summarized below in both molar and mass units:

Molar units, lb·mol/h:

Component	Feed	Distillate	Bottoms
C_3	4.762	4.762	0.000
iC_4	10.843	10.822	0.021
C_4	18.072	17.684	0.388
iC_5	11.651	0.809	10.842
C_5	20.388	0.292	20.096
	65.716	34.369	31.347

Mass units, lb/h:

Components	Feed	Distillate	Bottoms
C_3	210.0	210.00	0.0
iC_4	630.0	628.78	1.22
C_4	1050.0	1027.45	22.55
iC_5	840.0	58.36	781.64
C_5	1470.0	21.05	1448.95
	4200.0	1945.64	2254.36

f. Calculated values are now compared with the column specifications and evaluated.
- Maximum of 3 wt % iC_5 in distillate

$$\text{Calculated value} = \frac{58.36}{1945.64} \times 100$$
$$= 3.00 \text{ wt \%}$$

- Maximum of 1 wt % C_4 in bottoms

$$\text{Calculated value} = \frac{22.55}{2254.36} \times 100$$
$$= 1.00 \text{ wt \%}$$

Inasmuch as the revised separation is very close to that assumed initially, the relative-volatility data will not change. Therefore, the iterative calculation need not be repeated.

Distribution of Nonkey Components Graphically

In lieu of the above analytic expressions the Hengstebeck graphical approach can be used. The Fenske equation is restructured in terms of an arbitrary component i and the heavy-key reference component ($\alpha_{r,av} = 1.0$).

$$\log \left(\frac{d}{b} \right)_i = N_{\min} \log (\alpha_{i,av}) + \log \left(\frac{d}{b} \right)_r$$

On log-log paper the above equation is a straight line and can be located using the light and heavy key α and d/b values (see Fig. 9).

For a light component ($\alpha_i > \alpha_{mean}$) obtain $(d/b)_{light}$, and solve the material-balance equation for b_{light}:

$$b_{light} = \frac{f_{light}}{1 + (d/b)_{light}}$$
$$d_{light} = f_{light} - b_{light}$$

For a heavy component ($\alpha_i < \alpha_{mean}$), the necessary y pair of equations are:

$$d_{heavy} = \frac{f_{heavy} \, (d/b)_{heavy}}{1 + (d/b)_{heavy}}$$
$$b_{heavy} = f_{heavy} - d_{heavy}$$

EXAMPLE 12

Apply the Hengstebeck method to the results of Example 10. From that example the important results are:

	f, mol/h	α_{av}	d/b
C_3	4.762	4.90	(Very large)
iC_4	10.843	2.56	(541.15)
C_4 LK	18.072	1.97	45.1020
iC_5 HK	11.651	1.00	0.0738
C_5	20.388	0.84	(0.0141)

Plot $\log (d/b)$ vs. α_{av} for the two keys and draw a straight line through these points (Fig. 9). For the estimation of C_5, from the graph find d/b at $\alpha_{av} = 0.84$:

$$d/b = 0.0141$$

then
$$d = \frac{20.388 \,(0.0141)}{1 + 0.0141} = 0.284$$

$$b = 20.388 - 0.284 = 20.104$$

The number of cycles for the plot is at least six for the d/b terms in this example. This could be done by pasting sheets together. In this example the d/b values are so high it is simpler to calculate them as in Example 10.

Split Keys

When the separation is made between two groups of isomers, it is not clear which pair of components will characterize the separation. All that one can do by inspection is to select candidate pairs for the light and heavy keys. To find the proper pair, one must use the Fenske equation to evaluate each pair of components. The Fenske equation uses the desired separation and also the relative volatilities of the components. The relative volatility is a measure of ease of separation and is illustrated fully in previous examples. The Fenske equation estimates the minimum number of theoretical stages required for the desired separation. The aim of the key-selection search is to find the pair of components which yield the largest number of minimum stages. See Example 13.

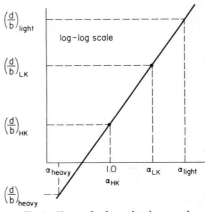

Fig. 9 Hengstebeck graphical approach.

The search requires predecision of the column operating conditions at top and bottom. With this information, calculate the relative volatility of the candidate pairs using a two-point (top and bottom) geometric average relative volatility:

$$\alpha_{av} = \sqrt{\alpha_{top} \times \alpha_{bottoms}} = \sqrt{\left(\frac{K_{LK}}{K_{HK}}\right)_{top} \left(\frac{K_{LK}}{K_{HK}}\right)_{bottoms}}$$

α_{av} is the average relative volatility of the light key (LK) with the heavy key as the reference. The Fenske equation is then used to determine the minimum number of equilibrium stages for each of the pairs.

EXAMPLE 13

A nine-component feed containing C_2 through C_8 is to be debutanized. The specifications are:
- A maximum of 0.05 wt % C_5's in the distillate (equivalent to 0.025 mole %)
- A maximum of 0.50 wt % C_4's in the bottoms (equivalent to 0.785 mole %)

The following table summarizes the feed and the separation of the nonspecification components in molar units.

	Flow rates, mol/h		
Component	Feed	Distillate	Bottoms
C_2	12.0	12.0	0.0
C_3	18.5	18.5	0.0
iC_4	3.5	$3.5 - w$	w
C_4	15.0	$15.0 - x$	x
iC_5	5.0	y	$5.0 - y$
C_5	10.2	z	$10.2 - z$
C_6	11.3	0.0	11.3
C_7	14.0	0.0	14.0
	100.0	$49.0 - w - x + y + z$	$51.0 + w + x - y - z$

The selection of the light and heavy keys is not as obvious as in Example 7. The candidates for the key components are light key iC_4 and C_4, heavy key iC_5 and C_5. With two possibilities for each of the keys, there are four possible combinations that must be evaluated:

Combination	Light key	Heavy key
(1)	iC_4	iC_5
(2)	iC_4	C_5
(3)	C_4	iC_5
(4)	C_4	C_5

To find the appropriate keys for this problem, we use the Fenske equation with each of the listed combinations. But to use the equation we need data on the relative volatility, which in turn implies knowledge of the compositions and conditions at the top and bottom of the tower. Using the techniques of Examples 7 and 8, the conditions at the top are found to be: temperature 125°F, pressure 265 lb/in² (abs). The specifications alone are not sufficient to determine bottom compositions from which a bubble-point temperature could be calculated. This is a trial-and-error situation. Assuming the following mole ratios and using the given product specifications, the product separation can be found:

$$\left(\frac{C_4}{iC_4}\right)_{bottoms} = \frac{x}{w} = 120.0$$

$$\left(\frac{iC_5}{C_5}\right)_{distillate} = \frac{y}{z} = 3.25$$

$$\text{Distillate specification} = \frac{0.025}{100} = \frac{y + z}{49.0 - w - x + y + z}$$

$$\text{Bottoms specification} = \frac{0.785}{100} = \frac{w + x}{51.0 + w + x - y - z}$$

Solving these four linear relations for w, x, y, and z:

$$w = 0.003$$
$$x = 0.400$$
$$y = 0.009$$
$$z = 0.003$$

Using these values, the separation in the debutanizer is estimated (and summarized):

	Flow rates, mol/h		
Component	Feed	Distillate	Bottoms
C_2	12.0	12.0	0
C_3	18.5	18.5	0
iC_4	3.5	3.497	0.003
C_4	15.0	14.600	0.400
iC_5	5.0	0.009	4.991
C_5	10.2	0.003	10.197
C_6	11.3	0	11.3
C_7	14.0	0	14.0
C_8	10.5	0	10.5
	100.0	48.609	51.391

Using the techniques shown earlier, assume the bottom pressure is 275 lb/in²(abs). Then a bubble-point calculation yields a temperature of 430°F. With these values the equilibrium K can be found from Fig. 1 for each component. The equilibrium data are summarized in the following table using C_5 as the reference compound:

Component	125°F and 265 lb/in²(abs)		430°F and 275 lb/in²(abs)		
	K	α	K	α	Avg α
C_2	2.75	19.64	8.09	5.09	10.00
C_3	1.03	7.36	4.76	2.99	4.69
iC_4	0.50	3.57	3.28	2.06	2.71
C_4	0.36	2.57	2.78	1.75	2.12
iC_5	0.17	1.21	1.79	1.13	1.17
C_5(ref)	0.14	1.00	1.59	1.00	1.00
C_6	0.06	0.43	1.03	0.65	0.53
C_7	0.02	0.14	0.62	0.39	0.23
C_8	0.01	0.07	0.46	0.29	0.14

In the above table, C_5 is the implied heavy key. The average values from the table apply to combinations (1) and (3). For the other combinations, the ratio of the average α is used. The next table summarizes the relative-volatility data:

Light key	Heavy key	α_{LK-HK}
iC_4	iC_5	2.316
iC_4	C_5	2.710
C_4	iC_5	1.812
C_4	C_5	2.120

Each set of candidates is evaluated using the Fenske equation. For the first pair:

$$N_{min} = \frac{\ln\left[(3.497/0.009) \times (4.991/0.003)\right]}{\ln 2.316} = 15.93$$

N_{min} for the other pairs is calculated in the same way. The results are tabulated below. The pair with the greatest number of stages is the key pair appropriate for the problem.

Light key	Heavy key	N_{min}
iC_4	iC_5	15.93
iC_4	C_5	15.24
C_4	iC_5	13.98
C_4	C_5	15.61

The first pair in the above table are chosen as the keys. This analysis is contingent upon having chosen the correct values for the product ratios: $(C_4/iC_4)_{bottoms} = x/w$ and $(iC_5/C_5)_{distillate} = y/z$. The ratios are checked by using the Fenske equation and the componental material balance $f = d + b$; the distribution of the nonkey components is calculated to see if the product specifications have been met. If not, these ratios must be rechosen and the entire process must be repeated. The method is demonstrated below.

The assumed reference was C_5, but the calculated reference was found to be iC_5. All the α_{av} are recalculated in this case by dividing all the old α_{av} values by 1.17, the α_{av} for iC_5. The results are summarized:

Component	Flow rates, mol/h			
	Feed	Distillate	Bottoms	Avg
C_2	12.0			8.55
C_3	18.5			4.01
iC_4(LK)	3.5	3.497	0.003	2.32
C_4	15.0			1.81
iC_5	5.0	0.009	4.991	1.00
C_5	10.2			0.85
C_6	11.3			0.45
C_7	14.0			0.20
C_8	10.5			0.12

As before, we find α_{mean}:

$$\frac{\alpha_{LK} + \alpha_{HK}}{2} = \frac{2.32 + 1.00}{2} = 1.66$$

Using the Fenske-Underwood forms developed earlier, calculate the distribution of the nonkey components: For $\alpha_i > \alpha_{mean}$, the situation for C_2 and C_3.

For C_3: $\qquad b = \dfrac{18.5}{1 + (0.009/4.991)(4.01/1.00)^{15.93}} = 2.53 \times 10^{-6} = 0.000$

Then $d = f - b = 18.5 - 0.000 = 18.5$ mol
For C_2, by inspection: $d = 12.000$ and $b = 0.000$ mol. For $\alpha_{LK} > \alpha_i > \alpha_{HK}$, C_4 only in this case; since α for $C_4 > \alpha_{mean}$, treat C_4 as a light component

$$b = \frac{15.0}{1 + (0.009/4.991)(1.81/1.00)^{15.93}} = 0.626 \text{ mol/h}$$
$$d = 15.0 - 0.626 = 14.374 \text{ mol/h}$$

For $\alpha_i < \alpha_{HK}$, C_5, C_6, C_7, and C_8.

For C_5: $\qquad d = \dfrac{10.2}{1 + (0.003/3.497)(2.32/0.85)^{15.93}} = 0.001 \text{ mol/h}$
$$b = 10.2 - 0.001 = 10.199 \text{ mol/h}$$

For C_6: $\qquad d = \dfrac{11.3}{1 + (0.003/3.497)(2.32/0.45)^{15.93}} = 5.93 \times 10^{-8} = 0.000$
$$b = 11.3 - 0.0 = 11.3 \text{ mol}$$

C_7 and C_8 by inspection distribute totally into the bottoms. Summarizing:

Component	Flow rates, mol/h		
	Feed	Distillate	Bottoms
C_2	12.0	12.0	0.0
C_3	18.5	18.5	0.0
iC_4	3.5	3.497	0.003
C_4	15.0	14.374	0.626
iC_5	5.0	0.009	4.991
C_5	10.2	0.001	10.199
C_6	11.3	0.0	11.3
C_7	14.0	0.0	14.0
C_8	10.5	0.0	10.5
	100.0	48.341	51.619

The products calculated must meet the specifications. For the C_5's in the distillate:

$$\text{Mol \%} = \frac{0.001 + 0.009}{48.381} 100 = 0.021$$

Specification is 0.025.
For the C_4's in the bottoms:

$$\text{Mol \%} = \frac{0.003 + 0.626}{51.619} 100 = 1.219$$

Specification is 0.785.
Since the bottoms specification is not met, the calculations must be repeated using as the new ratio assumptions:

$$\left(\frac{C_4}{iC_4}\right)_{bottoms} = \frac{x}{w} = \frac{0.626}{0.003} = 208$$

$$\left(\frac{iC_5}{C_5}\right)_{distillate} = \frac{y}{z} = \frac{0.009}{0.001} = 9.0$$

The Winn Equation[65]

The Fenske equation has a weakness—as the relative-volatility difference between column top and column bottom increases, the estimated minimum stages get increasingly too small. In 1958, Winn proposed a new equation which would account for temperature variation of equilibrium K values but which would have a form similar to the Fenske equation. Unfortunately Winn's method does not fit well into hand calculations and must be considered as part of a computerized shortcut procedure.

The relation proposed by Winn relates the equilibrium K of component i and reference heavy key as

$$K_i = \beta_i(K_r)^{\Theta_i}$$

where β and Θ are constants at a fixed pressure. The value of the two constants β and Θ must be determined for each component relative to the reference component (again the heavy key). There are two constants per component, which can be determined using the equilibrium data associated with the equilibrium calculations performed at the column top (dew point) and the column bottom (bubble point).

Determination of β and Θ for each component requires the preliminary determination of the constants A and B, by which the equilibrium data for the average pressure of the column are correlated with temperature. The equation has the structure of a modified Antoine equation and is

$$\ln(PK_i) = A_i + \frac{B_i}{T + 460}$$

where P is the average column pressure and T is the temperature (°F). The values of A and B for each component are determined using the equilibrium K values at the top and bottom of the column. The following pair of equations are used to obtain A and B:

$$B_i = \frac{\ln(PK_i)_{\text{top}} - \ln(PK_i)_{\text{bottom}}}{[1/(T_{\text{top}} + 460)] - [1/(T_{\text{bottom}} + 460)]}$$

$$A_i = \ln(PK_i)_{\text{top}} - \frac{B_i}{T_{\text{top}} + 460}$$

With the constants A and B defined, the Winn constants β and Θ can be obtained:

$$\Theta_i = \frac{B_i}{B_r} = \frac{\ln(PK_i) - A_i}{\ln(PK_r) - A_r}$$

$$\beta_i = \exp(A_i - \Theta_i A_r)(P)^{\Theta_i - 1}$$

The use of natural-base logarithms is dictated by their more common use on computing equipment including hand calculators. Use of the common base 10 logarithms can be made with no loss of generality.

The Winn equation for the minimum number of stages requires the use of mole fractions and is as follows:

$$N_{\min} = \frac{\ln\left[\left(\dfrac{x_D}{x_B}\right)_{\text{LK}}\left(\dfrac{x_B}{x_D}\right)_{\text{HK}}^{\Theta_{\text{LK}}}\right]}{\ln \beta_{\text{LK}}}$$

While the Fenske equation had the property of direct substitution of consistent units for the key components in the distillate and the bottoms, the Winn equation does not. Using molar flow rates instead of mole fractions, we get the following equation for $N_{\min}$:

$$N_{\min} = \frac{\ln\left[\left(\dfrac{d}{b}\right)_{\text{LK}}\left(\dfrac{b}{d}\right)_{\text{HK}}^{\Theta_{\text{LK}}}\left(\dfrac{B}{D}\right)^{1 - \Theta_{\text{LK}}}\right]}{\ln \beta_{\text{LK}}}$$

The Winn equation reduces to the Fenske equation at $\Theta = 1.0$, with β thus being akin to the relative volatility α.

With the number of stages $N_{\min}$ required at total reflux available, the Winn equation molar-flow form can be combined with the column-component material balance to estimate the fractionation of the nonkey components. The appropriate equations are given as follows:

$$d_i + b_i = f_i$$

$$\left(\frac{b}{d}\right)_i = \beta_i^{-N_{\min}}\left(\frac{b}{d}\right)_{\text{HK}}^{\Theta_i}\left(\frac{B}{D}\right)^{1 - \Theta_i}$$

$$\left(\frac{b}{d}\right)_i < 1 \qquad d_i > b_i \qquad \text{(light component)}$$

$$b_i = \frac{f_i(b/d)_i}{1 + (b/d)_i}$$

$$d_i = f_i - b_i$$

$$\left(\frac{b}{d}\right)_i > 1 \qquad b_i > d_i \qquad \text{(heavy component)}$$

$$d_i = \frac{f_i}{1 + (b/d)_i}$$

$$b_i = f_i - d_i$$

EXAMPLE 14

Using the final conditions from Example 11, find the minimum number of stages and the product distribution using the Winn equation. Subscript R identifies the reference component, i.e., the heavy key (HK).

a. The Winn equation has the following form:

$$K = \beta(K_R)^\Theta$$

with β and Θ defined in terms of A and B, which are related to the K value by

$$\log_{10}(\pi K) = A + \frac{B}{T + 460}$$

The values of A and B are obtained from the K values at the top and bottom of the column using the following formulas:

$$B = \frac{\log(\pi K)_{\text{top}} - \log(\pi K)_{\text{bottom}}}{[1/(T_{\text{top}} + 460)] - [1/(T_{\text{bottom}} + 460)]}$$

$$A = \log(\pi K)_{\text{top}} - \frac{B}{T_{\text{top}} + 460}$$

With A and B evaluated, compute Θ and β as follows:

$$\Theta = \frac{B}{B_R}$$

$$\beta = (10.0^{A - \Theta A_R})(P^{\Theta - 1})$$

The following K data are used to develop the Winn-equation parameters:

Component	K data	
	Column top	Column bottom
C_3	2.425	4.50
iC_4	1.175	2.55
C_4	0.87	2.05
iC_5	0.425	1.075
C_5	0.34	0.94

The values of A, B, Θ, and β are tabulated below. Note that β was evaluated at the average column pressure, 125 lb/in²(abs).

Component	B	A	Θ	β
C_3	−1263.27	4.55059	0.68001	4.31134
iC_4	−1563.77	4.73016	0.84177	2.40694
C_4(LK)	−1721.67	4.85935	0.92677	1.91988
iC_5(HK/R)	−1857.71	4.95582	1.0	1.0
C_5	−2028.41	4.95582	1.09189	0.86704

With the parameters established, determine the minimum number of stages from:

$$N_{\text{min}} = \frac{\log\left[(x_D/x_B)_{\text{LK}}(x_B/x_D)_{\text{HK}}^{\Theta_{\text{LK}}}\right]}{\log \beta_{\text{LK}}}$$

$$N_{min} = \frac{\log \left[\left(\frac{17.684/34.369}{0.388/31.347} \right) \left(\frac{10.842/31.347}{0.809/34.369} \right) \right]^{0.92677}}{\log 1.91988}$$

$$= 9.533$$

b. For the nonkey components, estimate the fractionation in the column using the following pair of equations:

$$f = d + b$$

$$\beta^{N_{min}} = \left(\frac{d}{b} \right) \left(\frac{b}{d} \right)_{HK}^{\theta} \left(\frac{B}{D} \right)^{1-\theta}$$

where

$$N_{min} = 9.533$$

$$\left(\frac{b}{d} \right)_{HK} = \frac{10.842}{0.809} = 13.402$$

$$\frac{B}{D} = \frac{31.347}{34.369} = 0.91207$$

For the nonkey components the product to be solved for is, by definition, the smaller numeric value. Therefore, based on the value of β_{av},

$$\beta_{av} = \frac{\beta_{LK} + \beta_{HK}}{2}$$

the following equations can be used:

$$\frac{d}{b} = \frac{\beta^{N_{min}}}{(b/d)_{HK}^{\theta}(B/D)^{1-\theta}}$$

$\beta > \beta_{av}$, light component $(d > b)$

$$b = \frac{f}{1 + d/b} \qquad d = f - b$$

$\beta < \beta_{av}$, heavy component $(b > d)$

$$d = \frac{f}{1 + b/d} \qquad b = f - d$$

Propane: $f = 4.762$, $\beta = 4.311$, $\Theta = 0.680$. This is a light component.

$$\frac{d}{b} = \frac{4.311^{9.533}}{(13.402)^{0.680} \times (0.91207)^{0.320}}$$

$$= 1.976 \times 10^{5}$$

$$b = \frac{4.762}{1 + 1.976 \times 10^{5}} = 0.0 \text{ lb} \cdot \text{mol/h}$$

$$d = f - b = 4.762 \text{ lb} \cdot \text{mol/h}$$

Isopropane: $f = 10.843$, $\beta = 2.407$, $\Theta = 0.842$. This is a light component.

$$\frac{d}{b} = \frac{2.407^{9.533}}{(13.402)^{0.680} \times (0.91207)^{0.320}}$$

$$= 494.15$$

$$b = \frac{10.843}{1 + 494.15} = 0.022 \text{ lb} \cdot \text{mol/h}$$

$$d = 10.843 - 0.022 = 10.821 \text{ mol/h}$$

Pentane: $f = 20.388$, $\beta = 0.867$, $\Theta = 1.092$. This is a heavy component.

$$\frac{d}{b} = \frac{0.867^{9.533}}{(13.402)^{0.680} \times (0.91207)^{0.320}} = 1.495 \times 10^{-2}$$

$$\frac{b}{d} = \left(\frac{d}{b} \right)^{-1} = 66.896$$

$$d = \frac{20.388}{1 + 66.896} = 0.300 \text{ lb} \cdot \text{mol/h}$$

$$b = 20.388 - 0.300 = 20.088 \text{ mol/h}$$

The separation calculated using the Winn equation is summarized below:
Molar units, lb · mol/h:

Component	Feed	Distillate	Bottoms
C_3	4.762	4.762	0.000
iC_4	10.843	10.821	0.022
C_4	18.072	17.684	0.388
iC_5	11.651	0.809	10.842
C_5	20.388	0.292	20.096
	65.716	34.368	31.348

Mass units, lb/h:

Component	Feed	Distillate	Bottoms
C_3	210.0	210.00	0.00
iC_4	630.0	628.72	1.28
C_4(LK)	1050.0	1027.45	22.54
iC_5(HK)	840.0	58.36	781.64
C_5	1470.0	21.05	1448.95
	4200.0	1945.59	2254.41

Direct comparison of the above summaries with those of Example 11 indicates excellent agreement.

Calculation of Minimum Reflux—The Underwood Equation[57]

Many investigators have presented methods for predicting the minimum reflux rate. By definition, minimum reflux implies an infinite number of equilibrium stages, the infinite number occurring at the zone of constant composition, or the pinch zone. As has been seen from the discussion on binary distillation, this pinch zone occurs in the vicinity of the feed, simultaneously in the rectifying and stripping section, and is true for most systems. Systems like ethanol-water illustrate the other common case of tangency between equilibrium and operating lines.

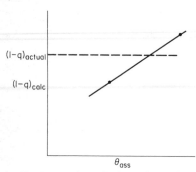

Of the developed methods for predicting the minimum reflux rate, all contain one or more assumptions and are in various degrees time-consuming in application. All the methods are approximate, but the Underwood method yields apparently good estimates and requires minimal computational effort.

Fig. 10   Auxiliary plot for Underwood.

It should be noted that the rigorous evaluation of the minimum reflux rate cannot be accomplished owing to the presence of infinite contact stages. The only exception is that of the ideal binary system having a constant relative volatility and adhering to the McCabe-Thiele assumptions.

The equations developed by Underwood are based on two assumptions: (1) constant molar overflow, and (2) knowledge of the component relative volatilities at the pinch zone. For a multicomponent system the following equations will yield an estimate of the reflux rate.

Based on the degree of feed vaporization, the value of Θ is solved for using the equation

$$\sum_i \frac{(\alpha_{i,\mathrm{av}})(z_{i,\mathrm{feed}})}{(\alpha_{i,\mathrm{av}}) - \Theta} = 1 - q$$

For the case of adjacent key components Θ is bounded by $\alpha_{\mathrm{HK}} < \Theta < d_{\mathrm{LK}}$. The z_i values are the molal compositions of the feed.

The value of $1 - q$ is the fraction of the feed that is vapor and is defined as

$$1 - q = \frac{H_f(\text{dew point}) - h_f(\text{feed condition})}{H_f(\text{dew point}) - h_f(\text{bubble point})}$$

The specific values needed to evaluate q, or rather $1 - q$, are the enthalpy values of the feed at its dew point, actual feed condition, and its bubble point. The determination of these thermal conditions requires that appropriate equilibrium calculations be made on the feed at the feed-stage pressure. For preliminary calculations, that is, until the actual number of trays or decks required has been determined, the average column pressure should be used.

$$P_{\text{feed stage}} = \frac{P_{\text{column top}} + P_{\text{column bottom}}}{2}$$

When the column stage/tray (latter implies an estimate of the tray efficiency) requirements have been established, the actual feed stage can be estimated and the calculation can be refined if necessary.

Evaluating Θ is a trial-and-error procedure. Assume a Θ about midway between α_{HK} and α_{LK} to start and then use an auxiliary plot to speed the convergence (see Fig. 10).

Use the Θ so found in the equation

$$R_{\min} + 1 = \sum_i \frac{(\alpha_{i,\text{av}})x_{iD}}{(\alpha_{i,\text{av}}) - \Theta}$$

If desired, it is possible to calculate minimum boilup $\bar{\beta}_{\min} = \bar{V}_{\min}/B$ from

$$-\bar{\beta}_{\min} = \sum \frac{(\alpha_{i,\text{av}})x_{i,B}}{(\alpha_{i,\text{av}}) - \Theta}$$

As pointed out previously, do not calculate both $R_{\min}$ and $\bar{\beta}_{\min}$ and then use them. Rather, calculate only one and then find the other via an energy balance so as to ensure consistency of calculation.

When the light key (LK) and the heavy key (HK) are not adjacent, the condition of distributed components exists. For the case of a single distributed component, designated via the subscript δ, two values of Θ are obtained from

$$\sum \frac{(\alpha_{i,\text{av}})(z_{i,\text{feed}})}{(\alpha_{i,\text{av}}) - \Theta} = 1 - q$$

with the relative-volatility data of the keys bounding the values of Θ,

$$\alpha_{HK} < \Theta_2 < \alpha_\delta < \Theta_1 < \alpha_{LK}$$

In order to determine the value of the minimum reflux $R_{\min}$, the following pair of equations must be solved simultaneously:

$$R_{\min} + 1 = \sum_{i \neq \delta} \frac{(\alpha_{i,\text{av}})x_{iD}}{(\alpha_{i,\text{av}}) - \Theta_1} + \frac{(\alpha_{\delta,\text{av}})x_{\delta D}}{(\alpha_{\delta,\text{av}}) - \Theta_1}$$

$$R_{\min} + 1 = \sum_{i \neq D} \frac{(\alpha_{i,\text{av}})x_{iD}}{(\alpha_{i,\text{av}}) - \Theta_2} + \frac{(\alpha_{\delta,\text{av}})x_{\delta D}}{(\alpha_{\delta,\text{av}}) - \Theta_2}$$

When there are two or more distributed components, the above procedure is simply extended by determining an additional value of Θ and having to solve an additional simultaneous equation involving the minimum reflux and the distillate composition of the distributed component.

Pseudo-Binary

Minimum-reflux estimation using the Underwood method can be time-consuming. For some problems, the work can be reduced a great deal if the multicomponent system is considered as a binary. This idea can be used with reasonable safety if the light and heavy keys together constitute 90% or more of the feed mixture.

To reduce a multicomponent system to a binary, either ignore the nonkey components or lump these components with the keys. Then revise the feed, distillate, and bottoms compositions to the new, keys only, binary. As usual, evaluate the feed, bubble-point, and dew-point temperatures at the estimated feed-stage pressure. The associated enthalpies are then calculated so that it is possible to estimate the feed q factor. Using the procedures for binary mixtures developed earlier, graphically find the operating line for minimum reflux. If the relative volatility is a constant for the mixture, the problem can be solved analytically for the intersection of the q line and the equilibrium line.

$$\text{Slope} = \left(\frac{L}{V}\right)_{\min} = \frac{R_{\min}}{R_{\min}+1} = \frac{y_D - y}{x_D - x} = \phi < 1.0$$

or

$$R_{\min} = \frac{\phi}{1-\phi} = \frac{L_{\min}}{D}$$

The minimum boilup ratio $\bar{\beta}$ can be evaluated from the slope of the operating line in the stripping section.

$$\text{Slope} = \left(\frac{\overline{L}}{\overline{V}}\right)_{\max} = \frac{\bar{\beta}_{\min}+1}{\bar{\beta}_{\min}} = \frac{y - y_B}{x - x_B} = \psi > 1.0$$

$$\text{or } \bar{\beta}_{\min} = \frac{1}{\psi - 1} = \frac{\overline{V}_{\min}}{B}$$

$R_{\min}$ and $\bar{\beta}_{\min}$ are not independent but are related through an energy balance. If both are calculated, the work must be checked for consistency using the energy balance

$$Fh_f + Q_R = Q_c + DH_D + Bh_B$$
$$Q_c = (R_{\min} + 1)D\lambda_D$$
$$Q_R = \bar{\beta}_{\min}B\lambda_B$$

In actual practice either $R_{\min}$ or $\bar{\beta}_{\min}$ is evaluated directly. The other is determined by use of the energy balances shown above. Note that λ_D and λ_B are latent heats and must be evaluated using bubble points, dew points, and enthalpy data of the components.

Stage-Reflux Correlations

A knowledge of minimum stages and minimum reflux in a column is of little interest unless these can be related to the actual number of stages and the actual reflux required.

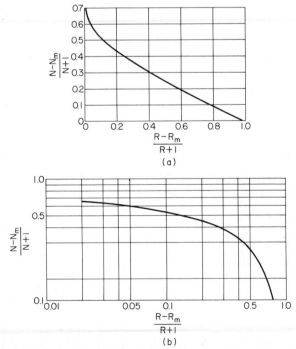

Fig. 11 The Gilliland stage-reflux correlation (a); on log-log scale (b). [*Reprinted with permission from* Ind. Eng. Chem., **32,**1220 (1940). *Copyright The American Chemical Society.*]

The two most widely accepted correlations are the Gilliland correlation and the Erbar-Maddox correlation. These are graphical and are reproduced as Figs. 11 and 12. Each relates the minimum column operating limits to the reflux and stages actually required.

The *Gilliland*[23] *correlation* is based on the Fenske-Underwood equation combined

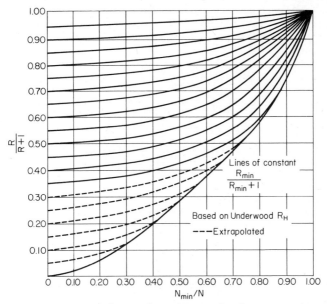

Fig. 12 The Erbar-Maddox stage-reflux correlation. [*Reprinted with permission from* Pet. Refiner, **40**(5), 184 (1961).]

with the Gilliland minimum-reflux equation and hand-computed stage-to-stage results to get stage-reflux values (see Fig. 11).

The *Erbar-Maddox*[18] *correlation* is based on the Fenske-Underwood minimum-stage equation and the Underwood minimum-reflux equation. A computer was used with rigorous stage-to-stage calculations to develop a set of consistent stage-reflux values (see Fig. 12).

The correlation recommended is the Erbar-Maddox. In order to exploit the full potential of this correlation, the economic relationship between investment and operating costs must be introduced. Results of previous economic designs have been compiled, compared, and reduced to a relationship of R to R_{min} or, equivalently, N to N_{min}, based on the

TABLE 3 Economic R/R_{min} and N/N_{min} vs. Heat-Sink Source

Method of condensing the reflux	$\dfrac{R}{R_{min}}$ ratio	$\dfrac{N}{N_{min}}$ ratio
Low-level refrigeration (-300 to $-150°F$)	1.05–1.10	2.0–3.0
High-level refrigeration (-150 to $50°F$)	1.10–1.20	1.8–2.0
Cooling water—at cost of circulation and limited treating	1.2–1.5	2.0–1.8
Air cooling	1.4–1.5	1.6–1.8

type of heat sink employed. Tabulated in Table 3 are economic ranges of reflux to minimum reflux and the resulting values of stages to minimum stages. The ratio of reflux ratio to minimum reflux ratio is referred to by process designers as the "reflux factor," and to a lesser degree the ratio of stages to minimum stages is termed the "tray factor."

Feed Location

Location of the proper feed for binary and pseudo-binary systems can be obtained directly from the $x-y$ diagram. For multicomponent systems which cannot be reduced to a pseudo-binary system, the feed stage can be estimated using either the Kirkbride equation or a modification of the Fenske equation.

The *Kirkbride equation* yields the ratio of the number of theoretical stages in the rectifying section m to the number of theoretical stages in the stripping section p. With the ratio m/p evaluated and the sum known, $m + p = N$, the feed stage can be determined. The Kirkbride equation is

$$\frac{m}{P} = \left\{ \frac{B}{D} \left(\frac{z_{HK}}{z_{LK}} \right)_{feed} \left[\frac{(x_{LK})_{bottoms}}{(x_{HK})_{distillate}} \right]^2 \right\}^{0.206}$$

The bottoms and distillate molal product rates B and D are required in addition to the compositions (mole fractions) of the key components in the feed and product streams.

Fenske Equation

As an alternate, the *Fenske equation*[20] can be used to estimate the minimum number of theoretical stages in the rectifying and stripping sections. The basis of the analysis is the separation of the distillate and the feed, and the feed and bottoms. The necessary equations are

$$(N_{min})_{rect} = \frac{\ln \left(\frac{\text{moles } LK_D \text{ moles } HK_F}{\text{moles } HK_D \text{ moles } LK_F} \right)}{\ln \left[\frac{(\alpha_{LK})_{av(rect)}}{(\alpha_{HK})_{av(rect)}} \right]}$$

$$(N_{min})_{strp} = \frac{\ln \left(\frac{\text{moles } LK_F \text{ moles } HK_B}{\text{moles } HK_F \text{ moles } LK_B} \right)}{\ln \left[\frac{(\alpha_{LK})_{av(strp)}}{(\alpha_{HK})_{av(strp)}} \right]}$$

The relative-volatility values that should be used are those representative of the rectifying and stripping sections and may have to be specially calculated. If a two-point geometric average was used to determine the minimum number of theoretical stages in the column, there exists only a single relative volatility describing the rectifying section and the stripping section:

$$(\alpha_i)_{av(rect)} = (\alpha_i)_{av(top)}$$
$$(\alpha_i)_{av(strp)} = (\alpha_i)_{av(bottom)}$$

For the case of a three-point geometric average being used for the column's minimum number of theoretical stages, there is an additional relative-volatility value representative of the conditions in the middle of the column, and thus:

$$(\alpha_i)_{av(rect)} = \sqrt{[(\alpha_i)_{av(top)}][(\alpha_i)_{av(middle)}]}$$
$$(\alpha_i)_{av(strp)} = \sqrt{[(\alpha_i)_{av(middle)}][(\alpha_i)_{av(bottom)}]}$$

It should be noted that the minimum number of theoretical stages in the respective column rectifying and stripping sections do not sum to the minimum number of theoretical stages for the desired separation because of variations in the relative-volatility data within the column. In general:

$$(N_{min})_{rect} + (N_{min})_{strp} \geq N_{min}$$

EXAMPLE 15

Using the separation established in Example 11 for the illustrative problem, calculate the minimum reflux via the Underwood equation. The column feed is a saturated liquid (i.e., at the bubble point). In addition, calculate the actual number of stages for a reflux factor of 1.14, along with the feed-stage location.

  a. Using the column feed, the following design equation is solved for Θ:

$$\sum \frac{\alpha_i z_i}{\alpha_i - \Theta} = 1 - q$$

The parameters needed are the feed mole fraction z_i, the relative volatilities α_i, and the feed condition q. For a bubble-point feed $q \equiv 1.0$

Component	mol/h	Mole fraction	α_i
C_3	4.762	0.0725	4.88
iC_4	10.843	0.1650	2.56
$C_4(LK)$	18.072	0.2750	1.98
$iC_5(HK)$	11.651	0.1773	1.00
C_5	20.388	0.3102	0.84
	65.716	1.0000	

The solution to the equation is a trial-and-error procedure, with the desired value of Θ being bounded by the relative-volatility data for the light and heavy keys.

$$1.98 > \Theta > 1.0$$

The secant method to find the value of Θ is shown below. Initial values of Θ are assumed, and the truncated expansion of $f(\Theta)$ is solved for the next trial value for Θ.

$$f(\Theta) = -(1 - q) + \sum \frac{\alpha_i z_i}{\alpha_i - \Theta} = 0.0$$

$$f(\Theta) = f(\Theta_1) + \frac{df}{d\Theta} \Delta\Theta$$

$$\frac{df}{d\Theta} = \frac{f(\Theta_1) - f(\Theta_2)}{\Theta_1 - \Theta_2}$$

$$\Delta\Theta = \Theta_1 - \Theta$$

$$\Theta = \Theta_1 - \frac{f(\Theta_1)}{df/d\Theta}$$

$$f(\Theta) = 0.0 + \frac{4.88 \; 0.0725}{4.88 - \Theta} + \frac{2.56 \; 0.1650}{2.56 - \Theta} + \frac{1.98 \; 0.2750}{1.98 - \Theta} + \frac{1.00 \; 0.1773}{1.00 - \Theta} + \frac{0.84 \; 0.3102}{0.84 - \Theta} = 0.0$$

$$\Theta_1 = 0.05 + 0.95(1.98) = 1.931$$

$$\Theta_2 = 0.05(1.98) + 0.95 = 1.049$$

Use the expanded $f(\Theta)$ with these values of and find:

$$\Theta_1 = 1.931 \quad \text{and} \quad f(\Theta_1) = 11.474$$
$$\Theta_2 = 1.049 \quad \text{and} \quad f(\Theta_2) = -3.908$$

Then $\Theta_3 = \Theta_1 - \dfrac{f(\Theta_1)}{[f(\Theta_1) - f(\Theta_2)]/(\Theta_1 - \Theta_2)} = 1.273$

and $f(\Theta_3) = -0.055$
Repeating the procedure:

$$\Theta_4 = 1.276 \quad \text{and} \quad f(\Theta_4) = -0.039$$
$$\Theta_5 = 1.283 \quad \text{and} \quad f(\Theta_5) = -0.004$$
$$\Theta_6 = 1.284$$

The final value of Θ is 1.284, since

$$|\Delta\Theta| = |\Theta_5 - \Theta_6| = 0.001$$

b. With the value of Θ determined, use the following design equation to calculate R_{min}, the minimum reflux ratio:

$$R_{min} = \sum \frac{\alpha_i x_{D,i}}{\alpha_i - \Theta} - 1.0$$

The parameters needed are the distillate mole fractions $x_{D,i}$, relative volatilities α_i, and the final value of Θ.

Component	mol/h	Mole fraction	α_i
C_3	4.762	0.1386	4.88
iC_4	10.822	0.3149	2.56
C_4	17.684	0.5145	1.98
iC_5	0.809	0.0235	1.00
C_5	0.292	0.0085	0.84

With $\Theta = 1.284$, $R_{min} = 1.185$.

c. With the operational bounds of stages and reflux established, use the Gilliland and Erbar-Maddox correlations to find the finite number of stages corresponding to a reflux factor of 1.14. The parameters needed are N_{min}, R_{min}, and R = factor × R_{min}.
Gilliland correlation (Fig. 11):

$$f(R) = \frac{R - R_{min}}{R + 1} = \frac{1.355 - 1.185}{1.355 + 1.0} = 0.072$$

$$\phi(N) = 0.565$$

$$N = \frac{N_{min} + \phi}{1.0 - \phi} = \frac{9.391 + 0.565}{1.0 - 0.565} = 22.9 \text{ or } 23 \text{ stages}$$

Erbar-Maddox correlation (Fig. 12):

$$\frac{L_0}{V_2} = \frac{R}{R + 1} = \frac{1.355}{1.355 + 1} = 0.575$$

$$\left(\frac{L_0}{V_1}\right)_{min} = \frac{R_{min}}{R_{min} + 1} = \frac{1.185}{2.185} = 0.542$$

$$\frac{N_{min}}{N} = 0.41$$

$$N = \frac{9.391}{0.41} = 22.9, \text{ or } 23 \text{ stages}$$

d. The feed stage is located using either the Fenske equation or the Kirkbride equation.
Using the *Fenske equation,* the number of stages in the rectifying and stripping sections can be determined under the condition of total reflux.
  ▪ Rectifying section (use the key components).

$$\alpha_{C_4} = \sqrt{2.05 \times 1.97} = 2.01$$

$$\alpha_{C_5} = 1.00$$

Since the Fenske equation uses ratios, any consistent set of flow rates or componential fractions can be used.

$$(N_{min})|_{rect} = \frac{\ln [(1027.45/58.36) \times (840.0/1050.0)]}{\ln (2.01/1.00)} = 3.789$$

  ▪ Stripping section (use the key components).

$$\alpha_{C_4} = \sqrt{1.97 \times 1.91} = 1.94$$

$$\alpha_{C_5} = 1.00$$

Using mass flow rates as before:

$$(N_{min})|_{strip} = \frac{\ln [(1050.0/840.0) \times (781.64/22.55)]}{\ln (1.94/1.00)} = 5.687$$

Distribute the stages between the two sections by ratio.

$$\text{Rectifying section} = 23 \times \frac{3.789}{3.789 + 5.687}$$

$$= 9.2 \quad \text{or } 9 \text{ stages}$$

$$\text{Stripping section} = 23 \times \frac{5.687}{3.789 + 5.687}$$

$$= 13.8 \quad \text{or } 14 \text{ stages}$$

The feed is introduced on the t.
  The *Kirkbride equation* may also be used.

$$\log \frac{m}{p} = 0.206 \log \left[\frac{B}{D}\left(\frac{X_{HK}}{X_{LK}}\right)_F \left[\frac{(X_{LK}) \text{ bottoms}}{(X_{HK}) \text{ distillate}}\right]^2\right]$$

where m = stages in the rectifying section
      p = stages in the stripping section

$$m + p = 23$$

$$\log \frac{m}{p} = 0.206 \log \left[\frac{31.347}{34.369}\left(\frac{11.651/F}{18.072/F}\right)\left(\frac{0.388/31.347}{0.809/34.369}\right)^2\right]$$

$$= -0.163$$

$$\frac{m}{p} = 0.688$$

$$m = 0.688\,p$$

$$\text{Stripping section} = p = \frac{23}{0.688 + 1.0}$$
$$= 13.4 \quad \text{or 14 stages}$$
$$\text{Rectifying section} = m = 0.688p$$
$$= 9.4 \quad \text{or 9 stages}$$

As before, the feed is introduced on the tenth theoretical stage from the top.

Determination of Condenser and Reboiler Energy Requirements

From the preliminary calculations the number of equilibrium stages and the associated reflux ratio and boilup ratio have been found. From these ratios the energy requirements of the column can be obtained.

Figure 2 illustrates the three basic energy input/output locations within the fractionation system. The energy associated with the feed is Q_F. This energy requirement must be consistent with the degree of feed vaporization ($1-q$; minimum-reflux determination) and is obtained directly by the following enthalpy balance:

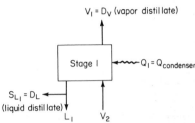

$$Q_F = F(H_F - h_F)$$

The values of enthalpy used are H_F (column entrance point) and h_F (battery limits) in

Fig. 13 Stage 1 condenser-reflux separator.

units of Btu per mole or Btu per pound of feed.

The condenser duty Q_c is obtained by writing an energy balance around the condenser/distillate drum-reflux accumulator. Since there are three possible conditions, the value of the condenser duty is defined for each using the following nomenclature: H_V, enthalpy of the vapor entering the condenser; H_D, enthalpy of the vapor leaving the drum/accumulator; and h_D, enthalpy of the liquid leaving the drum/accumulator.

$$L = RD$$
$$V = D(R + 1)$$

Case I: All-liquid distillate.

$$Q_c = D(R + 1)(H_V - h_D)$$

Case II: All-vapor distillate.

$$Q_c = D(R)(H_V - h_D) + D(H_V - H_D)$$

Case III: Distillate is liquid/vapor ($D = D_{\text{vapor}} + D_{\text{liquid}}$).

$$Q_c = (DR + D_{\text{liquid}})(H_V - h_D) + D_{\text{vapor}}(H_V - H_D)$$

With the condenser duty calculated, the reboiler duty Q_R can be obtained via an overall energy balance. See Fig. 13.

$$Q_R = Q_c + (D_V H_D + D_L h_D) + B h_B - F H_F$$

If the boilup ratio $\bar{\beta}$ is used as the design parameter, the reboiler duty can be obtained depending on whether an equilibrium thermosiphon/kettle or a recirculating thermosiphon is used. Thus, depending on which basic type of energy input is used, the appropriate reboiler duty can be calculated.

Case I:
Equilibrium-stage reboiler (kettle-type reboiler or a "once"-through thermosiphon). See Fig. 14.

The stream $\bar{V}$ is in equilibrium with the bottoms B, and the input $\bar{L}$ is at its bubble point.

$$\bar{V} \equiv \bar{\beta}B$$
$$\bar{L} = B(\bar{\beta} + 1)$$

$$Q_R = \overline{V}H_V + Bh_B - \overline{L}h_L$$
$$= \overline{\beta}BH_V + Bh_B - B(\overline{\beta} + 1)h_L$$

Case II:

Partial-equilibrium-stage reboiler. See Fig. 15. (30% vaporization has been recommended so as to yield uniform stripping vapor generation.)

The stream $\overline{V} + \overline{L}'$ is an equilibrium stream having 30 mol % vapor (per design

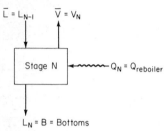

$\overline{L} = L_{N-1}$ $\overline{V} = V_N$

Stage N $Q_N = Q_{reboiler}$

$L_N = B = $ Bottoms

Fig. 14 Equilibrium-stage reboiler.

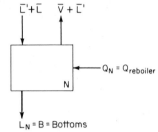

$\overline{L}'+\overline{L}$ $\overline{V} + \overline{L}'$

N $Q_N = Q_{reboiler}$

$L_N = B = $ Bottoms

Fig. 15 Partial-equilibrium-stage reboiler.

recommendation). Stream $\overline{L}$ is at its bubble point and has the same composition as the bottoms product B.

$$\overline{V} \equiv \overline{\beta}B$$
$$\frac{\overline{L}'}{\overline{V}} = \frac{1-f}{f}$$

where $f = 0.30$

$$\overline{L} = \overline{V} + \overline{L}'$$
$$Q_R = (\overline{V}H_V + \overline{L}'h_L') - \overline{L}h_{\overline{L}}$$

It should be noted that the process calculations associated with this type of reboiler require that a flash calculation be performed on the reboiler effluent where, even though the liquid-vapor split is known, the temperature is an undetermined quantity. Since the

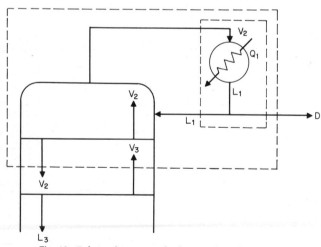

Fig. 16 Relationship at top of column with total condenser.

stream enthalpies are temperature-dependent, the reboiler energy requirement becomes a trial-and-error problem.

Using the boilup ratio $\bar{\beta}$ to obtain the reboiler duty, the condenser duty Q_C can be obtained via an overall energy balance

$$Q_C = FH_F + Q_R - (D_V H_D + D_L h_D) - B h_B$$

EXAMPLE 16

To complete the preliminary or shortcut calculations, energy balances must also be made. The results of Examples 7 through 13 will be used.

a. The following table of enthalpy data, obtained from J. B. Maxwell, "Data Book on Hydrocarbons," will be used in the subsequent calculations.

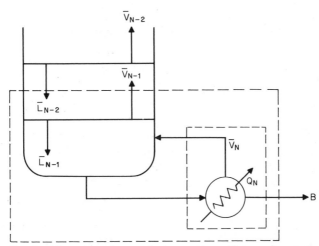

Fig. 17 Relationship at bottom of column with equilibrium stage reboiler.

Vapor enthalpy, Btu/lb at 125 lb/in²(abs) (avg):

Component	100°F	200°F	300°F
C_3	307.5	355.0	407.0
iC_4	277.5	326.0	379.0
C_4	298.0	346.0	400.0
iC_5	288.0	328.0	382.0
C_5	289.0	336.0	379.0

Liquid enthalpy, Btu/lb:

Component	100°F	200°F	300°F
C_3	170.0	228.0	288.0
iC_4	154.0	212.5	272.5
C_4	160.0	223.0	285.0
iC_5	148.0	211.0	277.5
C_5	152.5	215.0	280.0

b. With the basic column feed being available as a subcooled liquid at 100°F, the amount of feed preheat must be determined. As discussed in Example 13, the feed condition at the column entrance is that of a saturated liquid.

For the enthalpy associated with the 100°F liquid feed, use the equation

$$h_{feed} = \Sigma f_i \times h_i$$

Component	Feed, lb/h	Enthalpy, Btu/lb
C_3	210.0	170.0
iC_4	630.0	154.0
C_4	1050.0	160.0
iC_5	840.0	148.0
C_5	1470.0	152.5

$$h_{\text{feed}} = 0.6492 \times 10^6 \text{ Btu/h}$$

In order to calculate the energy requirement for the preheater, the feed bubble point must be determined (feed entering column is at the bubble point). The feed-stage pressure (tenth stage from the top) is 125 lb/in²(abs). Summarized below is the final iteration of the feed bubble-point calculation. At 125 lb/in²(abs) and 172°F:

Component	Feed, mol/h	K
C_3	4.762	2.90
iC_4	10.843	1.475
C_4	18.072	1.125
iC_5	11.651	0.55
C_5	20.388	0.45
	65.716	

$$\Sigma K_i x_i = \frac{65.7169}{65.716} = 1.000$$

The column feed at the 172°F bubble point is found by linear interpolation of feed enthalpy at 100°F (see part *a* above), and the enthalpy at 200°F. This last enthalpy is found using the feed rates of part *a* and the liquid enthalpy at 200°F,

Temp, °F	h_{feed}, 10^6 Btu/h
100	0.6492
172	0.8364
200	0.9092

$$\text{Feed preheat} = h_{\text{feed},172} - h_{\text{feed},100}$$
$$= 0.1872 \times 10^6 \text{ Btu/h}$$

c. For the external reflux ratio of 1.355 calculate the condenser duty. For this, the enthalpy of the distillate liquid and the overhead vapor must be determined at their respective conditions (same flow rates).

$$Q_{\text{condenser}} = (H_{\text{vap},148} - h_{\text{liq},130}) \times (R + 1)$$

▪ Summary of enthalpy data (Btu/h). The distillate and vapor flow rates are from Example 11, the final set in part *e*. The enthalpy data at 100 and 200°F were given above.

°F	h_{dist}, 10^6 Btu/h	H_{vap}, 10^6 Btu/h
100	0.3088	0.5681
130	0.3444	
148		0.6128
200	0.4275	0.6612

As before, a linear interpolation was used.

$$Q_{\text{condenser}} = [(0.6218 - 0.3444) \times 10^6](1.355 + 1.0)$$
$$= 0.6321 \times 10^6 \text{ Btu/h}$$

d. An overall energy balance yields the reboiler duty, since

$$h_{\text{feed},100} + Q_{\text{preheat}} + Q_{\text{reboiler}} = h_{\text{dist},130} + Q_{\text{condenser}} + h_{\text{bottoms},245}$$

The bottoms enthalpy must be found before the reboiler duty can be determined. The bottoms flow rates are in part *e* of Example 11. The liquid enthalpies were given earlier, the data at 200 and 300°F being used since the bottoms are at 245°F.

°F	h_{bottoms}, 10^6 Btu/h
200	0.4817
245	0.5482
300	0.6294

$$Q_{\text{reboiler}} = (0.3444 + 0.6321 + 0.5482 - 0.6492 - 0.1872) \times 10^6$$
$$= 0.6883 \times 10^6 \text{ Btu/h}$$

   e. The amount of stripping vapor generated by the reboiler is now determined. In this analysis the reboiler is assumed to be an equilibrium stage with the Q calculated above.

   The initial information needed is the enthalpy and composition of the vapor generated in the reboiler. This vapor by definition is in equilibrium with the bottoms product. The molal flow rates and the K data for 245°F were found in Example 11, part *b*. From these data the vapor concentrations are found using:

$$y_i = \frac{K_i b_i}{B} \quad \text{and} \quad \Sigma y_i = 1.000$$

Component	Vapor mole fraction	mol wt
C_3	0.0	44.1
iC_4	0.002	58.1
C_4	0.025	58.1
iC_5	0.372	72.1
C_5	0.601	72.1

NOTE: Average mol wt = 71.722.

Using the enthalpy data for vapor given above at 200 and 300°F, the molal enthalpy is calculated from

$$H = \Sigma y_i \times (\text{mol wt})_i \times (H_{\text{vap}})_i$$

- In summary:

°F	H_{bottoms}, Btu/lb·mol
200	23,897
245	25,426 (by interpolation)
300	27,294

Since the majority of the energy supplied to the reboiler is used to generate the stripping vapor, the vapor rate can be estimated by calculating a pseudo-latent heat.

$$\lambda \simeq \text{enthalpy of vapor} - \text{enthalpy of liquid}$$
$$\text{Vapor enthalpy} = 25,426 \text{ Btu/lb·mol}$$
$$\text{Liquid enthalpy} = \frac{0.5482 \times 10^6}{31.347}$$
$$= 17,488 \text{ Btu/lb·mol}$$
$$\lambda \simeq 25,426 - 17,488 = 7938 \text{ Btu/lb·mol}$$
$$\text{Stripping vapor rate} \simeq \frac{Q_{\text{reboiler}}}{\lambda}$$

$$\overline{V}_{\text{est}} \simeq 0.6883 \times 10^6 = 86.709 \text{ lb·mol/h}$$
$$= 6219 \text{ lb/h}$$

This estimated rate is verified by doing an energy balance around the (equilibrium stage) reboiler (Fig. 14).

$$\overline{L} = \overline{V}_{\text{est}} + B$$
$$= 86.709 + 31.347 = 118.056$$

From the mole fractions found above and $\overline{V}_{\text{est}}$, calculate the molal flow rates in $\overline{V}_{\text{est}}$. Flow rates of B were found in Example 11, part *e*. Then, by addition, the overall composition is easily calculated (and is shown in the next table). The bubble point and the enthalpy of stream $\overline{L}$ are needed for the balance and are calculated as shown before. The pressure is 127.5 lb/in²(abs), and the initial assumed temperature is 245°F, the conditions on the reboiler liquid as found in Example 11, part *b*.

Component	lb·mol/h	$K(245°F)$	$K(242°F)$
C_3	0.0	4.50	4.40
iC_4	0.194	2.55	2.50
C_4	2.556	2.05	2.00
iC_5	43.098	1.075	1.05
C_5	72.208	0.94	0.92
	118.056		

At 245°F:
$$\Sigma K_i x_i = \frac{119.940}{118.056} = 1.016$$

$$K_{iC_5} = 1.058 \qquad T_{calc} = 242°F$$

At 242°F:
$$\Sigma K_i x_i = 117.281 = 0.993$$

Using the molal flow rates, the molecular weights, and liquid enthalpies at 200 and 300°F (as tabulated earlier), calculate the mixture enthalpy at the extremes and interpolate linearly. In summary:

T, °F	Enthalpy, Btu/lb·mol
200	15,336
242	17,310
300	20,036

Combine the material and energy balances around the reboiler to reestimate V.
Material balance:

$$\overline{L} = \overline{V} + B \qquad \text{and} \qquad B = 31.347 \text{ lb·mol/h}$$

Energy balance:

$$\overline{L} \times 17,310 + Q_{reboiler} = \overline{V} \times 25,426 + B \times h_{bottoms}$$

and $Q_r = 0.6883 \times 10^6$ Btu/h

$$B \times h_{bottoms} = 0.5482 \times 10^6 \text{ Btu/h}$$

Solving simultaneously,

$$\overline{V} = 84.120 \text{ lb·mol/h}$$

Recalculate the stripping vapor rate using the mole fractions determined at the start of part *e*. Recalculate the reboiler feed rate, bubble point, and enthalpy. The flow rates, K data, and enthalpy data are presented below:

Component	lb·mol/h	$K(242°F)$	Btu/lb $(242°F)$
C_3	0.0	4.40	253.2
iC_4	0.189	2.50	237.7
C_4	2.491	2.00	249.0
iC_5	42.135	1.05	238.9
C_5	70.652	0.92	242.3
	115.467		

At 242°F:
$$K_i x_i = \frac{114.696}{115.467} = 0.993$$

Temperature unchanged.

At 242°F: Enthalpy = 17,310 Btu/lb mol

Recalculate the stripping rate in the same way as done earlier and find

$$V = 84.120 \text{ lb·mol/h}$$

This is the final value of the stripping rate from which the boilup is calculated.

$$\text{Molal boilup} = \frac{84.120}{31.347} = 2.684$$

ESTIMATION OF COLUMN SIZE

The procedures presented up to this point will yield results sufficient for preliminary sizing of a column. However, this section does not cover the design of equipment and the reader is referred to Secs. 1-7, 1-8, and 3-2 for details on the design of columns. This preliminary sizing may be required in the design of vacuum units where pressure drop across the column is critical to column performance.

RIGOROUS-SOLUTION METHODS

The rigorous analysis of a multicomponent fractionating column requires the simultaneous solution of the basic relationships associated with each stage. From a computational point of view each stage is considered to be an equilibrium stage—the vapor and liquid streams leaving the stage being in equilibrium with each other. The equations associated with each stage are the MESH equations: (1) material balances, (2) equilibrium relationships, (3) sum relations, and (4) energy (heat) balances. To minimize the number of variables with each stage, the stage pressure normally is preset via the results of the preliminary calculations.

Each equilibrium stage can be represented by Fig. 1. This figure incorporates all the features of a general stage: entering and leaving liquid and vapor streams, vapor and liquid sidestreams, an external feed, and provisions for external heat input/output. Excluding the sidestream drawoff (to make the development somewhat more simple), the restricting equations involved in the equilibrium stage model are:

1. Component material balances around stage n,

$$l_{i,n-1} - l_{i,n} - v_{i,n} + v_{i,n+1} + f_{i,n} = 0$$

where $l_i = Ly_i, v_i = Vy_i$, and $f_i = Fx_{Fi}$

2. Equilibrium relationship,

$$y_{i,n} = K_{i,n} x_{i,n}$$

3. Sum relations on the mole fractions,

$$\sum_i x_{i,n} = 1.0 \qquad \text{and} \qquad \sum_i y_{i,n} = 1$$

4. Energy balance around stage n,

$$L_{n-1} h_{n-1} - L_n h_n - V_n H_n + V_{n+1} H_{n+1} + F_n h_{Fn} + Q_n = 0$$

where h and H are the liquid- and vapor-phase enthalpies and Q_n is the energy input to stage n.

The top and bottom ends of the system are shown in Figs. 13 and 14. The MESH equations for the ends follow the same patterns and notation as for the general stage. Note in particular that the condenser-reflux separator is called stage 1 while the reboiler is called stage N so that stages are identified starting at the top of the tower. The condenser stage is considered an equilibrium stage when stream D_V is not zero (a partial condenser). The reboiler stage usually is considered an equilibrium stage (a kettle-type reboiler or a once-through thermosiphon).

For computerized calculations the number of variables considered cannot be over- or underspecified. Formal procedures for counting variables are given by Kwauk[32] and Hanson.[68] If one sets the stage pressure, the feed, sidestream drawoff, and the external heat load (condenser duty, reboiler duty, and the associated reflux ratio are all related), the number of variables associated with each stage is $2C+3$, where C is the number of components. The stage variables of interest are the vapor- and liquid-phase compositions, the vapor and liquid flow rates, and the temperature. For systems where the vapor-liquid equilibrium constant has a weak dependence on composition, the number of variables can be reduced by elimination of the vapor-phase composition using the relationship $y = Kx$. The stage variables are reduced from $2C+3$ to $C+3$, with the thermodynamic data being temperature and pressure-dependent.

Historically, the solution of the set of stage equations has been governed by the level of sophistication of the computational aids available to the process engineer. The methods used initially involved the reduction of a multicomponent system into a binary system.

The binary solution was then adjusted for the presence of the ignored components. In the early 1930s, Lewis and Matheson as well as Thiele and Geddes published methods for solving the set of equations. Both methods required that constant molal overflow be assumed (thus negating the enthalpy balance) and that the equations be solved in a decoupled form (that is, stage by stage). Because desk-top mechanical calculators and slide rules were the only machines available, the procedures were very time-consuming and repetitious. It was not until the advent of digital computers in the late 1950s that these methods were fully implemented.

With the coming of the large digital computer and the FORTRAN language alternate methods were developed for solving the complete set of equations. The emphasis was on using more advanced mathematical techniques and simultaneously extending the capabilities of the various steady-state programs that had been prepared. Chronologically the development started with the use of linear-algebra techniques to solve the equations in a decoupled form as before. The question of convergence soon arose, that is, what methods are best used so that a final solution is developed. In the study of the convergence methods applied to the solution of the decoupled set of column equations it was noted that many cases experienced convergence problems even though the algorithm was satisfactory for a large number of problems. In the late 1960s, the Newton-Raphson algorithm was used successfully to solve all the column equations simultaneously. At the same time there were significant advances in the size of the computer core and in associated hardware and software.

The Lewis-Matheson Method

The first really successful method for multicomponent distillation was developed by Lewis and Matheson.[35] Before the advent of computers, this method was the most widely used approach, and a number of useful procedures were developed. A good workaday reference to these is Hengstebeck.[69]

The method is a stage-by-stage calculation using equilibrium calculations and material balances alternately. Figure 2 gives the overall layout for the discussion. The method (in its simplest form) depends on two facts:

1. If we know the composition of one stream leaving an equilibrium stage, we can calculate the composition of the other stream leaving the stage.

2. For any given stage, if we have all molal flow rates except for one stream, we can calculate the flow rates for that stream by a material balance. Later on we will use another fact:

3. For any given stage, if we can calculate the enthalpy for all streams except one, we can calculate the enthalpy of that stream by an enthalpy balance.

The original method assumed molal flows of vapor and liquid were constant in each section of the tower (above the feed tray and below the feed tray). This was not unreasonable for petroleum hydrocarbon mixtures, and of course the petroleum industry was the principal user of multicomponent distillation even as it is now.

The calculations start with a series of assumptions and specifications. As usual, start with the keys and the split of the keys required between the distillate and the bottoms. Also needed are the overhead and bottoms compositions, which are estimated using the shortcut methods given previously. The shortcut methods also yield estimates of the external reflux ratio, total stages required, and feed-tray location. We must use the reflux ratio directly in the calculations, but the number of stages and the feed-stage location will be used only as check points or indicators. The shortcut procedures set the pressure in the condenser, and in the usual simple application this pressure is considered constant throughout the tower.

Simple Procedure (Constant Pressure and Constant Molal Overflow)   From the shortcut methods we have an estimate of the compositions of D and B and the external reflux ratio R. At the top of the tower the mass balances are

Overall mass: $V_1 = L + D = D(R + 1)$
Component: $V_2 y_{i,2} = D(R + 1)x_{i,D}$

Since we assume a total condenser, all streams must have the same composition. Because the reflux ratio R is known along with D, we can calculate the masses of streams L and V_2. We now know the flow rates (both total and componential) and compositions of the vapor stream leaving and the liquid stream entering the top stage in the tower. (Refer to Fig. 13

with $V_1 = 0$ and $D_L = D$, or to Fig. 16). Using the composition of V_2 leaving stage 2, do a dew point to get the composition of L_2 and also the stage temperature. Now, using the outermost envelope on Fig. 16, write a new material balance:

or

$$V_3 y_{i,3} = L_2 x_{i,2} + D x_{i,D}$$
$$y_{i,3} = \frac{L_2}{V_3} x_{i,2} + \frac{D}{V_3} x_{i,D}$$

There is one such equation for each component. This set of equations applies above the feed stage. For the simplified Lewis-Matheson method constant molal overflow is assumed so that:

$$L_1 = L_2 = L_3 = \cdots = L, \text{ to the feed stage}$$
$$V_2 = V_3 = V_4 = \cdots = V, \text{ to the feed stage}$$

The general form for these equations is

$$y_{i,n+1} = \frac{L}{V} x_{i,n} + \frac{D}{V} x_{i,D}$$

Note that the flow rates and compositions are known for the liquid stream leaving stage 2 and entering stage 3 and also for the vapor stream leaving stage 3 and entering stage 2. But this is the same type of data available at the top of stage 2; so the operations are repeated for each stage downward until the feed stage is reached. Rather than calculate the flow rates L and V, one can use the following equivalent formula:

$$y_{i,n+1} = \frac{R}{R+1} x_{i,n} + \frac{1}{R+1} x_{i,D}$$

The calculations stop at the feed tray. The usual criterion is to compare the ratio of the keys on any calculated stage with the ratio of the keys in the feed stream (in like phases). Usually the streams do not match compositions. This is because the starting assumed distillate composition was in error. Do not change anything yet. Instead shift attention to the reboiler.

In the reboiler we again have a full set of (assumed) componential flow rates for stream B which leaves the tower. From an overall mass balance:

$$F = D + B \qquad \text{or} \qquad B = F - D$$

and around the feed stage:

$$F + V_{F+1} + L_{F-1} = V_F + L_F$$

but

$$L_{F-1} = L \qquad \text{and} \qquad V_{F+1} = \overline{V}$$

then

$$F + L_{F-1} - V_F = L_F - V_{F+1} = \overline{L} - \overline{V}$$

where $\overline{L}$ and $\overline{V}$ are constant molal flow rates below the feed stage. Thus we know $B, \overline{L}, \overline{V}$, and the composition of B. Referring to Fig. 17 and starting at the reboiler, use the composition of stream B and the bubble point procedure to determine the vapor composition leaving the reboiler (i.e., stream $\overline{V}$ or V_n). The material balance around the first stage up from the reboiler is:

$$B x_{i,B} + V_{n-1} y_{i,n-1} = L_{n-2} x_{i,n-2}$$

or

$$x_{i,n-2} = \frac{V_{n-1}}{L_{n-2}} y_{i,n-1} + \frac{B}{L_{n-2}} x_{i,B}$$

For constant molal overflow:

$$L_{n-1} = L_{n-2} = \cdots = \overline{L}, \text{ to the feed stage}$$

$$V_n = V_{n-1} = \cdots = \overline{V}, \text{ to the feed stage}$$

The general form for these equations is:

$$x_{i,m-1} = \frac{V}{L} y_{i,m} + \frac{B}{L} x_{i,B}$$

where m is any stage between the reboiler and the feed stage. Since we know B, $\overline{L}$, $\overline{V}$, and new liquid compositions, we have the same type of information as we had for the reboiler and can continue the calculations upward to the feed stage. As before, the feed stage is chosen when the ratio of the keys in the feed matches the ratio of the keys on the calculated stage.

At this point, usually, the calculated compositions for the feed stage (one from above and one from below) do not match the feed-stream composition. This means that the assumed D and B stream compositions must be corrected. The methods developed by Hengstebeck[69] are recommended. When the D and B compositions have been revised, repeat the entire procedure until the assumed D and B compositions and the feed all mesh together. Obviously this is time-consuming—but it can be done. Example 17 illustrates the procedure.

EXAMPLE 17

Apply the Lewis-Matheson method to Example 7.

The reflux ratio is taken from Example 15. The compositions of distillate and bottoms are given in Example 11, but one of the concentrations is too small in the bottoms stream. As can be seen from the material-balance equations, the D and B streams *must* have some amount of each component, however small. In hand/desk-top calculations and (deliberately) in this example, the number of significant figures is limited. Thus it is necessary to assume values for $x_{i,D}$ and $x_{i,B}$ which are usable but not necessarily correct. The following are either known or quickly calculated:

$$F = 65.716 \text{ lb} \cdot \text{mol/h}$$
$$D = 34.369 \text{ lb} \cdot \text{mol/h}$$
$$B = 31.347 \text{ lb} \cdot \text{mol/h}$$
$$L = 46.570 \text{ lb} \cdot \text{mol/h}$$
$$V = 80.939 \text{ lb} \cdot \text{mol/h}$$
$$\overline{L} = 112.286 \text{ lb} \cdot \text{mol/h}$$
$$\overline{V} = 80.939 \text{ lb} \cdot \text{mol/h}$$

The equations for the components above the feed point are:

	$y_{i,n} = \dfrac{R}{R+1} x_{i,n-1} + \dfrac{1}{R+1} x_{i,D}$	$x_{i,D}$
C_3	$y_n = 0.5754 x_{n-1} + 0.0588$	0.1385
iC_4	$y_n = 0.5754 x_{n-1} + 0.1337$	0.3149
C_4	$y_n = 0.5754 x_{n-1} + 0.2185$	0.5146
iC_5	$y_n = 0.5754 x_{n-1} + 0.0100$	0.02355
C_5	$y_n = 0.5754 x_{n-1} + 0.0036$	0.0085

Note here that a limited number of significant figures are used, as would be the case for desk-top calculations. In addition, the K values used in subsequent calculations were read from De Priester charts and reflect the problems normal to reading graphs. The graph-reading problem may be eased by remembering that the pressure usually is considered to be constant in Lewis-Matheson calculations. If the pressure is constant, the De Priester chart can be read for K_i at different temperatures and an auxiliary plot made of K_i vs. $T°F$.

Using Fig. 3 and working downward from the top, we use the format of Fig. 6 and the equations just presented. Counting the condenser as stage 1, the next stage is

Stage 2: $P = 122.5$ lb/in²(abs), $T = 148°F$

	y_i	K_i	α_i	y_i/α_i	x_i
C_3	0.13855	2.41	5.738	0.02415	0.05755
iC_4	0.31488	1.17	2.786	0.11303	0.26937
(LK) C_4	0.51433	0.87	2.0714	0.24839	0.59195
(HK) iC_5	0.02354	0.42	1	0.02354	0.05610
C_5	0.00850	0.34	0.8095	0.01050	0.02502
			$K_R = \Sigma\,(y_i/\alpha_i) =$	0.41961	1.00000

Stage 3: $P = 122.5$ lb/in²(abs), $T = 154°F$

	y_i	K_i	α_i	y_i/α_i	x_i
C_3	0.09194	2.55	5.657	0.01622	0.03545
iC_4	0.028869	1.26	2.800	0.10310	0.22532
(LK) C_4	0.55907	0.94	2.044	0.27346	0.59764
(HK) iC_5	0.04227	0.45	1	0.04227	0.09238
C_5	0.01801	0.37	0.800	0.02251	0.04919
			$K_R = \Sigma\,(y_i/\alpha_i) =$	0.45757	0.99998

For the next stages we condense the calculations somewhat.

	Stage 4, $T = 160°F$			Stage 5, $T = 168°F$		
	y_i	K_i	x_i	y_i	K_i	x_i
C_3	0.07923	2.70	0.02939	0.07573	2.77	0.02727
iC_4	0.26335	1.34	0.19661	0.24683	1.40	0.17719
(LK) C_4	0.56235	1.00	0.56260	0.54219	1.06	0.51407
(HK) iC_5	0.06315	0.48	0.13162	0.08573	0.52	0.16569
C_5	0.03191	0.40	0.7981	0.04953	0.43	0.11577

	Stage 6, $T = 172°F$			Stage 7, $T = 176°F$		
	y_i	K_i	x_i	y_i	K_i	x_i
C_3	0.07452	2.90	0.02578	0.07366	3.00	0.02450
iC_4	0.23566	1.47	0.15755	0.22436	1.53	0.14630
(LK) C_4	0.51426	1.11	0.46460	0.48580	1.17	0.41423
(HK) iC_5	0.10533	0.54	0.19560	0.12254	0.56	0.21830
C_5	0.07022	0.45	0.15648	0.09364	0.475	0.19668

	Stage 8, $T = 180°F$			Stage 9, $T = 184°F$		
	y_i	K_i	x_i	y_i	K_i	x_i
C_3	0.07293	3.09	0.02348	0.07234	3.17	0.02289
iC_4	0.21788	1.60	0.13543	0.21163	1.65	0.12863
(LK) C_4	0.45682	1.21	0.37550	0.43453	1.26	0.34586
(HK) iC_5	0.13560	0.59	0.22858	0.14152	0.61	0.23267
C_5	0.11677	0.49	0.23701	0.13998	0.52	0.26997

On stage 8, the ratio of the keys is 1.64275. On stage 9, the ratio of the keys is 1.48648. For the feed, the ratio is 1.5511. The calculations now switch to the reboiler. A new set of equations are needed, but they all have the same form:

$$x_{i,n} = \frac{\overline{V}}{\overline{L}}\, y_{i,n+1} + \frac{B}{\overline{L}}\, x_{i,B}$$

The individual component equations are:

	$x_{i,n} = \dfrac{\overline{V}}{\overline{L}}\, y_{i,n+1} + \dfrac{B}{\overline{L}}\, x_{i,B}$	$x_{i,B}$
C_3	$x_i = 0.72083\, y_i + 0.00001$	0.00004
iC_4	$x_i = 0.72083\, y_i + 0.00019$	0.00068
C_4	$x_i = 0.72083\, y_i + 0.00346$	0.01239
iC_5	$x_i = 0.72083\, y_i + 0.09656$	0.34588
C_5	$x_i = 0.72083\, y_i + 0.17897$	0.64108

Working upward from the reboiler using Fig. 3 and using the format of Fig. 6, the following stage values are calculated:

Stage reboiler: $P = 122.5$ lb/in²(abs), $T = 243°F$

	x_i	K_i	α_i	$\alpha_i x_i$	y_i
C_3	0.00001	4.52	4.264	0.00004	0.00005
iC_4	0.00067	2.56	2.415	0.00162	0.00172
(LK) C_4	0.01238	2.06	1.943	0.02406	0.02559
(HK) iC_5	0.34587	1.06	1	0.34587	0.36791
C_5	0.64108	0.94	0.887	0.56850	0.60473
			$\Sigma \alpha_i x_i$	= 0.94009	
			$K_R = 1/\Sigma$	= 1.064	

Stage $R + 1$: $P = 122.5$ lb/in²(abs), $T = 241°F$

	x_i	K_i	α_i	$\alpha_i x_i$	y_i
C_3	0.00004	4.47	4.257	0.00017	0.00018
iC_4	0.00143	2.52	2.400	0.00343	0.00360
(LK) C_4	0.02191	2.03	1.933	0.04236	0.04446
(HK) iC_5	0.36176	1.05	1	0.36176	0.37987
C_5	0.61488	0.93	0.886	0.54461	0.57187
			$\Sigma \alpha_i x_i$	= 0.95233	
			$1/\Sigma$	= 1.05006	

For the remaining stages we condense the results:

	Stage $R + 2$, $T = 239°F$			Stage $R + 3$, $T = 236°F$		
	x_i	K_i	y_i	x_i	K_i	y_i
C_3	0.00014	4.42	0.00062	0.00046	4.35	0.00201
iC_4	0.00279	2.485	0.00693	0.00519	2.43	0.01264
(LK) C_4	0.03552	1.995	0.07088	0.05455	1.95	0.10664
(HK) iC_5	0.37038	1.035	0.38344	0.37296	1.005	0.37577
C_5	0.59119	0.91	0.53812	0.56686	0.885	0.50294

	Stage $R + 4$, $T = 232°F$			Stage $R + 5$, $T = 227°F$		
	x_i	K_i	y_i	x_i	K_i	y_i
C_3	0.00146	4.25	0.00623	0.00450	4.13	0.01851
iC_4	0.00930	2.36	0.02203	0.01607	2.27	0.03633
(LK) C_4	0.08033	1.885	0.15199	0.11302	1.81	0.20376
(HK) iC_5	0.36743	0.97	0.35775	0.35443	0.93	0.32832
C_5	0.54150	0.85	0.46200	0.51200	0.81	0.41308

	Stage $R + 6$, $T = 218°F$			Stage $R + 7$, $T = 205°F$		
	x_i	K_i	y_i	x_i	K_i	y_i
C_3	0.01335	3.905	0.05221	0.03765	3.60	0.13554
iC_4	0.02638	2.135	0.05615	0.04066	1.915	0.07786
(LK) C_4	0.15033	1.68	0.25295	0.18580	1.50	0.27869
(HK) iC_5	0.33322	0.855	0.28535	0.30225	0.755	0.22819
C_5	0.47673	0.74	0.35224	0.43367	0.645	0.27971

	Stage $R + 8$, $T = 184°F$			Stage $R + 9$, $T = 154°F$		
	x_i	K_i	y_i	x_i	K_i	y_i
C_3	0.09771	3.12	0.30479	0.21971	2.51	0.39215
iC_4	0.05632	1.61	0.09066	0.06554	1.235	0.05775
(LK) C_4	0.20435	1.23	0.25129	0.18460	0.92	0.12422
(HK) iC_5	0.26105	0.61	0.15921	0.21132	0.45	0.16418
C_5	0.38059	0.51	0.19406	0.31885	0.49	0.26172

Figure 18 is a plot of stage number vs. mole fractions (of C_4, iC_5, and C_5 only) and stage temperature. The mismatch is obvious but not unusual, and the next steps required are revision of distillate and bottoms compositions and a repeat of the calculations.

If the calculations are being done manually, the distillate and bottoms compositions should be revised using the methods presented in Hengstebeck.[69] Using these revised compositions, the entire set of calculations are repeated and repeated until the component curves mesh together to form smooth profiles as shown in Fig. 19.

Similar profiles can be developed using the procedures described below, which may include computerized approaches with internally generated revisions using the generally accepted Θ convergence procedure as developed by Lyster et al.[36] and treated in detail by Holland.[70,89]

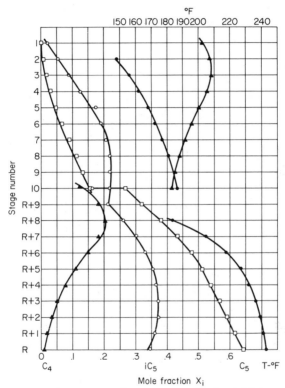

Fig. 18 Composition and temperature profile, Lewis-Matheson.

Lewis-Matheson with Nonconstant Molal Overflow

The simple method assumed constant molal overflow in each section of the tower. In general this is not realistic, especially when nonhydrocarbon systems are involved. The amount of labor increases drastically now because enthalpy balances must be made along with the material balances and equilibrium calculations. If such detail is needed, it is recommended that a computer approach be used. Time and programs can be rented from a service company, or a vendor of distillation equipment can be contacted. For this reason the procedure will only be outlined.

Starting at the top of the tower and working toward the feed stage, the following relations must be satisfied around each stage n:

$$y_{i,n} = g(x_{i,n}, T_n, P_n)$$
$$L_n + V_n = L_{n-1} + V_{n+1}$$
$$L_n x_{i,n} + V_n y_{i,n} = L_{n-1} x_{i,n-1} + V_{n+1} y_{i,n+1}$$
$$L_n h_n + V_n H_n = L_{n-1} h_{n-1} + V_{n+1} H_{n+1}$$

where in general $L_n \neq L_{n-1}$ and $V_n \neq V_{n+1}$. The procedure for hand calculations is multiple trial-and-error.

We know: $L_{n-1}, V_n, x_{i,n-1}, y_{i,n}, y_{i,n+1}, H_n, h_{n-1}, T_n, T_{n-1}$

We must find: $L_n, V_{n+1}, y_{i,n+1}, h_n, H_{n+1}, T_{n+1}$.

An unsophisticated procedure is:

Assume L_n; calculate $h_n, y_{i,n+1}, V_{n+1}$.

Using V_{n+1}, calculate $x_{i,n+1}, T_{n+1}$.

Using $T_{n+1}, V_{n+1}, y_{i,n+1}$, calculate H_{n+1}.

Compare this H_{n+1} to $(H_n + h_n - h_{n-1})$.

Revise L_n and repeat until the H_{n+1} values match.

In the procedure the following are used:

$$h_n = \Sigma L_n x_{i,n} h_{i,n}$$
$$V_{n+1} y_{i,n+1} = L_n x_{i,n} + V_n y_{i,n} - L_{n-1} x_{i,n-1}$$
$$V_{n+1} = \Sigma V_{n+1} y_{i,n+1}$$
Dew-point routine for $x_{i,n+1}$ and T_{n+1}
$$H_{n+1} = \Sigma V_{n+1} y_{i,n+1} H_{i,n+1}$$
$$H_{n+1} = H_n + h_n - h_{n-1}$$

When working upward from the bottom, the procedure is similar.

We know: $L_m, V_{m+1}, x_{i,m+1}, y_{i,m}, h_m, H_{al}, T_{al}, T_{al}$.

We must find: $L_{i,n-1}, V_n, x_a, T$.

One procedure is:

Assume V_a; calculate H, x_i, L.

Using L_{m-1}, calculate $y_{i,m}$ and T_m.

Using $T, L, and x_b$ calculate h_{m-1}.

Compare this h to $(H_n + h_n - H_{n-1})$.

Revise V_n and repeat until closure of the enthalpies.

The following relations are used:

$$H_n = \Sigma V_n y_{i,n} H_{i,n}$$
$$L_{n-1} x_{i,n-1} = L_n x_{i,n} + V_n y_{i,n} - V_{n+1} y_{i,n+1}$$
$$L_{n-1} = \Sigma L_{n-1} x_{i,n-1}$$
Point routine for αy_{in-1} and T_{n-1}
$$h_{n-1} = \Sigma L_{n-1} x_{i,n-1} h_{i,n-1}$$
$$h_{n-1} = H_n + h_n - H_{n+1}$$

Thiele-Geddes Method

Thiele and Geddes[55] proposed a method for multicomponent distillation that was significantly different from the Lewis-Matheson procedure in that the temperature profile of the tower was assumed first. This choice of independent variable implies that the number of stages in the tower is preset and hence that the shortcut procedures have been used. The usual other assumptions are column pressure, feed-plate location, feed condition, the external reflux ratio, and the internal reflux ratios L/V and $\overline{L/V}$.

As will be seen from the calculation procedure and by comparison with the computer methods discussed later, the Thiele-Geddes method is not well adapted to desk-top calculations but philosophically is akin to modern computer programs. In a sense, the Thiele-Geddes method was ahead of its time and needed computers for implementation. For these reasons, the method is only outlined below, and Refs. 70, 71, 72, 73, 74, and 76 can be used for greater detail. A number of computer programs exist which use the Thiele-Geddes approach.

The method proper usually employs composition ratios and not compositions. In the stripping section, the compositions are expressed as ratios to the bottoms composition, while in the rectifying section the distillate composition is used. The numerical values of the streams are not determined until the end of the calculation procedure. The method starts at either top or bottom and uses equilibrium relations and material balances alternately (as does Lewis-Matheson). For each component,

From the top:

$$y_{i,n} = K_{i,n} x_{i,n}$$

divide both sides by $x_{i,D}$ and rearrange to get

$$\frac{x_{i,n}}{x_{i,D}} = \frac{y_{i,n}}{K_{i,n}} \frac{1}{x_{i,D}} = \frac{y_{i,n}}{x_{i,D}} \frac{1}{K_{i,n}}$$

From the material balance,

$$Lx_{i,n} + Dx_{i,D} = Lx_{i,n} + (V - L)x_{i,D} = Vy_{i,n+1}$$

divide by $x_{i,D}$ and rearrange to get

$$\frac{y_{i,n+1}}{x_{i,D}} = \frac{L}{V}\left(\frac{x_{i,n}}{x_{i,D}} - 1\right) + 1$$

Starting from the bottom, the corresponding relations are

and

$$\frac{y_{i,m}}{x_{i,B}} = K_{i,m}\frac{x_{i,m}}{x_{i,B}}$$

$$\frac{x_{i,m-1}}{x_{i,B}} = \frac{\bar{V}}{\bar{L}}\left(\frac{y_{i,m}}{x_{i,B}} - 1\right) + 1$$

Calculations may start from either end using

$$\frac{y_{i,n}}{x_{i,D}} = \frac{y_{i,2}}{x_{i,D}} = 1.0 \qquad \text{or} \qquad \frac{x_{i,m}}{x_{i,B}} = 1.0 \qquad \text{as appropriate}$$

Consider the rectifying section. Use the assumed temperature for plate 2 to get $K_{i,2}$. With $y_{i,2}/x_{i,D}$ assumed equal to 1.0, calculate $x_{i,2}/x_{i,D}$ from the equilibrium relation. Use

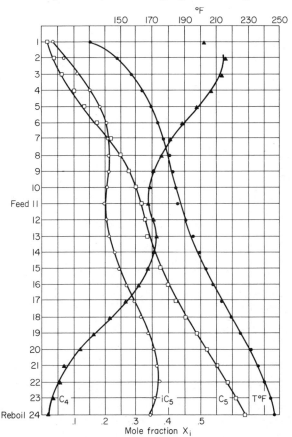

Fig. 19 Final column profiles.

this value and the known L/V with the material-balance equation to get $y_{i,3}/x_{i,D}$ and then repeat the process to the feed plate—ending with the term $y_{i,F}/x_{i,D}$.

From the bottom do the same thing but starting with $x_{i,m}/x_{i,B} = 1.0$ and ending with the term $y_{i,F}/x_{i,B}$.

Then calculate

$$\frac{y_{i,F}/x_{i,B}}{y_{i,F}/x_{i,D}} = \frac{x_{i,D}}{x_{i,B}}$$

From an overall material balance,

$$Dx_{i,D} + Bx_{i,B} = Dx_{i,D} + (F - D)x_{i,B} = Fx_{i,F}$$

Rearranging,

$$x_{i,F} = \frac{D}{F}(x_{i,D} - x_{i,B}) + x_{i,B}$$

Dividing each side by $x_{i,D}$ and rearranging,

$$x_{i,D} = \frac{x_{i,F}}{(D/F),(1 - x_{i,B}/x_{i,D})} + (x_{i,B}/x_{i,D})$$

Since $\Sigma x_{i,D} = 1.000$, the collection of $x_{i,D}$ terms can be solved for D or D/F. Note that the D estimated by the shortcut methods is the logical starting value. Once D is estimated, we find B from $B = F - D$. With both D and B estimated together with the values of $x_{i,D}$, find $x_{i,B}$ using

$$x_{i,B} = x_{i,D}\left(\frac{x_{i,B}}{x_{i,D}}\right)_F$$

Now find $x_{i,n}$ or $x_{i,m}$ on each plate for each component:

$$x_{i,n} = x_{i,D}\left(\frac{y_{i,n}}{x_{i,D}}\frac{1}{K_{i,n}}\right)$$

and

$$x_{i,m} = x_{i,B}\left[\frac{\bar{V}}{\bar{L}}\left(\frac{y_{i,m+1}}{x_{i,B}} - 1\right) + 1\right]$$

where the terms in brackets were calculated earlier.

For each plate find Σx_i. If the sum is not equal to 1.0000 ± 0.0005, normalize the estimated x_i by dividing by Σx_i:

$$x_{i,\text{new}} = \frac{x_i}{\Sigma x_i}$$

Using these new x_i on a given plate, find the bubble-point temperature for the plate and use this as the new estimated temperature.

If $\Sigma x_i = 1$ for all plates, the calculation is finished. If not, use the new temperature profile and repeat the calculations. After the last trial all the molal quantities can be determined—and indeed this is the only time they appear.

There are variations on the method which use ratios of moles of components instead of compositions; see Van Winkle.[76] For exploratory studies, the feed rate F can be set equal to 1.0 with some simplification in the formulas.

The Thiele-Geddes method, being suited to computers, usually includes enthalpy balances around each stage. The values of L and V (or $\bar{L}$ and $\bar{V}$) are adjusted to maintain the enthalpy balance, with latent heat effects predominating. The changes on any particular stage may be small, but the change in L/V or $\bar{L}/\bar{V}$ may be significant over a section of the tower.

Linear-Algebra Method

The application of linear algebra to the solution of problems in multicomponent distillation was proposed by Amundson and Pontinen[2] in 1958. In 1966, Wang and Henke.[61] applied the linear-algebra method only to the material-balance equations associated with the equilibrium relationships, but clearly illustrated how to handle multiple feeds and multiple sidestream drawoffs. Within this paper they called this technique the "tridi-agonal method."

For the sake of clarity, the following presentation will ignore intermediate sidestreams and will have a single feed stage. The computations start with the shortcut procedures developed earlier. This provides necessary input data, especially the number of stages, the feed-stage location, and the external reflux ratio. The actual computational procedure starts with the development of the internal flow map (vapor and liquid flow rates) and a temperature map. The easiest flow map assumes constant liquid overflow or constant stripping flow; the actual flow rates are functions of the distillate and bottoms flow rates, the feed rate, and the external reflux ratio. The simplest temperature map assumes a linear temperature distribution between overhead and reboiler temperatures.

The tridiagonal method still decouples the restricting equations. The application of linear algebra is confined to the solution of the mass and energy-balance equations for the respective compositions (liquid mole fractions) and internal stream flow rates (vapor-phase flow rates). The key to the tridiagonal method is development of standard forms. The stages are counted downward from the top with the condenser-reflux separator stage (Fig. 13) being stage 1. Neglecting sidestreams for simplicity, the material balance for any stage is written

$$L_n x_{i,n} + V_n y_{i,n} = L_{n-1} x_{i,n-1} + V_{n+1} y_{i,n+1} + F_n x_{i,nF}$$

The vapor-phase compositions are eliminated using $y_i = K_i X_i$ and the equation is regrouped into standard patterns:

$$L_{n-1} x_{i,n-1} - (L_n + V_n K_{i,n}) x_{i,n} + V_{n+1} K_{i,n+1} x_{i,n+1} = - F_n x_{i,nF}$$

Most stages will not have an F_n term and the right side of the equation will be zero.

Stage 1, the condenser-reflux separator, is a special stage with no external feed and can be written as

$$-(L_1 + D_L + D_v K_{i,1}) x_{i,1} + V_2 K_{i,2} x_{i,2} = 0.0$$

Stage N, the reboiler, is also special and is

$$L_{N-1} x_{i,N-1} - (B + V_N K_{i,N}) x_{i,N} = 0.0$$

The equations maintain a definite order in the subscripts for x_i. Only three groupings occur, and for ease in presentation these are defined as

$$a_n = L_{n-1}$$
$$b_n = -(L_n + V_n K_{i,n})$$
$$c_n = V_{n+1} K_{i,n+1}$$

For stages 1 and N,

$$a_1 = 0.0$$
$$b_1 = -(L_1 + D_L + D_v K_{i,1})$$
$$c_1 = V_2 K_{i,2}$$
$$a_N = L_{N-1}$$
$$b_N = -(B + V_N K_{i,N})$$
$$c_N = 0.0$$

For each component a matrix can be written in the form

$$
\begin{bmatrix}
b_1 & c_1 & \cdot & \cdot & \cdot & & & \\
a_2 & b_2 & c_2 & \cdot & \cdot & & & \\
\cdot & a_3 & b_3 & c_3 & \cdot & & & \\
& & & & & & & \\
\cdot & \cdot & \cdot & \cdot & & & & \\
& & & a_n & b_n & c_n & & \\
& & & \cdot & \cdot & \cdot & \cdot & \\
& & & & \cdot & \cdot & \cdot & \cdot \\
& & & & & a_{N-1} & b_{N-1} & c_{N-1} \\
& & & & & \cdot & a_N & b_N
\end{bmatrix}
\cdot
\begin{bmatrix}
x_{i,1} \\
x_{i,2} \\
x_{i,3} \\
\cdot \\
\cdot \\
x_{i,n} \\
\cdot \\
\cdot \\
x_{i,N-1} \\
x_{i,N}
\end{bmatrix}
=
\begin{bmatrix}
0 \\
0 \\
0 \\
\cdot \\
\cdot \\
F_n x_{i,nF} \\
\cdot \\
\cdot \\
0 \\
0
\end{bmatrix}
$$

This matrix is written for a tower with a single feed.

The $N \times N$ matrix is tridiagonal in form, and there is one such matrix for each component. All these matrix relations must hold true simultaneously, for they are stylized material balances. In each matrix, the values of the x terms are those of interest. In matrix operations, a good way to solve for the x terms is to "invert" the large $N \times N$ matrix and then premultiply each side of the given equation by the inverse. Symbolically, if $[ABC]$ is the original $N \times N$ matrix, $[ABC]^{-1}$ is the inverse. Then

$$[ABC]^{-1}[ABC][x_{i,n}] = [x_{i,n}] = [ABC]^{-1}[-F_n x_{i,nF}]$$

The tridiagonal matrix can be inverted by special methods, but more commonly it is inverted using a generalized matrix inversion program or a general simultaneous-linear-equations routine. Such routines normally are included in the various scientific subroutines packages prepared by computer firms. If such routines are not available, the solution can be obtained very efficiently using a gaussian elimination procedure. No matter what the technique used, each matrix is solved for the vector $[x_{i,n}]$ in which the $x_{i,n}$ are composition values for component i on each plate. Each component has its own tridiagonal matrix which must be solved—if there are j components, there are j sets of matrix equations.

After all the material balances are solved, the sum of the x_i values is found for each stage. If the flow streams and the temperature map are not correct, the sum of the mole fractions will not equal 1.0. In fact, any given stage may have a sum less than zero or greater than unity, and individual x_i values may show the same extremes.

The temperature map is corrected by determining bubble-point temperatures on each stage using the estimated compositions, but first these compositions must be corrected. (Direct use of the mole fractions obtained from the material-balance equations often leads to computational problems, in particular, inversion of temperatures on successive stages.) When x_i values as calculated are less than zero or more than unity, they must be corrected to the normal range. An arbitrary but workable correction is

- $x_i \geqslant 1.0$ $x_i = \dfrac{x_{i,\text{old}} + 1.0}{2.0}$

- $x_i \leqslant 0.0$ $x_i = \dfrac{x_{i,\text{old}} + 0.0}{2.0}$

The compositions on each stage are then normalized so that $\Sigma x_i = 1.0$:

$$x_{i,\text{new}} = \frac{x_{i,\text{old}}}{\Sigma x_{i,\text{old}}}$$

Bubble points are computed for each stage, as are the corresponding vapor compositions.

The internal flow map (of L values) is revised using a set of energy balances with the temperature map. This requires molal enthalpies if the systems are relatively ideal (petroleum mixtures) or partial molal enthalpies if the system is rather nonideal. Enthalpies of both liquid and vapor are required. If the solutions are ideal, the stage enthalpies are

$$H_n = \sum_i H_i(T_n) y_{i,n}$$
$$h_n = \sum_i h_i(T_n) x_{i,n}$$

Combining the energy balance with the overall material balance around each stage leads to the following set of relations:
Stage 1, condenser:

$$-V_1(H_1 - h_1) + V_2(H_2 - h_1) = - Q_c$$

Stage n, rectifying section:

$$-V_n(H_n - h_{n-1}) + V_{n+1}(H_{n+1} - h_n) = - D(h_n - h_{n-1})$$

Stage F, the feed stage ($\overline{h}_F$ is the molal feed enthalpy):

$$-V_F(H_F - h_{F-1}) + V_{F+1}(H_{F+1} - h_F) = - D(\overline{h}_F - h_{F-1}) - B(\overline{h}_F - h_F)$$

Stage m, stripping section:

$$-V_n(H_m - h_{m-1}) + V_{m+1}(H_{m+1} - h_m) = B(h_m - h_{m-1})$$

Stage N, the reboiler:

$$-V_N(H_N - h_{N-1}) = -Q_R + B(h_N - h_{N-1})$$

This set of equations is bidiagonal in form:

$$
\begin{bmatrix}
b_1 & c_1 & & & & \\
& \ddots & & & & \\
& & b_n c_n & & & \\
& & & \ddots & & \\
& & & & b_F c_F & \\
& & & & & \ddots \\
& & & & & b_m c_m \\
& & & & & & \ddots \\
& & & & & & & b_N
\end{bmatrix}
\cdot
\begin{bmatrix}
V_1 \\
\vdots \\
V_n \\
\vdots \\
V_F \\
\vdots \\
V_m \\
\vdots \\
V_N
\end{bmatrix}
=
\begin{bmatrix}
d_1 \\
\vdots \\
d_n \\
\vdots \\
d_F \\
\vdots \\
d_m \\
\vdots \\
d_N
\end{bmatrix}
$$

Stage 1:
$$b_1 = -(H_1 - h_1)$$
$$c_1 = H_2 - h_1$$
$$d_1 = -Q_c$$

Stage n:
$$b_n = -(H_n - h_{n-1})$$
$$c_n = H_{n+1} - h_n$$
$$d_n = -D(h_n - h_{n-1})$$

Stage F:
$$b_F = -(H_F - h_{F-1})$$
$$c_F = H_{F+1} - h_F$$
$$d_F = -D(\overline{h}_F - h_{F-1}) - B(\overline{h}_F - h_F)$$

Stage m:
$$b_m = -(H_m - h_{m-1})$$
$$c_m = H_{m+1} - h_m$$
$$d_m = B(h_m - h_{m-1})$$

Stage N:
$$b_{n = -\mp H_N} - h_N)$$
$$d_{n = -Q_R} + B(h_N - h_{N-1})$$

If the first stage is a total condenser, $V_1 = 0.0$. This matrix equation is solved as before and yields values of V_n. This new vapor-flow map is combined with the overall material balance to obtain a new liquid-flow map (L_n), and the first iteration is complete.

Similar trials are repeated until the computed Σx_i throughout the column differ from unity by a specified tolerance. Most computer programs also have an adjustable limit on the total number of iterations allowed and on the computational time allotted. This prevents waste of time should the program not converge.

The problem is not yet solved at this point—a particular case has converged. The print-out from the computer must now be studied for inconsistencies and to see if satisfactory results have been obtained. The distillate and bottoms compositions must be compared with the desired specifications. If the results are not satisfactory, additional computer trials are made in which the following parameters are altered (as based on personal experience):

- Total number of stages
- Number of stages in the rectifying section
- Number of stages in the stripping section
- Reflux or boilup
- Feed preheat

Often a matrix of computer runs is made to bound the parameters, after which a considered compromise is chosen for final trial.

NEWTON-RAPHSON METHOD

The past years have shown that while there are many methods available for multicomponent distillation calculations, all these methods, including the aforementioned methods, have convergence shortcomings. The Newton-Raphson method has the following advantages and disadvantages:

- Advantages

1. Absorbers, strippers, and reboiled absorbers can be calculated with the same algorithm used for distillation problems—the only adjustment being in the design constraints. (For example: The reflux ratio and the distillate rate can be replaced by the zero condenser and reboiler heat duties of conventional absorbers and strippers.)

2. The complexities of nonideality in the liquid and vapor phases can be accounted for rigorously as part of the overall computational procedure.

3. Stage inefficiencies can be included in a rigorous manner either by the Murphree equation or by the vaporization efficiency equation.

4. The number of feeds, sidestreams, and intermediate heat-exchanger systems is limited only by the number of stages.

5. The solution method is based on the linearization of the equations so that convergence accelerates as the solution is approached. This characteristic is not true of the Lewis-Matheson, Thiele-Geddes, or the linear-algebra methods, which have the tendency to approach the solution asymptotically.

- Disadvantages

1. The primary disadvantage of the Newton-Raphson method is the requirement of extensive computer storage for the first-order partial/total derivatives (even taking advantage of the overall structure of the linearized equations).

2. The necessity to establish the temperature and flow maps for the column can be rather troublesome, especially for separations involving extreme liquid-phase nonideality.

The equations which describe the continuous countercurrent multistage distillation column have already been presented. For the Newton-Raphson method the equations describing the nth stage (Fig. 20) are written as discrepancy functions, that is, a quantitative measure of the failure of the independent variables to satisfy the physical relationships of (1) material balances, (2) equilibrium relationships, and (3) energy balance.

The basic assumption is that the stage pressure is known (constant), and that the stage is

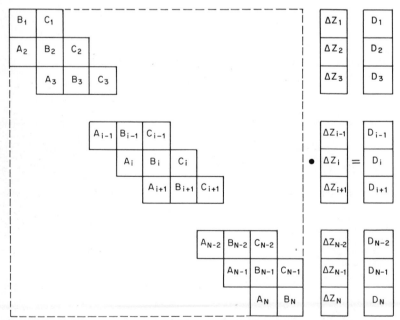

Fig. 20   Schematic representation of the column's linearized discrepancy functions.

adiabatic. Variable minimization while still retaining the maximum flexibility of problem definition is obtained by expressing the discrepancy functions in terms of componential flow rates and stage temperature, as contrasted to the methods previously discussed. The inclusion of nonideal vapor-liquid distribution factors (K values) along with the contact stage's departure from that of a theoretical stage maximizes the flexibility of this method.

The following equation set describes the general contact stage (n), with $n = 1$ being the condenser and $n = N$ being the reboiler. The condenser and the reboiler are considered to have a stage efficiency of 100% ($\eta = 1.0$). The equations account for sidestreams and intermediate coolers/heaters, as indicated on Fig. 20:

- Component material balances (moles/unit time):

$$F_1(i,n) = \left(1 + \frac{S_{L_n}}{L_n}\right) l_{i,n} + \left(1 + \frac{S_{V_n}}{V_n}\right) v_{i,n} - l_{i,n-1} - v_{i,n+1} - f_{i,n} = 0$$

- Equilibrium relationships, derived from the definition of the vapor phase Murphree stage efficiency (moles/unit time):

$$F_2(i,n) = \eta_n \frac{V_n}{L_n} K_{i,n} l_{i,n} - v_{i,n} + (1 - \eta_n) \frac{V_n}{V_{n+1}} v_{i,n+1} = 0$$

- Energy balance:

$$F_3(n) = (L_n + S_{L_n}) h_n + (V_n + S_{V_n}) H_n - L_{n-1} h_{n-1} + V_{n+1} H_{n+1} - F_n h_{fn} - Q_n = 0$$

The above discrepancy functions apply to all interior stages in the column, and also to the condenser and the reboiler if the heat duty is known (streams $n = 0$ and $n = N+1$ do not exist), which is true for absorber and stripper columns where the heat load is usually zero.

For the normal rating case of a distillation column, the distillate rate is fixed along with the external reflux ratio. With this pair of design constraints the energy balance discrepancy functions for the condenser ($n=1$) and the reboiler ($n=N$) are replaced by the following:

- Condenser ($n = 1$ and $\eta = 1.0$)

$$F_3(1) = \Sigma l_{i,1} - L_1 = 0$$
$$L_1 = \text{constant} = \text{reflux*distillate}$$

- Reboiler ($n = N$ and $\eta_N = 1.0$):

$$\Sigma l_{i,N} - L_N = 0$$
$$L_N = \text{constant} = \text{bottoms product}$$

Other specifications can be used in lieu of the above two specifications of the distillate rate and the external reflux ratio. Examples are the bottoms rate and boilup ratios; distillate rate and distillate drum temperature; etc.

The solution of the $(2C + 1)N$ discrepancy functions simultaneously cannot be accomplished directly because of the nonlinearities that are inherent in the functions. The Newton-Raphson method is based on representing each equation by a two-term Taylor's series, thus generating the set of $(2C+1)N$ linear equations, the unknowns being the corrections to the independent variables:

$$\Delta l_{i,n}, \Delta T_n, \text{ and } \Delta v_{i,n}.$$

The linearization of the discrepancy functions requires that the flow map (liquid and vapor) and the temperature map be initialized. The result of the linearization process is summarized for the general contact stage. The linearized discrepancy functions associated with the ends of the cascade ($n = 1$ and $n = N$) can be obtained easily, and hence are not included. It should be noted that the linearized equations are arranged so as to have the major temperature derivative on the diagonal of the Jacobian.

- Component material balances:

$$-\Delta l_{i,n-1} + \left(1 + \frac{S_{L_n}}{L_n}\right) \Delta l_{i,n} - \frac{S_{L_n}}{L_n^2} l_{i,n} \sum_K \Delta l_{K,n} + \left(1 + \frac{S_{V_n}}{V_n}\right) \Delta v_{i,n}$$
$$- \frac{S_{V_n}}{V_n^2} v_{i,n} \sum_K \Delta v_{K,n} - \Delta v_{i,n+1} = -F_1(i,n)$$

- Equilibrium relationships, including the Murphree stage efficiency:

$$\eta_n \frac{V_n}{L_n} K_{i,n} \Delta l_{i,n} - \eta_n \frac{V_n}{L_n^2} K_{i,n} l_{i,n} \sum_K \Delta l_{K,n} + \eta_n \frac{V_n}{L_n} l_{i,n} \sum_K \frac{\partial K_{i,n}}{\partial l_{K,n}} \Delta l_{K,n} + \eta_n \frac{V_n}{L_n} l_{i,n} \frac{\partial K_{i,n}}{\partial T_n} \Delta T_n$$

$$+ \eta_n \frac{1}{L_n} K_{i,n} l_{i,n} \sum_K \Delta v_{K,n} + \eta_n \frac{V_n}{L_n} l_{i,n} \sum_K \frac{\partial K_{i,n}}{\partial v_{K,n}} \Delta v_{K,n} - \Delta v_{i,n} + (1 - \eta_n) \frac{1}{V_{n+1}} v_{i,n+1} \sum_K \Delta v_{K,n}$$

$$+ (1 - \eta_n) \frac{V_n}{V_{n+1}} \Delta v_{i,n+1} - (1 - \eta_n) \frac{V_n}{V_{n+1}^2} v_{i,n+1} \sum_K \Delta v_{K,n+1} = -F_2(i,n)$$

- Energy balance:

$$-h_{n-1} \sum_K \Delta l_{K,n-1} - L_{n-1} \sum_K \frac{\partial h_{n-1}}{\partial l_{K,n-1}} \Delta l_{K,n-1} - L_{n-1} \frac{\partial h_{n-1}}{\partial T_{n-1}} \Delta T_{n-1} + h_n \sum_K \Delta l_{K,n}$$

$$+ (L_n + S_{L_n}) \sum_K \frac{\partial h_n}{\partial l_{K,n}} \Delta l_{K,n} + (L_n + S_{L_n}) \frac{\partial h_n}{\partial T_n} \Delta T_n + (V_n + S_{V_n}) \frac{\partial H_n}{\partial T_n} \Delta T_n + H_n \sum_K \Delta v_{K,n}$$

$$+ (V_n + S_{V_n}) \sum_K \frac{\partial H_n}{\partial v_{K,n}} \Delta v_{K,n} - V_{n+1} \frac{\partial H_{n+1}}{\partial T_{n+1}} \Delta T_{n+1} - H_{n+1} \sum_K \Delta v_{K,n+1}$$

$$- V_{n+1} \sum_K \frac{\partial H_{n+1}}{\partial v_{K,n+1}} \Delta v_{K,n+1} = -F_3(n) + F_n h_{F_n} + Q_n$$

The evaluation of the derivatives of the distribution factors (K values) and the enthalpy terms can be accomplished most readily numerically.

The solution of the linearized set of equations can be accomplished by grouping them stagewise,[79,80,82,83,84,85,86] or by grouping the material balances componentwise followed by the energy balance (so as to keep the maximum derivatives on the diagonal) and the equilibrium relationships.[81] For systems consisting of few components and many stages, as is common practice in the chemical and pharmaceutical industries, the former is the preferential grouping method. In addition, the stagewise grouping is also the preferred grouping method when there is componental dependency in the distribution factors and enthalpy.

The stagewise grouping of the linearized discrepancy functions results in a coefficient matrix that has a block tridiagonal structure as illustrated in Fig. 21. The solution of the $(2C+1)N$ set of equations can be done directly, but since the coefficient matrix (Jacobian) is rather sparse, the inherent structure is exploited to reduce both computational time and storage requirements. Using Gaussian elimination of matrices the set of corrections to the flow and temperature maps can be evaluated.

As an example of the Gaussian elimination of matrices the structure of A_n, B_n, and C_n should be noted (Fig. 21). The diagonal elements of A_n, which correspond to the component material balances, are -1.0. Thus with $2C+1$ entries to be evaluated (energy balance) the A_n matrix is generated implicitly, negating the storage requirement for A_n. The following Gaussian elimination procedure requires storage for a single B_n and $N-1$ C_n matrices plus the ND vectors.

- Forward substitution:

1. Invert B_1, perform $B_1^{-1} C_1$ and $B_1^{-1} D_1$, and replace C_1 and D_1 by the result of the matrix operations.

2. Eliminate A_n by performing $B_n - A_n C_{n-1}$ and $D_n - A_n D_{n-1}$, and replace B_n and D_n by the respective results of the matrix operation.

Invert B_n, perform the matrix operation $B_n^{-1} C_n$ and $B_n^{-1} D_{n-i}$, and replace B_n and D_n by the respective results.

3. Eliminate A_N by performing the operating $B_N - A_N C_{N-1}$ and $D_N - A_N D_{N-1}$, and replace B_N and D_N, respectively.

Invert B_N, perform the operation $B_n^{-1} D_n$, and replace D_N by the results.

- Back substitution:

1. Eliminate C_{N-1} by performing the operation $D_{N-1} - C_{N-1} D_N$ and then replacing D_{N-1} by the result.

2. Eliminate C_n by accomplishing the operation $D_n - C_n D_{n+1}$ and replacing D_n with the result.

3. Eliminate C_1 by performing $D_1 - C_1 D_2$ and replacing D_1 with the result.

The computed corrections are used to revise the componential vapor and liquid flow maps and the temperature map. The mechanism of interpreting the magnitude of the corrections, and imposing scaling factors (so that componential flows remain positive, for example), is generally regarded as proprietary. The correction of the flow and temperature maps is normally integrated with the overall convergence scheme.

Θ METHOD OF CONVERGENCE

The Θ method of convergence was developed by Lyster et al.[36] as a convergence procedure for the Thiele-Geddes type of calculation. The method has been expanded in scope over the years primarily by Holland and a number of coworkers. Because the details of the entire calculations are somewhat lengthy, the reader is referred to Refs. 70 and 89 and to two series of articles (Refs. 87, 90 to 107). This method is a way of adjusting distillate and bottoms compositions so as to satisfy the initial specifications on the tower. However, material balances, enthalpy balances, and equilibrium relations all are used to develop the calculated molal flow rates which are adjusted by the correction factor Θ. In addition, liquid and vapor flow rates and temperatures throughout the column are generated by the computer program. The procedure is iterative and may include forcing procedures although often these forcing procedures are not absolutely necessary.

Typical initial specifications are the number of plates in each section of the tower, column pressure, type of overhead condenser, reflux ratio, quantity and thermal condition of the feed, and the quantity of the distillate. Some of these specifications are developed using the hand procedures developed earlier. Note that the distillate composition is not preset (unlike the Lewis-Matheson procedure), which means that the final distillate composition calculated must be compared with the desired specifications. If the calculated composition does not meet the distillate specifications, the variables such as reflux ratio, the number of stages above or below the feed plate, and feed thermal condition must be changed and the whole calculation must be repeated. The type and extent of change of these variables required for a given problem is a matter of experience.

DESIGN OF EXISTING TOWERS

Whether an existing tower is suitable for a new use or not is a question that can arise. Obviously some factors are preset such as tower diameter, number and type of trays,

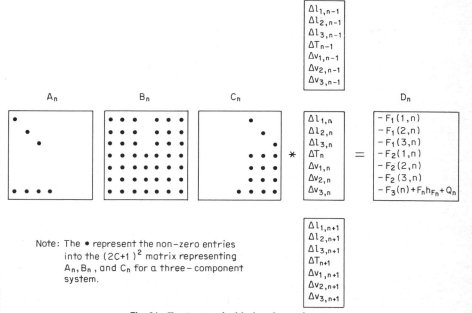

Fig. 21 Entries into the block tridiagonal matrix.

internal details, and the sizes of peripheral reboilers, condensers, and heat exchangers. The following procedures use the same general approach as for a new design but, of course, from a different viewpoint.

Structural Details

The physical strength and internal arrangement of an existing tower impose limits on operating pressures, temperatures, and flow rates of vapor and liquid. One must assemble copies of all the vessel drawings, including especially those with the internal details of the trays. The original design specifications on the units should be assembled, including the materials of construction. Tabulate for convenience the current pressure/vacuum rating on the shell, the tray characteristics (hole diameter and pitch, weir length, downcomer area, etc.), and details of construction on all the peripherals.

Utilities

Assemble and organize the current data on operating utilities. This should include available steam pressure, water or coolant available, and available flow rates. Determine the limitations on temperatures of coolant streams.

Start Shortcut Calculations

Use the normal shortcut calculation procedures to the extent that the column operating conditions can be estimated. The conditions at the top of the column are determined as before except that data on pressure drop through the condenser may be available or may be calculated. The pressure drop through the tower may be estimated with reasonable accuracy because the number of trays is known. As a guideline, see Kerns,[31] who gives a plot of ΔP/tray(bubble cap) vs. operating pressure.

Set Tentative Operating Pressure

The pressures calculated are compared with the structural limitations of the tower. The pressure calculated may lie within the design limits of the actual tower. If not, the tower may be structurally evaluated to see if field modifications are possible. Generally speaking, little can be done to raise the maximum operating pressure of a column. However, stiffening rings for vacuum service may be installed in the field. For some towers tray-support rings may act as stiffening rings. In either event the operating limits on the tower should be reexamined thoroughly and full consideration should be given to corrosion effects of previous and future service.

Establish Operating Parameters

Assuming structural problems are resolved, continue the shortcut calculations:
Determine minimum stages at total reflux (Fenske or Winn).
Determine minimum reflux (Underwood or Winn).
Locate the feed plate (Kirkbride or Fenske).
Then estimate the tray efficiency using, say, the O'Connell correlation (Fig. 16). Use the tray efficiency to calculate the equivalent number of theoretical stages in the actual column.

Using the N_{min}, N, and R_{min} just calculated, determine the reflux ratio R from the Gilliland or Erbar-Maddox correlations. With this value of R and the feed, distillate, and bottoms rates estimated earlier, calculate the liquid reflux and boilup rates at the top and bottom of the tower. These commonly are critical points for tower loadings.

Evaluate Tray-Performance Limits

With the rates already calculated, determine the vapor and liquid loadings at the top and bottom of the tower, and evaluate tray performances. The actual procedures for this are given in other sections of this handbook. (Refer to Sec. 3-2.)

The actual loadings may exceed allowable limits:
Beyond the upper load limit: flooding and blowing
Below the lower load limit: weeping and dumping
If the tower is overloaded, reduction of operating pressure and readjustment of feed preheat will tend to unload the tower. Sometimes changes in reflux ratio are possible. Such reductions in vapor/liquid loadings are specific to the case at hand. If such changes are not effective, redesign of the column internals can be considered. For example, in a

sieve-tray column with 6% hole area, the trays could be replaced with 8% hole area variable-orifice trays, thus increasing the upper load limit. If the tower is below the lower load limit, the column can be artificially loaded by increasing the boilup and reflux ratios, thus placing the column in a stable mode of operation. Since energy is expended to load the column artificially, an alternate to be considered is blanking off a portion of the tray to reduce the total hole area. Raising the column pressure will also increase the column loadings, but here the change is not large.

Check the Design

With a revised set of operating conditions and design parameters, a rigorous stage-to-stage evaluation of the column should be made to make sure the top and bottom stages are the critical load points. Normally the rectifying and stripping sections each will have a minimum and a maximum load point which should be examined. The computer print-out of the rigorous solution makes identification of these load points relatively easy, after which the tray-evaluation procedures are applied as before.

Review with Vendor

As a final step, the proposed loadings should be reviewed by the vendor who designed/fabricated the column internals. This is particularily important if changes in tray configuration are planned.

Evaluate Peripherals

With the flow rates, reflux ratio, boilup ratio, and feed preheat duty known, the heat-exchange peripherals can be evaluated. For each of the units calculate the heat duty and the ΔT. Estimate the needed transfer coefficients using the known flow rates and the physical arrangement of the exchangers, and then determine the transfer areas. If the units and duties do not match, it may be possible to change the tube-side flow pattern with baffles and improve the situation. If the area in the exchanger is too large, plugging some of the tubes may be sufficient. After a suitable configuration has been developed, the original vendor should be asked for an independent evaluation.

SPECIAL SEPARATIONS

When the components of a system have low relative volatilities ($0.95 < \alpha_{iR} < 1.05$), separation becomes difficult and expensive because a large number of trays are required and, usually, a high reflux ratio as well. Both equipment and utilities costs increase markedly and the operation can become uneconomic. If the system forms azeotropes, a different problem arises—the azeotropic composition limits the separation, and for a better separation this azeotrope must be bypassed in some way. Some systems may show an actual reversal of α values with the change in pressure from top to bottom in a tower. In other cases the composition of the azeotrope may show rapid change with pressure. Systems that have these characteristics are quite common in the drug and synthetic-chemicals industries and may occur in the petroleum industry.

For many such systems the vapor-liquid equilibrium of the feed-mixture components can be altered by deliberately adding a new material. The resulting system is more complex and the nonidealities are more pronounced. While the nonideality makes estimation of vapor-liquid equilibrium more difficult, the general procedures for multicomponent distillation still apply.

Two broad categories exist: azeotropic distillation and extractive distillation. In azeotropic distillation, the added entrainer forms a minimum-boiling azeotrope with one or more of the feed components and distills overhead. In extractive distillation, the solvent is a higher-boiling material that affects the heavier components in the feed and exits with the bottoms. Both types of operation are more complex and expensive than ordinary distillation because a foreign compound is added and because the entrainer or solvent involved may be special or relatively expensive compounds. If the materials to be separated are relatively cheap, recovery of the added agent can make or break the economics of the operation. Sometimes the added agent may be cheap—like water—but the required amount of added agent may so dilute one of the streams that the process becomes uneconomic from an energy viewpoint.

These special separations often may be considered as ternary systems. For troublesome binaries, such as ethanol and water with the well-known azeotrope, an added entrainer such as benzene forms a minimum-boiling ternary azeotrope which distills overhead. The overall system that must be considered is ternary and in general multicomponent. For mixtures such as butane-butenes an added solvent could be a furfural-water mixture. If the system contains other components as well, the added agent affects the relative volatilities of all the components, not just the desired ones, and some of the other components may have new relative volatilities in the range 0.95 to 1.05. The result is that the separation may remain poor. The usual cure for this is prefractionation of the proposed feed to limit the number of components or fractions. In a way, a many-component system is reduced to an effective binary and addition of solvent/entrainer changes the system to an effective ternary. Oliver[72] and Van Winkle[76] give specialized approaches for ternary systems which may be used instead of computer methods.

Design of azeotropic or extractive distillation systems requires significant preliminary work including:

- Choosing the solvent/entrainer
- Developing or finding necessary data
- Preliminary screening
- Computer simulation
- Small-scale testing to "prove out" the system

(The last step is *very* important.)

Solvent Selection

Solvent/entrainer selection still is more art than science. The requirements for the ideal solvent or entrainer include the following characteristics:

1. It must affect the vapor-liquid equilibrium relations of the key components (relative volatilities).

2. It should have a low latent heat.

3. It must be nonreactive with the other components in the feed mixture.

4. It must be easily separable from the components with which it leaves the column.

5. It should be noncorrosive and nontoxic.

6. It should be inexpensive.

7. It should remain soluble in the feed components and should not lead to the formation of two phases.

Naturally no single solvent or solvent mixture satisfies all the criteria, and compromises must be reached.

The selection of candidate solvents is not simple. Tassios[54] has presented techniques for rapid screening of solvents for extractive distillation using VLPC units. Ewell et al.[19] developed workable guidelines to identify chemical classes suitable for azeotropic distillation entrainers. These guidelines are discussed in some detail by Berg.[5] The basis for the classification of Ewell is the idea that liquid properties are related to the degree of bonding between molecules. The hydrogen bond is the most important of the bonds and serves as the most important criterion.

Class I. Liquids capable of forming three-dimensional networks of strong hydrogen bonds. This includes water, glycol, glycerol, amino alcohols, hydroxy acids, amides, etc.

Class II. Other liquids composed of molecules containing both active hydrogen atoms and donor atoms of oxygen, nitrogen, and fluorine. This includes alcohols, acids, primary phenols, oximes, primary and secondary amines, ammonia, hydrogen fluoride, hydrogen cyanide, nitro compounds with alpha-hydrogen atoms, etc.

Class III. Liquids containing donor atoms but no active hydrogen atoms. This includes ethers, aldehydes, esters, ketones, tertiary amines, and nitro compounds and nitriles without alpha-hydrogen atoms, etc.

Class IV. Liquids composed of molecules with active hydrogen atoms but no donor atoms. These include molecules with two or three chlorine atoms on the same carbon atom as a hydrogen atom, or one chlorine atom on the same carbon and one or more chlorine atoms on adjacent carbon atoms.

Class V. All other liquids—those without hydrogen-bond-forming capability. These include hydrocarbons, carbon disulfide, mercaptans, halohydrocarbons not in Class IV, and nonmetallic elements.

In mixtures, liquids from these classes may form hydrogen bonds or may break them and in so doing introduce deviations from Raoult's law. Remembering that + deviations from

Raoult's law may lead to minimum-boiling azeotropes (and vice versa), Table 4 is useful. Using the table, classes of compounds can be identified for screening tests. Not all classes will be effective, nor will all members of a given class be useful. Even when a class has been identified, care must be taken to avoid unwanted azeotropes. Methods for the prediction of azeotropes are not well established, and recourse should be made to the

TABLE 4   Deviations of Class Combinations from Raoult's Law

Classes	Deviations	H bonding
I + V	Always +	H bonds broken only
II + V	I + V frequently showing limited solubility	
III + IV	Always +	H bonds formed only
I + IV	Always +, I + IV limited solubility	H bonds both formed and broken (dissociation common)
I + I		
I + II	Usually +	H bonds both broken and
I + III	Some—giving maximum-boiling azeotropes	formed
II + II		
II + III		
III + III		
III + V	Quasi-ideal always	No H bonds involved
IV + IV	+ or ideal	
IV + V	Minimum azeotropes	
V + V	Minimum azeotropes (if any)	

Reprinted with permission from M. Van Winkle, "Distillation," McGraw-Hill, New York, 1967.

various tabulations of azeotropes.[13,62,78] Still, azeotropes may be desired, as can be seen from the following classification of separation possibilities:

Close-boiling compounds:

1. The entrainer forms a binary minimum-boiling azeotrope with only one component.

2. The entrainer forms binary minimum-boiling azeotropes with both compounds, but one azeotrope has a sufficiently lower boiling point than the other.

3. The entrainer forms a ternary minimum-boiling azeotrope with a boiling point sufficiently below that of any binary azeotrope. The ratio of the original feed components in the ternary must be different from the ratio in feed, and the ternary (preferably heterogeneous) must be separable in some way.

Minimum-boiling azeotropes:

1. The entrainer forms a binary minimum-boiling azeotrope which has a boiling point sufficiently below that of the original azeotrope.

2. The entrainer forms a ternary minimum-boiling azeotrope with a sufficiently low boiling point and with a different ratio of the original components from in the original azeotrope.

Finally, using binary data alone can be hazardous, and ternary data are required either from the literature or calculated from some correlation. A ternary (triangular) diagram is very useful in interpreting results of calculations and in considering possible distillation schemes. The following discussion is taken from Oliver.[72]

Figure 22 is a plot of bubble points of a hypothetical ternary somewhat similar to the methanol-acetone-chloroform system. Some knowledge of reading contour maps is helpful in analyzing this diagram. Contours (isotherms in this case) point up valleys and down ridges. The dot-dash line ATR is the crest of a ridge which rises in both directions from T. The dot-dash lines TS and TQ are the bottoms of valleys descending from T. Since T is the low point on a ridge, it is called a saddle or pass. Point T is a ternary "saddle-point" azeotrope. Points Q and S are minimum boiling azeotropes and point R a maximum boiling azeotrope.

Suppose one had data on the binary pairs only and wished to break the A-B azeotrope Q. He might propose to add C until the overall composition at 1 was reached, and distill to make pure B bottoms and A-C azeotrope S overhead. Point 1, however, is higher boiling than B, and the best separation that can be made is B overhead and point 2 bottoms. The crest of the ridge cannot be passed. Azeotrope S is more volatile than point 2, so S cannot be taken as the bottom product.

Again, suppose it is desired to break the A-C azeotrope S. On the basis of binary data one might propose to add A until the overall composition at 3 was reached. Distillation would be expected to give pure A overhead and B-C azeotrope R as bottoms. Point 4, however, is at the bottom of the valley and is more volatile than A. Hence A cannot be taken overhead.

Thus ternary data are necessary to insure that the barriers of ridges and valleys do not exist.

Dew points can also be plotted on the diagram, with tie lines to the bubble points. At azeotropes the dew and bubble point curves touch.

Extractive Distillation

In extractive distillation the solvent tends to associate with the higher-boiling component(s) and increases the relative volatility of the lighter components. The extractive

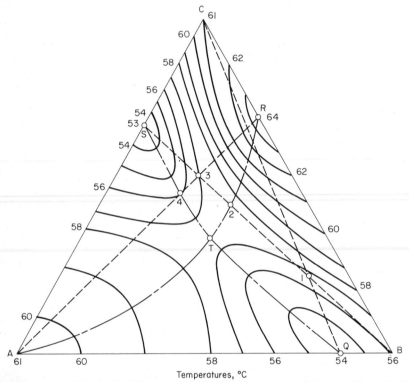

Fig. 22 Ternary diagram of bubble points. (*Reprinted with permission from E. D. Oliver, "Diffusional Separation Processes," Wiley, New York, 1966.*)

solvent usually is of low volatility, is added at or near the top of the column, is present throughout the tower, and exits with the bottoms. Later, the solvent must be separable from the original component(s) by some method (often distillation). If the solvent is not added at the top of the column, the plates from the top down to the solvent feed plate serve to "knock back" the solvent. This situation can arise when the solvent itself is somewhat volatile (it may even be one of the components of a ternary feed mixture). In such case, the solvent should be completely miscible with the feed components, because troubles can arise with immiscibility including foaming, reduction of relative volatilities, and operational instabilities.

The amount of solvent used is an important variable in effecting a separation. A common rule of thumb is: 1 to 4 mol of solvent per mole of feed—but this is only a way to start the problem. Each system has its own range which can be determined by experiment or by computer simulation.

The problem can be approached in several ways:

- Modified McCabe-Thiele
- Ternary diagram
- Multicomponent simulation

In the past, a modified McCabe-Thiele approach on a solvent-free basis has been used. The equilibrium curve is constructed using relative volatilities for the binary feed components in the presence of the solvent. An average α is developed for the rectifying section and another is developed for the stripping section. These α values need not be the same, and choosing good values may depend on experience or upon several trial calculations. See Fig. 23.

To minimize the work, it is useful to make an auxiliary plot of log α vs. mole fraction of the more volatile component on a solvent-free basis (the amount of solvent is prefixed). The variation in α is clearly seen on such plots. See Fig. 24.

The ternary diagram can also be used along with the delta-point or difference-point construction. The basic relations are derived from material balances and are summarized below:

$$\Delta = D = S + F + B$$
$$\Delta' = D - S = \Delta - S$$
$$\Delta' = F - B = F + \Delta''$$
$$\Delta'' = D - F - S = \Delta - F - S = B$$

The usual construction is shown in Fig. 25. As usual
  Δ applies from the condenser down to the first feed (solvent-addition plate).
  Δ' applies from the solvent-addition plate to the normal feed plate.
  Δ'' applies from the feed plate to the reboiler.

Finally, stage-by-stage calculations may be used. The Lewis-Matheson method is preferred for hand calculations, but the work is carried out from the reboiler upward to the condenser. This is done because the data refer to liquid compositions. The general form of the Lewis-Matheson equations is not changed—but there will be a third set of equations because of the second (solvent) feed.

Alternatively, a computer simulation can be done. The difficulty here lies in getting activity coefficients to reproduce the system. As pointed out earlier, prediction of ternary properties from binary data is not a simple task. However, recent advances and correlations have significantly improved the estimation of the needed ternary coefficients from various types of binary data.

Azeotropic Distillation

In an azeotropic distillation, the purpose of the added entrainer is to form a minimum-boiling azeotrope with one or more of the components. Entrainer selection is more difficult than choosing a solvent for extractive distillation simply because one of the feed components must be part of the azeotrope. In general, the entrainer should boil some 10 to 40°C below the feed, should form an azeotrope with a boiling point lower than any other boiling point in the system, and should be soluble in all components of the feed at all temperatures and conditions within the tower. In contrast to extractive-distillation solvents, the choice of azeotropic entrainers is very limited. Besides all this, a heterogeneous azeotrope is preferred so as to ease the problem of recovering the entrainer for reuse. The heterogeneous azeotrope splits into two phases upon condensation and cooling. Only one of the phases is returned to the column as reflux, which is another factor to complicate the situation.

The entrainer may be added with the feed or as a separate stream. If the entrainer has a volatility near that of the

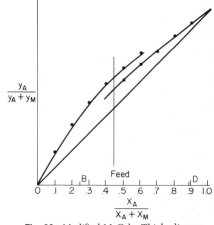

Fig. 23 Modified McCabe-Thiele diagram.

feed, it is mixed with the feed to form a single stream. If volatility is below that of the feed, the entrainer should be added above the feed.

Ideal operation would have no entrainer appear in the bottoms, but this seldom happens. In most cases entrainer is present throughout the column and can exert strong effects on relative volatilities throughout the column. As a result, the calculations become extremely dependent upon the data available, so that minor errors in data may have major impact on design. Because of this, experimental azeotropic distillations should be made before any design is finalized. The testing should be done in the largest unit available.

As with extractive distillation, the various shortcut methods (Fenske, etc.) do not really apply, so that the calculation methods are restricted to

- Modified McCabe-Thiele
- Ternary diagram
- Multicomponent simulations

The data may be plotted as curves of α vs. liquid composition or as vapor-composition curves overlaid on liquid compositions on ternary or triangular diagrams. For computer work the data must be fitted by one of the phase-equilibrium equations already mentioned (Renon-Prausnitz NRTL constants are widely used).

If the Lewis-Matheson method is used, the work closely parallels the extractive-distillation approach. Only the details of the problem vary, and the work remains a repetitious stage-by-stage construction—in this case with extremely good chances for unacceptable results.

Rather, the computer approach is recommended. A number of such programs are available at modest cost, and often the data requirements are quite minimal. However, use of a computer does not guarantee success either in simulation or in actual operation. In both cases, the computer print-out and available plant results should be studied carefully. As suggested earlier, ternary diagrams are of great value in considering the results of tests or simulations. Such studies may well predict failure for the proposed system, but they will not guarantee success. The final proof must be experimental or operational.

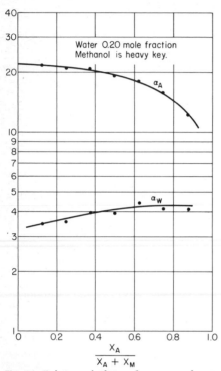

Fig. 24 Relative volatilities of acetone and water referenced to methanol at 1 atm and 0.2 mole fraction

Example 18 is meant to demonstrate the calculation procedures only and not to develop a workable design. An actual design would require a number of trials to develop acceptable solvent feed rates and solvent feed-point locations. The design should include enthalpy balances as well, because these special separations are basically nonideal and heat effects may be quite large. In the case of extractive and azeotropic distillations, the final decision is an economic one or one where environmental factors dictate such operations, whether economic or not.

EXAMPLE 18

A mixture of 45 mol/h of acetone (A) and 55 mol/h of methanol (M) is to be separated employing extractive distillation and using water (W) as the solvent. Water is to be used at the rate of 105 mol/h. The external reflux ratio is 4.0, and the feed is at the bubble point. The distillate is to contain 0.8 mol fraction A and 0.1 mol fractions of M and W. The bottoms contain 0.1 mol fraction A.

Using the ternary diagram (Fig. 26), locate the feed, distillate, and solvent points. Using the inverse-lever-arm rule or a material balance, find the bottoms composition:

$$A = 0.1$$
$$M = 0.3$$
$$W = 0.6$$

Also find that

$$D = 35.0 \text{ mol/h}$$
$$B = 170 \text{ mol /h}$$

Before going further, it is well to summarize the internal liquid and vapor flow rates (mol/h):

Condenser to solvent addition $L = 140$ $V = 175$

Solvent addition to feed $\overline{L} = 245$ $\overline{V} = 175$

Feed to reboiler $\overline{\overline{L}} = 345$ $\overline{\overline{V}} = 175$

From Ref. 24, p. 2350, calculate values of α_{acetone} and α_{water} referred to methanol as the heavy key. Assume that the water concentration in the tower is about 0.2 mole fraction (this can be checked later if necessary). The following values are calculated:

$\dfrac{x_A}{x_A + x_M}$	α_A	α_W
0.125	2.17	0.345
0.25	2.10	0.355
0.375	2.09	0.392
0.5	1.91	0.391
0.625	1.81	0.440
0.75	1.57	0.411
0.875	1.21	0.411

The calculated values also are plotted in Fig. 24.

   a. Construct a modified McCabe-Thiele diagram. For the lower concentrations of acetone, α_A varies from 2.2 to 2.0, with a geometric mean average of 2.098. For higher concentrations, α_A varies more widely from 2.0 at $x_A = 0.5$ to 1.15 at $x_A = 0.9$, with a mean value of 1.517. Calculating the equilibrium curve from the formula, the following values were found:

$\alpha_A = 2.098$		$\alpha_A = 1.517$	
x_A	y_A	x_A	y_A
0.1	0.189	0.4	0.503
0.2	0.344	0.5	0.603
0.3	0.473	0.6	0.695
0.4	0.583	0.7	0.780
0.5	0.677	0.8	0.858
0.6	0.759	0.9	0.932

Figure 23 is a plot of these results. Note the lack of continuity in the curve. The minimum reflux ratio is about 3.5. Using the given reflux ratio of 4, construct the operating lines and step off the stages. The preliminary estimate is:

Rectifying stages = 8.1
Stripping stages = 1.3

   b. Use the ternary diagram (Fig. 26). This particular diagram has liquid compositions on the triangular grid and vapor compositions (corresponding to the liquid) as a set of parametric curves overlaying the regular grid. This type of plot allows use of graphical methods as outlined below.

Assume constant molal overflow. On Fig. 25 locate points D, B, S, and F. Construct lines through D and S, and through F and B. At the intersection locate Δ'. The points Δ and Δ'' are located at D and B, respectively. Construction can be from the bottom or the top or both.

From the bottom, B is a liquid of known composition and V_B or V_{N+1} is in equilibrium with this liquid. From point B on the regular grid, read the curved vapor-composition grid to get the terms $y_{i,B}$.

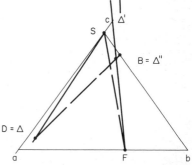

Fig. 25 Ternary construction—extractive distillation.

	$B = L_{N+1}$	$V_B = V_{N+1}$
A	0.1	0.41
M	0.3	0.37
W	0.6	0.22

Using these values, locate V_{N+1} on the triangular grid. The terms V_{N+1} and B can be used with the inverse-lever-arm rule to locate L_N, since $V_{N+1} + B = L_N$ or $V_{N+1} - L_N = B = \Delta''$. Note that L_N and V_{N+1}

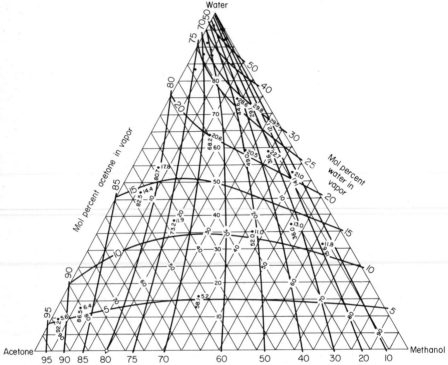

Fig. 26 Vapor-liquid equilibrium, acetone-methanol-water at 1 atm, mole fractions. (*From* Ind. Eng. Chem., **41**(10)(1949). *Copyright by the American Chemical Society.*)

correspond to $\bar{L}$ and $\bar{V}$, respectively. From L_N find $y_{i,N}$ and locate V_N. Point L_{N-1} is located on the line from B to V_N, and so on. The construction is continued until an equilibrium tie line crosses the line F to S, after which the Δ' point is used to locate the next stage.

From the top, the construction is similar, but to start a vapor-phase composition is known ($D = V_2$). Use the curved vapor-composition lines to locate L_2 (the top plate in the column).

	$x_{i,D} = y_{i,2}$	$x_{i,2}$
A	0.8	0.54
M	0.1	0.14
W	0.1	0.32

Then use the inverse-lever-arm rule to find V_3, since $V_3 = L_2 + D$ or $V_3 - L_2 = D = \Delta$. These $y_{i,3}$ values are read on the regular grid (it is a material balance) and then the values are used with the curved vapor lines to get $x_{i,3}$ or L_3:

	$y_{i,3}$	$x_{i,3}$
A	0.59	0.08
M	0.13	0.11
W	0.28	0.81

At some point along the way, the solvent S is added to a liquid stream on such a plate as to least upset the column. This may call for a series of trial points for the addition if the solvent has noticeable volatility. If the solvent is of low volatility, the addition stage is very near the top of the column. In this example suppose the solvent is added to the liquid on stage 4. By material balances:

$$V_4 = L_3 + D \quad \text{or} \quad V_4 - L_3 = D = \Delta$$
and
$$V_4 + S = L_3 + D \quad \text{or} \quad V_4 - L_3 = D - S = \Delta'$$

Using either of the relations, V_4 can be located. As usual the stepping process alternates equilibrium tie lines and material balances:

Locate L_4 from V_4.

Locate V_5 using $V_5 - L_4 = \Delta'$, and so on until the feed plate is passed.

   c. A Lewis-Matheson approach requires three sets of equations, but otherwise the procedure is unchanged. In the present example the following equations can be developed:

For the bottom section of the tower:

$$x_{m-1} = \frac{\overline{V}}{\overline{L}} y_m + \frac{B}{\overline{L}} x_B = \frac{175}{345} y_m + \frac{170}{345} x_B$$
$$x_{A,m-1} = 0.5072 y_m + 0.0493$$
$$x_{m,m-1} = 0.5072 y_m + 0.1478$$
$$x_{w,m-1} = 0.5072 y_m + 0.2957$$

For the top section:

$$x_{t-1} = \frac{V}{L} y_t - \frac{D}{L} x_D = \frac{175}{140} y_t - \frac{35}{140} x_D$$
$$x_{A,t-1} = 1.250 y_t - 0.2000$$
$$x_{m,t-1} = 1.250 y_t - 0.0250$$
$$x_{w,t-1} = 1.250 y_t - 0.0250$$

Note that in this approach, the ternary diagram gives the equilibrium compositions directly and the normal bubble- or dew-point calculation is not needed. However, often only α or K data are available and the full Lewis-Matheson approach must be used including stage-by-stage equilibrium calculations and heat balances. It is the authors' opinion that the graphical procedure should not be used except as a convenient check on the progress of a Lewis-Matheson approach.

REFERENCES

1. Abrams, D. S., and Prausnitz, J. M., *Am. Inst. Chem. Eng. J.*, **21**, 116 (1975).
2. Amundson, N. R., and Pontinen, A. J., *Ind. Eng. Chem.*, **50**, 730 (1958).
3. Amundson, N. R., Pontinen, A. J., and Tierney, J. W., *Am. Inst. Chem. Eng. J.*, **5**(3), 295 (1959).
4. Barbaudy, J., Sc.D. Thesis in Physical Sciences, University of Paris, 1925.
5. Berg, L., *Chem. Eng. Prog.*, **65**(9), 52 (1969).
6. Black, C., *Ind. Eng. Chem.*, **50**, 403 (1958).
7. Black, C., *Am. Inst. Chem. Eng. J.*, **5**, 249 (1959).
8. Bonham, M. S., M.S. Thesis in Chemical Engineering, MIT, 1941.
9. Cajander, B. C., Hipkin, H. G., and Lenoir, J. M., *J. Chem. Eng. Data*, **5**, 251 (1960).
10. Cook, E. J., Jr., M.S. Thesis in Chemical Engineering, MIT, 1940.
11. Cox, E. R., *Ind. Eng. Chem.*, **15**, 592 (1923).
12. Davies, J. A., *Ind. Eng. Chem.*, **39**, 774 (1947).
13. Dean, J. A. (ed.), "Lange's Handbook of Chemistry," 11th ed., McGraw-Hill, New York, 1973.
14. De Priester, C. L., *Chem. Eng. Prog. Symp. Ser.*, **49**(7), 1, (1953).
15. Derr, E. L., and Deal, C. H., *Int. Chem. Eng. Symp. Ser.*, **32**(3), 40 (1969).
16. Dodge, B. F., "Chemical Engineering Thermodynamics," McGraw-Hill, New York, 1944.
17. Dreisbach, R. R., "Physical Properties of Chemical Compounds," I, II, and III, *Am. Chem. Soc. Adv. Chem. Ser.*, nos. 15 (1955), 22 (1959), and 29 (1961).
18. Erbar, J. H., and Maddox, R. R., *Pet. Refiner*, **40**(5), 183 (1961).
19. Ewell, R. H., Harrison, J. M., and Berg, L., *Ind. Eng. Chem.*, **36**, 871 (1944).
20. Fenske, M. R., *Ind. Eng. Chem.*, **24**, 482 (1932).
21. Fredenslund, A., Jones, R. L., and Prausnitz, J. M., *Am. Inst. Chem. Eng. J.*, **21**(6), 1086 (1975).
22. Fredenslund, A., Michelsen, M. L., and Prausnitz, J. M., *Chem. Eng. Prog.*, **72**(9), 67 (1976).

23. Gilliland, E. R., *Ind. Eng. Chem.*, **32**, 1220 (1940).
24. Griswold, J., and Buford, C. B., *Ind. Eng. Chem.*, **41**(10), 2347 (1949).
25. Jordan, T. E., "Vapor Pressure of Organic Compounds," Interscience, New York, 1954.
26. Hala, E., Pick, J., Fried, V., and Vilim, O., "Vapor-Liquid Equilibrium," 3d ed. (2d English), Pergamon, New York, 1967.
27. Hala, E., Wichterle, I., Polak, J., and Boublik, T., "Vapor-Liquid Equilibrium Data at Normal Pressures," Pergamon, London, 1968.
28. Himmelblau, D. H., "Basic Principles and Calculations in Chemical Engineering," 3d ed., Prentice-Hall, Englewood Cliffs, N.J., 1974.
29. Holmes, M. J., and Van Winkle, M., *Ind. Eng. Chem.*, **62**(1), 21 (1970).
30. Kellog, M. W. Co., "Liquid Vapor Equilibrium in Mixtures of Light Hydrocarbons, Equilibrium Constants," New York, 1950; *Chem. Eng. Prog.*, **47**, 609 (1951).
31. Kerns, G. D., *Hydrocarbon Process.*, **51**(1), 100 (1972).
32. Kwauk, M., *Am. Inst. Chem. Eng. J.*, **2**, 240 (1956).
33. Kirkbride, C. G., *Pet. Refiner*, **23**, 32 (1944).
34. Leach, M. J., *Chem. Eng.*, **84**(12), 137 (1977).
35. Lewis, W. K., and Matheson, G. L., *Ind. Eng. Chem.*, **24**, 494 (1932).
36. Lyster, W. N., Sullivan, S. L., Billingsley, D. S., and Holland, C. D., *Pet. Refiner*, **38**(6), 221; (7), 151; (10), 139 (1959).
37. Margules, M., *Sitzungsber. Akad. Wiss. Wien. Math. Naturwiss. Kl.* II, **104**, 1243 (1895).
38. Maxwell, J. B., "Data Book on Hydrocarbons," Van Nostrand, Princeton, N.J., 1950.
39. Null, H. R., "Phase Equilibrium in Process Design," Wiley-Interscience, New York, 1970.
40. Palmer, D. A., *Chem. Eng.*, **82**(11), 80 (1975).
41. Perry, R. H. (ed.), "Chemical Engineers' Handbook," 5th ed., McGraw-Hill, New York, 1973.
42. Prausnitz, J. M., Eckert, C. A., Orye, R. V., and O'Connell, J. P., "Computer Calculations for Multi-Component Vapor-Liquid Equilibria," Prentice-Hall, Englewood Cliffs, N.J., 1968.
43. Prausnitz, J. M., "Molecular Thermodynamics of Fluid-Phase Equilibria," Prentice-Hall, Englewood Cliffs, N.J., 1969.
44. Prausnitz, J. M., Reid, R. C., and Sherwood, T. K., "Properties of Liquids and Gases," 3d ed., McGraw-Hill, New York, 1976.
45. Redlich, O., and Kister, A. T., *Ind. Eng. Chem.*, **40**, 341 (1948).
46. Reid, R. C., and Sherwood, T. K., "Properties of Liquids and Gases," 2d ed., McGraw-Hill, New York, 1966.
47. Renon, H., and Prausnitz, J. M., *Am. Inst. Chem. Eng. J.*, **14**, 135 (1968).
48. Rossini, F. D., et al., "Selected Values of Properties of Hydrocarbons and Related Compounds," API Project 44, Carnegie Press, Pittsburgh, Pa., 1957.
49. Scatchard, G., and Hamer, W. J., *J. Am. Chem. Soc.*, **57**, 1805 (1935).
50. Scheibel, E. G., and Jenny, E. F., *Ind. Eng. Chem.*, **37**, 80 (1945).
51. Smith, J. M., and Van Ness, H. C., "Introduction to Chemical Engineering Thermodynamics," 2d ed., McGraw-Hill, New York, 1959.
52. Stull, D., *Ind. Eng. Chem.*, **39**, 517 (1947).
53. Tassios, D., *Am. Inst. Chem. Eng. J.*, **17**, 1367 (1971).
54. Tassios, D. (ed.), "Extractive and Azeotropic Distillation," *Am. Chem. Soc. Adv. Chem. Ser.*, no. 115, 1972.
55. Thiele, E. W., and Geddes, R. L., *Ind. Eng. Chem.*, **25**, 289 (1933).
56. Thomson, G. W., *Chem. Rev.*, **38**, 1 (1945).
57. Underwood, A. J. V., *J. Inst. Pet.*, **31**, 111 (1945); **32**, 598, 614 (1946).
58. Underwood, A. J. V., *Chem. Eng. Prog.*, **44**, 603 (1948).
59. Van Laar, J. J., *Z. Phys. Chem.*, **185**, 35 (1929).
60. Van Ness, H. C., "The Classical Thermodynamics of Non-Electrolyte Solutions," Macmillan, New York, 1964.
61. Wang, J. C., and Henke, G. E., *Hydrocarbon Process.*, **45**(8), 155 (1966).
62. Weast, et al. (eds.), "Handbook of Chemistry and Physics," 56th ed., Chemical Rubber Company, Cleveland, Ohio, 1976.
63. Weissberger, A., Proskaners, E. D., Riddlick, T. A., and Toops, E. E., "Organic Solvents," 2d ed., Interscience, London, 1955.
64. Wilson, G. M., *J. Am. Chem. Soc.*, **86**, 127 (1964).
65. Winn, F. W., *Pet. Refiner*, **37**(5), 216 (1958).
66. Wohl, K., *Trans. Am. Inst. Chem. Eng.*, **42**, 215 (1946).
67. Wohl, K., *Chem. Eng. Prog.*, **49**, 218 (1953).
68. Hanson, D. N., Duffin, J. H., and Somerville, G. F., "Computation of Multistage Separation Processes," Reinhold, New York, 1962.
69. Hengstebeck, R. J., "Distillation," Reinhold, New York, 1961.
70. Holland, C. D., "Multicomponent Distillation," Prentice-Hall, Englewood Cliffs, N.J., 1963.
71. King, C. J., "Separation Processes," McGraw-Hill, New York, 1971.
72. Oliver, E. D., "Diffusional Separation Processes," Wiley, New York, 1966.

73. Robinson, C. S., and Gilliland, E. R., "Elements of Fractional Distillation," 4th ed., McGraw-Hill, New York, 1958.
74. Smith, B. D., "Design of Equilibrium Stage Processes," McGraw-Hill, New York, 1963.
75. Treybal, R. E., "Mass Transfer Operations," 2d ed., McGraw-Hill, New York, 1968.
76. Van Winkle, M., "Distillation," McGraw-Hill, New York, 1967.
77. Hummel, H. H., *Trans. A.1. Ch. E.*, **40**, 445 (1944).
78. Horsley, L. H., "Azeotropic Data,"Advances in Chemistry Series Nos. 6, 35, and 116, American Chemical Society, Washington, D. C.
79. Naphthali, L. M., Paper presented at the 56th National Meeting of the AIChE, San Francisco, May 1965.
80. Naphtali, L. M., and Sandholm, D. P., *AIChE J.*, **17**(1), 148 (1971).
81. Goldstein, R. P., and Stanfield, R. B., *Ind. Eng. Chem. Process. Des. Dev.*, **9**(1), 78 (1970).
82. Tomich, J. F., *AIChE J.*, **16**(2), 229 (1970).
83. Hirose, Y., Nagai, Y., and Tsuda, M., *Int. Chem. Eng.*, **18**(2), 258 (1978).
84. Roche, E. C., *Br. Chem. Eng. Process Technol.* **16**(9) 821 (1971).
85. Jelinek, J., Hlavacek, V., and Kubicek, M., *Chem. Eng. Sci.*, **28**, 1555 (1973).
86. Brannock, N. F., Verneuil, V. S., and Wang, Y. L., *Chem. Eng. Prog.*, **73**(10), 83 (1977).
87. Lyster, W. N., Sullivan, S. L., Jr., Billingsley, D. S., and Holland, C. D., *Pet. Refiner*, **38**(6), 221 (1959).
88. Fredenslund, A., Gmehling, J., and Rasmussen, P., "Vapor-Liquid Equilibria Using Unifac," Elsevier, Amsterdam-Oxford-New York (1977).
89. Holland, C. D., "Fundamentals and Modeling of Separation Processes," Prentice-Hall, Englewood Cliffs, N.J. (1975).
90. Lyster, W. N., Sullivan, S. L., Jr., Billingsly, D. S., and Holland, C. D., *Pet. Refiner*, **38**(7), 151 (1959).
91. Lyster, E. N., Sullivan, S. L., Jr., Billingsly, D. D., and Holland, C. D., *Pet. Refiner*, **38**(10), 139 (1959).
92. Lyster, W. N., Sullivan, S. L., Jr., McDonough, J. A., and Holland, C. D., *Pet. Refiner*, **39**(8), 121 (1960).
93. Hardy, B. W., Sullivan, S. L., Jr., Holland, C. D., and Bauni, H. L., *Hydrocarbon Process. Pet. Refiner*, **40**(9), 237 (1961).
94. Weisenfelder, A. J., Holland, C. D., and Johnson, R. H., *Hydrocarbon Process. Pet. Refiner*, **40**(10), 175 (1961).
95. Hardy, B. W., Holland, C. D., Canik, S. J., and Bauni, H. L., *Hydrocarbon Process. Pet. Refiner*, **40**(12), 161 (1961).
96. Dickey, B. R., Holland, C. D., and Cecchetti, R., *Hydrocarbon Process. Pet. Refiner*, **41**(2), 145 (1962).
97. McDonough, J. A., and Holland, C. D., *Hydrocarbon Process. Pet. Refiner*, **41**(3), 153 (1962).
98. McDonough, J. A., and Holland, C. D., *Hydrocarbon Process. Pet. Refiner*, **41**(4), 135 (1962).
99. Tomnie, W. J., and Holland, C. D., *Hydrocarbon Process. Pet. Refiner*, **41**(6), 139 (1962).
100. Holland, C. D., and Pendon, G. P., *Hydrocarbon Process. Int.*, **53**(7), 148 (1974.
101. Holland, C. D., and Eubank, P. T., *Hydrocarbon Process. Int.*, **53**(11), 176 (1974).
102. Holland, C. D., Pendon, G. P., and Gallun, S. E., *Hydrocarbon Process. Int.*, **54**(1), 101 (1975).
103. Holland, C. D., and Kuh, M. S., *Hydrocarbon Process. Int.*, **54**(7), 121 (1975).
104. Gallun, S. E., and Holland, C. D., *Hydrocarbon Process. Int.*, **55**(1), 137 (1976).
105. Hess, F. E., and Holland, C. D., *Hydrocarbon Process. Int.*, **55**(6), 125 (1976).
106. Hess, F. E., Holland, D. D., McDaniel, R., and Tetlow, N. J., *Hydrocarbon Process. Int.*, **56**(5), 241 (1977).
107. Hess, F. E., Gallun, S. E., Bentzen, G. W., Holland, C. D., McDaniel, R., and Tetlow, N. J., *Hydrocarbon Process. Int.*, **56**(6), 181 (1977).

Section **1.3**

Batch Distillation

R. W. ELLERBE, M.S.Ch.E., M.B.A. *Project Manager, The Rust Engineering Company; Registered Professional Engineer in Arkansas.*

NOMENCLATURE

L	Column liquid rate, moles/h
V	Column vapor rate, moles/h
S	Total moles of liquid in the still at any time
D	Distillate rate, moles/h
R	Reflux ratio L/D
x	Mole fraction of component in liquid phase
y	Mole fraction of component in vapor phase
α_{AB}	Relative volatility of component A with respect to component B
α_{BC}	Relative volatility of component B with respect to component C
t	Time, h
x, x_w, x_0	Ratio of mole fraction of lower-boiling component to mole fraction of higher-boiling component in liquid; subscript w refers to bottoms, subscript 0 refers to reflux composition
Y, Y_0	Ratio of mole fraction of low boiler to mole fraction of high boiler in vapor; subscript 0 refers to reflux
H	Column holdup, moles of liquid
J	Holdup per theoretical stage, moles
ϕ	Factor expressed in Fig. 11, as $n \ln$
θ	Time, h
t', t_H	Time for 90% approach to equilibrium (Fig. 10), and time to fill plates with dynamic holdup, h
n	Total theoretical stages

SUBSCRIPTS

A, B, C	Components in mixture
o	Initial quantity
D	Refers to distillate mole fraction
S	Refers to still mole fraction
W	Refers to still weight fraction
n	Refers to theoretical stage number
i	Refers to component i in the mixture
ss	Refers to steady state

INTRODUCTION

The process design of a batch distillation unit poses a challenging problem for chemical engineers. They are faced with a completely unsteady-state process. Compositions in the kettle and the distillation column change continuously with time. Consequently, rigorous mathematical solutions are not practical to make by hand calculations. Because of this, preliminary designs are ordinarily made by shortcut methods.

This section presents a concise, practical approach to the use of preliminary design methods. It also notes the limitations of these methods, so that process designers can make judgments regarding the accuracy of their designs. The designer who requires a rigorous solution of the batch distillation problem must use a digital computer. Several investigators (Huckaba and Danly,[9] Barb and Holland,[12] and Meadows[10]) have developed mathematical models and computer programs for solving the unsteady-state batch distillation problems.

It is well known that three complications increase considerably the difficulties in making rigorous calculations. They are:

1. *Unsteady state*—compositions throughout the unit change as product is withdrawn.

2. *Holdup*—the physical amount of material in the kettle, column, and auxiliaries has a marked effect on separation efficiency.

3. *Equilibration*—the length of time prior to beginning product withdrawal has a considerable effect on process economics.

During the past 70 years much effort has been directed toward developing design methods for batch distillation units. Most of the assumptions made in calculations for continuous distillations have been used in batch distillation theory. Despite the well-documented limitations of batch distillation design methods, these methods have been used for decades, and they do provide a sound engineering approach to the problem. Using them, the engineer can predict recovery efficiencies, cycle times, vapor-liquid rates, heat input, and auxiliary requirements.

Typically, the process designer must assemble vapor-liquid equilibrium data and, using these data, must choose the economic combination of kettle size, actual number of plates, and reflux rates to meet product specifications and product rates. This choice is never easy, because batch units characteristically operate with production cycles consisting of charging, equilibration, product drawoff, slop cut, product drawoff, dumping, and cleanup.

The number of product-drawoff periods depends, of course, on the number of components in the original mixture charged to the kettle. The size of the slop cut recycled to the next charge depends on the sharpness of separation between the products, and the sharpness of separation depends on holdup, relative volatility, reflux, and number of plates.

BATCH-STILL CHARACTERISTICS AND LIMITATIONS

The decision to use a batch unit for separating relatively pure components from a binary or multicomponent mixture usually depends on comparative economics between continuous and batch operation. However, there are separation problems for which the batch unit is a poor tool. Conversely, there are numerous situations which commend themselves to batch distillation.

Batch distillation is often preferable to continuous distillation where relatively small quantities of material are to be handled at irregularly scheduled periods. In many cases, the composition of the feed may vary widely from period to period. Furthermore, a general-purpose still is sometimes desired to be used in handling a number of different products.

Probably the most outstanding attribute of batch distillation is its flexibility. Little change is required when switching from one mixture to another. Reflux rate and throughput can be varied easily. No balance of feed and drawoff need be maintained. In a situation where the composition of the feed may change frequently or where completely different mixtures must be handled, the versatility of the batch unit is unexcelled.

The batch unit requires the least amount of capital for separating relatively pure components from a multicomponent mixture. Continuous separation requires a separate column and auxiliaries for each product (less one). Batchwise, one merely switches product receivers.

Batch Rectifying Unit

The conventional batch unit uses the column as a rectifying section. It is an efficient way to remove traces of high boiler from a single lower-boiling component. For this, it gives a sharp, clean-cut separation. The reason will be more apparent later, but suffice to say now, kettle holdup is a major factor in getting this sharp, clean-cut separation.

If only a few percent of a low boiler is to be removed and high-purity product is not required, the conventional batch unit can be used to take the light ends overhead. The rest of the charge can be recovered directly from the kettle in moderate purity. This saves considerable time over completing the distillation batchwise or running the same quantity of material through a continuous column. Such a distillation should not be expected to give a high product purity because of kettle-holdup implications.

Batch Stripping Unit

Sometimes the nature of the batch mixture may dictate the use of a "backward" still. Such a unit consists of a large accumulator placed beneath the condenser, a stripping column, and a reboiler. The initial charge is placed in the accumulator. The mixture is fed to the top of the column, and product is taken from the reboiler. This arrangement has reverse characteristics of a batch rectifier unit.

A backward still does a good job of removing traces of low boiler from a given component, but the distillation must be run to completion to accomplish this. The batch stripping still is a very unsatisfactory tool for recovering a high-purity product by removing a high-boiling impurity.

If only a few percent of high boiler is to be removed and a high-purity product is not required, this type of unit can be used to remove most of the high boiler by running only a part of the charge through, thus saving time.

Qualitatively, the usual characteristics of continuous and batch stills may be summarized as follows:

1. *Continuous still*—gives theoretical separation (after allowing for plate efficiency) at near minimum reflux ratios with intermediate cut (slop cut), equal to holdup per plate times the number of plates.

2. *Batch still*—gives less than theoretical separation at reflux ratios well above

minimum, large intermediate cut, and a poor yield of good material. Reflux is often equal to the number of plates in most practical distillations.

These characteristics are often dramatized by plant distillations. In one case, a batch unit had been used to remove traces of light ends and heavies from a by-product. Considerably increased capacity and equally good or better product was obtained by changing the operation. The unit was operated continuously as a stripping still to take the light ends overhead until the batch kettle was full. Then the feed was shut off, and the final separation was made by conventional batch distillation.

DISTILLATION CURVES

A batch still handling a mixture of A and B gives an intermediate fraction or slop cut containing both A and B. This cut is recycled and reloaded with the next charge. It is interesting to trace on a McCabe-Thiele diagram the progress of a distillation from one component through the intermediate fraction to the next higher-boiling component.

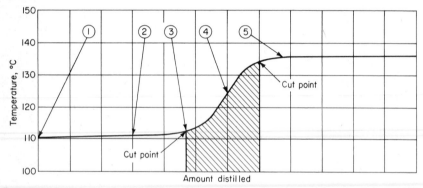

Fig. 1   Distillation curve for 50-50 mixture of components A and B. [*Adapted from* Pet. Refiner, **29**(2), 143 (1950).]

For illustrative purposes, assume a 50-50 mixture of A and B having an average relative volatility of 2.05. A has an atmospheric boiling point of 110°C, and B boils at 136°C. The distillation will be accomplished using a column of 10 theoretical plates and a 10:1 reflux ratio. Figure 1 shows an idealized distillation curve for such a mixture.

Figures 2 through 6 illustrate the relation of reflux ratio, plates, and composition at

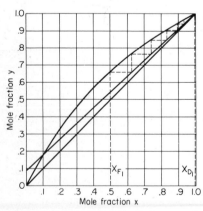

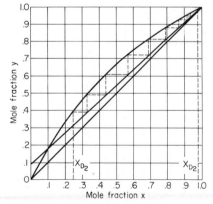

Fig. 2   Toluene–ethyl benzene system, first stage—50-50 mixture. [*Reprinted by permission from* Pet. Refiner, *29(2), 143 (1950).*]

Fig. 3   Toluene–ethyl benzene system, second stage—50-50 mixture. [*Reprinted by permission from* Pet. Refiner, *29(2), 143 (1950).*]

different stages of distillation. The numbers in the circles of Fig. 1 refer to the McCabe-Thiele figure representing that state of the distillation.

Figure 2 shows that at the beginning of the separation, when the kettle composition is 50-50, the 10 plates are enough to yield an overhead purity of 99.8 mol % x_A.

Figure 3 shows the construction when one-fourth of the charge has been distilled. The product purity has dropped to about 99%, and the kettle composition is about 25 mol % x_A.

Figure 4 shows the compositions when the distillation begins to enter the intermediate fraction. The product composition has dropped to 95% x_A; the kettle composition is a little over 10% low boiler.

Figure 5 shows the construction midway in the intermediate cut. At this point x_A has been reduced to about 5%.

Figure 6 shows that even after the intermediate fraction has distilled off and the product has only 5% low boiler, there is still about 0.5% low boiler in the kettle.

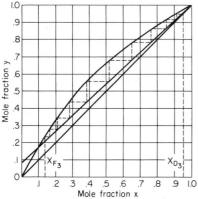

Fig. 4 Toluene–ethyl benzene system, third stage—50-50 mixture. [*Reprinted by permission from* Pet. Refiner, **29**(2), 144 (1950).]

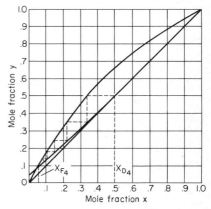

Fig. 5 Toluene–ethyl benzene system, fourth stage—50-50 mixture. [*Reprinted by permission from* Pet. Refiner, **29**(2), 144 (1950).]

These diagrams show that at a set reflux ratio, the overhead purity decreases slowly at first and then more rapidly as one approaches the break. As the distillation passes through the intermediate fraction, the overhead product composition only asymptotically approaches that of the pure high-boiling component. Using a constant reflux ratio throughout a particular distillation may require special justification in view of the soundly established practice of using low reflux along the plateaus of a distillation curve and high reflux at the breaks.

Curves shown in Fig. 7 are typical of batch-still temperature-recorder records. As shown, the column warms up and the top temperature rises above 110°C, the boiling point of A, but when reflux is applied the top temperature drops to 110°C. After a period of total reflux, equilibrium will be reached and the operator can withdraw product.

Figure 7 is a plot of top temperature and still temperature for the conditions of Fig. 1. Figure 7 is an excellent guide to the progress of a distillation. The kettle temperature gradually rises; it starts out between 110 and 136°C and will be above 136°C when all of A is removed. The difference as compared with overhead temperature is due to pressure drop in the column.

CALCULATION METHODS

General

The design of a batch distillation unit is based on material-balance and equilibrium-curve equations. These equations involve simplifying assumptions. Each assumption introduces a degree of inaccuracy into the final answer. Accordingly, the process designer must know the simplifying assumptions and must make allowances for the errors introduced by assumptions.

The preliminary design methods discussed in this section are based on the McCabe-Thiele graphical solution of the material-balance equations. The usual simplifying assumptions are:

1. *Equimolal overflow*—this assumption holds when the latent heats of vaporization of the components are nearly equal. It eliminates the need for tedious heat balances.

2. *Adiabatic operation*—this assumption holds for well-insulated columns. It poses the least problem of all assumptions.

3. *Negligible column holdup*—this assumption makes it possible to relate kettle and overhead compositions using straight operating lines. For most tray-type columns this assumption is reasonable. Some packed columns, however, have appreciable holdup, and this assumption could lead to serious design errors.

4. *Theoretical stages*—this assumption makes it possible to relate liquid and vapor compositions on each plate because for such stages they are in equilibrium.

5. *Constant relative volatility*—this assumption makes it possible to predict the shape of the equilibrium curve based on the equation

$$Y = \frac{\alpha x}{1 + (\alpha - 1)x}$$

The simplest case of batch distillation is one in which a multicomponent mixture is charged to a heated kettle fitted with a total condenser and product receiver. The mixture is distilled without reflux until a definite quantity of one of the components has been recovered or until a definite change in composition of the kettle contents has been effected. This process is called *batch differential distillation,* and the well-known Rayleigh equation expresses the system material balance:

$$\ln \frac{S_0}{S} = \int_{x_1}^{(x_A)_0} \frac{dx_A}{y_A - x_A} \tag{1}$$

Batch differential distillation has little commercial significance because it results in poor yields and low distillate purities. However, the principles led Smoker and Rose[5] to an extension of the theory to constant-reflux batch rectification operations. Bogart[6] developed the theory and equations for solving the variable-reflux batch rectification problem.

In batch rectification, one may be faced with the problem of determining the proper conditions for using an existing column with a fixed number of plates or designing a new column in which the selection of the number of plates must be made. In either case, two modes of operation should be considered:

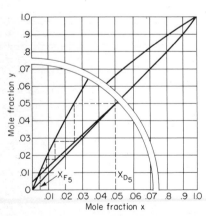

Fig. 6 Toluene–ethyl benzene system, fifth stage—50-50 mixture. [*Reprinted by permission from* Pet. Refiner, 29(2), 144 (1950).]

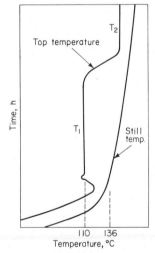

Fig. 7 Batch-still temperature-recorder chart. [*Reprinted by permission from* Pet. Refiner, 31(8), 95 (1952).]

1. *Constant reflux*—column has a fixed number of plates and a constant reflux rate, and overhead composition varies.
2. *Variable reflux*—column has a fixed number of plates and a variable reflux rate, and overhead composition remains constant.

Constant-Reflux Operation

In this mode of operation, the column has a fixed number of plates and operates with a constant reflux ratio and variable overhead-product composition. Rate of depletion of the kettle contents equals the rate of accumulation of the resulting distillate:

$$-\frac{dS}{d\theta} = \frac{dD}{d\theta} \tag{2}$$

where S is the moles of mixture in the still at time θ and D is the moles of distillate at time θ.

For any component:

$$-\frac{d(Sx_S)}{d\theta} = \frac{x_D \, dD}{d\theta} \tag{3}$$

where x_s is the mole fraction of the component in the kettle at time θ and x_D is the instantaneous mole fraction of the component in the distillate that is leaving the condenser at time θ.

Combining Eqs. (2) and (3) gives

$$\frac{dS}{S} = \frac{dx_S}{x_D - x_S}$$

Integrating this equation between indicated limits yields

$$\int_{S_0}^{S} \frac{dS}{S} = \int_{x_{S_0}}^{x_S} \frac{dx_S}{x_D - x_S}$$

$$\ln \frac{S}{S_0} = \int_{x_{S_0}}^{x_S} \frac{dx_S}{x_D - x_S} \tag{4}$$

where S_0 is the moles originally charged to the kettle and x_{S_0} is the mole fraction of the component in the kettle charge.

Equation (4) is similar to Eq. (1), the Rayleigh equation. Smoker and Rose have used the equation to estimate batch distillation curves for binary and ternary systems in which the mole percent of the components in the distillate is plotted against the percent of the charge distilled.

The right-hand side of Eq. (4) is integrated graphically by plotting $1/(x_D - x_S)$ vs. x_S. The area under such a curve, taken between x_{S_0} and x_S, is the value of the integral.

To establish the relation between x_D and x_S, the equilibrium curve is first plotted as shown in Fig. 8. Several values of x_D are selected, and operating lines having the same slope (L/V is constant) are drawn through the intersection of x_D and the diagonal. The diagonal has the equation $y = x$, and operating lines intersect this at x_D.

Once these lines are drawn, steps are drawn between the operating line and the equilibrium curve, as in the well-known McCabe-Thiele method. The kettle acts as a theoretical plate. The intersection of the last horizontal step (going down from x_D) with the equilibrium curve is the composition x_S of the liquid in the kettle.

Once the kettle composition is reduced to the desired concentration, the distillation is stopped. From the overall material balance and the fact that a constant reflux ratio was used, the total vapor produced by the kettle for the entire process can be calculated. The heat requirement consists of sensible heat added to reach the boiling point, latent heat added during the total reflux period, and latent heat of the vapor produced during product drawoff.

Variable-Reflux Operation

The column operates with a fixed number of theoretical stages, constant overhead composition, and varying reflux ratio. Overall material balance at any time is

$$S = S_0 \left(\frac{x_D - x_{S_0}}{x_D - x_S} \right) \tag{5}$$

Differentiating with respect to time gives

$$\frac{dS}{d\theta} = \frac{S_0(x_D - x_{S_0})dx_S}{(x_D - x_S)^2 d\theta} \tag{6}$$

Assuming equimolal vapor and liquid rates in the column at any instant, we find the rate of distillation from

$$\frac{dS}{d\theta} = (L - V) \tag{7}$$

Substituting in Eq. (6) and solving for the time gives

$$\theta = \frac{S_0(x_D - x_S)}{V} \int_{x_S}^{x_{S_0}} \frac{dx_S}{(1 - L/V)(x_D - x_S)^2} \tag{8}$$

This is the time required for distillation, exclusive of that required for charging the kettle, heating up, equilibration, shutting down, and cleaning up. Equation (8) is the same as that of Bogart.[6]

The vapor load V, moles per hour, can be calculated if the diameter of the column, allowable vapor velocity, operating pressure, and temperature are known. Conversely, if the time were fixed, the diameter of the column could be calculated using V obtained in Eq. (8).

To evaluate the integral term on the right-hand side of Eq. (8), the equilibrium curve is drawn as before. Several operating lines with different slopes L/V are drawn as in Fig. 9, each passing through x_D (which is constant) on the diagonal.

The steps equivalent to the total number of theoretical stages are drawn in the usual manner for each line. Again, the intersection of the last horizontal step with the equilibrium curve gives x_S. This is continued until x_S reaches the desired final value. The integral is then evaluated graphically by plotting $1/(1 - L/V)(x_D - x_S)^2$ vs. x_S and taking the area under the curve between x_{S_0} as before.

The minimum number of theoretical stages possible is that obtained when operating at total reflux between the specified x_D and the final kettle composition. All batch stills should be designed with sufficient plates to make the separation between the final kettle concentration x_S and the specified distillate purity x_D. By designing for this condition,

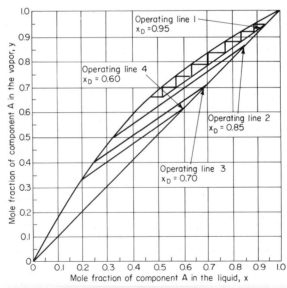

Fig. 8 Constant-reflux-ratio McCabe-Thiele diagram. *(Reprinted by permission from* Chem. Eng., *Jan. 23, 1961, p. 134.).*

more plates than necessary will be present during most of the distillation, allowing the column to operate near the minimum reflux ratio until the break is reached. If a column is designed to remove all but 1 or 2% of the more volatile component A, there will be little choice as to the number of plates and reflux ratio, because the equilibrium curve and the 45° line $y = x$ approach each other.

Total-Reflux Operation

Batch distillation columns are totally refluxed for a period of time prior to withdrawing product. This equilibration procedure assures that specification product is withdrawn when product recovery begins. However, the process designer must be concerned with

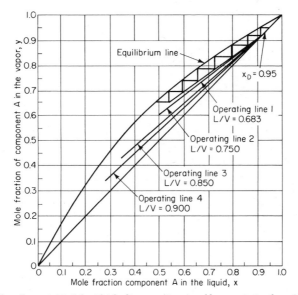

Fig. 9 Variable reflux ratio McCabe-Thiele diagram. *(Reprinted by permission from* Chem. Eng., *Jan. 23, 1961, p. 134.)*

the length of time required to achieve equilibrium. Equilibration time is a part of the total cycle time and thus affects the still production rate. The amount of liquid holdup in the condenser, column, and kettle is a major factor in the time requirement.

For a stage n, the basic material-balance equation is $J_n(dx_{i,n}/dt) = L_{n+1}x_{i,n+1} + V_{n-1}y_{i,n-1} - L_n x_{i,n} - V_n y_{i,n}$ where J_n is the molar holdup of liquid on the stage. Jackson and Pigford,[7] starting with this basic equation, developed a generalized method for estimating equilibration time under total-reflux operation.

Figure 10 shows a generalization of their results for the case of 90% approach to steady-state distillate composition. The ordinate is the ratio of time for reaching the 90% approach to the time required to fill the plates with their steady-state inventory of low-boiling material. Holdup enters the generalization at this point. The abscissa term includes the steady-state distillate content of low boiler.

Although Fig. 10 covers an efficiency range $E_{mv} = 0.5$ to 1.0, a range of ratios of kettle to plate holdup, and cases for different initial liquid compositions on the plates, it still must be regarded as approximate. Preferably, it should be used in connection with actual plant experience.

Berg and Ivor[8] developed three equations which allow the process designer to predict the time involved in reaching a given reflux composition. Their equations were experimentally verified. The first of the three is the most accurate and is applicable to systems with low volatility ratios. The other two are good approximations and are applicable when the value of α^n is high and solution of the first equation becomes cumbersome.

Following is their generalized expression of the relationship between relative volatility, column holdup, reflux flow rate, reflux and kettle compositions, theoretical stages, and time. The factor ϕ comes from their research and is expressed as $\phi = f(n \ln \alpha)$. Figure 11 is a graph of ϕ vs. $n \ln \alpha$ for use in the following equation:

$$1 - \frac{[Y_0/(X_w)(\alpha^{n+1})]}{1 - \alpha^{-n}} = \alpha^{-(\phi)(nL/H)(\alpha - 1/\alpha^n)t} \tag{9}$$

For values of $n \ln \alpha$ approaching 6, or where α^n approaches 400, the function ϕ becomes essentially equal to unity (see Fig. 11). The value of $1 - \alpha^{-n}$ also becomes very nearly equal to 1.0; so Eq. (9) reduces to

$$1 - \frac{Y_0}{X_w \alpha^{n+1}} = \alpha^{-(nL/H)(\alpha - 1/\alpha^n)t} \tag{10}$$

In cases where a fractionator with a large number of trays is being used with a system having a high relative volatility, in order to prepare a high-purity overhead product easily, the value of α^n becomes very large. Equations (9) and (10) are then somewhat difficult to handle; so Eq. (10) may be reduced to

$$\frac{Y_0}{X_w} = \frac{nL}{H}(\alpha - 1)(\alpha \ln \alpha) t \tag{11}$$

Equations (9), (10), and (11) are the three final forms of the equilibrium approach expression. Use of the proper equation is dictated by the value of α^n and/or the accuracy of the calculation desired.

There are two common cases where equilibration time lags become important. The first is the separation of components having relative volatility ratios approaching unity. The second is separations of components having relative volatility ratios in which other factors lead to excessive time lags.

Although the volatility ratio is a good indication of the difficulty of separation, and a general indication of the time required to closely approach equilibrium, the approach time is also strongly affected by the reflux and bottoms compositions and the column holdup. Even though the volatility may be high, the rate of approach to equilibrium may be quite slow if a relatively pure reflux stream is desired or if the column has excessive holdups.

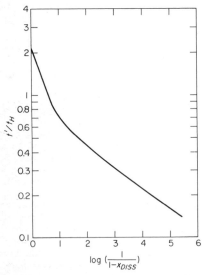

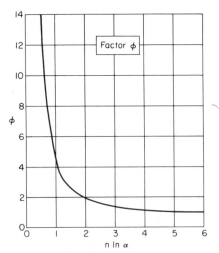

Fig. 10 Equilibration time for plate columns. (*Reprinted by permission from* Chem. Eng. *Apr. 22, 1968, p. 168.*)

Fig. 11 Factor ϕ employed in generalized equation for rate of approach to equilibrium. [*Reprinted by permission from* Chem. Eng. Prog., *44(4), 307 (1944).*]

The minimum diameter of the standard-construction bubble-cap columns used in petroleum-refining operations is about 30 in. When relatively small amounts of material are to be fractionated in a bubble-tray column, the column diameter as determined from vapor-load considerations may be less than 0.762 m (30 in). The fabrication of such columns has many mechanical limitations and may lead to the selection of an oversize column as an expedient to overcome the structural disadvantages of small-unit construction. Such a selection, while having sound fabricational and construction advantages, leads to excessive holdup. This results in operational and economic disadvantages.

Rigorous Methods

Much effort has been directed toward developing design methods for batch distillation columns. Most of the early work, beginning with the well-known Rayleigh equation for differential batch distillation and continuing through the many graphical and empirical hand methods, has been reduced to academic interest with the advent of computers.

Huckaba and Danly[9] presented the first comprehensive model of a batch distillation column. It employed heat and material balances but was developed for the binary case. Meadows[10] presented the first comprehensive model for multicomponent batch distillation. The model used heat and material balances as well as volume balances and was limited only by the assumptions of ideal plates, constant-volume holdup, adiabatic operation, and negligible vapor holdup.

Distefano[11] reported the results of a study devoted mainly to the analysis of various methods of numerically integrating the differential equations of a comprehensive mathematical model of multicomponent batch distillation. The model involved both heat and material balances and was limited only by the reasonable assumptions made by Meadows. It was concluded that the third-order Adams-Moulton-Shell predictor-corrector method was the most stable of the 11 methods tested.

Barb and Holland[12] have also developed a rigorous treatment of milticomponent mixtures. The integral-difference equations describing the material and energy balances are solved by numerical methods. The models are validated by comparing calculated distillation curves with experimental curves developed by Rose and O'Brien.[13] The agreement between calculated and experimental curves for the ternary system n-heptane–methylcyclohexane–toluene was extremely good.

Rigorous treatment of the batch distillation design problem is now possible if a computer is available. There should be little use for shortcut methods except for preliminary design studies.

Effect of Holdup

Just what is there about the arrangement of having the total charge contained in the kettle at the bottom of the column which handicaps the column and prevents it from giving the separation which might be expected by continuous operation? Is it the unsteady-state operation—the absence of equilibrium? This apparently is not the trouble, because, as will be pointed out below, when a column is operated batchwise with a small charge, quite sharp breaks are obtained even with liquids which are difficult to separate.

Evidence exists that the size of the charge in the kettle is of major importance in determining the sharpness of separation. For example, laboratory distillations give sharper separations when the break comes near the end, rather than early in the distillation. In other words, the separation is sharper when the amount of material in the kettle is small at the time of transition from one component to another.

The fundamental batch distillation equation has no terms which indicate a relation between the size of the intermediate fraction and the holdup per plate in the column. Holdup comes into the equation as a ratio between the size of the still charge and the column holdup. The equation indicates that sharper breaks obtain when the total charge is small. In general, this is borne out by experience.

Normal batch distillations give unsymmetrical "breaks" in the distillation curve. A more abrupt transition from good overhead to intermediate fraction is obtained on the lower half of the "break." As the distillation passes through the "break," the product approaches pure high boiler asymptotically. And, for difficult separations, considerable high boiler must be distilled over before substantially pure high boiler is obtained.

Experimental data and theory indicate that the larger the batch still, the larger will be the intermediate fraction. For example, if a 3790 L (1000-gal) batch containing 3032 L (800 gal) of the desired component gives only 379 L (100 gal) of the component with desired purity, a 7580 L (2000-gal) batch will not give 3411 L (900 gal) of good product. The larger batch gives about the same percentage of good product, namely, 758 L (200 gal).

One way to conceptualize the physical meaning of charge size on sharpness of separation is to regard the kettle as the first plate in the still—a plate with very large holdup. When the kettle charge is large, even a small fraction of a low-boiling impurity represents a sizable quantity of material. For example, if 3790 L (1000 gal) of a mixture remain in the kettle when the composition has been reduced to 0.5% low boiler, there still remains 18.95 L (5 gal) of low boiler to be removed from 3771 L (995 gal) of high boiler. If the holdup in the column is equal to 1 gal per equivalent plate, a thousand times as much material must interchange to remove the low boiler from the kettle as would be required to remove low boiler of the same concentration from one of the plates in the column.

For most practical applications, large columns have so low a ratio of holdup to charge that none of the effects of holdup is large for ordinary binary distillations. However, multicomponent distillation experience shows that the ratio of holdup to the amount of that component in the charge affects the sharpness of separation of that component. When the quantity of a component is small, the ratio will be large and the effects of holdup will likewise be large, even in industrial columns with large kettles and small columns. This also applies to binary mixtures with a low concentration of one component.

Failure of plant equipment to produce separations as good as laboratory and pilot-plant equipment has sometimes been due to differences in holdup.

TYPICAL DESIGN PROBLEM

Rose and Sweeney[14] describe the use of shortcut methods for making ternary batch distillation calculations for rectification of naphthalene tar acid oil. A commercial operation had to decide whether or not an existing batch column should be converted to continuous operation. This decision required a calculation of the column's annual output of naphthalene. The problem was complicated by the fact that naphthalene tar acid oil contains a number of components.

The following discussion focuses only on the constant-reflux separation calculations. Obviously, the process engineer would make similar calculations for the variable-reflux conditions in order to compare production rates. But such calculations require more information on the tower and its auxiliaries than was available. Accordingly, the following example is used to illustrate the application of preliminary design methods to a typical separation problem encountered in practice.

METHOD OF APPROACH

A ternary mixture of A, B, and C can be considered to consist of two binary mixtures. Therefore, two curves of distillate composition vs. percent distilled are calculated, one for the A-B binary and one for the B-C binary. These curves are then combined to give a similar curve for the ternary case.

Two precise laboratory distillations are required for the tar acid oil in order to determine relative volatilities and composition of the mixture. One distillation was made at 760 mmHg, the other at 100 mmHg. Figures 12 and 13 are the results of those distillations. Melting-point determinations were

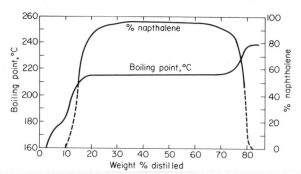

Fig. 12 Boiling-point curve at atmospheric pressure for ternary mixture. [*Reprinted by permission from* Ind. Eng. Chem. **50**(*11*), *1687 (1958).*]

made on the fractions to determine the naphthalene concentration. These analyses showed the constant-boiling material to contain 96% naphthalene. The distillation curves indicate the mixture can be broken into three pseudo-components (A, B, C). In addition, relative volatilities can be determined by making a Cox chart plot of the boiling points and pressures for the three components. Following is a tabulation of that information:

	Boiling point, °C		Wt %
	760 mm	100 mm	
A	180	114	16.0
B	216	144	61.8
C	239	166	22.2

The composition data (wt %) were computed by integrating areas under the distillation curves. From the boiling-point data, assuming Raoult's law holds, the relative volatilities were conservatively estimated to be $\alpha_{AB} = 2.2$ and $\alpha_{BC} = 1.6$ at atmospheric pressure.

Using these data, naphthalene recovery for the two binaries can be predicted for the existing still which is equivalent to 15 theoretical plates operating at 1 atm with a constant reflux ratio of 8:1 (reflux to distillate). The method is tested by making the two binary Rayleigh calculations and determining whether or not they meet together smoothly at a concentration of 95% of the intermediate component.

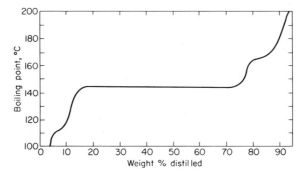

Fig. 13 Boiling-point curve at 100 mmHg for ternary mixture. [*Reprinted by permission from* Ind. Eng. Chem., **50**(*11*), *1687 (1958)*.]

CALCULATIONS

The distillation curve for the still can be predicted by the method outlined above. Table 1 summarizes the A-B binary results and Table 2 the B-C binary results.

The A-B binary consists of all the A and B in the charge. Thus, for 45 kg (100 lb) total charge A = 7.2 kg (16.0 lb) and B = 27.8 kg (61.8 lb). Therefore, the A-B binary composition will be A = (16.0/77.8) × 100 = 20.6%, and B = (61.8/77.8) × 100 = 79.4%.

A McCabe-Thiele diagram is plotted for $\alpha = 2.2$ and an 8:1 reflux ratio. The values of x_D for a still of 15 theoretical plates are found for various values of x_w. The Rayleigh equation is used to calculate x_D vs.

TABLE 1 A-B Binary Results Converted to Ternary Basis

A + B distilled, wt %	Total charge distilled, wt %	x_D	$1 - x_D$	Wt fraction naphthalene, $(1 - x_D)(0.96)$
0	0	0.999	0.991	0.00096
10	7.8	0.996	0.004	0.0038
14	10.9	0.805	0.195	0.187
18	14.0	0.530	0.470	0.451
20	15.6	0.433	0.567	0.544
30	23.3	0.123	0.877	0.842
36	28.0	0.0468	0.9532	0.914
40	31.1	0.0237	0.9763	0.936
43	33.5	0.0133	0.9867	0.947

SOURCE: Adapted from *Ind. Eng. Chem.*, **50**(11), (1958).

percent A + B distilled. Points are taken from this graph, x_D is converted to weight fraction actual naphthalene, and the percent A + B is converted to percent total charge distilled as follows:

At 20% of A + B distilled, $x = 0.433$
Weight fraction B (psuedo-naphthalene) $= 1 - x_D = 1 - 0.433 = 0.567$
Weight fraction pure naphthalene $= 0.567 \times 0.96 = 0.544$
$$\text{Total charge distilled} = \frac{9 \text{ kg (20 lb) distilled}}{45 \text{ kg (100 lb) A} + \text{B}} = 0.778 \frac{\text{lb A} + \text{B}}{\text{lb charge}} = 15.6\%$$

The B-C binary consists of all the B component and enough C to give an x_D close to 1.0. In this example, enough B was taken to make B + C equal 50% of the total charge. Thus for 45 kg (100 lb) of total charge:

B + C = 22.5 kg (50 lb)
B = 9.99 kg (22.2 lb)
C = 22.5 − 9.99 = 12.51 kg (50 − 22.2 = 27.8 lb)

Therefore, the B-C binary composition is

%B = (27.8 ÷ 50) × 100 = 55.6%
%C = 100 − 55.6 = 44.4%

Using the McCabe-Thiele and Rayleigh methods described earlier for constant-reflux operation, the results of Tables 1 and 2 are obtained. Plots of those results are made as shown in Fig. 14. Points are taken from the graph, and x_D is converted to weight fraction of actual naphthalene. Too, the percent of the last half of the charge is converted to percent of total charge distilled as shown below:

At 50% of the last half of charge distilled, $x_D = 0.740$
Wt fraction naphthalene = x_D (0.96) = 0.740 × 0.96 = 0.710
% total charge distilled = 50 + (wt % of last half of charge distilled) × 0.5 = 50 + (50 × 0.5) = 75%

TABLE 2   B-C Binary Results Converted to Ternary Basis

Wt % of last half of charge distilled	Wt % charge distilled	x_D	Wt fraction naphthalene in distillate x_D (0.96)
0	50	0.998	0.958
20	60	0.994	0.954
30	65	0.988	0.949
40	70	0.970	0.931
44	72	0.938	0.900
50	75	0.740	0.710
52	76	0.635	0.609
60	80	0.303	0.291
68	84	0.104	0.100
76	88	0.024	0.023

SOURCE: Adapted from *Ind. Eng. Chem.*, **50**(11), (1958).

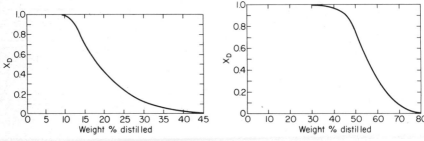

Fig. 14 Distillate compositions for A-B binary and B-C binary. [*Reprinted by permission from* Ind. Eng. Chem., **50**(11), 1687 (1958).]

Experimental Verification

The calculated results were experimentally verified by Rose and Sweeney.[14] Two laboratory distillations were made using the conditions set out for this example. The results are superimposed over the calculated curve in Fig. 15.

Analysis of Results

This method relies on the assumption that at some point in the distillation pure intermediate component B appears in the overhead. When this condition is met, all the heavy component in the charge remains in the kettle, and all the light component has been removed. Thus there is actually a binary in the kettle at this point and the usual binary batch distillation calculation procedure applies.

The first part of the distillation (A-B binary) cannot be calculated as rigorously as the

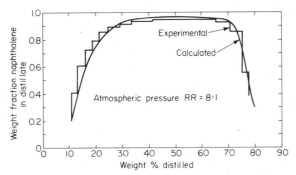

Fig. 15 Experimental and calculated results from ternary (A, B, C) distillation. [*Reprinted by permission from* Ind. Eng. Chem., **50**(11), 1687 (1958).]

second part. This is because x_D is not rigorously related to x_w, because the relationship between x_D and A/(A + B) is not independent of C. This is the only important approximation in the method. However, in most cases, the presence of C does not have a large effect on the relation.

For example, the top composition at 27.8% distilled (see Fig. 15) is 5% A and 95% B. The percent C in the bottoms at this point is

$$\%C = \frac{\%C \text{ in the charge}}{\% \text{ of charge remaining}} \times 100 = (22.2 \div 72.2) \times 100 = 30.8\%$$

Table 3 compares the results of a ternary plate-to-plate calculation and a binary calculation for the above top composition. There is no significant difference.

TABLE 3 Comparison of Plate-to-Plate Ternary Calculations with Binary

Composition	Ternary wt %	Binary wt %
Top composition:		
A	5.0	5.0
B	94.9	95.0
C	0.1	
Bottoms composition:		
A	0.36	0.43
B	70.26	99.57
C	29.38	
Bottoms composition on C-free basis:		
A	0.51	0.43
B	99.49	99.57

SOURCE: Adapted from *Ind. Eng. Chem.*, **50**(11), (1958).

This is an easy-to-use method for ternary batch distillation calculations without holdup. Since many chemical reactions have final reaction mixtures similar to the ternary just described, the method should have wide applications. The method is approximate, but accurate enough for practical use. Its success depends on one condition—at some time during the distillation the intermediate component must appear in the overhead with relatively little contamination by the other components. In the preceding example, the distillate attains a 99% purity, and very good results are obtained.

PROCESS CONTROL

Reflux Control

Most commercial batch distillations are designed for variable reflux control in order to maintain a constant overhead composition. Many batch stills operate with a variable boilup rate. Obviously, the use of product and reflux rotometers under such conditions requires too much operator attention. A better solution to the control problem is the use of a reflux splitter.

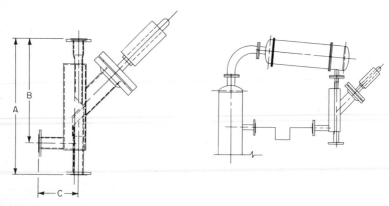

Fig. 16 Reflux splitter—offset design. (*Reprinted by permission from Chem-Pro Equipment Corporation.*)

Reflux splitters are on/off solenoid-operated devices actuated by a timer. The timer setting is equivalent to a particular reflux ratio. The solenoid operates a slide gate which diverts all the liquid flow to the column or to the product receiver. Figures 16 and 17 schematically represent reflux splitters for an offset condenser design and an in-column design.

In conventional reflux-splitter installations, the overhead condenser is located above the top tray of the distillation column. In such installations the liquid flows by gravity into the reflux splitter and onto the top tray of the column, or into the product receiver. Sometimes, however, the condenser may be well below the top tray. It is then necessary to install the reflux splitter on the discharge side of a reflux-product pump.

Figure 18 depicts a reflux-splitter installation under positive pump pressure. This particular installation permits gravity flow of reflux to the column and product to the receivers. To accomplish this, the splitter is installed at an appropriate elevation and a vent pipe is installed in the vapor space near the top of the reflux splitter to prevent a possible vacuum from developing at this point. Also, designing the reflux splitter and reflux line for maximum flow rate under gravity conditions (against a slight positive column pressure) ensures continuous flow. The flow rate to the reflux splitter is controlled by a level controller installed on the accumulator tank below the condenser.

When small columns are involved, with total overhead rates in the range of 37.9 to 379 L/h (10 to 100 gal/h), standard orifice-type instruments offer disadvantages for controlling reflux because of their expense and susceptibility to serious inaccuracies due to the small orifices. Experimental columns commonly use a partial condenser built into the top of the column to condense and return reflux, with a separate condenser installed for net product. Accurate control of such an arrangement is very difficult, however, particularly at high reflux ratios where only slight variations in the column boilup rate, or in the condensing rate, cause the net product to fluctuate widely. The reflux splitter is a much better solution to the reflux control problem.

Boilup-Rate Control

Block[19] has described several schemes for controlling the column vapor rate. Some indirectly control the true variable (vapor rate) and others directly control it. Figure 19 schematically illustrates several control methods.

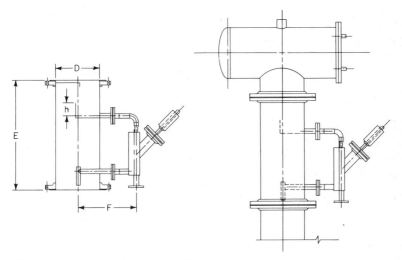

Fig. 17 Reflux splitter—in-column design. *(Reprinted by permission from Chem-Pro Equipment Corporation.)*

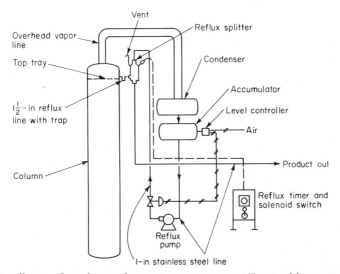

Fig. 18 Installing a reflux splitter under positive pump pressure. *(Reprinted by permission from Chem. Eng., Feb. 13, 1967, p. 180.)*

The simplest method for regulating boilup rate would be through a pressure controller on the inlet-steam line (Fig. 19a). This method works best for close-boiling mixtures where the heat-transfer rate does not change appreciably with concentration. If the mixture components are significantly different in boiling point and other physical properties, the vaporization rate gradually declines during the distillation, and total cycle time is longer than necessary.

This method can be improved by using a steam-flow controller as shown in Fig. 19b. The flow controller provides a constant steam flow to the coil. This system has the same disadvantage as steam pressure control. Boilup rate drops off as the kettle temperature rises during the distillation.

Steam flow can be regulated by differential temperature control. Thermocouples sense the steam and kettle vapor temperatures and a differential temperature controller operates the steam control valve as shown in Fig. 19c. A somewhat more elaborate arrangement cascades the differential temperature signal to a pressure controller as shown in Fig. 19d. These control schemes are intended to compensate for heat-transfer changes that take place during the course of a distillation. But they still indirectly control boilup rate.

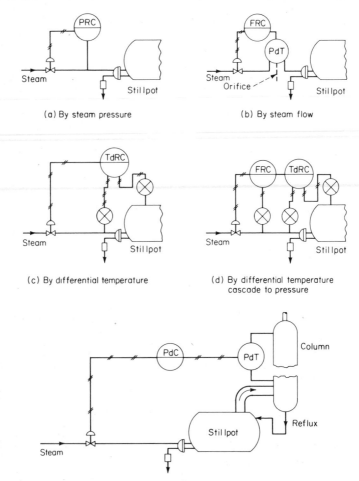

Fig. 19   Batch boilup control by several methods. (*Reprinted by permission from* Chem. Eng., *Jan. 16, 1967, p. 147.*)

A more direct approach is to have the vapor velocity in the column operate the steam controller. Column differential pressure can often be used to control the steam-flow rate automatically (Fig. 19e). The maximum steam and vapor flow rates will be limited by the heat-transfer capability of the steam coil. This method is particularly well suited for packed columns. It can also be used for bubble-cap or sieve-tray-column liquid loading, as the trays account for a large share of the pressure differential.

Column-Pressure Control

For operations above atmospheric, pressure can be controlled one of two ways, depending on whether noncondensables are present or are added for pressure-control purposes. If noncondensables are present, the pressure is controlled by regulating the waste-gas vent from the condenser. The amount of light ends so released is usually measured with a wet-gas test meter. If noncondensables are absent, the pressure is controlled by regulating the flow of coolant to the condenser coil.

For vacuum operation, column vacuum is controlled by a manostat. The manostat provides accurate control and has the flexibility necessary for even pilot-plant work.

START-UP AND TROUBLESHOOTING

Start-up

Mechanical Inspection A thorough mechanical inspection and preparation before start-up can eliminate many possible trouble spots. Most of the problems that occur during initial operation of a new unit are not the result of design errors. They are human errors that could be resolved during mechanical inspection and preparation for start-up. Mistakes in fabrication or in the assembly of internals in a distillation column cause operating problems that often require lengthy investigations, production losses, and plant shutdowns for repairs.

The real importance of a well-prepared inspection program before final tower closure cannot be overemphasized. This program is the only reliable basis for taking corrective measures during subsequent operation.

Operating engineers must become familiar with the unit design features, then plan the inspection program. They should prepare a checklist of conventional and special items in order to assure a complete and careful inspection. Such a list would include the following items:[21]

Materials of Construction. It is not uncommon to have changes in material classification at various sections in columns. In some cases, the material of the tray supports differs from that used in the tray itself.

Measurements of the Internals. All internal dimensions must be checked: weir heights, downcomer clearances, distances between tray, gaps at vessel walls, tray perforations, and total hole area. Note clearances between tray and outlets of distribution nozzles.

Location of Equipment. The location of each piece of equipment in relation to its function is very important. Antiswirl baffles, thermocouples in vapor or liquid space, and all inlet and outlet nozzles throughout the tower should be checked.

Tightness of Internals. Downcomers and trays must be pulled up tightly against their supports. Downcomer flexibility at trailing edges should be checked, particularly when inflow weirs are used, because the horizontal distance between the downcomer and weir is a critical dimension.

Cleanliness of Internals. Make sure no debris is left in nozzles, downcomers, catch pans, or on trays.

Flushing and Boilout After mechanical completion of the unit, a great deal of work remains before feed is introduced. These final preparation steps are made to ensure continuity of operation when the unit is started up. This flushing and testing procedure can be separated into the following categories:[21]

Inspection of Equipment. A piping and instrument checkout must be made against engineering flowsheets before any operations are carried out.

Cleaning of Equipment. Flushing of all process lines removes construction debris which can cause control and block valves to leak or even more serious damage to compressors and pumps. Adequate line velocities are important during the flushing procedure.

Final Pressure Testing. A final pressure test is sometimes recommended to be sure that all equipment disturbed by the flushing procedure has been securely headed up again.

Water Boilup. It is common practice to establish a water circulation through the plant and to boil up the columns on water as a final checkout of the operability of the equipment. This procedure allows operating shortcomings to be detected and corrected without purging, blanketing, and gas freeing some units.

Gas Blanketing of Equipment. Volatile chemicals often require gas blanketing of certain process equipment. When all preparation steps are complete and the unit is ready for operation, the operating engineer must decide how to blanket the system before introducing feed.

Troubleshooting

All aspects of distillation-equipment design and operation have not been reduced to an exact science. Although well concealed by reams of computer output, the spirit of the alchemist is still with us. It is therefore important to explore some of the troubleshooting techniques available to the process engineer.

Process designers should assume that the column will not function as expected and that they will be called to correct the problem. With this in mind, they should specify sample connections needed to obtain a satisfactory material and energy balance. In practice, one has to infer, deduce, and calculate what is going on internally from such indirect evidence as temperatures, sample connections, and pressure drops. Emphasis should be on having connections for leads or instruments if the need arises.

Figure 20 presents a tentative flow diagram for troubleshooting problems.[20] This should be considered only as a general guideline and partial checklist, rather than a complete logic diagram for solving all problems.

One of the first things to be determined is whether the problem is real or imaginary. This can be done with material and energy balances.

Next estimate the magnitude of the problem, at an early stage. Then isolate the problems with the column itself from those occurring with the environment in which the column operates. The proper approach is to determine what the column is capable of

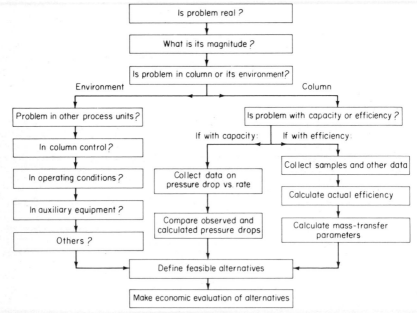

Fig. 20 Flowchart for column troubleshooting procedure. *(Reprinted by permission from* Chem. Eng., *June 1, 1970, p. 148.)*

doing. The troubleshooting task, therefore, consists of determining the actual performance characteristics of the column and then finding the most economical way to reconcile performance with design objectives.

The Environment Upstream units can affect the batch still. The question that needs to be asked is: "What conditions was this column designed for, and what conditions is it now expected to meet?"

Column control systems are frequently the root cause of operational problems. Problems of this nature are identified by starting with material and energy balances, extending them to detailed calculations of operating conditions, and checking calculated vs. apparent conditions at as many points as possible.

Improper operating conditions account for many process problems. Operating at reduced capacity and specifying more operating variables than the system has degrees of freedom are often the cause.

Auxiliary equipment may often cause problems that seem to originate in the column itself. For instance, reboilers, condensers, reflux pumps, motor valves, and other auxiliaries may be too small. If the energy flow—or its counterpart in reflux flow, steam flow, refrigerant flow, etc.—does not remain proportional to feed, a capacity limitation in the auxiliary equipment should be suspected.

Column-Capacity Problems Column-capacity problems are identified by pressure-drop measurements across various tray or packing sections. The variation of observed pressure drop with internal flows should be examined. Actual column capacity must be definitely established—i.e., at what rate or percent of design the column floods. Once maximum capacity has been determined, the section of the tower which floods first must be determined.

Column-Efficiency Problems A realistic appraisal of the need for exact tray efficiencies should be made before undertaking such a program. Collection of samples and data necessary to determine efficiency, plus the detailed analysis of such data, requires considerable time, effort, and money. If the simple qualitative fact of low efficiency has been established, it may be more pertinent to try to improve it rather than determine exactly how low it is.

If precise measurement is in order, considerable data must be collected. Valid composition samples from tray and other sample points are fundamental to the analysis. A good overall material balance is needed. Reflux flows and flow rates of heating and cooling mediums are essential.

Theoretical calculations must then be made by varying the assumed number of theoretical stages until the calculated and observed performance match.

Efficiency-testing programs should be a last alternative in the troubleshooting procedure.

REFERENCES

1. Kiguchi, S. T., *Ind. Eng. Chem.* **46**, 1363 (1954).
2. Fair, J. R., and Bolles, W. L., *Chem. Eng.*, **75**, 156 (1968).
3. Lloyd, L. E., *Pet. Refiner*, **29**(2), 133 (1950).
4. Coulter, K. E., *Pet. Refiner*, **31**(8), 95 (1952).
5. Smoker, E. H., and Rose, Al, *Trans. Am. Inst. Chem. Eng.*, **36**, 285 (1940).
6. Bogart, M. J. P., *Trans. Am. Inst. Chem. Eng.* **33**, 139 (1937).
7. Jackson, R. F., and Pigford, R. L., *Ind. Eng. Chem.*, **48**, 1020 (1958).
8. Berg, C., and Ivor, James, Jr., *Chem. Eng. Prog.*, **44**(4), 307 (1948).
9. Huckaba, C. E., and Danly, D. E., *Am. Inst. Chem. Eng. J.*, **6**, 335 (1960).
10. Meadows, E. L., *Chem. Eng. Prog. Symp. Ser.*, **59**(46), 48 (1963).
11. Distefano, G. P., *Am. Inst. Chem. Eng. J.*, **14**, 190 (1968).
12. Barb, D. K., and Holland, C. D., 7th World Congress, Mexico City, Apr. 2–8, 1967, Paper PD-16(4).
13. Rose, A., and O'Brien, V. S., *Ind. Eng. Chem.*, **44**, 1480 (1952).
14. Rose, A., and Sweeney, R. F., *Ind. Eng. Chem.*, **50**(11), 1687 (1958).
15. Grossberg, A. L., and Roebuck, J. M., *Chem. Eng.*, **54**(1), 132 (1947).
16. Galluzo, J. F., *Chem. Eng.*, **1967**, 180.
17. Badami, V. N., *Chem. Eng.*, May 8, 1967, 180.
18. Kelley, W. J., *Chem. Eng.*, Sept. 3, 1962, 154.
19. Block, B., *Chem. Eng.*, Jan. 16, 1967, 147.
20. McLaren, D. B., and Upchurch, J. C., *Chem. Eng.*, **76**(12), 139 (1970).
21. Murray, R. M., and Wright, J. E., *Chem. Eng. Prog.*, **63**(12), 40 (1967).

Steam Distillation/Stripping

R. W. ELLERBE, **M.S.Ch.E., M.B.A.** *Project Manager, The Rust Engineering Company; Registered Professional Engineer in Arkansas.*

INTRODUCTION

Steam distillation refers to a process in which live steam is in direct contact with the distilling system in either batch or continuous operation. This is called *open steam distillation.* More generally, a distillation conducted in the presence of any inert added component such as nitrogen, carbon dioxide, or flue gas is governed by the same fundamental relationship as steam distillation. Steam is widely used because of its energy level, cheapness, and availability. Steam distillation is commonly used in the following situations:

    1. To separate relatively small amounts of a volatile impurity from a large amount of material

    2. To separate appreciable quantities of higher-boiling materials

    3. To recover high-boiling materials from small amounts of impurity which have a higher boiling point

    4. Where the material to be distilled is thermally unstable or reacts with other components associated with it at the boiling temperature, e.g., glycerin

    5. Where the material cannot be distilled by indirect heating even under low pressure because of the high boiling temperature, e.g., fatty acids

    6. Where direct-fired heaters cannot be used because of danger, e.g., turpentine

This section deals with the specific case of liquids which are *immiscible* with water and those which, for all practical purposes, are immiscible with water. Some typical industrial steam-distillation processes are concentration and purification of essential oils for perfumery, purification of long-chain fatty acids, deodorization of fats and oils, recovery of absorption oils in by-product coke production, and aniline production.

The information in this section is intended to give the process engineer equilibrium

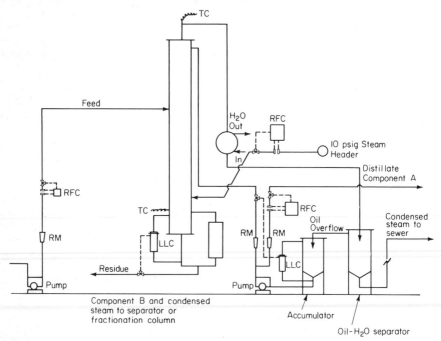

Fig. 1 Steam fractionation flowsheet. *(Reprinted by permission from* Pet. Refiner, **31**(11), *(November 1953).*

relationships for handling several diverse cases in steam distillation. Equations are presented for binary batch and binary continuous operations, as well as multicomponent batch operations. These equations permit the process engineer to compute (1) the quantity of steam required, (2) the time required to complete the process, (3) theoretical stage requirements, and (4) relative steam requirements for batch and continuous operation. These are the important factors affecting process economics.

Steam distillation of water-immiscible compounds may be practiced using a flowsheet similar to the one in Fig. 1. As noted, feed enters the column at the side and open steam at the bottom. Component A is taken overhead with water and component B as residue with water. The overhead is separated from the water and part is returned as reflux. The water is usually sewered. In some instances, where a slight solubility occurs, some water returns with the reflux, but it has no beneficial effect on separation. The residue may be separated from the water or fed to another column without separation. If a dry distillate or residue product is desired, a drying column or solid adsorbent can be used.

In a number of cases, water and organic compounds may not separate in the residues from the still. The presence of tars, polymers, inorganic salts, etc., may cause emulsions which cannot be separated by decantation or centrifuging. The oil industry has resorted to demulsifying agents and glass-wool filters to break the emulsions. In practice, it is sometimes best to feed troublesome residue emulsions to the following column and handle them in the same manner as a decanted residue. This eliminates the residue decanter and adds a large water load to the column, but if the residue is fed direct, without cooling or inventorying, the column heat load is not materially increased. The volatile material is distilled from the emulsified feed and usually separates from water in the distillate unless the impurities causing the emulsion are volatile.

Before plants are built using steam distillation, a good pilot plant should be operated using the feed proposed on the large unit. The equipment and controls should be designed with columns as models of the larger units. If good pilot-plant data are not obtained, emulsions, foaming, low plate efficiencies, and corrosion may make the large plant unsatisfactory.

THEORY

Conventional distillations involve separations of miscible liquids. Steam distillations permit a more complete separation of miscible liquids at lower temperatures for the same conditions of total pressure or vacuum. This is due to the unique vapor-pressure behavior of immiscible liquids, e.g., water and hydrocarbons. The vapor-pressure relations of two immiscible liquids are shown in Fig. 2. Each liquid exerts its own vapor pressure independently of the other. In such a system, at a given temperature, the total vapor pressure is the sum of the vapor pressure of the two liquids (if neither liquid dissolves in the other). Furthermore, the partial pressure for each component and the composition of the vapor phase, at constant temperature, are independent of the moles of liquid water or hydrocarbon present.

In the example shown, the total vapor pressure reaches 760 mmHg (atmospheric pressure) at 95°C. Boiling begins at that temperature, and both liquids distill together. When either one of the liquids distills away, the vapor pressure drops to that of the remaining liquid. Ordinarily bromobenzene would not boil at atmospheric pressure until its temperature reached 156.2°C, but the presence of a little water (or steam) makes bromobenzene boil at 95°C, a temperature that is easy to achieve even with low-pressure steam.

In 1918, Hausbrand published a vapor-pressure diagram that proved to be very useful in steam: distillation calculations. Figure

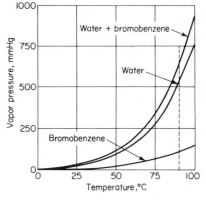

Fig. 2 Vapor pressure vs. temperature plot for system made up of water and bromobenzene. (*Reprinted from* Chem. Eng., *Mar. 4, 1974, p. 106.*)

3 shows that diagram. It plots $\pi - p_s$ at three system pressures (760, 300, and 70 mmHg) vs. temperature. These curves cut across the ordinary vapor-pressure curves of the materials to be distilled. The intersection of the water curve with the curves of the other materials gives the temperature at which steam distillations can take place.

Suppose an atmospheric still contains some impure toluene, and steam is blown into the still. Suppose also that the impurities are very-high-boiling compounds with negligible vapor pressure. Further, consider the case where the liquid in the still is heated solely by condensation of the steam. A water layer will therefore accumulate. As the temperature rises, the vapor pressures of the toluene and the water layers rise.

When the sum of the two vapor pressures equals 760 mmHg, the mixture begins to distill. At this point, the vapor pressure of the toluene is p_{im} mmHg, the vapor pressure of water is $\pi - p_{im}$ mmHg; Fig. 3 shows that this occurs at a temperature of about 84°C. In this case, the vapor passing over consists of toluene with a partial pressure of 350 mmHg, and water vapor with a partial pressure of 410 mmHg. The molar ratio of toluene to water would therefore be 350:410, or 85.4 parts toluene to 100 parts water.

The Hausbrand diagram makes possible a quick determination of the temperature at which a steam distillation takes place. It also presents graphically the molar richness of the vapor. Since most substances on the chart have a molecular weight considerably greater than that of water, the composition of the distillate by weight is much richer than would appear from the diagram.

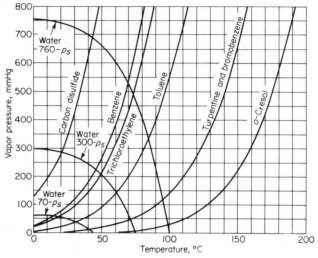

Fig. 3 Hausbrand diagram for various liquids at three system pressures. (*Reprinted by permission from* Chem. Eng., *Mar. 4, 1974, p. 108.*)

Batch Steam Distillation/Stripping

Carey has outlined an approach to developing a general equation for batch stripping. For an infinitesimal time interval $d\theta$, a constant sparging rate for the steam, and enough additional heat added to the charge so that only the steam and volatile component enter the vapor phase, the following equation describes a differential material balance for the volatile component:

$$-d(N_b X) = -N_b dX = S_b Y d\theta \tag{1}$$

where N_b = moles of inert component in the still charge
   X = instantaneous composition of the liquid in the still in moles of volatile component per mole of inert component
   S_b = steam rate in moles per unit time
   Y = instantaneous composition of the vapor leaving the still in moles of volatile component per mole of steam

The variables in Eq. (1) may be reduced to two by obtaining Y in terms of other variables. Carey notes that the actual partial pressure of the volatile component in the vapor is somewhat less than the theoretical because of mass- and heat-transfer resistance in the actual process. He therefore defines a vaporization efficiency as

$$E = \frac{p}{p^*} \tag{2}$$

where p = partial pressure of the volatile component in the vapor
   p^* = equilibrium partial pressure of the volatile component

Applying Dalton's law to the components in the vapor, Y is the ratio of the partial pressure of the volatile component p to the partial pressure of the steam $(\pi - p)$ where π is the total pressure maintained in the vapor space of the still. Then, by substitution of Eq. (2),

$$Y = \frac{p}{\pi - p} = \frac{Ep^*}{\pi - Ep^*} \tag{3}$$

Gibbs' phase rule, when applied to a three-component system with two phases present and with pressure specified, says that p^* is a function of two variables, e.g., temperature T and X, so

$$p^* = p^*(T,X) \tag{4}$$

Using this implicit equation for p^*, Eq. (3) and Eq. (1) lead to the general equation for batch stripping:

$$N_b \int_{X_r}^{X_f} \frac{\pi - Ep^*(T,X)}{Ep^*(T,X)} \, dX = \int_0^{\theta_b} S_b d\theta = S_b \theta_b \tag{5}$$

where X_r = composition of residue
   X_f = composition of feed
   θ_b = length of stripping period

Implicit equation (5) involves assumptions that are usually valid. If data for the influence of T and X on p^* and E for the process, usually assumed constant, are inserted into Eq. (5), an implicit expression for the total steam consumption results. When there are two liquid layers present in the still, the variables reduce to two, so that with pressure fixed only one degree of freedom remains. If X is the remaining independent variable, T is no longer subject to choice. So, when two liquid layers are present, p^* is a function of X only, once π has been set.

If all or almost all of the volatile component is distilled away, the partial pressure of that component becomes very small and the total pressure becomes essentially equal to the partial pressure of the steam, $\pi \cong p_s = P_s$. Therefore, in order to prevent condensation of the steam, its temperature should be selected to be higher than the saturation temperature of the steam at the pressure of the distillation.

Estimates of steam requirements for this batch process may be obtained by introducing some simple or approximate function for p^* into Eq. (5). Three such cases would include Raoult's, Henry's, or Lewis and Luke's laws. All three cases can be dealt with simultaneously, since the three laws have the same analytic form:

Raoult's law: $p^* = P \dfrac{X}{1 + X}$

Henry's law: $p^* = H \dfrac{X}{1 + X}$

Lewis and Luke's law: $p^* = K \dfrac{X}{1 + X}$

This leads to the general expression for Y in terms of X:

$$Y = \frac{X}{(k - 1)X + k} \tag{6}$$

where $K = \pi/EP$, π/EH, $1/EK$ when Raoult's, Henry's, and Lewis and Luke's laws, respectively, apply. Substituting this Y into Eq. (5) and integrating over the composition range specified gives the steam usage.

$$S_b \theta_b = N_b \left[(k) \left(\ln \frac{X_f}{X_r} \right) + (k - 1) (X_f - X_r) \right] \tag{7}$$

The moles of volatile component, vaporized, is $N_b(X_f - X_r)$; so the ratio of moles of steam used to moles of volatile component distilled is

$$\frac{S_b \theta_b}{N_b(X_f - X_r)} = \left[\frac{k}{X_{1m}} + (k - 1) \right] \tag{8}$$

where X_{1m} is the logarithmic mean value of X_f and X_r. Equations (7) and (8) are frequently used for batch steam-distillation approximations. For more exact calculations, the precise relationship for p^* required by Eq. (4) must be determined experimentally over the required operating range and then used in Eq. (5).

Equation (7), when expressed in mass rather than mole quantities, implies that the steam usage will be low when the ratio of the molecular weights of the volatile component to the inert component is low, and when the escaping tendency of the volatile component is high, i.e., when k is small. It also indicates the quantitative effect of operating at low pressures and high temperature to reduce steam usage.

Multicomponent Batch Steam Distillation

The most common situation confronted in multicomponent batch steam distillation is that in which n components are present. Some of these components are volatile and some nonvolatile. All components of the mixture are miscible in the liquid phase. Holland and Welch[5] derived the following equation to describe the process:

$$\sum_{i \neq s,r} \frac{L_c}{\beta i} \left[1 - \left(\frac{L_b}{L_b^0} \right)^{\beta_c^0} \right] + L_r^0 \ln \frac{L_b^0}{L_b} = \frac{E_b P_b}{\pi} (L^0 - L + S)$$ (9)

Where the nonvolatile components do not alter the mole fractions of the volatile components in the still, Eq. (9) reduces to

$$\sum_{i=s,r} \frac{L_i}{\beta_i} \left[1 - \left(\frac{L_b}{L_b^0} \right)^{\beta_i} \right] = \frac{E_b P_b}{\pi} (L^0 - L + S)$$ (10)

where $E_i = p_i/p^* =$ vaporization efficiency of i
$\quad L$ = moles of volatile components in still at any time
$\quad L = L^0$ at start of distillation
$\quad L_r^0$ = moles of nonvolatile material in still
$\quad P_i$ = vapor pressure of i at still temperature
$\quad p_i$ = partial pressure of i in vapor
$\quad \beta_i = (E_i/E_b) \, \alpha_{ib}$
$\quad \alpha_{ib}$ = relative volatility of i referred to $b = P_i/P_b$
$\quad \sum_{i \neq s,r}$ = summation of all components except steam and nonvolatiles
$\quad b$ = reference component
$\quad r$ = nonvolatile component
$\quad S$ = steam required, moles

Continuous Steam Distillation

In many cases, steam distillation or steam stripping is conducted as a continuous steady-state operation. The flow of steam may be countercurrent to the flow of feed or cocurrent with it. In the former case the vapor contacts the richest liquid in the system last and tends to be richer in the volatile component than in the cocurrent process.

Definitions of variables already used are modified to facilitate comparison of the different processes. Let N_c be the moles of steam admitted to the still per unit time in the countercurrent case. Assuming steady-state conditions, N_c is also the moles of inert component removed from the still per unit time. The volatile-component feed rate is $N_c X_f$, and the moles of this component leaving per unit time in the still bottoms is $N_c X_r$. The difference between these two quantities, $N_c(X_f - X_r)$, is the mole rate at which the volatile component passes into the vapor phase.

In the countercurrent process, the vapor last engages the feed liquid with composition X_f. For equilibrium operations, the vapor composition is predicted by the equilibrium properties of the system, e.g., $p^* = p^*(T, X_f)$. But the partial pressure of the volatile component in the vapor does not attain the full equilibrium value, only a fraction of it. Calling this fraction E, as in Eq. (2), the vapor leaving the still has the composition

$$Y_f = \frac{Ep^*(T, X_f)}{\pi - Ep^*(T, X_f)}$$ (11)

In the cocurrent-flow process, the effluent vapor last encounters liquid of composition X_r and it has the composition

$$Y_r = \frac{Ep^*(T, X_r)}{\pi - Ep^*(T, X_r)}$$ (12)

By making a material balance on the volatile component, the steam-usage equation is derived

$$S_c = N_c(X_f - X_r) \frac{\pi - Ep^*(T, X_f)}{Ep^*(T, X_f)}$$ (13)

$$S_p = N_p(X_f - X_r) \frac{\pi - Ep^*(T, X_r)}{Ep^*(T, X_r)}$$ (14)

The ratio of steam usage to volatile component removed is

$$\frac{S_c}{N_c(X_f - X_r)} = \frac{\pi - Ep^*(T, X_f)}{Ep^*(T, X_f)}$$ (15)

$$\frac{S_p}{N_p(X_f - X_r)} = \frac{\pi - Ep^*(T,X_r)}{Ep^*(T,X_r)} \tag{16}$$

Equations (13) and (14) are implicit exact expressions and are analogous to Eq. (5) for batch stripping. They and Eqs. (15) and (16) may be put in explicit but approximate form for the three cases previously considered for batch stripping. Using Eq. (6), we get

$$S_c = N_c \left[k \frac{X_f - X_r}{X_f} + (k - 1)(X_f - X_r) \right] \tag{17}$$

$$S_p = N_p \left[k \frac{X_f - X_r}{X_f} + (k - 1)(X_f - X_r) \right] \tag{18}$$

$$\frac{S_c}{N_c(X_f - X_r)} = \left[\frac{k}{X_f} + (k - 1) \right] \tag{19}$$

$$\frac{S_p}{N_p(X_f - X_r)} = \left[\frac{k}{X_r} + (k - 1) \right] \tag{20}$$

Depending on which solution law most closely applies, the appropriate value of k may be used in Eqs. (17), (18), (19), and (20).

Equation (8) for the batch process is very similar in form to Eqs. (19) and (20) for the continuous processes. The differences occur only in the first term of the bracketed quantity, $1/X_{lm}$ being replaced by $1/X_f$ and $1/X_r$ for the countercurrent- and parallel-flow processes, respectively. The similarities suggest a comparison of steam usage for equal amounts of stripping for the three processes.

Since $X_f > X_{lm} > X_r$, then $(1/X_f) < (1/X_{lm}) < (1/X_r)$. When applied to Eqs. (8), (19), and (20), these inequalities show that for the same operating conditions, but using different contacting processes, and assuming the same value of E for each case, the steam usage is largest for the parallel-flow and least for continuous countercurrent operation.

COMPARISON OF STEAM USAGE

It is interesting to compare the difference in the steam usage for the batch and countercurrent processes for the same operating conditions, since these are more widely applied. Calling this difference ΔS,

$$\Delta S = \frac{S_b \theta_b - S_c}{N_b(X_f - X_r)} = \frac{1}{X_f - X_r} \int_{X_r}^{X_f} \frac{\pi - Ep^*(T,X)}{Ep^*(T,X)} \, dx - \frac{\pi - Ep^*(T,X_f)}{Ep^*(T,X_f)} \tag{21}$$

Equation (21) can be expressed in approximate form for the three simple cases previously considered. Letting a be the ratio of final to initial concentration of the volatile component in the liquid,

$$X_r = aX_f \tag{22}$$

Then

$$\Delta S = \frac{k}{X_f} \left[\frac{\ln(1/a)}{(1 - a)} - 1 \right] \tag{23}$$

In Eq. (23) for ΔS, the bracketed term indicates the influence of the ratio of the final to initial concentration of the volatile component on the saving in steam usage. Equation (23) also implies that ΔS increases directly as π increases, and varies inversely with X_f. Temperature and escaping tendency also inversely affect ΔS. A quantitative picture of the effect of a and X_f alone on ΔS is given in Fig. 4, the other variables being held constant. This plot shows the pronounced advantage of the countercurrent process over the batch process. This is particularly true where the concentration of the volatile component in the residual liquid is reduced to a very low figure, as in deodorization of edible oils.

THEORETICAL STAGE REQUIREMENTS

If a plate stripping column is to be operated at a relatively high pressure, the pressure differential across the unit may be neglected in any calculations. For high-vacuum operations, the pressure drops cannot be neglected in design calculations without introducing large errors. With pressure drops of 1 to 4 mmHg per plate in industrial units, the pressure differential across the column may amount to several hundred percent more than the pressure maintained at the still head. Under these circumstances, the bottom plates are operating at higher pressures than the plates near the still head. These plates are therefore less effective, and the enriching of the vapor per plate drops off quickly as the column bottom is approached. This must be taken into account in the design of stripping columns or too few plates will be specified to reduce the product to the desired X_r when using a predetermined stripping-medium flow rate.

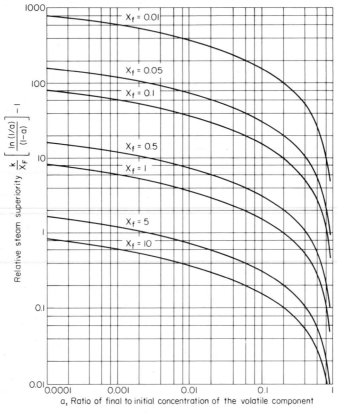

Fig. 4   Relative stripping superiority for batch vs. continuous countercurrent systems. [*Reprinted by permission from* Trans. Am. Inst. Chem. Eng., **39**, *121 (1943).*]

Figure 5 shows the construction of vapor-liquid equilibrium curves for each plate. This construction requires a knowledge of the termperature on each plate in the unit. For most engineering purposes involving preheated feed and superheated steam, the temperature differential across the column is small. So it is usually accurate enough to assume that the temperature of the liquid on each plate is about the same throughout the column.

The usual McCabe-Thiele procedure for binary mixtures is used to solve for the number of theoretical stages required. Since the pressure above each plate is different, Fig. 5 is constructed with a separate vapor-liquid composition for each plate. One curve applies to one plate only, as specified by the pressure at the still head plus the cumulative pressure drops down to the plate under consideration.

SAMPLE PROBLEMS

PROBLEM 1

An amount of 45,000 kg (100,000 lb) of food extract—approximate molecular weight is 450—containing 1.3 wt % hexane is to be subjected to steam distillation to reduce the hexane content to not more than 0.01 wt %. Assume $E = 0.7$, total pressure 1000 mmHg, $t = 100°C$. Determine the steam consumption for the continuous countercurrent process.

Solution

$$S_c = N_c \left[\frac{\pi}{EP} \frac{(X_f - X_r)}{X_f} + \left(\frac{\pi}{EP} - 1 \right) (X_f - X_r) \right]$$

Assume feed rate equals 450 kg/h (1000 lb/h)

$$N_c = \frac{1000 - 1000\,(0.013)}{450} = \frac{987}{450} = 2.19 \text{ mol/h}$$

$$X_1 = \frac{(1000)(0.013)/(86)}{2.19} = 0.069 \text{ mol/mol}$$

$$X_2 = \frac{0.0986/86}{2.19} = 0.00053 \text{ mol/mol}$$

$$\frac{\pi}{EP} = \frac{\pi}{E \left(\dfrac{X}{1 + X} \right) P_a} = \frac{1000}{0.7 \left(\dfrac{0.069}{1 + 0.069} \right) (1860)} = 11.9$$

$$S_c = 2.19(11.9) \frac{0.069 - 0.00053}{0.069} + (11.9 - 1)(0.069 - 0.00053)$$

$$S_c = 25.1(0.99) + 10.9(0.0685)$$
$$S_c = 25.5 \text{ mol steam/h}$$
$$W_c = 25.5(18) = 459 \text{ lb steam/1000 lb feed}$$

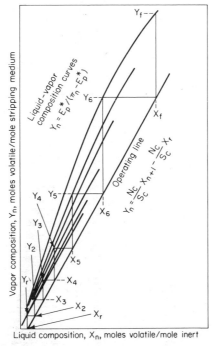

Fig. 5 Graphical determination of actual plates required for a stripping column at high vacuums. [*Reprinted by permission from* Trans. Am. Inst. Chem. Eng., 39, *121 (1943)*.]

PROBLEM 2

It is desired to separate three high-boiling materials from a nonvolatile material by batch steam distillation. At least 95% of the volatile material is to be removed. Determine the moles of steam required when the distillation is carried out at 100°C and 200 mmHg pressure. The data to be used are:

Initial mixture	Moles	P at 200 mmHg at and 100°C	E
Component 1	30	20	0.9
Component 2	25	14	0.9
Component 3	25	8	0.9
Nonvolatile	20		

Using component 3 as the base component,

$$\beta_3 = \beta_b = 1.0 \qquad \beta_1 = \frac{E_1 P_1}{E_b P_b} = \frac{0.9}{0.9} \times \frac{20}{8} = 2.5$$

$$\beta_2 = \frac{E_2 P_2}{E_b P_b} = \frac{0.9}{0.9} \times \frac{14}{8} = 1.75 \qquad \beta_r = \frac{E_r}{E_b} \times \frac{0}{8} = 0$$

L_t = total moles in still at end of distillation

$$L_t = L_1^0 \left(\frac{L_b}{L_b^0}\right)^{\beta_1} + L_2^0 \left(\frac{L_b}{L_b^0}\right)^{\beta_2} + L_3^0 \left(\frac{L_b}{L_b^0}\right)^{\beta_3} + L_r \left(\frac{L_b}{L_b^0}\right)^{\beta_r}$$

$$L_t = 30 \left[\frac{0.05(25)}{25}\right]^{2.5} + 25(0.05)^{1.75} + 25(0.05)^{1.0} + 20$$

$$L_t = 21.4$$

$$L = L_t - L_r^0 = 21.4 - 20 = 1.4$$

$$\sum_{i=s,r} \frac{L_i^0}{\beta_1}\left[1 - \left(\frac{L_b}{L_b^0}\right)^{\beta_1}\right] = \frac{30}{2.5}[1 - (0.05)^{2.5}] + \frac{25}{1.75}[1 - (0.05)^{1.75}] + \frac{25}{1}[1 - (0.05)^{1.2}]$$

$$= 12 + 14.3 + 23.75 = 50.05$$

$$L_r^0 \ln \frac{L_b^0}{L_b} = 20 \ln \frac{25}{(0.05)(25)} = 60.1$$

$$50.05 + 60.1 = \frac{E_b P_b}{\pi}(L^0 - L + S) = \frac{0.9(8)}{200}(80 - 1.4 + S)$$

$$110.15 - 0.036(78.602) = 0.036 \, S$$

$$\frac{110.15 - 2.83}{0.036} = S = 2980 \text{ mol steam}$$

REFERENCES

1. Ellerbe, R. W., Steam Distillation Basics, *Chem. Eng.*, May 4, 1974, pp. 105–112.
2. Coulter, K. E., Applied Distillation, Part III, *Pet. Refiner*, **31**(11), 156–158 (Nov. 11, 1952).
3. Potts, R. H., and White, F. B., Fractional Distillation of Fatty Acids, *J. Am. Oil Chem. Soc.*, **30** (2), 49–53 (February 1953).
4. Garber, H. J., and Lerman, F., Principles of Stripping Operations, *Trans. Am. Inst. Chem. Eng.*, **39**, 113–131 (1943).
5. Holland, C. D., and Welch, N. E., *Pet. Refiner*, May 1957, p. 251.

Continuous-Distillation Column Control

DAVID M. BOYD, JR. *Manager, Instrument Design & Service Department, UOP Process Division, UOP Inc.; Fellow, Instrument Society of America; senior member, Institute of Electrical And Electronics Engineers; member, American Chemical Society, American Institute of Chemical Engineers, and American Society of Mechanical Engineers; Chicago Technical Societies Council Award, 1965; the Sperry Award, 1966; Distinguished Engineering Award, University of Colorado, 1970; 160 patents in the field of instrumentation; author of numerous articles in instrumentation.*

INTRODUCTION

Taken as a whole, the control of a fractionating column would appear to be a very complex problem. However, like most complex problems it becomes much simpler when broken up into a consideration of its component parts. For this reason this section is in three parts: composition control, pressure control, and the control of fractionators making very-high-purity products with high recovery.

COMPOSITION CONTROL

Much material is to be found in the literature on the subject of the control of fractionators, but there is one fundamental principle at the root of it all. This may be simply stated: a material balance must be maintained at all times in the column by controlling the heat balance in such a manner that the desired separation will be obtained. Unfortunately, however, the control of fractionating columns is not simple, for no one answer is applicable to all problems, and various compromises become necessary.[1]

THREE FACTORS CONTROL FRACTIONATION

A typical fractionating column has a feed, consisting of components A and B, usually entering at about the midpoint. The two products are withdrawn from the column at the

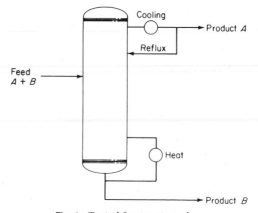

Fig. 1 Typical fractionating column.

top and bottom, respectively. Such a column is shown schematically in Fig. 1. The respective purities of the A and B streams are determined by three factors—the number of theoretical trays, the reflux ratio employed, and the material balance existing around the column.

The Number of Trays in the Fractionating Column

The number of theoretical trays must be determined during the design of the column. The number represents the result of engineering judgment in striking a balance between the cost of constructing the column and the anticipated cost of operating it. As the number of trays in a column increases, the operating cost to supply heat (in the feed and/or by reboiler) and to effect cooling of the reflux is decreased. This is because the required amount of reflux decreases as the number of trays increases in achieving specified purity for products A and B. On the other hand, the cost of erecting a fractionator increases as the number of trays increases. This is the first compromise.

A well-known procedure for determining the number of theoretical trays is that of McCabe and Thiele. This graphical method is shown in Fig. 2. The reflux ratio selected for the column determines the operating lines. When the mole fraction of the lighter component A in the liquid becomes equivalent to the mole fraction of A in the vapor ($x = y$), the slope of the operating line becomes 45° and the reflux becomes infinite. Consequently, as the operating lines approach infinite reflux, the number of trays becomes a minimum for any given separation. Again, compromise is necessary because at infinite reflux there can be no net overhead product.

Reflux Ratio

Control of the heat balance of the column involves control of the reflux ratio. This follows because, as previously stated, any fractionating column has a fixed number of trays and it is not practical to change the number being used as a means of control. Consequently, changing the reflux ratio used on a given column having a fixed number of plates necessarily will affect the quality of the products being made—A at the top and B at the bottom.

Consider a column containing 10 trays used to separate the feed into component A, the more volatile, and B, the less volatile. Assume that specifications require the top product, A, to contain a maximum of 2% of component B as an impurity in A. The bottom product has a specification of 0.5% A as an impurity in B. Such a column is shown at the left in Fig. 3. This specification can be made by using a reflux ratio of 3:1 on the overhead product.

If the reflux is now changed from 3:1 to 5:1 on the overhead product as shown in Fig. 3, the tower composition gradient would be so altered that the point where 2% of component B appears as an impurity in A will occur at tray 3, instead of at tray 1 as before. Also, the point where 0.5% of component A exists in B will occur at tray 8 instead of tray 10.

Consequently the overhead will contain 0.1% of B as an impurity in component A, and the bottom product will contain 0.05% of component A as an impurity in B. These compositions greatly exceed design quality requirements. Note that, in both examples, by using a given reflux ratio in the column and fixing the composition at one point in the column, the compositions at all other points in the column also have been substantially fixed.

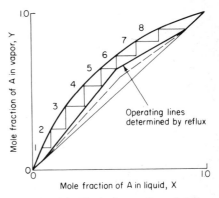

Material Balance

Any control system must be primarily concerned with the material balance around the column if the desired composition of the products is to be achieved—and maintained. That the desired material balance must be maintained is demonstrated by taking the following extreme case:

Fig. 2 McCabe-Thiele diagram for typical fractionating column.

Assume that the feed to a column is a mixture consisting of 100 units of component A and 100 units of component B. A reflux rate of 100:1 is employed with the intention of

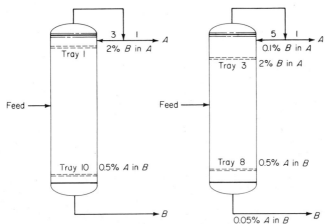

Fig. 3 Effect of reflux ratio on product purities.

securing 100% purity in the respective production of the two components. If, however, as shown in Fig. 4, the heat balance of the column is such that 101 units of overhead product is made, these purities cannot be obtained. Since there are only 100 units of component A in the feed when the volume of the overhead product is 101 units, there must be one unit of component B present in the overhead; the purity of the overhead then can only be 99% pure component A. The volume of the bottom product then must be only 99 units, but it will be essentially pure component B.

It is thus apparent that this column cannot be operated to produce specification products without proper attention to both material and heat balance. The proper reflux ratio can be established by means of a flow controller. A composition controller can be used to control the heat supplied to the reboiler so as to establish the composition gradient at the desired control point in the column—provided that the optimum composition control point is located in the bottom section of the column.

The most desirable place to control the composition often is found to be in the upper part of the column,[1] however. In this case the functions of the two controllers must be the reverse of the previous example: the flow controller controls the heat supply to the reboiler and the composition controller controls the reflux ratio to maintain the material balance of the column.

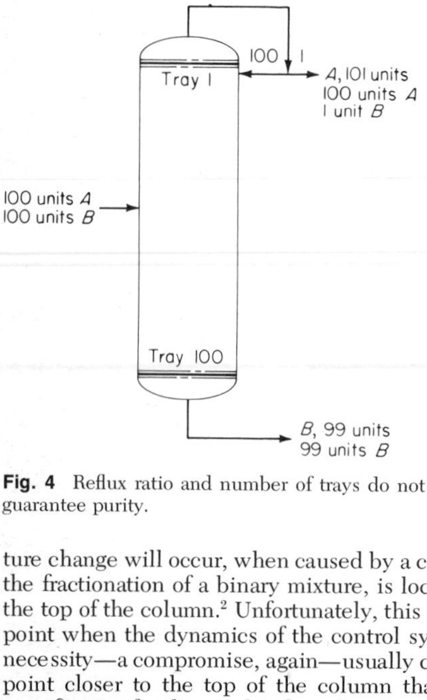

Fig. 4 Reflux ratio and number of trays do not guarantee purity.

The most popular type of composition controller used in fractionating columns is a temperature-sensing device. This is based on the fact that the boiling point of a binary mixture is proportional to its composition at any given pressure.

The actual location of the temperature control point in practice usually is the result of another compromise. If the required purity of the overhead product does not closely approach 100%, the temperature control point may be located in the overhead vapor line. When relatively pure products are desired, however, the temperature changes in the overhead vapor will be found to be too small for effective measurement and control; therefore, in practice, the control point must be moved down the column.

The point where the maximum temperature change will occur, when caused by a change in the purity of the overhead product in the fractionation of a binary mixture, is located halfway between the feed inlet tray and the top of the column.[2] Unfortunately, this point usually does not coincide with the ideal point when the dynamics of the control system are considered. Consequently, practical necessity—a compromise, again—usually calls for the location of the temperature control point closer to the top of the column than to the feed tray. Moreover, as the purity specification for the overhead product increases, the limitations of the single-temperature controller are quickly reached.

Control based upon the measurement of temperature at a single point in the column becomes impractical at high product purities because the boiling point of a binary mixture is proportional to composition (purity) only at a given pressure. Normal barometric pressure variations occurring within a short period of time alone can cause sufficiently large shifts in the boiling point that are not associated with a change in composition. These changes, if not taken into consideration, will cause the temperature controller to take such action that the allowable specifications for product purity will be exceeded.

DIFFERENTIAL TEMPERATURE CONTROL

The easiest way to compensate for pressure changes is to measure the temperature at two points, i.e., by employing a differential temperature controller.[3] A differential temperature controller assumes that the material on the top tray of the fractionating column is relatively pure. The temperature detected here, whether or not it is altered by barometric-pressure variations or by other pressure changes, is the reference point for the instrument. This controller constantly refers the temperature measured at some point nearer to the feed tray to the temperature of the top tray and uses the magnitude of the difference to achieve control of the quality of the product.

The differential temperature controller has many advantages in the control of fractionating columns to make a pure overhead product. However, it also has certain disadvantages which must be understood.

Consider the case of a column using differential temperature control which contains 30 trays and has the feed inlet at tray 15. The control temperature is measured on tray 8. The reference temperature is that on tray 3, so that it will not be affected by subcooling of the reflux. This arrangement is sketched in Fig. 5.

When the purity of the lighter component increases in the overhead product, the material on tray 8 will also have become purer in component A, and thus the temperature difference between this tray and tray 3 will decrease. However, a hydraulic gradient also exists between trays 3 and 8, characterized by a lower pressure at tray 3 than at tray 8. This pressure differential is the composite of the vapor velocity heads through the openings in the intervening trays and the respective liquid heads on these trays. A change in hydraulic gradient will appear to the differential temperature controller as a change in differential temperature, however, because of the corresponding changes induced in the respective liquid-boiling-point temperatures detected on these two trays.

Now two things have occurred to affect differential temperature: that ΔT caused by actual differences in composition between the respective materials on trays 3 and 8, and that ΔT resulting from pressure drop. These two ΔT's are represented by solid

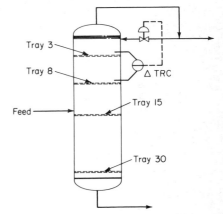

Fig. 5   Differential temperature control.

lines in the graph of Fig. 6. The instantaneous sum of these two effects—the resultant ΔT found by the differential temperature controller—is shown by the dashed line; it will be observed to possess a minimum point.

The differential instrument would tend to increase the reflux rate upon detecting an

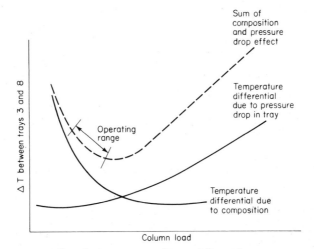

Fig. 6   Effect of column pressure drop on differential temperature control.

increase in the differential temperature existing between trays 3 and 8. This action would result in causing the controller to reduce the reflux and thereby maintain control of the fractionation. However, if the increase in reflux is sufficient to substantially affect the pressure difference between trays 3 and 8, the effectiveness of the controlled ΔT will be diminished.

Therefore, it is necessary to operate the column within the range shown in Fig. 6, that is, in such a manner as to stay to the *left* of the minimum point on the summation, or composite ΔT curve. Stated another way, the point where differential temperature reaches a minimum must not coincide with the desirable operating range for the column. The farther apart that these are separated, the better. One way to accomplish this is to avoid the use of trays characterized by high pressure drop in fractionating columns intended to produce high-purity products.

PRESSURE-COMPENSATED TEMPERATURE CONTROL

Fractionation problems are often encountered which cannot be solved, practically, by the use of differential temperature control. A good example of this is the requirement to fractionate a feed containing multiple components into an overhead product containing a number of components lighter than the light key and a bottom product containing the other heavier components. A variation in the lighter component (assumed to have a boiling point lower than that of the lighter key component) in the overhead product can, therefore, cause a change in the reference temperature.

It was previously pointed out that, in differential temperature control, the material on a tray in the upper section of the column is used as a reference to compensate for the effect of pressure variations on the boiling point (and therefore the purity) of the material being controlled. Since, under the new conditions in this example, the reference temperature can change disproportionally to the relation between the light and heavy key components, the use of differential temperature control should not be considered for this service.

The method of choice for a four-component application is the use of an absolute pressure transmitter output can be scaled so that at 0.875 kg/cm²(abs) [12.5 lb/in²(abs)] the controlled variable becomes that of maintaining vapor pressure constant at some selected point in the column.

To illustrate pressure-compensated temperature control, assume that the overhead product at desired purity boils at 120°C (250°F) when the pressure is 1.05 kg/cm²(abs) [15 lb/in²(abs)]. If the pressure rises to 1.4 kg/cm²(abs) [20 lb/in²(abs)] for any reason, the boiling point of the overhead will increase to 132°C (270°F) for the same purity. Both temperature and absolute pressure are measured on this tray. The transmitter outputs of both instruments will be separately fed to a computing unit and from it to a recorder-controller having a full-scale reading of 50 V. The arrangement is shown in Fig. 7.

The output of the temperature transmitter can be scaled so that at 116°C (240°F) the signal will be zero volts and at 138°C (280°F) it will be 50 V. Similarly, the absolute pressure transmitter output can be scaled so that at 0.875 kg/cm²(abs) [12.5 lb/in²(abs)] the signal will be zero volts and at 1.575 kg/cm²(abs) [22.5 lb/in²(abs)] 50 V.

Now, when the temperature is 120°C (250°F), the signal from the temperature-sensing instrument scaler will be 12.5 V; the pressure will be 1.05 kg/cm²(abs) [15 lb/in²(abs)] and the output of the absolute-pressure transmitter scaler will be 12.5 V also. The computer will be wired to always subtract the pressure signal voltage from the temperature signal voltage, and then to apply a 25-V bias to the result of this subtraction, thereby using the midscale point of the recorder-controller as the datum point, or set point. Under the circumstances given, the signal transmitted to the recorder-controller will be 25 V—the set point—and no compensating action will be initiated by the controller. Likewise, if the pressure and temperature increase to 1.4 kg/cm²(abs) [20 lb/in²(abs)] and 132°C (270°F) respectively, the outputs of each transmitter will be 37.5 V and the computer signal to the recorder-controller will remain unchanged at 25 V.

In the event that the temperature increases but pressure remains unchaged, the recorder-controller would sense that this measured variable is above the midscale set-point and take the appropriate corrective action. If pressure changes but temperature does not, the controller would sense that this measured variable is below the midscale set point and again take appropriate action.

Pressure-compensated temperature control is not a panacea for all fractionation problems, however, for if too much pressure compensation is used, the controller may appear to have reverse action.

In understanding this effect, it must be remembered that a substantial part of the pressure change being compensated for is the effect of the pressure drop resulting from vapor flow through the column. For a given pressure at the top of the column, an increase in vapor flow through the column will result in an increase in absolute pressure throughout the column and thus will affect the absolute-pressure transmitter.

Using the previous example of pressure compensation, as the absolute pressure increases from 1.05 (15) to 1.4 kg/cm² (20 lb/in²), the scaler output should go from 12.5 to

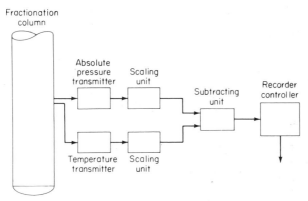

Fig. 7 Pressure-compensated temperature control.

37.5 V. But suppose the scaler was adjusted to overcompensate, so that when the pressure changed from 1.05 (15) to 1.4 kg/cm² (20 lb/in²) the scaler output voltage went from 12.5 to 40 V. Because the pressure-compensating signal is subtracted from the temperature signal, the recorder-controller would receive a signal which is below the set point; accordingly, it would increase the heat to the column via the reboiler.

The increase in heat input to the column would result in increased vapor flow. Since increased vapor flow will be recognized by the absolute-pressure transmitter as an increase in column pressure at the point of measurement, its signal voltage will increase still further and cause the recorder-controller to again call for more heat; ultimately, control will be lost as this cycle is repeated.[4]

The cure for this instability is simple, once the cause is recognized. The pressure compensator must never be set so that it will overcompensate. Undercompensation will not have the effect described.

PRESSURE CONTROL

The control of the composition in a fractionating column is, of course, of prime importance. Yet control of the pressure in the column has to be maintained if the tower is to stay at equilibrium condition. Also, many of the composition control schemes previously described require the measurement of a temperature for control. This temperature is related to composition only at one given pressure.

In the control of composition the speed of control or time constant depends for the most part upon the process itself, and usually little can be done about it. However, in controlling pressure, it is important to have as fast a control system as possible in order to make sure that pressure will not vary from its desired set point and that if it does deviate it can be brought back to the desired value at a rate at least ten times faster than the time constant of the composition control system. Therefore, it is necessary to choose a mechanical configuration of the column hardware which will give fast and adequate control of pressure.

The problem of controlling pressure is broken down here into three different types, each having many cases.

FRACTIONATOR PRESSURE CONTROL

Most fractionation control systems are based upon maintaining a constant pressure within the column. This pressure might be atmospheric, vacuum, or high pressure. Any variation from this set pressure will upset the control system by changing the equilibrium conditions on each tray. This constant pressure is obtained by the mechanical design of the equipment and by proper instrumentation. The choice of the latter is dependent upon the former.

Problems of fractionation pressure control usually can be separated into one of the following types: (1) atmospheric distillation, (2) vacuum distillation, and (3) pressure distillation including (*a*) large amount of uncondensables, (*b*) small amount of uncondensables (overhead product desired as liquid), (*c*) mixed uncondensables and vapor (overhead product desired as liquid), and (*d*) mixed uncondensables and vapor (overhead product desired as vapor).

This section will outline a pressure control system for each type of fractionation control problem.

Type 1: Atmospheric Distillation

Naturally, the usual type of atmospheric distillation will not require any sort of pressure control instruments. The only requirements are to locate a rather large vent on the receiver to equalize the pressure and to keep the column at substantially the same pressure as the receiver regardless of the load on the column. In the operation of a column of this type on very close fractionation service, it will be found that the quality will change with the barometric pressure even though the temperatures are held constant.

On fractionation problems requiring very close control of the tower, it is often necessary to install an absolute pressure controller and handle the tower as in 3*b*. The tower pressure is maintained just slightly higher than the highest atmospheric pressure expected in the area in order to permit the purging of uncondensables that might accumulate in the system.

Type 2: Vacuum Distillation (Fig. 8)

A vacuum-distillation column is essentially a low-pressure distillation unit with the pressure of the system controlled by the amount of uncondensables present. The majority of the overhead vapors are condensed, and the condensation rate in the condenser is inversely proportional to the amount of uncondensables that are blanketing the condensing surface. The uncondensables are pulled off by a vacuum source at a rate controlled by the pressure controller.

The most common vacuum source used on fractionating columns is the steam jet. It has no moving parts and requires very little maintenance. It is, however, sensitive to changes in steam pressure, and for this reason, it is advisable to install a pressure controller on the steam to keep it at the optimum steam pressure required by the jet.

It is not possible to control the system by controlling the steam to the jet, as the capacity curve of a given jet has a very abrupt break as the steam pressure goes below the critical value for the jet.

The recommended control system for this service uses a control valve to bleed air or gas into the line just ahead of the jet. This makes the maximum capacity of the jet available to handle any surges or upsets in the system. Placing the control valve in the vapor line leading to the jet is not recommended as the additional pressure drop across the control valve would greatly reduce the capacity of a given jet.

It is recommended that the receiver pressure be the controlled factor in a vacuum-distillation system, as this will involve the least time lag and, consequently, a very narrow throttling range can be used.

The pressure of the rest of the system will be proportional to the friction drop and will differ only by the pressure losses in the vapor lines, bubble trays, condenser, etc. It is necessary for good control (and operation) to reduce these losses to an absolute minimum in order to prevent pressure changes in the column with any change in the vapor-flow rate.

The pressure controller used on a vacuum control system should be provided with a pressure transmitter. This transmitter should be located above its pressure tap and have a slight bleed of air or gas into the line in order to keep it free of any condensate that might otherwise collect in this line and give a false reading.

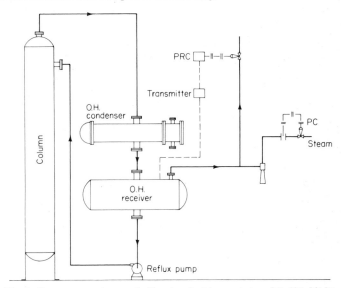

Fig. 8 Pressure control—type 2. *(Reprinted with permission of Gulf Publishing.)*

If the product from this still is to be of a high purity, it is essential that the transmitter be of the absolute-pressure type in order to eliminate fluctuations in quality with changes in atmospheric pressure.

Type 3: Pressure Distillation

Case a. Large Amount of Uncondensables (Fig. 9) This case is typified by an absorber. There is a large flow of gas that can be modulated by the control valve in order to maintain the proper pressure on the system.

The gas flow is large enough for the system pressure to respond quickly to a change in the gas flow, and yet the capacity of the system is large enough that the controller will not be unstable. These factors combined with a small time lag in the system make a very easy system to control.

The controller employed on this service can use a narrow throttling range and need be only a proportional controller not requiring automatic reset.

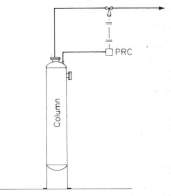

Fig. 9 Pressure control—type 3, case *a*. *(Reprinted with permission of Gulf Publishing.)*

Case b. Small Amount of Uncondensables (Product Desired as a Liquid) See Figs. 10 and 11. In this case the amount of uncondensable gas present is negligible and, consequently, it is not available for control purposes. This system is therefore controlled by the rate of condensation in the condenser.

The method of controlling the rate of condensation will depend upon the mechanical construction of the condensing equipment.

One method of control is to place the control valve on the cooling water from the condenser. This system is recommended only where the cooling water contains chemicals to prevent fouling of the tubes in the event of high temperature rises encountered in the condenser tubes. This method has very low maintenance costs because the valve is on the water line and will give very satisfactory service provided the condenser is designed properly.

The best condenser for this service is a bundle type with the cooling water flowing through the tubes. This water should be flowing at a rate of more than 4½ ft/s, and the water should have a sojourn time of less than 45 s. The shorter the sojurn time for the water, the better will be the control obtained, owing to the decrease in dead time or lag in the system.

With a properly designed condenser, the pressure controller need have only propor-

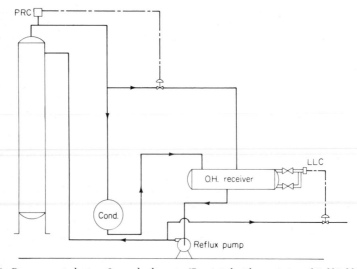

Fig. 10   Pressure control—type 3, case *b*, alternate. (*Reprinted with permission of Gulf Publishing.*)

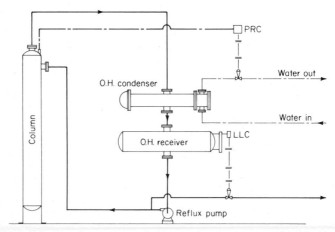

Fig. 11   Pressure control—type 3, case *b*. (*Reprinted with permission of Gulf Publishing.*)

tional control, as a narrow throttling range is required. Automatic reset is not necessary. However, as the sojourn time of the water increases, it will increase the time lag of the system; consequently, the controller will require a wider throttling range and will need automatic reset to compensate for the load changes. The results obtained by using a wide throttling range would not be satisfactory for precision-fractionation towers because of the length of time required for the system to recover from an upset, and this would interact with the composition control system.

It would be impossible to use this control system, for instance, on a condenser box with submerged tube sections. There would be a large time lag in the system because of the large volume of water in the box. It would take quite a while for a change in water-flow rate to change the temperature of the water in the box and finally the condensing rate.

In the presence of such unfavorable time lags, it becomes necessary to use a different type of control system, one which permits the water rate to remain constant and controls the condensing rate by controlling the amount of surface exposed to the vapors. This is done by placing a control valve in the condensate line and modulating the flow of condensate from the condenser. When the pressure is dropping, the valve cuts back on the condensate flow, causing it to flood more tube surface and consequently reducing the surface exposed to the vapors. The condensing rate is reduced and the pressure tends to rise. It is suggested that a vent valve be installed to purge the uncondensables from the top of the condenser, if it is thought that there is a possibility of their building up and blanketing the condensing surface.

Condenser below Receiver. A third possibility for this type of service is used when the condenser is located below the receiver. Many times this is done to make the condenser available for servicing and to save on steel work. It is the usual practice to elevate the bottom of the receiver 3 to 4.5 m (10 to 15 ft) above the suction of the pump in order to provide a positive suction head on the pump.

In this type of installation the control valve is placed in a bypass from the vapor line to the receiver. When this valve is open, it equalizes the pressure between the vapor line and the receiver. This causes the condensing surface to become flooded with condensate because of the 3 to 4.5 m (10 to 15 ft) of head that exists in the condensate line from the condenser to the receiver. The flooding of the condensing surface causes the pressure to build up because of the decrease in the condensing rate. Under normal operating conditions the subcooling that the condensate receives in the condenser is sufficient to reduce the vapor pressure in the receiver. The difference in pressure permits the condensate to flow up the 3 to 4.5 m (10 to 15 ft) of pipe that exists between the condenser and receiver.

A fourth possibility for this type of service is a modification of the flooded-condenser type. Instead of being placed on the outlet of the condenser, the control valve is placed in the product stream from the receiver. The receiver is run full and the pressure controller controls the overhead product-stream valve. The use of this control system is not recommended where the product stream feeds a succeeding column or where the feed contains any appreciable quantity of uncondensable gases.

Case c. Mixed Uncondensables and Vapor (Overhead Product Desired as a Liquid) See Fig. 12. This case is similar to Case *b*, but the problem is complicated by a higher percentage of inert gases.

The uncondensables have to be removed or they will accumulate and blanket off the condensing surface, thereby causing loss of control of the fractionator pressure.

The simplest method of handling this problem is to bleed off a fixed amount of gases and vapors to a lower-pressure unit, such as an absorption tower, if it is present in the system. If an absorber is not present, it is possible to install a vent condenser to recover all the condensable vapors possible from this purge stream.

It is recommended that the fixed continuous purge be used wherever economically possible; however, when this is not permitted, it is possible to modulate the purge stream. This might be desirable when the amount of uncondensable coming to the tower is subject to wide variation over a period of time.

As the uncondensables build up in the condenser, the pressure controller will tend to open up the control valve to maintain the proper rate of condensation. This is done by a change of air-loading pressure on the diaphragm control valve. This air-loading pressure could also be used to operate a purge control valve, as it passed a certain operating pressure. This could be done by means of a calibrated valve positioner or a second pressure controller.

Case d. Mixed Uncondensables and Vapor (Overhead Product Desired as a Vapor)   See Fig. 13. In this case the overhead product is removed from the system as a vapor. The pressure controller can consequently be used to modulate this flow of vapor, as it is a major factor, and the system will respond to changes in its flow.

A level controller is installed on the overhead receiver to control the cooling water to the condenser. It will condense only enough condensate to provide the column with reflux.

Here again this control system depends upon having a properly designed condenser in order to operate satisfactorily. The condenser requires a short sojourn time for the water as in Case *b.*

If the condenser is improperly designed for cooling-water control, it is recommended that the cooling-water flow be left at a constant rate and the level controller control a stream of condensate through a small vaporizer and mix it with the vapor from the pressure-control valve; or if the cooling water has bad fouling tendencies, it would be

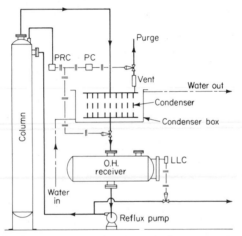

Fig. 12   Pressure control—type 3, case *c.* *(Reprinted with permission of Gulf Publishing.)*

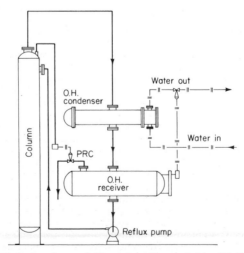

Fig. 13   Pressure control—type 3, case *d.* *(Reprinted with permission of Gulf Publishing.)*

preferable to use a system similar to the third possibility described under Case *b* and use the level controller to control the vapor bypass with the condenser located on the ground to make sufficient pressure drop available for proper control.

THE CONTROL OF FRACTIONATORS MAKING VERY-HIGH-PURITY PRODUCTS WITH HIGH RECOVERY

When it is desired to control the impurities of the products from a fractionator to the parts per million range, the following method is sometimes recommended. However, it should be emphasized that this method is not applicable if the impurities are going to be in the percent range. To understand this method's operation, it is necessary to take a new look at what occurs in a fractionating column when it is making high-purity products.

The natural composition profile in a fractionating column can be used to achieve tower control more accurately than most plant laboratory analyses by using uncomplicated, readily available temperature-measuring devices. Before discussing this type of process control, however, it is best to review just how much control can be imposed, and the limits placed upon such regulation by the design of the column.[5]

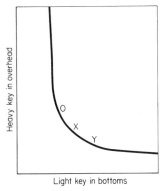

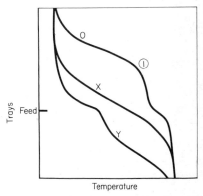

Fig. 14 Performance curve. (*Reprinted with permission of Chemical Engineering Progress.*)

Fig. 15 Typical temperature profile. (*Reprinted with permission of Chemical Engineering Progress.*)

Once the number of trays, the reflux ratio, the feed composition, and the feed-tray location have all been fixed, the column must operate along the curve of Fig. 14, which shows a fixed relationship between the impurities in the overhead and the impurities in the bottoms. Better purity of the overhead (light-key) product can be obtained by allowing some of the light-key product to appear in the bottoms product or vice versa. However, fixing the purity at one end of the column fixes the purity of the other product. The desired operating point is X, which gives the maximum separation of both ends under the conditions chosen.

Suppose we now consider the effect that moving along the curve in Fig. 14 has upon the temperature profile (or composition profile) of the column, shown in Fig. 15. Curve X corresponds to point X on Fig. 14, which gives the maximum separation at both ends of the column. Note that it starts down from the top of the column with very little temperature change from tray to tray. Then the rate of temperature change increases as the feed tray is approached, and this rate diminishes again as the bottom of the column is approached.

However, curve O is quite a different matter. Because there are some bottoms in the overhead, the temperature is slightly higher at the top tray of the column. Thus, because it does not take as many trays to make this purity as it does for curve X, the temperature increases much more rapidly as one goes down the column.

At point 1 on curve O, by vapor-liquid equilibrium, the same composition has been reached as is on the feed tray. Therefore, there is very little composition, or temperature, change between this point and the feed tray. This is sometimes referred to as the "pinch point."

As one moves down the column from the feed tray, the temperature approaches the bottoms temperature, giving very little temperature change in the bottom section. This, then, will give a higher bottom-product purity than the operation shown on curve X.

Curve Y is the opposite of curve O. The bottom temperature is slightly lower than that of curve X and the "pinch point" occurs below the feed deck. Above the feed deck the temperature profile rapidly approaches the top temperature, giving higher overhead purity than either curves X or O.

It would appear, therefore, that the area in the immediate vicinity of the feed tray, where such dramatic changes in temperature occur with changes in overhead or bottom composition, should hold much promise as a means of control instead of either the top or bottom of the column as is traditionally used. This section describes such a means.

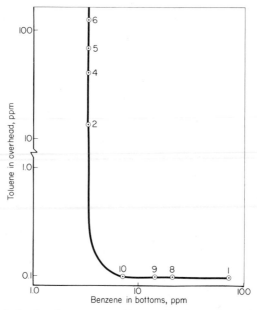

Fig. 16   Computer-calculated performance curve. (*Reprinted with permission of Chemical Engineering Progress.*)

Feasibility in Actual Fractionator

The column chosen for the tests was an actual column, feeding a solvent-extracted aromatic-hydrocarbon mixture and taking benzene as its overhead product. The bottom product is 50% toluene and the remainder mixed xylenes and ethylbenzene. It had 70 trays, with the overhead product drawoff on tray 5 (numbering from the top). It therefore had 65 real fractionating trays, the top five being mainly for drying. Feed was on tray 39. Using an assumed tray efficiency of 80%, a series of tray-to-tray calculations were run on a digital computer.

The resulting performance curve is shown in Fig. 16. Plotting the tray temperature vs. tray number for the 10 points shown results in the family of curves shown in Fig. 17. Note that each point in Fig. 16 defines a unique curve in Fig. 17.

The control system used on this operating column consisted of a differential temperature measurement between theoretical tray 5 and theoretical tray 15. (We would now use theoretical tray 5 and theoretical tray 20.) A second differential temperature was taken in the bottom section of the column, between theoretical tray 30 and theoretical tray 45. Thus we used two differential temperature-measuring circuits in each section of the column.

The two differential temperature signals were then fed into a subtraction relay and the difference between the two was the control signal. That is, we used the difference between the two differential temperature measurements for the control signal.

Figure 18 is a double plot, showing the effect of signal obtained by subtracting the lower ΔT from the upper ΔT upon the amount of toluene in the overhead product and the amount of benzene in the lower product. Several reflux-to-feed ratios are shown. Note that each point represents one given column operation. For example, 10 ppm of toluene in the overhead will give a control signal of $+10°F$, while 10 ppm of benzene in the bottoms will give a control signal of $-15°F$. Certainly this is a very large signal for such a small change in the purity of the products. In fact, the sensitivity was so great that some concern was felt for the stability of the control loop.

It can be seen in Fig. 19 that in going from 100.000 to 100.030 moles of overhead

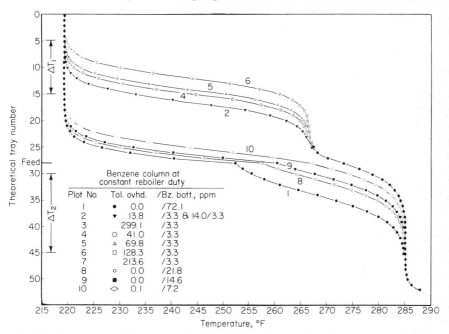

Fig. 17 Temperature profile vs. tray number (see Fig. 16). *(Reprinted with permission of Chemical Engineering Progress.)*

product the $\Delta\Delta T$ goes from -30 to $+10°F$. That is $40°F$ change in temperature for 3 parts in 10,000 change in overhead product.

It was apparent that an instrument with an extremely broad proportional band would be required to achieve stable control. Inasmuch as most controllers have a maximum proportional band of 500%, a feedforward system was devised by which, biasing the ratio setting on the feedforward system, we could multiply the proportional-band setting on the controller by another factor of 4. This would give a proportional band of 2000%.

However, again examining Fig. 17, there is some hope that it would be stable. Note that going from curve 2 to curve 10 would require a large change in the inventory on theoretical trays 15 to 25. Curve 10 has almost pure benzene on these trays, whereas curve 2 has a large amount of toluene to give the increase in temperature. Naturally, it will take time to change this inventory because it has to be accumulated from the feed to the column.

Feedforward Method Not to Be Used

The field tests proved this concern about stability to be unfounded, and the feedforward system was disconnected. The controllers were stable at a proportional-band setting of 20%. The reset rate was 5 min and the derivative was 1 min.

Therefore, in the future, the feedforward feature will not be used. However, the square-root extractor will be used on the controlled product drawoff to avoid the drastic change in gain produced by the nonlinear flow signal from a DP cell (Figs. 20 and 21).

The first installation of this system was made on benzene-toluene columns connected in series in the Trieste, Italy, refinery of Total Societia Italiana per Azoni (Aquila). The results were favorable, as shown by the chromatographic analyses in Table 1. The refinery chromatographic analyses were made on a new, special, high-sensitivity chromatograph in the refiner's research department, in order that a high sensitivity could be used.

The same samples that were analyzed on the refiner's research chromatograph were

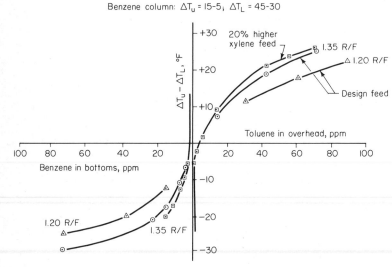

Fig. 18 Delta-delta (Δ-Δ) temperature vs. top and bottom composition. *(Reprinted with permission of Chemical Engineering Progress.)*

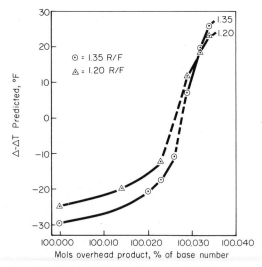

Fig. 19 Effect of overhead flow upon delta-delta (Δ-Δ) temperature (benzene column). *(Reprinted with permission of Chemical Engineering Progress.)*

TABLE 1 Effect of Double Differential Temperature Signal on Concentrations of the Key Components

Test	dd TRC, °C	Feed rate, % design	Toluene in benzene col. ovhd., ppm		Benzene in toluene col. ovhd., ppm		Xylenes in toluene col. ovhd., ppm		Toluene in toluene col. btms., ppm		Benzene in benzene col. btms., ppm	
			Ref.	UOP	Ref.	UOP	Ref.	UOP	Ref.	UOP	Ref.	UOP
1	0	80	0	14	273	1	0	50	0	93	350	1
2	+2	100	0	6	45	3	0	12	0	12		
3	+5	100	0	14	41	1	0	11				
4	+10	100	128	33	45	1	0	8				
5	+5	120	0	47	56	5	0	36				
6	+5	120	0	86	50	42	0	81	2.5	62	14.5	51

Basis: benzene-toluene splitter at Triéste. Analytical determinations were made at both the refinery and the UOP Research Laboratories.

col. = column, btms. = bottoms, ovhd. = overhead.

SOURCE: Reprinted with permission from *Chemical Engineering Progress*.

analyzed again at the UOP laboratory in Des Plaines, Ill., using a chromatograph equipped with a flame-ionization detector. The toluene in benzene was found to range from 6 to 14 ppm instead of zero (as suspected). However, the benzene in toluene was about 1 ppm instead of the 40 ppm reported by the refiner. (The design specifications permitted up to 500 ppm.)

Field Data Matched Original Computations

What is even more interesting is that when the plant data, using UOP's flame-ionization data, are plotted on the original computer calculation from this column, the plant data fall

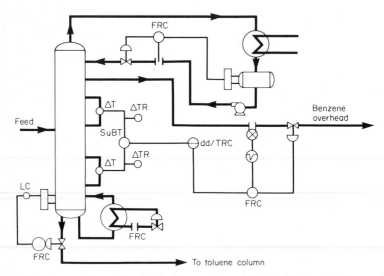

Fig. 20 Benzene column. *(Reprinted with permission of Chemical Engineering Progress.)*

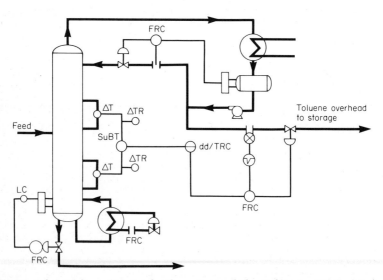

Fig. 21 Toluene column. *(Reprinted with permission of Chemical Engineering Progress.)*

on the original curve, as shown in Fig. 22. The original curve was for a reflux ratio of 1.35:1 on feed! The plant was run at 1:1. It must be concluded that, with the closer control, higher tray efficiency was being obtained, because the composition gradient was being held better than we have been able to in the past.

For high-purity fractionation, it appears that by running the computer calculations and then measuring the double differential temperature in the column, the impurities can be predicted more closely than the chromatographs available in the average plant control laboratory can determine.

The plant had been shut down by a power failure prior to the author's arrival and had

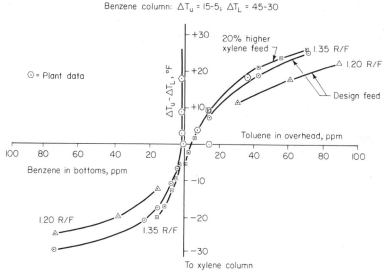

Benzene column: $\Delta T_u = 15\text{-}5$; $\Delta T_L = 45\text{-}30$

Fig. 22 Correlation of plant data with calculated data. *(Reprinted with permission of Chemical Engineering Progress.)*

just been brought back on stream. The benzene in the feed varied from 6 to about 14% over the next 2-day period and then drifted back to about 7.5% the rest of the time. (The design had been predicted on 20% benzene in the feed.)

The unit had been running at about 80% of design feed. However, charge was accumulated in the fractionator feed tank and was then increased to the design rate. Therefore, the samples for the last 3 days were at 100% of design rate. These are identified in Table 1 as tests 2 to 6.

Different settings of the control point of the double differential (dd) TRC were tried. Note that initially it was set at zero and the benzene in the toluene was 273 ppm. This was changed to +2°C, and the corresponding samples (test 2) were 45 ppm benzene in the toluene. The set point was next changed to +5°C, and the corresponding sample (test 3) showed 41 ppm benzene in the toluene. Therefore, the dd TRC setting was raised to +10°C. The corresponding sample (test 4) showed 128 ppm toluene in the benzene (UOP data, 33 ppm). It should be noted that tests 3 and 4 were run on the same day.

Naturally, the plant could not be left running in this condition. It was intended to run the column at 20% above the design rate, and the charge tank was becoming depleted. The set point was lowered from +10 to +5°C, the charge was set at 120% of design, and the steam rate to the reboiler was increased. This produced a big upset in the column, of course. However, the dd TRC cut the benzene to storage, which in effect put the column in total reflux. Then when the proper $\Delta\Delta T$ was established, it again increased the flow to storage. The samples taken 1 h later were on specification. These are shown as test 5.

Figures 23 and 24 show the effect of the changes made as well as the stability. They also show how the two controllers handled a partial steam failure the night of the third day. Note the bad swing in column temperature and that it cut the products going to storage until the column had lined out again. The samples taken the next morning were again on specification, the column operating at 120% of design.

It is also interesting to note the effect of the 20% increase in throughput on the upper and lower differential temperatures of the toluene column. Both increased because of the higher traffic. However, the difference between the two remained the same. This increase in differential temperature occurred because increased traffic increased the pressure drop through the column. At constant overhead pressure, the absolute pressure on each tray would increase. This would allow the products to go "off specs," sometimes seriously enough to cause a reversal in the action of the controller, thus completely upsetting the column.

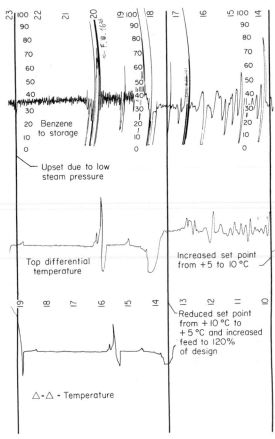

Fig. 23 Instrument records from benzene column. (*Reprinted with permission of Chemical Engineering Progress.*)

How the Controller Is Adjusted

In tuning the controllers on an installation of this system, it must be kept in mind that the double differential temperature controller is controlling the product to storage. However, in order for the temperature to be affected, the reflux at the point of measurement has to be changed.

In the case of the benzene column, this created no problems because the product to storage came from a trap tray in the column. However, in the case of the toluene column the product came from the overhead receiver. Therefore, to affect the temperature it was first necessary to change the level in the receiver, which would then change the reflux flow.

The level controller originally had a 50% proportional-band setting, which was much too slow. It was changed to 3% proportional band, and the performance improved. The double differential temperature controllers were each left with a proportional band of 20% 5 min reset, and 1 min derivative.

There has been a gradual evolution in the control of fractionating columns. Temperature has been the prime means of control. However, temperature will indicate composition only at a given pressure. When control for high purities is desired, even barometric-pressure changes can throw the product off specification.

A simple way to control product purity is to use differential temperature. One thermocouple is located in the relatively pure material and the other at the desired control point.

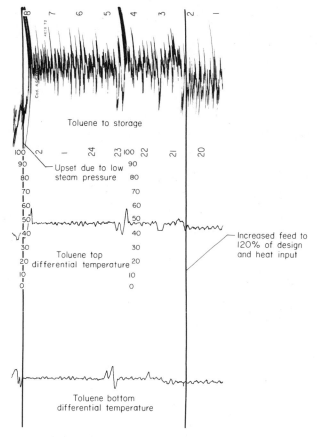

Fig. 24 Instrument records from toluene column. (*Reprinted with permission of Chemical Engineering Progress.*)

Therefore, as the pressure changes, both temperatures are affected, but the difference between the two remains the same.

As the column is loaded to improve composition, the differential temperature tends to decrease. However, the differential temperature tends to increase because of the pressure drop in the trays. The net result is that the measured differential temperature passes through a minimum and then increases again. This phenomenon generally confuses the operator of the column and very definitely limits the operating range.

With the new system, the differential temperature resulting from the pressure drop appears in the top differential temperature as well as in the bottom differential temperature. The bottom reading is subtracted from the top so that the effect on the reading is canceled out. For this reason, the differential temperatures are usually made over the same number of trays in the top and bottom section of the column.

IN CONCLUSION

It would appear that this method of control is a radical departure from the conventional method of controlling a fractionating column. Conventional methods control composition at one end of the column or the other by means of temperature or some analytical device.

This method, by taking advantage of the natural composition profile that exists in a fractionating column, controls near the feed tray instead of at either end of the column. By keeping the column out of "pinch" conditions around the feed tray, compositions can be maintained at either end of the column to an accuracy greater than that of most plant laboratory analyses. The measurements used are very simple temperature measurements that are well within the capabilities of available equipment and are very easy to maintain and keep in operation.

Field tests described here show the system to be extremely stable as compared with conventional systems and certainly well within the capabilities of the average operator to handle.

REFERENCES

1. Boyd, D. M., Jr., Fractionation, Instrumentation and Control, *Pet. Refiner,* **27**(10–11), (1948).
2. Uitti, K. D., The Effect of Control Point Location upon Column Control, presented before the Instrument Society of America, St. Louis, Mo., Sept. 12–16, 1949.
3. Fenske, E. R., and Broughton, D. B., Fractionation of Mixed Aromatics from Udex Extraction, *Ind. Eng. Chem.,* **47**, 714 (April 1955).
4. Cohen, R. G., Private communication on Pressure-Compensated Temperature Control.
5. Boyd, D. M., Fractionation Column Control, *Chem. Eng. Prog.,* **71**(6), (June 1975).

Section **1.6**

Solvent Recovery

JOHN W. DREW, Ch.E., P.E. *Director of Engineering, Chem-Pro Equipment Corp.; Member, American Institute of Chemical Engineers and American Chemical Society; patents on organic chemical processes; publications on costs of equipment and design of solvent-recovery systems; Licensed Professional Engineer in New York and New Jersey.*

INTRODUCTION

A solvent is a liquid agent used in the manufacture, purification, or application of a product. It dissolves, extends, or cleans part or all of the product during processing operation, but it is eventually removed for recovery or disposal. Recovery and reuse of organic solvents is generally practiced because of increased solvent costs and potential solvent shortages and because disposal often results in violation of air-, water-, or land-pollution regulations.

For pollution-control purposes recovery is becoming increasingly favored over incineration because of the expense of the added fuel usually required for incineration. This situation has forced the development of new recovery processes and better recovery equipment, some of which will be described in this section.

Water, the most widely used solvent, may also be considered for recovery. Where purification sufficient to allow its discharge is extremely difficult, where it is scarce, or where expensive pretreatment of fresh water is required, it is often more economical to recover and reuse water than to discharge it.

Recovery processes begin with the collection of the solvent from exhaust air, from waste water, or from processing operations. Following collection, purification of the solvent for reuse usually involves distillation.

SOURCES OF SOLVENT EMISSIONS

Listed in Table 1 are some major classes of products employing solvent in their manufacture. Beside each are the primary ways in which the solvent is discharged from the process. Where complete recycle of the solvent is an integral part of the process, the discharge is minimal, usually only from tank vents and spills. Otherwise recovery processes must be added.

TABLE 1   Sources of Solvent Emissions

Industry	Emission source
Synthetic fibers and films	Wastewaters, vapors in air
Paper coatings, graphic arts	
Printing	
Film coating, photographic and magnetic	Vapors in air
Protective coatings, industrial	
Architectural coatings	
Degreasing, dry cleaning	Contaminated liquids, vapors in air
Solvent refining of vegetable oils	Wastewaters, tank vents
Solvent extraction of pharmaceuticals from natural products	Wastewaters
Chemical manufacturing:	
Azeotropic and extractive distillations	Wastewaters, tank vents
Chemical reactions and crystallizations	Wastewaters, mixed liquids
Chemical drying	Vapors in air
Petroleum refining	Wastewaters, tank vents
Resin manufacturing:	
Reactor distillate	Wastewaters
Solvent quench	Concentrated vapors
Extractive metallurgy	Wastewaters
Nuclear-fuel reprocessing	Contaminated liquids

TABLE 2   Hydrocarbon Emissions from Stationary Sources

Source	Range of emissions, millions of tons/year
Organic-solvents usage:	
Degreasing	0.6–0.9
Dry cleaning	0.3–0.4
Graphic arts	0.2–0.3
Surface coatings	3.0–4.0
Petroleum refining and marketing	3.6–4.0
Chemical industries	1.0–1.4

Quantities of emissions from sources which have been reported by the EPA are shown in Table 2.[1]

RECOVERY PROCESSES

Solvents emitted in any of the classes listed in Table 1 may be recovered by the processes listed in the generalized flowsheet shown in Fig. 1. The best recovery path through this flowsheet depends primarily upon the nature of the solvents, the manner of emission, and the quality specifications of the recovered solvents. The best process would usually be that which produces the highest recovery of quality solvent at the lowest equipment and operating costs, while satisfying pollution-control and safety constraints.

COLLECTION

Flammable solvents discharging into ventilating air from dryers, coaters, and presses are generally kept at concentrations below 25% of the lower explosive limit (LEL) so as to avoid the possibility of fire or explosion in the ductwork and blowers. This means air volumes of the order of 280 m³ (10,000 ft³) of air per 3.79 L (gal) of solvent. Even where nonflammable solvents are used, air volumes are often of the same order of magnitude because of the high airflow requirements for workspace ventilation and product drying. Allowable concentrations of solvents for worker exposure, called *threshold-limit values,* are specified by the American Conference of Government and Industrial Hygienists. Upper or lower explosive limits of solvents in air are also published.[2]

Collection of solvent vapors from dilute mixtures in air can be accomplished by condensation (cooling or compression), absorption in water or oil, and adsorption on activated carbon or other solids. At present, adsorption on activated carbon is the only practical solution for high-percentage removal of common volatile solvents from dilute airstreams.

Condensation by cooling requires cooling to cryogenic temperatures. This involves

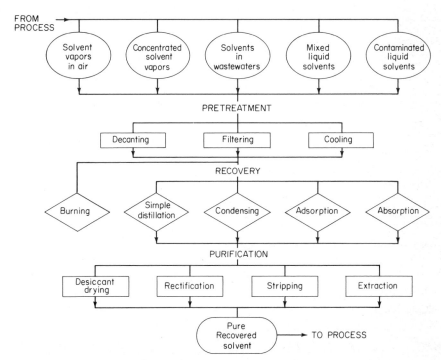

Fig. 1 Alternative recovery processes.

large expenditures of energy for air cooling, condensation, and freezing of atmospheric moisture plus complex and expensive equipment for refrigeration and ice removal.

Condensation by compression also consumes a great deal of energy because of the large volumes of gas and high pressures required to effect substantial removal. Recovery of mechanical energy from the solvent-free gas, although possible, is complex and costly.

Absorption of water-soluble solvents in water is often suggested. Most common solvents have rather high vapor pressure. Furthermore, their partial pressures from dilute aqueous solutions are 10 to 1000 times higher than from ideal solutions. This high partial pressure means that very large quantities of water have to be used to effect removal. As an example, to recover 90% of the acetone present in air at 25% of the lower explosive limit would require an absorption column of 9 theoretical stages using a minimum of 1500 kg (lb) of water per kilogram (pound) of acetone recovered. Distilling the recovered solution of 0.067% acetone would require a column of over 100 theoretical stages using 60 kg (lb) of steam per kilogram (pound) of acetone so as to produce 99% acetone overhead and water bottoms suitable for recycle to the absorber!

Absorption in water is practical, however, for high-boiling water-soluble solvents such as dimethyl formamide, glycols, glycol ethers, and glycol amines. However, some of these azeotrope with water, complicating final purification.

Absorption in oil is often more practical than absorption in water, since activity coefficients in oil tend to be closer to unity, especially for nonpolar solvents. Thus the ratio of oil to solvent need not be nearly as high as with water. However, losses of oil and stripping and recycling of the oil can be costly. The oil selected must have a low vapor pressure so as to minimize losses in the air leaving the absorber. Such oils have high molecular weights and viscosities well above that of water, which results in low stage efficiency in the absorber and low coefficients of heat transfer. Large absorption and heat-exchange equipment is therefore required. Absorption in oil is considered only where adsorption on carbon is not practical.

Activated Carbon in Solvent Recovery

Adsorption of organic vapors on activated carbon is very efficient. Although there are problems with its use, no other adsorbents have been found as desirable. Alumina and molecular sieves are common inorganic adsorbents, but these have greater affinity for water and low-molecular-weight gases than for the typical solvent molecule. Certain organic resins have been reported to be under successful development.

Ultimate retentivity of activated carbon for organic vapors can exceed 30% of its weight, and removal in excess of 98% is possible. In practice, however, most economical operation occurs at a solvent loading of 5 to 20%, while practical removal efficiencies are 80 to 95%. All organic vapors having molecular weights greater than that of air can be adsorbed; the higher the molecular weight the more readily adsorbed. Adsorption of mixed vapors is practical but some separation of components occurs in the bed, the adsorption taking place roughly as the inverse of the volatilities. This means a somewhat reduced capacity for the bed when compared with a single vapor.

Desorption and regeneration of the carbon is carried out by heating. Usually this is done by steam, resulting in a mixture of solvent and steam which is easily condensed. Hot gas may also be used. Since the quantity of hot gas needed is much less than that of the original air, a concentration of the vapor and transfer to another gas takes place. Desorption may also be carried out or aided by vacuum, but practical applications have been few, probably because of the added expense to make all equipment suitable for vacuum and the difficulty of transferring the necessary heat through the bed under vacuum. Hot-gas or vacuum regeneration processes also have the disadvantage of removing moisture from the bed. Moisture is an important factor in controlling bed temperature. Refer to Sec. 1.13 for a complete discussion and design information.

PURIFICATION

Vaporization

Solvent contaminated with nonvolatiles such as dirt, oils, greases, and resins can be recovered by simple distillation or vaporization. Prefiltration is used where particulate solids are present. Packaged stills are commercially available. They are of two types:

indirect-heated and direct-steam-injected. With indirect heat, the solvent is boiled by heating the walls of the still by steam, hot oil, or electricity. As the solvent boils away, dissolved impurities are concentrated in the boiler and tend to foul the heat-transfer surface. For this reason, scraper-type agitators are used. Keeping the walls clean and removing solvent-wet solids from the still are the major problems with the indirect-heated still.

The steam-injected still attempts to solve these problems. Steam fed into the still vaporizes the solvent by direct-contact heat exchange. The residue in the still becomes mixed with water, usually as a flowable slurry which may be readily discharged from the still. Residue removal can still be a problem, however, where resins form sticky masses.

Vaporization and recovery of water-immiscible solvents by steam injection is straightforward. Practically complete removal of the solvent from the residue is possible, and separation of the solvent from the water in the distillate is easily done by decantation. With partially or completely water-soluble solvents, the performance is not as good. Removal of essentially all solvent from the aqueous residue requires large amounts of steam, since the solvent concentration in the residue becomes more and more dilute. Batch operations permit more complete removal than continuous discharge of the residue. Also, since the distilled solvent is mixed with water, pure dry solvent must be obtained by fractionation in a distillation column.

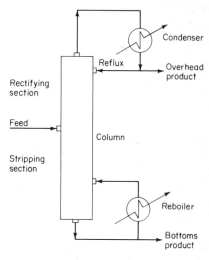

Fractionation of Binary Mixtures

In industrial practice, fractionation or fractional distillation refers to all column distillation operations including both rectification and stripping. In rectification, overhead product is fractionally distilled from a feed, either batchwise or continuously. In stripping, components of the feed are fractionally distilled from a bottoms product, usually continuously, in a column section below the feed.

In the purification of solvents by fractional distillation, a large number of processes and equipment arrangements are in use. This

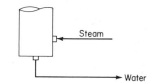

Direct steam injection where water is the bottom product

Fig. 2 Continuous fractional distillation.

results from the wide variation of relative volatilities, solubilities, and azeotropes existing among the many binary and multicomponent mixtures encountered. A few of these processes are quite common, and most problems can be solved by application of one or a combination of these. In many processes, distillation is supplemented by single or multistage extraction and decanting.

No Azeotropes Where no azeotropes are formed or azeotropes are acceptable as products, a continuous binary fractionation column (Fig. 2) is used. Where the bottoms product is or contains water, direct steam injection may be used in place of the reboiler. For small stills or for feed containing solids, a binary batch rectification column (Fig. 3) may be used. Examples of nonazeotroping mixtures are methanol-water, water-dimethyl-formamide, and hexane-toluene. A commonly accepted azeotrope is 95% ethanol.

Two-Phase Mixtures (Practically Insoluble) Here, an azeotrope is formed but the feed forms two liquid phases. When the solubility of one component in the other is less than the allowable contamination, the feed need only be decanted. Examples are toluene-water and heptane-water. For extreme purity requirements, these mixtures could be placed in the category below.

Partially Soluble Mixtures   Here, the allowable contamination is less than the solubility. If the azeotrope composition is well within the two-phase region, a combination of two stripping columns and a decanter may be used. This is shown in Fig. 4. Examples of such mixtures are ethyl acetate–water and butanol-water.

In the figure, the feed is assumed to be richer in water than the azeotrope and is shown entering the water column. If it is richer in solvent than the azeotrope, it is fed into the solvent column. No rectification sections are used. Rectification is not justified as long as a two-phase condensate is obtained, since purification by decantation is more efficient than by distillation. Dry solvent from the solvent column may be taken a few plates above the column bottom as a vapor side stream whenever high boilers or dirt may be present. A purge is then taken at the bottom of the solvent column.

Where only one component need be purer than the composition of the decanted layer, only one column should be used.

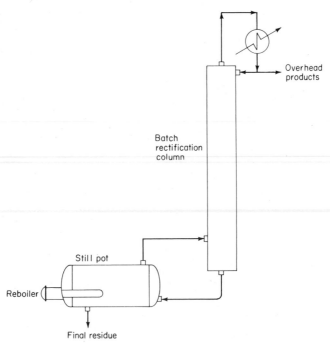

Fig. 3   Batch fractional distillation.

Separating Binary Single-Phase Azeotropes

Azeotrope compositions are a unique function of the pressure of distillation and of the components present. By changing either of these, the azeotrope composition may be shifted or eliminated and pure components obtained.

Pressure or Vacuum Distillation   The separation of MEK and water is illustrated in Fig. 5. The azeotrope contains 35% water at atmospheric pressure and 50% water at 7 kg/cm² (100 lb/in²). In this process, the azeotrope distilled off at atmospheric pressure is redistilled at 7 kg/cm² (100 lb/in²). Since the pressure azeotrope contains more water, all the water is removed with less than all the MEK, and dry MEK is produced. The pressure azeotrope is recycled to the feed.

Other examples of mixtures separated with one atmospheric and one column at another pressure are THF-water 7.6 kg/cm² (100 lb/in²), methanol–methyl ethyl ketone 7.6 kg/cm² (100 lb/in²), and methanol-acetone (200 torr).

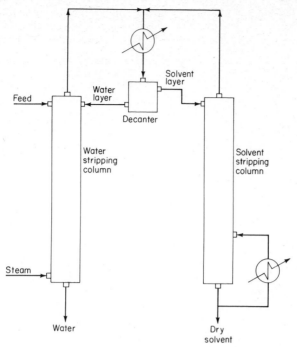

Fig. 4 Two stripping columns for partially soluble mixtures.

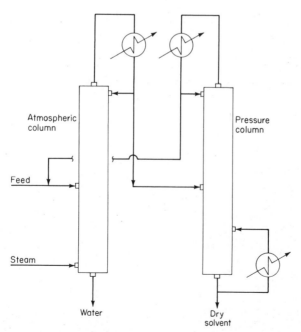

Fig. 5 Pressure distillation.

Azeotropic Distillation   In the most common case, azeotropic distillation involves adding a third component or entrainer, which forms a ternary azeotrope. The ternary azeotrope separates into two layers, one of which is enriched in one of the feed components. This layer is decanted while the other layer, containing most of the entrainer, is refluxed. Since the other feed component is taken from the bottom of the column, the entrainer must also be readily distilled from the bottoms product.

An example is the dehydration of ethanol, using benzene as entrainer. As shown in Fig. 6, 95% ethanol azeotrope from the binary column is fed to the azeotrope column. The water is taken overhead as part of the ternary azeotrope and is separated as the water layer in the decanter. Since it contains some benzene and ethanol, it is recycled to the binary column for recovery of the ethanol and benzene. Most of the benzene and ethanol are returned to the column in the benzene layer from the decanter. The stripping section of the azeotrope column can be operated so as to allow 2% benzene to exit with the anhydrous alcohol as denaturant.

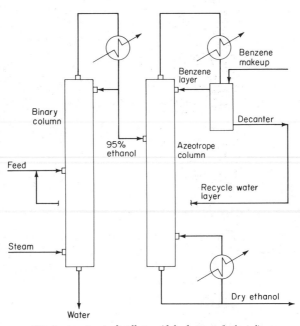

Fig. 6   Azeotropic distillation (dehydration of ethanol).

Extractive Distillation   In extractive distillation, a third component or solvent which does not azeotrope with either feed component is added to the column. It alters the relative volatility of the original pair, allowing one to distill overhead. It leaves the column with the bottoms product and is separated from it in a binary column.

A common example is acetone-methanol, shown in Fig. 7. Water as the solvent is added above the feed. It has a greater attraction for methanol, the more polar of the pair, and so the relative volatility of the methanol is decreased. This allows the acetone to distill overhead. The top column section serves to separate acetone from water, the middle section to extract methanol from the feed, and the bottom section to strip acetone out of the bottoms products.

Other examples are drying ethanol using ethylene glycol as solvent and drying THF using dimethyl formamide as solvent.

Salt Effect in Distillation   A salt added to an azeotrope will reduce the vapor pressure of the component in which it is more soluble. Thus an extractive distillation can be run in which a salt solution is the solvent. An example is the dehydration of ethanol using potassium acetate solution.[3] The advantage is that only a single distillation column need be used, although an evaporator is required. Figure 8 shows the system.

Liquid Extraction and Distillation Where two components of an azeotrope have different solubilities in a solvent, they can be separated by extraction. The solvent is chosen for high selectivity and so that it can be separated easily from each component by decantation or distillation.

An example is the methylene chloride–methanol azeotrope. The system is shown in Fig. 9. The methanol is extracted from the methylene chloride with water, after which the methanol and water are separated by binary fractionation. Methylene chloride and water are separated by decanting at the bottom of the extraction column. If necessary, the small amount of water dissolved in the methylene chloride product can be distilled off in a stripping column and recycled to the feed.

Another example is ethanol-toluene, where the solvent is also water. A comparison of azeotropic distillation, extractive distillation, and liquid extraction is given by Gerster.[4]

Chemical Action and Distillation Frequently the volatility of one component of an azeotrope can be reduced chemically. A common example is dehydration by chemical hydrate formation. As seen in Fig. 10, 95% THF is dehydrated by the use of solid sodium hydroxide. The water removed from the THF forms a 35 to 50% sodium hydroxide solution containing very little THF. That solution is insoluble in the THF and is removed as a liquid for evaporation or use elsewhere. The dried THF, containing about 0.5% water, may be used as is or further dried by distilling off the 95% azeotrope for recycle to the chemical drying column.

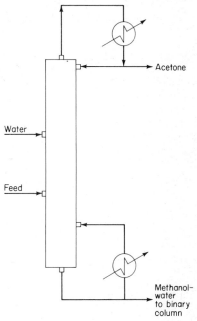

Fig. 7 Extractive distillation (separation of acetone and methanol).

Special Binary Mixtures

Acetic Acid–Water This pair does not azeotrope, but the relative volatility is so close to 1 that binary fractional distillation is usually not the most economical means of separation. Brown[5] compared the economics of nine separation processes which included fractional distillation, azeotropic distillation, extraction with low-boiling, and extraction with high-boiling solvents. The basis was recovery of 100 million lb per year of 99.5% acetic acid from solutions of 2 to 90% concentration. His correlation, reproduced in Fig. 11, showed that the most economical process varied with the concentration of the feed. Extraction processes gave the greatest return for 2 to 60% feeds and azeotropic distillation processes greater return at above 75% feed concentration. In no case was binary fractionation best.

Dilute Aqueous Solutions of High-Boiling Solvents Recovery of high-boiling solvents such as glycols, dimethyl formamide, and phenols from dilute solution in water is possible by binary distillation but not usually economical. Whether or not azeotropes are formed, a great deal of water must be vaporized in this process. Consequently, extraction with a low-boiling solvent and separation of the low- and high-boiling organics by distillation are commonly used. Various low-boiling chlorinated solvents, ketones, and hydrocarbon solvents are used. They are chosen for selectivity, low water solubility, safety, and compatibility with related processes. The process is shown in Fig. 12.

Multicomponent Mixtures

No Azeotropes Formed or Azeotropes Are Acceptable as Products In continuous fractionation, a maximum of two pure products can be withdrawn from a column, one

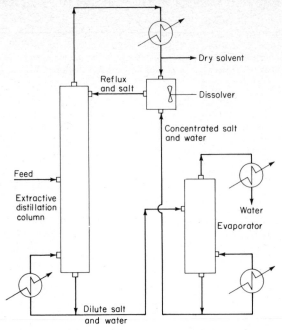

Fig. 8 Extractive distillation with salt.

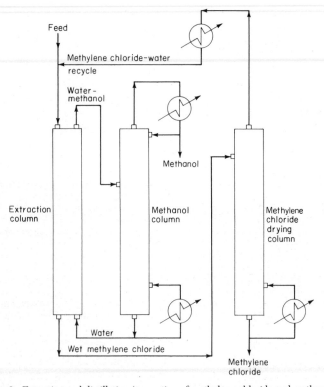

Fig. 9 Extraction and distillation (separation of methylene chloride and methanol).

product from each end. Therefore, in separating multicomponent mixtures, the number of columns in series must be one less than the number of components (see Fig. 13). For a large number of components it may be less expensive to use batch distillation since only one set of equipment and accessories is required for any number of products. Where batch times are at least 12 h, time losses during start-up and shutdown are not severe. Batch

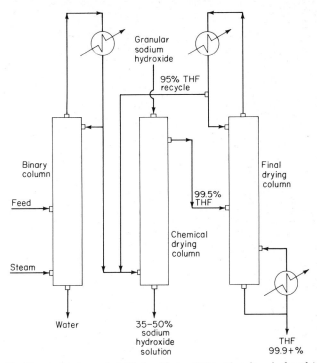

Fig. 10 Chemical drying and distillation (drying THF with sodium hydroxide).

stills of up to 10,000 gal capacity are quite common (see Fig. 11). Examples of such systems are THF-MEK-toluene, methanol-ethanol-propanol, and benzene-toluene-xylene.

Separating Azeotropes With multicomponent mixtures, several binary and multicomponent azeotropes frequently exist and two liquid phases are often possible. Problems with complete separations can be severe. In the worst cases, it may only be possible to take fractions by boiling-point range, reusing as much as possible and incinerating or selling the rest. Designing equipment for such separations is a real challenge to the distillation engineer. Often, sufficient data are not available for complete calculation so that the process must be developed in the laboratory or, at least, assumptions checked. Combinations of continuous and batch distillations, decanting, multistage extraction, and chemical treatment may often be combined in a successful process.

Of the many possible types of mixtures and recovery processes, a few common ones are described below.

a. One two-phase binary azeotrope and no ternary azeotropes are present. An example of such a system is methanol-toluene-water, and the process for separation is shown in Fig. 14.

Since the mutual solvent, methanol, is the lowest boiler and forms no azeotrope, it is distilled overhead. The toluene-water azeotrope is taken as bottoms, where it is decanted to yield pure toluene and water.

b. Two binary azeotropes and no ternary azeotropes are present. An example is MEK-toluene-water, shown in Fig. 15.

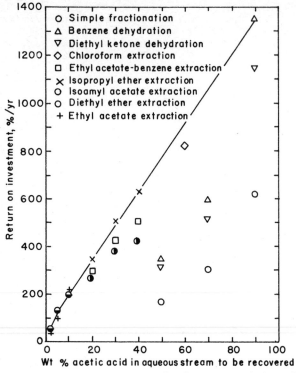

Fig. 11 Economics of acetic acid recovery. (*Reprinted with permission of* Chemical Engineering Progress.)

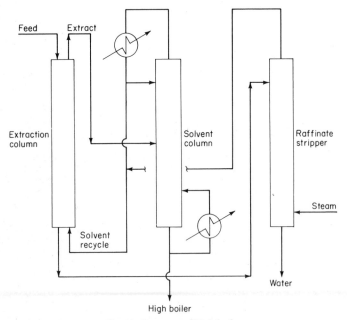

Fig. 12 Recovery of high boilers.

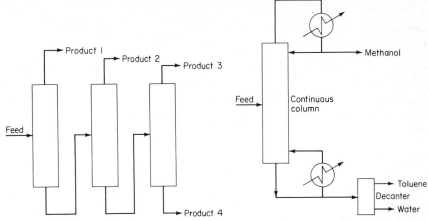

Fig. 13 Multicomponent mixtures.

Fig. 14 Three components—one azeotrope (separation of methanol, toluene, water).

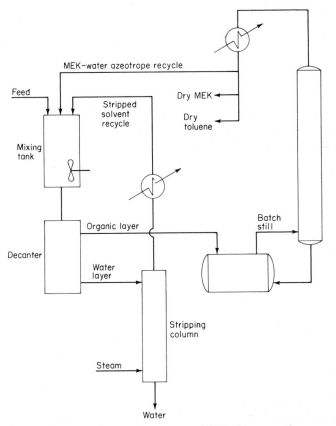

Fig. 15 Three components—two azeotropes (MEK, toluene, water).

This process combines single-stage extraction, continuous distillation, and batch distillation. Feed and recycle streams are mixed and decanted. Water layer from the decanter is stripped of its dissolved solvent in a continuous stripping column before passing to the sewer. Stripped solvent recycles to the mixing tank. The organic layer from the decanter, containing all the toluene and MEK plus some water, is sent to a batch still. Here, the solvent is dried by distilling overhead all the water as the MEK-water azeotrope and recycling it to the mixing tank. Pure dry MEK and toluene fractions are then taken off.

This process is useful in general where a large number of organic components are present as long as no organic azeotropes are formed.

c. Three binary azeotropes and/or a ternary azeotrope are present. An example is the separation of ethanol–ethyl acetate–water as shown in Fig. 16.

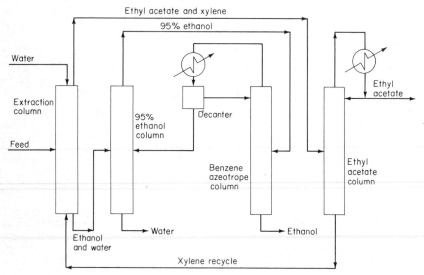

Fig. 16   Three components—three azeotropes (ethanol, ethyl acetate, water).

The process uses multistage fractional extraction, binary fractionation, and azeotropic distillation. Because of their differences in polarity, the two organics can be separated by partitioning between xylene and water. An added solvent is required because of the high mutual solubility of the three components. A high-boiling solvent, xylene, is selected so that a large amount can be circulated for extraction without its having to be distilled. Use of high-boiling solvent has the disadvantage that the small amount of water in the extract will distill over with the ethyl acetate. If this is a problem, a small drying column can be inserted before or after the ethyl acetate column. Alternatively, a low-boiling-point solvent such as methylene chloride can be used. The water-ethanol mixture leaving the extractor is dried using the conventional two-column benzene azeotropic distillation.

RECOVERY SPECIFICATIONS

Specifications for recovered products which can be met economically are:
   Water content of solvents, 0.05 and 0.5%
   Cross mixing of solvents, 0.5 to 5%
   Organic content of wastewater, 10 to 100 ppm
Usually, this means the recovered solvent will meet manufacturers' specifications for industrial-solvent grades.

RECOVERY EQUIPMENT

The unit operations involved in solvent recovery are often quite different from those found in the application of the solvent. Solvent-recovery equipment, therefore, is frequently installed as a packaged system, designed, skid-mounted, piped, and shipped to the user plant ready for start-up. This concept has the advantages of unit responsibility of one supplier and lower costs. Lower costs result since the equipment manufacturer is a specialist in the field and makes use of shop labor instead of the more expensive field labor for most of the erection and piping.

Condensers

Condensation of concentrated vapor mixtures by refrigeration is also carried out in packaged equipment systems. Units for recovery of gasoline vapors from tank vents in petroleum bulk terminals are commonly in use.

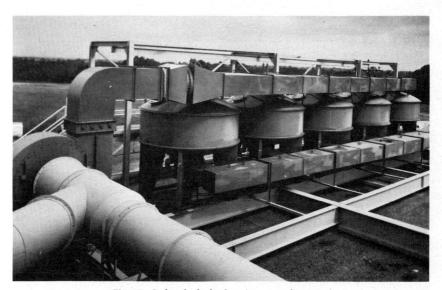

Fig. 17   Carbon-bed adsorber. *(Vic Manufacturing.)*

Absorbers

Towers for absorption of solvent vapors in oil or water are similar to towers for fractional distillation. Multistage contacting is used so as to carry out the recovery with a minimum amount of circulating absorbent. Ordinary wet scrubbers such as those used for dust removal are unsuitable, since they provide only one or two stages.

Adsorbers

A carbon-bed adsorber system is shown in Fig. 17. The carbon is installed in horizontal beds in the two horizontal tanks. Outlet valves and exhaust stacks are in the front of each tank. On the left is a cooling tower supplying water to the vapor condenser mounted between the tanks.

Vaporizers

Simple distillation for removal of up to 758 L/h (200 gal/h) of solvent from nonvolatile liquids and solids is carried out in small packaged vaporizers such as the one in Fig. 18.

These are heated indirectly with steam jackets or electrical heaters. Direct steam injection may be used with water immiscibles. Water is usually used for the condenser, although all electric units using the heat-pump principle are also manufactured in small sizes. Condensation by direct contact with chilled recovered solvent is often a feature of these units.

Fractionating Towers

Fractional-distillation towers containing packing or trays are arranged for continuous or batch operation. Single or multicolumn units together with reboilers, condensers, decanters, pumps, and instruments are packaged for capacities of up to 2250 kg/h (5000 lb/h) of recovered solvents. Associated extraction columns, where required, would also be included in the assembly. Figure 19 shows a typical multicolumn skid-mounted unit.

Corrosion

Organic solvents, whether wet or dry, can often be distilled in carbon-steel equipment without serious corrosion. However, organic and inorganic acids and salts may be present,

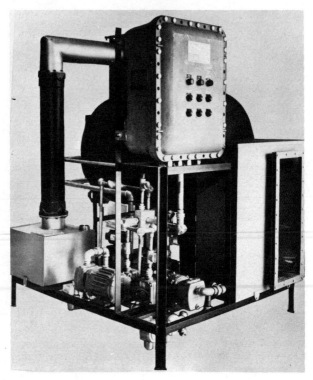

Fig. 18   Packaged vaporizer. *(Environmental Processing Systems, Inc.)*

and these corrode carbon steel rapidly at distillation temperatures. Esters and halogenated solvents in the presence of hot water or steam will hydrolyze to acids which can cause severe corrosion. Esters can normally be handled in stainless steel, but the mineral acid formed from halogenated solvents demands high-nickel alloys, glass, or resin-lined equipment.

Corrosion from trace impurities has wrecked many stills. For example, small quantities of acids, formed from decomposition of ketones in contact with activated carbon, can badly corrode subsequent condensers and stills. Inhibitors and catalysts from resin systems, trace contaminants from purchased solvent mixtures, and wash solvents can all find their way into recovery streams. A careful analysis of all streams, coupled with corrosion testing where indicated, makes good sense in selecting materials of construction.

Safety

The design and operation of solvent-recovery plants should follow the same safety practices as for chemical plants storing and handling flammable liquids in closed systems. Electrical requirements usually follow standards for Class I, Group D, Division 2 locations. Equipment normally operates at atmospheric pressure and is vented through direct connections to vented tanks and receivers. However, where steam and water at elevated pressures are connected to the equipment, they could, in the event of instrument failure or pluggage, raise the pressure in part or all of the equipment above the design pressure. Relief devices are therefore also used. They are located, for example, on the steam side of carbon beds, packed beds, and tray stacks. In addition, protection against cooling-water failure is always wise. A temperature sensor in a condenser outlet can be connected so as to sound an alarm and shut down steam in the event of undue temperature rise.

Fig. 19 Packaged multicolumn distillation unit.

ECONOMICS

Equipment Costs

Installed costs of carbon adsorption and distillation equipment have been presented in chart form.[6] These figures have been reproduced here with the costs updated to a Marshall and Stevens Process Industries Average Index of 451 (December 1975). In Fig. 20 the basic plant cost is found from the air rate and solvent rate. The basic plant consists of steel construction and comprises dust filter, air cooler, blower, adsorbers, condenser, decanter, automatic controls, and interconnecting piping. No building or collecting duct-work for the solvent-laden air is included. In Fig. 21, cost for a low-energy scrubbing system to preclean the air of particulates and high-boiling plasticizers may be calculated from the airflow rate and the point marked "scrubber." Also in this figure is a point marked "corrosion." Should a stainless-steel adsorber system be required, this point is used to calculate the cost to be added to the basic plant cost for stainless-steel or stainless-lined adsorbers and accessories.

Figure 21 also gives the cost of a fractional distillation plant based on solvent rate in gallons per hour. The cost read from the chart is approximately correct for a one-column packaged system, consisting of column, heat exchangers, pumps, piping, and instruments.

The total cost of the fractionating plant should be calculated by multiplying this figure by the number of distillation and extraction columns required. For stainless-steel construction, multiply the result by 1.5.

Operating Costs

The largest single operating cost is for steam. Steam usage per kilogram (pound) of solvent recovered may be estimated at 4 kg (lb) for the adsorber system and 2 kg (lb) for each column in the fractionating system. Cooling systems are sized to remove the heat input from the steam. Cooling-water circulating rates for a 10°C (20°F) rise through the condens-

Solvent recovery system

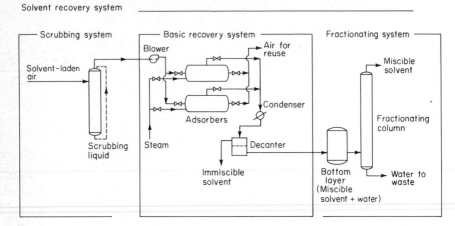

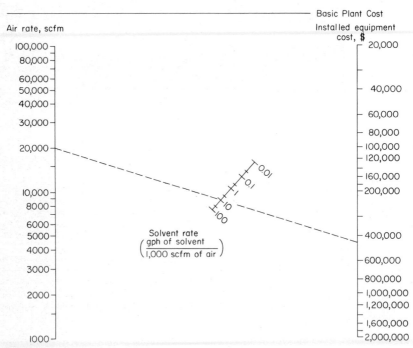

Fig. 20 Basic plant cost, solvent-recovery system.[6] (*Reprinted with permission from* Chemical Engineering.)

ers will be 0.379 L/min (0.1 gal/min) for each 0.45 kg (lb) per hour of steam input rate. For an automatic system, one operator per shift can handle all operations, including instrument monitoring, boiler and cooling-tower water treatment, and solvent-sample analysis. Variable and fixed overhead costs can be calculated in the usual way for fluids-processing plants.

Expected Recovery

In calculating solvent recovery, the recovery efficiency must be estimated. An adsorber system recovers greater than 95%. For the entire plant, however, losses from initial air

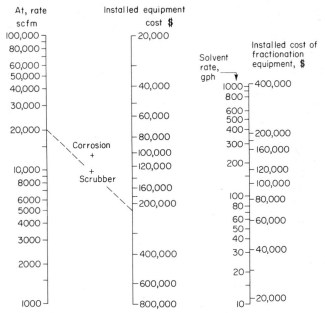

Fig. 21   Additional plant costs.[6] (Reprinted with permission from Chemical Engineering.)

collection, start-up, shutdown, occasional spills, and storage-tank vents reduce the overall annual recovery to about 75 to 80% of the solvent purchased. If recovery is less than this, improvement of the initial solvent-collection system may be warranted. If losses increase with time in an operating plant, reactivation of the carbon may be necessary.

REFERENCES

1. Hydrocarbon Pollutant Systems Study, EPA Contract EHSD 71-12, MSA Research Corporation, vol. 1, October 1972.
2. Flammability Characteristics of Combustible Gases and Vapors, U.S. Bur. Mines Bull. 627, 1965.
3. Furter, W. F., Salt Effect in Distillation: A Technical Review, Chem. Eng., June 1968, CE 173.
4. Gerster, J. A., Azeotropic and Extractive Distillation, Chem. Eng. Prog., 65(9), 43 (1969).
5. Brown, W. V., Economics of Recovering Acetic Acid, Chem. Eng. Prog., 59(10), 65 (1963).
6. Barnebey, H. L., and Davis, W. L., Costs of Solvent Recovery Systems, Chem. Eng., 65(26), 51 (1958).

Design of Packed Columns

JOHN S. ECKERT, P.E. *Consulting Engineer; Retired Director of Chemical Engineering, Chemical Process Products Division of the Norton Company; Formerly Director of Engineering R&D, U.S. Stoneware Company; Registered Professional Engineer in Ohio and in Texas; Certified Corrosion Engineer; Fellow, The American Institute of Chemical Engineers; Member, American Society of Mechanical Engineers, National Association of Corrosion Engineers, American Association for the Advancement of Science, and National Society of Professional Engineers.*

FUNDAMENTALS OF PACKED-COLUMN DESIGN

Types of Dumped Packings in Common Use in Distillation Towers

Raschig Ring The *Raschig ring* (Fig. 1*a*), the oldest of the high-efficiency tower packing shapes, was patented by Dr. Raschig of Germany in 1907. It is a hollow cylinder whose height is equal to its diameter.

Materials Available. These rings are commonly available in ceramics and metals; however, on rare occasions they may be supplied in plastic. The usual method of manufacture is to cut them from tubes or pipes, but in the case of metal they may also be rolled from metal strip. Table 1 gives the principal physical properties of this packing. At one time a great number of packed towers were filled with these rings or shapes closely

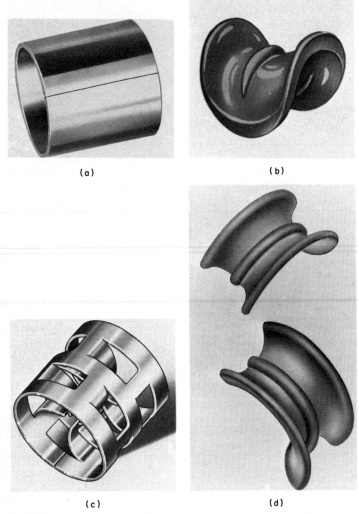

(a)   (b)

(c)   (d)

Fig. 1 (*a*) Raschig ring; (*b*) Berl saddle; (*c*) pall ring; (*d*) Intalox saddle. (*Norton Company.*)

related to them; but since the advent of more advanced packings, very few of them are used. Of the packings commonly used, the Raschig ring, because of its shape, is the most rugged and is recommended where a severe vibration or bumping condition may occur. It will have slightly poorer efficiency then either the Pall ring or Intalox saddle at pressure drops just below flooding (Fig. 5) and much poorer efficiency at lower pressure drops. Because of poor turndown and lack of capacity (Table 2), this packing should be used only for special service in distillation operations.

There have been some modifications of the Raschig ring which make it very useful for support structure on special occasions where corrosion makes material which can be fabricated for support plates very expensive. When used for this purpose, these rings are

TABLE 1 Tower Packing

Packing	Size (in)	Size (mm)	Pieces (ft³)	Pieces (m³)	Area (ft²/ft³)	Area (m²/m³)	Weight (lb/ft³)	Weight (kg/m³)	F_p (E*)	F_p (M†)	% voids	Wall (in)	Wall (mm)
Intalox ceramic	¼	6.3	117,500	8,149,770	300	984	54	865	725	220	75		
	⅜	9.5	49,800	1,758,744	240	787	50	801	330	100	76		
	½	12.7	20,700	731,028	190	623	45	721	200	61	78		
	¾	19.0	6,500	229,555	102	335	44	705	145	44	77		
	1	25.4	2,385	84,227	78	256	44	705	98	30	77		
	1½	38.1	709	25,039	59	194	42	673	52	16	80		
	2	50.8	265	9,358	36	118	42	673	40	12	79		
	3	76.2	53	1,872	28	92	37	593	22	7	80		
Super plastic	No. 1		1,200	42,378	77	253	39	625	60	18	79		
	No. 2		190	6,710	32	105	37	593	30	9	81		
	1	25.4	1,650	58,270	63	207	5¾	92	33	10	91		
	2	50.8	190	6,710	33	108	4¼	68	21	6	93		
	3	76.2	41	1,448	27	89	3½	56	16	5	94		
Pall-ring metal	⅝	15.9	5,950	210,124	104	341	37	593	70	21	93	0.018	0.46
	1	25.4	1,400	49,441	63	207	30	481	48	15	94	0.024	0.61
	1½	41.3	375	13,243	39	128	24	384	28	9	95	0.030	0.76
	2	52.4	170	6,004	31	102	22	352	20	6	96	0.036	0.91
	3½	88.9	33	1,165	20	66	17	272	16	5	97	0.048	1.22
Plastic	⅝	15.9	6,050	213,656	104	341	7	112	97	30	87		
	1	25.4	1,440	50,894	63	207	5½	88	52	15	90		
	1½	38.1	390	13,774	39	128	4¾	76	40	10	91		
	2	50.8	180	6,357	31	102	4¼	68	25	8	92		
	3½	88.9	33	1,165	26	85	4	64	16	5	92		
Berl saddles ceramic	¼	6.3	107,000	3,778,705	274	899	56	897	900	275	64		
	½	12.7	16,700	589,760	142	466	54	865	240	73	66		
	¾	19.0	4,950	174,809	87	285	49	785	170	52	71		
	1	25.4	2,180	76,987	76	249	45	721	110	34	73		
	1½	38.1	645	22,778	46	151	40	641	65	20	74		
	2	50.8	250	8,829	32	105	39	625	45	14	75		

*E = English units
†M = Metric units

TABLE 1 Tower Packing (Continued)

Packing	Size (in)	Size (mm)	Pieces (ft³)	Pieces (m³)	Area (ft²/ft³)	Area (m²/m³)	Weight (lb/ft³)	Weight (kg/m³)	F_p E*	F_p M†	% voids	Wall (in)	Wall (mm)
Raschig ring ceramic													
	¼	6.3	85,600	3,023,053	217	712	64	1025	1600	490	62	¹⁄₁₆	1.59
	⅜	9.5	24,700	872,280	147	482	65	1041	1000	305	67	¹⁄₁₆	1.59
	½	12.7	10,700	377,870	112	367	57	913	580	175	64	³⁄₃₂	2.38
	⅝	15.9	5,670	200,236	94	308	54	865	380	115	67	³⁄₃₂	2.38
	¾	19.0	3,090	109,123	74	243	50	801	255	78	72	³⁄₃₂	2.38
	1	25.4	1,350	47,675	58	190	44	705	160	47	74	⅛	3.18
	1¼	31.7	670	23,661	45	148	47	753	125	38	71	³⁄₁₆	4.76
	1½	38.1	387	13,667	37	121	42	673	95	29	73	³⁄₁₆	4.76
	2	50.8	164	5,792	28	92	42	673	65	20	74	¼	6.35
	3	76.2	50	1,766	19	62	37	593	36	11	78	⁵⁄₁₆	7.94
	4	101.6	20	706	14	45	37	593	30	9	80	⅜	9.52
Metal													
	¼	6.3	88,000	3,107,720	224	735	133	2130	700	215	72	¹⁄₃₂	0.79
	⅜	9.5	27,000	953,505	161	528	94	1506	390	120	81	¹⁄₃₂	0.79
	½	12.7	11,400	402,591	122	400	75	1201	300	91	85	¹⁄₃₂	0.79
	½	12.7	10,900	388,933	111	364	132	2114	410	125	73	¹⁄₁₆	1.59
	⅝	15.9	6,190	218,600	103	338	62	993	170	52	87	¹⁄₃₂	0.79
	¾	19.0	3,340	117,952	81	266	52	833	155	47	89	¹⁄₃₂	0.79
	¾	19.0	3,140	110,889	75	246	94	1506	220	67	80	¹⁄₁₆	1.59
	1	25.4	1,410	49,794	62	203	39	625	115	35	92	¹⁄₃₂	0.79
	1	25.4	1,310	46,263	56	184	71	1136	137	42	86	¹⁄₁₆	1.59
	1¼	31.7	725	25,603	48	157	62	993	110	34	87	¹⁄₁₆	1.59
	1½	38.1	390	13,776	39	128	48	769	83	25	90	¹⁄₁₆	1.59
	2	50.8	170	6,004	29	95	36	577	57	19	92	¹⁄₁₆	1.59
	3	76.2	51	1,801	20	66	25	400	32	10	95	¹⁄₁₆	1.59
HyPak metal	No. 1	30	850	30,018	54	177			42	13	96	0.0179	0.45
	No. 2	60	110	3,885	29	95			18	5.5	97	0.0239	0.61
	No. 3	90	30	1,059	18	59			15	4.5	98	0.0359	0.91

*E = English units
†M = Metric units

TABLE 2   Packing Factors[a] (Wet and Dump Packed)

Type of packing	Material	Nominal packing size, in										
		¼	⅜	½	⅝	¾	1	1¼	1½	2	3	3½
Super Intalox	Ceramic						60			30		
Super Intalox	Plastic						33			21	16	
Intalox saddles	Ceramic	725	330	200		145	98		52	40	22	
Hy-Pak rings	Metal						42			18	15	
Pall rings	Plastic				97		52		40	25		16
Pall rings	Metal				70		48		28	20		16
Berl saddles	Ceramic	900[b]		240[b]		170[c]	110[c]		65[c]	45[b]		
Raschig rings	Ceramic	1600[b,d]	1000[b,d]	580[e]	380[e]	255[e]	155[f]	125[b,a]	95[g]	65[h]	37[b,i]	
Raschig rings, 1/32 in wall	Metal	700[b]	390[b]	300[b]	170	155	115[b]					
Raschig rings, 1/16 in wall	Metal			410	290	220	137	110[b]	83	57	32[b]	

[a] $F \approx a/\epsilon^3$ obtained in 16-in and 30-in-ID tower.
[b] Extrapolated.
[c] Data by Leva.
[d] 1/16-in wall.
[e] 3/32-in wall.
[f] 1/8-in wall.
[g] 3/16-in wall.
[h] 1/4-in wall.
[i] 3/8-in wall.

stacked (Fig. 2) and may have one or more partitions in the interior (Fig. 3) to prevent dumped packings from falling through them. This type of construction is much more common in absorbers and regenerators than in distillation towers.

When the Raschig ring is diametrically parted in the interior with a single wall, it is commonly referred to as a Lessing ring or as a cross-partition ring if more than one

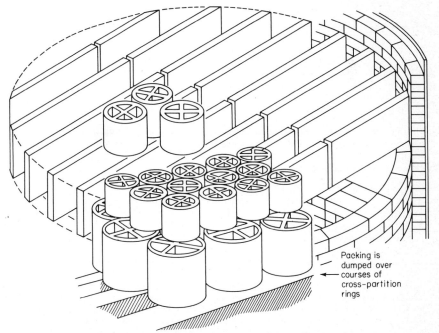

Packing is dumped over courses of cross-partition rings

Fig. 2 Stacked packing used as support. (*Norton Company.*)

partition wall is present. The Lessing type is available in metal or ceramic, while the partition type is available only in ceramic. Sometimes the partition wall in one of these rings may be spiraled to give a desired increase in efficiency, but this has become a rarity in recent years. Because of their special-purpose usage, partition rings are never offered smaller than 3 × 3 in in size. Some designers still use the Lessing ring as a dumped packing; so it may occasionally be found in sizes down through 1 in, but its use in this field may be said to be very minimal. The relative capacities of these packings may be found from the packing-factor table (Table 2) and using the relation

$$\frac{G_1}{G_2} = \left(\frac{F_2}{F_1}\right)^{0.5}$$

which is derived from the design chart (Fig. 4).

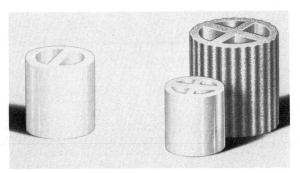

Fig. 3 Cross-partition rings. *(Norton Company.)*

Berl Saddle The *Berl saddle* (Fig. 1*b*) is the second major packing invention which has lasted through the years and like the Raschig ring has been gradually disappearing in recent years. Because of its shape, this packing can be easily manufactured only from ceramic.

Materials Available. Though there have been occasional offerings of this shape in both plastic and metal, the advantages to be realized over other shapes have not been sufficient to create a really viable market. This packing has properties similar to those of the Intalox® shape; so a detailed description particularly as to performance will not be given. It may be assumed that what is said about the Intalox saddle will also apply to the Berl saddle.

Intalox Saddle The *Intalox saddle* (Fig. 1*d*) may be regarded as the modern counterpart of the Berl saddle. Because it is a surface which may be generated by revolution, it is much more economical to produce in ceramic than the Berl saddle, and because of this general form it is also much more easily molded in plastic.

Materials Available. The Intalox saddle is available in both plastic and ceramic and probably accounts for over 90% of the market in ceramic packings at the present time. Its turndown property is not as good as that of the pall ring (Fig. 5) but is much superior to that of the Raschig ring.

Intalox saddles are commonly recommended for distillations when corrosive or temperature conditions are very severe such as occurs in aqueous or organic systems where water may be present. They should be considered in any environment where the corrosive condition is such that it can only be handled by expensive metal alloys and temperature conditions will not permit the use of plastics.

The plastic design of the Intalox saddle is finding increased use in low-temperature installations where an aqueous phase is present. Care should be exercised here in that some organic materials may attack the plastic where an anhydrous condition exists at some location in the tower or where severe oxidizing conditions exist.

Pall Rings *Pall rings* (Fig. 1*c*) represent a major improvement over the Raschig ring and in their old form or a newer form are the principal packing to be used in separations by distillation. This packing was invented in 1950 when Dr. Pfanmueller, knowing that the Raschig ring had a poor turndown performance largely because the interior of the ring was not actively engaging in mass transfer at low pressure drops, simply opened trapdoors in the side of this ring to make the interior available for mass transfer at low pressure drops. Thus he invented a packing that not only had good turndown properties but was also of greatly increased capacity. This packing not only has superior operating properties as compared with the Raschig ring, but it can also be manufactured somewhat more economically largely because it is not ordinarily offered in such heavy wall thicknesses as the Raschig ring.

Materials Available. Pall rings are available in metal and plastic. In Europe they have sometimes been offered in ceramic, but the demand for them has been quite low largely because the ceramic Intalox saddle is much more economical and somewhat more efficient.

The metal pall ring is the one commonly used in distillation in that it can be fabricated from just about any metal or alloy whose elastic limit is below 50,000 lb/in². The plastic version of the pall ring is sometimes used, but for distillation operations the plastic Intalox saddle is usually preferred because of better operating performance and the fact that it is offered in a wider range of plastic materials. The plastic pall ring is subject to about the same limitations physically and chemically as the Intalox saddle where the same plastic material is used.

Hy-Pak   In recent years an improved pall ring has been offered to the market under the trade name of Hy-Pak®. Basically this packing has been circumferentially corrugated to give it greater rigidity than the pall ring and the number of trapdoors or fingers has been doubled to increase its efficiency. This packing is sold in numbered sizes rather than in

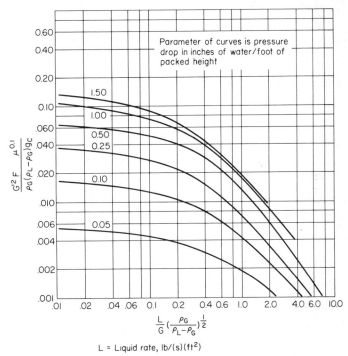

L = Liquid rate, lb/(s)(ft²)

G' = Gas rate, lb/(s)(ft²)

ρ_L = Liquid density, lb/ft³

ρ_G = Gas density, lb/ft³

F = Packing factor

μ = Viscosity of liquid, centipoise

g_c = Gravitational constant = 32.2

Fig. 4 Generalized pressure-drop correlation.

inch sizes because it has been set up in anticipation of the metric system; i.e., the No. 1 size is 30 mm, No. 2 is 60 mm, and No. 3 is 90 mm. Though these sizes are somewhat larger than the corresponding inch sizes, the increase in efficiency is such that they may be used interchangeably as far as efficiency goes. It will be noted from the table of packing

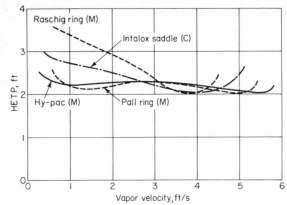

Fig. 5 Typical distillation performance of 2-in packings (octane-toluene, 3.75 IR.R, 740 mmHg absolute pressure).

factors (Table 2) that the capacity is somewhat greater than the corresponding inch size of the pall ring.

Materials Available. Up to the present writing, this packing is available only in metal and is limited as to metal or alloy to metals whose elastic limit is 3150 kg/cm² (45,000 lb/in²) or less, the slightly lower elastic limit being the result of lighter metal gage. Because of the increased rigidity and lesser metal thickness, this packing can be fabricated and sold for a lower price than the pall ring and is now rapidly becoming the accepted metal tower packing for distillations.

The foregoing discussion of types of packing is not intended to be all-inclusive. Many other types are available many of which are modifications to or improvements upon the types discussed. However, those listed are the ones most used at present.

Determining Packed Depth Required

Two general methods are used in calculating the height of packing required to effect a given separation.

Theoretical Plates   The more common and accepted method is that of the number of theoretical plates required.

Transfer Units   The more academic approach is to calculate the number of transfer units required, since a theoretical plate is somewhat a misnomer.

Theoretical Equilibrium Stage   The more proper term should be theoretical equilibrium stage, because this is the method used in arriving at this number, whereas a theoretical plate refers to a piece of equipment that does not really exist in that plates or trays tend to vary in efficiency, departing from 100% of theoretical depending on load, design, and possibly physical properties of the system. The theoretical-equilibrium-stage technique will be used here because at the present time most distillation separations are calculated with a computer on an equilibrium-stage basis using the fact that the enthalpy of the system will be essentially constant from the bottom to the top of the column (some small losses may occur as a result of heat radiation from the wall of the column).

Gas-Liquid Ratio Using Enthalpy. Knowing the enthalpy of the system, the designer can calculate the amount of gas and liquid at each equilibrium stage. This is important in that many separations such as acetone and water or methanol and water will show a considerable variation in reflux ratio between the top and bottom of the column because of the large difference in heat of vaporization of the water and the organic phase.

Dynamic-Equilibrium Data. It becomes very important to have the dynamic equilibrium (almost instantaneous) between the liquid and vapor phases of the system. This may be determined using the circulating-vapor technique (Fig. 6). Published vapor-liquid data will ordinarily be satisfactory; however, in those systems where the relative volatility is going to change considerably over the range of liquid composition desired, it is well worthwhile to run a check, particularly in the region between 95 and 100% of the light-boiling component, because in this region the relative volatility often approaches 1, requiring a very high reflux ratio to effect any further separation or purification.

Height Equivalent of a Theoretical Equilibrium Stage, or HETP. A great amount of experimental work[5,6,8] has proved that distillations with packed beds have efficiencies independent of physical properties of the commonly known systems. A 25-mm (1-in) pall ring will give a theoretical equilibrium stage (or HETP) for each 0.3 m (ft) of packed depth, a 38-mm (1½-in) pall ring will give 0.45 m (1½ ft) and a 50-mm (2-in) pall ring will give 0.9 m (2 ft) per theoretical equilibrium stage. These conclusions are based on results from very sophisticated laboratory equipment so the designer must consider the plant equipment will not have as perfect distribution[5] as that used in the laboratory. Therefore,

for design purposes a 30 to 50% factor of safety should be used at this point in the design. This means that the 25-mm (1-in) pall ring should be expected to practically develop a theoretical equilibrium stage or a theoretical plate for each 0.39 to 0.45 m (1.3 to 1.5 ft) of packed depth, a 38-mm (1½-in) pall ring 0.57 to 0.675 m (1.9 to 2.25 ft), and a 50-mm (2-in) pall ring 0.75 to 0.9 m (2.5 to 3.0 ft). If the reflux ratio required is high and a very good distributor has not been specified, it will be necessary to use greater depths of packing per theoretical plate.

The stripping section of the column quite often has a much higher liquid than gas rate; so the liquid may not be in complete turbulence in this zone. It is for this reason that the stripping section of the column is designed for the maximum amount of packing per theoretical plate or has turbulence equipment installed.

Sizing Tower for Required Diameter

Modified Sherwood Correlation The preferred method for sizing the tower descends from the original work of Sherwood, Shipley, and Holloway.[1] There have been considerable modifications through the years, but the basic variables controlling the size of bed required to effect a given separation at a given rate were established at this time. The first modification was by Lobo et al.[2] who, when reviewing the literature, determined that a/ϵ^3 (the area in square feet per cubic foot of packing divided by the cube of the voids) would not prop-

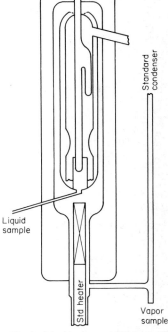

Fig. 6 Liquid-vapor equilibrium still.

erly define the gas-liquid handling capacity of a packed bed and proposed the use of an empirical number called a *packing factor*.

Packing Factor. Later, Leva[3] introduced parameters of constant pressure drop, and then Eckert[4] changed the packing factor of Lobo based on flooding to an averaged figure based on the operating range of packings. This was made possible after Eckert[5] determined that there were no lower and upper loading breaks in the pressure-drop curve (Fig. 7) of a packed bed but that the whole family of curves were smooth and a function of the gas loading on the packed bed depending initially on approximately the square of the gas mass velocity but increasing with the liquid holdup in the bed smoothly and progressively.

Further refinements were introduced by Eckert to provide for ambient liquid density[6] (the original work of Sherwood et al. was based on gases of very low density). Leva[7] had introduced a further correction for liquid density which later work showed to be inaccurate; so the factor was dropped.

Effect of Viscosity. It was found by Eckert[8] that the viscosity $(u^{0.2})$ was an excessive correction for liquids of low viscosity such as are ordinarily encountered in normal distillation and absorption practice; so this viscosity factor was reduced to $u^{0.1}$ for ordinary applications (when the packing factor rises, i.e., above 140-cP visocity, the exponent moves up to 0.2 and at very high packing factors, i.e., excess of 200 cP to as much as 0.3.)[8] With this background, the preferred design correlation (Fig. 4) has come to be the most accepted approach to the sizing of a tower for required area. The packing factors in Table 2 are those determined by the method of Eckert and are commonly accepted as being realistic for commercially available packings.

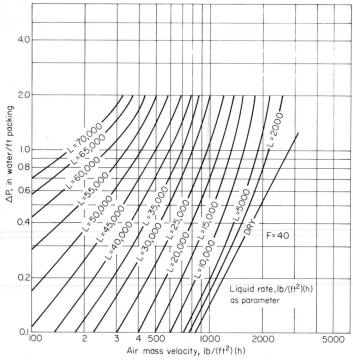

Fig. 7 Pressure drop vs. gas rate, 2-in Intalox saddles (porcelain). Column diameter = 30 in; packing height = 10 ft. (*Data of Eckert, Foote, and Walter.*)

Flooding. All dumped packings behave in a similar manner when operated in counter-current flow. At zero liquid rate, the pressure drop across the bed increases as approximately the square of the gas mass velocity (Fig. 7). When the bed is irrigated with liquid, the resistance to gas flow will increase faster than the square of the gas mass velocity owing to the reduction of void space in the bed caused by holdup of the liquid which increases as the gas rate increases. This holdup of liquid will continue to increase with the gas rate until it becomes about 62.5 mm (2½ in) per 0.3 m (ft) of packed depth, at which point the gas phase becomes discontinuous through the voids in the packing and will start to bubble through the liquid. Any further increase in gas rate will move the liquid up to the top of the bed, causing a flooded condition in and at the top surface of the bed. It should be noted from Fig. 7 that there are no breaks in the pressure-drop curves at the various liquid-rate parameters; in other words, there are no pressure-drop regions which could properly be described as the lower and upper loading regions of the bed, but the liquid holdup starts to increase as soon as gas is started through the bed and increases smoothly with increasing gas rate. The pressure drop at which the bed will flood is dependent not only upon the gas mass rate but also on the gas density, the liquid density, and the tendency of the liquid to foam.

Effect of Foam. Our design correlation takes all these factors into consideration except the foaming characteristic, which may be handled in either of two ways.

Empirical Correction. First, the usual technique involving a prior knowledge of the system is to make an empirical correction for foaming tendency; i.e., if the system is known to be a foamer, the capacity of the packing must be adjusted to take this foam into consideration (Fig. 8). A very severe foaming problem may be compensated for by designing to as low as 6.25 mm (0.25 in) pressure drop per 0.3 m (ft) of packed depth instead of the usual 18.75 to 25 (0.75 to 1.00). This will downgrade the gas-handling capacity of the bed from about 90% of ultimate to about 40% of ultimate capacity of the bed.

Quantitative Correction. The second method involves more time and expense but is also much more accurate. A 50-mm (2-in) column is packed with 6.25-mm (¼-in) Intalox

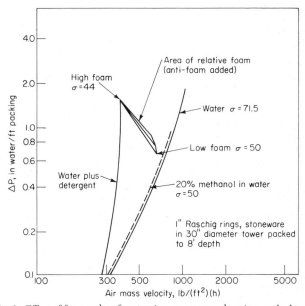

Fig. 8   Effect of foam and surface tension on pressure drop in a packed tower.

saddles or any other small packing and is calibrated with a loading of nitrogen gas and distilled water over a range of 12.5 mm (0.5 in) pressure drop per 0.3 m (ft) of packed depth to 25 mm (1.0 in) pressure drop per 0.3 m (ft) of packed depth, and from this a packing factor (Table 2) is calculated using the correlation (Fig. 4). This same column is then run under similar conditions with the process stream using process conditions, and a new packing factor is calculated. Then the design parameter of 1 in pressure drop is used and is downgraded using the following equation:

$$G_2 = G_1 \left(\frac{F_2}{F_1} \right)^{0.5}$$

When using this technique for designing distillation columns, it is usually advisable to check process conditions at the top as well as the bottom because the foaming conditions of some systems will vary considerably with concentration of individual components. One of the great advantages of using this latter technique is that when a foaming condition is found which is characteristic of a certain concentration range, it can be handled by increasing packing size in that zone of the column where the condition will prevail rather than downgrading the capacity of the entire column.

The first step in the design method is to determine the liquid-gas ratio of the process stream in the tower. For distillation operations, this becomes the desired operating internal reflux ratio for the rectification section of a given bed and the boil-up ratio for the stripping section of a bed. The liquid and gas density may be calculated (see the example below) and the value of the set of variables on the abscissa may be determined as being independent of the amount of the process stream.

Allowable Pressure Drop   Most distillations are relatively free from foaming. For most atmospheric or pressure separations the bed can be designed to operate at somewhere between 18.75 and 25 mm (¾ and 1 in) of pressure drop [18.75 mm (¾ in) for light liquids and 25 mm (1 in) for heavy liquids], which will be about 90% of the ultimate capacity of the bed (see Fig. 9). Reduced pressure distillations generate very high gas

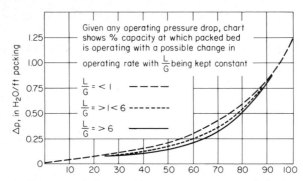

Fig. 9   Range of permissible operating rate with a packed bed.

velocities; so as the operating pressure becomes lower the maximum safe upper design limit for pressure drop becomes lower. High gas velocities generate mist because of mechanical shredding of the liquid surface. In other words, the work the gas is doing in the liquid is a function of $w = mv^2$, and under high vacuum where the gas density is very low, the velocity becomes quite high to generate a given pressure drop. Normally the allowable pressure drop in vacuum distillation design is determined more by a desired lower bottom of column temperature than an approach to flooding; however, the designer should be aware that at reasonably high vacuums (absolute pressure of 10 mm Hg or less) the column may behave transiently at a pressure drop as low as 0.2 in of water per foot of packed depth.

Other design methods have been proposed from time to time, but none of these has received sufficiently wide acceptance to be considered at present.

The packing factor (Table 2) which has been arrived at from the pressure-drop characteristics of the packing (Fig. 7) is an empirical number arrived at by calculation based on the average of the performance of a packing at 12.5, 25, and 37.5 mm (0.5, 1.0, and 1.5 in.) pressure drop per foot of packed depth. As may be seen in Table 2, these factors have been determined for all the commonly used packings. Because the packing factor is based on a range of pressure-drop parameters, it is subject to some variation. This variation is greater for the Berl saddle and the Raschig ring than for the Intalox saddle and the pall ring. The main reason for this is that the parameters of constant pressure drop (Fig. 4) have been located based on the operating characteristics of the latter two packings, rather than the other two, largely because the latter two shapes are those most commonly used in distillation columns. Distillation operations usually do not involve extremely high liquid-gas ratios. This variation of packing factor is therefore not very important for this type of operation except that if one of these extremes should be involved, it may be a good idea to recalculate the packing factor from an original pressure-drop chart such as Fig. 7 to be sure that the column will operate under the pressure-drop characteristics desired.

The viscosity exponent (Fig. 4) will increase with viscosity; however, only in exceedingly rare cases will a distillation be performed where the viscosity will be more than 6 cP, which is well within the range of exponent value $\mu^{0.1}$.

The tower must always be designed to handle the maximum expected operating condition. Because of its continuous nature, a fractionator will be subject to less variation in operating capacity than a batch still where usually a maximum reflux ratio and gas-handling capacity will be required at the fraction "cuts" and just a maximum gas-handling capacity during the removal of a fraction or a cut from the batch. Also, in batch distillations usually a much higher gas-handling capacity will be required at the beginning of the batch than at the end. For the above reasons, batch still columns tend to be larger than fractionators for the same 24-h throughput.

TYPICAL DESIGN

A typical fractionation design would be as follows: A feed stream consisting of 1895 L/min (500 gal/min) of plant wastes containing 10% methanol and 90% water is to be processed to recover the methanol at 95% purity or, better, leave a bottoms product of water containing not over 0.5% methanol. What size fractionator will be required if the maximum feed rate will reach 2274 L/min (600 gal/min) for short periods of ½ h and minimum rates of 948 l/min (250 gal/min) for periods up to 16 hr?

First, the column will have to be designed to handle the 2274 L/min (600 gal/min) rate because it will occur for a period of ½ h, which is more than a transient overload. Because this overload will exist for only ½ h and no foaming problem has been specified, it will be best to design the fractionator at this 2274 L/min (600 gal/min) load for a maximum pressure drop of 25 mm (1 in) H_2O per 0.3 m (ft) of packed depth. The turndown will be only a little over 50%, which is well within the range of the pall ring (Fig. 5c).

See Fig. 10 for vapor-liquid data on the system. To minimize the size of the fractionator, it will be assumed that both the feed and the reflux will enter the column at their bubble points.

With a feed consisting of 10% methanol, the density of the liquid will be 0.975 kg/L (8.21 lb/gal) at 60°F. The fractionator will have to be designed to handle the peak load of 2274 L/min (600 gal/min) of

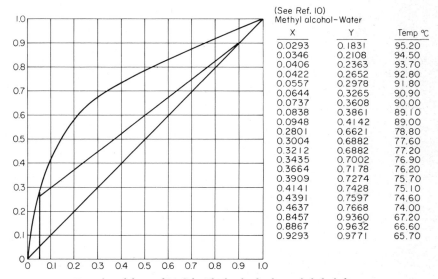

(See Ref. 10)
Methyl alcohol–Water

X	Y	Temp °C
0.0293	0.1831	95.20
0.0346	0.2108	94.50
0.0406	0.2363	93.70
0.0422	0.2652	92.80
0.0557	0.2978	91.80
0.0644	0.3265	90.90
0.0737	0.3608	90.00
0.0838	0.3861	89.10
0.0948	0.4142	89.00
0.2801	0.6621	78.80
0.3004	0.6882	77.60
0.3212	0.6882	77.20
0.3435	0.7002	76.90
0.3664	0.7178	76.20
0.3909	0.7274	75.70
0.4141	0.7428	75.10
0.4391	0.7597	74.60
0.4637	0.7668	74.00
0.8457	0.9360	67.20
0.8867	0.9632	66.60
0.9293	0.9771	65.70

Fig. 10 Vapor-liquid data and McCabe-Thiele plot for the methyl alcohol–water system.

solution in that these peaks have a duration of ½ h. To handle the peak load of 2274 L/min (600 gal/min) of solution:

$$2274 \times 0.975 \times 60 = 133{,}029 \text{ kg/h of solution} = 133{,}100 \text{ kg/h}$$
$$(600 \times 8.21 \times 60 = 295{,}560 \text{ lb/h of solution} = 296{,}000 \text{ lb/h})$$

The conditions of the feed will be
In metric units:

$$CH_3OH: 133{,}100 \times 0.10 = 13{,}310 \text{ kg/h} \div 32.04 = \frac{416 \text{ kg·mol/h}}{7064 \text{ kg·mol/h}} = 0.059 \text{ mole fraction}$$

$$H_2O: \quad 133{,}100 \times 0.90 = 119{,}790 \text{ kg/h} \div 18.02 = \frac{6649 \text{ kg·mol/h}}{7064 \text{ kg·mol/h}} = 0.941 \text{ mole fraction}$$

$$\text{Mol wt of feed} = 133{,}100 \div 7064 = 18.84$$

In English units:
$$CH_3HO: 296{,}000 \times 0.10 = 29{,}600 \text{ lb/h} \div 32.04 = 924 \text{ lb·mol/h} = 0.059 \text{ mole fraction}$$

$$H_2O: 296,000 \times 0.90 = 266,400 \text{ lb/h} \div 18.02 = \frac{14,784 \text{ lb} \cdot \text{mol/h}}{15,708 \text{ lb} \cdot \text{mol/h}} = 0.941 \text{ mole fraction}$$

Mol wt of feed $= 296,000 \div 15,708 = 18.84$
Boiling point of feed solution from Fig. 10 $= 91°C$
Vapor above feed 0.059 mole fraction, CH_3OH will contain 0.3 mole fraction CH_3OH
Mol wt of vapor $= 0.3 \times 32.04 + 0.7 \times 18.02 = 22.23$

The *overhead vapor* at 95% methanol will be as follows:
In metric units:

CH_3OH: $13,310 \times 0.95 = 12,645 \div 32.04 = 395 \text{ kg} \cdot \text{mol/h} = 0.913$ mole fraction
$\underline{666 \div 18.02 = 37 \text{ kg} \cdot \text{mol/h}} = 0.087$ mole fraction
$13,311 432 \text{ mol wt} = 30.83$

In English units:

CH_3OH: $29,600 \times 0.95 = 28,120 \div 32.04 = 878 \text{ lb} \cdot \text{mol/h} = 0.913$ mole fraction
$\underline{1,480 \div 18.02 = 82 \text{ lb} \cdot \text{mol/h}} = 0.087$ mole fraction
$29,600 960 \text{ mol wt} 30.83$

$$G = \frac{30.83}{359} \times \frac{273}{273 + 65} = 0.0694 \text{ lb/ft}^3 \qquad \text{boiling point} = 65°C$$

$$L = 0.81 \times 0.98 \times 62.4 = 49.5 \text{ lb/ft}^3$$

The *bottom liquor* at 0.5% methanol will be as follows:
In metric units:

CH_3OH: $13,310 - 12,645 = 665 \div 32.04 = 20.76 \text{ kg} \cdot \text{mol} = 0.00313$ mole fraction
H_2O: $119,790 - 665 = \underline{119,125 \div 18.01 = 6611 \text{ kg} \cdot \text{mol}} = 0.9969$ mole fraction
$119,790 6631.76 \phantom{\text{kg mol} = }1.0000$

In English units:

CH_3OH: $29,600 - 28,120 = 1,480 \div 32.04 = 46.2 \text{lb} \cdot \text{mol} 0.00313$ mole fraction
H_2O: $266,400 - 1,480 = \underline{264,920 \div 18.02 = 14.701 \text{ lb} \cdot \text{mol}} \underline{0.9969} \text{mole fraction}$
$266,400 14.747 \phantom{\text{lb} \cdot \text{mol}} 1.0000$

The slope of the equilibrium line near zero mole fraction methanol is $0.1831/0.0293 = 6.25$; therefore, the vapor composition over the reboiler will be $0.01956 \times 6.25 = 0.1223$, which is already greater than the amount of methanol in the feed. One theoretical separation would be a little delicate in that the reboiler may mist and not develop a complete equilibrium stage; therefore, two equilibrium stages of packed depth will be specified. Because of the large difference in heat of vaporization of the two components, an enthalpy balance will be needed. A McCabe-Thiele technique would not be entirely safe if many theoretical plates were required.

The enthalpy of the overhead vapor going to distillate would be

CH_3OH: $(28,120/32.04)\ 15,155 = 1.33 \times 10^7$ Btu/h
H_2O: $\underline{(1,500/18.02)\ 20,272 = 1.687 \times 10^6 \text{ Btu/h}}$
Total 960.8 lb $\cdot$ mol 1.499×10^7 Btu/h
or 1.591×10^4 Btu/lb $\cdot$ mol

If the column had equimolal flow, i.e., if the reflux line in a McCabe-Thiele diagram were a straight line, the minimum theoretical reflux ratio would be

$$L/D = \frac{X_d - Y_c}{Y_c - X_c} = \frac{0.87 - 0.3}{0.3 - 0.059} = 2.37$$

Normally columns are designed for not less than 1.25 times minimum theoretical reflux. A reflux ratio of 3:1 in this case would be 1.27 times theoretical; so it will be used. Now the total enthalpy of the column will be

$$4 \times 1.50 \times 10^7 = 5.995 \times 10^7 \text{ Btu/h}$$

The condition of the column at the top of the rectifying section will be

$G = 4 \times 960.8 = 3.84 \times 10^3$ lb $\cdot$ mol/h (mol wt 30.83) = 118,400 lb/h
$L = 3 \times 960.8 = 2.88 \times 10^3$ lb $\cdot$ mol/h (mol wt 30.83) = 88,800 lb/h
$G = 0.069$ lb/ft^3
$L = 49.5$ lb/ft^3

The loading condition at the bottom of the rectifying section will be

CH_3OH: $0.3 \times 15,160 = 4,548$ Btu
H_2O: $0.7 \times 20,450 = \underline{14,320 \text{ Btu}}$
$18,868$ Btu/lb $\cdot$ mol of vapor

The condition at the top of the column specifies an enthalpy of 6.00×10^7 Btu/h. Therefore, to maintain heat balance in the column, there must be $6.00 \times 10^7/1.887 \times 10^4 = 3180$ lb·mol gas/h at the bottom of the rectifying section. A material balance across this section will then be

$$- \text{ gas leaving top of column 3840 lb} \cdot \text{mol (mol wt 30.83)} = 118,400 \text{ lb/h}$$
$$+ \text{ liquid entering top of column 2880 lb} \cdot \text{mol (mol wt 30.83)} = 88,800 \text{ lb/h}$$
$$+ \text{ gas entering bottom of section 3180 lb} \cdot \text{mol (mol wt 22.23)} = 70,700 \text{ lb/h}$$

By differences:

$$\text{Liquid leaving bottom of section } \underset{\text{mol bal}}{\underline{2200 \text{ lb}}} \cdot \text{mol } \underset{\text{feed mol wt}}{[\text{mol wt 18.83 (calc)}]} = \underset{\text{wt bal}}{\underline{40,700}} \text{ lb/h}$$

The loading condition at the bottom of the rectifying section will be

$$L = 40,700 \text{ lb/h}$$
$$G = 70,700 \text{ lb/h}$$
$$L = 0.98 \times 0.97 \times 62.4 = 59.3 \text{ lb/ft}^3$$
$$G = \frac{22.23}{359} \times \frac{273}{273 - 91} = 0.046 \text{ lb/ft}^3$$

The loading condition at the top of the stripping section will be

$$L = 15,710 + 2200 = 17,910 \text{ lb} \cdot \text{mol (mol wt 18.83)} = 337,245 \text{ lb}$$
$$G = 3180 = 3180 \text{ lb} \cdot \text{mol (mol wt 22.23)} = 70,690 \text{ lb}$$
$$L = 59.3 \text{ lb/ft}^3$$
$$G = 0.046 \text{ lb/ft}^3$$

The slope of the equilibrium line at 0.059 mole fraction methanol, which is the composition of the liquid at the top of the stripping section, is approximately $0.298/0.0557 = 5.37$. So the composition of the liquid one equilibrium stage down from the top of the stripping-section bed will be $0.059/5.37 = 0.0110$ mole fraction methanol, which is already less than the theoretical vapor off the reboiler. Because the efficiency of the reboiler may be less than one equilibrium stage and the low number of stages required, packed depth for two stages should be specified. See bottom-liquor-condition analysis above.

The enthalpy balance at the bottom of stripping bed will be

$$\begin{array}{lll} \text{CH}_3\text{OH:} & 0.01 \times 15,160 & = 152 \text{ Btu} \\ \text{H}_2\text{O:} & 0.99 \times 1150 \times 18.02 & = 20,516 \text{ Btu} \\ & & 20,668 \text{ Btu/lb} \cdot \text{mol} \end{array}$$

It is necessary to provide a heat flow of 6.00×10^7 Btu/h up the column, which will amount to

$$6.00 \times 10^7/20,670 = 2903 \text{ lb} \cdot \text{mol of gas}$$

A material balance across the stripping section will give

$$\begin{array}{lll} \text{Liquid at top:} & 17,910 \text{ lb} \cdot \text{mol (mol wt 18.83)} = 337,245 \text{ lb} \\ \text{Gas at top:} & 3180 \text{ lb} \cdot \text{mol (mol wt 22.23)} = 70,690 \text{ lb} \\ \text{Gas at bottom:} & 2903 \text{ lb} \cdot \text{mol (mol wt 18.02)} = 52,312 \text{ lb} \\ \text{Liquid at bottom by difference:} & 17,695 \text{ (mol wt 18.02)} = 318,867 \text{ lb} \end{array}$$

$$L = 318,867 \text{ lb/h} \qquad L = 49 \text{ lb/ft}^3$$
$$G = 52,312 \text{ lb/h} \qquad G = \frac{18.02}{359} \times \frac{273}{373} = 0.0367 \text{ lb/ft}^3$$

Now return to the bottom of the rectifying section. The gas enters here with 0.3 mole fraction methanol. At the next equilibrium stage up in the bed the liquid would have 0.7 mole fraction water and 0.3 mole fraction methanol gas (see Fig. 10).

The tower should be sized at four places. A pressure drop of 1 in/ft of packed depth will be used because the situation occurs only transiently and the system is a nonfoamer.

	Top rectifying section	Bottom rectifying section	Top stripping section	Bottom stripping section
L	118,400 lb/h	40,700 lb/h	337,245 lb/ft³	318,867 lb/ft³
G	88,800 lb/h	70,700 lb/h	70,690 lb/ft³	52,312 lb/ft³
G	0.0694 lb/ft³	0.046 lb/ft³	0.046 lb/ft³	0.0367 lb/ft³
L	49.5 lb/ft³	59.3 lb/ft³	59.3 lb/ft³	49 lb/ft³
u	0.35 cP	0.30 cP	0.3 cP	0.3 cP

$$L/G \left(\frac{\rho_G}{\rho_L - \rho_G} \right)^{0.5} = \text{see abscissa, Fig. 4}$$

$X = 0.0500$	0.016	0.133	0.167

Using Fig. 4, move vertically from the calculated abscissa value above to the parameter of 1 in pressure drop, then horizontally to read the equivalent ordinate value:

$Y = 0.082$	0.10	0.065	0.060

Assuming a value for the packing factor F of 20, solve for the value of G':

$$G' = \left[\frac{Y_G(\rho_L - \rho_G)g}{Fu^{0.1}} \right]^{0.5}$$

$G = 0.708$	0.703	0.567	0.442

Tower area:

$A = 3.24 \text{ m}^2 \ (34.84 \text{ ft}^2)$	2.6 m² (27.95 ft²)	3.22 m² (34.63 ft²)	3.06 m² (32.88 ft²)

Diameter $\left[= \left(\frac{4A}{\pi} \right)^{0.5} \right]$:

$D = 2.03 \text{ m} \ (6.66 \text{ ft})$	1.82 m (5.96 ft)	2.02 m (6.64 ft)	1.97 m (6.47 ft)

The tower should be constructed 2 m (6 ft 8 in) in diameter and is limited by the condition at the top of the rectification section. A packing factor of 20 was chosen, which corresponds to a 50-mm (2-in) pall ring (Table 2). If some other packing were to be selected, a correction for area could be made from the relationship

$$\frac{G_1}{G_2} = \left(\frac{F_2}{F_1} \right)^{0.5} \qquad \text{or} \qquad G_2 = G_1 \left(\frac{F_2}{F_1} \right)^{0.5}$$

The HETP, or height equivalent to a theoretical plate, is dependent on packing size rather than on system properties;[8] so for the 50-mm (2-in) metal pall ring this will be 0.75 to 0.9 m (2.5 to 3.0 ft). The five separations needed in rectification can be obtained with $5 \times 0.825 = 4.125$ m ($5 \times 2.75 = 13.75$ or 14 ft) of 50-mm (2-in) metal pall ring packing.

The two stages in the stripping section can be obtained with $2 \times 0.9 = 1.8$ m ($2 \times 3.0 = 6$ ft) of 50-mm (2-in) metal pall rings. Because of the fall-off of required area in this lower section and the very high L/G ratio in this section, it would be desirable to install an in-bed wall wiper at 0.6 m (2 ft) above the bottom, which would choke the tower by 3.24 to $3.05 = 0.19$ m² (34.8 to $32.8 = 2$ ft²) or 5% to bring up the efficiency of this section, especially at low rates.

A distributor of the Fig. 15 type should be installed in the top. One bed of packing will be adequate in the top supported by a Fig. 11 support plate. The feed should be introduced through a Fig. 16 distributor in that it is of the same composition as the reflux at this point and there will be no need for a trap tray or a mixing redistributor. The Fig. 16 distributor will bring in the feed liquid with a minimum effect on the liquid already flowing the tower. The wall wiper should be of the Fig. 17 type, and it will not hurt the operation of the stripper too much if the wiper chokes the area of the tower 10% because it will produce only localized flooding. The stripping-section bed should be supported with another Fig. 11 support plate.

TOWER INTERNALS

Support Plates

A packing will not perform as well as expected unless it is properly supported and has both the gas and liquid evenly distributed over it. A support plate (Fig. 11) must have sufficient gas and liquid-handling capacity so that it will not flood or generate an excessive amount of pressure drop compared with the packing itself. Plates of a design similar to that in Fig. 11 will not generate more than about ¼ in of pressure drop under conditions which will be found in a distillation tower. It will be found that this cannot be done with a flat plate when packings over 1 in in size are used. An undulated plate of some type such as that shown in Fig. 11 is necessary to permit the liquid traffic to move through one set of openings and the gas traffic through another to avoid premature flooding of the packed bed at this point.

Redistributors

If it should be desirable to redistribute the liquid between beds in the tower, a device similar to Fig. 12 should be used. Omission of the liquid-distribution orifices in this plate will convert it to a trap tray or a drawoff tray for removal of liquid from the column at some intermediate point. Redistributors normally are not needed in distillation columns unless the reflux ratio is very high or the liquid rate if very low. Because of the low percentage of free area which can be built into them, they normally operate at too high a pressure drop to be used in vacuum distillation and still achieve any great degree of liquid redistribution.

TROUBLE SOLVING

Flooding at the Support Plate

Support-plate flooding can be easily detected by backing the burden off on the tower about 10%. If there is a sudden decrease in pressure drop over the tower (more than

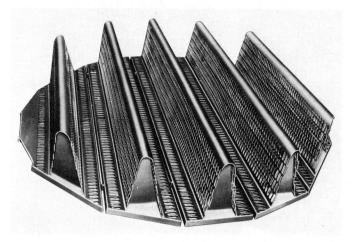

Fig. 11   Support plate. *(Norton Company.)*

would be predicted by Fig. 9), the support plate is flooded. If the loss of pressure drop falls on the curve in Fig. 9, the packed bed is flooded. If there is no noticeable loss of pressure drop, continue to reduce the burden on the tower until it starts to lose pressure drop and then do not attempt to operate above this rate until the tower is repacked.

Fouled Packing

Packed towers will lose capacity over a period of time if something in the liquid or the gas fouls the packing. This may be a mineral deposit like silica, calcium carbonate, etc., a tarry deposit resulting from thermal degradation of an organic material, or fibrous or sticky

Fig. 12   Redistributor. *(Norton Company.)*

material brought into the tower in suspension. A tower will also lose capacity if it is run too hot in the case of plastic packings or has been bumped or run flooded in the case of ceramic packings.

Damaged Packing

Ceramic packed towers, which may be subject to transient overloads or "bumps" in loading, are usually supplied with a holddown plate (Fig. 13) to keep the packing from

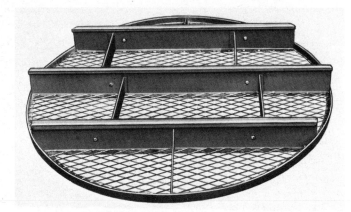

Fig. 13 Holddown plate. *(Norton Company.)*

fluidizing the top surface of the bed with consequent destruction of the packing which will filter down through the bed to the support plate, causing localized flooding.

Plastic packings present a different problem in that they tend to float out of the tower on either the liquid or the gas. Where this situation may prevail, it is best to install a bed limiter (Fig. 14) which will prevent the packing from floating out of the bed location in the tower.

Maldistribution

The desired efficiency of the packed bed can be obtained only if the liquid is evenly distributed over the top surface of the top bed. There is little or no evidence that the

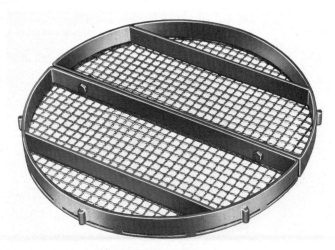

Fig. 14 Bed limiter. *(Norton Company.)*

efficiency of the bed will fall off because of maldistribution of the liquid within the bed as it moves down the tower. A good distributor (Figs. 15 and 16) is one that will lay the liquid evenly on the surface of the packed bed at a minimum of four points per square foot.[4] Extra packed depth will not correct faulty distribution even with the very best redistribution packings such as the Intalox saddle and the Berl saddle.

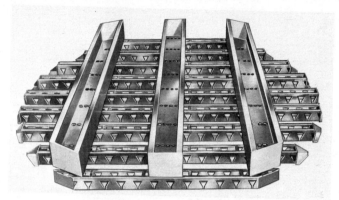

Fig. 15   Distributor. *(Norton Company.)*

Packing Size and Bed Depth

Normally, bed depths do not exceed 7.5 m (25 ft) in packed towers for reasons of support-plate strength and load concentration at the tower wall. In small-diameter towers these bed depths become much less because the geometry of the packing becomes large in relation to the size of the tower. Ordinarily, it is not desirable to have a Raschig ring in a tower that is larger than one-thirtieth the diameter of the tower; while a pall ring or a saddle can be from one-eighth to one-tenth the size of the tower. In any event, because of this decreasing geometric ratio of the packing size to the tower diameter, it becomes desirable to reduce the bed depth in small towers. In the case of Raschig rings, it is best not to exceed six tower diameters in bed depth, and in the case of pall rings and saddles no more than ten times the tower diameter.

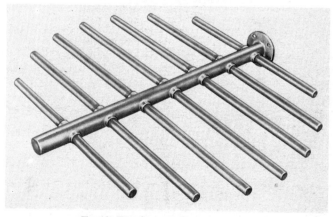

Fig. 16   Distributor. *(Norton Company.)*

Shorter bed depths in smaller towers make it necessary to increase the number of distribution points per square foot to as many as thirty for towers 150 mm (6 in) or less in diameter. The reason for this is the necessity of not losing any of the top of the bed to distribution of liquid flow in short beds.

Normally, perfect distribution of liquid is not as important at high liquid rates as it is at low liquid rates; however, it must be kept in mind that if the liquid rate at any spot in the cross section of the tower becomes less than minimum theoretical reflux, no separation will take place for the depth of bed this "pinch" area may extend.

Fig. 17   Wall wiper. *(Norton Company.)*

It follows that for those separations of low relative volatility, i.e., $\alpha = 1.2$ or less where a very high minimum theoretical reflux ratio is required, a very perfect distribution of liquid will be required, making it essential to assure uniform flow from all points of the distributor and even on occasion to use a distributor with more feed points. Also, because of the need for more perfect distribution, this is one instance where a liquid redistributor (Fig. 12) should be used if more than one bed is required in the column.

Importance of Pressure Drop

There is reason to believe that good agitation of the liquid is necessary to achieve a maximum rate of mass transfer, or in the case of distillation to most nearly approach equilibrium between the gas and liquid phases.[9]

In the rectification section of a fractionator, the amount of gas will always exceed the amount of liquid. There is no problem in achieving good agitation of the liquid either by pressure drop for dense gases (atmospheric and elevated-pressure distillation) or by gas velocity for light gases (vacuum distillation). The stripping section of the fractionator often presents a problem in achieving good liquid agitation because here the amount of liquid will exceed the amount of gas, so that quite often there is not enough gas to agitate the liquid adequately. Three general methods are available to correct this situation. First, a smaller-sized packing may be used which, because of a higher packing factor, will operate at a higher pressure drop and thus give better liquid agitation. Second, the tower may be reduced in diameter, thus giving a higher pressure drop because of increased loading. Third, in-bed wall wipers (Fig. 17) may be installed to cause the bed to flood regionally.

REFERENCES

1. Sherwood, T. K., Shipley, G. H., and Holloway, F. A. L., *Ind. Eng. Chem.*, **30**, 765 (1938).
2. Lobo, W. E., Friend, L., Hashmall, F., and Zenz, F. A., *Trans. Am. Inst. Chem. Eng.*, **41**, 693–710 (1945).
3. Leva, Max, "Tower Packings and Packed Tower Design," p. 40, The United States Stoneware Co., 1953.
4. Eckert, J. S., *Chem. Eng. Prog.*, **57**(9), 54 (1961).
5. Eckert, J. S., *Chem. Eng. Prog.*, **59**(5), 76 (1963).
6. Eckert, J. S., *Oil Gas J.*, Aug. 24, 1970.
7. Leva, Max, *Chem. Eng. Prog. Symp. Ser.*, **50**, 10, 51–59 (1954).
8. Eckert, J. S., Foote, E. H., Walter, L. F., *Chem. Eng. Prog.*, **62**(1) 59 (1966).
9. Eckert, J. S., Foot, E. H., Rollison, L. R., and Walter, L. F., *Ind. Eng. Chem.*, **59**(2), 41 (1967).
10. Ramalho, R. S., Tiller, F. M., James, W. J., and Bunch, D. W., *Ind. Eng. Chem.*, **53**, 895 (1961).

Section 1.8

High-Efficiency Low-Pressure-Drop Packings

PAUL G. NYGREN *Senior Staff Engineer, Union Carbide Corporation; Vice-chairman, Technical Committee and Technical Advisory Committee, Fractionation Research, Inc.; member, American Institute of Chemical Engineers.*

NOTATION

A, B, C Antoine equation constants, A and B are also compound names in design example
F factor Vapor loading, superficial flowing velocity times square root of vapor density

HELPD   High-efficiency, low-pressure drop
HETS   Height equivalent to a theoretical stage
P System pressure. With subscript, refers to pure-component vapor pressure
X Component mole fraction
x Mole fraction in liquid
y Mole fraction in vapor
T Temperature, degrees Celsius
α Relative volatility of light component to heavier component
ρ_v Vapor density, pounds per cubic foot

INTRODUCTION

High efficiency, low pressure drop means low height per theoretical stage and low pressure drop per theoretical stage. These are relative terms, and there is a wide variation in the performance of packings or devices which we classify as high-efficiency low-pressure-drop (HELPD). The low-pressure-drop characteristic is usually the factor which justifies use. This leads to high-vacuum designs, which are the principal emphasis of this section. However, the high-efficiency aspect can significantly affect the economics of a design whether or not pressure drop is important.

To put the subject in the proper perspective, we will briefly describe the three general classes of vapor-liquid contacting devices which are applicable in medium- to high-vacuum distillation. Refer to Fig. 1.

Trays are applicable down to 5 or 10 torr head pressure if design conditions permit a

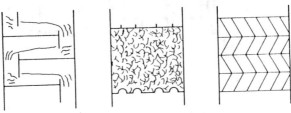

Fig. 1 Classes of contacting devices.

pressure drop of 3 to 5 mmHg per theoretical stage. This statement applies to conventional sieve trays, bubble-cap trays, and various proprietary devices including valve and sieve trays.

Dumped packings are applicable in the pressure-drop range from less than 1 to about 3 mmHg per theoretical stage. They include a great variety of designs of packings, most of which are configured as either rings or saddles.

Ordered Packings. HELPD packings fall in the ordered classification, although ordered packings are not necessarily low-pressure-drop or high-efficiency. They extend the range of pressure drop per theoretical stage in general down to about 0.25 mmHg, and in some cases as low as 0.1 mmHg.

GENERAL DESCRIPTION OF HELPD DEVICES

This section concerns industrial-scale equipment. Industrial scale is arbitrarily defined as 300 mm (12 in) diameter and larger. There are five devices marketed in the United States of widely differing construction and operating characteristics which can be classified as industrial-scale high-efficiency low-pressure-drop packings. These are Goodloe, Hyperfil, Neo-Kloss, and Koch-Sulzer packings and Leva film trays.

Goodloe and Hyperfil packings are formed from knitted and crimped wire mesh. The Goodloe mesh is spirally wound to form a circular layer to exactly fill the shell cross section. In Hyperfil packing, the mesh is folded back and forth upon itself into a circular assemblage that when bound will exactly fit the shell cross section. In both packings, the layers are stacked one upon the other in the column. In both packings capillary action generates a large liquid surface–liquid volume ratio.

In *Neo-Kloss* packing, fine-mesh screen is wound spirally to form a roll of the same diameter as the inside of the column shell; adjacent layers of the roll are separated by spacers. The rolls are stacked successively in the column shell. Capillary action disperses the liquid across the openings in the screen, and the vapor flows straight upward between the vertical layers of wetted screen.

Koch-Sulzer packing is formed from a woven fine-wire fabric. Parallel, corrugated strips of this fabric are assembled into a circular layer of the same diameter as the column shell. Successive layers are stacked in the shell. Capillary action spreads the liquid across the fabric; the upflowing vapor follows a tortuous path past the wetted surfaces.

Leva film trays are horizontal plates on spacings of about 50 mm (2 in) and have 50-mm- (2-in) diameter holes centered between holes in the adjacent plates. A venturi-shaped tube extends downward from each hole about halfway to the plate below. The rising vapor and the descending liquid flow countercurrently through these tubes. The liquid spreads on the tray below, flowing to adjacent holes. The film tray is the only one of the five devices in which capillary action is not the dominant factor in the generation of film surface. Although this is a multiple-tray device, it can also be considered an ordered packing.

These five packings are described and discussed in more detail in a later section. Table 1 also compares and provides additional information about the five devices.

OPTIMIZATION CONSIDERATIONS

HELPD packings are used where one or both of their two principal characteristics, low height of a theoretical stage and low pressure drop per theoretical stage, can be utilized to advantage. However, the decision to use such a packing and the choice of packing to purchase are often not obvious. The choices should be made from an optimization study of process conditions, the process alternatives, and the characteristics of the devices available.

HELPD packings are generally more expensive per unit of volume than the common devices. For the more costly ones, the difference can be a factor of 50 or 100. Therefore, there must be an advantage in one or more of the following factors:

1. The net cost of the equipment because of the higher volumetric efficiency
2. Lower pressure drop, which can provide several advantages
3. Reduced operating costs

Equipment Cost

Reduced equipment size may reduce the cost enough to justify the choice over the more conventional devices. The size reduction can result from lower packed height due to lower HETS and/or a higher allowable vapor F factor. Let us consider a simple example where the allowable F factor is the same as for a common packing. If the HETS is one-third that of the common packing and the volumetric cost is a factor of 3, the higher-efficiency packing can be justified by the saving in shell cost. However, if the packing-cost factor is 50, justification is highly unlikely.

Pressure Drop per Theoretical Stage

This is usually the primary consideration in the choice of HELPD packing. The approximate range where there is an advantage is from about 0.1 to 1.0 mmHg per stage. Above this value or even approaching the high end of the range we begin to overlap the capabilities of conventional packings.

Pressure drop can be important for several reasons:

1. To minimize kettle temperature. The objective is usually to avoid product degradation. However, limitations imposed by the heating source can also be a factor.
2. To generate more theoretical stages within available pressure drop. The improvement can be taken in the separation or the reflux ratio. The improved separation may provide either better recovery or higher product purity; the reduced reflux ratio can reduce energy consumption or increase production per unit of equipment cross-sectional area.
3. To permit operation at a higher head pressure. In existing equipment this can provide higher capacity. In new equipment, a column of smaller cross section is required. Occasionally, the higher resulting head temperature will provide more favorable condensing conditions.

A point to be emphasized is that the design loading is normally the highest allowable

TABLE 1 Comparison of Devices

	Goodloe packing	Hyperfil packing	Koch-Sulzer packing	Neo-Kloss packing	Leva film trays
General information:					
Type	Knitted multifilament	Knitted multifilament	Corrugated woven-wire fabric	Rolled screen with spacers	Multiple unsealed downcomer trays on close tray spacing
Approximate number of units 12 in and larger sold through 1975	610	90	500	43	120
Largest diameter sold to date	5 ft, 8 in [a]	5 ft [a]	11 ft [a]	6 ft [a]	14 ft, 6 in [b]
Materials in which available	[a]	[a]	[a]	[a]	[a]
Process and system considerations:					
Minimum head pressure, torr	1	1	0.5	0.5[i]	5
Liquid considerations:					
Minimum rates, gal/(min)/(ft²)[c]	0.016	0.1[i]	0.08	0.1[i]	0.05
Maximum rate, gal/(min)/(ft²)[c]	>4.9	4.7[i]	>8	3.8[i]	5
Maximum viscosity, cP	200				>100
Holdup, fraction of total volume, typical	0.07–0.12	0.1	0.04	0.03	
Sensitivity to uneven initial liquid distribution	Moderate	Moderate	Fairly low	High[d]	Moderate
Vapor F factor, based on internal cross-sectional area of shell, typical	0.5–1.5	1.0	1.8	2.5–3.0	
Maximum	1.7[c]	1.4	3.3	4.0	2.0
Minimum	low	0.14		0.16	0.25
HETS, in:					
Range	3½–8½	3½–9	4–10	4–18	12–24
Typical	5[c]	5[c]	7	8	18
Pressure drop, mmHg per foot of packed height	[e]	[e]	See Fig. 3 [g]	See Fig. 2 [g]	See Fig. 4 [g]
Pressure drop per theoretical stage	[f]	[f]			
Fouling considerations:					
Sensitive to particulate solids?	Yes	Yes	Moderately[h]	No	No
Sensitive to fouling by tarry substances?	Yes	Yes	Moderately[h]	Yes	No
Sensitive to fouling from polymer formation?	Yes	Yes	Moderately[h]	Yes	No
Mechanical considerations:					
Is the device furnished as a package including shell and internals?	Optional	Optional	Optional	Yes	Optional
Can it be installed through shell manholes?	Yes[j]	Yes	Yes	No	No
Can it be installed in an existing shell with only minor modifications?	Yes[j]	Yes	Yes		Yes[k]
Test facilities:					
Are pilot test facilities available?	Yes	Yes	Yes	Yes	Yes

[a] Any metal capable of being drawn into wire.

[b] Any metal which can be fabricated into the required shapes.

[c] These liquid rates are generally those claimed by the manufacturers. Very low liquid loadings [below 0.2 gal/(min)(ft²)] always require special attention to the design of the liquid-distribution system.

[d] Neo-Kloss packing requires highly precise initial liquid distribution because liquid cannot spread from one layer of screen to the next. Care is needed in the design of the distribution system (provided by vendor), in its installation, and in prevention of fouling.

[e] No general curves are available for estimating from F factor. See vendor bulletins for calculational methods.

[f] Vendor bulletins indicate that pressure drop per theoretical stage will be about 0.5 mmHg or less.

[g] For Koch-Sulzer packing, Neo-Kloss packing, and Leva film trays, a preliminary estimate can be made by dividing the pressure drop per foot of packing at a typical or expected F factor loading from Fig. 2, 3, or 4 by an assumed HETS (in feet).

[h] Relatively good irrigation properties minimize the potential for dry spots which promote fouling.

[i] Vendor considers those values to be extremes normally used but not absolute limits.

[j] Techniques have been developed to permit installation through a manhole. However, it is preferred and usually less costly to provide full shell flanges on either new or existing columns.

[k] Full shell opening required.

within the pressure-drop limitations, but below the flooding velocity of the packing. Examination of Figs. 3, 4, and 5 in conjunction with typical HETS shows that very low pressure drop may require very low vapor loading and therefore a large diameter.

Reduced Operating Costs

Lower costs result from the following factors:

1. A closer approach to minimum reflux ratio and the attendant reduction in energy consumption.

2. Higher recovery efficiency. The lower the kettle pressure, the lower the loss of light key in the bottom product for the maximum permissible kettle temperature.

3. The use of a one-column process where two columns might otherwise be required. There are many potential benefits such as savings in investment, energy, and operating labor.

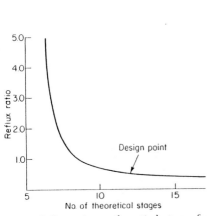

Fig. 2 Reflux ratio vs. theoretical stages for design example.

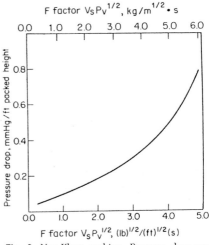

Fig. 3 Neo-Kloss packing. Pressure drop per foot of packed height for cis-trans-decalin at 5 torr head pressure. (*From Chem-Pro Bull. HV-510.*)

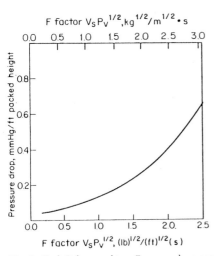

Fig. 4 Koch-Sulzer packing. Pressure drop per foot of packed height for cis-trans-decalin at 20 torr head pressure. (*From Koch Engineering Company Bull. KS-1.*)

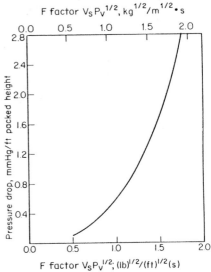

Fig. 5 Leva film tray. Typical pressure drop per foot packed height.

DESIGN CONSIDERATIONS

Design problems fit into two general categories:
1. All new designs
2. Use of existing equipment

The first category permits the greatest degree of freedom to optimize the design. The second category may be debottlenecking, improving performance, or adapting an existing column to a new service. Expensive internals may be an inexpensive solution to a problem if a complete new column can be avoided.

For either situation there are many considerations which must be taken into account with regard to the process, the equipment, and the physical layout. The following is a checklist which will cover the items which should be considered in developing the design for most installations. Several of the items are expanded upon in following sections.

Checklist of Design Considerations

1. *Reboiler temperature.* Determine the highest acceptable bulk temperature of the liquid in the reboiler. This is usually limited by the time-temperature sensitivity of the bottom product or by the temperature of the preferred heating medium and the temperature difference required for heat transfer.

2. *Base pressure.* Determine the base pressure corresponding to the bubble point for the desired bottom composition. Alternatively, determine the composition and bubble point corresponding to an assumed base pressure.

3. *Head temperature.* Determine whether there is a minimum head temperature based on condensing conditions. This might be cooling-medium temperature or physical properties of the distillate such as freezing point or high viscosity.

4. *Head pressure.* The final head pressure is the result of the optimized design, and is set by a combination of all the process requirements and equipment constraints.

5. *Theoretical stage vs. reflux ratio* relationship to achieve the required separation must be established by calculation or experimentation.

6. *Liquid holdup.* Is minimum liquid holdup required? The HELPD packings generally have very low liquid holdup as a fraction of the packed volume (see Table 1). The design of the base section, piping, and reboiler for low liquid holdup are the more important considerations.

7. *Fouling.* Is fouling expected? What is the nature of the fouling material? Can it be removed with in-place solvent cleaning?

8. *Physical limitations.* Establish dimensions and other characteristics of existing equipment which may affect the design. Establish layout limitations for new equipment.

9. *Liquid load.* Define the lowest liquid loading per unit of column cross section at the liquid distributor. Normally, the reflux is the lowest value. This is critical for many devices.

10. *Odor requirements.* This is a difficult if not impossible criterion to calculate. Plant or pilot experience is usually required.

11. *Corrosion.* Determine the acceptable materials of construction for the process conditions. With the possible exception of Leva film trays, the materials for HELPD packings must be completely resistant to corrosion.

12. *Flexibility.* Will it be necessary to provide for a range of process conditions, either for initial operation or for future changes?

VAPOR-LIQUID EQUILIBRIUM·

Accurate vapor-pressure data for the compounds to be separated is necessary for an accurate design. Uncertainties or inaccuracies in VLE data can cause overdesign, which is costly with respect to investment and may be wasteful of energy. The result can also be underdesign, which is even more costly when required capacity or separation capability is lacking.

In high-vacuum systems, ideal behavior can usually be assumed, and application of Raoult's and Dalton's laws provides the relationship where P_1 and P_2 are the pure component vapor measures at the system temperature and α is the relative volatility.

$$\alpha_{1,2} = \frac{P_1}{P_2} \tag{1}$$

The following relationship can be used to construct a binary equilibrium curve for use in graphical solutions if the relative volatility is known:

$$y = \frac{\alpha x}{1 + (\alpha - 1)x} \tag{2}$$

where x and y are the mole fraction of light component in the liquid and vapor, respectively.

If accurate vapor-pressure data are not available, relative volatilities of a binary system can be obtained from pilot-plant data by a method suggested by Lubowicz and Reich.[1] It is preferable that the work be carried out in a piece of equipment representative of the equipment to be used in the commercial plant.

Experimental Method for Determining Relative Volatility   Calibrate the pilot plant at total reflux with a well-defined test mixture such as cis-trans-decalin, chlorobenzene-ethylbenzene, or styrene-ethylbenzene. It is desirable to use a binary of ideally behaving components with a relative volatility of less than 1.5 to minimize errors in possible changes in relative volatility across the test composition range.

Fenske's equation[2] can be used to establish the efficiency of the column.

$$\log \left[\frac{(X_A/X_B)_1}{(X_A/X_B)_2} \right] = N \log \alpha \tag{3}$$

A and B are the components, A being the more volatile; 1 and 2 are the upper and lower sample points, respectively; X is concentration, mole fraction; N is the number of theoretical stages; α is the relative volatility of A to B. For the test conditions, the relative volatility is known, the concentrations are measured, and the number of stages N is calculated from Eq. (3). The calibration experiment should be done under conditions of vapor loading and operating pressure similar to those of the potential commercial column.

The next step is to run the system being studied in the same pilot column at total reflux. Since the number of stages has been determined in the calibration run, the relative volatility can be solved for in Eq. (3). Several runs may be needed to get relative volatilities over the concentration range. The authors refer to the relative volatilities obtained by this method as "pseudo-alphas," since they reflect the characteristics of the pilot device used.

The test described may also be helpful in obtaining information useful for scale-up to the commercial design. The HETS and pressure drop have been established from the pilot runs. The values thus obtained should be extrapolated with care and preferably with the advice of the proprietor.

REFLUX RATIO–THEORETICAL STAGE RELATIONSHIP

Many companies have sophisticated and rigorous tray-by-tray calculation programs available for determination of the reflux ratio–theoretical stage relationship. These should be used for final designs. There are several shortcut methods which are suitable for preliminary designs and sometimes for final designs. Some of these are:

The McCabe-Thiele[2] Graphical Method   This method does not require the assumption of constant relative volatility. The main disadvantage is the tedious multiple solutions required.

The Smoker Equation[3]   This method amounts to an analytical solution of a McCabe-Thiele diagram. The disadvantage is that it requires the assumption of constant relative volatility, although different values can be assumed for stripping and for rectifying. The advantage is rapid multiple solutions if a small computer is available.

The Gilliland Method[4]   An advantage is that it handles more than two components. A disadvantage is that the calculation shows only total stages and is not broken down between stripping and rectifying.

A *recommended shortcut approach* is to use the Smoker method using the geometric average of feed point and top of the column relative volatilities for the rectifying section and the feed point and bottom of the column for the stripping section. This should be used with caution, especially if the relative volatility varies widely across the column.

LIQUID DISTRIBUTION

Liquid distribution is an important consideration in all packed towers. The lower the loading per unit of area, the more important this factor becomes. The lower the absolute operating pressure, the lower this unit loading tends to become. Since HELPD packings are usually applied under high vacuum, liquid-distribution considerations are especially important. All the HELPD packing manufacturers recognize the importance and provide custom-designed distributors for their packings. Refer to Table 1 for a qualitative description of the sensitivity of the various packings to liquid considerations.

One should always obtain specific assurance from the proprietor that the liquid distributors, both reflux and feed, will function satisfactorily over the full range of required liquid loadings. The minimum liquid rate is not something to be negotiated as an afterthought just before an order is placed or in the approval-drawing stage. The minimum rates should be seriously considered when developing the inquiry specification and clearly spelled out therein. Do not arbitrarily ask for a lower rate than is necessary. A high turndown requirement may be difficult or impossible for a manufacturer to meet with a single distributor. The choice then is between installation of multiple liquid-distribution systems or a shutdown of the equipment to change distributors.

CONTROL

Good control is important for most distillation systems. It is especially important where a system has been closely designed for minimum investment, as should be the case for the relatively high cost HELPD packings. Good control is also essential to minimize energy consumption.

INSTRUMENTATION FOR TROUBLESHOOTING AND PERFORMANCE EVALUATION

Adequate instrumentation should be provided so that accurate heat and material balances can be made over the system for troubleshooting and for evaluation of equipment performance. The following paragraphs discuss the more important aspects.

1. Provide ability to measure the temperatures and determine the flow rates of all entering and exiting streams.

2. It is desirable to provide sufficient thermowells to define the column temperature profile. As a practical matter, it is often difficult to obtain in-bed temperatures. Liquid-phase measurements are preferred because of the difficulties of measuring vapor temperatures with assurance. As a minimum, provide measurement of column head, feed point, and base temperatures.

3. It is desirable to have enough sample taps to be able to get a column-composition profile. Representative in-bed liquid samples are difficult to obtain. Therefore, we are usually limited to reflux, bottom product, and any point where liquid is collected and redispersed. Frequently, the reflux from the rectifying section is collected and mixed with the feed. Liquid samples are preferred because it is difficult to ensure that liquid droplets are not included in a vapor sample. Vapor bubbles in liquid have small effect on a liquid sample, but liquid droplets in a vapor sample can have a large effect; the lower the absolute pressure, the greater the effect.

4. All heat-input values should be measurable. Especially important are reboilers, side heaters, and feed vaporizers. The heat out in condensers and coolers must often be calculated from heat balances. The exception is refrigerant flows, which are usually metered. Cooling-water flow is seldom measured and airflow for an air-cooled condenser is never known precisely.

OPERATING-PRESSURE OPTIMIZATION

A head pressure of 5 torr is a reasonable minimum goal. Below this value the required cross section increases very rapidly, since the area is proportional to the inverse of the square root of the vapor density. Also at a fixed reflux ratio the quantity of reflux tends to become very small and may fall below an amount which can be distributed evenly enough to assure good efficiency and the desirably low HETS. Certainly there are cases where pressures under 5 torr are warranted, down in the range of 1 torr. But be aware of the pitfalls and possibly give consideration to relaxing the bottom composition. Sacrificing a bit on recovery may be a more economic solution to the design problem at hand.

EVALUATION OF DEVICES AND VENDORS

There is a noticeable lack of unbiased published performance data for commercial-size HELPD packing installations. Many of the available data are from very small diameter experimental equipment and many are published by equipment manufacturers. Most of the data have been taken at total reflux under which the condition of liquid loading at the top of the packing is the most favorable possible. At a 1:1 reflux ratio, the liquid rate is only 50% of that for total reflux. Less than perfect liquid distribution for a low liquid rate can have substantial unfavorable impact upon the HETS.

The state of the art and the highly proprietary nature of the equipment makes necessary:

1. Close consultation with equipment vendors so that they understand all aspects of the separation problem.

2. Deep probing of vendors' general experience and knowledge of the type of design problem at hand. It is necessary to assure oneself that the supplier has the expertise to assure successful operation of the commercial plant-scale equipment.

DESIGN EXAMPLE

A feed stream of 4500 kg/h (10,000 lb/h) containing 3375 kg (7500 lb) of compound A and 1125 kg (2500 lb) of compound B must be separated into an overhead product of 99.5 wt % pure A and a 99% recovery of A. Time-temperature studies indicate that a maximum kettle temperature to prevent degradation is 186°C. Cooling water at 33°C is available for the overhead condenser. The vapor-pressure curves for the pure compounds can be represented by the Antoine equation,

$$\log P = A - \frac{B}{C + T}$$

	Compound A	Compound B
A	8.10488	7.31010
B	2517.9	2076.0
C	194.6	139.0
mol wt	150	194

P is in mmHg and T in degrees Celsius.

Since relatively high vacuum will be employed and the compounds are of a similar nature, we will assume that the vapor and liquid activity coefficients are 1.0 and that the perfect-gas law is followed for gas densities.

Step 1 *Establish the material balance.* The specified 99% recovery permits 75 lb of A in the bottom product. The 99.5% overhead product purity permits 37 lb of B in the overhead. In the tabulation below, the products are converted to moles and mole fractions for the later calculations.

	Feed,	Overhead product			Bottom product		
Compound	kg/h	kg/h	mol/h	Mole fraction	kg/h	mol/h	Mole fraction
A	3375	3341	22.3	0.996	33.75	0.225	0.038
B	1125	17	0.088	0.004	1108	5.713	0.962
	4500	3358	22.388	1.000	1141.75	5.938	1

	Feed,	Overhead product			Bottom product		
Compound	lb/h	lb/h	mol/h	Mole fraction	lb/h	mol/h	Mole fraction
A	7,500	7,425	49.500	0.996	75	0.500	0.038
B	2,500	37	0.191	0.004	2,463	12.696	0.962
	10,000	7,462	49.691	1.000	2,538	13.196	1.000

Step 2 *Establish the maximum base pressure.* From the vapor-pressure relationships and the composition established in step 1, calculate the base pressure. Assuming ideality, the base pressure is the sum of the partial pressures of the two components.

For T = 186°C,

	Pure component vapor pressure	Liquid phase mole fraction	Partial pressure
Compound A	30.85	0.038	1.17
Compound B	8.36	0.962	8.05
		Base presssure	9.22 torr

Step 3 *Establish the column head conditions.* From the base pressure of 9.2 torr calculated in step 2 and the general guidelines recommended in the section on operating-pressure levels, we will choose a tentative head pressure of 5 torr. This pressure can be adjusted as required by further calculations. The pressure drop available for the internals is 4.2 mmHg.

Step 4 *Establish the reflux ratio–theoretical stage relationship.* We shall calculate the relationship from the Smoker equation. Because of varying relative volatilities, we will use average values for the rectifying and stripping sections as shown. The temperature in the column at the feed point can be approximated by assuming an average column pressure for the feed composition bubble point.

	Pressure, torr	Temp, °C	α
Top	5.0	146	4.91
Feed	7.0	157	4.44
Bottom	9.2	186	3.69

$$\alpha_{A,B} \text{ (rectifying)} = (\alpha_T \cdot \alpha_F)^{1/2} = 4.67$$
$$\alpha_{A,B} \text{ (stripping)} = (\alpha_F \cdot \alpha_B)^{1/2} = 4.05$$

The result of the reflux ratio vs. stage requirements from the Smoker calculations is shown in Fig. 2. For this example we will choose a design reflux ratio of 0.52, which is 1.3 times the minimum reflux ratio of 0.4. This may appear to be a high multiple for today's energy-conscious environment. However, for this relatively "easy" separation the energy savings of a closer design are small. Also, a well-designed control system will permit the column to operate at the reflux ratio to accomplish the required separation.

Approximately 12 stages are required. They are split about evenly between rectification and stripping. The reboiler will provide about one theoretical stage, reducing the stages required from the packing to 11.

Step 5 *Establish a tentative tower diameter.* The loadings at the top of the tower are:

Vapor: 3385 × 1.52 = 5145 kg/h (7462 × 1.52 = 11,340 lb/h)
Liquid: 3385 × 0.52 = 1760 kg/h (7462 × 0.52 = 3880 lb/h)

For a high-vacuum tower, the top of the packing is usually limiting. An exception might occur if the feed is highly subcooled. The vapor required to heat the feed up to column temperature could place the maximum load in the stripping section. It is also possible that unusual composition effects might place the maximum load elsewhere. We will assume a bubble point feed condition.

Note also that the reflux requirement is liquid at the top of the packing rather than externally measured reflux. If the reflux has been subcooled to 50°C, only 1202 kg (2650 lb) of external reflux will generate the required internal reflux of 1760 kg/h (3880 lb/h). The effect of this is further discussed under step 7.

For a range of design F factors of 3.0 to 1.8, the required diameter varies from about 1.71 to 2.23 m (5.6 to 7.3 ft). Review of Table 1 indicates that Koch-Sulzer may satisfy our requirements. We will examine this packing in some detail to show the method. The same approach can be used to screen other packings.

For this preliminary sizing assume an HETS of 200 mm (8 in.). The packed height for the required 11 stages is 2.23 m (7.33 ft). Add 25% for safety factor and for such contingencies as feed mismatch, somewhat lower efficiency before the optimum liquid distribution is attained, a reheat zone for the subcooled reflux, etc. The safety factor could be reduced for designs with greater packed heights. This brings the total packed height to 2.80 m (9.17 ft). We will assume nine layers of 170-mm (6.7-in) thickness for the rectifying section and nine layers for the stripping section, for a total of about 3 m (10 ft) of packed height.

We have 4.2 mmHg pressure drop available. From Fig. 4 we could use a design F factor of 2.0. We will be a bit conservative and design for a value of 1.8 for a total drop of about 3.2 mmHg. The required diameter is 2.1 m (7 ft 3 in). For this diameter, the F factor at the bottom of the packing assuming a bubble point feed and constant molal overflow is 1.6.

Step 6 *Review head conditions.* For a base pressure of 9.2 torr and the estimated pressure drop of 3.2 mmHg, we can consider raising the head pressure to 6 mmHg. This will increase the vapor density by about 20% and would reduce the required column diameter to 2.15 m (7 ft). At this point we would prefer to keep the 1 mmHg as a design safety factor. Trimming the diameter can be reconsidered after review with the vendor.

Step 7 *Review the liquid loading.* The external reflux rate established in step 5 of 1202 kg/h (2650 lb/h) is 17.40 L/m (4.73 gal/min) based on a specific gravity of the liquid of 1.12. For the tower cross-sectional area of 3.84 m² (41.28 ft²), the unit-area liquid loading is 4.40 L/(m)(m²) [0.11 gal/(min)(ft²)]. This is approaching the minimum loading recommended by the proprietor. This column will not have much turndown ability. The situation would be helped somewhat by less subcooling of the reflux or even providing for reheat to the bubble point to give a higher quantity of reflux for distribution. Another possible adjustment of the design would be to use a higher reflux ratio with fewer theoretical stages, a tradeoff between diameter and height at the expense of energy usage. The higher reflux ratio will increase the required cross-sectional area. However, the liquid rate will increase more rapidly than the area.

Consider other packings which from a review of process requirements and equipment characteristics may be satisfactory. Do preliminary sizing as in steps 6 and 7 for all potential candidates.

Step 8 Prepare inquiry specifications for all potential suppliers, furnishing as much detailed information as possible.

DETAILED INFORMATION ON PACKINGS

Goodloe Packing

Vendor: Metex Process Equipment Corporation, 308 Talmadge Road, Edison, N.J. 08817.

Description: The packing is normally made of 0.0045-in-diameter wires, with 12 strands being knitted together to form a tube. The tube is then flattened to make a double-thickness ribbon approximately 6 in wide which is crimped, the creases of the crimping being at an angle to the centerline of the ribbon.

Two ribbons are then arranged in reversed relationship so that the creases cross each other and thereby determine the spacing of the adjacent ribbons. The two ribbons are rolled together until a cartridge is formed having enough layers to provide a diameter to fit the column snugly.

The vendor designs and supplies a line of custom liquid distributors.

Installation: The inner surface of the shell must be smooth. Care in installing the packing is essential.

Information for Preliminary Sizing: An information bulletin is available from the vendor for estimation of diameter, efficiency, and pressure drop.

Test Facilities: Tests can be arranged through the vendor if desired.

Hyperfil Packing

Vendor: Chem-Pro Equipment Corporation, 27 Daniel Road, Fairfield, N.J. 07006.

Description: Multifilament wire is knitted to form a tube, which is flattened into a ribbon and then crimped. A cartridge or a segment of a cartridge is then formed by folding the ribbon in parallel vertical layers, rather than being rolled into a spiral as with competing packings. The cartridges are stacked in the shell to the required height.

The vendor designs and supplies the liquid distributor. Two types of distributors are available: a weir type having about 24 distribution points per square foot and a capillary type with about 300 points per square foot. Use of the capillary type is claimed to decrease the HETS of the packing significantly. Wall-wiper liquid redistributors are placed at intervals of about 5 ft.

Information for Preliminary Sizing: An equation relating the pressure drop to the vapor mass velocity and molecular weight and the liquid density, and an equation for the maximum capacity, both at total reflux, are available from the vendor. Some data on HETS values can be supplied.

Test Facilities: Fairfield, N.J. (minimum absolute pressure: 0.1 torr; maximum temperature: 215°C).

Neo-Koss Packing

Vendor: Chem-Pro Equipment Corporation, 27 Daniel Road, Fairfield, N.J. 07000.

Description: Fine-mesh screen is rolled into 40-in-high cartridges. Spacer bands having $\frac{3}{16}$-in high upsets hold the successive turns $\frac{3}{16}$ in apart. The cartridges are stacked in the column shell to the desired height. The weight of each cartridge is carried on a supporting spider, from the center of which a thick bolt projects upward. The bottom spider of the assembly rests on lugs attached to the column shell. Uniform initial distribution of the entering liquid stream is essential since there is little or no opportunity for downflowing liquid to transfer from one layer of screen to an adjacent layer. A capillary-type distributor is supplied by the vendor as a part of the package.

Information for Preliminary Sizing: The vendor's *Bulletin HV-510* reports performance data at total reflux for cis-trans-decalin at 5 torr top pressure and dimethylphthalate/diethyphthalate at 1 torr. The data, pressure drop per foot of packing vs. F factor for the decalin system, are reproduced in Fig. 3 for use as a guide for preliminary sizing. Vapor-load F factors in the range of 4 to 5 have been demonstrated. However, beyond an F factor of about 2.5 rising pressure drop and HETS result in a rapid increase in pressure drop per theoretical stage. *Bulletin HV-510* provides pressure-drop equations for streamlined and turbulent flow based on vapor velocity, density, and viscosity.

Test Facilities: Fairfield, N.J. (minimum absolute pressure: 0.1 torr; maximum temperature: 215°C).

Additional Information: The greatest height of Neo-Kloss packing installed in one shell to date is 75 ft.

Koch-Sulzer Packing

Vendor: Koch Engineering Company, Inc., 4111 East 37th Street North, Wichita, Kans. 67220.

Description: Cylindrical or segmental packing sections are formed from parallel corrugated strips of woven-wire fabric. The corrugations are inclined with respect to the tower axis, and the direction of the corrugations is reversed on the adjacent strips. Two groups of parallel, crossed-flow passages of triangular shape are thus formed between adjacent strips. The packing sections, which are about 6.7 in thick, are stacked in the shell to the required height; each successive section is rotated 90°. A proprietary liquid distributor is furnished.

Information for Preliminary Sizing: The vendor recommends that preliminary sizing estimates be based on an F factor of 2.0 and HETS of 400 mm (8 in) and a pressure drop per theoretical stage of 0.40 mmHg. The vendor's *Bulletin KS-1* contains a generalized method for estimating the allowable vapor load for a given system at a specified operating pressure. Also contained in the bulletin are curves showing the relationship between the F factor and the HETS, pressure drop, and holdup for the system cis-trans-decalin at total reflux under an absolute pressure of 20 torr. Figure 4, reproduced from *Bulletin KS-1*, plots F factors vs. pressure drop per foot of packing for the cis-trans-decalin system at 20 torr. The chart can be used for preliminary design purposes. The vendor suggests an F factor of 3.3 as maximum. However, any loadings beyond a value of 2.0 should be used only upon specific advice of the proprietor.

Test Facilities: Wichita, Kans. (minimum absolute pressure 0.75 torr, maximum temperature 343°C).

Leva Film Trays

Vendor: Chemineer Mass Transfer, A Division of Chemineer, Inc., P.O. Box 1123, Dayton, Ohio 45401.

Description: Trays on spacings of 25 to 50 mm (1 to 2 in) have 50-mm (2-in) diameter holes on square pitch. The holes are centered between the holes on adjacent trays. Into each opening is fitted a venturi-shaped tube that extends downward about halfway to the tray below. The trays are assembled in bundles of, very roughly, 30 trays. Each bundle is held together by rods extending upward from a support grid; adjacent trays are separated by spacers. The top tray of each bundle is an inverted film tray modified to serve as a liquid distributor. An H-type pipe sparger is used to introduce reflux or liquid feed to a distributor tray. Each bundle must be level to ensure that even distribution of the liquid is established and maintained. Peripheral sealing rings between the trays prevent vapor from bypassing the entire bundle. The bottom bundle in the column rests on lugs welded to the column shell.

Information for Preliminary Sizing: Figure 5 is a generalized plot of pressure drop vs. F factor. Dr. Leva emphasizes that it should be used for preliminary estimates only and for diameters about 6 ft and larger. Note that the basis is feet of packed shell, including allowances for liquid distributors. However, the "typical" HETS of 0.45 m (18 in) also includes an allowance for distributors.

This device must operate at relatively low vapor F factors, usually in the range of 1.0 to 1.2, to achieve a pressure drop of 1 mmHg per theoretical stage for a typical HETS value. However, factors which make it competitive in some situations are favorable cost per unit of volume, lower sensitivity to fouling, and better resistance to corrosion.

Test Facilities: Pittsburgh, Pa. Additional testing facilities are located in Paris, France. Rental units are available for testing in a prospective customer's facilities.

Additional Information: The largest number of film trays installed to date in one column shell is 850. Dr. Max Leva is the inventor of the film tray and holds the patents on this device.

REFERENCES

1. Lubowicz, R. E., and Reich, P., *Chem. Eng. Prog.*, **67**(3), (March 1971).
2. McCabe, W. L., and Thiele, E. W., *Ind. Eng. Chem.*, **17**, 605 (1925).
3. Smoker, E. H., *Trans. Am. Inst. Chem. Eng.*, **34**, 165 (1938).
4. Gilliland, E. R., *Ind. Engl. Chem*, **32**, 1220 (1940).
5. Nygren, P. G., and Connolly, G. K. S., *Chem. Eng. Prog.*, **67**(3), (March 1971).

Liquid-Liquid Extraction

LANNY A. ROBBINS, Ph.D. *Associate Scientist, Dow Chemical USA, Midland, Mich.; member, American Institute of Chemical Engineers; member, Sigma Xi.*

NOMENCLATURE

A	Van Laar Constant
A^*	Activity of solute
E	Total flow rate of extraction
E'	Flow rate of extraction solvent in extract
F	Total flow rate of feed
F'	Flow rate of feed solvent in feed
K	Distribution coefficient, weight fractions
K'	Distribution coefficient, mole fractions
m	Distribution coefficient, weight ratios
N	Number of theoretical stages
P	Pressure
R	Total flow rate of raffinate
R'	Flow rate of feed solvent in raffinate
S	Total flow rate of extraction-solvent stream
S'	Flow rate of solute-free extraction solvent
W	Total flow rate of wash stream
W'	Flow rate of solute-free wash solvent
x	Weight fraction solute in raffinate

x^* Mole fraction solute in raffinate

X $\dfrac{\text{Weight solute}}{\text{Weight solute-free feed solvent in feed}}$

X'_r $\dfrac{\text{Weight solute in raffinate}}{\text{Weight solute-free feed solvent in feed}}$

X_f^B Pseudo X_f in Kremser equation, case B
X_f^C Pseudo X_f in Kremser equation, case C
y Weight fraction solute in extract
y^* Mole fraction solute in extract
y^{**} Mole fraction in vapor phase

Y $\dfrac{\text{weight solute}}{\text{Weight solute-free extraction solvent}}$

Y'_e $\dfrac{\text{Weight solute in extract}}{\text{Weight solute-free extraction solvent fed}}$

α Relative separation or selectivity
γ Activity coefficient
ε Extraction factor

SUBSCRIPTS

a Component a
b Component b
c Component c
e Extract
f Feed
r Raffinate
s Solvent
t Total
w Wash
$1,2,3, \ldots$ Stage number

THE EXTRACTION-PROCESS CONCEPT

Liquid-liquid extraction is a separation-technique which involves two immiscible liquid phases. It is an indirect separation technique because two components are not separated directly. A foreign substance, an immiscible liquid, is introduced to provide a second phase. This is in contrast to direct separation techniques, e.g., distillation, where heat is used to provide a vapor phase, or melt crystallization, where cooling is used to provide a solid phase.

A separation can be performed by liquid-liquid extraction whenever the ratio of one component to another is different in the two liquid phases. This is the same as performing a separation by distillation whenever the relative volatility is greater or less than 1.0. The simplest separation in extraction is when an immiscible solvent can be used to remove one component from a binary mixture. One example is solvent extraction of an impurity from wastewater. This is similar to a stripping or an absorption step in distillation where mass is transferred from one phase to the other.

In more complex separations, a binary mixture can be completely separated or fractionated by using a primary liquid to extract one of the components from a mixture and a secondary liquid to wash or scrub the extract free from the other component. This is similar to a fractional distillation scheme, where one component is enriched in one phase and another component is enriched in the other phase. The usual reason for using liquid-liquid extraction is to circumvent one difficult separation by using several relatively easy separations. For example, two close-boiling components that are difficult to separate by distillation may be separated easily by (1) liquid-liquid extraction, (2) solvent recovery, and (3) raffinate cleanup. The solvent-recovery and raffinate cleanup steps become an integral part of an extraction process.

Solvent recovery by distilling the solvent overhead is one of the most common schemes used in extraction processes. A typical example is the extraction of acetic acid from water with a low-boiling solvent (Fig. 1). In this case, the solute (acetic acid) is transferred from water into the solvent and the solvent is recovered as an overhead product from a distillation.

Solvent recovery by distilling the solute overhead can sometimes be achieved by selecting a high-boiling solvent.

Raffinate cleanup by stripping is a common practice in many extraction processes. In Fig. 1 the raffinate (water) is cleaned up by steam stripping residual solvent from the water. The relative volatility is enhanced by the low solubility of solvent in the raffinate.

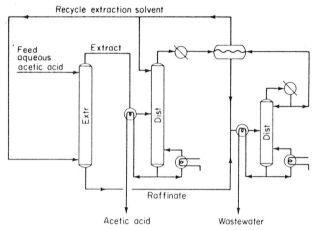

Fig. 1 Solvent extraction of acetic acid from water.

Brown[1] reviewed several recovery schemes for acetic acid recovery and reported the total operating cost to be about $6 per 1000 gal of water in 1963 when the acetic acid concentration was 2 to 10% and the solvents were either ethylacetate or diethylether. Lower operating costs can be achieved when less solvent can be used, e.g., in the Jones-Laughlin process for extracting phenol from water.[2] Even lower costs can be achieved if the solvent does not need to be recovered and the raffinate does not have to be cleaned up, e.g., the Phenex process for extracting 75 to 90% of the phenol from refinery water with light catalytic-cracking oil.[3]

Solvent recovery by extractive distillation is not an obvious technique to use in an extraction process, but it is used throughout the entire world in the Udex process (Fig. 2). The process, as originated by Dow Chemical, was developed by Universal Oil Products for separating aromatics from aliphatics.[4] In this process, the solvent is recovered by extractive distillation and the raffinate is cleaned up by a water wash. A light aliphatic fraction is used as the wash solvent in the extractor.

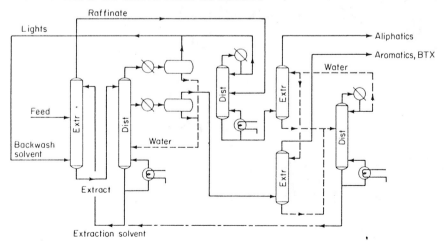

Fig. 2 Aromatics extraction from aliphatics with solvent recovery by extractive distillation.

Subsequent improvements on the process include the selection of solvents which would carry a higher loading of aromatics, e.g., sulfolane which is tetrahydrothiophene 1-1-dioxide,[5] triethylene glycol, and tetraethylene glycol.[6]

Solvent recovery by secondary extraction is a technique that can be used when high temperatures must be avoided or if lower-energy consumption can be realized. Somekh reported a Union Carbide Process[7] which used a secondary extraction for solvent recovery and a high-boiling wash solvent to reduce energy consumption. However, Kubek and Somekh[8] have reported that the best economics to date were achieved when tetraethylene glycol was used as the solvent, extractive distillation was used for solvent recovery, and mostly benzene was used as the wash solvent. Presumably, the light aliphatics were removed from the feed before the extraction and the water distillation was eliminated by using water from the lights decanter for final product washing.

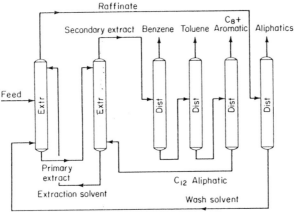

Fig. 3 Aromatics extraction from aliphatics with solvent recovery by secondary extraction. [*Somekh (Ref. 7) by permission.*]

The extraction process for purifying uranium[9,10] is another example where the solvent is regenerated by a secondary extraction (Fig. 4). This is typical of processes where the solutes are nonvolatile like metals or heat-sensitive like antibiotics.

These examples are presented here to show that the economic impact of an extraction process can be highly dependent on (1) the solvent that is selected and (2) the concepts used in the solvent recovery and raffinate cleanup. The selection and design of a specific extraction device can then be considered in proper perspective.

LIQUID-LIQUID EQUILIBRIUM

A separation of components by liquid-liquid extraction depends on the distribution of components (solutes) between two liquids (solvents). The two solvents must be immiscible or only partially miscible so that two liquid phases are present.

In the most simple case, one *solute* is present in a feed solvent. Then an immiscible extraction solvent is added and mixed with the liquid feed and the solute distributes between the two phases. Agitation of the two phases is continued until equilibrium is reached, and then agitation is stopped and the liquids are allowed to settle until both phases are clear. This constitutes one *equilibrium, theoretical,* or *ideal stage.* The liquid phase remaining that is rich in feed solvent is called the *raffinate.* The liquid phase remaining that is rich in extraction solvent is called the *extract.*

The *distribution coefficient* K is the ratio of the concentrations of the solute in each liquid phase at equilibrium:

$$K = \frac{y_a}{x_a} \quad \text{at equilibrium} \tag{1}$$

where y_a = weight fraction of solute a in the extract
x_a = weight fraction of solute a in the raffinate

The value of K is one of the main parameters used to establish the minimum solvent-feed ratio that can be used in an extraction process. For example, if the distribution coefficient is 4, a countercurrent extractor would require 0.25 or more pounds of solvent to remove all the solute from 1 lb of feed.

The *relative separation* or *selectivity* α can be described by the ratio of distribution coefficients for the two components:

$$\alpha(a/b) = \frac{K_a}{K_b} = \frac{y_a/x_a}{y_b/x_b} \tag{2}$$

This is similar to relative volatility in distillation. When two components are to be separated by liquid-liquid extraction, the ratio of one component to another must be different in each of the two phases.

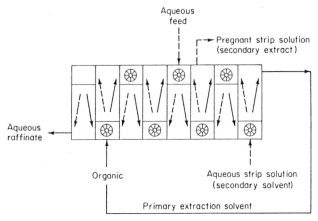

Fig. 4 Solvent extraction of uranium. *(Ref. 9.)*

The selection of a solvent for a liquid-liquid extraction process can often have a more significant impact on the process economics than any other design decision that has to be made. The liquid-liquid equilibrium for extraction involves almost entirely nonideal interactions because the starting point is to use two liquids which are so nonideal as to be immiscible or only partially miscible. Because of this, the selection of a solvent has usually been dependent on gathering an appreciable amount of experimental data from testing many solvents before narrowing down to one. But improvements are continually being made to understand the interactions between molecules and to predict the thermodynamic properties of mixtures, and this can be used to reduce the amount of experimental work required.

The *activity coefficient* of a solute in the liquid phase can be used to describe the thermodynamic basis for the distribution coefficient in liquid-liquid extraction. The activity A^* of a solute is the same in each phase at equilibrium [Eq. (3)].

$$A_c^* = \gamma_{cr}x_c^* = \gamma_{ce}y_c^* \tag{3}$$

where A_c^* = activity of component c
γ_{cr} = activity coefficient of component c in the raffinate phase
γ_{ce} = activity coefficient of component c in the extract phase
x_c^* = mole fraction of component c in the raffinate phase
y_c^* = mole fraction of component c in the extract phase

If the activity coefficient is the same in each phase, the solute will partition evenly between the two phases on a mole-fraction basis. But, if the activity coefficient is high in the feed solution and low in the extraction solvent, the distribution coefficient will be high.

EXAMPLE 1

The log of the activity coefficient of acetone in water was reported to be 0.89 at the infinite dilution based on vapor-liquid equilibrium data.[11] The activity coefficient of 0.022 mole fraction acetone in

1,1,2-trichloroethene was reported to be 0.732.[12] With acetone in water as a feed mixture, the distribution coefficient of acetone on a mole basis would be

$$K^* = \frac{y^*}{x^*} = \frac{\gamma_{cr}}{\gamma_{ce}} = \frac{(10)^{0.89}}{0.732} = \frac{7.76}{0.732} = 10.6$$

When this is converted to weight fractions at infinite dilution, K = 10.6 (18/133.4) = 1.43. This compares favorably with the experimental K value of 1.47 reported by Treybal.[12]

This extraction solvent, 1,1,2-trichloroethane, is probably one of the best available because it reduces the activity coefficient of the solute, acetone, in the extract layer but does not form a maximum-boiling-point azeotrope.

When a solvent does not interact with a solute, the activity coefficient of the solute will remain at unity and the escaping tendency of the solute into the vapor phase can be stated by Raoult's law; i.e., the partial pressure of the solute in the vapor phase is proportional to the pure-component vapor pressure times the mole fraction of the component in the liquid phase. If the escaping tendency is greater than predicted by Raoult's law, the deviation is considered to be a *positive deviation* and the activity coefficient is greater than 1.0. If the escaping tendency is lower than predicted by Raoult's law, the deviation is considered to be a *negative deviation* and the activity coefficient is less than 1.0. Positive and negative thus refer to the sign of the logarithm of the activity coefficient. A new extraction solvent will increase the partition coefficient if it can reduce the activity coefficient of the solute.

The primary source of activity-coefficient data is from vapor-liquid equilibrium measurements. When the vapor phase obeys the ideal-gas law, the activity coefficient can be calculated from Eq. (4).

$$\gamma_a = \frac{y^{**}_a}{x^*_a} \frac{P_t}{P_a} \qquad (4)$$

where y^{**}_a = mole fraction of component a in the vapor phase
P_t = total system pressure
P_a = vapor pressure of pure component a

EXAMPLE 2

Calculate the activity coefficient of 2.2 mole % acetone in 1,1,2-trichloroethane. Given[12,13]

Mole fraction acetone in liquid = 0.022
Mole fraction acetone in vapor = 0.080
System temperature = 113°C
System pressure = 1 atm
Acetone vapor pressure at 113°C = 5 atm

From Eq. (4):

$$\gamma = \frac{(0.080)(1)}{(0.022)(5)} = 0.727$$

Sometimes the azeotrope composition, temperature, and pressure are available for a binary system.[14] These data can be used to calculate activity coefficients if it is known whether the system forms a *homogeneous azeotrope* (one liquid phase) or a *heterogeneous azeotrope* (two liquid phases).

For a homogeneous azeotrope the vapor-phase composition is equal to the liquid-phase composition and the activity coefficient is proportional to the total system pressure divided by the pure-component vapor pressure, $\gamma_a = P_t/P_a$.

If a heterogeneous azeotrope is formed, the activity coefficient can be calculated from the azeotrope (vapor) composition and the saturated liquid composition at the azeotrope temperature in the same manner as any other vapor-liquid equilibrium data point. It should be emphasized, though, that the vapor composition is *not* equal to the liquid composition in a heterogeneous azeotrope. The heterogeneous-azeotrope data are especially helpful in evaluating the ease of stripping the extraction solvent from a raffinate stream by distillation. The activity coefficients and relative volatility can be calculated from the heterogeneous azeotrope composition and the saturated raffinate composition.[15]

Activity coefficients can also be calculated from the mutual solubility of two partially miscible liquids, if it can be assumed that the activity coefficient of the primary solvent is 1.0.

EXAMPLE 3

Calculate the activity coefficient of 1,1,2-trichloroethane in water from mutual-solubility data. Given[12]

Solubility of 1,1,2-trichloroethane in water is 0.44 wt %.
Solubility of water in 1,1,2-trichloroethane is 0.11 wt %.

Basis:

Feed solvent is water.
Extraction solvent is 1,1,2-trichloroethane.
Component b is 1,1,2-trichloroethane.

Calculate:

$x_b^* = 0.000596$, mole fraction trichlor in water phase.
$y_b^* = 0.9919$, mole fraction trichlor in trichlor phase.

Assume:

$\gamma_{bc} = 1.0$, activity coefficient of trichlor in trichlor phase.

Rearranging Eq. (3):

$$\gamma_{br} = \frac{\gamma_{bc}(y^*_b)}{x^*_b} = \frac{(1.0)(0.9919)}{0.000596} = 1664, \text{ activity coefficient of trichlor in water phase.}$$

Whenever one data point is available for calculating the activity coefficients in a binary system, the results can be extrapolated to other compositions using various equations. One of the most widely used sets of equations is the two-suffix Van Laar equations [Eqs. (5) to (8)]:

$$\log \gamma_a = \frac{A_{ab}}{(1 + A_{ab}x_a^*/A_{ba}x_b^*)^2} \tag{5}$$

$$\log \gamma_b = \frac{A_{ba}}{(1 + A_{ba}x_b^*/A_{ab}x_a^*)^2} \tag{6}$$

$$A_{ab} = \log \gamma_a \left(1 + \frac{x_b^* \log \gamma_b}{x_a^* \log \gamma_a}\right)^2 \tag{7}$$

$$A_{ba} = \log \gamma_b \left(1 + \frac{x_a^* \log \gamma_a}{x_b^* \log \gamma_b}\right)^2 \tag{8}$$

A_{ab} and A_{ba} are constants.

The constants A_{ab} and A_{ba} can be calculated from a single vapor-liquid equilibrium data point. The activity coefficients at infinite dilution are readily calculated from the limit where the concentration of that component goes to zero. A thorough discussion of the accuracy and limitations of the Van Laar equations is beyond the scope of this writing, but they have been evaluated elsewhere.[16,17]

In the absence of experimental data it is sometimes possible to estimate the activity coefficient at infinite dilution based on the molecular structures of the two compounds in solution. Pierotti, Deal, and Derr[18] presented a condensation of a large number of data points. More recent techniques involving the prediction of binary-component interactions include the method of Gilmont, Zudkevich, and Othmer,[19] the ASOG (analytical solution of groups) method,[20] and the NRTL (nonrandom two-liquid) method.[21]

Ewell, Harrison, and Berg[22] classified liquids into five groupings based on the possibility of a liquid being able to form hydrogen bonds. An initial attempt was made to describe the interactions among the various classes of solvents as to whether positive, +, or negative, −, deviations from Raoult's law were observed. Gilmont, Zudkevitch, and Othmer[19] subdivided some of the classes to give a total of nine groups which helped to develop a pattern of field-factor interactions which are useful for predicting binary vapor-liquid equilibrium data.

The present writer has combined these approaches into a table of interactions which can give a qualitative guide to solvent selection. Solvents were initially classified as in the Gilmont-Zudkevitch-Othmer tables. The number of maximum- and minimum-boiling-point azeotropes reported by Horsley[14] were then recorded for each binary interaction. A positive, +, deviation was assigned to a two-group interaction which produced a preponderance of minimum-boiling-point azeotropes. A negative, −, deviation was assigned for a preponderance of maximum-boiling-point azeotropes. A zero, 0, deviation was assigned to interactions which were seldom azeotropic or produced few of each type of azeotrope. Whenever subgroups provided the same interaction profile, they were grouped together. This led to the formation of a tenth group which did not fit into any other profile.

The table of interactions (Table 1) can provide a preliminary guide for the selection of the best solvents for an extraction process. For example, if a ketone were to be extracted from water, it would be desirable to use a solvent which reduced the activity coefficient of the ketone (solute). This would give a high distribution coefficient into the solvent. The two solvent groups which reduce the activity coefficients of ketones are 1 and 6. The solvent 1,1,2-trichloroethane is in group 6 and does reduce the activity coefficient of acetone below 1.0, as shown earlier.

TABLE 1   Solute-Solvent Group Interaction

		Solvent								
Group	Solute	1	2	3	4	5	6	7	8	9
1	Acid, aromatic OH (phenol)	0	−	−	−	−	0	+	+	+
2	Paraffin OH (alcohol), water, imide or amide with active-H	−	0	+	+	+	+	+	+	+
3	Ketone, aromatic nitrate, tertiary amine, pyridine, sulfone, trialkyl phosphate, or phosphine oxide	−	+	0	+	+	−	0	+	+
4	Ester, aldehyde, carbonate, phosphate, nitrite, or nitrate, amide without active-H, intramolecular bonding, e.g., *o*-nitrophenol	−	+	+	0	+	−	+	+	+
5	Ether, oxide, sulfide, sulfoxide, primary and secondary amine or imine	−	+	+	+	0	−	0	+	+
6	Multihalo paraffin with active H	0	+	−	−	−	0	0	+	0
7	Aromatic, halogen aromatic, olefin	+	+	0	+	0	0	0	0	0
8	Paraffin	+	+	+	+	+	+	0	0	0
9	Monohalo paraffin or olefin	+	+	+	+	+	0	0	+	0

Effect of interaction between groups on solute activity coefficient:
+   increases activity coefficient, minimum-boiling-point azeotropic
0   mild interaction, seldom azeotropic
−   decreases activity coefficient, maximum-boiling-point azeotropic

EXTRACTION SCHEMES

The equilibrium or theoretical stage in extraction was described earlier. In the following presentations it is shown as in Fig. 5. There are a number of multistage schemes that can be studied, but the main ones used in extraction are crosscurrent, countercurrent, and fractional extraction.

Crosscurrent extraction (Fig. 6) is a good laboratory procedure because each stage consists of an equilibrium stage where the two liquid phases are mixed together for a period of time until equilibrium is reached and then each phase is allowed to coalesce and they are separated by decanting. Fresh solvent is usually added to each progressive step. The extract (solvent) phase and the raffinate (feed) phase leaving the stage can each be analyzed to establish equilibrium data from each experiment. Also, the feasibility of extracting one or more components from the feed can be demonstrated with bench-scale experiments. However, a crossflow-extraction scheme is rarely used in a commercial process because of the large volume of solvent used and the low extract concentration which is generated.

Countercurrent extraction (Fig. 7) is used extensively in commercial processes. The solvent S enters the stage or the end of the extractor farthest from the feed point F, and the two phases pass countercurrent to each other. The objective is to strip one or more components from the feed liquid. If the contactor is an actual stage device, the phases will be separated before leaving each stage. But if the contactor is a differential device, one of one of the phases can remain as the dispersed (droplet) phase throughout the contactor as the phases pass countercurrent to each other. The dispersed phase is then generally allowed to coalesce at the end of the device before being discharged.

Dissociation extraction is a scheme that involves a nearly stoichiometric chemical reaction to bring about a liquid-liquid separation. A typical example is the use of caustic or a buffer at a high pH to preferentially extract acids or phenols from an organic solution.

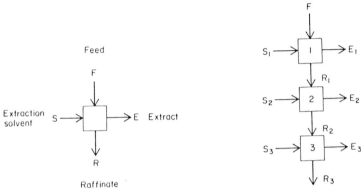

Fig. 5 Flow diagram of equilibrium (theoretical) extraction stage.

Fig. 6 Crosscurrent extraction.

The step of adding an acid to the aqueous solution to release the phenolic derivatives is referred to as *springing*. Similar schemes are used to extract amines or pyridines with acids; then a base is added to spring them back to free organics.[23]

Fractional extraction (Fig. 8) usually involves four components. Two immiscible liquids travel countercurrent through a contactor in order to separate at least two components from a feed mixture. One way of describing the operation is that the primary solvent S preferentially extracts (strips) one of the components from the feed F, and a wash solvent W preferentially scrubs the extract free from the unwanted solute. The washing in

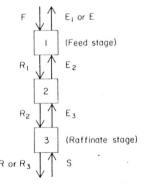

Fig. 7 Countercurrent extraction.

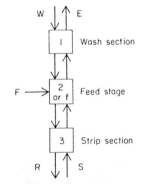

Fig. 8 Fractional extraction.

effect enriches the component being extracted. In this manner, two components in a feed mixture can be fractionated quite like stripping and rectification in continuous distillation.

Fractional dissociation extraction is a dissociation-extraction scheme where the mixture to be separated is fed near the center of the column. One of the fractionating solvents is an acidic stream and the other solvent is a basic stream.

LABORATORY EVALUATIONS

Exploratory-Solvent Selection

Liquid-liquid extraction can be used for many purposes in a laboratory, but the primary objective discussed here is development of a process. Bench-scale experiments are described for (1) selecting a solvent, (2) providing equilibrium data, and (3) providing data to help select a contactor.

In the exploratory stage of developing an extraction process there may be very few, if any, data in the literature that are relevant to the desired separation. One goal of the exploratory studies might then be to minimize the number of solvents tested and the number of analyses that have to be carried out.

TABLE 2   Typical Solvents Representing the Classes of Compounds

Class	Representative solvent
1	Phenol
2	Pentanol
3	Methylisobutylketone
4	Ethyl acetate
5	Ethyl ether
6	Methylene chloride
7	Toluene
8	Hexane
9	Butyl chloride

The table of interactions (Table 1) described earlier can be used to divide solvents into 9 classes. Generally the activity coefficient of a solute will stay within a factor of 2 or 3 for one class of solvents, but it may change by two or three orders of magnitude between classes. The relationship among classes of compounds has been correlated by Leo, Hansch, and Elkins[24] for comparing extraction data with the octanol-water system using several thousand data points. Treybal[13] and Francis[25] have also provided a collection of data which can help to evaluate solute-solvent group interactions.

When an experimental investigation is planned, typical solvents can be tested which represent the various classes; for example, see Table 2.

In the initial "shake-out" or equilibrium experiments one weight of solvent, say 20 g, can be contacted with an equal weight of feed mixture. In one study using ordinary laboratory separatory funnels[26] equilibrium was reached using about 50 inversions of the liquid layers during a 1- to 2-min period when the viscosities were as low as water. In some cases, e.g., at high temperatures, it may be desirable to use a stirred vessel. An agitator speed which will generate droplets of 1 to 2 mm in diameter is usually adequate. Violent agitation may be more detrimental in the coalescence and decantation steps that follow.

After the two liquid phases are mixed well and allowed to reach equilibrium, the agitation is stopped and the two liquid phases are allowed to coalesce and settle until there are no droplets left in either phase. The two phases are then sampled or separated by decantation for chemical analysis.

In exploratory experiments that do not require the ultimate precision, the feed sample x_f can be analyzed and then just the raffinate sample x_r which remains after contacting fresh feed with fresh solvent. If the solvents are relatively immiscible or symmetrically soluble and the solute concentration is low, the distribution coefficient can be calculated by material balance [Eqs. (9) and (1), Fig. 5].

$$Fx_f + Sy_s = Ey_e + Rx_r \tag{9}$$

$$K = \frac{y_e}{x_r} \tag{1}$$

If $F = S = E = R$ and $y_s = 0$,

$$K = \frac{x_f}{x_r} - 1 \tag{10}$$

This procedure can minimize the experimental work required, but considerable caution should be exercised because of the assumptions that are made and because of the error that can be associated with single observations. An analysis of both the extract and the raffinate layer is preferable.

If the intial exploratory experiments can identify the best two or three classes of solvents, subsequent experiments can be directed toward solvents within those classes. The distribution coefficients are dependent on the mole fraction of solute; so quite often a lower-molecular-weight solvent will improve the distribution coefficient on a weight-fraction basis. But at some point a lower-molecular-weight solvent may exhibit a high solubility in the raffinate and exhibit poor selectivity between the solute and the feed solvent. A high solubility of extraction solvent in the raffinate may have several effects on a commercial process, but not all are bad. Stripping of solvent from the raffinate may not be a serious problem if the relative volatility to the feed solvent is high, as discussed earlier for heterogeneous azeotropes. In wastewater extraction the more soluble solvents were found to be less toxic to fish.[27] Also a high solvent solubility can reduce the interfacial tension. This can mean lower agitation-intensity requirements in a commercial extractor so long as it is not so low as to create an emulsion problem.

In some cases it may be desirable to use a high-boiling solvent which does not have to be distilled overhead, so long as no degradation or undesirable azeotropes appear.

Providing Data for Process Design

After the choice of solvents has been narrowed down to one or two, reliable equilibrium data should be generated for design calculations and for evaluating the performance of extraction equipment. Expensive trial-and-error experiments in a pilot plant can often be minimized by generating complete equilibrium data before starting the pilot-plant investigation.

One important economic factor in a commercial process is the solvent loading that can be used, i.e., the amount of solute that can be carried with a solvent. The maximum concentration of solute is usually limited by the plait point in a Type I ternary-equilibrium diagram (Fig. 9). The solubility limits on a ternary diagram can often be determined by a simple titration procedure. A one-to-one mixture of feed solvent and extraction solvent can be stirred together while slowly adding solute until the solution becomes clear. The composition of the total mixture at that point establishes one point on the solubility line. Additional points can be determined by starting with other compositions, e.g., a two-to-one solvent ratio. The solubility of extraction solvent in a feed mixture can be determined by slowly adding solvent to a feed mixture until a few droplets or a cloudiness remains. The entire solubility line can usually be determined by these cloud-point and clear-point techniques.

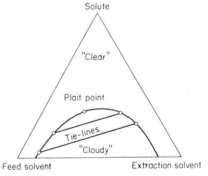

Fig. 9 Type I ternary-equilibrium diagram.

The consistency of equilibrium data can be visualized by using a correlation technique suggested by Hand.[28] The solute-to-solvent ratios are plotted on a log-log graph (Fig. 10). The ratio of solute to solute-free extraction solvent is on the Y axis and the ratio of solute to solute-free feed solvent is on the X axis. The use of solute-to-solvent ratios was first suggested by Bancroft.[29]

When a solubility limit point is determined by a cloud point (or clear point), the values of X and Y are the ratios of solute to solvents in the one-phase mixture.

When two liquid phases are contacted to reach equilibrium and separated for analysis, the Y value is the ratio of solute to solute-free "extraction solvent" in the extract layer and the X value is the ratio of solute to solute-free "feed solvent" in the raffinate layer. The Hand plot (Fig. 10) is very convenient for plotting liquid-liquid equilibrium data because the lines are nearly straight over one or two orders of magnitude. Also the equilibrium line can be extrapolated to intersect with the solubility line at the plait point.[12] The slopes of all three lines on a Hand plot are unity at low levels of solute, say below 0.1 wt % solute.

In a Type II ternary diagram (Fig. 11) the solute is not totally miscible in one of the solvents. Then the maximum loading of solute in the other solvent is not limited by a plait point. But for practical purposes the solute loading does not usually exceed 5 lb solute per pound of solvent. With neither X nor Y exceeding a value of about 5, the equilibrium line on a Hand plot is the same for Type I and Type II ternary systems. The difference is that the solubility line does not curve over to meet the equilibrium line in a Type II system.

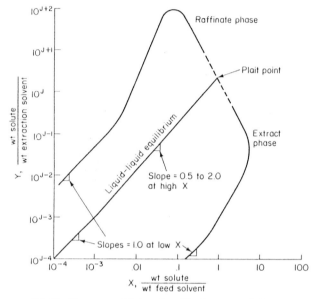

Fig. 10 Ternary-equilibrium data. Hand-type correlation.

By careful planning it may be possible to provide equilibrium data with the same experiments used to demonstrate feasibility of an extraction process at the bench scale. Usually there is some indication of what the distribution coefficient is from the exploratory experiments. So if a solvent-to-feed ratio S/F is selected to be about $1.5/K$, five crossflow equilibrium stages will remove about 99% of the solute from the feed and generate five equilibrium data points.

Other useful data can also be observed while performing the crossflow experiments. The density of the extract and raffinate streams is useful design information even if the

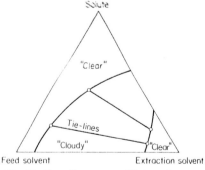

Fig. 11 Type II ternary-equilibrium diagram.

procedure is as simple as drawing a known volume into a pipette and weighing it into a sample bottle.

Other little observations can also be helpful, such as the stirrer size and r/min, i.e., tip speed, required to generate 1-mm-diameter droplets and the settling time required for coalescence. Any tendency for one phase to remain cloudy for a long period should be noted along with any formation of *rag*, i.e., solids or other buildup at the interface.

The viscosities of the solutions can affect the design and operation of an extractor and should be measured if the information is not available. A preliminary indicator can be the length of time required to drain a pipette as compared with the time for water.

The interfacial tension can also be an important design consideration in an extraction process. Treybal[30] observed a general correlation of interfacial tension with the amount of solute and dissolved solvent present in the layers (Fig. 12). This information may be helpful in decisions concerning solvent selection and type of extractor to be used in a commercial process.

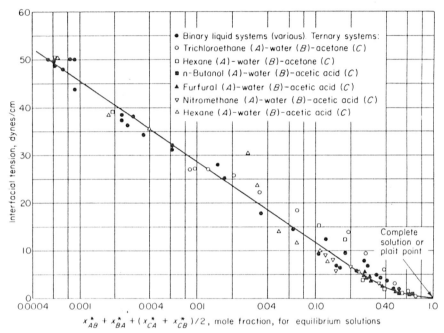

Fig. 12 A correlation of interfacial tension with mutual solubility for binary and ternary liquid mixtures. [*Treybal (Ref. 30) by permission.*]

Selection of an Extractor

Many different devices have been developed for liquid-liquid extraction, and the choice can involve many factors. These include the reliability of scale-up, the stages required, flow rates, capital, maintenance, materials of construction, floor space available, headroom available, turn-down flexibility, liquid residence time, emulsification tendencies, and volatility of the solvent. Treybal[30] and Logsdail and Lowes[31] have presented reviews on this subject. A decision network similar to the one proposed by Hanson[32] is presented here (Fig. 13). Generally the least complicated contactor which will perform the extraction with low maintenance is preferred for an industrial process.

The gravity columns without mechanical agitation require no moving parts and are utilized extensively in petroleum-refining applications which require only a few theoretical stages. But the height of the column can become excessive when a large number of theoretical stages is required. Mechanically agitated columns can generally attain a shorter height equivalent to a theoretical stage.

Mixer-settlers are used extensively in the mining industry where flow rates up to 10,000 gal/min are encountered. Mixer-settlers are also in common use for two-phase reactions or neutralizations which require a residence time of 0.5 to 2 min or longer to reach equilibrium.

Centrifugal extractors are used for extracting heat-sensitive antibiotics from fermentation broths with a very short residence time. Liquids which tend to emulsify severely can also be separated. But the maintenance tends to be high in a centrifugal extractor because of the fast-moving parts. Recently the reciprocating plate column has been used with a very low agitation intensity on systems which emulsify easily.

In order to exercise judgment concerning the ease or difficulty of dispersion (drop breakup), it is desirable to generate some experimental data. In general a system is easy to disperse if the interfacial tension is low and the density difference is small. But this also

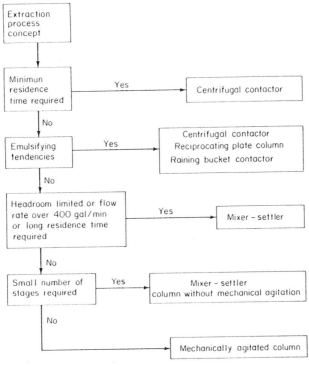

Fig. 13 Decision network for extraction selection.

can lead to slow coalescence and difficult phase separation. High column throughput can usually be attained only if the density difference is large and the continuous-phase viscosity is low. One problem is that the interaction among all these variables can be different for clean liquids than for actual plant streams. Consequently, in order for a person to develop a reliable design, the performance characteristics are usually obtained from an extractor which is operated with actual plant feedstock.

One column has recently been developed which can be operated over a very wide range of agitation intensities and can be used to characterize practically any extraction system with a small volume of liquid feed. This is a reciprocating-plate Karr column.[33] It has been scaled down to a 1-in-diameter laboratory column which requires only 10 to 50 gal of feedstock for a test.[34] The flooding characteristics have been correlated with a combination of most of the variables affecting an extraction system.[35]

Figure 14 shows the flooding curve for the xylene–acetic acid–water system, which is known to have a high interfacial tension, and for the MIBK–acetic acid–water system, which has a lower interfacial tension. The operating point for a commercial fermentation-broth system is also shown, which reportedly can emulsify quite easily.[36]

A Karr agitation intensity (r/min × stroke length/plate spacing) below 50 appeared to give a HETS (height equivalent to a theoretical stage) in the same range as other columns without mechanical agitation. The Karr column gave a HETS of 500 to 650 mm (20 to 26 in) with an agitation intensity of 0 to 50 in the MIBK (methylisobutyl ketone)–acetic acid–water system.[33] Similarly, a perforated-plate column gave a HETS of 550 to 900 mm (22 to 36 in) in the ethyl acetate–acetic acid–water system.[37]

The use of mechanical agitation in an extraction column can usually reduce the height that is required, especially if the interfacial tension is high. In the MIBK–acetic acid–water system the HETS was reduced from 650 mm (26 in) to 125 mm (5 in) by increasing the Karr agitation intensity to 160. For the high interfacial tension system, xylene–acetic acid–water, the HETS was reduced from 8750 mm (350 in) to 200 mm (8 in) by increasing the Karr agitation intensity to 250.

Mixer-settlers typically exert a high agitation intensity, i.e., tip speed of 171 m/mm (570 ft/min), and employ a settler throughput of 0.814 to 4.07m³/(h)(m²) [20 to 100 gal/(h)(ft²)] of interface area.[38] If such a low throughput and high agitation intensity are required in a Karr column to obtain a low HETS, possibly a mixer-settler should be considered.

The procedure for establishing the flooding characteristics in a Karr column, say 25 mm (1 in) in diameter by 0.6 m (2 ft) tall, consists of the following steps. First, the liquid feed and solvent are fed continuously into the column with the agitator turned off. The interface is maintained near one end of the column by controlling the bottom effluent. Then both flow rates are increased slowly until the droplets are touching each other as in fluidized bed. At higher feed rates the column will flood and a second interface will form at the end of the column where the dispersed (droplet) phase is being fed. After this flood point is determined with zero agitation intensity, the feed rates can be reduced to say 80% of flooding. Then the agitator can be started at a slow speed and gradually increased until flooding occurs again to establish another flood point. This procedure can be repeated for the entire range of column throughput. Then a similar flooding curve can be developed with the other phase dispersed, i.e., by starting with the interface at the other end of the column. The flooding points can be plotted on a graph similar to Fig. 14 for comparison with the known systems and in combination with Fig. 13 to help select an extractor for scale-up.

EXTRACTOR-PERFORMANCE AND THEORETICAL-STAGE CALCULATIONS

The first requirement for evaluating the performance or design of an extractor is reliable liquid-liquid equilibrium data. The procurement of these data was discussed earlier.

The main objective of theoretical-stage calculations is to evaluate the compromise between the ratio of solvent to feed and the number of stages, i.e., size of the equipment, required in an extraction process. The concept of a theoretical equilibrium stage was discussed earlier. The minimum solvent-to-feed ratio considered in a countercurrent extraction process is that which would create a pinch point in the design calculations. A pinch point requires an infinite number of theoretical stages to perform the required separation.

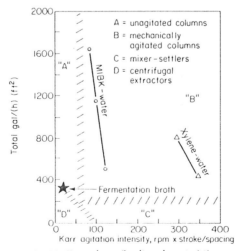

Fig. 14 Karr-column flooding characteristics.

The driving force for mass transfer in an extractor involves the solvent-to-feed ratio that is used relative to the minimum which would give a pinch point. This is usually expressed as the extraction factor ε [Eq. (11)].

$$\varepsilon = m \, \frac{S'}{F'}$$ (11)

where $m = Y/X$, distribution coefficient in Bancroft coordinates
S' = lb/h solute-free extraction solvent
F' = lb/h solute-free feed solvent

The extraction factor is similar to the stripping factor, absorption factor, or minimum reflux ratio used in vapor-liquid separations. The minimum solvent-to-feed ratio required to remove all of a specific component is $1/m$, that is, $\varepsilon = 1.0$.

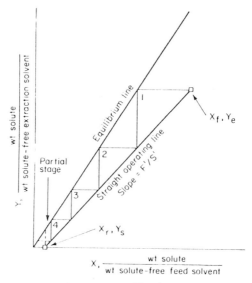

Fig. 15 Immiscible solvents.

As the driving force for mass transfer is increased, fewer theoretical countercurrent stages are required. In extraction, an increase in driving force for mass transfer means selecting a solvent with a higher distribution coefficient m or increasing the solvent-to-feed ratio S'/F'. An increase in solvent flow rate obviously implies an increase in solvent-recovery requirements. The continual expense of an increased solvent-recovery cost should be carefully compared against an increase in capital cost for larger equipment which would deliver more theoretical stages. The operating costs of an extractor are generally very low in comparison with the operating costs of a distillation column. Mechanically agitated extractions are in common use which give the equivalent of 5 to 8 theoretical countercurrent stages, with some designed to give 10 to 12 theoretical stages.

Countercurrent-Extraction Stage Calculations

The calculations for a countercurrent liquid-liquid extraction can be carried out in a number of ways using the concept of theoretical, ideal, or equilibrium stages. The equipment used for extraction may or may not actually mix and settle the phases as they pass countercurrent, but the performance of an extractor can be evaluated as stage efficiency in staged devices or height equivalent to a theoretical stage (HETS) in columns. The following calculations are concerned with computing the number of theoretical stages to bring about a desired separation. The stagewise computation of liquid-liquid extraction has much in common with the stagewise calculation of vapor-liquid separations, such as absorbers, strippers, and distillations. One helpful concept is to select the right parameters that will give straight operating lines in graphical or mathematical solutions. In vapor-liquid calculations, the concept of constant molar flow rates gave rise to the McCabe-Thiele method.

In liquid-liquid extraction, the concept of solute-free solvents passing countercurrent to each other can be used to simplify the calculations similar to the McCabe-Thiele method. The concentrations are then given as the ratio of solute to extraction solvent Y and the ratio of solute to feed solvent X. These concentrations will lead to a straight operating line on an X-Y diagram (Fig. 15). Equilibrium data in these concentrations have already been shown to be nearly a straight line on a log-log plot (see Fig. 10). The remaining problem then is to evaluate the primary ratio of solute-free solvents passing through the extractor.

Most extractions can be classified as having either (1) immiscible solvents, (2) partially miscible solvents with a low solute concentration in the extract, or (3) partially miscible solvents with a high solute concentration in the extract. The stagewise countercurrent-extraction scheme is shown in Fig. 16.

In *Case A* the solvents are immiscible so that the rate of solute-free feed solvent F' is the same as the solute-free raffinate rate R'. In like manner the rate of solute-free solvent S' is the same as the solute-free extract rate E'. The primary ratio of extraction solvent to feed solvent is therefore $S'/F' = E'/R'$. A material balance can be written around the feed end of the extractor down to any stage n. The material-balance equation can be rearranged to a McCabe-Thiele type of operating line with a slope of F'/S' [Eq. (12)].

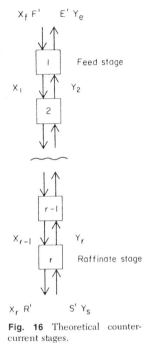

Fig. 16 Theoretical countercurrent stages.

$$Y_{n+1} = \frac{F'}{S'} X_n + \frac{S'Y_e - F'X_f}{S'} \qquad (12)$$

The end points of the operating line on an X-Y plot are X_r, Y_s, and X_f, Y_e. The equilibrium curve is taken from the Hand correlation shown earlier in Fig. 10. The number of theoretical stages can be stepped off graphically or calculated mathematically between the equation of the operating line and the equation of the equilibrium line.

In *Case B* the solvents are partially miscible and the miscibility is nearly constant throughout the extractor. This frequently occurs when all solute concentrations are relatively low. So, the feed stream is assumed to dissolve extraction solvent only in the feed stage and likewise the extraction solvent is assumed to dissolve feed solvent only in the raffinate stage. With these assumptions the primary "extraction solvent" rate is assumed to be S' and the primary "feed solvent rate" is assumed to be F'. But now the extract rate E' is less than S', so that the extract composition Y_e must be converted to the primary extraction solvent rate in order for it to fall on the operating line, that is,

$$Y'_e = \frac{E'Y_e}{S'} \qquad (13)$$

In like manner the raffinate rate R' is less than F', so that the raffinate composition X_r must be converted to the primary feed solvent rate in order for it to fall on the operating line, that is,

$$X'_r = \frac{X_r R'}{F'} \qquad (14)$$

The slope of the operating line is F'/S', and the end points are X'_r, Y_s, and X_f, Y'_e; see Fig. 17. The equation of the operating line is

$$Y_{n+1} = \frac{F'}{S'} X_n + \frac{S'Y'_e - F'X_f}{S'} \qquad (15)$$

For a graphical calculation using Case B the following is recommended:
1. Plot the equilibrium data on an X-Y plot.
2. Choose the design conditions X_f, X_r, X_s, Y_e.
3. Calculate the Y'_e and X'_r [Eqs. (13) and (14)].
4. Plot the ends of the operating line X'_r, Y'_s, and X_f, Y'_e and draw a straight line between them.
5. Plot the actual raffinate end conditions X_r, Y_s and step off the raffinate stage r only. See Fig. 17.

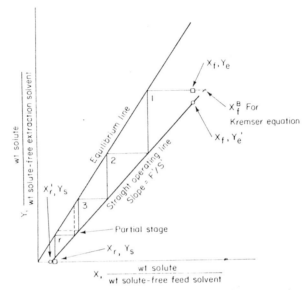

Fig. 17 Case B. Low solute concentration in extract. Mutual solubility constant through extractor.

6. Plot the actual feed end conditions X_f, Y_e and step off equilibrium stages until the raffinate stage is reached; see Fig. 17.

In *Case C* the solvents are partially miscible but the mutual solubility of solvent is relatively high at the feed end of the extractor and low at the raffinate end of the extractor. This occurs often when the solute concentration is high in the feed and extract but low in the raffinate. The feed stream is assumed to dissolve extraction solvent only in the feed stage just as in Case B. But the extraction solvent is assumed to dissolve a large amount of feed solvent in the feed stage when compared with the amount dissolved in the raffinate stage. With these assumptions the primary "extraction solvent" rate is assumed to be S' and the primary "feed solvent" rate is assumed to be R'. Again the extract rate E' is less than S', so that the extract composition Y_e must be converted to the primary extraction solvent rate in order for it to fall on the operating line, that is,

$$Y'_e = \frac{E'Y_e}{S'} \tag{13}$$

With similar reasoning the feed concentration would appear higher because so much feed solvent is lost in the feed stage; so the feed concentration X_f must be converted to the primary feed-solvent rate R' in order for it to fall on the operating line, that is,

$$X'_f = \frac{F'X_f}{R'} \tag{16}$$

The slope of the operating line on an X-Y diagram is R'/S' and the end points are X_r, Y_s and $X'_f Y'_e$; see Fig. 18. The equation of the operating line for Case C is

$$Y_{n+1} = \frac{R'}{S'} X_n + \frac{S'Y'_e - R'X'_f}{S'} \tag{17}$$

For a graphical calculation using Case C the following is recommended:
1. Plot the equilibrium data on an X-Y plot.
2. Choose the design conditions, X_f, X_r, Y_s, Y_e.
3. Plot the point of the raffinate end of the column X_r, Y_s.
4. Draw a straight operating line with a slope of R'/S' which goes through the raffinate point; see Fig. 18. The operating line should go through X_f', Y_e''.
5. Plot the point of the actual feed end of the column X_f, Y_e and step off the stages until the raffinate composition X_r is reached; see Fig. 18.

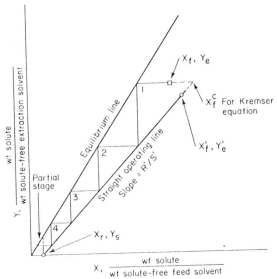

Fig. 18 Case C. High solute concentration in extract. Mutual solubility high in feed stage, low in raffinate stage.

EXAMPLE 4[10]

Forty-five kilograms (one hundred pounds) per hour of a 50% acetone–50% water solution is to be reduced to 10% acetone with 13.5 kg/h (30 lb/h) of 1,1,2-trichloroethane as solvent in a countercurrent multiple-contact operation at 25°C. Determine the number of stages and concentrations and weights of the various streams.

Solution Equilibrium data are from Treybal, Weber, and Daley,[12] Table 3, and Fig. 19. For first trial at calculating the mass balance, assume no mutual solubility of solvents.

TABLE 3   Equilibrium Data for Water–Acetone–1,1,2-Trichloroethane

Wt fraction in aqueous layer				Wt fraction in organic layer				
Trichlor	Water	Acetone	X	Trichlor	Water	Acetone	Y	Y(calc)
0.0052	0.9352	0.0596	0.0637	0.9093	0.0032	0.0875	0.0962	0.100
0.0054	0.9295	0.0651	0.0700	0.8932	0.0040	0.1028	0.1151	0.111
0.0068	0.8535	0.1397	0.1637	0.7832	0.0090	0.2078	0.2653	0.272
0.0102	0.7206	0.2692	0.3736	0.5921	0.0227	0.3852	0.6506	0.650
0.0160	0.6267	0.3573	0.5701	0.4753	0.0426	0.4821	1.014	1.02
0.0210	0.5700	0.4090	0.7175	0.4000	0.0605	0.5395	1.349	1.30
0.0375	0.5020	0.4605	0.9173	0.3370	0.0890	0.5740	1.703	1.68

$$\text{Exponent} = \frac{\log Y_2/Y_1}{\log X_2/X_1} = \frac{\log 1.014/0.1057}{\log 0.5701/0.0669} = 1.057$$

$$\text{Constant} = \frac{Y}{(X)^{1.057}} = \frac{1.014}{(0.5701)^{1.057}} = 1.84$$

$$Y(\text{calc}) = 1.84 \, (X)^{1.057}$$

$F' = 22.5$ kg/h (50 lb/h) water $= R'$
$S' = 13.5$ kg/h (30 lb/h) trichlor $= E'$
$X_f = 0.225$ kg (0.50 lb) acetone/0.225 kg (0.50 lb) water $= 1.00$
$X_r = 0.045$ kg (0.10 lb) acetone/0.405 kg (0.90 lb) water $= 0.11$

$$Y_e = \frac{F'X_f - R'X_r + S'Y_s}{E'} = \frac{(50)(1.0) - (50)(0.11) + 0}{30} = 1.48$$

Check Hand plot (Fig. 19) for mutual solubility.

At $Y_e = 1.48$ lb acetone/lb trichlor, the trichlor phase contains 7.5 lb acetone/lb water. Therefore, the water in the extract $= 1.48/7.5 = 0.197$ lb water/lb trichlor. At $X_r = 0.11$ lb acetone/lb water, the water phase contains 16.5 lb acetone/lb trichlor. Therefore, the trichlor in raffinate $= 0.11/16.5 = 0.0067$ lb trichlor/lb water.

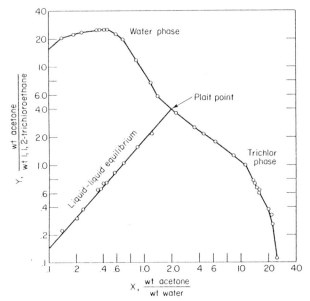

Fig. 19 Water–acetone–1,1,2-trichloroethane equilibrium and solubility [Hand correlation, (Ref. 28) Treybal et al. data (Ref. 12)].

For the second trial, note that very little trichlor is lost with the raffinate so that E' will be nearly equal to S'. The water in the extract will be $(0.197)(E') = (0.197)(30) = 5.91$ lb water/h; so the water in the raffinate will be the difference

$$R' = F' - 5.91 = 50 - 5.91 = 44.09 \text{ lb water/h}$$

The trichlor in the raffinate water will be $0.0067 R' = (0.0067)(44.09) = 0.30$ lb trichlor/h; so the trichlor in the extract will now be the difference

$$E' = S' - 0.0067 (R') = 30 - 0.3 = 29.7 \text{ lb trichlor/h}$$

The water in the extract would now be recalculated as $(0.197)(29.7) = 5.85$.
The extract composition can now be recalculated:

$$Y_e = \frac{(50)(1.0) - (44.1)(0.11)}{29.7} = 1.52$$

The water in the extract $= 1.52/7.3 = 0.208$ lb water/lb trichlor.

The mass balance has already converged within 5% in two trials. This is about the accuracy of graphical techniques, but a third trial will converge more precisely.

The water in the extract $= (0.208)(29.7) = 6.18$ lb water/h.

$$R' = 50 - 6.18 = 43.82 \text{ lb water/h}$$

The trichlor in the raffinate $= (0.0067)(43.82) = 0.29$ lb trichlor/h.

$$E' = 30 = 0.3 = 29.7 \text{ lb trichlor/h}$$

$$Y_e = \frac{(50)(1.0) - (43.8)(0.11)}{29.7} = 1.52 \text{ lb acetone/lb trichlor}$$

The weight percent acetone in the extract is

$$100y_e = \frac{(1.52)(100)}{1.52 + 0.208 + 1.00} = 55.72 \text{ wt \% acetone}$$

Compare the solubility of feed solvent in the extract with the solubility of the feed solvent in the extraction solvent leaving the raffinate stage. From the previous calculation, the solubility of water in the extract is 0.208 lb water/lb trichlor. But at the raffinate end of the extractor, the solubility of water in the trichlor is only 0.004 lb water/lb trichlor. Read trichlor solubility from Hand plot (Fig. 19) at raffinate composition $X_r = 0.11$, $Y_r = 0.165$, and solubility in the trichlor phase is 25 lb acetone/lb

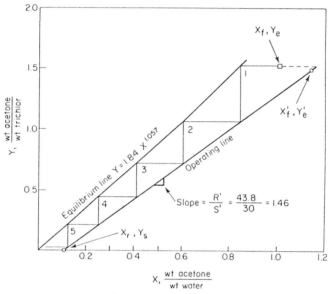

Fig. 20   Graphical stage calculations for Example 4.

water, so that the water in the trichlor leaving the raffinate stage is 0.11/25 = 0.0044 lb water/lb trichlor. This is much lower than the water leaving with trichlor in the extract stage; so use Case C assumptions.

Case C assumes the primary water rate going through the extractor is the raffinate water rate R'. The primary trichlor rate is assumed to be the solvent inlet rate S'. So, for consistency on the operating line, all concentrations must be in terms of S' and R'.

$$Y_s = 0$$
$$Y'_e = \frac{Y_r E'}{S'} = \frac{(1.52)(29.7)}{30} = 1.50$$
$$X_r = 0.11$$
$$X'_f = \frac{X_f F'}{R'} = \frac{(1.0)(50)}{43.8} = 1.14$$

On an X-Y diagram, the operating line is essentially a straight line going from X_r, Y_s, up to X'_f, Y'_e, with a slope of R'/S'; see Fig. 20. The equation of the operating line is as follows:

$$Y_{n+1} = \frac{43.8}{30} X_n + \frac{(30)(1.50) - (43.8)(1.14)}{30}$$
$$Y_{n+1} = 1.46 X_n - 0.164$$

Stepping off the theoretical stages can be done either graphically (Fig. 20) or mathematically (Table 4). The mathematical procedure is quite easy and is recommended for high recovery where the operating line and equilibrium line get very close. The equilibrium values of X and Y can be used from the Hand plot (Fig. 19) for each stage or calculated as from a straight-line log-log correlation: Calculations begin at $X_f = 1.0$, $Y_e = 1.52$ and proceed until $X_r = 0.11$.

Answer $r = 5.1$ theoretical stages.

The ease of converting this mathematical procedure into a programmed calculator routine is probably self-evident. The answer is in agreement with the 5.2 theoretical stages that Treybal obtained using a graphical technique on a triangular diagram.[30]

When the operating line is straight and the equilibrium line is straight, too, the countercurrent cascade of theoretical stages can be solved analytically to give the well-known Kremser equations.[39] See also McCabe and Smith[40] and Brian.[41] The Kremser equations are especially useful for studying the tradeoff between solvent-to-feed ratio and the number of stages required to achieve the desired raffinate concentration. This was one of the reasons for the complete development of linear operating lines for Cases A, B, and C.

TABLE 4   Mathematical Stage Calculations

$Y_e = Y_1 = 1.52$

$X_1 = (0.562) (1.52)^{0.946} = 0.835$

$Y_2 = (1.46) (0.835) - 0.164 = 1.06$

$X_2 = (0.562) (1.06)^{0.946} = 0.591$

$Y_3 = (1.46) (0.591) - 0.164 = 0.699$

$X_3 = (0.562) (0.699)^{0.946} = 0.401$

$Y_4 = (1.46) (0.401) - 0.164 = 0.421$

$X_4 = (0.562) (0.421)^{0.946} = 0.248$

$Y_5 = (1.46) (0.248) - 0.164 = 0.198$

$X_5 = (0.562) (0.198)^{0.946} = 0.121$

$Y_6 = (1.46) (0.121) - 0.164 = 0.0133$

$X_6 = (0.562) (0.0133)^{0.946} = 0.0094$

Step 6 has exceeded the raffinate level; so the fractional stage is calculated as follows:

$$\frac{0.121 - 0.110}{0.121 - 0.007} = \frac{0.011}{0.112} = 0.1$$

The case of immiscible solvents, Case A, was straightforward into the Kremser equations, but now the other cases, B and C, can also be used in the Kremser equations. However, a pseudo feed concentration X''_f or X^C_f must be used with the appropriate ratio of solvents in the calculations; see Figs. 17 and 18.

The main Kremser equations for calculating the number of theoretical extraction stages are as follows:

When $\varepsilon = 1$,

$$N = \frac{X_f - X_r}{X_r - Y_s/m} \tag{18}$$

When $\varepsilon \neq 1$,

$$N = \frac{\log \left[(1/U) (1 - 1/\varepsilon) + 1/\varepsilon \right]}{\log \varepsilon} \tag{19}$$

where ε = extraction factor (see specific case)

N = number of theoretical stages

$U = \dfrac{X_r}{X_f}$, fraction unextracted when pure extraction solvent is used

$U = \dfrac{X_r - Y_s/m_r}{X_f - Y_s/m_r}$, when recycle extraction solvent is used

The solution to these equations is presented graphically in Fig. 21.

The response of an extraction system can be evaluated by plotting the log of the raffinate concentration, vertically, vs. the solvent-to-feed ratio, horizontally, for a fixed number of theoretical stages. This will be the general operating characteristics of the commercial extractor in a process.

For Case A. Immiscible solvents. Use

$$\varepsilon = m \frac{S'}{F'} \tag{20}$$

X_r, Y_s, and X_f

For Case B. Low solute in extract. Use

$$\varepsilon = m \frac{S'}{F'} \tag{21}$$

X_r, Y_s, and X''_f

where

$$X_f^B = X_f + \frac{S' - E'}{F'} Y_e \qquad (22)$$

For Case C. High solute in extract. Use

$$\varepsilon = m \frac{S'}{R'} \qquad (23)$$

X_r, Y_s, and X_f^C
where

$$X_f^C = \frac{F'}{R'} X_f + \frac{S' - E'}{R'} Y_e \qquad (24)$$

The equations for the pseudo feed concentrations X_f^B and X_f^C were derived from the equations of the operating lines where $Y_{n+1} = Y_e$.

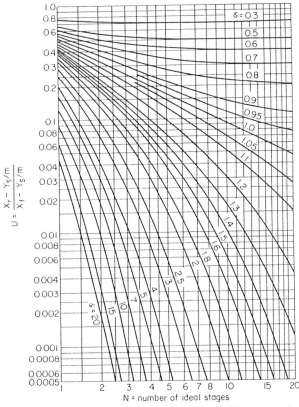

Fig. 21 Countercurrent multistage extraction with constant solvent/solvent ratio and constant distribution coefficient. [*Treybal (Ref. 30) by permission.*]

When the equilibrium line is not perfectly straight, Treybal[30] suggests using the geometric mean value of the distribution coefficients, i.e.,

$$m = \sqrt{m_r m_e} \qquad (25)$$

$$m_r = \frac{Y_r}{X_r} \qquad (26)$$

$$m_e = \frac{Y_e}{X_e} = \frac{Y_1}{X_1} \qquad (27)$$

EXAMPLE 5

Calculate the number of theoretical stages required in Example 4 using the Kremser equations. Basis: Case C, nonlinear equilibrium curve.

$$m_r = \frac{Y_r}{X_r} = \frac{1.84\,(0.11)^{1.057}}{0.11} = 1.62$$

$$m_e = \frac{Y_e}{X_e} = \frac{1.52}{0.835} = 1.82$$

$$m = \sqrt{m_r m_e} = \sqrt{(1.62)(1.82)} = 1.72$$

$$\varepsilon = m\,\frac{S'}{R'} = (1.72)\,\frac{30}{43.8} = 1.18$$

$$X_r = 0.11,\ Y_s = 0,\ X_f = 1.00$$

$$X_f^C = \frac{50}{43.8}\,1.00 + \frac{30 - 29.7}{43.8}\,1.52 = 1.15$$

$$U = \frac{X_r}{X_f^C} = \frac{0.11}{1.15} = 0.096$$

$$N = \frac{\log\left[(1/0.096)(1 - 1/1.18) + 1/1.18\right]}{\log 1.18} = \frac{0.387}{0.072}$$

$$N = 5.4\ \text{theoretical stages}$$

For preliminary design work where the distribution coefficient does not change drastically, the Kremser equations are recommended as long as the corrections are applied for mutual solvent solubility. For systems with highly nonlinear distribution coefficients, the graphical McCabe-Thiele type of method is recommended. When the distribution coefficients can be adequately described mathematically, the McCabe-Thiele type of method is also recommended for computer calculations. In all cases, the use of solute/solvent, i.e., Bancroft, coordinates is recommended for liquid-liquid extraction calculations.

Fractional Liquid-Liquid Extraction Stage Calculations

The most powerful liquid-liquid separation technique available is fractional extraction. Two solutes can be separated nearly completely by isolating one solute in the extraction solvent and the other solute in a wash solvent (see Fig. 22). The lower section of the extraction is much the same as a stripping extractor with the solvent S entering the bottom and stripping one of the components from the raffinate. As the extract travels above the feed stage, it is contacted countercurrently with a wash solvent W which scrubs the unwanted solute out of the extract, which in effect purifies the solute being extracted. With this concept in mind, the stripping section and washing section can be calculated individually with separate operating lines. The overall material balance must be satisfied at the feed stage.

For the simplest case, assume that the solvents are immiscible and that no solvents are present in the feed. In order for the extraction solvent to remove one of the components, a but not the other component, b, the extraction factor $\varepsilon = mS'/W'$ must be greater than 1 for component a and less than 1 for component b. For a symmetrical separation, the ratio of wash solvent to extraction solvent should be set equal to the geometric mean of the two distribution coefficients.[53]

$$\frac{W'}{S'} = \sqrt{m_a m_b} \tag{28}$$

This ratio of wash solvent to extraction solvent is the same in both sections of the column when no solvent is added with the feed. The desired separation can be selected, say 90% of component a in the extract and 90% of component b in the raffinate. Then the amount of feed can be selected so that the extract and raffinate loadings are reasonable, e.g., in the range of X_r or $Y_e = 0.01$ to 0.5. An overall material balance can be determined around the extractor. Then an X-Y plot can be constructed for each solute (see Figs. 23 and 24). The raffinate coordinates of the extractor X_{ar}, Y_{as} can be plotted for component a and the

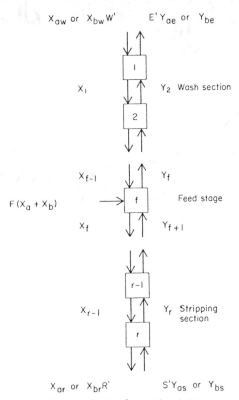

X_{aw} or $X_{bw} W'$ $E' Y_{ae}$ or Y_{be}

X_1 Y_2 Wash section

X_{f-1} Y_f

$F(X_a + X_b)$ f   Feed stage

X_f Y_{f+1}

X_{r-1} Y_r Stripping section

X_{ar} or $X_{br} R'$ $S' Y_{as}$ or Y_{bs}

Fig. 22   Fractional-extraction stages.

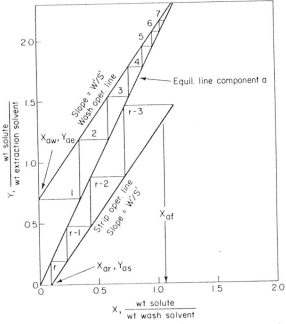

Fig. 23   Graphical calculation of fractional extraction, component a.

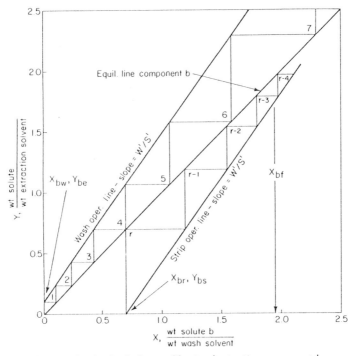

Fig. 24 Graphical calculation of fractional extraction, component b.

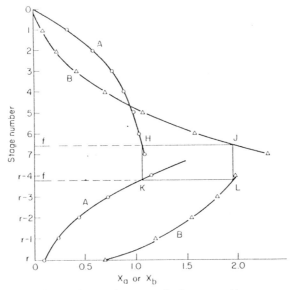

Fig. 25 Fractional-extraction-feed stage matching.

operating line can be drawn with a slope of W'/S'. The extract coordinates of the extractor X_{aus} Y_{ae} can also be plotted for component a with another operating line having the same slope W'/S'.

The theoretical stages can be stepped off in the conventional manner for each section of the extractor starting at the extract end, stage 1, and proceeding toward the feed stage f, then restarting at the raffinate end, stage r, and proceeding toward the feed stage f. The same procedure is repeated for component b (Fig. 24).

The feed-stage number can be determined by matching the concentrations and stage number at the point of feed introduction.[42] This is carried out by plotting the stage number vs. X for each component (see Fig. 25). The requirements are met when the rectangle $HJLK$ can be drawn as shown. The number of stages in the wash section plus the feed stage f is determined from the position of line HJ. The total number of stages r is calculated from the position of LK, which is also the feed stage.

It can be seen from both Fig. 23 and Fig. 24 that the solute concentrations are the highest at the feed stage and also the concentrations at the feed stage increase with the number of stages used. The amount of solvents used must be sufficiently high so that solute solubilities are not exceeded at the feed stage and so that pinch points are not created between the operating lines and equilibrium lines.

The ratio of wash solvent to extraction solvent can be optimized to give the lowest number of stages required. Final optimization is usually carried out experimentally in miniplant or pilot-plant equipment. The presence of solvent in the feed will change the slope of the operating line.

The use of solute reflux to the ends of the extractor can theoretically reduce the number of stages required by as much as a factor of 2.[41] But then higher solvent flow rates may be required to keep the concentrations down in the feed stage. Reflux can be especially beneficial in some Type II systems where the solute is not totally miscible with the solvent.

The reflux of pure solute in a Type I system, i.e., to the end of an extractor where total miscibility can occur, is nonproductive when the solvent is being recovered by distillation.[43,44] When secondary extraction is used for solvent recovery, however, the use of reflux can markedly reduce the solvent that has to be distilled.[45]

When the two solvents used in a fractional extraction are partially miscible, the assumptions of Case B discussed earlier can be used for calculations or other more detailed procedures described by Treybal[30] or Scheibel[46] should be used.

REFERENCES

1. Brown, W. V., *Chem. Eng. Prog.*, **59**(10), 65 (1963).
2. Lauer, F. C., Littlewood, E. J., and Butler, J. J., *Iron Steel Eng.*, **46**(5), 99 (1969).
3. Lewis, W. L., and Martin, W. L., *Hydrocarbon Process.*, **46**(2), 131 (1967).
4. Grote, H. W., *Chem. Eng. Prog.*, **54**(8), 43 (1958).
5. Deal, G. H., Jr., Evans, H. D., Oliver, E. D., and Papadopoulos, M. N., *Pet. Refiner*, September 1959, p. 185.
6. Somekh, G. S., and Friedlander, B. I., *Hyrocarbon Process.*, December 1969, p. 127.
7. Somekh, G. S., *Proc. Int. Solvent Extraction Conf. Soc. Chem. Ind. London*, **323** (1971).
8. Kubek, D. J., and Somekh, G. S., *Oil and Gas J.*, January 1974, p. 93.
9. Harrah, H. W., *Hydromet. Bull.* M4-B137, Denver Equipment Co., Denver, Colo.
10. Schreve, R. N., "Chemical Process Industries," 3d ed., McGraw-Hill, New York, 1967.
11. Gerster, J. A., in Perry, J. H., "Chemical Engineers' Handbook," 4th ed., p. 13-7, McGraw-Hill, New York, 1963.
12. Treybal, R. E., Weber, L. D., and Daley, J. F., *Ind. Eng. Chem.* **38**, 817 (1946).
13. Perry, R. H., "Chemical Engineers' Handbook," 5th ed., McGraw-Hill, New York, 1973.
14. Horsley, L. H., "Azeotropic Data III," American Chemical Society, Washington, D.C., 1973.
15. Robbins, L. A., *Chem. Eng. Prog.*, **71**(5), 4 (1975).
16. Prausnitz, J. M., "Molecular Thermodynamics of Fluid-Phase Equilibria," Prentice-Hall, Englewood Cliffs, N.J., 1969.
17. Null, H. R., "Phase Equilibrium in Process Design," Wiley-Interscience, New York, 1970.
18. Pierotti, G. L., Deal, C. H., and Derr, E. L., *Ind. Eng. Chem.*, **51**, 95 (1959).
19. Gilmont, R., Zudkevich, D., and Othmer, D. F., *Ind. Eng. Chem.*, **53**, 223 (1961).
20. Derr, E. L., and Deal, C. H., *Proc. Int. Symp. Dist. Inst. Chem. Eng. London*, 3, 40 (1969).
21. Renon, H., and Prausnitz, J. M., *Am. Inst. Chem. Eng. J.*, 14, 135 (1968).
22. Ewell, R. H., Harrison, J. M., and Berg, L., *Ind. Eng. Chem.*, 36, 871 (1944).

23. Wise, W. A., and Williams, D. F., Symposium on Less Common Means of Separation, Institute of Chemical Engineers, 1963.
24. Leo, A., Hansch, C., and Elkins, D., *Chem. Rev*, **71**(6), 525 (1971).
25. Francis, A. W., "Liquid-Liquid Equilibriums," Wiley-Interscience, New York, 1963.
26. Barry, G. T., Sato, Y., and Craig, L. C., *J. Biol. Chem.*, **174**, 209 (1948).
27. Kiezyk, P. R., and MacKay, D., *Can. J. Chem. Eng.*, **51**, 741 (1973).
28. Hand, D. B., *J. Phys. Chem.*, **34**, 1961 (1930).
29. Bancroft, W. D., *Phys. Rev.*, **3**, 120 (1895).
30. Treybal, R. E., "Liquid Extraction," 2d ed., McGraw-Hill, New York, 1963.
31. Logsdail, D. H., and Lowes, L., in Hanson, "Recent Advances in Liquid-Liquid Extraction," Pergamon, Oxford, 1971.
32. Hanson, C., *Chem. Eng.*, Aug. 26, 1968, p. 76.
33. Karr, A. E., *Am. Inst. Chem. Eng. J.*, **5**, 446 (1959).
34. Lo, T. C., and Karr, A. E., *Ind. Eng. Chem. Process Des. Dev.*, **11**(4), 495 (1972).
35. Baird, M. H. I., McGinnis, R. G., and Tan, G. C., *Proc. Int. Solvent Extraction Conf. Soc. Chem. Ind. London*, 251, 1971.
36. Wett, T. W., and Karr, A. E., *Chem. Process.*, December 1966, p. 68.
37. Mayfield, F. D., and Church, W. L., Jr., *Ind. Eng. Chem.*, **44**, 2253 (1952).
38. Shaw, K. G., and Long, R. S., *Chem. Eng.*, **64**(11), 251 (1957).
39. Kremser, A., *Nat. Pet. News*, **22**(21), 42 (1930).
40. McCabe, W. L., and Smith, J. C., "Unit Operations of Chemical Engineering," 3d ed., McGraw-Hill, New York, 1975.
41. Brian, P. L. T., "Staged Cascades in Chemical Processing," Prentice-Hall, Englewood Cliffs, N.J., 1972.
42. Treybal, R. E., "Mass-Transfer Operations," 2d ed., McGraw-Hill, New York, 1968.
43. Wehner, J. F., *Am. Inst. Chem. Eng. J.*, **5**, 406 (1959).
44. Skelland, A. H. P., *Ind. Eng. Chem.*, **53**, 799 (1961).
45. Scheibel, E. G., *Chem. Eng. Prog.*, **62**(9), 76 (1966).
46. Scheibel, E. G., *Pet. Refiner*, **38**(9), 227 (1959).

Section **1.10**

Commercial Liquid-Liquid Extraction Equipment

TEH CHENG LO *Technical Fellow, Development Department, Hoffmann-La Roche, Inc.; member, American Institute of Chemical Engineers, American Association for the Advancement of Science.*

ACKNOWLEDGMENTS: The author benefited from the many fruitful discussions on the topics of interest with Professor M. H. I. Baird, Chemical Engineering Department at McMaster University, Canada. His contribution to this work is gratefully acknowledged. Thanks are also extended to Francis Alechny and Cassius Brown for their assistance in the preparation of manuscript and drawings.

NOMENCLATURE

A_p	Pulse amplitude or reciprocating amplitude (distance of half stroke), ft
A	Interfacial surface area between liquids, ft²
a	Specific interfacial area, ft²/ft³ of active extractor volume
a'	Activity
c	A constant
C	Concentration, lb·mol/ft³
C^*	Solute concentration in a liquid phase in equilibrium with other liquid phases in the system
ΔC	Difference in concentration, lb·mol/ft³
C_o	Orifice coefficient [0.6 in Eq. (14), dimensionless]
D	Diameter of mixing vessel or extraction column, ft
D'	Diameter of mixing vessel or extraction column, cm
d_i	Rotor or impeller diameter, ft
d'_i	Rotor or impeller diameter, cm
d_o	Nozzle, perforation, or orifice diameter, ft
d'_o	Nozzle, perforation, or orifice diameter, in
d_p	Mean drop diameter, ft
d'_p	Mean drop diameter, cm
d_s	Diameter of baffle (stator-ring) opening, ft
d_p	Packing size, ft
D_v	Diffusivity of solute in dispersed phase, ft²/h
E	Axial diffusivity, axial dispersion coefficient, ft²/h
E'	Axial diffusivity, axial-dispersion coefficient, cm²/s
E_{Md}	Murphree dispersed-phase stage efficiency, fractional
E_o	Overall stage efficiency, fractional
f	Pulse frequency, cycles/h
F	Fraction of the flooding velocity at which the column is being operated
g	Acceleration of gravity, ft/h²
g_c	Gravitational conversion factor, 4.18×10^8 (lb mass) ft/(lb force) h²
g'_c	Gravitational constant, 32.2 ft/s²
H	Active height of extractor (height of packed section, etc.), ft
H_c	Height of compartment, or plate spacing, ft
h_c	See Eq. (36)
H'_c	Height of compartment, or plate spacing, in
H''_c	Height of compartment, or plate spacing, cm
HDU	Height of a diffusion unit, ft
HETS	Height equivalent to a theoretical stage, ft
HETS'	Height equivalent to a theoretical stage, cm
HTU	Height of a transfer unit, ft
HTU'	Height of a transfer unit, cm
K	Overall mass-transfer coefficient, lb·mol/(h)(ft²)(Δc)
Ka	Overall mass-transfer coefficient, lb·mol/(h)(ft³)(Δc)
K'	Overall mass-transfer coefficient, cm/s
k	Mass-transfer coefficient, lb·mol/(h)(ft²)(Δc)
$K'a$	Product of the interfacial area and specific-rate constant, as defined by Eq. (2) in Ref. 66.
l	A characteristic length, ft
m	Distribution coefficient, dimensionless
N	Rotational speed, r/h
N'	Rotational speed, r/s
N_f	Flux of mass transfer, lb·mol/h
N_{Po}	Power number = $P g_c / \rho N^3 d_i^5$, dimensionless
NTU	Number of transfer units, dimensionless
Npe	Peclet number for axial mixing = VH/E, dimensionless
P	Power per compartment, ft-lb force/h
Q	Total volumetric flow rate, ft³/h
N_{Re}	Impeller Reynolds number= $d_i^2 N \rho / \mu$, dimensionless
$S\phi$	Fractional free cross section or free space per plate
V	Superficial velocity, ft³/(h)(ft²) = ft/h
V'	Superficial velocity, cm³/(cm²)(s)
V_k	A characteristic velocity of drops, ft/h
V'_k	A characteristic velocity of drops, cm/s
V_o	$f A_p / S \phi$, ft/h
ε	Extraction factor = $V_k m / V_R$, dimensionless
δ	Differential operator
ε	Voids in a packed section, volume fraction

λ	A function
μ	Viscosity, lb/(ft) (h)
ρ	Density, lb/ft^3
$\Delta\rho$	Positive difference in density, lb/ft^3
$\Delta\rho'$	Positive difference in density, g/ml
σ	Interfacial tension, lb mass/h^2 = 28,700 (σ')
σ'	Interfacial tension, dynes/cm
ψ_f	Power function, ft^2/h^3
φ	Volume fraction of a liquid in vessel or extractor void volume
Z	Distance through the phases

SUBSCRIPTS

A	Phase A
av	Average
B	Phase B
c	Continuous phase
1	Model column
2	Objective column
d	Dispersed phase
E	Extract
f	Flooding
i	At interface
o	Overall
plug	Plug (unback-mixed) flow
R	Raffinate

INTRODUCTION

Industrial application of solvent extraction has increased rapidly over the past two decades. Simple mixer-settlers, packed columns, and unagitated plate towers were widely used in the process industries during the thirties and forties. The modern multistage continuous contactors, such as pulsed columns and mechanically agitated columns, which employ mechanical energy input to achieve a high rate of mass transfer, have been developed since the late forties and have found wide commercial application since the mid-fifties.

The unique ability of solvent extraction to achieve separation according to chemical type rather than according to physical characteristics (e.g., vapor pressure) allows a great variety of processes which range from nuclear-fuel enrichment and reprocessing to fertilizer manufacture, and from petroleum refining to food processing. Probably more types of contactors have been developed for solvent extraction than for any other chemical engineering unit operation. They have often been developed for specific processes, with which they then tend to become associated. As a result, selection of extractors can be quite bewildering to one who is contemplating a new process application.

Although extensive data on interphase dispersion and hydrodynamics in many contactors have been accumulated, pilot-scale testing still remains an almost inevitable preliminary to a full-scale contactor design for any new commercial process. In the calculation of stage efficiency (HTU or HETS) for a multistage differential contactor, the designer should take into account axial mixing.

Several texts[1-3] and reviews[4-10] have appeared in the past decade describing various types of extractors. This chapter describes the performance, features, applications, and design of those extractors proved in industrial operation, as well as new developments that appear interesting in extraction processes.

CLASSIFICATION OF COMMERCIAL EXTRACTORS

Contactors can be classified according to the methods applied for interdispersing the phases and producing the countercurrent flow pattern. Both of these can be achieved either by the force of gravity acting on the density difference between the phases or by applying centrifugal force. In the former method, they can be further classified according to the type of mechanical energy input applied. All continuous multistage contactors can be further divided into two broad categories according to the nature of their operation,

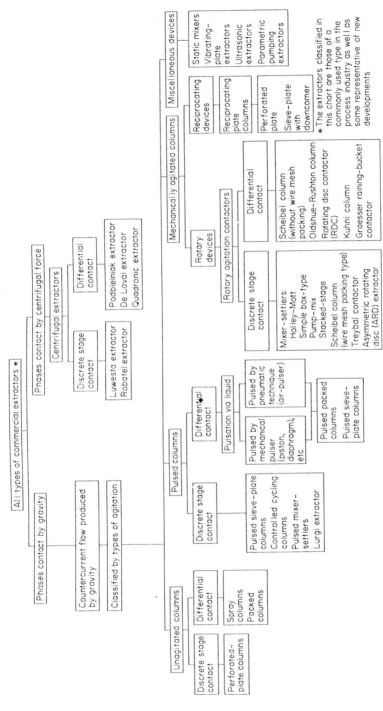

Fig. 1 Classification of commercial extractors. (Reprinted by permission from T. C. Lo, *Recent Developments in Commercial Extractors, a paper presented at Engineering Foundation Conference on Mixing Research at Rindge, N.H., 1975.*)

namely, discrete stagewise contact and differential contact. In the former, typified by mixer-settlers, two phases are equilibrated in one compartment and then separated in another chamber before being passed into another stage for equilibration. In differential contact, the two phases are brought into countercurrent contact in a tower, such as in a spray or packed column.

Figure 1 shows the classification of major types of commercial extractors according to the above description.[10]

BASIC PRINCIPLES OF EXTRACTORS

An ideal industrial extractor should be one which has (1) high throughput and low HETS or HTU, (2) low capital cost, and (3) low operating and maintenance costs. The effectiveness of an extractor can be measured by volumetric efficiency, which is obtained by dividing the throughput by HETS. The greater this number, the smaller the column volume required to do a given extraction job.

Most industrial processes require high-capacity multistage contacting to do the required separation. The design procedure for an extractor requires computation and evaluation of the following:

- The number of equilibrium stages or transfer units for the required separation
- The stage efficiency—HTU or HETS
- Throughput or volumetric flow capacity

The volumetric flow capacity of an extractor depends on type of agitation, degree of mechanical energy input to a column, and design of internals of a contactor. The design of

stagewise contactors is based on the number of equilibrium stages required for the separation and on the stage efficiency. In the *differential contactors*, the two phases are never at equilibrium, and it is necessary to know (1) the number of equilibrium stages required to effect a particular separation, (2) the mass-transfer rate at which equilibrium is established, and (3) the stage size necessary to attain equilibrium with a specified throughput. Design can be based on the number of equilibrium stages required and HETS.

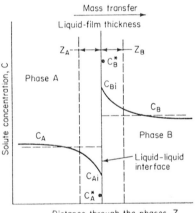

Fig. 2   Concentration profiles in interphase mass transfer. (*Reprinted by permission of the author.*)

Rate of Mass Transfer

The classical two-film theory has been used to explain the mechanism of mass transfer of a solute between two liquid phases. The theory suggests that the two phases are in equilibrium at interface, that there is consequently no resistance to transfer across the interface, and that the resistance to mass transfer occurs in the thin films on either side of interface. The concentrations in the bulk phases are assumed to be uniform as a result of combined eddy and molecular diffusion. The concentration gradients in interphase mass transfer are illustrated in Fig. 2. Considering the rate of mass transfer of a solute from a bulk solution of A to the bulk solution of B, the following equations may then be written

$$N_f = k_A A(C_A - C_{Ai}) = k_B A(C_{Bi} - C_B) \tag{1}$$
$$N_f = k_A A(C_A - C_A^*) = k_B A(C_B^* - C_B) \tag{2}$$

where k = film mass-transfer coefficient for a particular phase
A = interfacial surface area
C = solute concentration in the bulk of a phase
C_i = solute concentration in a phase adjacent to the interface
K = overall mass-transfer coefficient applicable to the two-phase system

$C*$ = solute concentration in a phase in equilibrium with other phases in the system

m = distribution coefficient

The overall mass-transfer coefficient is related to the individual film mass-transfer coefficients by the following equation:

$$\frac{1}{K_A} = \frac{1}{k_A} + \frac{1}{mk_B} \qquad (3)$$

The exact mechanism of interphase mass transfer in an extractor is much more complicated than the above simplified approach. The rate of mass transfer is greatly influenced by the complicated hydrodynamics of interfacial turbulence[11,12,13] and of droplet coalescence and redispersion.[14,16] Nevertheless, the two-film theory is still applicable to interpret the parameters of extraction that affect the rate of mass transfer.

Factors Affecting Rate of Mass Transfer

Mass-Transfer Coefficients. The mass-transfer coefficients are influenced by the following:

- Phase composition—governing diffusivity and causing interfacial turbulence
- Temperature—by affecting diffusion rates
- Degree and type of agitation—governing film thickness and interfacial turbulence
- Direction of mass transfer—depending on phase dispersed
- Physical properties of the system (densities, viscosities, interfacial tension, etc.)

Interfacial Area. The interfacial area depends on:

- Phase composition—by affecting the phase densities and interfacial tension
- Temperature—by affecting the interfacial tension
- Degree and type of agitation—by creating a more intimate dispersion of two phases
- Phase ratio
- Physical properties of the system—interfacial tension, etc.

Concentration Driving Force, ΔC. The concentration driving force depends on:

- The solute concentration of the bulk of the two phases
- Distribution coefficient—governing C_{ti} and C_{mi}
- Temperature-affecting distribution coefficient

Agitation or Mechanical Energy Input

From the considerations above, it is obvious that the interdispersion of the two phases by mechanical energy input into the system is an important parameter in determining mass-transfer rate. Therefore, the degree and nature of agitation can be expected to have an important influence on the rate of mass transfer by both decreasing the film thickness (thereby increasing the effective mass-transfer film coefficient) and increasing the interfacial area by creating a more intimate dispersion of the two phases. However, the increase in interfacial area falls as the agitation rate continues to rise. There are also other factors which become important beyond a certain point and tend to prevent a further rise in mass-transfer rate with increase in agitation. As the dispersed-phase droplet size decreases, the interfacial turbulence is decreased and fluid circulation within the drops is suppressed until the drops finally act as rigid spheres. All mass transfer within the drops takes place by the relatively slow process of molecular diffusion, causing the dispersed phase mass-transfer coefficient to decrease. Very fine droplets resulting from excessive agitation also reduce droplet interaction and, hence, decrease mass-transfer coefficients. All of these may compensate for the advantage gained by the increase in interfacial area. Therefore, mass-transfer considerations suggest the possibility of an optimum degree of agitation for the maximum overall mass-transfer rate.

Effect of Axial Mixing on Extraction Efficiency

Standard extractor designs are based on the assumption that the flow pattern is countercurrent with perfect plug flow of each phase. In practice, this assumption is rarely true, and deviations due to axial mixing have been found. The result of axial mixing is a reduction in the concentration driving force for interphase mass transfer below that assumed in the standard design procedure and, consequently, an increase in the HTU or HETS value. This is illustrated in Fig. 3.

The phenomenon of axial mixing or longitudinal dispersion in liquid-liquid extractors

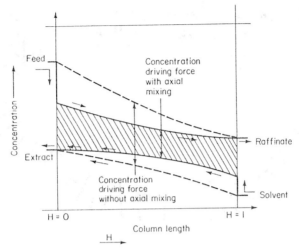

Fig. 3 Effect of axial mixing on concentration profiles in a countercurrent extraction column. *(Reprinted by permission of the author.)*

can be the result of a combination of factors which vary according to the type of contactor used and the liquid flow conditions within it. As discussed by Sleicher,[17] axial mixing in the continuous phase may result from the sum of the two effects:

1. The true turbulent and molecular diffusion in the axial direction. This may be caused by:

- Vertical circulation current
- Mixing in the eddies from the wakes of the dispersed droplets
- Entrainment of the continuous phase by the dispersed droplets
- Forced back-mixing action due to turbulence in the contactor or the effect of pulsation or vibration
- Channeling flow due to particular contactor or packing geometry

2. Nonuniform velocity and subsequent radial mixing or Taylor diffusion. This may predominate over eddy diffusion in contactors which have no or only a small degree of mechanical agitation, such as spray columns.

The dispersed phase can also be subjected to axial mixing, which also contributes to overall longitudinal mixing. However, the effect is generally not appreciable unless the extractor is operating near the flooding point. Axial mixing of the dispersed phase may result from:

- Local velocity eddies of the continuous phase which carry droplets countercurrent to the main direction of flow of the dispersed phase
- Forward mixing resulting from the size distribution of droplets
- The distribution of residence time in the droplet swarm which results from the distribution of droplet size and the distribution in rising velocities for droplets
- Channeling effect resulting from a particular contactor or packing geometry

Axial mixing occurs in all types of extraction equipment, but differential contactors are generally the most prone to it. In the case of some commercial extractors, 60 to 75% of effective height is accounted for by longitudinal mixing.[18] If allowance for the effect of axial mixing is not made in the design, the resultant plant would be seriously underdesigned. Axial mixing for various commercial contactors has been reviewed by Ingham.[18] Many mathematical methods for evaluation of axial mixing and methods for calculation of contactor efficiency which take into account longitudinal mixing have also been reviewed by Misek and Rod.[19]

In conclusion, modern extractors are developed according to the basic principles mentioned above with the aims of achieving high capacity, high mass-transfer rates, and low axial mixing.

PILOT-SCALE TESTS AND SCALE-UP CONSIDERATIONS

Pilot-Scale Tests

Because the factors relating mass transfer to fluid dynamics of the system in an extractor are extremely complex, particularly for mixed solvents and feedstocks of commercial interest, pilot-scale testing remains an almost inevitable preliminary to a full-scale contactor design for any new commercial process. The pilot tests provide the following qualitative and quantitative information for scale-up and design of extractors:

- Total throughput and agitation speed
- HETS or HTU
- Stage efficiency
- Hydrodynamic conditions—droplet dispersion, phase separation, flooding, emulsive-layer formation, etc.
- Selection of dispersed phase or direction of mass transfer
- Solvent-feed ratio
- Material of construction and its wetting characteristics
- Confirmation of desired separation (in cases where equilibrium data are not available)

For process chemistry studies, any laboratory-scale or bench-scale multistage units such as those described in the latter section will suit the purpose. However, for design of a large-scale commercial extractor, the pilot-scale extractor should be of the same type as that to be used on the large scale. Experimental data obtained from a mixer-settler or a pulsed column will provide little information for design of mechanically agitated columns or centrifugal extractors. Manufacturers generally provide pilot-plant facilities for scale-up tests, but for process development or a more detailed study, a pilot unit is generally designed and installed by the user.

Pilot tests should be carried out under simulated process conditions in an extractor of design and material of construction similar to those of a large extractor. It is important that actual process feed and solvent be used in the tests because the process streams usually contain impurities. The presence of surfactants, even in trace amounts, substantially reduces the mass-transfer rate by reducing the mass-transfer coefficients. Samples and measurements of extractor parameters are taken only after steady state has been reached in each run. Pilot-scale studies should cover a wide range of flow rates and phase ratios, and provision should be made for solvent recovery and recycle. Recycle of solvent can change solvent properties and significantly affect extractor performance. The solvent should be recycled for a reasonably long time to ensure that slow buildup of trace contaminants does not affect the performance of the extractor. If the contaminants do build up and eventually affect the performance, steps should be developed to remove them, and these steps should be incorporated in the process design flowsheet. Finally, extended runs should be carried out to confirm the optimal process.

Scale-up Considerations

Many factors affect scale-up. The important ones are listed below:

- Properties of the fluid system
- Total throughput
- Solvent-feed ratio
- Direction of mass transfer (from continuous phase to dispersed phase or vice versa)
- Phase of dispersion (e.g., aqueous phase or organic phase)
- Material of construction and its wetting characteristics
- Degree and type of mechanical agitation
- Size and size distribution of droplets
- Rate of droplet coalescence and redispersion
- Wall and end effects
- Axial mixing or longitudinal dispersion

Since mass transfer is controlled by the fluid dynamics of the system, direct scale-up requires that dynamic and geometric similarities be maintained in both small-scale and large-scale extractors. However, complete dynamic similarity cannot be maintained since Reynolds numbers ($lV\rho/\mu$), which control mass-transfer rate, and Weber numbers ($lV^2\rho/\sigma$), which control drop sizes and their distribution, are obviously unattainable on two scales for the same fluid system.[20,21] Scale-up by making the model an element or vertical slice of the large extractor has been reported[22] but is impractical and certainly not feasible for mechanically agitated contactors.

In the practical approach for the design of large-scale industrial extractors, one is often faced with the following problems:

1. Data on stage efficiency, HETS, and HTU appearing in the literature have generally been obtained on small scale under laboratory conditions using a pure solvent system. Flow-performance data on most types of large extractors are meager. Correlated data available for scale-up are generally based on limited liquid systems and a narrow scale range. When one is designing and scaling up a new industrial process, these data should be approached with some caution and judgment. For example, scale-up data are available for both a 3-in-diameter column and 1-ft-diameter column based on tests with MIBK–acetic acid–water, a typical testing system. However, when one is designing a 5-ft-diameter column for a new industrial process, data on a pilot test on the process should be obtained and scaled up judiciously.

2. Because of the time-consuming, tedious procedures requiring special skills and sophisticated instrumentation involved in obtaining data on drop sizes and size distribution and axial mixing, they are usually excluded in the pilot-test program. While these data are among the most important for the design of some extractors, axial mixing has been reported to depend more strongly on scale than any other factor in extractor design.[23]

3. Only overall volumetric mass-transfer coefficient Ka or average stage efficiency will be measured in the pilot-scale testing. While individual mass transfer coefficients, k's, and volumetric interfacial area, a, strongly depend on the composition and the hydrodynamics of the flow system in the column, they also generally vary strongly along the column. Without a full understanding of k's and a in the column, the scale-up problem appears to be formidable.

In view of the factors mentioned above, it seems unlikely that there will be direct scale-up methods to translate data from the pilot-scale model to the prototype[20] (large-scale device). Instead, a design method should be developed by a study of the components entering into the design. When these are known, it will be possible to design the large-scale extractors directly. Much research work still remains to be done in this area.

At the present state of knowledge, reliable scale-up for industrial large-scale extractors still depends on the correlations based on extensive performance data collected from both pilot-scale and large-scale extractors covering a wide range of liquid systems. Unfortunately, only limited data for few types of large commercial extractors are available in the literature. For the reasons mentioned above, the generalization given in the designs described in the following sections should be used only for approximate estimates, and a considerable design safety factor should be applied.

UNAGITATED COLUMNS

Three simple unagitated columns are shown in Fig. 4.

Spray columns are the simplest in construction but have very low efficiency because of poor phase contacting and excessive back mixing in the continuous phase. They generally give one or two stages.[24] A baffled-plate tower, 2.7 m (9 ft) in diameter and 24 m (80 ft) in height, for propane deasphalting of residue was reported to have only 3 to 3.5 theoretical stages.[5] Because of their simple construction, spray columns are still used in industry for simple operations such as washing, treating, and neutralization, often requiring no more than one or two theoretical stages. In recent years, considerable research on using the contactors for direct liquid-liquid heat transfer has been reported.[25]

Packed columns have better efficiency because of improved contacting and reduced back mixing. It is important that the packing material be wetted by the continuous phase to avoid coalescence of the dispersed phase. To reduce the effects of channeling, redistribution of the liquids at fixed intervals is normally done with long columns. Packing size should be less than one-eighth the column diameter to reduce wall effects and to achieve full packing density. Extensive mass-transfer data obtained in columns of various sizes up to 150 mm (6 in) in diameter using different types of packing and extraction systems have been summarized.[3] A packed tower, 3.6 m (12 ft) in diameter and 30 m (100 ft) in height, once used in furfural extraction plants for lubricating-oil manufacture, was reported to have the equivalent of about six theoretical stages.[5]

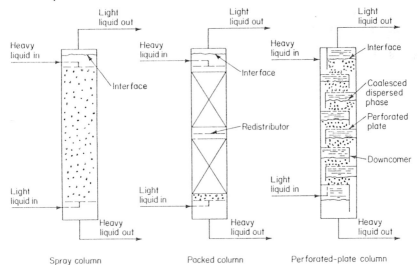

Spray column   Packed column   Perforated-plate column

Fig. 4 Unagitated column extractors. *(Reprinted by permission of the author.)*

Figure 5 shows a typical correlation for overall heights of transfer units vs. extraction factor for extraction of diethylamine from water to toluene.[26] Vermeulen et al.[27] have summarized their data on axial dispersion as shown in Fig. 6. In the scale-up for a commercial process, pilot data should be obtained for the same liquid system, packing, and method of operation. Overall heights of transfer units obtained from mass-transfer data should be corrected for axial dispersion on scale-up. For estimating the column size, detailed treatment is given in the standard text.[1]

Perforated-plate columns are semistagewise in operation. They are reasonably flexible and efficient. If the light phase is dispersed, the light liquid flows through the perforations of each plate and is dispersed into drops which rise through the continuous phase. The continuous phase flows horizontally across each plate and passes to the plate beneath through the downcomer as shown in Fig. 4. If the heavy phase is dispersed, the column is reversed and upcomers are used for the continuous phase. A perforated-plate column, 2.1 m (7 ft) in diameter and 24 m (80 ft) in height, used for the extraction of aromatics, was reported to have the equivalent of 10 theoretical stages.[5] Mass-transfer data obtained in various types of perforated-plate columns up to 225 mm (9 in) in diameter using different extraction systems have been summarized.[3] They generally are correlated in terms of overall heights of transfer units vs. flow velocities of continuous and dispersed phases for a specific column and system.[1,3] The following empirical equation[1] for estimating overall stage efficiency has been found to represent the available data reasonably well:

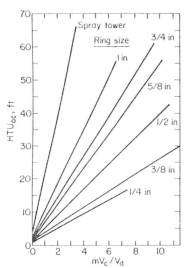

Fig. 5 Extraction of diethylamine from water to toluene. [*Reprinted by permission from Chem. Eng. Prog., 49, 405 (1953).*]

$$E_o = \frac{89,500 H_c^{0.5}}{\sigma_{gc}} \left(\frac{V_d}{V_c}\right)^{0.42} = \frac{0.9 (H_c')^{0.5}}{\sigma'} \left(\frac{V_d}{V_c}\right)^{0.42} \tag{4}$$

Krishnamurty and Rao[28] suggest that Eq. (4) can be improved if the right-hand side is multiplied by $0.1123/d_o^{0.35}$.

There are many variations of basic designs for perforated-plate (sieve-plate) columns. Detailed information is given in the literature.[1,3]

Because of the simplicity and low cost of packed and perforated-plate columns, they are still used in many commercial extraction processes, despite their low efficiency, particu-

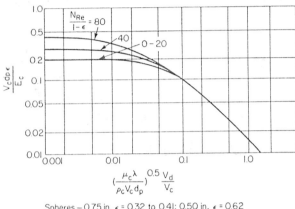

Spheres – 0.75 in, ϵ = 0.32 to 0.41; 0.50 in, ϵ = 0.62
Raschig rings – 0.50 in, ϵ = 0.62; 0.75 in, ϵ = 0.65
Berl saddles – 1.0 in, ϵ = 0.67

Fig. 6 Axial dispersion for the continuous phase in packed columns. [*Reprinted by permission from Chem. Eng. Prog.*, **62** *(9), 95 (1966).*]

larly for processes requiring few theoretical stages, and for corrosive systems where no mechanical moving parts are advantageous. A 12-m (40-ft)-diameter perforated-plate column with downcomers between each stage has been used for petroleum-refining processes.[29]

Features and applications of unagitated columns are given in Table 1.

MIXER-SETTLERS

Mixer-settlers are still widely used in the chemical process industry because of their many attractive features (Table 1). They are particularly practical and economical for the operations (e.g., washing and neutralization) that require high capacity and few stages (i.e., <3). Large, simple mixer-settlers with a throughput of up to 22,740 L/min (6000 gal/min) have been used in the mining industry. The obvious disadvantages of mixer-settlers are the space requirement and the inventory of material held up in the equipment. In the past decade, considerable effort has been expended on improving the designs, and many new devices have been reported.

Description, Application, and Performance of Mixer-Settlers

Agitated vessels are the simplest in construction and are still frequently used in industrial processes in either batch or continuous operations. There are many devices using this design principle. The *baffled agitated tank*, which uses a rotating impeller immersed in the liquid to accomplish mixing and dispersion, is the simplest single-compartment vessel for extraction or chemical reaction. Multiple-stage extraction can be carried out by crossflow batch extraction. In continuous extraction, a settler is incorporated with the mixing vessel, which may be sized to provide the residence time required. Agitated vessels are useful for liquids of high viscosity and systems requiring a high ratio of liquid flows.

TABLE 1   Summary of Features and Fields of Industrial Application of Commercial Extractors

Types of extractor	General features	Fields of industrial application
Unagitated columns	Low capital cost Low operating and maintenance cost Simplicity in construction Handles corrosive material	Petrochemical Chemical
Mixer-settlers	High-stage efficiency Handles wide solvent ratios High capacity Good flexibility Reliable scale-up Handles liquids with high viscosity	Petrochemical Nuclear Fertilizer Metallurgical
Pulsed columns	Low HETS No internal moving parts Many stages possible	Nuclear Petrochemical Metallurgical
Rotary-agitation columns	Reasonable capacity Reasonable HETS Many stages possible Reasonable construction cost Low operating and maintenance cost	Petrochemical Metallurgical Pharmaceutical Fertilizer
Reciprocating-plate columns	High throughput Low HETS Great versatility and flexibility Simplicity in construction Handles liquids containing suspended solids Handles mixtures with emulsifying tendencies	Pharmaceutical Petrochemical Metallurgical Chemical
Centrifugal extractors	Short contacting time for unstable material Limited space required Handles easily emulsified material Handles systems with little liquid density difference	Pharmaceutical Nuclear Petrochemical

SOURCE: Reprinted by permission from T. C. Lo, Recent Developments in Commercial Extractors, Engineering Foundation Conference on Mixing Research, Rindge, N.H., 1975.

The simple *box-type mixer-settler*,[30],[31] as shown in Fig. 7, is essentially a development of the first mixer-settler unit of Holley and Mott.[32] This design avoids all interstage piping by use of partitioned box construction and involves no interstage pumping. The driving force for the flow is derived simply from the density difference between the stages. The multistage unit is built as one box. The unit is simple and has proved very reliable in operation. It has been extensively used by the British Atomic Energy Authority for the separation and purification of uranium and plutonium.[33] The one major disadvantage of this type of contactor is its bulkiness, because the driving force for flow is proportional to the depth of the unit.

A *vertical type of mixer-settler*, as shown in Fig. 8, was developed by Hanson and Kaye[34],[35] in an attempt to reduce the space requirement and to increase throughput per unit volume without introducing pumping. Treybal[36] has also proposed a vertical mixer-settler whose design is aimed at minimizing back mixing.

A *pump-mix-type extractor* was first developed by Coplan et al.[37] to overcome the flow limitations of the simple gravity mixer-settler by introducing interstage pumping. This was primarily developed for and used in nuclear-fuel processing. This is a box-type design which can greatly reduce the flexibility of the unit; i.e., any variation in the phase flow rates or ratio has to be matched by a corresponding adjustment in the pumping unit to prevent one phase from being completely displaced from the settler with subsequent massive back mixing of the other phase. Hanson and coworkers[34],[38] proposed a design in which interstage flow is brought about by a simple lifting device instead of a pump (Fig. 9).

A new type of *IMI pump-mixer-settler* (Fig. 10a) has been developed by Israel Mining Institute[39,40] to handle high volumetric throughput. The unit has been widely used in many process industries. A unit with a capacity of 8338 L/min (2200 gal/min) has been used in phosphoric acid plants.[41] The unique part of this design is that the pumping device (Fig. 10b) is not required to act as the mixer and the two phases are dispersed by a

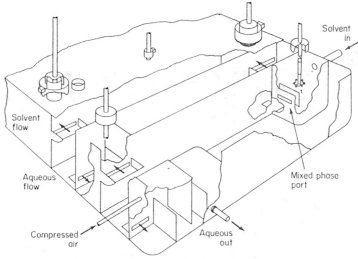

Fig. 7   Box-type mixer-settler. [*Adapted from Proc. 3d U.N. Conf. Peaceful Uses of Atomic Energy, Geneva*, 10, 224 (1964).]

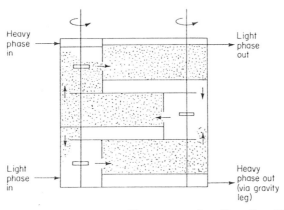

Fig. 8 . Vertical type of mixer-settler. [*Reprinted by permission from* Chem. Eng., 75 (18), 76 (1968).]

separate impeller mounted on a shaft running coaxially with the drive to the pump. Therefore, the mixing and pumping operations are separated and can be optimized individually. The settler is also a novel design. The dispersion is fed into the center of the settler, which is cylindrical in shape, and flows radially outward. Antiturbulence baffles are provided to increase the settling rate.

The *General Mills mixer-settler*[42] (Fig. 11) is another type of pump-mix unit designed by the manufacturer for metallurgical extraction. The unit has a baffled cylindrical mixer with a flat turbine which does both mixing and pumping for the incoming liquids. The settler is a shallow rectangular vessel designed for minimum holdup.

The *Davy Powergas unit*[43–45] (Fig. 12) is also a type of mixer-settler with the pump-mix approach. The liquids run through a draft tube and are mixed and pumped by an impeller running directly above the draft tube. The dispersion flows off the top of the mixer and down through a channel into a rectangular settler. Extensive tests have been reported, and a very large plant is in operation for copper extraction.[11]

The *Kemira mixer-settler*[46] developed in Finland and shown in Fig. 13 also adopted a pump-mix concept in which the mixing phase is drawn from a point in line with the mixing impeller. Only the heavy phase is pumped into the mixer, and the light phase is allowed to flow freely from the settler. A large auxiliary space is provided between the mixer and settler. This design, which differs from other types, has the following features:

1. It minimizes the probability of phase inversion in operation, which may reduce mass transfer and decrease phase separation.

2. It reduces start-up and shutdown periods and ensures that the desired dispersion is obtained on start-up. The unit has been successfully used in many metallurgical extraction and nitrophosphate fertilizer processes[47] and found to be particularly flexible when there are great variations in flow rate from stage to stage.

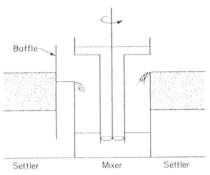

Fig. 9 Mixer-settler of Hanson and coworkers. [*Reprinted by permission from* Chem. Eng., 75(18), 76 (1968).]

Two types of *Lurgi mixer-settlers* have been developed in West Germany.

A *horizontal-type mixer-settler*[48] also uses an axial-flow impeller for both functions of mixing and pumping. The settler contains a series of horizontal trays to increase the throughput. The other type is the *Lurgi vertical mixer-settler*[49] where all settlers are arranged on top of one another in a round tower while mixing and phase transfer take place in pumps located outside the tower. The phases flow interstagewise by a complex

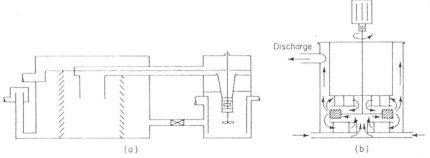

Fig. 10 IMI mixer-settler: (*a*) mixer-settler; (*b*) pump-mixer. [*Reprinted by permission from Mizrahi et al., Proc. Int. Solvent Extraction Conf. 1974,* 1, 141.]

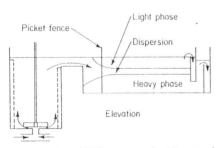

Fig. 11 General Mills mixer-settler. [*Reprinted by permission from* Chem. Eng., 83(2), 86 (1976).]

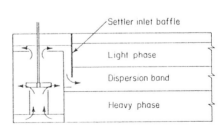

Fig. 12 Davy Powergas mixer-settler. [*Reprinted by permission from* Chem. Eng., 83(2), 86 (1976).]

arrangement of baffles within the settling zones. The settler is of multitray design with the outstanding features of small plot area, insensitivity to variations in throughput, and lower costs compared with conventional settlers. The Lurgi multitray settler is said to be suitable for many liquid-liquid extraction duties as long as the dispersion band is almost horizontal, although it was specifically developed for emulsions with low-settling-velocity problems. Figure 14 shows the new settler's characteristics in comparison with conventional ones. Equipment having a 1800 ton/h capacity has been commercially proved. Its design allows the circulation of either the light or the heavy phase. The unit is reported to have been used for separation of aromatic from aliphatic hydrocarbons (Fig. 15).

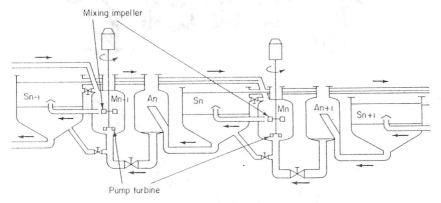

Mixing impeller

Pump turbine

M = mixing space, A = auxiliary space, S = settler

Fig. 13 The Kemira mixer-settler. [*Reprinted by permission from Proc. Int. Solvent Extraction Conf. 1974, 1, 169.*]

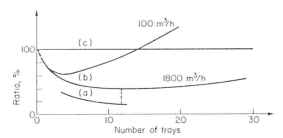

(a) Relative plot area requirements of LMTS/S for optimum design

(b) Relative cost of LMTS/S for two different throughput rates (100; 1800 m³/h)

(c) 100% = conventional settler S/S

Fig. 14 Characteristics of Lurgi multitray settler (LMTS) as compared with conventional settler. [*Reprinted by permission from Br. Chem. Eng., 16(7), 629 (1971).*]

A new mixer-settler has been designed and marketed by *Holmes and Narver, Inc.*[50] This unit incorporates multicompartment mixers and also has many other features.

The *Morris contactor*[51] (Fig. 16) is not designed for stagewise operation. One unique feature of this contactor is its capacity to handle solids in suspension. Used as a liquid-solid contactor, the solid particles move through the dispersed phase.

There are many mixers using mechanical agitation other than rotary-agitated types. Following are a few of them:

The *mixer-settler extractor described by Fenske and Long*[52] consists of compartments or trays stacked vertically, the contents of which are agitated by a reciprocating motion of a plate and are settled in each compartment. It has been found particularly useful as a laboratory device for evaluation of extraction processes because the stage efficiency is essentially 100%.

Large-scale industrial application has not been reported.

A *pulsed mixer-settler extractor* "*Honeycomb*" (Fig. 17) developed in the USSR is described by Karpacheva et al.[53] It has cocurrent pulsed column sections as mixing zones and arbitrarily arranged settling chambers. For any given system, there is a degree of mixing intensity corresponding to optimum total volume of mixing and settling chambers.

Based on the optimization studies on various sizes up to an industrial scale [16,020 L/h (4227 gal/h)], the authors have found that the settler volume, and hence solvent inventory, can be reduced by a factor of 4 or 5 compared with ordinary mixer-settlers.

An *air-pulsed mixer-settler*[54] developed in West Germany is used for reprocessing of nuclear fuels. The unit works without moving parts behind shielding and is particularly suitable for nuclear-fuel reprocessing. Scale-up from model to prototype has been reported.

Interdispersion of light and heavy liquids can be achieved in an *in-line mixer* which is fitted in the pipeline. There are many types, and only a few of the most important ones can be mentioned here.

Agitated Line Mixers. Flomix, a device which combines orifice mixers and agitators, is marketed by Nettco Corporation.[55] The *Lightnin Line Blender*,[56] using the impellers in separate stages shown in Fig. 18, is available in sizes to fit 100- to 500-mm (4- to 20-in) pipes.

Motionless in-line mixers utilize the energy for mixing and dispersing from pressure drop resulting from flow. Several devices using different designs are available in the market.

The *Static mixer*[57] of Kenics Corporation is an in-line, continuous mixing unit with no moving parts. It is constructed of a number of short elements of right- and left-hand helices. These elements are alternated and oriented so that each leading edge is at a 90° angle to the trailing edge of the one ahead. The element assembly is then enclosed within a tubular housing. Figure 19*a* shows the arrangement of the elements and the dispersion of the phase while liquids flow along the elements. Data on metallurgical extraction for copper using this unit compared with those employing agitated vessels have been reported.[58] The units can be arranged for countercurrent multistage extraction as shown in Fig. 19*b*.

Fig. 15 Lurgi vertical mixer-settler for extraction of aromatics from coke-oven benzene. (*Lurgi Mineraloltechnik GMBH, West Germany.*)

In addition to the mixer described above, several other motionless in-line mixers are available. They are different configurations to achieve the same goal.

The *ISG (Interfacial Surface Generator) mixer*[59] was developed by Dow Chemical Company and is manufactured by others under license.

The *LPD (Low Pressure Drop) mixer*,[59,60,61] also developed by Dow, is marketed by Charles Ross & Son Company.

The *Koch Sulzer Mixer*[59,62] is developed and marketed by Koch Engineering Company.

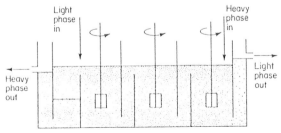

Fig. 16   Morris contactor. [*Reprinted by permission from* Chem. Eng. 75(18), 76 (1968).]

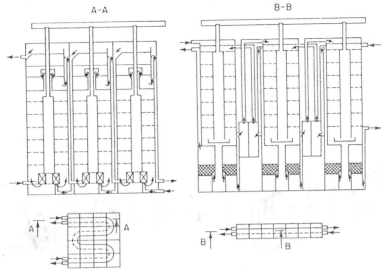

Fig. 17 Pulsed mixer-settler extractor "honeycomb." [*Reprinted by permission from Proc. Int. Solvent Extraction Conf. 1974, 1, 231.*]

Scale-up and Design of Mixer-Settlers

Mixer-settlers are relatively reliable on scale-up because they are practically free of interstage back mixing and stage efficiencies are at least 80% or higher. Because of limited space, only general rules and basic principles will be given here.

Mixers Results of many previous[20,63–66] studies in agitated vessels have led to the following general conclusions on the scale-up of mixers:

- The rate of extraction is a function of power input for both batch and continuous flow.
- Mixers can be reliably scaled up by geometric similitude at constant power input per unit mixer volume. However, this is not necessarily the most economically optimal scale-up.[20]
- The power input to the mixer can be calculated by using the Rushton correlation of power and Reynolds number:[67,68]

$$P = \frac{N_{P_0}}{g_c} \rho N^3 d_i^5 \tag{5}$$

For a baffled vessel, with or without an air-liquid interface, the density and viscosity for two-phase mixtures can be computed[69] by

$$\rho_{av} = \rho_c \varphi_c + \rho_d \varphi_d \tag{6}$$

$$\mu_{av} = \frac{\mu_c}{\varphi_c}\left(1 + \frac{1.5\mu_d\varphi_d}{\mu_d + \mu_c}\right) \tag{7}$$

From the results of studies in 150-, 300-, 500-, and 900-mm-diameter (6-, 12-, 20-, and 36-in) baffled tank mixers using Rushton turbines for uranium extraction of the Dapex process, Ryon et al.[66] demonstrated that the mixer can be scaled up, at least in ratios up to 200:1, by geometric similitude at constant power input per unit mixer volume, and the rate of uranium extraction is proportional to the cube root of power input for both the batch and continuous flow as shown in Fig. 20.

Warner also states that the method was successful in 100- to 200-fold scale-up of throughput.[70]

Settlers The processes taking place in a settler are complex and not quite understood. The flow capacity of settlers is characterized by a band of dispersion at the interface, the thickness of which is a measure of the approach to flooding. Properties of an emulsion band in a mixer-settler were studied by Pike and Wadhawan.[71] A plot of dispersion-band thickness vs. flow capacity (i.e., throughput per unit area of interface) is shown in Fig. 21 by Hanson et al.[11] The relation is often a power law, and two conclusions can be drawn from it:

1. There is a maximum specific throughput to use for design above which a small change can cause flooding.

2. The throughput per unit volume of dispersion decreases as the thickness of the dispersion band increases.

From the results of studies on an oil-water system in the simple gravity settlers, ranging from 0.15 to 1.8 m (0.5 to 6 ft) in diameter, Ryon et al.[66] found that the thickness of the band increases exponentially with increasing flow and showed that settlers can be scaled up by factors of up to 1000 on this basis. Figure 22 shows correlations of dispersion-band thickness vs. flow capacity of settlers in an oil-water system.

Basic designs can enhance the efficiencies of a settler:

- By minimizing turbulence in the settler
- By minimizing small drops in the mixer
- By maintaining linear velocities along the settler to a low value to avoid entrainment of small drops from the dispersion band

In large industrial mixer-settlers, settlers usually can easily represent 75% of the total volume of the units. There is therefore considerable incentive to find some practical means to increase the throughput and

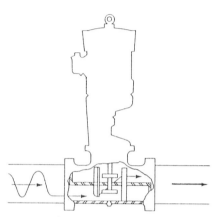

Fig. 18 Lightnin In-Line Mixer. (*Mixing Equipment Company.*)

(a)

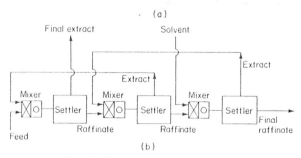

Final extract   Solvent

Extract

Extract

Mixer   Mixer   Mixer

Settler   Settler   Settler

Feed   Raffinate   Raffinate   Final raffinate

(b)

Fig. 19 Kenics static mixer. (*Kenics Corp.*)

hence to reduce the size of the settler and to lower the inventory of the solvent. One approach to the problem is to divide the overall dispersion into a number of thinner bands. Lurgi multitray settlers[18] and IMI compact settlers[40] use this approach to achieve a high throughput. Another approach is to introduce a coalescer to the settler. A woven-mesh packing incorporating materials that are both water- and organic-wetted has been developed, and excellent results with it have been reported.[45,72] Various devices such as cyclones, separating membranes, mechanical and electrical coalescers, and centrifuges have been used in the industry to increase settling rates.[3]

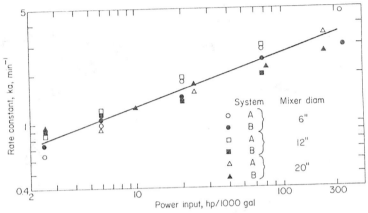

Fig. 20 Rate of extraction in continuous-flow mixers for uranium extraction of the Dapex process. [*Reprinted by permission from* Chem. Eng. Prog., **55**(10), 71 (1959).]

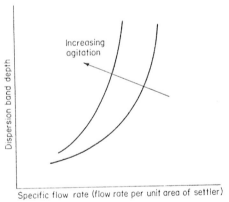

Fig. 21 Typical variation of dispersion-band depth vs. specific flow rate. [*Reprinted by permission from* Chem. Eng., **83**(2), 86 (1976).]

PULSED COLUMNS

The efficiency of simple sieve-plate or packed columns can be appreciably increased by the application of an oscillating pulse to the contents of the column as originally proposed by Van Dijck.[73] This is done by increasing both turbulence and interfacial areas. The mass-transfer efficiency is thus greatly improved over that of an unpulsed column, and there is a substantial reduction in HETS or HTU values. Because of the high efficiency of pulsed-plate columns and because the hydraulic pulsation can be activated from a safe area, pulsed perforated-plate columns have been widely used in the nuclear industry.

The features and fields of industrial application of pulsed columns are given in Table 1. Pulsed plate and packed columns are the most common types of column used in the chemical process industry.

The unique features of the pulsed column are low axial mixing and a relatively small increase in axial mixing with an increase in column diameter[74] (Fig. 23). These features are not generally found in mechanically rotary-agitated columns. In the past, because of the mechanical difficulties with the generation of pulsation, large-scale use of pulsed

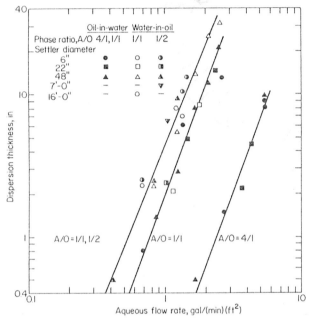

Fig. 22 Dispersion-band thickness vs. flow capacity of settlers in an oil-water system. [*Reprinted by permission of* Chem. Eng. Des., **55**(*10*), *71* (*1959*).]

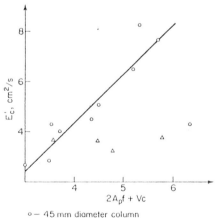

o — 45 mm diameter column
△ — 600 mm diameter column

Fig. 23 Axial-mixing coefficient vs. operating parameter. [*Reprinted by permission from Proc. Int. Solvent Extraction Conf. 1974, 3, 2339.*]

umns had been limited. However, in recent years, the industrial application of columns up to 2.7 m (9 ft) in diameter has been reported.[75]

ised Packed Columns

pulsed packed column consists of a vertical cylindrical vessel filled with packing. rious ordinary packings may be used, but it is most important to choose a material for packing that is wetted preferentially by the continuous liquid, thus ensuring that the ps of dispersed liquid will not be severely coalesced within the packed volume. Light l heavy liquids, either one of which is dispersed in the form of drops, pass countercurtly through the column. In the top or bottom of the column, the dispersed phase lesces at an interface layer. The liquids in the column are moved up and down by ans of a pulsating device connected to the bottom of the column.

Pulsed packed columns, up to 2.7 m (9 ft) in diameter, developed by DSM, Netherids, and manufactured under license by Bronswerk, Inc., have been installed for trochemical processing.[75] It is claimed that the unique design of the column is the lsation mechanism with its rotating valve. Figure 24 shows a pulsed packed column d its pulsation devices.

The results of studies on various sizes of DSM pulsed packed columns, ranging from 51 762 mm (2 to 30 in) in diameter, by Spaay et al.[76] have given the following conclusions d design procedures:

- Mass-transfer efficiency is largely determined by the degree of dispersion. The incipal factor in the design of a pulsed packed column is the Sauter average drop umeter of the dispersed phase.
- *Axial mixing* and *column efficiency* are relatively independent of column diameter the range investigated.
- The throughput or column diameter can be computed from the Sauter average drop umeter of the dispersed phase via the holdup of the dispersed phase. The holdup of the spersed phase related to superficial velocity can be expressed as[77].

$$\frac{V_d}{\varphi \epsilon} + \frac{V_c}{\epsilon(1-\varphi)} = V_k(1-\varphi) \tag{8}$$

he column diameter is so designed that the linear velocities of the two phases equal 65% the flooding velocities.

Thornton[78-81] showed that the holdup at flooding can be calculated from

$$\varphi_{df} = \frac{[(V_{df}/V_{cf})^2 + 8\,V_{df}/V_{cf}]^{1/2} - 3\,V_{df}/V_{cf}}{4(1 - V_{df}/V_{cf})} \tag{9}$$

Substitution of φ_{df} in Eq. (8) yields the values V_{df} and V_{cf}. The design velocities thus und are

$$V_d = \frac{65}{100}\,V_{df} \tag{10}$$

$$V_c = \frac{65}{100}\,V_{cf} \tag{11}$$

- The HTU and hence the column height can be computed from the Sauter average rop diameter via the specific interfacial area.

Since the principal resistance to mass transfer in a pulsed packed column is found in the ispersed phase, the following relation is expressed:

$$H = NTU_{od} \times HTU_{od} \tag{12}$$

For a given system, the number of transfer units (NTU$_{od}$) can be calculated from the quilibrium data and mass balance by means of the known procedures from Vermeulen nd Miyauchi.[27]

$$HTU_{od} = (HTU_{od})_{plug} + HDU_o \tag{13}$$

Vhen (HTU$_{od}$)$_{plug}$ denotes the height of a transfer unit for the dispersed phase under onditions where the two phases pass through the column in plug flow, (HDU)$_o$ is the eight of a diffusion unit, i.e., length to be added to (HTU$_{od}$)$_{plug}$ in order to compensate for he axial-mixing effect. For details of the design procedure, the work of Spaay et al.[76] hould be consulted.

Pulsed Perforated-Plate Columns

The column consists of a tower fitted with horizontal perforated plates (sieve plates) which occupy the entire cross section of the column with no downcomer as in an ordinary perforated-plate column. Small laboratory columns generally use 2.38 mm ($\frac{3}{32}$-in) holes and 25 mm (1-in) plate spacing, whereas larger columns use 3.18 to 6.35 mm ($\frac{1}{8}$- to $\frac{1}{4}$-in) holes with 2-in plate spacing. The total free area of the plate is about 20 to 25%. The columns are generally operated in frequencies of 100 to 260 cycles/min with amplitudes of 6.25 to 25 mm ($\frac{1}{4}$ to 1 in). Figure 25 shows the schematic of an operating pulsed perforated-plate column developed by the French Atomic Energy Commission.

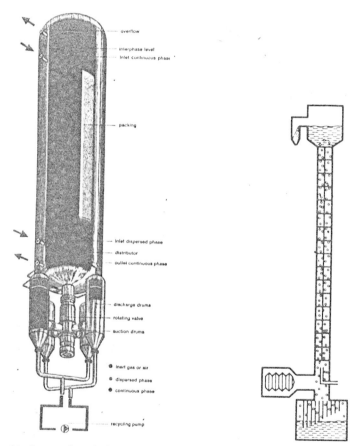

Fig. 24 Bronswerk pulsed packed column. (*Bronswerk—K.A.B., The Netherlands.*)

Fig. 25 Pulsed perforated-plate column. (*Reprinted by courtesy of ERIES, France.*)

The unique features of pulsed perforated-plate columns are their low axial mixing and high extraction efficiency which are due to uniform distribution of energy over a cross section of the column and, hence, uniform distribution of drops in the column as shown in Fig. 26. This type of column has been widely used in the nuclear industry because of its many advantages. Pulsed perforated-plate columns up to 0.9 m (3 ft) in diameter have been used in nuclear industry. Development of larger-scale columns is being undertaken in the nuclear-fuel reprocessing industry[71,82] because of the increasing demands for nuclear power.

Operating characteristics of the pulsed perforated-plate column have been described by Sege and Woodfield.[83] Several regions of operation can be distinguished depending on the flow rate and degree of pulsation, as illustrated in Fig. 27. The column basically can be operated in two different modes. At lower pulsed volume velocities (amplitude × frequency × column cross-section area), a discrete layer of liquid appears between plates during each reversal of the pulse cycle. At higher pulsed volume velocities, there is no coalescence, and fairly uniform dispersion of the dispersed phase occurs. The column will flood at a higher pulsed volume velocity.

From results of studies on hydrodynamics and efficiency of various sizes of pulsed perforated-plate columns 100, 300, and 600 mm (4, 12, and 24 in) in diameter, in uranium extractions, Rouyer et al.[71] have reached the following conclusions:

- Axial mixing is independent of diameter over the range studied as shown in Fig. 23.
- The pulsed plate column can be easily scaled up to at least 0.9 m (3 ft) in diameter without decrease in extraction efficiency measured in terms of height equivalent to a theoretical stage or height of a transfer unit for a given process.
- Throughput is independent of column diameter at least up to 600 mm (24 in) in diameter. A throughput as high as 1473 gal/(h)(ft²) has been achieved.

Fig. 26 Droplet dispersion in pulsed plate column. (*Reprinted by courtesy of ERIES, France.*)

Design The most comprehensive studies on flooding and mass transfer were made by Thornton et al.,[84,85,86] who developed empirical correlations based on experimental data on various sizes of column ranging from 50 mm (2 in) to 300 mm (12 in) in diameter using various liquid systems. Using dimensional analysis and multiple-regression techniques, Smoot et al.[87] have correlated the experimental flooding and mass-transfer data in literature for pulsed plate columns.

Power input is important in pulsed plate columns. Thornton[85] proposed the following correlation for the maximum power absorbed per unit mass of fluid. For a sine-wave pulse

$$\psi_f = \frac{\pi^2 (1 - S_\phi^2)(fA_p)^3}{2 S_\phi^2 C_o^2 H_c} \tag{14}$$

where C_o is the orifice coefficient.

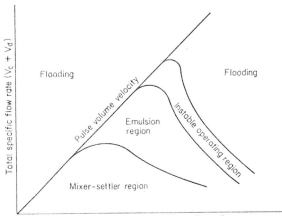

Fig. 27 Regions of operation of pulsed perforated-plate column. [*Reprinted by permission from* Chem. Eng. Prog., *50*, Symp. Ser. 13, 179 (1954).]

Column Throughput

$$\frac{(V_c + V_d)\mu_c}{\sigma} = 0.527 \left(\frac{V_c}{V_d}\right)^{-0.014} \left(\frac{\Delta\rho}{\rho_c}\right)^{0.63} \left(\frac{\psi_f\mu_c^5}{\rho_c\sigma^4}\right)^{-0.207} \left(\frac{d_o\sigma\rho_c}{\mu_c^2}\right)^{0.458} \left(\frac{g\mu_c^4}{\rho_c\sigma^3}\right)^{0.81} \left(\frac{\mu_d}{\mu_c}\right)^{-0.2}$$

(15)

The equation can be simplified where water is the continuous phase in which μ_c, ρ_c, and g are constants.

$$V_c + V_d = 3.2 \times 10^5 \frac{\Delta\rho^{0.63} d_o^{0.458} V_c^{0.014}}{\sigma^{0.144}\psi_f^{0.207}\mu_d^{0.2}V_d^{0.014}}$$

(16)

Once the total flooding velocity has been predicted for a given set of process conditions using Eq. (15) or (16), the required column diameter can be calculated from the relation

$$D = \left[\frac{4Q}{\pi F(V_c + V_d)_f}\right]^{1/2}$$

(17)

where F is the fraction of the flooding velocity at which the column is to operate and Q is the design capacity of the column in cubic feet per hour.

Column Height

$$HTU_{oc} = 10.4 \frac{V_c^{0.539}D^{0.317}H_c^{0.683}\sigma^{0.097}\Delta\rho^{1.04}}{D_r^{0.865}V_o^{0.434}d_o^{0.434}\rho_d^{2.34}\mu_d^{3.27}V_d^{0.636}}$$

(18)

$$HTU_{oc} = (HTU_{oc})_{plug} + HDU_o$$

(19)

$$H = NTU_{oc} \times HTU_{oc}$$

(20)

where $(NTU)_{oc}$ is calculated from the equilibrium data and mass balance by means of standard procedures.

Other Devices The rotary and pulsed forms of agitation have been combined to give what is claimed to be an improved performance over either form alone.[88,89] The agitators are rotated independently of the pulse generator, which leads to a more complicated design and equipment maintenance.

The controlled-cycling column[90,91] (Fig. 28) is an interesting recent development. The unit is essentially a perforated-plate column and is operated on a prearranged cycle involving introduction and removal of one phase followed by a coalescence period, then introduction and removal of the other phase, and finally, another coalescence period. Each phase is dispersed, in turn, in the other. It is claimed that the unit has a very much greater throughput than that achieved in a conventional pulsed column, but no large-scale application has been mentioned in the literature.

Pulsing Devices A reciprocating plunger or piston pump is a common mechanically pulsing device. However, the unit has the disadvantages that (1) corrosive liquid may contact the piston, and (2) high-speed pulsing may cause cavitation.

Bellows or diaphragms, constructed with stainless steel, Teflon, or some other special material, actuated by a reciprocating mechanism, are also used as pulse generators, as shown in Fig. 25.

Thornton[80] has reported an alternative arrangement using air pulses. The detailed design is given by Weech and Knight.[92]

The pneumatic pulser developed by DSM as shown in Fig. 24 is another unique design.[75] Generation of pulsation by the pneumatic technique, which has many advantages over conventional ways, has received increasing attention.[93,94] Figure 29 shows a schematic of an air-pulsing device.

MECHANICALLY AGITATED COLUMNS

Mechanically agitated columns can be divided into two main classes according to mechanical motion patterns, as shown in Fig. 1:

1. Rotary-agitated columns
2. Reciprocating- or vibrating-plate columns

For typical types, see Fig. 30.

Because of the mechanical advantages of rotary agitation, most modern multistage differential contactors employ this device. On the other hand, reciprocating- or vibrating-plate columns have been little explored, and these columns have gained commercial acceptance only in recent years.

Rotary-Agitated Columns

The interfacial area per unit volume within a column contactor can be increased by introduction of some form of mechanical agitation to bring about more efficient dispersion of two phases. To minimize the undesirable axial-mixing effect, some form of baffling is generally introduced. Various types of commercial rotary-agitated columns are given in Fig. 1. Of these, the Scheibel column, the rotating-disk contactor, and the Oldshue-Rushton multiple-mixer column are three well-known types of rotary-agitated columns that have been proved in many industrial installations. Table 1 shows features and applications of this type of column.

Scheibel Columns There are several design variations. The earlier model[95]

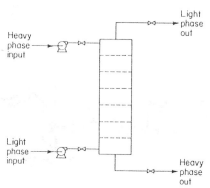
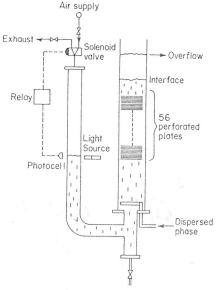

Fig. 28 Controlled-cycling type of column. [*Reprinted by permission from* Chem. Eng., 75(18), 76 (1968).]

Fig. 29 Pulsed perforated-plate column with pneumatic pulser. [*Reprinted by permission from Proc. Int. Solvent Extraction Conf. 1974, 2, 1571.*]

developed in 1948 is shown in Fig. 31. Alternate compartments are agitated with impellers, and the others are packed with an open woven-wire mesh having about a 97% void. This was the first agitated column to receive wide commercial application in the chemical process industry. Scheibel and Karr[85,95-97] presented capacity and mass-transfer data on 25-, 75-, and 300-mm-diameter (1-, 3-, and 12-in) columns using three different systems, including difficult (high interfacial tension, e.g., o-xylene–acetic acid–water) and easy (low interfacial tension, e.g., MIBK–acetic acid–water), to extract systems.

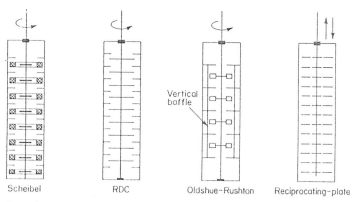

Fig. 30 Typical types of mechanically agitated columns. (*Reprinted by permission of the author.*)

The capacity of the column depends on the properties of the systems handled and varies from 14,247 to 24,452 L/(h)(m²) [350 to 600 gal/(h)(ft²)]. Table 2 shows a summary of the minimum HETS values and specific throughputs. The relationship between the size of the agitator, its speed, the height of the mixing compartment, and the physical properties of the system was investigated by Karr and Scheibel,[65] and they found that the Murphree-stage efficiency for a 12-in-diameter column can be correlated as

$$\frac{E_{Md}}{1-E_{Md}} = 1.09 \times 10^{-7} \left(\frac{Hc}{12d_i}\right)\left(\frac{Nd_i}{5}\right)^4 \left(\frac{\delta a'}{\delta C_d}\right)\left(\frac{\Delta\rho'}{\sigma'}\right)^{1.5} \tag{21}$$

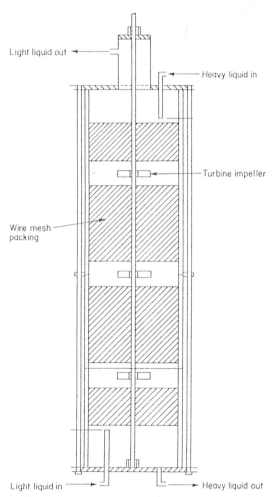

Fig. 31 First Scheibel extractor. [*Reprinted by permission of* Chem. Eng. Prog. **44**, 681 (1948).]

Scheibel[97] has stated that the HETS on this type of column varies as the square root of the diameter,

$$\frac{(HETS)_2}{(HETS)_1} = \left(\frac{D_2}{D_1}\right)^{1/2} \tag{22}$$

and that this correlation has been successfully extrapolated to the design of larger columns. Honekamp and Burkhart[98] have found that very little change occurs in the average drop size within wire-mesh packing and have determined extraction rates in the

TABLE 2   Performance Data on Scheibel Extractors[65,96,97]

Column type	Earlier extractor with wire-mesh packing and no horizontal baffle (Fig. 31)				Second Scheibel extractor with horizontal baffles only (Fig. 32b)	Second Scheibel extractor with horizontal baffles and wire-mesh packing (Fig. 32a)	
	Systems						
	MIBK–acetic acid–water	o-Xylene–acetic acid–water			o-Xylene–acetic acid–water	MIBK–acetic acid–water	o-Xylene–acetic acid–water
Column diameter, in	1	11½	1	11½	11½	11½	11½
HETS, in	2.5	9.2	4.2	15	5	3	6
Throughput, gal/(h)(ft²)	176	614				458	375

York wire-mesh packing. However, droplet behavior in this type of column has recently been investigated, and three distinct characteristics have been observed.[99] Gel'perin et al.[100] have measured axial mixing in a 50-mm-diameter (2-in) column of this type.

A newer type of column using horizontal baffles was developed in 1956.[97] It improves the HETS and permits a more efficient scale-up for large-diameter columns. There are two configurations of design; one is with wire-mesh packing and one is without packing between stages, as shown in Fig. 32a and b, respectively.

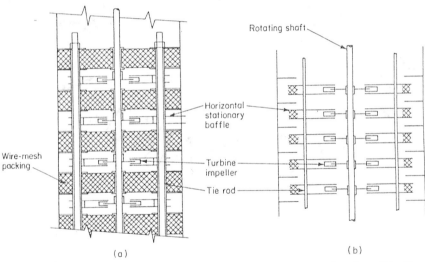

Fig. 32 Second Scheibel extractor with horizontal baffles: (a) with wire-mesh packing; (b) without wire-mesh packing between stages. [*Reprinted by permission of* Am. Inst. Chem. Eng. J., 2,74 (1956).]

Performance data for a 300-mm (12-in) column, with and without wire-mesh packing, are shown in Fig. 33, and the minimum HETS values are given in Table 2. From the results of studies in a 300-mm (12-in) column of this type, Scheibel[97] has stated that:

■ The efficiency of the mixing compartment can be correlated in terms of power input per unit volume of fluids flowing handled per compartment, and the ratio of the dispersed phase to the continuous phase. The power input can be calculated by the following correlation:

$$P = 1.85 \frac{d_i^5 \rho (N')^3}{g_c'} \tag{23}$$

■ Packing between the stages was found to increase the overall efficiency but to decrease the throughput of the column. For a difficult extraction system in which the

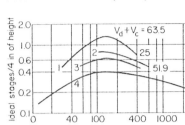

Fig. 33 Extraction in second Scheibel column [*From Scheibel, Am. Inst. Chem. Eng. J., 2, 74 (1956), with permission.*]

Curve	System
1,2*	Methyl isobutyl ketone–water–acetic acid
3*	o-Xylene–water–acetic acid
4†	o-Xylene–water–phenol
	Methyl isobutyl ketone–water–acetic acid
	o-Xylene–water–acetic acid

*Alternate mixing and packed sections.
†Packing omitted. Agitators in alternate and also every section.

interfacial tension is high, the benefit that can be derived from the packing would be insufficient to justify its use. On the other hand, in a system which has a low interfacial tension and is easy to extract, the greater amount of extraction obtained with the wire-mesh packing gives it an economic advantage over the use of additional mixing sections.

- The optimal height of the mixing section can be scaled up based on the square-root factor of diameter.
- The HETS varies with the square root of the diameter as shown in Eq. (22).

Although the column has been widely used in the past quarter century, relatively few data are available for the larger columns. Recently, data on scale-up and performance of a large-scale extractor have been reported by Lo,[10] as summarized in Table 3. From the results, the following general conclusions can be derived:

- The same stage efficiency can be maintained on scale-up.
- Total throughput can be increased by 3½ times at a compensation of higher HETS.

Scheibel columns are marketed by York Process Equipment Corporation, and sizes up to 2.55 m (8½ ft) in diameter are in service. For design details, the manufacturer should be consulted.

A third design, shown in Fig. 34, has been reported in recent years.[101] It is basically similar to the second design (Fig. 33b), but a pumping impeller instead of a turbine impeller is used in the mixing stage. It is claimed that the pumping impeller provides an improved agitated column structure and a relatively inexpensive structure for the large

TABLE 3   Summary of Data on Scale-Up and Performance of a 4 ft 9 in Diameter Scheibel Extractor

	Pilot-plant test	Performance	Scale-up factor
Column diameter	3 in	4 ft 9 in	19
Total throughput, gal/h	5	6000	1200
gal/(h)(ft²)	102	345	3.5
No. of actual stages	36	33	~1
Height per stage, in	1	8	8
Total height of extraction section, ft	3	22	7.3
Stage efficiency, %	22	20	~1
HETS, in	4.6	39	8.5
No. of theoretical stages	8	6.7	
Extraction efficiency, %	99+	99.5	
Volumetric efficiency, h⁻¹	35.6	14.2	0.4

SOURCE: Reprinted by permission from T. C. Lo, *Recent Developments in Commercial Extractors*, Engineering Foundation Conference on Mixing Research, Rindge, N.H., 1975.

columns. Columns up to 1.2 m (4 ft) in diameter are in service.[102] However, no performance data on this type of column are available.

Rotating-Disk Contactor (RDC)   This contactor, which was developed in Europe by Reman[103] in 1951, uses the shearing action of a rapidly rotating disk to interdisperse the phases. It consists of a number of compartments formed by a series of stator rings, with a rotating disk centered in each compartment and supported by a common rotating shaft, as shown in Fig. 35. RDC have been widely used throughout the world, particularly in the petrochemical industry, where they have become associated with furfural and SO_2 extraction, propane deasphalting, sulfolane extraction for separation of aromatics from aliphatics, and caprolactam purification. Columns up to 3.6 m (12 ft) in diameter are in service.

Because of the wide application of RDC, extensive research work has been done and performance data on industrial-scale extractors have been accumulated in the past twenty-five years, which provide more complete information in the literature for design of RDC than for any other contactors.

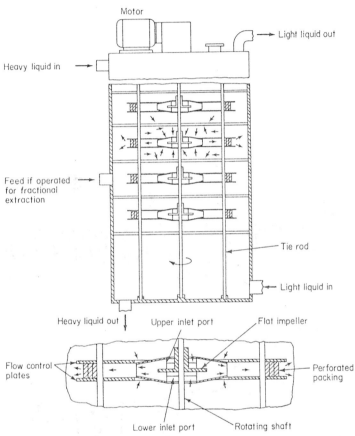

Fig. 34 Third Scheibel extractor. *(Adapted from U.S. Patent 3,389,970, 1968.)*

Throughput of the contactor is controlled by the size of the dispersed-phase droplets. According to Hinze's concept,[104] droplet size for a given system depends on only the power dissipated by the rotor disk per unit mass. The latter is characterized by

$$P \propto \frac{N^3 d_i^5}{H_c D^2} \tag{24}$$

Interfacial tension was found to be the most significant property controlling the drop settling velocity/power input relation when comparing different fluid systems. Therefore, the throughput of an RDC is a strong function of power input and interfacial tension of a liquid system, as illustrated in Fig. 36. However, the throughput of an RDC is not influenced by scale-up.

Extraction efficiency is reduced to some extent because the axial mixing is more intensive in a large column, as illustrated in Fig. 37. Effect of scale on axial mixing can best be illustrated on solutizer extraction reported by Reman,[5] as shown in Table 4. It should be noted that, when the naphtha phase is dispersed, axial mixing accounted for

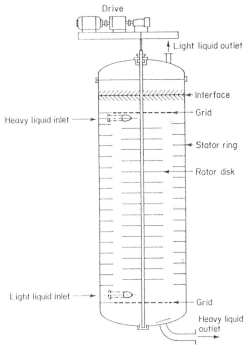

Drive

Light liquid outlet

Interface

Grid

Stator ring

Rotor disk

Heavy liquid inlet

Light liquid inlet

Grid

Heavy liquid outlet

Fig. 35   Rotating-disk contactor. (*Escher B.V., The Hague, The Netherlands.*)

80% of height of the column. It was found that the height of a theoretical stage for the overall mass transfer in the absence of axial mixing (HMS) is not size-dependent. In general practice, values of HMS have been found to range from about 10 in for easy-extraction systems to about 36 in for difficult-extraction systems.[105] Axial mixing for a

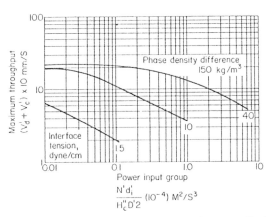

Fig. 36   RDC throughput vs. power input. (*Escher B.V., The Hague, The Netherlands.*)

continuous phase for RDC of 76 mm (3 in) up to 2.1 m (7 ft) in diameter can be correlated by[5,106]

$$E_c = 0.5H_cV_c + 0.12d_iNH_c\left(\frac{d_s}{D}\right)^2 \tag{25}$$

For the dispersed phase, the correlation has not been firmly established, but at high rotor speed, E_D varies from one to three times E_c.[106,107] An extensive review of axial mixing on RDC is given by Ingham.[19]

Reman[5] has reported that, for optimum capacity, stator- and rotor-to-diameter ratios can be kept constant in scale-up. Table 5 illustrates the design of compartment height for two systems having 2.1-m-diameter (7-ft) columns. For practical mechanical construction, the recommended minimum compartment height for a 2.1 m (7 ft) diameter is 200 to 300 mm (8 to 12 in).[5]

Misek[108] has compiled a design manual based on his own work and literature results. The extensive study reported by Strand et al.[109] has also provided an excellent theoretical framework for the scale-up. For detailed information on scale-up and design, their works should be consulted.

Oldshue-Rushton Column This column was developed[110] in the early fifties and has been used for a great variety of processes. It consists essentially of a num-

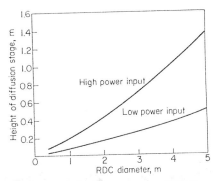

Fig. 37 Height of diffusion stage vs. RDC diameter. (*Escher B.V., The Hague, The Netherlands.*)

TABLE 4 Solutizer Extraction: Height of Transfer Units HTU$_0$

| RDC diameter | 3 in | 25 in | | 25 in |
	in	HTU$_0$,* in	$(E_dN_d + E_cN_c)$, in	HTU$_{plug}$, in
Naphtha dispersed	12	57	44	13
Solutizer dispersed	30	50	22	28

*HTU$_0$ = apparent overall height of transfer unit
HTU$_{plug}$ = overall height of transfer unit in plug flow
SOURCE: Reman, *Chem. Eng. Prog.*, 62(9), 56 (1966), with permission.

ber of compartments separated by horizontal stator-ring baffles, each fitted with vertical baffles and turbine-type impeller mounted on a central shaft. A schematic of the column is shown in Fig. 38. It is claimed that the column can also handle fluid containing suspended solids. Besides liquid-liquid extraction, the column is also used in mass-transfer operations as a gas absorber and solids dissolver, as well as a chemical reactor. Columns up to 2.7 m (9 ft) in diameter have been reported[111] to be in service for extraction.

TABLE 5 Calculation of Compartment Height, RDC Diameter: 7 ft

	Easily dispersed system	Difficultly dispersed system
Power-input group, ft²/s³	0.5	50
HTU$_{plug}$	8	20
Phase ratio	1	1
Dispersed-phase holdup, %	25	25
V_d, in/s	0.8	0.8
V_c, in/s	0.27	0.27

SOURCE: Reman, *Chem. Eng. Prog.*, 62(9), 56 (1966) with permission.

Performance Data. Although considerable research studies[9,19,100,110,112–119] on hydrodynamics, mass transfer, and axial mixing have been done with small-scale columns, data on large-scale columns are scarce in the literature. Oldshue and Rushton[110] have reported that a minimum HETS of 92.7 mm (3.7 in) using eight compartments, each 75 mm (3 in) high, at a throughput of 11,655 L/(h)(m²) [286 gal/(h)(ft²)] was achieved in a 150-mm-diameter (6-in) column. This corresponds to a stage efficiency of 81%. From the results of studies on a 150-mm-diameter (6-in) column, using water as the continuous phase and toluene or kerosene as the dispersed phase, Bibaud and Treybal[113] give the following correlations for the axial-dispersion coefficiencies.

For the continuous phase:

$$\frac{E_c \varphi_c}{V_c H_c} = -0.1400 + 0.0268 \frac{d_i N \varphi_c}{V_c} \tag{26}$$

For the dispersed phase:

$$\frac{d_i^2 N}{E_d} = 0.0393 \times 10^{-8} \left(\frac{d_i^4 N^2 \rho_c}{\sigma g_c}\right)^{1.51} \left(\frac{\rho_c}{\Delta\rho}\right)^{1.18} \left(\frac{d_i^2 N \rho_c}{\mu_c}\right)^{0.61} \tag{27}$$

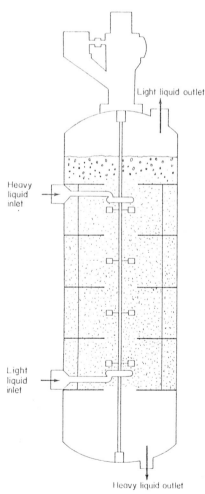

Heavy liquid inlet

Light liquid outlet

Light liquid inlet

Heavy liquid outlet

Fig. 38 Oldshue-Rushton column. (*Mixing Equipment Co.*)

Scale-up and Design Considerations. The column is said to have reliable predictability on scale-up.[120] Oldshue et al.[121] have recently suggested some general principles on scale-up and design. However, no detailed procedures for design are available.

1. *Column throughput* is scaled up based on the pilot testing data. The geometry conditions of the pilot scale are extrapolated to other geometry options with a design factor. The effect of stage openings on throughput is illustrated in Table 6.

2. *Stage efficiency* is scaled up according to the following procedures:

- Measure the stage efficiency in a pilot column and calculate the overall mass transfer Ka based on the dispersed phase.

- Calculate the Ka for large-scale columns from the data on the effect of scale-up, stage geometry, interstage opening, and impeller horsepower on Ka and interstage mixing.

- Calculate the interstage mixing; then obtain the overall stage efficiency.

As a general rule, because the stage height and the residence time increase on scale-up, the overall stage efficiency tends to increase as shown in Fig. 39.

A comparison of mixing factors for pilot scale vs. full scale is shown in Fig. 40.

Some scale-up calculation examples are shown in Table 7. For detailed information on design, the manufacturer should be consulted.

The Asymmetric Rotating-Disk (ARD) Contactor This column was developed by Misek and coworkers[122] in Czechoslovakia, and has been increasingly used in Western Europe in recent years. Its design is aimed at retaining the efficient shearing action of the RDC by using rotating disks to produce dispersion while using the coalescence-redispersion cycle produced in the separated transfer-settling zones to reduce back mixing. The principal features are shown in Fig. 41. The column consists of an asymmetrically located mixing zone and a transfer-settling zone partially separated from each other by means of a vertical baffle. The mixing zone contains a number of compartments, each of which is equipped with a disk-type mixing impeller mounted on a common rotor shaft. The transfer-settling zone consists of a series of compartments, separated by means of annular horizontal baffles.

TABLE 6   Oldshue-Rushton Column—Effect of Size of Opening between Compartments on Throughputs and HETS Values

SYSTEM: MIBK–acetic acid–water
COLUMN DIMENSION: 4 stages, 4-in stage height, 6-in diameter
EXTRACTION: Water to MIBK

Compartment opening, mm	Max stage efficiency	Min HETS, in	Flow rate, kg/(s)(m²)
			Constant flow rate
0	100	101	0*
54	83	122	2.9*
82	52	195	2.9
152	38	265	2.9
			At max efficiency
0	100	101	0*
54	83	122	2.9*
82	67	152	5.4*
152	38	265	6.0*

*Optimum flow rate.
SOURCE: Oldshue et al., *Proc. Int. Solvent Extraction Conf.*, **2**, 1651 (1974).

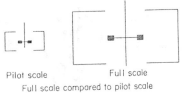

Pilot scale   Full scale

Full scale compared to pilot scale

Residence time	Higher
Blend time, undispersed	Longer
Interstage mixing, undispersed	Different
Interstage mixing, disp	Different
Concentration gradient, disp	Higher
Max impeller zone shear rate	Higher
Avg impeller zone shear rate	Lower
Avg tank zone shear rate	Lower
Turbulent shear rate	Different

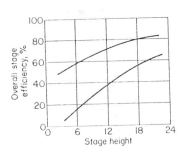

Fig. 39 Effect of stage height on stage efficiencies for Oldshue-Rushton column. [*Adapted from Proc. Int. Solvent Extraction Conf. 1974*, **2**, 1651.]

Fig. 40 Mixing factors compared for pilot and full scale. [*Adapted from Proc. Int. Solvent Extraction Conf. 1974*, **2**, 1651.]

The ARD extractor is used for the extraction of petrochemicals, caprolactam, pharmaceuticals, and phenol wastewater and propane deasphalting and furfural refining of oils, etc. Columns up to 2.4 m (8 ft) in diameter are in service.

Performance Data. Excellent research work has been done and extensive performance data in large-scale extractors have been accumulated by Misek et al.[108,122–124] which provide basic information for scale-up and design. The results of recent hydrodynamic studies[125] on columns of various sizes [i.e., 250 mm (10 in), 2.25 m (7.5 ft), 2.7 m (9 ft), and 3.9 m (13 ft) in diameter] have shown the importance of residence-time distribution in determining column performance, and that large ARD extractors operate practically under the same conditions as a mixer-settler. Figure 42 shows a correlation of the HETS, throughput, and rotor speed range with the extractor diameter, and the curves indicate maximal and minimal limits for various liquid systems. The ARD extractor is claimed to be satisfactory for handling throughput and efficiencies for systems with wide ranges of density differences, interfacial tensions, and solvent-feed ratio. The column is manufactured in a range of standard sizes from 0.6 m (2 ft) to 2.4 m (8 ft) in diameter, over which

TABLE 7   Examples of Scale-up Calculation on Oldshue-Rushton Column

	Pilot-scale data	Large-scale design
Example 1[121]		
Column diameter	6 in	10 ft
Stage height	3 in	6 ft
Stage efficiency	29%	85%
Ka	16.2	
Impeller power input		3 hp
Example 2[121]		
Column diameter	6 in	10 ft
Stage height	3 in	6 ft
Stage efficiency	4%	65%
Ka	4.6	
Impeller power input		4 hp
Example 3[10]		
Column diameter	6 in	6 ft
Throughput, gal/(h)(ft²)	230	258
Stage height	3 in	3 ft
Stage efficiency	20%	50%
HETS	15 in	6 ft

SOURCE: Oldshue et al., *Proc. Int. Solvent Extraction Conf.*, **2**, 1651 (1974).

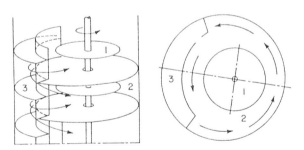

I. Rotating disk rotor;   2. mixing zone;   3. settling zone

Fig. 41   Asymmetric rotating-disk (ARD) extractor. (*Luwa A.G., Switzerland.*)

the HETS is claimed to vary from 0.6 m (2 ft) to 1.05 m (3½ ft) for average systems. The throughput of the column is in the order of 10,188 to 30,564 L/(h)(m²) [250 to 750 gal/(h)(ft²)] depending on the physical properties of the system.

Scale-up and Design. The scale-up procedure comprises a series of computations on flow characteristics in both extractors and corresponding determination of the throughput, rotor speed, and HTU or HETS of the large-scale extractor based on the optimal data obtained in the pilot tests.

The following correlations developed by Misek et al.[122,124] can be used for scale-up to a factor of throughput ratio of 250:1 or higher.

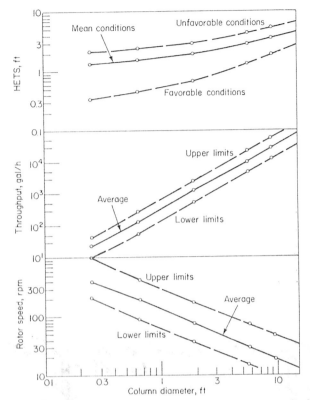

Fig. 42 Operating range of ARD extractor. (*Luwa A.G., Switzerland.*)

Scale-up of Throughput
1. Turbulent region:

$$\frac{V_2'}{V_1'} = \frac{\lambda_1}{\lambda_2}\left(\frac{D_2'}{D_1'}\right)^c \exp\,[0.00106(D_2' - D_1')] \tag{28}$$

2. Laminar region:

$$\frac{V_2'}{V_1'} = \frac{\lambda_1}{\lambda_2} \tag{29}$$

Scale-up of Rotor Speed
1. Turbulent region:

$$\frac{N_2'}{N_1'} = \frac{D_1'}{D_2'} \tag{30}$$

2. Laminar region:

$$\frac{N_2'}{N_1'} = \left(\frac{D_1'}{D_2'}\right)^c \tag{31}$$

Scale-up of HETS
1. Turbulent region:

$$\frac{(HETS')_2}{(HETS')_1} = \left(\frac{D_2'}{D_1'}\right)^c \frac{1 - \varepsilon_1}{1 - \varepsilon_2}\frac{\log \varepsilon_2}{\log \varepsilon_1} \tag{32}$$

2. Laminar region:

$$\text{HETS}' = \text{HTU}' \frac{\varepsilon_d \ln \varepsilon_d}{\varepsilon_d - 1} \tag{33}$$

$$\text{HTU}' \cong (\text{HTU}')_{\text{plug}} + \alpha \epsilon_d h_c \tag{34}$$

$$\frac{(\text{HTU}')_{\text{plug}2}}{(\text{HTU}')_{\text{plug}1}} = \left(\frac{D_2'}{D_1'}\right)^c \tag{35}$$

$$h_c = D'^{2/3}[0.645 + 5.08 \times 10^{-5}(D')^{1.116} \times N'/V_c'] \tag{36}$$

$$\varepsilon_d = V_c'/V_d' m \tag{37}$$

Back Mixing

1. Continuous phase:

$$\frac{1}{Npe_c} = \frac{E_c'}{V_c' H_c''} = \frac{c}{V_c'} \left(\frac{D'}{H_c''}\right)^{1/2} N'd_i' + 0.5 \tag{38}$$

2. Dispersed phase:

$$\frac{1}{Npe_d} = \frac{E_d'}{V_d' H_c'} = c \frac{N'd_i' S_{\phi}\varphi_d}{V_d'} + 0.7 \tag{39}$$

Height of a transfer unit

$$(\text{HTU}_{od}')_{\text{plug}} = \frac{V_d'}{K_d''a} \cong \frac{d_p' V_k' S_{\phi}}{6K_d'} \tag{40}$$

TABLE 8   The Results of a Design of a Commercial ARD Extractor for Extraction of Caprolactam

Column diameter, ft	5.9
No. of compartments	57
Compartment height, in	7.1
Overall compartment height, ft	33.8
Total throughput:	
gal/h	14,573
gal/(h)(ft²)	856
Solvent-feed ratio	0.24
No. of theoretical stages	8
HETS, ft	4.2
Caprolactam recovery, %	99.95
Volumetric efficiency, h⁻¹	27.2

SOURCE: Luwa AG (Zurich), technical report on ARD extractor prepared by Marek (1970), with permission.

Table 8 shows the results of a design for a caprolactam extraction scaled up based on the pilot-plant tests.

The Kuhni Contactor This mechanically agitated column[126] has gained considerable commercial application in recent years in Europe. The design of the mixing zone is similar to that of the second Scheibel column. The principal features of the column are the use of a shrouded turbine impeller to promote radial discharge within the compartments, and of variable hole arrangement to allow flexibility of design for different process applications. A typical compartment is shown in Fig. 43. In a particular application[10] on separation of aromatics from aliphatic hydrocarbons in the petrochemical industry, a 2.46-m-diameter (8.2-ft) column containing 52 actual stages gave a performance equivalent to about 15 theoretical stages, which is equivalent to 29% stage efficiency or an HETS value of

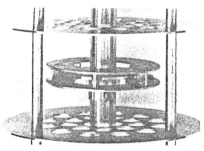

Fig. 43 Kuhni extractor. (*Kuhni A.G., Switzerland.*)

2.07 m (6.9 ft), at a total throughput of 30,565 L/(h)(m²) [750 gal/(h)(ft²)]. Columns up to 4.95 m (16½ ft) in diameter have been constructed.[11] Extensive studies on droplet size, holdup, and back-mixing characteristics for a 150-mm-diameter (6-in) column have been reported.[127,128]

Other Types of Rotary-Agitated Contactors   The *Treybal contactor*[129] was designed for semistagewise operation. The principal features of the column are shown in Fig. 44.

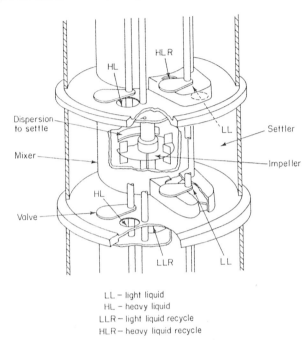

LL − light liquid
HL − heavy liquid
LLR − light liquid recycle
HLR − heavy liquid recycle

Fig. 44   Treybal extractor. (*U.S. Stoneware, Inc.*)

The unit is claimed to have a low HETS and a high-stage efficiency, 75 to 80% being maintained even with difficult extraction systems (e.g., toluene-water). A total throughput of 560 gal/(h)(ft²) has been quoted for the toluene-water system. No performance data on large-scale columns have been reported.

The *Wirz column*[130] also involves a coalescence-redispersion mechanism. A typical stage is shown schematically in Fig. 45. Experimental data have been reported by Leisibach[131] using columns from 100 mm (4 in) to 500 mm (20 in) in diameter. Values of HETS varied from 100 mm (4 in) to 300 mm (12 in) depending on the liquid system and the throughput.

The *Graesser raining-bucket contactor*[132] is of a horizontal design with the phases interdispersed by "water-wheel" arrangements. The basic features of the extractor are shown in Fig. 46. The unit is unique in that it is designed to disperse each phase into the other. The contactor was developed for handling the difficult settling systems found in the coal-tar industry, and it has proved attractive for other applications as well. Units have been built from 100 mm (4 in) to 1.8 m (6 ft) in diameter, giving a throughput of 8558 L/(h)(m²) [210 gal/(h)(ft²)]. Data on axial mixing have been reported.[133]

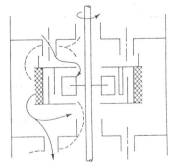

Fig. 45   Typical stage of Wirz column. (*Reprinted by permission of Brit. Chem. Eng., November 1968, p. 49.*)

Reciprocating-Plate Columns

Interdispersion can be achieved by reciprocating or vibrating plates, as well as by the mechanical rotary agitation mentioned above. In 1935, Van Dijck[73] proposed that the extraction efficiency of a perforated-plate column could be improved either by pulsing the liquid contents of the column or by reciprocating the plates. The latter idea lay dormant

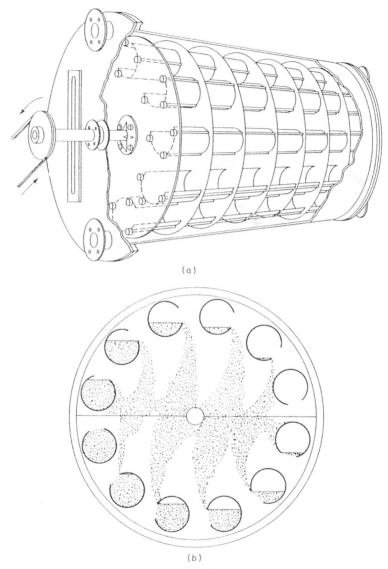

(a)

(b)

Fig. 46   Graesser raining-bucket contactor. [*Reprinted by permission of* Chem. Eng., 75(18), 76 (Aug. 26, 1968).]

until the late fifties, when Karr[134] reported data on a 75-mm-diameter (3-in) open-type perforated reciprocating-plate column. The column was further developed by Karr and Lo[135] by employing baffles for scale-up. A variant of this used for separation of zirconium-hafnium is described by Issac and DeWitte.[136] A simpler version has also been reported by Guyer.[137] Interest in research on this type of column has increased in the past decade.

Papers on various columns characterized by reciprocating perforated plates or packing have appeared in the literature.[136,138-141] Most of the data have lent support to the conclusion that reciprocating-plate columns generally have high volumetric efficiencies.

While pulsed columns are open to the criticism that considerable energy is required to pulse the entire liquid content of a column, particularly on a large-scale commercial extractor,[11] reciprocating the plates is an alternative solution to achieving uniform dispersion and similar mixing patterns using relatively much less energy. Some units that have been built for industrial use are described below. The general features and industrial applications of this type of column are given in Table 1.

Open-Type Perforated Reciprocating-Plate Column The column developed by Karr and Lo[135] consists of a stack of perforated plates and baffle plates which have a free area of about 58%. The central shaft which supports the plates is reciprocated by means of a reciprocating drive mechanism located at the top of the column. The amplitude is adjustable generally from 3 to 50 mm (⅛ to 2 in), and the reciprocating speed is variable up to 1000 strokes/min. Figure 47a and b shows the principal features of a 0.9-m-diameter (3-ft) reciprocating-plate column. Performance data on various column sizes—25, 75, 300, and 900 mm (1, 3, 12, and 36 in) in diameter—have been reported.[134,135,142-144] A minimum HETS of 153 mm (6.12 in) and a volumetric efficiency of 311 per hour were achieved in a 300-mm-diameter (12-in) column using the MIBK–acetic acid–water system. A minimum HETS of 500 mm (20 in) has been measured in a 900-mm-diameter (36-in) column using a relatively difficult extraction system, *o*-xylene–acetic acid–water. A summary of minimum HETS values for various column sizes at specific throughputs is given in Table 9. Figure 48 shows the effect of reciprocating speed on HETS.

The column possesses the following features:
- High throughput
- High mass-transfer rate and, hence, low HETS
- Great degree of versatility and flexibility
- Simplicity in construction

Research studies on hydrodynamics[145,146] and axial mixing[147] of reciprocating-plate columns have been reported by Baird et al. It is interesting to note that, in the reciprocating-plate columns, the presence of many closely spaced reciprocating impellers leads to conditions which approximate more closely the "uniform isotropic turbulence" basis of Kolmogoroff's theory. The experimental dimensionless constant in the equation for drop size is almost ten times greater than values obtained for stirred vessels. The more uniform distribution of energy dissipation in the reciprocating-plate columns gives high turbulence of interdispersing phases and low axial mixing, and hence results in a high mass-transfer rate and low HETS. Based on his studies on a 50-mm-diameter (2-in) reciprocating-plate column and a 150-mm-diameter (6-in) pulsed-plate column, Baird has reported that reciprocating-plate columns of small diameter have relatively low axial mixing and column diameter has insignificant effect. This is in agreement with the conclusion of Rosen and Krylov[148] that

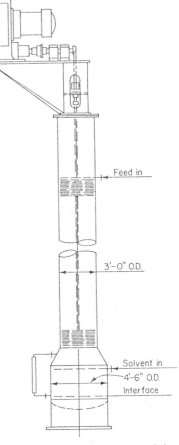

Fig. 47a Schematic arrangement of the 900-mm (36-in) reciprocating-plate column. [*Reprinted by permission from* Chem. Eng. Prog., *72 (9), (1976).*]

dispersion coefficients increase with column diameter only in large columns under two-phase flow conditions which give significant transverse nonuniformity in the flow pattern.

Scale-up Procedures. Based on the performance data obtained on various column sizes mentioned above, Karr and Lo[135,143,144] presented the following empirical equations for scale-up based on an experimental column of D_1 in diameter.

$$\frac{(\text{HETS})_2}{(\text{HETS})_1} = \left(\frac{D_2}{D_1}\right)^{0.38} \tag{41}$$

The corresponding reciprocating speed (SPM) at which the large-diameter column would be operated is given by

$$\frac{(\text{SPM})_2}{(\text{SPM})_1} = \left(\frac{D_1}{D_2}\right)^{0.14} \tag{42}$$

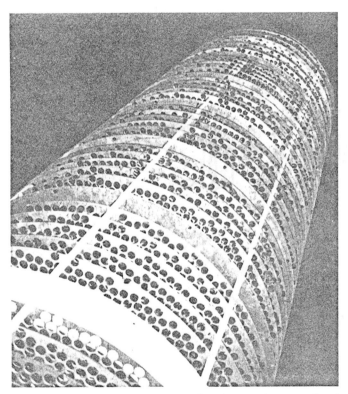

Fig. 47b   Open-type perforated-plates assembly. *(Chem-Pro Equipment Corp.)*

The above correlations are approximate, but they have been successful for scaling up columns of 25, 50, and 75 mm (1, 2, and 3 in) diameter to 900 mm (3 ft) diameter. It is expected that the scale-up correlation can be extended to columns larger than 900 mm (3 ft) in diameter.

The exponent for HETS as a function of a diameter for a difficult extraction system, which requires a higher operating speed and/or amplitude, would be expected to be somewhat greater than the exponent for a relatively easy extraction system, inasmuch as axial mixing is known to increase with reciprocating speed, especially in large-diameter columns. The exponent 0.38 in Eq. (41) is based on the relatively difficult extraction system of *o*-xylene–acetic acid–water. For MIBK–acetic acid–water, a relatively easy extraction system, the average exponent value is 0.25, although the exponent varies from

0.19 to 0.36, depending on which phase is dispersed and which phase is the extractant. For a safer design purpose, an exponent value for a relatively difficult extraction system should be chosen.

The following are the scale-up procedures recommended by the authors:

1. Data are obtained in a 25-, 50-, 75-mm-diameter (1-, 2-, or 3-in) column which need not have baffle plates;[134,142] 50- and 75-mm (2- and 3-in) pilot-plant columns are preferable for scale-up to columns exceeding 600 mm (2 ft) in diameter.

2. The optimum performance of the small-diameter column is determined. The criterion for optimum performance should be maximum volumetric efficiency.

TABLE 9   Summary of Minimum HETS Values and Volumetric Efficiencies for a Reciprocating-Plate Column

Column diam, in	Amplitude, in	Plate spacing, in	Agitator speed, strokes/min	Extractant	Dispersed phase	Min HETS	Throughput, gal/(h)(ft²)	Volumetric efficiencies $V_t/$HETS, h⁻¹
I. System: MIBK–Acetic Acid–Water								
1	½	1	360	MIBK	Water	3.1	572	296
			401			2.8	913	523
1	½	1	278	Water	MIBK	4.2	459	175
			152			8.1	1030	204
3	½	1	330	MIBK	Water	4.9	600	196
	½	1	245			6.3	1193	304
	½	2	355			7.5	1837	393
	½	1	320	Water	Water	4.3	548	205
	½	1	230			6.7	1168	280
	½	2	367	Water	Water	5.0	1172	376
			240			7.75	1707	353
12 (with baffle)	½	1	430	Water	MIBK	5.8	547	151
			285			5.7	1167	328
	½	1	244	MIBK	MIBK	4.4	599	218
			170			5.6	1193	342
	½	1	250	MIBK	Water	7.2	602	134
			225			7.2	1200	268
			150			14.0	1821	208
	½	1	225	Water	Water	7.0	555	127
			200			9.5	1170	197
			150			11.05	1694	246
	½	1	275	Water	MIBK	9.5	1179	199
	½	1	200	MIBK	MIBK	7.8	595	123
			150			6.2	1202	311
II. System: Xylene–Acetic Acid–Water								
3	1	1	267	Water	Water	9.1	424	75
3	½	1	537	Water	Water	8.2	424	83
3	¼	1	995	Water	Water	7.7	424	88
3	1	2	340	Water	Water	9.1	804	142
36	1	1	168	Water	Water	23.3	425	29*
36	1	1	168	Xylene	Water	20.0	442	36*

*Because of instrumentation limits, the maximum volumetric efficiencies have not been explored.
SOURCE: Karr and Lo.[134,135,142-144]

3. Suitable baffle plates in design and spacing are generally provided based on recommendations by the manufacturer.

4. The plate spacing is the same as in the small-diameter column.

5. The amplitude is the same as in the small-diameter column.

6. The total throughput per square foot is the same as in the small-diameter column.

7. The expected HETS is calculated by means of Eq. (41). A small upward/downward adjustment of the exponent may be justified if the degree of agitation required differs markedly from that of the present system.

8. The reciprocating speed required is estimated by Eq. (42).

Several dozen reciprocating-plate columns up to 900 mm (36 in) in diameter have been scaled up successfully by the above procedure for commercial applications.[143]

Commercial Applications. The reciprocating-plate columns have gained increasing industrial application in pharmaceutical, petrochemical, and wastewater-treatment industries.[144] They have been found to be suitable for processing mixtures with emulsifying tendencies (e.g., fermentation broth) and for liquids containing suspended solids. A 750-mm-diameter (2½-ft) column with many stages is used for complex fractional extraction (Fig. 49).

Small-diameter reciprocating-plate extraction columns have been successfully used for countercurrent and fractional liquid extraction in laboratory and pilot-plant process development and scale-up work.[142] A 50-mm-diameter (2-in) reciprocating-plate column was used in pilot-plant work leading to the development of tetraethylene glycol (TETRA) as

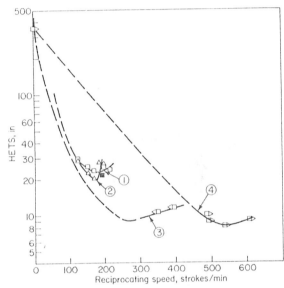

Fig. 48 Effect of reciprocating speed on HETS, *o*-xylene–acetic acid–water system. [*Reprinted by permission from* Chem. Eng. Prog., *72(9) (1976).*]

Symbol	Curve No.	Column diam, in	Phase dispersed	Phase extractant	Double amplitude, in	Plate spacing, in	Total throughput, gal/(h)(ft²)
□	1	36	Water	Water	1	1	425
△	2	36	Water	Xylene	1	1	442
◁□	3	3	Water	Water	1	1	424
□▷	4	3	Water	Water	½	1	424
■	Predicted minimum based on exponents of 0.36 in Eq. (41) and 0.14 in Eq. (42).						

an efficient solvent in the Udex process.[149] Figure 50 shows a typical 50-mm-diameter (2-in) pilot-scale extraction column.[30] A vacuum distillation column utilizing reciprocating-plate mixing of liquid and vapor was reported to produce 100% efficient operation.[150] This type of agitated column is found to be particularly suitable for gas-liquid reaction operations with suspended solid catalysts and is expected to be more widely applied to various mass-transfer operations[144,151] (e.g., gas absorption, gas-liquid reactions, liquid-liquid phase mass transfer with simultaneous chemical reaction).

Reciprocating Column—Perforated Plates with Segmental Passages Prochazka et al.[138] reported two new types of reciprocating-plate columns:

1. One type of column uses *perforated plates with segmental passages* for the continuous phase, with the dispersed phase passing through the relatively small perforations in the plates, as shown in Fig. 51. The authors have reported that, because the drops are formed from jets emerging from small openings, a substantially uniform dispersion is

Fig. 49 An industrial-scale reciprocating-plate column for multistage fractional extraction. (*Reprinted by permission from Hoffmann-La Roche, Inc.*)

Fig. 50 50-mm-diameter (2-in) pilot-scale reciprocating-plate column. (*Julius Montz, GMbH, West Germany.*)

produced. The drop size can be controlled by plate openings of suitable size and by the amplitude and frequency of the reciprocating motion. The perforated plate also gives uniform dispersion over the cross-sectional area of the column and thus reduces axial mixing. Largely because of the segmental passages for the continuous phase, which is a characteristic feature of the column, the throughput of the column is relatively higher than that of pulsed or other type of extractors.

Based on the performance data obtained on columns of various sizes up to 500 mm (20 in) in diameter for extraction of caprolactam in a commercial process (see Table 10), it has been reported that both the capacity and the HETS compare very favorably with other types of extraction columns. The axial-mixing data measured in three different columns having sizes ranging from 50 mm (2 in) to 500 mm (20 in) as shown in Table 11 indicate that the extractor has relatively low axial mixing and that the column diameter has minimal effect on axial mixing.

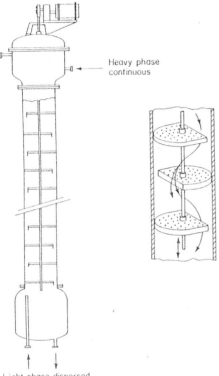

Heavy phase
continuous

Light phase dispersed

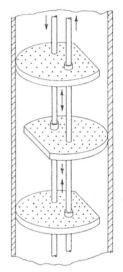

Fig. 51 Reciprocating column—perforated plates with segmental passages. [*Reprinted by permission of* Brit. Chem. Eng., **16**(*1*), 42 (1971).]

Fig. 52 Reciprocating-plate column with countermotion of plates. [*Reprinted by permission from* Brit. Chem. Eng., **16**(*1*), 42 (1971).]

TABLE 10   Performance Data for the Extraction of Caprolactam with Trichloroethylene in a Reciprocating-Plate Column

Direction of extraction	Aqueous → organic			Organic → aqueous		
Dispersed phase	Aqueous			Organic		
Column diameter, in	5.5	11.5	19.7	2.0	5.5	19.7
Effective extractor height, ft	11.8	18.0	19.7	9.8	11.8	26.2
Operating throughput, gal/(h)(ft²)	1792	1792	1470	2255	2210	1960
HETS, in	18.1	18.9	27.6	17.3	20.5	43.3
Volumetric efficiency, h⁻¹	159	152	85	210	173	73

SOURCE: Prochazka et al., *Brit. Chem. Eng.*, **16**(1), 42 (1971), with permission.

These data are consistent with the results which Baird found for the open-type perforated reciprocating column mentioned above.

The authors have also done considerable research work on hydrodynamic studies with this type of column.[152,153] Commercial applications of this type of column up to 20 in in diameter have been reported in Eastern Europe.

2. In the second design, the plates are carried by two shafts, being alternately attached to one and free to slide on the other, as shown in Fig. 52.

For the latter design, the authors claim that the drive has the advantage of allowing adjacent plates to move 180° out of phase with each other, thus spreading the load on the drive mechanism more uniformly over a cycle and reducing the stresses between the columns and the drive train. No commercial application has been reported.

Other Devices Several new developments using reciprocating devices, which have potential interest to the chemical process industry, are described below.

Extraction Column with Reciprocated Wire-Mesh Packing. Wellek et al.[140] described the performance of a new type of pulsating-liquid extractor in which the pulsating energy is imparted to the two countercurrently flowing liquid streams by a wire-mesh packing which is reciprocated vertically in the column at selected frequencies and amplitudes. Mass-transfer data were obtained on benzene–acetic acid–water, a relatively difficult

TABLE 11   Effect of Column Diameter on the Coefficient of Back Mixing

Tested at single-phase flow with water: 1227 gal/(h)(ft²) Amplitude: 0.065 in Frequency: 500/min			
Column diameter, in	2.0	5.5	11.5
Coefficient of back mixing in continuous phase	0.137	0.556	0.419

SOURCE: Prochazka et al., *Brit. Chem. Eng.*, **16**(1), 42 (1971), with permission.

extraction system, and on MIBK–acetic acid–water, a relatively easy one. These studies indicate that operating throughputs are significantly higher than those achieved with mechanically aided extractors, while maintaining high rates of extraction. No commercial application has been reported.

Multistage Vibrating-Disk Column. This column, which has been described by Takeba et al.,[151] consists of perforated plates which are reciprocated in the multistage compartments of the column. Axial mixing, mass transfer, and reaction data in gas-liquid contacting operations[140,154] have been reported.

CENTRIFUGAL EXTRACTORS

In this type of extractor, residence time can be reduced and phase separation accelerated by application of centrifugal force instead of gravity. Because of their precision construction, the capital cost of centrifugal extractors tends to be higher than that of other types of contactors, and they probably also have a greater maintenance requirement. However, they are compact, and relatively high throughput can be achieved in a small geometric space. The units are particularly useful when contact time must be short as for chemically unstable systems (e.g., extraction of antibiotics), when product inventory must be kept at a minimum, or when liquids tend to emulsify or are generally difficult to separate. General features and fields of industrial application are given in Table 1. Classification and types of commercial centrifugal extractors available are given in Fig. 1.

Differential Centrifugal Contactors

Podbielniak Extractor[155] This was the first centrifugal unit to gain commercial acceptance in the early fifties and is still probably the best known. The design consists essentially of a perforated-plate extraction column that has been wrapped around a shaft, which in turn, is rotated to create a centrifugal-force field that achieves a great reduction in the height and contacting time of a perforated-plate column. Figure 53 shows the basic features of the extractor. The light phase is fed under pressure through the shaft and is led via a channel through the perforated cylinders to the internal periphery of the rotating drum. The heavy phase is also introduced through the shaft to the center of the drum. The centrifugal force, acting on the density difference between the phases, causes the two liquids to flow radially countercurrent through the perforated cylinders. Interdispersion of the two phases takes place as liquids are being forced through the perforations. Performance data on this type of centrifugal extractor have recently been reported,[156] as illustrated in Table 12. The extractors have been widely used in the pharmaceutical industry (e.g., extraction of penicillin) and are increasingly used in other fields as well. Commercial units with throughput up to 98,540 L/h (26,000 gal/h) are quoted.

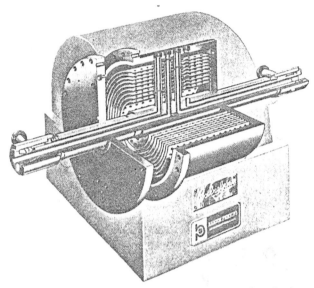

Fig. 53 Podbielniak centrifugal extractor. *(Baker Perkins, Inc.)*

TABLE 12   Performance Data on Podbielniak Centrifugal Extractor

Model	Rotor dimension, in			Holdup, gal	Total throughput, gal/h	Phase ratio, light/heavy	Interface location, R/R max	No. of theoretical stages
	ID	Width	r/min					
System: MIBK–Acetic Acid–Water; Extraction Direction: MIBK → Water								
Laboratory extractor A-1	7.0	1.0	10,000	0.13	20.7	1.37	0.40	5.4
			10,000	0.13	20.7	1.37	0.55	4.5
System: Kerosene-Butylamine-Water; Extraction Direction: Kerosene → Water								
B-10	22.8	9.8	3,000	15	1347	2.00	0.40	6.6
D-18	33.9	17.7	2,000	58	2906	2.0	0.4	5.0

SOURCE: Todd, *Proc. Int. Solvent Extraction Conf.*, 3, 2379 (1974).

The Alfa-Laval Extractor   The operating principle of this vertical centrifugal extractor[157] is shown in Fig. 54. Two phases are fed under pressure into the bowl through separate channels in the shaft—the light phase to the periphery of the drum and the heavy phase to the center. Under centrifugal force, the heavy phase moves toward the periphery, displacing the light phase to the center. Orifices located alternately at the top and the bottom of adjacent cylindrical baffles allow the liquids to flow between baffled zones. The shear force created by the countercurrent-flow conditions and the intimate mixing that takes place as the two phases pass in opposite directions through the holes leading from one concentric channel to the next create the conditions for mass transfer between two phases. The length of the contact path is about 25.5 m (85 ft) in the larger-sized units and approximately over half of this length in the smaller-sized units, and it is claimed that this can give up to 20 theoretical stages in one unit. The capacity of the standard unit depends on the systems being handled and ranges from 5685 to 21,224 L/h (1500 to 5600 gal/h).

The typical applications are for antibiotic extractions (e.g., recovery of penicillin, streptomycin, etc.) and for petrochemical processing.

Quadronic Extractor The extractor developed by Liquid Dynamics[158] is similar in several respects to the Podbielniak extractor, but different in internal structure. It is a horizontal cylindrical centrifuge with the heavy liquid flowing outward and light liquid moving inward within the bowl. A number of orificed-disk columns are mounted radially on the axis. These permit control of mixing intensity as the liquids pass radially through the drum. Figure 55 shows the essential features of the extractor. Ten theoretical stages can be achieved, and a throughput up to 83,380 L/h (22,000 gal/h) can be obtained in an extractor having a 1500-mm (60-in)-diameter rotor.

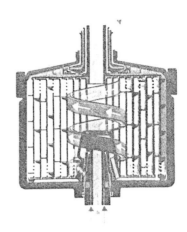

Fig. 54 The Alfa-Laval extractor. (*Alfa-Laval/DeLaval Company.*)

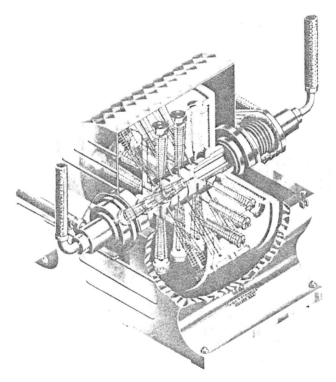

Fig. 55 Quadronic extractor. (*Liquid Dynamics Co.*)

Discrete-Stage Contact Centrifugal Contactors

The Westfalia Centrifugal Extractor This extractor[159,160] is also built on the vertical rotating principle shown in Fig. 56. The feed and solvent are mixed intensively by simultaneous flow through a number of parallel nozzles to establish extraction equilibrium, which is followed by a phase separation in a centrifugal separator bowl. Little information has been published on its performance. Its principal application is in the pharmaceutical industry. It has been reported that the capacity of the largest single-stage extractor is 49,270 L/min (13,000 gal/h).

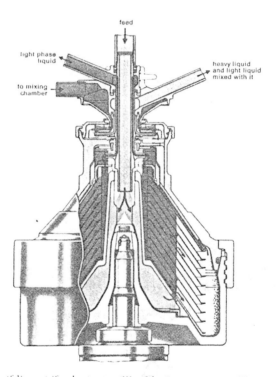

Fig. 56 Westfalia centrifugal extractor. (*Westfalia Separators A.G., West Germany.*)

Robatel Centrifugal Extractor Each stage is composed of a mixing chamber and a settling chamber. The high relative speed between the stationary frames of the agitating disk and the rotating frames of the mixing chamber create the shear force for interphase mixing in which extraction takes place. The mixed liquids, after entering the settling chamber, are separated by the centrifugal force, and then the light and heavy liquids pass via channels in the adjacent (above and below) stages.

The unit generally provides three to eight stages, and throughputs up to 6064 L/h (1600 gal/h) are quoted.

The extractors have found general application in the chemical, pharmaceutical, and petrochemical industries and have been used particularly in the nuclear industry. Technical and economic comparisons of the units with mixer-settlers and pulsed columns for nuclear extraction have been published.[161] Figure 57 shows an eight-stage Robatel nuclear extractor.

LABORATORY-SCALE EXTRACTORS AND
MISCELLANEOUS EXTRACTION EQUIPMENT

Several laboratory units which have proved useful in analysis, process control, process study and development, and obtaining basic data for extraction design are described below.

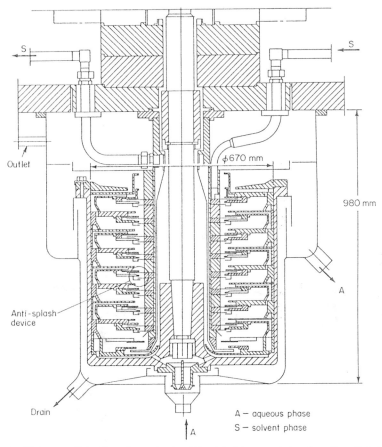

Fig. 57 Eight-stage Robatel nuclear extractor. [*Reprinted by permission from Proc. Int. Solvent Extraction Conf. 1974, 2, 1282.*]

The Craig Extractor

This apparatus consists of a series of specially designed glass tubes for a multiplicity of extractions and is an invaluable analytical tool[162] for:

- Separation of mixtures
- Identification of a substance
- Proof of purity of a substance
- Determination of distribution coefficients

Figure 58 shows an automatic Craig extractor with 500 contacting stages. Many machine models, which are capable of performing up to thousands of extractions and can operate automatically, are commercially available.[163]

The AKUFVE System

AKUFVE, a recent Swedish development, is a rapid and continuous system for measuring distribution coefficients in solvent extraction. The apparatus incorporates a mixer, a specially devised centrifuge for rapid and accurate separation of liquid phases, and on-line detectors, which are connected either in a once-through or in a closed-cycle arrangement. A schematic arrangement of the unit is shown in Fig. 59. This equipment, which was reported to be efficient for basic and applied research in solvent extraction:[164-166]

- Facilitates continuous and rapid measurement of distribution coefficients as a function of variations in simultaneously measured physical and chemical parameters.
- Is used for the determination of distribution and stability constants for various metal complexes, together with enthalpy and entropy values obtained from temperature-

Fig. 58   Craig extractor. *(Hoffmann-La Roche Inc.)*

dependency measurements, and the determination of reaction rates and activation energies.

- Provides data for evaluation and optimization of solvent-extraction processes.
- Is used for small-scale extraction for short-lived or unstable substances.

Bench-Scale Multistage Contactor

A continuous, bench-scale, multistage, countercurrent, liquid-liquid contactor was developed by Hanson et al.[167] The unit consists of a series of pump-mix mixer-settlers. Each stage is constructed individually as shown in Fig. 60. To obtain a multistage unit, the required number of individual stages are simply coupled together. Figure 61 shows a photograph of a multistage unit.

The apparatus provides stagewise contacting without back or forward mixing and complete flexibility in the number of stages used. An additional advantage is that it can be easily shut down and started up, without contamination of material between stages.

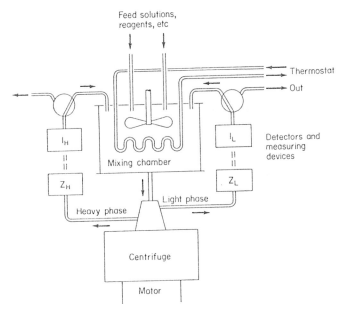

Fig. 59 Schematic arrangement of the AKUFVE. [*Reprinted by permission of* Chem. and Ind. (London), 488 (1970).]

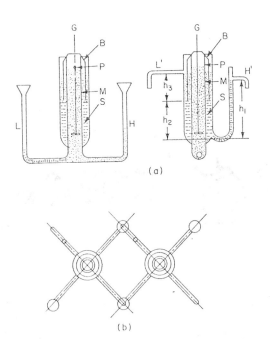

Fig. 60 Hanson's bench-scale mixer-settlers: (*a*) side view; (*b*) plan view. [*Reprinted by permission from* Chem. and Ind. (London), 9, 1090 (1969).]

Because the unit gives a known number of theoretical stages, it is particularly useful for preliminary laboratory work on development and optimization of new extraction processes which require practical confirmation of design flowsheets. It has been successfully used in a pilot plant for development of a nuclear-fuel processing installation.[168] A Pyrex unit is commercially available.[169]

Laboratory-Scale Continuous Mutlistage Extractor

A small, continuous, multistage, box-type, mixer-settler extractor has been reported by Rahn and Smutz.[170] The contactor uses the pump-mix principle in mixing chambers and individual interface control in settling chambers. This unit is claimed to have high-stage efficiency, flexibility of operation, and simplicity of control. The extractor was developed particularly for separation of rare earths. The apparatus is similar in principle to that of Hanson mentioned above.

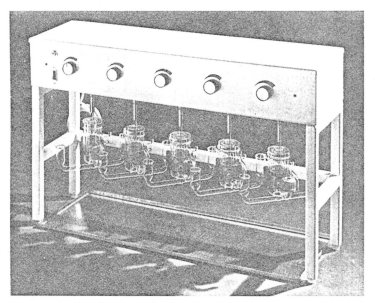

Fig. 61   Bench-scale mixer-settler. *(Horbury Technical Service, Ltd., England.)*

Laboratory-Scale Reciprocating-Plate Extraction Column

A 25-mm-diameter (1-in), open-type perforated reciprocating plate extraction column has been reported by Lo and Karr.[112] A minimum height of an equivalent theoretical stage (HETS) of 70 mm (2.8 in) and volumetric efficiencies of up to 532 per hour were achieved employing an MBK–acetic acid–water system. The column is simple in construction and versatile for laboratory process studies. The column has successfully been used for countercurrent and fractional liquid extraction in the laboratory. Figure 62 shows a 25-mm-diameter (1-in) reciprocating-plate extraction column and perforated-plate detail. Columns constructed of Pyrex with Teflon perforated plates are commercially available.[171]

SELECTION OF COMMERCIAL EXTRACTORS

Because of the great variety of commercial extractors available, the choice of a commercial extractor for a new process can be very bewildering. The following criteria should be taken into consideration when selecting a contactor for a particular application:[10]

- Stability and residence time
- Settling characteristics of the solvent system

- Number of stages required
- Capital cost and maintenance
- Available space and building height
- Throughput
- Experience with the type of contactor

Hanson[6] suggested a selection guide for use in choosing contactors (Fig. 63).

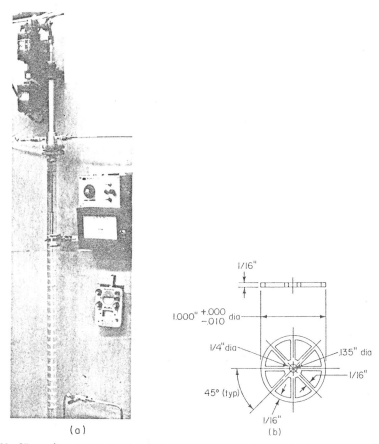

(a)   (b)

Fig. 62 25-mm diameter (1-in) reciprocating-plate extraction column: (a) column arrangement; (b) reciprocating-plate detail. [*Reprinted by permission of* Ind. Eng. Chem. Process Des. Dev., 11(4), 495 (1972).]

The preliminary choice of an extractor for a specific process is primarily based on consideration of the system properties and number of stages required for the extraction. A qualitative chart[124] of the economic operating range of various classes of extractor is shown in Fig. 64.

Choosing a contactor is still both an art and a science. It is largely based on one's experience. The vendor's experience, pilot-testing procedures, scaling-up methods, costs for capital equipment and maintenance, and reliability of operation should be considered and evaluated at an early stage before the pilot-plant tests are committed. Although cost ought to be a major balancing consideration, in many actual cases previous experience and practice are the deciding factors.

PROCESS CONTROL AND INSTRUMENTATION

An extraction plant should continue to operate at a steady state according to the design process flowsheet if the concentration and flow rate of all streams and the operating conditions are constant. However, this is not always the case in practical operation. Fluctuations in the feed streams can cause fluctuations in the products unless a sophisti-

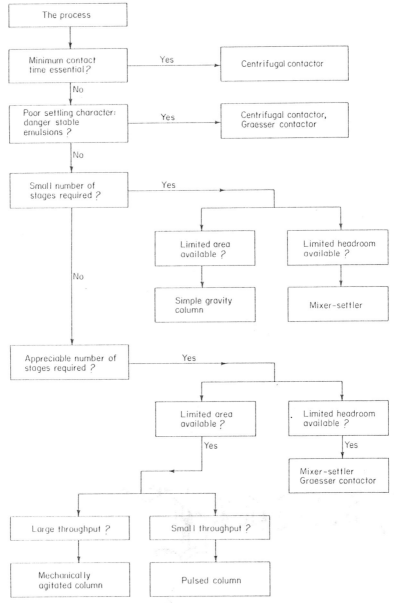

Fig. 63 Selection guide for use in choosing contactors. (*Reprinted by permission from Chem. Eng., Aug. 6, 1968.*)

cated system of feedforward control is used. Following are brief descriptions of general principles which are important to operating control.

Interface Control

Upset of column operation caused by flooding in the column always forces a shutdown of plant operation. Therefore, interface control could be most important in operation of an extraction plant. In general, certain steps should be taken by the operator who finds that the column is flooded. The following three methods are generally employed for interface control in a column at either the top or the bottom. The choice of method depends on the physical properties of the solvent system employed.

1. Hydrostatic pressure (e.g., DP cell)
2. "Float" (Archimedes' principle of buoyancy)
3. Dielectric-constant change (e.g., detected by capacitance type)

Automation and Instrumentation

Because of the interrelated and complex operations of the extraction process and the high sensitivity of the liquid-liquid contactor to fast flow fluctuation, automatic control requires a maximum of control loops. Based on previous experience, it is the author's opinion, which is shared by others,[172,173] that, at present, automatic process control would hardly be economically justified, and that it has not been found practical to rely totally on automatic control. The reasons are (1) investment and maintenance costs for an automatic on-line analytical system are high, (2) control rarely depends on a single analytical determination, and (3) complex dynamic responses always require lengthy studies before a decision can be made. It is more

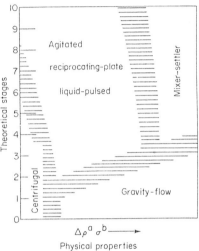

Fig. 64 Economic operating range of extractors. (*Luwa A.G., Zurich.*)

practical to design the plant based on (1) process decisions by trained technical personnel, (2) using off-line analysis and limited on-line automatic analysis, and (3) control panels equipped with manual/automatic control for motor speed, flow, interface level, temperature, etc. A process-control flow diagram[10] for an extraction column is shown in Fig. 65.

ECONOMICS OF EXTRACTION DESIGN

Although the economic picture is most important for any process, only few papers present the economics of solvent extraction.[174,175] The economics of a solvent extraction process may be divided into:

Capital costs:
1. Equipment
2. Inventory of material held up within plant

Prime costs:
3. Contactor operation
4. Solvent recovery
5. Solvent losses

Of these, solvent losses generally can be controlled to a satisfactory limit by design, and items 1 and 4 usually dominate the overall picture. The cost of solvent recovery is the most important economic question for a new process. Tables 13 and 14 illustrate the capital investment and operating cost for an extraction plant.[10] It is obvious that the capital investment for the extractor is only one quarter of the total investment. Reducing the height of the column, hence having fewer stages, would not give a significant cost saving. On the other hand, as shown in Table 14, solvent recovery has a greater impact on overall cost than does the number of contacting stages. Many processes would be more economical with a larger number of stages, so that the amount of solvent to be recovered would be less, and hence the cost for solvent recovery would be substantially reduced.

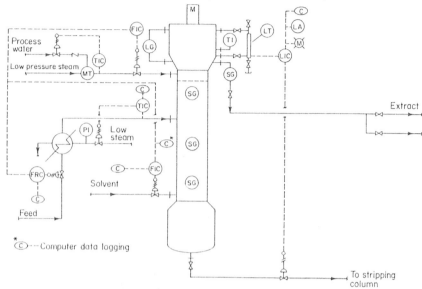

Fig. 65   Flow diagram of process control for an extraction column. (*Reprinted by permission from T. C. Lo, Recent Developments in Commercial Extractors, paper presented at Engineering Foundation Conference on Mixing Research at Rindge, N.H., 1975.*)

TABLE 13   Cost Estimation for Capital Process Equipment of an Extraction Plant

Item	Description	Cost, U.S.$, 1975
Extractor system:		
Extraction column	4 ft 9 in × 50 ft, s/s	117,000
Auxiliary units		45,500
Toluene flash evaporator units		182,000
Raffinate stripper units		45,840
Total:		
Uninstalled		390,340
Installed (f = 4.0)		1,560,000

SOURCE: T. C. Lo, Recent Developments in Commercial Extractors, Engineering Foundation Conference on Mixing Research at Rindge, N. H., 1975, with permission.

TABLE 14   Operating Cost for an Extraction Plant

	Cost, $/year (1975)
Labor: ½ worker/shift	30,000
Utilities:	
Steam: 77 million lb (evap. 28.5 million gal toluene)	245,000
Cooling-tower water	9,200
15 hp—power for column drive	2,525
Solvent makeup (0.1% loss)	16,400
Maintenance	35,000
Total	$338,125

SOURCE: T. C. Lo, Recent Developments in Commercial Extractors, Engineering Foundation Conference on Mixing Research at Rindge, N. H., 1975, with permission.

REFERENCES

1. Treybal, R. E., "Liquid Extraction," 2d ed., pp. 396, 463, McGraw-Hill, New York, 1963.
2. Logsdail, D. H., and Lowes, L., chap. 5 in C. Hanson (ed.), "Recent Advances in Liquid-Liquid Extraction," Pergamon, Oxford, 1971.
3. Perry, J. H., and Chilton, C. H., "Chemical Engineers' Handbook," 5th ed., p 12-1, McGraw-Hill, New York, 1973.
4. Akell, R. B., Chem. Eng. Prog., 62(9), 50 (1966).
5. Reman, G. H., Chem. Eng. Prog., 62(9), 56 (1966).
6. Hanson, C., Chem. Eng., 75(18), 76 (1968); 75(19), 135 (1968).
7. Mumford, C. J., Brit. Chem. Eng., 13, 981 (1968).
8. Warwick. G. C., Chem. Ind., May 1973, p. 403.
9. Oldshue, J., in Y. Marcus (ed.), "Solvent Extraction Reviews," Marcel Dekker, Inc., New York, 1971.
10. Lo, T. C., Recent Developments in Commercial Extractors, paper presented at Engineering Foundation Conference on Mixing Research, Rindge, N.H., The Engineering Foundation, New York, August 1975.
11. Bailes, P. J., Hanson, C., and Hughes, M. A., Chem. Eng., 83(2), 86 (1976).
12. Sawistowski, H., and Coltz, C. E., Trans. Inst. Chem. Eng., 41, 174 (1963).
13. Davies, J. T., Chem. Eng. Prog., 62(9), 89 (1966).
14. Sawistowski, H., chap. 9 in C. Hanson (ed.), "Recent Advances in Liquid-Liquid Extraction," Pergamon, Oxford, 1971.
15. Jackson, R., Chem. Eng. Prog., 62(9), 82 (1966).
16. Heertjes, P. M., and De Nie, L. H., chap. 10 in C. Hanson (ed.), "Recent Advances in Liquid-Liquid Extraction," Pergamon, Oxford, 1971.
17. Sleicher, C. A., Am. Inst. Chem. Eng. J., 5, 145 (1959).
18. Ingham, J., chap. 8 in C. Hanson (ed.), "Recent Advances in Liquid-Liquid Extraction," Pergamon, Oxford, 1971.
19. Misek, T., and Rod, V., chap. 7 in C. Hanson (ed.), "Recent Advances in Liquid-Liquid Extraction," Pergamon, Oxford, 1971.
20. Treybal, R. E., Chem. Eng. Prog., 62(9), 67 (1966).
21. Souders, M., Chem. Eng. Prog., 62(9), 103 (1966).
22. Johnstone, R. E., and Thring, M. W., "Pilot Plants, Models, and Scale-up Methods in Chemical Engineering," McGraw-Hill, New York, 1957.
23. Olney, R. B., and Miller, R. S., in "Modern Chemical Engineering," vol. 1, p. 89, Reinhold, New York, 1963.
24. Letan, R., and Kehat, E., Am. Inst. Chem. Eng. J., 13, 443 (1967).
25. Kehat, E., and Sideman, S., chap. 13 in C. Hanson (ed.), "Recent Advances in Liquid-Liquid Extraction," Pergamon, Oxford, 1971.
26. Lebison, I., and Beckman, R. B., Chem. Eng. Prog., 49(8), 405, (1953).
27. Vermeulen, T., Moon, J. S., Hennico, A., and Miyauchi, T., Chem. Eng. Prog., 62(9), 95 (1966).
28. Krishnamurty, R., and Rao, C. V., Ind. Eng. Chem., Process Des. Dev., 7(2), 166 (1968).
29. White, J. E., Exxon Europe, Inc., personal communication.
30. Williams, J. A., Lowes, L., and Tanner, M. C., Trans. Inst. Chem. Eng., 36, 464 (1958).
31. Lowes, L., and Larkin, M. J., Ind. Chem. Eng. Symp. Ser. 26, 111 (1967).
32. Holley, A. E., and Mott, D. E., British Patent 321,200, 1929.
33. Warner, B. F., Proc. 3d U.N. Conf. Peaceful Uses of Atomic Energy, Geneva, 10, 224, 1964.
34. Hanson, C., and Kaye, D. A., Chem. Process. Eng., 44, 27 (1963).
35. Hanson, C., and Kaye, D. A., Chem. Process. Eng., 44, 651 (1963).
36. Treybal, R. E., Chem. Eng. Prog., 60(5), 77 (1964).
37. Coplan, B. V., Davidson, J. K., and Zebroski, E. L., Chem. Eng. Prog., 50(8), 403 (1954).
38. Hanson, C., and Kaye, D. A., Chem. Process. Eng., 45, 413 (1964).
39. British Patent 1,117,959.
40. Mizrahi, J., Barnea, E., and Meyer, D., Proc. Int. Solvent Extraction Conf. 1, 141 (1974).
41. IMI Staff, Proc. Int. Solvent Extraction Conf. 2, 1386 (1971).
42. Ager, D. W., and Dement, E. R., Proc. Int. Symp. Solvent Extraction in Metallurgical Processes, p. 27 (Technologisch Instituut K. VIV, Antwerp, 1972).
43. Warwick, G. C. I., Scuffham, J. B., and Lott, J. B., Proc. Int. Solvent Extraction Conf. 2, 1373 (1971).
44. Warwick, G. C. I., and Scuffham, J. B., Proc. Int. Symp. Solvent Extraction in Metallurgical Processes, p. 36 (Technologisch Instituut K. VIV, Antwerp, 1972).
45. Jackson, I. D., et al., Inst. Chem. Eng. Symp. Ser., 42, paper 15 (1975).
46. Mattila, T. K., Proc. Int. Solvent Extraction Conf. 1, 169 (1974).
47. Niinimaki, L., and Orjans, J. R., Chem. Eng., 78(1), 63 (1971).
48. Stönner, H. M., and Wohler, F., Inst. Chem. Eng. Symp. Ser., 42, paper 14 (1975).
49. Mehner, W., Hochfeld, G., and Mueller, E., Proc. Int. Solvent Extraction Conf., 2, 1265, (1971).
50. Exhibition during International Solvent Extraction Conference ISEC '74, Lyon, 1974.
51. Morris, W. H., British Patents 885,503 and 974,829.

52. Fenske, M. R., and Long, R. B., U.S. Patent 2,667,407, 1954; *Chem. Eng. Prog.*, **51**(4), 194 (1955); *Ind. Eng. Chem.*, **53**, 791 (1961).
53. Karpacheva, S. M., et al., *Proc. Int. Solvent Extraction Conf.*, **1**, 231 (1974).
54. Johannisbauer, W., and Kaiser, *Proc. Int. Solvent Extraction Conf.*, **1**, 243 (1974).
55. Chase, U.S. Patent 2,183,859, 1939.
56. "Lightnin Mixers and Aerators," Mixing Equipment Company, Inc., *Bull.* B-528, Rochester, N.Y.).
57. "Kenics Static Mixers," Kenics Corp. *Bull.* KTEK-5, Danvers, Mass., 1972.
58. *Intermet Bull.*, **4**(3), 30 (April 1974).
59. *Chem. Eng.*, **80**(7), 111 (Mar. 19, 1973).
60. Schott, N. R., paper presented at Engineering Foundation Conference on Mixing Research, Rindge, N.H., August 1975 (The Engineering Foundation, New York).
61. Barbini, R., paper presented at Engineering Foundation Conference on Mixing Research, Rindge, N.H., August 1975 (The Engineering Foundation, New York).
62. Weinstein, B., paper presented at Engineering Foundation Conference on Mixing Research, Rindge, N.H., August 1975 (The Engineering Foundation, New York).
63. Miller, S. A., and Mann, C. A., *Trans. Am. Inst. Chem. Eng.*, **40**, 709 (1944).
64. Flynn, A. W., and Treybal, R. E., *Am. Inst. Chem. Eng. J.*, **1**, 324 (1955).
65. Karr, A. E., and Scheibel, E. G., *Chem. Eng. Prog. Symp. Ser.* **10**, 50, 73 (1954).
66. Ryon, A. D., Daley, F. L., and Lowry, R. S., *Chem. Eng. Prog.*, **55**(10), 71 (1959).
67. Rushton, J. H., and Oldshue, J. Y., *Chem. Eng. Prog.*, **49**(4), 161 (1953).
68. Bates, R. L., Fondy, P. L., and Fenic, J. G., chap. 3, p. 133, in Uhl, V. W., and Gray, J. B. (eds.), "Mixing-Theory and Practice," vol. 1, Academic Press, New York, 1966.
69. Vermeulen, T., Williams, G. M., and Langlois, G. E., *Chem. Eng. Prog.*, **51**(2), 85-F (1955).
70. Warner, B. F., "The Scaling-up of Chemical Plant and Processes," p. 44 Joint Symposium, London, 1957.
71. Pike, F. P., and Wadhawan, S. C., *Proc. Int. Solvent Extraction Conf.*, **1**, 112 (1971).
72. Davies, G. A., et al., *Chem. Eng.*, **266**, 392 (1972).
73. Van Dijck, W. J. D., U.S. Patent 2,011,186, 1935.
74. Rouyer, H., Lebouhellec, J., Henry, E., and Michel, D., *Proc. Int. Solvent Extraction Conf.* **3**, 2339 (1974).
75. Bronswerk, P. C. E. S., "The Bronswerk Technical Bulletin on Pulsed Packed Column," Amersfoort, The Netherlands.
76. Spaay, N. U., Simons, A. J. F., and ten Brink, G. P., *Proc. Int. Solvent Extraction Conf.* **1**, 381 (1971).
77. Gayler, R., Roberts, N. N., and Pratt, H. R. C., *Trans. Inst. Chem. Eng.*, **31**, 57 (1953).
78. Thornton, J. D., *Chem. Ind.*, 1581 (1954).
79. Thornton, J. D., *Chem. Eng. Sci.*, **5**, 201 (1956).
80. Thornton, J. D., *Chem. Eng. Prog. Symp. Ser.* **13** (1954).
81. Thornton, J. D., and Pratt, H. R. C., *Trans. Inst. Chem. Eng.*, **31**, 289 (1953).
82. Commissariat à l'energie atomique (Genas, France), Liquid-Liquid Extraction in C. E. A. Establishments, *Bull.* 25/74.
83. Sege, G., and Woodfield, F. W., *Chem. Eng. Prog.* **50**(8), 396 (1954).
84. Thornton, J. D., *Brit. Chem. Eng.*, **3**, 247 (1958).
85. Thornton, J. D., *Trans. Inst. Chem. Eng.*, **35**, 316 (1957).
86. Logsdail, D. H., and Thornton, J. D., *Trans. Inst. Chem. Eng.*, **35**, 331 (1957).
87. Smoot, L. D., Mar, B. W., and Babb, A. L., *Ind. Eng. Chem.*, **51**(9), 1005 (1959).
88. Angelino, H., Alran, C., Boyadzhiev, L., and Mukherjee, S. P., *Brit. Chem. Eng.*, **12**, 1893 (1967).
89. Pope, B. J., and Shah, N. R., *Proc. Int. Solvent Extraction Conf.* **1**, 699 (1971).
90. Szabo, T. T., et al., *Chem. Eng. Prog.*, **60**(1), 66 (1964).
91. Belter, P. A., and Speaker, S. M., *Ind. Eng. Chem. Process Des. Dev.*, **6**, 36 (1967).
92. Weech, M. E., and Knight, B. E., *Ind. Eng. Chem.*, *Process Des. Dev.*, **6**, 480 (1967); **7**, 156 (1968).
93. Cloete, F. L. D., and Streat, H., *Nature*, **200**, 1199 (1963).
94. Baird, M. H. I., and Ritcey, G. M., *Proc. Int. Solvent Extraction Conf.*, **2**, 1571 (1974).
95. Scheibel, E. G., *Chem. Eng. Prog.*, **44**(9), 681, 1948; U.S. Patent 2,493,265, 1950.
96. Scheibel, E. G., and Karr, A. E., *Ind. Eng. Chem.*, **42**(6), 1048 (1950).
97. Scheibel, E. G., *Am. Inst. Chem. Eng. J.*, **2**, 74, 1956; U.S. Patent 2,856,362, 1958.
98. Honekamp, J. R., and Burkhart, L. E., *Ind. Eng. Chem. Process Des. Dev.*, **1**(3), 176 (1962).
99. Jeffreys, G. V., Davies, G. A., and Piper, H. B., *Proc. Int. Solvent Extraction Conf.* **1**, 680 (1971).
100. Gel'perin, N. I., et al., *Theor. Found. Chem. Eng.* (in English, USSR), **1**, 552 (1967).
101. Scheibel, E. G., U.S. Patent 3,389,970 (1968).
102. Scheibel, E. G., personal communication, 1971.
103. Reman, G. H., *Proc. 3d World Pet. Cong.*, The Hague, Sec. III, p. 121, 1951.
104. Hinze, J. O., *Am. Inst. Chem. Eng. J.*, **1**, 289 (1955).
105. Escher, B. V., Rotating Disc Contactor for Solvent Extraction Processes, *Tech. Bull.*, The Hague, The Netherlands.
106. Stemerding, S., Lumb, E. C., and Lips, J., *Chem. Eng. Tech.*, **35**, 844 (1963).

107. Stainthorp, F. P., and Sudall, N., *Trans. Inst. Chem. Eng.*, **42**, 198 (1964).
108. Misek, T., "Rotating Disc Extractors and Their Calculation," State Publishing House of Technical literature, Prague, 1964.
109. Strand, C. P., Olney, R., and Ackerman, G. H., *Am. Inst. Chem. Eng. J.*, **8**, 252 (1962).
110. Oldshue, J. Y., and Rushton, J. H., *Chem. Eng. Prog.*, **48**(6), 297 (1952).
111. Mixer/Extraction Column, *Chemmunique*, June 1967.
112. Oldshue, J. Y., *Biotechnol. Bioeng.*, **8**(1), 3 (1966).
113. Bibaud, R., and Treybal, R., *Am. Inst. Chem. Eng. J.*, **12**, 472 (1966).
114. Haug, H. F., *Am. Inst. Chem. Eng. J.*, **17**, 585 (1971).
115. Ingham, J., *Trans. Inst. Chem. Eng.*, **50**, 372 (1972).
116. Dykstra, J., Thompson, B. H., and Clouse, R. J., *Ind. Eng. Chem.*, **50**, 161 (1958).
117. Gustison, R. A., Treybal, R. E., and Capps, R. C., *Chem. Eng. Prog. Symp. Ser.*, **58**(39), 8 (1962).
118. Gutoff, E. B., *Am. Inst. Chem. Eng. J.*, **11**, 712 (1965).
119. Miyauchi, T., Mitsutake, H., and Harase, I., *Am. Inst. Chem. Eng. J.*, **12**, 508 (1966).
120. Oldshue, J. Y., private communication, 1970.
121. Oldshue, J. Y., Hodgkinson, F., and Pharamond, J. C., *Proc. Int. Solvent Extraction Conf.*, **2**, 1651 (1974).
122. Misek, T., and Marek, J., *Brit. Chem. Eng.*, **15**, 202 (1970).
123. Marek, J., et al., paper presented at Society of Chemical Industry Symposium, Bradford, U.K., 1967.
124. Marek, J., Luwa AG (Zurich), technical report, March 1970.
125. Seidlova, B., and Misek, T., *Proc. Int. Solvent Extraction Conf.* **3**, 2365 (1974).
126. Fischer, A., *Verfahrenstechnik*, **5**, 360 (1971).
127. Ingham, J., et al., *Proc. Int. Solvent Extraction Conf.*, **2**, 1299 (1974).
128. Hody, D., *Chem. Rundsch.*, **28**, 9 (1975).
129. Treybal, R. E., U.S. Patent 3,325,255.
130. McEwen, C. K., *Specialties*, October 1966.
131. Leisibach, J., *Chem. Ing. Tech.*, **37**, 205 (1965).
132. Coleby, J., British Patents 860,880, 972,035, and 1,037,573.
133. Sheikh, A. R., et al., *Trans. Inst. Chem. Eng.*, **50**, 199 (1972).
134. Karr, A. E., *Am. Inst. Chem. Eng. J.*, **5**, 446 (1959).
135. Karr, A. E., and Lo, T. C., *Proc. Int. Solvent Extraction Conf.* **1**, 299 (1971).
136. Issac, N., and DeWitte, R. L., *Am. Inst. Chem. Eng. J.*, **4**, 498 (1958); *Dechema Monogr.*, **32**, 218 (1959).
137. Guyer, A., Guyer, A., Jr., and Mauli, K., *Helv. Chim. Acta*, **38**, 790, 955 (1955).
138. Prochazka, J., Landau, J., Souhrada, F., and Heyberger, A., *Brit. Chem. Eng.*, **16**, 42 (1971).
139. Elenkov, D., et al., *Khim. Ind. Sofia*, **4**, 181 (1966).
140. Wellek, R., et al., *Ind. Eng. Chem. Process Des. Dev.*, **8**, 515 (1969).
141. Tojo, K., Miyanami, and Yano, T., *J. Chem. Eng. Japan*, **7**, 123 (1974).
142. Lo, T. C., and Karr, A. E., paper presented at Engineering Foundation Conference on Mixing Research, Andover, N.H., Aug. 9–13, 1971; *Ind. Eng. Chem. Process Des. Dev.*, **11**(4), 495 (1972).
143. Karr, A. E., and Lo, T. C., *Proc. Int. Solvent Extraction Conf.*, 1977.
144. Karr, A. E., and Lo, T. C., *Chem. Eng. Prog.* (to be published).
145. Baird, M. H. I., McGinnis, R. G., and Tan, G. C., *Proc. Int. Solvent Extraction Conf.* **1**, 251 (1971).
146. Baird, M. H. I., and Lane, S. J., *Chem. Eng. Sci.*, **28**, 947 (1973).
147. Kim, S. D., and Baird, M. H. I., *Can. J. Chem. Eng.*, **54**, 81 (1976).
148. Rosen, A. M., and Krylov, V. S., *Chem. Eng. J.*, **7**, 85 (1974).
149. Somekh, G. S., *Proc. Int. Solvent Extraction Conf.*, **1**, 323 (1971).
150. Metcalfe, R. S., Ph.D. thesis, The Pennsylvania State University, 1970.
151. Takeba, K., preprint of the 10th General Symposium of the Society of Chemical Engineers, Japan, p. 124, 1971.
152. Prochazka, J., and Hafez, M. M., *Coll. Czechoslov. Chem. Comm.*, **37**, 3725 (1972).
153. Landau, J., Dim, A., and Shemilt, L. W., *Can. J. Chem. Eng.*, **53**, 9 (1975).
154. Myanami, K., Tojo, K., and Yano, T., *J. Chem. Eng. Japan*, **6**, 518 (1973).
155. Podbielniak, W. J., *Chem. Eng. Prog.*, **49**(5), 252 (1953).
156. Todd, D. B., and Davis, G. R., *Proc. Int. Solvent Extraction Conf.*, **3**, 2379 (1974).
157. Broadwell, E., paper presented at Society of the Chemical Industry Symposium, Bradford, U.K., 1967.
158. Doyle, C. M., Podbielniak Doyle, W. G., Rauch, E. G., and Lowry, C. D., *Chem. Eng. Prog.*, **64**(12), 68 (1968).
159. Eisenlohr, H., *Dechema Monogr.*, **19**, 222 (1951).
160. Anon., paper presented at Society of the Chemical Industry Symposium, Bradford, U.K., 1967.
161. Bernard, C., Michel, P., and Tamero, M., *Proc. Int. Solvent Extraction Conf.*, **2**, 1282 (1971).
162. Weissberger, A., "Technique of Organic Chemistry," Craig, L. C., and Craig, D., vol. 3, part 1, pp. 149–332; Scheibel, E. G., pp. 332–393, Interscience, New York, 1956.
163. Gaskins, P., *Am. Lab.*, **73**, October 1973.
164. Anderson, C., et al., *Acta Chem. Scand.*, **23**, 2781 (1969).

165. Reinhardt, H., and Rydberg, J., *Chem. Ind.*, 11, 488, (April 1970).
166. MEAB Metallextraktion AB (Sweden), "Brochure on AKUFVE 110."
167. Anwar, M. M., Hanson, C., and Pratt, M. W. T., *Chem. Ind.*, 9, 1090 (August 1969).
168. Naylor, A., Baxter, W., Duncan, A., and Scott, A. F. D., *J. Nucl. Energy* (pts. A/B), 18, 331 (1964).
169. Horbury Technical Services, Ltd. (Yorkshire, England), "Mixer-Settler MKIII—A Continuous Bench-Scale Multi-Stage Countercurrent Liquid Liquid Contactor."
170. Rahn, R. W., and Smutz, M., *Ind. Eng. Chem. Process Des. Dev.*, 8, 289 (1969).
171. Chem-Pro Equipment Co. (Fairfield, N.J.); Julius Montz, GMbH (Hofstrabe, West Germany).
172. I. M. I. Staff, *Proc. Int. Solvent Extraction Conf.* 2, 1386 (1971).
173. Warner, B. F., discussion on papers 9 and 260 during International Solvent Extraction Conference ISEC '74, Lyon, 1974.
174. Scheibel, E. G., *Chem. Eng.*, 64, 238 (November 1957).
175. Slater, M. J., *Chem. Ind.*, 10, 393 (April 1971).

Decantation

WILLIAM B. HOOPER, P.E. *Engineering Fellow, Process Design, Corporate Engineering Department, Monsanto Company; member, American Institute of Chemical Engineers.*

LOUIS J. JACOBS, Jr., M.S., P.E. *Engineering Specialist, Corporate Engineering Department, Monsanto Company; member, American Institute of Chemical Engineers*

DEFINITION AND ABBREVIATIONS

A "decanter" is defined as a vessel used to separate a stream continuously into two liquid phases using the force of gravity. Solid-liquid separations are covered in Sec. 4. Design methods for batch-type decanters will not be covered, since almost all industrial decanters are continuous. Of course, the basic principles still apply in designing a batch decanter.

Common English units are used throughout this section. However, most of the equations are given in nondimensional form so that any consistent set of units can be used.

A_I Area of the interface assuming flat interface, ft^2
A_L Cross-sectional area allotted to light phase, ft^2
A_H Cross-sectional area allotted to heavy phase, ft^2
d Droplet diameter, ft (1 ft = 304,800 μm)
D Decanter diameter, ft
D_h Hydraulic diameter = 4 (flow area)/wetted perimeter
g Local acceleration due to gravity (32.17 ft/s^2)
H_D Height of the dispersion band, ft
h Distance from center to given chord of a vessel, ft
I Width of the interface, ft
L Decanter length, ft
N_{Re} Reynolds number ($vD\rho/\mu$) in consistent units
Q Volumetric flow rate, ft^3/s
u Terminal settling velocity of a droplet, ft/s
v_C Continuous-phase crossflow velocity, ft/s
ρ_D Density of the fluid in the droplet, lb/ft^3
ρ_C Density of the continuous phase, lb/ft^3
μ_C Viscosity of the continuous phase, lb/(ft)(s) [1 cP = 6.72 × 10^{-4} lb/(ft)(s)]

DESIGN METHOD

The following steps should be used to design a decanter:
1. Calculate the settling velocity for droplets.
2. Pick a preliminary size based on overflow rates.
3. Estimate the time required for droplet coalescence.
4. Check the decanter size chosen for turbulence levels.
5. Specify feed and outlet geometry.

The theory for each of these steps will be discussed, and then two examples will be presented to illustrate application of the theory.

Settling Velocity

The basic equation is

$$u = \frac{gd^2(\rho_D - \rho_C)}{18\mu_C} \tag{1}$$

It is derived from Stokes' law using Newton's basic drag equation. A number of assumptions used in deriving Eq. (1) are not valid in most operating decanters. The assumptions most often violated are:
1. The continuous phase is a quiescent fluid.
2. The droplet is a hard sphere with no deformation or internal circulation.
3. The droplet moves in laminar flow (i.e., droplet $N_{Re} = ud\rho_D/\mu_C < 10$).
4. The droplet is large enough to ignore brownian motion.
5. The droplet movement is not hindered by other droplets or by walls.

Perry's "Chemical Engineers' Handbook"[1] gives more discussion of Eq. (1) and corrections for assumptions 2, 3, and 5. However, errors in these points are almost always small compared with assumption 1. Further, the uncertainty involved in choosing the design droplet diameter makes all other refinements trivial.

Design Droplet Diameter

The recommended *design droplet diameter* to use in Eq. (1) is 150 μm (d = 0.0005 ft). This is well below the sizes normally encountered in decanter feeds. For instance, the mean droplet diameter in an agitated vessel is normally in the 500 to 5000 μm range. Middleman[2] has a good summary of methods to predict droplet sizes in two-phase pipeline flow. His information confirms that droplets in turbulent flow in normally sized pipe run in the 200 to 10,000 μm range. The mean droplet size is normally over 1000 μm, and very little material is in droplets under 500 μm in diameter. In summary, one would expect to use a design droplet diameter above 300 μm. Because of the inadequacies of the design equation, however, decanters sized for droplets over 300 μm are usually too small to work well. Successful decanters have been designed using d's from 50 μm (0.00017 ft) to 300 μm (0.001 ft). The most popular choice is 0.0005 ft (150 μm or 0.15 cm). This is the size used in the API Design Method.[3]

Overflow Rate

The continuous phase must move vertically from the feed elevation to the outlet. Droplets of the dispersed phase must move opposite to this velocity to get to the interface. Ideally the continuous phase moves in a completely uniform plug flow with a velocity equal to the "overflow rate." The minimum-size dispersed droplet should then have a settling velocity a little greater than the overflow rate. The first step then is to pick an initial decanter size such that

$$\frac{Q_c}{A_I} < u_D \tag{2}$$

Of course, the actual flow of the continuous phase bears little resemblance to plug flow. However, there should be few if any droplets as small as the "design" droplet; so this is a good starting place.

Equation (2) requires identification of which phase is continuous and which is dispersed. Selker and Sleicher[4] give a useful correlation for predicting which phase is dispersed. Their correlation can be approximated using the following equation:

$$\theta = \frac{Q_L}{Q_H} \left(\frac{\rho_L \mu_H}{\rho_H \mu_L}\right)^{0.3} \tag{3}$$

θ	Result
< 0.3	Light phase always dispersed
0.3–0.5	Light phase probably dispersed*
0.5–2.0	Phase inversion probable, design for worst case
2.0–3.3	Heavy phase probably dispersed*
> 3.3	Heavy phase always dispersed

*Consult the article for critical designs.

Coalescence

The time required for the collected droplets to cross the interface often is the major limit on the size of the decanter. If the phases are not pure components, the droplets do not readily combine in the continuous phase. Therefore, most of them remain discrete until they reach the interface. When a droplet reaches the interface, it settles on the interface under the influence of gravity and makes a "dimple." Continuous phase is trapped between the droplet and the interface as a thin film. The film slowly drains around the droplet and back into the bulk of the continuous phase. When thinned sufficiently, the film ruptures and the droplet flows across the interface. Fragments of the film are often trapped on the wrong side of the interface as a fine "secondary dispersion" which must then get back across the interface. There is no simple equation to predict the time required for a drop to cross the interface. Typical values range from a fraction of a second up to 2 or 3 min. The time gets shorter as (1) the difference in the density of the phases increases, (2) the viscosity of the continuous phase decreases, (3) the viscosity inside the drop decreases, and (4) the interfacial tension increases. Notice that the first two factors are crudely corrected for when Eq. (1) is used to size the interfacial area.

Dispersion Band

The droplets waiting to cross the interface often back up to form a relatively deep band of dispersed droplets. Mizrahi and Barnea[5] report that the height of the dispersion band H_D is correlated by $H_D \propto (Q_c/A_I)^n$. Reported values of n range from 2.5 clear up to 7 for different systems. Since H_D is such a strong function of throughput Q_C, it should be kept small. If H_D becomes a significant fraction of the decanter height, a slight increase in feed will fill the decanter with dispersion band and separation will cease. About half of the space in the dispersion band is occupied by droplets of the dispersed phase. Therefore, the residence time of the dispersed phase in the dispersion band is $\frac{1}{2}H_D A_I/Q_D$. Good practice will normally keep $H_D < 10\%$ of decanter height and $\frac{1}{2}H_D A_I/Q_D > 2$ to 5 min.

Emulsion

The discussion so far has been limited to two pure liquids. Sometimes a surface-active agent or a dispersion of fine solids is present. Either can interfere with the coalescing process and lead to a stable emulsion. Indeed many decanters do accumulate a stable emulsion at the interface from time to time owing to traces of one of these agents. Since this "rag" interferes with proper operation of the decanter, sample nozzles should be provided to check for it. If rag is detected, facilities should be provided to drain it off and de-emulsify, or "break," it. Rag is like foam in vapor-liquid systems in that it is most often stabilized by very fine particles. Therefore, filtration will often break it. Other means used to break rags include heating, chemical addition, and reversing the phases.

Decanter Geometry

Both the overflow rate and coalescence are improved as the interfacial area is increased. Most decanters are cylinders for economic reasons. A horizontal cylinder has several times as much interfacial area as it would if it were vertical. It follows, then, that a horizontal tank should be used for the decanter unless there is some overriding extenuating circumstance. Some relationships in a horizontal drum are (see Fig. 1; if $A_L > A_H$, interchange subscripts)

$$I = 2(r^2 - h^2)^{1/2} \qquad A_I = IL$$

$$A_L = \frac{1}{2}\pi r^2 - h(r^2 - h^2)^{1/2} - r^2 \arcsin\left(\frac{h}{r}\right)$$

where arcsin is in radians [= degrees $(\pi/180)$]

$$A_H = \pi r^2 - A_L \qquad P = 2r \arccos\left(\frac{h}{r}\right)$$

$$D_L = \frac{4A_L}{I + P} \qquad D_H = \frac{4A_H}{I + 2\pi r - P}$$

An exception to the use of a horizontal drum is when the decanter would exceed 10 to 12 ft in diameter and would not be a pressure vessel. A vertical storage tank can then be used as a decanter. For best results it should be fed in a special center well and the products should be withdrawn uniformly around the periphery (as in a water clarifier).

If the decanter must be buried in order to serve a sewer system, it should be an API type or a proprietary plate separator.

Turbulence

In a horizontal decanter, the continuous phase is flowing perpendicularly to the settling of the droplets. This movement creates turbulence which interferes with the settling process. If the crossflow were fully laminar, there would be no problem, but this is usually impractical. The degree of turbulence is best expressed by the Reynolds number N_{Re} where

$$N_{Re} = \frac{vD_h\rho_C}{\mu_C} \tag{4}$$

The hydraulic diameter D_h should be calculated by four times the flow area allotted to the phase in question divided by the perimeter of the flow channel. Of course, v is the velocity down the flow channel.

The following guidelines pretty well summarize experience from successful decanters:

N_{Re}	Effect
Less than 5000	Little problem
5000–20,000	Some hindrance
20,000–50,000	Major problem may exist
Above 50,000	Expect poor separation

Interfacial Shear

It is good practice to proportion the decanter so that the velocities in both phases are similar. Normally the ratio should be less than 2:1. It is not necessary to make them equal, since Vijayan, Ponter, and Jefferys[6] report that some velocity difference seems to aid coalescing.

Feed Arrangement

Improper introduction of the feed is the most common limit to decanter performance. For optimum design, the feed should be:
1. Introduced uniformly across the active cross section of the decanter.
2. Done in such a way as to leave no residual jets or turbulence.

Often a feed is introduced through a nozzle at a velocity of 2 to 5 ft/s. When this is done,

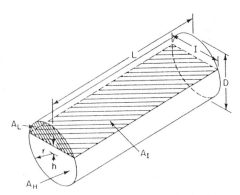

Fig. 1 Relationships in a horizontal cylinder.

the feed acts as a jet that shoots across the decanter. As the jet diffuses, it entrains the liquid around it and sets up major mixing in the decanter. The first step in feed design is then to limit the feed velocity. A practical approach is to limit the velocity head ρv^2 at the feed nozzle to 250 lb/ft-s² (or 0.002 atm). For water this is a velocity limit of 2 ft/s.

The next step is to make the feed undergo one or more direction changes as it enters the decanter. These two steps are adequate for some noncritical decanters. However, a decanter that is near a minimum size should also have a feed diffuser as the third step.

Feed Diffuser

The feed diffuser has two conflicting requirements. It should have a pressure drop many times higher than the pressure drop down the decanter to force distribution. However, the peak velocity leaving the diffuser should be only two to five times the mean velocity down the decanter. A good solution is to use two plates or rows of bars. The first plate or row of bars takes a high pressure drop. This requires that the open-flow area be 3 to 10% of the decanter cross-sectional area. The second plate or row of bars is placed to break up the jets from the first element. Further its open-flow area should be 20 to 50% of the decanter cross-sectional area. The spacing between the elements should be about the same as the openings in the elements. The energy in the jets will then be dissipated as turbulence in the volume between the elements.

Outlets

Attention should also be given to the outlets. A high outlet velocity must be avoided, since it can create a vortex that reaches up and sucks droplets out of the dispersion band. The outlet should draw from the clear liquid as uniformly as possible. Moreover, the velocity in the outlet should not exceed about ten times the velocity down the decanter.

Scale Model

If the decanter design is critical or if the design is unusual, you should seriously consider checking it with a scale model. The model can be made of any transparent material. Any convenient fluid and tracer can be used. The important thing is to check the flow patterns and amount of turbulence at the design Reynolds number. Notice that this means that the residence time in the model will be much lower and the overflow rate higher. For one such test, see Hooper.[7]

EXAMPLE 1

Design a decanter to separate

$$10,000 \text{ lb/h of oil, } \rho = 56 \text{ lb/ft}^3, \mu = 10 \text{ cP}$$
$$40,000 \text{ lb/h of water, } \rho = 62 \text{ lb/ft}^3, \mu = 0.7 \text{ cP}$$

Setup

$$Q_{oil} = 10,000/(56 \times 3600) = 0.050 \text{ ft}^3/\text{s}$$
$$\mu_{oil} = 10 \times 6.72 \times 10^{-4} = 60 \text{ lb/ft-s}$$
$$Q_{water} = 40,000/(62 \times 3600) = 0.18 \text{ ft}^3/\text{s}$$
$$\mu_{water} = 0.7 \times 6.72 \times 10^{-4} = 4.7 \times 10^{-4} \text{ lb/ft-s}$$

Check for dispersed phase From Eq (3), $\theta = \dfrac{0.05}{0.18}\left(\dfrac{56 \times 4.7 \times 10^{-4}}{62 \times 6.7 \times 10^{-4}}\right)^{0.3} = 0.12$. Since $\theta < 0.3$, the light

phase (oil) is always dispersed.

Find droplet settling rate Assume $d = 0.0005$ ft (150 μm); then from Eq (1):

$$u_{oil} = \frac{32.17(0.0005)^2(56\text{-}62)}{18 \times 4.7 \times 10^{-4}} = -0.0057 \text{ ft/s}$$

(Minus sign just means that it rises instead of settles.)

From overflow rate, pick size Assume l is 80% of D and $L/D = 5$; then from Eq. (2) $Q_c/A_1 < u_D$. Now A_1 $= lL = 0.8D(5D) = 4D^2$; so $Q_c/4D^2 < u_D$, or

$$D \geqslant \tfrac{1}{2}\left(\frac{Q_c}{u_D}\right)^{1/2} \tag{5}$$

then $D \geqslant \tfrac{1}{2}(0.18/0.0051)^{1/2} = 2.8$ ft. Use $D = 3$ ft and $L = 5D = 15$ ft for the first pass.

Set interface level Try holding the interface 1 ft below the top to ensure that the interface stays clear of the oil outlet. Then (see Fig. 1)

$$h = 0.5 \text{ ft and } r = 1.5 \text{ ft} \qquad l = 2(2.25 - 0.25)^{1/2} = 2.83 \text{ ft}$$
$$A_{oil} = \tfrac{1}{2}\pi(2.25) - 0.5(2.25 - 0.25)^{1/2} - 2.25 \arcsin(0.5/1.5)$$
$$= 3.534 \text{ ft}^2 - 0.707 \text{ ft}^2 - 0.765 \text{ ft}^2 = 2.06 \text{ ft}^2$$
$$A_{water} = \pi r^2 - A_{oil} = \pi(2.25) - 2.06 = 7.07 - 2.06 = 5.01 \text{ ft}^2$$
$$P = 2 \times 1.5 \arccos(0.5/1.5) = 3.69 \text{ ft} \qquad A_1 = 2.83 \times 15 = 42.4 \text{ ft}^2$$

Secondary settling Check the size of water drop (continuous phase) that can buck the oil overflow rate (dispersed phase) if it gets on the wrong side of the interface: $u_{water} \leqslant Q_{oil}/A_1 = 0.05/42.4 = 0.0012$ ft/s. From Eq. (1),

$$d = \left[\frac{18(6.7 \times 10^{-3})(0.0012)}{32.17(62 - 56)}\right]^{1/2} = 0.0009 \text{ ft (260 } \mu\text{m)}$$

This is acceptable *only* because oil will always be the dispersed phase, so there should not be any water in the oil phase.

Check coalescence time Assume $H_D = 10\%$ of D, or 0.3 ft. Then the time available to cross the dispersed band $= \tfrac{1}{2}(0.3 \times 42.4)/0.05 = 126$ s or >2 min. This is on the low side but is probably adequate for an oil-water separation.

Emulsion provision Provide sample nozzles at 4-in intervals above and below the interface to prove the location of the interface and to check for rag. Make them about 1 in size so that 2 or 3% of the feed can be removed as rag.

Check turbulence level

$$D_{oil} = 4(2.06)/(2.83 + 3.69) = 1.27 \text{ ft}$$
$$v_{oil} = 0.05/2.06 = 0.024 \text{ ft/s}$$
$$N_{Re} = 0.024 \times 1.27 \times 56/6.7 \times 10^{-3} = 250 \text{ (very good)}$$
$$D_{water} = 4(5.01)/(2.83 + 9.42 - 3.69) = 2.34 \text{ ft}$$
$$v_{water} = 0.18/5.01 = 0.036 \text{ ft/s}$$
$$N_{Re} = 0.036 \times 2.34 \times 62/4.7 \times 10^{-4} = 11,100 \text{ (fair)}$$

Check interfacial shear Velocity ratio = 0.036/0.024 = 1.5. This is less than 2; so it is satisfactory.

Set feed geometry

$$Q_{total} = 0.05 + 0.18 = 0.23 \text{ ft}^3/\text{s}$$
$$\rho_{mean} = 50,000/(3600 \times 0.23) = 60.7 \text{ lb/ft}^3$$

The economic pipe size for this flow is 2-in pipe. This gives a velocity of 0.23/0.0233 = 9.8 ft/s.

$$\rho v^2 = 60.7(9.8)^2 = 5830 \text{ lb/ft-s}^2 \text{ or} >> 250$$

Put a 4-in tee over the feed nozzle to give two 4-in feeds with a direction change at the inlet. Now velocity is $\frac{1}{2} \times 0.23/0.0884 = 1.3$ ft/s and $\rho v^2 = 103$ lb/ft-s^2 (fine).

Provide a diffuser Since N_{Re} for water is over 10,000, provide a two-element diffuser. For the distributor, put a 1.5-ft-high (D/2) baffle across the decanter. Since $h = 0.75$ ft for the baffle, the baffle area is

$$2 [0.75(1.5^2 - 0.75^2)^{1/2} + 2.25 \text{ arcsin } (0.75/1.5)] = 2 (0.97 + 1.18) = 4.3 \text{ ft}^2$$

Perforate the distributor with ½-in-diameter holes on a 1½-in triangular pitch. Open area = $(\pi\sqrt{3}/6)$ (diam/pitch)^2 = $(\pi\sqrt{3}/6)(\frac{1}{2}/1\frac{1}{2})^2$ = 0.101. Hole area = 0.101 × 4.3 = 0.43 ft^2. This is 6% of the 7.07 ft^2 decanter area (which is between 3 and 10%).

The baffle plate will also be perforated on a 1½ triangular pitch. However, it will be shifted 0.5 in to the side and 0.866 in up so that the jets from the ½-in holes impact on the solid web of the baffle. Use 1-in holes.

Hole area = $(\pi\sqrt{3}/6)(1/1\frac{1}{2})^2(4.3)$ = 1.7 ft^2, which is 25% of the decanter area. This meets the 20 to 50% criteria.

The gap between the elements should be about the same as the hole diameters, say 1 in. Note that two horizontal plates are required to "box in" the assembly and keep the feed from bypassing the distributor.

Look at outlets The water outlet should be about 10% of A_{water} or 5.01/10 = 0.50 ft^2. This is too big for an outlet nozzle; so provide an internal baffle. Try a plate placed as a chord 4 in from the water nozzle and open at both ends. $h = \frac{3}{2} - \frac{5}{12} = 1.17$ ft. Open area at each end is $\frac{1}{2}\pi(1.5)^2 - 1.17(1.5^2 - 1.17^2)^{1/2} - (1.5)^2$ arcsin (1.17/1.5) = 3.53 − 1.10 − 2.00 = 0.43 ft^2. Both ends open = 2 × 0.43 = 0.86 ft^2, which is greater than 0.50 ft^2 (fine). The nozzle itself should be about 0.86/10 = 0.086 ft^2. Use a 4-in nozzle (0.0884 ft^2). The line can be reduced to 2- or 3-in pipe once it is a foot or two away from the decanter. The width of the baffle plate should be around four times the nozzle diameter, or about 16 in. Use an 18-in width to be sure.

For the oil outlet, look at a baffle with a 2-in clearance. Area = 2 [3.53 − 1.33(1.5^2 − 1.33^2)$^{1/2}$ − 2.25 arcsin (1.33/1.5)] = 0.30 ft^2. This is more than 2.06/10 = 0.21 ft^2; so proceed. Try a 2-in nozzle, 0.0233/0.30 = 0.08, which is a little under the 10% guide. However, the actual velocity = 0.05/0.0233 = 2.1 ft/s, which is pretty low. Just to be sure, make the baffle width 12 in (6 diameters).

Summary The final design is shown in Fig. 2. It is a tight design. There may well be some "haze" or secondary dispersion in both phases. The oil-phase settling is marginal and the N_{Re} in the water phase is significant. If clarity of the products is critical, resize using $d = 0.0002$ ft (60 μm).

EXAMPLE 2

Design a decanter to separate

300,000 lb/h of product, $\rho = 56$ lb/ft^3, $\mu = 1.5$ cP
200,000 lb/h of solvent, $\rho = 61$ lb/ft^3, $\mu = 0.3$ cP

Setup

$$Q_p = 300,000/(56 \times 3600) = 1.49 \text{ ft}^3/\text{s}$$
$$\mu_p = 1.5 \times 6.72 \times 10^{-4} = 1.0 \times 10^{-3} \text{ lb/ft-s}$$
$$Q_s = 200,000/(61 \times 3600) = 0.91 \text{ ft}^3/\text{s}$$
$$\mu_s = 0.3 \times 6.72 \times 10^{-4} = 2.0 \times 10^{-4} \text{ lb/ft-s}$$

Check for dispersed phase From Eq. (3), $\theta = \dfrac{1.49}{0.91}\left(\dfrac{56 \times 2.0 \times 10^{-4}}{61 \times 1.0 \times 10^{-3}}\right)^{0.3} = 0.98$. This is so close to 1.0 that either phase could be dispersed, and so the decanter must be designed to work both ways.

Find droplet settling rate Assume $d = 0.0005$ ft (150 μm); then from Eq. (1):

$$u_p = 32.17(0.0005)^2 (56 - 61)/(18 \times 2.0 \times 10^{-4}) = -0.011 \text{ ft/s}$$
$$u_s = 32.17(0.0005)^2(61 - 56)/(18 \times 1.0 \times 10^{-3}) = 0.0022 \text{ ft/s}$$

From overflow rate, pick size Equation (5) gives for a horizontal cylinder:

$$D_p = \frac{1}{2}(1.49/0.0022)^{1/2} = 12.6 \text{ ft}$$
$$D_s = \frac{1}{2}(0.91/0.011)^{1/2} = 9.1 \text{ ft}$$

Clearly the size will be larger when the product is the continuous phase and the solvent is dispersed. 12.6 ft is large for a horizontal drum. Assuming the feed has a vapor pressure under 1 atm, adapt a vertical storage tank to serve as a decanter. Now $A_t = \frac{1}{4}\pi D^2$. From Eq. (2), $A_t > Q_c/u_D = 1.49/0.0022 = 677$ ft² or $D_p = 29$ ft. (Likewise $D_s = 10$ ft.) Use $D = 30$ ft for the first trial.

Set interface level   Start with the interface at the center of the tank, since we do not know which phase will be dispersed.

Check coalescence time   Assume the height of the tank is 10 ft. Further assume that $H_D = 10\%$ of 10 ft, or 1 ft. Then the available time is

$$\text{Solvent dispersed} = \frac{1}{2}(1 \times \frac{1}{4}\pi 30^2)/0.91 = 388 \text{ s } (6\frac{1}{2} \text{ min})$$
$$\text{Product dispersed} = \frac{1}{2}(1 \times 707)/1.49 = 237 \text{ s } (4 \text{ min})$$

So 10 ft is high enough to give adequate coalescence time.

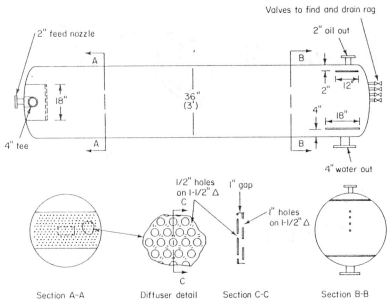

Fig. 2   Decanter for Example 1.

Emulsion provision   Provide five 2-in nozzles at 6-in spacing straddling the interface. These will permit verification of the true interface level and also removal of any "rag."

Check turbulence level   Calculate the Reynolds number at the mean diameter $D/\sqrt{2}$. Hydraulic diameter $= 4$ (flow area/wetted perimeter).

Flow area $= 5\pi 30/\sqrt{2} = 333$ ft² for both phases ·

Wetted perimeter $= 2\pi 30/\sqrt{2} = 133$ ft

$$D_h = 4 \times 333/133 = 10 \text{ ft}$$
$$v_p = 1.49/333 = 0.0045 \text{ ft/s}$$
$$N_{Re} = 0.0045 \times 10 \times 56/1.0 \times 10^{-3} = 2500 \text{ (good)}$$
$$v_s = 0.91/333 = 0.0027 \text{ ft/s}$$
$$N_{Re} = 0.0027 \times 10 \times 61/2.0 \times 10^{-4} = 8300 \text{ (fair)}$$

Check interfacial shear

$$\text{Velocity ratio} = 0.0045/0.0027 = 1.6 \quad (\text{OK}, < 2.0)$$

Set feed geometry

$$Q_{total} = 0.91 + 1.49 = 2.30 \text{ ft}^3/\text{s}$$
$$\rho_{mean} = 500,000/(3600 \times 2.30) = 58 \text{ lb/ft}^3$$

Normal practice would be to use an 8-in pipe for this mixture. This gives a velocity of 6.7 ft/s. The decanter must be well shielded from velocities this high. Start by bringing the feed into the closed end of a 24-in feed well. Velocity in the well is 2.30/3.012 = 0.76 ft/s. This gives $\rho v^2 = 58(0.76)^2 = 33$ lb/ft-s², or one-eighth of the limit of 250.

Provide a diffuser For distribution of the feed, cut twelve 4-ft-long slots in the side of the feed well. Make the slots ¼ in wide. The slot area is then $12 \times 4 \times \frac{1}{2}/12 = 1.0$ ft². Slot velocity is 2.30/1.0 = 2.30 ft/s, which is in the desirable 2 to 4 ft/s range. Break up the jets by mounting 3-in-wide baffle strips over each slot using a ½-in gap.

Look at outlets The product should be withdrawn using a weir around the top of the tank. Assume the weir is 9 in inside the wall. Then the weir length is $\pi(30 - 2 \times 9/12) = 89.5$ ft. An approximate formula for the height over a weir is $[1.7 \times Q/(\text{length}\sqrt{g})]^{2/3}$ or $h = [1.7 \times 1.49/(89.5\sqrt{32.17})]^{2/3} = 0.029$ ft (0.35 in). This is low enough that special care will be needed to make sure the weir is level.

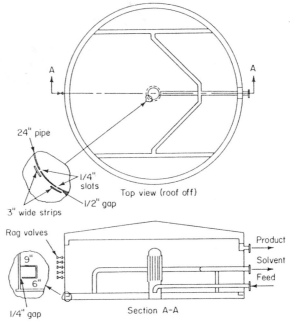

Fig. 3 Decanter for Example 2.

Assume the product leaves through an 8-in pipe. Velocity = 1.49/0.3601 = 4 ft/s. Normally the liquid will back up in the trough enough to overcome the entrance loss of the outlet pipe plus the acceleration loss. The entrance loss will be around one-half of a velocity head and the acceleration loss should be one velocity head. The velocity head expressed in feet of fluid is $\frac{1}{2}V^2/g$, or $\frac{1}{2} \times 4^2/32.17 = 0.26$ ft. The trough level should be about $1\frac{1}{2} \times 0.26 = 0.39$ ft, or about 40% of the available depth of 1 ft. This is on the high side; so provide an internal drawoff box about 1 ft wide and 1 ft deep.

The solvent should also be removed using an underflow weir. Limiting the area of the exit to 10% of flow area means that its height must be $0.1 \times 5 = 0.5$ ft. This is far too big to give good distribution. Therefore, use a ¼-in gap inside the outlet to control distribution. Gap area $= 30\pi\frac{1}{4}/12 = 2.0$ ft². Velocity = 0.91/2.0 = 0.46 ft/s. This may be high enough to give adequate distribution. Use four 8-in risers to help maintain distribution. The four lines will be brought to a single outlet nozzle using symmetrical internal piping.

Summary The design shown in Fig. 3 should give adequate performance.

PROPRIETARY DEVICES TO ENHANCE DECANTATION

Plate Separators

Hazen in 1904 first presented the concept of overflow rate Q/A_l as an important design criterion. He showed that the largest practical interfacial area per volume should be used to maximize separation. Decanters often have to be extremely large to meet the overflow-rate requirement. Hazen proposed parallel trays in settling basins to increase the settling

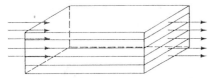

Fig. 4 Parallel-plate concept. *(The Pielkenroad Separator Co.)*

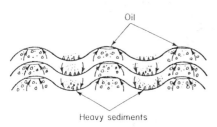

Fig. 5 Corrugated-plate flow. *(The Pielkenroad Separator Co.)*

area per volume. This increases the interfacial area by creating individual small separators operating in parallel (see Fig. 4). This simple concept was not widely practiced until the mid-1960s, when a parallel-plate settler was developed by the Royal Dutch/Shell group. Similar units are now in use throughout the world.

Parallel plates also attack the problem of turbulence in gravity settlers which often disrupts drop settling and causes breakup of the interface. Using a 18.75- to 37.50-mm (¾- to 1½-in) plate spacing, laminar flow can be obtained in a reasonably sized unit. That is, N_{Re} from Eq. (4) can be less than 2100.

With plates, the distance for any drop to travel to reach an interface is only about 18.75 mm (¾ in). Without the plates it would be several feet. Parallel-plate units are designed by the principles previously given. The plates just change the interfacial area and the hydraulic diameter.

Many variations of the parallel-plate settler exist commercially, though most use some variation of the corrugated design. The Corrugated Plate Separator (CPS) developed by Shell consists of corrugated plates spaced about 18 mm (¾ in) apart. The corrugations serve to concentrate the separated phase (see Fig. 5). The plates are usually in a rectangular bundle inclined at a 45° angle. Heil Process Equipment Corporation uses a standard pack with 47 plates at 18.75-mm (¾-in) spacing. Overall dimensions of the pack are 950 mm (38 in) high, 1050 mm (42 in) wide, and 1725 mm (69 in) long, which provides an effective area of about 74 m² (800 ft²).

When separating less than 1% light oil from water, countercurrent flow is used for oil removal, as shown in Fig. 6. Heavy sludge can be removed as well as light-phase material.

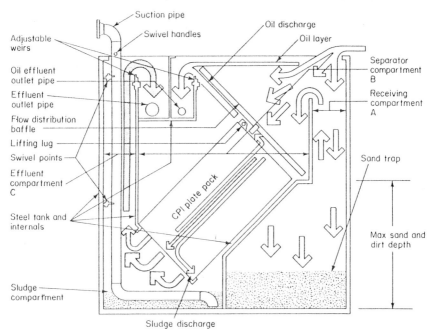

Fig. 6 Plate pack separator. *(Heil Process Equipment Corp.)*

With outlet oil concentrations of 20 to 50 ppm, design values of N_{Re} are usually about 100, with a maximum of about 400. Figure 7 provides a quick way to determine separation in a standard plate pack for drops settling in water.

EXAMPLE

Water at 70°F, specific gravity = 1.0; oil = 0.90 sp gr; and flow = 150 gal/min. From 150 gal/min on x axis proceed vertically to sp. gr. = 0.10; then proceed horizontally to the left into the temperature–particle size section and read the size particles which will be 100% removed at the operating temperature. In this case, at 70°F, all particles 60 μm or larger will be removed.

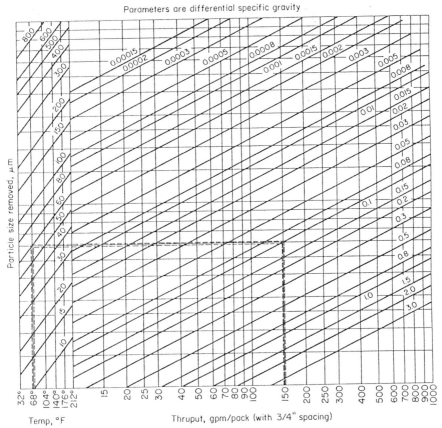

Fig. 7   Particle removal in plate pack. *(Heil Process Equipment Corp.)*

Inlet distribution and outlet collection are important in a good design. Most designs use some form of slotted inlet baffle to obtain uniform approach velocities to the plate sections. The dispersed-phase collection is usually done with gutters at the end of the corrugations. Some units with horizontal plates have holes in the plates to permit the dispersed phase to separate before reaching the end of the plate section.

The most common applications for the commercial units are in refinery wastewater cleanup and marine-bilge oil separation.

For large volumes of dispersed phase, as in chemical process separators, cocurrent flows are recommended. If high quality is needed from a system with a large dispersed-phase volume, a two-step process probably is required.

Coalescers

A coalescer is a vessel using any technique which accelerates the bringing together of two or more dispersed particles to form larger particles. The most common coalescers pass the phases through some type of solid bed, fiber mesh, metal screen, or membrane. The actual mechanism for coalescers using a solid surface is not completely understood. Most units are designed to have the dispersed phase preferentially wet the media. Dispersed-phase material is intercepted and held by the packing until large globules eventually disengage from the packing surface. A mesh pad or cartridge of fibers is typical. Metal wires or surfaces are wet by aqueous solutions while polypropylene, nylon, or polytetrafluorocarbon fibers are common for organic dispersed-phase wetting. Glass or glass-treated fibers are used and can be tailored for either aqueous- or organic-phase wetting. Fiber size or media pore size decrease with the need to handle smaller droplet size. Steady-state pressure drop for clean systems varies from less than 0.07 kg/cm^2 (1 lb/in^2) to as high as 1.4 kg/cm^2 (20 lb/in^2) for very difficult separations.

When a high concentration of each phase exists, it is not easy to determine which phase will be dispersed. Davies, Jeffreys, Ali, and Afzai[8] evaluated a bicomponent mesh pad using polymer and metal fibers intermeshed. Processing rates for kerosine-water dispersions were higher than obtained with either phase through single-component packing (see Table 1).

TABLE 1 Maximum Flow Rates of Dispersed Phase

Packing	Dispersed phase	Flooding velocity, ft/s
Stainless steel	Kerosine	0.235
	Water	3.38
Polypropylene	Kerosine	3.40
	Water	0.1615
Composite packing using both	Kerosine	4.43
	Water	4.43

If the wrong packing-dispersed phase combination is chosen or if phase inversion occurs, the rates will be different. The important feature of the bicomponent packing is the junction of the two materials. Bicomponent packing in the form of mesh pads is marketed by Knit Mesh, Ltd., of Great Britain.

Various single-material coalescers are available. Good decanter-design principles must be followed after passing the material through the coalescer elements. Such coalescers may be used to retrofit existing gravity settlers to improve efficiency or to allow increased capacity.

A big area of coalescer application is for removing low concentrations of droplets less than 20 μm in diameter as a polishing device following other separation techniques. Dense fiber beds are common. Spielman and Goren[9] have done an extensive literature review in this area. Spielman and Goren[10] later published experimental work.

Removal of 1 to 7 μm drops of oil through glass-fiber mats was independent of other drop sizes present and dispersed-phase viscosity. For these small particles and low concentrations continuous-phase wetted fibers were found to be more efficient, contrary to most coalescer applications. Smaller fiber diameters with equivalent porosity improve efficiency, as does lower superficial velocity. Outlet dispersed-phase globules ranged between 150 and 1000 μm, with larger fiber diameters giving larger globules.

Some commercial coalescers employ second-stage or separator elements which use a continuous-phase wetted medium of small pore size to reject and prevent entry of dispersed-phase droplets after previously being conditioned through a standard dispersed-phase wetted coalescer medium. Figure 8 is an example using both a coalescer and a separator-type element.

Coalescers are not without problems. Pore size in the elements is small, and the stream needs to be essentially free of solids. Most vendors recommend a prefilter to remove solids. Even then, especially with emulsions, finely divided solids are removed in the coalescing elements and require a periodic backflush or replacement. For situations where solids are significant, a packed bed similar to a deep-bed filter can be used with a periodic backflush. Madia, Fruh, Miller, and Beerbower[11] obtained good removal efficiency with dispersed-phase wetted packing.

Centrifugal Devices

Separation can be improved by applying additional force with centrifugal devices. Two basic types of centrifugal equipment are possible: (1) static equipment such as hydrocyclones or (2) mechanical centrifuges.

Hydrocyclones

Experimental work on liquid-liquid hydrocyclones was reported by Sheng, Welker, and Sliepcevich.[12] Efficiency per pass is not very great, and multibanks of units are required to

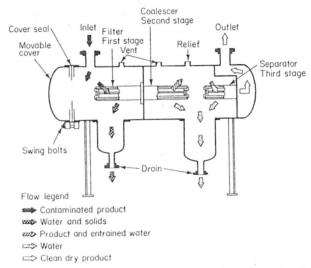

Fig. 8 Three-stage horizontal coalescer. [*Reprinted with permission from* Chem. Eng. Prog., 59(9), 89 (1963).]

exceed about 80 to 85% separation. Sheng et al. found a significant effect on efficiency by the degree to which the vortex tube was wetted by the light liquid phase. Liquid-liquid hydrocyclones are much less effective than those in solid-liquid service, since the complicated internal flow patterns and high shear can serve to break up liquid particles. We are not familiar with any widely used commercial liquid-liquid hydrocyclones.

Mechanical Centrifuge

The simplest liquid-liquid mechanical centrifuge is the tubular design, which is a high-speed, small-diameter bowl. Normally it is fed at one end with separated liquid removed at the other end through effluent take-offs at different radii. Machine capacity is limited to about 20 gal/min.

The most common design today is a disk-type centrifuge which is often operated to generate forces up to 9000 times that of gravity. Figure 9 is a typical configuration with vertical bowl and feed entering from the top on the axis. The inclined disks aid the separation by a mechanism similar to the plate settlers. The light liquid moves to the top of the disks and travels toward the center of the bowl. The heavy liquid and any solids move to the outer wall of the bowl. Both

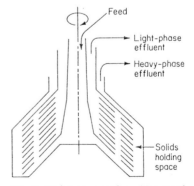

Fig. 9 Disk-type centrifuge. (*Reprinted with permission from "Chemical Engineers' Handbook."*[1])

liquid phases exit at the top in separate conduits. Some amount of solids can be handled along with the liquid separation. For small amounts of solids, accumulation at the bowl wall with periodic manual batch removal is acceptable. For more than 1% solids, an automatic solid discharge is required periodically, either by parting the bowl momentarily or opening nozzles to eject the solids.

Capacities of disk machines range from 20 to 1890 L/min (5 to 500 gal/min). Specific-gravity differences should be greater than 0.01 and droplets greater than 1 μm. Because of the high separating force available, the efficiency is very good. Some experimental results are reported by D. M. Landis.[13] Water was removed from aviation fuel down to 0 to 5 ppm even with as much as 15% water in the feed and at machine capacities of about 760 L/min (200 gal/min).

Flotation

Flotation is commonly associated with solid-liquid separation as in mineral processing and waste treatment. This is covered in Sec. 4 of this handbook. However, flotation is applied equally well to liquid-liquid separations and will briefly be discussed. Three

TABLE 2   Air Solubilities at Atmospheric Pressure

Temperature		Solubility,
°C	°F	lb/1000 gal
0	32	0.311
10	50	0.245
20	68	0.203
30	86	0.175
40	104	0.155
50	122	0.142
60	140	0.133
70	158	0.128
80	176	0.125
90	194	0.124
100	212	0.125

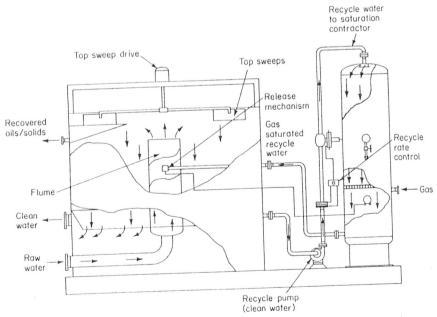

Fig. 10   Dissolved air flotation system. (*C-E Natco, Combustion Engineering, Inc.*)

basic flotation methods exist and differ in the mechanism of bubble formation: (1) dispersed air, (2) electrolytic gas formation, and (3) dissolved air.

Dispersed air systems involve use of agitators or gas spargers to form gas bubbles and often require the assistance of chemical addition. Bubble size is often large and not as efficient in contacting small liquid drops.

Electrolytic flotation uses a direct current between a cathode and anode in the liquid to generate small oxygen and hydrogen bubbles in water solutions.

Dissolved air flotation is the most common for liquid-liquid separations. Liquid is saturated with gas by various techniques, and on sudden letdown to lower pressure very small bubbles in the 30- to 100-μm range evolve. Cassel, Kaufman, and Matijevic[14] found optimum separations occurred with 50- to 60-μm bubbles. Boyd and Shell[15] report gas requirements of 0.0045 to 0.027 kg air/kg solid (0.01 to 0.06 lb air/lb solid). The lower range is probably more appropriate for liquid separation. They indicate gas released is equal to

$$S = Sg\left(\frac{fP}{14.7} - 1\right)$$

where S = gas released, lb/1000 gal
  Sg = gas saturation at atmospheric conditions, lb/1000 gal
  P = pressure at dissolving point, lb/in²(abs)
  f = system dissolving efficiency
Values of Sg can be obtained from Table 2.

Gas dissolving can occur by pressurizing the main stream, by a partial sidestream pressurization, or by using clarified recycle liquid (which is probably the most common). Recycle systems minimize disruption of the main flow and allow use of smaller dissolving equipment.

The dissolving is done either in a pressurized packed column (see Fig. 10), a pressure tank with sufficient retention time, or on the discharge of a pump with an eductor. The gas is usually dissolved at a pressure of 3.5 to 5.25 kg/cm² [50 to 75 lb/in²(abs)].

Common applications of flotation are in oil refineries cleaning up the effluent from API separators to meet water-effluent standards. Most of the free oil is removed prior to the flotation unit. Typical applications and efficiencies are shown in Table 3.

TABLE 3   Performance of Dissolved Air Flotation on Oily Waste

	ppm in	ppm out	% removal
Petroleum refinery	220	20	90
Hydrocarbon cracking	167	6	97
Oil tanker (ballast water)	35	7	80
Oil field (brine water)	87	7	92

SOURCE: Permutit Favair®, Mark II Dissolved Air Flotation Systems, *Bull.*, 1970.

Electrical-Charging Methods

Electrical-charging methods are used to remove dispersed aqueous materials and solids from organic fluids as in crude-oil desalting. Particles passed between electrodes either acquire induced dipoles which result in forces attracting particles, or particles having a net charge are attracted to an electrode where they coalesce with other particles. Waterman[16] discusses this phenomenon. Commerical equipment handling up to 7375 m³/day (125,000 bbl/day) of crude oil are in use from the Petrolite Corporation and others. Figure 11 shows a schematic of flow and electrical-field arrangements.

REFERENCES

1. Perry, R. H., and Chilton, C. H., "Chemical Engineers' Handbook," 5th ed., pp. 5–61 through 5–64, McGraw-Hill, New York, 1973.
2. Middleman, *Ind. Eng. Chem. Process Des. Devel.*, 13(1), 78–83 (1974).
3. API "Manual on Disposal of Refinery Wastes"—volume on liquid wastes, chap. 5, Oil-Water Separation Process Design, 1969.

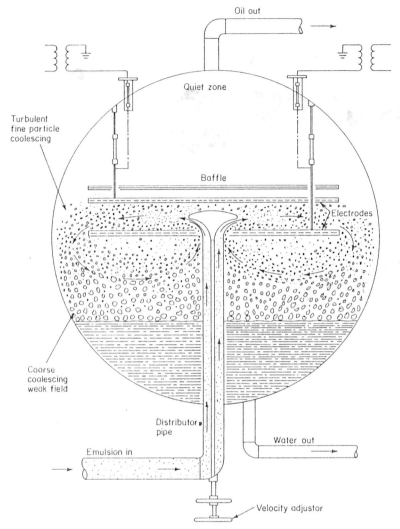

Fig. 11 Internal circulation and electric field, Petreco Cylectric coalescer (schematic). [*Reprinted with permission from* Chem. Eng. Prog., **61** *(10), 51 (1965).*]

4. Selker and Sleicher, *Can. J. Chem. Eng.*, **43**, 298 (1965).
5. Mizrahi and Barnea, *Process Eng.*, January 1973, pp. 60–65.
6. Vijayan, Ponter, and Jeffreys, *The Chemical Engineer J.*, **10**, 145–154 (1975).
7. Hooper, *Chem. Eng.*, **82**(16), 103 (1975).
8. Davies, Jeffreys, Ali, and Afzai, *The Chemical Engineer*, October 1972, pp. 392–396.
9. Spielman and Goren, *Ind. Eng. Chem.*, **62** (10), 10–53 (1970).
10. Spielman and Goren, *Ind. Eng. Chem. Fundam.*, **11**, (1) 66–72, 73–83 (1972).
11. Madia, Fruh, Miller, and Beerbower, Paper 79th National Meeting, American Institute of Chemical Engineers, March 1975.
12. Sheng, Welker, and Sliepcevich, *Can. J. Chem. Eng.*, **52**, 487–491 (1974).
13. Landis, D. M., *Chem. Eng. Prog.*, **61**, 58–63 (1965).
14. Cassel, Kaufman, and Matijevic, *Water Res.*, 9(12), (1975).
15. Boyd and Shell, *Proc. 27th Ind. Waste Conf.*, Purdue University, 1972.
16. Waterman, *Chem. Eng. Prog.*, 61(10), 51–57, 1965.

Ion-Exchange Separations

ROBERT E. ANDERSON *Senior Research Scientist, Functional Polymers Division, Diamond Shamrock Corporation, Redwood City, California*

INTRODUCTION

The basis of chemical separations is that components to be separated distribute unequally between separate or separable phases. In distillation the phases are liquid and vapor. In crystallization the phases are solid and solution. In separations using ion-exchange resins both phases can be considered liquid phases, one of which is confined in the form of a gel.

Ion exchange offers the possibility of removing one or more ionic species from one liquid phase and transferring them to another liquid phase via an intermediate solid. In many cases the transfer can be made on a selective basis and with good chemical efficiency.

The purpose of such a transfer may be:

1. Purification or modification of the original liquid phase
2. Concentration, isolation, and/or purification of one or more ionic components
3. Separation of mixed ionic species into two or more fractions

Synthetic ion-exchange resins have been in commercial production over thirty years. They are now used in such diverse fields as water treatment,[1,2] pharmaceutical manufacture,[3] trace analysis,[6,7] and hydrometallurgy.[5,10] Although the literature on ion exchange is extensive, a working knowledge of industrial ion-exchange processes is often lacking.

The basic description of ion-exchange interactions and column behavior will be developed on the basis of fully ionized resins. More complex partially ionized resin types will then be discussed as to differences from fully ionized resins, and the important possibilities for chemical separations they present.

Emphasis will be placed on ion-exchange applications in industrial separations. The huge field of analytical separations by ion exchange is covered in several books. Nonseparation uses for ion-exchange resins, such as in catalysis, will not be considered here.

Application of ion-exchange resins to practical separation problems can be enhanced by an understanding of the following factors: (1) nature of common types of ion-exchange resins, (2) types of distribution that occur between liquid and gel phases, and modifications of distributions, (3) rate-determining steps in moving components from one phase to the other, and (4) nature of column fronts and dependency on the distributions involved.

Theoretical aspects of each factor are available in the literature.[4] Methods for calculating column performance have been presented.[28] Unfortunately, this wealth of information is often more confusing than helpful. This section will describe the basic components of ion-exchange systems, give a qualitative overview of important variables in column performance, and present some semiquantitative techniques for screening proposed ion-exchange processes. In many cases, application of this information will allow the engineer to determine whether the required separation is possible by ion exchange, and the best ion-exchange method to use.

Laboratory column data should be obtained as the basis for further engineering once practicality has been shown. Laboratory work can be on a small scale with simple equipment. This phase of design can be shortened by application of general ion-exchange knowledge, since it will usually set practical limits on many of the process variables.

FULLY IONIZED RESIN SYSTEMS

All ion-exchange resins, whether cation or anion exchangers, strongly or weakly ionized, gel or macroporous, spherical or granular, can be viewed as solid solutions. Practically every observed ion-exchange behavior can be rationalized on the basis of a distribution of components between two solution phases, one of which is confined as a solid gel. The two phases can be referred to simply as the inside phase and the outside phase. Transfer of components takes place across the interface between the phases, which is the surface of the bead or granule.

Resin-Phase Components

The inside phase of an ion-exchange resin contains four necessary components:

1. A three-dimensional polymeric network
2. Ionic functional groups permanently attached to this network
3. Counterions
4. A solvent

Under certain conditions there may be other components inside the resin such as

5. A second solvent
6. Co-ions
7. Nonionic solutes

Each component will be considered as to interactions in the outside and inside phases.

Polymeric Network Most ion-exchange resins in commercial use are based on an organic polymer network, but inorganic polymers are also used. Regardless of the composition of the network, its primary function is to limit the solubility of the resin. The chemical nature of the polymer network is a major factor in determining the physical and chemical stability of the resin. However, it will have only secondary effects on the ionic distribution between the two phases.

With minor exceptions, strong-acid cation-exchange resins and strong-base anion-exchange resins contain styrene-divinylbenzene (DVB) as base polymer. Their structures are relatively well defined and are fully ionized over the entire pH range.

When styrene is polymerized, the resulting macromolecular chains are so long in relation to their diameters, they are essentially one-dimensional. When placed in an organic solvent the individual macromolecules are free to disperse. The plastic mass dissolves, and can be diluted infinitely.

If divinylbenzene (DVB) is added to the polymerization mixture, each molecule of this monomer becomes a part of two separate chains. As a result, the individual chains are interconnected at more or less regular intervals and the structure becomes three-dimensional or crosslinked.

If this crosslinked polymer is placed in an organic solvent, it will start to dissolve just as the linear polymer does. However, dispersion will stop when the osmotic force of solvation is balanced by the opposing force of the stretched polymer network. The higher the percentage of DVB in the monomer mixture, i.e., the higher the crosslinking, the smaller the amount of solvent picked up at equilibrium swelling.

The degree of crosslinking and average particle size are established during initial polymerization. The mixed organic monomers are suspended in water and polymerized as discrete spherical drops. The average particle size is determined by the rate of stirring.

Functional Groups The organic-swellable copolymer is converted to a water-swellable material by the introduction of functional ionic sites. Ionic sites, which give the polymer ion-exchange properties, are added to the styrene-DVB polymer network by one or more chemical reactions.

Strong-acid cation-exchange resins are prepared by sulfonating the benzene rings in the polymer:

Strong-base anion-exchange resins require two reactions: chloromethylation,

and amination.

Dry-Weight Capacity. The sulfonation reaction is relatively straightforward. With care, one sulfonate group, on the average, can be attached to each benzene ring in the copolymer bead, and it is very difficult to carry the reaction much beyond monosubstitution.

The sulfonic acid group on the benzene ring has essentially the same acid strength as the first hydrogen in aqueous sulfuric acid. It can be titrated with an aqueous base just like any acid, the only complication being that time must be allowed for the diffusion of ions in the gel phase.

Titration gives the number of ionic sites, usually expressed as hydrogen equivalents, in a given amount of resin. If the amount of resin is expressed as a weight of dry material, a dry-weight capacity is obtained, expressed as milliequivalents per dry gram (meq/g). The dry-weight capacity is a measure of the extent of functional group substitution in the resin. For sulfonated styrene-DVB resins having one sulfonic acid group per benzene ring, the dry-weight capacity is 5.2 ± 0.1 meq/g. Most commercial resins have dry-weight capacities of 5.0 ± 0.1 meq/g.

Strong-base resins are produced commercially with two types of functional groups. These are designated as Type I and Type II. Type I resins have three methyl groups substituent to the nitrogen in the quaternary ammonium structure. In Type II resins one of the substituents on the nitrogen is a hydroxyethyl group (Fig. 8). This minor structural variation results in important differences in the relative ionic selectivities and thermal stabilities of the two resins under certain conditions. Both Type I and Type II anion resins are fully ionized strong bases, essentially equivalent in base strength to sodium hydroxide in aqueous systems.

The anion resins derived from styrene-DVB copolymers do not show a tendency for monosubstitution on the benzene ring. Depending on the resin properties desired, the degree of substitution in these resins may be considerably less than, or somewhat more than, one functional group per benzene ring. Dry-weight capacities may range from under 2 to over 5 meq/g. In addition to variations in degree of substitution, there is the possibility of additional crosslinking during the chloromethylation reaction.

Swelling Volume. The ionized functional group converts the oil-swellable styrene-DVB copolymer to a water-swellable structure. A relatively low degree of functional substitution may be sufficient to give the copolymer some water swellability. The greater the degree of substitution, the more tendency there is for the bead to pull in water. The higher the crosslinking, the greater the elastic forces resisting this uptake of water. At equilibrium, the amount of water taken up by the bead is a function of the type and amount of ionic substitution, and the effective crosslinking of the polymer structure.

The degree of substitution obtained on sulfonating styrene-DVB copolymer is relatively constant, and no appreciable additional crosslinkage is formed in the reaction. The equilibrium water content of the resin and its swollen volume directly reflect the relative amount of DVB in the resin. It is common practice to indicate the amount of DVB in the base polymer of strong-acid resins by a "percent DVB" or an "X" number. Either number is an index of the relative water content of the resin, also referred to as porosity, or gel porosity.

Degree of substitution can vary widely in strong-base resins, and additional crosslinkage may be formed during chloromethylation. Designation of DVB content of the base polymer does not necessarily give an index to the porosity of the resin. Gel porosity is found by determining the water content or moisture-retention capacity of the fully swollen resin.

The relative amount of water in the swollen resin bead is an important property and has a direct bearing on equilibrium, kinetic, and mechanical characteristics. Water content is measured by oven-drying fully swollen beads which have been filtered or centrifuged to remove exterior water. Results are usually expressed as a weight percent of the fully swollen resin. An alternate method of expression is water regain, which is the weight of water per gram of dry solids. With the styrene-DVB resins, the water content of the swollen bead may vary from 20 to over 99%, but mechanical considerations limit this range to approximately 30 to 80% for most separation applications.

Maximum Capacity. Ion-exchange resins and processes are commonly rated on the basis of capacity per unit volume. If the bulk amount of resin titrated is expressed as a volume, a volume capacity is obtained. Fully ionized resins contain a fixed number of functional groups per unit weight, and therefore have a constant dry-weight capacity. The wet-volume capacity, however, varies with the water content and is inversely proportional to the swelling of the resin; the lower the crosslinking, the more the swelling, and the lower the total capacity on a volume basis.

For chemical process work, wet-volume capacity is most conveniently expressed as equivalents per liter. The water-conditioning industry has historically expressed capacity as kilograins of $CaCO_3$ per cubic foot. Conversion factors are given in Table 5.

Volume measurements of ion-exchange resins are subject to inherent experimental error. The normally accepted, relatively reproducible way to measure the volume of small quantities of resin is to place the resin with an excess of water in a graduated cylinder and to tap the cylinder gently until the resin reaches minimum volume, or tapped volume.

An alternate method is often used in column measurements. The resin is classified by fluidizing it with an upward stream of water. Flow is then stopped and the resin is allowed to settle. Water in the column is drained to the top of the resin bed and the height of resin is measured. Volume calculated from this height is called a backwashed, settled, and drained (BS&D) volume, and is somewhat greater than tapped volume.

In either case, the volume occupied by resin, or bed volume, includes the liquid in the interstices between resin beads. In a classified bed of spherical resin this volume, called the void volume, is 38 to 40% of the total bed volume.

Volume capacity is a measurement of the maximum usefulness of the resin in a stoichiometric exchange. When all the available exchange sites have been used, the process must be interrupted and the resin regenerated. Thus the relative ionic concentrations of resin and solution determine the maximum volume of solution that can be treated per cycle.

$$\frac{\text{Volume capacity (eq/L)}}{\text{Normality of ion(s) being removed (eq/L)}} = \frac{\text{max vol treated}}{\text{volume of resin}}$$

It is convenient to consider volume capacity as a normality. The standard strong-acid exchange resins have a normality of about 2.0 eq/L in sodium form. Strong-base resins have normalities of 1.0 to 1.4 eq/L in chloride form.

Ion-exchange processes in fixed columns are not usually attractive unless the volume of solution treated per cycle is at least several bed volumes. Ion exchange is thus normally limited to cases where the total ionic concentration of the components being removed is less than 0.2 to 0.3 eq/L. This restriction can be overcome in some cases by use of a continuous resin-solution contactor.

Counterions Fixed ionic sites in the resin structure must be balanced by a like number of ions of the opposite charge to maintain electrical neutrality. These ions are called counterions. True ion exchange is the transfer of counterions between the outside and inside phases. Such exchange is always on an equivalent basis; that is, the same number of ionic charges leave the inside phase as simultaneously enter it. Material balances can be established across a column on an equivalent basis. Once the liquid initially in the column has been displaced, the total ionic concentration of the effluent will approach that of the influent.

Equivalency of exchange may be obscured if the exchange is accompanied by chemical reaction. The most common case is exchange of an acid solution with the hydroxide form of a resin. The anion of the acid enters the resin in exchange for a hydroxide ion which reacts with hydrogen ion to form water. Other reasons for an apparent lack of equivalency of exchange include the formation of a precipitate which is filtered out on the resin bed, or the formation of a poorly ionized component which is physically adsorbed by the resin. The latter is common when organic acids and bases are present in solution or are generated in the exchange.

The nature of the counterion is a major factor in determining the osmotic forces tending to draw water into the resin. The water content, swollen volume, and volume capacity of the resin change as it is converted from one ionic form to another. The change in volume can be appreciable. Some typical relative-volume values are given in Table 1. The relative size of two ionic forms of a resin is also dependent on the crosslinking in the resin; the lower the crosslinking, the greater the volume difference for a given pair of ionic forms.

TABLE 1   Relative Resin Volumes for Various Ionic Forms

Strong-Acid Resin, 8% Crosslinked			
Li^+	1.07	NH_4^+	0.98
H^+	1.05	K^+	0.95
Mg^{2+}	1.01	Ca^{2+}	0.95
Na^+	1.00	Ba^{2+}	0.88
Type I—Strong-Base Resin, Moisture Content 45%			
OH^-	1.23	NO_2^-	1.03
F^-	1.19	Cl^-	1.00
CO_3^{2-}	1.15	Br^-	0.98
$CH_3CO_2^-$	1.11	NO_3^-	0.95
SO_4^{2-}	1.10	I^-	0.87
HCO_3^-	1.09		

Solvent The initial description of ion-exchange systems as consisting of two liquid phases, one of which is confined as a gel, presupposes water as the solvent in both phases. The solvent is free to move across the phase boundary. The amount of water in the inside phase is one of the equilibrium properties of the system. The water content is established by a balance between the osmotic forces of solvation of the ionized components and the elastic restraints of the polymer network.

The amount of water in the inside phase is an important factor in determining relative selectivity of the resin for different ions. The higher the water content of the resin phase, the more the inside phase starts to resemble the outside phase. As a general rule, as the water content increases, the difference in selectivity between ion species becomes less.

Any change in the solvent away from water changes the nature of the system. If part of the water in the system is replaced by a water-miscible liquid such as an alcohol, the two solvents equilibrate between the two phases. Usually the ratio of solvents in the inside phase will not be the same as that in the outside phase.

If the amount of the nonaqueous solvent is at all appreciable, the ion-exchange properties of the system are changed. If the ionic components inside are less strongly solvated by the mixed solvent, the resin will shrink. In a few cases, the second solvent may solvate the polymer structure and cause the resin to swell. Modification of the solvent may have marked effects on relative ionic selectivities and has been applied to a wide range of analytical separations. As the solvent becomes more nonpolar, the degree of ionization in both phases decreases and exchange becomes very slow or ceases.

Nonuniform distribution of two solvents between the inside and outside phases is the basis for the use of ion-exchange resins as desiccants. If a dry cation-exchange resin is placed in a nonpolar solvent containing a minor fraction of water, it will adsorb the water but essentially exclude the other solvent. Chlorinated hydrocarbons can be dried efficiently using the potassium form of a sulfonated styrene-DVB resin.[34] Resin desiccants are much more efficiently regenerated than conventional desiccants. Regeneration temperatures of 110 to 120°C are sufficient. As the solvent to be dried becomes more waterlike, the drying process becomes less efficient.

Co-Ions Ions having the same sign as the fixed ionic groups of the resin are called co-ions. When the total ionic concentration of the outside phase is low, the concentration of co-ions in the inside phase is negligible. The interface between the two phases acts as a semipermeable membrane which allows all the ionic species except the fixed functional groups to pass from one phase to the other and exhibits Donnan-membrane equilibrium behavior. Concentration of co-ions in the inside phase will be about one-tenth of the concentration in the outside phase. If the solution in contact with the resin is 0.1 N or less, the concentration of the co-ion in the resin due to ion invasion will be less than 0.01 N. This is less than 1% of the ionic concentration of the resin phase, and the effects will be negligible.

If the outside-phase solution has a concentration greater than 1.0 N, the concentration of invading co-ions may be 10% or more of the concentration of the fixed ionic groups. Since each invading co-ion and each fixed ionic group has an associated counterion, the effective ion-exchange capacity of the resin is greater than that due to fixed ionic groups and should be recognized when working with concentrated solutions.

The Donnan exclusion effect serves to sharpen the rate at which an electrolyte is rinsed out of a resin. Regeneration steps are commonly carried out at solution concentrations over 1.0 N, causing considerable invasion of the inside phase. As the rinse is introduced and ionic concentration of the outside phase drops rapidly, there is a strong force driving invaded electrolyte out of the inside phase to reestablish the Donnan exclusion equilibrium. This effect materially sharpens the rinse curve.

Nonionic Solutes Typical nonionic solutes, such as alcohols, ketones, and ethers, are not restricted from entering the resin by the Donnan exclusion effect. The difference in behavior between ionized and nonionized solutes is the basis of the ion-exclusion process. No net ion exchange is involved.[31]

Some nonionic solutes are preferentially adsorbed by the inside phase; others preferentially remain in the outside phase. Sometimes the distribution of such solutes can be changed by modification of the solvent or ionic form of the resin. Whenever two solutes show a differential distribution between the two phases, a chromatographic separation is possible and forms the basis for a number of analytical methods.[32]

Organic acids and bases may be either ionized or nonionized depending upon the pH of the system. When ionized, they behave like inorganic ions of like charge. When nonionized, they are often physically adsorbed in the resin phase.

Summary of Two-Phase System A generalized description of a strongly ionized ion-exchange bead is shown in Fig. 1. The inside phase is rendered insoluble by the presence of a three-dimensional polymer network which swells in water because of attached ionic functional groups. Each fixed ionic site is balanced electrically by an ion of opposite charge, or counterion, in its near vicinity.

This inside resin phase comes to equilibrium rapidly with the solution with which it is in contact. Water moves in or out of the resin phase so that the osmotic forces due to the solvation of the fixed charges and counterions are in balance with the elastic forces of the polymer network.

Counterions move in and out of the inside phase on an equivalent basis since electrical neutrality must be preserved. If two or more types of counterions are

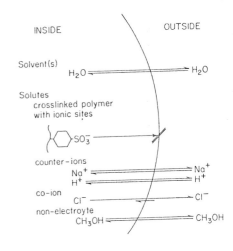

Stoichiometric Relations

Inside $\quad [\overline{Na^+}] + [\overline{H^+}] = [\overline{RSO_3^-}] + [\overline{Cl^-}]$

Outside $\quad [Na^+] + [H^+] = [Cl^-]$

Ionic Equilibria

Across
interface
$$[\overline{Cl^-}] \ll [Cl^-]$$
$$\frac{[\overline{Na^+}]}{[\overline{H^+}]} = K \frac{[Na^+]}{[H^+]}$$

Fig. 1 Schematic of typical component distribution in an ion-exchange resin system.

present in the system, they distribute themselves across the two phases. This distribution is dependent on the relative amount of counterions present and selectivity of the resin for one ion over another.

Co-ions are limited largely to the outside phase below an outside total ionic concentration of one normal. Those co-ions that are in the inside phase must be balanced by a like number of charges on counterions.

Other solvents or soluble nonionic species present in the system are distributed between the two phases in a way dependent upon the relative nature of these components and the resin.

All above factors are interrelated. Changes in the composition or nature of the outside phase produce a shift of the equilibrium distribution between phases. Every ion-exchange separation depends upon the differential distribution of two or more components between inside and outside phases.

Ionic Equilibria

Nonionized Products The fact that all the mobile components of an ion-exchange system can distribute across two phases makes various types of separations possible. However, a true ion-exchange separation can be defined as one in which the result is obtained by a net transfer of ions from the outside phase to the inside phase in exchange for an equivalent transfer in the other direction. Even though the difference in distribution of the two ions is quite small, column techniques allow the summation of many such transfers, and a useful separation may be realized.

Useful information can be obtained by considering ion-exchange reactions in simple mass-action terms. For instance, consider the reaction of the hydrogen form of a strong-acid resin with the base, sodium hydroxide.

$$\overline{H^+ \ R^-} + Na^+ \ OH^- \rightarrow \overline{Na^+ \ R^-} + HOH$$

In this reaction hydrogen ion leaving the inside phase immediately reacts with hydroxide ion to form water. The reaction will continue until there are either no sodium ions left outside, or no hydrogen ions inside, whichever occurs first. In the first case there will be no ions in the outside phase (except those few due to the ionization of water itself) to exchange with ions inside. In the second case, the sodium ions outside may continue to exchange with those inside, but there will be no further net exchange.

A similar case is the exchange of hydroxide-form anion-exchange resin with a mineral acid.

$$\overline{R^+ \ OH^-} + H^+ \ Cl^- \rightarrow \overline{R^+ \ Cl^-} + HOH$$

While these may appear to be trivial examples, the point is very important. If the product formed in the outside phase is nonionized, there is no tendency for the reaction to reverse. The reaction goes to completion and can be carried out as a single contact or batch reaction.

NOTE: All components of the inside phase will be designated by a superscript bar. The term $\overline{R^-}$ will be used to represent one equivalent unit of cation-exchange resin, and the term R^+ one equivalent unit of anion-exchange resin.

The product formed need not be water or even completely nonionized. If the product is a weakly ionized acid or base, there is little tendency to reverse the reaction.

$$\overline{R^+ \ OH^-} + NH_4^+ \ Cl^- \rightarrow \overline{R^+ \ Cl^-} + NH_3 + H_2O$$

$$\overline{H^+ \ R^-} + Na^+ \ C_2H_3O_2^- \rightarrow \overline{Na^+ \ R^-} + HC_2H_3O_2$$

$$\overline{H^+ \ R^-} + Na^+ \ HCO_3^- \rightarrow \overline{Na^+ \ R^-} + CO_2 + H_2O$$

Traces of iron or other metallic ions can be removed from concentrated organic acids such as acetic or glycolic very efficiently by hydrogen exchange on a strong-acid resin. In such cases, it may be necessary to have a small concentration of water in the acid treated to keep the resin from being dehydrated.

Another case in which a reaction is driven to completion is when the ion entering the outside phase is converted into a nonexchangeable form by complexing. An example is the exchange of the tetra sodium salt of ethylene diamine tetracetic acid (EDTA) with the copper form of a cation exchanger.

$$\overline{Cu^{2+}\ R_2^-} + (Na^+)_4 EDTA^{4-} \rightarrow \overline{2Na^+\ R^-} + (Na^+)_2 CuEDTA^{2-}$$

The copper loses identity as a cation when chelated into the EDTA complex.

Ionized Products *Ions of Identical Charge.* When the products of ion exchange are fully ionized, the reaction normally does not go to completion either right or left. Instead, an equilibrium is set up in which both reacting ions are found in both phases. An exchange between two ions of identical charge can be represented as

$$A + \bar{B} \rightleftharpoons \bar{A} + B$$

Ion A from the outside phase is displacing ion B from the resin. If the concentrations of reactants and products are expressed as moles per liter C_i, the concentration mass-action expression for this reaction is

$$K_{AB} = \frac{\dot{C}_A\ C_B}{\dot{C}_B\ C_A} \tag{1}$$

This concentration mass-action expression is not an accurate thermodynamic statement of the behavior of the system. Ion activity terms have been intentionally omitted. Ion activity terms for the inside phase cannot be independently determined. Ion-activity terms for the outside phase are reasonably close to unity for the dilute solutions in many ion-exchange processes. The selectivity coefficient K_{AB} is not constant for a given exchange. It may vary with the ionic form of resin and with minor variations in structure from one lot of resin to another. However, by using certain simplifications, the mass-action expression becomes a valuable tool in understanding ion-exchange behavior and in initial screening of proposed ion-exchange processes.

In all subsequent sections K_{AB} is the molar-selectivity coefficient as defined in Helfferich's equation (5-72) (Ref. 4, p. 154). The first letter in the subscripts indicates the ion entering the resin phase as the reaction is written, and the second letter indicates the ion leaving the resin phase. For all exchanges between ions of identical valence, the mass-action expression is given by Eq. (1). To use this relationship for screening ion-exchange processes, K_{AB} is considered constant for a given pair of ions exchanging on a given resin over a wide range of solution conditions.

Some consequences of mass-action behavior are seen more clearly if the relationship is put in terms of equivalent fractions for both phases. If the total ionic concentration (normality) of the solution phase is C eq/L, the equivalent fraction x_i of an ion of valence z_i present at concentration C_i is given by

$$x_i = \frac{z_i C_i}{C} \tag{2}$$

Concentrations in the resin phase are expressed in the same units as volume exchange capacity, equivalents per liter of bulk volume. If the total exchange capacity of the resin is $\bar{C}$, the equivalent fraction $\bar{x}$, of an ion of valence z_i, present in the resin phase at a concentration of $\bar{C}_i$ is given by

$$\bar{x}_i = \frac{z_i C_i}{\bar{C}} \tag{3}$$

When the values of C_A, C_B, $\bar{C}_A$, and $\bar{C}_B$ from Eqs. (2) and (3) are substituted in Eq. (1), the expression becomes

$$K_{AB} = \frac{\bar{x}_A\ x_B}{\bar{x}_B\ x_A} \tag{4}$$

Since only two counterions are present in each phase:

$$\bar{x}_A + x_B = 1$$
$$\bar{x}_A + \bar{x}_B = 1$$

and

Equation (4) can then be put in the form:

$$\frac{\bar{x}_A}{1 - \bar{x}_A} = K_{AB}\ \frac{x_A}{1 - x_A} \tag{5}$$

Equation (5) shows that the fraction of resin capacity occupied by a given ion at equilibrium will be dependent on the fraction of total ions in solution represented by the ion, and a selectivity coefficient. The total-concentration terms C and $\bar{C}$ do not appear in Eq. (5). Therefore, the distribution of ions between phases will not be directly dependent on total resin or solution concentrations. There may be secondary effects on relative ionic ratios, but these will be due to changing ionic activities.

Ions of Unequal Charge. When exchanging ions are of unequal charge, the total concentrations in the system have a major effect on the distribution of ions between phases.

For an exchange in which a divalent ion D displaces a monovalent ion B from the resin,

$$D + 2\bar{B} \rightleftharpoons \bar{D} + 2B$$

the molar mass-action expression is

$$K_{DB} = \frac{\bar{C}_D C_B^2}{\bar{C}_B^2 C_D} \tag{6}$$

Equation (6) can be converted to equivalent-fraction form by substituting values for $\bar{C}_D$, $\bar{C}_B$, C_D, and C_B derived from Eqs. (2) and (3). The equivalent fraction form of Eq. (6) becomes

$$K_{DB} \frac{\bar{C}}{C} = \frac{\bar{x}_D}{\bar{x}_B^2} \frac{x_B^2}{x_D}$$

$$\frac{\bar{x}_D}{(1 - \bar{x}_D)^2} = K_{DB} \frac{\bar{C}}{C} \frac{x_D}{(1 - x_D)^2} \tag{8}$$

Equations (7) and (8) may be used for an exchange between any divalent-monovalent ion pair on a fully ionized resin. While Eqs. (7) and (8) are expressed in equivalent fractions, K_{DB} is defined as the molar selectivity coefficient. If K_{DB} is taken as constant, this allows a direct estimation of the effect of solution concentration on distribution of ions in the system.

Equation (8) shows that in a divalent-monovalent exchange the equilibrium resin composition is dependent on a selectivity coefficient, the ratio of ions in solution, and the total ionic concentrations of solution and resin phases. The factor relating resin composition to solution composition is $K_{DA}\bar{C}/C$. For a given ion-exchange resin, the total capacity of the resin $\bar{C}$ is roughly constant for resins that are ionized over the entire pH range. However, in different applications or different steps of the same operation, total ionic concentration in solution may vary widely.

Since C is in the denominator of the factor, as solution concentration decreases the left-hand side of the equation becomes larger, or as solution in contact with resin becomes dilute, the resin is converted more to the divalent-ion form. Dependency of resin composition on solution concentration in divalent-monovalent exchange is inherent in the physical chemistry of the system, and holds for all resins of constant total capacity. In many cases dependency on solution concentration far outweighs the numerical value of the selectivity coefficient. This behavior permits changing the distribution behavior for a given pair of ions by changing total ion concentration of the solution phase and is important in certain ion-exchange processes by enabling both forward and reverse exchanges with reasonable efficiency.

Selectivity Coefficients *Cations.* A number of studies have been reported on selectivity coefficients in cation exchanges. Selectivity coefficients are not constants but vary somewhat as the equivalent fraction of the resin in a given ionic form changes. The pattern of variation is not regular from one ion pair to another. Selectivity coefficients in the literature are often given at a specified resin composition $\bar{x}_A$ or are integrated over the full resin composition. The numerical value of the selectivity coefficient will vary with the relative crosslinkage of the resin. Differences between two ions tend to become greater as crosslinkage increases, i.e., as the water content of the resin decreases. In a few cases, where the selectivity coefficient is close to unity, changing the crosslinkage of the resin may actually change the order of preference in a pair of ions.

There is not close agreement in the literature on numerical values for various selectivity coefficients. Approximate values are adequate for process screening. Bonner and coworkers developed a selectivity scale for most of the common monovalent and divalent cations on a strong-acid resin,[18] as given in Table 2. The numbers have been reduced to two significant figures to emphasize the approximate nature of calculations using these values. Bonner states that a selectivity coefficient for any pair of cations in the selectivity scale can be obtained by using the ratio of the values given. Thus the selectivity coefficient for potassium-hydrogen exchange $K_{K^+H^+}$ would be equal to 2.9/1.3 = 2.2. The value of $K_{Ca^{2+}Mg^{2+}}$ = 5.2/3.3 = 1.6; $K_{Ca^{2+}Na^+}$ = 5.16/2.0 = 2.6, and so on. The selectivity coefficient is valid as long as exchanging ions are of identical sign. When exchange is between divalent and monovalent ions, this procedure is questionable unless the mass-action relationship is expressed exactly as used by Bonner in deriving the selectivity scale.

The author has compared a number of divalent-monovalent selectivity coefficients derived from Table 2 with values calculated from experimental data in the literature and from laboratory work. The values from Table 2 are sufficiently in agreement with observed values to serve for process screening using Eqs. (5) and (8). Further, when working with divalent-monovalent exchange, the effect on ionic distribution of solution concentrations often outweighs any reasonable error in the value of the selectivity coefficient.

TABLE 2   Selectivity Scale for Cations on 8% Crosslinked Strong-Acid Resin

Li^+	1.0	Zn^{2+}	3.5
H^+	1.3	Co^{2+}	3.7
Na^+	2.0	Cu^{2+}	3.8
NH_4^+	2.6	Cd^{2+}	3.9
K^+	2.9	Be^{2+}	4.0
Rb^+	3.2	Mn^{2+}	4.1
Cs^+	3.3	Ni^{2+}	3.9
Ag^+	8.5	Ca^{2+}	5.2
UO_2^{2+}	2.5	Sr^{2+}	6.5
Mg^{2+}	3.3	Pb^{2+}	9.9
		Ba^{2+}	11.5

SOURCE: Ref. 18.

For cations of a given charge, the order of increasing selectivity parallels the order of decreasing hydrated ion size and is reflected by the fact that the swollen volume of the resin decreases in this same order. An ion of smaller hydrated size takes up less space inside the resin while satisfying the requirement of electrical neutrality. The stretching forces on the polymer network are reduced and the smaller ion is preferred. When no selectivity data are available for a pair of ions, the preferred ion can often be determined from the swollen volumes of the two forms of resin. The preferred ion will usually show a smaller swollen-resin volume.

Anions. Only limited data are available on selectivity coefficients on strong-base anion-exchange resins because the structure of anion resins is both more complex and less reproducible than in strong-acid resins. Moreover, many anions have both monovalent and divalent forms which exist in equilibrium in solution. There is also a greater tendency to form complex anionic species in solution than to form complex cationic species. Many complex anions show high tendencies to go into the inside phase.

Two types of functional structures are used in strong-base anion-exchange resins. Type I resins have three methyl groups in the quaternary nitrogen structure. Type II resins have two methyl groups and a hydroxyethyl group (Fig. 8). This small difference in structure has a strong influence on the selectivity between hydroxide ion and other anions. Measured relative to chloride ion, the hydroxide selectivity coefficients differ by at least a factor of 10.

Selectivity coefficients $K_{Cl^--OH^-}$ for Type I resins may range from 10 to over 30. This broad range of $K_{Cl^--OH^-}$ for Type I resins is due to the strong effect of relative water content of the resin on the selectivity coefficient. Figure 2 shows such dependence when water content of the resin is expressed as grams of water per equivalent of resin.

The difference in chloride-hydroxide selectivity for the two strong-base resin types has been interpreted as a difference in basicity. However, both resins are strong bases and will remove acids as weak as silicic acid quantitatively when in the hydroxide form, as long as the quaternary structure is intact. In practical terms the difference in chloride-hydroxide selectivity allows Type II resins to be regenerated to hydroxide form much more efficiently than Type I resins. In most cases, where a hydroxide cycle is used, increased chemical efficiency of Type II resins will outweigh the greater chemical stability of Type I resins.

Little data are available for direct comparison of the relative selectivities shown by Type I and Type II resins in ion pairs not including hydroxide ion. It appears that for most ion pairs the two types of resin show essentially the same selectivities. An approximate selectivity scale for a number of anions on a strong-base resin of medium moisture content is given in Table 3. Some of the values are only approximate, but they are adequate for process screening purposes using Eqs. (5) and (8).

Ion-exchange equilibria with divalent anions are interrelated with the relative degree of ionization of the acid from which the anion is derived. If the pH of the system is high enough, the acid remains completely in divalent form. The divalent anion equilibria are essentially the same as with the divalent cations. Under these conditions Eqs. (6) and (8) can be used to describe the approximate behavior of monovalent-divalent anion exchanges. Above a pH of 3 the exchange between sulfate and chloride is a typical monovalent-divalent exchange. Predominantly divalent anion behavior should be observed for carbonate above 10.5 pH, for sulfite above 8 pH, and for phosphate between 8 and 10 pH. No selectivity values as such have been presented for these divalent anions. The values in Table 3 are extrapolated from relevant experimental data.

Divalent sulfate, carbonate, and hydrogen phosphate anions are very low on the selectivity scale in relation to such monovalent anions as chloride or nitrate. This is opposite that observed with cations on a strong-acid resin where most divalent cations are slightly higher on the scale than sodium or potassium, another indication of the more complex behavior of anion-exchange resins.

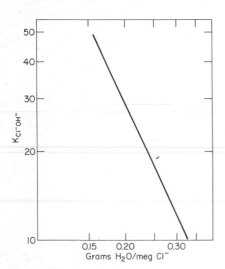

Fig. 2 Chloride-hydroxide selectivity for a Type I strong-base resin vs. moisture content.

When the pH of the system falls below the value where an acid anion is fully in divalent form, the outside phase contains both monovalent and divalent anions of the acid. The relative ratio of the two ions in solution will be determined by the total concentration of acid and salt and the pK_a of the acid. At equilibrium the ratio of monovalent to divalent anions in the resin phase will be a function of the ratio in the outside phase, the selectivity coefficient of the resin for the two anions, and the total concentration of the outside phase.

Equilibrium between sulfuric acid and a strong-base resin is a good example.[13] As the concentration of sulfuric acid decreases, the fraction of acid present as divalent sulfate

TABLE 3 Approximate Selectivity Scale for Anions on Strong-Base Resins

I^-	8	HCO_3^-	0.4
NO_3^-	4	CH_3COO^-	0.2
Br^-	3	F^-	0.1
HSO_4^-	1.6	OH^-(Type I)	0.05–0.07
NO_2^-	1.3		
CN^-	1.3	SO_4^{2-}	0.15
Cl^-	1.0	CO_3^{2-}	0.03
BrO_3^-	1.0	HPO_4^{2-}	0.01
OH^-(Type II)	0.65		

SOURCE: Ref. 24.

ions increases. A decreasing concentration of the outside phase also shifts the equilibrium in the system to favor divalent ions in the resin phase. As a result, the resin will be predominantly in sulfate form in very dilute acid and in bisulfate form in moderately concentrated acid. A strong-base resin that has been contacted with sulfuric acid more concentrated than 0.5 N will be largely in bisulfate form and will release sulfuric acid when rinsed with water. Prolonged rinsing will convert the resin completely to sulfate form.

$$2 \; \overline{R^+ \; HSO_4^-} \; \underset{dilute}{\overset{conc.}{\rightleftarrows}} \; \overline{R_2^+ \; SO_4^{2-}} + H_2^+SO_4^{2-}$$

Other inorganic acids such as phosphoric, sulfurous, and carbonic show similar behavior.

The chromate-dichromate pair of divalent anions presents a different case. In alkaline solutions hexavalent chromium exists in solution as the chromate ion CrO_4^{2-}. As pH drops below 6, chromate ion condenses to form dichromate ion $Cr_2O_7^{2-}$. Both ions appear to be held selectively over common monovalent anions. Dichromate ion shows a definite selectivity over chromate ion in relation to chloride ion. Maximum chromium loadings on a strong-base resin can be obtained by dropping the pH of the solution to approximately 5. A volume of resin of a given equivalent capacity can hold twice as much chromium as dichromate ion as it can hold as chromate ion. Half of the chromium held by the resin can be eluted merely by increasing the pH of the system.

$$\overline{(R_4N^+)_2, \; Cr_2O_7^{2-}} + 2NaOH \rightarrow \overline{(R_4N^+)_2, \; CrO_4^{2-}} + Na_2CrO_4 + H_2O$$

This reversibility is used in removing hexavalent chromium from cooling-tower blow-down and plating rinse waters.[23] While it is difficult to regenerate the resin efficiently from chromate form to another salt form, the capacity available from the pH swing makes the operation practical.

If a reaction is between a mixture of acids and the hydroxide form of a strong-base resin, all the acids will be exchanged onto the resin. As more solution is passed through the resin, acids will be displaced in order of increasing acid strengths.

When in the hydroxide-ion form, strong-base resins will remove even very weak acids from solution. Silicic acid can be removed to below detectable limits. However, weak acids are displaced by stronger acids moving down the column. The weakest acid will be the first component to appear in the effluent when the column becomes exhausted. Relative anion selectivities often do not apply when the component in outside phase is free acid instead of one of its salts. If there is a difference in degree of ionization of the acid of the anion on the resin and that of the anion in solution, the anion of the strongest acid will be preferred on the resin. For example:

$$\overline{R^+ \; CN^-} + H^+Cl^- \rightarrow \overline{R^+ \; Cl^-} + HCN$$
$$\overline{R^+ \; HCO_3^-} + H^+NO_3^- \rightarrow \overline{R^+ \; NO_3^-} + CO_2 + H_2O$$

Preference is due to the shifting of equilibrium by the nonionized weak acid.

Column Behavior

Isotherms and Separation Factor   Equilibrium between two exchanging ions is often shown as an ion-exchange isotherm as in Fig. 3. Each point on the isotherm gives the composition of the resin phase $\bar{x}_A$ that is in equilibrium with a given solution composition x_A. The separation factor α AB is defined as:

$$\alpha_{AB} = \frac{\bar{x}_A x_B}{\bar{x}_B x_A} \tag{9}$$

This definition in terms of equilibrium fractions holds regardless of the valence of the exchanging ions. In exchanges between ions of identical charge, the separation factor is identical with the molar-selectivity coefficient K_{AB} of Eqs. (1), (4), and (5).

When the exchange is favorable, K_{AB} and α_{AB} are greater than unity and the isotherm is convex above the diagonal. When the exchange is unfavorable, K_{AB} and α_{AB} are less than unity and the isotherm is concave below the diagonal. If K_{AB} is constant from $\bar{x}_A$ to 0 to $\bar{x}_A$ = 1, as assumed here, the isotherm is symmetrical. The shape and location of the isotherm are not greatly affected by changes in total solution concentration in exchanges between ions of identical charge.

In divalent-monovalent exchanges, the position of the isotherm is dependent on the total concentrations of the two phases as well as on the selectivity coefficient.[16] The separation factor for such an exchange may be greater than unity even though K_{DA} is less than unity. If the values of K_{DA}, $\check{C}$, and C are known, the isotherm for any divalent-monovalent exchange can be constructed by substituting values of $\bar{x}_D$ in Eq. (8) and calculating corresponding values of x_D. As long as K_{DA} is constant over the exchange, the shape and location of the isotherm will be constant for a given value of $K_{DA}\check{C}/C$. Figure 4

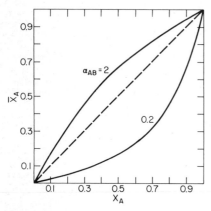

Fig. 3 Ion-exchange isotherms for the reaction $A + \bar{B} \rightleftharpoons \bar{A} + B$.

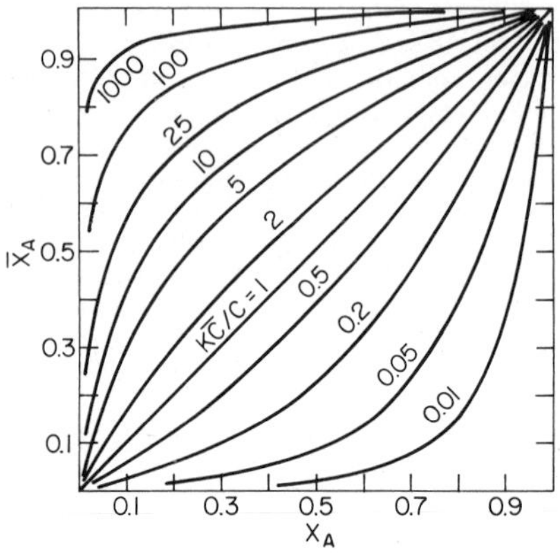

Fig. 4 Ion-exchange isotherms for the reaction $D + 2\bar{B} \rightleftharpoons \bar{D} + 2B$.

shows divalent-monovalent isotherms for values of $K_{DA}\check{C}/C$ from 0.01 to 1000. Even though K_{DA} is assumed to be constant, the isotherm is not symmetrical over the exchange. The separation factor changes as the resin converts from one form to the other. However, it is important to note that the separation factor will be greater than 1 as long as the value of $K_{DA}\check{C}/C$ is greater than 1, and vice versa.

The importance of dependence of separation factor on $K_{DA}\check{C}/C$ can be seen by considering an example. The scaling tendencies of certain natural waters can be reduced by removing their sulfate content by exchanging it for chloride on a strong-base resin.

$$SO_4^{2-} + 2\,\overline{Cl^-} \rightleftharpoons \overline{SO_4^{2-}} + 2\,Cl^-$$

The isotherms for the loading of sulfate on the resin and the regeneration of the resin with a concentrated sodium chloride solution can be approximated by calculating the appropriate values of $K_{DA}\check{C}/C$. The total capacity of a strong base is approximately 1.2 eq/L. The selectivity coefficient for the sulfate-chloride exchange $K_{SO_4^{2-}-Cl^-}$ is of the order of 0.15. If the water to be treated has a total concentration of 0.018 eq/L, the value of $K_{SO_4^{2-}-Cl^-}\check{C}/C$ will be $0.15 \times 1.2 \div 0.018 = 10$. The separation factor for loading sulfate on the resin is much greater than 1, and sulfate will be the preferred ion. A 10% solution of sodium chloride has a total normality of 1.8 eq/L. If this solution is used to regenerate the resin, $K_{SO_4^{2-}-Cl^-}\check{C}/C$ for the reaction as written is $0.15 \times 1.2 \div 1.8 = 0.1$. The separation factor will be much less than unity, and the resin will be selective for chloride, favoring the regeneration.

Ion-exchange processes usually must be cyclic to be practical. If resin is converted from B to A in the exhaustion step, it must be converted back to B, or regenerated, before the desired exchange can be carried out again. Unless there is some method of changing the nature of selectivity in the system, the separation factor for the reverse reaction will be the reciprocal of the forward separation. As a result, in monovalent exchange in fully ionized systems, one step of the cycle normally has a favorable separation factor while the reverse step of the cycle has an unfavorable separation factor. Ability to alter the value of the separation factor in exchanges between ions of differing valences is an important factor in efficient use of ion exchange.

Equilibrium Fronts   Perhaps the most important fact in understanding column operations is the cause-and-effect relationship between the numerical value of the separation factor and the nature of the exchange front in the column. The shape of the exchange front as it moves down a column is fundamentally different depending on whether the separation factor is greater or less than unity. The basis of this difference in behavior can be seen by following a series of equilibrium contacts in which solution containing ion A reacts with resin containing ion B.

Consider a series of stirred contactors, each of which contains one equivalent of resin in the B form. A quantity of solution containing one equivalent of A is added to the first contactor, and the exchange is allowed to come to equilibrium.

$$A + \bar{B} \rightleftharpoons \bar{A} + B$$

If the separation factor α_{AB} is known, the equilibrium compositions of the solution and resin can be calculated. The solution is drawn off, added to the equivalent of B-form resin in the second contactor, allowed to equilibrate, and drawn off. This is repeated until the solution has passed through each of the contactors. The equilibrium compositions are calculated after each contact.

A second increment of solution initially containing one equivalent of A is then passed through the series of contactors. Since the composition of resin in each contactor is known, a new series of equilibrium compositions can be calculated.

Solution compositions of 5 solution increments after each of 10 contacts are plotted in Fig. 5. In Fig. 5a the separation factor was taken as 5. In this case only a few contacts are required to convert the solution to almost pure B. The concentration histories of the successive solution increments tend to take on a repetitive pattern in which the entire exchange occurs in about five contacts. Once a solution increment has been converted to B it is not changed by subsequent contacts, nor are the later solution increments changed by passage through the first several reactors, since resin there has been converted to A form.

In Fig. 5b the separation factor was taken as 0.2. In this case the first increment of solution still contains nearly 10% of equivalents of A after 10 contacts. There is no tendency for the concentration histories of the successive solution increments to form a repetitive pattern. If the object is to produce pure B solution, it will obviously be much more practical when the separation factor is 5.0 than when it is 0.2.

Self-Sharpening Fronts. If an ion-exchange column is considered as a stack of contact units, it can be seen that the equilibrium factors determining the shape of the exchange zone that forms in the column will be similar to those in the series of batch contacts. When the separation factor for the forward reaction is greater than unity, the ion-exchange front attains and maintains a constant shape as it passes down the column. The front is said to be self-sharpening.

The shape of the breakthrough curve for component A at the bottom of the column is a direct reflection of the shape of the exchange front. As long as no part of the exchange front has reached the bottom of the column, the effluent will contain only B ions. When enough increments have passed through to move the exchange front to the bottom of the column, A will appear in the effluent. The concentration profile of the effluent will then mirror the

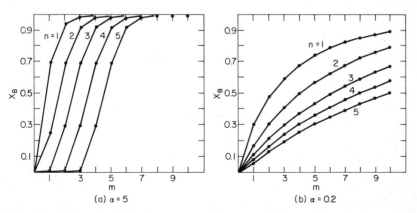

Fig. 5 Composition of solution increment n after m equilibrium contacts for the reaction $A + \bar{B} \rightleftharpoons \bar{A} + \bar{B}$.

shape of the exchange front. If the process is extended until the front has passed completely through the column, the effluent will have the same composition as the influent. All the resin in the column will then be in equilibrium with the influent. If the resin initially contains only B, and the influent only A, and the separation factor for A over B is greater than unity:

1. Exchange front will attain a constant shape.
2. Exchange zone will be narrow.
3. Effluent will be pure B solution up to a sharp breakthrough.
4. Resin will be converted cleanly to A form.

Nonsharpening Fronts. When separation factor is less than unity, the front does not attain a constant shape. Instead, the front becomes more diffuse as it passes through the column. The front is said to be nonsharpening. A nonsharpening front attains a proportionate pattern where the width of exchange zone increases in proportion of the distance traveled. The effluent will not contain the entering ion A until the exchange zone extends to the bottom of the column. However, since the exchange zone is constantly widening, it will reach the bottom of the column long before it will in the case of a self-sharpening front. The entering ion will appear in the effluent relatively early in the run and will increase in concentration slowly.

Equilibrium Effects in Column Performance   Equilibrium factors limit performance of the column. If exchange is unfavorable, no change in column variables such as height, resin size, temperature, or flow rate will give a sharp breakthrough curve. The consequences of the two drastically different types of exchange-front behavior must be considered for two ends: the preparation of a solution of a given ion, or free of a given ion; and the preparation of resin of a given ionic form, i.e., regeneration.

If the entering ion is preferred over the ion on the resin, $\alpha_{AB} > 1$, the exchange zone will be narrow and will occupy a relatively small proportion of total column at any one time. When the exchange reaches the bottom of the column ion A appears in the effluent. The run is usually stopped when the effluent concentration of A reaches some fraction of its concentration in the influent. Since the exchange zone is narrow, a high percentage of total resin in the column has been converted to A form by this time. Capacity of the column to the end point, or operating capacity, will approach equilibrium capacity of the column for A.

When the separation factor for the desired exchange is less than unity, the exchange zone will be broad and will continue to broaden as it moves down the column. Since the exchange zone may be spread over a considerable portion of total column length, net utilization of the resin capacity may be very low at the allowed breakthrough concentration. Operating capacity obtained will be much less than equilibrium capacity of the resin. Even though the separation factor is unfavorable, it will still be possible to obtain some product free of the influent ion; however, utilization of the resin capacity will be low.

Where the objective is to convert resin to a given ionic form, the same reasoning applies. If the entering ion is preferred, resin is converted cleanly to the entering ionic form behind a sharp exchange zone. Conversion is efficient and only a relatively small excess of regenerant is required to give a high degree of resin regeneration. If exchange front is nonsharpening, regeneration is inefficient. The exchange zone reaches the bottom of the column while a large portion of resin is unconverted and regenerant ion passes through the column without being exchanged. Resin can be converted to the form of the influent ion even if the exchange is unfavorable, but at the cost of an excess of regenerant ion.

Kinetic Effects in Column Performance   The degree to which actual column performance approaches limiting or equilibrium performance is dependent on kinetic and mechanical factors. Mechanical factors, such as axial dispersion and column-wall effects, tend to reduce the sharpness of exchange fronts. These effects can be minimized by good column design and operation.

Kinetic factors are associated with mass transfer of ions between the two phases. Exchange of ions is a coupled diffusional process. The rate-controlling step may be diffusion of the ions in the inside phase, through the interfacial film, or in the outside phase. The last is seldom of importance in ion-exchange reactions. A simple criterion has been devised for predicting whether particle diffusion or film diffusion will control overall rate of exchange (Ref. 4, p. 255). Factors included in the formulation are concentration of fixed ionic groups in the resin, concentration of the solution, radius of ion-exchange beads, ratio of the interdiffusion coefficients of ions in the resin phase and in the film, film thickness, and separation factor for the exchange. The criterion predicts that in an exchange between monovalent ions on a strongly ionized resin of standard size, the change-over from film-diffusion control to particle-diffusion control is at a solution concentration range of 0.1 to 1.0 N. Below this range film diffusion will tend to be the rate-controlling step. Above this range particle diffusion will be dominant.

A highly favorable separation factor will shift control toward film diffusion. When divalent ions are involved in the exchange, interdiffusion coefficients of different ions in the resin phase become an appreciable factor in determining rates. Exchange rates are usually much slower when divalent ions are involved.

When film diffusion is the rate-controlling step, exchange rate is proportional to interdiffusion coefficients of ions in the fluid film and the concentration of the solution. Exchange rate is inversely proportional to total capacity of resin and radius of the bead.

When particle diffusion is the rate-controlling step, exchange rate is proportional to the interdiffusion coefficient of ions in the inside phase, and inversely proportional to the square of the particle radius. Rate is independent of the concentration of the solution.

For monovalent exchanges on strongly ionized resins, half times for particle-diffusion-controlled exchange are on the order of seconds to minutes. In the case of film-diffusion control, half times can range from a few minutes in 0.1 N solution, to several hours in 0.001 N solutions.

The actual exchange front set up in the column will be determined by both equilibrium and kinetic factors. When the separation factor is favorable, a self-sharpening front will occur. The shape and width of this front will be determined by separation factor and exchange rate, and will vary depending on whether film diffusion or particle diffusion is the determining rate step. Film-diffusion control will tend to allow incoming ions to slip ahead of the equilibrium front. Particle-diffusion control will tend to cause tailing of the leaving ion. If exchange is favorable, the operating capacity and resin utilization obtained are largely independent of the magnitude of the separation factor, and may actually be limited by slippage due to film-diffusion control of the exchange rate.

The exchange zone of a self-sharpening front occupies a constant number of effective plates as it moves down the column. Lengthening the column will not increase sharpness of the exchange zone. However, the sharpness of the breakthrough curve in relation to the column volume will reflect average height of an effective plate in the column: the smaller the effective plate height the sharper the breakthrough curve. Effective plate height decreases with such factors as:

1. Decreasing linear flow rate
2. Decreasing ratio of solution concentration to resin concentration
3. Smaller resin particle size
4. Increasing exchange rates

With a self-sharpening front the sharpness of the breakthrough curve depends on how closely the actual front approaches the equilibrium front, and will be very dependent on column condition. However, even under relatively poor operating conditions, the exchange will be efficient and operating capacity will approach equilibrium capacity.

In case of a nonsharpening front, breakthrough is directly affected by the value of the separation factor: the smaller the separation factor, the more diffuse the breakthrough curve. Under similar column conditions, exchanges with nonsharpening fronts tend to run much closer to equilibrium than columns with self-sharpening fronts. As a result, the breakthrough curve obtained is less dependent on such factors as flow rate and the nature of the rate-limiting step.

In a number of common ion-exchange processes regeneration is the step in the cycle which is run with a nonsharpening front. Regeneration steps usually employ solutions with total concentrations of 1 N or higher. Under these conditions reduction in average particle size of resin usually results in a more efficient regenerant utilization.

To a certain extent, flow rate used in column processes is set in relation to the concentration of the component to be removed from the influent. Concentrated solutions must be passed through the column more slowly than dilute solutions. In some cases operating data have been correlated on the basis of a kinetic-load factor. The kinetic load is defined as the concentration of the influent ion times the volume flow rate:

$$\text{Kinetic load} = C_A \text{ (equivalent/liter)} \times \text{flow rate (hour}^{-1})$$

Kinetic-load factor is empirical. In some systems roughly equivalent operating capacities are obtained at equal kinetic loads over a range of concentrations and flow rates.

Summary   This is a qualitative overview of factors determining column performance in exchange systems where resins are fully ionized. A thorough treatment of ion-exchange equilibria and kinetics and relationship to column behavior is given by Helfferich,[4] secs. 5, 6, and 9.

A number of equilibrium and kinetic factors enter into performance of an ion-exchange process. However, as a generalization, in ion-exchange processes using fully ionized resins, the overall chemical efficiency of the process is dominated by an equilibrium-controlled factor. This is the unfavorable equilibrium (separation factor less than unity) for one direction of the reaction, usually the regeneration step.

PARTIALLY IONIZED RESIN SYSTEMS

Weakly ionized resins have the same necessary components as strongly ionized resins:
1. A three-dimensional polymer network
2. Permanently attached ionic, or potentially ionic, functional groups
3. Counterions to balance the fixed ionized sites
4. A solvent

Weakly ionized resins also may have the additional components of a second solvent, invaded electrolyte, and nonionic solutes.

Resin Structure

Most commercial strongly ionized resins are based on styrene-DVB polymer networks. A number of different networks are used in weakly ionized resins. Functional groups may be added to base polymers by subsequent chemical reactions as with strong-base resins, or may be formed by a chemical modification such as hydrolysis of a component in the base polymer. Finally, the functional group may be a part of one of the monomers in the polymerization mixture. Considerable detail on preparation and structure of a wide variety of ion-exchange resins is given by Wheaton and Hatch.[33]

Weak-Acid Resins   Currently produced weak-acid resins are essentially copolymers of divinylbenzene with acrylic acid or methacrylic acid. The actual manufacturing process does not normally use acrylic acid or methacrylic acid as such in polymerization. Some related compound, such as an ester or anhydride, is used and converted to the acid after polymerization is completed. As with styrene-based resins, divinylbenzene serves to crosslink the polymer into a three-dimensional network.

The basic structural unit of a polyacrylic acid-DVB weak-acid resin is:

$$\left[\begin{array}{c} -CH-CH_2- \\ | \\ C \\ HO \diagdown O \end{array} \right]_n \left[\begin{array}{c} -CH_2-CH_2- \\ \bigcirc \\ -CH_2-CH_2- \end{array} \right]_m$$

where m is much smaller than n. There are relatively few aromatic rings in the structure and the carboxylic acid group itself represents a high percentage of the total solids in the resin. Weak-acid resins have a high ion-exchange capacity per gram of solid and, if swelling is restricted to reasonable limits, a high capacity per unit volume.

The divinylbenzene content of the original copolymer represents the only major source of crosslinkage in these resins. Therefore, the swollen volume of the finished resin reflects the amount of DVB present, but in a more complex way than in the strong-acid resins.

Acrylic acid resins and methacrylic acid resins differ in relative acid strength. The acrylic acid group has an ionization constant about ten times that of the methacrylic acid group. The difference in acid strength directly affects the suitability of the two resins for certain applications.

Weak-Base Resins   At least four chemically different types of polymer networks are used in current weak-base resins. In some cases different types of functionality are used with one type of polymer network. The various types are given letter designations here purely for ease in future reference.

Type S. The styrene-DVB polymer network used in strong-base resins is also used for weak-base resins. Chloromethylation converts the polymer to a form that will react with an amine. If the intermediate is reacted with a secondary amine, such as dimethyl amine, the product formed is an essentially monofunctional resin, the unit structure of which can be represented by:

$$
\text{Type S-m} \quad \left[\begin{array}{c} -\text{CH}-\text{CH}_2- \\ \bigcirc \\ \text{CH}_2 \\ | \\ \text{N} \\ \diagup \quad \diagdown \\ \text{CH}_3 \qquad \text{CH}_3 \end{array} \right]_n
$$

There is a tendency in the amination for two chloromethyl groups to react with a single molecule of amine, resulting in a strongly basic group. Most Type S-m resins contain at least a few percent of strong-base capacity when manufactured. Strong-base capacity decreases on cycling of the resin. Total capacities of the monofunctional Type S-m resins are of the same order as those of strong-base resins, 0.9 to 1.4 eq/L. In Type S-m resins the amine groups are each isolated on a benzene ring, an important factor in determining the ion-exchange properties of these resins.

Another weak-base resin is produced by reacting the chloromethylated intermediate with an alkylene-polyamine. The resulting resin has a mixture of primary, secondary, and tertiary amine functionality and will be designated as Type S-p. Type S-p resins have a much higher fraction of amine groups to total resin structure than S-m resins. Total capacity of Type S-p resins may be two to three times that of S-m resins. However, since each nitrogen is in close proximity to one or two other nitrogens in the polyalkylene chain, interactions are likely to occur. In some cases interactions cancel the potentially higher exchange capacity under operating conditions.

Crosslinking in the Type S weak-base resins is very complex. In addition to the original DVB crosslinkage, crosslinking can arise during the chloromethylation and amination steps. The latter is quite pronounced in Type S-p resins. Properties of these resins have little direct correlation to the amount of DVB used in the original copolymer.

Type P. The first commercial synthetic anion-exchange resins were weak-base resins based on phenol-formaldehyde polymers. Weak-base functionality is of the polyalkyleneamine type. The polyamine may be either included in the original condensation polymerization or added to the polymer by subsequent reactions. Phenol-formaldehyde polymers are inherently three-dimensional and cannot be considered as linear chains with random crosslinks. Although the polymer structure is relatively rigid, the resins shrink and swell with changes in ionic form. They are probably the most resistant to physical breakdown from osmotic flexing of all ion-exchange resins. A high nitrogen content can be built into these resins. However, the polyamine functionality puts certain limitations on the utilization of high total capacity as in the Type S-p resins. The highly porous aromatic nature of these resins gives them useful sorptive properties as well as ion-exchange properties. The polymer for Type P resins is produced in bulk and ground to the desired size for use in ion exchange, giving granular rather than spherical resin.

Type E. A third type of weak-base resin is produced by the reaction of epichlorohydrin with a polyalkyleneamine.[11] The structure is highly branched and crosslinked but is more flexible than the phenol-formaldehyde structure. Functionality is again a mixture of primary, secondary, and tertiary amines with the same restraints on amine utilization as in the other polyamine resins. A very high percentage of nitrogen can be built into Type E resins, resulting in a total capacity as high as 2.5 to 3.0 eq/L. A high percentage of total capacity can be realized as operating capacity under some conditions.

Type A. Weak-base resins are also available in which the polymer network is a polyacrylate. There are several varieties of Type A weak-base resins, but in each case the functional amine is tied to the polymer network through an ester or amide structure. The amine functionality may be a mono-, di-, or polyamine. Type A resins do not contain aromatic rings unless DVB is used for crosslinking, and tend to be the most highly ionized weak-base resins.

Resin Properties

Exchange Capacity   It is in ion-exchange capacity that weakly ionized resins differ fundamentally from strongly ionized resins. Strongly ionized resins have a fixed exchange capacity per bead across the pH range. Size of the bead may change with pH so that the capacity per unit volume varies, but the actual number of exchange sites remains constant. Weakly ionized resins vary from negligible exchange capacity to a high exchange capacity depending on the pH of the solution with which they are in equilibrium.

Dependency of capacity on solution pH for two weak-acid resins is shown in Fig. 6. The resins have essentially zero exchange capacity below a solution pH of 5. Exchange capacity increases as pH increases and becomes constant above a pH of 11. At this point the carboxylic groups in the resin have been converted to ionized form and ion-exchange capacity is at a maximum.

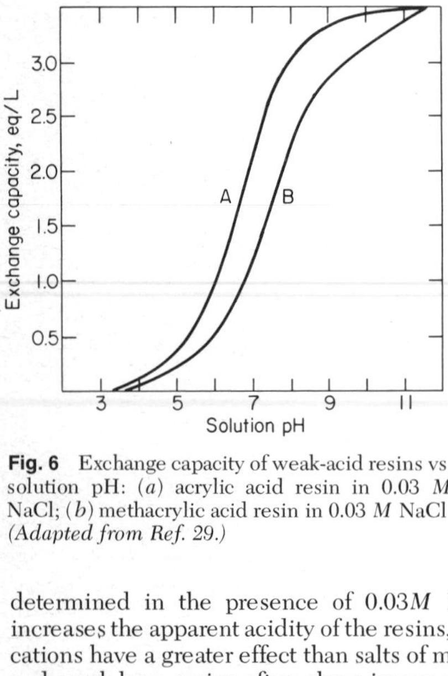

Capacity curve is dependent on both resin and solution compositions and will be shifted to right or left depending on the nature of functional groups in the resin. Methacrylic acid resin has an ionization constant about one-tenth of that of acrylic acid resin. It is thus a weaker acid and the capacity curve is shifted approximately one pH unit to the right of the capacity curve for acrylic acid resin.

Concentration of electrolyte in solution has a major effect on position of the capacity curve. Capacity curves at two different salt concentrations are shown in Fig. 7 for a Type S-m weak-base resin. As the salt concentration increases, the resin behaves as a stronger base and the curve shifts to the right. In the absence of salts the resin has little if any exchange capacity above pH 8. The same salt effect is seen with the weak-acid resins. The curves in Fig. 6 were determined in the presence of 0.03M NaCl. Increasing the concentration of NaCl increases the apparent acidity of the resins, and the curves shift to the left. Salts of divalent cations have a greater effect than salts of monovalent cations. One result is that weak-acid and weak-base resins often show improved performance when operating in a high-salt background.

Fig. 6 Exchange capacity of weak-acid resins vs. solution pH: (a) acrylic acid resin in 0.03 M NaCl; (b) methacrylic acid resin in 0.03 M NaCl. *(Adapted from Ref. 29.)*

Capacity curves for weak-base resins vary considerably in shape from one type to another, as shown by Fig. 7. Type S-m weak-base resins are nearly monofunctional, as shown by their relatively sharp change in ionization with pH. Other types of weak-base resins show somewhat different capacity curves. Type A resin is more highly ionized than Type S-m; so it shows more capacity at higher pH values. However, Type A resin is not as monofunctional as Type S-m; so the curve is less sharp.

Type E resins show very diffuse curves owing to the variety of amine structures and interaction between groups. Not all amine groups have equivalent basicities. In highly branched structures of Type E resins there is a mixture of primary, secondary, and tertiary amines. Also, there is a wide variety of ways in which the nitrogens are tied into the structure. These factors contribute to the energy required to protonate a particular nitrogen to its ionic form, and help account for the range of basicities shown in the sloping capacity curve. Functional groups are held in close proximity in polyamine resins. When one of the amines is converted to ionized form, an electrostatic field is set up. The field tends to reduce ionization of neighboring groups. As the resin becomes ionized, the remaining functional groups become progressively harder to ionize, reducing basicity.

Type S-p and Type P weak-base resins have curves similar to that shown for Type E resins, but tend to be somewhat lower in capacity and/or basicity. Capacity of Type E resins at pH 10 to 12 is due to the presence of strong-base capacity in the resin.

Accurate measurement of the capacity of weakly ionized resins is difficult. Capacity obtained is dependent on the conditions of equilibration, and these must be stated. The capacity will be different to different counterions. The presence of neutral salts in solutions usually increases the resin's capacity at a given pH. Even the way in which the sample is rinsed following equilibration will affect the result, since there is no certain way of removing excess solution without hydrolyzing acid from the resin. If an acid or base is too weakly ionized, it cannot exchange with a weakly ionized resin. Acid or base in solution and resin must be capable of existing in ionized forms at a common pH in order to form an ionic compound. Most weak-base resins show some capacity for bicarbonate, but none for borate or silicate. Acrylic-acid-type weak-acid resins have good capacity for ammonia and most aliphatic amines. Any anion or cation on a weakly ionized resin will be displaced by an anion or cation of a more highly ionized acid or base.

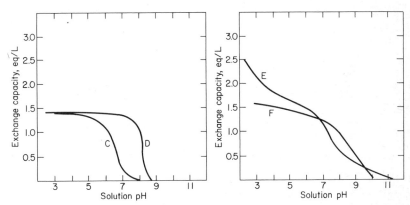

Fig. 7 Exchange capacity of weak-base resins vs. solution pH: (c) Type S-m resin in 0.009 m NaCl; (d) Type S-m resin in 0.5 M NaCl; (e) Type E resin in 0.5 M NaCl; (f) Type A resin in 0.5 M NaCl. (*Adapted from Ref. 30.*)

Swelling Volumes The swollen volume of a weak-acid or weak-base resin is determined by the same forces as in a strongly ionized resin: a balance between osmotic forces of solvation and elastic forces of the polymer network. In weakly ionized resins the shift in osmotic forces can be much greater. Nonionized forms of these resins tend to be less highly solvated than ionized forms. The volume increases as the resins are converted to salt forms. This is opposite to the behavior of strong-acid and strong-base resins which have their largest volumes in the hydrogen and hydroxide forms.

Typically, relative volumes of an acrylic acid resin in the H^+, Ca^{2+}, Mg^{2+}, NH_4^+, and Na^+ forms will be in the ratio of 1.00:1.15:1.55:1.80:1.90. A 90% increase in volume when the resin is converted from hydrogen form to sodium form can cause severe pressure-drop problems in column operations. Large volume changes are also a source of resin breakage during repeated cycling and can rupture lightweight laboratory columns.

It is not uncommon for weak-base resins to swell from 20 to 30% when converted from free-base form to chloride form. The sulfate form usually is less swollen than the chloride form.

Volume capacities in Figs. 5 and 7 are based on a volume of resin in nonionized form which is the accepted way of expressing capacities of weakly functional resins for most applications. Since resin swells during ionization, volume capacities based on equilibrium volume are less.

Acid-Base Reactions Strongly ionized cation and anion resins can be considered analogous to sulfuric acid and sodium hydroxide in neutralization reactions. Weakly ionized cation and anion resins can be considered analogous to acetic acid and ammonia.

$$\overline{\text{HOOCR}} + \text{NaOH} \rightarrow \overline{\text{NaOOCR}} + \text{H}_2\text{O}$$
$$\text{HC}_2\text{H}_3\text{O}_2 + \text{NaOH} \rightarrow \text{NaC}_2\text{H}_3\text{O}_2 + \text{H}_2\text{O}$$

$$\overline{\text{RNH}_2} + \text{HCl} \qquad \rightarrow \overline{\text{RNH}_3\text{Cl}}$$
$$\text{NH}_3 + \text{HCl} \qquad \rightarrow \text{NH}_4\text{Cl}$$

All four reactions are neutralizations. Because water does not appear as a product in the reaction of the weak-base resin as written, weak-base resins have sometimes been treated as acid adsorbers instead of as true ion exchangers. However, equilibrium and kinetic characteristics of weakly ionized resins can be rationalized consistently on the basis of ion-exchange mechanisms.

The most important result of dependency of ion-exchange capacity on pH is that it allows both forward and reverse reactions to be run at near 100% chemical efficiency. Since the exchange of an acid or base with these resins is a neutralization, the reaction will continue essentially to completion. Only a small excess of one of the reactants is necessary to convert the other completely, even in a batch contact. If carried out in a column, the effective separation factor will be quite large and a self-sharpening front will be set up.

The reverse reaction, regeneration, is also a very favorable reaction. An acid or base is added to the system to drive the pH to a range where the resin is no longer ionized:

$$\overline{\text{RNH}_3\text{Cl}} + \text{NaOH} \rightarrow \overline{\text{RNH}_2} + \text{NaCl} + \text{H}_2\text{O}$$

$$\overline{\text{NaOOCR}} + \text{HCl} \rightarrow \overline{\text{HOOCR}} + \text{NaCl}$$

These reactions can also be considered neutralizations. In regeneration of a weak-base resin a strong base is reacting with an acid salt to produce a neutral salt and water. Again, the reaction will go essentially to completion with only a small excess of one of the reactants. The reactions can be carried out batchwise. In a column the exchange front will tend to be self-sharpening.

Salt forms of weak-base resins undergo hydrolysis, as is normal with salts of weak acids or weak bases. A weak-base resin in its hydrochloride form will release hydrochloric acid slowly if rinsed with neutral water.

$$\overline{\text{R}_3\text{NH}^+\text{Cl}^-} \rightarrow \overline{\text{R}_3\text{N}} + \text{HCl}$$

A weak-base resin must be fully regenerated each cycle if acid-free effluent is to be obtained when it is put back on stream.

Applications of these resins tend to be removal of acidity or alkalinity from solution. Where operational characteristics allow use, they offer a major economic advantage over strongly ionized resins. They can be regenerated to free-base or free-acid form with an amount of acid or base only slightly greater than the equivalent amount of ion held on the resin. With strongly ionized resins regeneration to the hydrogen or hydroxide ion usually requires a large excess of regenerant.

Reactions with Neutral Salts It is commonly stated that weak-acid or weak-base resins in their free forms do not react with neutral salts to form acid or base. This is true for most practical purposes, but not in fact, and can be seen with a weak-acid resin as follows:

$$\overline{\text{HOOCR}} \rightleftharpoons \overline{\text{H}^+\text{-OOCR}}$$

An acrylic acid resin in equilibrium with pure water is about 5 M in carboxylate groups in the inside phase. Assuming that this polymeric acid dissociates to about the same extent as a monomeric acrylic acid (K_A approximately 1×10^{-4}), the concentration of hydrogen ions in the inside phase will be of the order of magnitude of 0.02 M.

Hydrogen ions cannot leave the inside phase, because electroneutrality must be maintained, and so do not affect pH in the outside phase. However, hydrogen ions in the inside phase are free to exchange with ions from the outside phase. If resin is placed in a solution of sodium chloride, an ion exchange will occur and hydrochloric acid will be formed in the outside phase.

$$\overline{\text{H}^+\text{ -OOCR}^-} + \text{Na}^+\text{Cl}^- \rightleftharpoons \overline{\text{Na}^+\text{ -OOCR}} + \text{H}^+\text{Cl}^-$$

The order of magnitude of the drop in pH which results from this exchange can be estimated. The resin probably does not show any appreciable selectivity between hydrogen ions and sodium ions on ionized sites, as there is no reason to differ from the sulfonic acid resin in this respect. If the degree of ionization of the resin remained constant, equilibration of resin with an equal volume of 0.01 M sodium chloride would drop the pH of the solution below 3. If the salt concentration is greater than 0.01 M volume, the pH drop will be greater. If the cation of the salt is more selectively held than sodium ion, the pH drop will be greater.

Apparent salt splitting with weakly ionized resins has sometimes been attributed to the presence of a small fraction of strong-acid or strong-base capacity in the resin. Many weak-base resins do contain strong-base capacity. However, resins which contain no strong or so-called salt-splitting capacity can produce a marked change in pH when a neutral salt is passed through them. Shift in pH is seldom observed with weak-base resins in practice, since the solutions fed to such resins are normally acidic and the object is neutralization. However, it is a common occurrence when weak-acid resins are used to remove alkalinity due to bicarbonates. The presence of neutral salts in the water being treated will give effluent containing mineral acidity if the weak-acid resin is regenerated fully. This phenomenon can also give undesired pH shifts when neutralizing process streams containing an appreciable electrolyte concentration.

Reactions with Salts of Weak Acids and Weak Bases   Consider the magnitude of the separation factor in the following three reactions:

$$\overline{\text{HOOCR}} + \text{NaCl} \rightleftharpoons \overline{\text{Na}^{+-}\text{OOCR}^-} + \text{H}^+\text{Cl}^-$$

$$\overline{\text{HOOCR}} + \text{Na}^+\text{OH}^- \rightarrow \overline{\text{NA}^{+-}\text{OOCR}^-} + \text{H}_2\text{O}$$

$$\left\{ \begin{array}{ll} \overline{\text{HOOCR}} + \text{Na}^+\text{HCO}_3 \rightleftharpoons \overline{\text{Na}^{+-}\text{OOCR}^-} + \text{H}^+\text{HCO}_3^- \\ \text{H}^+\text{HCO}_3^- \rightleftharpoons \text{H}_2\text{O} + \text{CO}_2 \end{array} \right.$$

In the first case hydrogen of the hydrochloric acid produced will suppress ionization of resin functionality. The overall reaction will be pushed strongly to the left. The separation factor for the forward reaction will be much smaller than unity. The numerical value of the selectivity coefficient in the exchange step will have little or no effect on the separation factor.

In the second case hydrogen ions are removed from the system as water is formed, allowing the resin to ionize completely. The overall reaction is dominated by formation of nonionized product and will go strongly to the right. The separation factor will be very large. Again, the selectivity coefficient in the ion-exchange step will have little effect on magnitude of separation factor and nature of exchange zone in the column.

In the third case one of the final products is weakly ionized. If the extent of ionization of the product is very similar to that of the resin, other factors may be decisive in determining the separation factor for the forward reaction. Based on column performance of acrylic weak-acid resins, the separation factor in this third set of reactions is less than unity, as a nonsharpening front is obtained. However, if the electrolyte is calcium bicarbonate instead of sodium bicarbonate, a self-sharpening front is obtained. The weak-acid resin shows a high selectivity for calcium over hydrogen vs. sodium over hydrogen. In the absence of a strong driving force due to a difference in reactant and product ionizations, the relative selectivity coefficients can determine the separation factor and thus the nature of the exchange front obtained. A similar situation should exist in the exchange between weak-base resins and solutions of ammonia or amine salts.

Reactions in Fully Ionized Systems   When weak-acid or weak-base resins are at a pH allowing an appreciable degree of internal ionization, they have most of the ion-exchange characteristics of strongly ionized resins and undergo equivalent exchange controlled by mass-action behavior. As long as the pH of the system does not change, the ion-exchange capacity of the resin will be relatively constant. Equations (5) and (8) can be used to estimate relative loadings. The stipulation that degree of ionization of the resin remains relatively constant eliminates hydrogen and hydroxide as counterions. Systems of interest are the exchange of cations on weak-acid exchangers in the pH range above 5 and anions on weak-base exchangers in the pH range below 6.

There is little selectivity data for weakly ionized resin systems. Differences between the various monovalent metallic cations are fairly small. The order of selectivity of alkali metals is reversed from that on sulfonic acid resins. Nitrate ion appears to be preferred on weak-base resins over chloride ion. Major differences from strongly ionized resins are in behavior with divalent ions. Weak-acid resins show appreciably greater differences in selectivity between divalent and monovalent ions, and within pairs of divalent ions, than sulfonic acid resins. Preliminary laboratory data on magnesium-sodium and calcium-sodium exchanges on an acrylic acid resin gave rough values of $K_{Mg^{2+}Na^+}$ of 5 and $K_{Ca^{2+}Na^+}$ of 55 at an equilibrium pH of 8. These values can be used in Eq. (8) to estimate equilibrium loadings over a total ionic concentration range of 0.01 to 2.0 N.

Recent work has shown that weak-base resins are more selective for sulfate vs. chloride than strong-base resins.[17] The extent of this selectivity varies with different types of weak-base resins. Behavior of weak-base resins to different ionic forms of multivalent acids is also quite different from that of strong-base resins. This reflects the greater sensitivity of weak-base resins to the different acid strengths of different ionized forms of the acids.

Exchange Rates   When exchange is between an ionized form of weakly ionized resin and an ionized solute, exchange rates are determined by the same factors as in exchange with strongly ionized resins. The selectivity coefficient for the exchanging ions is one of the factors affecting exchange rate. An appreciable suppression of rate will occur when a tightly held ion is displaced from the resin. Since selectivity coefficients tend to be relatively large when divalent ions are involved in the exchange, such slowing of rate is more likely to be a problem with weakly ionized resin systems.

Ion-exchange rates observed when the weakly ionized resins react with an acid or base are very different for forward and reverse reactions. Such reactions have been termed "ion exchange accompanied by reaction." Helfferich has given a theoretical discussion of the probable rate mechanisms in such exchanges.[19] Essentially all acid-base reactions with weakly ionized resins are rate-controlled by some form of ionic diffusion in the inside phase. Film diffusion can be ruled out as a rate-limiting step.

Three different particle-diffusion mechanisms have been postulated. The first mechanism occurs when an ionized form of resin is being converted back to the nonionized form, i.e., regeneration.

$$\overline{Na^{+-}OOCR^-} + H^+Cl^- \rightarrow \overline{HOOCR} + Na^+Cl^-$$

$$\overline{R_3NH^+Cl^-} + Na^+OH^- \rightarrow \overline{R_3N} + Na^+Cl^- + H_2O$$

Initial exchange is between a highly ionized resin and an ionized solution. However, as exchange proceeds from outside in, a shell of nonionized resin is formed. There are no fixed ionic sites in this shell to allow cross diffusion of exchanging counterions. However, there is no Donnan restriction on the invasion of electrolyte into the shell. The counterion and co-ion of the regenerant can diffuse jointly through the shell to the interface with unreacted core. At this point the exchange is of the same nature it was initially at the surface of the bead. Rate of exchange is dependent on the rate of diffusion of counterions and co-ions across the nonionized shell. Diffusion rate will depend on the concentration of invaded electrolyte which will in turn be directly dependent on concentration of the solution phase. The overall rate is proportional to solution-phase concentration, even though the rate-determining step is one of particle diffusion. Exchange rates are normally not a problem. Reagents used are relatively concentrated and small in volume in relation to volume of resin. Slow volume flow rates are used, but overall time required for the process is reasonable because of the small volume of solution required. When a divalent cation is being displaced from a weak-acid resin, considerable tailing may occur because of the slow particle-diffusion step involved.

Exchanges in which a nonionized form of the resin is reacting with an acidic or basic material tend to be slow. Half times in fully ionized exchanges are a few minutes or less. When the reacting resin is not ionized initially, the half time is usually at least an hour and may be on the order of days. Rate of exchange is qualitatively dependent on the acid or base strength of the resin. Two different rate mechanisms have been proposed for these exchanges. The first can be visualized by the reaction:

$$\overline{R_3N} + H^+Cl^- \rightarrow \overline{R_4NH^+Cl^-}$$

The counterion and co-ion of the acid diffuse into the bead jointly and react with a nonionized amine group. The resin is converted to the ionized form from the outside in. A distinct shell of ionized resin surrounds a nonionized core. There is little resistance to such a reaction initially. However, when the ionized shell attains a finite thickness, invasion of the co-ion of the acid is sharply restricted by Donnan forces. Rate of exchange of chloride ion will be controlled by concentration of hydrogen ion in the ionized shell which will be dependent on the concentration of the solution phase. Rate will also be dependent on any characteristic of the resin phase which allows more invasion. Rate should be relatively independent of the acid or base strength of the resin structure.

A third rate mechanism can be visualized by considering the reaction as taking place in two steps.

$$R_3N + H_2O \rightleftharpoons \overline{R_3NH^+OH^-}$$

$$\overline{R_3NH^+OH^-} + H^+Cl^- \rightarrow \overline{R_3NH^+Cl^-} + H_2O$$

The second step can be compared with the reaction of a strong-base resin with hydrochloric acid.

$$\overline{R_4N^+OH^-} + H^+Cl^- \rightarrow \overline{R_4N^+Cl^-} + H_2O$$

The two systems are quite similar in many respects, especially when a Type S-m weak-base resin is considered, since the strong-base and weak-base resins can have nearly identical polymer networks and water content. The self-diffusion rate of a given ion should be similar in both cases, yet the strong-base resin shows a half time of a few minutes, the weak-base resin a half time of about an hour. The half-time difference is due to internal concentration of hydroxide ion. In the electrically coupled diffusion process, each time a chloride ion enters the resin a hydroxide ion must leave. Rate is dependent on the relative concentrations of the leaving ion at the center and at the surface of the resin. Those concentrations fix the gradient for ionic diffusion across the radius of the bead. In strongly ionized resin concentration of leaving ion is high and the sharp concentration gradient gives a fast reaction rate. In weakly ionized resin concentration gradient is necessarily small and the exchange process slow.

In the third rate mechanism, exchange rate should be faster with resins having a higher internal ionization. An interesting example of the dependency of exchange rate on resin ionization was given by Bauman over 25 years ago:[15] "A resin containing —SO₃H, —COOH, and —OH in the same molecular structure, when titrated with NaOH will show almost instantaneous neutralization of the —SO₃H groups, a 2–4 hour drift to obtain a pH equilibrium with the —COOH groups, and a 24 hour drift with the phenolic groups." This interdiffusional rate mechanism does not give a sharp boundary as in the invasion mechanism, and the exchange becomes very slow near completion.

Both the invasion and the interdiffusional mechanisms occur with weakly ionized resins. Dominance will depend on degree of ionization of functional groups in the free acid or base form, and concentration of invaded co-ion. This in turn will be a function of the concentration of solution phase and physical structure of the resin. Hellfferich estimated that the switch from interdiffusion to invasion-rate control should occur in the concentration range of 0.001 to 0.01 M for a resin having an ionization constant of 1.5×10^{-5}.[19] Most experiments on rate of hydrochloric acid uptake on weak-base resins have shown that rate is independent of acid concentration and is faster with resins of higher basicity, at least at low solution concentrations. When an acid is taken up in a column operation, the acid concentration at the front of the exchange zone will be very low, and the interdiffusional mechanism should predominate in determining the shape of the initial breakthrough curve, regardless of the concentration of the entering solution.

It is generally found that exchange rates are inversely proportional to the square of the bead radius. All the rate mechanisms proposed for weak-acid or weak-base resins involve some type of particle-diffusion limitation and should be dependent on particle size. A problem in correlating rates with basicity in weak-base resins is that a polyfunctional resin will have a range of basicities. Only Type S-m resins approach monofunctionality.

Column Performance   When exchange starts with resin in a nonionized form, the column performance may be quite different from that in fully ionized systems. The fraction of total capacity of the resin, realized as useful operating capacity, will depend on both equilibrium and kinetic factors. Equilibrium capacity of resin is dependent on the nature of the acid or base, concentration, presence of neutral salts, and pH. Equilibrium capacity will be the limiting capacity the resin would show if run to an effluent composition equal to the influent composition.

Forward equilibrium is favorable when an acid or base is being neutralized. Separation factor will be large and there will be a tendency to set up a self-sharpening front. However, exchange rate in the reaction may be so slow that the effective number of plates in the column is low, and poor performance is obtained in contrast to fully ionized systems where the operating capacity may approach equilibrium capacity.

Operating capacity of a weakly ionized resin for acid or base removal is dependent on factors which affect the rate of exchange. Factors include flow rates, nature of the acid or base in solution, relative basicity or acidity of resin, temperature, and particle size of the resin. The smallest particles of resin that can be tolerated mechanically should be used. In acid removal, a higher operating capacity can sometimes be obtained by using a resin with a lower total capacity and a higher degree of ionization and thus a faster exchange rate. Type S-m resin has this characteristic. When solution concentrations are high and slow rates are mandatory, the higher total capacities of Type E resin can be utilized effectively. Type E resins give better performance to sulfuric acid than to hydrochloric acid.[11]

Performance of weakly ionized resins is much more sensitive to changes in temperature than performance of fully ionized resins. It is common to see poorer performance of weak-base resins in winter than in summer when treating surface waters. Equilibrium capacity of a weak-base resin is actually decreased by an increase in temperature. This behavior is the basis for the Sirotherm process for partial desalination using heat energy.[14] However, decrease in equilibrium capacity is more than compensated for by increase of exchange rate in standard column operations.

An appreciable fraction of a nonaqueous component in the process stream will usually make exchange more rate-sensitive. This is due to a suppression of the degree of ionization in both the inside and outside phase.

An interesting rate effect observed with weakly ionized resins is an improvement in performance due to the presence of a weakly ionized solute.[11] When a dilute mineral acid is being exchanged on a weak-base resin, the run is continued until mineral acid appears in the effluent. A higher capacity will often be obtained to breakthrough if a small amount of carbonic acid is present in the influent. The carbonic acid is exchanged, at least partially, onto the resin and then displaced by the advancing mineral-acid front. The resin in the bicarbonate form can exchange with the mineral acid much faster than the free-base resin can. The net effect is an increase in the number of effective plates in the column and a more efficient removal of the mineral acid.

The relationship among the equilibrium and kinetic factors controlling the performance of weakly ionized exchange resins is more complex and less well understood than in the case of fully ionized resins. As a generalization, in ion-exchange processes in which a poorly ionized form of a resin is involved, overall productivity of the process will be dominated by a kinetically controlled factor which is the inherently slow rate of exchange of the poorly ionized resin.

Specialty Resins

A number of resins have been prepared with functionality designed to have high specificity for an ion or a group of ions. Most highly specific resins have been based on compounds known to have chelating or complexing properties in solution. While none of the resins produced to date are specific for a particular ion, they can show much greater selectivity differences between certain groups of ions than conventional resins.

Iminodiacetate Resins A chelating resin containing iminodiacetate functionality has been made commercially since the 1950s. Similar resins are available from several resin manufacturers. Resin is usually prepared from the same type of chloromethylated styrene-DVB intermediate as the strong-base resins. The idealized resin structure is:

The functional group is the same as in ethylenediamine tetraacetic acid (EDTA). The iminodiacetate structure forms strong complexes with a wide variety of di- and trivalent cations.

$$R-N\underset{CH_2-\overset{O}{\underset{\parallel}{C}}-\ Na^+}{\overset{CH_2-\overset{O}{\underset{\parallel}{C}}{}^+\ Na^+}{<}}\ +\ Ca^{2+}Cl_2^-\ \rightarrow\ R-N\underset{CH_2-\overset{}{\underset{\parallel}{C^-}}}{\overset{CH_2-C^-}{<}}{-}{-}{-}{-}{-}{-}Ca^{2+}\ +\ NaCl$$

Selectivity of resin for various cations follows the same general pattern as the complex stability constants of EDTA in solution. However, incorporation of the functional group in a crosslinked polymer gel puts limitations on behavior. Synthesis of the material requires several steps, and it is difficult to obtain a monofunctional resin. Iminodiacetate resins made by various manufacturers differ in actual chelating ability.

There are two major restrictions on use of iminodiacetate resins because they are essentially weak-acid resins. They have little cation-exchange capacity below a pH of 5. This seriously restricts their use in hydrometallurgical processes, since many metal cations are not soluble above pH 5. Second, since they are weak-acid exchange resins, they have slow exchange rates in hydrogen form. Iminodiacetate resin offers some interesting possibilities when used in an ionized form above a pH of 5 to 6. Under these conditions the resin shows true ion exchange which is subject to mass-action rules, including normal monovalent-divalent ion behavior. However, because of the high selectivities shown by the resin for certain ions, successful removal of divalent ions over monovalent ions can be carried to a much higher solution concentration than with conventional resins. As an example, softening with the sodium form of a sulfonic acid is probably not practical above a total solution concentration of about 0.1 N. A conventional weak-acid resin has a selectivity coefficient for calcium over sodium about ten times that of a sulfonic acid resin. With a weak-acid resin in the sodium form softening is practical up to a total solution concentration of 0.5 to 1.0 N. Some iminodiacetate resins show calcium-sodium selectivity coefficients three to four orders of magnitude greater than those of strong-acid resins. Such resins can remove traces of calcium from saturated sodium chloride brine which has a total concentration of approximately 5 N.

Iminodiacetate resins also show greater selectivities between pairs of divalent ions than conventional cation resins do. Thus they can be used to remove such ions as zinc or lead from dilute streams where calcium is a major component. The pH of the stream must be high enough for the resin to have a useful exchange capacity. A comparison of relative selectivities of divalent cations on an iminodiacetate resin and a strong acid cation-exchange resin is given by Rosset.[25] Like conventional weak-acid resins, iminodiacetate resins show a large increase in volume when converted from hydrogen form to sodium form. Divalent-ion forms have intermediate volumes.

Polyisothiouronium Resins   Polyisothiouronium resins have a unit resin structure of the general form:

$$\left[\begin{array}{c} \overset{H}{\underset{|}{-C-CH_2^-}} \\ \bigcirc \\ CH_2 \\ | \\ S \\ | \\ H_2N\overset{N}{\diagup}\diagdown NH \end{array} \right]_2$$

Polyisothiouronium resins show very high selectivities for platinum-group metals and are used to extract the metals from dilute acid streams that occur in plating operations. Polyisothiouronium resins show good selectivity for several other heavy metals such as gold and mercury. They usually cannot be regenerated directly with acid or base. In some cases regeneration is possible with acid solutions of thiourea. In others, the metal value collected by the resin is great enough that the resins can be burned to isolate the metals.

Amidoxime Resins   Amidoxime resin has a functionality similar to polyisothiouronium resin. It is selective for cupric, ferric, auric, and a number of the platinum metal ions. It is active for some of these to below 2 pH. It can be regenerated with strong mineral acid. Amidoxime resin is currently used for removing unwanted components from certain plating baths and for the purification of fine chemicals.

n-Methylglucamine Resin   n-Methylglucamine resin has a functionality based on a hexose derivative. It has the ability to remove boron selectivity from solution. It can be regenerated with acid. It is currently used commercially to remove boron from magnesium chloride brine used as a raw material for magnesia ceramics.

Sulfhydryl Resins   A number of resins have been introduced in recent years which contain a sulfur-based functionality such as -SH. These and similar resins are used in Japan for removal of traces of mercury from various industrial-waste streams. Mercury levels can be reduced to below 0.5 ppb routinely. The resin is not usually regenerated but is retorted to recover mercury. They are used as a final polishing step following more conventional means of mercury removal. The level of mercury fed to them is very low and the cost is not prohibitive. Sulfhydryl resins also show selectivity for certain other metals such as lead.

Phenolic Resins   Phenol-formaldehyde polymers have ion-exchange activity above a pH of 10.5 to 11 where the phenolic group is in sodium-salt form. Under these conditions the resins have a good selectivity for cesium ion. This property is used for the isolation of cesium isotopes in certain nuclear processing steps. The resin may be used as the unaltered phenol-formaldehyde polymer or after the addition of carboxylic or sulfonic functionality.

Phenol-formaldehyde polymers when converted to Type P weak-base resins have a considerable capacity for mercury ion from solution. Mercury levels can be reduced to below 10 ppb. The mercury can be eluted with strong mineral acid and the cycle repeated. Similar resins are used in Scandinavia and Japan for treating waste streams from chlor-alkali production.

Inorganic Exchangers   A wide variety of inorganic compounds such as silicates and phosphates show ion-exchange activity. Some of the relative selectivities shown by inorganic compounds are significantly larger than shown by the organic-based exchangers. The naturally occurring mineral clinoptilolite is used to remove ammonium ion from tertiary sewage effluent.

RESIN STABILITY

Ion-exchange resins are organic compounds and are subject to chemical reactions which may alter their properties. In addition, they are subject to physical stress in use, which may change their physical structure. Chemical degradation of the resin may result from breaking of the polymer network, splitting off or modification of the functional groups, or fouling of the resin by a component from solution. Physical breakage may be due to excessive osmotic shrink-swell cycling or mechanical abrasion.

Strong-Acid Resins

Strong-acid resins are stable organic compounds, especially when in a salt form. It is not uncommon for strong-acid resins to have been in service for twenty years or more in a domestic softener and still perform adequately. Strong-acid resins are subject to oxidation by free chlorine, and it is recommended that the free-chlorine level be kept below 1 to 2 ppm. The hydrogen form of strong-acid resins is more subject to oxidation than the salt forms. The oxidative attack occurs on the polymer network; there is no loss of sulfur. Resins which have been oxidized show an increase in moisture retention and a drop in wet-volume capacity, but no change in dry-weight capacity. Mild oxidative attack does not appreciably reduce the resin's operating characteristics. Extensive oxidation increases the moisture content to the point where the resin becomes so soft that it compresses, causing excessive pressure drop. Chlorine in the feedwater can be eliminated by adding a small amount of sulfur dioxide. In addition to chlorine, severe oxidants are chromic acid, hydrogen peroxide, and hot concentrated nitric acid. Traces of iron and copper are catalytic in oxidations.

The sulfonic acid group shows an appreciable rate of hydrolysis above 150°C. The styrene-DVB network starts to depolymerize above about 200°C. Temperature limitations are rarely of concern in aqueous processing.

No irreversible fouling has been reported with strong-acid resins. Internal precipitation of calcium sulfate can occur if a resin highly in the calcium form is contacted with sulfuric acid above a few percent in concentration. Precipitate can be cleaned out by repeated washing with 6% hydrochloric acid free of calcium ion. A common type of fouling in softening applications is formation of a layer of hydrated ferric hydroxide on the surface of the beads. This can be minimized by proper backwashing. If ferric hydroxide fouling does occur, it can be cleaned out by treatment with a reducing agent.

Strong-acid resins are subject to breakage from osmotic shock, but this is seldom a serious problem. In some processes one or more dilution steps may be used in going across a large change in solution concentration to ease the transition. Opaque or macroporous forms of the resin are more resistant to osmotic shock.

There is little mechanical attrition in standard fixed-bed operations. The initial degree of cracking of the resin has only a secondary effect on the useful life of the resin in such units. In moving-bed systems good initial internal bead structure usually results in a longer service life.

Weak-Acid Resins

Weak-acid resins are extremely stable chemically. Fouling by internal calcium sulfate precipitation can occur as in the strong-acid resins but can be cleaned out with hydrochloric acid. Weak-acid resins undergo large cyclic volume changes between the free-acid form and monovalent-salt forms and are subject to osmotic-shock breakage in such cycles. Use of the opaque or macroporous forms will reduce osmotic-shock losses.

Strong-Base Resins

Where strong-acid resins primarily show attack on the polymer network, strong-base resins show modification of the functional group. The quaternary ammonium group is very stable below pH 7, where it is in a neutral-salt form. However, all quaternary-ammonium compounds tend to undergo a Hofmann degradation when in the hydroxide form, resulting in cleavage of one of the carbon-nitrogen bonds. Type I and Type II resins show different stability characteristics, both as to the rate at which degradation occurs and the nature of the change in resin properties.

Type I   In the Type I structure (Fig. 8a) the chance that one of the three methyl groups will be split off is roughly the same as that the entire amine will split away from the benzene ring. When a methyl group is lost, strong-base capacity is converted to weak-base capacity. When the amine group is split off, there is a loss in total ion-exchange capacity. The resultant change in ion-exchange properties is shown in Fig. 9a. Both total-exchange capacity and strong-base capacity decrease.

Rate of capacity loss is temperature-dependent. The half time, or time required to lose half of the strong-base capacity, has been shown to be of the order of 35 to 45 days at 90°C, but of the order of 5 years at 60°C. It is normally recommended that Type I resins not be exposed to temperatures much above 60°C when in the hydroxide form if a long service life is required.

The rate of loss of structure decreases as the resin ages. The quantity of organic con-

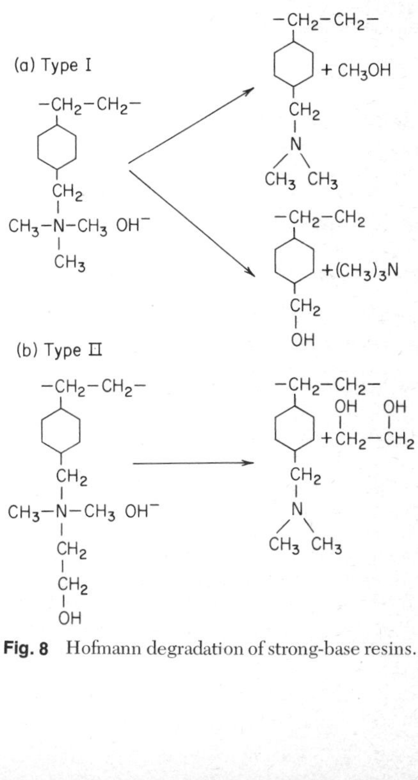

Fig. 8   Hofmann degradation of strong-base resins.

stituents lost from the resin is very small at any one time. Methanol or trimethylamine present no hazard at the levels at which they might be present. However, trimethylamine has an odor threshhold in air of a few parts per billion. This accounts for the fishy odor common to Type I resins.

Type II In Type II resins cleavage occurs almost entirely between the nitrogen and the ethanol group (Fig. 8b). Strong-base capacity is converted to weak-base capacity. There is no loss of nitrogen. Total ion-exchange capacity of the resin on a volume basis is essentially unchanged (Fig. 9b).

Rate of loss of strong-base capacity at a given temperature is faster in a Type II resin than in a Type I resin. However, the useful service life of a Type II resin may be 3 to 5 years where ambient temperatures are properly maintained. In many cases partial loss of basicity is not critical. Since there is no loss in total-exchange capacity, operating capacity of a Type II resin when used for neutralizing a mixed-acid stream may be relatively unchanged even though a considerable portion of the strong-base capacity has been converted to weak-base capacity. As long as there is strong-base capacity left in the resin, it will have some ability to pick up even very weak acids such as carbonic acid and silicic acid. A Type II resin which has lost half of its initial strong-base capacity will still show an operating capacity to a silica breakthrough of 85 to 90% of that of a new resin as long as the silica is no more than 5 to 10% of the total acids present in the influent. It is recommended that temperatures above 40°C be avoided with Type II resins at high pH.

The by-product of cleavage of the quaternary ammonium group in a Type II resin is odorless and nonvolatile. Type II resins therefore do not have any odor and are preferred when odor is unacceptable in the product stream.

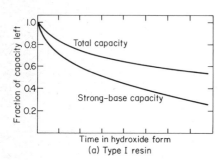

(a) Type I resin

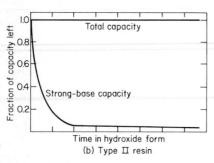

(b) Type II resin

Fig. 9 Change of capacity of strong-base resins from thermal degradation of hydroxide form (time scale dependent on temperature).

Oxidation of the polymer network is seldom observed in quaternary ammonium resins. When oxidative conditions exist, attack is predominantly on the functional group and results in a loss in strong-base capacity much as in the reactions described above.

Several types of fouling have been observed in strong-base resins. When hydroxide forms of strong-base resins are used to neutralize an acid stream, silicic acid is concentrated on the resin at the leading edge of the exchange front. There is a strong tendency for silicic acid to polymerize into large molecules that are present more as occluded inorganic solids than as exchangeable anions. Silicate ions which are present as such are easily eluted from the resin during regeneration with sodium hydroxide. To remove polymerized silica from resin, it must be depolymerized by holding the resin at a high pH. The depolymerization process is slow, particularly at ambient temperatures, and the use of elevated temperatures is restricted because of the instability of the quaternary group. If enough time at high pH is not allowed to dissolve silica from the resin each cycle, it will accumulate to a point where the resin cannot produce silica-free water. Serious silica fouling is difficult to clean out of a resin.

Certain types of complex organic acids which are common in surface waters tend to exchange and/or adsorb on strong-base resins somewhat irreversibly. The humic and fulvic acids which are by-products of decaying vegetable matter are typical fouling agents. One of the properties of resin most affected by organic fouling is rinse requirement following caustic regeneration. The acidic functionality of the adsorbed compounds is converted to sodium salts during regeneration of the anion resin. The sodium hydrolyzes off slowly as the resin is rinsed. Resin fouled in this way may require several hundred times the amount of rinse to achieve acceptable effluent quality than when new. If not too advanced, organic fouling can often be reversed by treating the resin with alkaline brine.

Organic fouling may also become evident as an inability to produce the required quality of water, or as a marked sensitivity of effluent quality to flow rate. This is usually only a problem in mixed-bed demineralizers when very high product purity is required.

Strong-base resins show very high selectivities for certain complexed metal anions. The cyanide and thiocyanate complexes of several of the transition metals are exchanged almost irreversibly. This can result in rapid deterioration of useful exchange capacity in hydrometallurgy applications if proper precautions are not taken. Resistance of strong-base resins to abrasion or osmotic-shock breakage is similar to that of strong-acid resins.

Weak-Base Resins

Each weak-base resin type has strengths and weaknesses where stability is concerned. All are subject to oxidation to some degree. Oxidation results in an increase in rinse requirement and/or an increase in rate sensitivity. Fouling with organic complexes also results in increased rinse requirements and often cannot be clearly differentiated from oxidation.

Type S-m resins are relatively stable to oxidation and fouling. Even though usually macroporous, S-m resins suffer osmotic-shock damage, especially in applications involving partially nonaqueous streams, or loading with organic acids. S-m resins have low particle density and care must be taken during backwashing to avoid loss from the unit. In some uses the resin has broken up badly because of silica fouling and loss of resiliency. Type S-p resins have been largely withdrawn from the market because of increased rinse requirement and loss of kinetic capacity due to autoxidation.

Type P resins are by far the most physically rugged of the weak-base resins. Type P resins are used in a number of chemical processing operations where osmotic shock is extreme. Type P resins tend to increase in rinse requirment under mild oxidative conditions. It is recommended that ammonia be used as the regenerant with Type P resins in place of sodium hydroxide to avoid a prolonged rinse.

Type E resins are resistant to autoxidative attack. Rinse requirements may increase from fouling. Type E resins are subject to osmotic-shock damage. Type A resins have high rinse requirements owing to the ester or amide link in their structure. Type A resins are only recommended where ammonia is used as the regenerant.

Macroporosity

Most types of ion-exchange resins are commercially available in gel and macroporous forms. These terms have come to be synonymous with clear and opaque. Gel resins are transparent, or at least semitransparent, and may be free of internal cracks, have a few cleavage lines, or be highly crazed. Cracking has no effect on the chemical properties of the resin but may be an important factor in determining a resin's ability to survive mechanical handling.

Opacity in resins is due to internal discontinuities which refract and reflect light. Opacity may be obtained without any appreciable degree of macroporosity in the bead. True macroporosity is obtained by adding a third component, or porogen, to the styrene-DVB mixture in the initial polymerization. Porogen is chosen to be soluble in the monomer mixture but insoluble in the polymer. As the polymer forms it precipitates internally, leaving the porogen as a separate phase. If enough DVB is included in the polymerization mixture, the polymer cannot collapse when the porogen is removed prior to converting the bead to an ion-exchange resin. Removal of porogen leaves discontinuities in the resin structure which may serve as diffusional channels and reduce the effective gel-diffusion path.

The degree of macroporosity obtained is dependent on the type and amount of porogen added, and the fraction of DVB in the monomer mix. If the porogen is a relatively small compound, 12 to 16% DVB must be used in the monomer mix to keep the polymer from collapsing when the porogen is removed. High crosslinking presents certain problems when the copolymer is converted to an ion-exchange resin, particularly in the case of strong-acid and strong-base resins. Sulfonation and chloromethylation both become more difficult at high DVB levels. As the result of the lower degree of functional substitution which results, and the lower density of the macroporous polymers, the strong-acid and strong-base resins obtained have lower total capacities than their gel counterparts.

A high DVB content also reduces the amount of water in the actual gel structure and increases the selectivity coefficients for all common metal cations over hydrogen, for calcium and magnesium over sodium, and for strong-acid anions over hydroxide. Water reduction tends to reduce the separation factor for regeneration steps in such important processes as softening, decationization, and acid neutralization and further decreases the efficiency of these already inefficient steps. The combination of a lower total capacity and a less efficient regeneration, plus their higher initial cost, makes the use of macroporous resins unattractive except in those cases where macroporosity per se gives a process advantage. The possible advantage of faster diffusion rates of macroporous resins appears questionable in the case of the strong-acid and strong-base resins. The limiting factor in almost all cyclic processes with fully ionized resin is the unfavorable equilibrium in either the forward or reverse step. Exchange rates are seldom controlling.

True macroporosity can have real value when the process is controlled by kinetic factors. Exchange rates tend to become very slow in nonaqueous or partially nonaqueous systems. The macroporous structure has proved to be much more effective in such cases. Advantages have been claimed for macroporous structures in certain cases where large ionic components are involved. One such case is removal of large colored ions from sugar and similar process streams. Gel resins are also used, and both types have strong advocates.

Weakly ionized resins are also available in macroporous forms. The relative advantages and disadvantages are much harder to evaluate. Type S-m resins are commonly macroporous. Type P resins are naturally macroporous. A variety of diffusional processes occur in weakly ionized resins, some of which are dependent on electrolyte invasion. The concept of a macroporous structure with electrolyte invasion uninhibited by Donnan forces is attractive. Kinetic studies on the rate of pickup of acid on a macroporous weak-base resin show that the rate is inversely proportional to the square of the radius as with gel resins. It is difficult to judge the relative contributions to reaction rates of electrolyte invasion and functional group ionization in complex systems.

Opacity and Osmotic-Shock Stability

Opacity, or rather the heterogeneity in the polymer which causes opacity, does give resin an increased resistance to fracturing. Ion-exchange resins may undergo large volume changes due to change in ionic form, change in solution concentration, or change from aqueous to nonaqueous conditions. Volume changes can result in severe physical breakage, especially during repeated cycling. The effect is much less with opaque beads. Opacity can be obtained in a resin with only a very nominal degree of heterogeneity in the polymer. Some of the so-called macroporous resins have little true macroporosity but have a high resistance to osmotic-shock breakage. The use of an opaque resin is recommended where the cycle can be expected to give considerable osmotic shock.

Highly macroporous resins show a unique type of attrition. The individual macroporous beads are made up of very tiny spherical particles held together by a relatively small amount of interconnecting polymer. As a result, they are quite brittle. If subjected to abrasive action, their surface tends to powder. While the beads do not break, they may gradually become smaller in use.

Useful Life

The question of useful resin life in an ion-exchange process is an important consideration as it has a direct bearing on the cost of the process. Unfortunately, it is a very difficult factor to predict except on the basis of past experience in similar processes. Short-term cyclic testing with solutions similar to those of the proposed process will point out any serious instabilities. Actual process solutions should be used if available. Occasionally a minor or overlooked component in the stream will give serious problems. This is also a good method of testing relative stability if several different resins are under consideration. Unfortunately, it is impossible with current knowledge to extrapolate such tests into useful resin life.

DESIGN OF ION-EXCHANGE PROCESSES

Initial Considerations

Definition of Problem In considering the application of ion exchange, the overall chemical process should be reviewed to find the most advantageous spot to obtain the desired result. This may be purification of the raw material going into the process, polishing of the final product solution, or treatment at one of the intermediate stages. Often the most favorable spot for ion-exchange treatment will be where the ionic component to be removed is present at its lowest concentration or where dilution of the process stream will be least objectionable.

Composition of the stream to be processed should be determined as completely as possible. Analysis should include all major components, both ionic and nonionic. Ionic components should be identified and concentrations determined in equivalents per liter. Cation and anion concentration should balance on an equivalent basis.

Any aggressive conditions such as oxidants or reductants should be checked. Temperature, pH, density, and clarity should be noted. Volume of the stream to be treated per hour or day should be specified.

In addition, the required change to be effected must be specified. Change may be in terms of purity obtained, percent of a component recovered, maximum waste volumes, etc. Other restrictions, such as source of the regenerant chemicals, capital limitations, and space requirements, should be stipulated.

Basis of Separation Once the problem has been defined, the chemical differences between the components of the system are examined to determine if there is a basis for a practical ion-exchange separation. In many cases the basis for separation is straightforward. In others much more complex relationships must be considered. The major bases for separation have been discussed in previous sections but are summarized below.

It is obvious that separation between ionized and nonionized solutes should be easy, as one will be subject to exchange and the other will not. Likewise, cationic components in solution are subject to exchange without affecting the composition of the anionic components, and vice versa. However, even in these cases the efficiency of the separation process is dependent on the nature of the ion-exchange reaction.

The largest separation factors occur when one of the products of the reaction has a low degree of ionization. These are most commonly acid-base reactions of some sort.

$$\overline{R_4N^+OH^-} + H^+Cl^- \rightarrow \overline{R_4N^+Cl^-} + H_2O$$

$$\overline{NH_4^+RSO_3^-} + Na^+OH^- \rightarrow \overline{Na^+RSO_3^-} + NH_3 + H_2O$$

$$\overline{H^+RSO_3^-} + Na^+CN^- \rightarrow \overline{Na^+RSO_3^-} + HCN$$

$$\overline{HOOCR} + Na^+OH^- \rightarrow \overline{Na^{+-}OOCR^-} + H_2O$$

$$\overline{R_3NH^+Cl^-} + Na^+OH^- \rightarrow \overline{R_3N} + NaCl + H_2O$$

All the above reactions go far to the right. The exchange rate, and therefore the column performance, will vary widely depending on the degree of ionization of the initial resin form.

Transformation of an ionic component to a nonionized component by a complexing agent in the outside phase will also drive the reaction far to the right.

$$\overline{Cu^{2+}(RSO_3^-)} + Na^+(EDTA^{4-})_2 \rightarrow \overline{2Na^+RSO_3^-} + Na_2^+(CuEDTA^{2-})$$

When all reactants and products are highly ionized, the separation factor in exchanges between monovalent ions is the selectivity coefficient of the resin for the particular ions involved. With simple monovalent ions most selectivity coefficients fall in the range of 0.1 to 10. As long as the selectivity coefficient for the forward reaction is greater than unity, the resin can be loaded efficiently with the preferred ion to almost its equilibrium capacity. However, if the preferred ion represents only a small fraction of the total ion content of the feed stream, the equilibrium capacity of the resin for this ion may be low. Equilibrium capacity can be estimated by a simple calculation which is an important step in determining process limits.

If the separation factor in an exchange between monovalent ions is greater than unity, the separation factor for the reverse reaction will be less than unity. The lower the reverse separation factor, the more chemically inefficient the reverse reaction will be. The more efficient monovalent-monovalent cyclic processes will be those in which the separation factor for the forward reaction is slightly greater than 1.0. This is true for the exchange between a strong-acid resin in the hydrogen form and any of the alkali metals except lithium.

In exchanges between ions of different valences the separation factor is dependent on the selectivity coefficient and the total ionic concentrations of the inside and outside phases. The exchange becomes more favorable for the ion of higher valence as the solution concentration becomes less. The exchange becomes less favorable for the ion of higher valence as the solution concentration increases. This offers the possibility of adjusting the separation factors in such exchanges for the forward and reverse reactions to optimize column behavior. The extent of this effect can be estimated by a simple calculation.

Any change which alters the nature of the inside or outside phase may shift the ionic equilibrium. Thus the selectivity coefficient between two ions may differ with resins of different water content or degree of functional substitution. Such differences are usually small. Changing the actual structure of the functional group can give marked changes in the selectivity coefficient. Such is the case with the hydroxide and chloride ions on Type I and Type II strong-base resins.

The presence of soluble complexing agents also affects ionic distribution. If the complex formed in solution is relatively nonionic, it will usually shift the distribution of the complexed ion toward the solution phase. If two ions are present which react with the complexing agent, the ion which forms the tightest complex will be preferred in the outside phase. This is an extremely powerful tool in the separation of certain groups of similar cations. The rare earths are purified on a commercial scale by a combination of ion exchange and chelate formation with ethylenediaminetetraacetic acid.[27] A similar method has been applied to the separation of isotopes and transuranium elements.

Some of the anionic complexes formed by various metals show very high affinities on anion-exchange resins. The composition and concentration of anionic complexes may be relatively insensitive to solution conditions over a fairly wide range of conditions. Examples are the cyanide complexes of many transition metals. Others may be almost transitory in nature, being very dependent on the concentration of complexing agent, pH, metal concentration, and temperature. The chloride complexes have been studied extensively on strong-base resins.[22] The observed behavior shows no consistent pattern from metal to metal. Iron is very strongly absorbed from strong hydrochloric acid as a complex anion on a strong-base resin. This behavior is used to purify commercial hydrochloric acid. Uranium forms complex anions in sulfate and carbonate solutions. The concentration of the complexes on a strong-base resin is a major step in many uranium-mining operations. Some nonionic compounds form metastable ionic complexes which can be used in ion-exchange separation techniques. Examples are aldehyde-sulfite complexes, sugar-borate complexes, and sugar-calcium salts. The latter are used on a commercial scale for the chromatographic separation of hexoses. Complex ions show mass-action ion-exchange behavior similar to that of simple ions. However, the concentration of the various complexes are so interrelated with solution conditions that identification and measurement of ionic species is difficult.

When conditions in the outside phase differ greatly from usual, resin selectivity may be altered. Very high ionic concentrations or replacement of water with an organic solvent are examples. In either case the resin may be largely dehydrated. Exchange rates are severely depressed in nonpolar systems. Resin selectivities may be enhanced or decreased.

Estimation of Process Limits

When a proposed process involves relatively common ions and an ion-exchange reaction can be written, certain limits can be estimated from equilibrium calculations.[12] The data needed are the compositions of the influent and/or regenerant solutions and some knowledge of the selectivity coefficient in the system. Only an estimate of the order of magnitude of the selectivity coefficient is sufficient in many cases to give useful answers.

If the calculated limits fail to meet the criteria set for a proposed process, the process can be dismissed, as the performance obtained will be poorer than the limits calculated.

The useful information that can be obtained includes approximate separation factors for the forward and reverse reactions and three types of process limits:

Maximum capacity

Maximum degree of regeneration

Initial leakage

The equivalent-fraction forms of the mass-action expression are the most convenient for calculation of separation factors and limits. Equation (5) is used for any exchange between ions of identical charge. Equation (8) is used for exchanges between monovalent and divalent ions.

In Equation (5) the separation factor is identical with the selectivity coefficient and must be obtained from the literature or by direct measurement. The separation factor is essentially independent of solution concentration. In Eq. (8) the separation factor is dependent on the term $K_{DA}\bar{C}/C$, which is inversely proportional to the solution concentration.

The term $\bar{x}_A$ defines the limit to which the resin can be converted to the A form when in contact with a solution of fractional composition x_A. Alternately, $\bar{x}_A$ is the maximum degree of regeneration that can be achieved with a regenerant of composition x_A. Or, if the resin at the bottom of the bed has the composition $\bar{x}_A$, the composition of the initial effluent can be calculated.

Limiting Capacity   The equilibrium capacity, or limiting operating capacity, is the maximum number of equivalents of a given ion that can be removed from solution per given volume of resin. If the influent contains only one species of counterion, the resin can be completely converted to that form if enough solution is passed through. In this case the limiting capacity of the resin is equal to total capacity. If the separation factor for the incoming ion is greater than unity, the operating capacity will approach limiting capacity.

If the influent contains two species of counterions, the limiting capacity for either will usually be less than the total capacity of the resin. If solution is passed through the resin until composition of the effluent is the same as composition of the influent, the resin will be in equilibrium with the influent. The approximate equilibrium composition of the resin can be calculated from the solution composition if some estimation of the selectivity coefficient is available. The limiting capacity for the more tightly held counterion will be equal to the fraction of the resin sites occupied at equilibrium, times the total capacity of the resin.

Once the limiting capacity has been found, the maximum number of bed volumes of solution that can be treated per volume of resin per cycle can be estimated. This is equal to the limiting capacity of the resin divided by the influent concentration of the ion being removed by the resin.

Limiting capacity is seldom realized in practice. Influent ions will appear in the effluent before equilibrium is reached and the run normally will be stopped. Further, if the reaction equilibrium is unfavorable, it may be impractical to approach the limiting value because of the inherent inefficiency of a nonsharpening front. Since use of the equilibrium expression requires estimation of a separation factor for the exchange, this will indicate the nature of the exchange front which will occur.

EXAMPLE 1

It is necessary to remove the ammonium ion from a recycled process stream of the following composition:

$$\begin{array}{ll} NH_4NO_3 & 0.05 \text{ eq/L} \\ NaNO_3 & 0.45 \text{ eq/L} \end{array}$$

What is the limiting capacity of a strong-acid resin for NH_4^+ from this stream? Assume the resin to have a total capacity of 2.0 eq/L.

$$K_{NH_4^+Na^+} = 2.55/1.98 = 1.3 \quad \text{(Table 2)}$$

$$x_{NH_4^+} = 0.05/0.50 = 0.10$$

$$\frac{\bar{x}_{NH_4^+}}{1 - \bar{x}_{NH_4^+}} = 1.3 \times 0.10/0.90 = 0.14$$

$$\bar{x}_{NH_4} = 0.11$$

The limiting capacity for NH_4^+ will be 0.11×2.0 eq/L = 0.22 eq/L. The maximum bed volumes of solution that can be treated per cycle are

$$\frac{0.22 \text{ eq/L}}{0.05 \text{ eq/L}} = 4.8$$

This is a relatively small volume of solution to treat per cycle. However, the separation factor is greater than unity and the exchange front will be sharp. Efficient regeneration can be obtained with a solution of sodium hydroxide converting NH_4^+ on the resin to weakly ionized ammonium hydroxide.

EXAMPLE 2

In order to avoid calcium sulfate precipitation in subsequent processing, it is necessary to remove either Ca^{2+} or SO_4^{2-} from a water with the following composition:

Ca^{2+}	1.5 meq/L	SO_4^{2-}	1.0 meq/L
Na^+	9.5 meq/L	Cl^-	10.0 meq/L
TDS	11.0 meq/L		9.0 meq/L

This can be done either by sodium softening on a strong-acid resin or by chloride softening on a strong-base resin. In either case NaCl would be used as the regenerant. And in either case the separation factor for the divalent ion on the resin is favorable. A direct comparison of the two systems can be made as shown below:

	Na^+ softening on strong-acid resin	Cl^- softening on strong-base resin
Resin capacity $\bar{C}$	2.0 eq/L	1.2 eq/L
Solution concentration, C	0.001 eq/L	0.011 eq/L
Equivalent fraction divalent ion	$x_{Ca^{2+}} = \dfrac{1.5}{11.0} = 0.14$	$x_{SO_4^{2-}}$
Selectivity coefficient	$K_{Ca^{2+} Na^+}$	$K_{SO_4^{2-}}$
$K_{DB}\bar{C}/C$	$2.6 \dfrac{2.0}{0.011} = 470$	$0.15 \dfrac{1.2}{0.011} = 16$
$\dfrac{\bar{x}_D}{(1 - \bar{x}_D)^2} =$	$470 \dfrac{0.14}{(0.86)^2} = 89$	$16 \dfrac{0.09}{(0.91)^2} = 1.74$
Fraction of resin in divalent form:	$\bar{x}_{Ca} = 0.90$	$\bar{x}_{SO} = 0.48$
Limiting capacity	0.90×2.0 eq/L = 1.8 eq/L	0.48×1.2 eq/L = 0.57 eq/L
Volumes treated per cycle	$\dfrac{1.8 \text{ eq/L}}{0.0015 \text{ eq/L}} = 1200$	$\dfrac{0.57 \text{ eq/L}}{0.001 \text{ eq/L}} = 570$

On the basis of volumes of influent treated per volume of resin, sodium softening would appear to have a clear advantage over chloride softening. However, in both cases the cycle length is reasonable, and the two processes should also be compared for regeneration effectiveness. Concentrated sodium chloride will be used as the regenerant. If the CaCl is used as a 3 N solution (approximately 15%) the nature of separation factors for the regeneration can be estimated.

Equation (8) for the divalent-monovalent exchange is written with the divalent ion as the entering ion. Regeneration is the reverse of this reaction; so the separation factor for regeneration will depend on the reciprocal of the quantity $K_{DA}\bar{C}/C$. The use of 3 N NaCl with the cation resin gives a value of $C/K_{DA}\bar{C}$ of $3/2.6 \times 2 = 0.6$. For the anion resin the value is $3/0.15 \times 1.2 = 17$.

On the basis of these numbers, regeneration of the cation resin will require an excess of salt. It will be difficult to get the bottom of the bed highly regenerated, which may result in calcium leakage. Regeneration of the anion resin to the chloride form will be efficient and the bed can be regenerated cleanly. In addition, there are fewer equivalents of sulfate to be removed per liter of feed than of calcium. Thus, although the anion resin system will have a shorter exhaustion time, the amount of salt required to treat a given volume of influent, and the regenerant wastes produced, will both be less than with the cation resin system. The initial costs for the anion system will be higher because a greater volume of more expensive resin will be required.

When calculating equilibrium resin compositions from the solution composition, the original state of the resin is immaterial. Any ions originally on the resin will have been displaced by ions from the solution by the time equilibrium is reached by definition. However, if the resin was not completely regenerated at the start of the exhaustion, the maximum operating capacity that can be realized will be the equilibrium capacity minus the unregenerated capacity on the resin.

The form of the mass-action expression used here is limited to systems with a single pair of competing ions. However, it is often possible to make meaningful estimates in multiple-ion systems by grouping similar ions. In dilute solutions, useful estimates on monovalent vs. divalent ion loadings can be made by considering all monovalent ions as one species and all divalent ions as the second species. The separation factor in such systems is often more dependent on solution concentration C than on the selectivity coefficient between the ions themselves.

As long as it is possible to estimate the separation factors between the various ionic components, the order in which the ions will occur in the effluent can be predicted. This may allow the estimation of the limiting capacity for an ion from a multicomponent system.

EXAMPLE 3A

A well water of the following composition is passed through a strong-base resin with a capacity of 1.2 eq/L until the effluent composition equals the influent composition. What is the approximate ionic composition of the resin?

Anion	Concentration, meq/L	Equivalent fraction x_i
HCO_3^-	4.9	0.66
SO_4^{2-}	1.1	0.15
NO_3^-	1.4	0.19
TDS	7.4	1.00

The fraction of the bed occupied by $SO_4^=$ at equilibrium can be estimated from Eq. (8) if the system is considered as having only two ionic components. For this purpose the NO_3^- and HCO_3^- are combined and considered as a single monovalent component B, requiring the assignment of a suitable selectivity coefficient for sulfate vs. component B. Since an approximate value is needed, the value can be taken as intermediate between the values derived from Table 3.

$$K_{SO_4^2 NO_3^-} \qquad\qquad 0.04$$
$$K_{SO_4^2 B} \qquad \text{estimated} \qquad 0.2$$
$$K_{SO_4^2 HCO_3^-} \qquad\qquad 0.4$$

The appropriate values are used in Eq. (8) to calculate the fraction of resin in the sulfate form at equilibrium, $\bar{x}_{SO_4^2}$.

$$\frac{\bar{x}_{SO_4}}{(1 - \chi_{SO_4^2})^2} = 0.2 \frac{1.2}{0.0074} \frac{0.15}{(0.85)^2} = 6.7$$

$$\chi_{SO_4^2} = 0.68$$

Approximately 68% of the resin sites will be in the divalent form at equilibrium. The remaining 32% of resin capacity will be shared by NO_3^- and HCO_3^-. Relative amounts of NO_3^- and HCO_3^- on the resin can be estimated by making the simplifying assumption that this 32% of resin sites are in equilibrium with a solution of NO_3^- and HCO_3^- of the same relative concentrations as in the feed. The equivalent fraction of NO_3^- in this solution will be 1.4/(1.4 + 4.9) = 0.22. The equivalent fraction of NO_3^- in the 32% of the monovalent-form resin, $\bar{x}_{NO_3^-}$ can be estimated from Eq. (5).

$$\frac{\bar{x}_{NO_3^-}}{1 - \bar{x}_{NO_3^-}} = 10 \times 0.22/0.78 = 2.8$$

$$\bar{x}_{NO_3^-} = 0.74$$

The fraction of the total resin capacity in the nitrate form will be 0.74 × 0.32 = 0.24, leaving 0.32 − 0.24 = 0.08, as the fraction of the resin in the bicarbonate form. The equilibrium composition is shown in Fig. 10a. Equilibrium capacity of the resin for sulfate will be 0.68 × 1.2 eq/L = 0.82 eq/L. Equilibrium capacity for nitrate will be 0.24 × 1.2 eq/L = 0.29 eq/L.

EXAMPLE 3B

What is the approximate operating capacity of the resin for nitrate from this water?

The relative positions of the three ions as they proceed down the column will be determined by the separation factors between the pairs. The separation factor for sulfate over nitrate will be related to $K_{SO_4^2 NO_3^-} C/C$, which has a value of 0.04 × 1.2/0.0074 = 6.5. $\alpha_{SO_4^2 NO_3^-}$ will be greater than unity and sulfate will tend to push the nitrate down the column. And $\alpha_{NO_3^- HCO_3^-} = K_{NO_3^- HCO_3^-} = 10$; so the nitrate will push the bicarbonate down the column. The sulfate will tend to concentrate in the top of the column. The order of appearance of the ions in the effluent will be HCO_3^-, NO_3^-, and finally SO_4^{2-}. (A similar three-component loading curve is shown in Fig. 13). The useful capacity of the bed for nitrate is greater than the equilibrium capacity of Example 3A. The maximum capacity occurs when the nitrate-bicarbonate exchange front reaches the bottom of the bed. At this point the upper portion of the bed contains the equilibrium mixture of three anions, while the lower portion of the bed contains an equilibrium mixture of nitrate and bicarbonate as represented in Fig. 10b. If the run is continued beyond this point, nitrate is progressively displaced from the resin by incoming sulfate.

Capacity for the intermediate ion, the nitrate ion in the above example, can be approximated by calculating the fraction of total column which is in equilibrium with feed solution at the time the intermediate ion breaks through. A pair of simultaneous equations is set up relating equivalents of ions

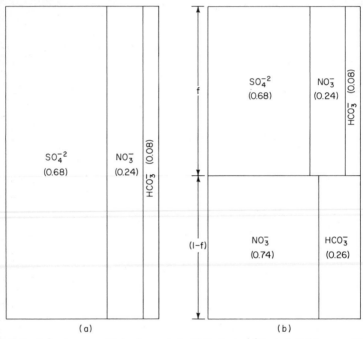

Fig. 10 Estimated resin compositions after contact with feedwater of Example 3: (a) composition after equilibration; (b) composition at nitrate-ion breakthrough.

passed through the resin and ions retained by the resin. Let f equal the fraction of the resin bed that is in equilibrium with the feed when the NO_3^- breakthrough occurs (Fig. 10b), and let v equal the liters of feed that have passed through each liter of resin at this point. All of the SO_4^{2-} introduced up to this point will be contained in the upper, or f, fraction of the bed.

The nitrate will be distributed between the upper portion of the bed which is in equilibrium with the feed, and the lower portion which contains only NO_3^- and HCO_3^-. The equivalent fraction of NO_3^- in this lower portion, $\bar{x}'_{NO_3^-}$, is assumed to be 0.74 as in the mixed monovalent resin in the upper portion.

The material balance for SO_4^{2-} will be:

$$\text{Equivalents on resin} = \text{equivalents fed}$$

$$\bar{C} \cdot \bar{x}_{SO_4^{2-}} \cdot f = C_{SO_4^{2-}} \cdot v$$

or 1.2 eq/L $\times$ 0.68f = 0.0011 eq/L $\cdot$ v

The material balance for NO_3^- will be:

$$\text{Equivalents in } f + \text{equivalents in } (1 - f) = \text{equivalents fed}$$

$$(\bar{C} \cdot \bar{x}_{NO_3^-} \cdot f) + (\bar{C} \cdot \bar{x}_{NO_3^-}) \cdot (1 - f) = C_{NO_3^-} \cdot v$$

or

$$(1.2 \text{ eq/L} \times 0.24f) + 1.2 \text{ eq/L} \times 0.74 (1 - f) = 0.0014 \text{ eq/L} \cdot v$$

Solving the simultaneous equations yields a value of 0.54 for f, and 400 L of solution per liter of resin for v. Approximately 54% of the column will be in equilibrium with the feed at the time of NO_3^- breakthrough. The capacity of the resin for NO_3^- to this point will be 400 L/L $\times$ 0.0014 eq/L = 0.56 eq/

L. If the column is run beyond this point, NO_3^- will be displaced by SO_4^{2-} until it is reduced to the 0.29 eq/L estimated in Example 3A.

Limiting Degree of Regeneration   The limiting degree of regeneration is the maximum degree to which the resin can be converted to the desired ionic form with unlimited quantities of regenerant solution of a given composition. If the regenerant is a pure solution of the desired ion, the resin theoretically can be regenerated completely to this form. If the regenerant contains another counterion, it will not be possible to convert the resin cleanly to the desired form no matter how much regenerant is used. The maximum degree of regeneration possible will be the equivalent fraction in the desired ionic form when the resin is in equilibrium with the regenerant solution.

EXAMPLE 4

A plant wishes to use a by-product nitric acid to regenerate strong-acid resin. What is the maximum degree of regeneration that can be obtained if the acid has the following analysis?

$$HNO_3 \qquad 6.0\%$$
$$NaNO_3 \qquad 1.1\%$$

$$\text{sp. gr. } 1.03$$

$$\text{Normality of NaNO}_3 = \frac{11g}{1000g} \times \frac{1030g}{1} \times \frac{1\ eq}{85g} = 0.13\ eq/L$$

$$\text{Normality of HNO}_3 = \frac{60g}{1000g} \times \frac{1030g}{1} \times \frac{1\ eq}{63g} = 0.98\ eq/L$$

$$x_{Na^+} = \frac{0.13}{0.13 + 0.98} = 0.12$$

$$\frac{\bar{x}_{Na^+}}{1 - \bar{x}_{Na^+}} = K_{Na^+H^+} \frac{Na^+}{1 - x_{Na^+}} = 1.5 \frac{0.12}{1 - 0.12} = 0.20$$

$$\bar{x}_{Na^+} = 0.17$$

Seventeen percent of the resin sites will still be in the sodium form even after exhaustive regeneration with the acid.

Initial Leakage   Leakage is the appearance of the influent ion in the column effluent. In the initial part of exhaustion, such leakage is not due to influent ions coming through the column without being exchanged. Rather, it is due to residual ions on the resin at the bottom of the column as a result of incomplete regeneration. Residual ions are displaced by ionic species coming down the column and appear in the effluent. As a first approximation, the concentration of leakage ions will be that generated by the equilibrium between the solution about to leave the column and the last layer of resin in the column. The extent of leakage is therefore dependent on the ionic form of the resin at the bottom of the column and the composition of the solution moving down the column. If the composition $(\bar{x})$ of the bottom of the bed is known, the composition (x) of the initial effluent can be calculated.

EXAMPLE 5

What will be the leakage of sodium ion from the regenerated column in Example 4 when treating the following water?

$$\begin{array}{ll} Ca^{2+}\ 1.6\ meq/L & Cl^-\quad 2.4\ meq/L \\ Mg^{2+}\ 0.8\ meq/L & SO_4^{2-}\ 1.6\ meq/L \\ Na^+\ 2.0\ meq/L & HCO_3^-\ 0.4\ meq/L \end{array}$$

The water leaving the column on the next loading will be in equilibrium with resin of composition $\bar{x}_{Na^+}$ = 0.17. The equivalent fraction of sodium in the effluent can be calculated.

$$\frac{x_{Na^+}}{1 - x_{Na^+}} = \frac{1}{K_H^{Na^+}} \frac{\bar{x}_{Na^+}}{1 - \bar{x}_{Na^+}} = \frac{1}{1.5} \times \frac{0.17}{1 - 0.17} = 0.136$$

$$x_{Na^+} = 0.12$$

The bicarbonates will be converted to carbonic acid and will not contribute to the ionic concentration of the effluent. Therefore, the effective total cation concentration in the effluent water will be that associated with the entering chlorides and sulfates, or 2.4 meq/L + 1.6 meq/L = 4.0 meq/L. Since the bottom of the bed was equilibrated with the regenerant acid, it will contain only hydrogen and sodium cations. The water leaving the column will be in equilibrium with this resin and so will contain only hydrogen and sodium cations. The sodium content of water will be 0.12 × 4.0 meq/L = 0.48 meq/L. This is equal to 11.0 ppm of the sodium.

In this case the resin has a uniform composition throughout the bed after regeneration. The leakage of sodium ion will be constant throughout the run up to breakthrough. In the more usual case where the bottom of the bed is less regenerated than the upper portions, the initial leakage will reflect the composition of the resin at the bottom of the column and then decrease up to breakthrough.

EXAMPLE 6

In many locations sea water is an inexpensive and readily available regenerant for softening operations. A large excess can be used; so the resin can be essentially equilibrated with the regenerant. The question is whether the resin will be regenerated enough to give the desired low hardness leakage on the next exhaustion cycle. Hardness leakage can be predicted from the composition of sea water and the water to be treated.

$$TH^{2+} \ 0.11 \ eq/L$$
$$Na^+ \ \ 0.49 \ eq/L \qquad or \qquad x_{TH^{2+}} = \frac{0.11}{0.60} = 0.18$$
$$\underline{TDS \ 0.60 \ eq/L}$$

In this case hardness ions are considered as a single divalent component TH^{2+}. There is no need to estimate a value for the selectivity coefficient for total hardness vs. sodium, as this value does not affect the level of leakage obtained. Since total resin capacity $\bar{C}$ and selectivity coefficient $K_{TH^{2+}Na^+}$ are the same for both regeneration and exhaustion steps, they can be carried through the calculation as constants. After exhaustive regeneration of the resin with sea water, the composition will be given by

$$\frac{\bar{x}_{TH^{2+}}}{(1 - \bar{x}_{TH^{2+}})^2} = K_{TH^{2+}Na^+} \frac{\bar{C}}{0.6} \frac{0.18}{(0.82)^2} = 0.45(K_{TH^{2+}Na^+})\bar{C}$$

The composition of water leaving the column on the next exhaustion will be in equilibrium with this resin composition and will have a composition given by

$$\frac{x_{TH^{2+}}}{(1 - x_{TH^{2+}})^2} = \frac{1}{K_{Na+}^{TH^{2+}}} \frac{C}{\bar{C}} \times 0.45 \ K_{Na+}^{TH^{2+}} \bar{C} = 0.45C$$

where C is the total ionic concentration of the water being softened.

When softening a water with a total solids content of 2 meq/L (100 ppm as $CaCO_3$), the hardness leakage will be

$$\frac{x_{TH^{2+}}}{(1 - x_{TH^{2+}})^2} = 0.45C = 0.45 \times 2 \times 10^{-3} = 9.0 \times 10^{-4}$$
$$x_{TH^{2+}} = 8.9 \times 10^{-4}$$
$$TH^{2+} = 8.9 \times 10^{-4} \times 2 \times 10^{-3} = 1.78 \times 10^{-6} eq/L$$
or
$$3.6 \times 10^{-2} ppm \ as \ Ca$$

For water containing 50 meq/L of total solids (2500 ppm) the hardness leakage will be

$$\frac{x_{TH^{2+}}}{(1 - x_{TH^{2+}})^2} = 0.45C = 0.45 \times 50 \times 10^{-3} = 0.023$$
$$x_{TH^{2+}} = 0.022$$
$$TH^{2+} = 0.022 \times 50 \times 10^{-3} eq/L = 1.1 \times 10^{-3} eq/L$$
or
$$22 \ ppm \ as \ Ca^{2+}$$

In the softening cycle, hardness leakage is dependent on the degree of regeneration of the bottom of the bed and the total ionic concentration of the water being treated. Low hardness leakage can be obtained even with a poor regeneration of the resin when treating low-solids water. Low leakage level can be realized with a high-solids water only if the bottom of the resin column is highly regenerated to the sodium form. Although illustrated by the softening cycle, this behavior holds true for monovalent-divalent exchanges in general.

Summary   If a solution is passed through a resin until the composition of the effluent is the same as the composition of the influent, no further net ion exchange occurs. The resin is in equilibrium with the influent solution. The equilibrium resin composition can be calculated from the solution composition and the selectivity coefficient for the ion-exchange reaction. Equilibrium resin composition can be interpreted directly as the limiting operating capacity on exhaustion, or the maximum degree of regeneration possible with a regenerant of known composition. Alternately, if the degree of regeneration is known, the initial leakage on the next exhaustion can be calculated.

Data required to make equilibrium-based estimates are the ionic compositions of the influent and regenerant solutions. Based on this information and some knowledge of the equilibria involved, a limiting operating capacity can be calculated. If this capacity is very low and only a small volume of influent can be treated per cycle, the process will not be very attractive, at least in conventional fixed-bed equipment. The same type of calculation will give the maximum degree of regeneration that can be obtained with a regenerant of given purity. Once the composition of the regenerated resin is known, the leakage on the next exhaustion can be calculated.

The equivalent-fraction form of the mass-action equation for monovalent-divalent exchange shows the powerful effect of the solution concentration on the equilibria in these exchanges. The effect of solution concentration often overshadows the effect of the actual selectivity coefficient. However, extrapolation of these calculations to highly concentrated solutions is questionable. True mass-action equations are properly based on ion activities and not concentrations. The two may differ widely as concentration increases.

Limit calculations are most applicable when working with the strong-base and strong-acid ion-exchange resins which are fully ionized over the entire pH range. Such calculations can be extended to weak-acid and weak-base resins in the case of an exchange between an ion in solution with the salt form of the resin at a constant pH. However, there are few meaningful selectivity-coefficient data for these systems.

It must be emphasized that the numbers obtained are limits and not predicted capacities. Ion-exchange columns are seldom run to "composition-out-equals-composition-in." And even though the equilibrium limit appears practical, it may be impractical to approach if the exchange front is nonsharpening.

Column Profiles and Leakage Patterns

Most cyclic processes with strongly ionized resins have one step in which the separation factor is less than unity. This is usually the regeneration step. The chemical efficiency, or regenerant utilization, is relatively poor because of the nonsharpening front, and becomes poorer as regeneration is carried to completion. For economic reasons, the resin is left partially in the unregenerated form in such processes.

In standard fixed-bed operations both regenerant solution and the stream being processed are passed down the column. During loading, the purity of the effluent is dependent on the composition of the resin at the bottom of the column. When regeneration is not carried to completion, the resin at the bottom of the column may still be appreciably in the unregenerated form, and leakage of unwanted ion results in the first part of the cycle. Initial leakage can be calculated from feed composition if the composition of the resin at the bottom of the column is known, using Eq. (5) or (8). However, as the run proceeds, the bottom layers of the column are actually regenerated by ions moving down from the fully regenerated portions of the column.

Figure 11a shows the approximate composition of a resin bed which has been regenerated to 40% using downflow regeneration, and a typical calcium-leakage curve from such

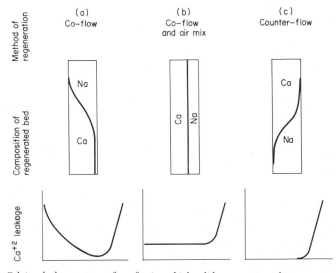

Fig. 11   Calcium leakage patterns for softening a high-solids water using a low regeneration level.

a bed when treating a high-solids water. The bottom of the resin bed is largely in the calcium form at the start of the exhaustion. The initial calcium leakage is high and then gradually decreases to a minimum value just before breakthrough. The practice, particularly in water treatment, is to increase the regenerant dosage until an acceptably low leakage is obtained. This results in inefficient regenerant utilization.

One method of reducing high leakage at the start of the run is to mix the resin after it has been regenerated and rinsed. The resin then has a uniform degree of regeneration throughout the length of the column. Leakage in the succeeding run will have a low constant value which may be estimated from the feed composition and the average degree of regeneration of the resin. The approximate resin composition and typical calcium-leakage curve for this method of operation of a sodium softener is shown in Fig. 11b. This technique may allow the use of a lower regeneration level without exceeding the leakage limit. The total leakage over the run will not be greatly different from that with the unmixed bed.

The highest chemical efficiencies and purest effluents can be obtained by passing the regenerant stream and process stream through the column in opposite directions.[8] This technique is called counterflow, or fixed-bed countercurrent. If the bed is regenerated upflow, the bottom of the bed is now the most highly regenerated portion and the

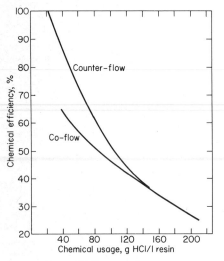

Fig. 12 Chemical efficiency of regeneration in a sodium-hydrogen exchange. Counterflow operation vs. co-flow operation.

unregenerated resin is at the top of the column. When the process stream is passed down through the bed, the last resin it contacts is highly regenerated and there is little or no leakage of unwanted ion. Figure 11c shows the approximate resin composition and a typical calcium-leakage curve for a softening cycle with counterflow regeneration.

For the counterflow system to be effective the resin must be prevented from fluidizing during the upflow step of the cycle. If fluidization is allowed, the resin mixing which occurs will tend to cancel some of the advantages of counterflow operation. For maximum purity, the bottom layer of the bed should never be exhausted, and the rinse water used following regeneration must be free of the unwanted ion.

Besides giving higher purities, counterflow operation gives high chemical efficiencies at low regenerant levels. Figure 12 shows the chemical efficiency, or regenerant utilization, realized for sodium-hydrogen exchange using co-flow and counterflow regeneration in a fixed bed. As regeneration level increases, chemical efficiencies of counterflow and co-flow operations become equal. However, counterflow operation maintains an advantage in product purity in the practical range of regenerant dosage.

Laboratory Verification

In many cases, consideration of the factors discussed in the previous sections will indicate when a proposed exchange process is within practical limits. If the proposed process closely parallels existing available commercial technology, laboratory data may not be required before design of a commercial unit. However, in chemical process applications it is advisable to evaluate the exchange in laboratory columns.

Laboratory Columns Excellent process data can be taken at laboratory scale. Data taken from 2.5 cm (1 in) ID columns can be scaled up directly if the depth of resin in the column is the same as used in the full-scale unit. If it is impractical to use this depth of bed in the laboratory, a bed height of about 1 m is adequate for most purposes.

Glass pipe is better for laboratory columns than ordinary glass tubing, since tubing may burst if the resin undergoes a large volume change in the process. Polyacrylate tubing makes excellent columns as long as no organic solvents are present in the system. The section of pipe should be at least 1.5 times as long as the height of resin used.

Underdrains can be prepared by using a rubber stopper with a short piece of glass tubing, covered by a layer of pebbles and then sand; or a wad of glass wool. The glass tubing should not extend above the top surface of the stopper or a stagnant volume will occur. More elaborate underdrains can be prepared from plastic granules cemented into a porous plug, or a layer of stainless-steel screen.

A rubber stopper with a short piece of tubing at the top of the column is adequate for introducing the feed. If dilution is to be minimized, a piece of tubing can be extended from the inlet to near the resin level. A small T on the bottom end of the tubing will prevent the incoming stream from disturbing the resin surface. A second opening in the upper stopper will allow introduction of rinse and backwash outlet.

Flow Rates Flow through the column can be maintained by gravity from overhead tanks, and the flow rate regulated by screw clamps. It is much more satisfactory to use a small peristaltic pump with a variable-speed drive, leaving the bottom outlet unrestricted. Master-Flex®* pumps have proved particularly suitable for laboratory column work.

Flow rate is a critical variable in most ion-exchange cycles, in either the exhaustion or the regeneration phase. The maximum flow rates that will give satisfactory performance should be determined as part of the laboratory study. Flow rates are then held constant in the scale-up.

Flow rate through an ion-exchange column may be expressed as either a linear velocity [m/h or gal(min)(ft²)] or a volume velocity [L/(L)(h) or gal/(min)(ft³)]. Volume velocity equals linear velocity divided by bed height. Both methods of expressing flow through a column have advantages and disadvantages. Linear velocity has a direct bearing on the height of an effective plate in the column. Volume velocity is more directly related to the total contact time between the solution and resin. If all scale-up is carried out at constant bed height, the terms are of course interchangeable.

In applications where the concentration of the exchange ion is less than 50 meq/L, flow rates are usually in the range of 15 to 80 bed volumes (BV) per hour. Flow rates as high as 400 BV/h are used in demineralizing low-solids waters in mixed beds. Regeneration flow rates are commonly held to the range of 0.5 to 5 BV/h to allow for a reasonable contact time with a small volume of concentrated solution. Factors which may require slower flow rates include high solution concentration, weakly ionized resin form, low solution temperatures, and the presence of nonpolar solvents. If divalent ions are involved in the exchange, slower flow rates may be required to obtain sharp breakthrough curves.

In general, breakthrough curves can be sharpened by adjusting column parameters that affect the rate of exchange as long as the separation factor is favorable. When the separation factor is unfavorable, adjusting column parameters will not be effective in sharpening breakthroughs.

Resin Handling The resin should be slurried in water to ensure that it is completely hydrated. A given volume of resin is measured in a graduated cylinder after tapping to a constant height. A volume of 460 mL of resin will fill a 2.5-cm (1-in) column to a height of 90 to 92 cm (36 in). If the resin is washed into a column which has some water in it, the bed will be fairly free of air pockets.

Before starting any processing steps, the bed should be backwashed to classify the resin. An efficient backwash consists of passing water up through the bed at a constant rate so that the resin holds steady at roughly 50% expansion. The bed should be allowed to backwash at equilibrium for 15 to 30 min. The water flow is then stopped and the resin allowed to settle to a packed bed before downward flow is started.

The resin as received from the manufacturers may not be in the proper ionic form for use in the evaluation. The ionic form should be indicated on the resin container. The forms most commonly supplied are:

Strong-acid resin:	sodium (or hydrogen)
Strong-base resin:	chloride
Weak-acid resin:	hydrogen
Weak-base resin:	free base

*Registered trademark, Cole-Parmer Instrument Company.

In any case, it is good practice to condition a new sample of resin through one or two acid-base cycles before starting process evaluation. This serves to remove traces of soluble impurities remaining from the manufacturing process. A suitable cycle is 2 or 3 volumes of 4% HCl, water rinse, a similar volume of 4% NaOH, water rinse. If the resin is in the sodium or free-base form, the HCl is passed through first. If the resin is in the hydrogen or chloride form, the NaOH is passed through first. A flow rate of about 5 volumes of solution per volume of resin per hour is satisfactory.

The resin is converted to the form desired at the start of the ion-exchange cycle. The conversion of weak-acid resin to the hydrogen form is usually complete when 4% HCl is passed through the resin until the effluent is strongly acidic. This may not be adequate if the resin is loaded with divalent ions. Similarly, a weak-base resin is readily regenerated with a small excess of 4% NaOH.

Complete conversion of a strong-acid or strong-base resin to the desired ionic form is more difficult. If the desired ion is not held selectively over the ion initially on the resin, a large excess of reagent may be required. If practical, regeneration should be continued until the concentration of the ion being stripped from the resin is less than 0.001 N in the column effluent. It is very important that the eluting solution be uncontaminated if a high degree of resin conversion is to be obtained. When stripping monovalent ions, 4% solutions are satisfactory. If the ions being stripped from the resin are di- or trivalent, the solution concentration should be increased to 3 to 5 N, if possible. Regenerants for fully ionized resins must be fully ionized themselves.

Conversions of fully ionized resins from one ionic form to another should always be done by passing a solution through a column of resin. The only time a batch contact is possible is when the resin is in the hydrogen or hydroxide form and is to be converted with a base or acid. Even in this case a column contact is preferable.

Following conversion the resin should be rinsed with deionized water until neutral. Ion-free water is recommended for rinses and regenerant makeup, at least in the initial stages of process evaluation, since it simplifies analyses and ionic material balances.

Saturation Loading Curve   A considerable amount of information can be obtained by determining a saturation loading curve. If possible, the actual process stream should be used for laboratory work. If not, a simulated process stream is made up as similar as possible to the eventual plant stream using stock chemicals. Simulated streams are usually satisfactory for the laboratory evaluation. As soon as the actual stream becomes available, it should be used for at least a couple of laboratory runs. Occasionally there are unknown or minor components in the plant streams that will affect the exchange step.

A loading curve is obtained by passing the process stream through the fully regenerated column, and collecting and analyzing the effluent. The column is deliberately overfed until the composition of the effluent approaches that of the influent. The total effluent is collected in successive fractions of known volume. Representative fractions are analyzed for ionic components. The ionic concentrations are plotted vs. the appropriate volume of effluent. Such a plot shows the nature of the sweetening-on and breakthrough curves.

It is convenient to plot component concentrations as normalities. As long as all the reactants and products of the exchange are fully ionized, once the water initially in the bed has been displaced the sum of the normalities out will equal the sum of the normalities in. Plotting the effluent data as normalities provides a check on the analytical data, or if one of the components is difficult to analyze, its concentration can be determined by difference.

The effluent volume should be plotted as multiples of the volume of the resin bed, commonly called "bed volumes." This yields a normalized curve which is essentially identical regardless of the size of the column.

Figure 13 shows a saturation loading curve for zinc on the hydrogen form of a strong-acid resin from a solution containing zinc chloride, sodium chloride, and hydrochloric acid. Hydrochloric acid appears in the effluent as soon as the interstitial water in the column has been displaced. This occurs at slightly under 0.5 bed volume of effluent.

The hydrochloric acid concentration increases rapidly until it is essentially equivalent to the total normality of the influent stream. When the resin shows a marked shrinking or swelling during this part of the cycle, the maximum concentration reached may be somewhat below or above the influent concentration because the resin gives water to or takes water from the outside phase.

When the sodium-hydrogen exchange front reaches the bottom of the column, the concentration of sodium chloride increases to approximately the sum of the sodium and zinc concentrations in the influent. The hydrochloric acid concentration falls to approximately its concentration in the influent.

When the zinc-sodium exchange front reaches the bottom of the column, the zinc breaks through. Zinc concentration increases and the sodium and hydrogen concentrations decrease until all components approach the influent levels as shown by the dotted lines in Fig. 13. When effluent concentrations are equal to influent concentrations, the resin is at equilibrium with the influent solution.

If the purpose of the exchange is to give an effluent free of zinc, the run is stopped at 18 bed volumes (point A, Fig. 13). The capacity of the resin for zinc to this point is equal to the bed volumes of solution passed through the column multiplied by the concentration of zinc in the influent, or approximately 18×0.067 eq/L = 1.2 eq/L.

If the purpose of the exchange is to load the resin with as much zinc as possible, the loading is extended to point B, increasing the loading to approximately 1.4 eq/L. The remaining capacity of the resin will be occupied by sodium and hydrogen ions.

The rather diffuse breakthrough of zinc chloride in Fig. 13 is due to kinetic factors. This experiment was run at a flow rate of 5 bed volumes per hour, a high flow rate for the exchange of a divalent ion from relatively concentrated solution. It should be possible to

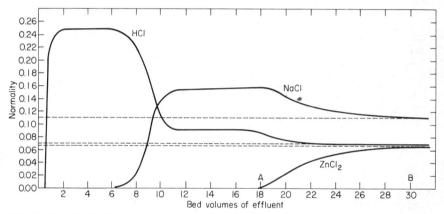

Fig. 13 Complete loading curve for a solution of $ZnSO_4$, Na_2SO_4, and H_2SO_4 on a strong-acid resin, H^+ form.

sharpen the breakthrough appreciably by using a slower flow rate, a smaller resin particle size, and/or higher temperature.

Complete loading curves give a basis for determining the volume of solution that can be processed per cycle and the nature of the effluent that will be obtained over the course of the loading. The character of the loading curve may be somewhat different after the process cycle is set. If the resin is not completely regenerated, there may be leakage of the influent ion early in the run, and the capacity realized will be somewhat lower.

Total-Elution Curve   A total-elution curve is obtained in a similar way by passing an excess of actual or simulated regenerant solution through the bed. A typical elution curve in which the separation factor is less than unity is shown in Fig. 14. The data collected in running a total-elution curve are used to choose a regeneration level that will give a workable compromise between resin conversion and regenerant efficiency.

Three quantities can be calculated from the individual ionic concentration and volumes of the regenerant effluent fractions.

1. The amount of regenerant that has passed through the column up to any point can be determined from the amount of co-ion that has appeared in the effluent up to this point. If the concentration of the co-ion in an effluent cut is multiplied by the volume of the cut, the product is the equivalents of regenerant chemical represented in the cut. Summing these values from the first to nth cut gives the equivalents of regenerant used to point n. Multiplying this by the equivalent weight of the regenerant and dividing by the bed volume gives the amount of regenerant required per volume of resin to point n.

Chemical requirement through cut n, g/L

$$= \frac{\Sigma(\text{co-ion conc., eq/L})(\text{cut volume, L})(\text{eq. wt., g/eq})}{\text{resin volume, L}}$$

2. The average degree of resin conversion after cut n can be determined from the amount of counterion eluted from the resin. If the concentration of the eluted counterion(s) in each of the cuts is multiplied by the volume of the cut and the products are summed, the total equivalents of the resin bed that have been converted to the desired form is obtained. If this sum is divided by the total capacity of the resin bed, the degree of conversion is obtained.

$$\text{Degree of conversion, after cut } n = \frac{\Sigma \,(\text{counterion conc., eq/L})(\text{cut volume, L})}{(\text{resin volume, L})(\text{resin capacity, eq/L})}$$

3. The overall chemical efficiency of the regenerant through the nth cut is obtained by dividing the sum of the counterions eluted by the sum of the co-ions eluted through cut n.

$$\text{Chemical efficiency through cut } n = \frac{\Sigma \,(\text{counterion conc., eq/L})(\text{cut volume, L})}{\Sigma \,(\text{co-ion conc., eq/L})(\text{cut volume, L})}$$

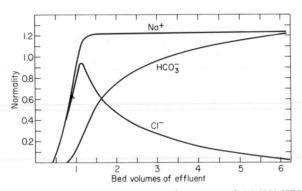

Fig. 14 Elution of Cl⁻ from a Type I strong-base resin with 1.3 N NaHCO₃.

Figure 15 shows the resin conversion and the chemical efficiency vs. chemical usage for the elution in Fig. 14. The degree of conversion is the average over the whole column. Actually, the top of the column may be almost completely regenerated while the bottom of the column is still largely unregenerated. It is the condition of the bottom of the column that determines the leakage in the succeeding run. Figure 15 shows that complete conversion of the resin in an unfavorable exchange can only be accomplished by a large excess of regenerant. Because of this inefficiency, the regeneration is seldom carried to completion in such cases. Instead, the amount of regenerant is chosen so that the column will be converted to a degree that will give the required quality of effluent on the next run and a reasonable run length. If the desired purity of product cannot be obtained at a reasonable chemical dosage, counterflow operation should be considered.

Choice of a logical chemical-dosage level for regeneration and an allowable breakthrough point establishes a tentative cycle. The column should be run through this cycle two or three times to stabilize the system. A cycle is then run with careful sampling across the total run. Analysis of these cuts will give a good indication of what can be expected in the full-scale column.

The effect on column performance of changing various parameters can be determined for optimization of the cycle. Parameters of interest include flow rates, feed volumes, chemical dosages and concentrations, rinse rates and volumes, advantages of air-dome vs. liquid-dome operation, etc.

Effluent Monitoring Rapid in-line monitoring of the column effluent is usually a necessity. Development of suitable methods should be an integral part of the laboratory evaluation. Rapid monitoring can also greatly reduce the analytical time required for process development and evaluation.

Physical methods used to follow effluent composition include:

Conductivity

pH

Specific gravity

Refractive index

Visible or ultraviolet spectra

These methods are easily run in the laboratory and can be scaled up to full-sized units with standard instrumentation. More sophisticated methods may be necessary if the components are very similar.

In many applications it is important to follow changes in composition rapidly to determine sweetening-on, breakthrough, and sweetening-off curves with a minimum of delay. In such cases the method of analysis need not be precise, but it should be in-line and direct-reading. In cases where the purity of the product is a major concern, considerable precision may be required. Most of these physical methods can be adopted to meet these needs.

Ion-exchange column operations, even though cyclic in nature, can be highly automated. As long as the entering streams have a reasonably constant composition, ion-exchange cycles are reproducible to a high degree. Cycles can be initiated by level controls and the feed streams and flow rates to the column programmed by a cycle timer. The product cut can be taken on the basis of a physical measurement on the effluent. If automation of the cycle is desirable, the scheme should be developed during the laboratory evaluation.

Scale-Up and Column Sizing

Ion-exchange data taken in 2.5-cm (1-in) columns can be scaled up directly to any diameter column if the height of the resin bed is kept constant. The most critical factor in this scale-up is that the feed be uniformly distributed over the resin surface and collected uniformly from across the bottom of the column.

In ion-exchange steps with favorable separation factors, the exchange zone usually occupies only a small part of the total column height. Scaling up the height in such a case will not change the shape of the effluent curve but will delay the breakthrough. Experi-

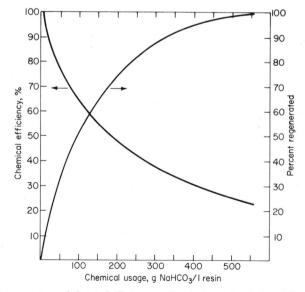

Fig. 15   Resin conversion and chemical efficiency vs. chemical usage for elution of Cl⁻ with NaHCO₃.

ence has shown that scaling up column height by a factor of 2 is more likely to give an improvement in useful performance over laboratory columns than to reduce performance.

The optimum flow rates determined in the laboratory are maintained in the full-sized column. As long as the volume velocity is held constant, the cycle time and concentration profile do not change, regardless of the column diameter. Only the volumes of liquid and resin change.

In ion-exchange units for chemical processing, columns should be sized on the factors of optimum flow and quantity of solution to be processed per unit time. The volume of resin that must be on stream is determined by dividing the volume of stream to be treated per unit time by the volume velocity.

$$\text{Resin volume on stream} = \frac{\text{solution to be treated (vol./hour)}}{\text{volume flow rate (vol./vol.-hour)}}$$

The diameter of a single column required on stream is equal to

$$2\sqrt{\frac{\text{resin volume}}{\text{resin height} \times \pi}}$$

Since the column must be taken off stream each cycle for regeneration, it will not be adequate unless the flow is intermittent. The unit can be increased in size to handle the flow at a faster rate allowing downtime for backwashing, regenerating, rinse, etc. If the exhaustion cycle is relatively long and the flow can be inventoried for short periods of time, this is a satisfactory arrangement. Again, if the capacity of the column can be designed to fit an 8- or 16-h work period, the column can be regenerated during an off shift when not needed. The column may be intentionally oversized to give a capacity that fits a work period.

When the requirement for treatment is sizable and/or continuous, two or more columns are preferred. This is also preferred if a single column would have to be much larger than 4 or 5 m in diameter. Using as few columns as possible reduces the cost of valves and instrumentation but gives less flexibility in operation of the process.

Liquid-Solid Contacting

Batch   The liquid-solid contact necessary for ion exchange can be carried out as a batch reaction in a stirred kettle. This is practical only when the reaction goes far to the right as in neutralization. Even then it is not usually convenient. The liquid must be separated from the resin at the end of the reaction, and usually a washing step is desired. Since this requires some type of filtering device in the kettle, it is usually more convenient to put the resin in a column and pass the solution through it.

The batch technique has been used on a small industrial scale where the ion exchange was run as a titration. Such a case could be the release of an amino acid from its strong-acid salt. Weak-base resin can be added to the solution in a stirred reactor until the desired pH end point is reached. The solution containing the neutral amino acid can be drawn off and the resin washed and regenerated for reuse. This technique avoids exposing the amino acid to a pH at which it would be picked up by the resin.

Fixed Bed   The majority of industrial ion-exchange processes are run in vertical columns and the process stream, rinses, and regenerants are passed down through the resin. The standard design is a vertical cylindrical tank equipped with a resin support-liquid collection system at the bottom and a distribution system above the resin level. Vertical columns are in use in diameters from a few centimeters to 6 m and with resin depths from less than 1 to more than 6 m. The most common bed depths are in the range of 1 to 3 m.

The distribution and collecting systems are critical design features, particularly as column diameter becomes greater than 1 m. The liquid must be delivered uniformly over the total surface of the resin bed without downward jetting which will disturb the level of the bed. The liquid leaving the bottom layer of resin must be collected uniformly across the bed with a minimum of mixing. These two steps are necessary to establish, maintain, and collect a sharp exchange zone. Both systems must be free of stagnant spots where solution can collect and contaminate subsequent steps.

A variety of resin support and collection systems have been used. One method is to use perforated laterals or spokes in the bottom of the column covered with graded layers of particulate matter. The particulate used must be inert to all the streams passing through the bed. Quartz is usually satisfactory. Graded anthracite has proved unsatisfactory where sodium hydroxide is involved, as it contains weak-acid functionality which adsorbs NaOH reversibly.

Instead of using a graded subfill, the laterals or spokes can be fitted with button distributors. These may be short sections of tubular Johnson®* weld screen, capped and threaded. Another arrangement uses plastic diffusers, either integrally cast screw-in units or layers of plastic disks held in place by a T bolt. The density and size of these outlets depend on the flow rate to be used. A common density is one to every 0.07 m² (0.8 ft²) of cross-sectional area.

In most systems it is usual practice to fill the dished end of the tank with concrete and build the collection system on the flat surface. The button distributors are normally mounted on the lower sides of the laterals, 45° from center. Even so, this leaves a potentially stagnant volume of liquid below the collection system which may not be acceptable if exchange fronts and rinse volumes are critical.

For chemical processing units with diameters up to 3 m where the collection system is critical, flat filtering surfaces which cover the cross section of the tank are preferred. Patented porous media, wire mesh, and perforated plates (Nevaclog) have been used successfully. When a flat surface is used, it must be firmly anchored so that it will not lift when the bed is backwashed.

The resin support must be fine enough that it will not pass resin particles. The most common size ranges for resin in chemical processing are −20 +50 mesh (0.8 to 0.3 mm) and −40 +80 mesh (0.4 to 0.18 mm). These sizes represent a practical compromise among chemical, kinetic, and mechanical properties.

Water Dome vs. Air Dome. The type of distribution system used depends somewhat on whether the column is run with a water dome or an air dome. In water-dome operations the column is kept completely filled with liquid at all times. In air-dome operations the liquid level is held a few inches above the resin level when liquid is being introduced into the resin.

In water-dome operation the feed distributor is located a few inches above the top of the bed when the resin is in its most expanded form. The distributor may be lateral pipes transversing the shell, or a spider or gridwork supported internally. It must be sturdy enough to withstand the pressure of the rising resin bed at the start of backwash. The distributor pipes are perforated on the sides or 45° from the vertical, in a pattern calculated to give a uniform flow of liquid over the entire surface of the resin.

If the solution being introduced into the column has a density even slightly greater than that of water, it will form a layer under a static water dome and proceed down through the bed. The amount of dilution will be primarily dependent upon the actual distance between the laterals and the top of the resin bed at the time the feed is started. If the resin undergoes a large shrink-swell cycle in the process, the distance between the distributor and the resin may be as much as 0.3 or 0.6 m (1 or 2 ft) when the resin is in its smaller form. In such cases the initial dilution will be much greater on this portion of the cycle.

When the desired volume of feed or regenerant has been introduced to the column, the solution in the lateral system is displaced with water. Rinse water is then introduced at the top of the column. The water dome is displaced down through the resin as rinse water. Any of the previous solution which had diffused into the space above the laterals will be rinsed down into the bed, leaving a fresh-water dome for the next introduction of solution.

The object of air-dome operation is to minimize dilution of the solution being introduced to the column. The liquid level in the column is adjusted to 25.4 or 50.8 mm (1 or 2 in) above the resin level prior to the introduction of the feed and again before the introduction of the rinse or regenerant. The adjustment is critical if minimum dilution is to be obtained and is difficult to instrument because of the varying resin level. Observation ports and direct operator control may be required.

The same type of laterals may be used as in water-dome operation. A unique type of distribution which has been used in air-dome operation is a full-cone spray from the top center of the tank. An inch or two of water on the resin prevents churning of the resin surface.

In addition to the problem of adjustment of liquid level before initiating feed streams,

*Registered trademark, Johnson Division, Universal Oil Products.

special precautions must be taken to keep the liquid level from falling below the resin level during the run. This may be done by pumping into the bed at a constant rate but controlling the flow from the bed by a liquid-level controller. Backwashing requires filling the dome with water and then draining it before the next cycle. Water-dome operation is much easier to instrument and control and is recommended unless air-dome operation is mandatory.

Flow through a water-dome column is usually supplied by a centrifugal pump with control of the flow rate at the bottom of the bed. Running the bed under a slight positive pressure helps to prevent the formation of gas pockets from feed supersaturated with air or from the release of carbon dioxide.

Dilution of solutions passed through an ion-exchange column cannot be eliminated, at least during introduction of the feed (sweetening on) and when the feed is rinsed out of the column (sweetening off). Forty percent of the volume of a bed of spherical resin is water. In addition, the resin proper is usually from 35 to 55% water by volume. Ionized solutes, which have little tendency to enter the resin because of Donnan exclusion, must displace approximately 40% of the volume of the bed before they appear in the effluent. Nonionized solutes or highly concentrated electrolytes which do penetrate the resin appreciably must displace from 40 to 70% of the volume of the bed before they appear in the effluent. A similar situation occurs during sweetening off. Mixing is inherent in these steps, and the effluent curves are S-shaped.

Dilution is most critical when treating concentrated solutions. The amount of dilution water added during sweetening on and sweetening off is dependent on size of bed and distribution of solutes between the two phases. It is independent of length of run. If total run length is several hundred bed volumes of solution, four to eight bed volumes of dilution during sweetening on and sweetening off is inconsequential. If the total run length is only 10 to 20 bed volumes, the same amount of dilution may represent a serious loss in concentration and is another reason ion exchange is often restricted to processing of dilute solutions.

Rinse Steps. Rinse steps are often required at the end of an exhaustion run (sweetening off) and after the regeneration step. In water conditioning, or when treating dilute solutions of minor value, no rinse step is required after exhaustion. If a concentrated or valuable stream is involved, the solution remaining in the column at breakthrough is displaced with water and recycled to feed.

It is necessary to use a rinse step after regeneration except in cases where the regenerant remaining in the bed will not be deleterious if mixed with the stream to be processed. In water conditioning usually the stream to be processed is used as rinse and the effluent is sewered or recycled until the necessary effluent quality is reached. In chemical processing the column is usually rinsed with water until the concentration of regenerant falls below some preset value as determined by pH or conductivity. It is important that the water used for rinsing be of sufficient quality that effluent requirements can be met without appreciably exhausting the resin bed.

Rinses should be carried out in two steps. The initial, or displacement, rinse is run at the same flow rate as the preceding step for one or two bed volumes. This serves to push regenerant or other solution through the bed at the flow set for the exchange. After initial rinse the flow rate may be increased to reduce the time required to reach an acceptable effluent before starting the next step of the cycle.

The relative volume of rinse needed to give the effluent quality required is an important characteristic of the system. If the rinse volume is excessive it is wasteful of water, consumes resin capacity, and extends the nonproductive time for the column. Rinse volumes are dependent on type of resin, nature of regenerant, design of the equipment, temperature, and flow rate. Rinse requirements tend to become greater as resin ages, particularly with anion-exchange resins. In strong-base resins increased rinse is most often due to irreversible adsorption of complex organic acids which form weak sodium salts when resin is contacted with sodium hydroxide. Weak-base resins also have a tendency to foul in this manner. Weak-base resins are also subject to mild oxidation, which often appears as an increase in rinse requirements. Rinse problems may also be due to the presence of precipitates in the resin bed or stagnant areas in the equipment.

Backwash. Ion-exchange resin beds are excellent filter media. Any suspended solids in the solutions passed through them will tend to be retained in or on the bed. If any appreciable layer of dirt or silt builds up on the surface of the resin during the run, operation of the bed may deteriorate and high pressure drop may develop. If the deposited layer is cohesive, it may shrink and either pull the resin away from walls of the tank at the upper surface or split and create a point of low resistance to flow. Shrinking or splitting results in channeling and premature breakthrough. It is recommended that solutions being treated in fixed-bed columns have a turbidity less than 10 ppm.

Fixed-bed columns are usually backwashed as a part of each cycle as protection against the accumulation of solids. It is customary to backwash between exhaustion and regeneration steps. Backwash is sometime used following regeneration with weak-base resins, which show large shrinkage when regenerated.

A second purpose of backwash is to reclassify the bed resulting in a gradual increase in particle size from top to bottom. Good classification of the resin bed helps to prevent channeling and assures that the exchange front will move uniformly down the column.

Backwash is carried out by introducing water through the bottom distributor system and allowing it to exit through openings at the top of the column. Exit openings are not screened since one of the purposes of backwashing is to remove debris and fine resin particles from the column. Loss of resin is avoided by controlling the backwash flow rate so that maximum expansion of resin bed is well within the free board allowed above the settled bed. Good backwashing requires holding the bed at approximately 50% expansion for 15 to 20 min. The expansion obtained is dependent on Stokes' law for falling spheres. The variables are:

1. The difference in density between the resin particles and the liquid
2. The particle size
3. The viscosity of the liquid
4. The linear flow rate of the liquid

Since both the density and viscosity of water are dependent on temperature, the expansion obtained is sensitive to water temperature. Backwash rates may have to be adjusted between winter and summer to avoid loss of resin. When the backwash period is over, flow is stopped. The resin should be allowed to settle before downward flow is started.

Mixed Beds.   Columns containing a mixed bed of hydrogen-form strong-acid resin and hydroxide-form strong-base resin give a degree of deionization difficult to achieve with individual resin beds in series. The presence of both resins in the same bed converts the deionization process to a reaction in which the only product is water. There is essentially no back reaction, and the process goes to completion.

$$\overline{H^+RSO_3^-} + \overline{R_4N^+OH^-} + NaCl \rightarrow \overline{Na^+RSO_3^-} + \overline{R_4N^+OH^-} + H_2O$$

Because there is no back reaction, the resins do not have to be completely regenerated to give high-quality water. It is necessary that the layer of resin at the bottom of the column contain an intimate mixture of the two resins, and that both resins are at least partially regenerated. The degree of resin mixing at the bottom of the bed is probably the most critical factor in determining the quality of the effluent.

When the column is exhausted, the two resins are separated into layers by backwashing. Good separation depends on the relative densities and particle-size ranges of the two resins. The resins are regenerated in place by introducing caustic and acid through headers placed above the resin bed and at the interface between the two resins. Resins are remixed by draining water in the column to a few inches above the resin and then introducing air through the bottom distributor.

Advantages of mixed-bed units are higher water quality and substitution of one column for two columns. The disadvantage is some loss in chemical and resin efficiency which results if resin separation is not complete or if the interface does not coincide with the central distributor.

Mixed-bed units may be used for the demineralization of solutions of nonelectrolytes such as sugar as glycerin and are particularly suited for final polishing where the ionic load is relatively low. In very large applications, such as condensate polishing in power-generating plants, the resins are often transferred to separate tankage for separation and regeneration.

Layered Beds   It has become common practice in water conditioning to use a combination of two different resins in the same shell. Usual combinations are a weak-acid resin layered above a strong-acid resin, or a weak-base resin layered above a strong-base resin. The purpose of such combinations is to increase chemical efficiency and capacity of the unit without the expense of a second shell.

Use of a weak-acid resin and a strong-acid resin in series results in an increase in overall regenerant efficiency if the stream being treated contains appreciable alkalinity. Weak-acid resin removes the cations equivalent to the alkalinity, and strong-acid resin completes the decationization. An excess of regenerant is required for the strong-acid resin to meet effluent requirements on the next run. Excess regenerant can be adjusted so that free acid in the regenerant effluent from the strong-acid unit is sufficient to regenerate the weak-acid unit. Regenerant efficiency can approach theoretical values and eliminate waste acidity.

Weak-base and strong-base resins in series are used when the stream contains a mixture of strong and weak acids. Both cation- and anion-exchange systems in series columns are in common use. Water to be treated is passed through weakly ionized resin and then through strongly ionized resin. Flow of regenerant is through strongly ionized resin and then through weakly ionized resin.

When two types of resins are used as layers in a single column, considerable flexibility is lost. Regeneration must be carried out upflow if strongly ionized resin is to be contacted first. Upflow regeneration fluidizes the resin bed to some degree unless some method of keeping the resin compacted is used, and column efficiency is lost. It is difficult to adjust the composition of effluent entering the upper layer, resulting in precipitation of calcium sulfate in weak-acid resins and silica in weak-base resins. Regeneration problems are minimized when the two resins are in separate shells, as the regenerant stream from the secondary unit may be diluted or fractionally discarded before introduction into the primary unit.

Layered units must be carefully designed both mechanically and in choice of resins. The capital savings must be carefully balanced against the added flexibility of separate units. Layered beds are not recommended for chemical processing where there are possibilities of resin instability.

Fixed-Bed Counterflow In conventional downflow operation a high regeneration level may be necessary to reduce the leakage of unwanted ion to the required level. When the exchange in the regeneration step is unfavorable, high regeneration levels are chemically inefficient.

By use of counterflow technique high product purity and high chemical efficiency can be obtained at low chemical usage. The regenerant is passed up through the bed and the feed down through the bed. Counterflow technique use is increasing, particularly in water treatment.

For the process to be truly effective the resin must be prevented from fluidizing during the upflow operation. A number of methods have been used to prevent fluidizing.[8] High product purity and high chemical efficiency can be obtained at the cost of somewhat more complex equipment and lower utilization of resin capacity by the use of counterflow operation. Pulsed ion-exchange equipment recently developed in Canada utilizes counterflow technique in shallow beds of fine-mesh resins.[20]

Continuous Ion Exchange A number of continuous or semicontinuous contactors have been developed for ion-exchange processing.[21] Exchange reactions in these contactors are said to be continuous countercurrent, meaning that the resin and solution pass countercurrent to each other. This is no different from standard fixed-bed operations. However, there are some potential advantages to continuous operation.

In a fixed bed only the volume of resin actually in the exchange zone is in use. Resin above the exchange zone is exhausted and inert. Resin below the exchange zone is not in use. Total resin inventory is considerably larger than needed at any one time. With continuous equipment steady-state zones can be set up as required for exhaustion, regeneration, and the required intermediate rinses. The volume of resin can be reduced to the sum of the working volumes plus the interconnecting resin transfer system.

When a steady-state exchange zone is set up, product can be taken off continuously and regenerant supplied continuously. The system can be subjected to McCabe-Thiele analysis. By adjusting the relative resin-liquid flow rates in each contact, chemical usage can be optimized. Rinse volumes can be minimized. However, there is no basic improvement in product purity as obtained with counterflow operation in fixed beds.

Continuous contactors can be divided into two major types: pulsed columns and fluidized beds. Either can give potential advantages, but each has disadvantages. Both types are in commercial use in water conditioning and chemical processing.

In the pulsed columns resin is moved up or down through the contacting zones by periodic application of pressure or vacuum. Solutions flow through the resins between pulses. Exchange takes place in a compacted bed with little loss of plate efficiency. Some loss of chemical efficiency due to partial reclassification of resin during the pulse occurs. Major problems with pulsed columns have been mechanical. A number of valves must be opened and closed each pulse. A pulse may be required as often as every 4 min to every 20 or 30 min. Some valves are quite large and some must work in a resin slurry. In addition to wear on the valves, considerable abrasion of the resin occurs. Resin-replacement rates as high as 30% per annum have been reported in water-conditioning applications.

In fluidized-bed systems the exchanges take place with noncompacted resin. The resin may fall down through a baffled column against the upflow stream, or the exchange may take place in stirred compartments or troughs where the resin is forwarded mechanically against the flow of solution. In either case, the expanded condition of the resin limits the number of effective plates to the number of discrete mechanical stages used. This is satisfactory when the reaction has a separation factor greater than unity and only a few contact stages are required. However, the contact method must be modified for regeneration steps where many plates are required. Continuous fluidized beds can be designed to eliminate valves working in the resin slurry, which reduces attrition of resins substantially.

Uniform distribution of feed solutions into and collection of product solutions from continuous contactors is a major problem when it is important to establish and maintain sharp exchange fronts. Uniform distribution is also a particularly difficult aspect of the system to scale up.

No clear design rules have been established as to when continuous contactors have an advantage over fixed-bed contactors. An area where continuous contactors will be required for successful application of ion exchange is processing of streams where the component being removed has a concentration much above 0.5 N. In such cases, the maximum volume of solution that can be treated per cycle in a fixed bed is two to three bed volumes. Dilution due to sweetening on and sweetening off becomes a major problem. With a continuous contactor the exchange processes and necessary rinse steps can be set up as standing waves. This technique should be applicable to certain hydrometallurgical processes.

Continuous contactors will be attractive whenever the total amount of material to be taken out of solution is large. One estimate has set 500 g/min as the point at which continuous ion exchange should be investigated. Even when working with dilute solutions, the size of a fixed-bed ion-exchange plant may become so large that the savings in resin inventory and space requirements offered by continuous contactors are economically significant.

Some continuous contactors have the ability to treat streams containing suspended solids which cannot be processed through conventional fixed beds without prefiltration. Continuous ion exchange should be of value in waste treatment and hydrometullurgical applications.

Mixed-bed deionization is carried out in continuous contactors. Exhausted resins are separated in a backwashing stage, regenerated and remixed, all on a continuous basis. Continuous mixed-bed units have an advantage in space requirements, resin inventory, and chemical efficiency over conventional units when large volumes of water are to be processed, but the quality of water produced tends to be lower than from conventional mixed-bed units. Output of the continuous unit may be polished by passage through a small conventional mixed bed.

Resin Availability

A wide variety of ion-exchange resins are available in commercial quantities. A list of manufacturers and their brand names is given in Table 4. Essentially all companies offer the four major types of resins. However, none of them offer all commercially available types when specialty resins are included. Several compilations of brand names for the different types of resins are in the literature.[9,28] A list of currently available resins can be obtained on request from the companies in Table 4.

Resin manufacturers are a prime source of engineering and performance data for ion-exchange processes. Most manufacturers maintain laboratories and technical-service staffs for consultation on ion-exchange applications. There are also a number of equipment manufacturers highly qualified in the design and construction of ion-exchange units.

TABLE 4   Ion-Exchange Manufacturers

Company	Location	Trade name
Akzo Chemie Ver Koopkantoor bv	Amsterdam, Netherlands	Imac®
Diamond Shamrock Corp.	Cleveland, Ohio	Duolite®
Dow Chemical Co.	Midland, Mich.	Dowex®
Farbenfabriken Bayer A.G.	Leverkusen, W. Germany	Lewatit®
Ionac Chemical Co.	Birmingham, N.J.	Ionac®
Mitsubishi Chem. Ind. Ltd.	Tokyo, Japan	Diaion®
Montecatini Edison Spa	Milano, Italy	Kastel®
Rohm and Haas Co.	Philadelphia, Pa.	Amberlite®
VEB Farbenfabrik Wolfen	Wolfen, GDR	Wofatit®

Economic Considerations

The economic justification of an ion-exchange process depends on a number of factors.[26]
Credit factors include:
1. Increased value of a product because of improved properties
2. Recovery of valuable components
3. Recovery of water for reuse
4. Better quality control due to elimination of process variability

5. Reduced waste-disposal costs due to elimination of objectionable components, or reduced volume.
6. Reduction of corrosion or scaling in subsequent equipment
Debit factors include:
1. Capital cost of equipment
2. Resin costs, which may be either capitalized or expensed, depending on resin life
3. Cost of chemical regenerants
4. Operating labor
5. Reconcentration costs
6. Disposal costs for regenerant wastes
If the purpose of the process is to recover or refine a particular chemical, the first requirement is that the value added be greater than the cost of regenerant chemicals used. If the chemical being processed is actually picked up on the resin, it must have a higher value per equivalent than the regenerant chemical used to displace it.

Equipment costs will be highly variable and dependent on such factors as materials of construction required and degree of automation desired. A fairly large selection of pre-engineered columns in various sizes is available. Engineering costs can be reduced if a process can be fit to an existing design. Materials of construction must be chosen to minimize corrosion. Where plastics are used for distributor and collector systems, the design must withstand mechanical stresses due to resin expansion and backwashing.

Resins are normally sold on a volume-in-place basis. Unit volume cost for strong-base resins has been three to four times that of strong-acid resins. Higher cost of strong-base resins is due to the considerably more complex manufacturing process required for the anion resins.

Weak-base resins are in roughly the same price range as strong-base resins. Weak-acid resins have run two to three times strong-acid resins in cost. Most chelating resins are more expensive than strong-base resins because of the more specialized manufacture and limited production. Special requirements such as narrow particle-size specifications will increase cost of any resin.

Operating labor requirement is fairly independent of size of the ion-exchange operation. If the operation is at least partially automated, part-time surveillance may be all that is required. Good instrument maintenance is mandatory in such cases.

TABLE 5   Conversion Tables for Ion-Exchange Processes

From	Multiply by	To
Volume:		
Gallons	3.78	Liters
Cubic feet	28.3	Liters
Flow:		
Gallons per minute	0.227	Cubic meters per hour
Linear velocity:		
Gallons per minute per square foot	2.44	Meters per hour
Volume velocity:		
Gallons per minute per square foot	8.02	(Bed volumes) per hour
Pressure drop:		
Pounds per square inch per foot of resin	0.23	Kilograms per square centimeter per meter
Bulk density:		
Pounds per cubic foot	16.0	Grams per liter
Solution concentration:		
Milligrams $CaCO_3$ per liter (ppm)	0.020	Milliequivalents per liter
Grains $CaCO_3$ per cubic foot	0.342	Milliequivalents per liter
Resin capacity:		
Kilograins $CaCO_3$ per cubic foot	0.0458	Equivalents per liter
Regenerant dosage:		
Pounds per cubic foot	16.0	Grams per liter
Regenerant concentration:		
Pounds per gallon	120	Grams per liter
To	Divide by	From

REFERENCES

Books

1. Applebaum, S. B., "Demineralization by Ion Exchange," Academic, New York, 1968.
2. Arden, T. V., "Water Purification by Ion Exchange," Plenum, New York, 1968.
3. Calmon, C., and Kressman, T. R. E. (eds.), "Ion Exchangers in Organic and Biochemistry," Interscience, New York, 1957.
4. Helfferich, F., "Ion Exchange," McGraw-Hill, New York, 1962.
5. Nachod, F. C., and Shubert, J. (eds.), "Ion Exchange Technology," Academic, New York, 1956.
6. Rieman, W., III and Walton, H. F., "Ion Exchange in Analytical Chemistry," Pergamon, New York, 1970.
7. Samuelson, O., "Ion Exchange Separations in Analytical Chemistry," Wiley, New York, 1953.

Articles and Reviews

8. Abrams, I. M., *Ind. Water Eng.*, January/February, 1973, p. 18.
9. Abrams, I. M., and Benezra, L., in "Encyclopedia of Polymer Science and Technology," vol. 7, p. 692, Wiley, New York, 1967.
10. Anderson, R. E., *Chem. Eng. Prog. Symp. Ser.*, **60**(48), 76 (1964).
11. Anderson, R. E., in Rembaum and Sélégny (eds.), "Polyelectrolytes and Their Applications," p. 265, D. Reidel Publishing Co., Dordrecht, Holland, 1975.
12. Anderson, R. E., *Am. Inst. Chem. Eng. Symp. Ser.*, **71**(152), 236 (1975).
13. Anderson, R. E., Bauman, W. C., and Harrington, D. F., *Ind. Eng. Chem.*, **47**, 1620 (1955).
14. Battaerd, H. A. J., et al., *Desalination*, **12**, 217 (1973).
15. Bauman, W. C., in F. C. Nachod (ed.), "Ion Exchange," p. 45, Academic, New York, 1949.
16. Bauman, W. C., and Eichhorn, J., *J. Am. Chem. Soc.*, **69**, 2830 (1947).
17. Boari, G., Liberti, L., Merli, C., and Passino, R., *Desalination*, **15**, 145 (1974).
18. Bonner, O. D., and Smith, L. L., *J. Phys. Chem.*, **61**, 326 (1957).
19. Helfferich, F., *J. Phys. Chem.*, **69**, 1178 (1965).
20. Hunter, R. F., Canadian Patent 834,256, Feb. 10, 1970.

21. *Ind. Water Eng.*, February 1970, p. 18.
22. Kraus, K. A., and Nelson, F., *Proc. Int. Conf. Peaceful Uses Atomic Energy Geneva*, **7**, 45 (1956).
23. Oberhofer, A. W., U.S. Patent 3,223,620, Dec. 14, 1965.
24. Peterson, S., *Ann. N.Y. Acad. Sci.*, **57**, 144 (1953).
25. Rosset, R., *Bull. Inf. Sci. Tech. Comm. à l'Energie Atomique*, no. 85, July–August 1964.
26. Seamster, A. H., and Wheaton, R. M., *Chem. Eng.*, Aug. 22, 1960, 115.
27. Spedding, F. H., and Powell, J. E., in Ref. 5, p. 359.
28. Vermeulen, T., Klein, G., and Hiester, N. K., in R. H. Perry, "Chemical Engineers' Handbook," 5th ed., sec. 16, McGraw-Hill, New York, 1973.
29. Weiss, D. E., et al., *Aust. J. Chem.*, **19**, 589 (1966).
30. Weiss, D. E., et al., *Aust. J. Chem.*, **19**, 561 (1966).
31. Wheaton, R. M., and Bauman, W. C., *Ind. Eng. Chem.*, **45**, 228 (1953).
32. Wheaton, R. M., and Bauman, W. C., *Ann. N.Y. Acad. Sci.*, **57**, 159 (1953).
    Dekker, Inc., New York, 1969.
34. Wymore, C. E., *Ind. Eng. Chem. Prod. Res. Dev.*, **1**, 173 (1962).

Section **1.13**

Activated-Carbon Systems for Separation of Liquids

ROY A. HUTCHINS *Technical Services Superintendent, Agricultural Chemicals Division, Texas Gulf, Inc., Aurora, N. C.*

ADSORPTION

Adsorption by activated carbon is an important unit process used to remove miscible liquids and dissolved solids from liquids. It is frequently the most economical method of separating such materials, particularly when the miscible liquids or dissolved solids are present at relatively low concentrations.

Adsorption occurs when the energy associated with a surface of a solid attracts molecular or ionic species from the liquid to the solid. The adsorbed material can form a layer on the surface of one to several molecules deep. The amount and properties of the surface and the environmental conditions at the surface will control adsorption.

An activated carbon might be described as a solid foam. It has extremely high internal surface area along the sides of an extremely intricate network of internal pores (holes). The ideal activated carbon would have the maximum amount of internal surface with ample volume in the pores to hold the maximum weight of adsorbed material. Because of its great porosity activated carbon is one of the few solids that can provide extremely high surface area and pore volume per unit weight or volume at relatively low cost.

Generally, the total surface area of activated carbons ranges between 450 and 1800 m^2/g with pore volumes of 0.7 to 1.8 mL/g. However, only the portion of that area and pore volume which is in pores of the proper size will be available for adsorption. Therefore, total surface area and pore volume data should not be used to rate the probable effectiveness of an activated carbon. For example, an activated carbon with an extremely high surface area may adsorb very rapidly, but its adsorptive capacity could be low because it lacks pore volume to hold the adsorbed material.

Little information is available about the adsorption of a wide variety of specific materials by activated carbon. This is presumably because considerable time and money are required to develop such information and pure-component adsorption is only occasionally related to the adsorption of complex mixtures.

Generally, adsorption of organics increases with increasing molecular weight (size) until the particle becomes too large to penetrate into the carbon pores. Adsorption increases with decreasing solubility. Molecules with more than three carbon atoms are generally adsorbed unless they are extremely soluble. Nonpolar organics are much more strongly adsorbed from water (which is polar) than polar organics.

Generally, adsorption increases as the solution pH decreases. Adsorption is reportedly more efficient when the organics in solution are near their isoelectric point and when the organics are neutral rather than charged particles.

Adsorption theory says that adsorption should decrease as temperature increases. However, decreasing viscosity and increasing molecular movement at higher temperatures apparently allow the organics to enter the carbon pores more easily, generally causing adsorption to increase as the temperature increases.

Inorganic materials are not, as a general rule, very adsorbable; but there are exceptions such as molybdates, gold chloride, mercuric chloride, silver salts, and iodine.

The type of activated carbon used will also affect what is adsorbed and the adsorption efficiency. For example, an activated carbon with a high concentration of small pores will tend to adsorb smaller molecules than a carbon with a high concentration of larger pores, which will tend to adsorb larger molecules. Also, differences in surface chemistry and in the ash constituents can affect adsorption.

The presence of other organics will greatly affect adsorption, since some organics are preferentially adsorbed.

Tests at probable plant-scale operating conditions are usually required to determine the adsorptive performance of a specific activated carbon for the removal of a specific material present as a pure component or in a complex mixture.

TYPES OF SEPARATIONS

A large number of important separations fall within the scope of adsorption by activated carbon. Among these are the removal of:

- Impurities causing color, taste, or odor from potable water and from chemical and food processing liquids.
- Oxidizable organics, colors, toxic organics, and/or metals from municipal or industrial wastewaters so environmental standards can be met or the water can be recycled.

• Small concentrations of products which may have no color, taste, or odor but which cause poor crystal yield or poor crystal habit.

• Color precursors which may have no original color but cause color to develop at a later stage of processing or after the product has been packaged and sold.

• Materials causing foam or other undesirable surface-active phenomena which interfere with processes such as evaporation, agitation, or air stripping.

• Impurities causing haze or turbidity, and haze or turbidity precursors.

• Trace quantities of water from water-immiscible organic solvents, or vice versa.

• Impurities from a liquid by extracting them with a trace quantity of immiscible solvent, and then adsorbing the resulting solution of solvent and impurities on carbon. This is useful where direct adsorption is unsuccessful.

• Trace quantities of ionic metals, by complexing with an adsorbable organic reagent and adsorbing the complex.

Two separations not concerned with the removal of impurities are:

• Concentration of valuable material from dilute solution by adsorption with recovery by elution.

• Scavenging of liquids or solutions dispersed in solids (e.g., staining oils in reclaimed rubber, drill-mud additivies in oil-well cements, herbicides, or insecticides in soils).

ACTIVATED CARBON

Types

Activated carbons can be divided in two classes: (1) gas-adsorbent carbons, which are used for purification applications in the vapor or gas phase such as solvent recovery, gas separation, or cigarette filter tips; (2) liquid-phase carbons, which are used to decolorize or purify liquids, solutions, and liquefiable materials such as waxes.

The main distinction between gas-adsorbing and liquid-phase carbons lies in the pore-size distribution (see Fig. 1). Gas-adsorbing carbons usually have the most pore volume in the micropore (3- to 50-Å radius) and in the macropore (1000 to 50,-000-Å radius) ranges, with little pore volume in the transitional pores (50- to 1000-Å radius). Liquid-phase activated carbons have significant pore volume in the transitional pore range, permitting ready access of liquids to the micropore structure and resulting in rapid attainment of adsorption equilibrium for smaller adsorbates. In addition accessibility to the internal pore structure improves subsequent adsorption for larger molecular and colloidal substances.[1]

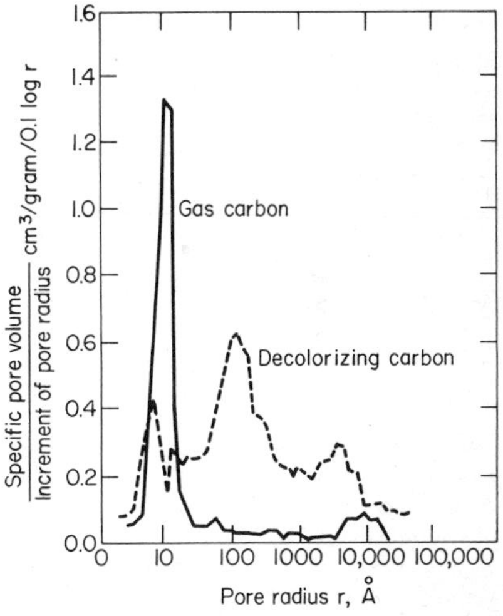

In general, liquid-phase activated carbons have about the same surface areas as gas-adsorbing activated carbons but larger total pore volumes.

Fig. 1 Pore-volume distribution of typical activated carbons. *(By permission, ICI United States, Inc.)*

All liquid-phase carbons are not equally well suited for all liquid-phase applications. Although all liquid-phase carbons have significant pore volume in transitional pores, the magnitude of the pore volume and surface area distributions are different. Figure 2 shows pore-volume distributions of three liquid-phase activated carbons made by different processes from different raw materials.[1]

Activated carbon can be made from a number of raw materials having a high carbonaceous content. The principal raw materials for commercially available liquid-phase carbons are:

Sawdust	Charcoal
Lignite	Bituminous coal
Fly ash	Petroleum coke

Physical Forms

Liquid-phase activated carbons are available in two physical forms, powdered and granular. Some adsorption problems are best handled by stirring powdered carbons with the liquid to be treated and then removing the carbon by filtration or settling. Others are handled most effectively and economically by passing the liquid through a bed of granular carbon.

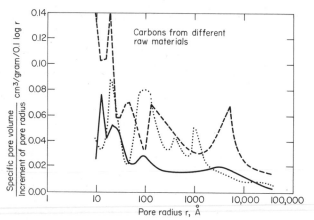

Fig. 2 Pore-volume distributions of liquid-phase activated carbons. *(By permission, ICI United States, Inc.)*

The choice between granular and powdered carbon for a given application depends not only on comparative economics; these factors must also be considered:
- Type of existing equipment
- The carbon usage rate
- Variability of flow rate, and the impurity concentration and composition
- Nature of the application
- Cleanliness
- Disposal

Type of Existing Equipment In many cases, equipment already exists which lends itself to powdered- or granular-carbon treatment. Examples of plants readily adaptable for powdered-carbon use are: (1) A chemical processing plant, which uses batch processing, may have tanks with agitators and pressure filtration equipment, or (2) an industrial wastewater-treatment plant may already have mix tanks (aeration basins) and clarifiers.

The Carbon-Usage Rate At low carbon-usage rates (less than about 400 lb/day), regeneration and reuse of granular carbon is not economical.[2] Since powdered carbons are generally less expensive than granular carbon, operating costs with powdered carbon could be lower.

Conditions which result in low carbon-usage rates are:
- Low impurity concentration
- High impurity concentration allowable in effluent
- Low impurity-removal requirements
- Low flow rates
- Low liquid volume

At high carbon-usage rates, granular carbon is usually preferred because it can be economically regenerated. However, regeneration methods for powdered carbon are now becoming commercially acceptable, which will permit its use at higher carbon-usage rates.

Variability of Flow Rate and the Impurity Concentration and Composition In cases where the flow rate and the impurity concentration and composition vary dramatically, granular carbon will usually be more suitable because adequate carbon would be present to compensate for the variations. Sophisticated instrumentation would be required if adequate powdered carbon were to be added as required to compensate for such variations.

Nature of the Application In choosing between powdered and granular carbon, it is necessary to consider whether all adsorbable material must be removed, or whether selective adsorption is required. In the treatment of beer, for example, carbon must remove the causes of chill haze, without removing flavor, color, or foam-causing constituents. Similar problems exist in treating other food products and pharmaceuticals. With powdered carbon, adsorption selectivity can be achieved by dosage control. With granular carbon, where a large mass of carbon is in contact with a relatively small volume of liquid at any one time, substantially all adsorbable materials can be stripped out.

Cleanliness Both granular and powdered carbons can be dusty if not properly handled. Powdered carbons generally present more of a problem, but dusting can be minimized. One method available to large users is bulk handling, receiving shipments in bulk railroad cars or trucks and conveying either dry or by slurry into storage by standard materials-handling methods. In preparing a slurry, the carbon should be added to the liquid as fast as it can be wetted and, if possible, in a closed slurry tank under slight vacuum so dust can be conveyed to an appropriate dust-collection system.

Disposal With powdered carbons, disposal of the spent carbon to a sewer of a municipal or industrial waste-treatment plant is usually permissible. In cases where disposal into sewers is not possible or permitted, disposal is usually by landfill or dumping. In cases where carbon has been used in liquids which are dangerous or has adsorbed dangerous materials, special disposal methods are required. Reactivation of powdered carbon which allows the carbon to be reused is now becoming commercially accepted. With reactivation disposal is no problem.

At carbon usage rates of above about 400 lb/day, granular carbon can be economically reactivated.[2] When granular carbon is not regenerated, it can be sewered, dumped, or used as landfill.

Properties

Table 1 shows the primary properties of powdered and granular carbons. These properties are those most directly concerned with cost and effectiveness of the carbon. Other properties could also be primary depending upon the application.[1,3,4,5] These properties could include particle size, moisture, ash, pH, and solubles.

Adsorptive Capacity Adsorptive capacity is the most important property with both powdered and granular carbons because it determines how many gallons of liquid can be treated per pound of carbon. Adsorptive capacity determines both the direct operating cost for carbon treatment and the sizing of equipment. For a given application, tests are required to determine the adsorptive capacity.

Surface area is not a primary property. High surface area does not mean high adsorptive capacity, for the following reasons:

- In adsorption only the wetted surface area is effective and the wetted surface area never equals the total surface area.
- In most applications, the material to be adsorbed is too large to enter the very small pores where the bulk of the surface area is.
- Data on surface area, pore volume, and surface nature usually have not been correlated with data on the material to be adsorbed. In many cases neither the identity nor the concentration of the material is known.

Filterability or Settling Characteristics of Powdered Carbon With powdered carbons, filterability is how fast a liquid at minimum pressure and maximum clarity can be filtered through a given cross section and depth of filter cake. This property is dependent

TABLE 1   Primary Activated-Carbon Properties

Powdered carbon	Granular carbon
Adsorptive capacity	Adsorptive capacity
Filterability or settling characteristics	Pressure drop or bed expansion
Bulk density (retention)	Abrasion resistance

upon the raw material, which determines the typical particle shape; and upon the grinding, which determines the particle size and size distribution. Poor filterability will cause short filter cycles and increased carbon-treatment cost. In cases where the powdered carbon will be removed by settling, the carbon must stay in suspension long enough for adequate adsorption to occur but settle rapidly enough so the liquid being decanted or overflowing will not contain carbon particles. This property is dependent not only upon the raw material, particle size, and size distribution, but also on particle density, which is a function of pore volume, and wettability, which is a function of the pore-size distribution.

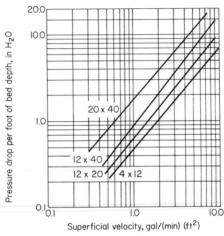

Fig. 3 Pressure-drop characteristics [water at 25°C (77°F)]. *(By permission, ICI United States, Inc.)*

Pressure Drop or Bed Expansion of Granular Carbon With granular carbons, pressure drop (head loss) in downflow and in confined upflow columns (see Fig. 3), or bed expansion in unconfined upflow columns (see Fig. 4) is of primary importance. The carbon's contribution to head loss is determined by particle size and size distribution, which is controlled by sieving and by particle shape controlled by raw material. Bed expansion is determined by particle size, shape and size distribution, and particle density. Head loss and bed expansion are design factors for carbon columns.

Since upflow and downflow pressure drop and bed-expansion characteristics can vary significantly for various commercially available granular activated carbons, this information should be obtained from the manufacturers of specific carbons.

Bulk Density of Powdered Carbon Bulk density is important when powdered carbon is removed by filtration because it determines how many pounds of carbon can be contained in a filter of given solids capacity and how much treated liquid is retained by the filter cake. When two carbons differing in bulk density are used at the same weight per gallon, more gallons can be filtered with the higher-density carbon before the available cake space is full. Since cleaning and restarting a filter is expensive, the denser carbon permits less filter downtime cost.

Even more important when valuable liquids are being treated is the cost savings in lower product retention by carbons of high bulk density. At the end of the filter cycle, the filter cake holds treated liquid. Some is held between the carbon particles and is easily recovered by displacement with steam, air, water, or other suitable fluid. A substantial amount, however, is retained within the pore structure of the carbon. This is not recoverable by displacement. Figure 5 shows the relationship between bulk density and the volume of liquid held in the pore structure per pound of carbon.[1] At 18 lb/ft³ the retention in a blown filter cake is 0.180 gal/lb, and at 30 lb/ft³ the retention is 0.086 gal/lb. The difference in retention is 0.094 gal/lb of carbon. If the retained liquid is valued at $1 per gallon, a 9.4 cents per pound higher cost results by using the less-dense carbon. With very costly liquids, the difference may be several times the purchase price of the carbon.

Bulk density is also important when the carbon is removed by settling. A carbon with a higher bulk density will provide a faster settling rate, and the volume of sludge which will have to be handled and/or dewatered will be less.

Attrition Resistance of Granular Carbon   Attrition resistance of granular carbons is frequently miscalled hardness. This property is important because of the harsh way granular carbons are handled and conveyed. Since the carbon is usually transferred in slurry form by pumping and dewatered by conveyors, resistance to attrition is important if carbon losses are to be minimized. Abrasion or hardness numbers devised for gas carbons are not reliable indicators of losses in a typical granular carbon system for treating liquids.

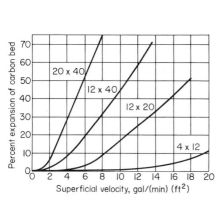

Fig. 4 Bed-expansion characteristics [water at 25°C (77°F)]. *(By permission, ICI United States, Inc.)*

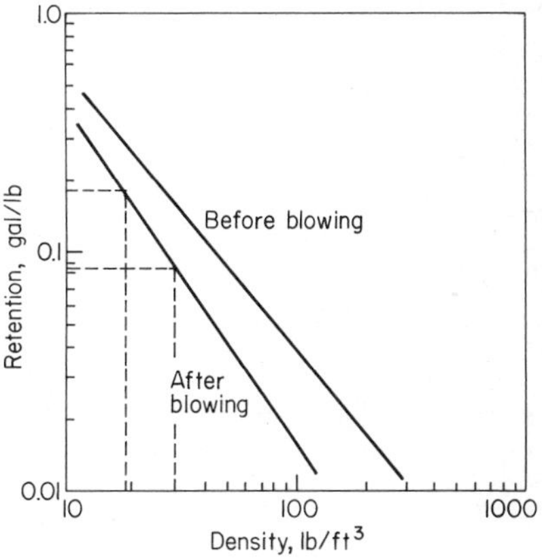

Fig. 5 Bulk density vs. filter-cake retention. *(By permission, ICI United States, Inc.)*

Of the analytical procedures currently in wide use, the National Bureau of Standards stirring-abrasion number appears to be the best indicator of relative handling and regeneration losses.[7,8,9] The test involves stirring a sample of carbon in a steel cup with an inverted T bar at 855 r/min for 1 h and measuring reduction in particle size. Granular carbons having NBS abrasion numbers of less than 25 (percent reduction in average particle size per millimeter of original average particle size) are suitable.

DEVELOPMENT OF DESIGN PARAMETERS

To make use of granular and powdered activated carbon economical at high carbon-usage rates (above 400 lb/day), the carbon must be regenerated for reuse.[2] Therefore, at high carbon-usage rates, data must be obtained which are suitable for designing both the adsorption and regeneration portions of the activated-carbon treatment system. At lower carbon-usage rates data for only the adsorption portion are required.

Powdered-Carbon Systems

Adsorption Portion   Powdered activated carbons can frequently be used in existing tanks, filtration, and/or settling equipment. In those cases, it is not necessary to develop design parameters, but it is necessary to determine how much carbon will be required to obtain the desired results so the operator will know how much carbon to use. The description and the development of design parameters for equipment required for powdered-carbon treatment, e.g., filters and clarifiers, are described in other sections of the handbook. However, several examples of possible powdered activated-carbon systems will give insight to other design considerations.

Batch Filtration System. Figure 6 shows a typical arrangement of the essential equipment for one type of powdered activated-carbon system. Many variations in arrangement, sizes, and types of the equipment are possible. Not included is a mechanical feeding device which could be used.

In the treating or mixing tank the carbon is thoroughly stirred with the impure liquid by means of a suitable mechanical agitator. Excessively rapid stirring should generally be avoided. Air beaten into the liquid may oxidize the impurity, changing its response to adsorption, or oxidize the liquid being purified. Where sensitivity to oxidation is known to be a serious factor, it may be necessary to conduct carbon treatment under vacuum or in an inert atmosphere.

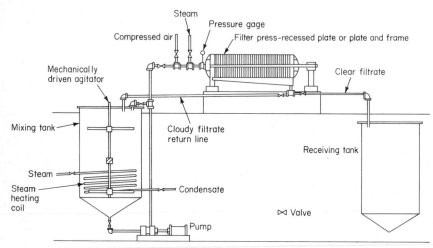

Fig. 6   Typical equipment for purifying liquids with powdered activated carbon.

Generally, carbon treatment should occur at as high a temperature as is compatible with the properties of the liquid. Heat can be supplied by built-in steam coils or by other means. The heating method chosen should provide adequate and economical heat transfer. Since powdered carbon is usually used at relatively low dosages (less than 1% by weight), plugging of heating coils is usually not a problem. Steam coils should be placed as low as possible so that during filtration the liquid will be kept at the desired temperature as the liquid level falls.

The stirring time (contact time) required to reach adsorption equilibrium is not generally the same as the time indicated by laboratory tests, since the stirring conditions are usually quite different. The laboratory contact time may be used as a first approximation for plant practice, and then revised as indicated by samples taken at suitably spaced time intervals from the treatment tank. Contact time is on the order of 20 to 30 min in most applications.

Filtration is begun after adsorption equilibrium is reached. During filtration, which usually requires an appreciable time, agitation should be continued to maintain a uniform suspension of the carbon remaining in the treatment tank.

Many types of filters can be used. Plate and frame, and vertical- or horizontal-leaf pressure filters are commonly used. Whatever type is used, the piping should be arranged so that the output can be returned to the treating tank. This is necessary because the first filtrate is usually cloudy and must be returned for refiltration. When the filtrate clears, it is directed to the final receiving tank.

Filter presses with bottom intake and top discharge are recommended because an effective and uniform precoat of the filter aid can be more easily attained.

Compressed air or steam lines connecting to the input line of the filter will permit "blowing" the cake to force out all free liquid held up in the cake and filter, thus minimizing retention losses. Most filters are equipped with backwashing facilities, permitting the filter cake to be washed with water to recover retained product. This diluted product is usually directed to a second tank.

A more detailed description of filters and the use of filter aids can be found in other sections of this handbook.

A pump is necessary to transfer the liquid-carbon mixture from the treating tank to the filter, to force under pressure the liquid through the filter, to recirculate the liquid, and to transfer the filtrate to the receiving tank. Both centrifugal and positive-displacement pumps are satisfactory. However, positive-displacement pumps are generally more expensive and check valves and moving metal parts in some types of positive-displacement pumps (such as gear or roller types) could require excessive maintenance, particularly in high-carbon-dosage applications.

With centrifugal pumps, filter pressure may be controlled by a valve on the discharge line. Such pumps should have sufficient clearance between impeller and housing to permit full throttling without stalling. They should also be designed to withstand the abrasiveness of carbon slurries.

With positive-displacement pumps, flow and pressure are controlled by a bypass valve in a pipe joining pump inlet and outlet, by adjusting the stroke or by a variable-speed drive. If a bypass is used, the pump always operates at full capacity, but part of its output is recirculated back to the suction side. As the bypass valve is opened, flow is diverted from the filter; so pressure against the filter medium is reduced. To maintain constant flow as filtration progresses, the bypass valve is gradually closed. When completely closed, maximum filter pressure is obtained.

In filtration there are two objectives of equal importance: maximum flow rate throughout the filtration cycle, and maximum clarity of filtrate.

Although carbon behaves as an incompressible solid in a filter cake, the liquid from which the carbon is being filtered frequently contains compressible solids. These are usually of a flocculant, colloidal, amorphous character. To minimize resistance to flow, pump pressure should be carefully regulated. At the beginning of filtration, the lowest pressure which will give a reasonable flow of liquid from the filter should be maintained. Pressure should be gradually increased to maintain this flow rate as the cake builds up.

The real filter medium in any filtration is the cake of deposited solids rather than the filter cloth, for example, which is a supporting surface for the cake. Until a cake is established, it is likely that fine particles will at first pass through the filter cloth. The liquid should be recirculated and refiltered until the filtrate runs clear. Once a cake begins to accumulate, even the finest particles are retained.

When the solids to be filtered out are fine and the filter medium is relatively coarse, the filtrate may not clear within a reasonable time. When products such as gelatin or pectin, which are colloidal, must be separated from carbon, the colloidal particles, or aggregates, carry the finest of carbon particles through even very retentive filters. When the solids to be removed are gelatinous, the pressure increase may be too rapid, causing short filter cycles. In each case, the problem can be lessened by precoating the filter surface with a 1/16- to 1/8-in layer of suitable retentive filter aid. Several weight percent of the filter aid could also be added to the carbon-treating tank just before filtration is begun.

Frequently, difficulty in clearing filtrate can be traced directly to holes in the filter cloth or screen. Support screens should be routinely checked and filter clothes closely inspected prior to use.

Flow rates in plant-scale filtration will vary from 3 to 4 gal/(ft²)(h) for viscous liquids to 30 to 40 gal/(ft²)(h) or more for low-viscosity liquids.

This is normally considered batch treatment; however, continuous operation can be obtained by duplication of equipment. Rotary vacuum-drum filters can be used for continuous operation.

Continuous System with Filtration. A diagram of a continuous system is shown in Fig. 7. The tank sizes are controlled by contact time and the flow rate of the liquid. If a 9 m³ (2400 gal/h) stream is to be treated for a 30-min contact time, the treatment-tank capacity is 4.5 m³ (1200 gal). To ensure complete treatment two 2.25-m³ (600-gal) tanks in series can be used, each equipped with an agitator and baffled to prevent hydraulic short circuiting. Both liquid and carbon slurry enter the lead tank continuously; treated solution with spent carbon leaves the second tank continuously. This continuous-treatment step should be followed by a continuous filtration to remove the spent carbon. Rotary vacuum-drum filters are suitable.

Continuous System with Settling. Another type of continuous system is shown in Fig. 8. In this case the carbon is added into a clarifier and allowed to settle while the liquid continuously flows into the unit and overflows. The settled carbon is periodically removed for regeneration, use in another stage of the treatment process, or disposal. Clarifier overflow rates with liquids with viscosities similar to that of water can vary from 0.976 to 8.44 m³/h [0.4 to 2.0 gal/(min)(ft²)] depending upon the type and design of the unit, and the activated carbon used. These overflow rates will generally permit the powdered carbon to settle while maintaining a reasonable size unit. If the application

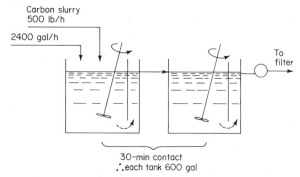

Fig. 7   Continuous tank treatment with powdered carbon. (*By permission, ICI United States, Inc.*)

requires that the product be free of suspended solids, this system will probably not give a satisfactory product unless clarification is followed by appropriate filtration.

Major Design Parameters. In general, the major design parameters for the design of the adsorption portion of a powdered-carbon system will depend on the type of adsorption system used.

For a mixing and treating tank with filter design for batch or continuous treatment, the major design parameters will be:

- The batch size. Volume of liquid to be treated.

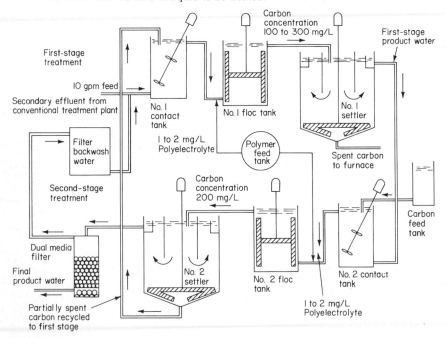

Fig. 8   Powdered-carbon two-stage countercurrent system. (*From Chem. Eng. Prog., Symp. Ser., Water 1970, by permission, American Institute of Chemical Engineers.*)

- The contact time. The time required with agitation to reach adsorption equilibrium.
- The filtration rate, m/h [gal/(min)(ft²)].
- The filtration time. This will be a function of the amount of carbon and filter aid used, the filter area, and allowable cake volume.

For a settling system, the major design parameters will be:
- The contact time
- The overflow rate
- Allowable solids in product

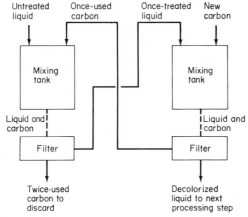

Fig. 9   Two-stage countercurrent carbon treatment. *(By permission, ICI United States, Inc.)*

Another consideration is the possibility of using the powdered carbon in countercurrent (Fig. 9) or divided-feed (Fig. 10) dosages. Both methods can reduce the carbon usage considerably, but more equipment is required.

Countercurrent treatment, by design, is continuous. Divided-feed treatment is designed for intermittent operation. However, continuous-treatment tanks can be used if

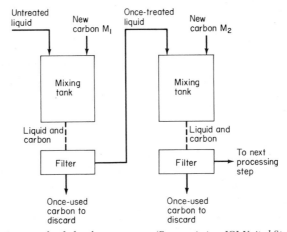

Fig. 10   Two-step divided carbon treatment. *(By permission, ICI United States, Inc.)*

it is important to keep the liquid at elevated temperature for the shortest possible time, or if space for equipment is limited.

In another method, the precoat or layer-filtration method, selected quantity of powdered carbon is mixed with treated liquid or with water to form a dilute slurry. The mixture is then pumped through the filter, forming a cake on the filter leaves. For liquids difficult to filter, a more porous cake can be developed by mixing filter aid with the carbon, often on a 50-50 basis. The filter cake should be uniform so channeling cannot occur. The liquid to be treated is then pumped once through or circulated through the filter cake, depending on the amount of treatment desired. Continuous treatment can be accomplished by multiple units.

TABLE 2   Adsorption-Isotherm Experimental Data

Carbon dosage, g/L	Concentration remaining, ppm C_f	Concentration reduction, ppm X	Reduction per unit carbon dosage, ppm/g/L $\dfrac{X}{M}$
0	400 (C_0)		
0.04	248	152	3800
0.08	164	236	2950
0.16	92	308	1920
0.32	34	366	1140

Regeneration (Reactivation) Portion   Regeneration (reactivation) of powdered activated carbon is just becoming commercially accepted. Numerous methods have been investigated,[11] many under the auspices of the Environmental Protection Agency. Most require that the excess water be removed from the carbon prior to thermal reactivation. In all cases, the major design parameter is the carbon-usage rate, e.g., the pounds of carbon which must be reactivated per day.

Design Procedure   Two design parameters, carbon-usage rate and the number of adsorption stages, for the adsorption and reactivation portions of a powdered-carbon system can be determined by a laboratory procedure, the adsorption isotherm. Other design parameters depend more on the type of equipment used in the system rather than the carbon and are discussed in other sections of this handbook.

Carbon Selection.   Numerous powdered activated carbons are commercially available which vary in many ways including particle size, adsorptive characteristics, bulk density, purity, and price. Although the properties which should be considered depend on the application,[3] generally the major considerations are the adsorptive capacity and delivered carbon price. Once the type of powdered-carbon system has been selected, various candidate carbons can be tested as described in the appropriate subsequent paragraphs. Generally, the powdered carbon giving the lowest operating cost on a price-performance basis will be the best.

Single-Stage System.   The carbon-usage rate or dosage can be determined by measuring the adsorptive capacity of the activated carbon with an adsorption isotherm.[1]

To run an isotherm, various quantities of carbon are added to a constant volume of liquid. The carbon-liquid mixtures are agitated vigorously for about 1 h. The carbon is then removed by filtration. The treated liquid is analyzed for the level of impurities and the amount adsorbed is found by the difference. The resulting data would appear as in Table 2. The first line applies to the blank untreated sample. The raw data are in the first two columns. X is calculated by subtracting C_f from C_0, which is the blank concentration, and X/M is computed. The adsorption isotherm is obtained by plotting X/M against C_f on logarithmic paper (Fig. 11). This isotherm is described by the Freundlich equation:

$$\frac{X}{M} = KC^{1/n}$$

X is equivalent to $C_0 - C_f$ (original concentration − final concentration). The equation of the isotherm then becomes

$$\frac{X}{M} = \frac{C_0 - C_f}{M} = KC^{1/n}$$

In logarithmic form the equation becomes

$$\log \frac{X}{M} = \log K + \frac{1}{n}\log C$$

which is of the form

$$Y = a + bx$$

So when X/M is plotted against C_f, on logarithmic paper, a straight line usually results. The line has an intercept K at $C = 1$ and a slope of $1/n$. From the plot, the dosage M

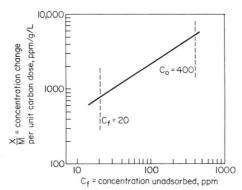

Fig. 11   Adsorption isotherm.

required to achieve a given desired residual concentration C_f can be calculated: at $C_f = 20$, the X/M value is read from the plot, e.g., 770. Since $X = C_0 - C_f$, in this case $400 - 20 = 380$, M is obtained by dividing X/M into X.

$$\frac{380 \text{ ppm}}{770 \text{ ppm/g/L}} = 0.49 \text{ g/L}$$

If 10,000 gal/day must be treated, the carbon-usage rate would be

$$0.49 \text{ g/L} \times \frac{3.785 \text{ L/gal}}{454 \text{ g/lb}} \times 10,000 \text{ gal/day} = 40.8 \text{ lb/day}$$

Countercurrent System. The adsorptive isotherm can be used to calculate countercurrent dosages or carbon-usage rates.[1] Figure 12 shows a very steep isotherm. The dosage calculation to reach 10% of the original concentration gives 306,000 ppm or 30.6% carbon on weight of liquid. This extremely high carbon dosage can be reduced by countercurrent treatment. Flowsheets for countercurrent treatment are given in Figs. 8 and 9.

The liquid is first treated with a quantity of once-used carbon which, after filtration, is discarded or regenerated. The once-treated solution is then given a second treatment with fresh carbon which, after filtration, becomes the once-used carbon for treating the next batch.

The isotherm in Fig. 12 shows that the once-used carbon meets solution having an impurity concentration of C_0 and leaves it with an intermediate concentration C_i. The carbon leaves the system at a loading corresponding to X/M at C_i instead of at the much lower loading indicated by X/M of C_f, the final concentration. The second treatment of the solution with new carbon takes the concentration from C_i to C_f and leaves the carbon after its first use loaded at the X/M value corresponding to C_f with reserve capacity for reuse to loading at X/M at C_i in its second use.

To determine what countercurrent dosage M_{cc} will achieve, the same final concentration C_f as the larger single-stage dosage M_s, these simultaneous equations derived from the Freundlich equation are solved:

$$\frac{M_{cc}}{M_s} = \left(\frac{C_f}{C_i}\right)^{1/n} \qquad \frac{M_{cc}}{M_s} = \frac{C_i - C_f}{C_0 - C_f}$$

They do not have an algebraic solution and must be solved graphically by trial and error. Figure 13 gives a general solution as a family of curves, each of which shows the variation of the ratio M_{cc}/M_s with the isotherm slope $1/n$ for a specific C_f value expressed as a fraction of C_0, the concentration before carbon treatment.

The first step is to measure $1/n$, the slope of the isotherm in Fig. 12. It is 1.60. On Fig. 13, go vertically from 1.6 on the abscissa to the intersection with $(100)(C_f/C_0) = 10\%$. The corresponding value on the ordinate is 18.5%, so that $M_{cc} = M_s \times 0.185 = 5.66\%$.

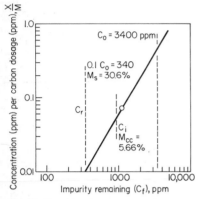

Fig. 12 Two-stage countercurrent-treatment isotherm. (*By permission, ICI United States, Inc.*)

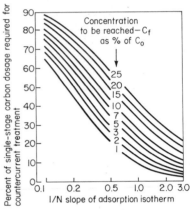

Fig. 13 Two-stage countercurrent carbon-dosage chart. (*By permission, ICI United States, Inc.*)

Three-stage countercurrent treatment will give even greater dosage reduction. A general solution is shown in Fig. 14. The three-stage countercurrent dosage is 3.27%. In general, as the number of stages is increased, the X/M effectiveness of the carbon after its final use approaches the X/M corresponding to C_0, which is the maximum obtainable. In Fig. 12 the three-stage value of X/M is shown by the circle on the line.

In most cases it is not practical to use more than three-stage treatment because the advantage gained by further stages does not justify the extra filtration.

Divided-Feed System. Countercurrent treatment is not practical when the liquid treated is available intermittently, because it is usually difficult to save or store partially used

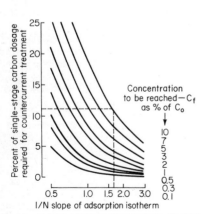

Fig. 14 Three-stage countercurrent carbon-dosage chart. (*By permission, ICI United States, Inc.*)

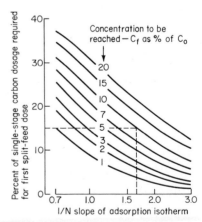

Fig. 15 Two-stage divided carbon treatment: Stage 1. (*By permission, ICI United States, Inc.*)

carbon. In such cases a sizable reduction in carbon dosage can be obtained with divided treatment as shown in Fig. 10.[1] Two treatments are used, each with fresh carbon. The minimum sum of the two treatments can be calculated from the Freundlich equation:

$$M_1 = \frac{C_0 - C_i}{KC_i^{1/n}} \qquad M_2 = \frac{C_i - C_f}{KC_f^{1/n}}$$

General solutions for M_1/M_s and M_2/M_s are in Figs. 15 and 16. From these figures and the isotherm in Fig. 12 the divided feed dosage can be calculated for a 10% residual impurity concentration as $M_1 + M_2 = 4.6 + 6.1 = 10.7\%$.

Precoat or Layer Filtration System. There is no convenient method for calculating the carbon-usage rate using the precoat or layer filtration method.[1] The relative effectiveness must be in actual pilot-plant or plant-scale simulation tests. However, in one application where liquid was filtered through a ⅜ in filter cake of powdered carbon to achieve 85% removal of impurities, the precoat method provided a 13% savings in carbon compared with single-stage batch treatment. The magnitude of the savings in any application will depend on the nature of the liquid, the carbon, and the treatment conditions used.

Granular-Carbon Systems

Adsorption Portion Granular carbons are most effectively used in columns where the liquid to be purified is passed through the bed of carbon. An example of what happens in a downflow column operation is shown in Fig. 17. The height of the carbon bed is h, the height of the adsorption zone is h_a, and the column diameter is d_c. The adsorption zone is defined as the portion of the carbon bed in which adsorption is occurring. It is the distance between the layer of exhausted carbon at the top of the zone and the layer of unused carbon at the bottom of the zone. Carbon at the top of the adsorption zone, which is always

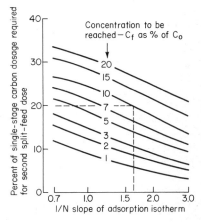

Fig. 16 Two-stage divided carbon treatment: Stage 2. *(By permission, ICI United States, Inc.)*

contacted by fresh, full-strength liquid, becomes exhausted (to the degree permitted by the operating conditions). Accordingly, after the lower boundary of the adsorption zone has progressed through the full depth of the bed and reached its bottom, the breakthrough has begun. As the adsorption zone moves out of the column, the breakthrough curve rises sharply. After the zone has moved all the way out, the entire bed of carbon is spent; the impurity concentration C_f of the effluent liquid is now the same as that of the feed C_0.

A number of factors determine the shape of the breakthrough curve and the height of the adsorption zone:

- Column dimensions
- Flow rate
- The carbon
- The composition and concentration of impurities in the feed liquid
- The temperature
- pH

Granular carbons can be used in two basic types of systems: fixed beds and pulsed beds.
Fixed Beds. Fixed bed systems can be operated as:
- A single column
- Multiple columns
- In parallel
- In series
- Combined parallel and series

A single-column system is indicated if: (1) the breakthrough curve for the carbon selected on the basis of laboratory tests is steep, (2) the carbon charge will last so long at the desired processing rate that the cost of replacing or regenerating it becomes a minor part of operating expense, (3) the capital cost of a second or third column cannot be justified because not enough carbon cost can be saved to pay for additional equipment, or (4) unusual temperature, pressure, or other controlling conditions must be maintained at the column (for example, to prevent crystallization or product deterioration).[13] Otherwise, a multiple-column system, which may be either series or parallel, should be designed.

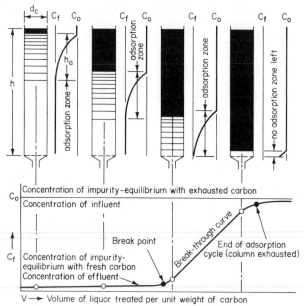

Fig. 17 What happens in a granular-carbon column. (*From Treybal, R. E., "Mass Transfer," McGraw-Hill, New York, 1955, by permission.*)

A multiple-column system is also indicated if: (1) the process cannot be interrupted for unloading, reloading, or regeneration, and a standby column is not available, and (2) the size or height of a single column will not fit existing space.[13]

A parallel-column system (a system in which the columns are placed on stream at evenly spaced time intervals, receive the same feed, and discharge into a common manifold—see Fig. 18) is indicated if pressure drop through the system is likely to be a problem. With this system, pumps can be smaller, power requirements lower, and

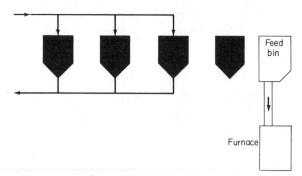

Fig. 18 Fixed-bed, parallel.

pressure specifications for columns and piping less stringent, especially if viscous materials will be treated or flow rates will be high.[13]

The carbon in a single column or in a parallel system which is removed for regeneration or discard usually is not completely spent. That is, the adsorption zone is still in the column when it must be removed.

A series-column system (in which the effluent from one column becomes the feed for the next—see Fig. 19) is indicated if: (1) the breakthrough curve is gradual and the highest possible effluent purity is desired, or (2) the combination of a gradual breakthrough curve and a high carbon requirement per unit of production makes it economically necessary to exhaust the carbon completely.[13]

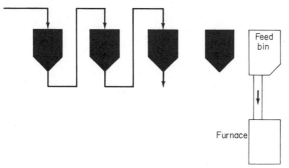

Fig. 19 Fixed-bed, series.

When the carbon in the lead column becomes spent, it is removed for discard or regeneration and a fresh column is put on stream at the end of the series. The second column in the series then becomes the lead column. The carbon removed has been more fully used than the carbon in the "downstream" columns because the adsorption zone has passed completely through it. The zone continues to pass through the "downstream" columns. The carbon in the column which was removed is completely exhausted while suitable effluent continues to be obtained from the last column in the series. Therefore, the operating costs for a series system would usually be lower than those for a single column or parallel system treating the same liquid.

A combined series-and-parallel system combines the high efficiency of series operation with the practicality of parallel design (see Fig. 20).

All of the foregoing are fixed-bed systems in which the carbon bed is confined and remains static. The liquid flow may be downward or upward.[13]

The best flow direction depends on the application. In general, with viscous liquids downflow or upflow with a restricted carbon bed (bed is not allowed to expand) is

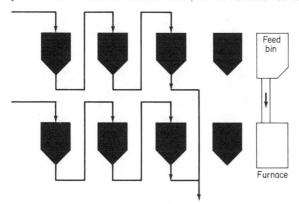

Fig. 20 Combined series-parallel.

preferred. The buoyancy of the carbon particles in a viscous liquid would result in the carbon's being floated out of an upflow adsorber with an unrestricted bed.

With liquids having viscosities similar to water, a major advantage to downflow operation is filtration. The suspended solids in the liquid will be removed by the finer carbon particles at the top of the bed. However, pumps must be sized to accommodate pressure

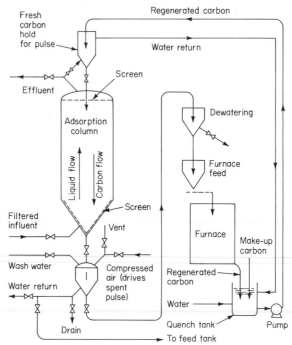

Fig. 21 Pulsed-bed system. (*From* Chem. Eng., *May 9, 1966, by permission, McGraw-Hill, Inc.*)

drop through the system, which could be substantial if suspended solids are present. When the pressure drop becomes unacceptable, the adsorber is backwashed. The backwashing frequency and volume of backwash water required can be high. Piping must permit downflow during the adsorption cycle and upflow during backwashing.

To operate upflow, the piping is simpler because the direction of the flow is the same during the adsorption and washing cycles. Pressure drops are low because the carbon bed expands with the flow. Only occasional washing is required to remove suspended solids which accumulate in the bed. Considerably less downtime and washwater are required. Upflow systems lend themselves to gravity feed. Although filtration occurs, upflow operation will not produce an effluent that is free of suspended solids or turbidity, and postfiltration is sometimes required.

For a given application, the sizes of the adsorption columns are identical; so the capital investment for major equipment would be similar.

Pulsed Beds. In pulsed beds, the carbon bed moves opposite to the flow of liquid. The liquid flows upward while spent carbon is removed from the bottom of the column and an equal volume of fresh or regenerated carbon is added to the top of the column. In effect, a pulsed-bed system is the same as a number of stacked fixed-bed columns operating in series. Removing a given volume of spent carbon from the bottom is similar to removing the spent carbon from the lead column in a series system. An example of a pulsed-bed system is shown in Fig. 21.

Pulsed beds are normally operated with the columns completely filled with carbon so there is no freeboard to allow bed expansion during operation or cleaning. Generally, it is desirable to maintain "plug" flow of carbon so a sharp adsorption zone will be maintained. Although some pulsed-bed units do permit bed expansion during operation and cleaning, the efficiency of the system suffers because mixing of the carbon disturbs the adsorption zone and the carbon withdrawn for regeneration could contain a mixture of spent and partially spent granules.

Pulsed-bed systems can be operated on a continuous or semicontinuous basis. Continuous operation means that spent carbon is continually withdrawn from the bottom of the column and virgin or regenerated carbon is continually added at the top. In a semicontinuous operation, which is the more common, a given volume of spent carbon is removed from the bottom of the column at given intervals, such as once every 8-h shift or once daily, and an equivalent volume of virgin or regenerated carbon is added to the top.

This process comes the closest to completely exhausting the carbon with a minimal capital investment (except perhaps for the instrumentation).

When to Use Pulsed- or Fixed-Bed Systems. Pulsed-bed systems and fixed-bed series systems are used to maximize the use of the activated carbon, thereby reducing operating costs. This is accomplished by regenerating only the carbon in the system which is the most completely used, that is, the carbon which is most nearly in equilibrium with the feed. This carbon has adsorbed more impurities than carbon farther "downstream" which has not been exposed to the impure feed.

Pulsed-bed systems are generally used when the feed is free of suspended solids and the carbon-usage rate is high. A pulsed-bed system cannot be used effectively when the feed is biologically active or contains high suspended solids. With such a feed, a given volume of spent carbon (a pulse) must be removed from the bottom of the column when the pressure drop builds up to an unacceptable level rather than when the carbon has been spent. Since the carbon has not been used to its fullest extent before regeneration, operating costs will be higher than they should be. Therefore, prefiltration of the liquid is recommended. An alternative is to use a pulsed-bed column with enough freeboard to permit cleaning of the bed (this can be done by passing water upflow through the bed at a rate significantly higher than the normal operating rate).

Fixed-bed systems are normally used when the liquid contains suspended solids or is biologically active and when carbon-usage rate is low.

Regeneration (Reactivation) Portion   To make the use of granular carbon economical at high carbon-usage rates [greater than about 180 kg/day (400 lb/day)], regeneration (reactivation) of the carbon is necessary.[2]

Regeneration Methods. The purpose of regeneration is to return the used carbon to its original adsorptive capacity by the removal of the adsorbed impurities. Regeneration can be by thermal, chemical, hot gas, solvent, or biological methods.

In thermal regeneration (reactivation), the carbon is heated in multihearth or rotary furnaces in the presence of steam to volatilize and carbonize sorbed material. The char residue is then activated. Gas-phase temperatures can range from 650 to 1040°C (1200 to 1900°F).

Chemical regeneration involves reacting the sorbed material with a regenerant to remove it from the carbon. For example, a carbon used for phenol removal can be regenerated with sodium hydroxide to produce water-soluble sodium phenolate.

Hot-gas regeneration can be used when a carbon has adsorbed a low-boiling organic material. Steam [to 120°C (250°F)], hot CO_2, air, nitrogen, etc., passing through the carbon causes the low-boiling material to be vaporized. The organic material is then removed from the regenerant by condensation or decantation.

Solvent regeneration involves passing a suitable solvent through the spent carbon to dissolve the sorbed material. The solvent can be recycled until it must be discarded or purified by distillation and reused.

Biological regeneration involves allowing aerobic, anaerobic, or both types of bacteria to remove adsorbed biodegradable material.

Sorbed materials can be recovered for reuse or sale in some cases where chemical, hot-gas, or solvent regeneration is used. However, the recovered sorbed material generally must be valuable and/or the regenerant and/or reaction products must be economically used in other parts of the plant, reprocessed for reuse, or otherwise economically disposed of to make these types of regeneration more economical than discarding the spent carbon.

Generally regeneration methods other than thermal will not be adequately effective if a mixture of organics has been adsorbed. Only a portion of the sorbed materials will be removed by a given solvent, hot-gas, chemical, or biological regeneration procedure. Therefore, the performance of the carbon will consistently decrease on successive regenerations, and after a few regenerations, the carbon usually has to be discarded. Thermal regeneration is the most versatile, and because most granular-carbon treatment systems remove a complex mixture of organics, it is by far the most widely used. For these reasons, only thermal regeneration will be discussed further.

Thermal Regeneration (Reactivation). Thermal regeneration (reactivation) is usually accomplished in either multihearth or rotary-tube furnaces.

Figure 22 shows a multihearth furnace. In a multihearth furnace, the removal of adsorbed organics is achieved by destructive distillation of adsorbed organic material and thermal oxidation of the char residue (from the pyrolysis).

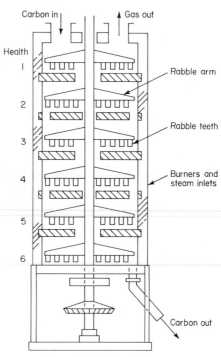

Fig. 22 Multiple-hearth furnace.

Moisture and volatile adsorbed materials are removed from the spent carbon in the upper hearths. The first hearth is operated with a 200 to 320°C (400 to 600°F) gas-zone temperature—carbon-bed temperature is probably 100 °C (212°F) in the first hearth because of the moisture evaporation.

On the next two or three hearths, the temperature is increased to about 430°C (800°F) and the adsorbed volatile material is driven off by destructive distillation. Only organic char and ash residue remains in the carbon pores.

On the bottom hearths, heat is supplied internally by gas- or oil-operated burners, and steam or water is injected. This is the activating zone. Temperatures can range from 650°C (1200°F) to 1040°C (1900°F) depending on the amount of sorbed material in the carbon, the residence time, and the type of carbon used. In this zone, the char is "burned" off and/ or activated. If the oxygen content in the furnace is not carefully controlled, excessive carbon losses will occur. The oxidation reaction ($C + O_2 = CO_2$) is exothermic. This favors the reaction to occur on the external surface and in larger pores near the outside of the carbon particle. Conversely, the steam-carbon reaction ($C + H_2O = CO + H_2$) is endothermic. This results in a much lower temperature gradient between the external and interior surfaces of the carbon particles. So development of smaller internal pores is favored by a steam atmosphere.

Unfortunately, part of the original carbon is burned along with the char residue. The amount of original carbon which must be burned to restore the carbon's adsorptive properties is proportional to the amount of adsorbed material inside the carbon pores. Reactivation conditions must be more stringent the more adsorbed material present.

After passing through the activating zone, the reactivated carbon falls into a quench tank (usually filled with water) and is transported to the adsorption system. A typical multi-hearth reactivation system is shown in Fig. 23.

Figure 24 shows a rotary-tube furnace. In a rotary tube the carbon tumbles through a sloped rotating tube while contacting the hot gases. There are two basic types of rotary-tube furnaces, internally and externally fired. Internally fired units generally have the

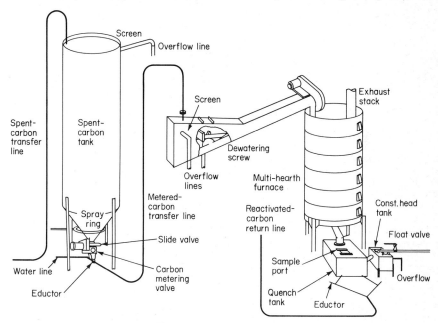

Fig. 23   Reactivation system.

burner in the carbon-discharge end of the tube. This is the activating zone and the hottest portion of the furnace. Gas-phase flow is countercurrent to the carbon flow, as it is in multihearth furnaces. Steam or water is sometimes injected into the activating zone. The mechanism for carbon reactivation is the same as that in a multihearth furnace.

Externally and internally fired rotary tubes are similar except that the heat source (burners) in externally fired rotary tubes is outside the tube. Both types of rotary furnaces are being used commercially to reactivate carbon. Most rotary furnaces presently in operation are internally fired.

Both types of rotary-tube furnaces and the multihearth furnaces appear to restore the carbon's adsorptive capacity with reasonable carbon losses.

Furnace manufacturers should be contacted for comparative assets and liabilities of each type of furnace.

Reactivation Control. Control of the reactivation is merely the control of carbon reaction rate in the furnace to produce carbon having essentially the same properties as the original carbon.

The primary variables affecting carbon reactivation are feed rate, temperature, residence time, and gaseous environment. These variables are dependent upon furnace design and the quantities of moisture and sorbed impurities in the spent carbon. Considerable variation in these factors is possible. Production scheduling or variations in carbon-usage rate may require an increase or decrease in carbon throughput rate. Therefore, analytical control of the reactivation process is required to maintain good quality in the reactivated carbon.

Measurement of adsorptive capacity for the impurity being removed in the plant determines the success of restoring a spent carbon. However, adsorption-isotherm tests are generally time-consuming and are not suitable as a rapid control technique. Quick adsorption tests with iodine and/or molasses are frequently used. Iodine adsorption gives an indication of the restoration of small pores, while molasses indicates large-pore restoration. Neither will necessarily correlate with the carbon's performance in a given application.

Measurement of carbon bulk density is the most commonly used procedure for reactivation control. As the carbon becomes exhausted in the adsorption process, it gains weight but its volume is not affected and its bulk density increases. If the reactivation process removes all the sorbed material, the carbon's bulk density will approximate that of the virgin carbon.

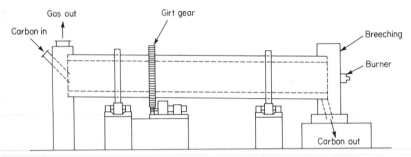

Fig. 24 Rotary-kiln reactivation furnace.

The measurement of carbon bulk density has an advantage over adsorption testing in that it is much simpler and quicker. It can be measured by pouring the carbon into a container of known volume and weighing the carbon. This technique suffers somewhat from lack of precision because of variable packing. A better method is to feed the carbon into the volume-calibrated container (graduated cylinder) from a prescribed height and standard rate using a vibrating trough. This assures greater precision in the measurement.

If the reactivation conditions result in a furnace product which is either over or under the desired reactivation control point, the feed rate and/or temperature should be changed. Feed rate is the ideal control variable for reactivation because the system responds more quickly to changes in feed rate than to any other variable change. The response time (time required to affect changes in product properties) is usually about 30 min.

Typically, feed rate is increased 5 to 10% if product is overreactivated or decreased the same amount if product is underreactivated.

Temperature is another controllable variable used to produce desired product properties. The principal drawback to this type of control is that 2 to 4 h is required after each temperature change before product properties reflect the change and are stabilized. If furnace product is underreactivated, temperatures should be raised by about 10°C (40°F), depending on the magnitude of underreactivation. Conversely, if carbon is being overreactivated, temperatures should be reduced by about 50°F.

Major Economic Factors   The major constant economic factors in the operation of a granular-carbon adsorption and regeneration system are depreciation, overhead, labor, and maintenance. These factors are established when a plant is built and will not change greatly in a plant operating at a "steady state" in a stable economic climate.[9]

The major variable economic factor is the carbon makeup costs. Carbon makeup costs are related to the percentage of carbon physically lost during an adsorption and reactivation cycle and the delivered price of the carbon. Normal losses range from 5 to 10%. Delivered prices for liquid-phase granular carbon range from $0.30 to $0.80/lb.

The factors which affect the carbon losses are:
- The design of the carbon adsorption, transporting, and reactivation system
- The operation of the reactivation furnace

- The frequency and duration of washing cycles necessary to keep the carbon beds reasonably clean
  - Quenching the reactivated carbon
  - The carbon-usage rate
  - The reactivatability of the carbon
  - The abrasion resistance of the granular carbon[9]

Of these the system design and the operation of the reactivation furnace have the greatest effects.

System Design. To minimize the amount of carbon lost as a result of attrition, the adsorption, reactivation, and carbon-handling system must be properly designed. System design will be discussed in subsequent paragraphs.

Operation of Reactivation Furnace. Poor operation of the reactivation furnace can contribute significantly to high carbon losses. In most cases where unusually high losses have occurred, the cause has been traced to poor furnace operation. Abnormally high losses in the furnace can be caused when:

- Reactivation rate too high because:
  - Residence time is too long.
  - Temperature is too high.
  - Inorganic catalysts are present.
  - Excessive oxidizing atmosphere is in furnace.
- Feed rate or moisture content of feed to the furnace too variable.
- Furnace draft too high, causing carbon to be blown into the afterburner or out the stack.
- Incorrect regeneration control point.[9]

Carbon Handling   Adsorbers may be filled with the carbon slurried or dry. Slurry loading is preferred, especially when water can be used.

Users of large quantities of carbon may receive carbon in bulk railroad cars or trucks. The most convenient unloading method is to dump the dry carbon by gravity into a storage tank below ground level. From that tank it can be slurried and hydraulically conveyed to the adsorbers. Another effective and clean method is to dump the carbon either directly into an eductor or a slurry tank and pump the slurry into the adsorbers. Eductors commonly used are 100 mm (4 in). Recessed impeller centrifugal slurry pumps are also suitable. Flexible, but not collapsible, 50- to 65-mm (2- to 2 1/2-in) hose is preferred. Water pressure to operate the eductors must be at least 4.2 kg/cm [60 lb/in² (gage)]. Higher pressures may be required, depending on the discharge head. If the trailers are so equipped, water or air pressure can be used in the trailer to assist in dumping the carbon.

It is also possible to unload bulk shipment dry by blowing the carbon directly into the adsorbers or storage tanks. However, this is generally not recommended because dust-control methods, sometimes elaborate, are required.

Users of smaller quantities of carbon may receive carbon in 18- to 27-kg (40- to 60-lb) bags or in pallets. Carbon to initially fill an adsorber or to provide makeup can be loaded into the reactivation-furnace quench tank to make a slurry, and hydraulically transferred to the adsorber. When an adsorber is being loaded with slurry, the excess water can be drained off simultaneously if it is equipped with a large enough drainage outlet or allowed to overflow. If allowed to overflow, the outlet should be screened or enough freeboard should be available to keep carbon granules from being carried out. The overflow procedure will help remove carbon fines. Wet loading lessens air entrapment.

When dry loading, several feet of water should be put into the adsorber so that screens or fittings will not be damaged. The carbon may be dumped manually into the top of the adsorber or transferred in by a conveyor loaded at ground level.

After the dry carbon has been charged, process liquid (or water) is slowly added, with the adsorber vented and the effluent valve closed, to prevent air entrapment and allow the carbon to settle. The charged adsorber should stand for at least 30 min.

When the adsorption cycle is started, the feed liquid should be introduced at a low rate and increased gradually.

Spent carbon can be removed from an adsorber by filling the column with water and applying low air or water pressure at the top of the carbon bed to help flush the carbon from the bottom of the adsorber.

When the spent carbon is reactivated, carbon handling is more critical than when it is discarded. The treatment the carbon receives in fixed-bed and pulsed-bed systems is basically the same (see Figs. 22 and 23) except that in the pulsed-bed system the excess product may be extracted from the spent carbon in a vessel separate from the adsorption column, while in the fixed-bed system this is done inside the column.

Pipelines for carbon transport can vary from 25 to 150 mm (1 to 6 in) diameter, depending on the size of the installation, the speed of handling required, and where they are in the system. They should contain sweeping bends rather than sharp right-angle elbows. Rubber elbows are frequently used. Piping is commonly made of black iron.

Eductors, blowcases (water or air pressure), diaphragm pumps, or recessed-impeller centrifugal slurry pumps should be used in the carbon-slurry transport system. Dry carbon can be transferred pneumatically in a properly designed system.

For pumping, the carbon-water ratio in a slurry should be about 0.9 kg (2 lb) of carbon per 3.79 L (gal) of water. A linear velocity of 0.9 to 1.8 m/s (3 to 6 ft/s) prevents the carbon from settling out and limits attrition. Pressure drops for carbon-water slurries approximate those for water alone.[13]

The "Process Design Manual for Carbon Adsorption" published by the U.S. Environmental Protection Agency, October 1973, has an excellent section on carbon handling.

When the spent carbon is to be reactivated, excess water should be removed before the carbon is put into the furnace. Dewatering systems for this purpose may employ vibrating screens, rotary vacuum screens, dewatering screws, or drainage hoppers which support the wet carbon on a fixed screen.

Gravity drainage will reduce the water content of the slurry to about 55% in about 12 h. A vibrating screen or rotary filter will not lower the water content much further but will reduce the drainage time. A 60-mesh screen is suitable for either method. The carbon can be fed to the reactivation furnace by a screw, vibrating, or belt conveyor.

As reactivated carbon emerges from the furnace, it falls into a quench tank. The quench tank is arranged so that the outlet from the reactivation furnace is always submerged in water. The makeup-water inlet and slurry outlet connections are installed to provide turbulent flow. The quench water, which is at ambient temperature, serves to cool the reactivated carbon. The resulting carbon-water slurry is then pumped into a storage hopper or an adsorber.[13]

Cleaning   Cleaning adsorbers is required in some applications to remove carbon fines, suspended solids which have accumulated during the adsorption cycle, or biological growth (in wastewater applications). Cleaning is normally accomplished with water. If flow is downward during the adsorption cycle, cleaning is usually accomplished by backwashing. If the flow is upward during the adsorption cycle, cleaning is usually accomplished by increasing the flow rate to allow the entrapped material to escape through the expanded carbon bed.

When an adsorber is first filled, some classification of the bed by density and particle size occurs. This may concentrate fine particles in sections of the bed, causing a high pressure drop or a low flow rate. When this happens, backwashing can remove the fine particles. If the adsorber effluent is filtered, cleaning also reduces the quantity of fines trapped by the filter in the effluent line.

The optimum cleaning rate is that which removes the fines but takes the least amount of useful carbon out of the bed. If a bed is cleaned at too low a rate, the fines not removed will compact on the top of the bed as soon as downward flow is started. Cleaning at too high a rate will reduce bed volume unnecessarily. Carbon particle size is the variable that determines the correct rate.

The optimum cleaning rate for granular carbon of 12 × 20-, 12 × 40-, and 20 × 40-mesh sizes, when the viscosity of the liquid approaches that of water, is 245, 131, and 61 L/(min)(m²) [6, 3.2, and 1.5 gal/(min)(ft²)], respectively. The recommended duration of backwash, when the rate is optimum, is 20 min per 3 m (10 ft) of bed height. If the rate is lower, the backwashing time should be longer.[13]

In wastewater applications cleaning rates are usually much higher because suspended solids and biological growth must also be removed. Cleaning rates vary from 326 to 734 L/(min)(m²)[8 to 18 gal/(min)(ft²)], depending on the application and the amount of free-board available. Cleaning is sometimes preceded by air scouring the carbon bed [passing air upward through the bed at 9 to 15 m³/m² (surface) (3 to 5 scfm/ft²)] to remove biological growth on the carbon granules and to inhibit anaerobic biological activity in the bed. Cleaning can be as frequent as twice daily.

Design Procedure The two major portions of many granular-carbon systems are the adsorption and reactivation portions.

The major design parameter for the adsorption portion is the contact time between the liquid and the carbon bed (the residence time in the carbon bed). Contact time (residence time) is based on the bed volume including the voids between the carbon particles but not including freeboard. When the residence time is established, the size of the adsorbers (or carbon bed) is established.

The major design parameters for the reactivation portion are (1) the carbon-usage rate or the reactivation rate required to maintain acceptable product, and (2) the residence time required in the furnace to obtain adequate restoration of adsorption capacity. The reactivation rate and the residence time in the furnace are the basic parameters used to size the reactivation system.

Once a reliable estimate of the combination of adsorption-system and reactivation-system size has been obtained, meaningful investment and operating costs can be estimated.

Carbon Selection. Numerous granular activated carbons are commercially available which vary in many ways including particle size, pore-size distribution, abrasion resistance, adsorptive characteristics, and price. In designing a granular-carbon system, however, the properties of virgin carbon are not as important as the properties of the carbon after it has been reactivated numerous times.

When granular carbon is regenerated, the total surface area, the pore volume and surface area in the small pores, and the surface area available for adsorption are significantly reduced.[9] This is attributed to ash plugging of or inadequate removal of impurities from the small pores. In cases where the small pores are used for adsorption, the performance will decrease as the number of regenerations increase. In applications where larger molecules are adsorbed, the performance of the carbon will not be greatly affected by regeneration.

In general, granular carbons, regardless of the manufacturer, which have undergone numerous regenerations, should have essentially identical pore structures and have essentially identical adsorptive characteristics.

It is extremely important that tests to develop design parameters be run with a carbon which has been regenerated numerous times to avoid the possibility of seriously undersizing the granular-carbon system. If a suitable regenerated carbon is not available, tests must be done with a virgin carbon but should include regeneration tests to determine how much the carbon's adsorptive properties will change on regeneration. A third alternative is to use a carbon in the test work which is not likely to change significantly on regeneration. Granular carbons made from lignite are likely to change less than carbons made from other raw materials.[4]

Particle size is another important consideration. Relatively coarse carbons (12 × 20 mesh, 8 × 30 mesh, etc.) are generally used in systems with downward flow, a high linear flow rate, relatively high suspended solids, relatively low impurity-removal requirements, and/or high-viscosity liquids. Finer carbons (12 × 40 mesh, 20 × 40 mesh, etc.) are generally used with upward flow, a low linear flow rate, relatively low suspended-solids levels, relatively high impurity-removal requirements, and/or low-viscosity liquids.

Feasibility Testing. Granular activated carbon is not suitable for solving all problems where dissolved solids or liquids must be separated from liquids. Therefore, before costly tests suitable for developing design data are conducted, preliminary feasibility tests should be made.

The first test to determine feasibility is usually the adsorption isotherm with pulverized samples of granular carbon. The adsorption isotherm has been discussed in preceding paragraphs.

The isotherm can give:
- A general idea of how effectively carbon will adsorb impurities present, and if the purification requirements can be obtained.
- An estimate of the maximum quantity of impurities a granular activated carbon will adsorb.
- "Ballpark" data to judge whether activated carbon may be an economic way to purify the liquid.

These tests cannot give definitive scale-up data for a granular-carbon system primarily because adsorption equilibrium is not obtained in a column, continuous flow causes preferential adsorption, and in wastewater treatment biological activity can occur in a column.

Column Testing. If the adsorption isotherm indicates a granular-carbon system may be suitable, granular-carbon column tests should be run to confirm feasibility and to develop design parameters.

Properly run column tests are operated using conditions which are expected to be used in the actual plant system. This will allow reliable results to be obtained directly or by extrapolation.

Improperly run tests can give results which are misleadingly poor and could cause unwarranted abandoning of granular carbon as an economical treatment method.

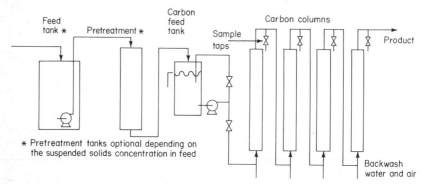

Fig. 25   Granular-carbon column test system. (*Pretreatment tanks optional, depending on the suspended-solids concentration in feed.)

Some undesirable column test practices are running tests:
- At linear flow rates or residence times unrealistic for plant-scale operations.
- In columns less than 1 in in diameter.
- For only a few hours or days to demonstrate that the desired impurity level can be reached, and not running long enough to obtain unacceptable effluent.
  - Using unrepresentative liquids and treatment conditions, such as temperature.
  - Using a feed which has not been pretreated as it will be in the plant.

To design a granular-carbon system, data must be obtained to size the adsorption and regeneration portions of the system.

One test uses a pulsed-bed column. The column is operated under conditions which ensure that the carbon intermittently and regularly removed from the bottom is completely exhausted (but not the carbon immediately above it) while acceptable effluent is maintained. The resulting residence time and carbon-withdrawal rate are the adsorption- and regeneration-system design parameters. This test is best for developing design parameters for a pulsed-bed system.

Another more conventional test system is shown in Fig. 25. It consists of at least four columns containing granular carbon operating in series. The more columns used, the better. A single test run in this system can provide data for pulsed-bed, series, and parallel system design.[14]

A column can be constructed from glass, plastic, reinforced fiberglass, or metal pipe. It should be constructed to allow loading of the carbon through the top.

For the downflow system, an influent line should be installed at the top of the column, with an effluent line at the bottom. To prevent the column from draining during operation (effluent leaving faster than influent entering), the effluent line should extend from the bottom of the column to the top of the carbon bed. This will keep the column filled with liquid at all times during operation.

If the column feed will contain suspended solids (to be avoided if possible), arrangements should be made to clean the carbon bed as required. This would involve the installation of a line from a source of clean water or column-treated process liquid to the bottom of the column.

The bed can be supported by glass wool, wire cloth, or other suitable retainer which will retain 100-mesh (U.S. Standard sieves) particles.

If the liquid is to be treated at an elevated temperature, the column should be heated by some suitable means such as electrical heating tape or jacket.[6]

If pretreatment of the liquid will be practiced in the plant system, the liquid should be pretreated before entering the carbon columns. The total bed depth of carbon and the linear flow rate should approximate that expected to be used in the plant system. Either up- or downflow can be used.

TABLE 3 Column-Exhaustion Data

		APHA color for column No.				
		1	2	3	4	5
Cumulative processing time, h*	Feed	Residence time, min				
	0	15.4	29.8	43.1	57.2	72.2
0	70,000			Start-up		
5	70,000	25,000	1,600	0	0	0
21	70,000	50,000	35,000	25,000	10,000	100
37	75,000	75,000	60,000	40,000	30,000	15,000
57	75,000		75,000	55,000	50,000	30,000
71	75,000			60,000	55,000	40,000
89	60,000			60,000	60,000	40,000

*4-in ID columns at 69L/(min)(m²) [1.7 gal/(min)(ft²)].

As liquid is passed through the system, instantaneous grab samples or daily or shift composite samples of the first column feed and of each column effluent are taken at regular intervals. The concentration of impurities in each sample is measured and recorded as in Table 3. Figure 26 shows these data, "APHA color" remaining plotted against processing time. The resulting curves are called "column-exhaustion curves." When the impurity concentration of the effluent from the last column in the series becomes unacceptable, the test can be stopped.

Sizing the Adsorption System. The adsorption system must be large enough to provide enough residence time to give suitable effluent. For example, greater than 1375 APHA

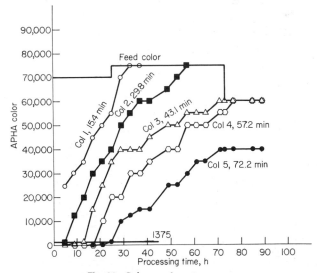

Fig. 26 Column-exhaustion curves.

color is considered unacceptable, in the application for which the data in Table 3 were developed. Suitable effluent was not obtained from Column 1 and was obtained from Column 2 for about 4.3 h. The total residence time in the first two columns was 29.8 min. Since suitable effluent was obtained, 29.8 min residence time is adequate. However, this procedure does not ensure that the carbon removed for discard or reactivation has been completely exhausted.

In a system designed for maximum performance, the carbon removed from a pulsed-bed system or the lead column in a series system should be completely exhausted. The residence time required so the carbon removed will be completely exhausted while suitable effluent is being maintained can be determined from the data in Table 3 and the

TABLE 4   Residence-Time Design Data

Column No.	Residence time,* min	Approx effluent color from first column at 1375 APHA color breakthrough	Color remaining †
1	15.4		
2	29.8	25,000	35.7%
3	43.1	35,000	50.0
4	57.2	45,000	64.3
5	72.2	55,000	78.6

*Based on volume of carbon bed including void volume.
†Feed color = 70,900 APHA.

curves in Fig. 26. The percentage color or impurity remaining in the effluent from the first column is calculated at the time the effluent from each column becomes unacceptable as in Table 4. Since the effluent from each column represents treatment at a specific residence time, the relationship between residence time and the color remaining in the effluent from the first column when breakthrough occurred at that residence time can be plotted as in Fig. 27. The residence time which gives suitable effluent while allowing the first column to be completely exhausted is the residence time when the curve or extrapolated curve crosses the 100% color-remaining line. In this example, the required residence time is about 93 min. Since this was a 5-column system, a 93-min residence time would give suitable effluent while permitting the first 1/5 of the total carbon bed to become completely exhausted.

For a pulsed-bed system where 1/20 of the bed would be removed periodically, the residence time could be somewhat less than 93 min.

If the plant-system liquid flow is not relatively constant, the adsorbers must be sized so suitable effluent will be produced during peak flow. If the peak flow is double the average flow, the adsorption-system size should be doubled to give adequate residence time at peak flow. The economics of using a surge or equalization tank with smaller adsorbers should be considered.

The configuration (height and cross-sectional area) of the adsorbers and the number of adsorbers in the adsorption system should be such as to allow the residence time required in the floor space and headroom available. The linear flow rate at peak flow should be the highest possible which will not cause intolerable pressure losses.

Typical linear flow rates for liquids with viscosities similar to water at average plant flows for parallel and single-column fixed-bed units are from 40.8 to 163L/(min)(m²)[1 to 4 gal/(min)(ft²)]; for series systems, from 122 to 286L/(min)(m²)[3 to 7 gal/(min)(ft²)]; and for pulsed-bed systems, from 204 to 367L/(min)(m²)[5 to 9 gal/(min)(ft²)]. For higher-viscosity liquids, lower linear flow rates are required. Once the residence time is established and a system configuration and linear flow rate are selected, the carbon-bed depths and adsorber-sidewall heights can be calculated.

The various manufacturers of granular-carbon adsorption equipment prefer specific system configurations, flow directions, materials of construction, adsorber linings, underdrains, and flow distribution and collection nozzles or launders depending upon the application. The manufacturers should be consulted to determine the assets and liabilities of the various alternatives.

Sizing the Regeneration (Reactivation) System. Size of the regeneration portion of the system is based on the carbon-usage rate and the residence time required in the furnace to

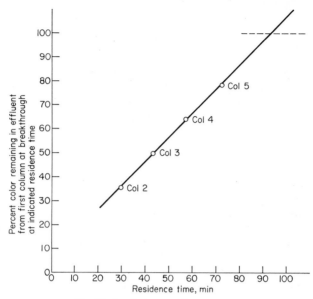

Fig. 27   Residence-time design curve.

remove the sorbed material adequately. To determine the carbon-usage rate, breathrough points are obtained from the column-exhaustion curves (Fig. 26). The "breakthrough point" is that point (processing time) on the column-exhaustion curve when the impurity concentration in any column effluent is no longer acceptable. Table 5 shows an example of column-breakthrough data. Figure 28 shows a plot of those data. The processing time at breakthrough should be a linear function of the residence time if the linear flow rate through the carbon bed and the impurity concentration and composition in the feed are reasonably constant throughout the test.[14]

The resulting curve shows how the breakthrough point (or adsorption wavefront) moves through the carbon bed. Since the processing time and residence time is a linear relationship, the curve can be described by the equation of a straight line.

TABLE 5   Residence Time–Processing Time Data*

Column No.	Residence time,† min	Processing time at breakthrough, h
1	15.4	0
2	29.8	4.3
3	43.1	12.9
4	57.2	17.4
5	72.2	24.6

*At 1.7 gal/(min)(ft²) linear flow rate. Average feed color 70,900 APHA and breakthrough at 1375 APHA.
†Based on volume of carbon bed including void volume.

The equation is

$$t = ar + b$$

where t = service time at breakthrough, h
$\quad r$ = residence time, min
$\quad a$ = slope
$\quad b$ = ordinate intercept

The slope a of the line is the numerical value in hours per minute of the amount of time required for the adsorption wavefront to move through a volume of carbon bed which provides a 1 min residence time under the test conditions used.

The reciprocal of the slope is the rate at which the carbon bed is spent. Multiplying this value by the apparent bulk density of the carbon, the expected cross-sectional area of the on-stream plant-scale adsorbers, and the expected plant-scale linear flow rate will give an estimate of the carbon-usage rate (or the rate at which the carbon must be regenerated) to produce acceptable product continuously.

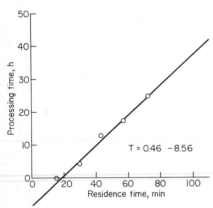

Fig. 28 Residence time–processing time curve.

The intercept of the abcissa (X axis) in Fig. 28 is the critical residence time (R_0). The critical residence time is defined as the minimum residence time required to obtain satisfactory effluent at time zero under the operating conditions used.

The intercept b of the ordinate (Y axis) in Fig. 28 is a measure of the rate of adsorption and is defined as the time required for the adsorption wavefront to pass through a volume of carbon equivalent to that which would provide a critical residence time R_0 under the test conditions used.

This equation describes the movement of an adsorption wavefront through a single fixed carbon bed. Therefore, using the equation directly to develop designs for systems other than single fixed-bed operation will give conservative design (oversized).

If a single fixed-bed column were removed for discard or reactivation when it began to produce unacceptable effluent, the carbon removed would not be completely exhausted (in equilibrium with the concentration of impurities in the column feed). Therefore, series or pulsed-bed systems are used to ensure that the carbon removed from the system has been more fully used. Ideally the carbon removed will be completely exhausted.

Pulsed-bed systems are typically designed so one-twentieth of the bed is removed for discard or reactivation at each pulse. Its performance is equivalent to that of a series system with 20 onstream columns operating at the same total residence time. Series systems normally contain from two to four on-stream columns.

A test comparing the carbon-usage rate in a pulsed-bed system with carbon-usage rate

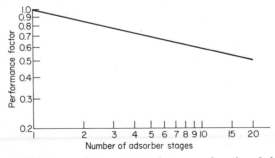

Fig. 29 Relationship between relative system performance and number of adsorber stages.

determined by the column test procedure indicated that the carbon-usage rate in the pulsed-bed system was one-half that predicted from the column test procedure. Therefore, a modification of the equation developed for a single-column fixed-bed system can be used as a basic design equation for series and pulsed-bed systems. Figure 29 shows the estimated relationship between the number of stages and the carbon-usage rate. The

TABLE 6 Cumulative Average Column-Exhaustion Data

Cumulative processing time, h	Feed	Cumulative average APHA color for column No.				
		1	2	3	4	5
		Residence time, min				
0	0	15.4	29.8	43.1	57.2	72.2
0	70,000			Start-up		
5	70,000	12,500	800	0	0	0
21	70,000	31,070	15,580	5,600	1,050	10
37	71,620	46,010	29,650	19,120	10,320	3,420
57	72,810		42,060	29,260	20,120	9,760
71	73,240			34,540	26,300	14,810
89	70,900			39,690	32,890	19,900

*4-in ID columns at 1.7 gal/(min)(ft²).

pulsed-bed carbon-usage rate (20-stage system) would be 0.5 of that obtained from the column test procedure. A four-stage series system would be 0.72 of that obtained from the column test procedure.

For an n-stage parallel system, the impurity concentration in the effluent would be the cumulative-average concentration of the effluents from n adsorbers. When a spent adsorber is removed from stream, it would be producing considerably poorer-quality effluent (the carbon is more fully spent) than it would if only one column were operating. The more columns in the parallel system, the closer to the desired cumulative-average impurity concentration the system can operate.

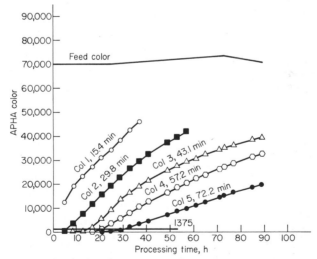

Fig. 30 Cumulative average column-exhaustion curves.

Column-exhaustion curves relating the cumulative-average effluent impurity concentration (or percentage removed or remaining) with volume of liquid processed (or processing time) are drawn from cumulative-average data (see Table 6) as in Fig. 30. From the exhaustion curves, breakthrough points are determined as in Table 7.

The processing time at breakthrough is then plotted against residence time (see Fig. 31) with breakthrough based on cumulative-average effluent concentration. The processing time to breakthrough should be a linear function of the bed depth if the linear flow rate and the feed-impurity composition and concentration are reasonably constant throughout the test.[14]

Since the processing time and residence time is a linear relationship, the curve can be described by the equation of a straight line.

TABLE 7   Residence Time–Processing Time Cumulative Average Data*

Column No.	Residence time,† min	Processing time at breakthrough, h
1	15.4	0
2	29.8	5.8
3	43.1	15.2
4	57.2	21.6
5	72.2	30.5

*At 1.7 gal/(min)(ft²) linear flow rate. Cumulative average feed color 70,900 APHA and breakthrough at cumulative average 1375 APHA.
†Based on volume of carbon bed including void volume.

The reciprocal of the slope of the line is the rate at which the carbon bed is spent. Multiplying this value by the apparent bulk density of the carbon, the expected cross-sectional area of the onstream plant-scale adsorbers, and the expected plant-scale linear flow rate will give an estimate of the carbon-usage rate (or the rate at which the carbon must be regenerated) to produce acceptable product continuously. This usage rate is valid for the feed and effluent concentration used for a parallel system with an infinite number of columns. Therefore, this procedure gives liberal rather than conservative values and becomes less accurate as the number of actual columns in the parallel system decreases.

Multihearth furnaces are available with various inside diameters and various numbers of hearths. The combination of diameter and number of hearths gives effective hearth areas. A common carbon-throughput rate of 17 kg/(h)(m²) [3.45 lb/(h)(ft²)] has been used to size multihearth furnaces in sugar applications and 23 kg/(h)(m²) [4.7 lb(h)(ft²)] in municipal wastewater applications. However, tests have shown that the reactivation time can vary widely from 5 to 125 min for a given activated carbon depending on the type and quantity of sorbed material.[2] Therefore, both the carbon-usage rate and the residence time are

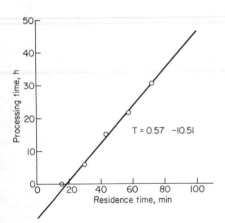

Fig. 31   Residence time–processing time curve based on cumulative-average data.

$T = 0.57 \quad -10.51$

required to size a regeneration furnace adequately.

Consultation with furnace manufacturers and possibly reactivation tests are required to establish the furnace residence-time parameter.

Cost Estimating. The total investment and operating cost estimate is a combination of estimates for the two separate portions—adsorption and reactivation.

Once the residence time in the adsorption system has been determined and the best configuration has been selected, the adsorption system can be sized. Once the carbon-usage rate has been determined for the best system configuration, and the residence time required in the furnace has been determined, the reactivation furnace can be sized.

When a reliable estimate of the adsorber size and configuration and the furnace size has been obtained, meaningful investment and operating costs can be estimated. These estimates can be compared with estimates of other treatment processes to select the most economical system.

Limitations and Alternatives. As with all design methods, the usefulness of this method is limited if the column feed does not represent the normal plant stream. Also, if chemical or biological activity occurs in the test system, it may not be equivalent to that in the plant-scale system.

To obtain the most useful design data, tests must be conducted at a reasonably constant linear flow rate and impurity concentration and composition in the feed.

Frequently, it is not possible to conduct tests under these circumstances. During the actual tests, inadvertent changes in flow rate, feed impurity composition and concentration, feed pH, etc., can occur. Therefore, tests should be conducted using the linear flow rate and bed depth (residence time) most likely to be used in the plant-scale system. Operation of the tests until unacceptable effluent is consistently obtained in the last column will give an estimate of the carbon-usage rate (reactivation rate) even with poor test conditions.

Even with very good data, the system design should also be based on experience and good judgment. It is usually advantageous, particularly in wastewater applications, to oversize carbon-adsorption and -reactivation systems to allow for:

- Unexpected surges in flow rate or high concentrations of impurities in the feed
- Possible future changes in the quantity, concentration, and composition of the feed
- Possible future tightening in the effluent-quality requirements
- Possible decreases in activated-carbon performance as a result of numerous regenerations (unless tests to develop design data were done with carbon which has been regenerated numerous times).

REFERENCES

1. "A Symposium on Activated Carbon," ICI United States, Inc., *Bull.* D-114, Wilmington, Del., 1968, revised 1970.
2. Hutchins, R. A., Thermal Regeneration Costs, *Chem. Eng. Prog.*, May 1975.
3. "Properties to Consider in Selecting Powdered Activated Carbon," ICI United States, Inc., *Bull.* D-96, 1965.
4. DeJohn, P. B., Carbon from Lignite or Coal: Which Is Better, *Chem. Eng.*, Apr. 28, 1975.
5. Hassler, J. W., "Purification with Activated Carbon," 3d ed., Chemical Publishing, New York, 1974.
6. "Evaluation of Granular Carbon for Chemical Process Applications," ICI United States, Inc., *Bull.* D-116, Wilmington, Del., 1971.
7. Glysteen, L. F., and Nickles, H. B., Hardness and Abrasion Values of Granular Activated Carbon in Relation to Regeneration Losses, paper presented at 7th Technical Session, Bone Char Research Project, Inc., and the National Bureau of Standards, Oct. 5–6, 1961.
8. Anonymous, "Activated Carbon Hardness Testing," Witco Chemical Corp., Tech. Bull. 55-3, New York.
9. Hutchins, R. A., Economic Factors in Granular Carbon Thermal Regeneration, *Chm. Eng. Prog.*, November 1973.
10. Berg, E. L., Villiers, R. V., Masse, A. N., and Winslow, L. A., Thermal Regeneration of Spent Powdered Carbon Using Fluid-Bed and Transport Reactors, *Chem. Eng. Prog., Symp. Ser.*, 1970.
11. Loven, A. W., Perspectives on Carbon Regeneration, *Chem. Eng. Prog.*, November 1973.
12. Treybal, R. E., "Mass Transfer," p. 499, McGraw-Hill, New York, 1955.
13. Fornwalt, H. J., and Hutchins, R. A., Purifying Liquids with Activated Carbon, *Chem. Eng.*, May 9, 1966.
14. Hutchins, R. A., New Method Simplifies Design of Activated-Carbon Systems, *Chem. Eng.*, Aug. 20, 1973.

Dialysis and Electrodialysis

ROBERT E. LACEY   *Southern Research Institute,*
Birmingham, Alabama.

DIALYSIS

Dialysis is the transfer of a solute through a membrane as a result of a transmembrane gradient in the concentration of the solute. It is always accompanied by osmosis, which is the transfer of a solvent through a membrane as a result of a transmembrane gradient in the concentration of the solvent. The direction of solute transfer in dialysis is opposite to that of solvent transfer in osmosis. The membranes used in dialysis must prevent convective mixing of the concentrated and dilute solutions being separated over a long enough time for economic operation. Thus the membranes must be "tight" enough to prevent flow of the solution itself and not its individual components, and the material from which the membranes are made must be stable enough to its environment to permit suitably long service lifetimes.

Since the driving force in dialysis is a concentration *gradient*, membranes should be thin and the transmembrane concentration difference should be large, if high fluxes are to be achieved.

Minimization of boundary-layer thicknesses is important in dialysis to minimize deleterious effects of osmosis and concentration gradients in the boundary layers.

Applications

The initial major use of dialysis was to recover NaOH from the "press liquor" generated in the manufacture of rayon. With the development of acid-resistant membranes and dialyzers by the Phelps Dodge Copper Company and The Graver Water Conditioning Company, dialysis found use as a means of separating acids from mixtures of metal salts and acids. Later the Asahi Glass Company, Ltd. and the Asahi Chemical Company, Ltd. in Japan developed dialyzers that made use of anion-exchange membranes. With these ion-exchange dialyzers, the dialysis rates of acids were as much as 60 to 80 times the rates of the metal salts, and extremely high separation factors were achieved.

These industrial applications of dialysis point out one generalization about the use of dialysis: dialysis should be considered when the concentration of the material to be separated is high. Another generalization about the applications of dialysis is that the process should be considered when either the material to be transferred or the materials to be retained can be damaged by harsher separation techniques. Dialysis has been used for years to remove undesirable metabolites from the blood of patients with renal failure because the gentleness of the separation process minimizes damage to blood components.

The gentleness of separation suggests that dialysis might be used for many separations in the food industry (e.g., removal of salts from solutions of larger components of foods without damage to structure or taste of the retained components), and the pharmaceutical industry (e.g., removal of metabolites that inhibit biological growth rates).

Mechanism of Dialysis

Mass transfer in dialysis occurs by diffusion. In many applications of dialysis the degree of convective mixing in the solution compartments is so low that the resistances of the nearly static boundary layers on each side of membranes are an appreciable part of the total resistance to mass transfer. With the assumptions that no volume changes occur on either side of the membrane and that mass-transfer coefficients are constant along the travel path of the solutions, Fick's law has been integrated over the total length the solution travels to give the following design expression[1] for industrial dialyzers:

$$W = U A \Delta C_{lm} \tag{1}$$

where W = weight of solute transferred through the membrane per unit time, g/min; U = overall dialysis coefficient, which includes the effects of the membrane and both boundary layers, cm/min; A = membrane area, cm²; and ΔC_{lm} = log-mean transmembrane concentration difference, g/cm³. For continuous dialyzers ΔC_{lm} is based on inlet and outlet values; for batch dialyzers ΔC_{lm} is based on initial and final values.

For approximations U can be estimated,[1] but for design purposes it is almost always necessary to determine U experimentally. Apparatus for these determinations is available from vendors of dialyzers.

The overall dialysis coefficient U is related to the coefficients of the membrane U_m and of the two boundary-layer films U_{l_1} and U_{l_2} by the usual expression for conductance terms in series.

$$\frac{1}{U} = \frac{1}{U_m} + \frac{1}{U_{l_1}} + \frac{1}{U_{l_2}} \tag{2}$$

Vromen[2] and Leonard[3] point out that the flux of solute W/A is not simply the diffusive flux of that solute through the membrane because both the flux of solvent due to osmosis and the fluxes of other solutes affect the flux of the solute in question. By the use of the thermodynamics of irreversible processes, Kedem and Katchalsky[4] and Katchalsky and Curran[5] show the effects of osmosis on solute flow. These authors also describe the conditions that result in anomalous osmosis and particularly *negative anomalous osmosis*, which is responsible for increases in the overall dialysis coefficients for acids with increases in concentration that is illustrated in Fig. 1 with data from Ref. 2. (Negative anomalous osmosis is a phenomenon in which solvent flows from a concentrated solution to a dilute solution. It occurs when the partition coefficient for the solute K_s is high and the dissipation of energy owing to friction between the solute and solvent and the solute and membrane is sufficiently low.)

Vromen[2] also observed that the dialysis coefficient of a strong acid can be much higher when the acid is dialyzed from a mixture of an acid and its metallic salt than when the acid is dialyzed by itself. For example, in the dialysis of a mixture of ferric nitrate and nitric acid, Vromen found the dialysis coefficient of HNO_3 was 41×10^{-2} cm/min, whereas the comparable value for pure HNO_3 was 1.6×10^{-2} cm/min. The reasons for this "common-ion effect" have been discussed by Vinograd and McBain[6] and by Kedem and Katchalsky.[4]

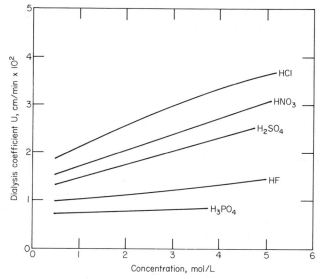

Fig. 1 Variation of dialysis coefficients with concentration.

In general, when a mixture of two electrolytes with a common ion [e.g., $Fe(NO_3)_3$ and HNO_3] is dialyzed, the movement of the anions [e.g., $(NO_3)^-$], which are present in greater numbers than either cation, exerts an electrostatic "pull" on the cations. The more mobile cation (H^+ in this case) will be speeded up by the electrostatic pull; the less mobile cation (Fe^{3+}) will move more slowly to maintain electroneutrality.

The effects of osmosis and of common ions can be seen clearly by consideration of the cross coefficients in phenomenological equations. However, the explanations are too lengthy for inclusion here. The reader is referred to Refs. 4 and 5 for dialyses in which these effects are important.

Design Considerations

In most industrial dialyzers the mixing that occurs in dialysis cells as a result of convection currents reduces the thicknesses of the boundary layers next to the surfaces of membranes, and thus increases the liquid-film coefficients of dialysis. In a dialysate compartment these convection currents are established because the osmotic transfer of solute into the compartment forms layers of dilute solution at the surfaces of the membranes that rise to the top of the compartment, since the dilute solution has a lower density than that of the bulk of solution. In a diffusate compartment the solute entering through the membranes forms layers of solution at the membrane surfaces that are more concentrated and have a higher density than that of the bulk of the diffusate solution. Therefore, the surface layers in diffusate compartments slide downward to establish convection currents. The effects of the convection currents on liquid-film coefficients are pronounced with concentrated dialysate solutions and dilute diffusate solutions. Since most industrial separations for which dialysis would be considered are for treatment of relatively concentrated dialysate solutions, no theoretical approach to prediction of dialysis coefficients has been adopted for industrial separations.

The general procedure in dialyzer design involves (1) experimentally evaluating the membrane and liquid-film dialysis coefficients for various solution flow rates and product-feed ratios, (2) determining membrane areas needed for different solution flow rates and

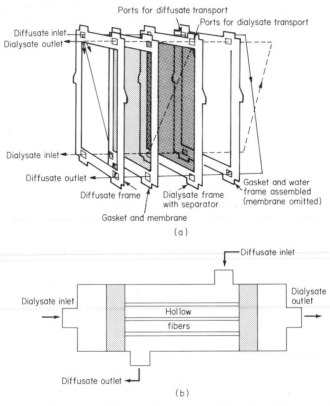

Fig. 2 Two types of dialyzers. (a) Plate-and frame dialyzer; (b) shell-and-tube type of hollow-fiber dialyzer.

product-feed ratios, and (3) determining the optimum ratios, flow rates, and areas based on economic considerations of capital investment, operating costs, and the value of recovered products.

As a result of his studies relating to dialysis for artificial kidneys, Leonard[3,7] developed three principles for dialyzer design, which have since been put into practice: (1) spacings between membranes should be minimized, (2) fluid path length should be short, and (3) some form of forced transverse mixing should be sought. The later mathematical developments of theory by Grimsrud and Babb,[8] Smith, Colton, Merrill, and Evans,[9] and Kaufman and Leonard[10] led to the design of several types of dialyzers that incorporate the three general principles. Among these are plate-and-frame dialyzers and hollow-fiber dialyzers, as shown in Fig. 2.

Examples of plate-and-frame dialyzers are Asahi, Brosites, Grover Hi-Sep, and Aqua-Chem dialyzers. In this type of dialyzer (top of Fig. 2), membranes are sandwiched between alternate dialysate and diffusate frames. The dialysate is fed to the bottoms of dialysate frames and withdrawn from the tops; diffusate is fed to the tops of diffusate frames and is withdrawn from the bottoms. Mixing within the dialysate compartments is enhanced by the flow of solution through meshlike spacers in the dialysate compartments.

Hollow-fiber dialyzers are made in a variety of sizes and configurations ranging from laboratory units with 1000 cm² of membrane area, or less, to units with 150 ft² of membrane area. Although there are several types of hollow-fiber dialyzers that differ in certain respects, the sketch at the bottom of Fig. 2 indicates features common to all types. The unit illustrated in Fig. 2 resembles a shell-and-tube heat exchanger, except that the "tubes" are in actuality a bundle of hollow fibers. The ends of the bundle of hollow fibers are potted with resin to form a header that can be sealed to the shell by O rings or other means. The dialysate is usually fed through the lumens of the fibers; the diffusate is fed through the shell.

Westbrook[11] presents expressions for analyzing the performance of hollow-fiber dialyzers operating at steady state in the continuous mode.

$$Q_F (C_{F_1} - C_{F_2}) = Q_D(C_{D_2} - C_{D_1}) = KA\Delta C_{lm} = Q_M\Delta C_{lm} \tag{3}$$

with countercurrent flow:

$$\Delta C_{lm} = \frac{(C_{F_1} - C_{D_2}) - (C_{F_2} - C_{D_1})}{\ln [(C_{F_1} - C_{D_2})/(C_{F_2} - C_{D_1})]} \tag{4}$$

If $Q_D = Q_F$,

$$\frac{C_{F_2}}{C_{F_1}} = \frac{Q_F}{Q_F + Q_M} \tag{5}$$

If $Q_D \neq Q_F$,

$$\frac{C_{F_2}}{C_{F_1}} = \frac{Q_D - Q_F}{\exp (Q_M/Q_F - Q_M/Q_D) Q_D - Q_F} \tag{6}$$

where Q_F = flow rate of feed solution, cm³/min
 Q_D = flow rate of dialysate solution, cm³/min
 C_{F_1} = concentration of solute in influent feed, mg/cm³
 C_{F_2} = concentration of solute in effluent feed, mg/cm³
 C_{D_1} = concentration of solute in influent dialysate, mg/cm³
 C_{D_2} = concentration of solute in effluent dialysate, mg/cm³
 K = dialysis coefficient, cm/min
 A = area of membranes, cm²
 Q_M = fictitious flow rate through membranes, cm³/min
The following example illustrates the use of Westbrook's equations.

EXAMPLE PROBLEM

Salt is to be removed from a solution containing 100 g/L of raffinose by dialysis in a shell-and-tube type of hollow-fiber dialyzer operating countercurrently. With a dialyzer having 1000 cm² area of membranes the dialysis coefficient for NaCl, K_s was determined to be 0.0415 cm/min, when the feed rate Q_F was 200 cm³/min, and the flow rate of the pure water used for diffusate Q_D was 500 cm³/min. If 90% of the NaCl is to be removed, what area of hollow-fiber membranes will be needed, if the same flow rates for feed and water are used?

$$C_{F_1} = 100 \text{ mg/cm}^3$$
$$C_{F_2} = (1-0.9)100 = 10 \text{ mg/cm}^3 \text{ for 90\% removal of salt}$$

The rate of salt removal must be

$$Q_F(C_{F_1} - C_{F_2}) = 200(100 - 10) = 18,000 \text{ mg/min}$$

Now, $C_{D_1} = 0$ (pure water) and $Q_D = 500$ cm³/min.

$$Q_D(C_{D_2} - C_{D_1}) = Q_F(C_{F_1} - C_{F_2}) = 18,000$$

$$C_{D_2} = \frac{18,000}{Q_D} = \frac{18,000}{500} = 36 \text{ mg/cm}^3$$

$$C_{lm} = \frac{(100 - 36) - (10 - 0)}{\ln [(100 - 36)/(10 - 0)]} = 29.1$$

$$A = \frac{Q_F(C_{F_1} - C_{F_2})}{K_s \Delta C_{lm}} = \frac{18,000}{(0.0415)(29.1)} = 14,904 \text{ cm}^2$$

ELECTRODIALYSIS

Electrodialysis is a process in which electrolytes are transferred through a system of solutions and membranes by an electrical driving force. With the conception of multiple-compartment electrodialysis in 1940, and the development of practical ion-exchange membranes in the late 1940s, electrodialysis became an important separation technique. As currently used, the term electrodialysis refers to multiple-compartment electrodialysis with ion-exchange membranes. The process can be used for separating electrolytes from nonelectrolytes, concentration or depletion of electrolytes in solutions, and the exchange of ions between solutions.

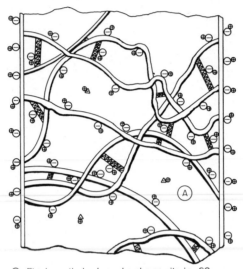

⊖ Fixed negatively charged exchange site; i.e., SO_3^-
⊖ Mobile positively charged exchangeable cation; i.e., Na^+
══ Polystyrene chain
⚁ Divinylbenzene crosslink

Fig. 3 Representation of an ion-exchange membrane.

As in any membrane process, selective membranes are the heart of electrodialysis. Therefore, the nature of ion-exchange membranes will be discussed first.

Ion-Exchange Membranes

Seemingly solid membranes are actually a jumble of polymer chains with void spaces between chains. Transferring species can pass through spaces between the intertangled or crosslinked polymer chains. Because of thermal motion, segments of chains between points of entanglement of crosslinking are in constant motion, and even the points of crosslinking have some motion. If these polymer segments are close together, transferring species must often push the segments apart to slide past them. Highly crystalline or highly crosslinked polymers are of this type. Commercially available ion-exchange membranes usually have relatively high crosslinking densities; thus the polymer segments between points of crosslinking are usually short. Part of the resistance to transfer results from the energy dissipated in collisions between polymer segments and moving species.

Ion-exchange membranes are selective in that they are permeable to positively charged ions (cations) but not to negatively charged ones (anions) or vice versa. The source of their selectivity is discussed with the aid of the sketch in Fig. 3. In Fig. 3, polymer chains are shown that have negatively charged groups chemically attached to them. The polymer chains are intertwined and also crosslinked at various points. Positive ions are shown dispersed in the voids between polymer chains but near a negative fixed charge. The fixed negative charges on the chains repulse negative ions that try to enter the membrane, and exclude them. Thus negative ions cannot permeate the membrane, but positive ones can (i.e., the membrane is a cation-exchange membrane). If positive fixed charges are attached to the polymer chains instead of negative fixed charges, positive ions cannot permeate the membrane but negative ions can, and the membrane is an anion-exchange membrane. This exclusion as a result of electrostatic repulsion is termed *Donnan exclusion.*

Selectivity by itself is not enough to make an ion-exchange membrane that is practical for low-cost processing. In addition, the resistance of the membrane to ion transfer must be low. To decrease the resistance, the crosslinking density may be decreased or the fixed-charge density may be increased, or both. Decreases in crosslinking density tend to increase the lengths of polymer segments that are free to move. Increases in fixed-charge density tend to increase the average distance between polymer chains because of the repulsion effects of like fixed charges. If the void spaces between polymer chains become too large owing to either decreased crosslinking density or increased fixed-charge density, there can be volumes in the center of the voids that are not affected by the fixed charges on the chains (as at point A in Fig. 3), since the repulsion effect of fixed charges falls off rapidly with the distance away from the charges. Such volumes that are unaffected by fixed charges result in ineffective repulsion of the undesired ions and lowered selectivity. A compromise between selectivity and low resistance must usually be made. In membranes now available, it has been possible to combine excellent selectivity with low resistance, high physical strength, and long lifetimes.

There are two general types of commercially available ion-exchange membranes: heterogeneous and homogeneous membranes. Heterogeneous membranes in which ion-exchange particles are incorporated in film-forming resins have been made by calendering mixtures of ion-exchange and film-forming materials in solutions of film-forming materials and allowing the solvent to evaporate, and by casting films of dispersions of ion-exchange material in partially polymerized film-forming polymers and completing the polymerization.

Homogeneous ion-exchange membranes have been made by graft polymerization of anionic or cationic moieties into preformed films, by casting films from a solution of a linear film-forming polymer and a linear polyelectrolyte, and allowing the solvent to evaporate, or by polymerization of mixtures of reactants that can undergo polymerization by either addition or condensation reactions. In the last method, one of the monomers must be either anionic or cationic.

Table 1 shows properties of some of the commercially available ion-exchange membranes, as reported by various manufacturers.

The Electrodialysis Process and Electrodialysis Stacks

In electrodialysis cation-exchange membranes are alternated with anion-exchange membranes in a parallel array to form thin solution compartments (0.5 to 1.0 mm thick). The entire assembly of membranes is held between two electrodes to form an electrodialysis stack, as shown in Fig. 4. A solution to be treated is circulated through the solution compartments. With the application of an electrical potential to the electrodes, all cations (positive ions) tend to transfer toward the cathode (negative electrode), and all anions (negative ions) tend to move toward the anode (positive electrode). The ions in the even-numbered compartments can transfer through the first membranes they encounter (cations through cation-exchange membranes, anions through anion-exchange membranes), but they are blocked by the next membranes they encounter, as indicated by the arrows in the diagram. Ions in the odd-numbered compartments are blocked in both directions. Ions are removed from the solution circulating through one set of compartments (even) and transferred to the other set of compartments (odd). Ion depletion is accomplished for one solution, and ion concentration is accomplished for the second solution.

Figure 5 is an exploded view of part of an electrodialysis stack that shows the main components. Component 1 in Fig. 5 is one of the two end frames, each of which has provisions for holding an electrode and introducing and withdrawing the depleting, the concentrating, and the electrode-rinse solutions. The end frames are made relatively thick and rigid to resist bending when pressure is applied to hold the stack components together. The inside surfaces of the electrodes may be recessed, as shown, to form an electrode-rinse compartment when an ion-exchange membrane, component 2, is clamped in place. Components 3 and 5 are spacer frames. Spacer frames have gaskets at the edges

TABLE 1   Reported Properties of Ion-Exchange Membranes*

Manufacturer and designation	Type of membrane	Area resistance, $\Omega \cdot cm^2$	Transference No. of counterion[a]	Strength	Approx thickness, mils	Dimensional changes on wetting and drying, %	Size available
ACI[b]		(0.5 N NaCl)		Tensile strength, kg/mm²			
CK-1	Cat-exch	1.4	0.85 (0.25/0.5 N NaCl)	2–2.4	9	15–23	44 × 44 in
DK-1	Cat-exch	1.8	0.85 (0.25/0.5 N NaCl)		9		44 × 44 in
CA-1	An-exch	2.1	0.92 (0.25/0.5 N NaCl)	2–2.3	9	12–18	44 × 44 in
DA-1	An-exch	3.5	0.92 (0.25/0.5 N NaCl)		9		
AGC[c]		(0.5 N NaCl)		Mullen burst, lb/in²			
CMV	Cat-exch	3	0.93 (0.5/1.0 N NaCl)	ca. 180	6		
CSV[d]	Cat-exch	10	0.92 (0.5/1.0 N NaCl)	ca. 180	12		
AMV	An-exch	4	0.95 (0.5/1.0 N NaCl)	ca. 150	6	<2	44-in-wide rolls
ASV[d]	An-exch	5	0.95 (0.5/1.0 N NaCl)	ca. 150	6		
IC[e]		(0.1 N NaCl)		Mullen burst, lb/in²			
MC-3142	Cat-exch	12	0.94 (0.5/1.0 N NaCl)	ca. 200	8		
MC-3235	Cat-esch	18	0.95 (0.1/0.2 N NaCl)	ca. 165	12	<3[g]	40 × 120 in
MC-3470	Cat-exch	35	0.98 (0.1/0.2 N NaCl)	ca. 200	8		
MA-3148	An-exch	20	0.90 (0.5/1.0 N NaCl)	ca. 200	8		
MA-3236	An-exch	120	0.93 (0.5/1.0 N NaCl)	ca. 165	12	<3[g]	40 × 120 in
IM-12	An-exch[a]	12	0.96 (0.1/0.2 N NaCl)[g]	ca. 145	6[g]	Not given	
MA-3475R	An-exch	11	0.99 (0.5/1.0 N NaCl)	ca. 200	14	Not given	
II[h]		(0.1 N NaCl)		Mullen burst, lb/in²			
CR-61	Cat-exch	11	0.93 (0.1/0.2 N NaCl)	115	23	Cracks on drying	18 × 40 in
AR-111A	An-exch	11	0.93 (0.1/0.2 N NaCl)	125	24		
TSC[i]		(0.5 N NaCl)	(by electrophoretic method in 0.5 N NaCl)	Mullen burst, lb/in²			
CL-2.5T	Cat-exch	3	0.98	ca. 80	6	Not given	40 × 50 in
CLS-25T	Cat-exch[d]	3	0.98	ca. 80	6	Not given	
AV-4T	An-exch	4	0.98	ca. 150	7	Not given	40 × 40 in
AVS-4T	An-exch[d]	5	0.98	ca. 140	7	Not given	
MOE[j]		(0.5 N NaCl)		Tensile strength, kg/mm²			
K-101	Cat-exch	3.3	Not given	2	1	Not given	15 × 30 in
CA-2	An-exch	2.3	Not given	2.2	1	Not given	15 × 30 in

*Properties are those reported by manufacturer, except for those membranes designated with footnote g.
[a]Calculated from concentration potentials measured between solutions of the two normalities listed.
[b]Asahi Chemical Industry, Ltd., Tokyo, Japan.
[c]Asahi Glass Co., Ltd., Tokyo, Japan.
[d]Membranes that are selective for univalent (over multivalent) ions.
[e]Ionac Chemical Co., Birmingham, N.J.
[f]Special anion-exchange membrane that is highly diffusive to acids.
[g]Measured at Southern Research Institute.
[h]Ionics, Inc., Cambridge, Mass.
[i]Tokuyama Soda Co., Ltd., Tokyo, Japan.
[j]M. O. Engineering Co., Ltd., Tokyo, Japan.

and ends so that solution compartments are formed when ion-exchange membranes and spacer frames are clamped together.

Usually the supply ducts for the various solutions are formed by matching holes in the spacer frames, membranes, gaskets, and end frames. Each spacer frame is provided with solution channels (at point E in Fig. 5) that connect the solution-supply ducts with the

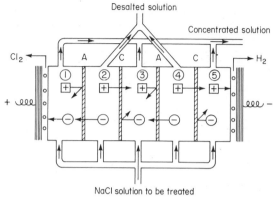

A = anion – permeable membrane
C = cation – permeable membrane

Fig. 4 Simplified representation of the electrodialysis process.

solution compartments. The spacer frames have mesh spacers or other devices in the compartment spaces to support the ion-exchange membranes to prevent collapse when there is a differential pressure between two compartments. These mesh spacers are selected or designed to aid in lateral mixing, which decreases the thicknesses of boundary layers.

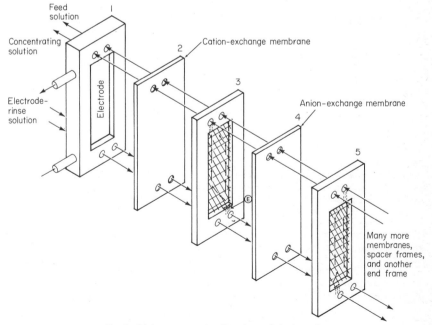

Fig. 5 Main components of an electrodialysis stack.

An electromembrane stack usually has many repeated sections, each consisting of components 2, 3, 4, and 5, with a second end frame at the end. These repeating units are termed *cell pairs*.

There are three basic types of electrodialysis stacks: tortuous-path, sheet-flow, and unit-cell stacks. In the tortuous-path stack, the solution flow path is a long, narrow channel as illustrated in Fig. 6, which makes several 180° bends between the entrance and exit ports of a compartment. The bottom half of the spacer gasket in Fig. 4 shows the individual narrow solution channels and the cross straps used to promote mixing, whereas the cross straps have been omitted in the top half of the figure so the flow path could be better depicted. The ratio of channel length to width is high, usually greater than 100:1. Spacer screens to support the membranes may or may not be used in tortuous-path stacks.

In sheet-flow stacks, spacer screens are almost always needed, since the width between gasketing devices is much greater than that in the usual tortuous-path stacks. The solution flow in sheet-flow stacks is in approximately a straight path from one or more entrance ports to an equal number of exit ports, as indicated in Fig. 5. As the solutions flow in and around the filaments of the spacer screens, a mixing action is imparted to the solutions to aid in reducing the thicknesses of diffusional boundary layers at the surfaces of the membranes.

Solution velocities in sheet-flow stacks are typically in the range of 5 to 15 cm/s, whereas the velocities in tortuous-path stacks are usually much higher, 30 to 100 cm/s. The drop in hydraulic pressure through a sheet-flow stack is normally lower than that through a tortuous-path stack because of the lower velocities and shorter path lengths.

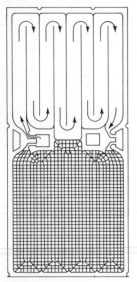

Fig. 6 Diagram of a tortuous-path spacer for an electrodialysis stack.

Unit-cell stacks were specifically developed for concentrating solutions. Each concentrating cell consists of one cation-exchange membrane and one anion-exchange membrane sealed at the edges to form an envelope-like bag, as shown in Fig. 7. Many of these concentrate cells are assembled with spacer screens between them to separate the cells so solutions being concentrated can flow between them. The entire assembly of alternating concentrate cells and spacer screens is held between a set of electrodes. When electric current flows through the stack, ions flow from the external solutions through the membranes to the insides of the concentrate cells, where they are trapped. Only osmotically and electro-osmotically transferred water flows through the membranes. Thus the maximum degree of concentration is effected. The concentrated solutions inside the concentrate cells flow through small tubes that lead from inside the concentrate cells to a plenum chamber arranged outside the stack.

Concentration Polarization

The detailed nature of ion transfer through a system of solutions and ion-exchange membranes warrants thorough discussion because it is not only the source of the depletion and concentration that occurs in electrodialysis but is also the source of excessive *concentration polarization,* which is responsible for most of the difficulties encountered in electrodialytic processing.

Figure 8 shows a cation- and an anion-exchange membrane mounted between two electrodes. A solution of an electrolyte flows through the compartments formed by the membranes and electrodes. With the passage of an electric current through the system of membranes and solutions, anions transfer toward the anode and cations transfer toward the cathode. These ion transfers are the way in which electric current is carried in an electrolytic medium. The fraction of the current carried by an ionic species is termed its *transference number* (t^+ or t^-).

The hydraulic characteristics of an electrodialysis solution compartment and the boundary layers of nearly static solution at the surfaces of the membranes can have a controlling effect on the current densities that can be used in electrodialysis, and therefore a controlling influence on the rate of demineralization. Because of the flow of solution through the center compartment formed by the two membranes in Fig. 8, there is a zone of relatively well-mixed solution near the center of the compartment. The velocity of the solution and thus the degree of mixing diminish as the surfaces of the membranes are approached. In Fig. 8, an idealization is used that there is a completely mixed zone in the

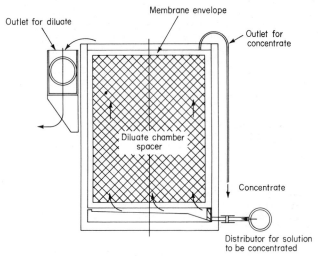

Fig. 7   Main structure of the unit-cell type of electrodialysis stack.

center of the compartment, and completely static zones of solution in boundary layers adjacent to the membranes. In the static boundary layers, ions are transferred only by electrical transfer and diffusion, but in the mixed zone ions are transferred electrically, by diffusion, and by physical mixing.

If the transference number of anions through the anion-exchange membranes (t_m^-) is 1.0 and that of anions through the solution (t_s^-) is 0.5, only half as many ions will be transferred electrically through the solution in the static boundary layer on the side of the membrane that anions enter as will be transferred through the membrane. The solution at the membrane interface will be depleted of ions. For the same reasons the static boundary layer on the other side of the membrane will accumulate ions. The interfacial concentrations will change until concentration gradients are established in the boundary layers that are sufficiently large to transfer by diffusion the ions not transferred through the boundary layers electrically, as indicated by the dashed lines in Fig. 8.

If the current density through the system is increased, the rate of electrical transfer of ions increases, and the diffusional transfer through the boundary layers must increase to supply the additional ions transferred electrically. Diffusional transfer can increase only if the boundary layers are made thinner, or if the differences between the ion concentrations at the membrane-solution interfaces and the concentrations in the well-mixed zones are increased. With any given thickness of boundary layer, a current density can be reached at which the concentrations of electrolytes at the membrane interfaces on the depleting sides will approach zero. At this current

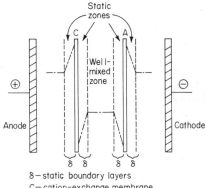

Fig. 8   Idealized representation of concentration gradients in electrodialysis.

density, called the *limiting current density*, H^+ and OH^- ions from ionization of water will begin to be transferred through the membranes. This continuous ionization of water is termed *water splitting*. The water splitting caused by exceeding the limiting current density has detrimental effects on the operation of electrodialysis, as discussed below.

Scaling of membranes by pH-sensitive electrolytes occurs at anion-exchange membranes because OH^- ions transfer through the membranes when water splitting occurs. The OH^- ions increase the pH within the membrane and at the interface on the concentrating side so that pH-sensitive substances, such as $CaCO_3$, precipitate.

Operation at or above the limiting current density causes the stack to have resistances higher than normal for the following reasons. When water splitting occurs, the rate of dissociation of water is increased because the H^+ and OH^- ions are continuously transferred away from the membrane interfaces, which are the locations of dissociation. An increased voltage is necessary to induce this continuous dissociation of water. Also, a thin film of highly depleted solution forms at the depleting sides of membranes. These films have high specific resistances, which are in series with the other resistances in an electrodialysis stack. Both the increased voltage needed for continuous ionization of water, and the high specific resistances of depleted films contribute to increased stack resistance.

Fouling of ion-exchange membranes by organic materials can also occur when excessive concentration polarization occurs. Fouling of anion-exchange membranes has been shown to be caused by large organic ions becoming attached to charged groups on membranes having the opposite charge.[12-17] References 16 and 17 are especially good if troubles are encountered with fouling of membranes.

Because of these detrimental consequences of excessive concentration polarization, the minimization of polarization is an important factor in the design of electrodialysis stacks.

Minimizing Concentration Polarization

As was implied in the discussion of Fig. 8, the changes in electrolyte concentrations in the solutions at the membrane-solution interfaces, which are responsible for the deleterious consequences of polarization, can be minimized by decreasing the thickness of the nearly static boundary layers.

With the idealization of completely static and completely mixed zones used in Fig. 5, the following expression has been derived to determine the thickness of boundary layers:[18,19]

$$\sigma = \frac{DFN}{i_{\lim}(t_m - t_s)} \tag{7}$$

where σ = thickness of boundary layers in depleting compartments, cm
  D = diffusivity of the salt being treated, cm/s
  F = the faraday, 96,500 C/eq
  N = concentration of the salt in the well-mixed zone, eq/cm³
  $i_{\lim}$ = limiting current density, i.e., the current density when the interfacial concentration becomes zero, mA/cm²
  t_m = transference number of the counterion in the membrane
  t_s = transference number of the counterion in solution

Equation (7) has been rearranged to a form in which the *polarization parameter* $i_{\lim}/N$ is expressed in terms of the above variables.

$$\frac{i_{\lim}}{N} = \frac{DF}{\sigma(t_m - t_s)} \tag{8}$$

The polarization parameter is more convenient to use in design and scale-up of electrodialytic equipment than the thickness of the boundary layer. It can be easily measured in small-scale stacks with a given value of bulk concentration N and used to predict limiting current densities in larger stacks at other values of N. (*Caution: In the literature, values of the polarization parameter are reported based on the influent concentration in the depleting stream, the effluent concentration, and the log mean of the influent and effluent concentrations. Therefore, the reader must be wary of comparisons of equipment based on the polarization parameter unless the values of polarization parameter are normalized to the same basis of concentration.*)

Although boundary-layer thickness at any point in a cell compartment depends on several factors,[20] the average boundary-layer thickness in Eq. (8) is primarily a function of the solution velocity in the depleting compartments and the mixing characteristics of spacer materials. Since the boundary-layer thickness varies with solution velocity, the polarization parameter does also. The relationships between solution velocity, the polari-

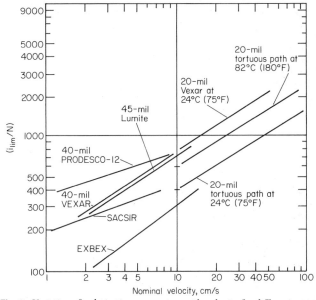

Fig. 9   Variation of polarization parameters with velocity for different spacers.

zation parameter, and the nature of ion-exchange membranes have been studied by different methods,[21–25] including determinations of changes in pH and stack resistances,[21,22] which indicate the limiting current density has been reached.

Values of i_{lim}/N may be measured in laboratory stacks assembled with the membranes to be tested. The solution of interest is circulated through the depleting compartments, and the stack voltage is adjusted to give a desired stack current. The flow rate and the influent and effluent concentrations of the solution being depleted are measured. The current is increased in increments, and at each increment measurements are made of the stack voltage E_s and stack current I_s. Values of E_s/I_s at each increment are plotted as a function of current density. At the limiting current density i_{lim} there will be a change in the slope of the curve. If plots of E_s/I_s vs. the reciprocal of current density are used, the break in the curve is more pronounced so that i_{lim} can be more closely defined. The polarization parameter is then calculated as i_{lim} divided by the log mean concentration N.

Figure 9 shows the change in the polarization parameter i_{lim}/N with nominal solution velocity for several commercial types of spacers, and for two thin-cell spacers. The data shown are from experiments in which the thicknesses of the spacers were carefully matched to the cell thicknesses, which is important in design of electrodialysis stacks because if the membrane surfaces are not in intimate contact with the filaments of the spacer, the spacer "floats" in the center of the cell, the mixing action is diminished, and values of i_{lim}/N decrease.

Because of the ill effects of exceeding the limiting current density and the fact that there can be stagnant or semistagnant areas in even well-designed electrodialysis stacks, most process designers use operating values of i_{lim}/N that are 50 to 70% of the limiting values.

Electrodes

The driving force for electrodialysis is an electric potential gradient. Since this involves the passage of a direct current from an external power supply, an anode and a cathode are required. The use of suitable electrode materials is important because electrodes are exposed to corrosive conditions.

At the cathode, electrons are transferred from the external circuit to ions in the solution by one or more of the following reactions:

$$M^{x+} + xe^- \rightarrow M^0 \qquad \text{metal deposition}$$
$$O_2 + 2H_2O + 4e^- \rightarrow 4OH^- \qquad \text{reduction of gaseous oxygen}$$
$$2H^+ + 2e^- \rightarrow H_2 \text{ (acidic solution)} \qquad \left.\right\} \text{evolution of}$$
$$2H_2O + 2e^- \rightarrow H_2 + 2OH^- \text{ (basic solution)} \qquad \text{gaseous hydrogen}$$

The last reaction is the most common in electrodialysis; so an electrode that is stable in the presence of base and hydrogen is required. Carbon steel or stainless steel are usually adequate materials.

At the anode, electrons are transferred from ions to the external circuit by one or more of the following reactions:

$$M^0 \rightarrow M^{x+} + xe^- \qquad \text{metal dissolution}$$
$$H_2 \rightarrow 2H^+ + 2e^- \qquad \text{oxidation of gaseous hydrogen}$$
$$2H_2O \rightarrow O_2 + 4H^+ + 4e^- \text{ (acidic solution)} \left.\right\} \text{evolution of}$$
$$4OH^- \rightarrow O_2 + 2H_2O + 4e^- \text{ (basic solution)} \text{gaseous oxygen}$$
$$2Cl^- \rightarrow Cl_2 + 2e^- \qquad \text{evolution of gaseous chlorine}$$
$$M^0 + xOH^- \rightarrow M(OH)_x + xe^- \qquad \left.\right\} \text{oxidation}$$
$$2M^0 + 2xOH^- \rightarrow M_2O_x + xH_2O + 2xe^- \text{of electrode}$$

Ideally, only reactions involving evolution of gases would occur at the anode of an electrodialysis stack. Oxidation or dissolution of the electrode will quickly destroy it or reduce its effectiveness. Platinum is good for use as anodes, but its cost is prohibitive. Platinum-coated titanium or tantalum is often used as anodes. Magnetite electrodes have been used in electrodialysis, but this material is fragile. Lead dioxide forms a durable and inexpensive electrode, which can be deposited on graphite substrates to which good connections to the external circuit can be made.

In most instances care must be taken to avoid reversal of polarity of the electrodes. Generally, good anode materials make poor cathodes and vice versa. Graphite appears to be the best material for reversible electrodes in electrodialysis stacks.

Reactions at both the anode and the cathode involve the evolution of gases. These gases must be removed as they are formed; otherwise they will mask the surfaces of the electrodes and cause increased electrical resistance or damage to the ion-exchange membranes. It is common practice to maintain a high velocity of rinse solution through the electrode compartments to flush out the bubbles. Because of the possibility of oxidation of membranes, oxidation-resistant membranes are often used adjacent to the anode. Their added expense is small, when spread over the hundred or more cell pairs in the usual stack.

Accumulation of basic precipitates in the cathode compartment can be a serious problem if the water being treated contains $Ca(HCO_3)_2$ or other pH-sensitive salts. Figure 10 illustrates a combination of techniques used to prevent precipitation in the vicinity of the cathode. These techniques include acidification of the catholyte, use of an isolating compartment, and softening of the catholyte.

Copper is usually the preferred material for conduction of electric current to the electrodialysis stack, but it is not used as an electrode. Therefore, somewhere in the circuit there must be a connection between copper and the electrode material, and this junction may be subject to severe corrosion, especially if it is damp. If possible, it is best to have a projection of the electrode material that passes through the end plate of the stack to an external connection that can be kept clean and dry. If the connection is internal, it must be well sealed with an insulation material that is impermeable to water.

Design Considerations

The viewpoint is taken here that most readers will be interested in selecting and specifying commercially available electrodialysis stacks and related equipment to effect a particular separation or concentration. Readers interested in criteria and methods for designing electrodialysis stacks are referred to Lacey,[26] Davis and Lacey,[27] and Shaffer and Mintz.[28] Readers interested in the advantages and disadvantages of various modes of operation of electrodialysis (e.g., batch, continuous, and feel-and-bleed) are referred to Mintz.[29]

In electrodialysis there are three categories of costs: (1) those that increase with current density (the cost of energy), (2) those that decrease with current density (the cost of membrane replacement and of amortization of capital investment), and (3) those that are

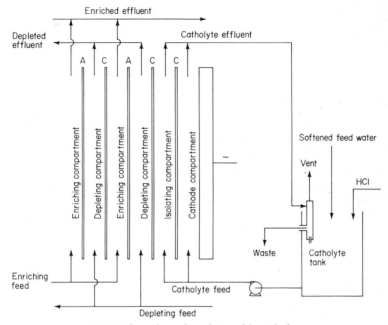

Fig. 10   Flow scheme for isolation of the cathode.

invariant or almost invariant with changes in current density (costs of chemicals, maintenance supplies, maintenance labor, operating labor, and payroll extras). Since two categories of costs vary in opposite directions with current density, there is an optimum current density. For many types of processes, the major design problem would be to determine the optimum current density and design at this current density, and parametric equations have been developed to accomplish this for electrodialysis.[30,31] However, for most applications of electrodialysis the limiting current density is lower than the optimal current density. Therefore, for demineralization of a solution by a continuous (rather than batch) operation, the first step in design is to determine the limiting current density and the polarization parameter for the solution to be treated. This can be done with bench-scale electrodialysis stacks as described previously. During the experiment to determine $i_{\lim}/N$, the data taken on voltage, current, effective membrane area, and number of cell pairs in the stack will enable the calculation of cell-pair voltage for each current density studied. By repeating the experiment with different feed concentrations N_f, curves of cell-pair resistance and cell-pair voltage can be developed for later use. The current efficiency to be expected in large-scale experiments can be calculated from data obtained during the experiments to determine limiting current density.

$$\eta_c = \frac{Q_s F (N_f - N_e)}{S_{cp} I_s} \tag{9}$$

where η_c = current efficiency
Q_s = volumetric flow rate through the stack, cm³/s
F = the faraday, 96,500 C/eq
N_f = feed concentration, eq/cm³

N_e = effluent concentration, eq/cm³
S_{cp} = number of cell pairs in stack, exclusive of any electrode-isolating compartments
I_s = current through the stack, A

With known values of i_{lim}/N, the operating value of i/N can be determined from the vendor's recommendations as to the best solution velocity for use and the percentage of i_{lim}/N to use as a safe operating value of the polarization parameter P_{oper}. The operating current density for use with any value of concentration N can then be calculated. If the log-mean concentration were used in the determination of i_{lim}/N, the operating current density for the first stage is calculated by Eq. (10).

$$(i)_{oper} = \frac{FqN_f}{A_{eff}\eta_c}\left[1 - \frac{1}{\exp(P_{oper}A_{eff}\eta_c/Fq)}\right] \tag{10}$$

where i_{oper} = operating current density, A/cm²
F = the faraday, 96,500 C/eq
q = volumetric flow rate to one depleting compartment = total flow to depleting compartments in the stack divided by the number of cell pairs
N_f = feed concentration, eq/cm³
A_{eff} = effective area of one membrane = total area minus portion blanked by space gaskets
η_c = current efficiency
P_{oper} = numerical value of i/N that is safe for operation

The effluent concentration N_e in the first stage can be calculated by Eq. (11).

$$N_e = N_f - \frac{i_{oper}A_{eff}\eta_c}{Fq} \tag{11}$$

The concentration of the effluent from the first stage is the feed concentration for the second stage.

The calculations of i_{oper} and N_e are repeated for subsequent stages until the concentration of the effluent is at the level needed for the given separation.

The cell-pair voltages E_{cp} for each stage can be estimated from curves of N vs. E_{cp} plotted from the data obtained during the determinations of i_{lim}/N. The log-mean concentration for each stage should be used.

With the procedures described above the number of stages needed, the total current in each stage and the cell-pair voltage can be determined. The number of stacks in parallel that will be needed to demineralize the solution at the desired rate is determined by dividing the desired total volumetric flow rate by the volumetric flow rate of the depleting solution through one stack. The amount of water transferred out of depleting compartments by osmosis and electro-osmosis will depend on the type of ion-exchange membranes used, the current density, and the differences in concentration between the depleting stream and the concentrating streams. Therefore, information about water transfer should be obtained from each vendor. For published information about water transfer see Lakshminarayanaiah,[32] Lacey,[33] and Shaffer and Mintz.[28]

The voltage needed for each stack in each stage may now be calculated (stack voltage = cell-pair voltage × number of cell pairs per stack), the dc power for each stack can be determined ($P_s = E_s I_s$), and specifications for rectifiers can be written.

The total ac power required is the sum of the ac power for pumping and to the rectifiers (efficiency of rectification is about 0.9). The power needed for pumping can be determined from the pressure drop across each stack at the solution velocity used (obtained from each vendor).

With the above information predesign estimates of capital and operating costs can be prepared to serve as a basis for operation with different types of stacks.

Capital costs include costs for pretreatment equipment (if needed), electrodialysis plant, and storage for feed and product. These costs may be estimated by standard chemical engineering methods.

Operating costs include costs for electrical energy (for stacks, pumping, lighting, and auxiliaries), membrane replacement, chemicals (acids for electrode streams), maintenance supplies, maintenance labor, operating labor, payroll extras, administrative overhead, amortization, taxes, insurance, and interest on working capital.

REFERENCES

Dialysis

1. Love, J. A., and Riggle, J. W., *Chem. Eng. Prog. Symp. Ser.*, **24**, 127 (1959).
2. Vromen, B. H., *Ind. Eng. Chem.*, **54**, 20 (1962).
3. Leonard, E. F., Dialysis, in R. E. Kirk and D. F. Othmer (eds.), "Encyclopedia of Chemical Technology," vol. 7, 1965.
4. Kedem, O., and Katchalsky, A., *J. Gen. Physiol.*, **45**(1), 143 (1961).
5. Katchalsky, A., and Curran, P., Membrane Permeability, chap. 10 in "Nonequilibrium Thermodynamics in Biophysics," Harvard, Cambridge, Mass., 1967.
6. Vinograd, J., and McBain, J., *J. Am. Chem. Soc.*, **63**, 2008 (1941).
7. Leonard, E. F., and Bluemle, L. W., Jr., *Trans. N.Y. Acad. Sci.*, **21**(7), 585 (1959).
8. Grimsrud, L., and Babb, A., *Chem. Eng. Prog. Symp. Ser.*, **62**, 20 (1966).
9. Smith, K., Colton, C., Merrill, E., and Evans, L., *Chem. Eng. Prog. Symp. Ser.*, **64**, 45 (1968).
10. Kaufman, T., and Leonard, E., *Am. Inst. Chem. Eng.*, **14**(3), 42 (1968).
11. Westbrook, G., *Chem. Eng. Prog. Symp. Ser.*, **68**, 283 (1971).

Electrodialysis

12. Olie, J. R., *Proceedings of the Milan Meeting of the Water Desalination Working Party of the European Federation of Chemical Engineers*, June 1965.
13. Solt, G. S., *Proceedings of the First International Symposium on Water Desalination*, Washington, D.C., October 1965, **2**, 13 (1967).
14. Mandersloot, W. C. B., *Proceedings of the First International Symposium on Water Desalination*, Washington, D.C., October 1965, **2**, 461 (1967).
15. Small, H., and Gardiner, R., *Office of Saline Water Research and Development Report* 565, 1970.
16. Korngold, E., deKorosy, F., Rahov, R., and Taboch, M., *Desalination*, **8**, 195 (1970).
17. Grossman, G., and Sonin, A., *Office of Saline Water Research and Development Report* 742, 1971.
18. Wilson, J. R., "Demineralization by Electrodialysis," Butterworth, London, 1960.
19. Cowan, D. A., and Brown, J. H., *Ind. Eng. Chem.*, **51**, 1445 (1959).
20. Sonin, A. A., and Probestein, R. F., *Desalination*, **5**, 293 (1968).
21. Rosenburg, N. W., and Tirrell, C. E., *Ind. Eng. Chem.*, **49**, 780 (1957).
22. Kressman, T. R. E., and Tye, R. L., *Trans. Faraday Soc.*, **55**, 1441 (1959).
23. Gregor, H., and Petersen, R., *J. Phys. Chem.*, **68**, 2201 (1964).
24. Cooke, B. A., *Electrochim. Acta*, **3**, 307 (1969); **4**, 179 (1961).
25. Mandersloot, W. G. B., and Koen, J., *Ind. Eng. Chem. Process Des. Dev.*, **4**, 309 (1965).
26. Lacey, R. E., *Office of Saline Water Research and Development Report* 228, 1967, National Technical Information Service, Operations Division, Springfield, Va., Cat. No. PB 210544.
27. Davis, T. A., and Lacey, R. E., *Office of Saline Water Research and Development Report* 710, Government Printing Office, Washington, D.C., 1970.
28. Shaffer, L. H., and Mintz, M. S., Electrodialysis, in K. S. Spiegeler (ed.), "Principles of Desalination," Academic, New York, 1966.
29. Mintz, M. S., *Ind. Eng. Chem.*, **55**, 18 (1963).
30. Mattson, M. E., Snedden, L. L., and Gugeler, J. E., *Proceedings of the First International Symposium on Water Desalination*, **3**, 265 (1965), U.S. Department of the Interior, Washington, D.C., 1966.
31. Booz, Allen Applied Research, Inc., *Office of Saline Water Research and Development Report* 488, National Technical Information Service, Operations Division, Springfield, Va., Cat. No. PB 203123.
32. Lakshminarayanaiah, N., "Transport Phenomena in Membranes," Academic, New York, 1969.
33. Lacey, R. E., *Office of Saline Water Research and Development Report* 343, National Technical Information Service, Operations Division, Springfield, Va., Cat. No. PB 206323.

Parametric Pumping

H. T. CHEN, Ph.D., P.E. *Professor of Chemical Engineering, New Jersey Institute of Technology, Newark, New Jersey.*

NOTATION

A Column cross-sectional area, cm²

b Dimensionless equilibrium parameter

C_1 $\dfrac{V_T}{Q(\pi/\omega)}$, fraction of dead volume of the top reservoir to displacement, dimensionless

C_2 $\dfrac{V_B}{Q(\pi/\omega)}$, fraction of dead volume of the bottom reservoir to displacement, dimensionless

h Column height, cm

k Partition coefficient, molar volumetric concentration in solid phase/molar volumetric concentration in gas phase

L Penetration distance, cm

M X/y

m Dimensionless equilibrium parameter

m_0 $0.5(m_1 + m_2)$

n Number of cycles of pump operation

P Product volumetric flow rate, cm³/s
p Pressure, lb/in² (abs)
Q Reservoir displacement rate, cm³/s
γ Purge volumetric flow rate/feed volumetric flow rate, dimensionless
T Temperature, °C
v_0 $Q/(A\epsilon)$, cm/s
v Interstitial velocity, cm/s
V_T Top dead reservoir volume, cm³
V_B Bottom dead reservoir volume, cm³
χ Concentration of solute in the solid phase, g·mol/g adsorbent
Y Mole fraction
y Concentration of solute in the liquid phase, g·mol/cm³
$<>$ Average value

GREEK LETTERS
ϕ Product volumetric flow rate/reservoir displacement rate, dimensionless
$\dfrac{\pi}{\omega}$ Duration of half cycle, s
ϵ Void fraction in packing, dimensionless
δ_s Density of adsorbent, g/cm³
α Separation factor, dimensionless

SUBSCRIPTS
0   Initial condition
1   Upflow
2   Downflow
BP   Bottom product
i Solute i
max   Optional or maximum condition
TP   Top product
B Stream from or to bottom of the column
T Stream from or to top of the column

INTRODUCTION

The process known as parametric pumping represents a new development in separation science. It has attracted considerable attention both because of its novelty and because it permits continuous operation in small equipment with very high separation factors.

  The basic principle of parametric pumping is to utilize the coupling of periodic changes in equilibrium conditions caused by periodic changes in some intensive variables (such as temperature, pressure, pH, or electric field) and periodic changes in flow direction to separate the components of a fluid (liquid or gas) which flows past a solid adsorbent. Techniques commonly used in the separation of fluid mixtures, including adsorption, extraction, affinity chromatography, and ion-exchange chromatography, might be adapted to parametric pumping. The adaptation could be made in principle in those situations in which a reversible differential shift in the distribution of components between a mobile and an immobile phase could conveniently and practically be brought about by variation of an intensive variable. In this section, we summarize the principle of parametric pumping and present several of its present implementations. Experimental methods are also discussed.

THERMAL PARAMETRIC PUMPING

The Batch Parametric Pump

The batch parametric pump shown in Fig. 1[37] consists of a jacked column packed with a particulate adsorbent and having reservoirs attached to each end. The mixture to be separated fills the column voids and the reservoirs. Let us assume that the mixture consists of a solution of two components—an adsorbable solute in an inert solvent. The reciprocating motion of pistons in the reservoirs causes the fluid to move up and down through the apparatus. At the same time that the flow direction in the column changes, the column temperature is changed by changing the temperature of the liquid in the jacket. Thus, for example, a hot liquid may be circulated through the jacket when flow in the column is upward, and a cold liquid may be circulated when the flow is downward.

During upflow in the heated column the solute desorbs and flows upward. During downflow in the cooled column the component is absorbed at higher position in the column than at the start of the previous period of upflow, and solvent depleted of solute flows downward. Each successive cycle of such operation results in movement of additional adsorbable material toward the top of the column. In the limit of a large number of cycles, substantially all of the solute will be removed from the lower reservoir and transferred to the opposite end of the apparatus. Thus the separation factor, defined as the ratio of the concentration of the solute in the upper reservoir to that in the lower reservoir, will be very large.

The theoretical analysis developed by Wilhelm[38] was reasonably successful in predicting the separations achieved but had the disadvantage of being so complex that the reasons for separation were obscured. The first advance after this was the equilibrium theory developed by Pigford.[23] The basic assumptions in the theory are local equilibrium throughout the adsorption column and absence of axial diffusion. The expressions predict that the batch pump should have the unusual capability of separating a two-component mixture into one fraction completely depleted in solute and another fraction enriched in solute by a factor of at most 2 or 3. The data of Wilhelm and Sweed[37] on the removal of toluene from toluene–n-heptane mixture, in which separation factors as great as 10^5 were readily attained, indicate that the theoretical limits which Pigford's expressions represent might be closely approached in practice. Later, Aris,[3] noting that the theory of Pigford is a special case of a more general theory, derived the general theory. Chen and Hill[8] then extended the general theory to include reservoirs of unequal size. Furthermore, they have derived mathematical models for the pump performance and showed that there are two possible regions of pump operation (Regions 1 and 3), depending on the relative magnitudes of penetration distances L_1 and L_2 and the column height h (Fig. 2),

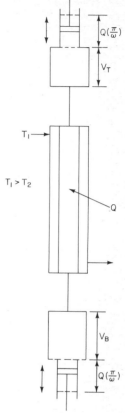

Fig. 1 Batch parametric pump.

Region 1: $L_2 \leqslant L_1$ and h
Region 3: L_1 and $L_2 > h$ (1)

The distance of penetration of the column is equal to the product of the slope of the characteristic curve and the half-cycle duration and is

$$L_1 = \frac{v_0}{(1 - b)[1 + 0.5(m_1 + m_2)]} \frac{\pi}{\omega} \qquad (2)$$

at the end of an upflow half cycle, and

$$L_2 = \frac{v_0}{(1 + b)[1 + 0.5(m_1 + m_2)]} \frac{\pi}{\omega} \qquad (3)$$

by the end of a downflow half cycle, where

$$m = \frac{\delta_s(1 - \epsilon)M}{\epsilon} \qquad (4)$$

and $M(= X/y)$ is the equilibrium distribution coefficient at temperature T. The quantity b is associated with a given two-phase system when operated at two specific temperatures and may be expressed as

$$b = \frac{0.5(m_2 - m_1)}{1 + 0.5(m_1 + m_2)} \qquad (5)$$

We will see from Table 1 that at steady state, solute removed from the lower reservoir is complete in Region 1 but only partial in Region 3. Figures 3 and 4 show some calculated results for concentration transients. The ordinate is the average reservoir concentration during the downflow half cycle divided by the initial liquid-phase concentration.

The upflow and downflow concentrations are identical in pumps with no reservoir dead volume. In pumps which have reservoir dead volume they are identical at steady state. Dimensionless concentrations greater than 1 are top reservoir concentrations, while those less than 1 are bottom reservoir concentrations. Figure 3 shows the effects of varying reservoir displacement volume in either reservoir. As long as the penetration distance L_2 is less than or equal to h (Cases 1 and 2), the separation factor approaches infinity as n becomes large. However, as L_2 is increased to the point where it exceeds h, the steady-state concentration in the lower reservoir abruptly switches to a finite value and the steady-state separation factor becomes finite. In Case 3, $L_2 = 120$ and $\alpha_\infty = 1.97$.

One can see from Fig. 3 and the expressions in Table 1 for the steady-state concentrations in the batch pump with no reservoir dead volume ($C_1 = C_2 = 0$) that not only can one obtain complete removal of solute from the lower reservoir—by making L_2 less than h—but at the same time one may obtain an arbitrarily high concentration in the upper reservoir by making $L_2 << h$, i.e., by making the column very long relative to the reservoir displacement volume.

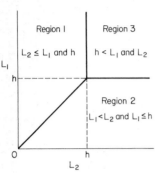

Fig. 2 Regions of parametric-pump operation.

Another way of achieving an infinite separation factor with arbitrarily high enrichment in the upper reservoir is with a pump with dead volume in the lower reservoir and no dead volume or less dead volume in the upper reservoir. Figure 4 shows concentration transients illustrating this point for a pump with $h = 100$ and a reservoir displacement volume corresponding to a downflow penetration distance of 100. In Case 1, there is no dead volume in either reservoir. The addition of dead volume to the lower reservoir equal to ten times the displacement volume produces an increase in steady-state top reservoir concentration roughly consistent with the ratio of the reservoir volumes. An addition of dead volume to the upper reservoir alone has the opposite effect. A bottom reservoir

TABLE 1 Steady-State Solutions

No.	Pump type	Region 1	Region 2
1.	Batch	$\dfrac{< yBP >\infty}{y_0} = 0$	
		$\dfrac{<yTP >\infty}{y_0} = 1 + \dfrac{2b}{1 + b}\dfrac{1}{1 + V_T/(Q\pi/\omega)} \times \left(\dfrac{h - L_2}{L_1 - L_2} + \dfrac{1 - b}{2b} + \dfrac{1 + b}{2b}\dfrac{V_B}{Q\pi/\omega}\right)$	
2.	Continuous top feed	$\dfrac{< yBP >\infty}{y_0} = 0$	$\dfrac{< yBP >\infty}{y_0} = \dfrac{(\varphi B - b)(\varphi T + \varphi B)}{\varphi B[(\varphi T + \varphi B) - b(1 + \varphi T\varphi B)]}$
		$\dfrac{< yTP >\infty}{y_0} = 1 + \dfrac{\varphi B}{\varphi T}$	$\dfrac{< yTP >\infty}{y_0} = \dfrac{(\varphi T + \varphi B)(1 - b\varphi B)}{\varphi T + \varphi B - b(1 + \varphi T\varphi B)}$
3.	Semicontinuous, continuous top feed during downflow, batch during upflow	$\dfrac{< yBP >\infty}{y_0} = 0$	$\dfrac{< yBP >\infty}{y_0} = \dfrac{(\varphi T + \varphi B)[\varphi B - b(2 + \varphi B)]}{\varphi B[(\varphi T + \varphi B) - b(2 + \varphi B - \varphi T)]}$
		$\dfrac{< yTP >\infty}{y_0} = 1 + \dfrac{\varphi B}{\varphi T}$	$\dfrac{< yTP >\infty}{y_0} = \dfrac{(\varphi T + \varphi B)(1 + b)}{\varphi T + \varphi B - b(2 + \varphi B - \varphi T)}$

arbitrarily large with respect to the top reservoir would lead to an arbitrarily large enrichment in the upper reservoir, but at the same time the transient time for depletion of solute from the lower reservoir would become very long.

The Continuous Parametric Pump

A continuous version of the parametric pump was first described by Chen and Hill.[8] A steady flow of feed is supplied to the top of the column and steady flows of top and bottom product streams are withdrawn from the apparatus. Figure 5 shows the schematic diagram of experimental apparatus. The equipment consists of a jacketed column packed with the adsorbent particles. The hot and cold baths are connected to the column jacket. Two three-way solenoid valves are wired to a dual timer so that hot water is always directed to the column jacket during upflow and cold water during downflow. Each half cycle is π/ω time units in duration and the reservoir displacement volume is $Q(\pi/\omega)$, where Q is the reservoir displacement rate. The pump has dead volumes V_T and V_B for the top and bottom reservoirs, respectively. The feed is directed to the top of the column at the flow rate $(\phi_T + \phi_B)Q$. The top-product flow rate is $\phi_T Q$ and the bottom-product flow rate is $\phi_B Q$, and ϕ_T and ϕ_B are the top- and bottom-product rates to the reservoir displacement rate.

In the batch pump the flow rates within the column in upflow and downflow are identical and are equal to the reservoir displacement rate Q. The column flow rates in the continuous pump may be determined by reference to Fig. 5. Material balances around the point of the bottom-product withdrawal show that the column flow rate in upflow must be $(1 - \phi_B)Q$ and in downflow $(1 + \phi_B)Q$.

By the equilibrium theory Chen and Hill have shown that for the continuous pump there are three regions of pump operation (Fig. 2):

Region 1: $\qquad$ $L_1 \geq L_2$ (or $\phi_B \leq b$) and $L_2 \leq h$
Region 2: $\qquad$ $L_1 < L_2$ (or $\phi_B > b$) and $L_1 \leq h$ $\qquad$ (6)
Region 3: $\qquad$ L_1 and $L_2 > h$

where
$$L_1 = \frac{v_0(1 - \phi_B)}{(1 - b)[1 + 0.5(m_1 + m_2)]} \frac{\pi}{\omega} \qquad (7)$$

and
$$L_2 = \frac{v_0(1 + \phi_B)}{(1 - b)[1 + 0.5(m_1 + m_2)]} \frac{\pi}{\omega} \qquad (8)$$

Region 3

$$\frac{<y_{BP}>\infty}{y_0} = 1 - \frac{h\left(1 - \frac{L_2}{L_1}\right)}{L_2(1 + C_2)} + (1 + C_1)(L_1 - L_2)\left(\frac{h}{L_1 L_2}\right)^2 \left[\frac{L_1 + \frac{h - L_1}{1 + C_2}}{\left(C_1 + \frac{h}{L_2}\right)\left(C_2 + \frac{h}{L_1}\right) - (1 + C_1)(1 + C_2)}\right]$$

$$\frac{<y_{TP}>\infty}{y_0} = 1 + \frac{h\left(\frac{1}{L_2} - \frac{1}{L_1}\right)\left(C_2 + \frac{h}{L_1}\right)}{(1 + C_1)(1 + C_2) - \left(C_1 + \frac{h}{L_2}\right)\left(C_2 + \frac{h}{L_1}\right)}$$

$$\frac{<y_{BP}>\infty}{y_0} = \frac{(\varphi T + \varphi B)\left(1 - \frac{h}{L_2}\right)}{\varphi T\left[1 - \frac{h}{L_1}(1 - \varphi B)\right] + \varphi B\left[1 - \frac{h}{L_2}(1 - \varphi T)\right]}$$

$$\frac{<y_{TP}>\infty}{-y_0} = \frac{\varphi T + \varphi B}{1 - \varphi T}\left\{\frac{1 + \varphi B - \frac{h}{L_1}(1 - \varphi B)}{\varphi T\left[1 - \frac{h}{L_1}(1 - \varphi B)\right] + \varphi B\left[1 - \frac{h}{L_2}(1 - \varphi T)\right]} - 1\right\}$$

$$\frac{<y_{BP}>\infty}{y_0} = \frac{(\varphi T + \varphi B)\left(1 - \frac{h}{L_2}\right)}{\varphi T\left(1 - \frac{h}{L_1}\right) + \varphi B\left[1 - \frac{h}{L_2}(1 - \varphi T)\right]}$$

$$\frac{<y_{TP}>\infty}{y_0} = \frac{\varphi T + \varphi B}{1 - \varphi T}\left\{\frac{1 + \varphi B - \frac{h}{L_1}}{\varphi T\left(1 - \frac{h}{L_1}\right) + \varphi B\left[1 - \frac{h}{L_2}(1 - \varphi T)\right]} - 1\right\}$$

When the pumps are operated in Region 1, the bottom-product concentration is

$$\frac{<y_{BP2}>_n}{y_0} = \frac{1-b}{1+b}\left[\frac{(1-b)/(1+b) + C_2}{1+C_2}\right]^{n-1} \tag{9}$$

At steady state ($n \to \infty$)

$$\frac{<y_{BP2}>_\infty}{y_0} = 0 \tag{10}$$

Thus, in this region, at the steady state, complete removal of solute from the bottom reservoir or bottom-product stream occurs, and the top-product stream must contain all of the solute supplied by the feed stream, i.e.,

$$\frac{<y_{TP2}>_\infty}{y_0} = 1 + \frac{\phi_B}{\phi_T} \tag{11}$$

When the pumps are operated outside Region 1 (i.e., Region 2 or 3), complete removal of solute from the product stream cannot be obtained.

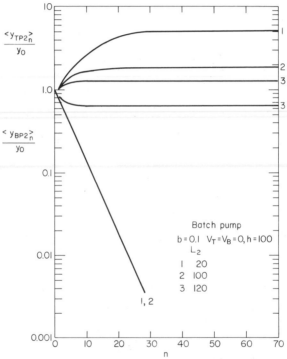

Fig. 3 Effect of reservoir displacement volume on concentration transients in batch parametric pump with no reservoir volume. [*Reprinted from* Sep. Sci., **6**, 427 (1971) *with permission of Marcel Dekker, Inc.*]

Figure 6 shows the comparison between the data and the calculated results based on the transient equations derived from the equilibrium theory (Chen[9]). One can see that in Region 1 the separation factor (defined as $<y_{TP2}>_n/<y_{BP2}>_n$) increases as n increases. It evidently will increase without limit as n becomes large, as theory predicts. On the other hand, the separation factors become finite in Regions 2 and 3.

A parametric pump with center feed into a central reservoir was studied by Horn and Lin.[17] The authors discussed the apparatus in general mathematical terms without any specific examples and experimental data. Chen and Hill investigated a continuous model

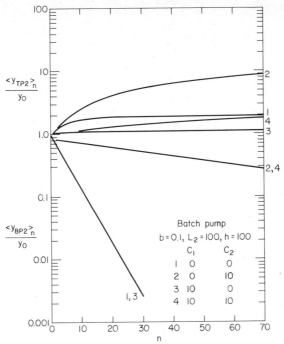

Fig. 4 Effect of reservoir dead volume on concentration transients in batch parametric pump operating at the switching point. [*Reprinted from* Sep. Sci. **6**, 428 (1971) *with permission of Marcel Dekker, Inc.*]

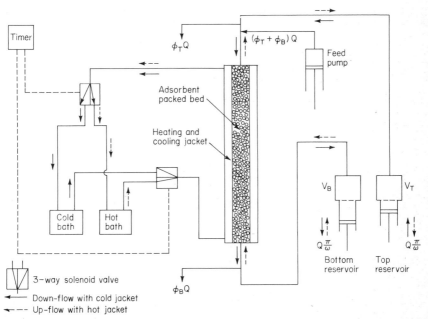

Fig. 5 Schematic diagram of experimental apparatus for thermal parametric pumping. [*Reprinted from* Sep. Sci. **9**, 36 (1974) *with permission of Marcel Dekker, Inc.*]

with center feed between an enriching column and a stripping column, and found this mode also capable of complete removal of solute from one product fraction and of arbitrarily large enrichment in the other.

The Semicontinuous Parametric Pump

The semicontinuous pump is operated batchwise during upflow and continuously during downflow (Fig. 7). The downflow penetration distance L_2 is the same for both the continuous and the semicontinuous pumps [Eq. (8)], because the feed and product streams are continuous during this half cycle. During the upflow half cycle, however, the

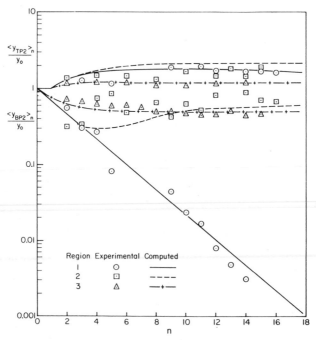

Fig. 6 Effects of L_1 and L_2 and h on product concentration for continuous thermal parametric pumping; the system used was toluene–n-heptane–silica gel. [*Reprinted from* Am. Inst. Chem. Eng. J., **18**, 357 (1972) *with permission of American Institute of Chemical Engineers.*]

penetration distance for the semicontinuous pump is the same as that for the batch pump [Eq. (2)].

In terms of the equilibrium theory, the semicontinuous pump, similar to the continuous pump, has three possible regions of pump operation (Fig. 2):

Region 1: $L_1 \geq L_2 \left(\text{or } \phi_B \leq \dfrac{2b}{1-b} \right)$ and $L_2 \leq h$

Region 2: $L_1 > L_2 \left(\text{or } \phi_B > \dfrac{2b}{1-b} \right)$ and $L_1 \leq h$ (12)

Region 3: L_1 and $L_2 > h$

Expressions for concentration transients, corresponding to all three regions, are available elsewhere (Chen[10]), and the steady-state solutions are presented in Table 1.

Figure 8 shows the comparison of the bottom-product concentration transients for the semicontinuous and continuous pumps. The performance characteristics of both pumps are similar in nature and approach infinite steady-state separation in Region 1 for both pumps. The main difference between the two pumps is the difference in the locus of switching points between Regions 1 and 2 (see Fig. 2). If, in a pump originally operated in

Region 1, L_2 is increased until it exceeds h, or L_1 becomes less than L_2, switching points are encountered and the steady-state behavior of the pump abruptly switches from a mode in which solute is completely removed from the lower reservoir to one in which solute removal is incomplete. Crossing the boundary, $L_1 = L_2$ can result from increasing the rate of bottom-product withdrawal, that is, increasing ϕ_B so that L_1 becomes less than L_2. For

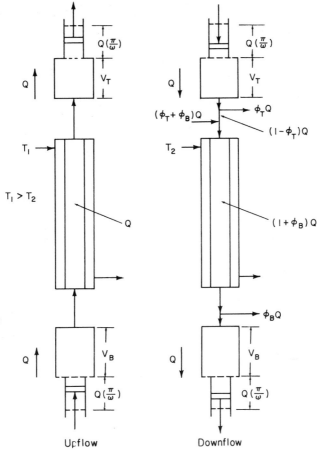

Upflow   Downflow

Fig. 7 Internal flow rates in semicontinuous pump. [*Reprinted from* Am. Inst. Chem. Eng. J., **19**, 590 (1973) *with permission of American Institute of Chemical Engineers.*]

the continuous pump, the switching points correspond to the condition $P_B/Q = \phi_B = b$. In the case of the semicontinuous pump the condition is $P_B/Q = \phi_B = 2b/(1-b)$. For the semicontinuous operation, the average flow rate during a complete cycle is $P_B/2$. Thus, in terms of the average flow rate, the switching condition is $(P/2)/Q = \phi_B/2 = b/(1-b)$. One can see that when both pumps are operating at their respective switching points the ratio of pure solvent produced by the semicontinuous pump to that produced by the continuous pump is $1/(1-b)$. It follows that for large values of b the bottom rate of production of pure solvent in the semicontinuous pump may be quite large relative to that of the continuous pump.

Figure 9 illustrates the steady-state concentrations for both the semicontinuous and continuous pumps. In the interval $0 < \phi_B < 0.22$, corresponding to $0 < \phi_B < b$, no solute appears in the bottom-product stream of either the semicontinuous or the continuous pump. Beyond the switching point of the continuous pump, that is, beyond $\phi_B = b$, solute appears in the bottom product of the continuous pump. However, no solute could appear in the bottom product of the semicontinuous pump until the switching point $\phi_B = 2b/(1 - b)$, that is, $\phi_B = 0.56$, is reached. This is beyond the present experimental range of $0 < \phi_B < 0.4$ because of the feed limitation $\phi_T + \phi_B = 0.4$. It should be pointed out that the same types of curves would result for the semicontinuous pump as for the continuous pump, provided the feed rate $\phi_T + \phi_B$ extended beyond 0.56. Over the inverval $0 < \phi_B <$

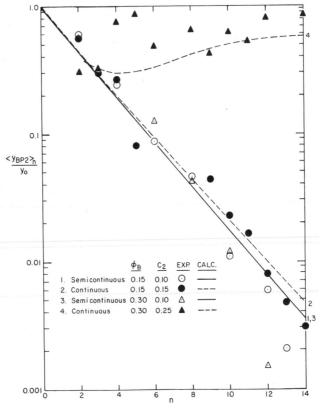

Fig. 8 Comparison of semicontinuous and continuous parametric pumps; the system used was toluene–n-heptane–silica gel. [*Reprinted from* Am. Inst. Chem. Eng. J., **19**, *594 (1973) with permission of American Institute of Chemical Engineers.*]

switching point, the top-product concentration increases according to Eq. (11). Beyond the switching point the top-product concentrations would decrease according to the appropriate expressions shown in Table 1.

Gregory and Sweed[15] have studied two open systems, nonsymmetric and symmetric. Both systems are semicontinuous in the sense that feed and products are withdrawn only during a portion of each cycle. However, they are different from the one described above. These two systems were then examined experimentally for the case of NaCl separation from H_2O, and the experiments were simulated using the STOP-GO theory (Sweed and Gregory[30] and Gregory and Sweed[16]). The STOP-GO model fit the data very well, while the equilibrium theory predicted more separation than was obtained experimentally.

Multicomponent Separations

The picture of parametric-pumping operation described above in connection with the separation of two-component mixtures may be extended to describe how a parametric pump will operate when treating a multicomponent mixture (Chen).[11] Assume that a mixture including s adsorbable components in an inert solvent is treated in a continuous or semicontinuous pump between two specified temperatures, and with a specified reservoir displacement flow rate Q, and also a specified product flow rate P_B. Let us assume further that the multicomponent mixture may be treated as s pairs of pseudo-binary systems. Each system includes one solute and the common inert solvent and could be characterized by a dimensionless equilibrium parameter b_i and corresponding values of the penetration distances of the hot and cold cycles L_{1i} and L_{2i}. L_{1i} and L_{2i} can be expressed in terms of $\phi_B(= P_B/Q)$ and b_i as shown in Eq. (2), (3), (7), and (8).

By treating the multicomponent mixture as a series of pseudo-binary systems, the multicomponent separation could be predicted by the existing mathematical expressions for binary systems.[8,9,10]

Now let us consider a mixture containing s solutes, each with its own b_i and

$$b_1 > b_2 > \cdots b_k \geqslant \phi_B > b_{k+1} \cdots > b_s \tag{13}$$

for the continuous pump, and

$$\left(\frac{2b}{1-b}\right)_1 > \left(\frac{2b}{1-b}\right)_2 \cdots \left(\frac{2b}{1-b}\right)_k \geqslant \phi_B > \left(\frac{2b}{1-b}\right)_{k+1} \cdots \left(\frac{2b}{1-b}\right)_s \tag{14}$$

for the semicontinuous pump. Also,

$$L_{2i} \leqslant h \qquad \text{where } i = 1, 2, \ldots, k \tag{15}$$

At steady state the components $i = 1, 2, \ldots, k$ for which the operations are indicated in Region 1 would appear only in the top-product stream, and the remaining components ($k + 1, \ldots, s$) would appear in both the top- and bottom-product streams. In the extreme case where $k = s$ the bottom-product stream would consist only of pure solvent. By proper adjustment of ϕ_B in Eq. (13) or (14), a solute split could be made which is analogous to that obtained by a multicomponent distillation column.

Figure 10 shows the separation of toluene-aniline-n-heptane by continuous parametric pumping (Chen et al.[11]). In the computation it was assumed that the system contains two pseudo binaries, each binary consisting of one solute as one component and the common solvent as the other component, that is, toluene–n-heptane and aniline–n-heptane. The switching point for toluene corresponds to the condition $\phi_B = b_{\text{toluene}} = 0.15$. In the case of aniline, the condition $\phi_B = b_{\text{aniline}} = 0.31$. Thus, when $\phi_B \leqslant 0.15$ (curve $3a$ and $3b$), the operation is in Region 1 for both toluene and aniline, and solute removal from the bottom-product stream may be complete at $n \to \infty$. If ϕ_B is increased to the interval range $0.15 < \phi_B \leqslant 0.31$ (curves $4a$ and $4b$), the operation switches to Region 2 for toluene and remains in Region 1 for aniline, and the bottom product could eventually contain only toluene. If ϕ_B is further increased, $\phi_B > 0.31 = b_{\text{aniline}}$, the operation is now in Region 2 for both toluene and aniline, and both toluene and aniline would appear in the bottom-product stream.

Figures 11 and 12 show the net direction of concentration fronts moving through the adsorption column as n

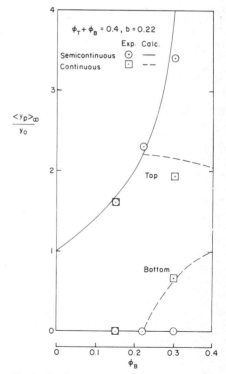

Fig. 9 Steady-state concentrations for semicontinuous and continuous parametric pumps; the system used was toluene–n-heptane–silica gel. [*Reprinted from* Am. Inst. Chem. Eng. J., **19,** 594 (1973) *with permission of American Institute of Chemical Engineers.*]

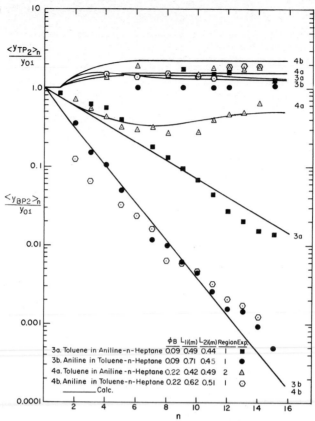

Fig. 10 Effects of operating conditions on product concentrations; the system used was aniline–toluene–n-heptane–silica gel. [*Reprinted from* Am. Inst. Chem. Eng. J., **20**, *309 (1974) with permission of American Institute of Chemical Engineers.*]

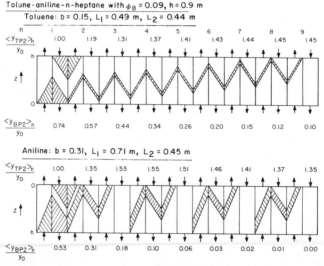

Fig. 11 Net movement of concentration fronts with aniline and toluene in Region 1. [*Reprinted from* Am. Inst. Chem. Eng. J., **20**, *309 (1974) with permission of American Institute of Chemical Engineers.*]

increases (or a function of time). The average top- and bottom-product concentrations relative to the feed concentrations are also given. In Fig. 11 the operation for both toluene and aniline is in Region 1, and since in the case $(L_{1i}/L_{2i}) \geqslant 1$, the net movements of the concentration fronts for both components are upward, the bottom-product concentrations show a steady decrease. Note that L_{1i}/L_{2i} for aniline is greater than that for toluene, and consequently aniline moves up the column faster than does toluene.

In Fig. 12, the operations for aniline and toluene are in Regions 1 and 2, respectively, that is, $(L_{1i}/L_{2i})_{\text{aniline}} \geqslant 1$ and $(L_{1i}/L_{2i})_{\text{toluene}} < 1$. The net movements of the concentration fronts of the two components are clearly in opposite directions. The net direction of

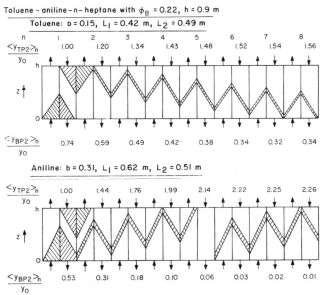

Fig. 12　Net movement of concentration fronts with aniline and toluene operated in Regions 1 and 2, respectively. [*Reprinted from* Am. Inst. Chem. Eng. J., **20**, *310 (1974) with permission of American Institute of Chemical Engineers.*]

aniline is upward and the bottom-product concentration $\langle y_{\text{BP2}} \rangle_n / y_{0i}$ decreases as n increases. The net direction of toluene is downward and $\langle y_{\text{BP2}} \rangle_n / y_{0i}$ approaches a limiting value. The net downward direction occurs because of an excessive bottom-product flow rate, that is, $\phi_B > b_{\text{toluene}}$. However, it must be noted that even though the net direction is upward, a modest separation will also occur if the reservoir displacement volume is excessive, that is, $L_{2i} > h$ (Region 3).

The system glucose-fructose-water on a cation exchanger (AG50W-X4, calcium form) was also studied by Chen and D'Emidio.[14] Agreement between experiment and the equilibrium theory was roughly equivalent to that obtained for toluene–aniline–n-heptane–silica gel.

The prediction of multicomponent separations by the simple method described above involves the assumption that solutes do not interact with one another. In practice, the concentration of solutes may be quite high and the high solute concentrations may cause competition for adsorption sites. However, this method does represent an upper bound on separation and provides a basis for commercial design.

The equilibrium theory has also been employed by Butts et al.[6] to study separations of multicomponent mixtures in a batch pump. The separation was achieved by using a nonsymmetric flow pattern. Later, Butts[7] applied a different approach to separate multicomponent mixtures of cations on Dowex 50 × 8 resin. In these experiments binary mixtures of K^+ and H^+ could be adjusted so that K^+ would migrate to the top reservoir and

H^+ to the bottom reservoir in a batch operation. Separation factors in the thousands were obtained.

Optional Performance of Equilibrium Pumps

The optional pump performance would be interpreted as the achievement of separation with maximum production of bottom product and complete removal of solutes $1, 2, \ldots, k$ [see Eqs. (13), (14), and (15)]. This would occur when:

1. $L_{2k} = h$ (that is, the pump is operated just on the verge of breakthrough of solute k from the top to the bottom of the column).
2. $\phi_B = b_k$ for the continuous pump.

$$\phi_B = \left(\frac{2b}{1-b}\right)_k \text{ for the semicontinuous pump}$$

Under these circumstances it can be shown that the maximum (or optional) reservoir displacement flow rate is

$$Q_{max} = Ah\epsilon \frac{1 + m_{0k}}{\pi/\omega} \quad \text{for the continuous pump}$$

$$= Ah\epsilon \frac{(1 + m_{0k})(1 - b_k)}{\pi/\omega} \quad \text{for the semicontinuous pump} \tag{16}$$

and the maximum bottom-product flow rate is

$$P_{Bmax} = Q_{max} b_k = Ah\epsilon \frac{(1 + m_{0k})b_k}{(\pi/\omega)} \quad \text{for the continuous pump}$$

$$= Q_{max}\left(\frac{2b}{1-b}\right)_k = Ah\epsilon \frac{(1 + m_{0k})2b_k}{\pi/\omega} \quad \text{for the semicontinuous pump} \tag{17}$$

Figure 13 shows the effects of Q and P_B on the product concentrations for the semicontinuous pump. As long as $Q \leq Q_{max}$ [or $Q(\pi/\omega) \leq Q_{max}(\pi/\omega)$] and $P_B \leq P_{Bmax}$ [or $P_B(\pi/\omega) \leq P_{Bmax}(\pi/\omega)$] $<y_{BP2}>_n/y_0$ decreases as n increases and will approach zero as theory predicts (curves 1 and 3). If, in a pump originally operated in a region where $Q < Q_{max}$ and $P_B < P_{Bmax}$, P_B is increased until it exceeds P_{Bmax} or Q becomes greater than Q_{max}, the steady-state behavior of the pump abruptly switches from a mode with infinite separation to one with finite separation. Hence Q_{max} and P_{Bmax} are the operating conditions necessary to accomplish separation with the maximum production of bottom product and infinite separation factors. It should be pointed out that the crossing of the boundary $Q = Q_{max}$ or $P_B = P_{Bmax}$ may also be thought of as switching from Region 1 to 3 or 1 to 2 (see Fig. 2).

Figure 14 shows a comparison of experimental and calculated concentration transients for $P_B < P_{Bmax}$ and $Q < Q_{max}$ with optimal theoretical results. One can see that variations in P_B, Q, and the mode of pump operation have no effect on the bottom-product concentration $<y_{BP2}>_n/y_0$. In every case, at the steady state ($n \rightarrow \infty$), solute removal from the bottom-product stream can be complete, and the top-product stream must carry away all the solute supplied by the feed stream. However, the pumps with $P_B = P_{Bmax}$ and $Q = Q_{max}$ (or the optimal operating conditions) are ideal separation devices in the sense that they can continuously or semicontinuously separate a system into one fraction completely free of solute and another fraction enriched with solute to the maximum degree.

Recuperative-Mode Separations

The form of parametric pumping described is the so-called direct mode in which the change in the intensive variable is applied uniformly over the entire bed. A second form, called the recuperative mode, is illustrated in Fig. 15. In terms of temperature as the intensive variable, bed temperature is changed by using a heat exchanger to change the temperature of fluid entering the column. The moving fluid carries not only the mixture components but also thermal energy into and out of the column as the flow direction alternates. The temperatures within the bed depend on fluid flow rate, temperatures of heat exchangers, heat capacity of both solid and fluid phases, and the rate of interphase heat transfer. Recuperative thermal parametric pumping was first introduced by Wilhelm[35] and was later studied in more detail by Wilhelm[38] and Rolke and Wilhelm.[25]

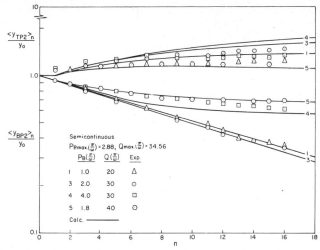

Fig. 13   Effects of P_B and Q on product concentrations; the system used was sodium nitrate–water–ion retardation resin. [*Reprinted from* Am. Inst. Chem. Eng. J., *20, 1021 (1974) with permission of American Institute of Chemical Engineers.*]

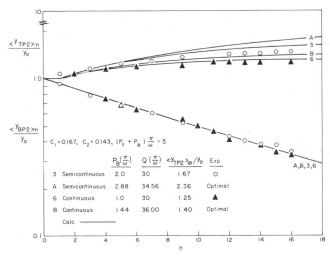

Fig. 14   Comparison of concentration transients for $P_B < P_{Bmax}$ and $Q < Q_{max}$ with optimal theoretical results; the system used was sodium nitrate–water–ion retardation resin. [*Reprinted from* Am. Inst. Chem. Eng. J., *20, 1022 (1974) with permission of American Institute of Chemical Engineers.*]

Experimental results for the $NaCl–H_2O$–ion-exchange resin system showed that the separation factors were substantially lower than that obtained by the direct mode. However, note that the recuperative mode has the advantage of better thermal contact between phases and of better heat recovery.

Separations of Gases

Much less work has been done on the thermal parametric pumping separation of gases than for liquids. Jencziewski and Meyers[18] studied separation of propane and ethane on activated carbon. A batch apparatus consisting of a single jacketed column equipped with piston pumps for reservoirs was used, which is identical to that used for liquid systems

(Fig. 1). Other separations accomplished by using batch—direct mode include argon–propane–active carbon and ethane–propane–active carbon (Jencziewski and Meyers[19]), and air–CO_2–silica gel (Patrick[22]). In these works the authors have shown that gases could be separated to a degree, but the difficulties with thermal parametric pumping were not mentioned. Because of the pressure-volume-temperature behavior of gases, if the operation is at a constant volume, the increase in temperature causes the pressure increases and an increase in adsorption, which clearly opposes the principle of thermal parametric pumping, i.e., the decrease of adsorption caused by the hot half cycle (or temperature increase). When the operation is at constant pressure, the change in temperature causes a change in volume, which in turn causes the increase in mixing and decrease in separation. Thus, while thermal parametric pumping has great promise for liquid separation, it does not appear to have potential for gas separations.

pH PARAMETRIC PUMPING

A batch pH parametric pumping system is shown in Fig. 16. The recuperative mode is used for pH control (direct mode could be envisioned using dialysis tubing to contain the mixture to be separated and surrounded by a cycling pH solution). Flow is upward during a high-pH half cycle and downward during a low-pH half cycle. Flow to and from the reservoirs during upflow and downflow is at a constant rate.

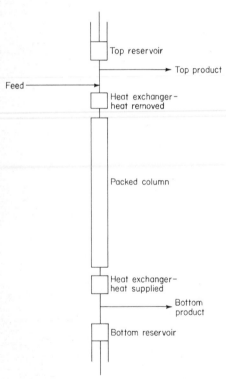

Fig. 15   Recuperative thermal parametric pump.

Prior to each run the entire system, including the interstitial column volume, the bottom reservoir, and the top dead reservoir volume, are filled with the feed mixture. Two constant-pH fields (i.e., high and low pH) are imposed periodically to the system by using a Bio-Fiber miniplant in each line entering the column. The reservoir are set to deliver about the interstitial column volume per half cycle. During the first half cycle the fluid in the bottom reservoir is pumped through the high-pH miniplant and into the bottom of the column. At the same time, solution that emerges from the column flows through the low-pH miniplant and fills the top reservoir. On the next half cycle the solution in the top reservoir flows back through the low-pH miniplant and passes through the column and then the high-pH miniplant to the bottom reservoir. This procedure is repeated in each of the succeeding cycles. Note that the miniplant uses the natural turbulence like a countercurrent heat exchanger to mix the buffer solution.

Shaffer and Hamrin[27] studied trypsin removal from α-chymotrypsin–trypsin mixtures by a parapumping system which is similar to the unit described above. Figure 17 shows the experimental results at pH gradients of 3 and 6. The parametric pumping runs were carried out using a 26 mm ID × 40 mm chromatographic column packed with Sepharose CVB CHOM. Separations were fitted to an equation of the form $\log_{10}(<y_B>_n/y_0) = \alpha n$, where α is a constant. Separations were much less than those predicted from the equilibrium data [Eq. (9)]. However, they do indicate that pH parametric pumping can be used for enzyme separations.

Sabadell and Sweed[26] used pH changes to remove K^+ and Na^+ from H_2O. The separation utilized the recuperative mode of semibatch operation with the low pH end being closed and the high pH end open. The maximum separation factor obtained for total K^+ + Na^+ was 1.84.

Recently, Chen[39] investigated a semicontinuous pH parametric pump for separating proteins. Various factors affecting separations were examined. It was shown that parametric pumping is capable of separating proteins with high separation factors.

HEATLESS PARAMETRIC PUMPING

The name "parametric pumping" was applied to the separation process by the inventor of the batch pump, the late R. H. Wilhelm of Princeton University. Since the time of that

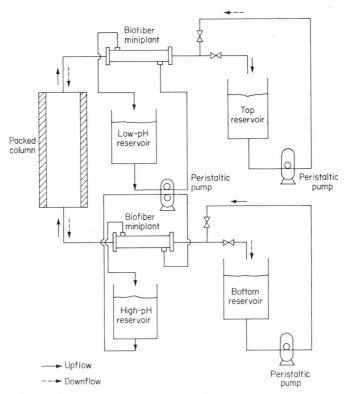

Fig. 16   Schematic diagram of experimental apparatus for continuous pH parametric pumping.

invention at least one preexisting industrial process has been recognized to operate on the parametric-pumping principle. One such process is the heatless fractionation process of Exxon (Sharstrom[28]). In this process two columns are used for the separation of gases. Adsorption of solute from a gas stream is done at high pressure and desorption is done at low pressure. This process has definitely been shown to be a form of parametric pumping by Shendalman and Michell.[29]

Figure 18 shows the schematic diagram of a heatless parametric pump. The unit includes two interconnected columns packed with an adsorbent. The heavy line depicts the gas flow during a half-cycle time π/ω. The high-pressure feed flows from the gas cylinder through a regulator and enters the bottom of column 1 at a constant flow rate F. During the adsorption half cycle of column 1, column 1 is maintained at the high pressure ρ_1. As the gas flows out the top of column 1 through an open solenoid valve, a portion of it leaves the system as the high-pressure top product at a flow rate q. The remaining portion of the gas flows through a pressure regulator which reduces it to a low pressure of ρ_2. This

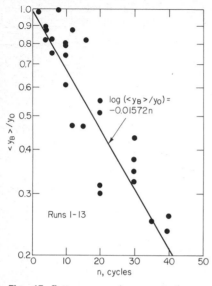

$$\log(<y_B>/y_0) = -0.01572n$$

Runs 1-13

Fig. 17 Bottom reservoir concentration vs. cycles; the system used was trypsin–α-chymotrypsin–Sepharose CVB CHOM; low pH =3, high pH = 6. [*Reprinted from* Am. Inst. Chem. Eng. J., **21**, 785 (1975) *with permission of American Institute of Chemical Engineers.*]

gas then flows into the top of column 2 at a low-pressure purge rate W during the desorption half cycle of column 2 and flows out of the column to the atmosphere. Column 2 is maintained at the low pressure ρ_2 during its desorption half cycle. This process continues until the end of the half cycle.

At the end of the half cycle the high-pressure feed is switched from column 1 to column 2 by closing the three-way solenoid valve at the bottom of column 1 and opening the one at the bottom of column 2, allowing high-pressure feed to flow into the bottom of column 2 at a rate of F. The pressure in column 2 then increases to ρ_1. A portion of gas from column 2 is drawn off as the top product (with a rate q) while the remaining part is reduced to low pressure ρ_2 by the regulator and flows into the top of column 1 at a rate W. This gas then exits column 1, which is maintained at pressure ρ_2, through the three-way solenoid valve to atmosphere. The flow continues until one cycle is complete. The procedure described is repeated in each of the succeeding cycles. Note that it is essential to set the half-cycle duration to be much longer as compared with the amount of time required to switch the column either from high to low pressure or from low to high pressure.

With the aid of the equilibrium theory and the concept of penetration distances Shendalman and Michell have shown that the conditions for achieving infinite separation are

$$L_1 = \frac{\epsilon v_1}{\epsilon + k(1-\epsilon)} \frac{\pi}{\omega} \leq h \quad \text{or} \quad F\frac{\pi}{\omega} \leq hA\left[\epsilon + k(1-\epsilon)\right] \tag{18}$$

$$\gamma = \frac{W}{F} = \frac{\text{purge volumetric flow rate}}{\text{feed volumetric flow rate}} \geq \left(\frac{\rho_1}{\rho_2}\right)^{\epsilon/\epsilon + k(1-\epsilon)}$$

and at the steady state ($n \to \infty$),

$$\frac{<Y_{TP}>_\infty}{Y_0} = 0$$

$$\frac{<Y_{BP}>_\infty}{Y_0} = \frac{1}{\gamma}\frac{\rho_1}{\rho_2} \tag{19}$$

Figure 19 shows the experimental results for the system CO_2–helium–silica gel. The straight line corresponds to the theoretically predicted exponential decay for $\gamma \geq 1$ and $k \geq 1$. Although the agreement was poor between the equilibrium theory and experiment, the theory does provide a qualitative or semiquantitative description of how a heatless parametric pump operates. Michell and Shendalman[21] later developed a theoretical model to include the effect of mass-transfer rate. Two versions of the model were constructed to bracket the effects of occurrences during pressure changes in the adsorption column. The results indicated these two versions did bracket the observations.

Weaver and Hamrin[34] studied the separation of hydrogen-deuterium mixtures by heatless parametric pumping. Alexis[1] discussed the application of heatless adsorption to purify low-grade hydrogen stream. Turnock and Kadlec[32] and Kowler and Kadlec[20] studied the separation of nitrogen and methane, and the optimization of the process. Despite the fact that heatless adsorption is considerably more complicated to model than the thermal

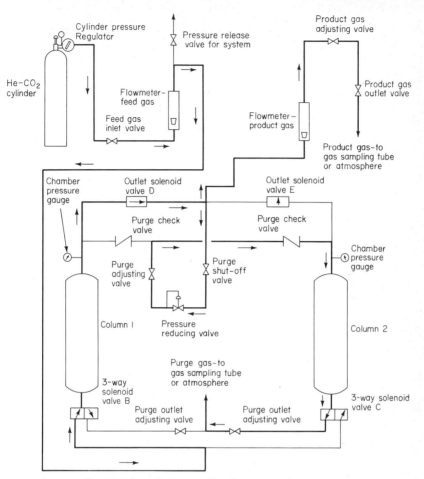

Fig. 18 Schematic diagram of heatless parametric pump.

adsorption which has been successfully modeled, heatless parametric pumping appears to have great promise for gas separations.

REFERENCES

1. Alexis, R. W., *Chem. Eng. Prog.*, **63**(5), 69 (1967).
2. Apostolopoulos, G. P., *Ind. Eng. Chem. Fundam.*, **14**, 11 (1975).
3. Aris, R., *Ind. Eng. Chem. Fundam.*, **8**, 603 (1969).
4. Booij, H. L., *Ind. Eng. Chem. Fundam.*, **8**, 231 (1969).
5. Booij, H. L., *Ind. Eng. Chem. Fundam.*, **14**, 281 (1975).
6. Butts, T. J., Gupta, R., and Sweed, N. H., *Chem. Eng. Sci.*, **27**, 855 (1972).
7. Butts, T. J., Sweed, N. H., and Camero, A. A., *Ind. Eng. Chem. Fundam.*, **12**, 467 (1973).
8. Chen, H. T., and Hill, F. B., *Sep. Sci.*, **6**, 411 (1971).
9. Chen, H. T., Rak, J. L., Stokes, J. D., and Hill, F. B., *Am. Inst. Chem. Eng. J.*, **18**, 356 (1972).
10. Chen, H. T., Reiss, E. H., Stokes, J. D., and Hill, F. B., *Am. Inst. Chem. Eng. J.*, **19**, 589 (1973).
11. Chen, H. T., Lin, W. W., Stokes, J. D., and Fabisiak, W. R., *Am. Inst. Chem. Eng. J.*, **20**, 306 (1974).
12. Chen, H. T., and Manganaro, J. A., *Am. Inst. Chem. Eng. J.*, **20**, 1020 (1974).
13. Chen, H. T., Park, J. A., and Rak, J. L., *Sep. Sci.*, **9**, 35 (1974).
14. Chen, H. T., and D'Emidio, V. J., *Am. Inst. Chem. Eng. J.*, **21**, 813 (1975).
15. Gregory, R. A., and Sweed, N. H., *Chem. Eng. J.*, **1**, 207 (1970).
16. Gregory, R. A., and Sweed, N. H., *Chem. Eng. J.*, **4**, 139 (1972).

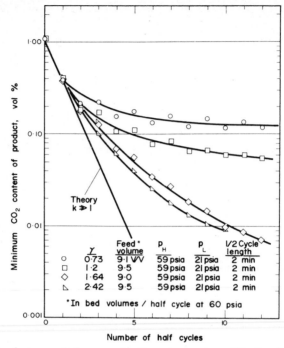

Fig. 19 Top-product concentration vs. cycles; the system used was CO_2–He–silica gel. [*Reprinted from* Chem. Eng. Sci., **27**, *1456 (1972) with permission of Pergamon Press.*]

17. Horn, F. J. M., and Lin, C. M., *Ber. Bunsenges. Phys. Chem.*, **73**, 575 (1969).
18. Jencziewski, T. J., and Meyers, A. L., *Am. Inst. Chem. Eng. J.*, **14**, 509 (1968).
19. Jencziewski, T. J., and Meyers, A. L., *Ind. Eng. Chem. Fundam.*, **9**, 316 (1970).
20. Kowler, D. E., and Kadlec, R. H., *Am. Inst. Chem. Eng. J.*, **18**, 1207 (1972).
21. Michell, J. E., and Shendalman, L. H., *Am. Inst. Chem. Eng Symp. Ser.* **69**(134), 25 (1973).
22. Patrick, R. R., Schrodt, J. T., and Kermode, K. J., *Sep. Sci.*, **7**, 331 (1972).
23. Pigford, R. L., Baker, B., and Blum, D. E., *Ind. Eng. Chem. Fundam.*, **8**, 144 (1969).
24. Rice, R. G., *Ind. Eng. Chem. Fundam.*, **14**, 280 (1975).
25. Rolke, R. W., and Wilhelm, R. H., *Ind. Eng. Chem. Fundam.*, **8**, 231 (1969).
26. Sabadell, J. E., and Sweed, N. H., *Sep. Sci.*, **5**, 171 (1970).
27. Shaffer, A. G., and Hamrin, C. E., *Am. Inst. Chem. Eng. J.*, **21**, 782 (1975).
28. Sharstrom, C. W., *Ann. N.Y. Acad. Sci.*, **72**, 75 (1959).
29. Shendalman, L. H., and Michell, J. E., *Chem. Eng. Sci.*, **27**, 1449 (1972).
30. Sweed, N. H., and Gregory, R. A., *Am. Inst. Chem. Eng. J.*, **17**, 171 (1971).
31. Tverdisl, V. A., Kleimeno, A. N., and Yakovenk, L. V., *Biofizika*, **18**, 251 (1973).
32. Turnock, P. H., and Kadlec, R. H., *Am. Inst. Chem. Eng. J.*, **17**, 335 (1971).
33. Wankat, P. C., *Ind. Eng. Chem. Fundam.*, **12**, 373 (1973).
34. Weaver, K., and Hamrin, C. E., *Chem. Eng. Sci.*, **29**, 1873 (1974).
35. Wilhelm, R. H., Rice, A. W., and Bendelius, A. R., *Ind. Eng. Chem. Fundam.*, **5**, 141 (1966).
36. Wilhelm, R. H., in K. B. Warren (ed.), "Intracellular Transport," p. 199, Academic, New York, 1966.
37. Wilhelm, R. H., and Sweed, N. H., *Science*, **159**, 522 (1968).
38. Wilhelm, R. H., Rice, A. W., Rolke, D. W., and Sweed, N. H., *Ind. Eng. Chem. Fundam.*, **7**, 337 (1968).
39. Chen, H. T., Hsieh, T. K., Lee, H. C., and Hill, F. B., *Am. Inst. Chem. Eng. J.*, **23**, 695 (1977).

Part 2

Liquids with Dissolved Solids

Section **2.1**

Membrane Filtration

Dr. M. C. PORTER *Vice President of Research & Development, Nuclepore Corporation, Pleasanton, California.*

NOMENCLATURE

A Area of the filtering surface
b Channel height

C	Concentration of membrane-retained species
C_B	Bulk stream concentration of species retained by membrane
C_F	Feed concentration
C_f	Concentration of solute in the filtrate
C_G	Gel concentration of retained species
C_0	Initial microsolute concentration
C_p	Concentration of solute in permeate
C_s	Concentration of retained species at surface of membrane
C_{total}	Total number of particles in the entire channel width
C_w	Concentration of water in the membrane
C_X	Number of particles per band
D, D_s	Diffusivity or diffusion coefficient of retained species
D_m	Solute diffusivity in the membrane
D_w	Diffusion coefficient of water in the membrane
D^*	Apparent diffusion coefficient in the presence of a shear field
d	Pore size
d_h	Equivalent hydraulic diameter
$F\left(\dfrac{r}{R}\right)$	Unspecified function of the radial position of the particle in the tube or channel
H	Henry's law constant (solubility of gas in water)
J, J_1, J_w	Water flux through membrane
J_0	Initial water flux
J_s	Solute flux to membrane surface $= J_w C_p$
K	Mass-transfer coefficient
K_M	Membrane permeability
k	Boltzmann constant
l	Pore length
L	Channel length
L^*	Distance in channel over which velocity profile is developing
m	Compaction slope
N	Pore density (number of pores per unit area)
P	Bubble-point pressure
ΔP	Transmembrane pressure drop
Q	Volumetric flow rate
R	Tube radius
R	Gas constant
R_C	Resistance to permeation due to cake or gel layer
Re	Reynold's number $= \dfrac{U d_h}{\nu}$
R_j	Retention or rejection of membrane fraction of solute in feed retained by the membrane
R_M	Resistance to permeation due to membrane
r	Radial position of particle in a tube or channel
r^*	Equilibrium radial position of the particle
r_p	Particle radius
s	Compressibility exponent of the cake (zero for perfectly noncompressible cake and unity for perfectly compressible cake)
Sc	Schmidt number $= \dfrac{\nu}{D}$
Sh	Sherwood number $= \dfrac{K d_h}{D}$
t	Membrane thickness
T	Absolute temperature
U	Fluid velocity down the channel or tube
U_0	Entrance velocity
V	Radial migration velocity
V_0	Initial volume
V_t	Volume of filtrate delivered or "throughput"
v_w	Partial molar volume of water
w	Channel width
w'	Weight of dry particulates per unit volume of filtrate
X	Distance from membrane surface
α'	Constant dependent on properties of filtration cake
γ	Surface tension of the liquid-gas interface
$\dot{\gamma}$	Fluid shear rate at the membrane surface
δ	Boundary-layer thickness

θ	Contact angle
μ	Fluid viscosity
ν	Kinematic viscosity $= \dfrac{\mu}{\rho}$
ξ	$\dfrac{J^3 bL}{6 U_6 D^2}$
π	Osmotic pressure
π_b	Osmotic pressure of bulk stream solution
π_s	Osmotic pressure at surface of membrane
ρ	Fluid density
σ	Fraction of total liquid flowing through the membrane passing through pores large enough to pass solute molecules

DEFINITIONS: MF, UF, AND RO

There is considerable confusion in the open literature as to the distinction between microfiltration (MF), ultrafiltration (UF), and reverse osmosis (RO). For example, ultrafil-

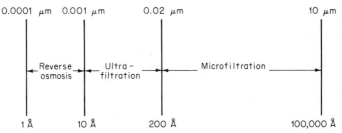

Fig. 1 Definitions of pressure-driven membrane processes based on the smallest particle (molecule) retained.

tration has been referred to as both microfiltration and reverse osmosis. Occasionally one will see it referred to by other names such as "hyperfiltration."

The most useful definition of these three processes is based on the smallest particles or molecules which can be retained by the various membranes. The present treatment will define each process by the rated pore size or molecular-weight cutoff of the membrane as follows (see Fig. 1):[1,2]

Microfiltration (MF) 0.02 to 10 μm (i.e., 200 to 10,000 Å)

Ultrafiltration (UF) 0.001 to 0.02 μm (i.e., 10 to 200 Å) or 300 to 300,000 mol wt based on globular proteins

Reverse osmosis (RO) 0.0001 to 0.001 μm (i.e., 1 to 10 Å) or <300 mol wt

Figure 2 is an indication of various particulates and molecules which may pass or be retained by any of the three processes considered. Thus MF membranes cannot be expected to retain macromolecules in solution (with the possible exception of very large globular proteins or carbohydrates) since the pores are too large. UF membranes, on the other hand, are ideally suited for this task. Proteins, carbohydrates (with a molecular weight over 300), or polymers may be concentrated, dewatered, or desalted along with colloids and suspended particles. The salts which pass through a UF membrane may be retained by RO membranes. As might be expected, the retention of these small ionic species means that a large osmotic pressure difference must be overcome (hence the name "reverse osmosis"). Thus RO is a relatively high-pressure process (300 to 1000 lb/in²) compared with MF and UF (1 to 100 lb/in²). In the latter case, the osmotic pressure of retained macromolecules, colloids, and particulates is negligible.

These three pressure-driven membrane processes should not be confused with other membrane processes such as dialysis or electrodialysis which utilize other driving forces such as concentration differences and electrical gradients.

The useful ranges of common separation processes are compared in Fig. 3. This chart, originally published in 1969, reflects the confusion in the literature among MF, UF, and RO. The three processes overlap on the chart, indicating some confusion in semantics. It is helpful to note that roughly speaking, microfiltration has the capability of performing separations equivalent to those obtained in a high-speed centrifuge (5000 to 10,000g). Likewise, ultrafiltration is equivalent to an ultracentrifuge (10,000 to 100,000g); but because centrifugal forces are not capable of separating ions out of water, there is no equivalent for reverse osmosis.

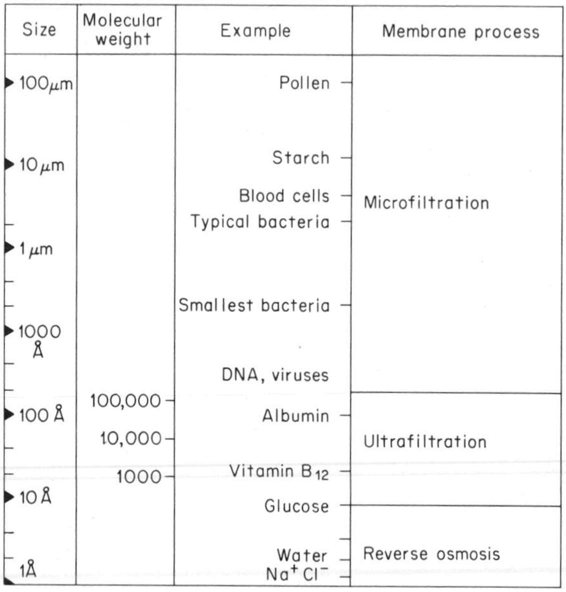

Size	Molecular weight	Example	Membrane process
▶ 100μm		Pollen ⊣	
▶ 10 μm		Starch ⊣	
		Blood cells ⊣	Microfiltration
		Typical bacteria ⊣	
▶ 1 μm			
		Smallest bacteria ⊣	
▶ 1000 Å			
		DNA, viruses	
▶ 100 Å	100,000 ⊣	Albumin	
	10,000 ⊣		Ultrafiltration
	1000 ⊣	Vitamin B$_{12}$ ⊣	
▶ 10 Å		Glucose ⊣	
		Water ⊣	Reverse osmosis
1Å		Na$^+$ Cl$^-$ ⊣	

Fig. 2   Species which may be retained by MF, UF, and RO.

MEMBRANE-RETENTION CHARACTERISTICS

RO Membrane Retention

A review of the rejection characteristics of a cellulose acetate RO membrane in Table 1 indicates that other factors besides size affect the rejection. Some organic molecules, which are larger than salts like sodium chloride, pass right through the membrane while sodium chloride is 95% retained. Granted that most species with molecular weights greater than 150 are well rejected, many low-molecular-weight nonelectrolytes are not. Certain substituted phenyl compounds such as phenol and phenol derivatives are actually negatively rejected (i.e., they are enriched because their permeation rate is higher than that of water). These data suggest that in addition to size, rejection will be influenced by the electrical charge of the molecule and by its "solution" in the polymeric membrane. The latter implies a "solution-diffusion" mechanism of transport (as in "perm-selective" membranes) rather than "pore-flow" alone (see Appendix D).

Data such as those from Table 1 have led to the postulation of the following rejection mechanism for RO membranes. Water is retained in the membrane in such a way that it still possesses the solubilizing properties attributable to its hydrogen-bonding capacity but has largely lost those properties attributable to its high dielectric constant. Small species whose solubilities in water are due partially or wholly to their hydrogen-bonding capacities (e.g., nonelectrolytes or hydrogen-bonding univalent ions) will tend to pass through the membrane unhindered. On the other hand, a small species whose water solubility is due primarily to the high dielectric constant of water (e.g., non-hydrogen-bonding univalent ions and all ions of valence greater than unity—regardless of hydro-

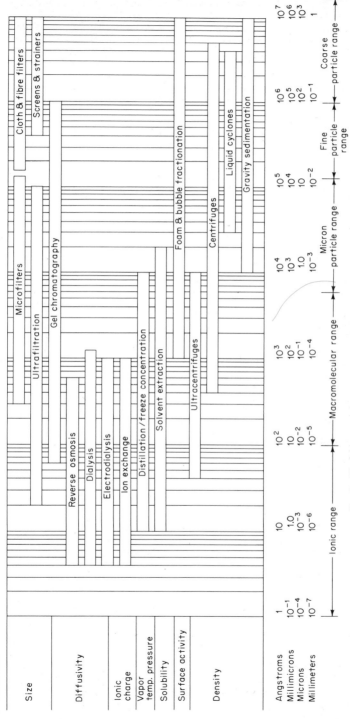

Fig. 3 Useful ranges of various separation processes. (*Dorr-Oliver, Inc.*)

gen-bonding characteristics) will tend to be rejected. Failure to recognize the many parameters which affect rejection has led to some costly mistakes.[3]

Another anomaly confusing to the layman has to do with retention of organisms by RO and UF membranes. It is well known that a 0.2-μm-pore-size MF membrane can provide "absolute retention" for the smallest known bacteria *(Pseudomonas diminuta)* at chal-

TABLE 1   Rejection Characteristics of an RO Membrane (Cellulose Acetate)

Solute	Molecular weight	Rejection, %
Dextrose	180	99
$MgCl_2$	203	99
$NaHCO_3$	84	98
NH_4Cl	54	97
$NaNO_3$	85	96
NaCl	58	95
H_3BO_3	62	56
Urea	60	27
NH_4OH	35	7
Aniline	93	4
Phenol	95	-9
2,4-Dichlorophenol	163	-34

TABLE 2   Removal of Bacteria by Reverse Osmosis

Time, h	Pseudomonas Reject ($\times 10^3$)	Product	Escherichia Reject ($\times 10^3$)	Product	Staphylococcus Reject ($\times 10^3$)	Product
$\frac{1}{4}$	5.9	0	5.8	0	3.2	0
$\frac{1}{2}$	7.7	0	9.4	0	4.1	0
1	8.9	0	6.6	0	3.6	0
2	12.7	10	6.1	0	2.8	0
4	14.5	8	7.4	0	2.1	0
6	21.8	22	4.3	0	1.8	0
12	19.4	17	8.9	0	2.2	0
24	17.4	12	10.6	0	1.4	0
Stopped 24 h						
$\frac{1}{4}$	7.3	0	7.4	0	2.4	0
$\frac{1}{2}$	21.4	4	8.3	0	1.3	0
1	18.6	2	7.2	0	1.1	0
2	17.0	0	10.4	0	1.8	0
6	14.3	1	6.2	0	1.4	0

SOURCE: Reprinted with permission from *Am. Lab.*, December 1973, p. 54.

lenge levels up to 10^8 to 10^{10} organisms per cm^2 of filter area. This is the maximum challenge possible short of clogging the filter and preventing further practical flow of liquid. Therefore, it would naturally be expected that UF and RO membranes with their smaller pore sizes would do as well. Such is not the case. Table 2 is a summary of data collected by Olten and Brown[4] using a DuPont Permasep B-9 reverse-osmosis module with relatively low challenge levels. Higher challenge levels of Staphylococcus also resulted in some passage. Similar results have been obtained with other types of UF and RO modules. The problem seems to be related to a wide pore-size distribution along with a few "defect holes." The presence of "defect holes" in these membranes has been confirmed by bubble-point tests (see Appendix A) which indicate maximum pore diameters above those of the MF membranes.

UF Membrane Retention

In general, UF and RO membranes seldom exhibit "absolute" retention for any species near their rated pore size. Ideally, the membrane should be capable of retaining *completely* all solutes or particles above some specified molecular weight or size and of passing completely all species below that size. Significant strides have been made in improving the "cutoff" characteristics of these membranes (see Fig. 4), but the perfect "sharp-cutoff" membrane has not yet appeared. Figure 5 and Table 3 present typical retention data for a series of ultrafiltration membranes and globular proteins. The diffuse cutoff of these membranes means that the pore-size rating or molecular-weight cutoff must be defined in terms of less than 100% retention and is typically chosen as the molecular weight of a globular protein that is 90% retained by the membrane.

The above is not always a satisfactory guide for membrane selection. The size, shape, charge, and deformability of molecules determine their retentivity. Molecular weight is only an indirect, albeit convenient, parameter to estimate size.

As a general rule, linear, flexible molecules in solution ("free-draining chains") tend to be retained to a lesser degree for a given molecular weight than do the more highly

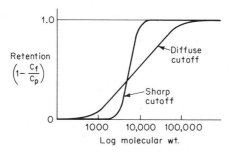

Fig. 4   Sharp vs. diffuse cutoff membranes.

structured molecules such as branched polymers. This tendency is illustrated in Table 4, where the relative ultrafiltration retentivities for globular proteins, branched polysaccharides, and linear flexible molecules of various molecular weights are shown.

In addition to the above, UF retention will be influenced by other factors. Fluid shear

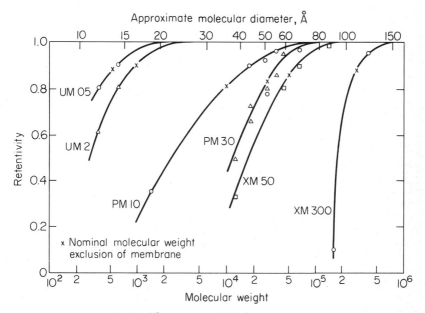

Fig. 5   Solute retention, Diaflo® membranes.

TABLE 3 Solute-Retention Diaflo® UF Membranes

| | | UM-05 | | Diaflo membranes | | | | | |
| | | pH 5 | pH 10 | UM-2 | UM-10 | PM-10 | PM-30 | XM-50 | XM-100 |
Solute	mol wt								
D-Alanine	89	80	15	0	0	0	0	0	0
DL-Phenyl-alanine	165	90	15–20	0	0	0	0	0	0
Tryptophan	204	80	15–20	0	0	0	0	0	0
Sucrose	342	90	80	60	10	5	0	0	0
Raffinose	594	75	90	80	50	35	0	0	0
Bacitracin	1,400	>95	75	60	50	20	0	0	0
Cytochrome C	12,400	>95	>95	>95	90	90			
Polyethylene glycol	16,000	~100	>95	>95	80	90		0	0
Myoglobin	17,800	~100	>95	>95	95	70	65	0	0
α-chymotrypsinogen	24,500	~100	~100	~100	>95	>95			0
Pepsin	35,000	~100	~100	~100	~100	~100	85	80	0
Ovalbumin	45,000	~100	~100	~100	~100	~100	>95	80	0
Hemoglobin	64,000	~100	~100	~100	~100	~100	~100	90	0
Albumin (human serum)	67,000	~100	~100	~100	~100	~100	~100	90	10
Dextran 110	110,000	~100	~100	~100	~100	30	20	10	0
Aldolase	142,000	~100	~100	~100	~100	~100	~100	>95	20
Gamma globulin	160,000	~100	~100	~100	~100	~100	~100	~100	90
Apoferritin	480,000	~100	~100	~100	~100	~100	~100	~100	~100

Percent retention figures below the heavy line indicate retention above 95% (average values, after 20 to 30 min continuous ultrafiltration). In some cases, retentivity is dependent on solute concentration and pressure.
   SOURCE: Amicon Corp.

stresses in the vicinity of the membrane pores will tend to uncoil free-draining chains and to reduce their effective cross-sectional area. The ionic strength and pH of the solution will affect the retention of polyelectrolytes. The more a polyelectrolyte is charged in solution and the lower the ionic strength of the medium, the larger the effective size of the polyelectrolyte for a given molecular weight.

When the solution to be ultrafiltered contains two or more solutes, the retention of the smaller of the two solutes may be increased dramatically. If the larger of the two solutes is retained by the membrane, it may form a "secondary-dynamic" membrane on the primary membrane surface. (This is discussed later under Concentration Polarization.) If the larger molecule is present in sufficient concentrations, this secondary membrane can be more

TABLE 4   Effect of Size and Shape of Molecules on Ultrafiltration Retentivity

	Solute class		
Membrane*	Globular proteins	Branched polysaccharides	Linear, flexible polymers
Diaflo XM50	Gamma globulin (160,000)† Albumin (69,000)		
Diaflo PM30	Pepsin (35,000)	Dextran 250 (236,000)	
Diaflo PM10	Cytochrome C (13,000) Insulin (5,700)	Dextran 110 (100,000)	Polyacrylic acid (pH 10; 50,000) Polyacrylic acid (pH 7; 50,000)
Diaflo UM10	Bacitracin (1,400)	Dextran 40 (40,000) Dextran 10 (10,000)	Polyethylene glycol (20,000)

*Molecules above a horizontal membrane line are completely retained by the membrane; below the line partial retention or complete clearance is observed.
†Number in parentheses denote molecular weights.

retentive than the primary membrane—hindering the passage of the smaller molecules, which are normally freely permeable.

Figure 6 shows the increase in rejection of ovalbumin (mol wt 45,000), chymotrysinogen (mol wt 24,500), and cytochrome C (mol wt 12,400) in the presence of albumin (mol

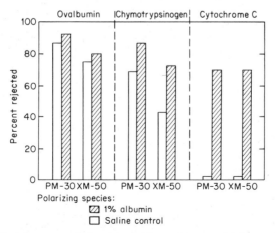

Fig. 6   Rejection of membrane-permeable solutes in binary mixtures containing membrane-retentive macrosolute. [*Reprinted from* Agr. Food Chem., **19(4)**, 589 *(July–August 1971) with permission.*]

wt 67,000) for two different membranes (molecular-weight cutoffs of 30,000 and 50,000, respectively). If the molecular weight of the smaller species is small enough (say below 500), little impedance is observed.

The latter result is illustrated by the data of Fig. 7, which compares the rejection of a series of proteins on a 50,000 mol wt cutoff membrane with their rejection in the presence of plasma (which contains higher-molecular-weight globular proteins). Species below 500 show less than 10% rejection even in the presence of the larger proteins. As a general rule, if there is a tenfold difference in molecular weight, fractionation may be accomplished.

As expected, the concentration of the larger retained species is a factor determining how much the rejection of the smaller species will be enhanced. Figure 8 shows the retention of cytochrome C (12,400 mol wt) on a 30,000 mol wt cutoff membrane as a function of the concentration of various larger solutes. It is noticed immediately that the concentration of

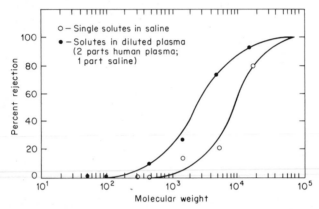

Fig. 7 Effect of plasma proteins on the rejection of solutes of various molecular weights. (Data from stirred cell, 1500 r/min, XM50 membrane.)

the larger solute has a dramatic effect on retention. It is also noted that the correlation between the molecular weight of the retained species and its ability to retain the smaller species is poor. Obviously, in addition to size the gel structure and molecular configuration of the retained species will determine the retention properties of the dynamic membrane formed.

One of the most coveted separations by membrane technologists is that between gamma globulin and albumin. It was largely for this application that an ultrafiltration

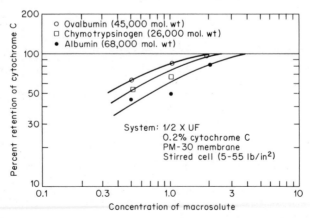

Fig. 8 Rejection of cytochrome C in presence of larger-molecular-weight solutes.

membrane with a 300,000 mol wt cutoff was developed. This membrane retained over 90% gamma globulin and less than 10% albumin when filtered separately. Unfortunately, as seen in Fig. 9, the presence of small quantities of gamma globulin increases the retentivity for albumin dramatically.

As will be seen in subsequent sections, the formation of a dynamic membrane on the primary membrane surface can be minimized by proper fluid-management techniques. This will sometimes facilitate the separation of binary mixtures. Alternatively, the binary mixture may be diluted down before fractionation, after which the two fractions may be reconcentrated separately. This of course increases the volume and time of filtration considerably, making it a rather cumbersome process.

MF Membrane Retention

Many of the remarks concerning UF membranes will also apply to microfiltration (MF). However, MF membranes are presently available with two radically different pore

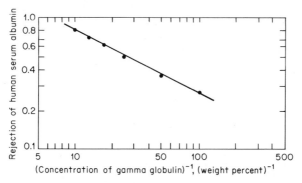

Fig. 9 The interference of gamma globulin with the passage of human serum albumin through a Diaflo® XM-300 membrane.

structures. One type, as typified by the cast cellulose ester membrane, has an open tortuous "spongelike" structure. The other, referred to in the literature as the "Nucle-pore" or "track-etch" membrane, resembles a sievelike structure with a narrow distribution of straight-through cylindrical pores. In addition, this membrane is typically one-twelfth the thickness of the spongelike membranes.

As might be expected, the retention of a given particle or solute molecule will be quite different on the two types of membranes even at the same rated pore size. The probability that a particle of a given size will find its way through a tortuous spongelike labyrinth of pores depends on the probability of finding a path which does not require passage through a hole smaller than the particle. Hence microporous membranes with a lower tortuosity and decreased thickness tend to show sharper cutoffs and enhanced separation factors. For example, Nuclepore membranes have been used to separate male- and female-determining sperm cells. The male-determining sperm has a Y chromosome and is smaller ($2~\mu m^+$) than the female-determining sperm ($3~\mu m^-$) which bears the X chromosome.

The difference in fractionation capability between the two membranes is best illustrated by the work of Davis et al.[11] at the Harvard Medical School. The retention for 0.05 μm and 0.005 gold colloids was measured on Nuclepore and cellulose ester membrane filters (see Table 5). Since the largest particle size is only one-half of the smallest pore size, a sharp-cutoff membrane should retain 0%. This is not the case for either of the membrane filters. Nevertheless, the effect of pore tortuosity, greater thickness, and much larger internal surface area (thirty times more) of the cellulosic membrane is dramatic.

The most severe test of an MF membrane's ability to fractionate various particle sizes occurs in *aerosol sampling*. Figures 10 to 12 present data of Spurny et al.[12] on the Nuclepore membrane. The effects of particle size (radius = r), pore size (radius = R), and flow rate (q = facial velocity) on particle retention (collection efficiency) are shown for specified particle densities S in grams per cubic centimeter.

Figures 10 and 11 show that the particle retention goes through a minimum when

TABLE 5   Percent Retention of Au Colloids Filtered through Cellulosic and Nuclepore Membranes

Colloid size, μm	0.05	0.005
0.1 μm Nuclepore	1.2	0.2
0.1 μm cellulosic	92.0	8.2
0.4 μm Nuclepore	1.3	0.2
0.45 μm cellulosic	46.9	12.2
1.0 μm Nuclepore	0.7	0.3
1.2 μm cellulosic	46.5	26.7
3.0 μm Nuclepore	0.4	0.2
5.0 μm cellulosic	59.3	17.9

SOURCE: Ref. 11.

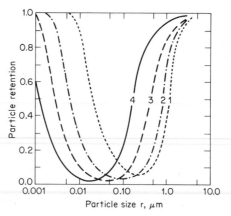

Fig. 10 Computed dependency of efficiency on particle size and face velocity. 1. $q = 0.1$ cm/s. 2. $q = 1.0$ cm/s. 3. $q = 5.0$ cm/s. 4. $q = 25.0$ cm/s. $R_0 = 4.0$ μm, $P = 0.05$. $s = 21$ g/cm^3. [*Reprinted from Environ. Sci. Technol., 3(5), 453 (May 1969) with permission.*]

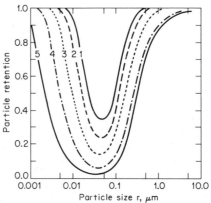

Fig. 11 Computed dependency of efficiency on particle size and pore size. 1. $R_0 = 0.4$ μm. 2. $R_0 = 0.5$ μm. 3. $R_0 = 1.0$ μm. 4. $R_0 = 2.5$ μm. 5. $R_0 = 4.0$ μm. $q = 7.5$ cm/s. $s = 21$ g/cm^3. [*Reprinted from Environ. Sci. Technol., 3(5), 453 (May 1969) with permission.*]

plotted against particle size. This phenomenon is not usually observed in liquid filtration, the reason being that particle capture by "diffusion" is insignificant owing to the large viscosity of the carrier fluid. However, in the case of aerosols, for very small particles, the diffusivity may be high enough (diffusivity is inversely proportional to particle size and viscosity) to permit particle diffusion to the pore wall and capture before the particle can pass through the pore. Higher pore velocities (as shown in Fig. 10) will decrease the residence time in the pore and thereby decrease collection by diffusion.

For relatively large particles, which are still smaller than the pore size, the predominant capture mechanism is "inertial impaction." In this case, the particles are small enough to pass through the pores but large enough that their inertial mass will prevent their following the fluid streamlines around corners (such as at the entrance to a pore). Thus, if the mass or velocity is great enough, the particles will impact on the surface surrounding

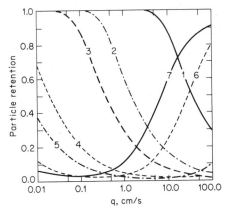

Fig. 12 Computed dependency of efficiency on flow rate. 1. $r = 0.001 \, \mu$m. 2. $r = 0.005 \, \mu$m. 3. $r = 0.01$ μm. 4. $r = 0.05 \, \mu$m. 5. $r = 0.1 \, \mu$m. 6. $r = 0.5 \, \mu$m. 7. $r = 1.0 \, \mu$m. $R_0 = 4.0 \, \mu$m. $P = 0.05$. $s = 2.2 \, \text{g/cm}^3$. [*Reprinted from* Environ. Sci. Technol., 3(5), 453 (May 1969) *with permission.*]

the pore or on the pore wall near the pore entrance. Figure 10 shows that an increased velocity and particle size (mass) lead to a higher retention value. Again, this effect will not be so pronounced with liquids owing to the higher viscosity.

Naturally, if the particle is larger than the pore size, collection is by interception (100%). Retention in this case is simply by a sieving mechanism. This is the primary mechanism for retention of particulates in liquids as well.

Figure 11 shows that at a fixed facial velocity (7.5 cm/s), the retention will be greatest for the smallest pore size regardless of whether collection is occurring via inertial impaction or diffusion. The smaller pores mean smaller distances over which diffusion must take place and sharper bends in flow streamlines.

Similarly, Fig. 12 shows that as the particle size increases, the minimum in the retention curve will occur at lower air velocities. The increased size means a lower diffusivity and a larger inertia.

As seen above, membrane-retention characteristics depend upon the mean pore size, the breadth of the pore-size distribution, the tortuosity, the membrane thickness, and the pore configuration. It will also be evident that membranes are inherently susceptible to internal plugging or fouling by particles or solute molecules whose dimensions lie within the pore-size distribution of the membrane. This fouling will result in a corresponding change in the rejection spectrum of the membrane.

Fouling will be most pronounced with solutes whose dimensions lie in the lower third of the pore-size distribution, since these will have the least difficulty in entering the structure but the greatest likelihood of lodging in pore constrictions. Since the lower end of the pore-size distribution is inactivated in the plugging process, most of the residual flow is forced to take place in the larger pores. Consequently, the plugged filter is often less retentive and size-discriminatory than it was before plugging. However, if the pore-size distribution is narrow, as pore obstruction proceeds, the retention may increase.

(In practice, the internal fouling problem may be minimized by selecting a membrane whose mean pore size is well below that of the dimensions of the particle or solute to be retained.)

Thus an understanding of the retention and fractionation capability will depend on a knowledge of the membrane's pore structure.

MEMBRANE STRUCTURE AND PRODUCTION

MF Membrane Structure and Production

Tortuous-Path Membranes   Historically, MF membranes evolved earlier than RO or UF membranes. In 1855, Fick cast nitrocellulose membranes on ceramic thimbles from an ether solution, but the properties of these membranes were not studied systematically until 1907, when Bechhold[5] discovered that he could vary the pore size by altering the collodion-solution concentration. In Bechhold's work he determined pore size by hydraulic-flow experiments using Poiseuille's law and by measuring the air pressure required to displace water from the pores of a wetted membrane (against the surface tension of the water in the pore). This "bubble-point" test is described in Appendix A. In 1918, Zsigmondy and Bachmann in Göttingen, Germany, developed methods for producing the collodion membranes commercially. Sartorius-Werke Aktiengesellschaft began to manufacture and market the membrane in 1927, but it was not until World War II that MF membranes began to receive widespread attention. In Germany, these membranes (0.5 μm) were used for *Escherichia coli* determinations in bomb-damaged water-supply systems.

Soon after the war (1947) researchers like Dr. Alexander Goetz, a technical consultant for the Technical Industrial Intelligence Branch of the U.S. Department of Commerce, studied the German membrane technology and developed methods for preparing the membranes domestically. Companies in both the United States and England made use of the German technology to begin manufacturing MF membranes by a phase-inversion process.

A "phase-inversion" membrane is a solvent-cast structure which owes its porosity to immobilization of the gel prior to complete solvent depletion. This is accomplished by not allowing the membrane to evaporate to dryness before its structure is set; partial solvent loss is effected so that the solution separates into two interspersed liquid phases. One of these phases represents the voids or vacuoles in the finished membrane. As evaporation continues, the polymer begins to precipitate and the gel structure is set.

Elford[15] was one of the first (1931) to employ the phase-inversion process in making collodion membranes. He found that amyl alcohol–ethyl alcohol or acetone–ethyl alcohol were excellent solvents for the nitrocellulose, but when both were employed together, the amyl alcohol and acetone were found to be mutually antagonistic in their solvent action. Therefore, as evaporation of acetone was more rapid than the amyl alcohol, gelation occurred around the dispersed phase of amyl alcohol. As expected, increasing the amount of amyl alcohol increased the permeability of the membrane. The result of this process is a semiopaque film with a porosity of roughly 80% and average pore sizes ranging between 0.1 and 1.0 μm.

In recent years, similar membrane structures have been produced from a variety of polymers, in addition to the cellulose ester, cellulose nitrate materials. These include nylon, polyvinyl chloride, copolymers of PVC and acrylonitrile, and even polytetrafluoroethylene (PTFE). Obviously, not all these membranes are made by a phase-inversion process.

A porous PTFE membrane is made by a process entirely different[16,17] from the phase-inversion process described above. A "paste-extruded" film is first prepared by extruding PTFE resin with an organic lubricant (such as mineral spirits). The film is then fed onto a heated roll to preheat the film to a temperature (200 to 300°C) at which it can be expanded (i.e., stretched biaxially or uniaxially at rates of 400 to 500% per second) to produce a porous film. The film is subsequently sintered at 370°C in the stretched condition, causing "amorphous locking" within the film, and cooled rapidly. The porous microstructure of the expanded material consists of nodes interconnected by very small fibrils. In the case of uniaxial expansion the nodes are elongated (the longer axis of a node being oriented perpendicular to the direction of expansion). The fibrils interconnecting the nodes are oriented parallel to the direction of expansion. Average pore sizes ranging from 0.1 to 15.0 μm can be produced, though the pore-size distribution is quite broad.

Another process which involves stretching polypropylene also produces a porous film. In this case the pores are below 0.1 μm.

All MF membranes described above may be classified as "tortuous" or "spongelike" membranes. The processes described are by no means exhaustive but indicate the variety of methods which may be used to produce microporous membranes of this type from a number of polymeric materials.

Track-Etch Membranes In the late sixties, a new MF membrane was developed by the General Electric Company which has a unique pore structure uncommon to other MF membranes. The new membrane, called Nuclepore, has straight-through cylindrical pores. In this case, since the pores are straight-through cylinders, the pore opening is a good measure of the pore size and permits the use of the electron microscope for quality control.

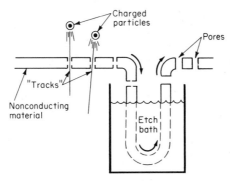

Fig. 13 The track-etch process for manufacturing Nuclepore membrane filters. (*Nuclepore Corporation.*)

The Nuclepore membrane, now produced by Nuclepore Corporation, an independent affiliate of General Electric, is made by a patented process which involves two independent stages and is called "track etch." Figure 13 schematically diagrams the process. The first stage consists of bombarding a dielectric film with massive energetic nuclei which, upon passing through the film, produce narrow trails of radiation-damaged material called "tracks." These tracks are not visible even with an electron microscope until the film is etched. This second-stage etching process selectively dissolves the damaged material, leaving cylindrical straight-through pores.

Since the two stages (of Fig. 13) are run separately and independently, Nuclepore membranes may be tailor-made with respect to pore size and density. The pore density, number of pores per square centimeter, is determined by the power of the nuclear reactor and the residence time of the film in the reactor. The pore size is determined by the temperature and residence time in the etch bath. Obviously upper limits exist whereby high densities and large pore sizes will result in significant losses in strength of the film. Table 6 shows the variation of pore density with pore size specified to maintain good strength (average tensile strength of 7000 lb/in²) commensurate with high flow rates for gases and liquids.

In the case of "standard" Nuclepore, the film is poly-bisphenol-A-carbonate which is irradiated in a collimated beam of U^{235} fission fragments. The fragments pass through the film in order to produce tracks across the entire thickness of the film, up to 15 μm maximum thickness. The irradiated film is then etched in warm caustic, and subsequent treatments render the membrane hydrophylic, partially hydrophobic, totally hydrophobic, and/or antistatic.

Theoretically,[18] any dielectric material may be track-etched whether on inorganic crystal or a long-chain polymer. In the crystal, the atoms along the path of an energetic highly charged particle are ionized and their mutual electrical repulsion causes them to explode into the surrounding lattice. At ordinary temperatures, these displaced ions never return to their original positions and the disordered cylindrical region is highly distorted and easily attached by a strong reagent. In the amorphous polymer, an energetic charged particle excites as well as ionizes molecules, in either case breaking chemical bonds between atoms in the long-chain molecules. The chain ends rarely unite at room temperature but usually react with oxygen or other gas dissolved in the polymer, forming chemically reactive species along the particle trajectory.

Because of the random nature of the bombardment process during irradiation, there are occasional pore overlaps. To retain the absolute retention characteristics of the membrane, the angle of particle impingement on the film is allowed to vary between 0 and 29°. This ensures that the probability of pore coincidence throughout the thickness of the film is negligible (see Fig. 14) except in the case of the large pore sizes, where the length-diameter ratio approaches unity. (Incidentally, there is a difference in surface roughness between the "shiny" and "matte" sides of a Nuclepore membrane.)

There are two other important differences between the cellulose ester MF membranes

TABLE 6   Standard Nuclepore Membrane Specifications

Specified pore size, μm	Pore-size range, μm	Nominal pore density, pores/cm²	Nominal thickness, μm	Typical flow rates at 10 lb/in² (gage), ΔP, 70°F	
				Water, gal/(min)(ft²)	N₂, ft³/(min)(ft²)
8.0	6.9–8.0	1×10^5	8.0	144.0	138.0
5.0	4.3–5.0	4×10^5	8.6	148.0	148.0
3.0	2.5–3.0	2×10^6	11.0	121.0	128.0
1.0	0.8–1.0	2×10^7	11.5	67.5	95.0
0.8	0.64–0.80	3×10^7	11.6	48.3	76.0
0.6	0.48–0.60	3×10^7	11.6	16.3	33.0
0.4	0.32–0.40	1×10^8	11.6	17.0	33.0
0.2	0.16–0.20	3×10^8	12.0	3.1	8.9
0.1	0.08–0.10	3×10^8	5.3	1.9	5.3
0.08	0.064–0.080	3×10^8	5.4	0.37	2.6
0.05	0.040–0.050	6×10^8	5.4	1.12	1.3
0.03	0.024–0.030	6×10^8	5.4	0.006	0.19

SOURCE: Nuclepore Corporation.

and the Nuclepore MF membrane. A Nuclepore membrane has a much lower porosity (nominally 10%) than a cellulosic membrane (nominally 80 to 85%), but the thickness is only one-twelfth that of the cellulosic. Thus, even though the Nuclepore porosity is lower than that of the cellulosics, the thickness and tortuosity difference compensate to give flow rates equal to or greater than the cellulosics.

The differences in porosity and thickness have important ramifications for various applications of the membrane. As pointed out earlier, the available area for adsorption of species on the membrane is much higher for the cellulosic type—thirty times more surface area per unit facial area than exists with Nuclepore (as determined by BET adsorption measurements). For small-batch filtrations of proteins or viruses, often most of the product will be lost by adsorption on the membrane; 2 to 5% yields are not uncommon. The same filtration conducted with Nuclepore results in very little loss: 95 to 100% yield.

On the other hand, the tremendous surface area available within the cellulosics should enhance their dirt-loading capability and increase the liquid throughput prior to plugging. Indeed, for some applications, this is the case. It is, however, a curious fact that the Nuclepore pore structure sometimes yields a throughput equal to or even higher than that of the equivalent-pore-size cellulosic (see Fig. 15). It is possible in cases like this that the particulates accumulate on the surface of the

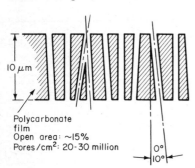

10 μm

Polycarbonate film
Open area: ~15%
Pores/cm²: 20-30 million

0°
10°

Fig. 14 Angle of particle impingement and pore overlap (up to 29° in Nuclepore membrane.

Nuclepore membrane, providing a prefilter cake which enhances the dirt-loading capacity of the membrane. In the case of the cellulosic membranes, the particles may tend to become entrapped in the pore constrictions within the membrane.

Both the cellulosics and Nuclepore have relatively low "throughputs" compared with the so-called "depth filters." Prefilters of the depth type are often used in conjunction with these membranes to enhance their dirt-loading capability.

The thinness and low porosity of the Nuclepore membrane also permit folding and pleating of the membrane which cannot be done with a normal cellulosic material (unless reinforced) without cracking the film. This is helpful in fabricating pleated cartridges (to

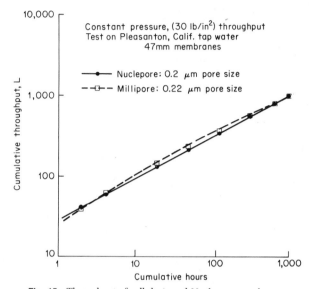

Fig. 15 Throughput of cellulosic and Nuclepore membrane.

incorporate a large membrane area in a small package) and membrane-encapsulated systems. The Nuclepore material can also be heat-sealed or solvent-sealed.

RO Membranes

Historically, the next membrane to be developed after MF (disregarding dialysis or electrodialysis membranes) was the RO (reverse-osmosis) membrane. The development had heavy OSW (Office of Saline Water) support in pursuit of a membrane capable of desalting sea water.

Reid and Breton[19] of the University of Florida in 1957 to 1959 screened a number of potential candidates and found that cellulose acetate was capable of rejecting electrolytes from aqueous solutions and that its semipermeability far exceeded that of any other film tested. However, they also concluded that such membranes could not be made thin enough (practically) to achieve reasonable flux values.

Loeb-Sourirajan Membrane Loeb and Sourirajan (1958–1961)[20] at UCLA worked, in the same period, first with preshrunk Schleicher and Schuell cellulose acetate filter membranes (allegedly having pores of 50 Å or less), and later developed their own technique for producing what was to become a major breakthrough in the RO and UF membrane field—the "anisotropic" membrane. This membrane (sometimes referred to as a Loeb-Sourirajan-type RO membrane) has an asymmetric structure consisting of a 0.2-0.5-μm-thick, dense layer, supported by a 50- to 100-μm-thick porous substructure. The substructure has pores 0.1 to 1.0 μm in diameter whereas the dense layer, or skin, has pores estimated to be of the order of 10 Å. This type of structure was a breakthrough to the state of the art since the thin skin offers minimal resistance to flow along with the necessary solute-retention characteristics. The porous substructure offers no resistance to flow but provides integral support for the thin skin, which could not be handled otherwise. This structure has the capability of reasonable flux with high salt rejection.

The original Loeb-Sourirajan recipe was as follows:

Cellulose acetate (acetyl content 39.8%)	22.2%
Acetone	66.7
Water	10.0
Magnesium perchlorate	1.1

This casting solution was cast in a cold box at 0 to $-10°C$ on a glass plate with an 0.010-in gap. The more volatile solvent (acetone) was allowed to evaporate at cold-box temperatures for 3 to 4 min. The glass plate was then immersed into ice-cold water for 1 h.

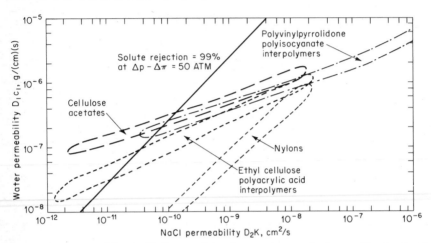

Fig. 16 Water permeability vs. NaCl permeability for several types of polymers. *(Reprinted with permission from "Theory and Practice of Reverse Osmosis and Ultrafiltration," chap. 8, Wiley Interscience, New York.)*

Subsequently, after complete gelation, the film was peeled off the glass plate and heat-treated (5 min in water at 65 to 85°C) to improve the retention characteristics. It was found that the membranes had to be kept in a very humid atmosphere, preferably under water, lest they dry and the pore structure collapse (see Appendix B).

Although the original recipe and subsequent modifications were deeply rooted in empiricism, the process will be recognized as a phase-inversion process. After the casting solution has been cast on the glass plate, the volatile components (e.g., acetone) of the solvent system evaporate, doing so more rapidly from the solution-air interface than from the interior of the film.

Owing to solvent loss, the polymer density increases at the film surface and a skin layer forms while the underlying solution is still fluid. This accounts for the degree of anisotropy sometimes observed with MF membranes. In the case of RO (or UF) membranes, the skin is more pronounced owing to immersion in a nonsolvent medium like water with more rapid dissolvation at the solution-liquid interface. In addition, water rapidly diffuses into the interior of the film where its concentration soon exceeds the solubility of the polymer with consequent gelation. During the posttreatment or annealing step, the membrane shrinks and the porosity decreases.

The asymmetry of such membranes can be rationalized in terms of membrane-precipitation kinetics by analogy to crystallization from supersaturated solutions. At the outermost surface, where the cast membrane solution is in direct contact with the water of the precipitation bath, the degree of supersaturation is extremely high and the density of nuclei and the rate of their growth are also high. This results in a finely dispersed structure which corresponds to the final membrane skin. As the precipitation front moves farther into the film, its composition becomes progressively richer in solvent, and water has to diffuse through the already formed membrane skin into the precipitation zone. Since the degree of supersaturation is significantly lowered, the precipitate becomes increasingly coarser. Thus the average pore size increases from top to bottom of the membrane.

Even though the anisotropic cellulose acetate membrane has poor chemical resistance (it is hydrolyzed below pH 4 and above pH 8) and inadequate thermal resistance (hydrolyzed above 38°C), it is a curious fact that extensive funding by OSW and investigations by many researchers over a 15-year period failed to uncover or develop a superior membrane of this type. Cellulose acetate combines three essential requirements for a practical RO membrane: (1) high permeability to water, (2) low permeability to salts, and (3) excellent film-forming properties. Figure 16 shows data on the first two characteristics for a number of polymers; cellulose acetates occupy the most favorable position on the chart.

This is not to underestimate the potential of Loeb-Sourirajan type membranes made from other polymer-solvent systems. Membranes have been prepared from polyacrylonitrile, an aromatic polyamide and a polyimide using dimethylacetamide (DMAc), dimethylformamide (DMF), n-methylpyrrolidone (NMP), or dimethyl sulfoxide (DMSO) as basic solvent with water as the precipitation agent.[23] Lithium chloride or formic acid have been used as additives to the casting solution. Although some of these membranes exhibited high rejection values for NaCl (99%), they suffered when compared with cellulose acetate membranes in overall performance (e.g., water permeability).

Hollow-Fiber Membranes   Increasing the active membrane area can compensate for a low water permeability provided this can be done inexpensively. The use of hollow fine fibers has permitted the use of improved polymers which have lower water permeabilities than cellulose acetate. In this configuration, the outside of the fiber is pressurized such that salt-free permeate is produced at the end of the hollow-fiber bundle (Fig. 17). The small diameter of these fibers (typically 85 μm OD, 42 μm ID) makes possible large surface-area packing (12,000 ft²/ft³) and resistance to pressure collapse [even at 600 to 1000 lb/in²(gage)]. DuPont has pioneered in this field, developing an asymmetric aromatic polyamide fiber which has a flux of 1.8 gal/(ft²)(day) at 400 lb/in² (gage) and an NaCl rejection of 93%. Even though the flux is considerably below that of cellulose acetate film [(10 to 15 gal/(ft²)(day)], the inexpensive, compact area of the hollow-fiber geometry makes it competitive.

Dow Chemical Co. has also developed a cellulose triacetate hollow-fiber membrane for brackish-water desalination. After 1460 h of operation, modules of these fibers had fluxes averaging 1.5 gal/(ft²)(day) and salt rejections of 98% at 52% recovery and 600 lb/in²(gage). Cellulose triacetate membranes normally exhibit lower water permeabilities than cellulose "diacetate" (more commonly used for desalination membranes) but have higher salt rejections and greater resistance to biological attack.

Although not yet commercialized, a number of other polymeric materials have been used as hollow-fiber membranes (see Table 7). The advent of PBI (polybenzimidazole) hollow fibers is illustrative of strides being made to achieve higher temperature stability (over 100°C with PBI) and enhanced chemical resistance (from pH 2 to concentrated caustic). The most prominent morphological feature is the presence of elongated radial pores or voids. Nevertheless, longitudinal views of unsectioned samples show a void-free surface, suggesting that it is covered by a Loeb-type skin.

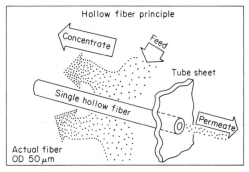

Fig. 17   RO hollow-fiber configuration.

Composite Membranes   More recently, researchers[25,26,27] have sought for greater flexibility in polymer selection by developing so-called composite membranes. These are formed by casting a dense, thin film (200 to 8000 Å thick) on another microporous substrate. Thus the composite membrane is structurally similar to the asymmetric Loeb-Sourirajan membrane but provides greater flexibility in tailoring membrane properties. Thin-film thickness can be controlled, perhaps down to levels not attainable by the Loeb-Sourirajan method. Most important, the thin films can be prepared with relative ease from materials that are not amenable to asymmetric membrane formation. The thin film and porous support can be prepared from different polymers, allowing each to be optimized for its specific function.

Some of the most successful "composite-membrane" developments have been made at North Star Research and Development—among which are the NS 100 and NS 200 membranes. The NS 100 membrane[27] is a composite comprised of a microporous polysulfone support film coated with polyethylenimine (PEI) and crosslinked with tolylene-2,4-diisocyanate (TDI). This membrane is fabricated by coating PEI onto the microporous surface of the polysulfone, followed by an interfacial reaction with TDI for the crosslink-

TABLE 7   Several RO Hollow Fibers in Development

Polymer	Company	Flux, gal/ft²/day	Rejection, % NaCl at 3000 to 5000 ppm NaCl	Pressure, lb/in²
Aromatic polyamide	Du Pont	1.8	93	400
Cellulose triacetate	Dow	1.5	98	600
Cellulose acetate	Monsanto	3.0	97	250
Cellulose acetate	Hercules	4.6	94.5	600
Polybenzimidazole	Celanese	2.9	96.7	600

ing step. Heat curing is then necessary to achieve the final rejection and water-flux properties.

A schematic diagram of the NS 100 fabrication process is shown in Fig. 18. The *polysulfone support film* is a standard asymmetric membrane of the Loeb-Sourirajan type formed by casting a solution of polysulfone in dimethylformamide followed by precipitation in water. The resulting membrane has pores generally less than 200 Å on one side and up to 20,000 Å on the other and is basically a UF membrane. It can be air-dried without loss of permeability and is relatively resistant to pressure compactions.

The second zone is the *polyethylenimine* (PEI) coating, which becomes water-insoluble after heat curing. In the uncrosslinked state, this zone exhibits high fluxes but low salt rejection (70% at 1500 lb/in²).

The third zone, a high-salt-rejecting barrier, is produced by reaction of the *PEI with tolylene-2,4-diisocyanate (TDI)*. The PEI-TDI coating thickness as determined by

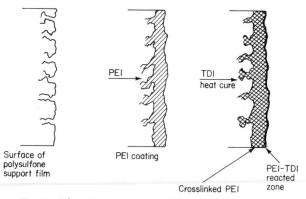

Fig. 18 Schematic representation of the NS 100 membrane.

infrared penetration is apparently less than 2100 Å, but as determined gravimetrically it is 6000 Å. This simply means that the pores of the polysulfone support film contain a significant amount of the coating.

At 1500 lb/in², using synthetic sea water, the NS 100 membrane has a flux of 18 gal/(ft²)(day) with a 99.4% NaCl rejection.[27] It has excellent stability over a broad pH range (pH 2 to 13) and shows a 95.5% cyanide rejection for zinc cyanide plating rinse waters[26] compared with only 28% for aromatic polyamide hollow fibers (Permasep B-5 TM).

The NS 200 membrane consists of a sulfonated polyfuran film on the microporous polysulfone membrane. Furfural alcohol, sulfuric acid, and polyethylene oxide are blended in a water-isopropanol solution and coated on the support. A slow heat cure at 150°C causes the alcohol to condense into a highly crosslinked, insoluble polyfuran. Part of the sulfuric acid is incorporated into the resin as hydrophilic sulfonic acid groups. Apparently the polyethylene oxide helps to maintain the openness of the structure, permitting higher water-flux values. At its best performance, the membrane removed 99.8% of the salt from sea water at a rate of 45 gal/(ft²)(day).

A fringe benefit of the "composite membrane" approach is improved resistance to "compaction" (see Appendix C). It is well known that even with clean water (no fouling), the water flux through anisotropic membranes decreases with time. This phenomenon is apparently the result of a creep process in which the skin grows in thickness by amalgamation with the finely porous substrate immediately beneath it. At constant pressure, compaction effects seem to be more pronounced with higher flux membranes and at higher temperatures. Preliminary results with composite membranes indicated they are much more resistant to compaction.

It may be presumed that composite membranes will be more expensive than the standard Loeb-Sourirajan membrane, since in the latter the skin and substructure are integrally made in one operation whereas in the former the two components are made separately.

Composite Hollow-Fiber Membranes   Gulf South Research Institute[28] and FRL (subsidiary of Albany International)[29] have developed polysulfone hollow fibers for use as a porous substrate to make NS 100 or NS 200 hollow-fiber membranes.

The FRL process consists of forcing the polysulfone solution through an orifice around a needle into the coagulation bath. Fluid passing through the needle prevents fiber collapse while the coagulation bath draws solvent from the solution, causing the polymer to form. In the NS 200 membrane, the sulfonated polyfuran coating is applied to a second bath and quickly cures into shiny black fibers. FRL reports a rejection of over 99% for a 2000 ppm NaCl solution at 800 to 1000 lb/in² and for zinc cyanide as well. Fluxes of 2 to 4 gal/(ft²)(day) at 800 lb/in² are common.

Dynamic Membranes   During the period 1966 to 1969, Johnson and coworkers[30] at the Oak Ridge National Laboratory found that they could form a "dynamic membrane" by flowing a feed solution containing 50 to 100 mg/L of an inorganic membrane-forming material tangential to a clean porous surface at velocities from 5 to 50 ft/s under pressures from 500 to 1200 lb/in². This was a natural consequence of "concentration polarization" (see below). Within minutes after start-up, a significant salt rejection was achieved, and membrane formation was usually complete within an hour. The most successful dynamic membranes were those formed by depositing a layer of polyacrylic acid (PAA) on top of a hydrous metal oxide like zirconium oxide. Since then the characteristics of Zr (IV)–PAA membranes have been studied extensively with salt-solution feeds, dye waste, primary sewage effluent, kraft-paper pulp-mill and bleach-plant wastes, food wastes, low-level radioactive wastes, and laundry wastes.

The porous support tubes have ranged all the way from carbon tubes to porous plastic films on porous stainless-steel tubes.

The dynamic membrane is best described by outlining the procedure for its formation:

1. Adjust to pH 4 with HCl a solution of $0.05M$ NaCl and $10^{-4}M$ hydrous Zr(IV) and introduce to a circulating loop.

2. Start pressurizing and circulating pumps and adjust pressure at 900 to 1000 lb/in².

3. Monitor chloride rejection until it is in the range of 30 to 50%; then rinse the remaining Zr (IV) from the loop with pH 4 adjusted water.

4. Introduce a $0.05M$ NaCl solution (pH 2) containing 50 ppm PAA to the loop and circulate for ~ 30 min.

5. Add NaOH at 30-min intervals to raise pH a unit or so at a time until pH 7.

6. Rinse excess PAA from the loop with distilled water.

The dynamic membranes formed by the above process are reported to have a rejection for $0.05M$ NaCl between 90 and 95% with enormous fluxes [100 to 200 gal/(ft²)(day) at 950 lb/in²(gage)]. In addition, such systems can operate at 65°C or more with no loss of rejection. From an analysis of the amount of metal oxide found on the substrate, it appears that the salt-rejecting layer is often no more than a few hundred angstroms thick. Apparently, the carboxyls on the PAA are converted to the salt form without displacement of the organic polyelectrolyte from the hydrous oxide. Increases in rejection parallel the increase in cation-exchange capacity expected from the neutralization of the acid. The advantages of the dynamic-membrane concept are that high water fluxes can be achieved at moderate rejection levels and the membranes can be generated and regenerated as often as necessary in situ, thereby avoiding the costs of module replacement.

UF Membranes

Loeb-Sourirajan Membranes   The anisotropic membrane structure developed by Loeb and Sourirajan was the breakthrough needed for practical ultrafiltration membranes as well. UF membranes with a more homogeneous structure akin to the isotropic MF membranes were available previously from Sartorius (collodion bags and cellulose nitrate membranes) and others. The new asymmetric membrane provided a higher flux (due to the thinness of the membrane skin) and superior resistance to plugging by retained solutes. The latter feature is merely a consequence of the fact that a molecule barely large enough to penetrate the pore opening will find an ever-increasing pore diameter as it passes through to the more open microporous substructure and finally to the pore exit. Thus for molecules or particles which enter a pore, there is a high probability they will pass through to the other side without becoming trapped within the membrane. This antifouling characteristic is important in achieving life of the membrane for industrial applications. It is, however, not to be confused with surface fouling (discussed later), which occurs on the surface of the membrane but not within the membrane. After the development of the anisotropic membrane, UF membranes were made in the early sixties from a variety of polymers including cellulose acetate (like Loeb-Sourirajan) but extending to polymers more resistant to bases, acids, and high-temperature environments. The membranes are cast using a phase-inversion process almost identical to that used for RO membranes but with larger pore sizes in view.

Asymmetric UF membranes have been cast from a number of polymers including cellulose acetate, polycarbonate, polyvinylchloride, polyamides, modacrylic copolymers (e.g., PVC, acrylonitrile, and styrene–acrylic acid copolymers), polysulfones, halogenated polymers (such as polyvinylidene fluoride), polychloroethers, acetal polymers, acrylic resins, and various polyelectrolytes. Membrane users are constantly looking for more solvent-resistant membranes, but the nature of the process makes it clear that the "universal membrane" will never be made by this process since the polymer must be soluble in at least one solvent and be a film former in order to be cast. Nevertheless, membranes with remarkable solvent resistance, pH resistance, and thermal stability are on the market today.

A series of membranes with various retentivities (molecular-weight cutoffs) can be formed from the same polymer by adjusting the casting dope composition in accordance with the principles discussed above. Figure 19 shows how a small change in the water flux (porosity) can dramatically affect the retention for a series of polyelectrolyte complex membranes.

Changes in porosity can also be effected by alternating the solvent for a given polymer system. Figure 20 shows the various porosities obtained from a 15% solution (by weight) of cellulose acetate in different solvents. The effect has been correlated with the solubility parameter of the solvent—the water content increasing with the solubility parameter. This appears to be a general rule independent of the nature of the polymer.[31] For example, tetrahydrofuran always gives very dense membranes and dimethylsulfoxide or dimethylformamide always gives very porous membranes. This fits the model given previously that membrane porosity may be largely governed by the rate of diffusive exchange of solvent and water during the precipitation of the polymer. In general, water will enter the solution layer faster if the solvent solubility parameter is closer to that of water (i.e., high 23.4), and this results in rapid precipitation and a higher water content or porosity.

Table 8 shows a series of membranes produced by Amicon Corporation. The prefix letters UM, PM, and XM refer to different polymers. (For example, only the PM membrane can be steam-autoclaved.) The digits refer to the nominal molecular-weight cutoff. It will be noticed that retentivity is not always related to flux even within the same polymer series. For example, the PM 30 membrane is more open than the PMIO but has a lower flux value. Similarly, a different polymer system may exhibit different flux values for the same cutoff. For example, the XM 50 membrane has a flux only one-half that of the PM 30, though its cutoff is higher. Anomalies such as these may be explained concep-

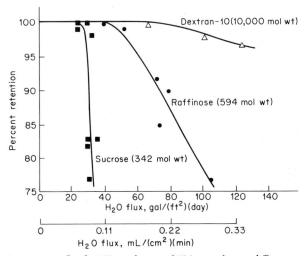

Fig. 19   Retention vs. flux for UF membranes of UM series having different porosities.

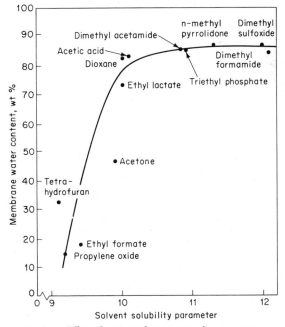

Fig. 20   Effect of casting solvent on membrane porosity.

tually on the basis that larger numbers of smaller-sized pores are available to yield higher fluxes for relatively "tight" membranes.

Hollow-Fiber Membranes   Hollow-fiber UF membranes[10] have also been made with the skin both on the outside and on the inside of the fiber. As will be seen under Concentration Polarization, there are advantages to passing the feed stream through the fiber lumen (provided the skin is on the inside) to provide improved fluid management. Since UF can be a relatively low pressure process (say 25 lb/in² or less), fibers having an inside diameter of 0.018 in with burst strengths approaching 100 lb/in² can be pressurized internally so that the fiber "weeps" permeate.

Inorganic Membranes   The technology utilized in producing dynamic membranes (see RO membranes) has been utilized by Union Carbide to commercialize an inorganic membrane with UF retention capability. The membrane is sold in module form only, known as Ucarsep®, which consists of an unspecified inorganic coating on the inside of

TABLE 8   Amicon Ultrafiltration Membranes and Filters

Diaflo®	Nominal mol wt cutoff	Apparent pore diam, Å	Water flux, gal/ft²/day at 55 lb/in²
UM 05	500	21	10
UM 2	1,000	24	20
UM 10	10,000	30	60
PM 10	10,000	38	550
PM 30	30,000	47	500
XM 50	50,000	66	250
XM 100A	100,000	110	650
XM 300	300,000	480	1300

SOURCE: Amicon Corp.

carbon tubes. The major advantage cited for these systems is better chemical resistance (pH 1 to 14) and thermal stability (up to and over 149°C) than with organic polymeric membranes. (However, some polymeric systems will come close to this performance.) Even if the process stream is relatively "mild," it can be a significant advantage to clean a membrane with hot caustic.

It is interesting that the first commercial exploitation of the dynamic-membrane technology does not take advantage of in situ generation (see RO membranes). Instead, modules are prepared by the manufacturer (Union Carbide) ready to use in the field. Presumably, this reflects the inconvenience and poor reproducibility of dynamic-membrane formation in situ.

CONCENTRATION POLARIZATION

Reverse osmosis and ultrafiltration would probably not be feasible commercial processes today had it not been for the advent of the asymmetric membrane. Prior to this development, membranes could be made that had sufficient rejection for species like salt, but the flux through these membranes was too low to be of any practical utility. The flux could be increased by decreasing the membrane thickness, but in this case the membrane was so fragile it could not be handled. The anisotropic structure with the thin (0.2 to 0.5 μm) dense barrier layer, supported by a 50- to 100-μm-thick porous substructure, reduced the resistance to flow and maintained membrane strength. Further, the graded (almost conical) pore structure in these asymmetric membranes results in excellent "internal fouling resistance," since the small pore opening virtually ensures no entrapment of solutes or particulates within the pore network.

However, the development of anisotropic membranes is only half the battle. More often than not, the membrane is not the limiting resistance to flow, but rather the buildup of retained solutes and particulates on the membrane. As we have seen, dynamic membranes can be formed by depositing a film of particulates on a porous substrate; the dynamic membrane becomes the limiting resistance to flow and offers a higher retention capability than the porous support.

The convection of retained solutes and particulates to the membrane surface by the flux of carrier fluid through the membrane (see Fig. 21) is known as "concentration polarization."[32] The term simply refers to the fact that any membrane system with a flux greater than zero will show a higher concentration of retained species adjacent to the membrane surface than in the bulk stream.

Concentration Polarization in Microfiltration

For any type of pressure filtration, the filtration rate per unit area or flux J will be proportional to the pressure drop ΔP across the membrane or filter (i.e., the driving force)

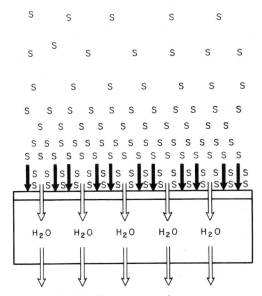

Fig. 21 Concentration polarization.

divided by the resistance to flow. The resistance term consists of two parts: the resistance of the cake which accumulates on the upstream surface of the membrane R_c and the resistance contributed by the membrane itself R_m:

$$J = \frac{\Delta P}{R_c + R_m} \tag{1}$$

The resistance of the membrane R_m is easily determined by the resistance to flow observed with ultrapure water and should be a constant for a given membrane operating with a specified fluid and temperature.

The resistance of the cake of accumulated particulates R_c is more complicated. Indeed, in conventional filtration it is a variable which increases as filtration proceeds, resulting in a progressively lower filtration rate at constant pressure. This is due to the continually increasing thickness of the cake and its compaction under the pressurized conditions of filtration. If R_m is defined as above, R_c must also include the effect of pore plugging within the membrane.

In conventional filtration of particulates:

$$R_c = \frac{\alpha' w' V_t (\Delta P)^s \mu}{A} \tag{2}$$

where α' = constant dependent on properties of the cake
w' = weight of dry particulates per unit volume of filtrate
V_t = volume of filtrate delivered, or "throughput"
ΔP = pressure drop

s = compressibility exponent of the cake (s is zero for a perfectly noncompressible cake and unity for a perfectly compressible cake; normal values range between 0.1 and 0.8 for commercial slurries)

μ = viscosity of the filtrate

A = area of the filtering surface

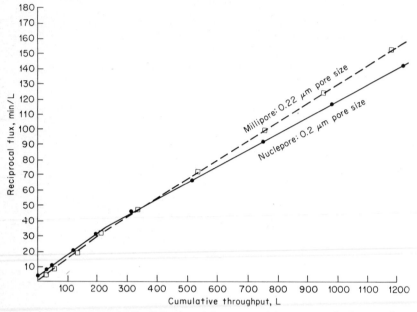

Fig. 22   Constant-pressure (30 lb/in²) filtration of Pleasanton, Calif., tap water with 47-μm membranes.

Combining Eqs. (1) and (2),

$$J = \frac{\Delta P}{\alpha' w' V_t (\Delta P)^s \mu / A + R_m} \tag{3}$$

Thus the flux declines as the throughput increases. Inverting Eq. (3),

$$\frac{1}{J} = \frac{\alpha' w' V_t (\Delta P)^{s-1} \mu}{A} + \frac{R_m}{\Delta P} \tag{4}$$

Data from a constant-pressure filtration may be plotted as a straight line if the reciprocal of the flux is plotted vs. the accumulated filtrate. That this applies in microfiltration is seen in Fig. 22. Knowing w', μ, A, and ΔP means that α' and s are independent constants which can be determined from the slopes of two sets of data taken at different pressure drops.

If one assumes that the limiting resistance to flow is that due to accumulated particulates on the membrane or within the pores, Eq. (3) becomes

$$J = \frac{A \Delta P^{1-s}}{\alpha' w' V_t \mu} \tag{5}$$

It is interesting to note that for a "perfectly compressible" cake, the flux becomes independent of pressure. Higher pressures simply increase the resistance to flow of the cake enough to offset increases in the flow rate due to the higher driving force ΔP.

Polarization Control by Reducing Flow Rate   Rearranging Eq. (5) gives us the dependence of throughput on other variables.

$$\frac{V_t}{A} = \frac{\Delta P^{1-s}}{\alpha' w' J \mu} \tag{6}$$

Equation (6) shows that as a first approximation, the throughput (total volume processed per square foot of membrane area) will be inversely proportional to the flow intensity or flux [in gal/(min)(ft²)]. Figure 23 shows that Eq. (6) is a good approximation for at least one case. In other cases (see Fig. 24), the throughput seems to increase more sharply for a decrease in flow intensity. This may be due to an inertial-impaction phenomenon, which Eq. (6) does not take into account. Particles smaller than the rated pore size may be captured on the membrane at high velocities, whereas at low velocities they may follow the flow streamlines more easily and pass through the membrane without capture. Disputing this hypothesis is the fact that small-particle retention appears to be invariant with flow intensity. Another possible explanation is that particles impacting on the

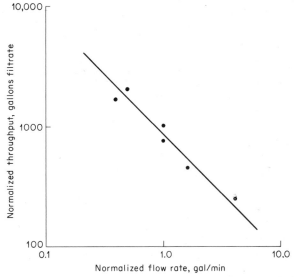

Fig. 23   Variation of throughput with flow intensity for Nuclepore membranes with Pleasanton, Calif., tap water.

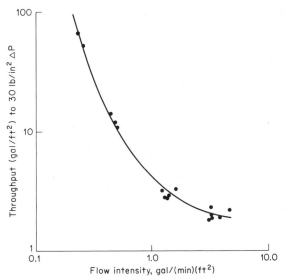

Fig. 24   Variation of throughput with flow intensity.

membrane at higher velocities may show a higher effective compressibility exponent than those collected at low velocities. According to Eq. (6), this would reduce the throughput to a given ΔP.

The variation of throughput with flow intensity has far-reaching implications on the most economical way to run conventional membrane filtration processes. For example, if a fixed volumetric flow rate (gal/min) must be filtered, the total volume (in gallons) which may be processed before plugging (run to a set ΔP) is proportional to the square of the membrane area [J in Eq. (6) has units of gal/(min)(ft²)]. Thus increasing the fixed capital investment (in housings, etc.) by two times (to increase the membrane area by a factor of 2) will increase the volume processed (to membrane exhaustion) by a factor of 4. Consequently, replacement costs of membranes will be reduced by a factor of 2.

Polarization Control with Prefiltration   Microfiltration (MF) is generally used in a "dead-ended" configuration where all the liquid is forced through the membrane (Fig. 21). In this case, the particulates retained on the membrane will accumulate on the surface and in the pores of the membrane until the resistance of the cake R_c reduces the filtration rate to an unacceptable level at a given pressure drop. One way of decreasing the resistance of the cake R_c is to distribute the accumulated particulates within a depth filter upstream of the membrane. If the depth filter has a high surface area available for absorption such as will be obtained with a fine fibrous media, the dirt-loading capacity of the prefilter will extend the life of the membrane manyfold.

Referring to Eqs. (3) to (6), the area term A is really the area over which the dirt is collected as opposed to the superficial area on which the flux J is based. Thus an effective depth filter with an internal surface area 100 times that of the superficial area would be expected to increase the throughput a hundredfold.

Polarization Control with Backwashing   If the accumulation of particulates takes place on the surface of the membrane rather than in the pores, there is a good chance that the particulates can be removed from the surface by intermittent backwash pulses which blow the particulates off the surface and out of the system. In general, the Nuclepore type of membrane is much more amenable to "backwashing" than the thicker reticulated structures associated with the other MF membranes. Particles do not become entrapped in pore constrictions as they do with the reticulated pore structures. Accumulation on the surface of Nuclepore, unless gummy and gelatinous, is more easily removed than that accumulated on the spongelike structure of the other MF membranes. Further, particles

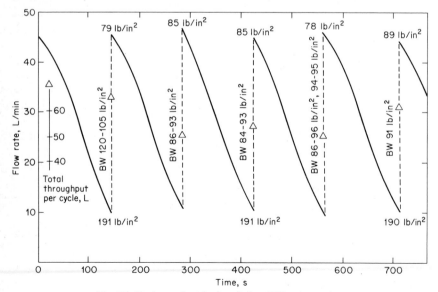

Fig. 25   Nuclepore beer backwash data 0.85-μm pore size.

within the straight-through cylindrical pores are more easily blown out than those entrapped in the reticulated structures.

Figure 25 presents data from a backwash test on beer with Nuclepore 0.85-μm-pore-size membranes. Because of the nature of this on-line test, the pressure increases as the flow rate decreases. The backward pulse reduces the pressure drop from 190 to about 85 lb/in² and restores the flow rate from 10 to 45 L/min.

It is important that the backwashed debris be removed from the system; otherwise, when normal forward filtration is resumed, the backwashed debris will simply reaccumulate along with the normal debris from new feed to result in an even more rapid decline in flux. This is illustrated in Fig. 26.

In these data, a five-plate, backwashable system (Fig. 27) was installed in a 26-gal/min

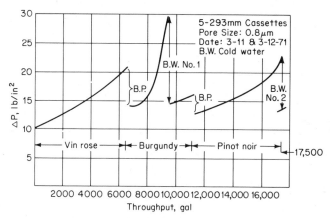

Fig. 26 Results of a 2-day wine-filtration run.

line. The five Nuclepore membranes (0.8-μm pore diameter) amounted to approximately 3 ft² of membrane area. After completion of the vin rosé filtration, a bubble-point test (see Appendix A) was made which resulted in some recovery of pressure drop (Fig. 26). The pressurization of gas on the downstream side of the membrane until wine is displaced from the largest pores resulted in a mini-backpulse. However, since particulates were not flushed from the system, they merely reaccumulated on the membrane. Note that an extrapolated vin rosé plot tends to intersect the upper end of the burgundy plot. It is only in a fully flushed backwash (No. 1) that significant and lasting recovery is achieved.

A word of caution: Not all fluids deposit particulates which can be backwashed as readily as the data of Figs. 25 and 26 would indicate. The "backwashability" depends on the morphology and size of the particulates. Gelatinous particulates are particularly difficult to backwash.

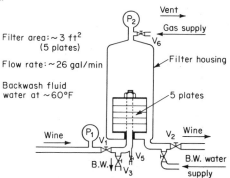

Fig. 27 Backwashable system for wine filtration.

Backwashing has also been done with UF and RO membranes with reasonable success (see below). The author has restored the decaying flux of hollow-fiber UF membranes operating on Lexington tap water with a daily backpulse. This procedure was successful in restoring flux until the end of the test—over a period of several months. However, this technique must be used with caution on anisotropic membranes; a sizable backpulse with significant reverse pressure drop can blow the skin of an asymmetric membrane right off its substrate.

Polarization Control with Crossflow   The accumulation of particulates on the surface of membranes can be minimized by flowing the feed stream over the surface (see Fig. 28) to literally sweep away part of the accumulated layer. This technique of fluid management is referred to as "crossflow" filtration. It requires recirculation of the process stream at rates that may be one or more orders of magnitude larger than the rate of filtration. The result is greatly extended membrane life, which reduces operating costs (membrane-replacement costs). This must be balanced against the increased capital investment (pumps and system) and additional power (required for fluid recirculation).

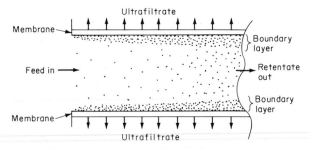

Fig. 28   Schematic of crossflow fluid management.

In the Nuclepore laboratories, the author has been able to concentrate yeast suspensions from 0.2 to 8% by weight in a crossflow device (see Fig. 29),[1] a feat impossible with conventional dead-ended filtration. In fact, an extrapolation of the data indicates that in large systems a concentration of 50% is possible.

The effectiveness of crossflow fluid management is readily seen in the data of Fig. 30, which show an 80% increase in flux for a similar increase in recirculation rate. (The recirculation rate is a measure of the fluid velocity tangential to the membrane surface.)

Figure 31 shows crossflow filtration data from two runs designed to dewater and remove the cellular debris from a fermentation broth. The data show that higher long-term flux values are obtained with ultrafiltration membranes (XM 300) than with microfiltration membranes (0.2 μm). This is undoubtedly due to the anisotropic structure of the UF

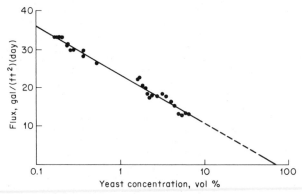

Fig. 29   Concentration of yeast with 0.2-μm Nuclepore membrane in RF 10 (radial flow–crossflow device with 10-mL channel).

membrane, which tends to prevent internal fouling of the pores. The tortuous-path MF membranes tend to be more prone to internal fouling than the straight-through cylindrical pores of the Nuclepore MF membrane. From these data, one might conclude that UF membranes are preferred for separations of this type unless some sort of fractionation capability is desired.

The usefulness of a crossflow fluid-management technique with microfiltration is further illustrated in plasmapheresis—the fractionation of blood into the cellular components (red cells, white cells, and platelets) and the plasma. For example, when obtaining blood from a blood donor it is often desirable to be able to return the cellular components to the donor so that more frequent bleedings can be made (every 2 weeks instead of every 2 months). At present, most plasma is obtained via a two-stage batch centrifugation process. Until 1970, microfiltration of blood was considered unfeasible because of membrane plugging and high concentrations of hemoglobin (hemolysis) in any plasma

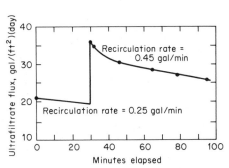

Fig. 30 Twofold concentration of fermentation broth by ultrafiltration.

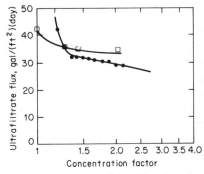

Fig. 31 Removal of cellular debris from fermentation broth by ultrafiltration.

obtained. In 1970, the crossflow technique was utilized by researchers at Amicon Corporation to produce relatively hemoglobin-free plasma.[34] This work showed that a reasonable plasma flux could be obtained in a laminar flow, crossflow device with tolerable hemolysis levels. Subsequent work showed that the smooth surface of a 0.6-μm-pore-size Nuclepore membrane was effective in reducing the degree of hemolysis even further, permitting up to 70% plasma recovery in a single pass (no pumps to reduce hemolysis) with a hemolysis level of less than 0.25%. The use of single-pass crossflow filtration for plasmapheresis raises the question of how other variations of crossflow compare with the more common recirculating mode at relatively high velocities. Figure 32 compares the filtrate volume obtained after various times with crossflow filtration of "turbid" bovine serum (clarification) through a 0.6-μm MF membrane. In order of increasing effectiveness above dead-ended filtration, stirred cells, thin-channel single-pass, and thin-channel recirculating systems are best. One also notices the improvement effected by going to thinner flow channels to increase the shear against the membrane.

Polarization Control with Ultrasonic Energy Still another means of reducing concentration polarization with MF membranes is the application of ultrasonic vibrations to the membrane or the fluid adjacent to the membrane. This tends to prevent the accumulation of particulates on the membrane by ultrasonically dispersing them back into the bulk solution.

Semmelink[36] applied the ultrasonic energy directly to the membrane by coupling it solidly with an ultrasonic transducer. However, better results were obtained when the transducer was not connected but placed close to the membrane.

In addition, Semmelink investigated several different types of filtration media including sintered stainless steel, woven screens, photoetched screens, and Nuclepore membranes. He found that the length and the shape of the pores in the filter have an important bearing on the increase in flux with ultrasonic agitation. The filter with the least tortuosity gave the best results. For example, the photoetched material showed the largest increase in flux but was destroyed by cavitation.

Concentration Polarization in Ultrafiltration

The accumulation of solutes or particulates on the surface of an anisotropic UF membrane also leads to an apparent fouling of the membrane which some workers have interpreted as pore blockage. As mentioned earlier, the asymmetric pore structure of these membranes practically eliminates internal fouling of the pores. This is readily shown by easy restoration of the initial flux by simply washing the membrane surface.

From the early days of UF membrane development, it was recognized that control of the boundary layer was crucial to the success of the process. On a laboratory basis this was done with magnetically driven stirring bars suspended above the membrane. On an industrial basis, it has been done by flowing the feed stream over the surface of the membrane (crossflow) to literally sweep away part of the accumulated layer as in Fig. 28, crossflow filtration.

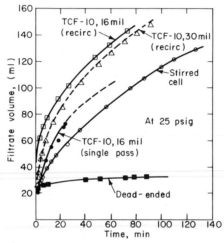

Fig. 32 Filtration of "turbid" bovine serum using Diapor 0.6-μm filter.

On a qualitative basis, thin-channel crossflow fluid management improves the flux by a factor of 2 to 20 times over that of stirred cells. Presumably this is due to the greater shear at the membrane surface in a controlled thin-channel system. However, the author has shown that when the stirring bar is converted into a wiper blade which sweeps across the membrane surface, even higher flux values are obtained.

Charm and Lai[37] made a comparison of various UF systems available for use in the laboratory in the ultrafiltration of enzymes. Typical results from their study are shown in Fig. 33. Again, crossflow fluid management produced the highest flux values. Thin-channel units were more effective than turbulent crossflow units. Vibrating units were least effective. The surprising thing in this study was that catalase was more stable in the thin-channel system; inactivation after 3 h of exposure was 7%, 26%, and 46% for the thin-channel, turbulent-flow, and vibrating units, respectively.

Considerable theoretical and experimental work has made it possible to predict the concentration polarization of *dissolved solutes* and its effect on the flux through the membrane. This has been more difficult with *particulates* and *undissolved colloidal* species, as will be shown below.

Referring to Fig. 34, as permeation takes place through the membrane, the convective transport of solute to the membrane surface results in a higher concentration of solute at the membrane C_s (since it is retained by the membrane) than in the bulk solution C_b. Steady state is reached when the back-diffusive transport of solute away from the membrane (due to a concentration-gradient driving force) is just equal to the convective transport of solute to the membrane:

$$J_w C - D \frac{dc}{dx} = 0 \qquad (7)$$

where D = local solute diffusivity in solution

x = normal distance from the membrane surface

Of course, steady-state conditions can be satisfied only if the solute concentration in the layer of solution adjacent to the membrane sufrace C_s is higher than that in the bulk solution C_b.

The trivial solution of Eq. (7) results in

$$\frac{C_s}{C_b} = \exp \frac{J_w \delta}{D} \tag{8}$$

where δ is the stagnant boundary-layer thickness. Thus the polarization modulus C_s/C_b

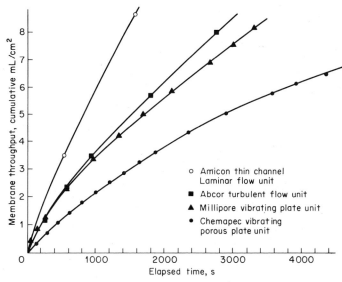

Fig. 33 Comparison of membrane throughput for various ultrafiltration systems. Catalase solution (1 g/L) PM 30 membrane. [*Reprinted from* Biotechnol. Bioeng., **13**, *189 (1971) with permission.*]

increases exponentially with the transmembrane flux and with the boundary-layer thickness, and decreases exponentially with increasing solute diffusivity. This means that polarization is particularly severe with high-permeability membranes and high-molecular-weight solutes (as in UF). The boundary-layer thickness is uniquely determined by the fluid flow adjacent to the membrane surface.

Effect of Concentration Polarization on Retentivity   The polarization modulus C_s/C_b will affect the retention of a molecule by the membrane. The rejection or retention R_j of a solute may be defined as the fraction of the solute present in the feed solution which is retained by the membrane.

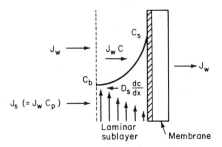

Fig. 34   Polarization in ultrafiltration.

$$R_j = \frac{C_b - C_f}{C_b} = 1 - \frac{C_f}{C_b} \tag{9}$$

where C_b = concentration of solute in the bulk solution upstream of the membrane
C_f = concentration of solute in the filtrate
The retention R_j may also be viewed as

$$R_j = 1 - \sigma \tag{10}$$

where σ is the fraction of the total liquid flowing through the membrane passing through pores large enough to pass the solute molecules. Thus σ is a function of the pore structure of the membrane (average pore size, pore-size distribution, etc.). Comparing Eqs. (9) and (10),

$$\sigma = \frac{C_f}{C_b} \tag{11}$$

However, as the concentration at the membrane surface C_s is increased above the bulk concentration by the polarization mechanism, the solute which is carried through the larger membrane pores is then at a higher concentration

$$\sigma = \frac{C_f}{C_s} \tag{12}$$

Thus, if the retention is defined by Eq. (9), as the polarization modulus increases, C_f will increase, resulting in a lower value of the rejection.

Figure 35 presents data from ultrafiltration of dilute dextran solutions (1%) of various molecular weights in a stirred cell using a 50,000 mol wt cutoff membrane. As the pressure increases, the flux J increases and the polarization modulus C_s/C_b also increases [see Eq. (8)]. Baker[38] has even suggested that this phenomenon could be used to obtain a series of different molecular-weight fractions from the same membrane. The smallest molecules would be eluted at the lowest pressures; on raising the pressure, the next higher molecular-weight fraction would be eluted, etc.

If two solutes are present, concentration polarization may result in a secondary dynamic membrane formed of the larger solute. As mentioned earlier, this secondary membrane can be more retentive than the primary membrane, hindering passage of the smaller molecules, which are normally freely permeable. Therefore, an understanding of the parameters affecting dynamic-membrane formation is important to optimizing membrane rejection characteristics as well as membrane flux. With proper polarization control, the fractionation capability of the membrane can be exploited if the concentration of dissolved solutes is not great.

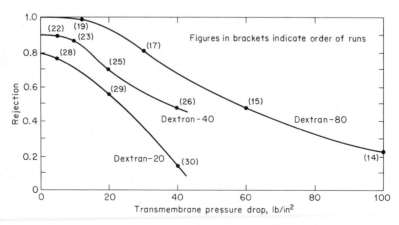

Fig. 35   Rejection as a function of pressure for 1% dextran solution.

Gel-Polarization Model Equation (8) holds for both RO and UF membranes. For UF, the higher polarization modulus means that the concentration of solute at the membrane surface C_s quickly increases to the maximum value obtainable; i.e., the solute precipitates or forms a thixotropic gel. Gel formation on the membrane is analogous to cake formation in conventional filtration (discussed earlier). Thus, under these circumstances, Eq. (1) applies and the cake or gel resistance R_c may again be more significant than the membrane resistance R_m. Figure 36 is analogous to Fig. 34 but is specific to UF. When polarization occurs at the membrane surface, the solute concentration at that surface C_s may be considered to be a constant which is virtually independent of the bulk solute concentration C_b, operating pressure, fluid-flow conditions, or membrane characteristics (e.g., R_m). It is sometimes referred to as the gel concentration C_g. It will depend upon the chemical and morphological properties of the solute and may vary from 10 to 75% by volume.

The gel layer may be thought of as a secondary dynamic membrane which is hydraulically permeable to solvent. Often the hydraulic permeability of the gel layer is considerably smaller than that of the membrane itself, i.e., $R_c \ggg R_m$. Thus the gel layer becomes the limiting resistance to flow and the flux obtained is independent of the membrane permeability unless, of course, the pore size is so large that the solute is no longer retained by the membrane.

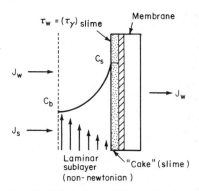

Calculated values of the cake resistance R_c for the ultrafiltration of albumin in a relatively tight (10,000 mol wt cutoff) UF membrane are shown in Fig. 37. It will be seen that at high stirrer speeds and low protein concentrations the cake resistance may be reduced below that of the membrane ($R_m = 350$ in Hg, min/cm). However, at protein concentrations of over 6.5%, the cake resistance is always higher than

Fig. 36 Polarization in UF membranes.

that of the membrane. For more open UF membranes, the membrane resistance will be reduced considerably (e.g., about 80 in Hg, min/cm for a 30,000 mol wt cutoff membrane) and the cake resistance will increase owing to higher convection rates to the membrane surface (lower membrane resistance).

It is also interesting to note (Fig. 37) that a twofold increase in pressure (from 32 to 64 in Hg) increases the flux by only 18% at a protein concentration of 6.5%. The cake resistance has increased by almost twofold; this is due to the increase in thickness of the gel layer because of the increased convection of protein to the membrane surface.

Equation (7) refers to the steady-state condition; the convective transport of solute toward the membrane $J_w C$ must be equal to the diffusive or convective transport of solute away from the membrane surface back into the bulk solution $D(dc/dx)$. Equations (1) and (7) explain why the solvent flux often becomes invariant with increasing transmembrane pressure drop ΔP. Solvent flux J_w increases with pressure until the concentration at the membrane surface reaches some critical solute concentration which represents the close packing of colloidal particles or the incipient point of gel formation. Since the concentration cannot increase further, the gel layer begins to thicken or compact, increasing the resistance to flow R_c until the convective transport of solute toward the membrane is reduced to a value just equal to the diffusive back transport of the solute away from the membrane into the bulk solution. Any finite increase in ΔP will cause the gel layer to thicken or compact by an equal amount, resulting in a transmembrane solvent flux invariant with the transmembrane pressure drop.

For the gel-polarized region where the flux is invariant with pressure, Eq. (7) may be integrated to yield Eq. (8), which may be rewritten as

$$J_w = K \ln \left(\frac{C_g}{C_b} \right) \tag{13}$$

where K is a mass-transfer coefficient equivalent to the diffusivity D over a boundary-layer thickness δ and the concentration at the membrane surface C_s becomes a constant which is the gel concentration C_g. Equation (13) expresses a surprising, but experimentally confirmed, conclusion: the ultrafiltration rate J_w is a dependent variable governed solely by the solute concentration C_b and the mass-transfer conditions K which exist adjacent to the membrane surface.

Stirred-cell data[43] demonstrate another consequence of the limiting flow resistance of the gel layer—that flux is independent of membrane permeability. Three membranes of

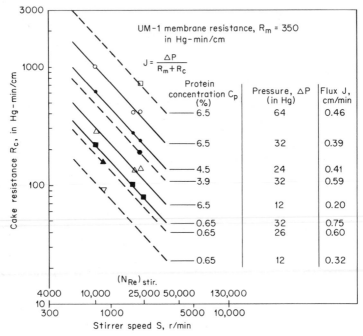

Fig. 37   Characterization of caking in ultrafiltration (M 50 cell).

widely differing permeabilities show virtually the same flux when operated above the threshold transmembrane pressure.

Thus experimental data confirm that, as predicted by Eq. (13), above some threshold pressure, membrane flux will be invariant with transmembrane pressure drop or membrane permeability and is fixed by the solute characteristics C_g and the mass-transfer coefficient K for back diffusion from the membrane surface. Under gel-polarized conditions, Eq. (13) indicates that the variation of flux with concentration C_b can be plotted as a straight line on semilog paper. Figure 38 shows that this is indeed the case for both colloidal suspensions and macromolecular solutions. Equation (13) also indicates that the intercept of the straight line on the abscissa should be at the gel concentration C_g. It is interesting to note on Fig. 38 that the gel concentration for colloidal suspensions (60 to 70%) is higher than that for protein solutions (25%). This is not fortuitous, since in the case of colloidal suspensions the "gel layer" is expected to resemble a layer of close-packed spheres having 65 to 75% solids by volume whereas many protein solutions are known to gel at around 25% solids.

According to Eq. (13), improved mass-transfer conditions adjacent to the membrane (higher K) will result in higher slopes on the J vs. ln (C_b) plots. This is illustrated by the data of Figs. 39 and 40, which show the dependence of flux on recirculation rate, which is a measure of the fluid velocity across the membrane surface.

Evaluation of Mass-Transfer Coefficient in Laminar Flow. The mass transfer–heat transfer analogies well known in the chemical engineering literature make possible an evaluation of the mass-transfer coefficient K of Eq. (13) and provide insight into how membrane geometry and fluid-flow conditions can be specified to optimize flux.

The Graetz or Lévêque solutions[43,44] for convective heat transfer in laminar-flow channels, suitably modified for mass transfer, may be used to evaluate the mass-transfer coefficient for the case where the laminar parabolic velocity profile is assumed to be established at the channel entrance but where the concentration profile is under development down the full length of the channel. It can be shown that for all thin-channel lengths of practical interest, this solution is valid.[41] Lévêque's solution gives

$$\text{Sh} = 1.62 \left(\text{Re Sc} \, \frac{d_h}{L} \right)^{0.33} \tag{14}$$

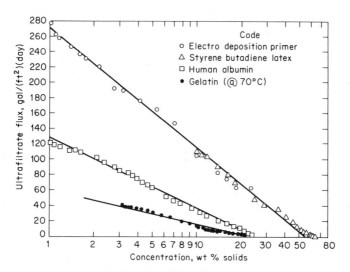

Fig. 38 Variation of flux with concentration.

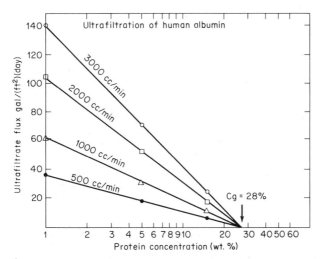

Fig. 39 Cross plot of flux vs. recirculation rate.

for $100 < \mathrm{Re\ Sc}\dfrac{d_h}{L} < 5000$

where $\mathrm{Sh} = \text{Sherwood number} = \dfrac{Kd_h}{D}$

$\mathrm{Re} = \text{Reynolds Number} = \dfrac{Ud_h}{\nu}$

$\mathrm{Sc} = \text{Schmidt number} = \dfrac{\nu}{D}$

$d_h = \text{equivalent hydraulic diameter}$
$L = \text{channel length}$
$U = \text{average velocity of fluid}$
$\nu = \text{kinematic viscosity} = \dfrac{\mu}{\rho}$

$\mu = \text{viscosity of fluid}$
$\rho = \text{density of fluid}$

or
$$K = 1.62\left(\frac{UD^2}{d_hL}\right)^{0.33} \tag{15}$$

For flat rectangular channels, where $d_h = 2b$ (b = channel height),

$$K = 0.816\left(\frac{6UD^2}{bL}\right)^{0.33} \tag{16}$$

or
$$K = 0.816\left(\frac{6QD^2}{b^2wL}\right)^{0.33} \tag{17}$$

where $Q = \text{volumetric flow rate}$
$w = \text{channel width}$
More generally,

$$K = 0.186\left(\frac{\dot\gamma}{L}\,D^2\right)^{0.33} \tag{18}$$

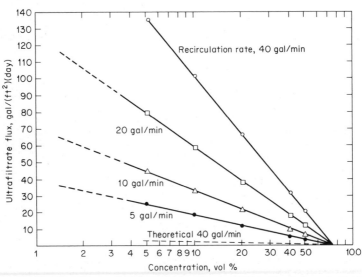

Fig. 40 Ultrafiltration of styrene butadiene polymer latex in LTC-1, 15-mil channels, XM 50 membrane, 60 lb/in² (gage) average transmembrane pressure.

where $\dot{\gamma}$ = fluid shear rate at the membrane surface

$$= \frac{6U}{b} \text{ for rectangular slits}$$

$$= \frac{8U}{d} \text{ for circular tubes}$$

Reviewing Eqs. (14) to (18), it is clear that the flux (or mass-transfer coefficient) may be increased by increasing the channel velocity (U or Q) or by decreasing the channel height b. In more general terms, any fluid-management technique which increases the fluid shear rate $\dot{\gamma}$ at the membrane surface will increase the flux.

Indeed, Eq. (18) shows that in laminar flow, at a fixed bulk stream concentration C_B, the flux should vary directly as the cube root of the wall shear rate per unit channel length. This has been confirmed for a large number of solutions ultrafiltered in a variety of channel geometries.

In a few cases, a higher power dependence of flux on recirculation rate (channel flow rate) than that indicated by Eq. (17) has been noted. When faced with anomalous data such as these, one is tempted to use correlations such as that presented by Gröber et al.[45] for the special case in which velocity and concentration profiles are both developing down the full channel length.

$$\text{Sh} = 0.664 \left(\text{Re} \, \frac{d_h}{L} \right)^{0.5} \text{Sc}^{0.33} \tag{19}$$

or

$$K = 0.664 \left(\frac{U}{L} \right)^{0.5} \frac{D^{0.66}}{\nu^{0.17}} \tag{20}$$

or

$$K = 0.664 \left(\frac{Q}{bwL} \right)^{0.5} \frac{D^{0.66}}{\nu^{0.17}} \tag{21}$$

Although these correlations fit some anomalous data better than Eq. (17), they cannot be justified, since Gröber[45] points out that the velocity profile is completely developed at a distance from the channel inlet L^* given by the following equation:

$$\frac{L^*}{d_h} = 0.029 \, \text{Re} \tag{22}$$

For thin channels, where $d_h = 2b = 0.060$ in maximum, L^* maximum $= 0.029$ $(2000)(0.060) = 3.48$ in, which is not an appreciable fraction of the channel length.

Evaluation of Mass-Transfer Coefficient in Turbulent Flow. Perhaps the best-known heat-transfer correlation for fully developed turbulent flow is that due to Dittus and Boelter:[46]

$$\text{Sh} = 0.023 \, \text{Re}^{0.8} \text{Sc}^{0.33} \tag{23}$$

$$K = 0.023 \, \frac{U^{0.8} D^{0.67}}{d_h^{0.2} \nu^{0.47}} \tag{24}$$

It can be argued that any turbulent-flow correlation should not be applied for Re $<10,000$. However, in the case of current thin-channel ultrafiltration devices, the entrance geometry is such that fully developed turbulent flow occurs at much lower Reynolds numbers. Measurements of fluid velocity vs. pressure drop show a definite transition from laminar to fully developed turbulent flow at Re $= 2000$.

For flat rectangular channels where $d_h = 2_b$, Eq. (24) becomes

$$K = 0.02 \, \frac{U^{0.8} D^{0.67}}{b^{0.2} \nu^{0.47}} \tag{25}$$

or

$$K = 0.02 \, \frac{Q^{0.8} D^{0.67}}{bw^{0.8} \nu^{0.47}} \tag{26}$$

As in the case of laminar flow, the flux (or mass-transfer coefficient) may be increased by increasing the channel flow rate Q and decreasing the channel height b. The effect is, however, much more dramatic in turbulent flow—the flux varying with flow rate to the 0.8 power and inversely with channel height. Furthermore, the flux is independent of channel length, since both the velocity and concentration profiles are established rapidly in the first section of the channel.

Again, as in the case of laminar flow, most of the ultrafiltration data taken on solutions in turbulent flow are in good agreement with theory, i.e., with the Dittus-Boelter correlation.

Quantitative Calculation of Flux from Mass-Transfer Coefficients for Macromolecular Solutions. The success of the Lévêque and Dittus-Boelter relationships in indicating the variation (power dependence) of ultrafiltrate flux with channel geometry and fluid velocity for macromolecular solutions is gratifying. The more crucial test of the theory, of considerable interest to the design engineer, is whether these relationships can be used to calculate *quantitatively* the ultrafiltrate flux knowing the channel geometry, fluid velocities, and solute characteristics.

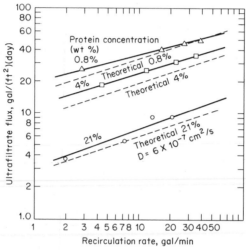

Fig. 41   Ultrafiltration of human albumin [LTC-1, XM 50, 15-mil channel, 30 lb/in² (gage) average transmembrane pressure drop]. Laminar-flow data.

The accuracy of the Lévêque and Dittus-Boelter relationships has been verified in linear thin-channel tubular equipment ultrafiltering protein solutions.

Figure 41 presents laminar-flow data from 15-mil channel tubes compared with theoretical values using a diffusivity of 6×10^{-7} cm²/s and a gel concentration of 45% as determined from Fig. 42. Again, experimental and theoretical values agree within 25%— although the theoretical values are consistently lower.

In Fig. 42, the departure from the linear semilog plot at low concentrations is due to incomplete gel polarization. The high recirculation rates have reduced the wall concentration to a point where it is no longer the limiting resistance to flow. Higher pressures could be used to increase the flux up to the limiting gel-polarized flux. It will be noted that the lower recirculation-rate data with less efficient polarization control are fully gel-polarized even at the lowest concentration.

The ability of the Lévêque solution to describe adequately the variation in flux with diffusivity of the retained solute is illustrated in Fig. 43. Here albumin data were compared with whole-serum data. The larger globulins in whole serum have a lower diffusivity ($D = 4 \times 10^{-7}$ cm²/s)[47] and are strongly polarizing. The theoretical curves are 15 to 20% below the experimental data in both cases. Thus the dependence of flux on diffusivity to the 0.67 power is confirmed.

As a final test of the theory for macromolecular solutions, data from the literature[48,49] on the ultrafiltration of polyethylene glycol (Carbowax 20M) solutions in 1-in-diameter tubular membranes operating in turbulent flow are presented, along with theoretical flux values, in Fig. 44. A diffusivity of 5×10^{-7} cm²/s (Refs. 48, 49, 50) and a gel concentration of 7.5% (see Fig. 45) were assumed. The agreement between theoretical and experimental values is within 14 to 27% at the higher and lower Reynolds numbers, respectively.

Quantitative Calculation of Flux from Mass-Transfer Coefficients for Colloidal Suspensions The agreement between theoretical and experimental ultrafiltration rates for *macromolecular solutions* can be said to be within 15 to 30%. For colloidal suspensions, experimental flux values are often one to two orders of magnitude higher than those indicated by the Lévêque and Dittus-Boelter relationships.

Equations (17) and (26), for laminar and turbulent flow, respectively, both indicate that the mass-transfer coefficient should vary with the diffusivity of the retained solute to the 0.67 power. That this is the case was shown in Fig. 43.

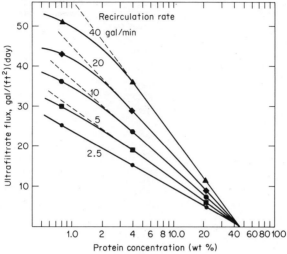

Fig. 42   Ultrafiltration of human albumin [LTC-1, XM 50, 15-mil channel, 30 lb/in² (gage) average transmembrane pressure drop]. Gel concentration.

In accordance with the Stokes-Einstein relation for diffusivity,

$$D = \frac{kT}{6\pi\mu r_p} \tag{27}$$

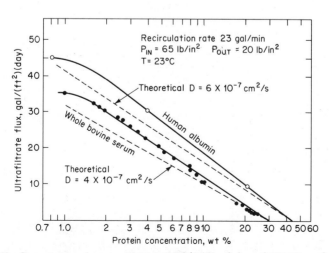

Fig. 43   Concentration of serum proteins in LTC lS (15-mil channel, PM 10 membrane).

where k = Boltzmann constant (i.e., the molar gas constant divided by the Avogadro number)

= 1.380×10^{-16} erg deg^{-1}

T = absolute temperature, K

μ = viscosity, P

r_p = radius of the particle diffusing, cm

Macromolecules of higher molecular weight will have larger particle dimensions and lower diffusivities. For example, protein diffusivities of interest are as follows:

Protein	mol wt	$D_{20°C}$, cm²/s
Albumin	65,000	6×10^{-7}
γ globulin	170,000	4×10^{-7}
Collagen (gelatin)	345,000	0.7×10^{-7}

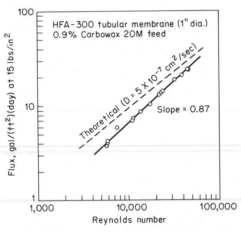

Fig. 44 Ultrafiltration flux in turbulent flow.

Thus these proteins should exhibit decreasing flux rates in the order given; this can be seen in Figs. 38 and 43.

On this basis, one would expect that the ultrafiltration rate from whole blood, including the 8-μm red cells, should be considerably less than that from the plasma above. That such is not the case is illustrated in Fig. 46.

Likewise, the polymer-latex data of Fig. 38 have a much higher slope (mass-transfer coefficient) than would be expected from the latex-particle diffusivity. It is well known that monodisperse polystyrene latices have both suspension viscosities and sedimentation coefficients in good agreement with the predicted Stokes-Einstein diffusivity.[51,52] The Stokes-Einstein diffusivity was calculated for a carboxylic modified styrene-butadiene

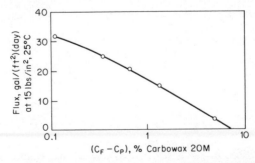

Fig. 45 Ultrafiltration of polyethylene glycol–Abcor tubular HFA 300 membrane.

copolymer latex which had an average particle size of 0.19 μm and was found to be 2.3 $\times$ 10^{-8} cm²/s. This is in good agreement with other values reported in the literature.[51,52]

The gel concentration of this material has been determined to be 75% in numerous thin-channel runs in both tubular (see Fig. 40) and spiral-flow equipment.

Using the values cited above for the diffusivity D and the gel concentration C_G, the theoretical flux is plotted for the 40 gal/min recirculation rate in Fig. 40 and found to be one-thirty-eighth of the experimental thin-channel tube value. Even if the calculated diffusivity were an order of magnitude larger, the theoretical flux would still be less than one-eighth of the experimental value.

Flux vs. recirculation rate data for a thin-channel tubular unit are plotted in Fig. 47. Theoretical values are also plotted assuming laminar and turbulent flow. Most of the data on this plot (below 30 gpm) are taken at Reynolds numbers below 2000 and therefore

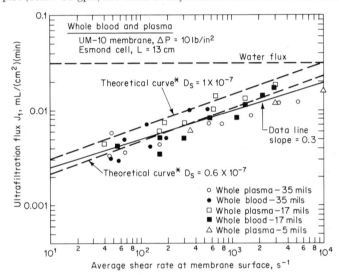

Fig. 46 Ultrafiltration of whole blood and plasma.

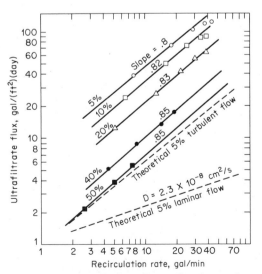

Fig. 47 Flux vs. recirculation rate—ultrafiltration of styrene-butadiene polymer latex in LTC-1 [15-mil channel, XM 50, 60 lb/in² (gage) average transmembrane pressure drop].

assumed to be in laminar flow. The discrepancy between the experimental-flux value and theoretical laminar-flow value is a factor of 38 at 40 gal/min and a factor of 15 at 5 gal/min. If the assumption is made that these data were taken in turbulent flow (completely unwarranted), the slope is more nearly that indicated by theory, but the discrepancy is still a factor of 7.5 at all recirculation rates.

The gross discrepancy between theoretical and experimental flux values also exists in data (Fig. 48) obtained from spiral-flow thin-channel equipment (see Fig. 49). In these data, the transition from laminar flow is clearly seen (change in slope at Reynolds numbers only slightly below 2000. However, there is no abrupt change in flux, as

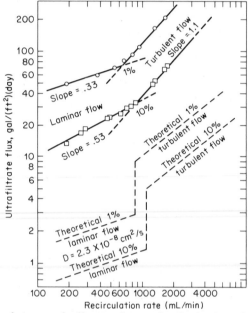

Fig. 48 Flux vs. recirculation rate ultrafiltration of styrene-butadiene polymer latex TC-1D [30-mil channel, PM 30 membrane, 40 lb/in²(gage) average pressure transmembrane pressure drop].

calculated from theory, suggesting that the data labeled "turbulent flow" may be in a more nebulous transition region between laminar and turbulent flow. The experimental slopes tend to be higher than the theoretical slopes in both laminar and turbulent flow except for the laminar-flow data at 1% concentration, which follows the predicted one-third power dependence. Again, the gross failure of the theory in estimating experimental values is evident. The experimental laminar-flow values are a factor of 19 to 29 higher than the theoretical values, whereas the experimental turbulent-flow values are a factor of 8 to 15 higher than the theoretical values.

An evaluation of more than 40 different colloidal suspensions in the author's laboratories has indicated that the diffusion coefficient calculated from the ultrafiltrate flux using the Lévêque or Dittus-Boelter relationships is generally from one to three orders of magnitude higher than the theoretical Stokes-Einstein diffusivity.

It is evident from the above that minor adjustments in molecular parameters such as diffusivity, kinematic viscosity, or gel concentration are incapable of resolving order of magnitude discrepancies.

Similar discrepancies were noted by Blatt, Dravid, Michaels, and Nelson[41] for colloidal suspensions such as skimmed milk, casein, polymer latices, and clay suspensions. They state that "actual ultrafiltration fluxes are far higher than would be predicted by the mass transfer coefficients estimated by conventional equations, with the assumption that the proper diffusion coefficients are the Stokes-Einstein diffusivities for the primary particles." They concluded that either (1) the "back-diffusion flux" is substantially augmented over that expected to occur by Brownian motion or (2) the transmembrane flux is not limited by the hydraulic resistance of the polarized layer. They favored the latter possibility, arguing that closely packed cakes of colloidal particles have quite high permeabilities. The present author rejects this hypothesis for the following reasons:

1. If the gel layer is not the limiting resistance to flow, the flux will be invariant with concentration—which is clearly not the case (see Figs. 38, 40, 47, and 48).

2. If the gel layer is not the limiting resistance to flow, the membrane must be, and the flux should be proportional to the transmembrane pressure drop. Experimental data deny this—showing threshold pressures above which flux is independent of pressure (see Fig. 50).

3. If the gel layer is not the limiting resistance to flow, the layer will continue to grow until the channel is completely full of 75% solids material. If this were the case, we would see a drop in recirculation rate with time even though the feed concentration and channel pressure drop are constant. Thin-channel tubes and spiral-flow modules running continuously at constant latex feed concentration and pressure drop for periods approaching 1 year have shown no decreases in recirculation rate or accumulation of polymer latex in the channels.

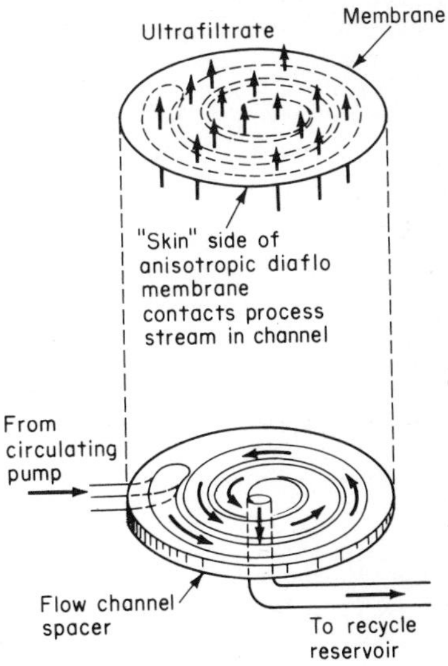

These observations lead to the conclusion that the back-diffusive transport of colloidal particles away from the membrane surface into the bulk stream is substantially augmented over that predicted by the Lévêque or Dittus-Boelter relationships. *In other words, for colloidal suspensions, mass transfer from the membrane into the bulk stream is driven by some force other than the concentration gradient.* It is the author's contention that the so-called "tubular-pinch effect" is responsible for this augmented mass transfer.

Fig. 49 Spiral-flow thin-channel plate.

Tubular-Pinch Effect. It has become increasingly apparent in the last few years that there are many colloidal suspensions of large and small particles which exhibit less frictional pressure drop than would be expected from the fluid viscosity. The apparent viscosities of such suspensions vary with tube radius, length, and flow rate.[53] To account for such anomalies, it has been postulated that there exists a lubricating particle-depleted ("plasmatic") layer at the wall of vessels in which there is a nonuniform shear field. For example, it has been determined that blood flowing through fine glass capillaries reaches an equilibrium state in which the red-cell concentration in the tube is less than that in the inflowing or outflowing blood—presumably the result of axial drift of red cells and their consequent faster average transit than plasma.[54] Palmer [55] was able to skim off a plasma-rich layer at the wall through fine branches and was able to measure increases in hematocrit from near the wall to the axial region.

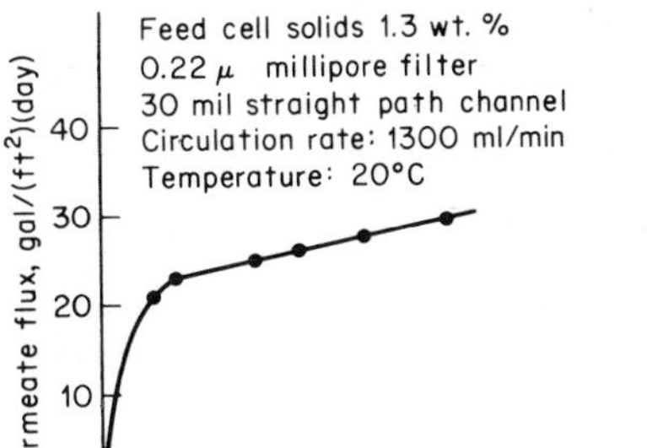

Fig. 50 Dependence of flux on transmembrane pressure ($C_{av} = 5.0\%$; $Q = 20$ cm³/S).

Segré and Silberberg,[56] working with dilute suspensions of rigid spheres, were the first to publish their observations of the tubular-pinch effect whereby the particles migrated away both from the tube wall and the tube axis, reaching equilibrium at an eccentric radial position. At this position, the spheres became regularly spaced in chains extending parallel to the tube axis. The observations of Segré and Silberberg have spawned a number of theoretical and experimental studies investigating the effect and analyzing the cause.

As reported by Karnis, Goldsmith, and Mason,[57] the tubular-pinch effect has also been observed experimentally for suspensions of rubber disks, carbon black, polystyrene spheres, PVA spheres, aluminum-coated nylon rods, elastomer filaments, aluminum particles, insoluble salts, glycerol, and silicone oil in various continuous-flowing media.

One of the more spectacular visual studies was made by Brandt and Bugliarello.[58] They made direct photographic observations of small Dylite spherical beads suspended at concentrations of 1.7 to 5% in a glycerin-water solution and flowing through long rectangular channels made of Plexiglass. The ratio of the channel width to the bead diameter was 25.6. The Reynolds number was varied between 402 and 1640.

Figure 51 gives the average half-channel particle distribution (obtained photographically) expressed as the ratio of the number of particles per band C_X to the total number of particles C_{total} in the entire channel width. Brandt and Bugliarello found that the process was accelerated by increases in flow rate and delayed by increases in average concentration.

It would appear that the physical cause of the tubular-pinch effect is still somewhat controversial. Several analyses have been made, resulting in surprisingly similar expressions for the migration velocity away from the wall.

Radial-Migration Velocity Due to "Slip-Spin" Magnus Force. It is well known that particles flowing in a shear field, in laminar or turbulent flow, spin because the fluid velocity is higher on one side than on the other. Rubinow and Keller[59] postulated that radial migration of particles in small channels occurred because of a transverse force which arises from "slip-spin" force akin to the Magnus force used to explain phenomena such as the curve-ball effect. They derived an expression for the radial-migration velocity:

$$V = \frac{1}{9} U \text{ Re} \left(\frac{r_p}{R}\right)^4 \frac{r}{R} \tag{28}$$

where V = radial-migration velocity
U = average fluid velocity down the channel or tube
Re = Reynolds number
r_p = particle radius
R = tube radius
r = radial position of the particle in the tube

Although there is nothing wrong with this analysis, it is also known that *nonrotating* spheres migrate. Oliver[60] conducted experiments in a vertical tube with spheres whose centers of mass were eccentrically located, to prevent rotation. Again, migration from the wall was observed. Theodore[61] measured the inward radial-migration force on a nonrotating sphere directly by connecting the sphere in a horizontal tube via a thin vertical wire to an analytical balance.

Radial-Migration Velocity Due to Inertial Effects ("Slip-Shear Force"). Saffman predicted the radial migration of particles from solutions of the Navier-Stokes equations retaining the *inertial* terms.

$$V = 0.86U \text{ Re} \left(\frac{r_p}{R}\right)^4 \frac{r}{R} \tag{29}$$

which, except for the numerical coefficient, is similar to Eq. (28). Unlike the Rubinow-Keller[59] theory, which depends critically on particle rotation, Saffman's "slip-shear" lift force is independent of the angular velocity of the sphere; i.e., the particle would migrate even if it was prevented from rotating. Since Saffman's prediction is based on inertial forces, it is expected to occur to a greater extent at higher Reynolds numbers and to occur for turbulent as well as laminar flow.

The theories given above, while able to predict qualitatively the radial migration away from the wall, are unable to account for the *two-way* migration of neutrally buoyant particles observed by Segré and Silberberg.[62] A full treatment of the problem of a freely rotating rigid sphere has been attempted by Cox and Brenner.[63] They obtained the first-order solution of the Navier-Stokes equation and computed the lateral force required to maintain the sphere at a fixed r. The resulting migration velocity, from application of Stokes' law, was for the neutrally buoyant case

$$V = \frac{1}{2} U \text{ Re} \left(\frac{r_p}{R}\right)^3 F \left(\frac{r}{R}\right) \tag{30}$$

where $F (r/R)$ is a function of the radial position of the particle in the tube or channel.

Equation (30) has the same form as the *empirical equation* used by Segré and Silberberg[62] to correlate their data:

$$V = 0.17U \text{ Re} \left(\frac{r_p}{R}\right)^{2.84} \frac{r}{R} \left(1 - \frac{r}{r^*}\right)$$

(31)

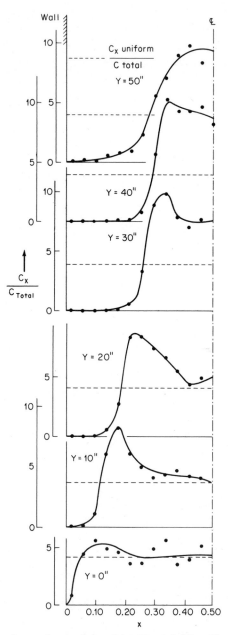

Fig. 51   Flow of monolayers of suspended particles. [*Reprinted from* Trans. Soc. Rheol., **10**(1), 250 (1961) *with permission.*]

where r^* is the equilibrium radial position of the particle, which was found to decrease as r_p/R increased.[57]

Deformable particles, such as liquid drops or elastomer filaments, always migrated to the tube axis ($r^* = 0$) in contrast to rigid particles.

Thin-Channel Ultrafiltration and the Tubular-Pinch Effect. The tubular-pinch effect has the potential of explaining much of the anomalous ultrafiltration data obtained with colloidal suspensions in thin-channel equipment. The migration of particles away from the membrane wall will certainly augment the back-diffusive mass transfer described by the Lévêque and Dittus-Boelter relationships.

Since the migration effect predicted by Saffman[64] and Cox and Brenner[63] is inertia-dependent, it is expected to be strongly affected by the fluid velocity. Equations (28) to (31) all predict a radial-migration velocity V increasing as the *square* of the bulk flow velocity U. King[65] has pointed out that theoretical and experimental mass-transfer coeffi-

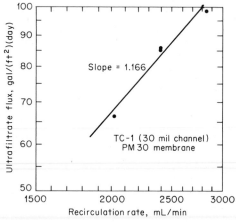

Fig. 52   Ultrafiltration of 7% solids electrocoat paint in turbulent flow.

cients for Newtonian fluids in turbulent flow exhibit exponents on velocity ranging from 0.75 to 1.0 but that *an exponent greater than 1.0 on velocity cannot be predicted by any accepted theory of diffusive convective transport in turbulent flow.* (Of course, this does not include the nondiffusive tubular-pinch effect.) Yet Figs. 52 and 53 (electrodeposition paint data) clearly show a dependence of ultrafiltration rate on channel velocity at exponents greater than unity. Indeed, Fig. 54 shows a 1.33 exponent in *laminar flow!* As far as the author knows, the highest exponent predicted on the basis of diffusive mass transfer in laminar flow is 0.5—for the case where velocity and concentration profiles are both developing simultaneously [see Eq. (21)]. Exponents higher than 0.33 to 0.50 were also noted in Fig. 47, which exhibited an exponent of 0.8 to 0.85 in laminar flow.

Figure 55 shows a flux vs. channel-velocity plot for plasma and whole blood exhibiting slopes of 0.33 and 0.6, respectively. The Lévêque solution would have predicted flux for whole blood considerably lower than that for plasma but with the same 0.33 power dependence on velocity. It is the author's contention that the tubular-pinch effect tends to depolarize the membrane surface of red cells, yielding a flux similar to that obtained with plasma alone. It is also conceivable that the migration of red cells away from the membrane wall could augment the mass transport of plasma proteins in the same direction by means of the relatively large-scale motions of the red cells. The particle migration and particle rotation would be expected to generate secondary flows by virtue of local drag effects in the continuous medium.

Along similar lines, Forstrom and Blackshear[66,67] have shown how the combined rotational and translational motion of red cells augments the apparent diffusion coefficients of various constituents of blood. Their analysis leads to the following relationships:

$$\frac{D^*}{D} - 1 = 3 \times 10^{-8} \frac{1}{D} \dot{\gamma} \tag{32}$$

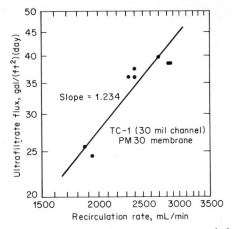

Fig. 53   Ultrafiltration of 15% solids electrocoat paint in turbulent flow.

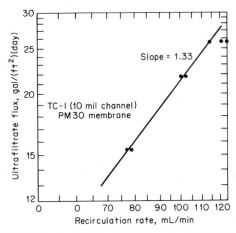

Fig. 54   Ultrafiltration of 15% solids electrocoat paint in laminar flow.

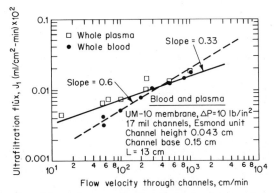

Fig. 55   Effect of flow velocity on ultrafiltration rate for plasma and whole blood.

where D^* = apparent diffusion coefficient
$\quad\quad D$ = diffusivity in stagnant plasma
$\quad\quad \dot{\gamma}$ = local shear rate
or for values of $D^*/D \gg 1$,

$$D^* = 3 \times 10^{-8} \, \dot{\gamma} \tag{33}$$

Blackshear has shown Eq. (32) to be reasonably accurate for the axial diffusion of plasma proteins in blood for values of D^*/D from 10 to 20.

Bixler and Rappe[68] patented a process (assigned to Amicon) for improving the ultrafiltrate flux by introducing solid particulate materials into the process stream. It was found that this technique augmented the back diffusion of retained solutes from the membrane surface and consequently increased the flux. They used spherical glass and methylmethacrylate (MMA) beads ranging in size from 29 to 100 μm.

The present author's interpretation of these data is as follows:

1. The lower-density (0.94 g/cm^3) MMA beads show less tendency to settle onto the membrane surface and may never get there—resulting in a relatively inefficient use of the beads.

2. The high-density (2.50 g/cm^3) glass beads are more efficient—provided the beads are not too small. In the latter case, Eqs. (28) to (31) predict a lower migration velocity away from the wall. The transport of glass beads to the membrane surface is driven by gravity and convective transport via the solvent flux through the membrane.

3. The decrease in flux at higher concentrations may be due to the decreased migration velocity at higher concentrations as observed by Brandt and Bugliarello.[58] It will be noted that this effect is seen only with the higher flux membrane (UM2), where the convective transport of beads to the surface is greater.

Examination of Eqs. (28) to (31) indicates that the migration velocity should be proportional to the fluid velocity U raised to the second power and the ratio of the particle size to the channel dimension r_p/R raised to powers ranging between 2.84 and 4.0. *This would indicate that small channel sizes are advantageous in depolarizing the membrane surface via the tubular-pinch effect.* This effect is over and above the predicted increase in back-diffusive mass transfer by virtue of the increased fluid shear rate at the membrane surface $\dot{\gamma}$ and may explain the larger discrepancies between experimental and theoretical flux values in 15-mil channels (see Fig. 47) than those in 30-mil channels (see Fig. 48).

King[65] has also suggested that the secondary flows generated by virtue of radial particle migration and the inertial effects of the particles themselves could conceivably cause the transition from laminar to turbulent flow to occur at a Reynolds number below 2000. If so, this could aid in explaining the data from Figs. 47 and 48. In the ultrafiltration of the paint, several cases have been found where the apparent transition between laminar and turbulent flow in thin-channel tubular units has occurred at a Reynolds number as low as 1000 to 1500.

Unfortunately, most of the theoretical analysis and experimental work on the tubular-pinch effect has been done in laminar flow. Although the effect should be greater in turbulent flow because of the larger inertial effects, we are lacking in quantitative expressions for the radial-migration velocity. Even in laminar flow, with the current state of knowledge, it is almost impossible to predict quantitatively the enhanced mass transfer of particles away from the wall due to the tubular-pinch effect. For the present, all that can be said is that the radial-migration velocity will increase with the Reynolds number, fluid velocity, and particle size but will decrease with increasing channel dimensions. One can envision that the effect of velocity on this depolarizing mechanism will be different for different types of colloidal material, depending, for example, on how deformable the particles are and their shape. Eirich[53] has shown that deformable particles are even more prone to move away from the wall than rigid particles. This could account for the fact that the observed flux-velocity dependence is different for different colloidal suspensions.

Concentration Polarization in Reverse Osmosis

Much of the treatment of concentration polarization for UF also applies to RO. The control of the boundary layer adjacent to the membrane is just as important.

One fundamental difference, however, must be taken into account: the difference in driving force due to the osmotic pressure of retained low-molecular-weight species. In the case of MF and UF, these low-molecular-weight species were freely permeable to the membrane and the osmotic-pressure difference between the feed solution and the permeate was negligible. With the tighter RO membranes exhibiting 99% retention for NaCl, this is no longer the case.

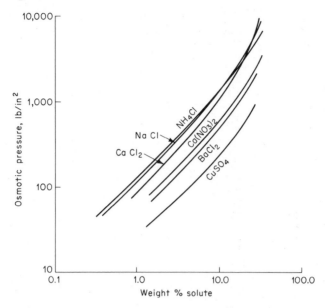

Fig. 56   Osmotic pressure for various salts in water at 25°C.

Figure 56 shows that for 3.5% NaCl solutions ($\sim$ sea water) the osmotic pressure is close to 400 lb/in². The osmotic pressure π of any species may be approximated for dilute solutions by the well-known van't Hoff equation:

$$\pi = iCRT \tag{34}$$

where C = molar concentration of the solute
  R = gas constant
  T = absolute temperature
  i = number of ions formed if the solute molecule dissociates (e.g., for NaCl, i = 2; for BaCl$_2$, i = 3)

Thus Eq. (1) must be modified to take the osmotic-pressure difference $\Delta\pi$ between the feed solution and the permeate into account:

$$J_w = \frac{\Delta P - \Delta\pi}{R_c + R_m} \tag{35}$$

For sea water, it is obvious that the imposed hydraulic pressure ΔP must exceed 400 lb/in² to obtain permeation. In practice, the pressure required is much greater, owing to concentration polarization.

Effect of Concentration Polarization on Osmotic Pressure   Since the osmotic-pressure difference $\Delta\pi$ is associated with the concentration of solutes at the membrane surface, if the boundary layer is not controlled, a high $\Delta\pi$ will result, reducing the permeate flux J for any applied pressure drop ΔP. Thus concentration polarization affects the flux through an increase in the apparent osmotic pressure which must be overcome rather than through the formation of a gel layer or cake on the membrane surface.

Referring back to Eqs. (7) and (8), the polarization modulus is lower for RO membranes than for UF membranes. The membranes have a lower permeability (a higher resistance to flow R_m) and the species retained are smaller with higher diffusivity [see Eq. (27)].

Therefore, in the case of RO, the concentration of retained solutes C_s at the membrane surface seldom exceeds the solubility limit or forms a thixotropic gel. When it does occur, the literature often refers to the phenomenon as "fouling" of the membrane. Fouling has been found to arise from precipitation of iron and manganese oxides or calcium carbonate and calcium sulfate. In addition, organic macromolecules are severe foulants in the processing of food products or municipal sewage. Generally, colloids and particulates are removed with appropriate pretreatment steps such as prefilteration, maintaining iron and manganese oxides in the reduced forms by excluding air or adding reducing agents, adding precipitation inhibitors such as sodium hexametaphosphate, controlling pH to prevent inorganic precipitates, and passing the feed through an activated-carbon column to remove organics. Thus, for the general case (good pretreatment), Eq. (35) can be reduced to

$$J_w = \frac{\Delta P - \Delta \pi}{R_m} \tag{36}$$

since $R_c = 0$.

Unlike UF in the gel-polarized region, Eq. (36) indicates that the flux through an RO membrane is dependent on pressure and the permeability characteristics of the membrane. In addition, the concentration of solute at the membrane surface will affect the flux through an increase in the osmotic-pressure difference.

Referring back to Eq. (8),

$$\frac{C_s}{C_b} = \exp \frac{J_w \delta}{D} \tag{8}$$

it is clear that the polarization modulus can be decreased by a reduction in the boundary-layer thickness δ. Equation (8) has been solved for RO with a variety of conditions in both laminar- and turbulent-flow regimes by Brian and co-workers.[69-71] The solutions are quite complex, in spite of several simplifying assumptions. The water flux decreases down the channel because $\Delta \pi$ constantly increases as water is removed and the boundary layer develops. Solute flux increases (i.e., the rejection goes down) for the same reason. As a result of these complexities, the results do not appear in closed form, at least in the laminar-flow case, and finite-difference methods have been used.

Polarization in Laminar Flow   In laminar flow, the concentration gradient develops over the entire height of the flow channel. As we move down the channel, a large fraction of the water may be removed before the boundary layer is fully developed. Thus we must contend with two different laminar-flow regimes: an "entrance region" where the boundary layer is under development and an "asymptotic region" where the boundary layer is fully developed. In this region, the concentration gradient extends across the entire flow channel and does not change shape as additional water is removed.

Near the entrance region of the channel formed by two parallel plates, an approximate solution has been derived by Dresner[72] and confirmed by experiment

$$\frac{C_s}{C_b} = 1 + \xi + 5[1 - \exp(-\sqrt{\xi/3})] \tag{37}$$

where $\xi = \dfrac{J^3 b L}{6 U_0 D^2}$ \hfill (38)

U_0 = entrance velocity

For downstream in the channel, as the polarization modulus increases, the solution reduces to

$$\frac{C_s}{C_b} = 1 + \frac{J^2 b^2}{6 D^2} \frac{JL}{U_0 b} \tag{39}$$

where $JL/U_0 b$ is the water recovery over the length of the channel (Brian has given the exact solutions to the above.[70,71])

One of the interesting things about Eq. (39) is that for a fixed fractional water recovery, the polarization modulus varies with channel height b and flux J, but not with feed velocity U_0. This curious result is explained by the fact that even though the feed velocity U_0 affects the polarization at each point along the length of the channel, the length of the channel required for a given recovery increases with increasing velocity, resulting in a constant average polarization modulus.

It is obvious from Eqs. (37) to (39) that the maximum polarization modulus is found in the asymptotic region. For this region, hardware design is often predicated on maintaining an "entrance region" condition at all points. This can be done if multiple passes are combined with low water recoveries per pass or by inducing mixing periodically down the length of the channel.

The feasibility of RO in the laminar-flow regime can be estimated from Eqs. (37) to (39), assuming that the polarization modulus cannot exceed 2.0 (requiring a minimum ΔP of 850 lb/in² for sea water). For a 50% water recovery per pass, most of the recovery will occur in the "far-downstream region." Assuming a typical flux of 10 gal/(ft²)(day) [4.7 × 10^{-4} cm³/(cm²)(s) and a solute diffusivity of 1.6×10^{-5} cm²/s (NaCl)], the required channel height may be calculated from Eq. (40):

$$ b = \frac{D}{J}\sqrt{12} = \frac{1.6 \times 10^{-5}}{4.7 \times 10^{-4}}(3.46) = 0.12 \text{ cm, or 46 mils} \tag{40} $$

Polarization in Turbulent Flow The polarization problem is less complex in the turbulent-flow regime because the boundary layer is fully developed in the first few hydraulic radii and there is not a significant entrance region. In this case the solution to Eq. (8) is straightforward. Using the Dittus-Boelter relation developed in the UF case,

$$ K = 0.023 \frac{U^{0.8}D^{0.67}}{d_h^{0.2}\nu^{0.47}} \tag{24} $$

Thus the polarization modulus becomes

$$ \frac{C_S}{C_B} = \exp \frac{J_W d_h^{0.2}\nu^{0.47}}{0.023U^{0.8}D^{0.67}}. \tag{41} $$

Several implications are clear from Eq. (41). The geometry of the feed channel is of little importance, but water flux, feed velocity, and solute diffusivity are all important. This is to be contrasted with the laminar-flow case where the channel height was all-important.

Figure 57[71] shows the thoretical average concentration polarization modulus for tubular membranes with a 50% water recovery and a flux of 10 gal/(ft²)(day). Flow was assumed to be laminar for values of the Reynolds number at the channel *inlet* less than 2100 and turbulent for values at the channel *outlet* greater than 2100. Since half the water is removed through the membrane, there is a range representing a factor of 2 in inlet velocity which separates these two regions; in this transition zone a dashed line is drawn to connect the laminar- and turbulent-flow solutions.

Note that since the water recovery is fixed at 50% in Fig. 57, the polarization modulus is

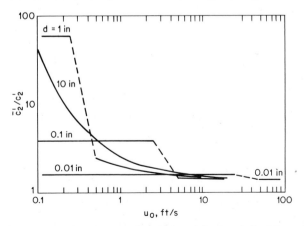

Fig. 57 Theoretical average concentration polarization modules for tubular RO membranes with a 50% water recovery and an average flux of 10 gal/(ft²)(day) as a function of inlet channel velocity.

independent of the inlet channel velocity U_0 for laminar flow but decreases with inlet channel velocity in turbulent flow. Further, polarization decreases as the tube diameter decreases in either laminar or turbulent flow, but the effect is generally more pronounced in the laminar-flow regime.

Laminar-flow operation avoids the high viscous-flow pressure drops required to maintain turbulent flow but requires thin-channel equipment of increased complexity. One exception to this is the hollow-fiber configuration. However, flow inside the lumen of a hollow fiber is generally not feasible at the high pressures required for reverse osmosis.

Effect of Concentration Polarization on Rejection   Regardless of whether the flow of solvent and solute across a membrane occurs by pore flow or by a solution diffusion model (see Appendix D), the solute flux is proportional to the upstream concentration of solute adjacent to the surface of the membrane C_s. Thus for a given bulk stream concentration C_b the rejection will decrease as the polarization modulus increases.

FLUX STABILITY AND DECAY

In many commercial RO and UF plants, the flux remains fairly constant over a number of months and sometimes years. Indeed, there are some plants with which the author is acquainted where the flux has actually increased slightly with time. This flux stability has an important bearing on the operating costs of a given plant. Membrane replacement is usually dictated by the point at which the output of the plant (filtrate) drops to a value which is unacceptable.

Constant flux over a period of months is possible only with an effective fluid-management technique such as crossflow filtration. Concentration polarization must be controlled. In a conventional dead-ended system, the cake or gel layer on a membrane will continue to grow in thickness as filtration proceeds until the flux drops to a negligible value. In the gel-polarization model we saw that the steady-state equilibrium is achieved when the transport of retained species to the membrane by pressure-activated convective transport (in the solvent) is just balanced by the concentration-gradient-activated back diffusion away from the membrane surface. If the fluid-velocity component tangential to the membrane is zero, the magnitude of back diffusion will be almost zero [see Eqs. (15) and (24)] and the flux will be zero as well.

Likewise, in RO, without crossflow fluid management or a similar technique, the polarization modulus will continue to rise until the solute precipitates out at the membrane surface (gel polarization) or until the osmotic-pressure difference $\Delta\pi$ equals the hydraulic pressure difference ΔP, yielding zero flux.

However, even though the theory predicts a steady-state value of the flux for a constant feed-stream composition, in most real-life situations the flux does decline over a long period of time. The rate of this decline will determine the frequency of membrane replacement and the magnitude of the operating costs. (Membrane-replacement costs are often of the same order of magnitude as power costs and amortization of the plant.)

In the case of reverse osmosis, or any membrane process running at high pressure drops across the membrane, there will always be the phenomenon of "membrane compaction" (discussed in Appendix C) to contend with. This is apparently the result of a creep process in which the thin skin of an anisotropic membrane grows in thickness by amalgamation with the finely porous substrate immediately beneath it. As shown in Appendix C, the flux usually decreases logarithmically with time and may be expected to reduce the flux after 1 year to a value between 50 and 85% of the initial value depending on the applied pressure (500 to 1500 lb/in²). Since the membrane resistance is often significant in RO, this can be the controlling factor.

In the case of low-pressure microfiltration, membrane compaction is not a problem, but internal pore fouling is. Since most microfiltration membranes are not anisotropic, particulates can penetrate into the membrane where they become lodged in a pore constriction but cannot be swept away by tangential flow. This was illustrated by the data of Fig. 31 where the flux of an MF membrane dropped below a lower-permeability UF membrane due to internal pore fouling.

In the case of anisotropic membranes, internal pore fouling does not appear to be a significant problem, but surface fouling can be. We have already spoken of fouling of RO membranes due to precipitation of iron oxide, manganese oxide, calcium carbonate, and calcium sulfate. As pointed out earlier, pretreatment can often be utilized to prevent precipitation of these foulants. However, even if they do precipitate, the flux decline can often be minimized if not eliminated by higher recirculation rates (higher tangential velocity) or better hardware design (channel height, membrane configuration, etc.). The real difficulty is associated with species which attach themselves to the membrane

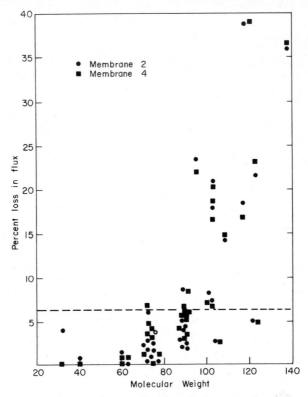

Fig. 58   Relationship between flux loss and solute molecular weight.

irreversibly and are no longer subject to back diffusion from the membrane surface. In this case, the theory behind the gel-polarization model becomes invalid. The fundamental assumption on which the theory is based is that species are freely transported to and away from the membrane. We have already seen how other phenomena such as the tubular-pinch effect may augment the back diffusion and increase the flux above those values predicted from theory. When we speak of irreversible fouling, we are referring to an imposed restriction on back diffusion from the membrane surface. As the gel layer continues to grow in thickness due to irreversible attachment of "fouling species" to the membrane, the flux will continue to drop.

Organic macromolecules, colloids, and various organisms appear to be among the worst offenders in terms of fouling. Slime-producing bacteria may grow on the surface of the membrane, resulting in catastrophic reductions in flux. This is an acute problem in plants which are turned off over the weekend without the addition of biostats or other disinfectants.

Duvel et al.[73] have looked at 38 different organic solutes and their effect on the flux through a tubular cellulose acetate membrane. Figure 58 shows the flux decline as a function of the molecular weight of the organics. Each run utilized a 0.01M test solution. Figure 59 shows the effect of the number of carbon atoms in the solute molecule and the concentration of the organic.

The nature of the membrane may also affect the rate of flux decline. Presumably, this is due to an interaction between the membrane and the foulant due to polarity or charge effects built into the membrane. For example, anionic electrodeposition paints exhibit relatively stable flux values with noncellulosic UF membranes, but a cationic paint suffered irreversible catastrophic flux decline. Switching to a cellulose acetate membrane solved the problem with a very stable flux.

The techniques used to control flux decline involve either minimizing the formation of deposits on the membrane or removing deposits already formed.

Effect of Pretreatment on Fouling

The treatment of sewage effluents by RO is illustrative of the effect of pretreatment on flux decline. Figure 60[74] shows typical results for a series of sewage effluents. The flux-decline rates for secondary sewage effluent are at least a factor of 5 greater than the decline observed due to compaction of the membrane under nonfouling conditions. It is clear that some of the flux decline on raw sewage is due to the deposition of suspended particulates on the membrane, since the removal of these by chemical clarification (primary sewage) is helpful. The digestion of dissolved organic matter to produce secondary sewage is even more beneficial. Removal of other organics by carbon treatment results in still further improvement. Thus it is clear that both dissolved and suspended organic matter contribute to flux decline.

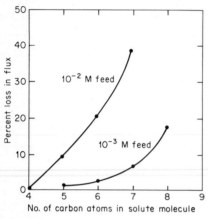

Fig. 59 Relationship between flux loss and number of carbon atoms in solute molecule.

Figure 61[75] shows the flux decline on settled and chemically treated sewage. The greater effectiveness of chemical clarification is pronounced. It will also be noted that a plot on log-log paper of flux vs. time again produces a straight line.

Pretreatment is needed if the foulant(s) cause irreversible fouling or if the cleaning frequency is unacceptable. There are several different approaches that can be used to minimize the problem of membrane fouling:

1. The influent stream can be pretreated to remove the fouling materials before they reach the membrane. This pretreatment may be by precipitation, filtering, activated carbon, etc.

2. The influent stream can be pretreated to *dissolve* the fouling materials. The most common example is the injection of an acid to prevent the formation of calcium carbonate.

3. The influent stream can be pretreated to "suspend" the fouling materials. This pretreatment is exemplified by the injection of a dispersant such as hexametaphosphate.

Deep well waters are generally easiest to treat because their composition is relatively constant and they usually do not contain colloidal solids. Municipally treated waters are more difficult, but by far the most difficult class of water to treat is surface water. It is characterized by seasonal shifts in particulates and colloids. It usually contains very fine clay minerals which not only deposit on the membrane but also act as adsorption and nucleation sites for other foulants. When these other foulants are humic acid-type organics, the most severe fouling occurs.

The need for pretreatment and the type needed can often be indicated by a silting-index test (ASTM F52-65T). A modification of the silting-index test, a so-called plugging-factor test, is used by a large supplier of RO equipment for screening work in the field[76] and involves the following procedure:

The water to be treated is filtered, under a fixed inlet pressure [30 lb/in²(gage)] through a 0.45-μm cellulose acetate membrane, and the *time* required to pass the "initial" 500 (or 100) mL of water is measured. The water is then continuously passed through the same membrane under the same pressure for a period of 15 (or 5) min, and then the *time* required to pass a "final" 500 (or 100) mL at the end of this period is measured. In the case where the water contains almost no foulant matter, a relatively small increase in the filtration times (initial vs. final, after the 15- or 5-min periods) is observed. When a water contains significant foulant, the filtration rates are often dramatically slowed by passage of that water through the filter during the interim period. From these purely temporal data

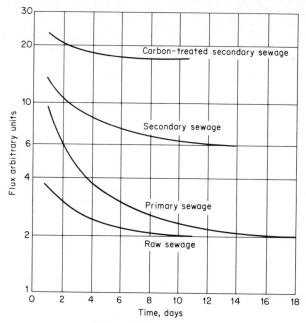

Fig. 60   Flux decline in sewage effluents.

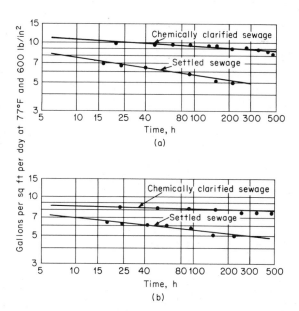

Fig. 61   Flux decline on settled and chemically treated sewage: (*a*) membrane type 300; (*b*) membrane type 500.

then, a "plugging factor" or "percentage plugging" can be calculated and used as an index of water quality at any point in a multiple pretreatment process.

$$\text{Plugging factor} = 100 - 100 \times \frac{t \text{ initial seconds}}{t \text{ seconds after 15 (or 5 min)}} \quad (42)$$

For example, a plugging factor (15 min using a 100-mL sample) of 80 will tend to foul a membrane rapidly, while a factor of 20 is conducive to relatively long periods of effective module operation between cleanings.

Fouling in RO is due to diverse mechanisms and cannot often be simply characterized by fouling index alone.

Table 9 summarizes recommended pretreatment steps for various foulants.

TABLE 9   Recommended Pretreatment Steps for Various Foulants

Foulant	Recommended pretreatment
Fe, Mn, S (solubles oxidized to insolubles)	Chlorination or air oxidation followed by detention and filtration
Small amounts of Fe ($< \frac{1}{2}$ ppm) or Al	Use acid feed pH 5 or below
Fe, Cu, C, or metal oxides	Filtration after coagulant feeding
Insolubles (clay, sand, silica, algae, colloids)	Filtration with or without coagulant
$CaSO_4$, $CaCO_3$ scaling	Use hexametaphosphate
High sulfates	Feed HCl instead of H_2SO_4 for pH correction, utilize weak-acid cation dealkalizer, and chloride anion dealkalizer with zeolite softening
Biological growth	Continuous chlorine feed for CA membranes, periodic formaldehyde disinfection for polyamide membrane

Effect of Pressure on Fouling

As might be expected, high pressure results in a more rapid rate of flux decline (see Fig. 62). This is due to compaction of the gel layer or increased thickness to compensate for the higher flux (back diffusion is controlling).

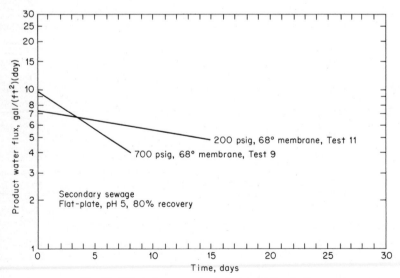

Fig. 62 Effect of pressure on product water flux decline.

Effect of Fluid Shear Rate on Fouling

The beneficial effect of high tangential velocities in thin channels for control of concentration polarization in achieving high flux has already been noted. If these high fluid shear rates are beneficial in minimizing the gel layer (the boundary layer) adjacent to the membrane, it is logical that they should also reduce the irreversible attachment of foulants and arrest flux decay.

Figure 63[78] illustrates the effect of tangential (axial) velocity on flux decay using three

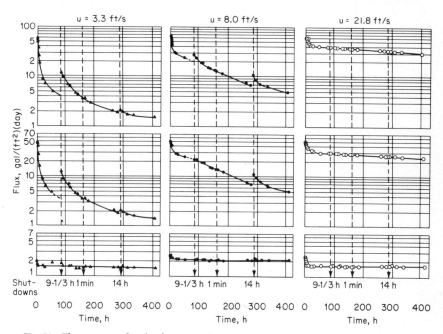

Fig. 63 Flux recovery after shutdowns as a function of axial velocity and initial membrane flux.

different cellulose acetate membranes (having different water-flux values) on primary sewage. It will be noted that on the same membrane, the final flux after 400 h of operation was considerably higher for the higher velocities. In fact, a ninefold increase in axial velocity resulted in a fifteenfold increase in final flux. Further, for the two high-flux membranes (having relatively high initial flux values), the initial flux seemed to have no bearing at all on the final flux. However, for the low-flux membrane, at higher velocities, the final flux was limited by the membrane resistance (not so with lower velocities).

The effect of high fluid shear rates in ultrafiltration is illustrated in the case of cheese whey where fouling by proteins is a severe problem. Scanning electron microscopy has been used[79] to observe various forms of surface fouling materials and their progressive buildup during ultrafiltration. It was suggested that small amounts of relatively large-sized protein structures and microorganisms readily settle on the membrane and provide anchor points or collection points for the more abundant constituents. Some of the latter, especially β-lactoglobulin, have covered the entire membrane with a thick deposit. If the fluid shear rates are high enough, these protein molecules may be kept in suspension.

Effect of Hardware Configuration on Fouling

The high pressure drops and shear forces associated with thin-channel units are advantageous in preventing curd formation within the channel. However, it is also true that a more open tubular system might be less subject to fouling albeit with a somewhat lower flux.

Schatzberg et al.[82] conducted a study on four different types of modules in ultrafiltration of oil-water mixtures. The tubular configuration used was the UCARSEP inorganic membrane on porous carbon tubes (0.25 in ID and 40 in length).

The spiral-wound module was similar to the design shown in Fig. 64. Figure 65 is a schematic drawing of how the unit is assembled. A single folded assembly is termed a "leaf" and is comprised of an envelope of membrane glued on the edges, attached to the axial product-water tube and enclosing a porous material which serves as the product-water channel. Some spiral-wound modules have two or more leaves to cut down on the product-water resistance. The feedwater spacer, which is generally a meshlike material, allows the feedwater to flow axially between the leaves under pressure. Permeate flows through the membrane into the product-water channel where it is conducted spirally to

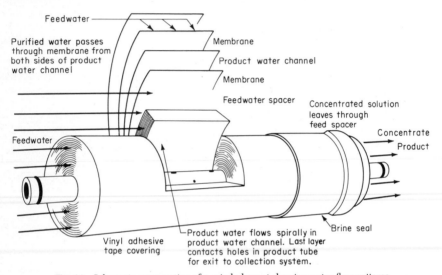

Fig. 64 Schematic cross section of a spiral element showing water flow patterns.

the product-water tube around which the leaves are wound. The whole unit is then placed in a pressure vessel.

The major advantage of a spiral-wound membrane element is its compact, preassembled, economical, ready-to-use shape. It is capable of providing a large number of square feet of membrane surface per cubic foot of module and provides more flux per dollar than the tubular design. On the debit side, spiral elements are not ideally suited for handling feed solutions with high solids contents owing to the thinness of the feed channels and the stagnant areas associated with the cross members of the feedwater spacer (mesh). The author has worked with corrugated feedwater spacers which were much less susceptible to plugging but which cut down on the total membrane area which could be wound in a module.

Tubular systems offer the ability to handle feed solutions with high solids contents. Although they present a relatively small membrane area per unit volume of feed solution, their very structure is hydraulically "clean" and straight-through. They do, however, represent some of the highest cost per unit membrane area of any module design available.

The Schatzberg study used hollow-fiber configurations. In hollow-fiber configuration I, the fiber ID was 0.017 in and the membrane cutoff was 10,000 mol wt, whereas in configuration II, the ID was 0.020 in and the cutoff was 80,000 mol wt. The skin is on the inside of both fibers, and the lumen of the fiber is the feedwater channel. The fibers are bundled together and potted at the ends so that the feedwater may be directed into the fiber lumen. The permeate is then collected on the shell side of the bundle. Hollow fibers represent the least expensive membrane area in terms of dollars per square foot but do not necessarily represent the greatest flux per dollar. This is because of the burst strength of the fiber, which limits the inlet feed pressure and consequently the lumen velocity down the fiber. Further, hollow-fiber units are, in general, more susceptible to plugging than either spiral-wound or tubular designs—particularly at the header inlet. More will be said about this below.

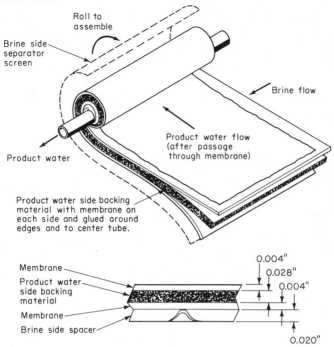

Fig. 65   Spiral-wound module concept, unrolled.

If a flux-decline parameter is calculated similar to the compaction slope m in Eq. (Cl), page 2-92, the results are as follows (between 15 and 45 h):

Spiral wound	-1.13
Tubular	-1.09
Hollow fiber II	-1.06

Undoubtedly, the reason for similar flux decay has to do with the fact that the oil-water mixture did not have suspended solids or constituents other than oil.

In RO, the two most common hardware configurations are spiral-wound (as in Fig. 64) and hollow-fiber. RO hollow-fiber units are almost always run with the feed stream on the outside of the fiber (as in Fig. 17) rather than in the lumen of the fiber because of the high pressures (greater than 400 lb/in²) required. The fibers are generally much smaller in diameter (see Fig. 66) than UF hollow fibers and can withstand the compressive stresses due to the external pressure without collapse.

The Du Pont or Dow hollow-fiber units are similar in general configuration (see Fig. 67). The module design uses the fibers in a half loop with both free ends potted in epoxy resin, much like the tube sheet in a shell-and-tube heat exchanger. The loop end is contained in a pressure vessel. Feedwater is introduced around the outside of the hollow fibers—as mentioned above. Desalted water passes through the fiber walls and flows up the bore to the open-fiber end at the epoxy head. Brine is discharged from the opposite end of the pressure vessel. Because the fluid management of the feed stream through the fiber mass is poor, hollow-fiber units of this type are much more susceptible to plugging than are spiral-wound elements, but the polyamide hollow fiber is less susceptible to attack by cleaning agents than is cellulose acetate.

Table 10 compares the fouling characteristics of two types of hollow-fiber modules with the spiral-wound type.[84] The fouling index FI is similar to the plugging index in Eq. (42). Spiral-wound modules will tolerate feed streams with a fouling index as high as 15, with no pretreatment, whereas hollow-fiber units cannot tolerate feedwaters with an index over 3. In terms of turbidity, the spiral-wound units can tolerate turbidities of 1 JTU whereas the hollow-fiber units must stay under 0.5 JTU.

Since spiral-wound modules have higher flux values [15 to 18 gal/(ft²)(day)] than do hollow-fiber units [1.5 to 2.0 gal/(ft²)(day)], the propensity to fouling of hollow-fiber units is even more dramatic. In the spiral-wound units, every square foot of membrane area is exposed to ten times as much permeating water as for the hollow fibers.

In UF, if there are no large particulates (over 50 μm in diameter) in the feed stream, or if these can be removed by prefiltration, it is often advantageous to go to a high-shear-rate thin-channel design. This is readily demonstrated even in the ultrafiltration of pseudo-plastic or thixotropic colloidal suspensions.

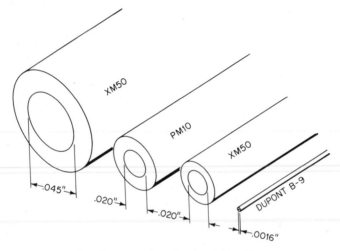

Fig. 66   Dimensions of various hollow fibers.

A disposable linear thin-channel cartridge has been used to obtain very high flux values with low fouling characteristics for feed streams of the type mentioned above. These cartridges employ sixty tubes in a shell-and-tube configuration. The individual tubes (0.5 in in diameter) contain splined cores which direct the process stream into thin channels adjacent to the membrane. Ultrafiltrate then passes through a braided support which gives the tube a high pressure capability with virtually no hydraulic resistance to filtrate flow. These cartridges have been used to concentrate fermentation broths, polymer latices, clay

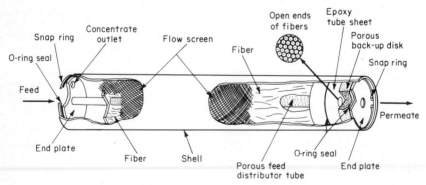

Fig. 67   Permasep permeator.

suspensions, and secondary sewage sludge to 20 to 65% solids. However, if large particulates (particularly fibers) are not removed from the feed stream, they will very rapidly foul the inlet header.

Surprisingly enough, even a "leaf module" design like that shown in Fig. 68 can show severe fouling by fibrous materials which collect along the leading edge of the leaves.

TABLE 10   Comparison of Three Major Types of Reverse-Osmosis Modules

	Tri-acetate hollow fibers	Polyamide hollow fibers	Cellulose acetate spiral-wound
Module sizes and flow, gal/day at 400 lb/in²	5 × 48 in, 4000 gal/day; 10 × 48 in, 20,000 gal/day	4 × 48 in, 4200 gal/day (1); 8 × 48 in, 14,-000 gal/day (1)	4 in × 21 ft (6), 4200 gal/day; 8 in × 21 ft (6) 24,-000 gal/day
Recommended operating pressure, lb/in²	400	400	400
Flux, permeate rate, gal/day/ft²	1.5	2	15–18
Seals, pressure	2	2	12
Recommended max operating temp, °F	86	95	85
Effluent quality (guaranteed % rejection)	90	90	90
pH range	4–7.5	4–11 0.1 > pH 8.0	4–6.5
Chlorine tolerance	0.5–1.0	0.25 > pH 8.0	0.5–1.0
Influent quality (relative—FI*)	FI–<4	FI<3	FI<15
Recommended influent quality	FI–<3	FI<3	FI<3
Permeate back pressure (static), lb/in²	75	75	0
Biological attack resistance	Resistant	Most resistant	Least resistant
Flushing cleaning	Not effective	Not effective	Effective
Module casing	Epoxy-coated steel	FRP	Epoxy-coated steel and FRP
Field membrane replacement	Yes	No (future yes)	Yes

*FI = fouling index.² (1) Initial flow. (6) Six modules per 4200 gal/day.
SOURCE: Ref. 84.

The 1-in-diameter tube design (also shown in Fig. 68) is probably least subject to fouling of all the UF module designs. However, this does not mean it is most cost-effective. Relatively clean feeds can often be processed more economically through high-performance thin-channel equipment or through hollow fibers.

Effect of Membrane Cleaning on Flux Restoration

No pretreatments are needed if foulants can be removed from the membrane by periodic flushing or cleaning. This in turn is dependent on the type of foulant, the membrane configuration, and the membrane's resistance to the cleaning agents. For example, one advantage of the inorganic membranes is that they may be cleaned in hot caustic.

As an example of the various types of cleaning which may be effective, we will take the case of asymmetric UF hollow-fiber systems having fibers similar to those pictured in Fig. 66. Breslau et al.[83] has designated two types of cleaning which can be used with these hollow fibers (where the feed stream runs through the lumen of the fiber): cleaning by back flushing and cleaning by recycling (see Fig. 69).

The back-flushing mode serves to clean the membrane surface by forcing permeate or other fluid back through the fiber, which loosens and lifts off the cake accumulated on the inside wall of the fiber. This operation requires a reservoir to accumulate filtrate. It is important that the back-wash fluid contain no suspended matter which might foul the outer spongelike structure of the fiber.

The "recycling" mode is accomplished by simply closing off the permeate ports. As the feed stream continues to recirculate through the fibers, permeate is continuously produced, resulting in a buildup of pressure in the cartridge shell until it reaches an equilibrium value which is very close to the average of the inlet and outlet pressures of

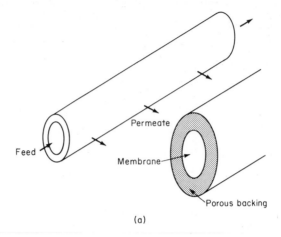

(a)

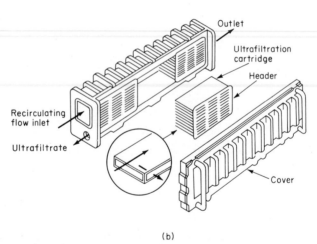

(b)

Fig. 68 Various membrane geometrics: (*a*) open tube (½ to 1 in diameter); (*b*) membrane-coated porous plates (0.125-in spacing).

the feed stream. If the inlet pressure is 25 lb/in² and the exit pressure is 10 lb/in², the cartridge shell should equilibrate at 17.5 lb/in². As shown in Fig. 70, this means that the first half of the fiber length has an inside pressure greater than that in the shell. In the last half of the fiber length, the reverse is true. Thus permeate will flow from the inside of the fiber to the shell side in the first half of the length and then back into the fiber in the last half. This recycle back-flushing action is coupled with the high shear sweep of fluid across the inside wall of the fiber, effectively removing material loosely adhering to the membrane surface. Often this material can be seen extruding out of the fiber ends like spaghetti.

Back-flushing and recycling operations are often more effective when carried out with cleaning agents. For example, these membranes can withstand pH environments from 1.5 to 13. This means that sodium hydroxide can be used.

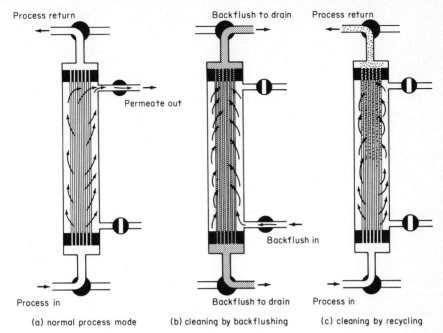

Process return ← | Backflush to drain → | Process return ←

Permeate out →

Backflush in ←

Process in → | Backflush to drain → | Process in →

(a) normal process mode | (b) cleaning by backflushing | (c) cleaning by recycling

Fig. 69 Schematic of hollow-fiber cartridge showing three modes of operation used in systems design.

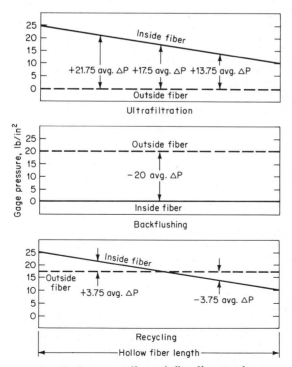

Inside fiber

+21.75 avg. ΔP +17.5 avg. ΔP +13.75 avg. ΔP

Outside fiber

Ultrafiltration

Outside fiber

−20 avg. ΔP

Inside fiber

Backflushing

Gage pressure, lb/in²

Inside fiber

Outside fiber +3.75 avg. ΔP

−3.75 avg. ΔP

Recycling

Hollow fiber length

Fig. 70 Pressure profile in a hollow-fiber cartridge.

Figure 71 shows the flux decline seen on tap water and the restoration in flux accomplished by back flushing with 1% Clorox. The difference in well water and surface water is seen in Fig. 72.

Figure 73 shows flux data before and after fouling with a mixture of 2% cultured buttermilk in skim milk and letting the mixture set overnight at 22°C. Back flushing with water partially restored the flux, but back flushing with 0.2% NaOH resulted in complete flux restoration.

Other examples of chemical regeneration of membranes may be helpful. Notable among these are ultrafiltration of fermented enzyme streams and electrophoretic paint.

For poorly filtered fermented enzyme streams, the organism cells progressively accumulate, resulting in a lowering of flux. The probable cause is that the cells progressively

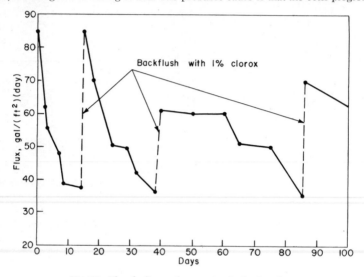

Fig. 71 Flux decline and restoration by backwashing.

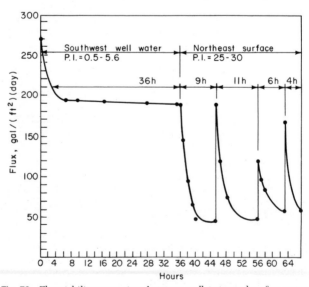

Fig. 72 Flux stability comparison between well water and surface water.

die leaving a tenacious, slimy deposit. It has been found[85] that batch concentrations of the enzyme stream over 8 h followed by 1 to 2 h cleaning with alkali completely restores the flux.

In the processing of electrophoretic paint, the accidental introduction of tap water led to precipitation of calcium and magnesium compounds on the membrane with sudden, catastrophic flux decline. In one case, the flux had been stable at 20 gal/(ft²)(day) for many months and suddenly dropped to 1 gal/(ft²)(day). Alkali regeneration with chelating agents rapidly restored the flux to 20 gal/(ft²)(day).

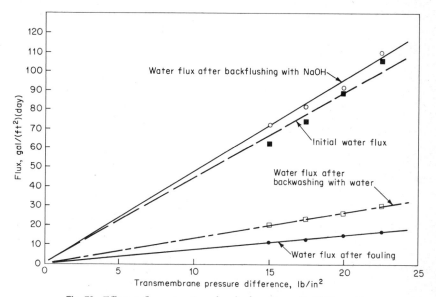

Fig. 73   Effect on flux restoration when backwashing with 0.290 caustic.

ECONOMICS

As seen above, fouling of the membrane can be a significant factor in determining membrane process economics. In calculating the operating cost, membrane replacement, utilities (pumping power), fixed operating costs (including pretreatment amortization, property taxes, and insurance), labor (including maintenance), and overhead must all be considered. Operating-cost estimates are helpful to the user in making a decision among various membrane modules and are often the determining criterion for plant configuration (single-pass or recirculating mode, and the number of recycle stages). In addition, the optimum operating conditions may be specified from this information.

Figure 74 presents the final results from an optimization study for one of the simplest UF plant configurations—a single recycle stage with 80% recovery. The cartridge flow rate (recirculation rate in crossflow) is selected as the single most important operating parameter; it determines the flux through the membrane and consequently the membrane area required to handle a given process stream. As we have seen, an increase in recirculation rate (fluid velocity across the membrane) increases the flux and reduces the membrane-area requirement, thereby reducing the membrane-replacement costs (assuming an annual replacement). In addition, the size of the plant will be reduced, thereby decreasing amortization, labor, and overhead. On the other hand, increased recirculation rates will increase the utility costs due to increased pumping power.

To construct Fig. 74, pilot-plant data are needed on the variation of membrane flux with recirculation rate at the required concentrations as well as life data on the membrane. Often these may be determined from a test of short duration by extrapolating the flux-decay curve on log-log paper.

The minimum in the total operating-cost curve is the optimum design point for operation. In the case of Fig. 74, this amounts to 36 gal/min per cartridge. It will be noticed, however, that small variations in recirculation rate around this point produce little change in the total operating cost.

As an illustration of how operating costs may be useful in selection of the most appropriate module, consider the comparison of operating costs shown in Figs. 75 and 76. These figures are based on calculations made in 1972, and are now out of date. Nevertheless, they illustrate the tradeoffs which should be considered. For simplicity, only the major components of the total operating cost (membrane replacement, amortization, and power cost) are considered.

Pumping and membrane costs may be considerably reduced by utilizing hollow-fiber units. The parasite drag due to the core in a tubular thin-channel system increases the frictional pressure drop. Further, the membrane area is much less expensive in the hollow-fiber case (one-fifth that of the linear thin-channel system) because of inexpensive

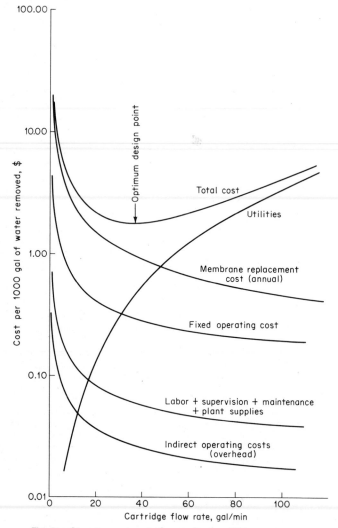

Fig. 74   Operating costs—single recycle stage, 80% recovery.

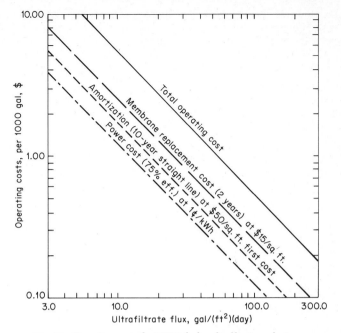

Fig. 75 Operating costs for LTC tubular ultrafiltration plants.

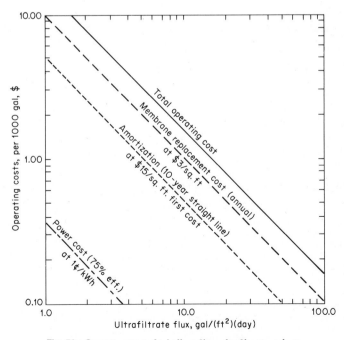

Fig. 76 Operating costs for hollow-fiber ultrafiltration plants.

fabrication. However, the hollow-fiber burst pressure limits the inlet pressure [typically 30 lb/in²(gage) limit], meaning that for a given fiber length and fluid viscosity, the velocity through the fiber is also limited. Thus hollow-fiber cartridges have ultrafiltration rates one-half to one-fifth those exhibited by the LTC tubular cartridges; the latter are also less prone to fouling, resulting in less frequent replacement (estimated 2 years instead of annual replacement).

When all these factors are considered, it will be seen that both systems give equivalent operating costs when the flux of the tubular cartridge system is approximately three times greater than that for the hollow fibers. This is often the case, and under these circumstances the choice between hollow fibers and tubular cartridges is often made on the basis of other criteria—such as whether pretreatment is required.

Figure 76 shows that for hollow-fiber systems, power costs are insignificant compared with membrane-replacement costs. This represents an unoptimum situation, and reflects the necessity for improving fiber burst strength and life. If these goals are achieved, membrane-replacement costs should drop considerably by virtue of extended life or higher operating pressures. The latter will permit a higher membrane flux at the expense of some additional power costs. The net effect will be to reduce the total operating cost, permitting a more favorable comparison with tubular cartridge systems.

Curves similar to those presented in Fig. 74 have been prepared for RO by the various manufacturers. Lacey[89] presents plots of RO operating costs as a function of the permeate-to-brine flow rate. As expected, total operating costs are not as strong a function of the brine flow rate across the membrane as in UF. The lower costs for the system with highly permeable membranes are attributable to both the lower system pressure (720 lb/in² compared with 960 lb/in²) and the lower membrane area required.

APPLICATIONS

A several-volume encyclopedia could be written on the numerous applications for MF, UF, and RO. In this treatment, we will confine ourselves to industrial applications of broad usage or interest. In some cases, combinations of MF, UF, or RO have been used effectively. Nevertheless, we will classify them with respect to each membrane process.

Applications for Reverse Osmosis

Electronics Water   The applications for RO equipment are extremely varied. Most of the systems now in use produce water which is fed to deionizers with the ultimate destination a semiconductor manufacturing line. Semiconductor fabrication requires ultrapure water of high resistivity (18 MΩ · cm) and low levels of particulate. Ionic impurities can alter the current-carrying properties of a critical semiconductor surface in a variety of ways. Ionic impurities may diffuse into the surface of an integrated circuit, rendering nonconductive channels conductive.

Historically, distillation was first used to remove the inorganic salts; but even triple-distilled water (in quartz) can produce only 3 MΩ · cm resistivity water. Today the combination of reverse osmosis (RO) with ion exchange (IX) is capable of producing 18 MΩ · cm resistivity.

Ion-exchange (IX) systems can treat moderate (550 ppm) to low TDS (total dissolved solids, expressed as ppm CaCO₃) water economically. For large volumes, a succession of cation- and anion-exchange columns are easiest to regenerate and most economical. However, since ion-exchange reactions are reversible, the removal of dissolved ions is less than total. Successful production of 18 MΩ · cm water usually requires a mixed bed where breakthrough can be eliminated. Unfortunately, these mixed-bed exchanges are not easily regenerated and are more costly.

Reverse-osmosis (RO) systems can economically treat high TDS (3000 ppm) to moderate TDS (500 ppm) water. Thus, if RO precedes IX, the incoming TDS can be reduced by 90 to 95% before being treated by IX. This reduces the work load on the IX columns drastically, but by itself RO cannot produce 18 MΩ·cm resistivity economically. RO is, however, very effective in removing many of the organics found in natural water supplies. Water-soluble organic compounds with molecular weights greater than 300 are almost totally removed and compounds between 100 and 300 mol wt are generally 90% removed. It is extremely beneficial to remove these organic compounds before they enter the IX columns, since the resins are easily fouled by high-molecular-weight compounds like humic acid.

It is clear from the above that the combination of RO and IX is much more effective than either one alone. Quinn et al.[90] have evaluated and compared IX and RO systems on both a technical and economic basis to determine the optimum combination of the two. Figure 77 graphically compares the economics of the two individual processes with two hybrid systems (RO + IX). Total cost is plotted against influent TDS levels for a 1 million gal/day product plant of each system. In these data, a 7-year operating period is assumed with the plant amortized over 3 years at 7.5% interest. It will be noted that RO operating costs (per gallon) are insensitive to changes in the influent TDS; this is a good assumption for TDS values below 1500 ppm since the osmotic pressure is not significant. On the other hand, IX operating costs increase proportionately with increases in the TDS. Thus, at low

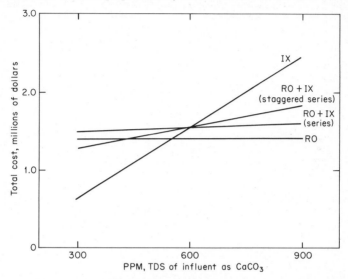

Fig. 77   System cost comparison.

influent TDS levels, IX operating costs are lower than those for RO. The crossover point is about 560 ppm TDS for a 7-year operating period and will shift to higher TDS values for shorter operating periods (710 ppm for a 3-year period). This is because of the higher capital costs associated with RO plants compared with those for IX plants.

The two hybrid systems are shown schematically in Figs. 78 and 79. In both cases the RO permeators precede the IX resin beds. The combination of IX and RO in series costs approximately 10% more than RO alone for TDS values below 1000 ppm. The series hybrid system costs more than IX alone at low influent TDS values but is more economical at high TDS values. Even at low TDS values, the improved quality of the effluent water may justify the additional costs. The effluent is thirty times more pure than that from RO alone, and ten times more pure than that from IX alone. For example, with a 600 ppm TDS influent, the RO effluent is 60 ppm TDS and the IX effluent is 18 ppm TDS, but the RO + IX series configuration produces effluent on only 1.8 ppm. In addition, the RO + IX series incorporates the advantage of organic removal which is not only beneficial in final product-water quality but may extend the life of the strong-base anion-exchange resin by eliminating fouling.

The alternate combination shown in Fig. 79 (staggered series) blends the effluent from the RO unit with the influent to produce a TDS level equal to approximately 50% of that of the influent. The primary advantage of the staggered system is its flexibility. If a given process has periods of large demand during which the water quality is not critical, the RO and IX systems may be operated in parallel.

Boiler Feedwater and Cooling-Tower Blowdown   RO is also being used extensively to remove organics, salts, and silica ahead of deionizers in boiler feed systems. Its first application for boiler feedwater was reported in 1971 by H. Rowland of the Burbank, Calif., Public Service Department. The Burbank plant was relatively small (10,000 gal/day) and lacked operating experience (e.g., data on membrane life). Nevertheless, they reported savings of over 33% as a result of a five- to tenfold increase in DI water between regenerations, decreased regenerative chemical cost (by 90 to 95%), extended life of the ion-exchange resins, improved water quality, and lower manpower and maintenance requirements. These initial impressions have been proved to be correct.

Butterworth and Bass[91] cite economic data from a central electric generating station located in the southwest. Condenser cooling water was supplied from lake water which, owing to evaporation losses and scarce rainfall, contained increasing TDS. As a result, the ion-exchange demineralizer output decreased dangerously. Regeneration was required every 60,000 gal instead of every 175,000 gal (design basis). Installation of RO upstream of the IX system immediately increased throughput above 1,000,000 gal and the demineralizer effluent decreased in conductivity from 50 to 75 to less than 6 μS/cm. The mixed bed showed a corresponding improvement.

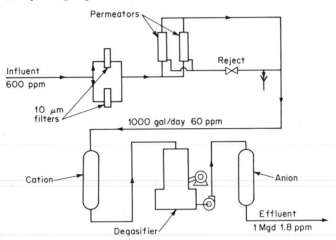

Fig. 78   Hybrid system series.

Butterworth[91] reports that the chemical operating cost of the IX unit (for regeneration) dropped from $3.30 per thousand gallons down to $0.25 per thousand gallons. Based on regenerant chemical-cost savings of $3 per thousand gallons, the utility realized on equipment payback for the RO within the first year of operation.

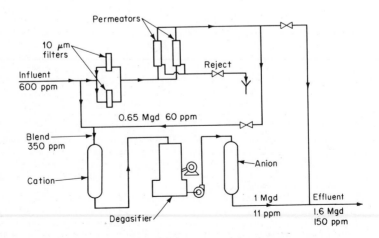

Fig. 79   Hybrid system—staggered series.

In addition to these cost savings,

1. The large increase in demineralizer output permitted the shutdown of some evaporators.

2. Chemical restoration of anion resins (formerly an annual project) has not been required since because of elimination of organic fouling.

3. The cleaning of the internal collectors of the cation unit to remove precipitated calcium sulfate every third or fourth regeneration is no longer required.

Not unrelated to the boiler-feedwater application is the use of RO on recirculating cooling-tower blowdown. During recirculation, the concentration of some dissolved solids such as $CaSO_4$ reaches supersaturation levels and precipitates as scale deposits. To minimize scale buildup, some fraction of the cooling water is continuously discharged as blowdown and fresh water is added to make up the loss. The waste discharged as blowdown is high in TDS and contains 5 to 20 ppm of chromate, phosphate, and chlorinated hydrocarbons used as corrosion inhibitors and algicides. Present RO membranes are capable of demineralizing the cooling-tower blowdown for reuse but are inadequate for removing chromates.

Potable Water RO was initially looked upon as a process for the demineralization of sea water and brackish water to produce fresh water for municipal and industrial purposes. High desalting costs associated with the demineralization of sea water have so far prevented large-scale application. Distillation still has the competitive edge for large plants. However, the lower TDS of brackish water has permitted broad use of RO, particularly for smaller plants.

Recent surveys have indicated that a large number of communities are suffering rapid deterioration in the quality of their municipal water supplies. A decline in well-water quality through overpumping and resorting to deeper wells has become a major public concern. The 1962 Public Health Service Drinking Water Standards recommended that a TDS level of 500 ppm not be exceeded. Groundwater supplies of high salinity are prevalent throughout the southeastern United States, the Middle West, and many coastal areas. For example, the state of Florida is virtually surrounded by sea water, with an abundance of brackish groundwater supplies. As of this writing, many of the largest RO plants for potable water in the United States are located in Florida.

The two leading membrane processes used for desalting brackish water are RO and ED (electrodialysis). Table 11[92] shows operating results for six plants in Florida. Note that the salt passage from the ED plants is much higher than that from RO. It is apparent that RO treated waters can meet drinking-water standards of less than 500 ppm TDS even with a feedwater TDS of 7000 ppm. The ED plants were not normally successful in producing product water of less than 500 ppm TDS.

Table 12 compares the cost of the two processes. It will be noted that the RO operating costs tend to be lower but are comparable with ED costs. In data of this type, it must be recognized that amortization is a fixed annual cost which remains the same whether the plant is operating 5 or 75% of the time. Therefore, a plant operating at a low load factor is penalized in terms of the amortization cost allocation per 1000 gal of product water.

Both ED and RO operating costs are less than $1 per 1000 gal, which cannot be matched by processes such as distillation, at least for plants of comparable size. A better comparison between ED and RO can be made by calculating the cost of desalting per 100 ppm of TDS removed per 1000 gal of product water. The last column in Table 12 shows that the unit costs based upon TDS removal are considerably lower for RO plants than for ED plants

TABLE 11 Effect of Treatment on Water Quality

City	Type	Typical feed TDS	Typical product TDS	Salt passage, %	Rejection, %
Buckeye	ED	1782	568	32	68
Siesta Key	ED	1319	739	56	44
Gillette	ED	2600	720	28	72
Greenfield	RO	2250	142	6	94
Ocean Reef	RO	6912	421	6	94
Rotonda	RO	6332	380	6	94

SOURCE: Ref. 92.

(Siesta Key may not be representative because of membrane-fouling problems).

As wells become more heavily utilized and feedwater quality decreases, RO plants are better able to produce a good-quality product water below the PHS drinking-water standard of 500 ppm TDS. There is less "margin of safety" for ED plants to meet these standards.

Waste Water   *Municipal Wastes.*   Any water-resource management technique designed with water reuse as a factor must take into consideration treatment of municipal wastes for reuse. Conventional waste-treatment methods are inadequate since they are designed primarily for removal of biological and biodegradable organic compounds. They do not remove inorganic salts, phosphorus, or nitrogen compounds, the concentrations of which will increase with reuse. Therefore, any municipal water-reuse concept must include a demineralization step, particularly if the "zero population by 1985" standards are to be met.

Pilot-plant tests on municipal waste processing by RO have been conducted at Pomona, Calif., for the last 6 years under the sponsorship of the Environmental Protection Agency. The tests were designed to determine how effectively RO would operate on treated

TABLE 12   Cost of Desalination by ED and RO

City	Type	Capacity, million gal/day	Plant capital cost	Operating and amortization cost		¢/100 mg/L removed/ 1000 gal
				¢/1000 gal	% load factor	
Buckeye	ED	0.65	$283,000	72.6	30	6.0
Siesta Key	ED	1.83	605,000	92.1	62	15.9
Gillette	ED	1.0	465,500	94.0	26	5.0
Greenfield	RO	0.15	94,346	86.2	48	4.1
Ocean Reef	RO	0.63	300,000	42.5	73	1.4
Rotonda	RO	0.5	257,000	65.1	34	1.5

SOURCE: Ref. 92.

municipal waters. Fouling of the membrane along with methods for prevention were studied and evaluated.

Results of these tests[93] were encouraging; the permeate of the RO unit had:

1. Less than 7% of the dissolved solids originally present in the feed.
2. No detectable turbidity, color, or suspended solids.
3. Less than 1% of the feed phosphate content.
4. Less than 10% of the feed ammonia nitrogen content.
5. Less than 25% of the feed nitrate content.
6. Less than 7% of the feed chemical oxygen demand (COD) content when operating on secondary effluent.
7. Less than 35% of the feed COD when operating on primary effluent. However, tests indicated that the remaining COD could be lowered to approximately 10% of the feed by aeration.

We have already seen that one of the primary problems in RO processing of municipal waters is membrane fouling (see Figs. 60 to 63). In the Pomona tests, average product-water fluxes in the range of 8 to 10 gal/(ft²)(day) were maintained by regular chemical cleaning. Three cleaning solutions in which the active constituent was either the enzyme detergent BIZ®, ethylene diamine tetraacetic acid (EDTA), or sodium perborate performed well in the field tests. Test results indicated that enzymes per se were not the specific active materials in the detergent-based compound. To prevent membrane damage due to exposure to high pH, all cleaning solutions were adjusted to a pH less than 8.0. Periodic depressurization also aided in maintaining the product-water flux.

After the Pomona tests, RO plants have been installed in Escondido (150,000 gal/day) and Kirkwood Meadows (a skiing and summer resort). More recently (1977), California's Orange County Water District has started up a mammoth 5 million gal/day RO installation on municipal wastewater in Fountain Valley. The unit is operating on tertiary wastewater at 85% recovery with 95% removal of impurities.

Metal-Plating Wastes.   After metal plating, "drag-out" of plating solution from the plating tank necessitates rinsing, which is done most efficiently in a countercurrent rinse line (see Fig. 80). The concentration of metal in the plating tank may vary from 0.5 to 5.0%, whereas the concentration in the waste stream usually varies from 25 to 1000 mg/L. Many plating operations have a rinse stream of about 25,000 gal/day. The metals most commonly encountered are chromium (hexavalent), copper, cadmium zinc, tin, and nickel. Many of these are plated out of a cyanide solution, which is often encountered in the plating wastes. Traditionally, the plating wastes have been treated chemically—to reduce C_r^{6+} to C_r^{3+}, to precipitate hydrated oxides, or chlorination to convert the cyanide to cyanate or nitrogen and carbon dioxide.

RO is especially suited to operate on the rinse water. The purified permeate is recycled back to the rinse line and the concentrated reclaimed salts are recycled to the plating tank (see Fig. 80). In cases where the concentrated salts cannot be recycled to the plating tank, there may be other savings associated with the reduced amount of water which must be discarded.

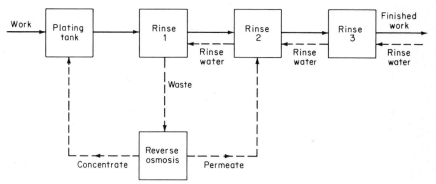

Fig. 80 Typical plating line with RO to reclaim salts.

Okey[95] has presented values of recovered products and payout times (see Table 13). His data are somewhat conservative, being based on RO operating costs of $2 per 1000 gal. Nevertheless, when the metal values exceed $3 per lb, the installation of RO appears to be economically attractive.

The most severe limitation to be overcome in RO treatment of plating wastes is the pH resistance of the membrane, since the pH of many rinse waters will accelerate hydrolysis of the cellulose acetate membrane. For example, in zinc cyanide plating wastes, the pH cannot be lowered below 9 because of precipitation. Cellulose acetate membranes are most stable to hydrolysis in the pH range of 3 to 8.

TABLE 13 Economic Relationships in Treating Metal-Plating Wastes

Basic metal cost, $/lb	Value of metal, $/day	Value of water	Total recovered value $/day	Total recovered value $/year	Payout time, 8% per year
1	25	$7.50	32.50	11,900	
2	50	7.50	57.50	21,000	
3	75	7.50	82.50	30,800	3.8
4	100	7.50	107.50	39,300	1.9
5	125	7.50	132.50	48,400	1.2
6	150	7.50	157.50	59,500	1

System flow: 30,000 gal/day
Metal concentration: 100 mg/L
Cost of water: $0.25/1000 gal
Initial cost of membrane treatment plant: $30,000 [flux 10–20 gal/(day)(ft²)]
Metal lost per day: 25 lb
Operating cost: $2.00/1000 gal (annual = $21,900)
Net dollars applied to capital investment = total recovered $/year – operational cost

$$\text{Payout time} = \frac{\text{capital cost}}{\text{annual net savings adjusted for interest requirement}}$$

Other Industrial Wastes. Considerable attention has been focused on the use of RO to alleviate various pollution problems associated with industrial wastes. Where possible, the recovery of valuable by-products from the waste stream has been sought to offset the additional processing costs.

There are a number of water-reuse and by-product-recovery possibilities in the textile industry. Dyeing wastes are rejected well by RO membranes in general. Desizing wastes are also of interest. Dynamic membranes have been investigated for this use, with some success. Brandon and Porter[98] investigated four commercial membrane module configurations for this application and concluded that two of the commercial units would process 2 million gal/day for a capital cost of $1.65 million and a net operating saving of 43 cents per 1000 gal (including credit for water recycle, chemicals, and energy).

Applications for Microfiltration

Analytical   As we have seen, World War II brought microfiltration membranes to prominence. In Germany, in cities that had suffered heavy bomb damage, Dr. G. Mueller and others applied membrane-filter techniques to the bacteriological examination of water. Microfiltration membranes of about 0.5 μm pore size were used for *E. coli* determination in bomb-damaged water-supply systems.

In 1957, the U.S. Public Health Service officially accepted the use of the membrane-filter technique in the bacteriological examination of water. In this technique, a known sample volume of the water to be tested is passed through a membrane filter that retains organisms of a given size at or near its surface. The membrane pore structure is such that nutrients can freely diffuse upward through the membrane, from a nutrient medium, in order to supply nutrition for the retained bacterial cells. As a result, the organisms grow and form colonies on the surface of the membrane, which may be counted with the naked eye (or low magnification), providing an accurate measurement of the number of organisms present in the original sample. For example, by varying the incubating temperature and the type of medium used, the technique can be used to test water for the presence of coliform, enterococci, or fecal coliform. A total aerobic bacterial count can also be taken. Microbiological analysis of soft drinks, wine, pharmaceuticals, and other liquids represents a big market for microfiltration membranes.

In addition to microbiological analysis, there are numerous analytical techniques using MF membranes which are routine in clinical and medical research laboratories. For example, cytology, the study of cells from the body, has been greatly facilitated by the simple technique of filtering body fluids through the membrane and concentrating the exfoliative cells suspended in the fluid onto the surface of the filter.[99] The cells can then be stained and viewed through a microscope (some MF membranes like the Nuclepore track-etch membranes are nonstaining). Cancerous cells can then be readily identified.

Further, the use of MF membranes in the detection and analysis of particulate contamination has increased in importance. Analysis and control of micrometer and submicrometer particulates in such diverse fluids as aviation fuels, intravenous solutions, and aerosols has become mandatory. Again, the membrane-filter technique makes possible the collection of the particulates in a large volume of fluid on a small planar area for observation under a microscope, and for chemical and gravimetric analysis. Automated particle-analysis systems have been introduced to lessen the burden of manual counting through a microscope.

The list of analytical laboratory applications for MF is almost endless; however, our primary concern in this treatment is for large-scale industrial applications for MF membranes.

Electronics Water   We have already seen the application of RO producing ultrapure water of high resistivity to be used in the electronics industry in the manufacture of semiconductors. Contaminated high-purity water is one of the major threats to production yields in microelectronics processing. In integrated-circuit (IC) manufacturing, nearly every operation ends with a deionized water rinse. Impurities in this water can alter the current-carrying properties of the finished IC through insulation breakdown, internal arcing, and shorts. The greatest concern to electronics processors has always been ionic impurities which may diffuse into the surface of an integrated circuit, making nonconductive channels conductive. Thus the use of RO and/or ion exchange has been mandatory in producing the required 18 M$\Omega \cdot$ cm resistivity water.

In the last few years, electronics processors have also become concerned about other contaminants not reflected in the usual resistivity measurements. For example, particulate contamination in a photoresist coating gives rise to impurities in the finished solid-state device. The particles block and scatter light during exposure and image formation, resulting in a pattern defect and a flawed circuit. Conductor widths and spacing are often in the micrometer range; this means that 0.5-μm particles can affect production yields. In the case of IC, particles cause pinholes in the various thin layers, leading to open or shorted circuits. Similarly, in cathode-ray-tube manufacture, particulate contamination causes pinholes in the phosphors and sometimes degrades their response characteristics. Removal of these particulates down to 0.5 μm is essential to high production yields, and indeed some electronics manufacturing plants are talking about removal down to 0.1 μm. Only a membrane filter can provide this removal reliably. It has been proved over and over that MF filtration can easily pay for itself and achieve considerable cost savings if the reject rate is reduced by even 1%. In some cases, 75% of the rejects can be attributed to particulate contamination.

Ironically, the ion-exchange columns and carbon beds which are used to eliminate dissolved contaminants contribute significant particulates to the product water (e.g., bits of carbon, DI resin fines). In addition, ion-exchange columns are notorious breeding grounds for microorganisms. Although the RO unit preceding the IX columns should retain many of the microorganisms in the incoming water, the retention is not absolute— as discussed earlier. The microorganisms which leak through have all the disagreeable attributes of a particle plus the ability to multiply. Under ideal conditions, one can become a million in a matter of hours. These organisms thrive in the rich accumulation of organics in the carbon beds and ion-exchange resins. Unless checked, they can overrun a distribution system, clog filters, and foul ion-exchange resins. The organisms tend to clump and bridge together, making them even more troublesome. Moreover, when alive, the bacteria often give off small quantities of metabolic by-products which are in essence "dissolved organics" which tend to form films upon evaporation of the water. Further, if the bacteria are destroyed, their cell walls rupture and spew out organic substances which adversely affect adhesion of the various layers of an IC or of the phosphors in a cathode-ray tube and can lead to uneven thickness of plated or vacuum-deposited layers.

Frequent regeneration of the IX columns with strong acids and alkalies will kill many bacteria. Ironically, one of the benefits of RO is that it reduces the required frequency of regeneration. The problem is particularly acute in mixed-bed DI columns where the intervals between regeneration cycles are longest.

The use of MF membranes as the final process step, after the ion-exchange columns in Figs. 78 and 79, can remove microorganisms and particulates down to the required size. MF membranes rated at 0.4 μm remove most waterborne bacteria, but species of Pseudomonas, common to many water systems, require a 0.2-μm rating. This will also remove the smaller colloidal particles of iron and silica.

It is often important to install the MF membranes as close as possible to the point of use. Transit of the water through the distribution piping with exposure to valves and storage tanks is likely to pick up additional particles and microorganisms. Therefore, installation downstream of all valves with a minimum run of pipe to the use point is prudent.

The user must also be cautious concerning the nature of the membrane utilized. Some media tend to shed particulates downstream of the membrane or leach soluble materials into the product water. The latter is sometimes reflected in a prolonged "conductivity rinse-down." That is, when the membrane cartridge is first put on stream, the effluent from the cartridge may be considerably lower than the 18 M$\Omega \cdot$ cm resistivity water coming into the cartridge housing. Depending on the leachables in the cartridge, the resistivity may rise within a few minutes or over a period of hours.

Sterile Fluids *Pharmaceutical.* A great many of the drugs and solutions used in hospitals must be sterile and relatively free of particulates—especially if the product is to be injected into the bloodstream. Contamination is almost unavoidable when a drug is mixed, reconstituted, transferred, or otherwise exposed to the environment of a hospital pharmacy or of a pharmaceutical manufacturing plant.

If a membrane filter (MF) is autoclaved in its housing to kill all bacterial downstream of the membrane, a sterile product can be produced. However, one must be careful to ensure that pressure differentials are not allowed to exceed 20 to 30 lb/in^2 or to fluctuate in order to prevent "pinholing" of the media. The bubble-point test (see Appendix A) can often be used to check the membrane integrity nondestructively.

Intravenous solutions, tissue-culture media, vitamin solutions, bacteriological-broth media, ophthalmic solutions, antibiotics, vaccines, vegetable or mineral oil, serum, plasma, phage and virus suspensions, albumin, and allergins are commonly sterilized by MF membrane filtration.

MF membranes are also used in the sterilization of gases for fermentation processes in the pharmaceutical industry. Before the introduction of MF membranes, microorganisms were removed in conventional depth filters comprised of asbestos, acrylically bonded glass fibers, or a combination of the two. However, because these media were not absolute, contamination often occurred. MF has provided reliable sterilization at a reasonable cost.

Food and Beverage. Microfiltration has often been used in the biological stabilization of beer, wine, and soft drinks. Yeasts, molds, and other spoilage organisms can be effectively removed with pore sizes of 0.6 μm and below. Product stabilization by membrane filtration eliminates the need for pasteurization or chemical sterilization, and offers the additional advantage of high product clarity. For a time, brewers were enamored with membrane sterilization to produce "draft beer." Unlike pasteurization, membrane filtration does not remove the lighter esters that give draft beer its distinctive flavor.

Lamoureux and Peterson[100] concluded that membrane filtration was more economical than either pasteurization or chemical sterilization.

Applications for Ultrafiltration

As in microfiltration (MF), there are numerous laboratory (small-scale) applications for UF. Many of these are analytical, but most attempt to concentrate, dewater, purify, desalt, or fractionate various components. Macromolecules such as proteins, carbohydrates, or other polymers, along with colloids of various types, may be processed in this way with UF.

On an industrial scale, many of the applications, although technically feasible, have not proved economically attractive and have turned out to be little more than a pipe dream. There are, however, at least five applications which have proved to be economically attractive and useful:

1. Electrocoat paint recovery
2. Latex recovery
3. Polyvinyl alcohol recovery
4. Oil-water separations
5. Miscellaneous pharmaceutical and biological separations

Electrocoat Paint Recovery   Perhaps the first truly large industrial market for UF was found in the electrocoat paint industry.[101] Electrophoretic painting was introduced in the early sixties and has been rapidly accepted as a means of priming mass-produced metal parts such as automotive bodies and parts, appliances, and office furniture. Its prime advantage lies in its high throwing power, guaranteeing a coating in areas not normally accessible, resulting in a superior corrosion resistance to that provided by direct dip coating.

The paint is based on an aqueous solution of a resin, solubilized in ionic form by an amine or other alkali such as potassium hydroxide, with pigment and organic-solvent additives. During electrodeposition the resin, together with associated pigment, is deposited on the substrate (made the anode in the process), and alkali is liberated at the cathode. After removal from the tank, undeposited paint is washed off in a spray station, and the part is then cured in a stoving oven during which the solvent aids secondary flow of the film, partly eliminating irregularities caused by the coating taking the contours of the substrate during deposition.

One of the prime disadvantages of the process has been the drag-out loss of undeposited paint, normally amounting to more than 50%. The removal of this in a water-spray station produces large volumes of effluent; and treatment processes such as chemical flocculation and settling are expensive to install and operate, and are not always reliable. Problems relating to local authority acceptance of final effluent are not uncommon. Since drag-out loss is higher at high paint concentration, there has been a tendency to reduce, at the expense of throwing power, the applied paint concentration to minimize drag-out and effluent problems.

Being an electrolytic-deposition process, electropainting is adversely affected by ionic impurities. Careful control is necessary but contaminants can, and do, arise from a variety of sources including carryover of materials such as chromic acid and zinc phosphate from pretreatment stages, the carbonation of the alkali by carbon dioxide in the atmosphere, electrolytic decomposition reactions of the resin, and the accumulation of the alkali liberated at the cathode. Quality-control problems are variable, depending on the nature of the paint and rate at which it is turned over. The electrodialytic control method, frequently called the dialysis control method, can be used to liberate the base in isolated cells so that this particular problem can be successfully controlled.

Another related problem is that of the water balance or content of the paint in the dip tank. As the paint is used up, it is brought up to strength by adding paint concentrate, which adds additional water to the system. Water can also be added to the system through the introduction of wet substrates, but normally steps are taken to avoid this. Although evaporation helps to control the water content, in some installations it is necessary to discharge, to waste, some diluted paint and reestablish the concentration by adding paint concentrate.

All these problems in one way or another involve the paint being associated or contaminated with an excess of water or low-molecular-weight impurities, making it very amenable to an ultrafiltration solution.

An ideal electropainting system would keep the dip tank in constant chemical balance and would not discharge any pigment or resin material—avoiding paint loss, as well as pollution of the plant effluent. While flocculation and flotation methods may be used to remove paint residue from plant effluent, this renders the paint unusable.

Several industries—primarily in the automotive and appliance fields—have found membrane ultrafilters economically effective for removal of excess water (and of impurities such as phosphates) from the electrocoating primer bath and recovery of paint from the rinse water. Hundreds of such systems are presently in use.

Figure 81 illustrates loop arrangements in which the water from the postdip rinse (which carries excess paint from the rinsed metal part) is allowed to return to the main tank. The tank is continually dewatered by ultrafiltration, while simultaneously purging the tank of excess salts. A small fraction of the paint-free ultrafiltrate stream is discarded to maintain constant chemical balance while most of it is recycled as rinse water. Such a system offers multiple benefits:

- The savings in paint alone (typically $100 to $800 per day) will pay for the ultrafiltration.
- The organic solvents present in the ultrafiltrate are more efficient in the removal of drag-out paint than is pure water.
- Because of the recycle, there is no need for solvent makeup in the paint tank.
- The demand for spray water is minimized to a final rinse.
- The ultrafiltrate bleed solves water and purification requirements in the same facility.
- Except for the bleed stream, all pollution is eliminated.

Performance data on UF modules operating on ED paint tanks vary considerably. High-performance thin-channel units may offer flux values of 30 to 50 gal/(ft²)(day) compared with 10 gal/(ft²)(day) on 1-in-diameter tubes. However, the thin-channel units are much more susceptible to fouling, and ED paint tanks are commonly filled with debris from paper cups, cigarette wrappers, etc.

Latex Recovery   Closely akin to electrodeposition primers are polymer latices. Polymer-latex wastes from polymerization-kettle washdown and other sources are often discharged directly into rivers and streams or into a sewage-treatment plant. Municipal treatment plants are becoming increasingly reluctant to handle latex wastes because of the problems they cause: reduction of biochemical activities, plugging, and odor.

Ultrafiltration can provide an inexpensive way to clean up wastewater from latex-polymer manufacture, while recovering the latex for recycling. Latices concentrated in the author's laboratories include styrene butadiene, acrylic, styrene maleic anhydride, polyvinylidene chloride, methyl methacrylate–butyl acrylate copolymer, butyl acrylate–diacetone acrylamide copolymer, and polybutadiene latices. In many cases, the latex may be concentrated up to 60 to 70% by weight. Figure 40 presents typical data for the concentration of a carboxylic modified styrene-butadiene copolymer latex from 5 to 50% solids in 30-mil channels. If desired, the concentrated latex can be further dewatered by

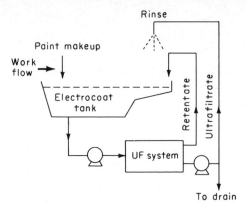

Single-rinse system

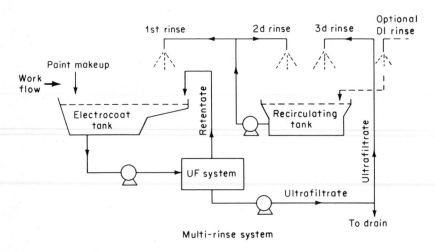

Multi-rinse system

Fig. 81 Ultrafiltration applied to electrocoat paint operations.

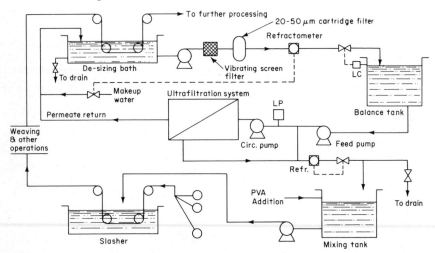

Fig. 82 PVA recovery system.

chemical coagulation. The UF permeate contains no suspended solids and therefore may be recycled back to the process.

Polymer latices may also be purified from unwanted salts or surfactants by diafiltration (see Appendix G). This process also permits replacement of one surfactant by another.

Polyvinyl Alcohol Recovery Polyvinyl alcohol (PVA) is a sizing agent for warp yarn which replaces starch and natural gum on synthetic fibers. Because PVA is not fully biodegradable and is more expensive than starch, its recovery and reuse offer important environmental and economic advantages. Figure 82 shows a schematic of the overall operation with UF installed. This system concentrates the 0.8 to 1% PVA solution from the desizing bath to an 8% solution for recycle to the sizing system. The permeate, containing less than 0.1% PVA, is returned to the desizing bath. Because of the high viscosities of concentrated PVA solutions, operating temperatures of 80 to 90°C are used to reduce the viscosity to an acceptable level (150 cP at 10%).

Oil-Water Separations Many waste streams consist of an oil-in-water emulsion formed by either mechanical emulsification or spontaneous emulsification. In either case, oil-droplet particle sizes are rarely as small as 0.1 μm. Most of the time, they are stabilized in the presence of surfactant. Thus most oil droplets can be retained by a UF (even an MF) membrane, presuming there is little coalescence of the droplets. In the ideal case, only the oil freely soluble in water, a very low value, will be seen in the permeate. In the real world, some "unstabilized tramp oil" will be present. If the membrane is oil-wet, it may weep this oil into the permeate. At the very least, the unstabilized oil will spread on the membrane, resulting in significant fouling.

In practice, membranes with a high surface-free energy (e.g., cellulose acetate) will outperform those of lower free energy.[104] This effect is particularly pronounced at low tangential velocities, where noncellulosic materials rapidly become wetted with oil and quickly lose both their flux and their rejection characteristics. On the other hand, membranes with a low free surface energy (e.g., hydrophobic membranes) often have vastly improved chemical and thermal resistance and are thereby the only possible choice. Careful pilot evaluation is mandatory for the selection of the most appropriate membrane.

There are three general categories of oil-water emulsions: metalworking lubricants, metal-finishing bath wastes, and mechanical oil-water emulsions.

Metalworking Lubricants. Oil-in-water emulsions are commonly used in the metalworking field to cool and lubricate tools and dies during the cutting, forming, and grinding of metals. UF becomes commercially important in the metalworking emulsion field after the emulsions have been removed from the metal-forming machinery and are ready to be discarded. At this point, the oil content of the emulsion is generally about 2%. The object of UF is to concentrate this oil waste to an oil content of approximately 30%, at which point it will support combustion. While the reuse of oily waste is possible in principle, it has seldom been attempted in waste from metal-cutting operations. Generally, the products of UF are a sewerable permeate and a combustible concentrate.

The cost of UF is a strong function of the flux and fouling rate. Low-lubricity emulsions (with a high "synthetic" oil content) are most amenable to concentration by UF. They are characterized by high flux rates, low fouling rates, and permeates with less than 5 ppm oil.

Table 14 shows the composition of the feed and permeate with raw and acidified

TABLE 14   Composition of Feed and Permeate Samples for Oil Emulsion from Metalworking Operation

Approx concentration ratio	Feed samples		Permeate samples		
	Oil and grease, mg/L	Total solids, %	Oil and grease, mg/L	Total solids, %	Notes
1x	16,600	1.69	506	0.67	
2x	37,300	2.6	764	0.75	
4x	100,700	3.82	693	0.80	No acid added,
8x	88,200	9.35	405	0.86	pH 10.2
15x	156,400	13.6	742	0.94	
20x	227,900	18.4	331	0.93	
2.5x	31,900	4.1	205	0.99	
6x	80,600	7.43	442	1.03	Acidified sample,
10x	242,500	12.1	159	0.74	pH 5.4
17x	371,300	16.7	211	0.81	

samples from a metalworking operation. It is to be noted that the oil and grease content of the permeate is essentially independent of the feed concentration, since the solubility of organics in the aqueous phase of a two-phase system should be independent of the relative volumes of the two phases. The fact that acidification greatly reduced the oil and grease content in the permeate is presumed to be due to coagulation of soluble anionic surfactants which would be retained by the membrane. This was confirmed by acidifying permeate from the nonacidified sample which developed substantial turbidity.

Metal-Finishing Bath Wastes. UF systems have also been employed for cleaning up metal-finishing baths which are used to wash oil from steel and aluminum coils prior to roll coating, painting, or stamping and forming operations. The objective of the UF system is to separate and concentrate the emulsified oil from water while returning permeate (containing detergents) to the cleaner bath. The permeate is hot (54°C), requiring reasonable thermal stability of the membrane. The recovered oil can be reused provided acid treatment or some other means can be used to clean the oil of the accompanying dirt.

Bansal[102] has published retention data for inorganic membranes (UCARSEP) operating on a metal-finishing bath. The membrane rejection for oil was over 99%. Since the alkaline detergents passed through the membrane, the pH of the permeate was higher than that of the feed. In certain tests, the dilute concentrations of feed solution were not well emulsified and severely fouled the membrane. The addition of Triton X-100 (nonionic surfactant) emulsified the oil and greatly reduced the fouling.

Bansal's cost estimates indicate that the operating costs for a 50,000-gal/day unit would vary between $2.00 and $3.60 per 1000 gal of feed processed.

Mechanical Oil-Water Emulsions. By far the largest volume of oil-in-water emulsions are mechanical emulsions which are largely free of surfactant. They are commonly found in the petroleum and petroleum transportation industry and in ship bilges. Mechanical separation is normally the first means used to coalesce and separate these emulsions. In some cases, stray surfactants or other emulsifying agents do stabilize the system and UF is effective in making the separation.[82]

Pharmaceutical and Biological Separations   There are numerous applications in the pharmaceutical and biologicals industry for UF.[8,10,37] The isolation and recovery of many biologically active materials in low concentrations is often difficult. The sensitive nature of biological specimens to thermal and chemical environments limits the selection of the separations technique employed. Vacuum evaporation generally cannot be used for concentration of thermally labile materials because of the loss of biological activity in the product. Other techniques such as solvent precipitation, crystallization, or solvent extraction may sometimes denature the product owing to phase change. Dewatering with lyophilization may also denature proteins because of changes in ionicity and pH. Other techniques such as ultracentrifugation and liquid chromatography are costly and time-consuming.

A number of characteristics are associated with membrane ultrafiltration which make it ideal for concentration and purification of labile biologicals.

1. Althermal–low-temperature operation if desired (thus, by ultrafiltering at cold room temperatures, −5 to +5°C, loss in activity due to exposure to high temperature is avoided and "autodigestion" can be minimized).
2. No phase change required.
3. Operation at low hydrostatic pressure.
4. Gentle and nondestructive.
5. No chemical reagents required.
6. Simultaneous concentration and purification if desired.
7. Maintenance of constant ionic strength and pH of the concentrate—avoiding inactivation of enzymes.
8. Economical.

Other Applications   There is a considerable body of literature covering other applications for UF. Many of these show great potential but have not yet been accepted because of economic considerations or other technical problems. The reader is referred to the survey literature for further information.[1,2,8,10,35,39,40,81,88,99,103,115,116,117]

A PERSPECTIVE ON THE FUTURE

The hydraulic-pressure-activated membrane processes of MF, UF, and RO are now in wide use for cold sterilization, particulate removal, water demineralization, wastewater treatment, recovery of valuable products and/or by-products from waste streams, and the isolation and purification of biologicals and other macromolecules.

Increasing use for waste treatment is inevitable as the standards for zero discharge are enforced. At present, waste treatment with membranes can seldom be justified unless valuable products and/or by-products can be recovered to offset process costs.[88]

Water recycling and reuse will become mandatory as our water requirements continue to grow. This will tend to make uneconomical membrane processes economical as water costs rise and can be applied as a credit to processing costs.

Sea-water desalination with RO membranes should become more competitive with evaporation processes as energy costs rise and fresh-water resources become exhausted.

Increased use of membranes in a "kidney" function on continuous process lines should become more commonplace. Removal of impurities from electrocoat paint tanks with containment of the valuable paint is an example of what could be applied more broadly. We have seen the extension of this concept to metal-degreasing lines and metal-plating wastes.

The use of membranes for novel processing concepts such as membrane fermentors and enzymatic reactors will become more commonplace.[10]

The production of single-cell protein to feed the world's billions will be cumbersome on a batch basis. The use of MF or UF membranes to allow continuous removal of toxic metabolites and other by-products while containing the cellular biology will allow quantitative conversion of feed (substrate) to product (SCP) on a continuous basis.

Membrane technology is in a state of rapid development. New membranes capable of handling severe thermal and chemical environments with greater fouling resistance are on the horizon. Novel hardware concepts, now on the drawing boards, should drop membrane module costs (dollars per square foot), making possible applications hitherto uneconomical.

REFERENCES

1. Porter, M. C., Selecting the Right Membrane, *Chem. Eng. Prog.*, **71**(12), 55 (December 1975).
2. Porter, M. C., What, When and Why of Membranes, MF, UF AND RO, *Am. Inst. Chem. Eng. Symp. Ser., Filtration*, 1977.
3. Merson, R. L., and Morgan, A. I., Jr., Juice Concentration by Reverse Osmosis, *Food Technol.* **22**(5), 97 (May 1968).
4. Olten, G., and Brown, G. D., Bacteria and Pyrogens in Water Treatment, *Am. Lab.*, December 1973, p. 54.
5. Bechhold, Z. *Phys. Chem.*, **60**, 257 (1967).
6. Method of Test for Bubble Point, ASTM D 2499.
7. Meltzer, T., and Meyers, T. R., The Bubble Point in Membrane Characterization, *Bull. Parenter. Drug Assoc.*, **25**, 165 (July–August 1971).
8. Porter, M. C., and Nelson, L., Ultrafiltration in the Chemical, Food Processing, Pharmaceutical, and Medical Industries, *Recent Dev. Sep. Sci.*, **2**, 227 (1972).
9. Blatt, W. F., Membrane Partition Chromatography—A Tool for Fractionation of Protein Mixtures, *Agr. Food Chem.*, **19**(4), 589 (July–August 1971).
10. Porter, M. C., Applications of Membranes to Enzyme Isolation and Purification, *Biotechnol. Bioeng. Symp.*, no. 3, 115–144 (1972).
11. Davis, M. A., Jones, A. G., and Trindade, H., A Rapid and Accurate Method for Sizing Radio Colloids, *J. Nucl. Med.*, **15**(11), 923–928.
12. Spurny, K. R., Lodge, J. P., Jr., Frank, E. R., and Sheesley, D. C., Aerosol Filtration by Means of Nuclepore Filters—Structural and Filtration Properties, *Environ. Sci. Technol.*, **3**(5), 453 (May 1969).
13. Spurny, K. R., Stober, W., Ackerman, E. R., and Lodge, J. R., Jr., A Note on the Sampling and Electron Microscopy of Asbestos in Ambient Air by Means of Nuclepore Filters, Paper 74-47, presented to the 67th APCA meeting, Denver, Colo., June 9–13, 1974.
14. Kesting, R. E., "Synthetic Polymeric Membranes," McGraw-Hill, New York, 1971.
15. Elford, W. J., A New Series of Graded Collodion Membranes Suitable for General Bacteriological Use, Especially in Filterable Virus Studies, *J. Pathol. Bacteriol.*, **34**, 505 (1931).
16. Gore, R. W., Process for Producing Porous Products, U.S. Patent 3,953,566, Apr. 27, 1976.
17. Gore, R. W., Very Highly Stretched Polytetrafluoroethylene and Process Therefore, U.S. Patent 3,962,153, June 8, 1976.
18. Flesicher, R. L., Price, P. B., and Walker, R. M., Nuclear Tracks in Solids, *Sci. Am.*, **220**(6), 30 (June 1969).
19. Reid, C. E., and Breton, E. J., *J. Appl. Polym. Sci.*, **1**, 133 (1959).

20. Loeb, S., and Sourirajan, S., in SeaWater Research, Department of Engineering, University of California, Los Angeles, *Repts.* 59-3 (1958), 59-28 (1959a), 59-46 (1959b), 60-5 (1960a), 60-26 (1960b), 60-60 (1961).
21. Strathmann, H., Scheible, P., and Baker, R. W., A Rationale for the Preparation of Loeb-Sourirajan Type Cellulose Acetate Membranes, *J. Appl. Polym. Sci.*, **15**, 811–828 (1971).
22. Lonsdale, H. K., Theory and Practice of Reverse Osmosis and Ultrafiltration, chap. VIII in Lacey and Loeb, "Industrial Processing with Membranes," pp. 125–178, Wiley Interscience, New York, 1972.
23. Strathmann, H., Saier, H. D., and Baker, R. W., The Formation Mechanism of Asymmetric Reverse Osmosis Membranes, *4th Int. Symp. Fresh Water from the Sea*, **4**, 381–394, 1973.
24. Model, Frank S., and Lee, L. A., PBI Reverse Osmosis Membranes: An Initial Survey, from H. K. Lonsdale and H. E. Podall (eds.) "Reverse Osmosis Membrane Research," pp. 285–297, Plenum, New York, 1972.
25. Riley, R. L., Milstead, C. E., Wrasido, W. J., Grabowsky, R. L., Hightower, G. R., Lyons, C. R., and Togami, M., "Research and Development of Composite Membrane Technology," Office of Saline Water, no. 851, May 1973.
26. Rozelle, L. T., Cadotte, J. E., Nelson, B. R., and Kopp, C. V., "Ultrathin Membranes for Treatment of Waste Effluents by Reverse Osmosis," Applied Polymer Symposium no. 22, pp. 223–239, Wiley, New York 1973.
27. Rozelle, L. T., Kopp, C. V., Jr., Cadotte, J. E., Cobian, K. E., NS-100 Membranes for Reverse Osmosis Applications, *Trans. ASME, J. Engl. Ind.*, February 1975, pp. 220–223.
28. Cabasso, I., Klein, E., and Smith, J. K., Polysulfone Hollow Fibers I Spinning and Properties, *J. Appl. Polym. Sci.*, **20**, 2377–2394 (1976).
29. Davis, R. B., Allegrezza, A. E., Coplan, M. J., Charpentier, J. M., Hollow Fiber Composite Membranes of Furan Resins on Polysulfone: Chemistry, Stability, and Processing Conditions, paper presented at First Chemical Congress of North American Continent, Mexico City, December 1975.
30. Johnson, J. S., Jr., Minturn, R. E., and Wadia, P., Hyperfiltration XXI Dynamically Formed Hydrous Zr (IV) Oxide-Polyacrylate Membranes, *J. Electroanal. Chem.*, **37**, 267–281 (1972).
31. So, M. T., Eirich, F. R., Strathmann, H., and Baker, R. W., Preparation of Asymmetric Loeb-Sourirajan Membranes, Polymer Letters Edition, John Wiley & Sons, Inc., **11**, 201–205 (1973).
32. Michaels, A. S., New Separation Technique for the Chemical Process Industries, *Chem. Eng. Prog.*, **64**(12), 31–43 (December 1968).
33. Henry, J. D., Jr., Cross Flow Filtration, *Rec. Dev. Sep. Sci.* **2**, 205 (1972).
34. Blatt, W. F., Agranat, E. A., and Rigopulos, P. N., Blood Fractionating Process and Apparatus for Carrying Out Same, U.S. Patent 3,705,100, Dec. 5, 1972.
35. Porter, M. C., and Michaels, A. S., Membrane Ultrafiltration—Part 4. Application in Processing Vegetable Foods and Beverages, *Chem. Technol.*, **1**, 633 (October 1971).
36. Semmelink, A., Ultrasonically Enhanced Liquid Filtering, *Ultrasonics Int. Conf. Proc.*, Session 1, 1973.
37. Charm, S. E., and Lai, C. J., Comparison of Ultrafiltration Systems for Concentration of Biologicals, *Biotechnol. Bioeng.*, **13**, 185–202 (1971).
38. Baker, R. W., Methods of Fractionating Polymers by Ultrafiltration, *J. Appl. Polym. Sci.*, **13**, 369–376 (1969).
39. Porter, M. C., and Michaels, A. S., Membrane Ultrafiltration, *Chem. Technol.*, January 1971, pp. 56–63.
40. Porter, M. C., Ultrafiltration of Colloidal Suspensions, *Am. Inst. Chem. Eng. Symp. Ser.*, **68**(120), 21–30 (1972).
41. Blatt, W. F., Dravid, A., Michaels, A. S., and Nelson, L., Solute Polarization and Cake Formation in Membrane Ultrafiltration: Causes, Consequences, and Control Techniques, from Flinn, J. E. (ed.), "Membrane Science and Technology," Plenum, New York, 1970.
42. Porter, M. C., Concentration Polarization with Membrane Ultrafiltration, *Ind. Eng. Chem. Prod. Res. Dev.*, **11**, 234 (September 1972).
43. Graetz, L., Über die Wärmeleitungsfähigkeit von Flüssigkeiten, *Ann. Phys. Chem.*, **18** (1883).
44. Lévêque, M. D., Les Lois de la Transmission de Chaliur por Convection, *Ann. Mines*, **13** (April 1928).
45. Gröber, H., Erk, S., and Grigull, U., "Fundamentals of Heat Transfer," p. 233, McGraw-Hill, New York, 1961.
46. Dittus, F. W., and Boelter, L. M. K., *Univ. Calif. Berkeley, Publ. Eng.*, **2**, 443 (1930).
47. Sober, H. A., "Handbook of Biochemistry," Chemical Rubber Co., 1968.
48. DeFilippi, R. P., and Goldsmith, R. L., Application and Theory of Membrane Processes for Biological and Other Macromolecular Solutions, from Flinn, J. E. (ed.) "Membrane Science and Technology," pp. 33–46, Plenum, New York, 1970.
49. Goldsmith, R. L., *Ind. Eng. Chem. Fundam.*, **10**, 113 (1971).
50. Brandup, J., and Immergul, E. H., "Polymer Handbook," John Wiley & Sons, 1975.
51. Cheng, P. Y., and Schachman, H. K., *J. Poly. Sci.*, **16**, 19 (1955).
52. Dunning, J. W., and Angus, J. C., *J. Appl. Phys.*, **39**, 2479 (1968).
53. Eirich, R. R., "Rheology, Theory, and Applications," vol. 4, chap. 2, Academic, New York, 1967.

54. Fahraeus, R., *Physiol. Rev.*, **9**, 241–227, (1929).
55. Palmer, A. A., *Am. J. Phys.*, **209** (1965).
56. Segré, G., and Silberberg, A., *Nature*, **189**, 209 (1961).
57. Karnis, A., Goldsmith, H. L., and Mason, S. G., *Can. J. Chem. Eng.*, **44**, 181 (1966).
58. Brandt, A., and Bugliarello, G., *Trans. Soc. Rheol.*, **10**(1), 229–251 (1961).
59. Rubinow, S. I., and Keller, J. B., *J. Fluid Mech.*, **11**, 447 (1961).
60. Oliver, D. R., *Nature*, **194**, 1269 (1962).
61. Theodore, L., Eng. Sci. D. Dissertation, New York University, New York, 1964.
62. Segré, G., and Silberberg, A., *J. Fluid Mech.*, **14**, 136 (1962).
63. Cox, R. G., and Brenner, H., forthcoming publication.
64. Saffman, P. G., *J. Fluid Mech.*, **1**, 540 (1956).
65. King, C. J., Department of Chemical Engineering, University of California. (Berkeley), personal communication, 1971.
66. Forstrom, R. J., Voss, G. O., and Blackshear, P. L., Jr., Fluid Dynamics of Particle (Platelet) Deposition for Filtering Walls: Relationship to Atherosclerosis, *J. Fluids Eng.*, **96**, 168–171 (1974).
67. Forstrom, R. J., Barlelt, K., Blackshear, P. L., Jr., and Wood, T., Formed Element Deposition onto Filtering Walls, *Trans. Am. Soc. Artif, Int. Organs*, **21** (1975).
68. Bixler, H. J., and Rappe, G. C., Ultrafiltration Process, U.S. Patent 3,541,006, Nov. 17, 1970.
69. Sherwood, T. K., Brian, P. L. T., Fisher, R. E., and Dresner, L., Salt Concentration at Phase Boundaries in Desalination by Reverse Osmosis, *Ind. Eng. Chem. Fundam.*, 4(2), 113–118 (May 1965).
70. Brian, P. L. T., Concentration Polarization in Reverse Osmosis Desalination with Variable Flux and Incomplete Salt Rejection, *Ind. Eng. Chem. Fundam.*, 4(4), 439–445 (November 1965).
71. Brian, P. L. T., Mass Transport in Reverse Osmosis, chap. 5 in Merten, U. (ed.), "Desalination by Reverse Osmosis," The M.I.T. Press, Cambridge, Mass, 1966.
72. Dresner, L., Boundary Layer Buildup in the Demineralization of Soft Water by Reverse Osmosis, *Oak Ridge Natl. Lab. Rept.* ORN L-3621, May 1964.
73. Duvel, W. A., Jr., Helfgott, T., and Genetelli, E. J., Flux Loss in Reverse Osmosis Due to Dispersed Organics, paper presented at 69th national meeting of American Institute of Chemical Engineers, Cincinnati, Ohio, May 17–19, 1971.
74. Sammon, D. C., and Stringer, B., The Application of Membrane Processes in the Treatment of Sewage, *Proc. Biochem.*, March 1975, pp. 4–12.
75. Rex Chainbelt, "Amenability of Reverse Osmosis Concentrate to Activated Sludge Treatment," U.S. Environmental Protection Agency, 17040 EUE, 07/71.
76. Beach, W. A., and Epstein, A. C., Summary of Pretreatment Technology for Membrane Processes, *Ind. Water Eng.*, August–September 1975.
77. Aerojet General Corp., "Reverse Osmosis Renovation of Municipal Wastewater," FWQA-ORD 17040EFQ 12/69.
78. Thomas, D. G., Gallaher, R. B., and Johnson, J. S., Jr., "Hydrodynamic Flux Control for Wastewater Application of Hyperfiltration Systems," Environmental Protection Agency. EPA-RZ-B-228, May 1973.
79. Lee, D. N., and Merson, R. L., Prefiltration of Cottage Cheese Whey to Reduce Fouling of Ultrafiltration Membranes, *J. Food Sci.*, **41**, 403–410 (1976).
80. Marshall, P. G., Dunkley, W. I., and Lowe, E., Fractionation and Concentration of Whey by Reverse Osmosis, *Food Technol.*, **22**(8), 969 (1968).
81. Porter, M. C., and Michaels, A. S., Membrane Ultrafiltration—Part 2—Applications in the Processing of Dairy and Poultry Products, *Chem. Technol.*, April 1971, pp. 248–254.
82. Schatzberg, P., Harris, L. R., Adema, C. M., Jackson, D. F., and Kelly, C. M., Oil-Water Separation with Noncellulosic Ultrafiltration Systems, 1975 Conference on Prevention and Control of Oil Pollution, San Francisco, Calif., Mar. 25–27, 1975.
83. Breslau, B. R., Agranat, E. A., Testa, A. J., Messinger, S., Cross, R. A., Hollow Fiber Ultrafiltration—A Systems Approach for Process Water and By-Product Recovery, *Chem. Eng. Prog.* **71**(12), 74–80 (December 1975).
84. Crits, G. J., Some Characteristics of Major Types of Reverse Osmosis Modules, *Ind. Water Eng.*, December 1976–January 1977, pp. 20–23.
85. Forbes, F., Considerations in the Optimisation of Ultrafiltration, *Chem. Eng.*, January 1972, pp. 29–34.
86. Sourirajan, S., "Reverse Osmosis," Academic, New York, 1970.
87. Goldsmith, R. L., Roberts, D. A., and Burre, D. L., Ultrafiltration of Soluble Oil Wastes, *J. Water Pollut. Control Fed.*, **46**(9), 2183–2192 (September 1974).
88. Porter, M. C., Membrane Ultrafiltration for Pollution Abatement and By-Product Recovery, *Water, 1972, Am. Inst. Chem. Eng. Symp. Ser.*, 1972, pp. 100–122.
89. Lacey, R. E., The Costs of Reverse Osmosis, chap. IX, from Lacey and Loeb, "Indsustrial Processing with Membranes," pp. 179–189, Wiley-Interscience, New York, 1972.
90. Quinn, R. M., Hamilton, R. S., Anderson, J. R., and Weiss, C. O., Some Economic Factors Relating to the Use of Reverse Osmosis and Ion Exchange for the Treatment of Water and Waste Water, *Water, 1971, Am. Inst. Chem. Eng. Symp. Ser.*, **124**(68), 262–269 (1971).

91. Butterworth, D. J., and Bass, W. C., Performance Histories of Boiler Feedwater Plants Highlight the Advantages of Reverse Osmosis, paper presented at TAPPI Engineering Conference, Toronto, September 1975.

92. Skrinde, R. T., and Tang, T. L., Operating Results of Electrodialysis and Reverse Osmosis Municipal Desalting Plants, *J. NWSIA*, **1**(1), 15–22 (July 1974).

93. Cruver, J. E., Waste-Treatment Applications of Reverse Osmosis, *Trans. ASME*, February 1975, pp. 246–251.

94. Kremen, S. S., Reverse Osmosis Makes High Quality Water Now, *Environ. Sci. Technol.*, **9**(4), 314–318 (April 1975).

95. Okey, R. W., The Treatment of Industrial Wastes by Pressure-Driven Membrane Processes, chap. XII, pp. 249–277, from E. Lacey and S. Loeb, (eds.) "Industrial Processing with Membranes," Wiley Interscience, New York, 1972.

96. Golomb, A., Application of Reverse Osmosis to Electroplating Waste Treatment, from S. Sourirajan, (ed.), "Reverse Osmosis and Synthetic Membranes," pp. 481–494, National Research Council Canada, Ottawa, 1977.

97. Bansal, I. K., and Wiley, A. J., Application of Reverse Osmosis in the Pulp and Paper Industry, pp. 459–480, from S. Sourirajan (ed.), "Reverse Osmosis and Synthetic Membranes," National Research Council Canada, Ottawa, 1977.

98. Brandon, C. A., and Porter, J. J., "Hyperfiltration for Renovation of Textile Finishing Plant Wastewater," EPA 600/2-76-060, March 1976.

99. Porter, M. C., "A Novel Membrane Filter for the Laboratory, *Am. Lab.*, November 1974.

100. Lamoureux, P. E., and Peterson, J., Wine Stabilization and Quality Control with Millipore Filters, presented at the Canadian Society of Genologists, Montreal, Apr. 21, 1966.

101. Forbes, F., Ultrafine Filtration for Electrophoretic Painting, *Prod. Finish.*, **23**(11), 24–29 (November 1970).

102. Bansal, I. K., Concentration of Oily and Latex Wastewaters Using Ultrafiltration Inorganic Membranes, *Ind. Water Eng.*, October/November 1976, pp. 6–11.

103. Mir, L., Eykamp, W., and Goldsmith, R. L., Current and Developing Applications for Ultrafiltration, *Ind. Water Eng.*, May/June 1977, pp. 14–19.

104. Eykamp, W., Ultrafiltration of Aqueous Dispersions, paper presented at 79th national meeting American Institute of Chemical Engineers, Houston, Tex., Mar. 18, 1975.

105. Goldsmith, R. L., Roberts, D. A., and Burre, D. L., Ultrafiltration of Soluble Oil Wastes, *J. Water Pollut. Control Fed.*, **46**(9), 2183–2192 (September 1974).

106. Dunkley, W. L., Concentrating and Fractionating Whey, Symposium cosponsored by National Canners Association and USDA, Albany, Calif., Jan. 23, 1969.

107. Klein, C. L., and Wong, C., "Study Reveals Potential for Profit in Waste Recovery through Desalination," news release, OSW, Department of Interior, Nov. 2, 1969.

108. Horton, B. S., Goldsmith, R. L., Hossain, S., and Zall, R. R., Membrane Separation Processes for the Abatement of Pollution from Cottage Cheese Whey, presented at the Cottage Cheese and Cultured Milk Products Symposium, University of Maryland, College Park, Md., Mar. 11, 1970.

109. Hayes, J. F., Dunkerley, J. A., Muller, L. L., and Griffin, A. T., Studies on Whey Processing by Ultrafiltration II Improving Permeation Rates by Preventing Fouling, *Aust. J. Dairy Technol.*, September 1974, pp. 132–140.

110. Blatt, W. F., Feinberg, M. D., Hopfenberg, H. P., and Saravis, C. A., Protein Solutions: Concentration by a Rapid Method, *Science*, **150**, 224 (1965).

111. Wang, C. C., Sonoyama, T., and Mateles, R. I., Enzyme and Bacteriophage Concentration by Membrane Ultrafiltration, *Anal. Biochem.*, **26**(2), 277 (1968).

112. Wang, D. I. C., Sinskey, A. J., and Sonoyama, T., Recovery of Biological Materials through Ultrafiltration, *Biotechnol. Bioeng.*, **11**, 987 (1969).

113. Wang, D. I. C., Sinskey, A. J., and Butterworth, T. A., Enzyme Processing Using Ultrafiltration Membranes, in J. E. Flinn (ed.), "Membrane Science and Technology," Plenum, New York, 1970.

114. Ericksson, K. E., and Rzedowski, W., Extracellular Enzyme System Utilized by the Fungus Chrysosporium Lignorum for the Breakdown of Cellulose, *Arch. Biochem. Biophys.*, **129**(2), (1969).

115. Porter, M. C., Schratter, P., and Rigopulos, P. N., By-Product Recovery by Ultrafiltration, *Ind. Water Eng.*, June/July 1971.

116. Porter, M. C., and Michaels, A. S., Membrane Ultrafiltration—Part 3—Applications in the Processing of Meat By-Products, *Chem. Technol.* July 1971, pp. 440–445.

117. Porter, M. C., and Michaels, A. S., Membrane Ultrafiltration—Part 5—A Useful Adjunct for Fermentative and Enzymatic Processing of Foods, *Chem. Technol.*

118. Kesting, R. E., A Novel Technique for the Study of Dry Process Membranes during their Nascent Phases, *Ion Exch. Membranes*, **1**, 197–204 (1974).

119. Merten, U., Lonsdale, H. K., Riley, R. L., and Vos, K. D., Reverse Osmosis Membrane Research, *U.S. Off. Saline Water Res. Dev. Prog. Rept.* 208, General Dynamics, General Atomic Division, 1966.

120. Breton, E. J., Jr., Water and Ion Flow through Imperfect Osmotic Membranes, *Office of Saline Water, R & D Prog. Rept.* 16, PBI 61391, 1957.

APPENDIX A   Description and Theory of Bubble-Point Test

The "bubble-point test" is based on the fact that a specified gas pressure is required to displace liquid from the pores of a wetted membrane; this may be calculated from the equation

$$P = \frac{4\gamma \cos \theta}{d} \tag{A1}$$

where P = bubble-point pressure

γ = surface tension of the liquid-gas interface

θ = contact angle between the liquid and the pore wall

d = pore size

As the pressure is raised, the first gas bubbles to appear through the membrane represent the displacement of water from the largest pore. The pressure at which this occurs is referred to as the bubble point and corresponds to the maximum pore size in the membrane.

It should be recognized that any of the three types of membranes discussed in this section (MF, UF, or RO) may exhibit lower bubble points than that corresponding to the rated pore size, particularly if large areas are incorporated in a single module or cartridge. This is presumably due to "defect holes" introduced in the manufacturing operation (e.g., in a pleating operation) and is confirmed by organism passage through the membrane. In other cases, where integrity and absolute retention have been maintained, the lower bubble points have been explained by poor wetting or by gas diffusion through the liquid-filled pores with desorption on the downstream side of the membrane. Both phenomena are enhanced by a large membrane area.

Figure 83 represents a series of calculations to estimate the diffusion of N_2 gas across a water-wetted Nuclepore membrane at *pressures below the bubble point.*

It is realized that even though all pores are filled with water, gas will dissolve at the upstream face of the membrane, diffuse through the pores in solution, and come out of solution at the lower pressures downstream of the membrane. The question is what kind of gas-permeation rate is expected and whether is it detectable. Obviously, the porosity of the membrane and its thickness will limit the amount of diffusion that takes place.

The values on the graph were calculated from the following equation:

$$J = \frac{N\pi d^2}{4} (DH) \frac{\Delta P}{l} \tag{A2}$$

where J = permeating rate, $g \cdot mol/(s)(cm^2)$

N = pore density, number/cm²

d = pore diameter, cm

D = diffusivity of the gas in water [for N_2 in water at 20°C

$D = 1.64 \times 10^{-5}$ cm²/s)]

H = solubility of the gas in water [for N_2 in water at 20°C

$H = 6.9 \times 10^{-7} g \cdot mol/(atm)(cm^3)$]

ΔP = pressure differential across the membrane, atm

l = pore path length across the membrane, cm. (This has been assumed equal to the thickness of our membrane.)

For example, membrane material having a pore size of 0.27 μm, a pore density of 6 × 10⁷ pores/cm², and a thickness of 10^{-3} cm has a diffusion rate of

$$\frac{J}{\Delta P} = \frac{6 \times 10^7 (0.785)(0.27 \times 10^{-4})^2(1.64 \times 10^{-5})(6.9 \times 10^{-7})}{4 \ 10^{-3}}$$
$$= 3.89 \times 10^{-10} g \cdot mol/(s)(atm)(cm^2)$$

Using the gas constant and converting to more convenient units, this becomes

$$\frac{J}{\Delta P} = 0.0355 \ mL/(min)(lb/in^2)(ft^2)$$

Thus, for a 15-ft² cartridge tested at 30 lb/in², the permeating rate is about 16 mL/min.

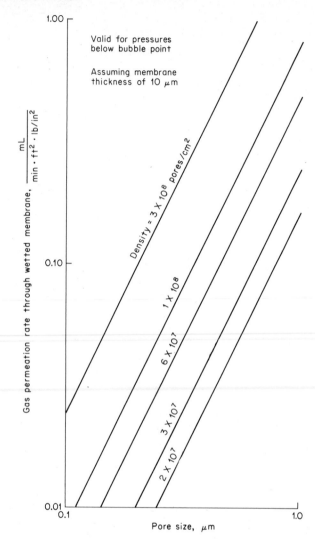

Fig. 83 Diffusion of nitrogen across a wetted Nuclepore membrane.

APPENDIX B   Pore Collapse during Drying of RO and UF Membranes

"Pore collapse" is a common phenomenon in both RO and UF membranes. The fine pore structure of these membranes means that the capillary stresses present during the drying operation may well exceed the yield stress of the polymer, bringing the opposing sides of the cell wall in contact with one another and causing irreversible collapse of many of the smaller pores (see Fig. 84).

On the surface it would appear that a membrane which can be dried without collapse must either be non-water-wettable or be made from a polymer with a very high yield strength. The former is usually undesirable, since once dried, it is very difficult to resaturate. The latter approach has been used very successfully in the author's experience in producing dryable UF membranes with relatively large pores (i.e., mol wt cutoff above 50,000).

The membrane may also be rendered semidryable by posttreating in the wet state. It may be saturated with a nonvolatile liquid like glycerin or glycol, which remains behind after drying. Alternatively, the water may be replaced with a low-boiling organic-like alcohol or acetone which has a low surface tension and can be evaporated without creating large capillary stresses.

Kesting[118] has been successful in developing a RO membrane with a dry process in which solvent is lost solely by *complete* evaporation of the solvent system. It is a variation in the Loeb-Sourirajan procedure in that no aqueous gelation bath is employed to precipitate the membrane skin. It is similar to the process used to produce MF cellulose ester membranes. A casting solution consisting of a polymer, solvent(s), and swelling agent(s) is cast onto a suitable substrate. Immediately after the solution is cast, solvent starts to evaporate. Since solvent is lost more rapidly from the casting solution/air interface than from the interior of the solution, solvent is rapidly depleted at this interface and polymer concentration increases—rapidly rising to a level where the remaining solvent is insufficient to maintain the polymer solution. The polymer then comes out of the solution and forms the skin layer. After the skin has been formed, desolvation is slowed considerably, so that the rate of gelation in the substructure can be considerably slower than it was

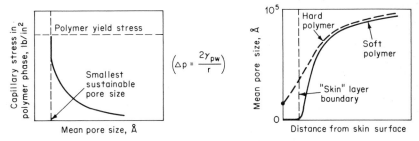

Fig. 84 Capillary consolidation stresses in microporous membranes.

in the skin. Eventually, however, remaining solvent power becomes insufficient to maintain all the components in a homogeneous true solution. Since the solvent is usually more volatile than the swelling agent, a large amount of the latter remains even after much of the solvent has been lost. The swelling agent now separates from the continuous phase as tiny droplets of a dispersed phase. The polymer molecules aggregate about these droplets because of surface activity and because insufficient solvent is available to keep them in the continuous phase. As more solvent is lost, the polymer-coated droplets of swelling agent approach one another more closely. Eventually, as both solvent and swelling agent depart, the polymer coating of the droplets is stretched too thin and ruptures, leaving behind an open-celled substructure.

The nature of the "dry" phase-inversion process is such that the substructure has a larger void size, 1 to 2 μm, compared with 0.1 to 1.0 μm for the "wet" process membranes. This could explain the greater tendency of the wet-process membranes to collapse when dried. However, it is claimed that the dry-process membranes have "wet-dry reversibility." It is not clear why the pores in the skin structure do not collapse unless one thinks of the skin as having no pores. In this case, permeability would occur by a solution-diffusion mechanism.

APPENDIX C   Membrane Compaction

At constant operating pressure, the clean-water flux through an asymmetric membrane decreases with time. It is not noticeable at low pressures or with open membranes in UF or MF. In these cases, any flux decline is usually due to fouling or plugging of the membrane.

With RO membranes of the asymmetric type, the phenomenon of compaction is apparently the result of a creep process in which the thin skin grows in thickness by amalgamation with the finely porous substructure.

It has been found empirically that if flux is plotted vs. time on log-log paper, the result is a straight line. For example, Fig. 85 shows a series of cures at different operating pressures. The water flux J at any time t can be calculated from the expression

$$J = J_0 t^m \qquad\qquad\qquad\text{(C1)}$$

where J_0 is the initial flux and m is the compaction slope. The compaction slope m will be a function of applied pressure (as shown) and membrane heat-treatment temperature as well as operating temperature. In addition, compaction effects are more pronounced with higher flux membranes but less so with "composite" membranes. It is almost impossible to predict what any membrane will do unless previous data are available or unless a few days of data can be obtained. The line can then be extrapolated to a flux value one year hence or any other time.

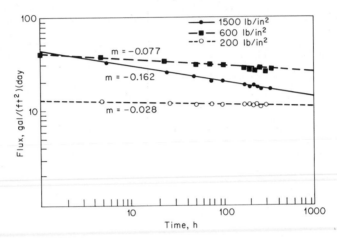

Fig. 85 Effect of time and pressure on flux of Eastman membrane RO 89 to 106°F.

Compaction phenomena are often confused with membrane fouling. The compaction data represent the maximum flux possible at any given time.

Typical compaction slopes and their effect on water flux for Loeb-Sourirajan membranes annealed at 85°C are given below.[119]

Applied pressure, lb/in²	Typical compaction slope	Flux at 1 year/ flux at 1 day
500	−0.03	0.84
1000	−0.06	0.70
1500	−0.09	0.59

APPENDIX D   Solute-Solution Transport in Membranes

There are several theoretical models in use to describe solute-solvent transport in membranes. Two of these models are based on pore flow and a solution-diffusion mechanism, respectively. The distinction between pore flow and solution-diffusion is sometimes blurred. That pore flow is the predominant mechanism in MF membranes is obvious. Flow rate and retention are governed by porosity, pore-size distribution, and other specific interactions within the pore fluid. The second model portrays the membrane as a nonporous diffusion barrier. All molecular species dissolve in the membrane in accordance with phase-equilibrium considerations and diffuse through the membrane under the influence of a concentration gradient. The solution-diffusion model is useful for describing what happens in an RO membrane. Both models have been used to describe various phenomena which occur in UF membranes.

Diffusional membranes separate by virtue of different diffusion coefficients and solubilities of the solute in the membrane. Since even quite similar molecules often differ markedly in their solubility and diffusion coefficients, separations are often possible between molecules of approximately the same size. For example, organic molecules similar in size to salt pass through cellulose acetate RO membranes much more readily than salt. On the other hand, as the molecular size and weight increase, so does the energy of activation. When the backbone length of a molecule exceeds about 10 atoms, the energy of activation becomes so high as to reduce the flux of the molecule by a diffusional process to negligible proportions.

In principle, experimental data on the effect of such parameters as pressure on the flux of solute and solvent through the membrane should be helpful in distinguishing between the two mechanisms of flow. In a solution-diffusion process, the rate of transport of water and salt is proportional to the chemical potential gradient of each species across the membrane. For both water and salt, this gradient is governed by the differences in both pressure and concentration in the two solutions. For water, the relative magnitude of these two contributions is expressed by Δp and $\Delta \pi$ as seen in Eq. (35). In the case of good salt rejection, however, the ratio of upstream to downstream salt concentration is very large, and the contribution of the concentration terms to the chemical-potential gradient for salt dominates the pressure term completely. Thus the *solute flux Js* should be independent of pressure in a solution-diffusion process.

$$J_s = H \frac{D_m}{t} (C_s - C_f) \tag{D1}$$

where H = solute-distribution coefficient between membrane and solution
D_m = solute diffusivity in the membrane
t = membrane thickness
In a pore-flow process, the *solute flux* should be proportional to pressure

$$J_s = \sigma J_w C_s = \frac{\sigma(\Delta P)C_s}{R_m + R_c} \tag{D2}$$

where σ = fraction of the total liquid flowing through the membrane passing through pores large enough to accommodate solute molecules
The solvent flux (J_w for water) is directly dependent on pressure in both solution-diffusion and pore-flow transport.

Obviously, the upstream concentration of solute C_s at the surface of the membrane will be a function of the polarization modulus C_s/C_b. Therefore, some caution must be exercised in interpreting experimental data to separate out the effect of pressure on the polarization modulus and on the solute flux.

Data like those presented in Fig. 35 where the polarization modulus is low appear to indicate a pore-flow mechanism for solute transport. The author believes this is prevalent for most UF membranes.

On the other hand, data like those presented in Fig. 86 seem to fit a solution-diffusion transport mechanism.[120] Increasing the pressure increases the water flow while leaving the salt flow nearly unchanged in accordance with Eq. (D1). This dilutes the salt concentration in the filtrate, resulting in an increase in the salt rejection.

APPENDIX E   Effect of Temperature on Membrane Flux and Solute Retention

It has been found experimentally for a large number of membrane systems (including MF, UF, and RO) and feed streams that the permeation rate is inversely proportional to the fluid viscosity. Since the viscosity of water decreases by about 3% for every °C rise in temperature, membrane researchers often refer to the 3% rule (that flux increases 3% per °C) as a rough rule of thumb.

Pure-Water Transport   These results are not unexpected for the transport of pure water through porous membranes. Poiseuille's law predicts that flow J through any porous media should be described as follows:

$$J = \frac{N \pi d^4 \Delta P}{128 \mu l} \tag{E1}$$

where N = number of pores per square foot
$\quad d$ = pore diameter (average)
$\quad \Delta P$ = pressure drop
$\quad \mu$ = fluid viscosity
$\quad l$ = pore length (including a turtuosity factor)
The only variable in Eq. (E1) which is temperature-dependent is the viscosity.

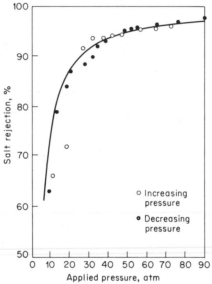

Fig. 86 Effect of pressure on semipermeability to sodium chloride 40% acetyl cellulose acetate membrane, 0.1 M NaCl feed.

Likewise for solution-diffusion membranes, the experimental results correspond quite well with the 3% rule. Lonsdale[22] has pointed out that in solution-diffusion membranes, where the difference in water concentration across the membrane is small,

$$J_w = \frac{D_w C_w v_w (\Delta P - \Delta \pi)}{RTt} \tag{E2}$$

where D_w = diffusion coefficient of water in the membrane
$\quad C_w$ = concentration of water in the membrane
$\quad v_w$ = partial molar volume of water
$\quad R$ = gas constant
$\quad T$ = absolute temperature
$\quad t$ = membrane thickness
Since the diffusion coefficient of water in the membrane D_w can be approximated by a relationship similar to the Stokes-Einstein diffusivity

$$D = \frac{kT}{6\pi\mu r_p} \tag{27}$$

Substitution into Eq. (E2) indicates that

$$J \propto \frac{C_w v_w (\Delta P - \Delta \pi)}{\mu} \tag{E3}$$

neglecting the variation of the partial molar volume and osmotic pressure (no solute present in pure water) with temperature again leads to an inverse dependence on viscosity.

Figure 87[86] is a plot showing the increase in flux with temperature and the invariance of the product of flux and viscosity with temperature.

RO Water Flux with Solute Present   In the case of reverse osmosis (a solution-diffusion membrane), so long as the solute concentration is low, Eqs. (E2) and (E3) are still valid. However, the osmotic-pressure difference across the membrane can mean a significant reduction in driving force for water transport across the membrane. Since the osmotic pressure for a given concentration of solute is a function of temperature in accordance with Eq. (34),

$$\Pi = iCRT \tag{34}$$

Thus an increase in temperature increases the osmotic pressure and tends to decrease the flux, whereas the decrease in viscosity tends to increase the flux. The latter effect is more prominent since the ΔP term is larger than the $\Delta \Pi$ term. Thus, in Fig. 87, even though $0.5M$ NaCl is present, the flux is still inversely proportional to viscosity.

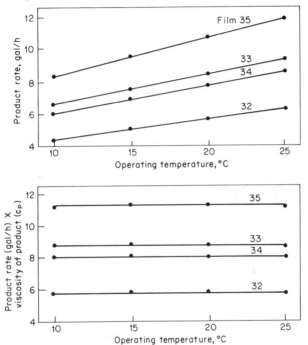

Fig. 87   Effect of operating temperature on the product rate characteristics of S & S porous cellulose acetate membranes for the system NaCl-water.

UF Water Flux with Gel Polarization   In the case of gel polarization, the membrane flux is given by Eq. (13)

$$J_w = K \ln \left(\frac{C_g}{C_b} \right) \tag{13}$$

Independent of the variation of the mass-transfer coefficient K with temperature, it would be expected that the gel concentration might vary with temperature. For example, proteins would be expected to be more soluble at elevated temperatures provided denaturation does not take place.

Figure 88 is illustrative of the change in gel concentration with temperature in the ultrafiltration of a high-protein meal.[4]   The intercept with the abscissa is not to be construed as the true gel-concentration value, since the total solids content was comprised not only of proteins retained by the membrane but also of small-molecular-weight species such as salts, peptides, and low-molecular-weight carbohydrates which were freely permeable to the membrane. Nevertheless, these data illustrate a lower gel concentration at the lower temperature (4°C), as would be expected qualitatively. (Incidentally, the recirculation rate in the low-temperature experiment was also due to the increased viscosity of the process stream.)

In other protein systems, the gel concentration does not appear to be a function of temperature.

Laminar-Flow Dependence. It will be noted that the dependence of flux on temperature in the data of Fig. 89 roughly corresponds to the inverse-viscosity rule. For example, the viscosity of water at 17°C is 1.083 cP and at 37.5°C is 0.686 cP. The ratio of these two viscosities is the same as the flux ratio at the two temperatures indicated. For example, at 0.2% protein, the flux ratio = 47/30 = 1.6. If the inverse viscosity rule is valid, this would mean that the gel concentration would have to be invariant with temperature since, at that

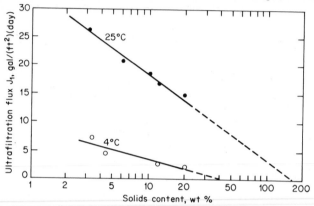

Fig. 88 Variation of flux with concentration and temperature.

point, the flux is zero by definition—regardless of temperature. In other words, the data of Fig. 88 represent a radical departure from the inverse-viscosity rule.

The data of Fig. 89 were taken in laminar flow for which

$$K = {}^{\iota}0.816 \left(\frac{6UD^2}{bL} \right)^{0.33} \tag{16}$$

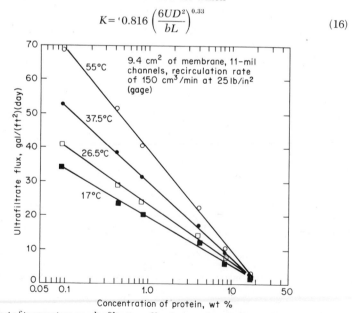

Fig. 89 Effect of temperature on ultrafiltration of bovine serum in TCF-10 with PM 30 membrane.

Assuming that the channel velocity is controlled to remain constant (it will tend to increase at elevated temperatures)

$$K \propto D^{0.66} = \left(\frac{KT}{6\pi\mu r_p}\right)^{0.66} \tag{E4}$$

where μ, the viscosity of the solvent, has a temperature dependence of

$$\mu = \mu_0 \exp\frac{\Delta E_0}{RT} \tag{E5}$$

and where ΔE_0 is the activation energy (3750 cal/g·mol for water).

For laminar flow, the key question is how the diffusivity varies with temperature. Differentiating:

$$\frac{dD}{dT} = \frac{K}{6\pi r_p}\frac{\mu - T(d\mu/dt)}{\mu^2} \tag{E6}$$

or

$$\frac{dD}{dT} = D\left(\frac{1}{T} - \frac{1}{\mu}\frac{d\mu}{dt}\right) \tag{E7}$$

Recalling Eq. (E5),

$$\frac{d\mu}{dT} = -\mu_0\left[\exp\left(\frac{\Delta E_0}{RT}\right)\right]\frac{\Delta E_0}{RT^2} \tag{E8}$$

or

$$\frac{d\mu}{dT} = -\mu\frac{\Delta E_0}{RT^2} \tag{E9}$$

Substituting Eq. (E9) into Eq. (E7),

$$\frac{dD}{dT} = D\left(\frac{1}{T} + \frac{\Delta E_0}{RT^2}\right) \tag{E10}$$

Therefore, assuming that C_G is independent of temperature as Fig. 89 indicates

$$\frac{dJ_w}{dT} = \ln\left(\frac{C_G}{C_b}\right)\frac{dK}{dT} \tag{E11}$$

From Eq. (16),

$$\frac{dK}{dT} = 0.816\left(\frac{6U}{bL}\right)^{0.33}0.66D^{-0.33}\frac{dD}{dT} \tag{E12}$$

or

$$\frac{dK}{dT} = K\frac{(0.66)}{D}\frac{dD}{dT} \tag{E13}$$

Substituting Eq. (E13) into Eq. (E11),

$$\frac{dJ_w}{dT} = \frac{J_w(0.66)}{D}\frac{dD}{dT} \tag{E14}$$

or from Eq. (E10),

$$\frac{dJ_w}{dT} = J_w(0.66)\left(\frac{1}{T} + \frac{\Delta E_0}{RT^2}\right) \tag{E15}$$

This expresses the variation of flux with temperature in laminar-flow gel polarization. Experimentally, the flux seems to vary inversely with the viscosity, or

$$\frac{1}{J_w}\frac{dJ_w}{dT} = \frac{-1}{\mu}\frac{d\mu}{dT} = \frac{\Delta E_0}{RT^2} \tag{E16}$$

The question, then, is whether Eq. (E15) is similar in magnitude to Eq. (E16), or is

$$0.66\left(\frac{1}{T} + \frac{\Delta E_0}{RT^2}\right) \qquad \text{equal to} \qquad \frac{\Delta E_0}{RT^2}$$

In the range of interest, $\Delta E_0 = 3750$ cal/g·mol and $R = 1.99$ cal/g·mol/K. At room temperature $T \cong 300$ K

$$\frac{\Delta E_0}{RT^2} = \frac{3750}{(1.99)9 \times 10^4} = 0.021$$

and

$$0.66\left(\frac{1}{T} + \frac{\Delta E_0}{RT^2}\right) = 0.66\left(\frac{1}{300} + 0.021\right)$$
$$= 0.002 + 0.014 = 0.016$$

which is close (within 25%).

One consequence of the empirical relationship given in Eq. (E16) is that, if correct, the

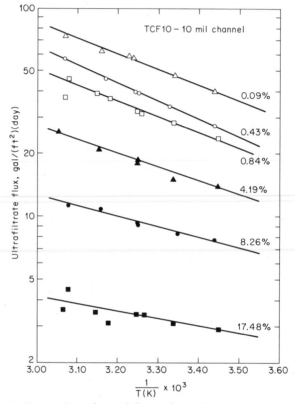

Fig. 90   Dependence of albumin flux on temperature.

flux should have an activation energy equal to the viscous activation energy of water. Figure 90 is an attempt to calculate the flux activation energy from data similar to those presented in Fig. 89, but for human albumin. It would appear that the activation energy is a function of concentration ranging from 1310 to 3920 cal/g·mol. The flux activation energies calculated from Fig. 89 ranged from 2360 to 4270 cal/g·mol.

Turbulent-Flow Dependence. Next consider the turbulent-flow case under gel polarization. Simplifying Eq. (25),

$$K = \frac{0.02\rho^{0.5}U^{0.8}D^{0.66}}{b^{0.2}\mu^{0.5}} \tag{E17}$$

Thus:

$$\frac{dK}{dT} = \frac{0.02\rho^{0.5}U^{0.8}}{b^{0.2}}\left[\frac{\mu^{0.5}(0.66/D^{0.33})(dD/dT) - D^{0.66}(0.5/\mu^{0.5})(d\mu/dT)}{\mu}\right] \tag{E18}$$

or
$$\frac{dK}{dT} = K\left(\frac{0.66}{D}\frac{dD}{dT} - \frac{0.5}{\mu}\frac{d\mu}{dT}\right) \tag{E19}$$

Substituting in Eq. (E7),

$$\frac{dK}{dT} = K\left[0.66\left(\frac{1}{T} - \frac{1}{\mu}\frac{d\mu}{dT}\right) - \frac{0.5}{\mu}\frac{d\mu}{dT}\right] \tag{E20}$$

$$\frac{dK}{dT} = K\left(\frac{0.66}{T} - \frac{1.16}{\mu}\frac{d\mu}{dT}\right) \tag{E21}$$

remembering Eq. (E9),

$$\frac{dk}{dT} = K\left(\frac{0.66}{T} + 1.16\frac{\Delta E_0}{RT^2}\right) \tag{E22}$$

and thus, assuming C_G is invariant with temperature,

$$\frac{dJ_w}{dT} = J_w(0.66)\left(\frac{1}{T} + \frac{1.75\,\Delta E_0}{RT^2}\right) \tag{E23}$$

This expresses the variation of flux with temperature in turbulent-flow gel polarization. It will be noted that Eq. (E23) bears a striking resemblance to the laminar-flow equation (E15). Again, how close are we to the inverse-viscosity relationship (at 300 K)

$$0.66\left(\frac{1}{T} + \frac{1.75\,\Delta E_0}{RT^2}\right) = 0.66(0.0033 + 0.037) = 0.026$$

which is again within 25%.

Table 15 presents turbulent-flow flux data for the ultrafiltration of dilute Syton 2X silica sol. Again the product of flux times viscosity is relatively constant.

Figure 91 presents data taken on the ultrafiltration of a water-oil emulsion in turbulent flow through a 1-in-diameter tube.[87] Increasing the temperature from 15 to 38°C increases the flux by a factor of 2.25. The ratio of the viscosities at these temperatures is only 1.14/0.68 = 1.67.

Using Eq. (E23), with $T = 300$ K, the predicted increase in flux should be $(0.026)(23) = 0.6$, or 60%, which is very close to the actual viscosity (as opposed to theoretical). The additional increase in flux at elevated temperatures may be due to the augmented diffusion of oil droplets via the tubular-pinch effect, since the radial-migration velocity should increase as the viscosity decreases [see Eq. (30)].

TABLE 15   Effect of Viscosity on Flux Rate

Temp, °C	Flux, IGFD*	Water viscosity, cP	Flux × viscosity
17	30.6	1.083	32.9
20	34.3	1.000	34.3
27	39.4	0.8545	33.7
28.2	39.4	0.836	32.8
30.2	42.2	0.801	33.8
32	43.0	0.768	33.0
34	45.6	0.737	33.4
35.5	47.7	0.715	34.2
37.5	49.8	0.686	34.2
40.0	51.2	0.656	33.5
42.0	52.8	0.631	33.4
43.0	52.2	0.621	32.4
45.5	55.4	0.593	32.8
49	58.4	0.559	32.6
53	62.8	0.523	32.8
57.5	69.0	0.487	33.4
62.5	71.6	0.452	32.2
63.5	75.4	0.44	33.4
69.5	81.5	0.408	33.4

*Imperial gallon per square foot per day [gal/(ft²)(day)].

Solute Transport   The activation energy for salt transport in RO membranes is dependent on the nature of the salt, but it generally exceeds that for water transport so that solute rejection improves with decreasing operating temperature. For example, the activation energy for NaCl transport is about 7.4 kcal/mole. In one case, increasing the temperature from 0 to 35°C increased the flux from 4 to 12.5 gal/ft²/day (a threefold increase) (the viscosity ratio is 1.8/0.8 = 2.25), but the rejection for NaCl fell from 97.8 to 96.8%.

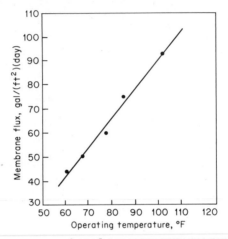

Fig. 91   Membrane flux vs. operating temperature.

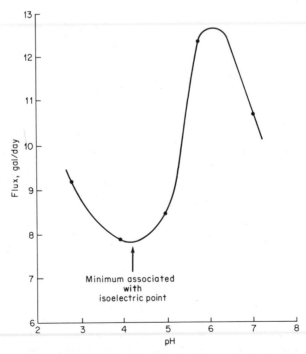

Fig. 92   Effect of pH on flux (cheshire-cheese whey).

APPENDIX F   Effect of pH on Gel Concentration

If the gel-polarization model has any correspondence to reality, the gel concentration C_G should vary as the solubility of the species retained by the membrane. As pointed out in Appendix E, it is surprising that C_G is often invariant with temperature, and one begins to wonder about the nature of this constant.

The effect of pH on membrane flux can be rather dramatic, as seen in Fig. 92. The minimum is associated with the isoelectric point of the proteins present in the cheese whey. At about pH 4, there is minimum solubility of the proteins present. The decrease in flux at pH 7 and above is logical, since these proteins denature in alkali and calcium and/or magnesium salts are precipitated. Unfortunately, no flux vs. concentration data are available to derive the gel concentration at each pH. However, it may be surmised that the

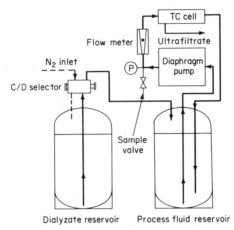

Fig. 93   Batch diafiltration.

gel concentration shifts with pH since the mass-transfer coefficient is not expected to be affected significantly.

APPENDIX G   Diafiltration

Diafiltration differs from conventional diaylsis in that the rate of microspecies removal is not dependent on concentration but is simply a function of the ultrafiltration rate (membrane area) relative to the volume to be exchanged or dialyzed.

Reduction or alteration of microsolutes in a solution is generally obtained by dialysis, a time-consuming and relatively inefficient procedure at low concentration. The same effect can be achieved in a fraction of the time in an ultrafiltration system. Repeated or continuous addition of fresh solvent, called diafiltration, flushes out or exchanges salts and other microspecies efficiently and rapidly.

Referring to Fig. 93, a batch diafiltration is carried out on a process-fluid reservoir by adding dialyzate or wash solvent to the reservoir at a rate equal to removal of ultrafiltrate. (Practically, this can be achieved by using level controllers in the reservoir or a pressure-balance system such that the liquid volume in the reservoir remains constant.) Under this arrangement, the reduction in microspecies, which are freely permeable to the membrane, can be computed from

$$\ln \frac{C_0}{C_f} = \frac{V_t}{V_0}$$

where C_0 = initial microsolute concentration in the reservoir
$\quad\quad C_f$ = microsolute concentration after volume V_f of wash solution has passed through the cell
$\quad\quad V_t$ = volume of ultrafiltrate or wash solution added to the cell from the reservoir

V_0 = volume of initial preparation added to the cell (kept constant during run) When salt or microsolute is to be added, exchanged, or "washed in" to the macromolecular preparation, the following is applicable:

$$\ln \frac{C_f}{C_f - C} = \frac{V_t}{V_0}$$

where C_f = concentration of microsolute in the feed reservoir
C = cell concentration of microsolute after volume V_f has passed through the cell

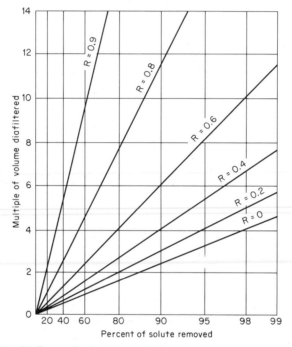

Fig. 94   Diafiltration for the partition of solutes not retained by the membrane.

Figure 94 may be used to predict passage through a membrane of partially rejected solutes as a function of the total volume diafiltered. With a nonrejected permeable solute ($R = 0$) as the impurity, effective clearance (> 99%) will be obtained when filtrate volumes five times the original cell volume have been collected. With partial solute retardation, filtrate volumes must be accordingly increased. For a partially rejected contaminant, e.g., $R = 0.4$, about 8 volumes of solvent must pass to obtain 99% removal of this species.

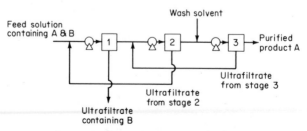

Fig. 95   Continuous countercurrent diafiltration.

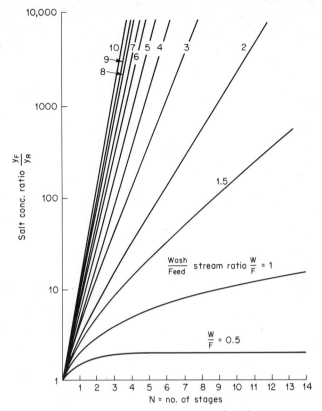

Fig. 96 Salt-concentration ratio vs. number of stages.

The excessive volume of dialyzate (wash solvent) required for complete purification in batch diafiltration can be reduced with a staged-process continuous countercurrent diafiltration (see Fig. 95). The addition of extra stages permits a lower wash solvent W to feed stream F ratio to achieve the same degree of purification (see Fig. 96).

Hydrometallurgical Extraction

GORDON M. RITCEY *Head, Extractive Metallurgy Section, Mineral Sciences Laboratory, Canada Centre for Mineral and Energy Technology, Department of Energy, Mines and Resources, Ottawa, Ontario.*

ALLAN W. ASHBROOK *Manager, Research and Development Division Eldorado Nuclear, Ltd. Ottawa, Ontario.*

NOMENCLATURE

Antagonism: Antonym of synergism.

A/O ratio: Volume-phase ratio of aqueous to organic phases; may also be expressed as O/A.

Aqueous feed: The aqueous-solution feed to the extraction stage which contains the metal or metals to be extracted.

Back mixing: Deviation from an ideal (plug) flow pattern in a contactor.

Contactor: A device for dispersing and disengaging immiscible solution mixtures; it may be single or multistage.

Continuous phase: The coherent phase in a contactor.

Countercurrent extraction: Extraction in which the aqueous and organic phases flow in opposite directions.

Crud: The material resulting from agitation of an organic phase, an aqueous phase, and fine solid particles that form a stable mixture. Crud usually collects at the interface between aqueous and organic phases.

Diluent: The organic liquid in which an extractant and modifier are dissolved to form a solvent.

Dispersed phase: The phase, in a contactor, which is discontinuous. Generally the dispersed phase is in the form of droplets.

Distribution: The apportionment of a metal (solute) between two phases.

Distribution coefficient: See *Extraction coefficient.*

Distribution isotherm: See *Extraction isotherm.*

Equilibration: Treatment of the solvent prior to its entering the extraction stage.

Equilibrium: The position when the chemical potentials of both aqueous and organic phases are equal.

Equilibrium constant: The equilibrium constant of a specified distribution reaction expressed in terms of thermodynamic activities.

Extract: Used as a verb to describe the transfer of a metal from one phase to another.

Extractant: The active organic component of the solvent primarily responsible for the extraction of a metal.

Extraction: The operation of transferring a metal from an aqueous to an organic phase.

Extraction coefficient E: The ratio of total concentrations of metal (in whatever form) after contacting an aqueous and an organic phase under specified conditions:

$$E = \frac{\text{concentration of metal in organic phase}}{\text{concentration of metal in aqueous phase}}$$

Extraction isotherm: The graphical presentation of isothermal equilibrium concentrations of a metal in the aqueous and organic phases over an ordered range of conditions in extraction.

Extraction raffinate: The aqueous phase from which a metal (or metals) has been removed by contacting with an organic phase.

Flooding: The discharge of mixed phases from one or both exit ports of a contactor.

Load: To transfer a metal from an aqueous to an organic phase.

Loaded solvent: The organic solvent containing the maximum concentration of a metal for the conditions under which extraction occurred.

Loading capacity: Refers to the saturation limit of a solvent for a metal or metals.

Maximum loading: See *Loading capacity.*

Mixed solvent: A solution of more than one extractant in an organic diluent.

Modifier: A substance added to a solvent to increase the solubility of the extractant, salts of the extractant, or of the extracted metal species, during extraction or stripping. Also added to suppress emulsion formation.

Phase inversion: The change in a solvent-extraction system when the dispersed phase becomes the continuous phase, or vice versa.

Phase ratio: See *A/O ratio.*

Raffinate: The aqueous phase from which the metal has been removed by extraction; generally a waste stream from a solvent-extraction circuit.

Scrubbed solvent: The organic phase after removal of contaminants by scrubbing.

Scrub raffinate: The aqueous phase after contacting the loaded solvent.

Scrub solution: The aqueous solution used to contact the loaded solvent for the removal of contaminants.

Scrubbing: The selective removal of a metal or impurities from a loaded solvent prior to stripping. Also removal of solvent degradation products and nonstrippable complexes from the solvent, usually after stripping.

Separation factor: The ratio of the extraction coefficients of two metals being compared.

Settling: Separation of dispersed immiscible phases by coalescence or sedimentation.

Solvent: A mixture of an extractant, diluent, and in some cases a modifier. The organic phase which preferentially dissolves the extractable metal species from an aqueous solution.

Solvent extraction SX: Separation of one or more solutes from a mixture by mass transfer between immiscible phases in which at least one phase is an organic liquid.

Solvent inventory: The total quantity of solvent in the process.

Stage: A single contact (dispersion and disengagement). Also refers to a theoretical stage, which is a contact that attains equilibrium conditions in a particular system.

Strip isotherm: Similar to extraction isotherm but for stripping.

Strip liquor: The aqueous solution containing the metal recovered from a loaded solvent by stripping.

Strip solution: The aqueous solution used to contact the loaded (or scrubbed) solvent to recover the extracted metal.

Stripped solvent: The solvent after removal of extracted metal by stripping.

Stripping: The removal of extracted metal from the loaded solvent. Selective stripping refers to separate removal of specific metals from a solvent containing more than one metal.

Stripping coefficient S: The reciprocal of the extraction coefficient.

Synergism: The cooperative and beneficial effect of two or more extractants or reagents that exceeds the sum of the individual effects.

INTRODUCTION

The distribution of a solute (metal) between two immiscible phases was first placed on a scientific footing in 1872 when Berthelot and Jungfleisch[1] enunciated a law describing this phenomenon. This was followed in 1891 by the better-known Nernst equation.[2] The technique of solvent extraction of metals developed from this time, and has been used for many years as an analytical tool.

From a practical point of view, solvent extraction (liquid-liquid extraction) first became an industrial process with its application, in the 1940s, to the processing of uranium and radioactive materials. The 1950s saw the application of this approach to processing to the treatment of base-metal solutions, and since that time it has continued to gain favor in the hydrometallurgical field. Today, some 100 plants of varying size are in operation throughout the world in the processing of some 25 different metals.

Basically, the process of solvent extraction as applied in hydrometallurgy is a simple one, depending essentially on shifting the reaction equilibria of a system, usually by adjustment of the aqueous-phase pH. Indeed, this basic simplicity has a particular attraction for the application of solvent extraction to the separation and isolation of metals from aqueous media in metallurgical processing. But while processes are never as simple as we would like them to be, nevertheless the solvent-extraction process provides an elegant method for the separation of metals in a high state of purity.

In this section we shall endeavor to provide an overview of the solvent-extraction process as it applies to the hydrometallurgical processing of many materials. The subject will be treated in a systematic manner, starting with the basic requirements through process development, operating processes, and then through analytical and environmental aspects to economic considerations.

THE GENERAL SOLVENT-EXTRACTION PROCESS

The general process of solvent extraction as it is applied in hydrometallurgy is shown in Fig. 1. In the extraction stage the feed solution, containing the metal of interest, and the solvent phase are contacted, usually countercurrently, and then allowed to disengage (settle). The metal-barren aqueous phase (raffinate) is either processed further or goes to waste. The metal-rich solvent phase (loaded solvent) may be fed to a second contactor (scrub stage) where it is mixed with a scrub solution to remove unwanted metals. After settling, the aqueous scrub raffinate is recycled, and the solvent (scrubbed solvent) goes to another stage for stripping. A scrub stage may not be required, depending on the particular system. In the strip stage the scrubbed (loaded) solvent is contacted with a strip solution by which the metal is stripped from the solvent. The strip liquor then goes for further processing to recover the metal, and the stripped solvent is recycled to the extraction stage. Depending on the system, the stripped solvent may be treated (equilibrated) prior to return to the extraction stage.

GENERAL PRINCIPLES

The most important factors involved in a solvent-extraction system are the conditions prevailing in the aqueous phase and the composition of the solvent phase. In distinguishing among solvent-extraction systems, since it is impossible to do this by a consideration of the aqueous phase, such systems are classified according to the extractant type. There are several classification systems,[3] and the one used here is as follows:

1. Those systems which involve compound formation
2. Those systems which involve ion association
3. Those systems which involve solvation of the metal ion by the extractant

Our knowledge of the theoretical aspects of solvent extraction decreases from 1 through 3. It is useful, therefore, to consider the theory of solvent extraction in this order.

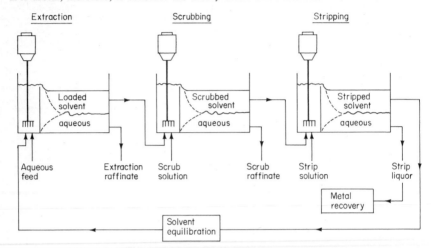

Fig. 1 The general process of solvent extraction.

Extractants involved in compound formation are the chelating and acidic types (Table 1), and the simplest equation describing metal extractions with these reagents is

$$M^{n+} + \overline{nHA} \rightleftharpoons \overline{MA_n} + nH^+ \tag{1}$$

where M^{n+} is the metal ion (e.g., Cu^{2+}, Fe^{3+}), HA is the extractant (e.g., LIX64N), n represents the oxidation state of the metal ion, and the bar represents the solvent phase. Thus these systems involve cation exchange.

As written, Eq. (1) is an equilibrium reaction, and for the extraction of a metal we require the equilibrium to be shifted to the right. For stripping a metal from the solvent phase, the equilibrium must be shifted to the left, and as suggested by Eq. (1), this can be done using acids (H^+).

The degree to which a metal is extracted by a solvent can be readily determined by analysis of the aqueous and solvent phases after they have been contacted. This is expressed as a ratio:

$$\frac{\text{Total concentration of metal in organic phase}}{\text{Total concentration of metal in aqueous phase}} = E \tag{2}$$

where E (or D) is the extraction coefficient for the particular system under the experimental conditions obtained at equilibrium. The stripping coefficient is the inverse of Eq. (2).

For a system described by Eq. (1) we can define an equilibrium constant K_E:

$$K_E = \frac{[\overline{MA_n}][H]^n}{[M^{n+}][\overline{HA}]^n} \tag{3}$$

Now we know that $[\overline{MA_n}]/[M^{n+}] = E$, then

$$K_E = \frac{E[H]^n}{[\overline{HA}]^n} \quad \text{or} \quad E = K_E \left(\frac{HA}{H}\right)^n \tag{4}$$

Thus the extractability of a metal E *under the experimental conditions used* is propor-

TABLE 1   Commercial Extractants

Extractant	Type	Structure	Supplier
Di(2-ethylhexyl) phosphoric acid	Acidic	$\left(\begin{array}{c}CH_3(CH_2)_3CHCH_2O\\ \quad\quad\quad\vert\\ CH_3CH_2\end{array}\right)_2 POOH$	Union Carbide
Naphthenic acids			Shell Chemicals
Versatic acids			Shell Chemicals
LIX 63	Chelating	$\underset{\underset{HON\ \ OH}{\Vert\ \ \vert}}{CH_3(CH_2)_3CHC\ \ CHCH(CH_2)_3CH_2}$ (with $_5H_2C$ and C_2H_5 groups)	General Mills
LIX 64		+LIX 63	General Mills
LIX 64N		LIX 65N + LIX 63 (~1 vol %)	General Mills
LIX 65N			General Mills
LIX 70	Chelating	+LIX 63	General Mills
LIX 71		LIX 70 + LIX 65N	General Mills
LIX 73		LIX 70 + LIX 65N + LIX 63	General Mills
Kelex 100			Ashland Chemicals
SME 529			Shell Chemicals

TABLE 1   Commercial Extractants (Continued)

Extractant	Type	Structure	Supplier
Acorga P17, P50 (R = $CH_2C_6H_5$; H)		C_9H_{19} ... $R-\overset{}{\underset{HON}{C}}$—⟨benzene ring⟩—OH	Acorga, Ltd.
Primary amines, secondary amines, tertiary amines, quaternary ammonium halides	Ion association	$R\cdot NH_2$ (R = C_{12}–C_{14}) R_2NH (R = C_{10}–C_{12}) R_3N (R = C_8–C_{10}) $(R_3N^+CH_3)Cl^-$ (R = C_8–C_{10})	Rohm & Haas, Ashland Chemicals, General Mills
Tri-n-butyl phosphate, methyl isobutyl ketone	Solvating	$(CH_3(CH_2)_3O)_3P{=}O$ $(CH_3)_2CH_2CH_2\overset{\overset{O}{\|\|}}{C}CH_3$	Ashland Chemicals

tional to the nth power of the extractant concentration, inversely proportional to the hydrogen-ion concentration of the aqueous phase, and independent of metal concentration. Equation (4) says nothing, however, about the kinetics of the system.

Another way of expressing metal extractability in a given system is as percent metal extracted P:

$$P = \frac{100E}{E + v/\bar{v}} \tag{5}$$

where v and $\bar{v}$ are the volumes of the aqueous and solvent phases, respectively. For the case where $v = \bar{v}$, then $v/\bar{v} = 1$; but for all other conditions,

$$P = \frac{100(Ev/\bar{v})}{(Ev/\bar{v}) + 1} \tag{6}$$

From (4) and (5),

$$E = \frac{P}{100 - P} = K_E \left(\frac{HA}{H}\right)^n \tag{7}$$

The above equations apply generally to all three of the extractant systems listed above, and are used widely in process-development studies. It must be stressed that the use of E provides *only an overall assessment* of an extraction system at equilibrium, *under the experimental conditions used,* and provides no insight into the reactions and mechanisms occurring during the attainment of the equilibrium state.

EXTRACTANTS, DILUENTS, AND MODIFIERS

The solvent phase in solvent-extraction processing consists of an extractant dissolved in a diluent, and in some cases includes a modifier (see Nomenclature for definitions).

Extractants

Typical commercial extractants available today are listed in Table 1. In Tables 2 and 3 are listed some diluents and modifiers, respectively, which are available for use in solvent-extraction processing.

Chelating extractants are those which form metal chelates, as for example, with the LIX extractants:

$$MSO_4 + 2 \ldots \rightleftharpoons \ldots \quad (8)$$

$$+ H_2SO_4$$

All the chelating extractants available today have been designed specifically to extract copper from acid solutions.[4] By variation in the pH of the system, other metals may also be extracted.[5-7]

Systems involving acidic extractants (Table 1), such as di(2-ethylhexyl) phosphoric acid (EHPA), are generally described by the same qualitative considerations as for chelating extractants, but mechanisms of extraction and composition of the extracted species are less predictable.

The order of metal extraction as a function of pH is the same for all acidic extractants, with the odd exception, and follows the order of metal hydrolysis constants. For example, EHPA extracts metal in the order $Fe^{3+} > Zn^{2+} > Cu^{2+} > Co^{2+} > Ni^{2+} > Mn^{2+} > Mg^{2+} > Ca^{2+}$ with increasing pH. A similar order is found with carboxylic acid extractants.

Solvent-extraction systems involving ion association are more difficult to analyze than those involving compound formation. This arises mainly because of the need for high-ionic-strength solutions and an increasingly complex chemistry of such systems. Most extractants employing ion association are long-chain amines (Table 1). Such systems are also classified as anion-exchange systems, for obvious reasons. Basically, the extraction of a metal M^{n+}, which forms anionic complexes with an anion A^- in the aqueous phase, by an amine salt R_3NHB can be represented as an anion-exchange process:

$$MA_m^{(m-n)-} + \overline{(m-n)R_3NHB} \rightleftharpoons \overline{[(R_3NH)_{m-n}MA_m]} + (m-n)B^- \qquad (9)$$

For example, the extraction of uranium from a sulfate liquor by a tertiary amine sulfate:

$$\overline{(R_3NH)_2 {}^*SO_4} + UO_2(SO_4)_2^{2-} \rightleftharpoons \overline{[(R_3NH)_2UO_2(SO_4)_2^{2-}]} + {}^*SO_4^{2-} \qquad (10)$$

The order of metal extraction by tertiary amines from hydrochloric acid media is $Zn^{2+} > Pb^{2+} > Fe^{3+} > Cu^{2+} > Co^{3+} > Mn^{2+} > Cr^{3+} > Ni^{2+}$ with increasing acid concentration.

The fact that nickel does not form anionic chloro complexes, whereas cobalt and copper do, is the basis for the separation of cobalt and copper from nickel by tertiary amines.

Solvating extractants (Table 1) are those which can solvate inorganic molecules or complexes and solublize them in organic solutions. These are two major extractant groups here: those containing oxygen bonded directly to carbon, such as ethers, esters, ketones, and alcohols; and those containing oxygen bonded to phosphorus, such as alkyl phosphate esters. Generally, the extractive power of phosphorus-containing extractants increases with increase in C-P bonds over the series: phosphate-phosphonate-phosphine oxide. Because of the complex chemistry involved in metal extraction by solvating extractants, we know relatively little about the actual extraction mechanisms.

The best known of this type of extractant is tri-n-butyl phosphate (TBP), mainly because of its wide use in the nuclear field.[8,9] Generally, the higher the metal oxidation state the greater the extraction by TBP. Thus at $5M$ nitric acid concentrations, the order of metal extraction is $U(VI) > Th(IV) > Bi(III) > Fe(III) > Cu(II) \approx Zn(II) \approx Ni(II) \approx Co(II)$.

Alcohols are used mainly for the extraction of phosphoric acid produced from dissolution of phosphate rock. Methyl isobutyl ketone is used in the separation of zirconium and hafnium.[10]

Diluents

Those diluents which have been considered for, or actually used in, solvent-extraction processes are shown in Table 2, together with some of their properties.[11] Diluent compo-

TABLE 2 Commercially Available Diluents

Diluent	Flash pt., °F	Component analysis Aromatic	Component analysis Paraffin	Component analysis Naphthenes	sp. gr. (20°C)	BP, °F	Kauri butanol	Solubility parameter	Viscosity at 25°C
Isopar L[a]	144	0.3	92.7	7.0	0.767	373	27	7.2	1.60
Isopar E[a]	<45	0.05	99.94		0.723	240	29	7.1	
Isopar M[a]	172	0.3	79.9	19.7	0.782	405	27	7.3	3.14
Norpar 12[a]	156	0.6	97.9	1.1	0.751	384			1.68
Esso LOPS[a,f]	152	2.7	51.8	45.4	0.796	383			2.3
DX3641[a]	135	6.0	45	49	0.793	361	33.6	7.7	1.165
Shell 140[b]	141	6.0	45	49	0.785	364	32		
Napoleum 470[c]	175	11.7	48.6	39.7	0.811	410	33		2.10
Escaid 110[a]	168	2.4	39.9	57.7	0.808	380			2.51
Shell livestock spray[b]									
Mentor 29[a]	270	15	48	37	0.819	512			
Shell Parabase[b]	280	15	48	37	0.800	500			
Escaid 100[a]	210	20	56.6	23.4	0.788	428			1.78
NS-144[a]	168	16	42	42	0.790	376			
NS-148D[a]	140	4.5	38.6	43			30.9		1.603
Solvesso 100[a]	160	98.9	1.1		0.876	315	92	8.8	
Solvesso 150[a]	112	97.0	3.0		0.895	370	90	8.7	1.198
Xylene[a]	151	99.7	0.3		0.870	281	98	8.9	0.62
HAN[a]	80	88.5	4.1	6.8	0.933	357	105	8.9	1.975
Cyclohexane[e]	105			100				8.2	
Chevron 40L[d]	141	78			0.886	360	76		
Chevron 370[d]	127	0			0.758	346	27		
Chevron 44L[d]	154	69			0.893	366	73		
Chevron 3[d]	145	98			0.888	360	88		
Chevron 425[d]	142	2			0.787	360	26		
Chevron LOS[d]	130	0			0.779	350	33		
Chevron 25[d]	115	99			0.875	316	94		

[a] Esso or Exxon.
[b] Shell.
[c] Kerr-McGee.
[d] Chevron.
[e] Others.
[f] Now Escaid 200.

sition can range from essentially aliphatic (e.g., Isopar E) to essentially aromatic (e.g., Solvesso 100), and all are fractions from petroleum distillation.

While in the past diluents have been thought not to enter into the chemistry of the solvent-extraction process, recent studies have shown this to be far from the truth.[11] Indeed, the choice of diluent can have a considerable effect on the extraction system, not only from the chemical reactions involved but also from the physical point of view, such as in phase separation and emulsion-forming tendencies.

A priori choice of a diluent cannot be made, since we understand very little of the reasons for diluent effects on solvent-extraction systems. Thus diluent selection must be made after experimental work on the system.

TABLE 3   Modifiers

	sp. gr.	BP, °F	Flash pt., °F (TOC)
2-Ethylhexanol	0.834	365	185
Isodecanol	0.840	428	220
Tri-*n*-butyl phosphate	0.973	352	380
p-Nonyl phenol	0.94		

Modifiers

Modifying reagents are usually employed as third-phase inhibitors. The action here is to increase the solubility of the extracted species in the solvent phase. Modifiers can also help in inhibiting the formation of emulsions. Like diluents, however, a priori choice of a modifier cannot be made. Further, modifiers can influence the extraction and stripping characteristics of a solvent.[11] The most popular modifiers used today in commercial solvent-extraction processes are given in Table 3.

DISPERSION AND COALESCENCE

Of major importance in solvent-extraction processing are those aspects concerned with the physical mixing and separation of the aqueous and solvent phases. In dispersing one phase in the other we can define two extreme situations, one a temporary dispersion from which the dispersed phase rapidly coalesces and separates from the nondispersed or continuous phase, and one in which a stable emulsion is produced.

Ideally the rapid coalescence of the dispersed phase after mixing is desirable. However, this may provide for poor mass transfer of metal because of an inadequacy of mixing or dispersion. Mass-transfer rates across liquid-phase boundaries depend in part on the drop-size distribution, that is, on the interfacial surface area of the dispersed phase. As a result, the type of agitation and energy input provided in mixing becomes important. Too fine a dispersion (very high interfacial area) may also result in poor mass-transfer rates because very small bubbles tend to behave like rigid spheres.

Dispersion characteristics of a system usually vary depending on whether the aqueous or the solvent phase is the dispersed phase. Thus while an organic-continuous system may behave best for an extraction stage, the opposite may be true for the stripping stage. But whatever the system and conditions chosen, the best results are invariably the result of a compromise among the various operating conditions to provide maximum mass transfer, coalescence rates, and throughput in the system.

In phase coalescence or disengagement, two distinctly separate stages can be defined: a primary break of the phases, followed by a secondary break. While the primary break provides the major separation of the phases, secondary break time is of importance because if this is long, entrainment of one phase in the other will result, which can prove costly in terms of solvent losses or in a decrease in purity of the recovered metal.

The process of drop coalescence is shown in Table 4 for a binary dispersed system.[12] The main factors affecting coalescence of a dispersed phase are drop size, secondary drop formation, phase viscosity and density, interfacial tension, presence of solids, and temperature. While considerable data, and theory, are available on dispersion and coalescence in the absence of mass transfer, the situation is decidedly more complex in the presence of mass transfer as it occurs in solvent-extraction processes, and these phenomena are not well understood. Accordingly, the physical properties of any solvent-extraction process have to be determined and optimized by empirical methods.

TABLE 4   The Process of Drop Coalescence in a Binary Dispersed System[12]

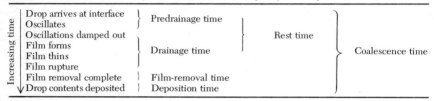

SOURCE: Hanson, C. (ed.): "Recent Advances in Liquid-Liquid Extraction," Pergamon, New York, 1971, with permission.

DEVELOPMENT OF A SOLVENT-EXTRACTION PROCESS

In the development and design of a solvent-extraction process, several stages are to be considered, and data to be acquired, such as:

1. Extractant selection and solvent composition
2. Equilibrium data, for both extraction and stripping
3. Kinetics of extraction and stripping
4. Calculation of theoretical stages (extraction and stripping) for maximum process efficiency
5. Retention times for mixing and settling
6. Size of equipment for desired throughput, mixing and settling volumes, and areas
7. Product purity, and raffinate composition
8. Solvent losses

In the choice of a suitable solvent for a particular system, certain preliminary screening tests can be carried out in order to narrow the range of reagents to be tested. The solvent should have a very low solubility in the aqueous medium, and no degradation on recycle; it should not form a stable emulsion when mixed with the aqueous phases, and should be relatively inexpensive. These screening tests or shake-outs are normally carried out in separatory funnels in the laboratory. When the choice of possible extractants has been narrowed, additional shake-outs are performed to determine the following: how specific the extractant is for a particular metal, the loading capacity, the pH and temperature dependency, kinetics of extraction and stripping, the requirements for modifier or special diluents, and the scrubbing and stripping characteristics.

In the development of a flowsheet upon which plant design and operating cost estimates may be made, considerable information is required. Normally, most use is made of the extraction coefficient E, which as stated earlier, is a measure of the extractability of a metal.

The active organic dissolved in the inert diluent is made to a certain concentration and therefore is capable of holding, at saturation, a limited amount of a metal. This "saturation loading" value for the system is therefore quite important, since it governs the volume of solvent flow necessary to contact a fixed tonnage of feed solution. By variation of the ratio of aqueous and organic volumes in bench-scale shake-outs, or by repeated contact of organic with fresh feed solution, the saturation loading for a particular system may be determined. These data on saturation can also be utilized to construct an *extraction isotherm* or *distribution isotherm,* as shown in Fig. 2.

Each process of mixing and separation of the organic and aqueous phases is considered a *stage*. Depending upon the number of *theoretical stages,* and therefore the type of equipment to be used, the feasibility of the process may be determined.

The extraction isotherm becomes of additional value when it is used to calculate the number of theoretical stages of extraction (or stripping) necessary to achieve the required organic saturation with a minimum loss of metal to the raffinate. This is known as a McCabe-Thiele diagram. In the construction of such a diagram, the extraction isotherm is first plotted. Then a second line, known as the *operating line* (slope = phase ratio), is drawn. The operating line is based on the mass balance of the system, and therefore the concentrations of metal in the solvent entering and in the aqueous raffinate leaving any stage are coordinates of points on the operating line. Similarly, the concentrations of metal in the aqueous feed entering and in the extract leaving any stage are coordinates of such points. A typical diagram is shown in Fig. 3. The theoretical stages are stepped off by extending a horizontal line from the upper end of the operating line to intercept the operating line. This is continued until the lower end of the operating line is intersected. Each step is called a theoretical stage.

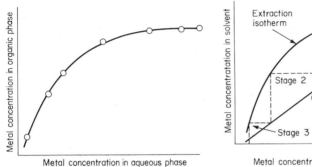

Fig. 2   Typical extraction isotherm.   **Fig. 3**   McCabe-Thiele diagram.

In the example shown, slightly over three theoretical stages would be required. In actual practice, if mixer-settlers were to be considered for extraction, a cascade of four contactors would be required. When multistage contactors such as columns or centrifuges are used, in which an extraction stage is not evident, the McCabe-Thiele diagram permits the calculation of extraction and stripping efficiency and the correlation of this efficiency with such operating variables as flow rates, contact time, mixing, and temperature. It is usually found that the speed with which equilibrium is attained is affected by variables governing the rate of mass transfer, such as the turbulence or degree of mixing, the temperature, and the viscosity of the system.

It has been the practice in bench-scale testing to use a series of cascading mixer-settlers. These are usually very small, perhaps up to 100 mL in capacity, and in some cases are ideal for testing the process in the laboratory on a continuous basis. One can probably safely say that all the design data for the copper and uranium solvent-extraction plants were obtained initially by small-scale mixer-settler units. Neat and compact designs are available from suppliers.[13-15]

After sufficient data have been obtained to construct a conceptual flowsheet, the next stage in process development is to construct and operate a pilot plant. Typical data obtainable from pilot-plant operations are:

1. Aqueous and solvent flow rates, liters/min (gal/min) (phase ratio)
2. Number of extraction stages required; stage volumes (size of mixer)
3. Number of stripping stages required
4. Contact times (min)
5. Settling areas, $m^2/(L)(min)[ft^2/(gal)(min)]$
6. Temperature requirements
7. Recycle requirements, liters/min (gal/min)
8. Analysis of aqueous and organic phases, solvent losses, product purity
9. Metal-separation efficiency
10. Power requirements

Pilot-plant operations should be continued until sufficient data and experience are gained to scale-up the process confidently to plant requirements.

EQUIPMENT

In order to arrive at the proper decisions in the choice of contactor for the solvent-extraction process, one must have an understanding of the physical aspects of the process.

These are primarily concerned with the dispersion of the two phases on mixing, the size of droplet, and the rate and completeness of the coalescence process. The rate of mass transfer is a function, among other variables, of the drop-size distribution or interfacial area between the phases. The settling area required is dependent upon the rate of coalescence, which is dictated by the droplet size. Also the kinetics of the system, together with the dispersion and coalescence, will influence the choice of the contactor. Systems with slow kinetics, for example, demand greater dispersion and contact time, so that mixer-settlers could probably be used; whereas in a fast kinetic system the choice of contactors becomes wide.

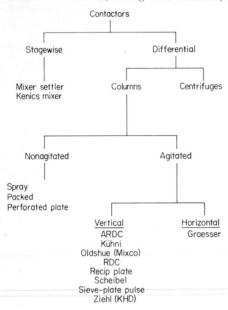

Fig. 4   Contactor classification.

The most common type of contactor used during the early years of solvent extraction was the mixer-settler; columns were also used in the nuclear field. Mixer-settlers have undergone modifications during the last two decades; today there are various designs, as well as a multitude of column types and other contactor designs. Thus with so many types of contactors presently available, a clear understanding of the chemical and physical aspects of the process, as well as what the contactors will do, is necessary in the selection of the most suitable contactor for a particular process. The economics of the choice must be considered with respect to particular conditions of solutions and solvent type, throughput, kinetics and equilibrium, dispersion and coalescence, solvent losses, corrosion, number of stages, and available area.

The major emphasis in the development and optimization of equipment for the solvent-extraction process has been to increase throughput while maintaining the efficiency, and to increase efficiency without reduction in throughput. However, with all the equipment that is available, and the amount of engineering data produced on synthetic systems, very little can be predicted regarding scale-up to plant size without first running a pilot plant.

It should be emphasized that there is no "universal contactor" which is suitable for all processes, because of the many variables already mentioned. Therefore, selected contactors are usually evaluated on the process feed solution, and based on this assessment subsequent pilot-plant tests will optimize the process variables and the most suitable contactor for scale-up to plant size.

Mass transfer in solvent-extraction processing is usually accomplished in continuous countercurrent contacting equipment, which can be divided into two main classes according to whether their mode of operation is stagewise or differential. The classification of types of contactors is shown in Fig. 4.[16] The differential column contactor can be further subdivided into nonagitated and agitated. The simplest nonagitated contactors such as spray, packed, or perforated-plate columns achieve both mixing and countercurrent flow. Some of the major advantages and disadvantages of the various contactors are shown in Table 5.[16]

Section 1.10 describes some of the more common contactors.

MATERIALS OF CONSTRUCTION

Because many alternatives are often available in the choice of materials of construction for the solvent-extraction plant, each plant designer and contractor must choose carefully. The ultimate selection can greatly affect the plant costs—from both a capital and an operating standpoint, because of corrosion problems, frequency of maintenance, and often poor metallurgy resulting from corrosive or degradation problems. Where existing plants are in operation, the knowledge and experience derived from those plants can be invaluable.

With the usually corrosive conditions existing in most solvent-extraction plants, owing to both the solvent and the aqueous feed solution, severe restrictions are imposed on the choice of materials of construction. Synthetic materials such as polyvinyl chloride and reinforced polyester resins have been common in recent years, although coated mild steel

TABLE 5   Equipment Performance[16]

	Mixer-settlers	Nonagitated differential	Agitated differential	Centrifugal
Advantages	Good contacting of phases Handle wide-range flow ratio (with recycle) Low headroom High efficiency Many stages Reliable scale-up Low cost Low maintenance	Low initial cost Low operating cost Simplest construction	Good dispersion Reasonable cost Many stages possible Relatively easy scale-up	Handle low gravity difference Low holdup volume Short holdup time Small space requirement Small inventory of solvent
Disadvantages	Large holdup High power costs High solvent inventory Large floor space Interstage pumping may be necessary	Limited throughput with small gravity difference Cannot handle wide flow ratio High headroom Sometimes low efficiency Difficult scale-up	Limited throughput with small gravity difference Cannot handle emulsifying systems Cannot handle high flow ratio Will not always handle emulsifying systems, except perhaps pulse column	High initial cost High operating cost High maintenance Limited number of stages in single unit, although some units have 20 stages

and stainless steel are still common. A comparison of costs for tanks constructed of different materials is shown in Table 6.[19] Such a reduction in tank costs can result in lower plant construction costs provided that the choice of material is compatible with the particular process.

There appears to be a gradual evolution of materials of construction in solvent-extrac-

TABLE 6   Comparison of Tank Costs[52]
(Covered tanks, without nozzles in dollars)

Volume, m³	Steel	Rubber-lined steel	Opanol-lined steel	PVC	PVC + fiberglass-reinforced polyester	Fiberglass-reinforced polyester	Stainless steel 304
1	100	229	263	183	246	80	314
10	685	1257	1429		1114	829	2,057
70	4200	6343	6971		4400	3686	12,686
120	5857	8829	9714		7457	6857	17,571

tion plants. Present-day copper and uranium plants bear little resemblance to the early plants designed for the recovery of uranium from mill leach solutions, all using a type of mixer-settler design which also evolved over the years. Mild steel or sheet iron covered with a plastic coating were frequently used in the early mixer-settler plants, both in construction of the mixer and the settler, as well as for storage and feed tanks. Piping was frequently polyvinyl chloride and the valves were of numerous types. Later, the tanks used were steel-lined with a polyester reinforced with fiberglass. Some of the mixing vessels were mild steel lined with polypropylene, while Type 316 stainless steel and fiberglass tanks were in use for settlers and storage and feed tanks. Subsequently, concrete and compartmented mixer-settler tanks were introduced, lined with fiberglass or fiberglass-reinforced plastic. Piping, valves, and transfer pumps were of 316SS or 316LSS, while the latter were also plastic-lined. Storage and feed tanks were constructed of reinforced polyester resin or fiberglass-lined steel. Impellers and shafts were often of reinforced bisphenol polyester.

The early copper plants appeared to be influenced by the experience gained by the uranium mills. The Rancher's Bluebird mill in Arizona, the first large copper solvent-extraction plant, had the mixing chambers constructed of steel lined with isophthalic reinforced fiberglass and 316SS impellers.[20] Three layers of the same resin covered the concrete settlers. Most of the piping was reinforced fiberglass or polyvinyl chloride, although some consisted of 316SS. Pumps and valves were of 316SS or Alloy 20. When the next large copper plant, Bagdad, Ariz. was constructed, 316SS was used extensively throughout the plant for the mixers, settlers, various tanks, and piping.[21]

Solvent-extraction plants employing multistage column contactors are usually stainless steel 316, 304L, or 316L or in some cases glass, depending upon the corrosive conditions. The uranium-refining plants using the nitric acid system have generally employed 304LSS in the sieve-plate pulse columns,[22] as was also the case in a recent nitric acid zirconium circuit for the valves, piping, and Mixco columns.[23] A more complete summary of materials of construction for various plants is to be published.[17]

As stated earlier, although capital costs and installation can be effectively reduced by choice of certain materials of construction, the maintenance costs may be considerably different for the various materials and these costs can be increased because of corrosion, brittleness, or some other factor. For example, in one system, the plastic liner of a carbon-steel mixer-settler degraded with continual contact with the solvent, resulting in subsequent corrosion of the tanks by the acid feed solution. Liners of epoxy are often hard and brittle, and if undamaged can give long service. Once the coating is damaged, the underlying steel or concrete is readily corroded. In the case of concrete tanks lined with epoxy, the concrete is rapidly attacked by acid once the coating is damaged, often resulting in crud problems due to the chemical attack of the concrete by the acid solution. Polyesters are more resistant to strong oxidants than are the epoxies, and while fiberglass-reinforced coatings are often brittle they are easily repaired. Stainless steel, although more expensive, is used for many applications where both corrosion and shock resistance are required.

The choice of materials of construction is greater in alkaline extraction systems than in acidic systems. In the stripping stage, if acid is used, the corrosion conditions are similar to those found in acidic systems.

The reader is referred to an excellent article by G. J. Bernard of Bechtel Engineering on materials selection and performance, in which various materials for possible use in a copper solvent-extraction plant are discussed and evaluated.[24]

PROCESSES

This discussion could be dealt with extensively, but because of the restricted length and scope of this section, only a few processes will be cited. Thus examples of processes employing solvating, ion association, chelating, and acidic extractants will be briefly described.

Solvating

Tributyl phosphate (TBP) has been used extensively in the uranium-refining industry since the first solvent-extraction plants were put into service in the early 1950s. More recently, this extractant has been used for other metals such as iron from chloride systems and zirconium from nitrate solutions. The recent zirconium circuit of Eldorado Nuclear is briefly discussed.[25]

The feed solution at 100 g Zr/L, and containing up to 7 g Si/L and about 2 g Hf/L, with a solution specific gravity of 1.4 is fed to a 9-m-high (30 ft), 750-mm-diameter (30-in) Mixco column. This feed solution is contacted, organic continuous, and organic/aqueous (O/A) of 4/1 with a 50% solution of TBP in kerosine which had previously been equilibrated with nitric acid. Following extraction the loaded solvent is scrubbed with $4.5N$ HNO_3 to remove coextracted hafnium. The scrubbed solvent then passes to a third column for stripping with water to recover zirconium nitrate, which is subsequently evaporated and recrystallized. The resultant product contains <200 ppm Hf on a Zr basis. Zirconium tetrafluoride can be produced from this concentrated strip solution by the addition of hydrofluoric acid.[26] The flowsheet is shown in Fig. 5.

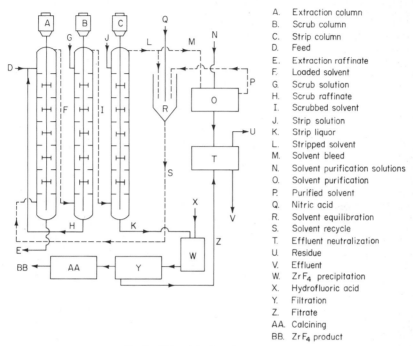

A.	Extraction column
B.	Scrub column
C.	Strip column
D.	Feed
E.	Extraction raffinate
F.	Loaded solvent
G.	Scrub solution
H.	Scrub raffinate
I.	Scrubbed solvent
J.	Strip solution
K.	Strip liquor
L.	Stripped solvent
M.	Solvent bleed
N.	Solvent purification solutions
O.	Solvent purification
P.	Purified solvent
Q.	Nitric acid
R.	Solvent equilibration
S.	Solvent recycle
T.	Effluent neutralization
U.	Residue
V.	Effluent
W.	ZrF_4 precipitation
X.	Hydrofluoric acid
Y.	Filtration
Z.	Fitrate
AA.	Calcining
BB.	ZrF_4 product

Fig. 5   Eldorado's zirconium process.

Ion Association

After the initial uranium-refining operations using TBP, the next plants were essentially those for the recovery of uranium from mill leach solutions. The extractant was almost exclusively the tertiary amine, Alamine 336, and in recent years, also Adogen 364, although EHPA also had limited use. The processing of uranium ores has been usually by sulfuric acid leaching followed by solvent extraction and precipitation, although alkaline carbonate leaching can be a satisfactory lixiviant for certain ores, but no satisfactory extractant has been developed for that medium. The reader is referred to many particularly good publications on uranium-ore processing, such as "The Extractive Metallurgy of Uranium" by Merritt, "Uranium Ore Processing" by Clegg and Foley, and the numerous proceedings of international uranium agencies. One of the recent uranium mills is briefly described below.

The Highland Uranium operation in Casper, Wyo., commenced production in October 1972.[27] Ore, grading approximately 0.0017 kg U_3O_8/kg [3.5 lb U_3O_8/ton] ore, is mined by open-pit methods. After grinding to 80% minus 35 mesh, the ore, at 35% solids, is leached in eight 5.4-m-diameter (18-ft) wooden tanks, using sulfuric acid and sodium chlorate. Following CCD separation and clarification the feed to solvent extraction contains 0.5 to 0.7 g U_3O_8/L. Extraction takes place in four mixer-settlers, using a solvent mixture consisting of 2.5% Alamine 336 and containing 1.8% isodecanol. The final raffinate, after extraction, contains less than 0.001 g U_3O_8/L. Four stages of stripping, using 135 g/L $(NH_4)_2SO_4$ at pH 4.5, and at a temperature of 85°F, recover a strip solution containing about 27 g U_3O_8/L. Uranium is recovered by precipitation with anhydrous NH_3 at pH 7.0 in two stages, achieving a barren solution of less than 0.01 g U_3O_8/L. This barren solution is recycled to stripping, with a bleed to control the buildup of $(NH_4)_2SO_4$. The flowsheet is shown in Fig. 6.

The tertiary amine extractant is also used in chloride media, and Falconbridge Nickel, Kristiansand, Norway, processes a solution resulting from leaching nickel matte in HCl.[28] The leach solution, after removal of iron with TBP, is contacted with a tertiary amine to coextract the Cu and Co, which are selectively recovered by waterstripping. The nickel is recovered from the raffinate by crystallization.

Chelating

Following the development of the uranium technology, the first copper plants were put into production in the 1960s. Many more have gone on stream in recent years. Chelating extractants, LIX64N and LIX65N, have been the common reagents, although Kelex 100 is another comparatively recent development, as is Shell SME 529 and Acorga.

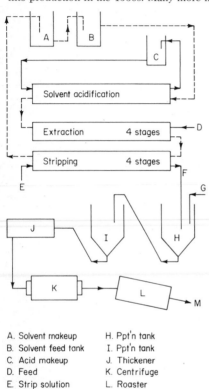

The first plant to use a dump leaching–solvent extraction–electrowinning sequence was the Ranchers Bluebird Mine at Miami, Ariz.[29] Production began in 1968, treating 6.8 m³/min (1800 gal/min) of solution. Subsequently the Bagdad Copper Company, also in Arizona, went on stream.[30] Here the total flow of 12 m³/min (3200 gal/min) solution at 1.4 g Cu/L is fed to the circuit, where it is split and fed to four banks of mixer-settlers in parallel. The solvent is LIX64N in Napoleum 470. The last-stage raffinate contains about 0.2 g Cu/L and is recycled to dump leaching. Copper is recovered from the solvent by stripping with 130 g H₂SO₄/L, resulting in a concentration of about 56 g Cu/L, containing 130 g free H₂SO₄/L in the strip solution. Cathode copper is produced by electrowinning. The flowsheet is shown in Fig. 7 for the production per day of about 18,000 kg (40,000 lb) of electrolytic copper in the tankhouse.

A larger copper plant went into production in 1974, also using the same extractant. This is the Nchanga Consolidated Copper Mine in Zambia,[31] treating 30,000 tons/day of ore to produce about 250 tons/day of cathode copper, at a feed rate of about 15,-000 gal/min containing 2.5 g Cu/L. The total flow of 20% LIX64N solvent and aqueous solution is about 32,000 gal/min. Plants in Zaire and Chile are expected to be even larger.[31]

A. Solvent makeup
B. Solvent feed tank
C. Acid makeup
D. Feed
E. Strip solution
F. Strip liquor
G. Ammonia
H. Ppt'n tank
I. Ppt'n tank
J. Thickener
K. Centrifuge
L. Roaster
M. Product

Fig. 6 Highland uranium operation.

The Anaconda Company at Twin Buttes, Ariz., began operation in 1975 to treat 10,000 tons/day of oxide ore by sulfuric acid leaching, followed by contact of the clarified solution, containing 2 to 3 g Cu/L, with LIX64N and subsequent recovery of 80 tons/day of cathode copper by electrowinning.[32]

From alkaline ammonium sulfate solution the SEC Corporation in El Paso, Tex., recovers nickel using LIX64N at pH 8, after initially extracting copper from the original acid liquor, also with LIX64N.[33] In another alkaline system, the Anaconda Arbiter process of alkaline leaching is followed by solvent extraction using LIX65N, and electrowinning to recover high-purity cathode copper.[34] The use of Kelex 100 for the extraction and separation of Cu, Ni, Zn, and Co from ammonical solutions containing sulfate or carbonate has recently been described.[35]

Acidic

The acidic extractants, although not as selective as the chelating, have been commonly used for the extraction of such metals as uranium, vanadium, rare earths, cobalt, and nickel, as well as other metals[36] where the aqueous medium has been sulfate, nitrate, or chloride. The alkyl-phosphoric acids, such as EHPA, have been used most commonly in acidic solutions, while the carboxylic acids are more suitable in alkaline solutions in the presence of a high salt content.[37] The use of EHPA for the extraction and separation of cobalt and nickel from solutions resulting from the dissolution of ores and concentrates in nitric or sulfuric acids, or mixtures of the two, is briefly described below.

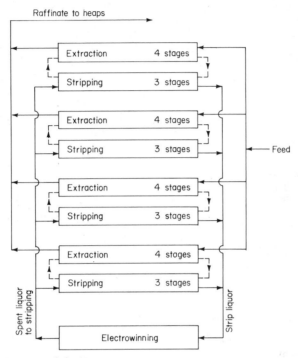

Fig. 7   Bagdad solvent extraction–electrowinning circuit for copper.

The process, developed by Eldorado Nuclear, Ltd., resulted from the requirement to treat cobalt ores and concentrates, also containing Ni, As, Cu, Bi, Fe, in order to recover high-purity products of cobalt and nickel. Following leaching and liquid-solid separation, the filtrate is treated in a solvent-extraction circuit, or by cementation, to remove the copper. Then an oxidation-neutralization stage removes the iron, and the filtrate pH is adjusted to about pH 5. The filtrate is sent to the solvent-extraction circuit for the separation of cobalt and nickel. The diagrammatic flowsheet for the process is shown in Fig. 8, and the reader can refer to the literature for the integrated circuit and complete process description.[38] The extraction and scrub circuits were designed for the use of sieve-plate pulse columns, while mixer-settlers were adequate, and often desirable, for stripping and solvent equilibration.

Extraction is effected using 20% DEHPA in an aliphatic diluent, also containing 5% TBP as a modifier. The solvent is previously equilibrated with ammonia to enable the system to be maintained at the desirable pH range of 5.0 to 5.5 during extraction. Increased rate of coalescence was achieved by operating between 40 and 55°C. The loaded solvent, containing about 20 g Co/L and 0.2 g Ni/L, is contacted in a scrub column with a cobalt sulfate scrub solution at pH 5 to remove any coextracted nickel. A bleed of the scrub recycle stream is returned to the extraction circuit as the nickel content is increased because of scrubbing. Cobalt is recovered from the scrubbed solvent by stripping with sulfuric, nitric, or hydrochloric acid, depending upon the refined final product, and results in a liquor containing 100 g Co/L. If sulfuric acid stripping is used, the resultant cobalt sulfate solution is returned to the scrub circuit to maintain the necessary cobalt balance in that circuit. If nitric acid is used for stripping, the cobalt nitrate strip solution is evaporated, denitrated, and dried at 500°C to produce a cobalt oxide product. Reduction of the oxide at 800°C under a hydrogen atmosphere provides cobalt metal powder.

The raffinate, containing the nickel, is evaporated and crystallized to produce nickel ammonium disulfate. Further treatment by heating these crystals to 400°C in steam,

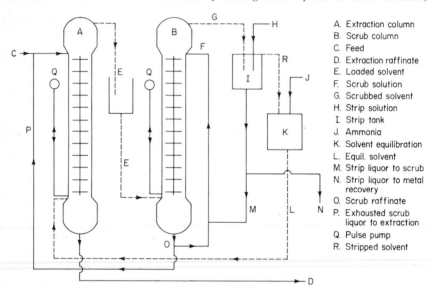

A. Extraction column
B. Scrub column
C. Feed
D. Extraction raffinate
E. Loaded solvent
F. Scrub solution
G. Scrubbed solvent
H. Strip solution
I. Strip tank
J. Ammonia
K. Solvent equilibration
L. Equil. solvent
M. Strip liquor to scrub
N. Strip liquor to metal recovery
O. Scrub raffinate
P. Exhausted scrub liquor to extraction
Q. Pulse pump
R. Stripped solvent

Fig. 8   Separation of cobalt and nickel using EHPA as extractant.

drying to remove water of hydration at 110°C, and finally heating to 400 to 500°C converts the product to nickel sulfate.

The process was piloted on a relatively large scale, where 550-mm-diameter (22-in) columns were used to treat solutions containing 6 to 21 g Co/L and ranging from Co-Ni ratios in the feed of 3:1 to 1:3. It was found that, depending on the cobalt-nickel ratio, there were definite advantages in the choice of the continuous phase. Purity of the cobalt product was also dependent upon the amount of scrubbing used, and a ratio of Co to Ni of 1000:1 could be readily achieved.

The patent[39] for the process has been licensed to Inco, Canada, and a plant subsequently built. The process is also under license to a major mining company in Japan.

SOLVENT LOSSES AND ANALYTICAL ASPECTS

The loss of solvent in a solvent-extraction process cannot be avoided, since all solvent phases are soluble in an aqueous phase to some extent. However, by suitable choice of solvent and other conditions, such losses can be minimized.

Solvent loss arises from essentially three causes: solubility, entrainment, and volatilization. Little can be done about soluble losses, which are governed by the reagents used and the operating conditions chosen for maximum efficiency of metal extraction and separation. Losses due to entrainment can vary widely, and result from such as poor equipment choice or design, poor solvent composition, too much energy input into the extraction stage, systems upsets, and so on. Volatile losses depend on the solvent constituents, temperature, and how well the system is enclosed.

The determination of solvent losses can be achieved in two ways, by inventory and by analysis. Inventory provides only long-term losses, and includes all losses. Analysis provides fast information on solvent losses. However, the analysis of raffinates for solvent components is difficult, and most losses reported from processes are inventory losses. If analytical methods are available, they should be used, especially on pilot-plant runs, because such runs are usually not of a sufficient duration to provide good estimates of solvent losses.

Extractant losses reported for some processes are given in Table 7; they are probably typical of those particular operations.

One area which has not received much attention to date is that of solvent degradation, which provides another cause of loss of solvent constituents.

Actually, little is known about solvent losses. This is surprising, because the solvent-

TABLE 7   Extractant Losses Reported for Some Processes

Metal extracted	Extractant	Extraction pH	Loss, ppm	References
	Naphthenic acid	4.0	90	73
Ni	Naphthenic acid	6.5	900	74
	Versatic 911	7.0	900	75
			300	Shell Chemicals Data
Co	Versatic 911	7.7	100	70
Rare earths	D2EHPA	2.0	7	76
Co	D2EHPA	5.5–6.5	30	71
U	Tertiary amines	1.5–2.0	10–40	77–80
Cu	LIX 64	1.5–2.0	4–15	81–84
U	TBP	2.0	25–40	76
Cu	Kelex 100	1–2	1–10	85,86
Hf	MIBK	1–5M HCl	20,000	87

extraction process is dependent on the economics associated with the solvent, which comprises a large part of the system's inventory costs.

Analytical requirements in solvent-extraction processing fall into two groups, those required for analysis of the aqueous phase and those for the solvent phase. While methods for the inorganic components of both phases are generally available, major problems occur in the analysis of the organic components. Simple and reliable methods are required for the determination of solvent components in both phases, and especially so in raffinates. Here, while some methods are available for the determination of extractants, few are available for diluents and modifiers, especially in the presence of each other. Similarly, while the determination of extractant concentration in the solvent phase is usually quite straightforward, methods for other components are sadly lacking. A recent review of analytical chemistry in solvent-extraction processing has been published, and the reader is referred to this for detail.[55]

Another major problem associated with analysis, as with all other areas of analysis, is that of sampling. Because of the surface-active nature of many of the reagents used in solvent extraction, it is of prime importance that the sampling aspects be thoroughly investigated if the analyses done are to be meaningful. Some aspects of the sampling of raffinates and solvents have been published.[55]

The choice of sampling method and analytical procedure is of considerable importance, and is usually not easy. Compromises usually result between speed and sophistication, and costs of analysis. It should be pointed out, however, that good pilot-plant or plant work can be ruined by insufficient attention being paid to sampling and analysis.

ENVIRONMENTAL ASPECTS

Little has been done or said about environmental pollution resulting from solvent-extraction processes. Since such processes do have solvent losses, and these can be discharged to the environment, the effect on aquatic life becomes important. In order to develop proper water-quality criteria, biodegradability and toxicity must be determined for solvent-extraction reagents.

In the evaluation of toxicity to fish due to chemicals, four fish types are usually used: fathead minnows, blue gill, goldfish, and guppies. Table 8 shows some toxicity tests, at 25°C, for some organic compounds when tested on these fish.

The significance of the acute toxicity data given in Table 8 can be determined only through a clear understanding of the conditions under which the compounds tested are normally used. This will include a knowledge of dosage rates, treatment frequencies, residual concentration in the effluent, and the composition of the local fish population.

Because the results given here were derived from short-term toxicity tests, all values should be viewed as relative toxicities only. Much more study is required before the impact of solvent-extraction reagents on aquatic (and other) life in the environment can be properly assessed.[57]

To conclude this section, a word about the recovery of solvent from effluents. Entrained solvent losses have been minimized in some processes by the installation of such equipment as skimmers, centrifuges, coalescers, settlers, flotation cells, and foam-fractionation units, and by activated carbons.[58] Only the last of these, carbon, is suitable for the

TABLE 8   Median Tolerance Limits for Some Petrochemicals[89]

| Compound | TL$_m$ 96, mg/L | | | |
	Fatheads	Bluegills	Goldfish	Guppies
Benzene	33.47	22.49	34.42	36.60
Chlorobenzene	229.12	24.00	51.62	45.53
o-Chlorophenol	11.63	10.00	12.37	20.17
3-Chloropropene	19.78	42.33	20.87	51.08
o-Cresol	12.55	20.78	23.25	18.85
Cyclohexane	32.71	34.72	42.33	57.68
Ethyl benzene	48.51	32.00	54.44	97.10
Isoprene	86.51	42.54	180.00	240.00
Methyl methylacrylate	159.1	311.0	232.2	368.1
Phenol	34.27	23.88	44.49	39.19
Styrene	46.61	26.05	64.74	74.83
Toluene	34.27	24.00	57.68	59.30

recovery of soluble solvent. While such equipment and processes may prove costly, they may be necessary in order to meet environmental specifications.

In conclusion, it is evident that considerably more research into solvent losses to the aqueous phase, and the effects on downstream processes and the environment, must be undertaken because of both economic and environmental constraints.

ECONOMICS

The concentration of metals in the solution, the value of the metal being recovered, together with the flow throughput, may singly or in combination contribute to the decision on the possible use of solvent extraction. Also, the upstream and downstream aspects of the overall process have to be considered.

Economic considerations may be divided into capital investments and operating costs:

Capital Cost

Capital cost is primarily related to the size of equipment necessary for a given throughput. Usually the building required to house the process is an important cost factor, as is the solvent inventory.

Operating Cost

Operating costs may be divided into several areas:
1. Preparation of the feed solution
2. Pretreatment of the solvent if required
3. Scrubbing of undesirable metals if necessary
4. Stripping costs
5. Solvent losses
6. Labor and maintenance costs
7. Loss of some of the desired metal to the raffinate

It is evident that one of the factors involved in operating costs is the chemical requirements of the system, such as the neutralization or preparation of the feed solutions prior to the solvent-extraction process. This is demonstrated in Figs. 9 and 10, which show the relationship between feed-metal concentration and the processing costs for systems covering both the acidic and basic range of extraction using acidic, basic, neutral, and chelating extractants.[36] The costs are similar for the same extraction system. Figure 11 shows the effect of variation of flow rate and feed concentration on the operating cost, indicating decreased cost at the higher throughputs.[36] Similar effects, shown in Fig. 12, illustrate operating cost as a function of the daily copper-production rate.[59]

Other factors affecting the cost of the process are solvent concentration, viscosity, agitation speed, continuous phase, number of stages, and flow ratio. The solvent-concentration effect has been reported by Robinson,[60] and is shown in Fig. 13.

Labor requirements can be optimized and minimized by automation, and a plant producing 18,000 kg (30,000 lb) of copper per day from a solution containing 4.5 g Cu/L reported their circuit would require four operators and three laborers to cover the 24-hour, 7-day per week operation.[20]

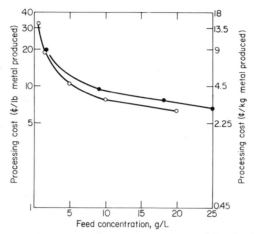

Fig. 9   Processing cost as a function of feed concentration and pH; cobalt and zinc-EHPA systems: ●— zinc, pH 3; ○—cobalt, pH 5.5. Both at 100 gal/min flow.

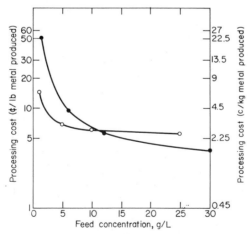

Fig. 10   Processing cost as a function of feed concentration and extractant for cobalt and copper ammoniacal systems. ●—Cu-LIX63; ○—Co-V911; both at 100 gal/min flow rate.

One of the items that must be mentioned as affecting costs is the type of equipment used, which ranges from stirred tanks and mixer-settlers to centrifugal contactors and various types of columns. The economics of equipment selection must be considered with respect to the type of contactor to suit particular conditions of a given throughput, solution and solvent type, kinetics and equilibrium, dispersion and coalescence, solvent losses, capacity and stage requirements, solvent residence time, phase flow ratio, physical properties, direction of mass transfer, holdup, presence of solids, available area, corrosion, and overall performance. Table 9 shows an economic comparison based on extraction data derived from the evaluation of several contactors, together with estimated costs of equip-

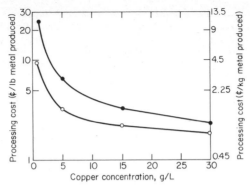

Fig. 11   Variation of feed concentration and flow rate, effect on processing cost; Kelex-copper system: ●—100 gal/min; ○—1000 gal/min.

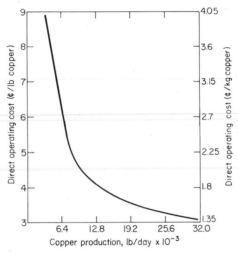

Fig. 12   Direct operating costs vs. copper production.

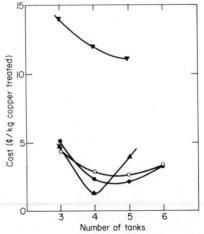

Fig. 13   Cost vs. LIX64N concentration. ▼— 10v/o; ▲—14v/o; ○—18v/o; ●—22v/o LIX64N.

ment, solvent, and building requirements.[61] The data are based on the extraction of 5 g Cu/L with Kelex 100, from an ammoniacal ammonium sulfate solution containing copper, zinc, and nickel, at about pH 8.

Finally, another cost factor to be considered in the design of the solvent-extraction process is that of environmental constraints. As mentioned elsewhere, it may become mandatory to ensure that effluents entering watercourses contain little or no quantities of detrimental organic compounds or metal-organic species. Table 10 shows the estimated costs for removal and recovery of amine from a process effluent using activated carbon.[91]

TABLE 9   Economic Evaluation of Contactors for Extraction[94]

Contactor	Equipment diam, ft	Required length, ft		Equipment cost, $ × 1000	After settler required,[a] $ × 1000	Solvent cost,[b] $ × 1000	Building space,[c] $ × 1000	Total cost, $ × 1000
Mixer-settler[d]	2			60		11.2	80.0	151.2
Mixco[e]	3	5	16	100	10	59.2	77.5	246.7
Pulse	1	5	60	160		14.0	87.5	261.5
Kenics	3	2	28	230	10	15.1	81.0	336.1
Podbielniak	3-D36			300		6.3	71.7	378.0
Graesser	15	5	3.0	88	5	20.0	200.0	308.0

[a]External settlers required for phase disengagement, capacity of each 600 ft³.
[b]Solvent cost used was $7.00 per imperial gallon of mixed solvent.
[c]Equipment spaced with 10-ft clearance. Cost at $40.00/ft² on first floor and $20/ft² on other floors.
[d]Mixers: 150 gal capacity; settlers: 150 ft² × 4 ft deep, with 9-in depth of solvent.
[e]Mixco run only at room temperatures. All other equipment evaluation at 50°C.

TABLE 10   Estimated Costs for Treatment of Uranium Solvent-Extraction Raffinates by Carbon Adsorption[91]

	¢/lb amine	¢/lb U	¢/1000 gal treated
Steam*	1.2	0.1	0.4
Carbon†	8.6	0.4	3.0
Labor‡	19.1	0.8	6.7
Capital§	0.1	0.01	0.1
	29.0	1.31	10.2

*Steam consumption: 1.5 lb steam/lb carbon at $1.00/1000 lb steam.
†Carbon loading: 0.14 lb amine/lb carbon, carbon loss of 5% at a cost of 12¢/lb carbon.
‡Labor: 12 h at $4/h.
§Capital: Installed cost of $100,000 10-year straight line depreciation. Mill: 3000 tons/day, 0.1% U_3O_8, 500 gal/min, 35 mg/L amine.

Some estimated plant costs for the extraction of a single metal, or the separation of several metals from complex solution mixtures, are shown in Table 11 (page 2–128).[17] These demonstrate the wide variation that can occur with each particular system because of the concentration, chemical costs, capital cost, etc. No allowance was made for environmental costs, which could vary widely from plant to plant—even for the same process installation.

REFERENCES

1. Berthelot, M., and Jungfleisch, J., *J. Ann. Chim. Phys.*, **26**, 396 (1872).
2. Nernst, W., *Z. Phys. Chem.*, **8**, 110 (1891).
3. Marcus, Y., and Kertes, A. S., "Ion Exchange and Solvent Extraction of Metal Complexes," Wiley-Interscience, New York, 1969.
4. Ashbrook, A. W., *Coord. Chem. Rev.*, **16**, 285 (1975).
5. Agers, D. W., and DeMent, E. R., LIX 64 as an Extractant for Copper-Pilot Plant Data, presented at the Joint Meeting of SME of AIME, September 1967.
6. Anon., LIX 70, A Major Advance in Liquid Ion Exchange Technology, presented at the annual AIME Meeting, New York, February 1971.
7. Ritcey, G. M., and Lucas, B. H., Some Aspects of the Extraction of Metals from Acidic Solution by Kelex 100, *CIMM Bull.*, February 1974.
8. Blake, C. A., Brown, K. B., and Coleman, C. F., USAEC, *Rept.* ORNL-1930, 1955.

TABLE 11 Approximate Solvent-Extraction Costs for Recovery of Some Metals (flow rate of 100 gal/min)

Metal	Feed conc., g/L	Aqueous system	Organic extractant	A/O ratio	Equipment type	Acid, adj.	Processing costs, ¢/lb metal produced						
							Equipment and installation*	Solvent treatment	Scrubbing	Stripping	Estimated§ solvent losses	Labor and maintenance	Total costs
Co-Ni	Co 10, Ni 10	H_2SO_4, pH 5-6	20% D2EHPA	1:1	Columns		1.4	3.1	0.2	1.8	0.6	1.3	8.4
Cu	Cu 2	H_2SO_4, pH 2	10% LIX 64N	2:1	Mixer-settlers		3.3			1.5	1.5	7.0	13.3
Cu	Cu 15	H_2SO_4, pH 0.5-1.5	20% Kelex 100	1:1	Mixer-settlers		0.3			1.5	0.9	1.1	3.8
R.E.† and Y	Y 0.10	H_2SO_4, pH 1.5-2.0	3.3% D2EHPA	20:1	Centrifuge and columns	18.0‡	0.6			30.4	7.0	25.0	81.0
U	U 2.0	H_2SO_4, pH 1.5-2.0	5% Adogen 364 or Alamine 336	2.5:1	Mixer-settlers		2.5	0.5		1.2	3.0	2.5	9.7
Zr-Hf	Zr 95, Hf 1.5	HNO_3, 7.5N	50% TBP	1:3	Columns	16.0	0.20	20.0	4.0		0.8	0.03	41.0
Hf-Zr	Zr 100, Hf 2	HCl-HCNS (1M) (3M)	MIBK	1:2	Columns	6.0	4.6	2.8	3.9	5.2	3.9	30.0	56.4

*10-year straight line depreciation.
†Assume a separation of each of the rare earths and yttrium is being made.
‡Neutralization and pH adjustment in the circuit divided among the rare earths and yttrium.
§Estimated total of extractant, diluent, and modifier.

9. McKay, H. A. C., Healy, T. V., Jenkins, I. L., and Naylor, A. (eds.) "Solvent Extraction of Metals," Macmillan, London, 1967.
10. Technical Editor, *Eng. Min. J.*, **161**, 78 (1960).
11. Ritcey, G. M., and Lucas, B. H., *Proc. Int. Solvent Extraction Conf., Soc. Chem. Ind., London*, 1974, p. 2437.
12. Jeffreys, G. V., and Davies, G. A., in Hanson, C. (ed.), "Recent Advances in Liquid-Liquid Extraction," p. 495, Pergamon, New York, 1971.
13. Rydberg, J., and Reinhardt, H., *Chem. Ind.*, 1970, p. 488.
14. Croda Scientific Co., Ltd., Reid Works, Horbury Bridge, Wakefield, Yorkshire, England.
15. Bell Engineering, Tucson, Ariz.
16. Ritcey, G. M., *Proc. Am. Inst. Chem. Eng. Solvent Ion Exchange*, Tucson, Ariz., May 11, 1973.
17. Ritcey, G. M., and Ashbrook, A. W., "Solvent Extraction in Mining and Metallurgy," to be published by Elsevier, Amsterdam.
18. Wellek, R. M., Ozsoy, M. U., Carr, J. J., Thompson, D., and Konkle, T. V., *Ind. Eng. Chem. Process Des. Dev.*, October 1969, p. 515.
19. Israel Mining Industries, *Proc. Int. Solvent Extraction Conf.*, 1971, pp. 1386–1408.
20. Miller, A., in Kibby, R. M. (ed.), "The Design of Metal Producing Processes," AIMI, 1967.
21. McGarr, H. G., *Chem. Eng.*, Aug. 10, 1970, pp. 82–84.
22. Burger, J. C., and Jardine, J. M., *Proc. 2d UN Int. Conf. Peaceful Uses of Atomic Energy*, Geneva, **4**, 3 (1958).
23. Private communication to G. M. Ritcey.
24. Bernard, G. J., Material Selection and Performance for Solvent Extraction and Electrowinning Plants, paper presented at the Canadian Region Eastern Conference of the National Association of Corrosion Engineering, Montreal, Sept. 27–29, 1971.
25. Ritcey, G. M., and Conn, K., Liquid-Liquid Separation of Zirconium and Hafnium, Eldorado Nuclear Ltd., *R & D Rept.* 767-7. Ottawa, 1967.
26. Craigen, W. J., Joe, E. G., and Ritcey, G. M., *Can. Met. Quart.*, CIM, **9**(3), 485–492 (1970).
27. Abramo, J. A., and Lowings, S. W. A., *Proc. Symp. Solvent Ion Exchange*, AIChE, Tucson, Ariz., May, 1973.
28. Thornhill, P. G., Wigstol, E., and VanWeert, G., *J. Metals*, **23**(7), 13–18 (1971).
29. Power, K. L., *Proc. Int. Solvent Extraction Conf.*, 1971, pp. 1409–1415.
30. McGarr, H. J., Berlin, N. H., and Stolk, W. F. A., *Eng. Min. J.*, **110**, 66 (1969).
31. Private communication to G. M. Ritcey.
32. *Chem. Week*, Apr. 2, 1973, p. 44.
33. Eliasen, R. D., and Edmunds, E., Jr., SEC Nickel Process, *CIM Bull.*, **67**(742), 82–86 (February 1974).
34. Kuhn, M. C., Arbiter, N., and Kling, H., Anaconda's Arbiter Process for Copper, *CIM Bull.* **67** (742), 62–79 (February 1974).
35. Ritcey, G. M., and Lucas, B. H., Extraction and Separation of Copper, Nickel, Zinc and Cobalt from Ammoniacal Solution Using Kelex 100, *CIM Bull.*, February 1975.
36. Ritcey, G. M., Some Economic Consideration in the Recovery of Metals by Solvent Extraction Processing, *CIM Bull.*, June 1975.
37. Ritcey, G. M., and Lucas, B. H., *Proc. Int. Solvent Extraction Conf.*, 1971, p. 463.
38. Ritcey, G. M., Ashbrook, A. W., and Lucas, B. H., Development of a Solvent Extraction Process for the Separation of Cobalt from Nickel, *CIM Bull.*, January 1975.
39. Ritcey, G. M., and Ashbrook, A. W., U.S. Patent 3,399,055, August 1968.
40. Fletcher, A. W., and Flett, D. S., in McKay, H. A. C. (ed.), "Solvent Extraction Chemistry of Metals," p. 359, Macmillan, London, 1966.
41. Fletcher, A. W., and Hester, K. D., *Trans. AIME*, **229**, 282 (1964).
42. Ashbrook, A. W., *J. Inorg. Nucl. Chem.*, **34**, 1721 (1972).
43. Private communication to the authors.
44. Tunley, T. H., and Faure, A., The Purlex Process, paper presented at the annual AIME meeting, Washington, 1969.
45. Tremblay, R., and Bramwell, P., *Trans. Can. Inst. Min. Metall.*, no. 62, 1959, p. 44.
46. Bellingham, A. I., *Proc. Australas. Inst. Min. Metall.*, no. 198, 1961, p. 85.
47. Lloyd, P. J., *J. S. Afr. Inst. Min. Metall.*, **62**, 465 (1961).
48. Dasher, J., and Power, K. L., *Eng. Min. J.*, **172**(4), 111 (April 1971).
49. Rawling, K. R., *World Min.*, **22**, 34 (1969).
50. McGarr, H. G., *Eng. Min. J.*, **171**(10), 79 (October 1970).
51. Agers, D. W., House, J. E., Swanson, R. R., and Drobnick, J. L., *Trans. Soc. Min. Eng. AIME*, **235**, 191 (1966).
52. Ritcey, G. M., *Proc. 2d. Ann. Meeting, Canadian Hydrometallurgists*, CIMM, October 1972, p. 11.
53. Hartlage, J. A., Kelex 100—A New Reagent for Copper Solvent Extraction, paper presented at Soc. Min. Engrs., AIME Fall Meeting, Salt Lake City, September 1969.
54. Flett, D. S., *Min. Sci. Eng.*, **2**, 17 (1970).
55. Ashbrook, A. W., *Talanta*, **22**, 327 (1975).
56. Pickering, Q. H., and Henderson, C., *J. Water Pollut. Control Fed.*, September 1966, p. 1419.

57. Ritcey, G. M., Lucas, B. H., and Ashbrook, A. W., *Proc. Int. Solvent Extraction Conf.*, 1974, p. 2873.
58. Ritcey, G. M., Lucas, B. H., and Ashbrook, A. W., *Proc. Int. Solvent Extraction Conf.*, p. 943.
59. Palley, J. N., and Paige, P. M., *EIM J.*, July 1972, pp. 94–96.
60. Robinson, C. G., *NIM Rept.* 1085, November 1970, Johannesburg.
61. Ritcey, G. M., and Lucas, B. H., *Proc. Hydrometallurgy Symp.*, Institute of Chemical Engineers, Manchester, 1975.
62. Burkin, A. R., *Proc. 1st Hydrometallurgy Meeting*, CIM, Mines Branch, Ottawa, Oct. 28–29, 1971.

Section **2.3**

Evaporation

R. C. BENNETT *Manager, Swanson Division of Whiting Corporation*

INTRODUCTION

Evaporation is the removal of a solvent from a solution by vaporization. Most commonly the solvent is water and the solute is an inorganic or organic solid with very low vapor pressure at the temperature of evaporation. This distinguishes evaporation from distillation where all or most of the components in the liquid phase have an appreciable vapor pressure and appear in the overhead vapors. In evaporation the overhead vapor is primarily solvent contaminated only by small amounts of liquid-phase material which are called entrainment.

Evaporation is used in a wide variety of chemical processes for concentrating solutions containing dissolved solids or liquids of extremely low vapor pressures. In many applications some of the dissolved solids exceed their solubility limit in one or more of the evaporating stages and are precipitated as solid-phase material in a saturated solution. These vessels are called evaporative crystallizers, and the common examples are evaporators employed for the precipitation of evaporator-pan salt (NaCl) and sugar.

MODERN DESIGNS

Most modern evaporators in use today are of the long-tube vertical (LTV) or forced-circulation (FC) variety (Figs. 1 to 3). The most notable exception is the sugar industry, which still uses the short-tube vertical calandria evaporator for both concentration and crystallization.

The LTV evaporator is the usual choice where high-capacity evaporation from solutions of low viscosity and low boiling point elevation is required. It can be produced in an almost unlimited range of capacities and materials of construction. The FC evaporator is used with viscous materials or in cases where scaling or precipitation of solids occurs.

Both the LTV and FC evaporators built today are capable of high economy, simple operation, long operating cycles, and a wide range of capacities without a significant change in operating economy. In addition to the concentrated solution or slurry which is their product, they deliver condensate often containing less than 100 ppm of solids. All uncondensed gas or vapor leaving an evaporator is passed through a vacuum pump and/or scrubber, and therefore these devices find ready application in the solution of pollution problems where there can be zero discharge from a system.

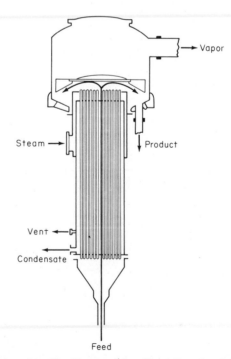

Fig. 1   LTV evaporator—rising film. (*Reprinted from* Chem. Eng., *Dec. 9, 1963, with permission.*)

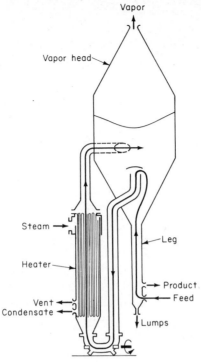

Fig. 2 Forced-circulation evaporator—submerged-tube type. (*Reprinted from* Chem. Eng., *Dec. 9, 1963, with permission.*)

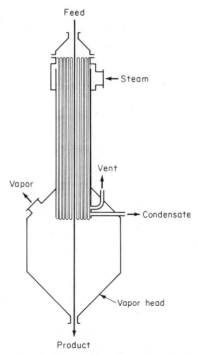

Fig. 3 Falling-film evaporator. (*Reprinted from* Chem. Eng., *Dec. 9, 1963, with permission.*)

A number of special types of evaporators are made in addition to the ones described above, but the most common encountered are the wiped-film evaporator, which is primarily for highly viscous materials which cannot be properly circulated in the LTV or FC types, and the falling-film evaporator, which is a variation of the LTV type and is used primarily in applications where very short retention time of the process fluid at high temperature is required for process reasons.

DEFINITION OF TERMS

For purposes of this section, it is useful to define the precise meaning of certain terms related to evaporator use and design.

Heating Surface

This refers to the heat-transfer surface, which in almost all modern evaporators consists of tubes ranging in diameter from 19 to 63 mm (¾ to 2½ in). In almost all modern evaporators the flow of liquid is through the tubes and steam or evaporator vapor is condensed on the outside of the tubes. Overall heat-transfer relations for evaporators are calculated from the fundamental heat-transfer formula $Q = UA \, \Delta T$. The area may be taken as the inside or outside area of the tubes. In natural-circulation evaporators, the ΔT is the overall steam temperature to equilibrium liquid temperature at the vapor pressure in the vapor head. The overall coefficient varies greatly with changes in ΔT (see Figs. 9 and 10). For forced-circulation evaporators, the ΔT is taken to be the log mean ΔT between the steam temperature in the heat exchanger and the incoming and outgoing liquid temperatures at the heat exchanger.

Boiling-Point Elevation

The boiling-point elevation is the difference in temperature between the saturated-vapor temperature (corresponding to the vapor pressure above the solution) and the temperature of the solution. This boiling-point elevation represents a thermal gradient that is largely unavailable for heat transfer.

With most inorganic salts this value is about 1.6 to 5.5°C (3 to 10°F). For high concentrations of acid or alkali this difference can be as much as 44 to 55°C (80 to 100°F). Since these data are so important in their effect on the heat-transfer-area calculations, data are needed throughout the range of concentrations and temperatures in which the evaporator will function so that its effect can be properly anticipated. These data are commonly correlated by means of a Düring chart, which is a plot of the boiling point of the solution vs. the boiling point of water at the same pressure. It results in a straight line over a considerable span in temperature.

Economy

The economy of an evaporator is the pounds of water evaporated per pound of steam used. As a rule of thumb it is roughly 0.8 for a single effect, 1.6 for a double effect, 2.4 for a triple effect, etc. The term efficiency is not properly used with evaporators, since unlike mechanical equipment there is no loss of heat energy within the evaporator circuit except by radiation and typically such losses are only 2 or 3%. The economy of the evaporator is influenced not only by the number of effects (the major variable) but also by feed temperatures, crystallization heat effects, flow of condensate, and heat-recovery streams.

Effect

The term effect is applied to each body of an evaporator along with its related circulating piping, vapor piping, and feed piping. Numbering on evaporators starts with the body into which the incoming steam is placed. Bodies are numbered in the direction of the vapor flow from this first effect. The term evaporator applies to one or more effects. Multiple-effect evaporators have been built with as many as twelve bodies so interconnected.

Forward Feed

In this type of evaporator the steam and the feed solution enter the first effect and then progress to the succeeding effects.

Backward Feed

In a backward-feed evaporator the feed enters the last stage and progresses toward the first stage. Mixed-feed sequences are also possible where the feed enters at an intermediate body and progresses toward the last stage and then returns, bypassing the bodies in which it has been processed, and moves toward the first stage.

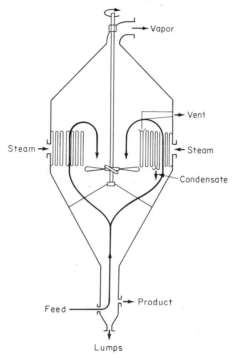

Fig. 4   Propeller calandria evaporator. (*Reprinted from* Chem. Eng., *Dec. 9, 1963, with permission.*)

Natural Circulation

This terms refers to movement of liquid within the evaporator body. In natural-circulation evaporators the thermosiphon effect of the boiling liquid moving through the tubes causes high tube wall velocities which lead to improved rates of heat transfer. As vapor forms within the tubes, the density of the liquid column in the tube is reduced and more liquid fills its place from the body of liquid within the vessel itself. Sometimes this liquid is conducted from the vapor head to the lower portion of the heat exchanger through an external pipe, and sometimes it is supplied through internal pipe(s) as in a calandria evaporator.

Calandria

Also known as the short-tube vertical, STV, or standard evaporator, the calandria contains a heating element normally with a large central downtake (about equal in cross-sectional area to the tubes) and vertical tubes from 0.9 to 1.5 m (3 to 5 ft) in length. Tube diameters vary, but usually are 38 to 75 mm (1½ to 3 in). Early designs depended entirely on natural circulation. More recent units usually have propellers to improve the natural circulation (see Fig. 4). They may be used for concentration with or without solids precipitation; however, for the former, the FC is preferred.

Forced Circulation

These are evaporators wherein the liquid movement through the heat exchanger is controlled by an internal or external circulating pump. Such pumps are usually low-head type such as axial-flow or mixed-flow designs. Normally, they are used when solids are precipitated.

LTV

Refers to the long-tube vertical evaporator, which is a natural-circulation evaporator similar in principle to the calandria except that tube lengths are typically 4 to 9 m (13 to 30 ft). At present it is probably the most common and most economical of the large-scale evaporators used for processing materials which are not heat-sensitive and during the concentration of which no solids precipitate.

MSF

MSF evaporators or multistage flash evaporators are generally horizontal-tube machines of a compact design used primarily when the ratio of evaporation to feed is relatively low. Feed is heated in surface condensers from the last stage to the first without evaporation and then flashed from one effect to the next. Common applications are for the concentration of sea water, brackish water, or alumina waste liquors.

MVR

Mechanical vapor-recompression evaporators use a mechanical vapor compressor to increase the pressure of the water evaporated from the solution to compress it to a pressure where condensation can take place at a temperature high enough to be condensed in the heat exchanger and sustain the process. Evaporators of this type are used where electric power is relatively inexpensive compared with steam.

TVR

Thermal vapor-recompression evaporators are used where high-pressure steam is available at a cost no greater than waste steam. The thermal compressor is a large ejector which receives high-pressure steam and entrains some of the water evaporated from the vapor head, compressing it to a temperature and pressure such that it can be condensed in the heat exchanger. The disadvantage of this scheme is that the high-pressure steam condensate is mixed with contaminated vapor which has been boiled out of the evaporator and condensed.

Condenser

Condensers are used to condense the water evaporated and separated in the vapor head. They may be of the tube-and-shell type or the direct-contact type.

Venting

All steam and process solutions contain small traces of noncondensable gases such as air and carbon dioxide which must be removed from the heat-transfer surface in order that the transfer rates may be sustained. This is done by venting this gas at concentrations of roughly 10% to some suitable lower-pressure source such as the condenser or ejector.

Salting

This is the growth on the body walls or heat-transfer surface of material having a solubility that may either decrease or increase with temperature. The salting occurs because of localized high supersaturation caused by the evaporation of liquid or by cooling or heating of the liquid (depending on the nature of the solubility curve). Large deposits can build on the walls of evaporators, precipitating solids, and these must be periodically removed or they will interfere with the liquid circulation and vapor separation occurring in the evaporator.

Scaling

This is the deposition and growth on the body walls or on the heat-transfer surfaces of material undergoing an irreversible chemical reaction or decrease in solubility as a result of the processing step. The scaling is typically of materials having relatively small mass, but interfering with heat transfer by precipitation on the tubes or heat-transfer surfaces. Often the rate of precipitation of such material is reduced by artificially increasing the slurry density of the scaling material.

Fouling

This is a general term referring to the formation of deposits other than salt or scale on heat-transfer surfaces or on body walls which interfere with the operation of the equipment. Examples are corrosion on the inside or outside of the tubes or the entrapment of solid material entering with the feed which sticks or coagulates to the heat-transfer surfaces or body walls.

Entrainment

This term refers to the presence in the condensate of materials other than the condensed vapor. Typically evaporator condensate contains about 0.1 to 0.5% liquid of a composition equal to that in the body from which the vapor was generated. Special means (mesh) can be employed to reduce this entrainment to very low values, often in the range of 10 to 100 ppm. Entrainment does not refer to materials which are volatile and are present by virtue of their vapor pressure.

Foaming

Since almost all evaporators operate under vacuum at least in some effects, the presence of foamy materials can have a serious effect on the operation of the equipment, particularly by increasing the entrainment. Antifoamants or defoamants are commonly added, but in some cases it is impossible to control foaming when boiling vapor from a free liquid surface. For such applications there are specialized designs that are successful in many cases in breaking up the foam so that entrainment can be reduced. The LTV evaporator with its internal deflector is relatively successful in this regard.

EVAPORATOR DESIGNS

There are so many kinds of evaporators in existence and being sold for a variety of applications that it is pointless to try to include a description or sketches of all of them. Some styles such as the calandria evaporator exist largely because they have been used historically in certain industries such as the sugar industry and up until now there has been little incentive to switch to more modern designs. Changing materials of construction and new construction techniques have caused other styles such as the horizontal-tube evaporator to largely disappear in recent years. A few modern or at least frequently encountered evaporator styles are shown in Figs. 1 through 6 to illustrate the general concepts underlying evaporation equipment and to review their nomenclature.

Figure 1 shows an LTV evaporator, which is probably the most widely used type being produced today. This evaporator is built in a wide variety of sizes and consists of a vertical heat exchanger topped by a vapor-disengagement vessel called the vapor head. Immediately above the tubes is a deflector to act as a primary liquid-vapor separator and foam breaker. Typically the feed entering the bottom liquor chamber is heated in a single-pass flow in the lower portion of the tubes, and as it rises at velocities generally around 0.15 m/s (½ ft/s) or less, vapor bubbles start to form as the liquid moves up the tubes. These vapor bubbles increase the velocity of the two-phase flow through the tubes and heat transfer per unit area increases causing more vaporization. Liquid leaving the end of the tubes is a two-phase mixture traveling at high velocity. Disengagement between the bulk of the liquid and the vapor occurs at the deflector, with the liquid leaving the lower portion of the vapor head where it may be partially recycled or advanced to the next effect. Vapor leaving the vessel may go through an entrainment separator or in many cases leaves via a vapor pipe to the heat exchanger of the succeeding body or to a condenser.

Figure 2 shows an outside heating element forced-circulation evaporator, which is widely used for the handling of materials where scaling or salting occurs. Liquid is pumped from the vessel through the suction line to a circulation pump and then through a single- or multiple-pass heat exchanger. Here the temperature of the liquid is increased roughly 2.2 to 4.4°C (4 to 8°F) and the liquid reenters the vessel either axially or tangentially as indicated. Most commonly the inlet of such a recirculation line is submerged so that boiling occurs only after the heated fluid entering the body has been mixed with cooler body liquid. Vapor is evolved from the liquid surface and leaves for the condenser or for the next body as in the case of the LTV. Recirculated liquor or slurry is returned to the suction of the circulating pump. If solids are formed during the evaporation process, an elutriation leg for collecting these solids is often used such as in the precipitation of sodium chloride. Depending upon the service the suction pipe may or may not contain a separation device designed to prevent solids which may build up on the walls of the vessel from entering the circulation loop. If the precipitation of solids occurs in such a unit, it could be termed a crystallizer, and depending upon the nature of the solids, the style and the speed of the circulating pump could have a strong influence on the size of the product discharged from the leg. Vessels of this type have been built in

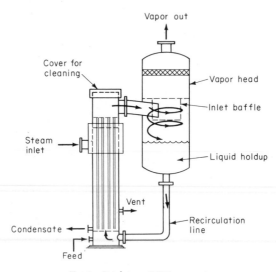

Fig. 5   Batch-type LTV evaporator.

diameters ranging up to 12 m (40 ft) and containing either single or multiple circulation systems as shown.

In Fig. 3 is shown a falling-film evaporator which in many ways is similar to the rising-film evaporator (LTV) except for the location of the heat exchanger. In this style of exchanger the boiling at the inlet of the tubes is not suppressed and therefore the heat-transfer coefficient shows very little change with ΔT. In order to operate properly a significant amount of liquid must be circulated through the tubes, and this requires a substantial energy demand to recirculate liquid from the bottom of the vapor head to the inlet of the tubes. In addition, careful attention must be paid to the method of distributing the liquid at the tube inlets. This type of unit has been extensively considered for the concentration of sea water; however, current practice favors the rising-film machine for this application. This design is useful because of its very short liquid retention at the steam temperature. Its greatest field of application is with heat-sensitive materials.

A *calandria evaporator* is shown in Fig. 4, and although this style has been in use for over 75 years it still finds some application in the sugar industry and in other services where the use of circulating pumps is objectionable because of leakage or operating considerations such as in radioactive-waste concentration. It is capable of handling solids precipitation but has relatively short operating cycles when used for this service. It requires much lower headroom than other styles of evaporators, and it therefore finds some use in applications requiring small equipment. As shown in Fig. 4, this type has often been supplied with large propeller circulators to aid in the natural circulation caused by vapor formation in the tubes.

Figure 5 is a *recirculating LTV* having a substantial liquid holdup. Such vessels find some usefulness in applications where a batch concentration is done on solutions having a relatively high evaporation-feed ratio. This general style is also useful where frequent cleaning of the heat exchanger is required and, therefore, placing the exchanger external to the body (rather than internal as with the LTV shown in Fig. 1) makes the design attractive. This style is often used in processing food materials.

Shown in Fig. 6 is a *wiped-film evaporator* wherein the treated liquid is moved at high speed on a heated surface by means of mechanical wiper arms. This type of equipment can handle materials of very high viscosity, in some cases as high as 100,000 cP. Retention time at the liquor temperature corresponding to the heated wall is relatively short, often a matter of seconds. For very high vacuum operation, this type has the advantage that there is almost no hydraulic head to prevent the heated liquid from achieving equilibrium vapor pressure. Some styles of this equipment have been developed that are capable of discharging dried solids. The main disadvantage is the relatively high cost per unit of evaporation.

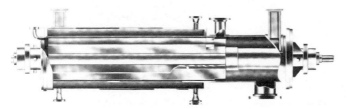

Fig. 6   Thin-film mechanically aided evaporator. *(Artisen Industries.)*

THEORY

From the foregoing descriptive material, it is obvious that the main elements of any evaporator are a heat-transfer surface, a vapor-liquid separator, and a vapor condenser. The vapor-separation area in all early evaporators was determined primarily by the mechanical design, since the heat-exchange surface was formed integrally with the separation zone or vapor head. This is true in most horizontal and calandria evaporators and to a large degree in wiped-film evaporators. It is often found in such machines that there is ample liquid-disengaging space in a given sized machine of this style. In evaporators where the heat-transfer surface is independent of the vapor-disengaging space such as in the FC, the falling-film, and the LTV designs, entrainment is often the limiting factor in determining the vapor head size. The entrainment is a function not only of the vapor velocity, ΔT, and inlet style but also of the physical properties of the fluid being concentrated such as the viscosity and surface tension. The droplets which form when the liquid and vapor are separated are frequently in a size range which will be supported by the outgoing vapor velocity even if the velocities are relatively low. Some general guidelines regarding this sizing as a function of entrainment have been published by Standiford,[5] and these are shown in Fig. 7. These values must be used only as a general guide and then with some caution, since entrainment is also strongly influenced by mechanical design and this is considered proprietary information by most manufacturers. Many liquids foam during concentration, particularly under vacuum and foaming can greatly increase entrainment even to the point where operation cannot be economically sustained. Defoamants are often used to reduce or eliminate this problem. Antifoam agents are commercially available such as silicones, vegetable oils, and fatty acids. Foam can also be reduced or at least its harmful effects negated by mechanical designs which break up the foam in or near the disengaging space.

In vessels similar to the FC shown in Fig. 2 which contain an outside heating element where solution or slurry is pumped through the tubes, heat-transfer rates can be treated with conventional heat-transfer calculations; however, the presence of air in the successive heat exchangers of multiple-effect evaporators causes a marked lowering of the theoretical coefficient. The theoretical coefficients U frequently mentioned in the literature such as those shown in Fig. 8 are often difficult to achieve in practice because of venting problems, particularly on large-scale equipment. Test data taken on small-sized

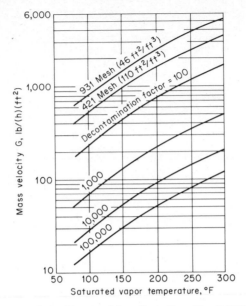

Fig. 7   Allowable mass velocity. (*Reprinted from* Chem. Eng., *Dec. 9, 1963, with permission.*)

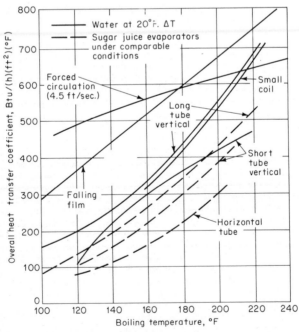

Fig. 8   Comparison of heat-transfer coefficients. (*Reprinted from* Chem. Eng., *Dec. 9, 1963, with permission.*)

exchangers tend to give higher measured coefficients than full-scale equipment owing to the improved venting which can be achieved on small-scale equipment.

For almost all evaporator work the overall coefficient is commonly used, and this is defined as

$$Q = UA \, \Delta T$$

where Q = overall rate of heat transfer, Btu/h or kcal/h
U = overall heat-transfer rate based on the liquid side of the tubes, Btu/(h)(ft²)(°F)
ΔT = difference between steam temperature and liquid temperature leaving the vapor head for natural-circulation machines and log mean ΔT for forced-circulation machines

The heat-transfer coefficient in forced-circulation evaporators is a function of the thermal conductivity, viscosity, specific heat, in-tube liquid velocity, and the steam film

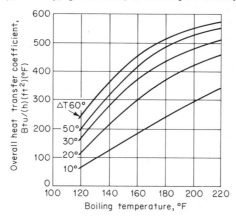

Fig. 9 Typical coefficients for water in short-tube LTVs having steel tubes.

coefficient. It is relatively hard to anticipate the steam film coefficient, and this is often so much higher than the liquid film coefficient, particularly at low temperatures or with high-viscosity materials, that the liquid-side film coefficient controls heat transfer. In cases where the tubes remain relatively clean and free from scale and viscosities are low, very high overall coefficients are sometimes obtained. In these cases the steam film coefficient can be a significant part of the overall heat-transfer resistance. Heat transfer with various types of evaporators is shown in Fig. 8.

With natural-circulation evaporators wherein there is two-phase flow the calculation of the overall heat-transfer coefficient is not normally done by theoretical calculations. Both the short-tube vertical and the long-tube vertical machine coefficients are strongly influenced by ΔT, which is a driving force causing the natural circulation. Shown in Figs. 9 and 10 are heat-transfer coefficients for the concentration of sea water in short-tube vertical and long-tube vertical evaporators. These values are measured in actual equipment operating under laboratory conditions with clean heat-transfer surfaces. For practical application these values must be reduced considerably, and if the liquids being treated are more viscous than water, the coefficients obtained in natural-circulation evaporators of this type can be far below those shown. It is particularly desirable with natural-circulation equipment to have actual test data taken on units of the same tube length.

CALCULATION TECHNIQUES

The calculations necessary for design of or analyzing the operation of evaporators can be summarized best in the form of an energy balance and mass balance around each stage or effect of the evaporator. For a single-effect evaporator the case of the material balance may be trivial in that the feed entering the evaporator differs from the discharge from the evaporator only by virtue of the overhead vapors which are condensed or emitted to the atmosphere. For most evaporators the overhead vapor is essentially pure water containing a few tenths of a percent liquid from the evaporator. For purposes of the material balance, this entrainment is normally ignored.

The heat balance is best made by assuming the reference temperature to be the temperature in the evaporator; and the feed-stream sensible heat difference, the heat of crystallization (if there is any), and the heat required to vaporize the water evaporated make up the total heat requirements of the stage. This heat is supplied by the steam entering the heat exchanger, and in some cases the heat energy resulting from the circulating pump is also considered. To this steam rate is added between 1 and 5% for the effects of radiation.

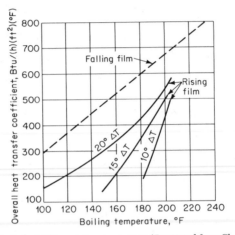

Fig. 10   Actual coefficients for LTV sea-water evaporators. (*Reprinted from* Chem. Eng., *Dec. 9, 1963, with permission.*)

In a multiple-effect evaporator the heat to the succeeding stage is that contained in the vapor evaporated in the first or preceding stage. In addition, when condensate flashing is used, the heat from this source must be added to the heat from the condensing vapor. The heat balance around each successive stage is made in the same way. Although these calculations can be made rigorously by solving the algebraic equations written around the system, they are commonly solved by hand by successive approximations by assuming the evaporation per stage, so that the sensible heat effects can be calculated, and assuming a steam rate to the first stage of the evaporator. After a trial balance is made, the values are compared with the assumptions and the overall scheme is iterated until successive approximations give an indicated evaporation rate and change in concentration per effect close to or identical with the assumptions. This same iterative-solution method is employed by most digital-computation methods.

It is convenient to think of each evaporator effect representing a series resistance to the flow of heat through the system. For this purpose the resistance is thought of as the reciprocal of $U \Delta T$, where the ΔT for the overall system is the difference in temperature between the steam and the cooling water leaving the condenser. The U or coefficient varies with the physical properties of the solution and with the liquor temperature and the type of evaporator in use. The ΔT used in this description refers to the overall ΔT and includes all the temperature differences for heat transfer, the losses of temperature corresponding to pressure drops in the transfer of vapor, as well as the boiling-point elevation of the liquid at each stage. It is conventional in evaporator practice to calculate the heat of condensation as that occurring at the pressure above the liquid rather than the liquid temperature (which is higher in liquids having boiling-point elevation). This ignores any superheat in the vapor, which is generally small for liquids of small boiling-point elevation. For liquids containing very high boiling-point elevations, a more complex correction is required.

In preparing an evaporator design, it is common to assume lower U for each of the succeeding effects in a multiple-effect evaporator, and sometimes to maintain the heat flux the lower U values are compensated for by increased ΔT values in the last stages. In other cases the compensation is made by increasing the heat-transfer surface in the higher-numbered stages.

In an actual evaporator during operation the ΔT per stage adjusts depending upon the U actually obtained and the condition of the heat-transfer surface. There is only control of the steam pressure or temperature in the first-stage heating element and of the condensing temperature or pressure in the last stage. If control of the individual bodies were necessary, it could be achieved only at the expense of throttling vapor between stages, which would increase the unavailable ΔT and thereby reduce the capacity of the evaporator.

It is obvious that to increase the amount of water evaporated per pound of steam used it is necessary to increase the number of effects in the evaporator. As each additional effect is added, the boiling-point elevation of the feed or of the intermediate liquid is added and thereby some of the available ΔT is lost. Therefore, if stages are added, the amount of heat-transfer surface must be increased so that the capital cost of the evaporator increases. In the case of certain materials the boiling-point elevations are so high that only two or three stages are possible. In other cases such as the evaporation of sea water a very large number of stages are possible, sometimes as many as 11 or 12.

Another factor having strong influence on the steam usage in an evaporator is how the condensate from the first-effect heating element and the successive heating elements is handled. Frequently the condensate from the first stage is sent back directly to the boiler even if it is above atmospheric pressure. In other cases it is flashed to atmospheric pressure and the vapor generated by this flashing is directed to one of the evaporator effects to increase the economy. The condensate from each effect's heating element can also be flashed to the succeeding heater to increase the economy at the expense of a combined condensate of lower temperature. Since the condensed water from evaporators is frequently used for process purposes, this scheme may or may not be attractive depending on the situation.

The feed sequence of an evaporator is often a function of the process design, and systems fed as forward-feed evaporators, backward-feed evaporators, and mixed-feed evaporators are all known and have specific advantages. If the greatest overall economy is the only requirement, it is common to feed the evaporator at the stage whose liquid temperature most closely approximates the feed temperature. The feed is passed forward to the succeeding stages and then returned to the stage preceding the feed stage to be fed backward to the first stage. In sea-water evaporators the feed sequence normally dictates feeding to the first stage, not because of heat-balance requirements but because scaling due to heating more concentrated liquid from the successive stages is a factor. When the liquid fed to an evaporator must be heated considerably to reach the temperature of the stage into which it is injected, preheating using condensate or vapor from the appropriate intermediate stages is normally employed.

In recompression evaporators some of the vapor boiled out of the evaporator is compressed to a higher pressure (recompressed) where it can be used to transfer heat to the heat exchanger by condensing at the higher pressure. The energy used for compressing the vapor can be derived from any convenient source of power, either turbine, electric motor, or diesel engine. Normally it is advantageous to do this only in cases where electrical energy or mechanical energy is available at far lower cost than heat energy because of generation of hydroelectric power or other special situations. In the case of compressing the evaporated water with a mechanical compressor, such as a centrifugal or axial-flow type, the inefficiency compared with isentropic compression is normally about 60 to 80% depending on the compressor design, and the inefficiency results in superheat of the compressed vapor. This inefficiency is recovered by spraying condensate into the superheated vapor to bring it back to saturation at the pressure corresponding to the compressor discharge. The work that must be done to compress the vapor consists of pumping the required quantity through a differential pressure corresponding to the difference between the vapor pressure in the body and the discharge pressure corresponding to the heating-element pressure. This can be thought of as an overall ΔT

corresponding to the boiling-point elevation plus the ΔT in the heat exchanger. To maintain a power requirement as low as possible, therefore, most recompression evaporators operate with extremely low overall ΔTs, often in the range of 6 to 8°C (10 or 15°F). Recompression finds its greatest use in those applications where the boiling-point elevation is relatively low, such as in the concentration of sea water or saline waters, in sodium sulfate solutions, and to a limited degree in the production of salt from its saturated brine.

There is also a thermorecompression evaporator. However, it is limited to those cases where either high volumes must be handled because of the low operating pressure and temperature requirements of the process or cases where high-pressure steam is available at the same cost as low-pressure steam. In such cases the thermocompressor, which is merely a large ejector, throttles high-pressure steam through a nozzle and into a diffusion section designed to entrain some low-pressure steam from the evaporator and compress it to the discharge pressure, which would be the heating-element pressure. In such cases the discharge vapors may be desuperheated like a mechanical compressor, although the quantity is different.

EXAMPLE CALCULATIONS

In order to keep track of the effects of boiling-point elevation and varying latent heats of vaporization and specific volume as temperatures change throughout the evaporator, most multiple-effect evaporators are presented in calculation form as a temperature frame like that in Table 3. Before these calculations can be completed, a material balance must be made as shown in Tables 1 and 2. Refer to Fig. 11. Commonly both sets of calculations are made starting with the assumed temperature frame for the effects in question; then material-balance calculations for the design rate are required, and finally calculation of the heat-transfer surface and vapor-head diameters in accordance with the summary at the bottom of the temperature frame. In this example a triple-effect evaporator has been considered using thermorecompression on the first stage.

In the example calculations which follow, the first step is to prepare a sketch showing all known liquid, vapor, and condensate flows similar to that shown in Fig. 11. It is useful to designate the stream numbers at this time and carry the same nomenclature through on the material balance which will

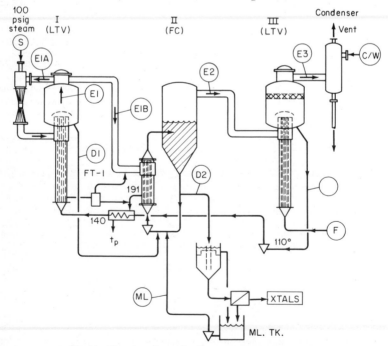

Fig. 11 Flowsheet for triple-effect recompression evaporator.

follow. An overall material balance similar to that shown in Table 1 is useful in establishing the desired evaporation rate and the loss of water with the product and purge stream if there is one. In this particular case, since one of the bodies is precipitating solids, the density of the underflow stream containing the slurry fed to the settler is also indicated as well as the required mother-liquor return path.

The next step is to prepare the temperature frame, listing the temperatures anticipated for the bodies and the physical properties of the vapor under the different pressure and temperature conditions under consideration. In this case, since the first effect is a recompression evaporator, a very small ΔT across the heat-transfer surface was chosen. Generally the recompression evaporator will operate at or close to atmospheric pressure, since the specific volume of steam is more favorable for vapor recompression under these conditions than at higher vacuums. The crystallizing body was chosen as the second stage so the slurry would be high enough in temperature to enter the dryer above its wet-bulb temperature. If the crystallizing body had been the first stage and operated at the atmospheric boiling point, there would be a greater tendency for the slurry to blind the centrifuge screen. The third body has a much greater ΔT than the other two, since cooling water is available at a temperature which permits operation of that body at a liquor temperature of 45°C (113°F). By increasing the ΔT, the heat-transfer surface in that body is reduced. In the crystallizing body, the ΔT chosen was 14°C (25°F) because the material precipitated is sodium sulfate, which has an inverted solubility curve and requires a limited ΔT in the heat-exchange surface to prevent scaling. Once these temperature selections are made, all the

TABLE 1   Overall Material Balance for Triple-Effect Recompression Evaporator

	Total	Na_2SO_4	Water
Feed	75,000	5,000	70,000
Evaporation	69,754		69,754
Crystals	5,246	5,000	246

D2 assumed to be 25% solids (wt).

items in the temperature frame down to and including the latent heat of vaporization can be written down from physical properties or assumptions.

The next step is to make a trial heat balance by assuming an economy and thereby making a selection for the steam rate to the first effect. Since this is a recompression evaporator, the entrained steam ratio is now fixed by the design of the thermocompressor, and in this case the value selected gives an entrained vapor-steam ratio of 1.05.

By making an assumption as to the evaporation which will be achieved in the third stage, the heat balance can be started by summing the heat effects of all streams around each body as shown in Table 4. At the end of the calculation, the total evaporation is computed and compared with that shown in the required overall material balance in Table 1. If the value is below or above the required value, the steam is adjusted accordingly. With a little practice, a close material balance can be achieved by the second or third trial assumption. In this case the economy is 3.53 lb evaporation per lb high-pressure steam. It should be noted here that without recompression the triple-effect evaporator economy would be only 2.66 for these conditions.

The heat balance around the preheater must be checked to be certain that there is a useful ΔT as well as sufficient heat to confirm the assumption that the feed was heated from 43 to 60°C (110 to 140°F). In this case, that assumption was fulfilled.

Based on the evaporation rates selected and the balance obtained as shown in Table 4, a detailed material balance can now be made as shown in Table 2. In this case the feed is entered on the top line and each successive processing step and its influence on the concentration is computed as shown. It is

TABLE 2   Detailed Material Balance for Triple-Effect Recompression Evaporator

lb/hr	Total	Na_2SO_4	Water	% solids	Temp, °F
F	75,000	5,000	70,000	6.67	100
E3	15,899		15,899		
D3	59,101	5,000	54,101	8.46	110
E1			36,786		
D1	22,315	5,000	17,315	22.41	197
ML	14,682	4,415	10,267	30.1	166
E2	17,069		17,069		
D2*	19,928	9,415	10,513	Slurry	166
Crystals	5,246	5,000	246	95.3	166

*Solubility assumed at 43 parts Na_2SO_4/100 parts H_2O = 30.1%.

clear from this balance that the discharge stream D1 leaving the first-effect LTV is still unsaturated. If it were supersaturated or extremely close to saturation, it would be necessary to alter the flow scheme or the relative evaporation rates per stage in order to prevent overconcentration in the LTV, which cannot operate for a reasonable period of time when precipitating solids.

Based on the heat and material balance shown in Table 4, it is now possible to enter the evaporation rate and heater heat-flux rate in the temperature frame in Table 3. From this the heat-transfer surface and vapor-head sizing may be readily calculated.

If an existing design were being checked rather than the preparation of a new design as shown in this illustration, certain additional information would be available which could be used for calculation of the heat-transfer coefficients and for checking the accuracy of the overall material balance. An excellent review of the techniques for doing this on a number of different evaporator styles is given in the AIChE Evaporator Testing Procedure.[4] In addition to this, an article reviewing the test procedure code was written by H. H. Newman.[6]

TABLE 3   Temperature Frame for Triple-Effect Recompression Evaporator

Effect	I	II	III
Steam pressure, in Hg abs	30.0	19.45	9.20
Steam temp, °F	212	191	158
ΔT, °F	15	25	45
Liquor temp, °F	197	166	113
BPE, °F	5	7	3
Vapor temp, °F	192	159	110
Vapor pressure, in Hg abs[a]	19.84	9.42	2.59
LHV, Btu/lb[b]	982	1,003	1,032
Evaporation rate, lb/h[c]	36,786	17,069	15,899
Q/1000, Btu/h[d]	39,155	16,580	17,119
U, Btu/(h)(ft²)[e]	380	400	225
LMT, °F		22.4	
HS, ft²	6,870	1,850	1,690
$U\Delta T$	5,700		10,125
Temperature out of heater		170	
Circulation rate, gal/min		8,290	
Vapor head diam	15	12	14

[a]A 1°F loss is assumed due to friction loss in the vapor pipe.
[b]The latent heat of this vapor is taken at saturation.
[c]In I, the entrained-steam ratio is 1.05 for the conditions chosen.
[d]Heat to be transferred in the heat exchangers.
[e]Based on previous experience.

TABLE 4   Heat Balance for Triple-Effect Recompression Evaporator

(S)	HP steam 19,780	Recompressed vapor (E1A)	
(EIA)	Entrained 20,769† ⟵		
	(40,549)(985.4*)(0.98)	$= +$ 39,155,794	
(D3)	Sens. ht. (59,101)(197-140)(0.9)	$= -$ 3,031,830	20,769
	EVAPORATION IN I	$+$ 36,123,964 $\div$ 982 = 36,786	
(E1B)	To II effect	$+$ 15,728,806	16,017
	Cond. fls. (40,549)(212-191)(1)	$= +$ 851,529	
	To II heater	$+$ 16,580,335	
(D1)	Sens. ht. of FD (22,314)(197-166)(0.78)	$= +$ 539,553	
	EVAPORATION IN II	$+$ 17,119,887 $\div$ 1003 = 17,069	
(F)	Ht. feed in III (75,000)(110-100)(0.95)	$= -$ 712,500	
	EVAPORATION IN III	16,407,387 $\div$ 1032 = 15,899	

Total evaporation = 36,786 + 17,069 + 15,899 = 69,754

$$\text{Economy} = \frac{69,754}{19,180} = 3.53 \frac{\text{lb evaporation}}{\text{lb steam}}$$

Preheater Heat Balance
$(75,000 - 15,899)(140-110°)(0.95) = (16,017 + 40,549)(191-t_p)$
$t_p = 161° > 140°$ assumed liquor outlet temperature

†From the manufacturer's data 1.05 lb entrained/lb steam.
*Determined from difference between steam and condensate streams for I heater.

The typical calculations required for a mechanical recompression evaporator concentrating sodium carbonate solutions are shown in Tables 5, 6, and 7. The basic vapor and liquid flows are shown in Fig. 12. The overall material balance indicates a required evaporation of 36,957 lb/h. In this example, steam is compressed from a suction pressure equivalent to saturation at 93°C (200°F) to a pressure equivalent to steam saturation at 110°C (230°F). This is a 1.8 compression ratio and could be handled by a single-stage centrifugal compressor.

Because the feed is substantially colder than the operating temperature, a small amount of makeup steam is required. Vapor evaporated from the solution during processing is compressed from 587.5 mm (23.5 in)/Hg abs to 1057.5 mm (42.3 in)/Hg abs. Some condensate would be recycled as indicated to desuperheat the vapor leaving the compressor before entering into the heat exchanger. Assuming 75% compressor efficiency (over the isentropic path) approximately 65.3 Btu is required to compress 1 lb of vapor.

From the heat balance shown in Table 7, an energy-use ratio can be computed wherein the Btu equivalent to the evaporation rate are divided by the Btu put into the system by the compressor and by the makeup steam. In this case, the ratio is 11:87, which corresponds to a much higher economy than

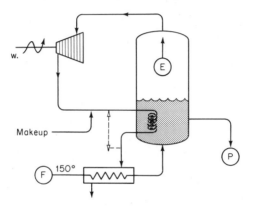

Fig. 12   Single-effect recompression evaporator.

could be obtained with normal multiple-effect evaporators. Bear in mind, however, that this is attractive only when the solution has a low boiling-point elevation such as shown in this example and when the vapor is noncorrosive so that it can be handled in compressors readily available. Hot condensate leaving the heat exchanger is used to preheat the feed from 66 to 92°C (150 to 199°F).

TABLE 5   Material Balance for Single-Effect Recompression Evaporator

	Total	Na_2CO_3	H_2O	Conc.
(F) Feed	50,000	3,000	47,000	6%
(E) Evaporation	36,957		36,957	
(P) Prod.	13,043	3,000	10,043	23% (Unsat.)

TABLE 6   Temperature Frame for Single-Effect Recompression Evaporator

SP	in Hg abs	42.3		H_v at discharge = 1195 (isentropic)
ST	°F	230		H_v at suction = 1146 (isentropic)
ΔT	°F	18		ΔH_v = 49
LT	°F	212		
BPE	°F	10		
VT	°F	202		Say 75% eff. ∴ $\dfrac{49}{0.75}$ = 65.3 Btu/lb
PFL	°F	2		
Suction	°F	200		$\dfrac{36,957\ (65.3\ \text{Btu/lb})}{2544}$ = 949 hp at
VP	in Hg abs	23.5		compressor

(side labels: 1.8 comp. ratios ; 30° ΔT)

In each case, the overall economy is limited by the fluid processed and is a function of the heating surface chosen and the mechanical layout. In any given case under consideration the larger the heating surface selected for the heat exchanger (reducing the ΔT required for heat transfer) to reduce the power consumption, the higher the capital investment in heat-exchanger surface. An economic selection must always be made where the savings in energy for compression over a selected equipment life are offset by higher capital charges on the initial installation. Other costs which must be considered but are less objectively defined are the required investment in spare parts (which is much greater in recompression evaporators than in multiple-effect evaporators), penalty for interrupted operation during repairs, and part-load response.

The choice of the drive and the ΔT in the heat exchanger lead to logical engineering choices, but the part-load response is the more complex question because the efficiency of the compressor chosen decreases as the capacity is reduced. Depending on the initial heat-transfer-area selection, however, the effect of excess heating surface at part-load

TABLE 7   Heat Balance for Single-Effect Recompression Evaporator

Vapor from compressor	$(36{,}957)(1195 - 198)(0.988)$	$= +\ 36{,}109{,}206$
Sens. heat	$(50{,}000)(198 - 212)(0.92)$	$= -\quad 644{,}000$
Evaporated water	$(36{,}957)(976.6)$	$= -\ 36{,}092{,}206$
		$-\quad 627{,}000$
Makeup	$(654)(958.8)$	$= +\quad 627{,}000$
		0

$$\text{Energy-use ratio} = \frac{\text{Btu evaporation}}{\text{Btu input}} = \frac{(36{,}957)(976.6)}{(36{,}957)(65.3) + 654(958.8)} = 11.87$$

Preheater Balance

$$(50{,}000)(t - 150)(0.92) = (36{,}957 + 654)(230 - 170)(1)$$
$$t = 199 \quad \text{(close check)}$$

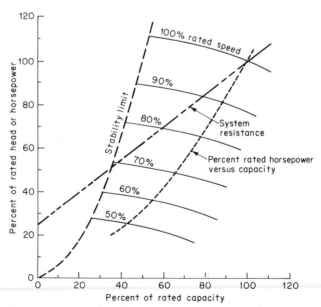

Fig. 13   Recompression evaporator compressor and system characteristic curves.

conditions tends to reduce the relative compression range required and thereby leads to lower power requirements per pound of vapor handled. Another important consideration with centrifugal and axial-flow compressors is that there is a stability limit at approximately 55% of design flow at design speed below which the system becomes mechanically inoperable. As shown in Fig. 13, this stability limit restricts the operating area of the system considerably. The major head against which the compressor pumps is proportional to the boiling-point elevation plus the ΔT for heat transfer. In forced-circulation evaporators this heat transfer ΔT directly is proportional to the evaporation rate. This means that the system response (neglecting pipe-friction loss) is essentially a straight line intersecting the head curve at some finite level equivalent to the boiling-point elevation. To operate at loads other than the design load, therefore, it is necessary either to reduce the compressor speed or to change the operating pressure so that the compressor sees an increase in specific volume of the vapor and hence a lower operating temperature. Depending on the choice of the compressor driver and conditions, this may impose limits of operation that are fairly restrictive. It also illustrates the advantages in operating such a compressor with a variable-speed drive, but this type of drive is not easily achieved with electric power.

REFERENCES

1. Perry, J. H.: "Chemical Engineers' Handbook," 3d ed., sec. 7, McGraw-Hill, New York, 1952.
2. Perry, J. H.: "Chemical Engineers' Handbook," 4th ed., sec. 11, McGraw-Hill, New York, 1963.
3. Perry, J. H., and Chilton, C. H., "Chemical Engineers' Handbook," 5th ed., sec. 11, McGraw-Hill, New York, 1973.
4. "Evaporators," A.I.Ch.E. Equipment Testing Procedure, American Institute of Chemical Engineers, 1961.
5. Standiford, F. C.: Evaporation, *Chem. Eng.*, Dec. 9, 1963, pp. 157–176.
6. Newman, H. H.: How to Test Evaporators, *Chem. Eng. Prog.*, **64**(7), 33–38 (July 1968).
7. The Versatile Evaporator, *Brit. Chem. Eng.*, September 1957, pp. 473–488.
8. Beagle, M. J.: Recompression Evaporation, *Chem. Eng. Prog.* **58**(10), 79–82 (October 1962).
9. Ruths, D.: Plate Evaporator System, *Ind. Eng. Chem.* **57**(6), 47–49 (June 1965).
10. Parker, N. H.: How to Specify Evaporators, *Chem. Eng.*, July 22, 1963, pp. 135–140.
11. Beard, James L.: private communication, May 16, 1977.
12. Whitt, F. R.: Performance of Falling Film Evaporators, *Brit. Chem. Eng.*, **11**(12), 1523–1525 (December 1966).
13. Agitated Thin Film Evaporators, *Chem. Eng. Prog.*, Sept. 13, 1965, pp. 175–190.

Section **2.4**

Crystallization from Solutions

GYANENDRA SINGH *Section Head, Food Products Research, The Procter & Gamble Company, Cincinnati, Ohio*

ACKNOWLEDGMENTS: The author is grateful to the Procter & Gamble Company for permission to publish this.

NOTATION

a, b	Kinetic parameters
A	Crystal-surface area
A, B, K	Constants
C	Concentration
D	Diffusivity
G	Free energy
ΔG_v	Volume free-energy change per unit volume of the crystal
ΔG_c	Activation-energy barrier to nucleation
i	Nucleation sensitivity parameter
J	Nucleation rate [nuclei/(volume) (time)]
K_v, K_1	Geometric shape factors
L	Crystal size
n	Crystal-size distribution
M	Mass of crystals in suspension
N	Cumulative number oversize
r	Growth rate
R	Universal gas constant
S	Supersaturation
t	Time
T	Temperature, K
v	Volume
U	Velocity
V_0	Molar volume
$W(X)$	Dimensionless weight fraction function
X	L/r_r, mean residence time
Z	Constant

GREEK LETTERS

ρ	Density
σ	Interfacial free energy per unit area of the nuclei
τ	Residence time
δ	Film thickness

INTRODUCTION

Crystallization is the process of production of crystals from vapor, melt, or solution and can be used to separate solids from liquids, liquids from liquids, or solids from solids. The most important process industrially is the crystallization of dissolved solids from solutions, which is the subject of this section.

If a solution in equilibrium between the solid and the liquid phase is altered in such a way (whether by cooling the solution or evaporating some of the liquid) that the amount of dissolved solids exceeds the equilibrium concentration, the system will seek to attain equilibrium by getting rid of this excess solids concentration. The resulting process is called *crystallization from solutions* and the concentration-gradient driving force is called *supersaturation*.

Crystallization from a solution consists of two distinct phenomena: crystal formation and crystal growth. The former is also called *nucleation*. Both nucleation and crystal growth require that the solution be supersaturated, but the effect of supersaturation is different in the two processes.

Supersaturation is a measure of the quantity of solids actually in solution as compared with the quantity that would be present if the solution were kept for a long period of time with the solid phase in contact with the solution (equilibrium solubility). Supersaturation is generally expressed as a coefficient:

$$S = \frac{\text{parts solute/100 parts solvent}}{\text{parts solute at equilibrium/100 parts solvent}}$$

Solutions vary greatly in their ability to withstand supersaturation without crystallization. The most common everyday examples are sugar and salt; while sugar solutions can withstand a great deal of supersaturation without crystallizing (supersaturation coefficients of 1.5 to 2.0), salt solutions crystallize with so little supersaturation that it is difficult to even measure it.

CRYSTALLIZATION THEORY

Crystallization Equilibria

Crystallization equilibria of any system may be defined in terms of its solubility and supersolubility curves. The supersolubility curve differs from the solubility curve in that its position is not a property of the system only but also depends on other factors like the cooling rate, the degree of agitation, and the presence of foreign particles.

Under specific conditions, however, the supersolubility curves for a given system are definite and reproducible and represent the maximum supersaturation the system can tolerate, at which point nucleation occurs spontaneously. The solubility curve describes the equilibrium between the solute and the solvent and represents the locus of conditions at which solute crystals and the mother liquor coexist in thermodynamic equilibrium. The solubility and supersolubility curves divide the entire concentration-temperature field into three zones: the undersaturated region to the right of the solubility curve, the metastable region between the two curves, and the supersaturated or labile region to the left of the supersolubility curve. A typical crystallization-equilibria diagram is shown in Fig. 1.

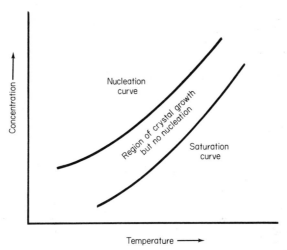

Fig. 1   Crystallization equilibria: Miers' solubility-supersolubility curves.

According to Miers' original theory, in the undersaturated region crystals of solute will dissolve, crystal growth will occur in the metastable region, and nucleation will occur instantaneously in the supersaturated or labile region.[1] Subsequent workers found, however, that factors other than supersaturation also affect nucleation.[2,3] Crystal-surface area, rate of cooling, and mechanical shock have been shown to affect the position of Miers' curves. An implication of the subsequent work has been that for real systems the supersolubility curve is nearly always a "band" or a region and not a line. Nevertheless, if the process conditions are kept constant, the supersolubility curves for a given system are definite and reproducible and represent the maximum supersaturation the system can withstand without homogeneous or heterogeneous nucleation. The solubility curve, on the other hand, differs from the supersolubility curve in that its position is a property of the crystal system alone and does not depend on process variables.

Nucleation

Nucleation basically occurs by three different mechanisms: homogeneous, heterogeneous, or secondary nucleation. *Homogeneous* nucleation refers to nucleation in the bulk of a fluid phase without the involvement of a solid-fluid interface. *Heterogeneous* nucleation is nucleation in the presence of surfaces other than that of the crystal itself such as that of the container, agitator, or foreign particles. *Secondary* nucleation refers to nucleation due to the presence of the crystals of crystallizing species itself.[4]

According to the homogeneous-nucleation theory of Gibbs, molecular aggregates form in solution which grow into nuclei if they exceed a certain size. Aggregates below this critical size are called embryos and above the critical-size nuclei. Two energy effects result from the formation of a nuclei: the volume free energy of the system is reduced and the surface free energy increased. The activation-energy barrier to crystallization is the sum of these two effects. It can be shown that

$$G_C = \frac{32\sigma^3 V_0^3}{R^2 T^2 \ln^2 (S + 1)} \tag{1}$$

Thus the activation-energy barrier to nucleation is directly proportional to the interfacial free energy and inversely proportional to supersaturation. For any given supersaturation a nucleus of critical size exists, which is in unstable equilibrium with the supersaturated solution. The higher the supersaturation, the smaller is the critical size.

Volmer-Weber-Becker-Doring developed rate expressions for nucleation from the vapor phase. In all cases the rate expression is of the form

$$J = A \exp\left(-\frac{\Delta G_C}{KT}\right) \tag{2}$$

or

$$J = A \exp\left(-\frac{32\sigma^3 V_0^2 N}{(RT)^3 \ln^2 (S + 1)}\right) \tag{3}$$

or

$$J = A \exp\left(-\frac{Z}{\ln^2 (S + 1)}\right) \tag{4}$$

where Z is a term that consolidates all the terms which are constant for a specific crystal system and temperature.

In the case of nucleation from a liquid phase (solution or melt) diffusion effects become significant, as one now has to deal with closely packed molecules of a liquid phase instead of the free-flowing molecules of a vapor phase. The diffusion effects are taken into account by introducing a term in the rate equation for nucleation, as follows:

$$J = B \exp\left(-\frac{\Delta G_D}{RT} - \frac{\Delta G_C}{RT}\right) \tag{5}$$

If the free energy of activation for diffusion can be assumed constant over the temperature range of interest, the above equation reduces to the preceding one.

From the above equations it can be seen that nucleation rate is directly proportional to the supersaturation and inversely proportional to the crystal-solution interfacial free energy. Further, the nature of this relationship is exponential, which makes it extremely sensitive to even small changes in these two variables.

Several authors have asserted that the exponential term in Eq. (2) can be adequately represented by a power function of supersaturation.[5,6,7]

$$J = \frac{dN^0}{dt} = K_N S^b \tag{6}$$

Further, the effects of secondary nucleation can be accounted for by including a crystal-surface-area term in the kinetic model:

$$J = K_2 S^b A^j \tag{7}$$

where S is the supersaturation ($C - C_{eq}$) and A is the crystal-surface area of the suspension.

The kinetic models represented by Eqs. (6) and (7) have often been called the "engineering models" because of their overwhelming application in engineering design.

Crystal Growth

Crystal growth can be represented as occurring in two steps:
1. Diffusion of the solute to the suspension-crystal interface.
2. Surface reaction to absorb the solute into the crystal lattice.

If the surface reaction is first-order, the overall process may be represented by

$$\frac{dM}{dt} = \frac{1}{\delta/D + (1/K_i)(C/C_{eq})} = K_j AS \qquad (8)$$

where M = total mass of crystals in suspension and δ = effective film thickness.

As $\delta \to 0$, the growth rate is reaction-controlled; as $K_i \to \infty$, the growth rate is diffusion-controlled. We can express the mass and surface of the crystal in terms of its size:

$$M = K_v L^3$$
$$A = K_A L^2 \qquad (9)$$

where K_v and K_A are geometric shape factors and L is a characteristic crystal dimension. Equation (8) can now be reduced to

$$r = \frac{dL}{dt} = K_g S \qquad (10)$$

where S = supersaturation = C/C_{eq}. This relationship has been experimentally verified for several systems.

In a classical series of experiments, McCabe[8] observed that for a wide variety of substances in aqueous solution the growth rate of a crystal along a linear dimension is independent of the size of the crystal:

$$\frac{dr}{dL} = 0 \qquad (11)$$

This law is generally known as McCabe's ΔL law, and when it can be assumed to hold, it results in considerable simplification in the mathematical analysis of crystallization processes. In real systems, deviations from the ΔL law have been observed. Further, the linear relationship of growth rate with supersaturation has often been shown not to hold true. Therefore, several workers have suggested the power-law model:

$$r = K_g S^a \qquad (12)$$

where the power a may determine a kinetic dependency[9] close to or significantly different from unity. The proportionality constant is a function of temperature, environment, and crystal size.

Crystal-Size Distribution

Crystal-size distribution (CSD) is the single most important factor in the analysis of crystallization systems, and has the greatest interaction with all crystallization problems: habit (i.e., the external appearance and shape of the crystal), purity, scale-up, stability, and liquid-solid separation. However, it remains the single most difficult and least-understood problem of crystallization.

Conventional mass and energy balances per se are not sufficient for the analysis of CSD, as they tell us nothing about the particle-size distribution. From mass and heat balances, one can, of course, predict the total production rate of a crystallizer but without any clue as to whether this production comes out, say, as a very fine powder or as large coarse crystals. However, by incorporating a population balance in addition to mass and energy balances, it is possible to predict the crystal size distribution as well as yield and production rate.

Population balance is based on the principle that, like mass and energy, the number of discrete particles must also be conserved in any dispersed system, and that given the proper representation of birth and death rates, all particles can be accounted for. Such an accounting, called a population balance, results in a mathematical expression characterizing the size distribution. The size distribution is dependent on the birth rate, the growth rate, and when applicable, the death rate or the rate of disappearance.

When a crystallizer is operated under certain conditions which can be readily achieved in practice, the differential equations characterizing the resultant size distribution can easily be solved to yield nucleation and growth rates. Such a unit is called the mixed-suspension, mixed-product-removal (MSMPR) crystallizer, and the size distribution predicted by population balance is an exponential expression. The constraints on such a crystallizer are:

- Well-mixed suspension
- Mixed-product removal
- Negligible breakage
- No crystals in the feed
- Steady state
- McCabe's ΔL law

The implication of McCabe's ΔL law is that the linear growth rate is independent of size. Note that it is not necessary to restrict the analysis to the exponential form of the CSD in order to solve for growth and nucleation rates, as once the population density is known, the characterizing equations can be solved directly using numerical techniques. Once the basic population distribution is known, one can calculate several quantities of interest from different moments of the distribution, e.g., the cumulative or the distributive weight distributions, the mass-weighted mean size, population-weighted mean size, dominant size, surface area, solids concentration in suspension, and other pertinent properties.

Population Balance for Continuous Crystallizers

To formulate the population balance, it is necessary to define a continuous variable to represent the discrete distribution. Such a function can be defined as the number of crystals per unit volume in a given size range n called the population density. Thus

$$\lim_{\Delta L \to 0} \frac{\Delta N}{\Delta L} = \frac{dN}{dL} = n \tag{13}$$

By definition of the population-density function, the number of particles in the range of size from L to $L + dL$ found in a volume of liquid V (solids-free basis) is given by

$$dN = Vn(L)\, dL \tag{14}$$

The units of population density n are (ft^{-3}), (ft^{-1}), or (ft^{-4}). To find the number of crystals per unit volume of liquid in some finite range L_1 to L_2, it is only necessary to integrate n over the range of L. Thus,

$$\int_{L_1}^{L_2} n\, dL = \text{number of crystals with sizes from } L_1 \text{ to } L_2 \text{ per unit volume}$$

Consider now a well-stirred crystallizer with a constant-composition feed as shown in Fig. 2. A well-mixed slurry with unclassified product leaves the crystallizer continuously. It is further assumed that no particles enter with the feed. Assuming steady state, the population balance for an arbitrary size range L_1 to L_2 and time interval Δt is as follows:

Let r = linear particle growth rate
Q = volumetric feed and discharge rate

Then,   Input =   Output
or $Vr_1 n_1 t = Vr_2 n_2 t + Q\bar{n}Lt$
i.e.,

Number of crystals growing into range over the time interval	=	number of crystals growing out of range over the time interval	+	number of crystals in the size range removed from the crystallizer

or

$$\frac{V(r_2 n_2 - r_1 n_1)}{L} + Q\bar{n} = 0 \tag{15}$$

In the limit as $\Delta L \to 0$,

$$V\frac{d}{dL}(rn) + Qn = 0 \tag{16}$$

Letting V/Q equal τ, the hold time, and assuming McCabe's ΔL law applies, i.e., growth is independent of size, we get

$$\frac{dn}{dL} + \frac{n}{r\tau} = 0 \tag{17}$$

If n^0 is the population density of the zero-size particles (which can be assumed to be the nuclei), integration yields

$$\int_{n^0}^{n} \frac{dn}{n} = \int_{0}^{L} \frac{dL}{r} \tag{18}$$

or

$$n = n^0 \exp\left(-L/r\tau\right) \tag{19}$$

Equation (19) gives the functional relationship between L and n and characterizes the size distribution. Therefore, knowing n as a function of L and τ, the hold time in the crystallizer, it is possible to calculate r, the growth rate, and n^0, the population density of the nuclei.

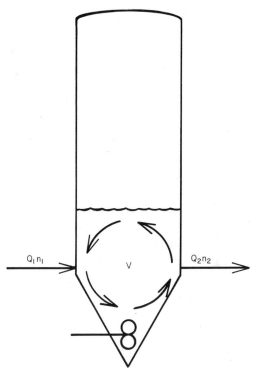

Fig. 2   Mixed-suspension, mixed-product-removal crystallizer.

Nucleation and Growth Kinetics

The population density can be calculated from the CSD obtained and plotted on a semilog paper, i.e., log n vs. L. The resulting plot forms a straight line with slope related to $-1/r\tau$ and intercept n^0 (Fig. 3). Knowing τ, the growth rate can be calculated. To calculate the nucleation rate, we use the following:

$$n^0 = \frac{dN}{dL}\bigg|_{L=0} \qquad r = \frac{dL}{dt} \tag{20}$$

$$\text{Nucleation rate} = \frac{dN}{dt}\bigg|_{L=0} = \left(\frac{dN}{dL}\bigg|_{L=0}\right)\frac{dL}{dt} \tag{21}$$

The above analysis suggests that by changing the hold time $\tau = V/Q$ (by changing the flow rate under identical crystallizer conditions) different growth and nucleation rates resulting from the different supersaturations will be obtained. Thus, if the existing steady-state supersaturation can be measured, one can relate the kinetic rates to supersaturation and develop kinetic models.

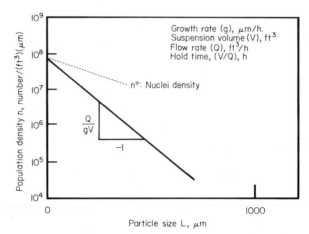

Fig. 3 Typical population-density curve for MSMPR crystallizer.

Nucleation and growth may be viewed as parallel kinetic reactions, the former determining the rate of particle formation, the latter determining the rate of deposition of solute on existing crystals. Therefore, the final crystal-size distribution obtained can be controlled through varying the relative rate of nucleation as compared with growth; if it is desired to increase the mean crystal size, nucleation must be suppressed relative to growth, resulting in fewer but larger crystals. Quantitatively this can be expressed as

$$R = \frac{\text{nucleation rate}}{\text{growth rate}}$$
$$= \frac{K_N S^b}{K_g S^a}$$
$$= K_R (S)^{b/a}$$

For two different levels of supersaturation S_1 and S_2, $S_2 > S_1$,

$$\frac{R_1}{R_2} = \left(\frac{S_1}{S_2}\right)^{b/a}$$

If $b > a$, $R_2 > R_1$; that is, at the higher level of supersaturation, nucleation rate has increased relative to growth rate. The result will be a crystal-size distribution with greater percentage of fines.

If $b < a$, $R_2 < R_1$; that is, at the higher level of supersaturation, growth rate has increased relative to nucleation rate. The result will be a crystal-size distribution with a larger mean crystal size.

If $b = a$, $R_2 = R_1$, and the crystal-size distribution will be invariant.

The design of crystallizers should be based on increasing growth rate relative to nucleation rate. If growth rate is of higher order than nucleation, operating at as high a supersaturation as possible is desirable; therefore, a batch- or a plug-flow crystallizer is preferred to a back-mix crystallizer. If one is restricted to a back-mix crystallizer, a shorter-residence-time unit will give better results as compared with one with a longer residence time. However, if the growth rate is of lower kinetic order than nucleation rate, decreasing supersaturation will decrease growth rate less than the corresponding decrease in nuclea-tion rate. Therefore, operating at low supersaturation is desirable and a back-mix crystal-lizer with a long residence time should be designed for such cases.

ANALYSIS OF A CRYSTALLIZATION SYSTEM

In analyzing a crystallization system, we seek to answer the following questions:

1. What will be the yield under different operating conditions? What material and energy inputs-outputs can be expected?

2. What are the optimum operating conditions?

3. What will be the supersaturation under different process conditions?

4. At what rate will crystal nucleation and growth take place, and how are they related to operating conditions?

5. What is the effect of impurities and additives on the crystal-size distribution?

6. What crystal-size distribution can be expected?

7. What is the optimum crystallizer design—geometry as well as configuration?

8. What is the dynamic behavior of the process? What is the effect of transient conditions, as at start-up, shutdown, or system upsets caused by such things as changes in the feed rate, temporarily shutting off product discharge, etc.?

The first three questions can be answered from the knowledge of the crystallization equilibria, the solubility-supersolubility relationships for the system; the fourth from the crystallization kinetics, the relationship between nucleation and growth rates and the driving forces for crystallization; the fifth requires a knowledge of the effect of the chemical environment on the crystal system—i.e., how does the presence of impurities and additives affect the crystal-size distribution? The sixth and seventh require the knowledge of crystallizer residence-time distributions in addition to the above; the last relates to crystallizer dynamics, i.e., system response to external and internal disturbances.

Crystallization Equilibria for the Given Crystal System

The concentration driving force for crystallization—supersaturation—that will be present in the system for any given set of process conditions must be determined. Different modes of operation are used to create supersaturation in a crystallization system, the net effect of all of which is to create the driving force—supersaturation—and a generalized analysis is applicable to all of these. The choice of the specific mode of operation for a particular application depends upon the slope of the solubility curve for the crystal system. The solubility curves for several inorganic salts in water are shown in Fig. 4. Depending on whether the slope of the solubility curve is positive, negative, or flat, indirect or vacuum cooling or evaporation may be employed. It should be noted that crystallization-equilibria phase diagrams, such as Miers' solubility-supersolubility curves for a two-component system, or a system that can be described as such, describe only the primary-nucleation effects in the system.

However, the effect of other crystallization mechanisms must also be considered. Primary nucleation, which includes homogeneous and heterogeneous nucleation, is sometimes not as significant a factor as is secondary nucleation—nucleation generated by the presence of crystals themselves. Once primary nucleation has occurred, secondary nucleation often proceeds at a rapid rate under much lower supersaturations than that required for either homogeneous or heterogeneous nucleation. The obvious question then is: Of what value are crystallization-equilibria relationships which primarily show the effect of primary nucleation?

The answer is twofold: first, crystallization-equilibria information is needed to carry out material and energy balances and yield calculation; second, it can be used to optimize the process conditions for crystallization.

Regardless of the dominant mechanism, whether it be primary or secondary nucleation, controlling supersaturation means controlling nucleation. This should be no surprise in light of many studies which have shown that nucleation rate almost always correlates with supersaturation, even though for some systems better correlations can be obtained when a crystal-suspension surface area or a mechanical-energy term is included to account for secondary nucleation. Therefore, if the effect of process variables such as cooling rate, solvent concentration, agitation rate, or feedstock concentration on the supersaturation curves is known, the former can be adjusted so as to optimize nucleation for desired process results. Staging of the process can be carried out to ensure operations at levels of supersaturation which maximize throughput. In fact, in many processes it is possible to characterize the crystal suspension at the point where nucleation first occurs to control and optimize the process.

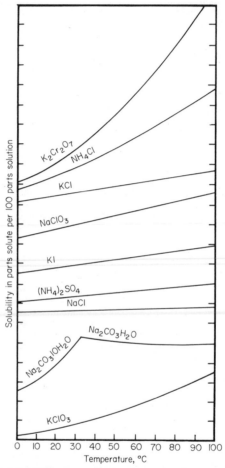

Fig. 4 Solubility curves of various inorganic salts in water.

Crystallization Kinetics for Nucleation and Crystal Growth

While the dominant nucleation mechanism may be either primary or secondary, both depend on supersaturation, and therefore in most cases the power-law "engineering model" of nucleation kinetics is sufficient. Also, recent studies have shown that power-law nucleation kinetics observed in mixed-suspension crystallizers apply equally well to those systems where secondary nucleation is the dominant mechanism.[10,11]

Crystal growth can be either diffusion-controlled or particle-integration (surface-reaction)-controlled. In the case of diffusion-controlled growth, crystal growth rates become size-dependent. However, in a well-mixed crystal suspension growth rate is generally particle-integration-controlled and therefore independent of size. Consequently, in most systems it is sufficient to check that the growth rate is in fact particle-integration-controlled, and empirical correlations of the power-law form suffice to characterize it adequately.

The population-balance theory and the mixed-suspension, mixed-product-removal (MSMPR) crystallizer concept make it possible to obtain quantitative nucleation and growth-rate data and predict crystal-size distributions. A number of predictive CSD studies have been carried out and kinetic models developed for several crystal systems.[12]

The Effect of Chemical Environment

The chemical environment, i.e., the presence of relatively low concentrations of substances other than the crystallizing species as either impurities or additives, plays an important role in optimization of a crystallization system. Its role is important for several reasons. First, all materials are impure to a certain extent or contain trace impurities introduced during processing. Random variation of impurities is undesirable. Their effect on the crystallizing species must be understood if satisfactory control over the crystallization system is to be achieved. Second, and more importantly, it is possible to influence the output and control of a crystallization system or change the crystal properties by the addition of small amounts of carefully selected additives. Thus, by adding certain types and amounts of additives it is possible to tailor the crystal size, crystal-size distribution, crystal habit, and purity of the product crystals.

The chemical environment is one of the three key variables that can be used to influence nucleation and growth kinetics, and hence the crystal-size distribution, the other two being supersaturation and residence-time distribution. Indeed, in certain crystallization systems it can be more influential than the other two; therein lies its unique value in industrial practice of crystallization. The chemical environment can be appropriately varied to:

1. Significantly alter the crystallization kinetics and hence the crystal-size distribution.

2. Gain better control of a crystallizer.

3. Improve product quality and/or yield by producing a desired type of product crystal.

4. Produce very pure crystals of certain materials in which impurities are unacceptable as they distort the crystal lattice and affect its physical properties.

Garrett has summarized many examples from industrial practice.[13] A more recent review by Botsaris et al. presents a list of impurities that influence the crystallization of certain materials.[14]

Crystallizer Residence-Time Distribution

Ideally, the residence-time distribution for both the liquid and the solid phases should be known and controllable. However, where one of the two is dominant, it may be sufficient to know the residence-time distribution of the dominant phase only.

The crystallizer residence-time distributions depend primarily on the crystallizer geometry and process configuration (seeding, product classification, fines removal, clear-liquor advance, staging, etc.). The crystallizer residence-time distributions, together with the crystallization kinetics, determine the crystal-size distribution that will be obtained.

The process configurations used to bring about changes in the crystallizer residence-time distributions are as follows:

Classified Product Removal   Crystals of a desired size-distribution range are preferentially removed by a separating device, the most common of which is an elutriating leg even though other devices such as hydrocyclone or a wet screen are also used, and the undersized particles are recycled to the crystallizer. In an elutriation leg, clarified mother liquor can be circulated through the bottom of the leg to fluidize the crystals prior to discharge and selectively return the smaller crystals to the crystallizer. The classification of crystals is controlled by the flow rate of the fluidizing liquid. The net result is a shift in the crystal-size distribution toward a smaller average particle size within a narrower size range, as shown in Fig. 5.

Fines Removal/Destruction   Particles smaller than a specified size are separated from the crystal suspension, leaving fewer crystals in suspension on which solute deposition can take place. This causes an increase in supersaturation leading to increased average crystal size due to higher growth rates. The effect of fines removal, which can be either internal or external, on the population density of the crystal suspension is shown in Fig. 6.

The net effect is equivalent to reduction in the effective nucleation rate relative to that in the absence of fines destruction. Fines removal must be carefully controlled, since an increase in supersaturation helps only up to a limit; at excessive growth rates poorly formed and impure crystals result or alternately excessive homogeneous nucleation results once supersaturation exceeds the metastable limit, which negates the effect of fines destruction.

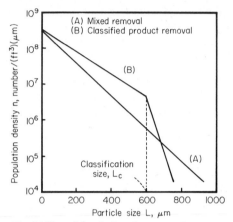

Fig. 5 Effect of classified product removal on population density.

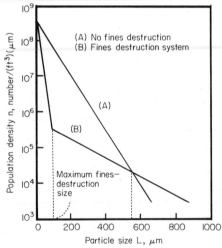

Fig. 6 Effect of fines destruction on population density.

Clear-Liquor Advance   For increasing slurry density from weak feed liquors or for producing large crystals at low supersaturations, clear liquor may be withdrawn from the top of the crystallizer through a settler. Simultaneously, the product stream is drawn off from the mixed underflow at a much lower rate, so as to increase the residence time of the solids in the system, causing an increase in the average crystal size. The net effect of clear-liquor advance is to lower the supersaturation while raising the solids concentration and increasing the slurry density. Clear-liquor advance is limited by secondary-nucleation effects at high solids concentration or an increase in slurry density to the point that the crystal suspension becomes unstable.

Seeding   Fine crystals are deliberately added as crystal nuclei on which further crystal growth is to take place. Seeding is generally unnecessary in continuous systems because of secondary-nucleation mechanisms but can be very necessary in some batch-crystallization processes, particularly where the system can withstand a high degree of supersaturation without nucleating—as is the case in sugar crystallization. Where seeding is the only source of nuclei or where very little additional nucleation takes place, a narrow crystal size distribution can be produced.

Staging   The best way to produce a narrow crystal-size distribution in continuous systems is through a number of crystallizers in series. As the number of stages in series increases, the residence-time distribution of the liquids in the system approaches that of a plug-flow crystallizer. If nucleation can be confined to the first stage, a very narrow CSD will be produced under conditions analogous to a seeded batch crystallizer. While this is difficult to achieve in practice, it is possible to carefully control the process to maximize nucleation in the first stage and hence optimize the CSD obtained.

Classified-Bed Crystallizers   For crystal systems with size-dependent growth rates, exceedingly large crystals can be produced in classified-bed crystallizers by contacting the large particles near the bottom with the freshly supersaturated liquor. This results in the larger particles being exposed to higher supersaturation than the smaller particles at the top of the crystal suspension, and hence growing at a relatively faster rate. The only limitation is the maximum supersaturation that the liquor can withstand without homogeneous nucleation.

Crystallizer Dynamics

System dynamics are somewhat more complex for crystallization than other unit operations owing to an inherent process instability arising from internal feedback loops.[15] Therefore, both the system response to outside disturbances (transients) and this process instability must be taken into account when analyzing the dynamic behavior of a crystallization system. Transient oscillations decay with time, whereas instability causes undamped cyclic oscillations. A basically stable crystallizer will oscillate as an underdamped system when subject to upsets or perturbations.

Transient CSD calculations can be used to evaluate the likely effect of major process upsets, as reasonable agreement between experimental observations and theoretical calculations has been obtained.[16] For most crystallization systems, a transient period of from 6 to 10 retention times is necessary to reach steady state.

The inherent instability of crystallization systems is due to the dependence of nucleation and growth rates, which determine CSD, on supersaturation, which in turn is dependent on the crystal area and hence the CSD. Recent studies of secondary nucleation indicate that other feedback loops may also exist due to the mechanisms of secondary nucleation. The two feedback loops shown in Fig. 7 provide a qualitative explanation for this type of cyclic instability. Linear-stability analysis of crystallization systems[17] indicates that the nucleation/growth rate sensitivity parameter, defined as

$$i = \frac{d \ (\log \text{ nucleation rate})}{d \ (\log \text{ growth rate})}$$

must exceed 21 before the system becomes unstable. In practical terms, it means that only nucleation discontinuities can cause system instability, as, for example, a system with fines removal forcing supersaturation into a region of homogeneous nucleation (nucleation discontinuity) because of excessive fines dissolution. Recent studies, however, indicate that process instability can also occur at low values of the nucleation-sensitivity parameter in certain systems such as those with product-classification schemes. Thus, overregulation of crystallization systems should be avoided.

An overview of the crystallization system is provided in Table 1, which shows the system inputs, responses, and outputs, It is necessary to understand the relationship between the system inputs and responses to be able to control the responses and hence the output.

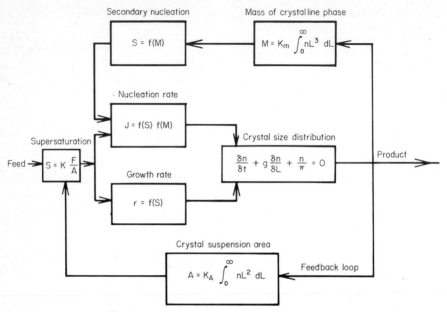

Fig. 7 Natural feedback loop in continuous crystallizers.

CRYSTALLIZATION EQUIPMENT

Because of the widely varying requirements for different applications, it is rare for units on different sites to be identical, even though many basic principles apply to the design of each individual system. These basic principles are:

1. Controlling the level of supersaturation to correspond to low rates of nuclei formation.

2. Maintaining a sufficient number of seed crystals in suspension so that there is adequate suspension surface area to relieve supersaturation by the deposition of the solute.

3. Contacting the seed crystals with the supersaturated liquor as soon as possible to avoid losses due to time decay.

4. Removing excess nuclei as soon as possible after their formation.

5. Minimizing secondary nucleation by keeping mechanical energy input and crystal attrition as low as possible.

6. Maintaining as high a magma density as possible. In general, the higher the magma density, the larger the average crystal size.

7. Minimizing solids buildup by eliminating localized heat and mass-transfer gradients, unnecessary restrictions to fluid flow, and operating at as low temperature gradient or supersaturation as possible.

8. Providing a chemical environment, i.e., impurities and additives, which favors the desired crystal shape or growth.

Two schemes have been broadly employed to classify crystallization equipment according to (1) the method of generating supersaturation or (2) the method of suspending the growing crystals.

There are five basic methods of generating supersaturation:

1. *Evaporation.* By vaporization of the solvent.

2. *Cooling.* By cooling a solution through indirect heat exchange.

3. *Vacuum Cooling.* Flashing the feed solution adiabatically to a lower temperature and inducing crystallization by simultaneous cooling and evaporation of the solvent.

4. *Reaction.* By chemical reaction.

5. *Salting.* By the addition of a third substance to change the solubility relationship.

TABLE 1   The Crystallization System

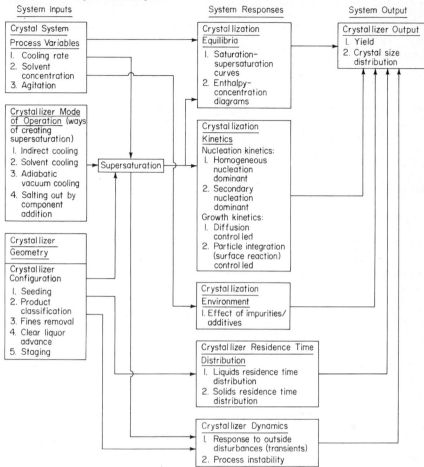

However, classification according to the method of generating supersaturation is not entirely satisfactory as different means of generating supersaturation may be employed in the same type of equipment, e.g., Oslo (Krystal) crystallizers in which either of the first three methods, evaporation, cooling, or vacuum, may be employed. Therefore, classification according to the method of suspending the growing crystals is increasingly coming into vogue. Under this classification scheme, four basic types of crystallizers may be identified.

1. *Circulating Magma.* All the growing crystals are circulated through the zone of the crystallizer where supersaturation is generated. This may be accompanied by mixed or classified product removal with or without fines destruction.

2. *Circulating Liquor.* Only the liquor is circulated, with the bulk of the growing crystals not being circulated. Supersaturation is imparted to the liquor in one part of the equipment; then the liquor is circulated to another part of the crystallizer where it gives up this supersaturation to the growing crystals, and is then recirculated again. This type of equipment is also available with or without fines destruction.

3. *Scraped Surface.* Crystallization is induced by direct heat exchange with a cooling medium at a heat-transfer surface which is continuously scraped or agitated to minimize fouling/solid deposition.

4. *Tank Crystallizers.* Crystallization is induced by cooling the feed solution in

either static or agitated tanks by natural convection/radiation or by surface cooling through coils in the tank or a jacket on the outside of the tank.

An overview of crystallization equipment showing both the classification schemes is provided in Table 2. The table shows the various types of crystallizers available, criteria for specific applications, and examples of known applications. Both classification schemes are shown, as both are likely to be encountered in practice, even though only the latter is used here.

Circulating-Magma Crystallizers

Circulating-magma crystallizers are by far the most important type of crystallizers in use today; both the forced-circulation (FC) and draft-tube-baffle (DTB) belong to this class.

The FC crystallizer is shown in Fig. 8. The feed enters the circulating pipe below the product discharge at a point sufficiently below the free-liquid surface to prevent flashing. The crystal suspension together with the feed is circulated through a heat exchanger, where its temperature is raised. The deposition of solute on the heat-exchanger tubes can be minimized by heating under pressure to prevent vaporization. The heated magma enters the crystallizer near the liquid surface, raising the temperature near the point of entry to cause boiling. The vaporization of the solvent and the consequent cooling result in the supersaturation necessary for nucleation and crystal growth. The product is continuously withdrawn from the circulating pipe.

The critical design parameters in FC crystallizers are the internal recirculation rate and velocity, crystallizer hold volume, and the speed of the circulating pump. At internal circulation rates less than the optimum, excessive flashing can occur at the boiling surface, causing excessive supersaturation. If all this supersaturation cannot be relieved by deposition of the solute because of the lack of adequate crystal-surface area in the suspension, excessive nucleation will occur, causing excessive fines in the CSD and solids buildup on the crystallizer walls. In general, recirculation rates aim at restricting the flashing at the boiling surface to about 1.65 to 4.4°C (3 to 8°F), while a magma-density range of 20 to 30% is typical even though the exact optimum depends on the particular crystal system. The

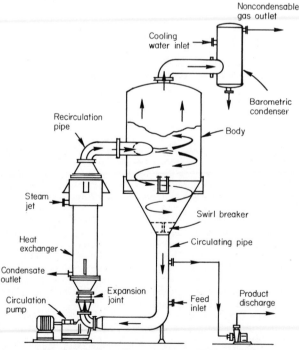

Fig. 8 Forced-circulation (evaporative) crystallizer.

pump design should be picked to permit the lowest speed possible to minimize the mechanical-energy input to the system. If the speed of the pump or propeller is too high, severe attrition and secondary-nucleation effects can cause a smaller average crystal size and excessive fines in the crystal-size distribution. The crystal sizes typically produced by FC crystallizers range from 30 to 60 mesh. Some improvement in the performance of the FC crystallizers can be obtained by improving the liquid distribution over the boiling surface, such as using a radial rather than tangential recirculation inlet, as in conispherical magma crystallizers.

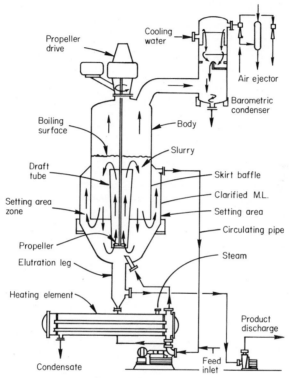

Fig. 9 Draft-tube-baffle (DTB) crystallizer.

The draft-tube-baffle (DTB) crystallizer gives significantly larger crystals than the FC crystallizers under equivalent conditions. As shown in Fig. 9, it consists of a closed vessel with an inner baffle forming a partitioned settling area, inside which is a tapered vertical-draft tube surrounding a top or bottom entering agitator located close to the bottom. The agitator is of the axial-flow type and operates at low speeds. An elutriating leg can be fitted to the bottom cone if further classification of crystals is desired. The draft tube is centered by support vanes to prevent body swirl and minimize turbulence in the circulating magma. Supersaturation may be generated by either evaporation or cooling. A modified form of the above is the draft-tube (DT) crystallizer in which the baffle is omitted and the internal agitator is sized to achieve low circulation rates at which secondary nucleation in the suspension does not become a problem.

In DTB crystallizers, the feed enters the base of the crystallizer and is directed into the upward-draft tube. The axial-flow agitator circulates the liquid and crystals from the bottom to the top liquor surface. Mother liquor is continuously removed from the top, which causes an upward liquor flow through the settling area. The larger crystals settle down into the active volume of the crystallizer, while fine crystals below a certain size are removed by the liquor flow. The magma density can be regulated by controlling the liquid

TABLE 2 Crystallization Equipment

Crystallizer type	Crystallizer name	Basic criteria for application	Known applications
Classification by the method of creating supersaturation:			
Evaporative	Tank crystallizer, strike pan, forced-circulation evaporator, Oslo (Krystal) evaporator crystallizer, draft-tube/draft-tube baffle evaporative crystallizer	Where solubility coefficient is very small or negative (solubility decreases with temperature). In such cases, crystallization can be effected only by evaporation	Complex inorganic salts from diluted ammonia solution, organic substances from aqueous solution by dilution with methanol, isopropanol, or acetone, crystallization of substances from high-boiling solvents under reduced pressure, nickel sulfate from the electrolytic copper refining liquor, sodium chloride from the electrolytic cell caustic soda liquor, crystallization of sodium chloride, anhydrous sodium sulfate, sodium carbonate monohydrate, ammonium sulfate, ammonium perchlorate, sodium nitrate, citric acid, etc.
Cooling	Tank crystallizer, forced-circulation cooling crystallizer, Oslo (Krystal) cooling crystallizer, draft-tube/draft-tube baffle cooling crystallizer, scraped-surface crystallizer, Votator	Where solubility coefficient is positive, i.e., solubility increases with temperature	Inorganic salts such as Glauber salt, Epsom salt, copperas, hypo, copper sulfate, sodium potassium tartate, organic compounds such as p-xylene, fatty acids, marine oils, chlorobenzene, benzoic, butyric, sorbic, boric acids, etc.
Vacuum	Swenson vacuum crystallizer, Oslo (Krystal) vacuum crystallizer, draft-tube vacuum crystallizer	Where continuous operation is necessary or where corrosive conditions exist necessitating low-temperature operations	Sodium nitrate; mono-, penta-, and octa-hydrates of barium hydroxide, sodium sulfate, sodium thiocyanate, silver nitrate, nickel sulfate, tartaric acid, borax, boric acid, potassium nitrate, di- and trisodium phosphate, aniline salt, sodium citrate; inorganic sulfates such as sodium, zinc, magnesium, nickel, copper; iron, manganese, ammonium; organic acids such as tartaric, citric, fumaric, oxalic, adipic
Reaction	Tank crystallizer, forced-circulation crystallizer	Where a combination reaction/crystallization operation is to be carried out	Molecular-sieve zeolites
Salting out	Tank crystallizer, forced-circulation crystallizer	Where the addition of a third component changes the solubility relationship	Glauber's salt from sodium sulfate brines

Classification by the method of suspending growing crystals:

Circulating magma	Forced-circulation crystallizers (evaporative, cooling, surface-cooled, etc.). Draft-tube crystallizers (evaporative, cooling, direct-contact-cooling, etc.). Draft-tube-baffle crystallizers (evaporative, cooling, surface-cooled, vacuum, etc.)	For the production of large crystals	Ammonium sulfate, sodium/potassium chlorides, borax, trisodium phosphate, potassium nitrate, sodium ferrocyanide, sulfamic acid, sodium tetraborate, etc.
Circulating liquor	Oslo (Krystal) crystallizers (evaporative, cooling, surface-cooled and vacuum)	For the production of smaller, more uniform crystals	Gypsum, molecular-sieve zeolites, sodium sesquicarbonate, silver nitrate, sodium nitrate, ammonium nitrate, urea, sodium sulfate, ammonium sulfate; inorganic sulfites, thiosulfates, chlorates and dichromates
Scraped surface	Swenson-Walker crystallizer, Votator, Armstrong continuous scraped cooling crystallizer	For organic compounds or where scaling is a major problem	p-Xylene, fatty acids, marine oils, chlorobenzene, organic dyes, organic acids such as benzoic, butyric, sorbic, boric, sebacic, propionic, adipic; urea, naphthalene, formates, cyclamates
Tank	Static or stirred-tank crystallizers (surface-cooled or evaporative)	For batch operations, or where crystals are not the major product	Fatty acids, vegetable oils, sugar, molecular-sieve zeolites

circulation. Further classification and thickening of slurry discharge is carried out by the elutriating flow, which carries small crystals back to the body for further growth. A fines-destruction feature can also be added in systems where a substantial density difference exists between the crystals and mother liquor. The coarse crystals are separated from the fines by gravity sedimentation, the slurry containing the fines withdrawn from the settler and heated to dissolve the fines prior to being recirculated back to the crystallizer.

The DTB crystallizer has two advantages over the FC units in that it reduces supersaturation at the boiling surface down to 0.55 to 1.1°C (1 to 2°F) and it brings growing crystals to the boiling surface where the supersaturation is most intense. The internal circulation within the crystallizer is considerably higher than in FC units and is achieved against extremely low heads of a few inches of water, requiring much less energy input than is possible in an external circulating loop with the same amount of circulation. Because of the large quantity of seed surface present at the boiling surface, the supersaturation is easily relieved by crystal growth; hence fresh nucleation is minimized. Both the heavy slurry density at the boiling surface and concentration of boiling in the center of the body reduce the overall amount of fouling on the vessel walls. This increases the operating cycles, giving lower operating costs. Using DTB crystallizers, it is possible to obtain more uniform-quality product and crystals ranging from 8 to 30 mesh.

For systems where it is preferable to generate supersaturation by indirect cooling by heat exchange across a surface, a forced-circulation tube-and-shell heat exchanger may be combined with a DTB crystallizer as shown in Fig. 10.

In some cases, direct-contact cooling is employed to generate supersaturation by mixing the coolant directly with the feed. The coolant is then separated from the crystal magma after crystallization. Obviously, for successful operation of such systems the coolant must be easily separable for recycling into the crystallization system. This type of crystallizer is used for systems where (1) the allowable temperature difference between the coolant and slurry is very small [less than .36°C (5°F)] so that heat exchange across a surface becomes

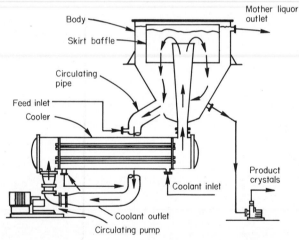

Fig. 10 Forced-circulation-baffle surface-cooled crystallizer.

impractical owing to the large surface area required and (2) the slurry viscosity is so high that the mechanical energy required for circulating the magma is excessive. Other side benefits of direct cooling are the elimination of fouling on cooling surfaces and a reduction in the process-energy requirements for crystallization. However, the overall process-energy requirements may or may not be reduced depending on the energy requirements for the separation and recycling of the coolant.

Circulating-Liquor Crystallizers

This type of crystallizer is called the circulating-liquor type because the crystals are retained in the suspension vessel and only the liquor is circulated. The crystals are maintained in suspension by the upflowing solution whereby each face of the growing crystal is exposed to the fresh supersaturated liquor, resulting in the growth of large, regular well-defined crystals. The output and crystal-size distribution are controlled by varying the liquor flow rate.

This type of crystallizer is unique in that the same basic design and operating principles are used whatever the method of achieving supersaturation—indirect cooling, vacuum cooling, or evaporation. The crystallizer consists of three interrelated parts; a suspension of crystals where most of the crystal growth takes place, a vessel or means of producing supersaturation under controlled conditions, and a means of circulating the liquor throughout the crystallizer. The saturated mother liquor is drawn off from the suspension vessel and mixed with incoming feed liquor. The mixture is then passed through the supersaturation zone to build up the supersaturation before being circulated through the crystal suspension to relieve the supersaturation via crystal growth.

The best-known example of this type of crystallizer is the Oslo/Krystal crystallizer of which three versions are shown in Fig. 11. The supersaturation is developed in one part of the system by any of the three methods: vaporization, surface cooling, or vacuum cooling. The mother liquor leaving the supersaturation zone is only slightly supersaturated so as to stay within the metastable zone to minimize fresh nucleation. A stream of mother liquor is withdrawn from the top of the suspension vessel for fines removal and destruction to obtain larger crystal sizes.

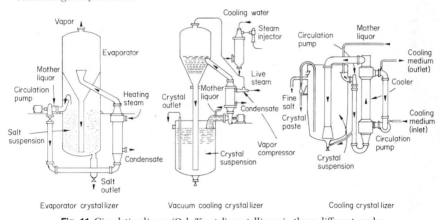

Fig. 11 Circulating-liquor (Oslo/Krystal) crystallizers in three different modes.

Scraped-Surface Crystallizers

High-viscosity solutions are best handled in scraped-surface crystallizers and are unsuited for circulating-magma or circulating-liquor crystallizers. The supersaturation is generated by direct heat exchange between the slurry and the coolant or heat exchange across a jacket or double wall containing a cooling medium. The cooling medium circulates through the shell side at velocities high enough to give good shell-side heat-exchange coefficients, and an internal agitator fitted with spring-loaded scrapers scrapes the heat-transfer surface to prevent solids buildup and maintain the heat-transfer coefficient at the highest possible level. The equipment can be operated in either continuous or recirculating-batch manner. While its capacity is generally smaller than the other types of crystallizers, being limited by the heat-transfer area available, it is cheaper to install, as it does not require expensive installation or supporting structures.

Examples of scraped-surface crystallizers are the double-pipe scraped-surface crystallizer (Votator and Armstrong) and the Swenson-Walker crystallizer. The first is essentially a double-pipe heat exchanger fitted with scrapers to wipe the inner pipe walls. The units are typically built in 12-m (40-ft) sections, which can be mounted in parallel or in series to give the desired crystallizer hold time. At typical temperature differentials of 16.5°C (30°F) or higher, heat-transfer coefficients in the range of 0.004 to 0.02 cal/(s)(cm²)(°C) [30 to 150 Btu/(h)(ft²)(°F)] have been reported.[18]

The Swenson-Walker crystallizer is a continuous-cooling crystallizer which consists of an open round-bottomed trough 600 mm (24 in) wide by 3 to 12 m (10 to 40 ft) long, jacketed for circulating the coolant, with a ribbon mixer throughout the length of the crystallizer which turns slowly at about 3 to 10 r/min. The blades of the mixer have a narrow clearance from the crystallizer wall, typically about 1.5 mm (1⁄16 in). Several of these units may be connected together with the agitators being driven from a single shaft. For even greater capacity, a second set of crystallizers can be set up at a slightly lower level and fed by overflow from the end of the first set. During operations, the hot feed solution enters continuously at one end and the crystal slurry overflows at the other end. The agitator blades scrape the crystals from the cold walls, giving them a forward motion through the solution. This maximizes crystal growth while keeping nucleation to a minimum in the bulk of the slurry. The main limitation to the capacity of these units is their heat-transfer-surface requirement. Typically, the heat-transfer coefficients range from 0.001 to 0.0034 cal/(s)(cm²)(°C) 3L/cm [10 to 25 Btu/(h)(ft²)(°F)] for a heat-transfer surface ratio of about 1.1 m²/m (3.5 ft²/ft) of crystallizer length, at a crystallizer holdup ratio of about 3 L/cm (25 gal/ft) of length. Even though the unit can be operated at large temperature differences, as the heat-transfer surface is continuously scraped, it is not attractive economically at heat-transfer rates exceeding 200,000 Btu/h.

Tank Crystallizers

This type of equipment is generally used for smaller capacities and with materials of normal solubility. Both static and stirred-tank crystallizers are used. Applications include fine chemicals, pharmaceuticals, food products, and systems in which the liquor and not the crystals is the desired product. Supersaturation is generated by either natural convection and radiation, surface cooling through coils in the tank, or circulating a coolant through a jacket on the outside of the tank. Agitation improves the heat and mass-transfer characteristics of the crystallizer, eliminates localized gradients, and reduces the solids buildup on cooling surfaces. At the same time, as agitation increases so do secondary-nucleation effects so that an optimum agitation rate exists. For agitated systems, the overall heat-transfer coefficient is typically in the range of 0.0027 to 0.027 cal/(s)(cm²)(°C) [20 to 200 Btu/(h)(ft²)(°F)].

In static tanks, nucleation is difficult to control or predict and cooling rate varies considerably. Because of the absence of agitation, localized supersaturation levels rise to very high values, resulting in the formation of dendritic crystals and considerable occlusion of mother liquor. Either very large singular crystals or a "slush" consisting of large quantities of fines with occluded mother liquor can result. Therefore, the operation of static-tank crystallizers is generally more difficult and expensive as compared with stirred-tank crystallizers and is limited to areas where the labor costs are low relative to capital costs.

DESIGN OF CRYSTALLIZERS

The crystallizer is only part of a crystallization system, as is evident from the schematic diagrams of two typical crystallization systems shown in Fig. 12. Therefore, the design of a crystallization system requires complete specifications regarding the chemical to be crystallized, the crystallizer and accessories necessary to operate it, and the site where it is to be installed and operated. A checklist of design specifications for crystallization systems is provided in Table 3.

For the design of typical accessories such as heat exchangers, condensers, vacuum equipment, refrigeration equipment, pumps, centrifuges, and filters, standard references concerned with such equipment may be consulted. The design procedure for crystallizers is outlined below and illustrated by the examples that follow:

    1. Select the crystallizer type based upon the criteria outlined in Table 2.

    2. Select operating temperatures from the solubility-supersolubility curves.

    3. Calculate the material and energy flows from material and energy balances.

    4. Determine the average residence time necessary to produce the desired CSD from the nucleation and growth-rate data.

    5. Size the crystallizer on the basis of average residence time or vapor-release area.

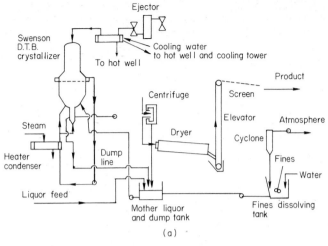

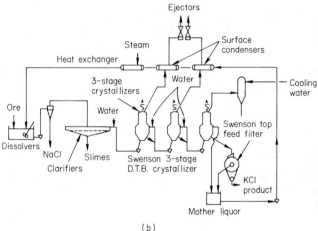

Fig. 12 Typical crystallization systems: (*a*) crystallization of ammonium sulfate from Nylon-6 or acrylonitrile by-product liquors and (*b*) crystallization of potassium chloride.

EXAMPLE 1

Given the kinetic models

$$J = 31.91 \times 10^6 r^{1.7}$$

and

$$r = 0.052 s^{0.8}$$

Where J and r are the nucleation and growth rates, respectively, with the units number of crystals per hour and millimeters per hour. Calculate the mean residence time required for an ammonium sulfate crystallizer given the requirements that more than 90% of the product crystals be greater than 100 mesh and the saturated feed and mother-liquor temperature should be 109 and 54°C (200 and 130°F) respectively.

From Fig. 4, the solubility of ammonium sulfate at 109 and 54°C (200 and 130°F) is seen to be 36 and 32.6%, respectively. Therefore, supersaturation $s = 36/32.6 = 1.1$ and growth rate $r = 0.056$ mm/h.

The size distribution of solids can be calculated from Eq.(19).

$$W(x) = 1 - e\left(\frac{x^3}{6} + \frac{x^2}{2} + \frac{x}{1} + 1\right)$$

where $W(x)$ is the weight fraction up to size L and $x = L/r\tau$, where x is the mean residence time. (Values of $W(x)$ as a function of x are given in the Appendix.)

The size distribution that can be expected for a given mean residence time can now be calculated and the mean residence time selected according to the product size requirements as shown below.

For mean residence times 1 and 2 h, respectively:

Mean residence time, h	Screen size mesh	L, mm	X	$W(x)$	Cumulative % retained $100[1-W(x)]$
1	20	0.833	14.88	0.998	0.2
	28	0.589	10.52	0.995	0.5
	35	0.417	7.44	0.939	6.1
	48	0.295	5.26	0.770	23.0
	65	0.208	3.72	0.510	49.0
	100	0.147	2.63	0.271	72.9
2	20	0.833	7.44	0.939	6.1
	28	0.589	5.26	0.770	23.0
	35	0.417	3.72	0.510	49.0
	48	0.295	2.63	0.271	72.9
	65	0.208	1.86	0.119	88.1
	100	0.147	1.31	0.044	95.6

Therefore, the theoretically expected size distribution from a 2-h mean-residence-time crystallizer would result in 95% of the product being larger than 100-mesh size. The nucleation rate under those conditions is:

$$J = 31.91 \times 10^6 r^{1.7}$$

or
$$J = 31.91 \times 10^6 \, 0.056^{1.7}$$

or
$$J = 0.2376 \times 10^6/h$$

or
$$J = 66/s$$

TABLE 3   Design Specifications for a Crystallization System

A. Specifications concerning the chemical
  1. Chemical formula, name, and purity of the product to be crystallized
  2. Output required; quantity, purity, and particle size
  3. Composition of feed liquor including impurities and their effect on the crystal system
  4. Solubility and supersolubility curves
  5. Crystal-size distribution and/or crystallization kinetics
  6. Physical properties:
    a. Specific heats of solutions and crystals
    b. Specific gravity of solutions and crystals
    c. Viscosity of solutions
    d. Boiling-point elevation of saturated solutions
    e. Heat of crystallization
    f. Settling velocity of crystals of average size in mother liquor
    g. Vapor pressure of solvent
    h. Decomposition temperature of crystals
    i. Whether the crystals are deliquescent, hygroscopic, efflorescent, toxic, corrosive, inflammable, explosive, etc.
B. Specifications concerning the crystallizer
  1. Batch or continuous process
  2. Crystallizer type and special design features, e.g., fines removal, classification, seeding
  3. Materials of construction
  4. Accessories, i.e., pumps, valves, centrifuges, dryers, filters, instruments, etc.
C. Specifications concerning the site
  1. Location, area, height available, site elevation and access—obstacles, narrowest load, lowest bridge, etc.
  2. Utilities costs and availability (steam, oil, gas, water, refrigerant, etc.)
  3. Regulations and methods of disposal of effluents and wastes
  4. Manpower costs and availability

EXAMPLE 2

Design an ammonium sulfate crystallizer for the above example given the following:
  Production rate: 10,000 lb/h
  Wash water added on the centrifuge: 20 lb/100 lb of product
  Cooling water for the condenser: 86°F
  Boiling-point elevation: 24°F
  Heat of crystallization: 100 Btu
  Slurry density: 1.66 lb/gal
  1. A forced-circulation type of crystallizer is indicated from the relatively "flat" solubility curve.
  2. The temperature frame is as follows:
  Steam: 212°F
  Liquor: 130°F
  Boiling-point elevation: 24°F
  Vapor: 106°F
  Cooling-water inlet: 86°F
  Cooling-water temperature rise: 14°F
  Cooling-water outlet: 100°F
  Latent heat of vaporization: 1033 Btu/lb
  Special volume: 296 ft³/lb
  3. *Material Balance:* Basis 100 lb feed.

Stream	Total	Water	Ammonium sulfate
Feed	100.00	64.00	36.00
Purge	10.00	6.74	3.26
Product	32.74		32.74
Evaporation	57.26	57.26	

Ratio up to 10,000 lb/h
  Feed (100/32.74) 10,000 = 30,500 lb/h
  Evaporation (57.26/100) 30,500 = 17,465 lb/h
  Purge (10/100) 30,500 = 3050 lb/h
  4. *Heat Balance:* Assume that the latent heat of the vapor is equal to that of steam at the pressure in the vessel and the heat input from the circulating pump is negligible.
  Sensible heat (30,500)(70)(0.62) = 1,323,700 Btu/h
  Heat of crystallization (10,000)(100) = 1,000,000 Btu/h
  Latent heat of vapor (17,465 + 2000)(1033) = −20,107,345 Btu/h
  Steam (19,915)(940)(0.95) = 17,783,645 Btu/h
  5. *Crystallizer hold volume:* The crystallizer hold volume can be calculated from the required mean residence time and the slurry density.

$$\frac{10{,}000 \text{ lb/h}(2 \text{ h})}{1.66 \text{ lb/gal}} = 12{,}050 \text{ gal}$$

  6. Body size: Maximum vapor velocity.

$$U_m = K_v \sqrt{\frac{\rho_1 - \rho_2}{2}}$$

  where K_v = constant
    ρ_1, ρ_2 = density of the liquid and vapor respectively

$$U_m = 0.06 \sqrt{(72.2 - 0.0034)/0.0034} = 8.75 \text{ ft/s}$$

$$\text{Minimum diameter} = \frac{(19{,}465 \text{ lb/h})(296 \text{ ft}^3/\text{lb})}{(3600 \text{ s/h})(8.75 \text{ ft/s})}$$
$$= 183 \text{ ft}^2 = 13.5 \text{ ft diam}$$

Hence, for 12,050 gal volume a crystallizer vessel 15 ft in diameter by 10-ft liquid level should be used.
  7. Accessories: The heat exchanger, condenser and vapor pipe, vacuum system, and pumps must now be designed. Any standard reference may be consulted. The mechanical design of the crystallizer should be in accordance with the ASME code for pressure vessels.

CRYSTALLIZER OPERATIONS

The key operating problems in crystallization systems are:
  1. Control of crystal-size distribution

2. Minimizing fouling (extension of operating cycle)
3. Maintaining crystallizer stability
4. Separation of solids from the mother liquor

The keys to the control of CSD are the rate of nucleation and magma density. The rate of nucleation required to produce crystals of a given size decreases exponentially with increasing size of the product. For example, 1 lb of 150-mesh seed crystals is sufficient to produce 1 ton of 14-mesh product crystals. Therefore, careful control of nucleation is necessary in order to produce large-sized crystals, particularly since nucleation rates are relatively fast as compared with crystal growth rates and even a small upset can produce a large number of unwanted nuclei. Careful attention must be exercised to prevent prenucleation in the incoming feed stream or unwanted nuclei being introduced in the crystallizer through recycle streams. Careful control of the magma (slurry) density is essential to the control of crystal size in many systems. In such systems, an increase in the magma density increases the product size through a reduction in the nucleation rate and increased residence time of the crystals in the growing bed. Conversely, a reduction in the magma density will generally increase the nucleation rate and decrease the particle size.

The method of controlling CSD depends on the type of crystallizer. In FC units seeding, classification, or controlling the slurry density are the primary control variables. In DTB units, the flow in the circulating pipe is regulated so as to withdraw a portion of fines from the crystallizer. The exact quantity of solids removed depends on the size of the product crystals and the capacity of the fines-dissolving system. At steady state, the quantity of solids overflowing should be relatively constant, with some solids always appearing in the stream being withdrawn from the crystallizer. However, too high a circulation rate or too high a slurry density in the crystallizer will cause a large quantity of crystals to appear in the overflow, which can disable the fines-destruction system. On the other hand, too low a flow will destroy an insufficient number of fines and result in smaller product crystal size than desired. In classifying crystallizers, control of the fluidizing flow is required in addition to the control of the fines-removal stream. The flow required to fluidize the crystal suspension varies with the particle-size distribution and hence must be varied whenever the CSD is transient as during start-up/shutdown and system upsets. Thus, from the standpoint of complexity of operations, the FC units are the simplest, followed by DTB and classifying crystallizers.

All crystallizers, no matter how well operated, eventually undergo fouling, causing system shutdown for a washout. Typical washout cycles range from 4 h to several days, resulting in valuable production time being lost. The fouling problem is perhaps the second most important operating problem, after the control of particle size, in most installations. While fouling cannot be completely eliminated, it can be controlled to a great extent by determining the optimum magma density and circulation rates. Again, there is a tradeoff between more nucleation by attrition, change in habit to a rounded shape, and increased pumping costs vs. the length of operation that can be achieved before washout.

The areas most likely to foul are the vapor-release space, boiling liquid-vapor interface, and inlets to circulation lines and circulating downcomers. Two basic designs are used to remove heat at a boiling interface: a separate vapor-release area for the clear liquor or direct vaporization above the mixed magma. In general, the latter is preferred as the driving forces are generated in the presence of the full crystal surface, which allows supersaturation levels to decay immediately and not continue to exist for a long time. This reduces the likelihood of homogeneous nucleation at the boiling interface and hence minimizes fouling. On the other hand, crystallizers with a separate, clear-liquor vapor-release section can be forced into the zone of homogeneous nucleation, since adequate crystal-surface area may not be available for supersaturation to be relieved, leading to excessive fouling if not controlled.

It should be noted that while in the beginning fouling may be a function of the rate of nucleation on the vessel walls, subsequently it proceeds by growth and nucleation on the crystals already deposited on the fouled surface. Thus a combination of both molecular growth and heterogeneous nucleation is responsible for building up fouling deposits. Impurities/additives can have a pronounced effect on the extent and rate of fouling and should be employed to minimize fouling wherever possible. Likewise, variations in mother-liquor purity affect fouling and must be controlled to the extent possible. Steady operations are relatively more important to crystallization as compared with most other unit operations. This is due to the relatively slow rates of crystal growth, which typically require 2 to 6 h of crystallizer mean residence time to produce crystals of the desired particle size. Since four to six times the mean residence time is generally required to damp out the effects of a system upset, 8 to 36 h may elapse before steady operations are resumed. Unusually long recovery periods characterize crystallization systems. The problem is further complicated by the inherent process instability of crystallization, which as discussed earlier can cause long periods of cycling during which the CSD undergoes wide swings ranging from very coarse to very fine crystal sizes.

The final step at the end of any crystallization operation is the separation of the crystals from the mother liquor, which provides a focal point of interaction between crystal habit, CSD, and crystallizer operation. Therefore, the solid-liquid separation step often becomes the controlling step in the entire operation. The product stream from crystallizers typically contains less than 30 wt % solids, which is usually thickened to 50% or greater solids prior to the solids-liquid separation step, for which either conical settlers or hydrocyclones can be used depending upon the settling characteristics of the particles. The final solids-liquid separation is generally accomplished via filters or centrifuges. Filters are preferred for high-volume applications and wherever a separate washing step is necessary, while centrifuges are preferred for a more thorough separation of solids and mother liquor. The choice between centrifuges and filters is not clear-cut, however, as it depends on many factors, e.g., solids concentration, liquid and solids throughput, necessity of wash-liquor separation, breakage of crystals, dryer capacity, and maintenance costs, and should be determined by pilot testing for each individual operation.

CRYSTALLIZER COSTS

While it is difficult to provide an accurate picture of crystallizer costs in general, the circulating-liquor type of crystallizer is the most expensive followed by the DTB, forced-circulation, and tank crystallizers. The installed costs of the four types can range from 4:1 to 3:1 as compared with tank crystallizers, depending upon the capacity of the equipment. Typically, crystallizer cost ranged from $4000/(ton/day) to $1000/(ton/day) in 1975 economic values.

REFERENCES

1. Miers, H. A., *J. Inst. Metals*, **37**, 331 (1927).
2. Ting, H. H., and McCabe, W. L., *Ind. Eng. Chem.*, **26**, 1201 (1934).
3. Preckshot, G. W., and Brown, G. G., *Ind. Eng. Chem.* **44**, 1314 (1952).
4. McCabe, W. L., and Smith, J. C., "Unit Operations of Chemical Engineering," McGraw-Hill, New York, 1967.
5. Randolph, A. D., and Larson, M. A., "Theory of Particulate Processes," Academic, New York, 1971.
6. Timm, D. C., and Cooper, T. R., *Am. Inst. Chem. Eng. J.*, **17**, 639 (1971).
7. Randolph, A. D., *Chem. Eng.*, **77**(10), (1970).
8. McCabe, W. L., *Ind. Eng. Chem.*, **21**, 112 (1929).
9. Van Hook, A., "Crystallization," Reinhold, New York, 1961.
10. Ottens, E. P. K., Janse, A. G., and DeLong, E. J., *J. Cryst. Growth*, **13/14**, 500 (1972).
11. Bennet, R. C., Fiedleman, H., and Randolph, A. D., *Chem. Eng. Prog.*, **69**(7), 86 (1973).
12. Moyers, C. G., and Randolph, A. D., *Am. Inst. Chem. Eng. J.*, **19**, 1089 (1973).
13. Garrett, D. E., *Chem. Eng.*, **4**, 673 (1959).
14. Botsaris, G. D., Denk, D. G., Epsun, G. S., Kirwan, D. J., Ohara, M., Reid, R. C., and Tester, J., *Ind. Eng. Chem.*, **61**, 72 (1969).
15. Randolph, A. D., and Larson, M. A., *Am. Inst. Chem. Eng. J.*, **424** (1965).
16. Timm, D. C., and Larson, M. A., *Am. Inst. Chem. Eng. J.*, **14**, 452 (1968).
17. Randolph, A. D., *Am. Inst. Chem. Eng. J.*, **11**, 42 (1965).
18. Garrett, D. E., and Rosenbaum, *Chem. Eng.*, **65** (16), 127 (1958).

APPENDIX: Dimensionless Weight-Fraction Function, $w(x)$

$$w(x) = \frac{1}{6}\int_0^x e^{-p}p^x dp$$

$w(x)$	x	$w(x)$	x	$w(x)$	x	$w(x)$	x
0.000	0.01	0.002	0.51	0.020	1.01	0.067	1.51
0.000	0.02	0.002	0.52	0.020	1.02	0.068	1.52
0.000	0.03	0.002	0.53	0.021	1.03	0.070	1.53
0.000	0.04	0.002	0.54	0.022	1.04	0.071	1.54
0.000	0.05	0.003	0.55	0.022	1.05	0.072	1.55
0.000	0.06	0.003	0.56	0.023	1.06	0.073	1.56
0.000	0.07	0.003	0.57	0.024	1.07	0.075	1.57
0.000	0.08	0.003	0.58	0.024	1.08	0.076	1.58
0.000	0.09	0.003	0.59	0.025	1.09	0.077	1.59
0.000	0.10	0.003	0.60	0.026	1.10	0.079	1.60
0.000	0.11	0.004	0.61	0.027	1.11	0.080	1.61
0.000	0.12	0.004	0.62	0.027	1.12	0.082	1.62
0.000	0.13	0.004	0.63	0.028	1.13	0.083	1.63
0.000	0.14	0.004	0.64	0.029	1.14	0.084	1.64
0.000	0.15	0.005	0.65	0.030	1.15	0.086	1.65
0.000	0.16	0.005	0.66	0.030	1.16	0.087	1.66
0.000	0.17	0.005	0.67	0.031	1.17	0.089	1.67
0.000	0.18	0.005	0.68	0.032	1.18	0.090	1.68
0.000	0.19	0.006	0.69	0.033	1.19	0.092	1.69
0.000	0.20	0.006	0.70	0.034	1.20	0.093	1.70
0.000	0.21	0.006	0.71	0.035	1.21	0.095	1.71
0.000	0.22	0.006	0.72	0.036	1.22	0.096	1.72
0.000	0.23	0.007	0.73	0.036	1.23	0.098	1.73
0.000	0.24	0.007	0.74	0.037	1.24	0.099	1.74
0.000	0.25	0.007	0.75	0.038	1.25	0.101	1.75
0.000	0.26	0.008	0.76	0.039	1.26	0.102	1.76
0.000	0.27	0.008	0.77	0.040	1.27	0.104	1.77
0.000	0.28	0.008	0.78	0.041	1.28	0.106	1.78
0.000	0.29	0.009	0.79	0.042	1.29	0.107	1.79
0.000	0.30	0.009	0.80	0.043	1.30	0.109	1.80
0.000	0.31	0.010	0.81	0.044	1.31	0.110	1.81
0.000	0.32	0.010	0.82	0.045	1.32	0.112	1.82
0.000	0.33	0.010	0.83	0.046	1.33	0.114	1.83
0.000	0.34	0.011	0.84	0.047	1.34	0.115	1.84
0.000	0.35	0.011	0.85	0.048	1.35	0.117	1.85
0.001	0.36	0.012	0.86	0.049	1.36	0.119	1.86
0.001	0.37	0.012	0.87	0.050	1.37	0.120	1.87
0.001	0.38	0.013	0.88	0.052	1.38	0.122	1.88
0.001	0.39	0.013	0.89	0.053	1.39	0.124	1.89
0.001	0.40	0.014	0.90	0.054	1.40	0.125	1.90
0.001	0.41	0.014	0.91	0.055	1.41	0.127	1.91
0.001	0.42	0.015	0.92	0.056	1.42	0.129	1.92
0.001	0.43	0.015	0.93	0.057	1.43	0.131	1.93
0.001	0.44	0.016	0.94	0.058	1.44	0.132	1.94
0.001	0.45	0.016	0.95	0.060	1.45	0.134	1.95
0.001	0.46	0.017	0.96	0.061	1.46	0.136	1.96
0.001	0.47	0.017	0.97	0.062	1.47	0.138	1.97
0.002	0.48	0.018	0.98	0.063	1.48	0.139	1.98
0.002	0.49	0.018	0.99	0.064	1.49	0.141	1.99
0.002	0.50	0.019	1.00	0.066	1.50	0.143	2.00
0.145	2.01	0.245	2.51	0.355	3.01	0.466	3.51
0.147	2.02	0.247	2.52	0.357	3.02	0.468	3.52
0.148	2.03	0.249	2.53	0.360	3.03	0.470	3.53
0.150	2.04	0.251	2.54	0.362	3.04	0.472	3.54
0.152	2.05	0.253	2.55	0.364	3.05	0.474	3.55
0.154	2.06	0.255	2.56	0.366	3.06	0.476	3.56
0.156	2.07	0.258	2.57	0.369	3.07	0.478	3.57
0.158	2.08	0.260	2.58	0.371	3.08	0.481	3.58
0.160	2.09	0.262	2.59	0.373	3.09	0.483	3.59
0.161	2.10	0.264	2.60	0.375	3.10	0.485	3.60

$w(x)$	x	$w(x)$	x	$w(x)$	x	$w(x)$	x
0.163	2.11	0.266	2.61	0.377	3.11	0.487	3.61
0.165	2.12	0.268	2.62	0.380	3.12	0.489	3.62
0.167	2.13	0.271	2.63	0.382	3.13	0.491	3.63
0.169	2.14	0.273	2.64	0.384	3.14	0.493	3.64
0.171	2.15	0.275	2.65	0.386	3.15	0.495	3.65
0.173	2.16	0.277	2.66	0.389	3.16	0.498	3.66
0.175	2.17	0.279	2.67	0.391	3.17	0.500	3.67
0.177	2.18	0.282	2.68	0.393	3.18	0.502	3.68
0.179	2.19	0.284	2.69	0.395	3.19	0.504	3.69
0.181	2.20	0.286	2.70	0.398	3.20	0.506	3.70
0.183	2.21	0.288	2.71	0.400	3.21	0.508	3.71
0.185	2.22	0.290	2.72	0.402	3.22	0.510	3.72
0.187	2.23	0.293	2.73	0.404	3.23	0.512	3.73
0.189	2.24	0.295	2.74	0.406	3.24	0.514	3.74
0.191	2.25	0.297	2.75	0.409	3.25	0.516	3.75
0.193	2.26	0.299	2.76	0.411	3.26	0.518	3.76
0.195	2.27	0.301	2.77	0.413	3.27	0.520	3.77
0.197	2.28	0.304	2.78	0.415	3.28	0.522	3.78
0.199	2.29	0.306	2.79	0.418	3.29	0.525	3.79
0.201	2.30	0.308	2.80	0.420	3.30	0.527	3.80
0.203	2.31	0.310	2.81	0.422	3.31	0.529	3.81
0.205	2.32	0.313	2.82	0.424	3.32	0.531	3.82
0.207	2.33	0.315	2.83	0.426	3.33	0.533	3.83
0.209	2.34	0.317	2.84	0.429	3.34	0.535	3.84
0.211	2.35	0.319	2.85	0.431	3.35	0.537	3.85
0.213	2.36	0.321	2.86	0.433	3.36	0.539	3.86
0.215	2.37	0.324	2.87	0.435	3.37	0.541	3.87
0.217	2.38	0.326	2.88	0.437	3.38	0.543	3.88
0.219	2.39	0.328	2.89	0.440	3.39	0.545	3.89
0.221	2.40	0.330	2.90	0.442	3.40	0.547	3.90
0.223	2.41	0.333	2.91	0.444	3.41	0.549	3.91
0.226	2.42	0.335	2.92	0.446	3.42	0.551	3.92
0.228	2.43	0.337	2.93	0.448	3.43	0.553	3.93
0.230	2.44	0.339	2.94	0.450	3.44	0.555	3.94
0.232	2.45	0.342	2.95	0.453	3.45	0.557	3.95
0.234	2.46	0.344	2.96	0.455	3.46	0.559	3.96
0.236	2.47	0.346	2.97	0.457	3.47	0.561	3.97
0.238	2.48	0.348	2.98	0.459	3.48	0.563	3.98
0.240	2.49	0.351	2.99	0.461	3.49	0.565	3.99
0.242	2.50	0.353	3.00	0.463	3.50	0.567	4.00
0.569	4.01	0.659	4.51	0.736	5.01	0.799	5.51
0.570	4.02	0.661	4.52	0.738	5.02	0.801	5.52
0.572	4.03	0.663	4.53	0.739	5.03	0.802	5.53
0.574	4.04	0.664	4.54	0.741	5.04	0.803	5.54
0.576	4.05	0.666	4.55	0.742	5.05	0.804	5.55
0.578	4.06	0.668	4.56	0.743	5.06	0.805	5.56
0.580	4.07	0.669	4.57	0.745	5.07	0.806	5.57
0.582	4.08	0.671	4.58	0.746	5.08	0.807	5.58
0.584	4.09	0.673	4.59	0.747	5.09	0.808	5.59
0.586	4.10	0.674	4.60	0.749	5.10	0.809	5.60
0.588	4.11	0.676	4.61	0.750	5.11	0.811	5.61
0.590	4.12	0.678	4.62	0.751	5.12	0.812	5.62
0.592	4.13	0.679	4.63	0.753	5.13	0.813	5.63
0.593	4.14	0.681	4.64	0.754	5.14	0.814	5.64
0.595	4.15	0.682	4.65	0.755	5.15	0.815	5.65
0.597	4.16	0.684	4.66	0.758	5.16	0.816	5.66
0.599	4.17	0.686	4.67	0.758	5.17	0.817	5.67
0.601	4.18	0.687	4.68	0.759	5.18	0.818	5.68
0.603	4.19	0.689	4.69	0.761	5.19	0.819	5.69
0.605	4.20	0.690	4.70	0.762	5.20	0.820	5.70

APPENDIX: Dimensionless Weight-Fraction Function, w(x) (Continued)

$$w(x) = \frac{1}{6} \int_0^x e^{-p} p^x \, dp$$

w(x)	x	w(x)	x	w(x)	x	w(x)	x
0.607	4.21	0.692	4.71	0.763	5.21	0.821	5.71
0.608	4.22	0.694	4.72	0.765	5.22	0.822	5.72
0.610	4.23	0.695	4.73	0.766	5.23	0.823	5.73
0.612	4.24	0.697	4.74	0.767	5.24	0.824	5.74
0.614	4.25	0.698	4.75	0.768	5.25	0.825	5.75
0.616	4.26	0.700	4.76	0.770	5.26	0.826	5.76
0.617	4.27	0.701	4.77	0.771	5.27	0.827	5.77
0.619	4.28	0.703	4.78	0.772	5.28	0.828	5.78
0.621	4.29	0.704	4.79	0.773	5.29	0.829	5.79
0.623	4.30	0.706	4.80	0.775	5.30	0.830	5.80
0.625	4.31	0.707	4.81	0.776	5.31	0.831	5.81
0.626	4.32	0.709	4.82	0.777	5.32	0.832	5.82
0.628	4.33	0.710	4.83	0.778	5.33	0.833	5.83
0.630	4.34	0.712	4.84	0.780	5.34	0.834	5.84
0.632	4.35	0.713	4.85	0.781	5.35	0.835	5.85
0.634	4.36	0.715	4.86	0.782	5.36	0.836	5.86
0.635	4.37	0.716	4.87	0.783	5.37	0.837	5.87
0.637	4.38	0.718	4.88	0.784	5.38	0.838	5.88
0.639	4.39	0.719	4.89	0.786	5.39	0.839	5.89
0.641	4.40	0.721	4.90	0.787	5.40	0.840	5.90
0.642	4.41	0.722	4.91	0.788	5.41	0.841	5.91
0.644	4.42	0.724	4.92	0.789	5.42	0.842	5.92
0.646	4.43	0.725	4.93	0.790	5.43	0.842	5.93
0.648	4.44	0.727	4.94	0.791	5.44	0.843	5.94
0.649	4.45	0.728	4.95	0.793	5.45	0.844	5.95
0.651	4.46	0.729	4.96	0.794	5.46	0.845	5.96
0.653	4.47	0.731	4.97	0.795	5.47	0.846	5.97
0.654	4.48	0.732	4.98	0.796	5.48	0.847	5.98
0.656	4.49	0.734	4.99	0.797	5.49	0.848	5.99
0.658	4.50	0.735	5.00	0.798	5.50	0.849	6.00
0.850	6.01	0.889	6.51	0.919	7.01	0.941	7.51
0.851	6.02	0.890	6.52	0.919	7.02	0.942	7.52
0.852	6.03	0.890	6.53	0.920	7.03	0.942	7.53
0.852	6.04	0.891	6.54	0.920	7.04	0.942	7.54
0.853	6.05	0.892	6.55	0.921	7.05	0.943	7.55
0.854	6.06	0.892	6.56	0.921	7.06	0.943	7.56
0.855	6.07	0.893	6.57	0.922	7.07	0.944	7.57
0.856	6.08	0.894	6.58	0.922	7.08	0.944	7.58
0.857	6.09	0.894	6.59	0.923	7.09	0.944	7.59
0.858	6.10	0.895	6.60	0.923	7.10	0.945	7.60
0.858	6.11	0.896	6.61	0.924	7.11	0.945	7.61
0.859	6.12	0.896	6.62	0.924	7.12	0.945	7.62
0.860	6.13	0.897	6.63	0.925	7.13	0.946	7.63
0.861	6.14	0.897	6.64	0.925	7.14	0.946	7.64
0.862	6.15	0.898	6.65	0.926	7.15	0.946	7.65
0.863	6.16	0.899	6.66	0.926	7.16	0.947	7.66
0.863	6.17	0.899	6.67	0.927	7.17	0.947	7.67
0.864	6.18	0.900	6.68	0.027	7.18	0.948	7.68
0.865	6.19	0.901	6.69	0.928	7.19	0.948	7.69
0.866	6.20	0.901	6.70	0.928	7.20	0.948	7.70
0.867	6.21	0.902	6.71	0.929	7.21	0.949	7.71
0.867	6.22	0.902	6.72	0.929	7.22	0.949	7.72
0.868	6.23	0.903	6.73	0.930	7.23	0.949	7.73
0.869	6.24	0.904	6.74	0.930	7.24	0.950	7.74
0.870	6.25	0.904	6.75	0.930	7.25	0.950	7.75
0.871	6.26	0.905	6.76	0.931	7.26	0.950	7.76
0.871	6.27	0.905	6.77	0.931	7.27	0.951	7.77
0.872	6.28	0.906	6.78	0.932	7.28	0.951	7.78
0.873	6.29	0.907	6.79	0.932	7.29	0.951	7.79
0.874	6.30	0.907	6.80	0.933	7.30	0.952	7.80

$w(x)$	x	$w(x)$	x	$w(x)$	x	$w(x)$	x
0.874	6.31	0.908	6.81	0.933	7.31	0.952	7.81
0.875	6.32	0.908	6.82	0.934	7.32	0.952	7.82
0.876	6.33	0.909	6.83	0.934	7.33	0.953	7.83
0.877	6.34	0.910	6.84	0.934	7.34	0.953	7.84
0.877	6.35	0.910	6.85	0.935	7.35	0.953	7.85
0.878	6.36	0.911	6.86	0.935	7.36	0.953	7.86
0.879	6.37	0.911	6.87	0.936	7.37	0.954	7.87
0.880	6.38	0.912	6.88	0.936	7.38	0.954	7.88
0.880	6.39	0.912	6.89	0.936	7.39	0.954	7.89
0.881	6.40	0.913	6.90	0.937	7.40	0.955	7.90
0.882	6.41	0.913	6.91	0.937	7.41	0.955	7.91
0.883	6.42	0.914	6.92	0.938	7.42	0.955	7.92
0.883	6.43	0.915	6.93	0.938	7.43	0.956	7.93
0.884	6.44	0.915	6.94	0.939	7.44	0.956	7.94
0.885	6.45	0.916	6.95	0.939	7.45	0.956	7.95
0.885	6.46	0.916	6.96	0.939	7.46	0.957	7.96
0.886	6.47	0.917	6.97	0.940	7.47	0.957	7.97
0.887	6.48	0.917	6.98	0.940	7.48	0.957	7.98
0.888	6.49	0.918	6.99	0.941	7.49	0.957	7.99
0.888	6.50	0.918	7.00	0.941	7.50	0.958	8.00
0.958	8.01	0.970	8.51	0.979	9.01	0.985	9.51
0.958	8.02	0.970	8.52	0.979	9.02	0.985	9.52
0.959	8.03	0.971	8.53	0.979	9.03	0.986	9.53
0.959	8.04	0.971	8.54	0.979	9.04	0.986	9.54
0.959	8.05	0.971	8.55	0.980	9.05	0.986	9.55
0.959	8.06	0.971	8.56	0.980	9.06	0.986	9.56
0.960	8.07	0.971	8.57	0.980	9.07	0.986	9.57
0.960	8.08	0.972	8.58	0.980	9.08	0.986	9.58
0.960	8.09	0.972	8.59	0.980	9.09	0.986	9.59
0.960	8.10	0.972	8.60	0.980	9.10	0.986	9.60
0.961	8.11	0.972	8.61	0.980	9.11	0.986	9.61
0.961	8.12	0.972	8.62	0.981	9.12	0.986	9.62
0.961	8.13	0.973	8.63	0.981	9.13	0.987	9.63
0.962	8.14	0.973	8.64	0.981	9.14	0.987	9.64
0.962	8.15	0.973	8.65	0.981	9.15	0.987	9.65
0.962	8.16	0.973	8.66	0.981	9.16	0.987	9.66
0.962	8.17	0.973	8.67	0.981	9.17	0.987	9.67
0.963	8.18	0.973	8.68	0.981	9.18	0.987	9.68
0.963	8.19	0.974	8.69	0.982	9.19	0.987	9.69
0.963	8.20	0.974	8.70	0.982	9.20	0.987	9.70
0.963	8.21	0.974	8.71	0.982	9.21	0.987	9.71
0.964	8.22	0.974	8.72	0.982	9.22	0.987	9.72
0.964	8.23	0.974	8.73	0.982	9.23	0.987	9.73
0.964	8.24	0.975	8.74	0.982	9.24	0.988	9.74
0.964	8.25	0.975	8.75	0.982	9.25	0.988	9.75
0.965	8.26	0.975	8.76	0.982	9.26	0.988	9.76
0.965	8.27	0.975	8.77	0.983	9.27	0.988	9.77
0.965	8.28	0.975	8.78	0.983	9.28	0.988	9.78
0.965	8.29	0.975	8.79	0.983	9.29	0.988	9.79
0.966	8.30	0.976	8.80	0.983	9.30	0.988	9.80
0.966	8.31	0.976	8.81	0.983	9.31	0.988	9.81
0.966	8.32	0.976	8.82	0.983	9.32	0.988	9.82
0.966	8.33	0.976	8.83	0.983	9.33	0.988	9.83
0.966	8.34	0.976	8.84	0.983	9.34	0.988	9.84
0.967	8.35	0.976	8.85	0.984	9.35	0.989	9.85
0.967	8.36	0.977	8.86	0.984	9.36	0.989	9.86
0.976	8.37	0.977	8.87	0.984	9.37	0.989	9.87
0.967	8.38	0.977	8.88	0.984	9.38	0.989	9.88
0.968	8.39	0.977	8.89	0.984	9.39	0.989	9.89
0.968	8.40	0.977	8.90	0.984	9.40	0.989	9.90

APPENDIX: Dimensionless Weight-Fraction Function, *w(x)* (*Continued*)

$$w(x) = \frac{1}{6} \int_0^x e^{-p} p^x \, dp$$

$w(x)$	x	$w(x)$	x	$w(x)$	x	$w(x)$	x
0.968	8.41	0.977	8.91	0.984	9.41	0.989	9.91
0.968	8.42	0.978	8.92	0.984	9.42	0.989	9.92
0.968	8.43	0.978	8.93	0.984	9.43	0.989	9.93
0.969	8.44	0.978	8.94	0.985	9.44	0.989	9.94
0.969	8.45	0.978	8.95	0.985	9.45	0.989	9.95
0.969	8.46	0.978	8.96	0.985	9.46	0.989	9.96
0.969	8.47	0.978	8.97	0.985	9.47	0.989	9.97
0.970	8.48	0.979	8.98	0.985	9.48	0.990	9.98
0.970	8.49	0.979	8.99	0.985	9.49	0.990	9.99
0.970	8.50	0.979	9.00	0.985	9.50		

Foam Separation Processes

YOSHIYUKI OKAMOTO, Ph.D. *Professor of Chemistry, Polytechnic Institute of New York, Brooklyn, New York.*

ENG J. CHOU, Ph.D., P.E. *Sr. Development Engineer, Research and Development, Research Cottrel Inc., Somerville, N.J.*

INTRODUCTION

The foam separation process is an adsorptive bubble separation method and a chemical engineering process that selectively separates surface-active compounds in a solution, collecting them at the interface between the liquid and gas, and thereby concentrating and separating them. This process is especially effective for separation of materials at low concentrations. Surface-inactive compounds (colligens) can be removed from solution if an appropriate surface-active material (collector) is added to unite with the compound so that it can be adsorbed at the bubble surface. Foam separation may be conveniently categorized into foam fractionation and froth flotation (Fig. 1). Foam fractionation separates dissolved material, while froth flotation separates insoluble material.[1] Froth flotation may be divided into several branches depending on the substance, which may be molecular, colloidal, or macroparticulate.

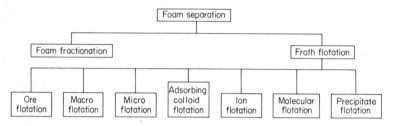

Ore flotation: A special case of froth flotation for the separation of minerals from their ores.
Macroflotation: A separation of macroscopic particles by foaming.
Microflotation: The separation of microscopic particles by foaming, especially colloids or microorganisms.
Absorbing colloid flotation: The separation of dissolved material by adsorption on colloidal particles which are surface-active.
Ion flotation: The separation of surface-inactive ions by foaming with a surfactant (collector) which yields an insoluble product, especially if removed as a scum.
Molecular flotation: The separation of surface-inactive molecules by foaming through the use of a surfactant which yields an insoluble product.
Precipitate flotation: The process in which a precipitate is removed by a surfactant which is not the precipitating agent.

Fig. 1 Classification of the foam separation process.

PRINCIPLE

Of all the foam separation methods, foam fractionation is the one for which separation theory is most advanced. Some of this theory may also be applied to other foam separation processes. Under equilibrium conditions of dilute solutions (assuming activity to be unity), adsorption of surface-active species from bulk solution at a gas-liquid interface can be quantitatively described by the Gibbs equation:[2]

$$\frac{\Gamma}{C} = -\frac{1}{RT}\frac{d\gamma}{dC} \qquad (1)$$

where Γ is the surface excess of the adsorbed solute (i.e., concentration at the surface, $g \cdot mol/cm^2$), C is the bulk equilibrium concentration, Γ/C can be considered a distribution factor, γ is the surface tension, R is the gas constant, and T is the absolute temperature. The value of $d\gamma/dC$ may be readily determined from the slope of γ-C plots. A hypothetical surface tension vs. surface-active species concentration curve is shown in Fig. 2. In the very low concentration region, i.e., below a in Fig. 2, little adsorption can occur, since few surface-active species are present and the surface tension is close to that of the solvent. The surface excess is then close to zero and separa-

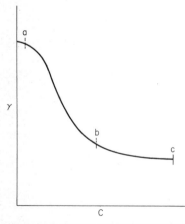

Fig. 2 Hypothetical surface tension vs. concentration for a surface-active agent in a solution.

tion occurs only to a small extent. At intermediate concentrations (between a and b), γ decreases with increasing concentration of surface-active material; i.e., a negative slope occurs. Γ/C becomes constant and separation of surface-active material from the bulk solution can be achieved. The average concentration of surface-active material in this region is 10^{-7} to 10^{-3} M depending on various conditions. In the region above concentration b, the slope becomes close to zero. This is the region in which micelles form, and the point at which the curve levels off is called the critical micelle concentration (CMC). According to Eq. (1), the distribution coefficient should become close to zero and no removal should occur by foam separation processes. Practically, foam is nevertheless formed in the micelle region and separations can be successfully carried out; however, better separation would occur below the CMC.[3,4]

Surface-inactive species can be removed from solution if an appropriate surface-active material is added to unite with this material so that it can be adsorbed at the bubble surface. This can occur through the formation of a chelate, electrostatic attraction, or some other mechanism. For example, if it is desired to remove a surface-inactive univalent anion, a cationic surfactant (S^+X^-) is added to the solution. Thus the anion exchanges with the anion from the surfactant. The surfactant acts as a mobile ion exchanger, and the reaction can be written as

$$(A^-)_b + (X^-)_s \rightleftharpoons (A^-)_s + (X^-)_b \tag{2}$$

where A^- and X^- denote the anion being removed and the anion of the surfactant added, respectively, and subscripts b and s denote bulk and surfactant phases, respectively. The exchange constant is then

$$K_{ex} = \frac{[A^-]_s[X^-]_b}{[A^-]_b[X^-]_s} \tag{3}$$

The K_{ex} values will depend on the relative affinities of A^- and X^- for the cationic surfactant and their relative solubilities in the solution. When the surface tensions of S^+A^- and S^+X^- are similar, the extent of removal will depend on the exchange constant K_{ex}. The more favorable the formation of the ion pair, the better will be the removal. A similar principle can be applied using a chelating surfactant when a metallic ion can coordinate with the surfactant to form a complex ion which is also surface-active. Consider a solution of two different metallic ions A and B and a chelating surfactant S. When they form complexes, the equilibria are

$$S + A \rightleftharpoons SA \qquad K_A = \frac{C_{AS}}{C_s C_A} \tag{4}$$

$$S + B \rightleftharpoons SB \qquad K_B = \frac{C_{BS}}{C_s C_A} \tag{5}$$

where K_A, C_S, C_A, and C_{SA} are the chelate-formation constant, the concentrations of free surfactant, free A, and complex A and S, respectively. A relative distribution coefficient (selectivity) α for A and B removal may be defined as the ratio of their individual distribution factors:

$$\alpha_{AB} = \frac{\Gamma_{AS}/C_{AS}}{\Gamma_{BS}/C_{BS}} \tag{6}$$

For a Gibbs adsorption isotherm, Eq. (6) may be rewritten as

$$\alpha_{AB} = \frac{\partial \gamma_{AS}/\partial C_{AS}}{\partial \gamma_{BS}/\partial C_{BS}} \tag{7}$$

Thus, in consideration of the chelation equilibrium in foam separation, Eq. (7) can be modified to give Eq. (8):

$$\alpha_{AB} = \frac{\partial \gamma_{AS}/\partial C_{AS}}{\partial \gamma_{BS}/\partial B_S} \times \frac{1 - C_A/C_{AO}}{1 - C_B/C_{BO}} \tag{8}$$

where C_{AO} and C_{BO} are the total concentration of A and B, respectively. When K_A and K_B are relatively large and the concentration of surfactant added is much greater than the sum of their concentrations, the second term in Eq. (8) becomes unity and the equation is reduced to the form of Eq. (7). For example, Cd and Cu ions can be removed almost quantitatively from aqueous solution using chelating surfactants such as 4-dodecyldiethylenetriamine. The chelate-formation constants log $K_1 K_2$ for Cd and Cu are 15.1 and 20.3, respectively.[5] Thus, when the total concentration of Cd and Cu ions is smaller than that of the surfactant, practically all those ions are complexed with the surfactant and there is no competition for complex formation. Selectivity in this instance is controlled by the respective surface tensions of the complexes. Since the Cd-surfactant complexes have consistently lower surface tensions than the corresponding Cu complexes, Cd can be removed preferentially. Conversely, when the total concentration of Cd and Cu ions is larger than the surfactant concentration, there is competition for complex formation. The chelate-formation constant for Cu is much larger than that for Cd; therefore, the Cu competes more favorably for complex formation and thus Cu ion can be removed selectively.

FOAM SEPARATION OPERATION

In foam separation processes, a variety of techniques can be used for operation of the foam column, depending on the specific requirements. However, most of the arrangements reported fall into one of the following categories.

Batch-Type Foam Separation System

In a batch foam separation system, foamate is continuously removed and the original solution is continuously depleted of the foaming agent. Solutions giving both stable and unstable foam may be treated in a batch-type apparatus.[5-7]

If the volume of collapsed foam (foamate) collected is small compared with the liquid-pool volume, a mass balance carried out on the collapsed foam produces the following relationship:[8]

$$\frac{\Gamma}{C} = \left(\frac{C_s}{C_b} - 1\right)\frac{ld}{6} = (E - 1)\frac{ld}{6} \tag{9}$$

where l, d, and E are the foam ratio in mL liquid/mL foam, the average bubble diameter, and the enrichment ratio C_s/C_b, respectively.

Continuous Foam Separation System

In a continuous apparatus, feed solution is continuously added to the column while foamate and residue are withdrawn continuously. All continuous foam separation systems have the same basic configuration as is shown in Fig. 3. Recovery (percent removal) and the grade of the product (percent product in foamate) can be increased by using a stripping mode and an enriching mode, respectively. In the stripping mode the feed is introduced into the foam so that some amount of separation takes place while it is descending through the foam itself. In the enriching mode a certain amount of reflux is achieved by feeding part of the foamate back to the top of the column for increasing the separation. If a foam column is sufficiently tall, the performance can be expressed by Eqs. (10) and (11), if the column contains an enricher mode:[9]

$$C_D = C_W + \frac{GZ\Gamma_W}{D} \tag{10}$$

$$C_W = C_F - \frac{GZ\Gamma_W}{F} \tag{11}$$

and Eqs. (12) and (13), if it contains both an enriching mode and a stripping mode:

$$C_D = C_F + \frac{GZ\Gamma_F}{D} \tag{12}$$

$$C_W = C_F - \frac{GZ\Gamma_F}{W} \tag{13}$$

where Z, G, F, W, and D are the ratio of bubble surface to bubble volume, the flow rates of gas, feed, residue, and net overflow, respectively. C_F, C_W, and C_D are the corresponding concentrations of feed, residue, and net overflow, and Γ_F and Γ_W are the corresponding surface excesses of C_F and C_W at equilibrium.

The number of theoretical stages in the foam can be found by a graphical method as in a distillation process. The number of transfer units in the foam based on the rising stream for an enriching column which is concentrating a surfactant of constant Γ while operating at a reflux ratio of R is expressed by Eq. (14):[9]

$$\text{NTU} = R \ln \frac{RGZ\Gamma(F - D)}{(R + 1)GZ\Gamma(F - D) - (R + 1)FD(C_D - C_F)} \tag{14}$$

Foam separation, like other separation schemes, can be conducted in a series of single stages or in one multistage countercurrent column. One possible arrangement of a series of single-stage, continuous columns is shown in Fig. 4. In this example, the purpose is to remove as much as possible of the solutes in order to recover a pure residue. If the

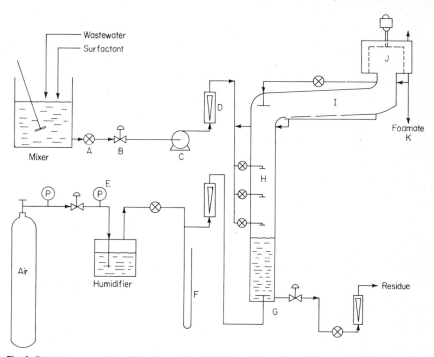

Fig. 3 Continuous operation of foam separation. (A) Valve; (B) regulator; (C) pump; (D) flowmeter; (E) pressure gage; (F) Hg monometer; (G) bubbler; (H) foaming tower; (I) drainage section; (J) foam breaker; (K) to surfactant recovery unit.

operation is to remove a surface-inactive material by complex formation with a suitable surfactant, the formate (complex) may be broken by a chemical reaction to form an insoluble compound of the removed surface-inactive material, thus regenerating the surfactant. The insoluble compound is filtered from solution and the regenerated surfactant is recycled. It has been demonstrated experimentally that surfactant-Hg or -Cd complexes removed as foam could be broken with H_2S or Na_2S and the HgS or CdS precipitated and filtered. The regenerated surfactant could then be recycled.[10]

FACTORS IN FOAM SEPARATION

Many factors affect the performance and efficiency of a foam separation system, the relative importance of each depending on the specific conditions. These include basic variables such as concentration of the surfactant and auxiliary reagents, pH, ionic strength, temperature, viscosity, and other operating variables like gas flow rate, feed rate, reflux ratio, foam height, foam density, bubble size, and equipment design.

Effect of Surfactant Concentration

A primary advantage of foam separation is the small volume of foamate necessary to perform separations which would require much larger volumes by other techniques. The production of a large amount of foam is both unnecessary and undesirable. An excess of collector can reduce the separation by competing against the collector-colligen complex for the available surface. It can also reduce the separation by foaming micelles in the bulk solution which adsorb some of the colligen, thus keeping it from the surface. For example, in the removal of mercury ion using a chelating surfactant, the removal rate increased with an increase in the mercury ion–surfactant ratio until a particular ratio was reached and above which the efficiency of removal decreased, as is shown in Fig. 5.[10] For ion flotation, it was reported[11] that the most effective removal of Y^{3+} ion using 2-sulfohexadecanonic acid as the collector is achieved when the ratio of collector to colligen is 3:1.

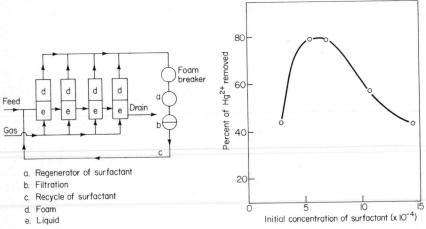

a. Regenerator of surfactant
b. Filtration
c. Recycle of surfactant
d. Foam
e. Liquid

Fig. 4 A brief sketch for a series of continuous single-stage columns.

Fig. 5 The effect of surfactant concentration (4-dodecyldiethylenetriamine); initial concentration of Hg^{2+} = 10 ppm, pH = 9.2, gas flow rate = 200 cm³/min, foaming time = 3 h.

Effect of Auxiliary Reagents

Various auxiliary reagents are being used successfully in foam separation techniques for improved separation. The effects are, in some cases, due to the flocculation of the particulates or the activation of collector for enhanced adsorption. Most commonly used flocculation agents in foam separation are alum, ferric salts, and organic polyelectrolytes. For example, the removal of phosphate and suspended solids from wastewater was enhanced by adding these flocculation agents.[12] Frequently, in a foam separation process, an activator which promotes selective adsorption of the collector on a particular mineral is used.

Sulfate and phosphate ions can activate ilmenite $(FeTiO_3)$ and it can be floated with dodecylammonium acetate.[13] Fe^{3+} can be floated by anionic surfactants such as sodium oleate. In this case, the addition of Ca^{2+} enhanced the flotation.[14] A selective complex agent can also be used in the foam separation process. For example, in $1N$ HCl, Hg^{2+} forms a strong anionic chloride complex, whereas many other metal ions such as Fe^{3+} and Co^{3+} do not. Thus mercury can be separated from these other metals in $1N$ HCl using a cationic surfactant.[15] Clearly this approach can be applied to many other types of complexes. It is worth pointing out that the information developed for separation in ion exchange using this approach can be directly applied for prediction of separation in a foam separation process.

Effect of Solution pH

The pH of the solution will determine the sign and the magnitude of the charge on a variety of inorganic particulates. Therefore, the extent of removal of the particulates by foam separation techniques will be controlled by the solution pH. Excellent separations of minerals from one another is achieved in practice by choosing appropriate pH conditions.[16]

In the foam separation of metal ions, metal ions and collector are attracted by single ion-pair formation or by weak coordination. With hydrolyzable metals such as aluminum(III), lead(II), and zinc(II) and the transition metals, the mechanism of removal depends to a large extent upon the pH of the medium. When these metal salts are added to water, they will dissociate and the ions will become hydrated and also react with water in some cases. The exact distribution of the various species formed depends primarily on the solution pH. For example, when a lead salt is dissolved in water, the following hydrolytic reactions take place:[17]

$$Pb^{2+} + H_2O \rightleftharpoons PbOH^+ + H^+$$
$$PbOH^+ + H_2O \rightleftharpoons Pb(OH)_2(aq) + H^+$$
$$Pb(OH)_2(aq) \rightleftharpoons HPbO_2^- + H^+$$

Therefore, when lead is separated using an anionic surfactant such as sodium lauryl sulfate at low pH, maximum removal can be expected because of the high concentration of Pb^{2+} and $PbOH^+$. However, actual experimental results showed that the removal rate of lead at pH 1.5 was very slow and the results were accounted for by H^+ competition and instability of the lead–lauryl sulfate complex in this pH region. Thus the removal reached a maximum at pH 8.2 because the stoichiometry of collector to metal is most favorable in the presence of the $PbOH^+$ ion. Removals decrease above pH 8.2 because of formation of the soluble lead hydroxide [$Pb(OH)_2$ and $HPbO_2^-$].[18]

In certain metal ions, the initial pH of the solution will determine whether the process to be used should be foam fractionation or ion flotation. For example, the removal of zinc(II) ion with the anionic surfactant sodium lauryl sulfate can best be accomplished below pH 8, where the zinc exists primarily as the Zn^{2+} and $Zn(OH)^+$ ions and can be removed by foam fractionation. However, above pH 8, the insoluble hydroxide is produced, which can be removed by ion flotation.[19]

Effect of Ionic Strength

The effects of increasing the ionic strength in different foam separation techniques bear little relationship to each other. It is generally agreed that the presence of neutral salts decreases the efficiency of ion flotation and that this arises because of competition for collector between ions originally present and the added ions. The efficiency also depends on the nature of the ions; for example, in ion flotation of dichromate ($Cr_2O_7^{2-}$) using a cationic collector the interference with flotation was found to follow the order $PO_4^{3-} > SO_4^{2-} > Cl^-$.[20] Orthophosphate (that is, HPO_4^{2-}) and phenol ions have been foam fractionated using a cationic surfactant: ethylhexadecyldimethylammonium bromide. Reduced fractionation of both orthophosphate and phenol was observed in the presence of sulfate and chloride ions. Sulfate was also more detrimental than chloride, owing primarily to competition with HPO_4^{2-} and $C_6H_5O^-$ for the monovalent cationic surfactant.[21] However, the flotation of copper hydroxide with sodium dodecylsulfate was found to be unaffected by an increase in the ionic strength of the solution.[22]

Effect of Temperature

Temperature has been suggested as an operating variable for cases where the foam stability of surface-active components is different at different temperatures. In the case of froth flotation of minerals, surfactant adsorption and hence flotation could be expected to decrease with an increase in temperature if the binding of the collector to the mineral surface is due to physical adsorption. If the adsorption is due to chemical forces between the surfactant and the mineral particles, opposite effects could be expected.[23] When floating hexacyanoferrate(II), [$Fe(CN)_6$]$^{4-}$, with dodecylpyridinium chloride, the efficiency of the floating is reduced by a factor of 2, with an increase in the temperature from 5 to 30°C. It is attributed to the fact that as adsorption is an exothermic process, an increase in temperature leads to a decrease in the amount of surfactant on the bubbles, and thus the efficiency is reduced.[24] However, in many cases, temperature was found to have little affect in ion flotation[25] and foam fractionation processes.[26]

Effect of Gas Flow Rate

Gas flow rate strongly affects the rate of removal of dissolved substances without significantly affecting steady-state removals. The removal of dissolved substance involves their distribution or partition between gaseous and aqueous phases. An increase in interfacial area, as occurs with increasing gas flow rate, causes an increase in removal at any given time. However, low gas flow rate is in general beneficial for separation and for high enrichment.[27] There must, of course, be sufficient gas flow to maintain the foam height that is essential for good separation, the optimum flow rate being determined by the concentration of the surfactant and the transiency of the foam.

Effect of Foam Drainage

When a foam separation has been carried out, the extract must be concentrated to as small a volume as possible. Foam drainage is commonly carried out by passing the foam upward through a length of column of expanded diameter.[28] It can also be arranged as shown in Fig. 3 (see I), in which the foam from the separation process travels nearly horizontally. This type of drainage section has two very considerable advantages over the vertical system. In the first place, not only is the vertical component of gas velocity reduced to zero, but the distance through which the liquid must drain can be made uniform and set at any desired value. Second, the drainage that takes place may be predicted from laboratory experiments with static foam.

Effect of Other Physical Variables

Among the other physical variables, bubble-size distribution, agitation, column length, etc., do not produce any primary effects on the ultimate separation of materials.[29] However, it was reported[30] that the foam height had a significant effect on the separation of albumin, and the effect was very pronounced near the foam-liquid interface. A change in foam height from 3 to 17 cm produced a drastic change in the foam stream. At a foam height of 3 cm, the volume of the solution carried away in the form of foam was 24 mL/min, while at 17 cm of foam height a volume of 10 mL/min was obtained. Further, it was observed that an increase in foam height produced a small decrease in the efficiency of the process. There is little variation in values of the height of a transfer unit (HTU) with column length at good operation conditions for countercurrent lengths of 10 to 28 cm. If the liquid distribution is not uniform at the feed point, inefficient countercurrent contact for a distance below the feed point may result, with efficient contact below this region.[28]

Effect of Overflow Rate

For a fairly dry foam, the results of overflow rate for a vertical column are given by Eq. (15):[110]

$$D = \frac{G^2\mu}{Ag\rho d^2} \phi \frac{\mu^3 G}{\mu_s^2 g\rho A} \tag{15}$$

where A is the cross-sectional area of the empty column, g is the acceleration due to gravity, ρ is the liquid density, μ is the viscosity of liquid, and μ_s is an effective surface viscosity. The importance of μ_s as a factor for foam drainage is well established and the value of μ_s is on the order of 10^{-4} dyn · s/cm.[111,112] The function ϕ depends on the detailed relationship between interstitial downflow and bulk foam upflow. Figure 6 shows a theoretical relationship for overflow.[110] For foam containing more than a few percent of a liquid on a volumetric basis, D from Eq. (15) should be multiplied by the quantity $(1 + 3D/G)$.

EXAMPLE OF THE DESIGN OF A FOAM SEPARATION PROCESS

An aqueous waste stream flowing at the rate of 5000 L/h contains a trace of a metallic ion which needs to be removed. When using a simple-mode continuous foam fractionation column with average bubble diameters of 0.1 cm in the liquid pool, if it is desired to reduce the concentration of this ion to 5% of its

initial concentration while producing no more than 300 L/h of collapsed foam, an important consideration is the air flow rate and column diameter required. Assume that Γ/C is constant and is equal to 5×10^{-3} cm.

Solution   Substituting the given information, $d = 0.1$ cm, $\Gamma/C = 5 \times 10^{-3}$ cm, $C_W = 0.05\ C_F$, and $Z = 6/d$ into Eq. (11) yields

$$G = \frac{(C_F - C_W)F}{(6/d)\Gamma_W} = \frac{(C_F/C_W - 1)Fd}{6\Gamma_W/C_W}$$

$$= \frac{(1/0.05 - 1)(5000)(0.1)}{6 \times (5 \times 10^{-3})} = 5280\ \text{L/min}$$

Figure 6 is now used to find the column diameter. In the absence of physical-property data, the viscosity μ is estimated to be equal to that of water, which is 10^{-2} dyn·s/cm², that is, 10^{-5} g/(s)(cm).

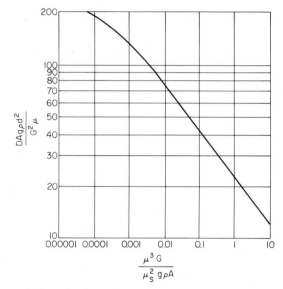

Fig. 6   Theoretical relationship for predicting the rate of foam overflow from a column. *From Fanlo and Lemlich, Am. Inst. Chem. Eng. Symp. Ser., no. 9, 1965.)*

Similarly, $\rho = 1$ g/cm². The surface viscosity μ_s is taken as the typical value 10^{-4} dyn·s/cm, i.e., 10^{-7} g/s. From Eq. (15),

$$\frac{DAg\rho d^2}{G^2\mu} = \frac{(300 \times 1000/3600) \times A \times 980 \times 1 \times (0.1)^2}{(5280 \times 1000/60)^2 \times 10^{-5}} = 0.01055A$$

and

$$\frac{\mu^3 G}{\mu_s^2 g\rho A} = \frac{(10^{-5})^3 \times (5280 \times 1000/60)}{(10^{-7})^2 \times 980 \times 1 \times A} = 8.98/A$$

where A is in cm².

Relating the ordinate to the abscissa by means of the curve, and solving for A by trial and error or by auxiliary plot yields $A = 14{,}000$ cm². This corresponds to a column diameter of $(4A/\pi)^{1/2}$, or about 135 cm.

EXAMPLES OF FOAM SEPARATION

Foam separation techniques have been used for removal or recovery of minerals, surfactants, enzymes, proteins, microorganisms, organic compounds, and various metallic ions.
Table 1 summarizes some of the work published. There are many other examples of foam
separation processes in the literature.

TABLE 1 Examples of Foam Separation

Substance	System and remarks	Reference
	Separation of Metallic Ions	
Al	Ion flotation using potassium stearate	31
Ag	Ion flotation of complex anions (silver thiosulfate) with ethylhexadecyldimethylammonium bromide at pH 4.5	32
Ag	Using dimethylbenzyllaurylammonium bromide	33
As	Ion flotation using sodium oleate	34
As	Ion flotation using colophony amine acetate	35
Au	Ion flotation using dodecylbenzene sulfonate	33
Ba	Foam fractionation using potassium laurate	36
Be	Ion flotation using dodecylsulfate	37
Bi	Ion flotation using cationic surfactants	38
Ca	Foam fractionation using potassium tridecanoate	36
Cd	Colloid flotation using hexadecyltrimethylammonium chloride	39
Cd	Foam fractionation using sodium dodecylbenzene sulfonate at a pH less than 4	6
Cd	Foam fractionation using chelating surfactant	5, 10
Ce-144	Colloid flotation at pH between 7.8 and 8.5 using cetyltrimethylammonium bromide	40
Ce-144	Foam fractionation using N-dodecylbenzyldiethylenetriaminetriacetic acid	41
Co	Ion flotation using potassium laurate	42
Co	Ion flotation using cetyltrimethylammonium bromide	43
Co	Ion flotation using cetylpyridinium chloride	33
Co	Foam separation using hexadecyltrimethylammonium bromide	44
Cr	Foam separation using hexadecyldimethylbenzylammonium chloride	45
Cr	Ion flotation using sodium diisobutylnaphthalene sulfonate	46
Cu	Ion flotation using potassium laurate	47
Cu	Ion flotation using sodium α-sulfolaurate	48
Cu	Foam fractionation using sodium dodecylbenzene sulfonate	6
Cu	Foam fractionation using 4-dodecyldiethylenetriamine	5
Eu	Ion flotation using sodium tetradecylsulfate	49
Fe	Ion flotation using potassium stearate	31
Ge	Ion flotation using cetylpyridyl acetate	50
Ge	Ion flotation using laurylamine	51
Hg	Foam fractionation using hexadecyltrimethylammonium bromide	52
Hg	Foam fractionation using 4-dodecyldiethylenetriamine	10
In	Precipitate flotation using gallein	53
La	Ion flotation from aqueous solution as citrate complexes using a cationic collector, cetylpyridinium bromide	54
Mg	Foam fractionation using potassium myristate	36
Mn	Ion flotation using colophony amine acetate	35
Mo	Foam separation using hexadecyldimethylbenzylammonium chloride	45
Mo	Ion flotation using dimethyllaurylbenzylammonium bromide	33
Ni	Ion flotation using potassium laurate	42
Ni	Ion flotation using cetylpyridinium chloride	33
Ni	Ion flotation using cetyltrimethylammonium bromide	43

Substance	System and remarks	Reference
	Separation of Metallic Ions	
Pb	Ion flotation with Fe (III) and hexadecyltrimethylammonium bromide at pH > 3 for wastewater treatment	55
Pb	Foam fractionation using sodium dodecylbenzene sulfonate at pH < 4	6
Pd	Ion flotation using dimethylbenzylcetylammonium chloride	33
Pm-147	Colloid flotation at pH between 6.3 and 8.2 using cetyltrimethylammonium bromide	40
Pt	Ion flotation using dimethylbenzylcetylammonium chloride	33
Re	Foam separation using hexadecyldimethylbenzylammonium chloride	45
Re	Ion flotation using diethylcetylammonium hydrochloride	33
Ru	Froth flotation using alkyltrimethylammonium chloride as collector	56
Ru	Ion flotation using dodecylamine	57
Sb	Ion flotation using colophony amine acetate	35
Sn	Ion flotation using long-chain ammonium compounds	38
Sr	Foam fractionation using potassium laurate	36
Sr	Foam fractionation from radioactive wastes using n-dodecylbenzyldiethylenetriaminetriacetic acid	41
Tm	Frothless solvent ion flotation from aqueous solution using citric acid at pH 10.5	58
Tm	Ion flotation using cetylpyridinium bromide	49
UO_2^{2+}	Ion flotation using cetyltrimethylammonium bromide	59
V	Ion flotation using lauryltrimethylammonium chloride	60
V	Ion flotation using aminate colophony	61
V	Foam separation using hexadecyldimethylbenzylammonium chloride	45
W	Foam separation using hexadecyldimethylbenzylammonium chloride	45
W	Ion flotation using octadecylamine	62
Y-91	Ion flotation using dodecyl sulfate	63
Yb	Ion flotation using cetylpyridinium bromide	49
Zn	Foam fractionation using sodium dodecylbenzyl sulfate	64
Zn	Ion flotation using sodium monostearyl phosphate	65
Zn	Precipitate flotation using sodium lauryl sulfate	19
	Separation of Anions	
CrO_4^{2-}	Ion flotation with the cationic surfactant, ethylhexadecyldimethylammonium bromide, from aqueous solution	66
$Cr_2O_7^{2-}$	Ion flotation using ethylhexadecyldimethylammonium bromide	67
ClO_3^-	Foam fractionation using ethylhexadecyldimethylammonium bromide	68
CN^-	Precipitate flotation with Fe (II) and ethylhexadecyldimethylammonium at pH 6.0	69
CN^-	Ion flotation using ethylhexadecyldimethylammonium bromide at an iron-cyanide ratio of about 0.35 and pH 7 or less	70
F^-	Colloid flotation using sodium laurylsulfate	71

TABLE 1 Examples of Foam Separation (*Continued*)

Substance	System and remarks	Reference
	Separation of Anions	
I^0	Foam fractionation using ethylhexadecyldimethylammonium bromide	72
MoO_4^{2-}	Colloid flotation using rosin amine acetate	73
NO_3^-	Foam fractionation using ethylhexadecyldimethylammonium bromide	72
	Foam fractionation using ethylhexadecyldimethylammonium bromide	74
PO_4^{2-}	Continuous foam fractionation with ethylhexadecyldimethylammonium bromide at pH 5.4	75
PO_4^{2-}	Colloid flotation using ethylhexadecyldimethylammonium bromide	20
SCN^-	Foam fractionation within pH range 5.6 to 5.8 using ethylhexadecyldimethylammonium bromide	68
$S_8O_3^{2-}$	Ion flotation from aqueous solution using ethylhexadecyldimethylammonium bromide	76
	Separation of Colloids	
Apatite	Using dodecylammonium chloride at pH > 6 and sodium dodecylsulfonate at pH < 6	77
Barite	Using cetyltrimethylammonium bromide	78
Calcite	Using fatty acid soap at pH 8 to 9.5	79
Cellulose	Microflotation using lignosulfonate or xylan	81
Cellulose fiber	Separated by froth flotation	82
Chalcocite	Floated by ethyl xanthates, fatty acids, and long-chain sulfates and sulfonates	80
Clay	Using cetyltrimethylammonium chloride	83
Galena	Using fatty acids and long-chain sulfonates	80
Hematite	Froth flotation using hexamate	84
Kyanite	Floated using petroleum sulfonates and alcohol frothers at pH 3	85
Lignin	Ion flotation using cetyldimethylbenzylammonium chloride	86
Lignin	Ion flotation using cetol	87
Magnesium hydroxide	Using sodium dodecylbenzene sulfonate	88
Magnetite	Using ethylhexadecyldimethylammonium bromide best above pH 10.5	89
Magnetite	Using sodium oleate and n-dodecylamine	90
Martite	Froth flotation using sodium oleate	91
Pyrite	Using ethyl xanthates or fatty acids in acid solution	80
Stannic oxide	Using cetyltrimethylammonium bromide	92
Titanium dioxide	Froth flotation using n-dodecylamine	90
	Colloid flotation using sodium laurylsulfate	93
	Separation of Dyes and Organic Acids	
Bromothymol blue	Using laurylpyridinium chloride	94
Crystal violet	Using sodium lauryl sulfate	95
Citratogermanic acid	Using rosin amine acetate	96
Fatty acid	With saponin	97
Humic acid	Using rosin amine acetate	98
Phenol	Foam fractionation using ethylhexadecyldimethylammonium bromide	72
Tartratogermanic acid	Using rosin amine acetate	96

Substance	System and remarks	Reference
	Separation of Proteins	
Albumin	Separated from dilute solution containing organic and inorganic materials	30
Apple protein	Proteins concentrated in foam	99
Catalase	Purified by foaming from amalase	100
Diastase	Distase concentrated in foam, lipase left in residue	101
Tyrosinase	Concentrated in foam	102
Urease	Separated from catalase	103
	Separation of Miscellaneous	
Acid mine drainage	Foam fractionation using anionic surfactants	104, 105
Benzene amyl alcohol	Froth flotation using saponin	97
Polyvinyl alcohol	Selective concentrated in the foam of syndiotactic polyvinylalcohol	106
Polyvinyl alcohol	Foam separation using alkyltrimethylammonium chloride	107
Polystyrene	Froth flotation using sodium lignosulfonate	108
Polystyrene sulfonate	By adsorbing particle flotation	109

REFERENCES

1. Karger, B. L., Grieves, R. B., Lemlich, R., Rubin, A. J., and Sebba, F., *Sep. Sci.*, **2**, 401 (1967).
2. Gibbs, J. W., "Collected Works," Longmans, New York, 1928.
3. Newson, I. H., *J. Appl. Chem. (London)*, **16**, 43 (1966).
4. Sebba, F., *Am. Inst. Chem. Eng. Symp. Ser.*, **1**, 14 (1965).
5. Okamoto, Y., and Chou, E. J., *Sep. Sci.*, **11**, 79 (1976).
6. Huang, R. C. H., and Talbot, F. D., *Can. J. Chem. Eng.*, **51**, 709 (1973).
7. Jorne, J., and Rubin, E., *Sep. Sci.*, **4**, 313 (1969).
8. Rubin, E., and Gaden, E. L. Jr., Foam Separation in Schoen, H. M. (ed.), "New Chemical Engineering Separation Techniques," Wiley, New York, 1962.
9. Lemlich, R., Principles of Foam Fractionation and Drainage in Lemlich, R., (ed.), "Adsorptive Bubble Separation Techniques," Academic, New York, 1972.
10. Okamoto, Y., and Chou, E. J., *Sep. Sci.*, **10**, 741 (1975).
11. Rose, M. W., and Sebba, F., *J. Appl. Chem. (London)*, **19**, 185 (1969).
12. Garrott, D. E., Phosphate Removal by Foam Separation, presented at 3d AIChE-IMIQ Meeting at Denver, Colo., August 1970.
13. Nakatsuka, K., Matsouka, I., and Shimeiizaka, J., *Proc. 9th Int. Process. Cong.*, Prague, 1970.
14. Koyanaka, Y., *Nippon Genshiryoku Gakkaishi*, **7**, 621 (1965).
15. Karger, B. L., Poncha, R. P., and Miller, M. W., *Anal. Letters*, **1**(7), 438 (1968).
16. Aplan, F. F., and Fuerstenau, D. W., in Fuerstenau, D. W. (ed.), "Froth Flotation," 50th Anniversary Volume, p. 170, AIME, 1962.
17. Fuerstenau, M. C., and Atak, S., *Trans. AIME*, **232**, 24 (1965).
18. Rubin, A. J., and Lapp, W. L., *Anal. Chem.*, **41**, 1133 (1969).
19. Rubin, A. J., and Lapp, W. L., *Sep. Sci.*, **6**, 357 (1971).
20. Grieves, R. B., Wilson, T. E., and Shih, K. Y., *Am. Inst. Chem. Eng. J.*, **11**, 820 (1965).
21. Grieves, R. B., and Bhattacharyya, D., *Sep. Sci.*, **1**, 81 (1966).
22. Rubin, A. J., Johnson, J. D., and Lamb, J. C., *Ind. Eng. Chem. Process Des. Dev.*, **5**, 368 (1966).
23. Mitrofanov, S. I., in Almquist and Wiksell, *Progress in Mineral Dressing*, **46**, 1 (1957) Stockholm.
24. Pinfold, T. A., *Sep. Sci.*, **5**, 379 (1970).
25. Grieves, R. B., and Crandall, C. J., *Water & Sewage Works*, **112**, 432 (1966).
26. Schoen, H. M., and Mazzella, G., *Ind. Water Wastes*, **6**, 71 (1961).
27. Lemlich, R., and Lavi, E., *Science*, **134**, 191 (1961).
28. Haas, P. A., and Johnson, H. F., *Am. Inst. Chem. Eng. J.*, **11**, 319 (1965).
29. Somasundaran, P., *Sep. Sci.*, **10**(1), 93 (1975).
30. Ahmad, S. I., *Sep. Sci.*, **10**(6), 673 (1975).
31. Tadzhikbaev, P., and Adilov, T. A., *Khim, Redk. Tsvetn. Met.* (T. D. Artykbaev and Sh. Z. Khamudkhanova, eds.,) p. 33, 1975.
32. Bhattacharyya, D., and Grieves, R. B., *Am. Inst. Chem. Eng. J.*, **18**, 200 (1972).

33. Charewicz, W., and Walkowiok, W., *Sep. Sci.*, **7**, 631 (1972).
34. Mukai, S., Nakahiro, Y., and Shirakawa, Y., *Suiyokai-Shi*, **18**, 51 (1974).
35. Amanov, K. B., Skrylev, L. D., and Paderin, V. Ya., *Izv. Akad. Nauk Turkm. SSR, Ser. Fiz.-Tekh. Khim. Geol. Nauk*, **1973**(2), 108, 1073.
36. Skrylev, L. D., and Dashuk, L. A., *Izv. Vyssh. Uchebn. Zaved., Khim. Khim. Tekhnol.*, **18**, 366 (1975).
37. Pustovalov, N. N., Chupin, V. V., Fominykh, V. E., and Pushkarev, V. V., *Izv. Vyssh. Uchebn. Zaved., Khim. Khim. Tekhnol.*, **17**, 767 (1974).
38. Makarenko, V. K., and Pyatykh, N. V., *Uch. Zap., Tsentr. Nauchno-Issled. Inst. Olovyannoi Prom-sti*, **2**, 58 (1974).
39. Kobayashi, K., *Bull. Chem. Soc. Japan*, **48**, 1750 (1975).
40. Kepak, F., and Kriva, J., *Sep. Sci.*, **7**, 433 (1972).
41. Schoen, H. M., *Ann. N.Y. Acad. Sci.*, **137** (art 1), 148 (1966).
42. Amanov, K. B., and Skrylev, L. D., *Izv. Akad. Nauk Turkm. SSR., Ser. Fiz.* Tehk., Khim. Geol. Nauk, *1975(1)*, 96 *(1975)*.
43. Jurkiewicz, K., and Waksmundzki, A., *Rocz. Chem.*, **49**, 187 (1975).
44. Karger, B. L., and Miller, M. W., *Anal. Chim. Acta*, **48**, 273 (1969).
45. Charewicz, W., and Grieves, R. B., *J. Inorg. Nucl. Chem.*, **36**, 2371 (1974).
46. Selecki, A., and Kawalac-Pietrenko, B., *Pr. Inst. Inz. Chem. Politech. Wars.*, **1**, 363 (1972).
47. Skrylev, L. D., and Amanov, K. B., *Zh. Prikl. Kim. (Leningrad)*, **46**, 819 (1973).
48. Kobayashi, K., *Bull. Chem. Soc. Japan*, **48**, 1180 (1975).
49. Szeglowski, Z., Bittner-Jankowska, M., and Mikulski, J., *Nukleonika*, **18**, 299 (1973).
50. Seifullina, I. I., Pozhariskii, A. F., Skrylev, L. D., Belousova, E. M., and Kozina, N. V., *Iav. Vyssh. Uchebn. Zaved., Khim. Khim. Tekhnol.*, **17**, 973 (1974).
51. Klassen, V. I., and Shrader, E. A., *Tr. Inst. Obogashch. Tverd. Goryuch. Iskop.*, **1**, 156 (1971).
52. Miller, M. W., and Sullivan, G. L., *Sep. Sci.*, **6**, 553 (1971).
53. Minczewski, J., Trybulowa, Z., and Krzyzanowska, M., *Chem. Anal. (Warsaw)*, **20**, 630 (1975).
54. Szeglowski, Z., Bittner-Jankowska, M., Mikulski, J., and Machej, T., *Nukleonika*, **17**, 307 (1973).
55. Ferguson, B. B., Hinkle, C., and Wilson, D. J., *Sep. Sci.*, **9**, 125 (1974).
56. Berezyuk, V. G., Evtyukhova, O. V., and Pushkarev, V. V., *Radiokhimiya*, **17**, 103 (1975).
57. Kepak, F., and Kriva, J., *Sep. Sci.*, **55**, 385 (1970).
58. Stachurski, J., and Szeglowski, Z., *Sep. Sci.*, **9**, 313 (1974).
59. Shakir, K., *J. Appl. Chem. Biotechnol.*, **23**, 339 (1973).
60. Evtyukhova, O. V., and Berezyuk, V. G., *Tr. Ural'sk. Politekhn. in-ta, Sb.*, **222**, 7 (1974).
61. Skrylev, L. D., Borisov, V. A., and Grinchenko, L. I., *Izv. Vyssh. Uchebn. Zaved. Tsvetn. Metal.*, **2**, 18 (1975).
62. Novgorodova, I. Z., Kuz'kin, S. F., Nebera, V. P., Yudin, G. G., and Filobok, E. D., *Izv. Vyssh. Uchebn. Zaved. Tsvet. Metal.*, **16**, 78 (1973).
63. Pustovalov, N. N., and Pushkarev, V. V., *Radiokhimiya*, **16**, 240 (1974).
64. Onda, K., Takeuchi, H., and Takahaski, M., *Kogyo Kagaku Zasshi*, **74**, 1721 (1971).
65. Yamada, K., Koide, Y., and Tanoue, S., *Nippon Kagaku Kaiski*, **10**, 1900 (1972).
66. Grieves, R. B., Bhattacharyya, D., and Ghosal, J. K., *Sep. Sci.*, **8**, 501 (1973).
67. Grieves, R. B., and Schwartz, S. M., *Am. Inst. Chem. Eng. J.*, **12**, 746 (1966).
68. Grieves, R. B., and The, P. J. W., *J. Inorg. Nucl. Chem.*, **36**, 1391 (1974).
69. Grieves, R. B., and Bhattacharyya, D., *Sep. Sci.*, **4**, 301 (1969).
70. Grieves, R. B., and Bhattacharyya, D., *Sep. Sci.*, **3**, 185 (1968).
71. Clarke, A. N., and Wilson, D. J., *Sep. Sci.*, **10**, 417 (1975).
72. Moore, P., and Phillips, C. R., *Sep. Sci.*, **9**, 325 (1974).
73. Skrylev, L. D., Amanov, K. B., Borisov, V. A., and Seifullina, I. I., *Zh. Prikl. Khim. (Leningrad)*, **48**, 1432 (1975).
74. Grieves, R. B., Bhattacharyya, D., and The, D. J. W., *Can. J. Chem. Eng.*, **51**, 173 (1973).
75. Primiani, S., Nguyen, Y. V., and Phillips, C. R., *Sep. Sci.*, **9**, 211 (1974).
76. Grieves, R. B., and Bhattacharyya, D., *Eng. Bull. Purdue Univ. Eng. Ext. Ser.*, **140**, (pt. 2), 902 (1973).
77. Somasundaran, P., *J. Colloid Interface Sci.*, **27**, 659 (1968).
78. Arafa, M. A., Boulos, T. R., and Yousef, A. A., *Inst. Min. Metall. Trans.*, **84**, C38 (1975).
79. Lee, O., R12744 USBM, 1926.
80. Taggart, A. F., "Handbook of Mineral Dressing," pp. 12–55, Wiley, New York, 1964.
81. Roberts, K., and Barla, P., *J. Colloid Interface Sci.*, **49**, 75 (1974).
82. Ng, K. S., Mueller, J. C., and Walden, C. C., *Pulp Pap. Mag. Can.*, **75**, 263 (1974).
83. Horikawa, Y., *Clay Sci.*, **4**, 281 (1975).
84. Noblitt, H. L., *Can. Mines Branch Invest. Rep.*, IR71-51, 1971.
85. Thom, C., in Fuerstenan, D. W. (ed.), "Froth Flotation," 50th Anniversary Volume, p. 343, AIME, New York, 1962.
86. Wang, M. H., Grenstrom, M. L., Wilson, T. E., and Wang, L. K., *Water Resour. Bull.*, **10**, 283 (1974).
87. Wilson, T. E., and Wang, M. H., *Eng. Bull. Purdue Univ. Eng. Ext. Ser.*, **137**(pt.2), 731 (1974).

88. Kalman, K. S., and Ratcliff, G. A., *Can. J. Chem. Eng.*, **49**, 626 (1971).
89. Grieves, R. B., and Bhattacharyya, D., *J. Appl. Chem.*, **18**, 149 (1968).
90. Pope, N. I., and Sutton, D. I., *Powder Technol.*, **5**, 101 (1972); **7**, 271 (1973); **9**, 273 (1974).
91. Emel'yanov, M. F., and Maksimov, I. I., *Obogashch. Rud*, **19**, 37 (1974).
92. Grieves, R. B., and Bhattacharyya, D., *Am. Inst. Chem. Eng. J.*, **11**, 274 (1965).
93. Rubin, A. J., and Haberkost, D. C., *Sep. Sci.*, **8**, 363 (1973).
94. Sebba, F., *Nature,* **188**, 736 (1960).
95. Karger, B. L., and Rogers, L. B., *Anal. Chem.*, **33**, 1165 (1961).
96. Seifullina, I. I., Belousova, E. M., Skrylev, L. D., and Pozharitshii, A. F., *Zh. Prikl. Khim. (Leningrad),* **48**, 1311 (1975).
97. Howe, R. H. L., *Prog. Hazardous Chem. Handl. Disposal, Proc. 3d Symp., 1972, p. 130.*
98. Skrylev, L. D., Purich, A. N., and Gorodentsev, Yu. S., *Zh. Prikl. Khim. (Leningrad).* **48**, 685 (1975).
99. Davis, S. G., Fellers, C. R., and Esseler, W. B., *Food Res.,* **14**, 417 (1949).
100. Bader, R., and Schutz, F., *Trans. Faraday Soc.,* **42**, 571 (1946).
101. Ostwald, W., and Mischke, W., *Kolloid Z.,* **90**, 205 (1940).
102. Dognon, A., and Dumontet, H., *C. R. Soc. Biol.,* **135**, 884 (1941).
103. London, M., Cohen, M., and Hudson, P. B., *Biochem. Biophys. Acta,* **13**, 111 (1954).
104. Bikerman, J. J., Hanson, P. J., and Rose, S. H., *Water Pollu. Control Res. Ser.,* no. 14010 DEE, FWQA, Department of Interior, 1970.
105. Hanson, P. J., *Water Pollut. Control Res. Ser.,* no. 14010 FUI, EPA, 1971.
106. Imai, K., and Matsumoto, M., *Bull. Chem. Soc. Japan,* **36**, 455 (1963).
107. Fukunaga, K., Japan Kokai 74-51344 and 74-48746, 1974.
108. Ismui, S., and Tanaka, H., Ger. Offen. 2,407,953, 1974.
109. Yamasaki, T., Shimoizaka, J., and Sasaki, H., *Nippon Kogyo Kaishi,* **79**, 313 (1963).
110. Fanlo, S., and Lemlich, R., AIChE-ICE (London) Symp. Ser., no. 9, pp. 75, 85, 1965.
111. Davies, J. T., and Rideal, E. K., "Interfacial Phenomena," 2d ed. Academic, New York, 1963.
112. Osipow, L. I., "Surface Chemistry, Theory and Industrial Applications," ACS Monograph, ser. 153, Reinhold, New York, 1962.
113. Lemlich, R. (ed.), "Adsorptive Bubble Separation Techniques," Academic, New York, 1972.
114. Karger, B. L., and DeVivo, D. G., General Survey of Adsorptive Bubble Separation Processes, *Sep. Sci.,* **3**, 393 (1968).
115. Wadsworth, M. E., Separation with Foams, in Berl, W. G. (ed.), "Physical Methods in Chemical Analysis," Academic, New York, 1961.

Part 3

Gas (Vapor) Mixtures

Section **3.1**

Gas-Phase Adsorption

J. LOUIS KOVACH, D.Ch.E.   *President, Nuclear*
Consulting Services

CONCEPT OF ADSORPTION

Molecules or atoms interacting individually or in a group are considered a *system*. A system which consists of physically homogeneous parts is referred to as a *phase*. A system consisting of a single phase is called *homogeneous*, while those consisting of two or more phases are called *heterogeneous systems*. When we are dealing with a heterogeneous system, the boundary between the phases is termed an *interface*. These heterogeneous systems may exist in both equilibrium and nonequilibrium states. When the phase composition stays constant with time, it is considered to be at equilibrium. If the phase composition or its thermodynamic parameters change with time, the system is nonequilibrium. Naturally, a spontaneous change in the state of a nonequilibrium system goes toward the establishment of the equilibrium state.

In any heterogeneous system consisting of atoms, molecules, or ions, the interaction between the phases begins with chemical or physical interaction at the phase interface. When a molecule under kinetic motion hits a surface (as an example, a phase boundary) from a random direction, it can either bounce back from the surface elastically with the angle of reflection equaling the angle of incidence, or the molecule may stay at the surface for a period of time and come off in a direction unrelated to that from which it came. Generally, the latter is the case, the residence time depending on the nature of the surface and the molecule, the temperature of the surface, and the kinetic energy of the molecule.

This phenomenon, molecular or atom interaction at the surface, can be observed at interfaces between gas and solid, gas and liquid, liquid and solid, two liquids, and in rare circumstances, two solid phases. In cases where the residence time of a molecule at the surface is very short, it is difficult to determine the residence time by conventional physical or chemical means. However, existence of the phenomenon can be proven by the fact that, gas can be heated by a hot surface with which it is in contact. The interaction at the interface involving the transition of a molecule from one phase to another can be considered the sorption of the molecule by the given phase. The type of interaction between the phases of a heterogeneous system depends on the properties and composition of all the components of the system. From the standpoint of the distribution of the substances to be sorbed, two types of sorption can occur: adsorption and absorption. Adsorption takes place when the molecules or atoms sorbed are concentrated only at the interface, while absorption takes place when the molecules or atoms are distributed in the bulk of the interacting phases. As a rule, adsorption takes place when one of the phases consists of a solid.

Adsorption of atoms or molecules on a solid from the gas phase is, therefore, a spontaneous process, and as such it is accompanied by a decrease in the free energy of the system. The process involves loss of degrees of freedom. Therefore, there is also a decrease in entropy.

Thus,

$$\Delta F = \Delta H - T\Delta S \tag{1}$$

The adsorption process is always exothermic regardless of the type of forces involved.

Gas-Phase Adsorption

Almost all air pollution problems where adsorption is considered as a unit operation involve gaseous contaminants. Gas- (the term gas will denote both gas and vapor) phase adsorption is treated in detail.

As mentioned in the introduction, the number of molecules which are present at the surface is dependent on the number which reach the surface and on the residence time of these molecules on the surface. According to de Boer,[1] if n molecules strike a unit area of a surface per unit time, and remain there for an average time t, then σ number of molecules are present per unit area of surface:

$$\sigma = nt \tag{2}$$

Using square centimeter as unit surface and second as unit time, n is number of molecules falling on 1 cm²/s. The number n thus denotes the number of molecules striking each square centimeter of the surface every second, and this number can be calculated using the Maxwell and the Boyle–Gay-Lussac equations. The number n is directly related to the speed of the molecules within the system. It is important to realize that the velocity of the molecules is not dependent on the pressure of the gas, but the mean free path is inversely proportional to the pressure.

Thus,

$$n = 3.52 \times 10^{22} \times \frac{p}{\sqrt{MT}} \tag{3}$$

where p = pressure, mmHg
M = molecular weight
T = absolute temperature, K

From this equation at 20°C and 760 mmHg pressure, the following values can be obtained:

H_2: $n = 11.0 \times 10^{23}$ molecules/(cm²)(s)
N_2: $n = 2.94 \times 10^{23}$ molecules/(cm²)(s)
O_2: $n = 2.75 \times 10^{23}$ molecules/(cm²)(s)

The molecule residence time t on the surface is not as easy to determine as the number n. Reflection experiments can indicate the residence time on a smooth surface because, if a molecule is retained on the surface for any finite time, the angle of removal will be random.

Early attempts to directly determine t by Holst and Clausing[2] were not very successful. However, the exchange of thermal energy between a molecule and the surface it strikes is indicative of the residence time. In the case of elastic rebound from the surface, there is hardly any degree of exchange of energy. Clausing[3] later used a method involving the estimation of the velocity with which molecules of a gas pass through narrow capillaries when the pressure is low enough that no mutual collisions of the gas molecules take place. Experimenting with argon in glass capillaries, Clausing obtained the following results:

At 90 K: $t = 3.1 \times 10^{-5}$ s
At 78 K: $t = 75 \times 10^{-5}$ s

Same order-of-magnitude results were obtained also for nitrogen on glass, but for neon the value of

$$t = 2 \times 10^{-7} \text{ s} \qquad \text{at 90 K}$$

was obtained.

It is evident from the argon residence time results at two different temperatures that t is greatly dependent on temperature. According to Frenkel,[4]

$$t = t_0 e^{Q/RT} \tag{4}$$

where t_0 is the time of oscillation of the molecules in the adsorbed state, and Q is the heat of adsorption. It can be seen from Eq. (4) that the adsorption energy (heat of adsorption) Q is the all-determining factor for the magnitude of t.

Adsorption Forces

It has to be strongly emphasized that there are no special adsorption forces. Forces causing adsorption are the same ones which cause cohesion in solids and liquids, and which are responsible for the deviation of real gases from the laws of ideal gases. The basic forces causing adsorption can be divided into two groups: intermolecular, or van der Waals, forces and chemical forces, which generally involve electron transfer between the solid and the gas. Depending on which of these two force types plays the major role in the adsorption process, we distinguish between physical adsorption, where van der Waals or molecular interaction forces are in prevalence, and chemisorption, where heteropolar or homopolar forces cause the surface interaction. Thus, if in the process of adsorption, the individuality of the adsorbed molecule (adsorbate) and of the surface (adsorbent) are preserved, we have physical adsorption, If between the adsorbate and the adsorbent any electron transfer or sharing occurs, of if the adsorbate breaks up into atoms or radicals which are bound separately, then we are presented with chemisorption.

Although the theoretical difference between physical and chemical adsorption is clear in practice, the distinction is not as simple as it may seem. The following parameters can be used to evaluate an adsorbate-adsorbent system to establish the type of adsorption:

1. The heat of physical adsorption is in the same order of magnitude as the heat of liquefication, while the heat of chemisorption is of the same order as of the corresponding chemical reaction. It has to be pointed out here that the heat of adsorption varies with surface coverage because of lateral interaction effects. Therefore, the heat of adsorption has to be compared on corresponding levels.

2. Physical adsorption will occur under suitable temperature-pressure conditions in any gas-solid system, while chemisorption takes place only if the gas is capable of forming a chemical bond with the surface.

3. A physically adsorbed molecule can be removed unchanged at a reduced pressure at the same temperature where the adsorption took place. The removal of the chemisorbed layer is far more difficult.

4. Physical adsorption can involve the formation of multimolecular layers, while chemisorption is always completed by the formation of a monolayer. (In some cases physical adsorption may take place on the top of a chemisorbed monolayer.)

5. Physical adsorption is instantaneous (the diffusion into porous adsorbents is time-consuming), while chemisorption may be instantaneous, but generally requires activation energy.

Adsorption Rate in Porous Adsorbents

As is shown in the previous discussion, the boundary layer is most important in the phase interaction. Therefore, to achieve a high rate of adsorption, it is expedient to create the maximum obtainable surface area within the solid phase. High surface area can be produced by creating a large number of microcapillaries in the solid. All commercial adsorbents, such as activated carbon, silica gel, alumina, etc., are prepared in this manner.

While adsorption is nearly instantaneous, the passage of molecules through capillaries (pores) may involve some time. There have been observations that in rare cases it took several days to reach adsorption equilibrium.

The movement of molecules into the pores is a diffusion process. Regardless of the mechanism used, there is a correlation between σ, the number of molecules adsorbed per square centimeter at a given pressure and temperature, and time t_r required to complete the adsorption process:

$$\sigma_t = (1 - e^{-k_d t_r}) \tag{5}$$

where σ_t = value of σ at the time t_r

σ_∞ = the equilibrium value of σ

k_d = a constant

The constant k_d is inversely proportional to the square of the distance which the molecules have to travel, and proportional to the diffusion constant D, which can be obtained from the following equation:

$$D = D_0 e^{-Q_d \div RT} \tag{6}$$

where D_0 = a constant

$\quad Q_d$ = the energy of activation

The temperature dependence of the diffusion constant D, and the temperature dependence of constant k_d in Eq. (5), and of the whole phenomenon of the rate of reaching adsorption equilibria is a function of the activation energy Q_d. The molecules may proceed in three major manners within the pore structure. First, they may collide with the wall of the capillary and bounce off immediately according to the cosine law, collide again, etc. In this case there is no activation energy. Second, the molecules may collide with the wall, stay there for time t, reevaporate, collide again, etc. The energy of activation will then be equal to the heat of adsorption. Third, the molecules may migrate along the wall of the capillary during a sufficiently long time t, or they might make hopping movements as described by de Boer.[5] The energy of activation is then given by the fluctuations in the heat of adsorption ΔQ_a. As an example, the rate of adsorption of ethane on activated carbon gives an energy of about 3 kcal/mol, which is roughly half of the value of the heat of adsorption.

Adsorption Equilibrium

Both in nature and in technological processes, the solid-gas phase interaction may occur under two conditions. One involves the random mixing of the phases; the other, their direct relative motion. Thus, static adsorption occurs when the adsorption process takes place in relative rest, or random mechanical mixing of the phases of the solid-gas system takes place, and ends in the establishment of an adsorption equilibrium among the interacting phases. Dynamic adsorption represents a sorption process accomplished under conditions of direct relative motion of one or both phases. Although in air pollution control most applications involve dynamic conditions where adsorption equilibrium is not reached, it is essential to survey the equilibrium conditions because their modified effect is of major importance in the dynamic nonequilibrium systems.

Adsorption equilibrium is defined when the number of molecules arriving on the surface is equal to the number of molecules leaving the surface to go into the gas phase. The adsorbed molecules exchange energy with the structural atoms of the surface and, provided that the time of adsorption is long enough, will be in a thermal equilibrium with the surface atoms. In order to leave the surface, the adsorbed molecule has to take up sufficient energy from the fluctuations of thermal energy at the surface so that the energy corresponding to the vertical component of its vibrations surpasses the holding limit.

Adsorption Isotherms

When studying the adsorption of a particular gas on the surface of a particular adsorbent, one may start by inserting into Eq. (2) the generalized values for n from Eq. (3) and for t from Eq. (4), to get

$$\sigma = \frac{Np}{\sqrt{2\pi MRT}} \times t_0 e^{Q/RT} \tag{7}$$

because the adsorbent and the adsorbate are defined M and t_0 are constants, thus:

$$\sigma = \frac{k_0 p}{\sqrt{T}} e^{Q/RT} \tag{8}$$

in which

$$k_0 = \frac{Nt_0}{\sqrt{2\pi MR}} \tag{9}$$

If the temperature is considered, then the simplest form of adsorption isotherm is

$$\sigma = kp \tag{10}$$

which shows that the amount of gas will be directly proportional to the pressure.

From Eq. (1), it can be seen that isotherms show essentially the free energy change as a function of the amount adsorbed. Under real conditions where the mutual interaction of the adsorbed molecules takes place, and at increasing coverage, the adsorption energy Q changes, deviation from Eq. (10) takes place.

Isotherms as measured under existing conditions can yield qualitative information about the adsorption process and also give indication to the fraction of the surface coverage, and thus, with certain assumptions, to the surface area of the adsorbent.

In Fig. 1, the five basic types of adsorption isotherms are presented as classified by Brunauer.[6] The Type I isotherm represents systems where adsorption does not proceed

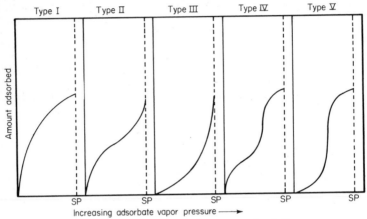

Fig. 1 Adsorption isotherm classification according to Brunauer.

beyond the formation of a monomolecular layer. Such an isotherm is obtained when oxygen is adsorbed on carbon black at $-183°C$. The Type II isotherm indicates an indefinite multilayer formation after completion of the monolayer. As an example, the adsorption of water vapor on carbon black at $30°C$ results in such a curve. Type III isotherm is obtained when the amount of gas adsorbed increases without limit as its relative saturation approaches unity. The convex structure is caused by heat of adsorption of the first layer being less than the heat of condensation due to molecular interaction in the monolayer. This type of isotherm is obtained when bromine is adsorbed on silica gel at $20°C$. The Type IV isotherm is a variation of Type II, but with a finite multilayer formation corresponding to complete filling of the capillaries. This type of isotherm is obtained by the adsorption of water vapor on active carbon at $30°C$. The Type V isotherm is a similar variation of Type III obtained when water vapor is adsorbed on activated carbon at $100°C$.

Although a large number of equations developed to date have been based on theoretical considerations, none of them can be generalized to describe all systems. A critical evaluation of these equations can be found elsewhere[7,8]; therefore, only the two most important methods will be presented here.

The potential theory of adsorption was introduced by Polanyi in 1914,[9] and because of its thermodynamic character, it is still regarded as fundamentally sound. The Polanyi concept for a typical cross section of a gas-solid phase boundary system is shown in Fig. 2. The forces holding a molecule to the surface diminish with distance and the multimolecular adsorbate film is lying in the intermolecular potential field. The force of attraction is conveniently measured by the adsorption potential E, defined as the work done by the adsorption forces in bringing a molecule from the gas phase to that given point. The adsorbate volumes included in these equipotential fields $E = E_1, E_2, E_3, \ldots, E_i, \ldots, 0$ are $\Phi_1, \Phi_2, \Phi_3, \ldots, \Phi_i, \ldots, \Phi_{max}$, the last quantity denoting the volume of the entire adsorption space. As ϕ increases from 0 to Φ_{max}, E decreases from its maximum value E_0 at the adsorbent surface to 0 to the outermost layer. The process, therefore, is represented by the curve

$$E = f(\Phi) \tag{11}$$

which in reality is a distribution function. The adsorption potential is postulated to be independent of the temperature; therefore, Eq. (11) is the same for a given system at all temperatures. The usefulness of the potential theory is that it is not explicitly an isotherm equation. The characteristic curve replaces the normal p vs. v curve. The application of the potential theory requires the empirical determination of a single isotherm, the calculation of the characteristic curve, and the prediction of the isotherms at other temperatures from this curve. As an example, adsorption isotherms of carbon dioxide determined at 0°C were found to correspond empirically to the calculated values from

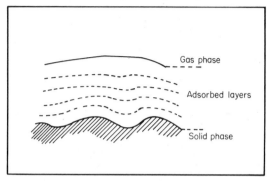

Fig. 2 The structure of the adsorbed phase according to the Polanyi potential theory.

about −110 to +120°C. Polanyi[10] and Berenyi[11,12] examined the relationship between the characteristic curves of different gases on the same adsorbent. For activated carbon, they found empirically that for any two gases

$$\frac{E_{0_1}}{E_{0_2}} = \left(\frac{a_1}{a_2}\right)^{1/2} \tag{12}$$

where a is the van der Waals constant.

The actual determination of the work done when the adsorbed layers are assumed to be liquid is

$$E_i = RT \ln \frac{P_0}{P_s} \tag{13}$$

where P_s = equilibrium vapor pressure in the gas phase
P_0 = vapor pressure in the adsorbed phase

The value ϕ_i corresponding to E_i is x/d_T where x is the weight of the adsorbed film and d_T is the density of the liquid at temperature T. The Polanyi equation is very important in chemical engineering design calculations because of its ability to predict isotherms for any gas, given a single isotherm for one gas on the same solid. Dubinin et al.[13−16] published a large number of adsorption correlation results using the Polanyi equation. Lewis et al.[17] improved the Polanyi-Dubinin method by substituting fugacity in place of pressure and replacing V, the molar volume of the liquid at temperature T, by V^1, the molar volume of the liquid at a temperature where its vapor pressure is equal to the equilibrium pressure p.

The Polanyi potential theory of adsorption basically describes the volume filling of micropores in a typical adsorbent structure. The theory is valid primarily below the critical temperature of the adsorbate. The applicability of this theory to industrial adsorbents is expressed by the equation

$$W = W_0 \exp(-kE^2) \tag{14}$$

where W = volume of the adsorbate as liquid
W_0 = volume of adsorbent filled when E decreases to zero, which in most cases is the total pore volume of the adsorbent
k = constant

The plot of W/W_0 against E, the characteristic curve representing the relationship between volume of the available adsorption volume and the adsorption potential, is basically a statistical relationship expressing the fraction of the pore volume filled at different adsorption potential E values.

The filled volume of the adsorbent is

$$W = av_m \tag{15}$$

where a = constant and
$\qquad v_m$ = molar volume of the adsorbate

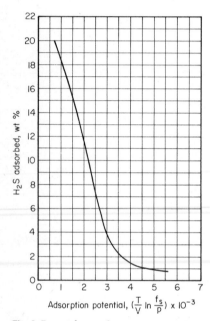

Fig. 3 Potential curve for H_2S on 13× molecular sieve.

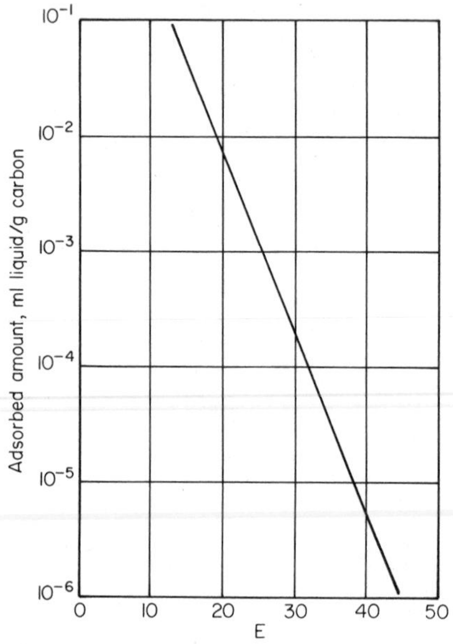

Fig. 4 Experimental characteristic curve for 1000 to 1100 m²/g coconut-shell activated carbon.

Equation (12) indicated the temperature invariance of the adsorption potential, and

$$\frac{E_1}{E_2} = \beta \approx \frac{v_{m_1}}{v_{m_2}} \approx \frac{|P_1}{|P_2} \tag{16}$$

where β = the affinity coefficient or the relative molar work of vapor adsorption compared with a standard substance (often for benzene = 1.0), and
$\qquad |P$ = parachor for the bulk liquid phase

Therefore, with these equations it is possible to calculate the adsorption isotherm of nearly every substance at any temperature from a single measured adsorption isotherm at one temperature.

A typical potential curve is shown on Fig. 3 for W.R. Grace 13X molecular sieve, and a generalized form is shown on Fig. 4 in conjunction with Table 1 for 1000 to 1100 m²/g steam-activated coconut-shell carbon (Nusorb CC 1000 grade). The E values of Table 1 are given for 20% of the ACGIH threshold limit values.

Langmuir[18] presented an ideal monolayer adsorption isotherm

$$V = \frac{V_m bp}{1 + bp} \tag{17}$$

where V = volume of gas (0°C, 760 mmHg) adsorbed per unit mass of adsorbent
V_m = volume of gas (0°C, 760 mmHg) adsorbed per unit of adsorbent with a layer one molecule thick
b = empirical constant in reciprocal pressure unit which has limited practical application

Brunauer, Emmett, and Teller[19] expanded the Langmuir isotherm to include multilayer adsorption

$$V = \frac{V_m C x}{(1 - x)[1 + (C - 1)x]} \tag{18}$$

where V_m and C are empirical constants and $x = p/p_s$. The constant C is derived from the heat of adsorption and V_m represents the volume of gas required to cover the surface with a monomolecular layer.

Although the BET equation has limitations (such as the assumption that the heat of adsorption is constant over the entire surface coverage of the monolayer and that the

TABLE 1   E Values for Use With Potential Plot, Fig. 4
(E values are for 20% of the ACGIH threshold limit values)

Contaminant	E	Contaminant	E
Acetone	18	Formaldehyde	75
Acetaldehyde	27	Hydrogen	418
Acetylene	48	Hydrogen chloride	81
Allyl alcohol	23	Hydrogen fluoride	166
Ammonia	54	1-Hexene	10
Amyl alcohol	7	n-Hexane	10
Benzene	16	Hexamethylcyclotrisiloxane	5
n-Butane	17	Hydrogen sulfide	63
Butene-1	19	Isopropyl alcohol	15
cis-Butene-2	18	Isobutyl alcohol	11
trans-Butene-2	19	Methylene chloride	22
n-Butyl alcohol	10	Methyl chloroform	12
Butraldehyde	11	Methyl ethyl ketone	14
Butyric acid	8	Methyl isopropyl ketone	10
Carbon disulfide	22	Methyl alcohol	31
Carbon monoxide	110	3-Methyl pentane	10
Chlorine	55	Methane	43
Chloroacetone	13	Monomethylhydrazine	20
Chlorobenzene	9	Methyl mercaptan	35
Chloropropane	16	Nitric oxide	107
Cyclohexane	11	Nitrogen tetroxide	32
Cyclohexanol	8	Nitrous oxide	42
Cyanamide	18	Propylene	28
1.1-Dimethylcyclohexane	7	Isopentane	13
trans-1,2-Dimethylcyclohexane	7	n-Pentane	13
2.2-Dimethylbutane	10	Propane	24
1,4-Dioxane	16	n-Propylacetate	9
Dimethylhydrazine	23	Propyl mercaptan	13
Ethyl alcohol	20	Phenol	9
Ethyl acetate	12	Sulfur dioxide	49
Ethylene dichloride	17	Toluene	9
Ethylene	42	Trichloroethylene	13
Ethylene glycol	7	Tetrachloroethylene	10
trans-1, me-3, ethylcyclohexane	6	1,1,3-Trimethylcyclohexane	6
Ethyl sulfide	10	Tetrafluoroethylene	30
Ethyl mercaptan	23	Freon-21	23
Freon-11	22	Valeric acid	4
Freon-12	22	Vinyl chloride	27
Freon-22	28	Vinylidene chloride	19
Freon-23	40	m-Xylene	7
Freon-114	14	o-Xylene	8
Freon-114 unsymmetrical	16	p-Xylene	7
Freon-125	25		

monolayer is completed before the formation of the secondary layer begins, with a heat of adsorption equaling that of the heat of liquefaction), it is very useful because it permits the numerical determination of surface area. Knowing the area occupied by a single molecule of the adsorbent and the number of molecules needed to form a monolayer, it is possible to express the surface area of the adsorbent in cm²/g or m²/g.

Surface areas of commonly used adsorbents determined in this manner are:

Activated alumina: 50 to 250 m²/g
Silica gel: 200 to 600 m²/g
Molecular sieve: 800 to 1000 m²/g
Activated carbon: 500 to 2000 m²g

Adsorption isotherms are most commonly used to select the adsorbent or even the adsorption process as a unit operation for the adsorptive separation of gases.

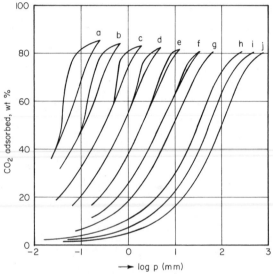

Fig. 5 Isotherms for the adsorption of carbon dioxide on steam-activated coconut shell carbon, 1800 m²/g. (a) 130.4 K, (b) 136.0 K, (c) 142.1 K, (d) 148.3 K, (e) 155.3 K, (f) 162.8 K, (g) 169.0 K, (h) 181.6 K, (i) 155.3 K, (j) 196.3 K.

If the adsorption isotherm shape is Type I, II, or IV, adsorption can be used to separate the adsorbate from the carrier gas. If it is Type III or V, adsorption will probably not be economical for the separation.

The shape of the adsorption isotherm for each temperature is a function of both the properties of the adsorbate and the adsorbent. As an example, Figs. 5 and 6 show CO_2 and N_2O adsorption isotherms on the same 1800 m²/g coconut-shell carbon. These are pure-gas isotherms where concentration is expressed as absolute adsorbate pressure. The CO_2 adsorption at low temperature shows significant hysteresis (higher sorption capacity on desorption than on adsorption). As the temperature increases the hysteresis loop diminishes, and above 169 K the adsorption is reversible. The saturation adsorption is constant in the reversible range but increases with the development of hysteresis toward low temperatures. The adsorption of N_2O on the same carbon shows that hysteresis is much smaller and occurs in a different shape at low and high temperatures. Also at high temperatures a rise in the saturation capacity is observed.

Figure 7 shows the CO_2 adsorption isotherm on a structurally different 1400 m²/g coal carbon, which obviously has larger pore diameters also. The low-temperature hysteresis loop disappears with increasing temperatures and the saturation capacity increases with increasing temperature. The data not only stress the importance of determining the adsorption isotherm on the exact adsorbent used, but also the importance of knowing

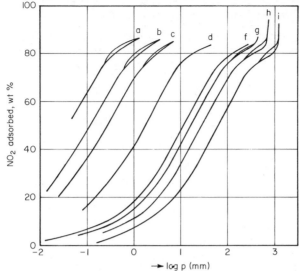

Fig. 6 Isotherms for the adsorption of nitrous oxide on steam-activated coconut-shell carbon, 1800 m²/g. (a) 130.3 K, (b) 136.5 K, (c) 141.5 K, (d) 155.2 K, (e) 172.9 K, (f) 176.4 K, (g) 183.5 K, (h) 192.8 K.

where adsorption and desorption isotherms deviate from each other. Additional adsorption isotherms are shown in Figs. 8 to 10.

Figure 11 indicates the variation in the shape of the benzene adsorption isotherm on three increasing-surface-area activated-coconut-shell carbons. As the carbon surface area is increased, the average micropore diameter also increases. This increase results in an improvement of adsorption capacity at high benzene concentrations; however, there is a commensurate decrease in adsorption at low concentrations. Thus the higher surface area adsorbent is not necessarily the best adsorbent at all concentrations if the average pore diameter also increases with the increasing surface area.

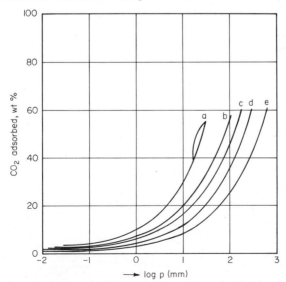

Fig. 7 Isotherms for the adsorption of carbon dioxide on steam-activated coal carbon, 1400 m²/g. (a) 162.7 K, (b) 173.2 K, (c) 178.2 K, (d) 183.1 K, (e) 193.2 K.

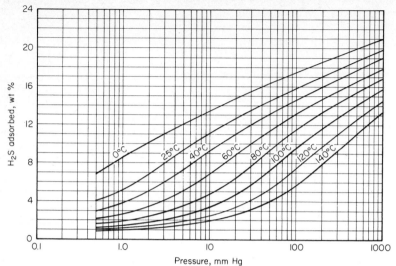

Fig. 8 Hydrogen sulfide adsorption isotherms on 13× molecular sieves.

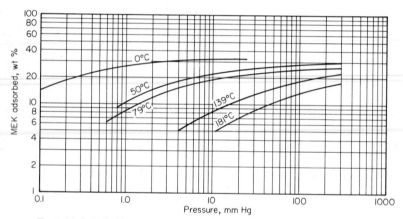

Fig. 9 Methyl ethyl ketone adsorption isotherms on Union Carbide 45 carbon.

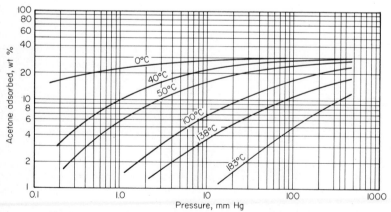

Fig. 10 Acetone adsorption isotherms on Union Carbide 45 carbon.

Water adsorption isotherms on activated carbon have a unique behavior in that the shape and location of both the adsorption and desorption branches change with aging (exposure to oxygen) of the activated carbon (Fig. 12). This is because the amount of surface oxide complexes changes on the carbon surface, influencing the polarity of the surface and significantly increasing the number of active sites where water adsorption

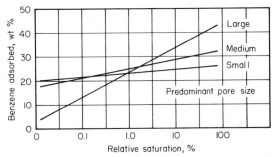

Fig. 11 Benzene adsorption isotherms on 750, 1000, and 1600 m²/g coconut-shell carbons, showing effect of increasing micropore structure.

initiates. While the carbon underwent this "aging" process, there was no structural change in the carbon when the nitrogen adsorption isotherm was used to measure pore size distribution.

This increased water adsorption capacity of activated carbons due to the development of oxygen-containing surface complexes will influence the residual adsorption capacity of

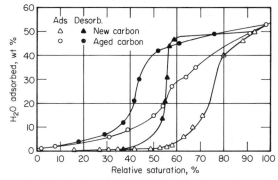

Fig. 12 Water vapor adsorption isotherm at 25°C as a function of aging (slow oxidation) in humid air. coconut-shell carbon: 1050±50 m²/g.

carbons in regenerative systems when the adsorbate is also polar. The shift in the water adsorption isotherm was not the only change observed; the pH of the water extract of the carbon also decreased from 10.5 to 8.7, indicating that the resulting surface oxides were acidic in character. The development of the acidic surface oxides can also have catalytic effects in cross-reactions between different adsorbates.

Adsorption Isobars

If we keep the pressure constant in Eq. (8), a direct correlation exists between the number of molecules adsorbed and the temperature producing the equation for adsorption isobars:

$$\sigma = k_2 \frac{e^{Q/RT}}{\sqrt{T}} \qquad (19)$$

The effect of $\sqrt{T}$ is small compared with the e power function. Therefore, the number of adsorbed molecules at constant pressure will be exponential with increasing temperature. In a plot of σ as a function of T, the isobar at a high pressure lies above the isobar at lower pressure. When a disruptively sharp increase of σ is observed with increasing T, this indicates a change from physical adsorption to chemisorption. In this case, the new σ value will decrease exponentially with increasing temperature. This type of phenomenon is also caused when the adsorbate penetrates into the structure of the adsorbent from the surface and bulk absorption takes place.

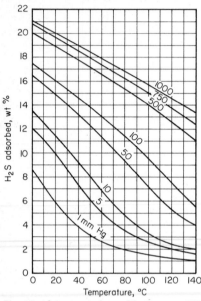

Fig. 13 Adsorption isobars of hydrogen sulfide on 13× molecular sieves.

A family of adsorbtion isobars of H$_2$S on 13X Grace molecular sieves is shown on Fig. 13. Adsorption isobars are very useful expressions of data when thermal regeneration of the adsorbent is considered. As an example from Fig. 13, if adsorption takes place at an H$_2$S partial pressure of 10 mmHg at 40°C and desorption occurs at 120°C, even if the concentration of the sweep gas is not changed the theoretical operating capacity is 6% by weight.

Adsorption Isosteres

In considering Eq. (8), if we keep σ constant, a relationship between pressure and the temperature corresponding to the same degree of surface coverage is obtained. This type of relationship representation is very useful when a certain amount of contaminant has to be adsorbed; however, some degree of freedom is available in selecting temperature and pressure conditions of the adsorption. The equation for adsorption isostere is

$$p = k_3 \sqrt{T} e^{-Q/RT} \tag{20}$$

For approximation, the effect of $\sqrt{T}$ can be neglected and

$$\ln p = -\frac{Q}{RT} + B_a \tag{21}$$

is obtained, where B_a is constant. The plot of $\ln p$ against $1/T$ for a constant value of σ gives a straight line on arithmetic-coordinate paper. The other important feature of adsorption isosteres is that the slope of this straight line corresponds to the heat of adsorption Q.

Very often the operating experience exists for a particular contaminant, and ranges of operating capacities are known; however, the exhaust flow of the operation is too high and either cooling or dilution with clean air is considered. With adsorption isosteres, the selection of direct cooling versus dilution can easily be evaluated. A typical family of adsorption isosteres is given in Fig. 14.

When the heat of adsorption is calculated from the slope of the adsorption isostere, it also gives relative values for the strength of adsorption as indicated by Eq. (4). Therefore, the higher the heat of adsorption, the longer the residence time of the particular adsorbate on the surface in the adsorbed stage. It is important to know the heat of adsorption of the adsorbate on the adsorbent—particularly at high adsorbate concentrations—because the temperature rise due to the decrease in free energy by decrease in the vapor pressure, and the decrease in entropy by the decrease in mobility of the adsorbate will lower the adsorption capacity of the adsorbent, and may involve safety considerations also.

ADSORBENTS

Activated Carbons

Activated carbons are produced either by thermal decomposition of carbonaceous materials and subsequent controlled oxidation (activation) using steam, CO_2, air, or by a mixture of these to burn a network of pores into the grains. Another method of one-step decomposition and activation is by thermal treatment of cellulosic materials with phos-

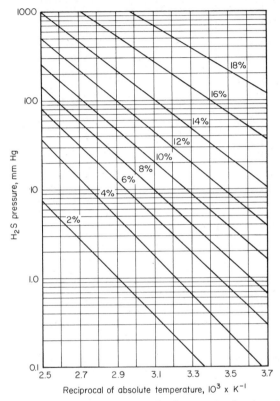

Fig. 14 Isosteres of hydrogen sulfide on 13× molecular sieve. Loading in weight percent.

phoric acid or zinc chloride. The structure development is different during the thermal decomposition and the activation step. Generally materials which have no initial pore structure cannot be activated by either process. Natural grain materials typically have little porosity in >300-Å pores (macropores), while the reconstituted carbons produced by fine-grinding raw materials and then forming them into distinct shapes have 2 to 3 times larger macroporosity at equal microporosity (<300-Å pores).

Typical values of micropore volume and surface area for a series of steam-activated coconut-shell carbons is given in Figs. 15 and 16 at various CCl_4 adsorption capacities, which are one of the industry standards for gas phase carbon grading.

Activated carbons are less polar than other adsorbents but it is incorrect to call them nonpolar because the surface oxides render them slightly polar. The polarity generally increases with aging of the carbon in humid air.

Carbons are generally used from 4 × 6 mesh (coarse) to 12 × 20 mesh (fine) range in gas phase application. Apparent densities vary between 25 and 35 lb/ft³. The hardness of the carbon is an important parameter particularly when in-place regeneration is used, resulting in thermal stresses in addition to the packing and fluid dynamic stresses.

Nearly all carbons contain ash (or other noncarbonacous material such as sulfur, which is not reported by manufacturers as ash) which influences potential reactions between the adsorbed molecules. Such catalytic effects should be investigated before the adsorbent material is selected, and the evaluation should be performed under both the adsorption and the regeneration condition.[20]

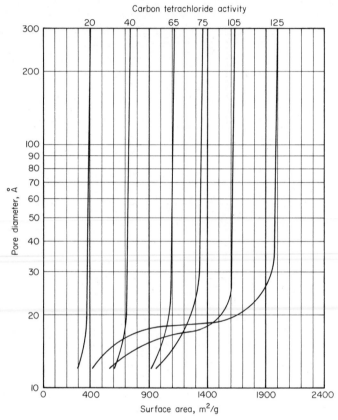

Fig. 15 Cumulative surface area distribution of Nusorb steam-activated coconut carbons.

Activated Aluminas[21]

Activated alumina and alumina gel are porous forms of aluminum oxide of high surface area. They adsorb liquids, vapors, and gases without any change of form or properties. Activated alumina and gel are inert chemically to most gases and vapors, are nontoxic, and will not soften, swell, or disintegrate when immersed in water. High resistance to shock and abrasion are two of their important physical characteristics. The adsorbed material may be driven from the alumina desiccant by suitable choice of reactivating temperature, thus returning it to its original highly adsorptive form.

The aluminas are used to dehydrate gases and liquids and to control humidity in air-conditioning units.

Activated alumina is used in maintenance programs for transformer and lubricating oils. It will neutralize acids and adsorb moisture from these oils, thus preventing the deterioration of the oil and formation of sludge.

Activated alumina and alumina gel will also adsorb gases and liquids other than water. Important components of mixtures may be recovered or impurities removed from the mixture by taking advantage of the preferential adsorption of alumina.

Activated alumina and alumina gel may be reactivated to their original adsorptive efficiency by employing a heating medium at any temperature between 177°C and 316°C. For thorough reactivation, the temperature of the reactivating gas on the exit side of the bed should reach at least 177°C. The alumina desiccants may be reactivated an almost unlimited number of times without seriously lowering the efficiency of adsorption.

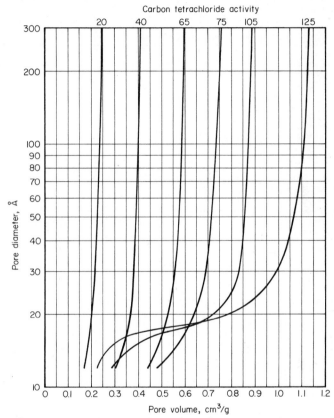

Fig. 16 Cumulative pore volume distribution of Nusorb steam-activated coconut carbon.

There are many types of alumina desiccants manufactured by thermally treating granules of hydrated alumina and by forming spheres of alumina gels.

Typical grades of one manufacturer (Alcoa) are shown in Table 2.

Molecular Sieves[22,23]

Molecular sieves are crystalline metal aluminosilicates with a three-dimensional interconnecting network structure of silica and alumina tetrahedra.

The tetrahedra are formed by four oxygen atoms surrounding a silicon or aluminum atom. Each oxygen has two negative charges and each silicon has four positive charges. This structure permits a neat sharing arrangement, building tetrahedra uniformly in four directions. The trivalency of aluminum causes the alumina tetrahedron to be negatively charged, requiring an additional cation to balance the system. Thus, the final structure has sodium, potassium, or calcium cations in the network. These "charge balancing" cations are the exchangeable ions of the zeolite structure.

In the crystalline structure, up to half of the quadrivalent silicon atoms can be replaced by trivalent aluminum atoms. By regulating the ratios of the starting materials, it is possible to produce zeolites containing different ratios of silicon to aluminum ions and different crystal structures containing various cations.

Type A Molecular Sieves   In the most common commercial zeolite, Type A, the tetrahedra are grouped to form a truncated octahedron with a silica or alumina tetrahedron at each point. This structure is known as a sodalite cage. It contains a small cavity which is of no practical significance, since the largest openings through the six-sided faces of the octahedron are not large enough to permit the entrance of even small molecules.

TABLE 2   Typical Properties of Alcoa Alumina Desiccants

	F-1	F-5	F-6	H-151
Typical properties				
Al_2O_3, %	92	77	88	90
Na_2O, %	0.90	0.80	0.86	1.6
Fe_2O_3, %	0.08	0.06	0.07	0.13
SiO_2, %	0.09	0.09	0.09	2.2
Loss on ignition (1100°C) (after reactivation), %	6.5	12.4	8.5	6.0
$CaCl_2$, %		9.3		
$CoCl_2$, %			2.0	
Form, %	Granular	Granular	Granular	Ball
Surface area, m^2/g	210			390
Bulk density, loose, lb/ft³	52	57	54	51
Bulk density, packed, lb/ft³	55	60	57	53
Specific gravity	3.3			3.1–3.3
Static sorption at 60% RH, %	14–16	18–23	13–15	22–25
Crushing strength	55			75
Sizes available	¼–½ in, ¼-in–8 mesh, 8/14 mesh, 14/28 mesh, and finer	¼–½ in	¼-in–8 mesh, 8/14 mesh	⅛-in spheres ¼-in spheres

When sodalite cages are stacked in simple cubic forms, the result is a network of cavities approximately 11.5 Å* in diameter, accessible through openings on all six sides. These openings are surrounded by eight oxygen ions. One or more exchangeable cations also partially block the face area. In the sodium form, this ring of oxygen ions provides an opening "window" of 4.2 Å diameter into the interior of the structure. This crystalline structure is represented chemically by the following formula:

$$Na_{12}(AlO_2)_{12}(SiO_2)_{12} \cdot X\,H_2O$$

The water of hydration which fills the cavities during crystallization is loosely bound and can be removed by moderate heating. The voids formerly occupied by this water can be refilled by adsorbing a variety of gases and liquids. The number of water molecules in the structure (the value of X) can be as great as 27, making the water in the saturated formula 28.5% of the weight of the anhydrous zeolite.

The sodium ions which are associated with the aluminum tetrahedra tend to block the openings, or conversely may assist the passage of slightly oversized molecules by their electrical charge. As a result, this sodium form of the molecular sieve, which is commercially called 4A, can be regarded as having uniform openings of approximately 4 Å diameter.

Because of their base exchange properties, zeolites can be readily produced with other metals substituted for a portion of the sodium.

Among the synthetic zeolites, two modifications have been found particularly useful in industry. By replacing a large fraction of the sodium with potassium ions, the 3A molecular sieve is formed (with openings of about 3 Å). Similarly, when calcium ions are used for exchange, the 5A (with approximately 5 Å openings) is formed.

Type X Molecular Sieves   The crystal structure of the Type X zeolite is built up by arranging the basic sodalite cages in a tetrahedral stacking (diamond structure) with bridging across the six-membered oxygen atom ring. These rings provide opening "win-

*An angstrom unit Å is one hundred millionth of a centimeter (1×10^{-8} cm).

dows" 9 to 10 Å in diameter into the interior of the structure. The overall electrical charge is balanced by positively charged cation(s), as in the Type A structure. The chemical formula that represents the unit cell of Type X molecular sieve in the soda form is shown below:

$$Na_{12}(AlO_2)_{86}(SiO_2)_{106} \cdot XH_2O$$

As in the case of the Type A crystals, water of hydration can be removed by moderate heating and the voids thus created can be refilled with other liquids or gases. The value of X can be as great as 276, making the water in this type of molecular sieve 35% of the weight of the anhydrous zeolite.

The typical product types of one manufacturer (Union Carbide) are shown in Table 3.

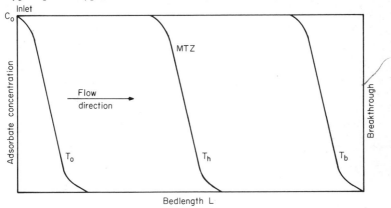

Fig. 17 The formation and movement of the MTZ within an adsorber bed until breakthrough is reached.

Silica Gel[23]

Silica gel is a porous amorphous form of SiO_2 manufactured by chemical reaction between sodium silicate and sulfuric acid. The purity is typically 99.7% SiO_2 and the primary trace impurities are Al_2O_3, CaO, and Na_2O. The typical properties of W. R. Grace silica gel are given in Table 4.

ADSORPTION DYNAMICS

In most adsorptive separation cases the process takes place in a dynamic system. The adsorbent is generally used in a fixed bed and the contaminated air is passed through the adsorbent bed. Depending on the concentration and market value, the contaminant is either recovered or discarded when the loading of the adsorbent requires regeneration. Although isotherms are indicative of the efficiency of an adsorbent for a particular adsorbate removal, they do not supply data to permit the calculation of contact time or the amount of adsorbent required to reduce the contaminant concentration below the required limits.

At this point, it is necessary to evaluate the dynamic capacity in a little more detail. When a contaminant-containing fluid is first passed through a bed of adsorbers, most of the adsorbate is initially adsorbed at the inlet part of the bed and the fluid passes on with little further adsorption taking place. Later, when the adsorber at the inlet end becomes saturated, adsorption takes place further along the bed. The situation in the bed while it is in normal operation may therefore be represented by Fig. 17, which shows the building up of a saturated zone starting at the inlet end of the bed.

As more gas is passed through, the adsorption proceeds, the saturated zone moves forward until the breakthrough point is reached, at which time the exit concentration begins to rise rapidly toward the inlet concentration in the fluid. If the passage of the fluid is continued on still further, the exit concentration continues to rise until it becomes substantially the same as the inlet concentration. At this point the bed is fully saturated. While the concentration at saturation is a function of the material used and the tempera-

TABLE 3 Typical Properties of Union Carbide Type X Molecular Sieves

Basic type	Nominal pore diameter, angstroms	Available form	Bulk density, lb/ft³	Heat of adsorption (max), Btu/lb H₂O	Equilibrium H₂O capacity, % wt	Molecules adsorbed	Molecules excluded	Applications
3A	3	Powder $\frac{1}{16}$-in pellets $\frac{1}{8}$-in pellets	30 44 44	1800	23 20 20	Molecules with an effective diameter < 3 Å, including H₂O and NH₃	Molecules with an effective diameter > 3 Å, e.g., ethane	The preferred molecular sieve adsorbent for the commerical dehydration of unsaturated hydrocarbon streams such as cracked gas, propylene, butadiene, and acetylene. It is also used for drying polar liquids such as methanol and ethanol.
4A	4	Powder $\frac{1}{16}$-in pellets $\frac{1}{8}$-in pellets 8 × 12 beads 4 × 8 beads 14 × 30 mesh	30 45 45 45 45 44	1800	28.5 22 22 22 22 22	Molecules with an effective diameter < 4 Å, including ethanol, H₂S, CO₂, SO₂, C₂H₄, C₂H₆, and C₃H₆	Molecules with an effective diameter > 4 Å, e.g., propane	The preferred molecular sieve adsorbent for static dehydration in a closed gas or liquid system. It is used as a static desiccant in household refrigeration systems; in packaging of drugs, electronic components and perishable chemicals; and as a water scavenger in paint and plastic systems. Also used commercially in drying saturated hydrocarbon streams.
5A	5	Powder $\frac{1}{16}$-in pellets $\frac{1}{8}$-in pellets	30 43 43	1800	28 21.5 21.5	Molecules with an effective diameter < 5 Å, including n-C₄H₉OH,† n-C₄H₁₀,† C₄H₈ to C₂₂H₄₆, R-12	Molecules with an effective diameter > 5 Å, e.g., iso compounds and all 4-carbon rings	Separates normal paraffins from branched-chain and cyclic hydrocarbons through a selective adsorption process.
10X	8	Powder $\frac{1}{16}$-in pellets $\frac{1}{8}$-in pellets	30 36 36	1800	36 28 28	Iso paraffins and olefins, C₆H₆, molecules with an effective diameter < 8 Å	Di-n-butylamine and larger	Aromatic hydrocarbon separation.
13X	10	Powder $\frac{1}{16}$-in pellets $\frac{1}{8}$-in pellets 8 × 12 beads 4 × 8 beads 14 × 30 mesh	30 38 38 42 42 38	1800	36 28.5 28.5 28.5 28.5 28.5	Molecules with an effective diameter < 10 Å	Molecules with an effective diameter > 10 Å, e.g., (C₄F₉)₃N	Used commercially for general gas drying, air plant feed purification (simultaneous removal of H₂O and CO₂), and liquid hydrocarbon and natural gas sweetening (H₂S and mercaptan removal).

ture at which the unit is operated, the breakthrough capacity is dependent on the operating conditions, such as inlet concentration, fluid flow rate, and bed depth. The dependence on inlet concentration and fluid flow rate arises from heat effect and mass-transfer rates, but the dependence on bed depth, as can be seen from the above description, is dependent on the relative sizes of saturated and unsaturated zones. The zone of the bed where the concentration gradient is present is often called the *mass-transfer zone*, or MTZ. In most published data on dynamic adsorption, results are expressed in terms of the dynamic capacity or breakthrough capacity at given inlet concentration, temperature and flow rate condition of the bed, and bed dimensions. It is extremely important that the adsorber bed should be at least as long as the transfer zone length of the key component to be adsorbed. Therefore, it is important to know the depth of this mass-transfer zone.

TABLE 4   Typical Properties of W. R. Grace Silica Gel

Property, (unit)	Regular	Intermediate
Pore volume, mL/g	0.43	1.15
Average pore diameter, Å	22	140
Surface area, m²/g	750–800	340
Volatile at 1750°F, % wt	5–6.5	4.5
Specific heat, Btu/(lb)(°F)	0.22	0.22
True density, lb/ft³	137	137
Apparent density, lb/ft³	45	25
Thermal conductivity, Btu/(ft²)(h)(°F)(in)	1	1

The following factors play the most important role in the length and rate of movement of the MTZ:

1. The type of adsorbent
2. The particle size of an adsorbent (may depend on maximum allowable pressure drop)
3. The depth of the adsorbent bed
4. The gas velocity
5. The temperature of the gas stream and the adsorbent
6. The concentration of the contaminants to be removed
7. The concentration of the contaminants not to be removed, including moisture.
8. The pressure of the system
9. The removal efficiency required
10. Possible decomposition or polymerization of contaminants on the adsorbent.

While the adsorbate molecule is carried along by the carrier gas flow and diffusion in the direction of pressure and concentration gradients, the gas flows among the adsorbate particles, and a thin film, which is practically immobile, adheres to the particle surface. The transfer of the carrier gas and adsorbate molecules is by diffusion through this film. Adsorption to the interior adsorbent surface also takes place by pore diffusion. Additionally, diffusion in the bulk packing will also occur, i.e., external, bulk diffusion in the longitudinal direction and transverse to the carrier gas flow. Therefore all the parameters which affect any of these diffusion steps will also influence the reaching of equilibrium conditions within the bed and determine the shape of the mass-transfer zone.

The early determination of the MTZ length from empirical data was performed by Shilov[24] and Mecklenburg.[25] Both of these theories predict that the MTZ moves with parallel transfer, and for different column lengths the penetration curves are parallel and independent of the column length maintaining the MTZ shape. This condition is maintained only after the MTZ achieves a steady state by traveling several times its length along the column in the direction of the carrier gas flow.

It is also important to realize that molecules of the carrier gas also travel through the bed by adsorption and desorption steps and a delay of each carrier gas molecule is proportional to t of Eq. (2). Thus in strict terms the carrier gas itself is an adsorbate. The degree of interference of the carrier gas or the adsorption of the contaminant is dependent on similar adsorbate-adsorbent relationships which control the adsorption of any other component.

These early empirical breakthrough equations are in the form

$$t_b = \frac{as}{C_0 V}(L - h) \tag{22}$$

where t_b = time to breakthrough

  a = volumetric adsorption capacity at C_0 concentration, cm³/cm³

  s = cross section of the adsorbent bed, cm²

  V = volumetric flow rate, cm³/min

  L = length of the adsorbent bed, cm

  h = half-length of the MTZ, cm

  C_0 = adsorbate volume fraction, cm³/cm³

In this type of expression h is approximately 1 half-length of the MTZ, and the equation only approximates the dynamic adsorption conditions. For the exact determination of the MTZ length, more precise equations are required. One of these was developed by Michaels.[26]

The Length and Rate of Movement of the MTZ

As the C_0 concentration adsorbate containing carrier gas is introduced into the adsorber bed, the formed S-shape MTZ will travel along the adsorber bed at a velocity u. The determination of u is made by determining the time t elapsed to the appearance of $C_0/2$ adsorbate concentration in the outlet of the column,

$$u = \frac{L}{t} \tag{23}$$

Graphic representation of this condition is shown on Fig. 18.

As is shown, the $C_0/2$ concentration can be used only if the breakthrough curve is symmetrical; if not, the areas above and below the curve have to be integrated, and equal areas above and below will determine time t.

The exact determination of the MTZ can be performed by monitoring the rate of movement of the MTZ either in the adsorbent bed or as it exits from the column in the concentration region 0 to C_0. The time period from the measurement of the initial detectable concentration to the inlet concentration C_0 is expressed as Δt (the time required for the MTZ to travel its own length) in Fig. 18.

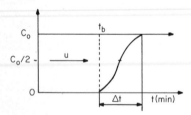

After u and Δt are determined, the length of the MTZ, M, can be calculated by

$$M = u\,\Delta t \tag{24}$$

Fig. 18 The illustration of the Δt and u in an adsorber bed.

For the experimental determination of the u and t values it is important that the experimental column length be sufficiently long for the MTZ to reach steady-state conditions. Data generated by Illes,[27] shown in Fig. 19, demonstrate the formation of steady-state conditions.

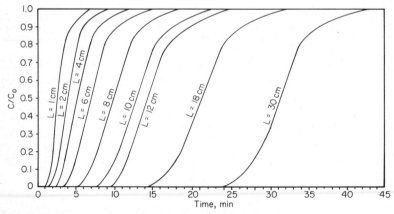

Fig. 19 Breakthrough curves for different bed depths under identical inlet conditions, showing the establishment of steady-state conditions.

By preparing a material balance during adsorption in an L length, S cross-section adsorber containing m weight of adsorbent with a void volume of E into which C_0 volume fraction concentration adsorbate is introduced with a carrier gas flow rate of V, and with MTZ assumed to be symmetrical on $C_0/2$ outlet concentration, then for the appearance of this outlet concentration the time is t and the quantity in the adsorbed equilibrium phase is $V_0(\text{mL/g})$:

$$VC_0t - (LS)(EC_0) = V_0LSm \tag{25}$$

The first expression on the left-hand side of the equation describes the quantity of adsorbate fed into the column, and the second part describes the quantity of adsorbate in the void volume of the adsorbent bed. The right-hand side shows the quantity of adsorbate adsorbed.

The adsorbed quantity per unit weight (g) is:

$$V_0 = \frac{VC_0t}{LSm} - \frac{EC_0}{m} \tag{26}$$

and with Eq. (23) substituted into Eq. (26),

$$V_0 = \frac{VC_0}{uSm} - \frac{EC_0}{m} \tag{27}$$

Rearrangement of Eq. (26) results in

$$t = LS \frac{V_0m + EC_0}{VC_0} \tag{28}$$

from which the rate of movement of the MTZ can be determined as follows:

$$u = \frac{L}{t} = \frac{V}{S} \frac{C_0}{(V_0m - EC_0)} \tag{29}$$

and by substitution of linear velocities v (based on empty cross section of the adsorber, S) and v_t (the true velocity based on ES),

$$u = v \frac{C_0}{V_0m + EC_0} \tag{30}$$

Even in the case of mild adsorption capacity of the adsorbent toward the adsorbate, the quantity of adsorbate in the void volume will be negligible; thus

$$u = v \frac{C_0}{V_0m} = v^t \frac{1}{q} \tag{31}$$

where $q = V_0m/C_0 =$ the distribution function of the adsorbate between the adsorbent and the gas phase (slope of the adsorption isotherm).

Thus, in the case of a linear adsorption isotherm, the rate of movement of the MTZ is a linear function of the true gas velocity. In case of a Type I isotherm, if Eq. (31) is used only in a narrow low-concentration range, the assumption that q is constant can be made with close but inaccurate approximation. Figure 20 shows the rate of movement of MTZ of propane at 40 and 80 cm/min superficial gas velocities as a function of C_0.[27] Figure 21 shows the effect of superficial gas velocity on the rate of MTZ movement vs. the superficial carrier gas velocity at four different concentrations.

The above derivation gives a more accurate description of the MTZ than those using the time to breakthrough in the mass balances due to the partial saturation of the MTZ.

Mass transfer in packed beds is analogous to heat transfer in packed beds, and Klotz[28] described an equation for the MTZ length based on earlier work by Hougen et al.[29] However, the value of capacity was taken for t_b, and other diffusional problems apparent in the original heat transfer equations were also uncorrected.[30] However, the equation describes factors which influence the MTZ length M, or more correctly $M/2$, because of the above mentioned mass-balance selection. This value [which is identical with h in the Mecklenburg equation, Eq. (22)] is designated as I and is shown to be composed of two separable factors I_t relating to external diffusion and I_r relating to internal pore diffusion. This equation is valid only in turbulent flow and is incorrect at low superficial velocities.

$$I = I_t + I_r \tag{32}$$

where

$$I_t = \frac{1}{f}\,\mathrm{Re}^{0.41}\mathrm{Sc}^{0.67}\ln\frac{C_0}{C_b} \tag{33}$$

$$I_r = kv^t \ln\frac{C_0}{C_b} \tag{34}$$

and f = external surface of adsorbent, cm²/cm³
   Re = Reynolds number
   Sc = Schmidt number
   C_b = exit concentration at breakthrough
   k = constant
   Empirical correlation evaluating the length and rate of movement of the MTZ with the equation

$$M = u\Delta t \tag{35}$$

can be relatively easily performed for the proposed adsorbent under the conditions expected, and correlation to other conditions can be made with Eq. (31).
   In general the following individual effects have to be evaluated.

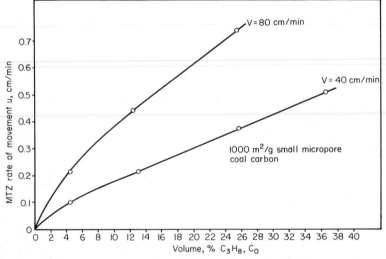

Fig. 20 The rate of movement of the MTZ of propane as a function of concentration.

Type of Adsorbent

Most industrial adsorbents are capable of adsorbing both organic and inorganic gases. However, their preferential adsorption characteristics and other physical properties make each one more or less specific for a particular application. As an example, activated alumina, silica gel, and molecular sieves will adsorb water preferentially from a gas phase mixture of water vapor and an organic contaminant. This is a considerable drawback in the application of these adsorbents for organic contaminant removal. Activated carbon preferentially adsorbs nonpolar organic compounds. Recently, through evaluation of the type of surface oxides present on activated carbon surfaces, it was found that the preferential adsorption properties of carbon can be partially regulated by the type of surface oxide induced on the carbon.
   Silica gel and activated alumina are structurally weakened by contact with liquid droplets; therefore, direct steaming cannot be used for regeneration.
   In some cases, none of the adsorbents have sufficient retaining adsorption capacity for a particular contaminant. In these applications, a large surface area adsorbent is impreg-

nated with inorganic or, in rare cases, with a high-molecular-weight organic compound which can chemically react with the particular contaminant. Iodine-impregnated carbons are used for removal of mercury vapor, and bromine-impregnated carbons are used for ethylene or propylene removal. The action of these impregnants is either catalytic conversion or reaction to a nonobjectionable compound, or to a more easily adsorbed

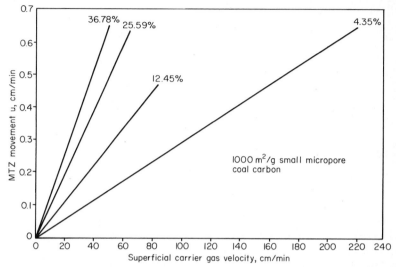

Fig. 21 The rate of movement of the MTZ of propane as a function of carrier gas velocity.

compound. It has to be noted here that the general adsorption theory does not apply anymore on the gross effects of the process. For example, the mercury removal by an iodine-impregnated carbon proceeds faster at a higher temperature, and a better overall efficiency can be obtained than at a low-temperature contact. An impregnated adsorbent is available for most compounds which under the particular conditions are not easily adsorbed by nonimpregnated commercial adsorbents.

As was previously mentioned, adsorption takes place at the interphase boundary; therefore, the surface area of the adsorbent is an important factor in the adsorption process. Generally, the higher the surface area of an adsorbent, the higher is its adsorption capacity for all compounds, however, the surface area has to be available in a particular pore size within the adsorbent. At low partial pressure (concentration) the surface area in the smallest pores into which the adsorbate can enter is the most efficient. At higher pressures the larger pores become more important. At very high concentrations, capillary condensation will take place within the pores, and the total micropore volume is the limiting factor. Figure 22 shows the relationship between maximum effective pore size and concentration for the adsorption of benzene vapor at 20°C. It is evident that the most valuable information concerning the adsorption capacity of a certain adsorbent is its surface area and pore volume distribution curve in different-diameter pores. Figure 23 shows the characteristic distribution curves for several different adsorbent types. As Fig. 11 indicates, the relationship between adsorption capacity and surface area in optimum pore sizes is concentration dependent; thus, it is very important that any evaluation of adsorption capacity is performed under actual concentration conditions.

The action of molecular sieves is slightly different from that of other adsorbents in that selectivity is determined more by the pore size limitations of the molecular sieve. In selecting molecular sieves, it is important that the contaminant to be removed is smaller than the available pore size, and that the carrier gas is not adsorbed, or the not-to-be-removed component is larger, and thus not adsorbed.

The Effect of Particle Size *d*

The dimensions and shape of particles affect both the pressure drop through the adsorbent bed and the diffusion rate into the particles. The pressure drop is lowest when the adsorbent particles are spherical and uniform in size. However, the external mass transfer increases inversely with $d^{3/2}$ and the internal diffusion rate inversely as d^2. The pressure drop will vary with the Reynolds number, being roughly proportional to velocity and varying inversely with particle diameter. It is evident that, everything else being equal, adsorbent beds consisting of smaller particles, although causing a higher pressure drop, will be more efficient. Therefore, a sharper and shorter mass-transfer zone will be obtained.

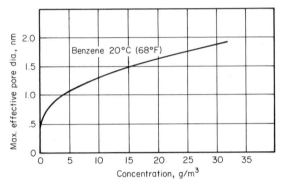

Fig. 22 Maximum effective pore size vs. concentration.

Figure 24 indicates the possible effects of change in particle size, depending upon the rate-controlling steps. If the rate-controlling step in the process is solely a surface reaction or the adsorption step by itself, then there is no variation in MTZ length M vs. particle size as shown by line A. If the process is both diffusion-controlled and surface-reaction- (or adsorption-) rate controlled, the effect of particle size change is shown by line B. If the sole rate-controlling step is diffusion, the effect of particle size is indicated by line C.

An important fact must be noted regarding generation of experimental data as a function of the ratio of particle diameter to bed diameter. To eliminate wall effect in the test bed, the bed diameter should always be a minimum of 12 times the average particle diameter.

The Depth of the Adsorbent Bed, L

The effect of bed depth on adsorption mass transfer is twofold. First, it is important that the bed is deeper than the length of the transfer zone. Second, any multiple of the minimum bed depth gives more than proportional increased capacity. In general, it is advantageous to size the adsorbent bed to the maximum length which is allowed by pressure-drop considerations.

If the steady-state MTZ length is more than double the adsorbent bed depth (or because of polymerization high boilers, etc., it becomes longer), the initial efficiency of the system will be less than 100% and immediate breakthrough will occur, i.e., if bed depth is less than h from the Mecklenburg equation.

Gas Velocity (V) Effects

The primary effect of gas velocity is on the rate of movement of the MTZ; however, as is obvious from Eq. (35), gas velocity will affect the length of the MTZ also.

When the superficial gas velocity is changed, and the rate-controlling step is thereby changed (change from laminar- to turbulent-flow regime), the length of the MTZ can become shorter. Such is the case when in laminar flow the rate-controlling step is bulk diffusion, while in the turbulent-flow region the rate-controlling step is pore diffusion. At very low velocities the forward diffusion of the adsorbate can be faster than the rate of movement of the MTZ because of carrier gas velocity effects, resulting in very long MTZ lengths. For many adsorbents the transition from laminar to turbulent flow takes place approximately at a modified Reynolds number of 10.

In Fig. 25, data produced by Illes is shown giving the correlation between the length of the MTZ and its rate of movement. The origin of the curve is not zero, and the curve slope flattens out with the increased velocity.

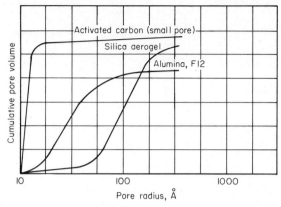

Fig. 23 Pore volume distribution curve of several adsorbent types.

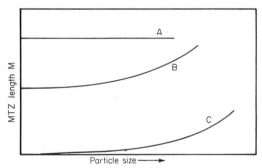

Fig. 24 The effect of adsorbent particle size on the MTZ length for various mechanisms of adsorbate removal.

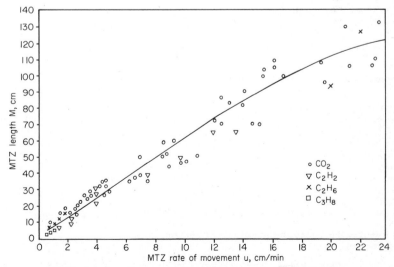

Fig. 25 Correlation between MTZ rate of movement u and its length m.

Theoretically there is no minimum or maximum velocity for the adsorption process to take place; however, there are velocity ranges at high pressures where the adsorbent material will be crushed by hydrodynamic pressure. Manufacturers of adsorbents have data available relating to these limiting "crushing" velocities.

The effect of carrier gas velocity on the rate of movement of the MTZ is shown in Fig. 21.

Temperature Effects

As was discussed in the *Adsorption Equilibrium* section, the physical adsorption capacity of any adsorbent (σ, V_0, etc.) will decrease with increasing temperature. At the same time, all diffusion rates will increase and the MTZ will be shorter as the temperature increases while its rate of movement increases. In gas-phase adsorption, when surface reaction per se is not part of the rate-controlling step, there is no overall advantage in increasing temperature because the time to breakthrough and the adsorption capacity are decreased with increasing temperature. The reason for this is that the effect of the shorter MTZ (M) is overridden by the higher rate of its movement (u). This is often not true in liquid-phase adsorption, where diffusion rates are very slow for large-molecular-weight adsorbates.

In gas-phase adsorption, refrigerated systems are common for separation or purification of rare gases, nitrogen, etc. In the case of refrigerated systems, care has to be taken that the adsorption of the carrier gas is not increased to the point where it becomes a competing adsorbate on the surface.

It was also shown that the adsorption process is always exothermic. As the adsorption front moves through the bed, a higher-temperature front due to the heat of adsorption will also move in the same direction, and some heat is imparted to the carrier gas stream. After the carrier gas leaves the MTZ, the heat exchange will reverse and the gas will impart heat to the adsorbent bed. This increase in temperature of the bed ahead of the MTZ, during adiabatic operation of the adsorber, will decrease the equilibrium adsorption capacity of the adsorbent. (This effect is one of the reasons for the MTZ coming to steady-state shape only after it has had a chance to travel its length several times.) This adiabatic temperature rise in the adsorber can be calculated by assuming that there is a thermal equilibrium between the gas and the adsorbent bed and that the temperature of the outlet gas stream is essentially the same as that of the adsorbent bed downstream of the MTZ:

$$Q = (Gc_a + mc_c + Vc_v)\,\Delta t \tag{36}$$

where G = weight of adsorbed adsorbate, kg
c_a = specific heat of adsorbate, kcal/(kg)(°C)
m = weight of adsorbent, kg
c_c = specific heat of adsorbent, kcal/(kg)(°C)
V = volume of carrier gas passed through bed, m^3
c_v = specific heat of carrier gas, kcal/m^3
Δt = temperature rise of the bed, °C
Q = heat of adsorption, kcal/kg

Adsorbate Concentration (C_0) Effects

Increasing the adsorbate concentration of the influent gas into an adsorber will increase the rate of movement of the MTZ while decreasing the length of the time for the MTZ to travel its own length, the Δt, and in case of curvature in the isotherm, the length of the MTZ itself. The effect of increasing concentration is twofold, while the breakthrough curve becomes steeper: the trailing part (nearing the inlet concentration C_0 in the outlet stream) elongates because of heat of adsorption and the overall change in Δt and M is not as large as it would be under true isothermal conditions. The time to breakthrough, as on the leading part of the breakthrough curve, shapes more significantly with the increasing concentration. Therefore, mass balance calculations which use t_b are increasingly less correct with increasing adsorbate concentration, and for precise data the use of time to $C_0/2$ or the time to equal area above and below the MTZ becomes more important.

The Effect of the Concentration of Adsorbates not to Be Removed

It is important again to stress that all gases and vapors are adsorbed to some extent on all adsorbents. Because these gases (including carrier gases) compete for the available surface area and/or pore volume, their effect will be the lowering of the adsorption capacity for the particular adsorbate which is to be removed. Under ambient conditions, very little air is being adsorbed on commercial adsorbents; however, even O_2 and N_2 adsorption competes with Kr adsorption, the relative capacities being 1.5:10, while for xenon adsorption the relative capacities are 1:100. Water vapor, carbon dioxide, and any other larger molecular weight or large-concentration low-molecular-weight adsorbate will influence the adsorption capacity for the design adsorbate.

Although activated carbon is less sensitive to moisture interference than the more polar adsorbents (molecular sieve, silica gel, aluminas, etc.), at high moisture contents or inlet relative humidities its adsorption capacity will be considerably lower than for adsorption from a dry gas stream. This interference from moisture is particularly strong during the adsorption of low-molecular-weight polar adsorbates.

TABLE 5   Effect of High Boilers* on Adsorbent-Adsorbate Behavior

Wt % Oil on adsorbent	MTZ rate of movement, cm/min	MTZ length, cm	Equilibrium capacity, $N \cdot mL/g$
0	.32	2.5	2.51
5	1.01	7.1	0.09
10	1.07	8.0	0.63
20	1.35	11.1	0.48
35	2.53	25.3	0.26

*All other conditions equal.

The presence of moisture increases the rate of movement of the other adsorbate MTZ and increases its length; any displacement of adsorbed water by organic adsorbates (displacement adsorption) is not obvious in short bed depths, such as in air-cleaning adsorption panels or gas mask cartridges, which are typically less than 1 or 2 MTZ lengths deep. Naturally both preconditioning and test humidity will influence the adsorption. Often high boilers can also accumulate on the bed, and the effect of these contaminants can permanently influence both the rate of movement and the length of the MTZ. Table 5 indicates the effect of artificially loaded compressor oil vapor on the MTZ length and rate of movement change for methyl acetylene on a propietary adsorbent. In this case very little displacement of the compressor oil would take place.

However, in some cases the presence of another adsorbate on the surface of the adsorbent is advantageous because of adsorbate-adsorbate interaction or adsorption of one adsorbate in another which is already adsorbed on the surface, or chemical reaction on the surface, where the surface acts as a catalyst. Such an example is indicated on Fig. 26, where the quantity of SO_2 converted and adsorbed as H_2SO_4 on activated carbon, based on its moisture content, is shown.

Pressure Effects

Generally, the adsorption capacity of an adsorbent increases with increasing pressure, if the partial pressure of the contaminant increases. However, at high pressures a decrease in capacity will be observed because of retrograde condensation, a decrease in the fugacity of the more easily adsorbed compound, and increased adsorption of the carrier gas.

The Removal Efficiency of the System

All breakthrough determinations are naturally subject to the sensitivity of the instrument used to detect the adsorbate in the effluent from the adsorber. If very high (99.99%) removal efficiencies (or decontamination factors) are required, the sensitivity of the detection system has to be also very high, and breakthrough has to be designated at C/C_0 < 0.0001, while if a less critical adsorbate is being removed, detection may be made at C/C_0 > 0.01. Naturally such differences in detection limits will influence the length of the MTZ, and to a lesser extent, its rate-of-movement measurement; the sizing of the adsorber bed will have to be commensurate with the obtained MTZ length.

Decomposition and Polymerization of the Adsorbate

Some solvents or compounds may decompose, react or polymerize when in contact with adsorbents. The decomposed product may be adsorbed at a lower capacity than the original substance or the decomposition product may have different corrosion, etc., properties. As an example, in an airstream, NO is converted to NO_2 when in contact with activated carbon. Polymerization on the adsorbent surface will significantly lower its adsorption capacity and render it nonregenerable by conventional methods. An example is the adsorption of acetylene on activated carbon at higher temperatures. Decomposition may also take place in regenerative systems during direct steam stripping or heating of the adsorbent bed.

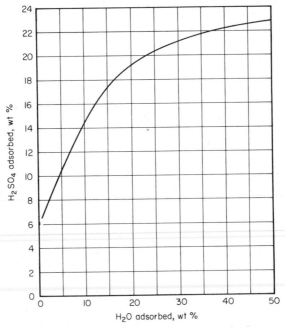

Fig. 26 Dependence of the quantity of sulfuric acid formed on activated carbon on moisture content.

Intermittent Operation of the Adsorber

Very often, the adsorbers are operated periodically, or the concentration of the contaminant varies greatly, depending on the periodic discharge of contaminants. The performance of the adsorption system is impaired under these conditions. This is caused by the variation of adsorbate concentrations with bed height. In Fig. 27, an MTZ diagram of a system is shown where under normal operation an MTZ curve (A) is obtained. The continued circulation of the carrier gas in the absence of contaminant causes the adsorbate to diffuse through the bed by the process of desorption into the carrier and readsorption until the elongation of the MTZ takes place, as represented by curve B (the dashed line).[31]

It may be concluded that short periods of intermittent operation do not affect greatly the overall capacity of an adsorption system if the bed depth equals several MTZ lengths, but long periods of intermittent operation, particularly in an undersized system, will cause a serious capacity drop.

Multicomponent Adsorbate Systems

If a carrier gas containing two or more adsorbates is introduced into an adsorbent bed, two or more MTZs form and travel along the adsorber at different rates. The adsorbate least well adsorbed travels at the highest rate, producing a zone containing all of the adsorbates at the inlet side and zones of decreasing number of adsorbates in the direction of carrier gas flow. The zone separation is a function of relative adsorption capacities of the adsorbent. However, after the above-described initial formation phase, displacement effects can take place if one of the components is very strongly adsorbed in relation to another. In such case the zone near the inlet of the adsorbent bed will become richer in the most strongly adsorbed component at the expense of the less strongly adsorbed components, and the least-well-adsorbed component may be present only in the gas phase in the macropores of the first zone.

If the relative adsorption affinity of the adsorbates is close together, coadsorption and blending of the zones will occur, resulting in combined MTZ. The type of zone formation is not only a function of the adsorbates but is dependent on the same factors discussed for single adsorbates, i.e., adsorbent type and operating conditions. In case of coadsorption, although the zones may be merged, the concentration gradient forms still exist, but for determination of the exact adsorbate MTZ, specific detectors are required.

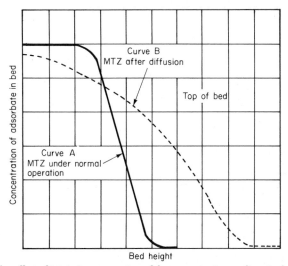

Fig. 27 The effect of intermittent operation of the concentration gradient in the adsorber.

Typical multicomponent separation is shown in Fig. 28.[32] Complete separation, where only one component is present in any zone, is impossible in continuously fed multiadsorbate systems. Such separation is possible only if the feed is pulsed in a chromatographic manner.

Regeneration Conditions

When an adsorbent reaches its capacity (whether to breakthrough or saturation), the adsorbate can be removed from the adsorbent by several techniques. The most common ones are:

1. Displacement of adsorbate
2. Desorption of adsorbate
3. Combustion of adsorbate
4. Decomposition or radioactive decay

On large adsorbers the typical regeneration practices are (1) and (2) or their combination. Combustion is generally practiced when the adsorbate is very strongly adsorbed. Care has to be taken that the adsorbent itself, if combustible, is protected by keeping the oxygen partial pressure low or using protective reducing gases in combination with the oxidizing gases. When combustion-type removal of the adsorbate is practiced, the operation is generally referred to as reactivation, while when displacement and/or desorption is practiced, the term regeneration is used.

Several considerations have to be made when establishing the conditions of regeneration for an adsorber system. Very often, the main factor is an economical one, i.e., to establish that an in-place regeneration is or is not preferred to the replacement of the entire adsorbent charge. Aside from this factor, it is important to establish that the recovery of the contaminant is worthwhile, or only the regeneration of the adsorbent is required. If recovery is the main problem, the best design can be based on a prior experimental test to establish the ratio of the sorbent fluid to the recoverable adsorbent at the different working capacities of the adsorbent. A well-designed plant, for example, will have a steam consumption in the region of 2 to 6 lb of steam per pound of recovered solvent. For activated carbons under most conditions, a direct steam regeneration is the most efficient. The steam entering the adsorbent bed not only introduces heat, but adsorption and capillary condensation of the water take place, which supply additional heat and displacement for the desorption process. The following factors must be considered when designing the stripping process:

1. The length of time required for the regeneration should be as short as possible. If continuous adsorption and recovery are required, multiple systems have to be installed.

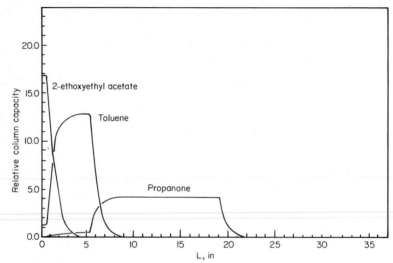

Fig. 28 Multicomponent MTZ as it occurs in continuously fed adsorber system.

2. The short regeneration time requires a higher steaming rate and thus increases the heat duty of the condenser system.

3. The steaming direction should be opposite to the direction of the adsorption to prevent the possible accumulation of polymerizable substances, and also to permit the shortest route for the desorbed contaminant.

4. To make possible fast stripping and efficient heat transfer, it is necessary to sweep out the carrier gas from the adsorber and condenser system as fast as possible.

5. A larger fraction of the heat content of the steam is used up to heat the adsorber vessel and the adsorbent, thus it is essential that the steam condenses quickly in the bed. The steam should contain only a slight superheat to allow condensation.

6. Whether it is advantageous to use a medium- or a low-retentivity carbon to allow the adsorbate to be stripped out easily. When empirical data are not available, the following heat requirements have to be taken into consideration:

    (*a*) Heat to the adsorbent and vessel.

    (*b*) Heat of adsorption and specific heat of adsorbate leaving the adsorbent.

    (*c*) Latent and specific heat of water vapor accompanying the adsorbate.

    (*d*) Heat in condensed, indirect steam.

    (*e*) Radiation and convection heat loss.

Since the adsorbent bed must be heated to regeneration temperature in a relatively short time, it is necessary that the regeneration steam rate calculation be increased by some factor which will correct for the nonsteady state heat transfer.

During the steaming period, condensation and adsorption of water will take place in the adsorbent bed, increasing the moisture content of the adsorbent. Also, a certain portion of the adsorbate will remain on the carbon. This fraction is generally referred to as "heel" or residual charge. To achieve the minimum efficiency drop for the successive adsorption cycles, it is important that the adsorbent bed be dried and/or cooled before being returned to the adsorption cycle. The desired state of dryness will depend on the physical properties of the adsorbate and on the concentration of the adsorbate in the carrier stream. For high adsorbate concentrations, it may be desirable to leave some moisture in the adsorbent so that the heat of adsorption may be used in evaporating the moisture from the adsorbent, thus preventing any undue temperature rise of the adsorbent bed.

It is also necessary to establish the materials of construction on the basis that several compounds, particularly chlorinated hydrocarbons, will undergo a partial decomposition during regeneration, forming hydrochloric acid.

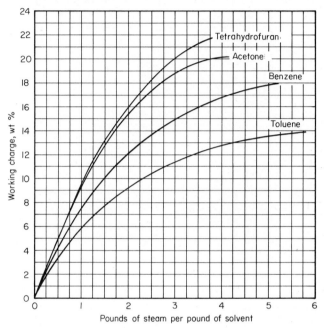

Fig. 29 Relationships between working charge and steam/solvent ratios with a medium activity, low metals, low ash content, coconut-shell activated carbon; 1200 m²/g, 6×12 mesh.

Some safety considerations have to be weighed also in designing a regeneration system, to ensure that the adsorber is not used at temperatures higher than the self-ignition point of the contaminant. Experiments[33] show that carbon does not lower the ignition temperatures of solvents; for example, solvent adsorbed on carbon ignites at the same temperature as the solvent vapor alone. However, the adsorbate may decompose on carbon, resulting in lower-ignition-temperature compounds.

Empirical determination of the regeneration efficiency can be made by collecting and analyzing the condensate in fractions during a pilot regeneration test. Such data are shown in Fig. 29 for tetrahydrofuran, acetone, benzene, and toluene. As the quantity of steam is increased, the working charge (capacity obtainable between regenerations) asymptotically approaches the breakthrough capacity. Depending on the adsorbent-adsorbate characteristics, the slope of this working charge vs. steam/solvent ratio will change. The slope decreases with increasing molecular weight, increasing adsorption strength (retentivity), and adsorbent pore diameter and with decreasing inlet adsorbate concentrations.

These non-adsorption-related considerations are used in determining the arbitrary operational steam/solvent ratio selection. If the recovered solvent is nonmiscible with water, typically higher steam solvent ratios are used because further separation is by decantation only, while if distillative separation is used, the quantity of water in the recovered solvent may be a limiting factor. If the solvent tends to decompose or oxidize in the adsorbed state, it is preferential to strip the adsorbate more efficiently to result in a lower heel.

In the United States, the general practice developed of using "high-retentivity" carbons for regenerative operation. This often results in high heels and high steam/solvent ratios, and except at low inlet concentrations does not result in economical operation.

In the case of multicomponent adsorbate systems, one or more components may be completely removed from the adsorbent during regeneration and the heel may consist of only the more strongly adsorbed components. Figure 30 shows the desorbed effluent composition vs. time during steaming of a multiadsorbate-containing adsorber.

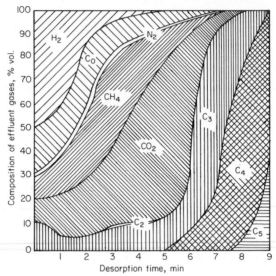

Fig. 30 Desorbed effluent composition vs. time.

Lukchis[34] analyzed the method of regeneration direction for thermal-swing cycles which includes steam-regenerated units. In thermal-swing cycles, an additional consideration involves the direction of flow during the cooling step. Purge flow during the heating step can be cocurrent or countercurrent to the adsorption flow. For each of these two cases, the cooling flow can be either cocurrent or countercurrent. A set of X-L diagrams that show qualitatively the general shape of the residual loading gradients after the cooling step for each of the four cases is presented in Fig. 31. Some qualitative observations are:

1. Countercurrent heating and cooling can provide the lowest residual loading at the effluent end of the bed in the adsorption step if the cooling fluid is free of sorbable component. Countercurrent cooling would be highly detrimental to subsequent adsorption if the cooling fluid contained significant quantities of sorbable component, because the sorbable component from the cooling purge would be adsorbed as the bed was cooling. This would increase the residual loading, which would be highest at the end of the adsorber that becomes the adsorption-step effluent.

2. Cocurrent heating and cooling always leave a relatively high residual loading at the effluent end of the bed. Both in cocurrent heating and cooling and in countercurrent heating and cooling, desorption continues during the cooling step, because the movement of concentration and thermal gradients is not reversed when cooling commences. The cooling fluid, after being heated upstream, acts as a hot purge and continues to do some desorbing as it flows through downstream sections of the bed.

3. Countercurrent heating and cocurrent cooling makes the residual low at the effluent end of the bed in the subsequent adsorption step. This is a preferred mode of operation in many drying and purification operations because adsorption is not afterwards harmed by the sorbable component in the cooling fluid. In many commerical processes, the sorbable component may be present in the regeneration fluid during start-ups and periods of emergency operation.

4. Cocurrent heating and countercurrent cooling is not a common mode of operation.

It is used occasionally when feed fluid is the closed-cycle hot purge, and purified product is the cool purge—for example, to minimize dilution of the desorbed component.

It is obvious from the previous discussion that the adsorption capacity of a regenerative system will be less than that of an unused bed. Even under proper regeneration condi-

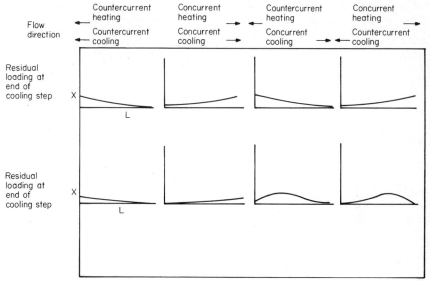

Fig. 31 Regeneration condition effect on residual adsorbate loading.

tions, the adsorbate will be present in the inlet side of the bed, and any residual moisture or temperature gradient in the adsorber will influence the rate of movement of the MTZ when the adsorption cycle begins. The residual charge (heel) in some systems may be as high as the working charge of the system; in such cases, the MTZ is already established (although in elongated form) in the adsorber, and the first step after reintroduction of the adsorbate is the achievement of a steady-state MTZ front and thereafter the normal movement of the MTZ under the effects described in the previous section.

The cycling of an adsorber between adsorption and regeneration cycles should be controlled by instrumentation determining breakthrough rather than by timers alone. When timers are used the adsorber is normally regenerated before breakthrough occurs, and because the quantity of steam (or other regenerate) required to heat up and sweep the adsorber is constant, the steam/solvent ratio will be higher than for the breakthrough operation case. Without exception the steam/solvent ratio will be inversely proportional to the quantity of the adsorbate on the adsorbent.

If the steam or other regenerant gas is introduced into the adsorber at a very low velocity, the pressure drop of the adsorbent bed may not be sufficiently high to assure uniform steam distibution at the regenerant inlet side of the adsorber. Such operation can leave pockets of high concentration of the adsorbate on the bed, enhancing adsorbate decomposition or polymerization. The use of steam distributors is recommended for low-velocity steaming of adsorbers.

Limiting temperature conditions (temperature at which structural damage or pore deformation would take place) for regenerate are:

Activated carbons	Approximately 900°C (in nonoxidizing atmoshere)
Molecular sieves	475°C
Silica gels	310°C
Alumina activated	175 to 300°C
Alumina gel	670°C

Even if only a low quantity of nonremovable high boilers are present in the inlet stream, or if polymerization takes place only to a very small extent in the adsorbers, the working charge will decrease to a point where very high steam/solvent ratios are required to maintain operation at acceptable working charges. In such cases it is more economical to remove the adsorbent from the vessels and burn off the "plugging" component. This reactivation of the adsorbent costs approximately one-third to one-half the new carbon price, and the yield (including screening out of fines generated) is typically 80 to 90% by weight. The makeup adsorbent should be fresh material and not other reactivated product. This type of operation will result in eventual complete replacement of the adsorbent, while discarding the structurally weakened particles. The requirement for such reactivation and the quality of the reactivated product should be monitored by, at least, performance of MTZ length and rate-of-movement studies using the actual adsorbate. Tests using "simulant" adsorbates such as CCl_4 are only rarely meaningful for determination of residual capacity of the adsorbent for other adsorbates.

Vacuum regeneration of the adsorbates is also being practiced to a lesser extent to avoid the corrosion and thermal reactions on the adsorbent during steam or other high-temperatur stripping. However, with conventional adsorbents the vacuum regeneration still requires heat, and heat transfer in adsorbent beds under vacuum is greatly limited. The use of modified adsorbents where the porosity is tailor-made for vacuum regeneration offers better economics for the process.

SYSTEM DESIGN DESCRIPTIONS

Process Evaluation

The first step is to determine if an adsorbent-adsorbate pair can be found which results in Type I, II, or IV adsorption isotherm shape. If such a relationship exists, then adsorption as a unit operation will probably be economical. If the adsorbent-adsorbate relationship selection results only in Type III or V isotherm shapes, then adsorption will very rarely be economical.

If the adsorbate-containing carrier gas is at elevated temperature or the adsorbate is weakly adsorbed and cooling of the gas stream will be required, the determination of an adsorption isobar (adsorption capacity vs. temperature determination) should be made to establish the conditions of cooling or refrigeration required for economically optimum adsorption conditions. Once an operating adsorption temperature is selected, further evaluation is performed at this temperature. If thermal-swing regeneration is considered, at this time the range of temperatures used for the adsorption-desorption range can be established by extending the range of temperature in which the adsorption isobar is determined.

After the potential of using adsorption is established by determining the isothermal relationship of C_0-0 adsorbate concentration range, the value of q can be determined for Eq. (31) and a close approximation of the MTZ rate of movement vs. velocity can be obtained. In case of multicomponent adsorbates, the actual measurement of u (rate of MTZ movement) and M (MTZ length) can be established by measuring u and Δt at least for the least-well-adsorbed component (which would break through first), but preferably for all components. If significant coadsorption takes place, the measured u and M values may represent multiple adsorbates.

On the basis of these results, the cross-sectional area and the superficial gas velocity can be established and the adsorption capacity to breakthrough can be calculated. The adsorption cycle life can also be established if no in-place regeneration is used, or—if very low heel is expected—an approximation of the time between regeneration can be estimated.

Once the adsorbent, operating temperature, superficial gas velocity, and approximate breakthrough times are selected, a column experiment can be performed using the actual adsorbate composition. This experiment is confirming for sizing; however, it is conducted only to breakthrough. (Typical column size is 5.0 cm ID and 4 to 5 MTZ length.) At this time the regenerant steam is introduced, the adsorber effluent is conducted through a condenser, and the condensate is collected in fractions.

The fractions are analyzed for water and solvent content until the predetermined steam/solvent ratio is reached. At that time the adsorber is cooled and/or dried and put back on stream. Several cycles are conducted in this manner if the adsorbed contaminant is not suspected of possible decomposition or polymerization on the adsorbent. If polymerization is suspected or expected, several hundred cycles have to be run and for several cycles the adsorption experiment is conducted to $C = C_0$ in the outlet to determine MTZ rate of movement and length changes.

The collected condensate is also analyzed for potential decomposition (acid, etc.) products. On the basis of these experiments, the working charge–steam/solvent ratio curves are prepared, and if necessary the bed depths are adjusted to give the desired adsorption cycle time, and the quantity of steam required for the regeneration and the quantity of heel are determined.

While these regeneration experiments are being performed, it is advantageous to insert planned-material-of-construction coupons both above and within the adsorbent bed and monitor their corrosion rate. Without exception, corrosion rates will be higher in contact with the adsorbent because oxidation and other decomposition can be catalyzed when the adsorbate is on the surface. Depending on the adsorbent and the material of construction, electrolytic corrosion also can occur in adsorber beds. The utilization of the adsorption experiments for corrosion measurements also can shorten the time required for completion of the design experiments.

If acids are generated during the adsorption-desorption process (chlorinated hydrocarbons, ketones, acetates), then a neutralization system also has to be included in the process as soon after condensation as possible to prevent corrosion of downstream components.

The carrier gas stream should also be sampled for both liquid and particulate aerosol droplets. The obtained particle size spectra should be converted to contaminant-content particle size distribution on a mass basis. On the basis of the obtained data, the requirement for any aerosol cleanup (filtration, wet scrubbing, etc.) can be established. If filtration is used it should be performed ahead of any heat exchangers to prevent fouling of the heat-exchanger surfaces.

Once the process design experiments are completed, preliminary process and instrumentation diagrams and operational logic diagrams are prepared. Typical operational parameters are as follows:

	Range	*Design*
Superficial gas velocity	20 to 50 cm/s	40 cm/s
	(40 to 100 ft/min)	(80 ft/min)
Adsorbent bed depth	3 to 10 MTZ	5 MTZ
Adsorption time	0.5 to 8 h	4 h
Temperature	−200 to 50°C	
Inlet concentration		
Adsorption base	100 to 5000 vppm	
LEL base	40%	
Adsorbent particle size	0.5 to 10 mm	4 to 8 mm
Working charge	5 to 20% wt	10%
Steam solvent ratio	2:1 to 8:1	4:1
Adsorbent void volume	38 to 50%	45%
Steam regeneration temperature	105 to 110°C	
Inert gas regenerant termperature	100 to 300°C	
Regeneration time	½ adsorption time	
Number of adsorbers	1 to 6	2 to 3

A typical flow diagram for a dual-bed adsorber is shown in Fig. 32.

Equipment Design

The design of the major components of the hardware is performed in the following manner.

Adsorber Vessels and Internals The adsorber vessels for small carbon-based solvent-recovery systems (< 2.500 ft³/min) are typically vertical vessels. For larger airflows, the vessel is in a horizontal position, the adsorbent being held on supports horizontally in the adsorber. However, particularly in hydrocarbon processing, larger vertical beds with 8 to 10 ft ID are also used. High-pressure adsorbers are also typically vertical adsorbers. Annular beds have been designed in both vertical fixed-bed and horizontal rotating-bed systems, but so far have resulted in a larger percentage of the carbon or other adsorbent breaking down. Vertical annular-bed units also show more severe corrosion due to the increased stress acting on the carbon which settles toward the lower end of the adsorber while the bed expands and contracts in successive desorption-adsorption cycles.

Figure 33 shows a conventional vertical-bed system. Figure 34 shows a conventional system assembled from dual polymer-lined flanged vessels. Figure 35 shows the rotary single-bed system and Fig. 36 shows a vertical stacked-bed adsorber.

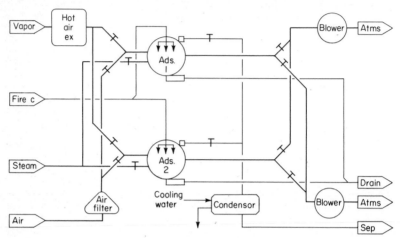

Fig. 32 Typical flow diagram for two-adsorber system.

The most critical part of the adsorber is the support holding the adsorbent bed. The design has to result in both weight-supporting rigidity and sufficient open area to prevent the development of large pressure drops. The direct adsorbent bed support can consist of perforated metal, expanded metal, or perforated plastics. These direct bed-holding screens are supported on "subway grating." Both of the supports are held on ledges within the adsorber bed.

The European practice of using a gravel bed both for bed support and as a heat recovery device, while useful for the recovery of stable compounds such as toluene, xylene, and hexane, will have disastrous effects in the recovery of easily decomposing or polymerizing compounds such as ketones, aldehydes, and some esters. The heat retained in the gravel bed is released at the time when the adsorbent bed is most vulnerable to overtemperature-caused fires.

Regardless of the type of support used, the design has to include the potential weight increase in the adsorbent during regeneration and when the vessel is flooded for fire control. Particularly horizontal vessels have been known to become slightly oval, and the internals tend to drop to the bottom of the adsorber after contraction and expansion from the adsorption-desorption cycle, when the ledge (on which the bed support sets) is inadequately designed. Even if the consequences are not as serious, loss of adsorbent through the support grid will result in damaged valve seats, particularly if Teflon is used.

The direct-adsorbent-supporting material should always be one grade more corrosion resistant than the vessel material. The use of titanium was found to be acceptable for all commonly recovered solvents using activated-carbon adsorbents.

Adsorber vessels should be equipped with manholes or inspection ports both above and below the adsorbent bed for inspecting any irregularity or to permit repairs on the system.

Adsorbers should be equipped with isolatable ΔP-indicating devices to monitor pressure drop across the adsorbent bed. Lowering of pressure drop will indicate creation of voids or movement of the adsorbent, while increasing pressure drop will indicate contamination of the adsorbent with particulate matter or degradation of the adsorbent.

In some cases where the entry of high boilers into the adsorber cannot be prevented, it is advantageous to divide the adsorbent bed into two sections. The inlet section should be sized to completely remove the high boilers (10 to 20 cm deep) and prevent migration of the high boilers into the main adsorber section. This manner of operation permits periodic removal and reactivation of only this narrow section instead of the entire bed. This

Fig. 33 Vertical-bed adsorber system. *(Barnebey-Cheney Co.)*

Fig. 34 Horizontal adsorber. Assembled from dual plastic-coated vessels. *(Westates Carbon Inc.)*

Fig. 35 Rotating horizontal annular bed adsorber. *(Sutcliffe-Speakman Ltd.)*

Fig. 36 Vertical multiple stacked bed adsorber. *(Sutcliffe-Speakman Ltd.)*

"sacrificial" bed should not be placed into a nonregenerable section of system if the adsorbate can decompose easily in the adsorbed state (ketones, aldehydes, acetates, etc.).

When adsorbent media sampling is planned, ports should be placed to permit removal of adsorbent at different positions along the direction of the carrier gas flow. The rate of "poisoning" or high-boiler loading of the adsorbent media can be then monitored in a quantitative manner.

Adsorbent loading and unloading can be well accomplished by vacuum techniques, rather then pressure loading. The use of vacuum will tend to dedust the adsorbent material, which is desirable whether loading or unloading is used. Carbon-type adsorbents can also be safely loaded in water-slurry form. This type of loading will result in sufficient retention of water in the adsorbent to prevent high-temperature spots developing as a result of the heat of adsorption of the first cycle.

If long adsorbers are used, the introduction of the adsorbate-containing gas should be through distributors to assure uniform gas flow distribution across the adsorber. In case of doubt, a pitot tube can be inserted and moved to various positions in the adsorber bed, close to the outlet face of the bed, to measure uniformity of gas flow.

Valves and Valve Seats Most inlet and outlet valves used in adsorbers are butterfly valves. If polymer seating or gasketing components are used, the compatibility of the polymers should be evaluated for the adsorbate-water mixture, both at operating and regenerating temperature. A very common ailment of adsorber systems is caused by stuck or leaking valves, resulting in high exit concentrations, excessive steam losses, and elongated MTZ, particularly if regenerant in-leakage occurs during adsorption cycle. The leakage of the adsorbate-containing carrier-gas boundary valves during desorption or cooling can result in fires in the adsorbers.

Drains Adsorber vessel low points should be drained to prevent condensate-accumulation-caused corrosion within the adsorbers.

Steam Quality The heat transfer economics for the carbon-bed systems is best when low-pressure saturated steam is used. The effect of steam conditioning chemicals on the poisoning of the adsorbent bed by deposition or cross-reaction should be evaluated. If boiler capacity is inadequate, the steam introduction valve to the adsorber should be a slow-opening type to prevent "blowing down" of the boiler into the adsorber. The boiler water generally contains chlorides and such blowdown into the adsorber will have disastrous corrosion effects on the support screens of the adsorber, particularly if 304 or 316 SS construction is used.

If recovered condensate is used for boiler feed, it should not contain inorganic or organic compounds which interfere with the adsorption.

Neutralization Systems Neutralization of acidic recovered condensate is frequently required (halogenated hydrocarbon, ketone, aldehyde, acetate, etc., recovery) immediately downstream from the condenser. Such systems should be designed to neutralize the acidity not only generated to the then-existing level, but should assure neutralization of acidic compounds which may be generated subsequent to that point in time. Particularly ketone decomposition products are slow in converting to acids, and pH adjustment has to be made to excess alkalinity to neutralize the slowly released acids.

Blowers Often carrier-gas blowers are designed to overcome the pressure drop of the new system only. Typically there will be an increase in pressure drop of a system with operating life. It is prudent to oversize blowers and use variable inlet vanes to control carrier-gas flow at the desired level. The use of dilution air should be practiced only if the adjustment and control of the LEL is required. It has to be remembered that strictly from a adsorption standpoint, the higher the adsorbate concentration, the higher is the adsorption capacity and the working charge at constant steam/solvent ratios.

Controls The use of adsorbate, temperature, humidity, adsorbate decomposition level, etc., sensors should be carefully evaluated for location and failure mode to prevent off-design conditions developing into hazards. The use of microprocessors or minicomputers is highly economical and permits the evaluation of the readiness of the system for the next operating step before the switchover is made, therefore preventing the pyramiding of occurrences and resulting gross system failures.

Any safety instruments should have backup power available.

Utility Requirements Typical utility requirements are shown on Table 6 for various adsorbate and total flow systems.[35]

SAFETY CONSIDERATIONS

General

Adsorption systems are often plagued by corrosion, fire, and adsorbent poisoning problems. All of these failures can be attributed to unsatisfactory design, caused by either insufficient information supplied to the designer or lack of understanding of adsorption principles on the part of the hardware supplier. Many systems are being built identically, regardless of the adsorbate type and property influences and requirements on the system. The blame should be equally shared by the buyer and the supplier for not buying and

TABLE 6   Typical Utility Requirements of Solvent Recovery Systems

ACFM	SLA Temperature, °F	Solvent	Solvent lb/h	Electrical, kW	Steam, lb/h	Cooling water, gal/min
10,000	95	Acetone	950	75	2,200	420
10,000	130	Toluene and methanol	280 60	80	1,150	400
60,000	300	Toluene and ethanol	2000 200	210	6,000	1300
100,000	130	Toluene and mixed hydrocarbons	1200	510	5,200	1050
85,000	105	Toluene	1600	400	4,800	500
120,000	115	Toluene and xylene	3000	350	10,750	1160
20,000	180	Methanol	680	100	2,580	270
24,000	100	Acetone	1600	65	4,710	520
30,000	200	Toluene and miscellaneous	1500	130	3,800	825
75,000	100	Xylene	2600	225	6,360	700
60,000	150	Toluene	2600	245	5,000	1100

supplying a designed system, but only equipment. The cost of modifying, repairing, and maintaining even well-built but badly designed systems is much more than the cost of performing the proper input source sampling, design experiments, and pilot plant work, which would result in a smoothly operating system.

A particular example of the problem is activated carbon-base ketone recovery systems, where past experience indicates corrosion failure within 24 h for some systems and disastrous fires upon start-up for others. The systems which failed were often built using the same adsorbent and the same configuration as those built to recover toluene or some similar material which is a noncorrosive, nondecomposing solvent.

It is not claimed that the recommendations presented here are the preventive solutions to all safety problems or that if these steps are followed, operational safety can be achieved for all types of adsorbate-adsorbent pairs, but there will be less chance of failure and/or accidents if these steps are followed.

After a preliminary process and instrumentation diagram (P&ID) is prepared, a detailed failure analysis of components should be made to assure that a single equipment failure will not result in major damage to the system. On the basis of this analysis, the P&ID should be revised to assure safe operation.

A secondary review should be made after the "as built" drawings are made, but before start-up. Often modifications made during erection of the system are not carefully evaluated as to whether they result in unsafe consequences, and a second safety review is a must to avoid compromise of the safe operation of the system.

Adsorbent Selection

Adsorbents containing impurities which catalyze decomposition or polymerization of the adsorbate should not be used. As an example, metallic salts accelerate the decomposition of halogenated hydrocarbons and ketones. The resulting compounds are more corrosive than the undecomposed adsorbate, and the decomposition reactions can be exothermic, resulting in spot overtemperatures and fires. Some of the adsorbates, even in the absence of metallic catalysts, decompose much more easily in the adsorbed state because of their lower degrees of freedom. Not only is the selection of adsorbents not containing these catalysts required, but care has to be taken that the adsorbents do not become "impregnated" by these catalysts as a result of vessel or grating corrosion during operation. Many carbon types contain oxygen-containing surface compounds which protect the adsorbent from further oxidation. In a fire, these compounds are removed and the resulting fresh carbon surface is subject to slow oxidation from the humid air. If their oxidation is too fast and the heat of the exothermic oxidation is not carried away, the result is another fire.

The same components which cause decomposition hazards in halogenated hydrocarbon and ketone adsorption can be beneficial for other adsorbates. Carbons containing metallic oxides (particularly iron) will decompose and prevent the formation of peroxides formed in ethers including tetrahydrofuran.

Other adsorbents, although reported by the manufacturer to contain less than 1.0 wt % ash, do contain 2 to 4% sulfur. Even in slightly oxidizing carrier-gas streams, part of the sulfur will bleed off as sulfur dioxide or in humid streams as sulfur trioxide, which can cause both corrosion and adsorbate cross-reactions on the surface.[20,36] These carbons should not be used with nickel alloys, particularly at high temperatures.

Thus adsorbents selected have to be evaluated for non-adsorption-related factors also, for safe operation of the system.

OPERATIONAL PARAMETERS

If the generation of high temperatures is to be prevented both during adsorption and regeneration steps (desorption/cooling), certain precautions can be taken. These are aimed primarily at preventing rather than extinguishing fires in the adsorbers.

The carrier gas, containing easily oxidizable adsorbates, should enter the adsorber bed at as low a temperature as possible. As an example, ketones decompose slightly but detectably even at ambient temperature; however, in air, above 80°C, the decomposition rate becomes exponential. The order of decreasing decomposition rate is cyclohexanone, methyl ethyl ketone, methyl isobutyl ketone, acetone; no specific chemical or physical property has been found to correlate with decomposition rate in the adsorbed state. Thus some decomposition cannot be avoided even in normal operation. However, steps can be taken to prevent the decomposition from reaching disastrous levels.

Activated-carbon beds, where the fire hazard is the most serious, ignite in the high-velocity airstreams at 340 to 430°C, depending on the type of impurities in the carbon. However, the ignition temperature is both velocity and bed-depth dependent.[37-39] The carrier-gas velocity always has to be sufficiently high to remove the heat produced by an oxidation step which cannot be prevented.

Although the presence of humidity in the carrier gas is slightly detrimental for the adsorption process itself, if water is adsorbed to some extent on the carbon bed during adsorption, the introduction of steam will not result in as high a temperature rise above the steam temperature as would be the case for dry carbon beds. As an example, introducing 135°C steam into a dry carbon bed can raise carbon-bed temperature, in a zone where the water adsorption takes place, above 190°C.[42] In the same manner, at the end of the desorption cycle, the cooling air should not dry out the carbon bed and it should be introduced at a sufficiently high velocity to remove heat from the carbon bed to prevent the acceleration of residual solvent decomposition to a fire.

The water left in the adsorbent bed, which will be displaced by the adsorbing solvent during the adsorption cycle, will counteract the heat of adsorption of the solvent. Thus the decision whether cooling or drying or their combination is used should be made not only from the adsorption but also from the safety standpoint. Often a compromise has to be made at the expense of adsorption efficiency to assure safe operation.

If quantitative evaluation is made, the heat of adsorption (or desorption) of water for the particular adsorbent can be calculated from adsorption isosteres.

Solvent decomposition by oxidation results in both CO and CO_2 generation long before either the ketone-decomposition product or the ketone and carbon actually ignite; thus by proper monitoring of CO and CO_2 (or their differential amount if both exist in the inlet stream) the oxidation process can be detected and fire prevented.

Monitoring the adsorbent bed itself by placing temperature-sensing devices in the adsorbent bed is an unsafe practice. The thermal conductivity of the adsorbents is low and any hot spot developing downstream of the temperature sensor will not register. Temperature sensors should be placed in the inlet and outlet of the adsorber; between the vessel and the valves is a preferred location.

If excessive oxidation is detected, the adsorber has to be isolated to prevent introduction of additional oxygen into the system and cooled down to stop the oxidation. It is important to realize that the oxidation processes are autoinerting and strongly temperature dependent. After any such occurrence, the next step should be an extensive regeneration and cooling before the system is put back on stream. If adsorbers are shut down for extended periods, double or triple regeneration step should be used, and the bed should be cooled and isolated. The re-start-up step should also be regeneration.

Mechanical Parameters

All effort has to be made to detect off-design conditions and prevent their becoming safety hazards by not only selecting proper process parameters but also by proper selection of mechanical components. It also has to be assured that particularly safety-related equipment will also have emergency backup power available, and that adverse weather conditions will not prevent operation of safety-related equipment. (For example, freeze-up of steam-traced controls when boilers are shut down, use of humid air for instrument air, etc.)

Particular safety-related equipment parameters are:

Valve seating should prevent air and/or steam leaks into or out of the system.

Loss of electric or pneumatic power should result in isolated adsorbers.

Humidity and CO_2 sensors should be protected from malfunction due to adsorbate-caused fouling or lack of proper ambient-temperature protection (i.e., protection from freeze-up).

Fast cooling of the adsorber, which indicates adsorbate decomposition, should be accomplished before a fire starts. Water-spray lines for this purpose should be protected from freeze-up.

Safety pressure-relief valves should be installed in each adsorber.

Flame arresters should be used at each interface between ambient air and solvent-containing lines.

If flooding is considered, the possibility of non-water-miscible solvent floating on top of the water layer—and reaching the fire before the water—should be considered.

ACKNOWLEDGMENTS

The contribution of material by the following organizations is gratefully acknowledged: Union Carbide Corporation, Aluminum Company of America, W. R. Grace & Co., C&I/Girdler Inc., Hoyt Manufacturing Corporation, Vic Manufacturing Company, Vulcan-Cincinnati, Inc., Lurgi Apparate-Technik GmbH., Calgon Corporation, Westates Carbon, Inc., Ray Consulting Company Inc., Bamag Verfahrenstechniks GmbH., Barnebey-Cheney Company, Sutcliffe-Speakman Ltd., and Oxy-Catalyst Div. of Research Cottrell.

REFERENCES

1. de Boer, J. H., "The Dynamical Character of Adsorption," Clarendon, Oxford, 1953.
2. Holst, G., and Clausing, P., *Physica*, **6**, 48 (1926).
3. Clausing, P., *Ann Phys.*, **7**, 489 (1930).
4. Frenkel, J. Z., *Physica*, **26**, 117 (1924).
5. de Boer, J. H., *Z. Elektrochem.*, **44**, 488 (1938).
6. Brunauer, S., "The Adsorption of Gases and Vapors," Oxford, London, 1953.
7. Adamson, A. W., "Physical Chemistry of Surfaces," Interscience, New York, 1960.
8. Young, D. H., and Crowell, A. D., "Physical Adsorption of Gases," Butterworths, London, 1962.
9. Polanyi, M., *Verh. Dtsch. Phys. Ges.*, **16**, 1012 (1914).
10. Polanyi, M. *Z. Elektrochem.*, **26**, 371 (1920).
11. Berenyi, L., *Z. Phys. Chem.*, **94**, 628 (1920).
12. Berenyi, L., *Z. Phys. Chem.*, **105**, 55 (1923).
13. Dubinin, M. M., and Zaverina, E. D., *Acta. Physiochim. URSS*, **4**, 647 (1936).
14. Dubinin, M. M., and Timofeev, D. P., *C. R. (Dokl.) Acad. Sci. URSS*, **54**, 701 (1946).
15. Dubinin, M. M., and Timofeev, D. P., *C. R. (Dokl.) Acad. Sci. URSS*, **55**, 137 (1947).
16. Dubinin, M. M., and Timofeev, D. P., *Zh. Fiz.-Khim.* **22**, 133 (1948).

17. Lewis, W. K., et al., *Ind. Eng. Chem.*, **42**, 1362 (1950).
18. Langmuir, I., *J. Am. Chem. Soc.*, **40**, 1361 (1918).
19. Burnauer, S., et al., *J. Am. Chem. Soc.*, **60**, 309 (1938).
20. Deuel, C. L., NASA report 1666F, Analytical Research Lab., 1971.
21. Alcoa technical information.
22. Union Carbide technical information.
23. W. R. Grace technical information.
24. Shilov, N. A., *Zh. Russ. Khim. Ova.*, **51**, 1107 (1929).
25. Mecklenburg, W., *Zh., Elektrochem.*, **31**, 488 (1925).
26. Michaels, A. S., *Ind. Eng. Chem.*, **44**, 1922 (1952).
27. Illes, V., MAFKI Report 233, 1961.
28. Klotz, I. M., *Chem. Rev.*, 241 (1946).
29. Hougen, O. A., and Watson, K. M., *Ind. Eng. Chem.*, **35**, 529 (1943).
30. Kovach, J. L., Heat Transfer in Packed Carbon Beds, NUCON 017, August 18, 1972.
31. White, P. A. F., and Smith, S. E., "Inert Atmospheres," Butterworths, London, 1962.
32. Juhola, A. J., USEPA report EPA-R2-73-202, 1973.
33. Woods, F. J., and Johnson, J. E., U.S. Naval Research Laboratory report 6090, 1964.
34. Lukchis, G. M., *Chem. Eng.*, July 9, 1973.
35. Vulcan-Cincinnati technical information.
36. Kartyan, D. K., USAEC report TR3750, 1957.
37. Kovach, J. L., and Green, J., *Nucl. Saf.*, **8**, 41 (1966).
38. Brotz, W., *Chem.-Ing.-Tech.*, **23**, 408 (1951).
39. Bratzler, K., *Chem.-Ing.-Tech.*, **28**, 569 (1956).

Section **3.2**

Design of Gas Absorption Towers

F. A. ZENZ *Vice-President Engineering, the Ducon Company, Mineola, N.Y.*

INTRODUCTION

There exist today literally hundreds of varieties of hardware available for promoting contact between a gas and liquid stream. The business of thereby removing components from a gas stream by dissolving or absorbing them in a liquid from which they can afterwards be more easily recovered and concentrated is currently mushrooming under the impact of pollution restrictions on certain gaseous emissions. Academically, the process design of certain types of such equipment has attained a level of sophistication and idealization that surpasses the capabilities of the most careful mechanical designs and tries the patience of the majority of the vast number of people who suddenly find themselves charged with the responsibility for detailed specifications and installations. With these considerations in mind, the author intends the following pages to provide ready reference to most, if not all, the calculational tools, design guides, and details of mechanical equipment necessary for choosing and achieving an operable installed gas absorber.

Accomplishment of absorption depends on intimate contact of gas and liquid under conditions wherein the forces promoting solution will always be the most favored. Attaining such contact involves maximizing the exposure of surfaces of gas and/or liquid to each other; conceptually, this can take the form of breaking the liquid up into tiny drops dispersed through a volume of gas (as should occur in spray towers), or breaking the gas up into little bubbles passing through a volume of liquid (as should occur in trayed towers), or breaking the liquid up into a multiplicity of slow-flowing films which form and re-form through a volume of gas (as should occur in packed towers). These concepts are pictured in their idealization in Fig. 1. In practice, none conforms entirely or exclusively to its conceptual classification and each has certain advantages and disadvantages. Where a choice is not critical on process grounds, speed of installation, versatility, or simply investment cost may dominate.

PACKED TOWERS

Nomenclature

a	Square feet of packing surface in a cubic foot of packed volume
F	Fraction void volume in packing
μ_L	Liquid viscosity, cP
V	Flooding gas rate, cfm per square foot of tower cross section
Q	Flooding liquid rate, gpm per square foot of tower cross section
ρ_G	Gas density, lb/ft³
ρ_L	Liquid density, lb/ft³
ΔP_F	Pressure drop per unit height of packing at incipient flooding

a/F^3	Packing factor
K_Ga	Overall mass-transfer coefficient, $lb \cdot mol/(h)(ft^3)(atm)$
K_ga	Gas film mass-transfer coefficient
K_la	Liquid film mass-transfer coefficient
U_o	Superficial gas velocity up tower, ft/s
g	32.2
L/G	Pounds of liquid downflow per pound of gas upflow
D_T	Tower diameter, ft
G_F	Gas rate at incipient flooding
ΔP_o	Pressure drop per unit height of packing at design operating point
J_D	Mass transfer number defined in Fig. 10 (dimensionless)
Re	Reynolds number defined in Fig. 10 (dimensionless)
S	Square feet of surface of one piece of packing
G	Specific gas rate
S_c	Schmidt number, $\mu_G/\rho_G D_G$ (dimensionless)
D_G	Diffusivity of solute gas through bulk gas
P	Total pressure, atm
M	Molecular weight of bulk gas stream
HETP	Height equivalent to a theoretical plate

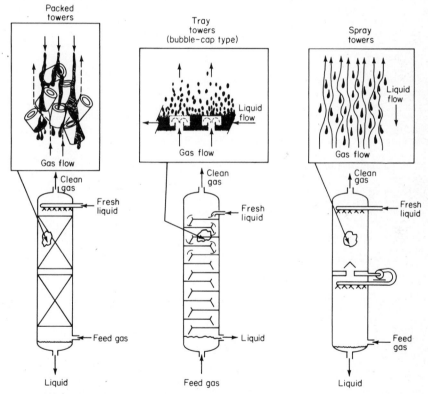

Fig. 1 Operating principles of packed, tray, and spray towers. (*Reprinted from* "Kirk-Othmer Encyclopedia of Chemical Technology," *2d ed., Wiley-Interscience.*)

Advantages

In number of installations, if not in capacity, packed towers appear to predominate. This popularity stems from a number of advantageous features:

Minimum of Fabrication   As opposed to a multiplicity of trays and their supports, a packed tower relies on one packing support and one liquid distributor perhaps only every 10 ft along its height.

Versatility   The packing material can be changed simply by dumping it out of the column and replacing it with styles giving better efficiency, lower pressure drop, or higher capacity; the depth of packing can also relatively easily be changed if efficiency turns out to be less than anticipated or if feed and/or product specifications change.

Handling Corrosive Fluids   Ceramic packing is common and in many instances preferable to metal or plastic because of its corrosion resistance. When packing does deteriorate, it is easily replaced with a minimum of shutdown time.

Low Pressure Drop   Unless a tower is operated at very high liquid rates, where the liquid becomes the continuous phase as its films thicken and merge, the pressure drop per lineal foot of packed height is usually less than in a trayed tower.

Range of Operability   Though efficiency varies with the gas and liquid feed rates, the range of operability, without fluid bypassing gas contact, is broader than in a tray tower, unless the tray tower is conservatively designed with high-pressure-drop trays.

Low Investment   Where plastic packings are satisfactory or where towers are less than about 3 to 4 ft in diameter, investment comparisons usually favor the packed tower.

Capacity Limits (Tower Diameter)

A first step in sizing a packed tower involves the prediction of how large a shell diameter is necessary to pass the desired gas and liquid rates such that the gas will not simply blow the liquid back out, or in other words, "flood." The capacity will depend on the resistance or tortuosity imposed by the packing and the liquid film it bears. Liquid viscosity; liquid density; gas density; and packing size, type, surface, and free space become the determining variables. The pioneer work in this field was carried out in the late 1930s and early 1940s. The correlation, shown in Fig. 2a, derived theoretically by Sherwood et al.,[1] incorporated all the conceivable variables and in many instances still prevails in its original form.

The scatter of the data in Fig. 2a led Lobo et al.[2] to investigate the variables and the experimental procedures more closely. They quickly discovered that the values of a and F, representing the packing material, varied depending on how the packing was placed in the tower. Just as a bed of sand can be loosely poured into a vessel or shaken down more compactly, so packing can be loose or more densely packed, thereby creating greater or lesser restriction to the flow of fluids. Lobo et al., therefore, suggested the combination of a and F into a single variable termed the "packing factor" and recommended values of this term based on the method of packing.

In extending the study of flow through packings to incorporate pressure drop, it was noted that many of the data plotted by Sherwood et al. and by Lobo et al. corresponded to rates which fell short of actual flooding and were reported by various investigators as "flooding" simply because they were approaching this condition and liquid accumulation within the packing became visually more pronounced. Eliminating these so-called loading points resulted in the more well defined curve of Fig. 2b based on truer flooding rates and in-tower packing factors.[3]

The correlation of Fig. 2b persisted till the early 1960s when it was noted that a simple rearrangement would result in a far more useful correlation,[4] easier to comprehend and to extrapolate. Plotting the square root of the ordinate vs. the product of abscissa and square root of ordinate, in volumetric units, transformed Fig. 2b into that shown in Fig. 2c, which with incorporation of the term $\sqrt{a/F^3}\,\mu_L^{0.2}$, representing the restriction imposed to flow by the tower internals, as a parameter, is represented in Fig. 3. Figure 3 is the Sherwood et al. correlation replotted in its simplest form. It has the virtues of least error in extrapolation, of requiring no trial-and-error procedure to calculate the gas rate which would flood a given tower, and of establishing the flooding curve for a new packing from only a test of the maximum liquid rate able to pass through without building up a liquid head atop the bed. It can be represented by an equation of the form:

$$\left(\frac{28.6V}{\sqrt{\rho_G/\rho_L}}\sqrt{\frac{2}{F^3}\,\mu_L^{0.2}}\right)^{1/3} + \left(\frac{Q}{7.481}\sqrt{\frac{a}{F^3}\,\mu_L^{0.2}}\right)^{1/2} = 18.91$$

which for a given gas-liquid system[5] simplifies to:

$$V^{1/3} + Q^{1/2} = \text{constant}$$

permitting calculation of all the combinations of flooding rates without resort to a chart once a single such condition has been established.

Pressure Drop

The significance of Fig. 3 can best be realized in relation to what might be described as an operating diagram—a plot of gas rate vs. pressure drop with liquid rate as a parameter. Such a diagram, typical of any specific packing and gas-liquid system, is represented schematically in Fig. 4a. Note that at constant gas rate, an increase in liquid throughput, which takes up more room in the packing (increased holdup) and hence affords less room for the gas (representing greater restriction), is accompanied by an increase in pressure drop until the flooding liquid rate is reached, at which point any slight excess which cannot pass through remains atop the packing, building up a deeper and deeper head or pressure drop, hypothetically reaching an infinite value. Similarly, at constant liquid downflow, increasing gas flow is again accompanied by increasing pressure drop until the flooding rate is reached, whereupon the slightest increase will cause a decline in permissible liquid throughput causing the remainder again to accumulate atop the packing so that pressure drop again increases infinitely.

The shapes or slopes of the curves in Fig. 4a appear identical for all packings and systems. Early investigators treated such curves as two distinct lines, interpreting the point of change in slope, illustrated in Fig. 4b, as a so-called loading point. Subsequent detailed studies with all available data have shown no justification[6] for such treatment, and instead, by analogy at constant liquid rate to critical (flooding) and subcritical flow of gases through orifices, the studies suggest the generalized correlation[3] given in Fig. 5. This analogy visualizes an irrigated packing as a restriction, or orifice, and as such it must exhibit a characteristic equivalent area, represented by the term $\sqrt{(a/F^3)}\mu_L^{0.2}$, and a characteristic critical velocity, represented by the curves of Fig. 3, at which the critical pressure ratio ΔP_F limits further throughput. The packing factor a/F^3 and the pressure drop at flooding ΔP_f are basic characteristics of the packing and are reasonably constant if care is taken to consistently maintain standards in the method of packing. Typical values for a variety of popular packing styles and sizes are presented in Table 1. These two packing characteristics in conjunction with Figs. 3 and 5 permit the prediction of the entire operating diagram of Fig. 4a for any gas-liquid system passing through the packing. A typical specification sheet for establishing tower diameter and pressure drop is shown in Fig. 6.

Absorption Rate and Efficiency (Packed Height)

Having established an operable tower diameter for the desired fluid throughputs with a chosen packing material, we ask how high a tower will be necessary. Barring for the moment the considerations surrounding feed and drawoff ends, consider first the required height of packing, since this will be interrelated with the fluid rates, the choice of packing, and the design tower diameter.

If a mixture of gases is brought into contact with a liquid in which only one of the components is soluble (or the solubility of the other negligibly small) and the mixture kept in contact at a fixed temperature, an equilibrium is reached at which no further solution occurs. Carried out at various concentrations of the soluble gas in the total gas mixture, such experiments result in a so-called equilibrium curve, which is usually represented in the manner illustrated in Fig. 7. Obviously, if the concentration of the soluble gas in the total gas, in contact with its solvent liquid, is greater than the concentration which would be in equilibrium with the liquid, there will be a migration of solute gas molecules absorbed into the liquid so that the concentration in the liquid increases and the concentration in the gas decreases to reach a point on the equilibrium curve. The so-called driving force for this migration is the degree to which the solute gas concentration (partial pressure) in the total gas exceeds its concentration in equilibrium with the liquid at any point in time. If a water-soluble gas at concentration y_B enters the bottom of a packed absorption column, passing upward countercurrent to downflowing fresh water which enters at the top of the packing, it is obvious that the gas will decrease in soluble component and the liquid will increase. If gas and water rates and feed concentrations are known, specifying the exit gas concentration, or degree of solute gas removal, establishes

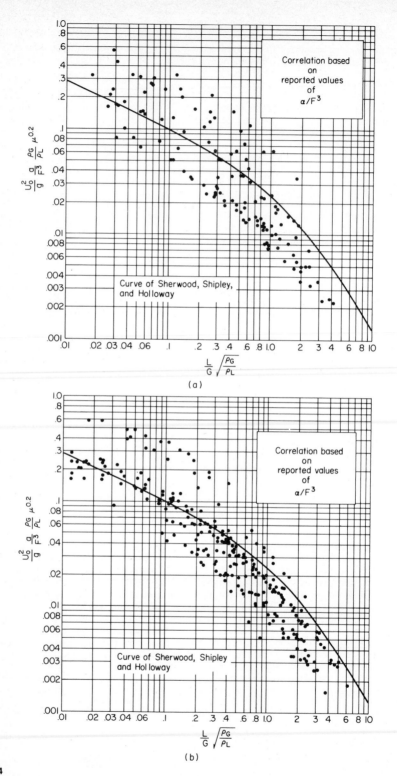

(a)

(b)

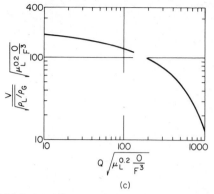

(c)

Fig. 2 Development of flooding correlations. [(*a*) *From* Trans. A.I.Ch.E., **41**, *693–710 (1945)*. (*b*) *From* Chem. Eng. Prog., **43**, *415–428 (1947)*. (*c*) *From* Petrol. Refiner, **40**, *130–132 (1961)*.]

TABLE 1 Typical Packing Characteristics
These figures vary among manufacturers, material (such as ceramic, metal, or plastic), and method of packing, and should be regarded here only as typical.

Packing type	Nominal size, in	Design packing factor	ΔP_F air-water, in H_2O
Raschig rings	3	34	2.5
	2	62	2.5
	1½	90	4.0
	1¼	115	4.0
	1	160	3.0
	¾	230	3.0
	⅝	281	2.5
	½	500	3.5
Pall rings	3½	16	2.1
	2	25	2.0
	1½	32	2.0
	1	52	2.0
Berl saddles	2	45	2.2
	1½	65	2.2
	1	110	2.5
	½	240	2.0
Intalox saddles	2	40	3.0
	1½	53	2.5
	1	98	2.1
	¾	140	2.1
Heilex	3	14	1.8
	2	21	1.5
Broken stone	¾	630	
	½	1190	
	¼	2190	

by material balance the exit absorbing liquid concentration or the so-called operating line as illustrated in Fig. 8. The vertical distance between equilibrium and operating lines defines the driving force for solute gas migration at any point along the packed height.

The rate at which the migrating solute molecules are absorbed is termed the "interfacial mass-transfer rate" and is dependent on the amount of liquid surface exposed. The latter is not only directly related to the packing material's size and shape, but to the liquid rate, its distribution over the packing surface, its potential to blind flow areas and surfaces, its wetting tendency, and a host of similar details. In all likelihood, all of these effects will for practical reasons never be academically completely resolved for all possible operating conditions. At best, performance of a packing can be rated against cases where surfaces were 100% coated or wetted and where blinding was minimal. Such cases can be represented via short-term runs by passing, for example, dry air over beds of water-

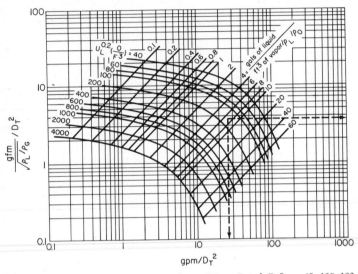

Fig. 3   Simplified packed tower flooding correlation. [*From* Petrol. Refiner, **40**, *130–132 (1961).*]

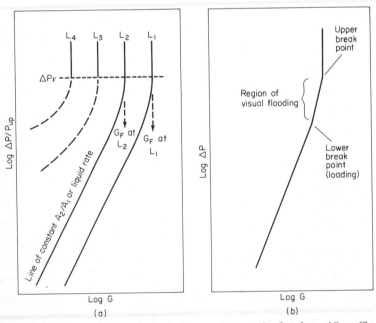

Fig. 4   (*a*) Typical pressure drop curve predicted by analogy to orifice flow theory. [*From* Chem. Eng. Prog., **43**, *415–428 (1947).*] (*b*) Typical pressure drop curve suggested by early investigators.

saturated spheres and cylinders and measuring the rate of humidification or mass transfer of water molecules at the interface of packing and air. At zero liquid downflow no blinding can occur, and for short-term runs all possible surface is actively wetted and calculable in area. The results of such investigations by Gamson et al.[7] were found to be correlated by the familiar mass-transfer number when plotted against Reynolds number, as shown in

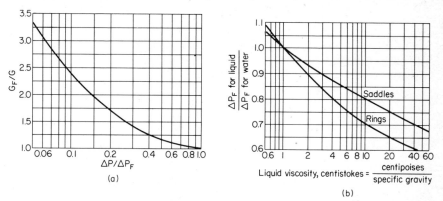

Fig. 5   Generalized pressure drop correlation for packed towers. [*From* Chem. Eng. Prog., 43, 415–428 (1947).]

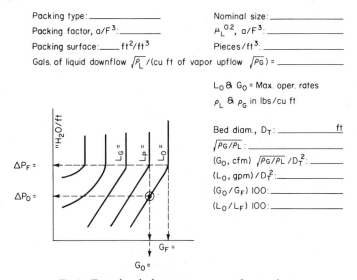

Fig. 6   Typical packed tower process specification sheet.

Fig. 9. The curve of Fig. 9 can be regarded as an upper limit, the case of 100% wetting of the packing, or in other words, the best one can expect to achieve.

Any overall comparisons of counterflow packed-tower performance on the basis of Fig. 9 neglect resistance to migration or diffusion of the solute gas molecules through the absorbing liquid. Such resistance in the liquid is a less common situation and can be found treated at length in the literature.[8] Other mechanistic treatments of the absorption process (HETP, HTU, penetration theory) have also been proposed, but in all instances require again some empirically determined parameters which have in no case been experimentally explored to the extent devoted to mass-transfer coefficients. It must be appreciated that treatment based solely on Fig. 9 assumes that the overall rate of mass transfer $K_G a$ is independent of any resistance in the liquid, or that $k_L a$ is large relative to

K_ga, and the liquid resistance is therefore negligible. Expressed analytically as a sum of resistances in series, this can be represented as:

$$\frac{1}{\text{overall rate}} = \frac{1}{\text{gas side rate}} + \frac{1}{\text{liquid side rate}}$$

$$\frac{1}{K_Ga} = \frac{1}{k_ga} + \frac{1}{k_1a}$$

$$K_Ga = k_ga \qquad \text{when } k_1a \text{ is large}$$

When the results of packed-tower tests, with actual counterflow of liquid, are plotted on Fig. 9, the exactitudes of the idealization appear to be overcome by the physical vagaries

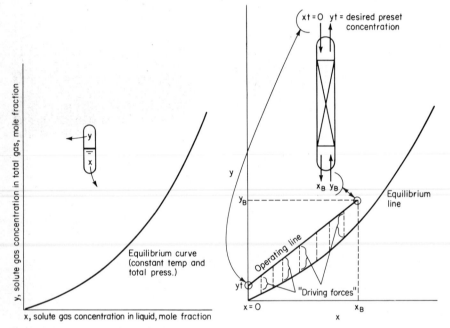

Fig. 7 Conventional representation of solute gas and solvent equilibrium.

Fig. 8 Graphical illustration of operating line and absorption driving forces.

of practicality, as evidenced in Fig. 10. Figure 10 represents "best-fit" straight lines drawn through the data of many investigators of different packings at a variety of liquid rates.[8] If the results are carefully sorted out, they show a trend with packing size and liquid rate as indicated in Fig. 11. The degree to which the lines fall short of the upper curve for 100% wetting may be regarded in terms of efficiency of contact. The higher the liquid rate, the greater the degree of wetting of the dry packing and hence the closer the approach to the maximum possible rate. The larger the packing size or the higher the gas rate, the less the degree of surface blinding, and therefore, again the closer the approach to the maximum possible rate.

Various authors have sorted out the data in Fig. 10 and fitted equations to the lines, resulting in recommended analytical expressions for specific packings of specific size at various liquid rates.[9] Typical results are represented by expressions such as given in Table 2. Developing such relationships for large-size packing requires large-diameter columns and hence relies solely on data gathered from industrial rather than laboratory equipment. The virtue of Fig. 10 is that it allows us to derive from only one data point a more general relationship simply by plotting the point and drawing through it a line whose slope is approximated from its neighbors. The resulting relationship, of the form of those in Table 2, is the most reasonable guide for extrapolation and interpolation to other conditions and other systems.

By analogy to distillation, a common alternate approach to estimation of packed height treats absorption as a stagewise process and refers to the height of packing required to achieve the calculated degree of absorption in one such stage as the height equivalent to a theoretical plate (HETP). This procedure is no more accurate than that based on the mass-transfer coefficient $K_G a$, but is frequently simpler to apply. Best-fit curves drawn through

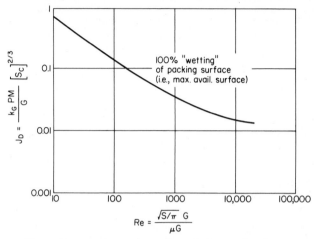

Fig. 9 Mass-transfer correlation for perfectly wetted packing.

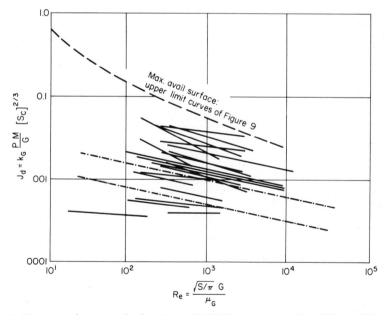

Fig. 10 Experimental mass-transfer data. (*From* "Kirk-Othmer Encyclopedia of Chemical Technology." *2d ed., Wiley-Interscience.*)

data for various packings are given in Fig. 12. An example of such stage calculations is given in the discussion of tray absorption towers.

The design sequence to this point involves the steps outlined in Table 3 with the understanding that step 2 represents the simplest case of physical absorption (or solution) without chemical reaction and with negligible resistance in the liquid phase. It should also be obvious that if the solution is accompanied with an evolution of heat, this must be simultaneously taken into consideration when evaluating the equilibrium values to correct the driving forces along the tower height.

TABLE 2   Typical Analytical Representations of the Curves in Fig. 10.
Typical average values of the coefficients in the analytical representation of the curves of Figure 10 as given by $J_D = A(L)^B/(Re)^c$, where J_D and Re are the ordinate and abscissa and L is the liquid downflow rate.

Packing	A	B	C
½-in Raschig rings	0.00216	0.3	0.100
1-in Raschig rings	0.0108	0.3	0.225
1½-in Raschig rings	0.0252	0.3	0.28
3-in Raschig rings	0.0320	0.3	0.32
1-in saddles	0.0123	0.3	0.34
2-in saddles	0.0270	0.3	0.39
3-in saddles	0.0350	0.3	0.40

TABLE 3   Design Sequence

Step	Given	Using	Find
1.	Gas and liquid properties, gas and liquid rates	Figure 3 in conjunction with a practical packing chosen from Table 1	Operable practical tower diameter
2.	Results of Step 1 and the solute gas-liquid equilibrium curve	Figure 10 or relationships derived therefrom as in Table 3	Required packing volume and hence ideal packed height
3.	Results of Steps 1 and 2	Practical considerations	Overall tower design

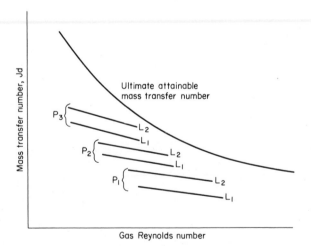

Fig. 11   Schematic representation of effect of packing diameter and liquid rate on gas-film mass-transfer correlation. (*From* "Kirk-Othmer Encyclopedia of Chemical Technology," *2nd ed., Wiley-Interscience.*)

Practical Design Considerations

An interrelation between steps 1 and 2 results from the operable tower diameter and the calculated active volume both being dependent upon the packing initially chosen in step 1. The choice is neither infinite nor one requiring extensive economic comparisons. Exact rules are also not always justifiable, but for approximation purposes if the gas rate is greater than about 14 m³/min actual (500 acfm), a nominal packing size smaller than 25 mm (1 in) would probably not be practical, and similarly about 56 m³/min actual (2000 acfm) sizes smaller than 50 mm (2 in) would likely also be impractical. Such "rules" depend on many other considerations of allowable pressure drop, possible height restrictions, liquor rate, etc., and must, therefore, be left principally to the designer's discretion with the only generally practical limitation that the nominal size of the packing never exceed about one-twentieth the tower diameter in order to maintain the greater voids at the interface of vessel wall and packing to a negligibly small fraction of the total voids in the packing.

Approach to Flooding

In the strictest interpretation, Fig. 3 yields the ultimate capacity limits or the minimum tower diameter. The design diameter must be somewhat larger to allow for uncertainties in the exact packing dimensions (e.g., thicknesses of metal or ceramic rings), the vicissitudes of the crew engaged to put the packing in place, the probability of unexpected surges in either flow rate, allowance for possible later changes in packing material if product specifications tighten, and the limitations on pressure drop, which decreases rapidly as rates fall below the flooding mark. Again, the choice of how close the design diameter should approach the minimum cannot be precisely defined. It is usual practice to design so that the operating gas rate is 75% of the rate which would cause flooding; this is equivalent to a tower diameter 15% greater than calculated from Fig. 3.

There are three possible definitions of "degree of approach to flooding." With reference to Fig. 13, A is the allowable increase in gas rate at constant liquid rate, B is the allowable increase in liquid rate at constant gas rate, and C is the allowable increase in both gas and liquid rates with their ratio kept constant. A is the usual approach where $G/G_F = 0.75$, but again the choice may depend on the intended operating procedure.

Packing Procedures

The wide variations in performance evidenced from any careful study of data such as represented in Fig. 10 are most frequently a result of a lack of appreciation of the importance of design details. The maximum possible attainment of the idealization of Fig. 10 (i.e., the minimum perfect-contact-packed volume) depends heavily upon minutely uniform dispersion of both gas and liquid streams thoroughly wetting the packing throughout its volume. Such is never completely attainable, but can be reasonably approached if given every opportunity. A first requisite is to pack "wet," when possible, and drain just prior to start-up, establishing the liquid downflow rate before introducing gas. Packing "wet" requires the column supports and shell to be designed for full hydraulic head and may be hampered by the overhead liquid-feed distribution devices if they were installed prior to packing. If packing "wet" is impractical, it would at least be desirable to fill the column slowly with liquid after packing and prior to start-up.

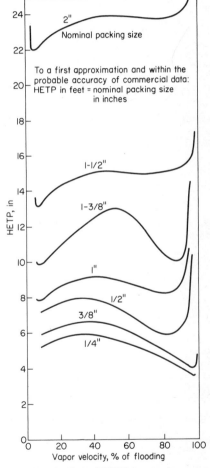

Fig. 12 Typical HETP test results.

To avoid damage to the packing support plates, liquid redistributors, or the packing material itself, a common practice involves dumping the packing from wire buckets lowered into the column from above. This procedure cannot be employed with plastic packing lighter than water. Plastic packings, more than any other, require upper hold-down plates and can be prewet by filling the tower with liquid after packing but before

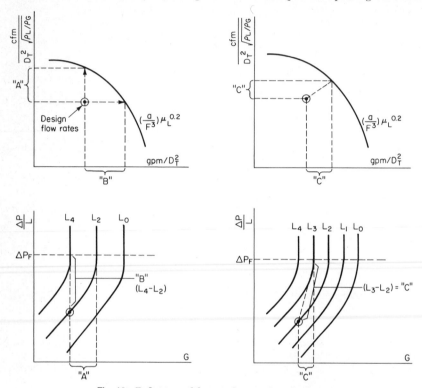

Fig. 13 Definitions of degree of approach to flooding.

start-up. The buckets of packing, lowered into the column usually through a manhole, should be emptied at several points especially around the periphery of the tower to avoid undesirable preferred packing orientations. This is particularly the case, for example, with Raschig ring-type packings which, if dumped continuously in the center of the tower, would build up a conical heap from which rings could only reach the tower periphery by tumbling down its slopes and in so doing would tend to roll on their sides, resulting in preferentially horizontally oriented packing lying at the vessel walls. As opposed to a completely random orientation, a tower packed in this fashion would show a pronounced poorer performance.

Feed Liquid Distribution

Maintaining good gas and liquid dispersion throughout the packing is of the greatest importance, and yet all too frequently is given insufficient attention. Packing will not of itself adequately distribute liquid which is simply poured on its surface at one point. However irregular the path a liquid stream must take, it still flows down by gravity over a relatively small cross section. Ideally, it should be fed to the top of the packing over an infinite number of points. One such means would be spraying from one or more full-cone nozzles covering the entire tower cross section. Though not an uncommon practice, this requires more pumping horsepower, is less flexible in accommodating changing liquid loads, can be troublesome if the absorbing liquid is recycled and might accumulate dust or other precipitated solid particulate matter, and usually makes an overhead mist eliminator mandatory to control entrainment of the finest droplets. As a practical compromise, other types of feed distributors are used. These distributors are normally supported on internal lugs welded to the tower shell, or on narrow flanged beams spanning the tower cross section. Some can serve the dual purpose of a packing holddown plate, with or without an expanded metal mesh on the underside, if purposely so positioned and properly supported.

One distributor type disperses the liquid feed through small perforations and allows the upflowing gas to pass out through the larger-diameter chimneys. This type is restricted to relatively clean liquids and a narrow range of liquid loads. Too low a load could allow some perforations to run dry and instead pass gas, thus providing liquid to only the remaining cross section. Too high a liquid rate could overflow the gas chimneys and lead to flooding if the chimneys are not properly oversized. Dirty liquid could lead to pluggage of the perforations. Another distributor type provides countercurrent passage of gas and

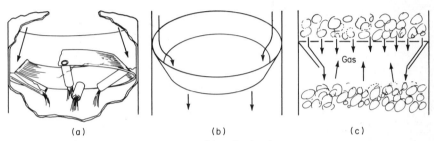

(a)   (b)   (c)

Fig. 14   Typical liquid redistributors.

liquid through notched chimneys. A weir-type notch directs the liquid down one edge of the chimneys, thus minimizing interference with the upflowing gas and, by widening, accommodates ever-increasing liquid rates if the level in the pan were to rise with increased feed. This distributor affords a lesser number of feed streams to the packing surface but handles more reliably a wider range of flow rates.

The most commonly used distributors are the so-called weir type and the perforated ring. The weir type is the least likely to give difficulties in operation and can handle a wide range of feed rates very effectively. The weir type must be carefully installed with precise level gages to assure overall equal overflow rates; the weir notches must be carefully cut and the troughs rigidly anchored.

The perforated-ring type is very popular because of its simple construction and ease of support. Perforations are usually on the underside of the rings directed down on the packing. It is not uncommon practice to submerge this type directly within the packing, thus using the packing absorption section to support the distributor and the packing layer above the rings as a separator or mist eliminator to collect any liquid droplets which might be entrained in the upflowing gas.

Liquid Redistribution

The greater the care taken to ensure random packing and well-dispersed liquid feed, the less the probability of subsequent maldistribution of liquid as it flows by gravity down over the packing obstructions. Inevitably, perfect random distribution will not prevail, particularly at the tower walls, which interface represents not only a discontinuity in the surface structure, but also a region of higher voidage or greater open area relative to the internal regions of the bed of packing, and hence an area from which liquid is less likely to leave once having randomly reached. Liquid cannot flow into or from the tower wall and thus it represents a place for liquid to accumulate. Means therefore must be provided to redirect liquid from the walls back into the central portions of the packing to maintain good distribution. Several designs of such devices are in common usage.

Figure 14a illustrates an arrangement in which liquid is intentionally collected and piped to a more central point. This particular form is most suitable for small towers in the range of 600 mm (24 in) diameter or less. Figure 14b represents a simple conical shell "wiper" which skims liquid off the walls to an inner diameter. If a distributor of this type is contemplated, it must be realized that if buried within packing, as in Fig. 14b, it also restricts tower capacity so that the inner diameter must be checked for approach to flooding. In the arrangement of Fig. 14c the distributor is no longer likely to represent a flooding hazard. However, if such a packing support is incorporated, then it would be even more desirable to employ a complete support grid design duplicating that at the bottom of the tower, where both gas and liquid distribution are incorporated. Figure 17 in Sec. 1.7 represents a more effective version of Fig. 14b in which the cone is sliced into a number of troughs or fingers directing liquid to various points along the tower radii. Here also care must be taken to assure no undue restriction which might induce flooding at the redistribution level.

When redistribution is to be considered, the question arises as to "how frequently?", or at what levels in the tower or within the packed height. Again, the necessity is dependent on how well the feeds are distributed initially. If redistribution were carried to the extreme of every 1 or 2 ft of packed height, this would approach in effect a trayed tower. One virtue of packed towers is the intended accomplishment of several trays' worth of separation within the distance between only one upper and one lower distribution device widely separated. Within the additional restraints of practicality, which vary from tower to tower, a reasonable rule of thumb is to space liquid distributors or redistributors at intervals no greater than 10 ft or 5 tower diameters, whichever is the smaller distance.

Feed Gas Distribution

The distribution of feed gases at the bottom of the packing is equally as important as the distribution of feed liquid at the top. It may be argued that if liquid ideally dripped

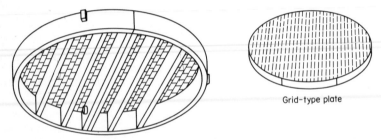

Grid-type plate

Fig. 15   Typical packing support plates.

uniformly from every piece of packing at the bottom of the tower, then gas could be admitted without specific direction and would find its own naturally equal distribution between the dropping liquid streams. Such is the hope when packing is supported on a flat mesh as illustrated in Fig. 15. Where a very coarse mesh, without support ribs beneath, would be preferred, or considered adequate, it is also common practice to use a few layers of larger-size packing atop the coarse mesh as a support for the main packing above. Reasonable assurance of distribution, beyond reliance on ideal randomness, is afforded with support plates of the type illustrated in Figs. 11 and 12 in Sec. 1.7. These provide slightly shorter preferred paths of lower pressure drop for the gas distributed over their cross section.

Packing Holddown Plates

To avoid gross fluid maldistribution which could result from packing physically being lifted, shifted in position near the top of the bed, or even broken as a result of abnormal flow rates or short surges, it is recommended practice to provide some form of holddown plate. Such plates normally consist of wire mesh on weighted rib supports, fabricated in sections which can be fitted through manholes and bolted together inside the tower. Typical forms are illustrated in Figs. 13 and 14 in Sec. 1.7. Since packing frequently settles with time, it is not uncommon to find support plates mounted so that they rest on the upper surface of the packing. Lugs protruding from the edges are fitted into mating slotted members welded to the inside of the tower shell, which are then sealed with a removable bolt or pin at the upper ends, so that though the holddown plate can ride up and down on the packing, it is ultimately limited in its upward travel.

Entrainment Suppressors

Composites summarizing the foregoing are shown in Fig. 16 to illustrate the simplest and least costly investment vs. alternates which would give the greatest assurance of meeting the desired design performance levels. This qualitative comparison alone should make it evident that scatter in performance data, of the degree experienced in Fig. 10, will never be resolved without detailed knowledge of column construction, and the near-impossible

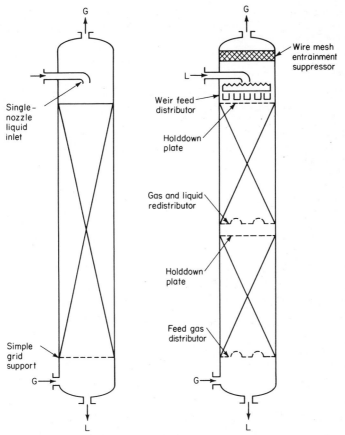

Fig. 16 Absorber internals assembly. (*a*) Simplest and least-efficient design. (*b*) Well-engineered absorber.

performance evaluation of what might be termed end effects which could even occur within entrainment-separating mesh. Entrainment is normally no problem in packed towers since below the flooding point, interfacial velocities are too low to significantly rift a liquid film. Only in an instance where even the least entrainment is a serious problem because of corrosive fluids, delicate downstream equipment, or insufficient disengaging space, is additional packing or a wire-mesh demister considered essential. Entrainment rate is more significant in plate towers for which correlations will be compared with data from aerated pools.

Random vs. Oriented Packing

Small packing poured randomly into a shell is certainly the more popular and commonly employed form of packed-tower design. However, it should be pointed out that in certain instances stacked or oriented packings have also been used and are specified where exceptionally low pressure drops and very high flow rates are involved. Only those packings of cylindrical shape and with a diameter larger than 3 in would be practical to install in a stacked form. The worker-hours involved with smaller pieces would be prohibitive. It is not uncommon to find stacked packings in square or even rectangular tower shells where their arrangement in triangular or square pitch is relatively straightforward; in cylindrical towers their stacking arrangement is generally from periphery to center, two layers of which appear in plan as shown in Fig. 2 of Sec. 1.7. In general, stacking reduces pressure drop and also mass-transfer efficiency to some extent. However, migration of the liquid to the walls is also greatly reduced, so that redistribution at intervals is not necessary. Good initial distribution of both gas and liquid feeds is, however, of paramount importance.

References

1. Sherwood, T. K., Shipley, G. H., and Holloway, F. A. L., *Ind. Eng. Chem,* **30**, 765 (1938).
2. Lobo, W. E., Friend, L., Hashmall, F., and Zenz, F. A., *Trans. AIChE,* **41**, 693, (1945).
3. Zenz, F. A., *Chem. Eng. Prog.* **43**, 415, (1947).
4. Zenz, F. A., and Eckert, R. A., *Petrol. Refiner,* **40**(2), 130 (1961).
5. Zenz, F. A., *Chem. Eng.,* **60**(8), 186 (1953).
6. Prahl, W. H., *Chem. Eng.,* August 11, 1969, p. 89.
7. Gamson, B. W., Thodos, G., and Hougen, O. A., *Trans. AIChE,* 39.1–35, (1943).
8. Kirk-Othmer, "Encyclopedia of Chemical Technology" (2d ed.), Vol. I, Wiley-Interscience, New York, 44–76, 1963.
9. Treybal, R. E., "Mass Transfer Operations," McGraw-Hill, New York, 1955.

TRAYED TOWERS

Nomenclature

σ_L	Liquid surface tension
μ_L	Liquid viscosity
S	Submergence or equivalent static depth of liquid on tray
V_h	Gas velocity through perforations, ft/s
ρ_G	Gas density, lb/ft^3
L_w	Weir length, in
H_w	Weir height, in
h_w	Height of liquid crest flowing over weir, in
TS	Tray spacing
A_o	Total hole area on tray
L/V	Liquid-to-vapor flow ratio
W	Weepage rate
U_o	Superficial gas velocity
E	Entrainment rate

Forms

There exists an almost infinite variety of conceived tray-type gas-liquid contacting towers; in operating principle, they could be divided into at least three general forms:

1. *Random counterflow* through weeping grids where gas and liquid flow through the same openings (as in Shell Turbogrid trays). See Fig. 17.

2. *Directed counterflow* either by means of a head differential or by single opening as in a baffle tray (exemplified here by the disk and doughnut arrangement) in which contact occurs principally by flow through a series of liquid curtains, as shown in Fig. 18.

3. *Directed counterflow* by piping liquid flow from tray to tray with contact occurring only on the tray proper, as practiced in the more conventional perforated-plate columns. Refer to Fig. 19.

The dimensional requirements for efficient operability of this general form of crossflow sealed downcomer design is the subject of this section.

Operating Characteristics

Basically this sealed-downcomer form of perforated-plate design involves bubbling gas up through a layer of liquid as the liquid flows over a plate having orifices or perforations through which the gas enters and froths the liquid layer. See Fig. 20.

Obviously there must be some limits on the gas and liquid flows. For example, the higher the gas velocity through the perforations, the more liquid droplets carried higher into the space above the frothed liquid layer, and hence the more liquid carried (entrained) by the gas through the perforations back into upper trays. This back-mixing is undesirable in most situations since it creates higher internal tray-to-tray circulating

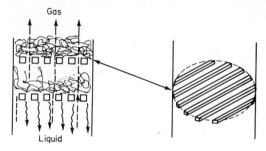

Fig. 17   Random weeping grid trays.

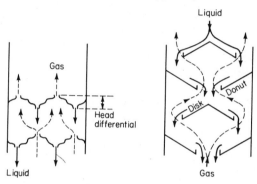

Fig. 18   Directed cascading liquid trays.

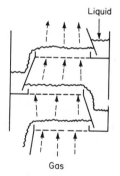
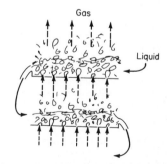

Fig. 19   Typical cross-flow trays.   **Fig. 20**   Bubbling contact on a cross-flow tray.

liquid loads and destroys the staging, or the temperature or concentration gradients, that should be established from bottom to top of tower. The same effect is caused by excessive liquid load. The higher the liquid throughput, or more correctly, the greater the liquid depth (calculated as though it were flowing across an unaerated plate as pure liquid) on the tray, the greater the droplet entrainment to the tray above (if for no other reason than the obvious decrease in space between the level of liquid and the tray above when the liquid depth is increased).

Just as it is undesirable to operate under conditions of excessive liquid entrainment (back-mixing) to upper trays, it is equally if not sometimes more undesirable to operate with insufficient gas velocity through the perforations such that excessive liquid drips or "weeps" through the perforations countercurrent to the gas flow, and thereby bypasses its anticipated residence time or the number of gas contact points existing along its flow path across the full tray length. The greater the liquid depth on the tray, the greater the required gas velocity through the perforations to prevent counterflowing liquid weepage.

These relationships between gas and liquid throughput, tray spacing, and weepage are best illustrated by reference to a typical operating diagram[1,2] for a given gas-liquid system and for a fixed perforation size and pattern as illustrated in Fig. 21.

The incipient weepage locus, shown as a dashed line in Fig. 21, represents the combinations of minimum vapor loads or maximum static liquid submergences at which this particular plate will just begin to weep. If at a given liquid load corresponding to a crest of 25 mm (1 in) flowing over a 25 mm (1 in) high weir [resulting in a static submergence of 50 mm (2 in)], the vapor load was such that the product of hole velocity and the square root of the vapor density (essentially the square root of the emerging vapor's kinetic energy) exceeded 8.8 (point A in Fig. 21), no liquid would weep through the perforations. If the vapor flow were reduced 30% (to point B in Fig. 21), or if instead the liquid load were increased to yield a static submergence of 100 mm (4 in) (point C in Fig. 21), then tray liquid would weep through the perforations at a rate of W_1 gallons of liquid per minute per square foot of hole area on the tray.

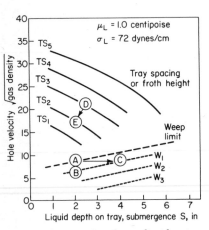

Fig. 21 Capacity chart for gas-liquid system on a perforated plate with ³⁄₁₆-in holes and 8.35% hole area.

Normally one would choose to design at some point well above the incipient weepage locus so that reduced load operation would also still fall above this locus. Ordinarily this is quite feasible, particularly since reduction in vapor load is usually accompanied with reduction in liquid load so that reduced load operation might move an operating point from a design condition at a point such as D in Fig. 21 to a condition corresponding to point E. However, there exist instances in which pressure drop considerations force the design point to be quite close to the incipient weepage locus, and in such instances partial-load operation can fall into the weepage region. The higher the ordinate (or the greater the abscissa), the greater the pressure drop through the operating tray. The former is simply a reflection of higher gas velocity through the holes and the latter simply a matter of more liquid head to be overcome.

Though Fig. 21 is qualitatively a representative so-called capacity chart for a perforated plate, it is not quantitatively universally applicable. The position of the weepage limit and the tray spacing (or equivalent entrainment) curves shift as a function of the liquid properties and the physical configuration of the trays; for example:

1. Lowering the surface tension of the tray liquid increases the weepage limit vapor rates (e.g., shifts the dashed curve of Fig. 21 upward on the ordinate).

2. Reducing the perforation diameters reduces the weepage limit (e.g., shifts the dashed curve of Fig. 21 downward on the ordinate). If the holes are small enough, it is possible to hold liquid on the tray without weepage even when no gas is flowing up through the holes (e.g., an ordinate of zero).

3. At constant hole velocity and submergence, smaller holes permit closer tray spacing (e.g., the solid curves of Fig. 21 shift diagonally up and to the right).

4. Decreasing the plate thickness raises the weepage limit (e.g., shifts the dashed curve of Fig. 22 upward—thin plates weep more easily than thick plates).

5. Increasing percentage free area (e.g., increasing the number of holes or the hole diameter or both) increases the weepage limit and increases the tray spacing at equivalent entrainment.

Capacity Charts

The capacity charts[3] shown in Figs. 22 through 28 are based mostly on experimental data and illustrate the magnitude of the effects referred to above.

In connection with the use of these charts, the following definitions are relevant:

1. Submergence S refers to the equivalent depth of unaerated liquid which would exist in flow across the tray if all the holes were plugged. The depth of this liquid would

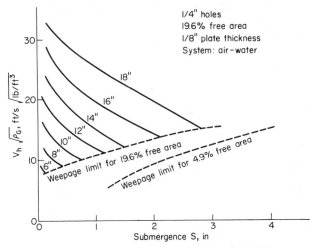

Fig. 22 Capacity chart for ¼-in-diameter holes and 19.6% free area.

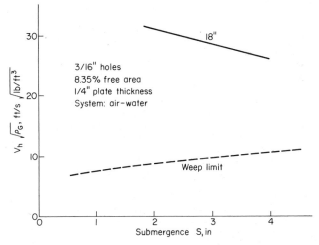

Fig. 23 Capacity chart for ³⁄₁₆-in-diameter holes and 8.35% free area.

be equal to the sum of the overflow weir (or dam) height H_w above the tray floor plus the height or thickness of the liquid crest h_w over the weir necessary to pass the volumetric flow. The weir height H_w is set mechanically (frequently the weir plate is bolted in place and provided with slotted bolt holes to permit later adjustment to higher or lower position if it is expected to operate periodically over widely different throughputs or to "tune" the tower to maximum efficiency). The flowing crest over the weir, h_w is calculable from the desired liquid throughput and the weir length via the Francis weir equation given below for conventional straight-edged weir plates.

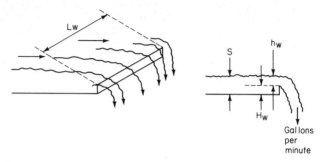

$$S = H_w + h_w$$
$$h_w = 0.48 \,(gpm/L_w)^{2/3} \quad \text{inches} \tag{1}$$

where L_w is the weir length in inches.

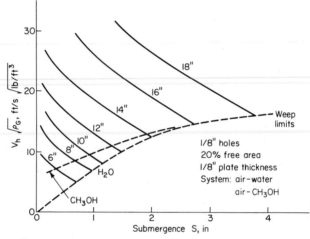

Fig. 24 Capacity chart for ⅛-in-diameter holes and 20% free area.

2. Percent free area refers to the so-called unit "cell" of the pattern of the perforated portion of the tray. For example, if the perforations consisted of 6 mm (¼-in) holes on 25 mm (1-in) centers in square pitch this would represent a free area of 4.9%. Percent free area does *not* mean hole area divided by plate area or hole area divided by tower cross-sectional area.

3. Tray spacing (TS) refers to the vertical distance from any tray to the next tray above. It is conventionally measured as the distance from perforated metal to perforated metal, regardless of the operating depth of liquid on the lower tray or any special configurations of the liquid downspouts.

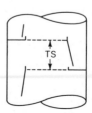

The use of the capacity charts in design consists simply of first arbitrarily choosing a hole size and percent free area (e.g., a particular capacity chart), and second selecting an appropriate design point such as the value of the ordinate about 20% above the weep limit at 50 mm (2 in) to 75 mm (3 in) of submergence. From this value of the ordinate and the desired volumetric throughput, one can determine the hole area and therefore the total

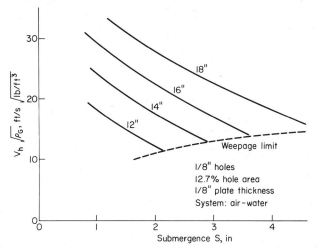

Fig. 25   Capacity chart for ⅛-in-diameter holes and 12.7% free area.

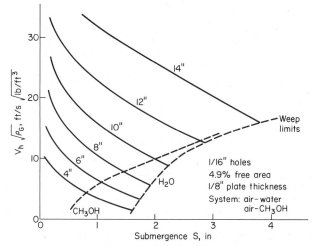

Fig. 26   Capacity chart for ¹⁄₁₆-in-diameter holes and 4.9% free area.

area of the perforated metal. This area is then multiplied by 1.6 (to account for liquid downflow areas and peripheral-hole-area losses) and taken as the cross-sectional area of the tower from which its diameter can be ascertained. A layout sketch is then made from which the weir length can be determined and therefrom the submergence [to make certain this does not exceed the original assumed 50 mm (2 in)] and the attendant required tray spacing.

EXAMPLE

Suppose, for example, that it is desired to size a tower to handle the following load:
   Liquid: water, 1516 L/min (400 gpm)
   Gas: essentially nitrogen, 319 m³/min actual (11,400 acfm); density 1.2 kg/m³ (0.075 lb/ft³)

As a first step arbitrarily choose 3 mm (⅛ in) holes on triangular pitch, three holes per 25 mm (inch), 12.7% hole area, which is the capacity chart of Fig. 25.

At a submergence of, say, 60 mm (2.4 in) of liquid and tray spacing of 350 mm (14 in), the ordinate gives $V_h\sqrt{\rho_G}$ of about 15.5, which is about 28% above the weep limit ordinate of 12 at this submergence. Therefore try this as a first estimate of a design point:

$$V_h\sqrt{\rho_G} = 15 = \frac{11,400\sqrt{0.075}}{60A_o}$$

Therefore

$$A_o = 0.32 \text{ m}^2 \text{ (3.46 ft}^2\text{)}$$
$$\text{Perforated area} = 3.46/0.127 = 2.56 \text{ m}^2 \text{ (27.2 ft}^2\text{)}$$
$$\text{Tower area} = 1.6 \times 27.2 = 4.1 \text{ m}^2 \text{ (43.5 ft}^2\text{)}$$
$$\text{Tower diameter} = \sqrt{\frac{43.5 \times 4}{\pi}} = 2250 \text{ mm (7 ft, 6 in)}$$

A trial layout gives the following:

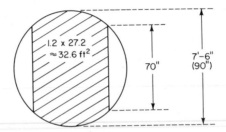

The crest over a 1750 mm (70 in) straightedge weir would amount to

$$h_w = 0.48\,(400/70)^{2/3} = 38.5 \text{ mm (1.54 in)}$$

therefore a 25 mm (1 in) high weir gives $S = 63.5$ mm (2.54 in), which would then correspond on Fig. 25 to a tray spacing of, say, 375 mm (15 in).

Obviously the above procedure can be repeated with any other hole size or pattern until the most economical tower size accommodating a desired range of loads is determined.

The table[4] of segmental functions in the Appendix to this section is extremely helpful in making tray layouts to determine weir lengths, chord areas, and similar necessary dimensions.

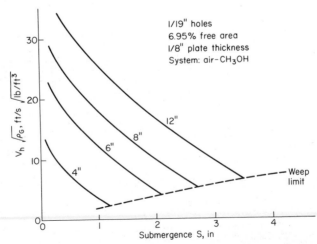

Fig. 27 Capacity chart for ¹⁄₁₉-in-diameter holes and 6.95% free area.

It should be obvious also that for any chosen hole pattern, a reduction in tower diameter would normally call for increased tray spacing, since a smaller tower accommodating an equal perforation area would be restricted to a shorter weir and hence a greater crest or submergence. Within reason this can be offset by changing the physical weir height H_W; however, it is not generally advisable to design for an H_W less than 25 mm (1 in) and

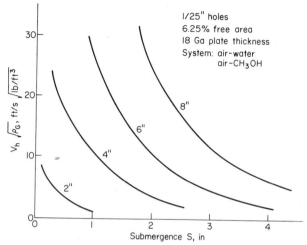

Fig. 28   Capacity chart for ½₅-in-diameter holes and 6.25% free area.

certainly not less than 12.5 mm (½ in). Only in extraordinary cases have such towers been designed without any overflow weir whatsoever (in which cases the submergence was a result of solely the liquid flow), and only in special cases such as air fractionation are perforated plates usually designed with weirs less than 6 mm (¼ in) in height [with tray spacings of 62.5 to 100 mm (2½ to 4 in) and holes of the order of 0.8 mm (¹⁄₃₂ in) in diameter].

Rapid Sizing Charts

By the reverse of the procedure described above, it is possible to calculate the liquid and vapor capacities of any diameter of tower (provided with any given weir length). If the weir height is assumed constant, then merely weir length and liquid capacity fix the abscissa of a capacity chart. Knowing the weir length, and assuming that the entire area between the weirs is perforated, it is a simple matter to calculate the available hole area for any diameter tower. The desired gas capacities in conjunction with this hole area fix the ordinate of a capacity chart. The ordinate and abscissa then yield the design tray spacing. In any one diameter of tower there can be fitted trays having a reasonably large variation in weir length; for example, in Fig. 29, B would accommodate a higher liquid rate and a lower gas rate than A, since it provides a longer weir but at the expense of less available perforation area. Similarly, C would accommodate a still higher liquid-to-gas flow ratio.

For fixed weir heights of 25 mm (1 in) and fixed ratios of weir length to tower (or tray) diameter, the above three configurations have each been rated for gas and liquid capacity over a reasonable range of tower diameters when provided with a perforation pattern of 12.7% free area using 3 mm (⅛ in) diameter holes. The results are presented graphically in Figs. 30, 31, and 32.

For illustrative purposes consider again the rapid size estimation of a tower to accommodate the following load:
Liquid: water 1516 L/min (400 gal/min)
Gas: essentially nitrogen, 319 m³/min actual (11,400 acfm); density 1.2 kg/m³ (0.075 lb/ft³)

$$V\sqrt{\rho_G} = \frac{11,400}{60}\sqrt{0.075}$$
$$= 52$$

Proceed as follows: Try the high L/V, Fig. 31. Locate the liquid load line for 400 gal/min and the gas load line of 52 as illustrated schematically below.

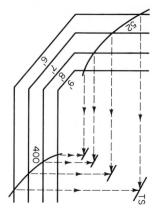

Note that the gas load line is intersected by tower diameters of 1.8, 2.1, 2.4, and 2.7 m (6, 7, 8, and 9 ft). Follow these diameter lines to the point where they intersect 400 gal/min and then draw the horizontal and vertical lines as illustrated in the sketch, and at their point of intersection read off (perhaps by

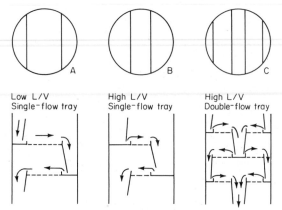

Fig. 29   Cross-flow tray and tower arrangements.

interpolation) the design tray spacings corresponding to essentially negligible entrainment. The following figures should be obtained:

Tower diameter		$V_h\sqrt{\rho_G}$	Tray spacing	
Meters	Feet		Millimeters	Inches
1.8	6	28	500	20
2.1	7	20	400	16
2.4	8	15	350	14
2.7	9	12	300	12½ (barely above weep limit)

Note that the previous solution gave a 2.25 m (7 ft, 6 in) diameter at 375 mm (15 in) tray spacing with a $V_h\sqrt{\rho_G}$ of 15.5. The choice of optimum design is now a matter of capital or operating expenditure. The 1.8 m (6-ft) tower at 500 mm (20-in) tray spacing costs less in steel than any of the larger sizes; however, it consumes the highest pressure drop (gas pressure drop is proportional to the square of $V_h\sqrt{\rho_G}$). Note also that in a highly competitive situation, a detailed layout is much to be preferred in the final preparation of a bid. Figures 30, 31, and 32 are principally helpful in making rapid "ball park" estimates.

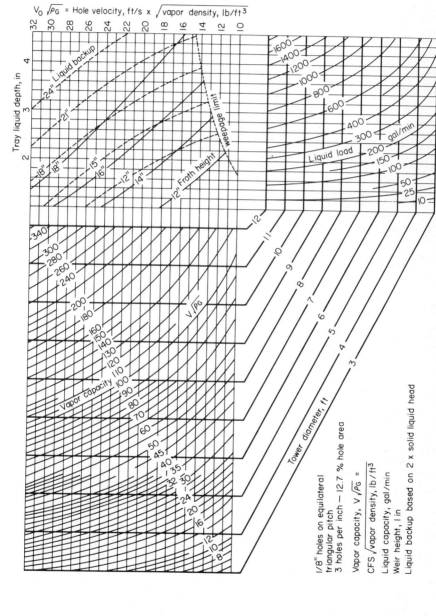

Fig. 30 Perforated-plate sizing chart, single-flow trays, low L/V.

1/8" holes on equilateral
triangular pitch
3 holes per inch — 12.7 % hole area

Vapor capacity, $V \sqrt{\rho_G}$ =
$\dfrac{CFS \sqrt{\text{vapor density, lb/ft}^3}}{}$

Liquid capacity, gal/min
Weir height, l in
Liquid backup based on 2 x solid liquid head

$V_0 \sqrt{\rho_G}$ = Hole velocity, ft/s x $\sqrt{\text{vapor density, lb/ft}^3}$

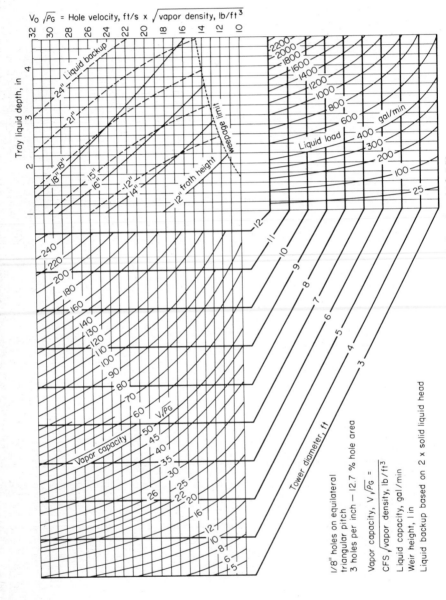

Fig. 31 Perforated-plate sizing chart, single-flow trays, high L/V.

1/8" holes on equilateral triangular pitch
3 holes per inch — 12.7 % hole area

Vapor capacity, $V\sqrt{\rho_G} =$
CFS $\sqrt{\text{vapor density}}$, lb/ft^3
Liquid capacity, gal/min
Weir height, l in
Liquid backup based on 2 x solid liquid head

$V_O\sqrt{\rho_G}$ = Hole velocity, ft/s x $\sqrt{\text{vapor density}}$, lb/ft^3

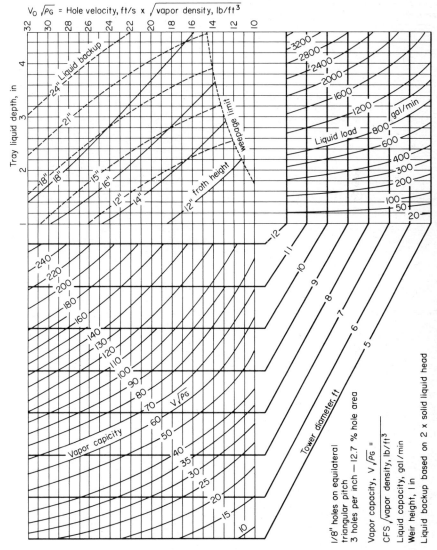

Fig. 32 Perforated-plate sizing chart, double-flow trays.

1/8" holes on equilateral triangular pitch
3 holes per inch — 12.7 % hole area

Vapor capacity, $V\sqrt{\rho_G} =$

CFS $\sqrt{\text{vapor density, lb/ft}^3}$

Liquid capacity, gal/min

Weir height, l in

Liquid backup based on 2 x solid liquid head

These capacity charts are based on conventional nonfoaming liquids such as water, light hydrocarbons, alcohols, liquid air, etc. They may be applied to liquids of greater foaming tendency such as ethanolamines, furfural, dilute detergents etc. by simply adding height [usually 150 mm (6 in) is ample] to the design tray spacings.

Gas Pressure Drop

The gas pressure drop across a perforated plate is the sum of the head required to overcome the equivalent liquid depth (the submergence) on the tray plus the gas resistance in passing through the perforations. The latter is calculable from the relationship

$$\Delta P_h = \frac{(V_h \sqrt{\rho_G})^2}{214} \tag{2}$$

where ΔP_h = gas ΔP through perforations, inches of water
$\quad\quad V_h$ = gas velocity through perforations, ft/s
$\quad\quad \rho_G$ = gas density, lb/ft³

The former (submergence) was given in Eq. (1). Note that if the tray liquid has a density significantly different from water (density = 62.4 lb/ft³), then S must be converted from inches of tray liquid to equivalent inches of water. The perforation loss ΔP_h in Eq. (2) is given in units of inches of water (of 62.4 lb/ft³ density).

When Eq. (2) is applied, V_h may be taken as the gas load in cubic feet per second divided by the total area of the perforations on a tray (which assumes that all the holes are blowing) as long as the design point or operating conditions lie above the weep limit.

Liquid Flow, Distribution, and Resistance

There are a number of variations possible in the manner of conducting liquid through the column from tray to tray. The most common takes the form of so-called segmental downcomers illustrated schematically in Fig. 33a, in which the tower shell forms one wall of the "downpipe." The liquid flows down through an area that has the shape of a segment of a circle, as shown in Fig. 33b.

Where it is simpler to accommodate the liquid flow through round pipes, rather than seal a downcomer plate to the shell wall, either method shown in Fig. 34 can be employed. In Fig. 34a the straight overflow weir length sets the crest height [via Eq. (1)], and in Fig. 34b the perimeter of the downpipes protruding above the tray set the crest height according to the following relationship:

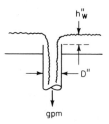

$$h_w = [\text{gpm}/10D]^{0.704} \quad\quad \text{inches} \tag{3}$$

where $h_w \leq D/5$. Exceeding the criterion of Eq. (3) can cause severe reduction in liquid throughput by a form of vapor lock. If the crest over the pipe lip is so great that the inflow from all points on the perimeter reaches the center of the downpipe, then gas is trapped in the downpipe (e.g., as though liquid were trying to pour into an unvented tank), and the liquid can only continue flowing in as the trapped gas "gulps" up out of the pipe sporadically. Under extreme conditions this state can cause the entire tower to "flood."

In instances where there is very low liquid load and it is desired to minimize the risk that, because of out-of-levelness, the liquid flow across the tray might occur predominantly (if not entirely, depending on the care in erection to avoid out-of-levelness) along one side of the tray, it is common practice to serrate the edge of the overflow weir as shown below.

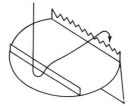

The height of liquid crest within the notch of the serration is calculable from Eq. (4):

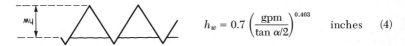

$$h_w = 0.7 \left(\frac{\text{gpm}}{\tan \alpha/2} \right)^{0.403} \quad \text{inches} \quad (4)$$

In addition to the "single flow" segmental and pipe downcomer arrangements illustrated in Figs. 33 and 34, so-called radial- and double-flow trays are relatively common. The radial-flow arrangement (Fig. 35) requires only every other tray to be sealed to the vessel shell and yields equal length of flow path to all elements of liquid. However, it is not too well suited to high liquid loads because of the limitations in weir or overflow perimeter on the center downflow trays.

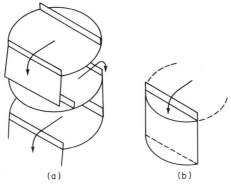

(a)   (b)

Fig. 33   Segmental downcomers. (*a*) Arrangement. (*b*) Downcomer segment.

The double-flow arrangement (Fig. 36) is the equivalent of two half-round towers side by side, with the wall between removed to allow gas and liquid mixing between trays. More exotic tray forms are beyond the intent of this discussion and of significance only in rare applications.

The introduction of liquid onto the top tray is carried out in a wide variety of arrangements. The main requirement is that the liquid feed be introduced calmly with a minimum of "splashing" or jetting and be distributed so as to flow over the perforated portion reasonably uniformly across the width of the tray. The simplest arrangement on single-flow trays is to provide an internal baffle at the feed nozzle, extend the nozzle with an elbow, or feed via a straight pipe down through the vessel head, as shown in Fig. 37.

On large-diameter towers, the splash baffle may be much wider or the elbow end in a distributor pipe may be slotted or perforated along its bottom. See Fig. 38.

Similarly, on a double-flow tray, it is most convenient to feed onto the center of an outer downflow tray with an internal pipe extension of the feed nozzle again slotted or perforated on its underside. See Fig. 39.

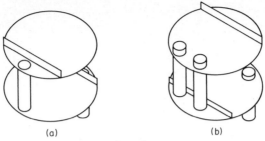

Fig. 34   Pipe downcomers.

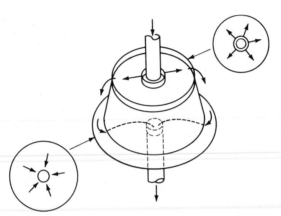

Fig. 35   Radial-flow trays.

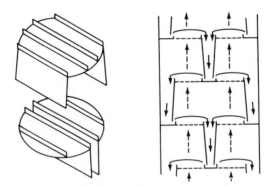

Fig. 36   Double-flow trays.

Liquid drawoff is usually accomplished via a so-called drawoff pot (or seal pot) or frequently by simply utilizing the bottom of the vessel for this purpose. The drawoff or seal pot is simply a closed-off downcomer.

The sizing of horizontal drawoff nozzles, seal pots (or so-called trap-out boots) is illustrated in Eq. (5) below.

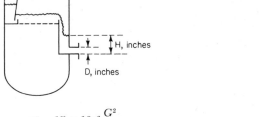

$$H = 15 \times 10^{-6} \frac{G^2}{D^4} \tag{5}$$

where $D = H_{\text{minimum}}$
     G = liquid flow, gal/h

The obviously similar arrangements for a double-flow tray follow the principles illustrated in Fig. 40 for single crossflow trays. The bottom tray of a double-flow tower would preferably be a center downflow tray where all liquid comes into one downcomer,

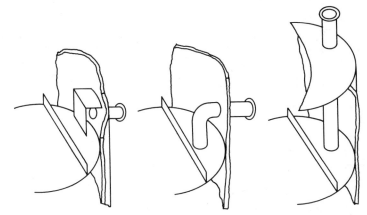

Fig. 37   Internal liquid feed nozzle arrangements.

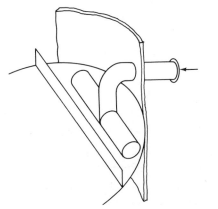

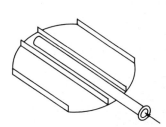

Fig. 38   Feed distributor pipe for a single-flow tray.

Fig. 39   Feed distributor pipe for a double-flow tray.

therefore requiring only one seal pot and one drawoff nozzle unless the tower bottom is utilized as the seal pot.

In an arrangement such as in Fig. 41b, the center downcomer must be contracted to a spout so that feed gas can get around to its opposite side unless two diametrically opposed gas feed nozzles are provided. If the downcomer in Fig. 41b is sealed to the tower shell,

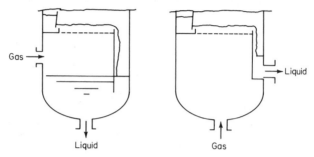

Fig. 40 Liquid draw-off arrangements for a single-flow tray.

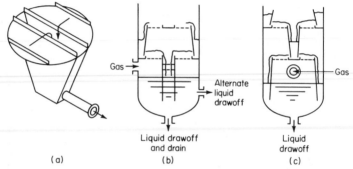

Fig. 41 Liquid draw-off arrangements for double-flow trays.

then it must be provided with a hole (e.g., a pipe sealed to each side as shown dotted), to allow feed gas access to both halves of the tower from a single gas-inlet nozzle.

In order to assure that gas passes up the column by flowing through the perforations on each tray rather than up through the liquid downpipes, it is necessary to create a liquid pool or seal at the bottom of every downcomer. This usually takes the form of an inlet weir or a depressed seal pan, as shown in Fig. 42. In either case the overflow weir provides additional seal and could alone provide sufficient seal under liquid flows adequate to immerse the bottom edge of the downcomer from the tray above. The arrangement in Fig. 42b (the depressed seal pan) is generally more costly and usually employed only when the liquid load is high and would create inordinate splashing or additional kinetic head on the tray were it allowed to flow over a conventional inlet weir.

Flooding

As even in the case of an empty vertical pipe, there exists a limit to the quantity of liquid and gas which can be simultaneously passed through the pipe in opposing directions of countercurrent flow. A perforated-plate tower with sealed downcomers can reach its ultimate or flooding capacity by either excess tray pressure drop, excess liquid entrainment from tray to tray, or vapor lock in the downcomers.

The sealed downcomer shown in Fig. 43 in the form of a pipe represents effectively one leg of a manometer. If there were no flow of gas up through the trays (assume that surface tension prevents the trays from leaking), then $P_1 = P_2$. If the clearance R between the bottom of the pipe and tray 1 offered no resistance to the liquid flow, as in Fig. 43a, then the level in the downcomer pipe would simply be that at A, reflecting the depth on the tray (with inlet weirs no shorter and no higher than tray outlet weirs).

If the clearance at R were restrictive, as in Figure 43b, so that the liquid would have to acquire a "head" or pressure drop to overcome the resistance at the desired liquid throughput, then the liquid level in the downpipe would rise to a level such that $B - A$

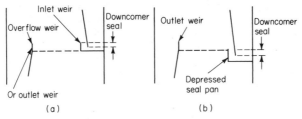

Fig. 42   Downcomer seal arrangements.

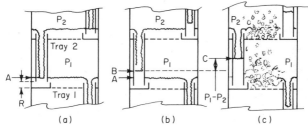

Fig. 43   Pressure drop determines liquid level in downcomer.

reflected the restrictive or "downcomer clearance" loss. This loss is calculable from the relationship

$$h_c = 0.56V_L^2 \tag{6}$$

where h_c = contraction loss of liquid flowing under downcomer, inches of tray liquid

V_L = liquid velocity through constricted area under or around downcomer, ft/s

Obviously in tray design it is desired to provide sufficient clearance to minimize this loss without undue sacrifice of a reasonable seal.

When gas is passed up through the trays (as in Fig. 43c), pressure P_1 will be greater than P_2 because of the gas pressure drop encountered in passing through tray 2. The pressure inside the downpipe is therefore P_2 and that on the outside P_1. Therefore the manometric difference $P_1 - P_2$ will be reflected by a rise in the downcomer liquid level. The excess in pressure at P_1 over that at P_2 will push down on the liquid outside the downpipe and force the level inside to rise to equal the differential.

The level inside the downcomer can therefore be represented by:

$$\underbrace{h_w + H_W} \quad + \quad \underbrace{h_c} \quad + \quad \underbrace{H_W + h_W} \quad + \quad \underbrace{\Delta P_h}$$

Level determined by weir height and crest (as in Fig. 44a) based on inlet and outlet weirs, whichever yields the largest combination	Additional head to overcome downcomer clearance loss (as in Fig. 43b)	Submergence head	Perforation resistance to gas flow
			Gas pressure differential between tray 1 and tray 2 (i.e., ΔP through tray as in Fig. 43c)

For the usual case where the tray inlet weir is lower or equal to the height of the tray outlet weir and equal or greater in length:

$$\left.\begin{array}{l}\text{Liquid height in}\\\text{downcomer, or}\\\text{``liquid backup''}\end{array}\right\} = \Delta P_h + h_c + 2h_w + 2H_W \tag{7}$$

The various terms in Eq. (7) are calculable from Eqs. (1), (2), and (6), and in practice all should be expressed in units of tray liquid head.

It is the usual practice in design not to allow liquid backup [Eq. (7)] to exceed 50% of the tray spacing. In other words, in operation the liquid level in the downcomers should ride midway between the trays. For example, if the level rose to a height equal to the tray spacing, then the tower would be at its flood point, since no increased liquid or gas could be fed. Obviously, increasing gas rate increases tray pressure drop and accordingly backs liquid up higher in the downcomers. Similarly, increasing liquid rate increases weir crest height, downcomer clearance loss, and tray gas-pressure drop (because of increased submergence), and hence also backs liquid up higher in the downcomers. The design figure of backup equal to 50% of tray spacing allows for some additional capacity and represents in effect an equivalent of the "percent of flooding" in packed-tower design. There are instances in which trays have been designed to operate with backups as high as 80% of tray spacing; it is the designer's option to decide how conservative this criterion is. It also affects the safety factor with regard to unpredictable liquid foaming tendencies when it is not known how rapidly the foam created on the tray will "break" during its residence time in the downcomer, should it be carried incompletely defoamed over the tray outlet weir.

Flooding can also occur as a result of excessive entrainment. The liquid entrained from any tray to the tray above returns via the downcomer from the tray above. Hence entrained liquid simply constitutes an internal liquid recycle, which effectively increases the tray liquid loads above the net feed rate. Its influence on flooding is then the same as that of an equal increase in net liquid feed as discussed above.

A less frequent (but nevertheless significant) source of flooding may lie in vapor locking of downcomers. If, for example, in Eq. (3) the crests meet in the pipe center, the liquid would bridge the pipe and could not get past the trapped gas bubble below (the limitation of $h_w \leq D/5$ is intended to avoid such vapor lock). Even in a segmental weir arrangement, the crest flowing over the weir should not have so great a "throw" as to reach the tower wall and thus trap gas below.

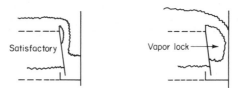

The table below gives values of the "throw" of water over sharp-edged weirs.

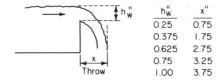

h_w''	x''
0.25	0.75
0.375	1.75
0.625	2.75
0.75	3.25
1.00	3.75

Weepage Rate

When (as a result of temporary reduced load or intentional design for a raining tray of low pressure drop) it is necessary to operate at a point below the weep limit, the amount of liquid weepage can be quantitatively estimated from the correlation[2] presented in Fig. 44 and Eq. (8).

$$6 \log_{10}(w + 1) = (V_h\sqrt{\rho_G})_{w\to 0} - (V_h\sqrt{\rho_G}) \tag{8}$$

In practice, prediction of the weepage rate through a tray involves a trail-and-error procedure. Given the total liquid flow to a tray, one assumes a weepage rate and by difference obtains the weir flow rate. From the Francis weir formula, one calculates the crest over the weir and this plus the weir height gives the liquid submergence. The liquid gradient on perforated plates may usually be neglected; it is in any event so small as not to affect the calculation beyond the limits of accuracy of the experimental weepage data. At this calculated liquid submergence, one obtains from a capacity chart, for the particular

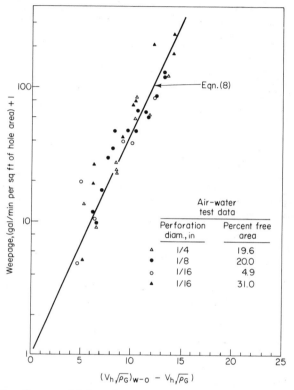

Fig. 44 Correlation of weepage data for various trays. [*From* Hydrocarbon Proc., **46**, *138–140 (1967)*.]

hole size and configuration, the minimum vapor load expressed as $(V_h\sqrt{\rho_G})_{w\to 0}$ at incipient weepage.

From the operating vapor capacity and Fig. 44 [or Eq. (8)], one can then calculate a weepage rate which must check the rate initially assumed.

EXAMPLE

As an illustration, consider a 1.8 m (6 ft, 0 in) diameter tower bearing single-flow trays having 4.7-mm ($\frac{3}{16}$-in) perforations on 20.3 mm ($\frac{13}{16}$ in) square pitch centers plus a hole centered in each square. The overflow weir is 1500 mm (60 in) long and 50 mm (2 in) high, and the total hole area on the tray amounts to 0.135 m² (1.45 ft²).

Tray liquid load: 707 L/mm (450 gal/min)
Tray vapor load: 30.66 m³/min actual (1095 acfm)
Vapor density: 1.76 kg/m³ (0.11 lb/ft³)

$$\text{Percent hole area} = 2\left[\frac{\pi(\frac{3}{16})^2}{4}\right]\frac{100}{(\frac{13}{16})^2} = 8.35\%$$

Assume weepage through tray = 56.85 L/min (15 gal/min)
Then net flow over weir = 1707 − 5685 = 1650.15 L/min (450 − 15 = 435 gal/min)
Weir crest = 0.48 (435/60)²ᐟ³ = 45 mm (1.8 in)
Submergence = 50 + 45 = 95 mm (2 + 1.8 = 3.8 in)

From Fig. 24: $(V_h\sqrt{\rho_G})_{W\rightarrow 0} = 10.5$

$$(V_h\sqrt{\rho_G})_{oper} = \frac{1095}{60 \times 1.45}\sqrt{0.11} = 4.18$$

Substituted in Eq. (8), these values yield

$$6\log_{10}(w + 1) = 10.5 - 4.18$$
$$w = 25.1 \text{ m}^3/\text{h} \ [10.3 \text{ gal/(min)(ft}^2)]$$

Calculated weepage rate = $10.3 \times 1.45 = 14.92$ gal/min which checks the rate initially assumed.

Entrainment

A universally accepted and completely generalized correlation of entrainment in perforated plate columns has not as yet been fully developed. It is not likely to be a matter of great concern in simple water scrubber and gas cooler design, but should an estimate be desirable, the work of Friend[5] et al. is that most generally accepted as the best available. By judicious cross-plotting, a reasonable estimate can be made for the effect of most variables over the practical ranges of design dimensions.

In 1960 Friend, Lemieux, and Schreiner[1] published entrainment data for the air-water system using plates with 4.68-mm, 9.38-mm, and 12.5-mm ($\frac{3}{16}$-in, $\frac{3}{8}$-in, and $\frac{1}{2}$-in) holes at 150-mm, 225-mm, and 300-mm (6-in, 9-in, and 12-in) tray spacings. Their data for 4.68-mm ($\frac{3}{16}$-in) holes are shown in Fig. 45. The dashed curve is intended to show comparison with

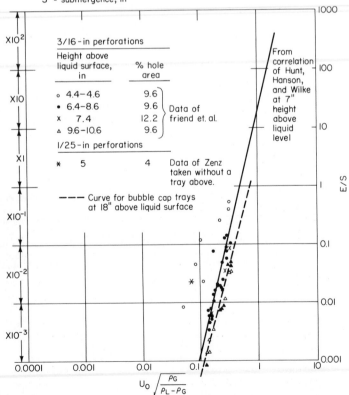

U_0 = superficial gas velocity, ft/s
ρ_G = gas density, lb/ft^3
ρ_L = liquid density, lb/ft^3
E = lb of liquid entrained per second per ft^2 of bubbling area
S = submergence, in

Fig. 45 Perforated-plate entrainment data showing tray spacing effects.

bubble cap tray entrainment data and the solid curve comparison with the correlation of Hunt, Hanson, and Wilke.[6] Figure 46 shows the data of Friend et al. for several different perforation diameters at essentially equal spacings and percent hole areas. Note that as perforation diameter decreases, the entrainment also decreases. In Figs. 45 and 46 there is also shown a point attributed to Zenz taken in a small unit measuring total entrainment at

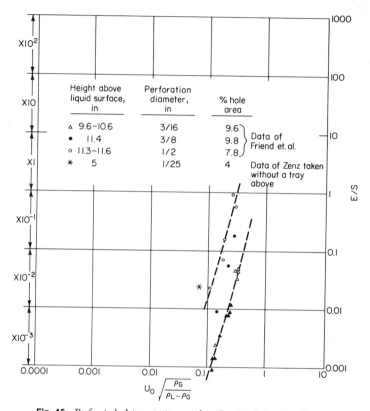

Fig. 46   Perforated-plate entrainment data showing hole size effects.

a distance of 125 mm (5 in) above a plate having 1 mm (1/25 in) diameter perforations. This point lies high because it represents total entrainment as opposed to entrainment collected atop an identical dry tray located above the test tray. This point is shown simply to illustrate the effective deentrainment action which a tray affords.

The data in Fig. 45 indicate that percent hole area (at least between 9.6 and 12.2 percent) has no detectable effect on entrainment. Cross-plotting the data in Figs. 45 and 46 yields the smoothed curves shown in Fig. 47a and b for perforated plates with 3 mm and 6 mm (1/8 in and 1/4 in) diameter holes.

Tray Efficiency

The estimation of tray efficiency is, in its rigorous procedures, extremely involved and complex. There are a number of published procedures which depend on path length, relative volatility, liquid viscosity, and tray details, no two of which will necessarily yield the same figure and among which it is not surprising to calculate extremes of 30 to 80% for a given tray and system.

For the case of simple gas cooling or absorption towers, operating with low-viscosity liquids and without weepage or excessive entrainment, it is suggested that the following figures be used for design purposes:

Perforation diameter	E, %
<1.5 mm (<¹⁄₁₆ in)	80
1.5 to 3 mm (¹⁄₁₆ to ¹⁄₈ in)	75
3 to 4.6 mm (¹⁄₈ to ³⁄₁₆ in)	70
6.25 to 9.4 mm (¹⁄₄ to ³⁄₈ in)	65

Since the most economical column is generally the one of smallest diameter, there is incentive to design for higher gas and liquid rates (i.e., smaller diameter columns) than would be indicated by Figs. 22 through 28 or 30 through 32. Obviously if the diameter is reduced, then entrainment increases and tray efficiency decreases. The loss in tray efficiency could presumably be offset by providing a few extra trays as long as their added

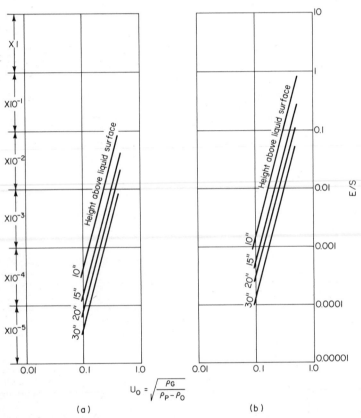

$$U_o = \sqrt{\frac{\rho_G}{\rho_P - \rho_O}}$$

(a)   (b)

Fig. 47 Smoothed entrainment curves from cross plots of Figs. 45 and 46, guided by the similar correlations for bubble-cap trays.

cost is less than the saving in total cost by virtue of the reduction in column diameter. If the idea is carried to its logical conclusion, there should exist a minimum-cost arrangement at the point where the cost of added trays exceeds the saving in tower diameter. This optimum point (probably not practical in small towers with only a few trays) has been derived for the case of bubble cap and perforated plates, yielding cost curves such as illustrated schematically in Fig. 48. The locus of minimum cost[1.7] e_m represents an entrainment rate in units of pounds of liquid entrained per pound of vapor upflow equal to 17½% of the ratio of liquid downflow to vapor upflow in the tower. Thus the minimum-cost tower is one pinched in diameter to approach an entrainment rate of 0.175 L/V.

Miscellaneous Mechanical Details

In large-diameter towers the trays are supported on channels or even I beams which span the tower diameter. In towers of 6 m (20 ft) or more the web of such a beam might be a significant determinant in selecting an operable tray spacing. Such large trays are made up in sections which must fit through a vessel manhole and which are bolted together inside the vessel with metal-to-metal seals or frequently with asbestos-rope gasketing. Leakage through such joints should not be excessive, but in moderation acts simply as more bubbling area, and compensates for the loss in perforated area along the unperforated (except for bolt holes) and usually overlapping edges of the tray sections. The peripheral edges of the tray sections bolted to the tray-support clips, spot welded to the tower shell, are also void of perforations, as are the edges to which the overflow and inlet weirs are bolted. These joint and peripheral losses partially account for the factor of 1.6 in the diameter estimate calculation given in the text under the heading of *Capacity Charts.* Typical tray- and beam-clamping arrangements are shown in Figs. 49 and 50. Figure 51 defines the common nomenclature.

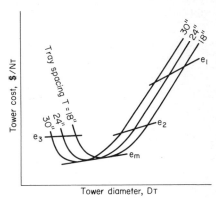

Fig. 48 Effect of diameter and tray spacing on absorption tower cost.

In large towers where it is not practical to remove an entire tray to ascend or descend through the tower for maintenance or replacement operations, it is common practice to provide at least one approximately 1000- × 1000-mm (24- × 24-in) section of the perforated area, as a separately bolted piece on each tray, oriented in vertical alignment. These manway sections can be removed to provide access for a worker to move through the column after having entered through a manhole in the shell above the top tray or below the bottom tray.

The perforations in the trays are normally punched in the tray plate metal rather than gang-drilled. The cost of punching is far less than the cost of drilling. As a rule of thumb, punching is feasible in those cases where tray thickness divided by hole diameter is less than unity. In other words, a hole of $L/D < 1.0$ can usually be punched whereas a hole of $L/D > 1.0$ must be drilled. This ratio may be slightly exceeded for the case of very soft metals or conversely might not be economically attainable with some of the harder stainless steels because of excessive tool failure. This should be taken into account when specifying tray metal thickness relative to tray support beams. Where possible, standard-pattern perforated sheets are best purchased from suppliers.

In order to permit full drainage of the tower on shutdown, it is common practice to drill weep holes in the tray inlet weirs flush with the tray floor. The number and size of these weep holes are usually based on a capacity of around 2% of the liquid throughput when operating at full load.

In instances where it is practical to flange the column head, or where the vessel diameter is too small to admit a welder, or where a specified desire exists to permit rapid change of internal tray details (even to the extent of possibly later changing to dumped packings), consideration should be given to the use of so-called cartridge-type construction. Though adaptable to any tray form, it is particularly attractive for radial-flow designs since only alternate trays need be sealed to the vessel shell. In cartridge construction, the trays are provided with a lip that conforms to the shell ID so that, when provided with an attached rope-asbestos, polyethylene, or similar gasket material, an entire stack of trays can be slid into the shell like a series of pistons. In a radial design, such as illustrated in Fig. 52, the bottom ends of the downcomers are provided with bolt "legs" which penetrate the tray below and are secured with a nut so that the entire stack can be pulled out or inserted in sections, yet the entire assembly is integral when in place.

The foregoing treatment has in many respects been focused on perforated plates because of their simplicity; economy; and ease of fabrication, design, and installation procedures, which are available to almost any shop. The variety of capped, valved, and other miscellaneous designs is legion when one considers such types as ballast, Varioflex, Jet, Flexitray, Kittel, ripple, Turbogrid, slot, disk and donut, bubble cap, tunnel, Linde, film, and Penchem trays to name but a few. As with random tower packings, there are

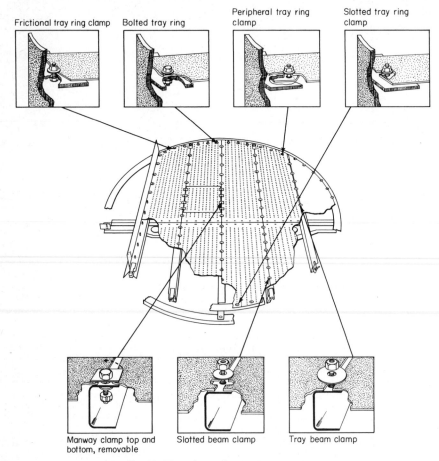

Fig. 49 Typical tray clamping arrangements.

innumerable patents and designs available, some of which have been found particularly attractive in specific applications.

Of these many variants, those most commonly finding practical industrial application in gas absorption are predominantly valve trays, followed by grid trays, tunnel cap trays, and on occasion some variation of a multiple-downcomer arrangement. More exotic varieties find greatest application in distillation of close-boiling fractions.

Valve Trays

As their name implies, valve trays present variable-area vapor openings through which gas bubbles into each tray. The available valves consist of numerous designs, some involving more than one floating element.

Though valve trays superficially appear more intricate and therefore more costly than a simple flat perforated plate, they are in many instances less costly. The number of holes in the tray floor is substantially less; the holes are large, and therefore the tray can be thicker,

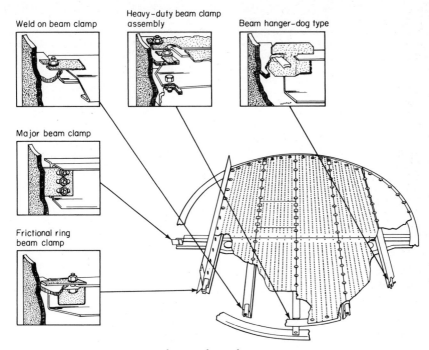

Weld on beam clamp

Heavy-duty beam clamp assembly

Beam hanger-dog type

Major beam clamp

Frictional ring beam clamp

Fig. 50 Typical support beam clamping arrangements.

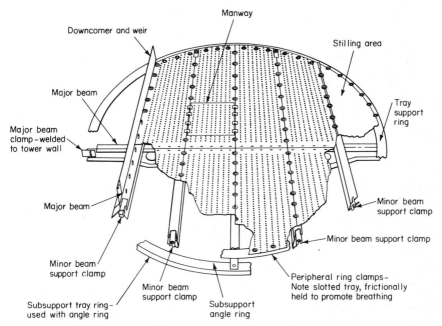

Manway

Downcomer and weir

Stilling area

Major beam

Tray support ring

Major beam clamp-welded to tower wall

Major beam

Minor beam support clamp

Minor beam support clamp

Minor beam support clamp

Minor beam support clamp

Peripheral ring clamps-Note slotted tray, frictionally held to promote breathing

Subsupport tray ring-used with angle ring

Subsupport angle ring

Fig. 51 Typical tray mechanical and structural components.

requiring less support structure, and the floating disks are relatively simple to punch out and insert. Valve trays compete rather well with custom-designed perforated plates, unless other uncompromisable restrictions such as very close tray spacings prevail. With judicious crimping, the sheets from which valve disks are stamped can be recovered as a form of expanded metal packing such as exemplified by the Glitsch grids illustrated in the section *Packed Columns.* Competitive forms of valve trays among the numerous vendors, fabricators, and innovators range from the simple single disk to multiple disks and hinged slats. Except for the details of valves, such trays are assembled and installed in column shells identical to methods illustrated in Figs. 49, 50, and 51.

Grid Trays

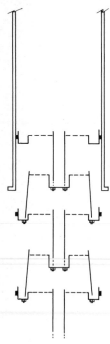

Fig. 52 Radial flow cartridge tray assembly.

Grid trays designed to circumvent the need, cost, inconvenience, and expense of downcomers were introduced with varying success over the past few decades. The Turbogrid, ripple, and Kittel forms found reasonably broad application. In general, opinion appears divided on the ability of such trays to maintain efficiency and controllability over wide ranges of operating loads, in view of the interdependence among tray liquid level, weepage or downflow rate, bubbling area, and gas or vapor throughput. Properly designed and with sufficient pressure drop, successfully operating units indicate that efficiency need not be an insurmountable problem. There are undoubtedly many unknowingly poorly designed conventional perforated trays with downcomers operating with excessive weepage and yet performing satisfactorily.

Partial control of liquid flow on downcomerless trays is achieved in the ripple and Kittel designs. In the ripple concept, ordinary perforated plate is corrugated so that some perforations are in valleys, some on peaks, and some on the slopes. Because of the differential in head between peaks and valleys, liquid preferentially gathers in the valleys and rains down to the tray below, whereas the rising gas follows the paths of least resistance and enters the tray liquid preferentially through the peaks. The ripple tray thus ideally constitutes a perforated plate from which liquid falls in a series of curtains.

The Kittel tray is made up of triangular pieces of expanded metal, butted together into hexagonal plates, with specially formed pieces around the edges making the plate circular. The slots on the tray are so arranged as to impart radial direction to the vapors leaving the plate. This causes the liquid on alternate trays to move either toward the tower shell or toward the center of the column, where it is accumulated by the directional force of the vapor entering the tray and rains down onto the tray below, simulating the liquid path on conventional radial-flow trays. To visualize how the openings on the trays are made, imagine a metal sheet pierced by a narrow slit. If one lip on the slit is pushed downward to form a triangular opening and the opposite lip is pushed upward in similar fashion, a vertically inclined rhombical opening results. Kittel plates are alternately spaced approximately 200 and 400 mm (8 and 16 in) apart. In diameters even as large as 2.4 m (8 ft), the plates are usually assembled with spacer rods and inserted as a cartridge into the shell proper. The very proprietary nature of grid trays coupled with the need for reasonably detailed and less conventional design know-how has decidedly limited their application relative to flat sieve or valve trays.

Capped and Multiple Downcomer Trays

In direct contrast to grid trays, there have been proposed a number of variations of multiple downcomer trays. Such designs are most applicable to high liquid loads and follow conventional design practice. Again, specific versions are structurally proprietary, and hence royalties coupled with the higher cost of all the added downcomer steel and greater weight erase nearly all their competitive attractions.

Seldom do absorption operations call for the characteristics associated with conventional round bubble-cap trays. However, a nearly equal and more economical form composed of channel irons arranged as so-called tunnel or channel caps has found frequent application; the design procedures are no different than those for other bubble-cap forms. See Fig. 53.

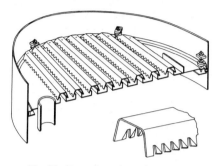

Fig. 53   Tunnel- or channel-cap trays.

REFERENCES

1. Perry, R. H. (ed.), Chemical Engineers' Handbook" (4th ed.), McGraw-Hill, New York, 1963, Chap. 18, p. 18–15.
2. Zenz, F. A., Stone, L., and Crane, M., *Hydrocarb. Proc.*, **46**(12), 138–140 (1967).
3. Zenz, F. A., *Petrol. Refiner*, **33**(2), 99–102 (1954).
4. Glitsch, F. W., and Sons, Inc., Bulletin 4900, pp. 23–25, 1961.
5. Friend, L., Lemieux, E. J., and Schreiner, W. C., *Chem. Eng.*, October 31, 1960, p. 101.
6. Hunt, C. A., Hanson, D. N., and Wilke, C. R., *AIChE J.*, **1**, 441 (1955).
7. Zenz, F. A., *Petrol. Refiner*, **36**(3), 179–181 (1957).

Appendix

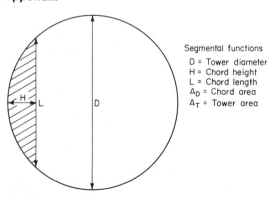

Segmental functions

D = Tower diameter
H = Chord height
L = Chord length
A_D = Chord area
A_T = Tower area

Tables giving H/D for this figure are on pages 3-94 to 3-96.

H/D From .0 to .1

H/D	L/D	A_D/A_T	H/D	L/D	A_D/A_T	H/D	L/D	A_D/A_T	H/D	L/D	A_D/A_T	H/D	L/D	A_D/A_T
.0000	.0000	.0000	.0200	.2800	.0048	.0400	.3919	.0134	.0600	.4750	.0245	.0800	.5426	.0375
.0005	.0447	.0000	.0205	.2834	.0050	.0405	.3943	.0137	.0605	.4768	.0248	.0805	.5441	.0378
.0010	.0632	.0001	.0210	.2868	.0051	.0410	.3966	.0139	.0610	.4787	.0251	.0810	.5457	.0382
.0015	.0774	.0001	.0215	.2901	.0053	.0415	.3989	.0142	.0615	.4805	.0254	.0815	.5472	.0385
.0020	.0894	.0002	.0220	.2934	.0055	.0420	.4012	.0144	.0620	.4823	.0257	.0820	.5487	.0389
.0025	.0999	.0002	.0225	.2966	.0057	.0425	.4035	.0147	.0625	.4841	.0260	.0825	.5502	.0392
.0030	.1094	.0003	.0230	.2998	.0059	.0430	.4057	.0149	.0630	.4859	.0263	.0830	.5518	.0396
.0035	.1181	.0004	.0235	.3030	.0061	.0435	.4080	.0152	.0635	.4877	.0266	.0835	.5533	.0399
.0040	.1262	.0004	.0240	.3061	.0063	.0440	.4102	.0155	.0640	.4895	.0270	.0840	.5548	.0403
.0045	.1339	.0005	.0245	.3092	.0065	.0445	.4124	.0157	.0645	.4913	.0273	.0845	.5563	.0406
.0050	.1411	.0006	.0250	.3122	.0067	.0450	.4146	.0160	.0650	.4931	.0276	.0850	.5578	.0410
.0055	.1479	.0007	.0255	.3153	.0069	.0455	.4168	.0162	.0655	.4948	.0279	.0855	.5592	.0413
.0060	.1545	.0008	.0260	.3183	.0071	.0460	.4190	.0165	.0660	.4966	.0282	.0860	.5607	.0417
.0065	.1607	.0009	.0265	.3212	.0073	.0465	.4211	.0168	.0665	.4983	.0285	.0865	.5622	.0421
.0070	.1667	.0010	.0270	.3242	.0075	.0470	.4233	.0171	.0670	.5000	.0288	.0870	.5637	.0424
.0075	.1726	.0011	.0275	.3271	.0077	.0475	.4254	.0173	.0675	.5018	.0292	.0875	.5651	.0428
.0080	.1782	.0012	.0280	.3299	.0079	.0480	.4275	.0176	.0680	.5035	.0295	.0880	.5666	.0431
.0085	.1836	.0013	.0285	.3328	.0081	.0485	.4296	.0179	.0685	.5052	.0298	.0885	.5680	.0435
.0090	.1889	.0014	.0290	.3356	.0083	.0490	.4317	.0181	.0690	.5069	.0301	.0890	.5695	.0439
.0095	.1940	.0016	.0295	.3384	.0085	.0495	.4338	.0184	.0695	.5086	.0304	.0895	.5709	.0442
.0100	.1990	.0017	.0300	.3412	.0087	.0500	.4359	.0187	.0700	.5103	.0308	.0900	.5724	.0446
.0105	.2039	.0018	.0305	.3439	.0090	.0505	.4379	.0190	.0705	.5120	.0311	.0905	.5738	.0449
.0110	.2086	.0020	.0310	.3466	.0092	.0510	.4400	.0193	.0710	.5136	.0314	.0910	.5752	.0453
.0115	.2132	.0021	.0315	.3493	.0094	.0515	.4420	.0195	.0715	.5153	.0318	.0915	.5766	.0457
.0120	.2178	.0022	.0320	.3520	.0096	.0520	.4441	.0198	.0720	.5170	.0321	.0920	.5781	.0460
.0125	.2222	.0024	.0325	.3546	.0098	.0525	.4461	.0201	.0725	.5186	.0324	.0925	.5795	.0464
.0130	.2265	.0025	.0330	.3573	.0101	.0530	.4481	.0204	.0730	.5203	.0327	.0930	.5809	.0468
.0135	.2308	.0027	.0335	.3599	.0103	.0535	.4501	.0207	.0735	.5219	.0331	.0935	.5823	.0472
.0140	.2350	.0028	.0340	.3625	.0105	.0540	.4520	.0210	.0740	.5235	.0334	.0940	.5837	.0475
.0145	.2391	.0030	.0345	.3650	.0108	.0545	.4540	.0212	.0745	.5252	.0337	.0945	.5850	.0479
.0150	.2431	.0031	.0350	.3676	.0110	.0550	.4560	.0215	.0750	.5268	.0341	.0950	.5864	.0483
.0155	.2471	.0033	.0355	.3701	.0112	.0555	.4579	.0218	.0755	.5284	.0344	.0955	.5878	.0486
.0160	.2510	.0034	.0360	.3726	.0115	.0560	.4598	.0221	.0760	.5300	.0347	.0960	.5892	.0490
.0165	.2548	.0036	.0365	.3751	.0117	.0565	.4618	.0224	.0765	.5316	.0351	.0965	.5906	.0494
.0170	.2585	.0037	.0370	.3775	.0119	.0570	.4637	.0227	.0770	.5332	.0354	.0970	.5919	.0498
.0175	.2622	.0039	.0375	.3800	.0122	.0575	.4656	.0230	.0775	.5348	.0358	.0975	.5933	.0501
.0180	.2659	.0041	.0380	.3824	.0124	.0580	.4675	.0233	.0780	.5363	.0361	.0980	.5946	.0505
.0185	.2695	.0042	.0385	.3848	.0127	.0585	.4694	.0236	.0785	.5379	.0364	.0985	.5960	.0509
.0190	.2730	.0044	.0390	.3872	.0129	.0590	.4712	.0239	.0790	.5395	.0368	.0990	.5973	.0513
.0195	.2765	.0046	.0395	.3896	.0132	.0595	.4731	.0242	.0795	.5410	.0371	.0995	.5987	.0517

H/D From .2 to .3

H/D	L/D	A_D/A_T	H/D	L/D	A_D/A_T	H/D	L/D	A_D/A_T	H/D	L/D	A_D/A_T	H/D	L/D	A_D/A_T
.2000	.8000	.1424	.2200	.8285	.1631	.2400	.8542	.1845	.2600	.8773	.2066	.2800	.8980	.2292
.2005	.8007	.1429	.2205	.8292	.1636	.2405	.8548	.1851	.2605	.8778	.2072	.2805	.8985	.2298
.2010	.8015	.1434	.2210	.8298	.1642	.2410	.8554	.1856	.2610	.8784	.2077	.2810	.8990	.2304
.2015	.8022	.1439	.2215	.8305	.1647	.2415	.8560	.1862	.2615	.8789	.2083	.2815	.8995	.2309
.2020	.8030	.1444	.2220	.8312	.1652	.2420	.8566	.1867	.2620	.8794	.2088	.2820	.8999	.2315
.2025	.8037	.1449	.2225	.8319	.1658	.2425	.8572	.1873	.2625	.8800	.2094	.2825	.9004	.2321
.2030	.8045	.1454	.2230	.8325	.1663	.2430	.8578	.1878	.2630	.8805	.2100	.2830	.9009	.2326
.2035	.8052	.1460	.2235	.8332	.1668	.2435	.8584	.1884	.2635	.8811	.2105	.2835	.9014	.2332
.2040	.8059	.1465	.2240	.8338	.1674	.2440	.8590	.1889	.2640	.8816	.2111	.2840	.9019	.2338
.2045	.8067	.1470	.2245	.8345	.1679	.2445	.8596	.1895	.2645	.8821	.2116	.2845	.9024	.2344
.2050	.8074	.1475	.2250	.8352	.1684	.2450	.8602	.1900	.2650	.8827	.2122	.2850	.9028	.2349
.2055	.8081	.1480	.2255	.8358	.1689	.2455	.8608	.1906	.2655	.8832	.2128	.2855	.9033	.2355
.2060	.8089	.1485	.2260	.8365	.1695	.2460	.8614	.1911	.2660	.8837	.2133	.2860	.9038	.2361
.2065	.8096	.1490	.2265	.8371	.1700	.2465	.8619	.1917	.2665	.8843	.2139	.2865	.9043	.2367
.2070	.8103	.1496	.2270	.8378	.1705	.2470	.8625	.1922	.2670	.8848	.2145	.2870	.9047	.2372
.2075	.8110	.1501	.2275	.8384	.1711	.2475	.8631	.1927	.2675	.8853	.2150	.2875	.9052	.2378
.2080	.8118	.1506	.2280	.8391	.1716	.2480	.8637	.1933	.2680	.8858	.2156	.2880	.9057	.2384
.2085	.8125	.1511	.2285	.8397	.1721	.2485	.8643	.1938	.2685	.8864	.2161	.2885	.9061	.2390
.2090	.8132	.1516	.2290	.8404	.1727	.2490	.8649	.1944	.2690	.8869	.2167	.2890	.9066	.2395
.2095	.8139	.1521	.2295	.8410	.1732	.2495	.8654	.1949	.2695	.8874	.2173	.2895	.9071	.2401
.2100	.8146	.1527	.2300	.8417	.1738	.2500	.8660	.1955	.2700	.8879	.2178	.2900	.9075	.2407
.2105	.8153	.1532	.2305	.8423	.1743	.2505	.8666	.1961	.2705	.8884	.2184	.2905	.9080	.2413
.2110	.8160	.1537	.2310	.8429	.1748	.2510	.8672	.1966	.2710	.8890	.2190	.2910	.9084	.2419
.2115	.8167	.1542	.2315	.8436	.1754	.2515	.8678	.1972	.2715	.8895	.2195	.2915	.9089	.2424
.2120	.8174	.1547	.2320	.8442	.1759	.2520	.8683	.1977	.2720	.8900	.2201	.2920	.9094	.2430
.2125	.8182	.1553	.2325	.8449	.1764	.2525	.8689	.1983	.2725	.8905	.2207	.2925	.9098	.2436
.2130	.8189	.1558	.2330	.8455	.1770	.2530	.8695	.1988	.2730	.8910	.2212	.2930	.9103	.2442
.2135	.8196	.1563	.2335	.8461	.1775	.2535	.8700	.1994	.2735	.8915	.2218	.2935	.9107	.2448
.2140	.8203	.1568	.2340	.8467	.1781	.2540	.8706	.1999	.2740	.8920	.2224	.2940	.9112	.2453
.2145	.8210	.1573	.2345	.8474	.1786	.2545	.8712	.2005	.2745	.8925	.2229	.2945	.9116	.2459
.2150	.8216	.1579	.2350	.8480	.1791	.2550	.8717	.2010	.2750	.8930	.2235	.2950	.9121	.2465
.2155	.8223	.1584	.2355	.8486	.1797	.2555	.8723	.2016	.2755	.8935	.2241	.2955	.9125	.2471
.2160	.8230	.1589	.2360	.8492	.1802	.2560	.8728	.2021	.2760	.8940	.2246	.2960	.9130	.2477
.2165	.8237	.1594	.2365	.8499	.1808	.2565	.8734	.2027	.2765	.8945	.2252	.2965	.9134	.2482
.2170	.8244	.1600	.2370	.8505	.1813	.2570	.8740	.2033	.2770	.8950	.2258	.2970	.9139	.2488
.2175	.8251	.1605	.2375	.8511	.1818	.2575	.8745	.2038	.2775	.8955	.2264	.2975	.9143	.2494
.2180	.8258	.1610	.2380	.8517	.1824	.2580	.8751	.2044	.2780	.8960	.2269	.2980	.9148	.2500
.2185	.8265	.1615	.2385	.8523	.1829	.2585	.8756	.2049	.2785	.8965	.2275	.2985	.9152	.2506
.2190	.8271	.1621	.2390	.8529	.1835	.2590	.8762	.2055	.2790	.8970	.2281	.2990	.9156	.2511
.2195	.8278	.1626	.2395	.8536	.1840	.2595	.8767	.2060	.2795	.8975	.2286	.2995	.9161	.2517

Spray Towers

Although there exist in the published literature a reasonable number of papers dealing with experimental investigations of gas absorption in spray towers, no one has yet succeeded in establishing an indisputable generalized design, or sizing, correlation. As might be expected, the number of variables, the method of operation, the differences in mechanical details and the effects of the size of the test equipment easily lead to what usually appear to be hopelessly scattered data.

The classical spray tower consists of nothing more than an empty cylindrical shell through which gas passes upwards against a downflowing spray or rainfall of absorbing liquid. The liquid is the dispersed phase and the gas is the continuous phase. Presumably, mass transfer occurs at the liquid droplet surface, and hence published correlations for single drops, single particles, and even clouds of particles might be expected to apply. However, beyond the academic realm of single drops of very dilute sprays, the practical means necessary to produce the sprays frequently introduce quantitatively uninterpretable effects.

The liquid is in practice introduced through high-pressure spray nozzles. In order to obtain good contact and avoid bypassing any upflowing gas, the spray must cover the entire column cross section. This is nearly impossible without having some portion of the liquid spray, after leaving the point-source nozzle, impinge on the vessel wall and stream down as a film that no longer affords drop-type contact. By the same token, no spray nozzle produces perfectly uniform-size drops, and hence some inevitably catch up with others in falling down through the column. The drop size, velocity, residence time, etc. therefore vary considerably with height or location in the tower. In large-diameter columns, with overlapping sprays that fully cover the cross-sectional area, the unpredictable heterogeneity is even greater.

Despite these practically imposed resistances to correlation in terms of fundamentals, spray towers still find application in areas where less effective tower volume is a small price to pay for the ability to handle difficult fluids.

Areas of Application

Spray towers are of particular interest in instances where one or more of the following limitations are involved:

1. The gas to be scrubbed is corrosive either alone or in its dilute solution in the absorbing liquid.

2. The gas contains dust, soot, or some form of fine particles which would cause interstitial blockages if a conventional packed tower were used.

3. Only one or at most two theoretical stages of contact are required.

4. The gas-side pressure is at a premium and the pressure drop must therefore be kept to a minimum.

Its disadvantages lie in the cost of power for pressuring the absorbing liquid through the spray nozzles, in the absolute necessity of installing entrainment suppressors or mist eliminators to control the inevitable entrainment, and in the substantial height necessary to achieve a theoretical stage of absorption.

Capacity Limits

Capacity limits, analogous to flooding in packed and tray columns, are not meaningful with respect to sizing spray columns. The countercurrent gas-liquid flooding correlations for empty towers are based on a continuum of liquid as opposed to dispersed droplets. The criterion in spray towers is simply that the upflowing gas velocity be low enough to allow droplets to fall freely. The terminal or free-fall velocity of liquid droplets as a function of their size and the physical properties of the liquid and its surrounding gaseous medium is given by Fig. 54. Nominally superficial gas upflow velocities in spray towers range from 0.5 to 2 ft/s. The denominators of abscissa and ordinate of Fig. 54 are, for example, 0.00108 and 1.83 ft/s, respectively, for water droplets in atmospheric air so that a typical 400-μm spray-nozzle droplet would have a free-fall velocity of about 5.5 ft/s, which is well above the usual nominal operating gas velocity. A common guide is a maximum of 20% of the free-fall velocity of the mean droplet diameter characteristic of the nozzles at the desired operating conditions; by this rule a superficial velocity of $5.5 \times 0.2 = 1.1$ ft/s would be specified as a basis for determining the tower diameter in this instance.

Entrainment

In view of the multitude of operational variables, and the broad spectrum of available droplet sizes and distributions from available nozzle designs, the correlation of entrainment has never been satisfactorily resolved. Since a prime area of application of spray towers lies in corrosive acid or caustic absorption, even relatively minor entrainment is

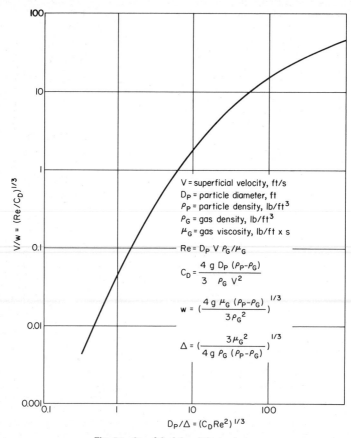

Fig. 54 Simplified free-fall correlation.

usually undesirable, if not intolerable. It is therefore usual practice to install a shallow bed of ceramic packing or, more commonly, a wire-mesh mist separator in the area above the nozzles. The capacity or sizing of such a knitted wire-mesh demister pad, which need not necessarily cover the entire tower cross section, is generally quoted on the basis of an allowable superficial gas velocity given by

$$V = K \sqrt{\frac{\rho_L - \rho_G}{\rho_G}} \qquad \text{ft/s}$$

where K would characteristically be 0.4 to 0.45 at flooding against a downflow (or collected) liquid rate of about 200 lb/(h)(ft²). At the recommended operating gas rate $K = 0.3$, which corresponds to 65 to 75% of flooding. This coefficient is also a function of the density of packing of the wire mesh and presumably of the viscosity of the tower liquid phase. It therefore relates the mesh characteristics to an equivalent packing factor, as discussed and defined under *Packed Towers*. A mesh characterized by a K of 0.4 to 0.45 at a liquid counterflow of 200 lb/(h)(ft²) (if the liquid is water) would have an equivalent ordinate and abscissa in Fig. 3 of 20 and 0.31, respectively, indicating an equivalent packing factor of 60.

Liquid Feed and Distribution

Absorbing liquid is simply fed to a spray tower via a manifold of nozzles. With the exception of the exit-gas demister pad, these nozzles frequently represent the only tower internals, obviating the economic simplicity of this form of absorber.

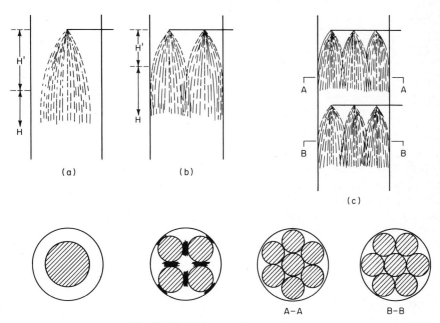

Fig. 55   Nozzle layout and spray coverage.

As mentioned earlier, the disposition of the nozzles should be such as to produce a nonagglomerating "rain" over the entire tower cross section. This is theoretically impossible, as demonstrated by the arrangements depicted in Fig. 55. Obviously, the arrangement in Fig. 55a suffers from incomplete coverage at the periphery, although with a smaller column or wider nozzle this would approach the theoretical perfect coverage. Figure 55b illustrates the more common multiple-nozzle arrangement where again areas of incomplete coverage and agglomerating-droplet overlap distort the ideally desirable flow patterns. A more desirable arrangement is illustrated in Fig. 55c where theoretically incompletely covered areas at section A-A are offset by rotating the nozzle arrangement 90° at the lower level, as in section B-B. The theoretical losses in area coverage illustrated in Fig. 55b and c are in practice never so distinct, since in an upflowing countercurrent gas stream the droplet fall pattern is inevitably distorted to a broader though still imperfect coverage.

Nozzle Characteristics

Nozzles providing a wide variety of spray patterns are commercially available. Those most relevant to absorption towers are the so-called hollow-cone and full-cone patterns illustrated in Fig. 56. As depicted in Fig. 56, the full-cone nozzles are those normally specified and those intended in design correlations from which required tower volumes are derived. Manufacturer's bulletins provide a wide variety of specifications concerning spray patterns and droplet sizes as functions of nozzle orifice size, flow rate, and pressure. Typical figures are shown in Table 4.

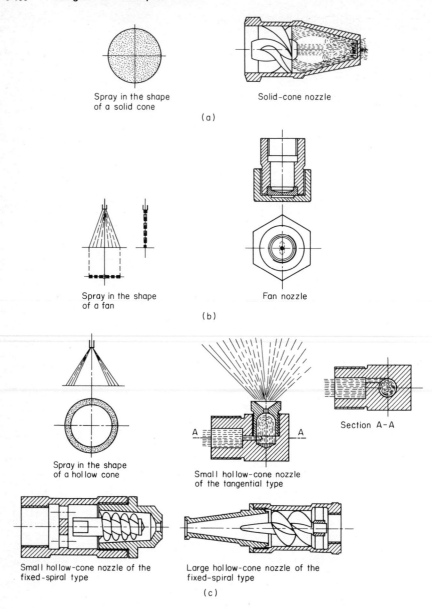

Spray in the shape
of a solid cone

Solid-cone nozzle

(a)

Spray in the shape
of a fan

Fan nozzle

(b)

Spray in the shape
of a hollow cone

Small hollow-cone nozzle
of the tangential type

Section A-A

Small hollow-cone nozzle of the
fixed-spiral type

Large hollow-cone nozzle of the
fixed-spiral type

(c)

Fig. 56 Conventional spray nozzle designs and characteristics.

TABLE 4   Discharge Rates and Included Angle of Spray of Typical Pressure Nozzles

Nozzle type	Orifice diameter, in	Discharge, gpm, and included angle of spray, degrees							
		10 lb/in²		25 lb/in²		50 lb/in²		100 lb/in²	
		Discharge	Angle	Discharge	Angle	Discharge	Angle	Discharge	Angle
Hollow	0.046			0.10	65	0.135	68	0.183	75
cone	.140	0.535	82	0.81	88	1.10	90	1.50	93
	.218	1.25	83	1.88	86	2.55	89	3.45	92
	.375	7.2	62	11.8	70	16.5	70		
Solid	.047			0.167	65	0.235	70	0.34	70
cone	.188	1.60	55	2.46	58	3.42	60	4.78	60
	250	3.35	65	5.40	70	7.50	70	10.4	75
	.500	17.5	86	27.5	84	38.7	73		
Fan	.031	0.085	40	0.132	90	0.182	110	0.252	110
	.093	0.70	70	1.12	76	1.57	80	2.25	80
	.187	2.25	50	3.70	59	5.35	65	7.70	65
	.375	9.50	66	15.40	74	22.10	75	30.75	75

Transfer Rates and Tower Volume

Variations in nozzle pattern and droplet size with flow rate exacerbate the inevitable difficulties in correlating performance. As illustrated in Fig. 55, the volume within height H', which varies also with nozzle number and orientation, introduces additional ambiguities in correlation, particularly between laboratory-scale single-nozzle results and industrial-scale multinozzle installations.

To propose a generalized correlation for the mass-transfer coefficient in any spray tower is essentially impossible; however, industrial experience with solid-cone nozzles, over a maximum height of 4 ft from the specified range of sizes of nozzles [i.e., $H + H$ of Fig. 55 = 1.2 m (4 ft) maximum], has conservatively suggested the empirical relationship shown in Fig. 57. Mean droplet diameters are usually obtainable from nozzle manufacturers.

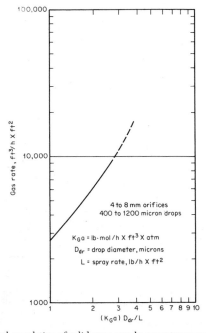

Fig. 57   Empirical correlation of solid cone nozzle spray tower performance tests.

Overall Arrangement

The basic elements of a conservative two-stage spray tower design are illustrated in Fig. 58. In order to minimize distortion of the spray pattern, it is advisable to provide some form of feed-gas distributor. This can take the form of a conventional packed-tower

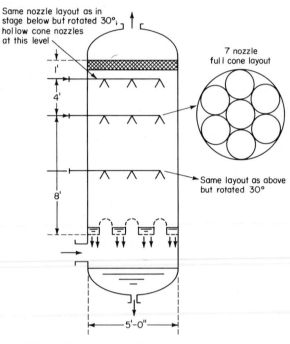

Fig. 58   Basic elements of a two-stage spray tower.

distributor of the type illustrated in Fig. 15*b* and *c*. In instances of very high *L/G*, a feed-gas distributor might not be absolutely necessary to achieve the calculated mass transfer.

Figure 58 illustrates the use of hollow-cone nozzles (just below the demister pad) as entrainment reducers. Such nozzles are frequently employed without a superimposed demister, and even more frequently the demister pad is employed alone without the hollow-cone nozzles. The inherent simplicity of the spray tower makes field trials a convenient and more positive design determinant than calculations based on plots such as Fig. 57.

SAMPLE CALCULATIONS

To illustrate the use of some of the foregoing material in arriving at a required vessel diameter or height, several types of sample problems are presented on the following pages. They are directed principally to process sizing aspects rather than mechanical design, and are in some instances illustrative of rapid approximation methods and in others of more detailed analysis.

1. PACKING FACTOR FROM FLOODING DATA

Figure 2 on p. 5 of the F.W. Glitsch Co. Bulletin No. 7070 pertaining to "Glitsch Grids" (a punched-metal tower packing) gives flooding data in terms of countercurrent volumetric air and water flows through a 4-ft ID test tower as follows:

You wish to consider this packing for application in other gas-liquid systems and therefore need a generalization of its flooding characteristics. What is the effective packing factor (a/F^3) of this material?

Superficial air velocity, ft/s	Water rate, gal/min
16	Approx. 0
15	125.8
14	252
13	377
12	504
11	629
10	755

Water density = 62.4 lb/ft³
Water viscosity = 1.0 cP
Air density = 0.076 lb/ft³

The procedure involves converting the above-tabulated flooding data into the coordinates of Fig. 3, plotting the data on this Fig. 3 and by interpolation, or cross-plot extrapolation, determining the value of the labeled parameter. This procedure will also clarify the consistency of the data. If the data do not describe a curve paralleling those in Fig. 3, then the data might be questionable. The conversions are performed in the following fashion:

$$\left(\begin{matrix}\text{Superficial air}\\ \text{velocity, ft/s}\end{matrix}\right)\left(\frac{60}{1}\right)\left(\frac{\pi}{4}\right)\sqrt{\frac{0.076}{62.4}} = \text{ordinate}$$

$$\text{gpm}/4^2 = \text{abscissa}$$

which yields the figures below

$\dfrac{cfm/\sqrt{\rho_L/\rho_G}}{D_T^2}$	$\dfrac{gpm}{D_T^2}$
26.31	0
24.67	7.86
23.02	15.75
21.38	23.56
19.73	31.50
18.09	39.31
16.45	47.19

When plotted on Fig. 3, these points describe a higher parallel curve which upon cross-plotting may be considered to have a parametric value of 20. Since the liquid viscosity was 1 cP, this means (a/F^3) or the equivalent packing factor of this material has a value of 20, which may now be applied generally to calculate the countercurrent flooding rates through this material for any gas-liquid system.

2. PREDICTION OF FLOODING RATES FOR UNKNOWN PACKING CHARACTERISTICS

You are confronted in the field with a 36 in ID packed tower of unknown internal packing size or type, and are told that flooding was noted once at a liquid rate of 6380 gal/h with a countercurrent gas rate of 56,500 ft³/h. The tower is currently operating at the 6380 gal/h rate with a countercurrent gas rate of 38,000 ft³/h. The operator wants to know whether the liquid rate can be increased to 8000 gal/h without making an alteration of the gas flow necessary. Is this proposed increase in liquid rate possible without flooding?

Since the tower, the packing, the fluids, and the operating conditions are constant, the solution simply involves determining whether the sum of the cube root of the intended volumetric gas flow and the square root of the intended volumetric liquid flow exceeds the similar sum under the previously noted flooding conditions. Under flooding conditions:

$$\left[\frac{56,500/60}{\pi(3)^2/4}\right]^{1/3} + \left[\frac{6380/60}{7.481\pi(3)^2/4}\right]^{1/2} = 6.525$$

Under the operating gas flow:

$$\left[\frac{38,000/60}{\pi(3)^2/4}\right]^{1/3} + \left[\frac{L_F/60}{7.481\pi(3)^2/4}\right]^{1/2} = 6.525$$

from which $L_F = 13,335$ gal/h, and hence the operator can safely expect the tower to operate at the increased liquid rate of 8000 gal/h without flooding.

3. NUMBER OF TRAYS IN A GAS COOLER

It is proposed to use a packed or a tray tower to cool a gas stream. The gas volume is 78,300 cfm at 14.7 lb/in²(abs) and 153°F saturated with water vapor and has the following composition:

	Moles
Carbon dioxide	16.65
Nitrogen	114.45
Water vapor	47.40

This gas is to be cooled to 105°F with 90°F feedwater leaving the tower at 120°F. Given the enthalpies, vapor pressures, and latent heats in the first six columns of the table below, calculate the number of theoretical trays required and the amount of water condensed in cooling the gas.

The solution consists of drawing operating and equilibrium curves for the system and stepping off the required number of equilibrations, as is typical of such operations, best exemplified by the McCabe-Thiele graphical solution for the number of trays required in a separation of two components by distillation. In this instance the equilibrium curve is best represented by plotting the enthalpy of the water-saturated dry gas (as for example Btu per pound of dry gas) vs. temperature. The operating line would represent the enthalpy of the gas stream vs. the cooler liquid temperature existing at the same point in the tower. The temperature difference represents the driving force. The calculation proceeds in the following manner. From the given dry gas composition:

	Molecular weight	*lb·mol/min*		*lb/min*	*Pounds per pound of dry gas*
CO_2	44	× 16.65	=	732	0.19
N_2	28	× 111.45	=	3120	0.81
	30	128.10		3852	1.00

In the table below, col. 8 = 0.19 (col. 1) + 0.81 (col. 2) + (col. 7) (col. 3 + col. 4), and mole fraction water in total gas = vapor pressure/14.7.

T, °F	Enthalpy above 32°F, Btu/lb			Water's latent heat of vaporization, Btu/lb	H_2O vapor pressure lb/in²(abs)	Mole fraction H_2O in total gas	Pounds H_2O per pound of dry gas	Btu/lb of dry gas at saturation
	CO_2 (1)	N_2 (2)	H_2O (3)	(4)	(5)	(6)	(7)	(8)
90	11.51	14.42	58.00	1042.1	0.69	0.0470	0.0294	46.24
95	12.52	15.66	62.95	1039.3	0.80	0.0545	0.0345	53.08
100	13.54	16.91	67.97	1036.4	0.94	0.0640	0.0408	61.32
105	14.54	18.15	72.96	1033.6	1.19	0.0809	0.0525	75.51
110	15.57	19.40	77.94	1030.9	1.26	0.0855	0.0559	80.56
115	16.60	20.64	82.93	1028.1	1.45	0.0985	0.0653	92.36
120	17.63	21.89	87.91	1025.3	1.67	0.1133	0.0764	106.15
125	18.67	23.13	92.90	1022.4	1.92	0.1305	0.0899	122.45
130	19.71	24.38	97.89	1019.5	2.20	0.1495	0.1050	140.65
135	20.75	25.62	102.89	1016.6	2.51	0.171	0.1232	162.74
140	21.80	26.87	107.88	1013.7	2.88	0.196	0.1460	189.95
145	22.85	28.11	112.88	1010.8	3.26	0.222	0.1708	219.15
150	23.90	29.36	117.87	1007.8	3.70	0.252	0.2010	254.35
155	24.96	30.60	122.87	1005.0	4.18	0.284	0.2370	296.54
160	26.02	31.85	127.87	1002.0	4.70	0.320	0.2810	347.75

The equilibrium or saturation curve represented by col. 8 plotted vs. temperature is shown in Fig. 59. The operating line is determined by a heat balance. The heat gained by the water must equal the heat lost by the gas so that

$$L(C_p)(120 - 90) = 3852(280 - 75.5)$$

where L = required lbs of 90°F feedwater
C_p = specific heat of water, Btu/(lb)(°F)
$(120 - 90)$ = temperature rise in the water, °F
3852 = gas flow, lb/min dry
$(280 - 75.5)$ = difference in enthalpy of the gas between 153 and 105°F, Btu/lb of dry gas

Thus, with a specific heat of 1.0 Btu/(lb)(°F), the required feedwater flow L is 26,240 lb/min. Again, equating the enthalpy change in the gas stream to the enthalpy change in the liquid stream defines the operating line:

$$(280 - H)\,3852 = L\,(C_p)(120 - t)$$

Thus, with $C_p = 1.0$, the above relationship allows plotting H in Btu/lb of dry gas vs. t, shown as the operating line in Fig. 59. The number of theoretical trays is obtained graphically as 1.5.

The amount of water condensed out of the feed gas stream is simply the dry gas flow rate times the difference in water content between feed and exit gas streams.

t, °F	Pounds H_2O per pound of dry gas
153	0.2220
105	−0.0525
	0.1695 (3852) = 653 lb water condensed

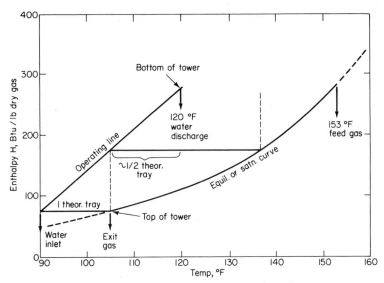

Fig. 59   Graphical solution of Example 3.

4. OPTIMIZATION OF PACKED ABSORPTION TOWER DESIGN

The following example illustrates the optimization of tower proportions for a fixed tower packing material with no limitations on space, water availability, or allowable pressure drop.

What is the probable optimum diameter and height of a 2-in polypropylene pall-ring packed tower for the absorption of a highly water soluble contaminant gas in an 80,000 acfm air stream if isothermal operation at 16°C and 1 atm is assumed? The contaminant gas has a molecular weight of 38. Its concentration in the feed air stream is 0.02 grains/ft³; its concentration in the gas leaving the absorber must not exceed 0.001 grains/ft³.

The optimization procedure consists of calculating absorber heights and diameters for various water rates. For illustrative purposes it will be assumed that the only information available on the specified packing consists of the data shown in Figs. 60 and 61, as they appear in the manufacturer's bulletin.

If the densities of air and water are taken as 0.076 and 62.4 lb/ft³, respectively, the packing factor can be calculated as in the case of Example 1 by noting in Fig. 60 that flooding has essentially been reached at a pressure drop of 2 in of water per foot. Values of gas, G_F, and liquid, L_F, flooding rates read from Fig. 60 can be converted to the coordinates of Fig. 3 as follows:

$$\frac{L_F \times 7.481(\pi/4)}{60 \times 62.4} = \frac{L_F}{637.5} = \frac{\text{gpm}}{D_T^2}$$

$$\frac{G_F(\pi/4)}{60 \times 0.076}\sqrt{\frac{0.076}{62.4}} = \frac{G_F}{166} = \frac{\text{cfm}/D_T^2}{\sqrt{\rho_L/\rho_G}}$$

From Fig. 65		Calculated Coordinates of Fig. 3	
L_F, lb/(h)(ft²)	G_F, lb/(h)(ft²)	$\dfrac{\text{gpm}}{D_T^2}$	$\dfrac{\text{cfm}}{D_T^2}\sqrt{\dfrac{\rho_L}{\rho_G}}$
70,000	440	110	2.65
50,000	630	78.5	3.80
30,000	1300	47.0	7.85
10,000	2400	15.7	14.5

Plotting the last two columns of figures on the generalized flooding correlation of Fig. 3 shows the data falling directly on the curve labeled $\mu_L^{0.2}$ $(a/F^3) = 40$. Since $\mu_L = 1.0$ cP, the 2-in polypropylene pall rings have an effective packing factor of 40.

Figure 61 gives K_Ga values which must also be converted to generalized coordinates in order to

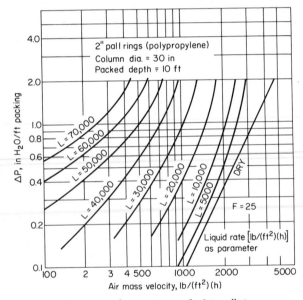

Fig. 60   Pressure drop vs. gas rate for 2-in pall rings.

allow extrapolation and interpolation to other operating conditions. The generalized mass-transfer correlation is based on J_D vs. Re as in Fig. 10 and 11, where

$$J_D = \frac{k_G PM}{G}\,(\text{Sc})^{2/3}$$

and

$$\text{Re} = \sqrt{\frac{\text{ft}^2 \text{ of surface/piece}}{\pi}}\;\frac{G}{\mu_G}$$

Since $(\text{Sc})^{2/3}$ for CO_2 in air is essentially unity:

$$J_D = \frac{k_Ga \text{ from Fig. 3}}{31}\,\frac{29}{500} = \frac{k_G a}{535}$$

where 31 is the manufacturer's reported square feet of pall ring surface in a cubic foot of packed volume, and

$$\text{Re} = \sqrt{\frac{31/180}{\pi}}\;\frac{500}{0.0182 \times 2.42} = 2660$$

where 180 is the manufacturer's reported number of these pall rings contained in a cubic foot of packed volume and 2.42×0.0182 is the viscosity of air in units of lb/(ft)(h). The data in Fig. 3 therefore reduce to the following values of the generalized coordinates:

From Fig. 3		Coordinates of Fig. 10	
L	$k_G a$	J_D	Re
5,000	1.9	0.00355	2660
10,000	2.25	0.00420	2660
20,000	2.6	0.00486	2660
40,000	2.9	0.00541	2660
60,000	3.0	0.00560	2660

Plotting the above values of J_D vs. Re on Fig. 10 and drawing lines parallel to analogous packings gives a basis for obtaining gas-phase mass-transfer coefficients at any gas and liquid rates in any system for 2-in pall rings. Note that $K_1 a$ has been assumed to be negligible so that $k_G a$ is essentially the overall coefficient $K_G a$.

With generalized mass-transfer and flooding data, it is now a relatively routine procedure to calculate packed depths and tower diameters for the conditions of the problem at various water rates.

Instead of randomly choosing a series of liquid rates, it is more methodical to select values along a curve of Fig. 3 parallel to that for the packing factor of 40 (the 2-in polypropylene rings), but displaced therefrom by a factor of 0.75 along any diagonal of constant L/G. Any operating point along such a curve would require the corresponding gas and liquid rates to both be increased by a factor of 1/0.75 before the tower would flood, so that the tower diameters calculated from the displaced curve would be operating at 75% of flooding, which is a reasonable design basis. Since the gas density in this case is essentially 0.071 lb/ft³, the liquid density is 62.4 lb/ft³, and the gas flow is 80,000 acfm, selected intervals along the abscissa of the 75% of flooding curve yield the following figures:

From Fig. 3 at 75% of Flooding

$\dfrac{\text{gpm}}{D_T^2}$	$\dfrac{\text{acfm}}{D_T^2}\sqrt{\dfrac{\rho_G}{\rho_L}}$	$\dfrac{\text{gpm}}{\text{acfm}}$	gpm	D_T, ft	L, lb/(h)(ft²)	G, lb/(h)(ft²)
100	1.4	2.405	192,400	44.0	63,700	226
60	4.0	0.505	40,400	26.0	38,200	645
30	7.6	0.133	10,630	18.9	19,100	1220
10	12.5	0.027	2,160	14.7	6,370	2012
6	15.0	0.0135	1,080	13.5	3,820	2420
3	16.5	0.00612	490	12.8	1,910	2660
1	19.0	0.00177	142	12.0	637	3060

The corresponding required packing depths or tower heights are obtained from the J_D-vs.-Re curves generated from the data cited in Fig. 3 as described above. The last column in the table above leads to the values of Re in the table below, and hence to the values of J_D at the corresponding liquid rates. Assuming that the Schmidt number Sc is again equal to unity allows calculation of the predicted mass-transfer coefficient $k_G a$ from the gas rates G and the values of J_D, as shown in col. 3 of the table below, where $k_G a$ has the units of pound-moles of contaminant transferred from the gas to the liquid per hour per cubic foot of packed volume per atmosphere of driving force. The driving force is taken as the log mean of the difference between the actual partial pressure of the contaminant in the gas stream and the partial pressure it would exhibit if the gas were in equilibrium with the liquid adjacent to it as they exist at the bottom and top of the tower. The log mean of these differences is satisfactory as long as the equilibrium curve (partial pressure in the vapor vs. concentration in the adjacent liquid) is a straight line. This is usually the case at low concentrations. In this instance, with no equilibrium data and a contaminant which is highly soluble, it will be assumed that the equilibrium partial pressure is zero at all contaminant concentrations in the liquid over the range of interest.

Since the 38 molecular weight contaminant gas is to be stripped out of the 80,000 acfm airstream at an inlet concentration of 0.02 grains/ft³ to an outlet concentration of 0.001 grains/ft³ the transfer rate amounts to

$$\frac{80,000(0.02 - 0.001)60}{7000 \times 38} = 0.342 \text{ lb} \cdot \text{mol/h}$$

The mole fractions (or partial pressures, in atmospheres) of the contaminant in the 0.071 lb/ft³ density air stream amount to

$$\frac{0.02 \times 29}{7000 \times 0.071 \times 38} = 0.0000308 \text{ atm} \qquad \text{in the feed gas}$$

$$\frac{0.001 \times 29}{7000 \times 0.071 \times 38} = 0.00000154 \text{ atm} \qquad \text{in the exit gas}$$

so that the log mean driving force (with zero equilibrium partial pressures) amounts to

$$\frac{(0.0000308 - 0) - (0.00000154 - 0)}{\ln \dfrac{(0.0000308 - 0)}{(0.00000154 - 0)}} = 0.00000975 \text{ atm}$$

From the definition of k_Ga, and its predicted values based on the J_D-vs.-Re correlation, the corresponding packed volumes in the table below are obtainable as

$$\text{Packed volume required, ft}^3 = \frac{0.342}{0.00000975 k_Ga}$$

These volumes, and the previously calculated diameters, yield the corresponding packed heights.

To Strip 80,000 acfm
from 0.02 to 0.001 grains/ft³

Re	J_D	k_Ga	Packing, ft³	Packed height, ft³	From table above D_T, ft	gpm
1,200	0.0068	1.65	21,300	14.0	44.0	192,400
3,430	0.0051	3.52	10,000	18.9	26.0	40,400
6,500	0.0041	5.22	6,720	23.9	18.9	10,630
10,700	0.0028	6.02	5,830	34.2	14.7	2,160
12,890	0.0024	6.20	5,660	39.6	13.5	1,080
14,160	0.0018	5.14	6,840	53.0	12.8	490
16,300	0.0012	3.92	8,950	79.1	12.1	142

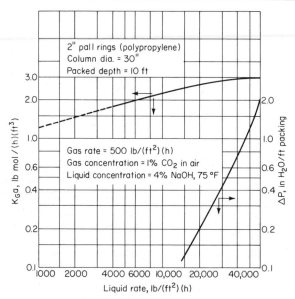

Fig. 61   Mass-transfer coefficients of 2-in pall rings.

The optimum design, using the specified packing with no other restrictions, such as limited available pressure drop or water supply, etc., would fall in the neighborhood of the 13.5-ft-diameter tower with a minimum of 39.6 ft of packed height which represents essentially the minimum packed volume in this instance.

Part 4

Solid-Liquid Mixtures

Section **4.1**

Filtration Theory

CLIFFORD W. CAIN, Jr. *Senior Research Engineer, Filtration and Minerals Research, Johns-Manville; Associate Member, American Institute of Chemical Engineers; Member, Filtration Society.*

INTRODUCTION

Filtration processes can be separated into three broad categories which must be dealt with separately. These three categories are: cake filtration, depth filtration, and surface filtration.

Cake filtration occurs when a liquid containing solid particles is forced through a porous filter medium which is open enough to allow the passage of the liquid, but tight enough to retain the solid particles. As more and more liquid is forced through the medium, the solids form a thicker and thicker filter cake. The main characteristic of cake filtration is that the cake which is formed must be porous enough to permit continued fluid flow through it as filtration progresses.

In depth filtration, the liquid containing solid particles is forced through a bed of porous material. The solid particles are trapped within the relatively coarse interstices of the bed, allowing relatively clear liquid to pass through. Sand filtration and cartridge filtration exemplify depth filtration. As solid particles continue to accumulate within the filter bed, there comes a time when either fluid flow is restricted below acceptable limits, or solid particles are forced through the bed into the filtrate. At this time the bed must be regenerated, or the cartridge replaced.

Surface filtration is essentially a straining mechanism where the solid particles are stopped by a matrix of controlled pore size. This case differs from cake filtration in that the flow rate decreases because of plugging of the matrix pores. The matrix becomes plugged before any significant cake thickness is attained.

CAKE FILTRATION

In order to develop the necessary flow equations for the case of cake filtration, we must assume that as solid particles are deposited on the filter cake, they remain in place without migration. If we also assume that the filter cake is hard and incompressible, we can write the following differential flow equation using the modified D'Arcy equation[1] to describe the fluid flow through the filter cake:

$$\frac{dV}{A\,d\theta} = \frac{K'\Delta P}{\mu L}$$ (1)

where V = liquid volume, mL
K' = filter cake permeability, darcies
ΔP = pressure drop across the cake, atm
A = filtration area, cm²
θ = filtration time, s
μ = liquid viscosity, cP
L = filter cake thickness, cm

As filtration proceeds, the value of L increases. The value of L can be calculated if the total volume of filtered liquid is known, and the concentration of filterable solids can be determined:

$$L = \frac{VC}{\rho_c A}$$

where C = solids concentration, g/mL
ρ_c = filter cake density, g/cm³

Combining these two equations, we obtain the differential flow equation for the case of an incompressible filter cake:

$$V\,dV = K'A^2\rho_c\,\frac{\Delta P\,d\theta}{\mu C}$$ (2)

This flow equation can be integrated for whatever flow conditions are maintained for a given filtration.

Constant-Rate Filtration

Many industrial filtrations can be approximated by the special case of constant-rate filtration. For this case, the value of $dV/d\theta$ remains constant at V/θ. The differential flow equation reduces to

$$\Delta P = \frac{\mu CV^2}{\theta K'A^2\rho_c}$$ (3)

The specific flow rate is defined as

$$Q = \frac{V}{A\theta}$$

Substituting into Eq. (3) we obtain

$$\Delta P = \frac{\mu CQ^2\theta}{K'\rho_c}$$ (4)

For the case of an incompressible filter cake with a constant filtration rate, a plot of ΔP vs. θ should result in a straight line. Actual practice, however, shows nearly all constant-rate cake filtrations to have some curvature to the plot of ΔP vs. θ. This is probably because all filter cakes exhibit some compressibility, and allow some particle migration during the filtration run. Nevertheless, we can use Eq. (4) for approximation purposes if the filter cake is relatively hard, and if ΔP is held to a fairly low level, say 2.8 kg/cm² (40 lb/in²).

Inspection of Eq. (4) indicates that the filtration time is inversely proportional to the specific flow rate squared. A small change in the flow rate of a constant-rate filtration will result in a significant change in the observed filtration time. This fact can be used to advantage in designing a filtration system. By designing the filtration area on the high side, we can assure that the filtration cycles we obtain in practice will not be excessively short, requiring excessive cleaning or downtime.

Thus far in our analysis, we have not taken into consideration the pressure drop caused by equipment piping and filter medium resistance. For a constant-rate filtration, the pressure drop caused by piping and medium resistance will be constant and simply add to the pressure drop across the filter cake. For most filtrations, the resistance of the equipment and medium are quite small when compared with the resistance of the filter cake.

Constant-Pressure Filtration

While a good number of industrial filtrations can be approximated by constant-rate filtration, it is most convenient to gather laboratory data using a constant-pressure setup. In a constant-pressure filtration, the pressure drop across the system is maintained at a constant level. Liquid flow starts at a high rate, but decreases as the filter cake builds. If we integrate Eq. (2) for the case of constant pressure, we obtain the following equation:

$$V^2 = \frac{2K'A^2\rho_c\,\Delta P\theta}{\mu C} \tag{5}$$

A plot of the logarithm of V vs. the logarithm of θ should result in a straight line having a slope of 1/2 and an intercept of $2K'A^2\rho_c\Delta P/\mu C$. This analysis again assumes that the equipment and medium resistance to fluid flow is negligible. Equation (5) gives the flow equation for the filter cake alone.

In practice, however, the equipment and medium resistance are appreciable at the beginning of the test, and decrease considerably as the filtration rate decreases because of cake buildup. Figure 1 illustrates the effect of equipment and medium resistance on the log-log plot of V vs. θ. Note that the plot of V vs. θ approaches a straight line having a slope of 1/2 as an asymptote. We can determine the cake permeability if we run the constant-pressure filtration test long enough to graphically estimate the position of the straight line based on the asymptote of our filtration data points. Once the filter cake permeability is known, we can use this value to calculate filtration cycle length using Eq. (1) or an integrated form of Eq. (1), depending upon the designed flow conditions for the filter.

In some cases, there will be an appreciable bleed of solid material through the filter medium at the start of a constant-pressure filtration until such time as the solid particles bridge across the medium openings. The effect of this behavior upon the log-log plot of V

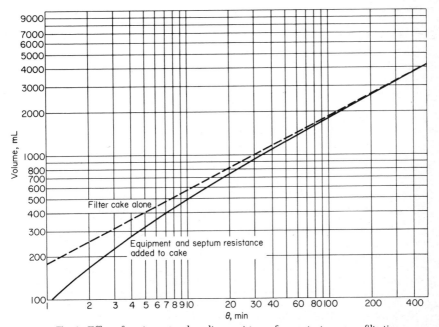

Fig. 1   Effect of equipment and medium resistance for constant-pressure filtration.

vs. θ is shown in Fig. 2. Again, notice that this line also approaches a straight line having a slope of 1/2 as an asymptote. It is important to run a constant-pressure filtration long enough to overcome the effects of filter medium and equipment resistance and solids bleed-through.

Thus far in our discussion, we have dealt with the case of an incompressible filter cake. Most filter cakes, however, exhibit some degree of compressibility, so we must consider methods for handling compressible filter cakes. Two techniques will be dealt with.

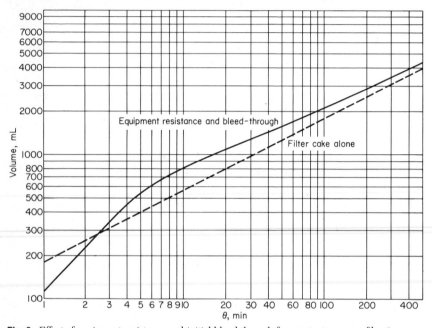

Fig. 2 Effect of equipment resistance and initial bleed-through for constant-pressure filtration.

Compression-Permeability Cells

Probably the most rigorous technique for analyzing a compressible filter cake is to measure the permeability of the filter cake under various applied pressures using a compression-permeability cell. When the data are gathered, the flow equation (1) can be integrated by graphical or computer methods to obtain a plot of ΔP vs. time for the required flow conditions (usually dictated by the characteristics of the pump used). This technique has been described by Grace and Tiller.[2,3,4]

In most cases, however, it is very difficult to obtain a large enough sample of filter cake to fill a compression-permeability cell. Compressible filter cakes are slow filtering, and even if enough cake could be obtained, the manipulation of the solids would probably change their character to some extent. The use of compression-permeability cells also requires special techniques to correct for side-wall friction effects. For these reasons, gathering data is quite tedious, so other techniques are normally employed.

Use of Filter Aids for Compressible and Extremely Fine Turbidity Solids

Filter aids are a class of highly porous, inert powders which can be added to a liquid being filtered to increase the permeability of the filter cake. The function of the filter aid is to trap turbidity solids and to maintain a porous filter cake when filtering compressible and very fine turbidity solids. The size and quantity of the pores that the filter aid provides, therefore, determine the effectiveness of the filter aid for a given system. Figure 3 shows the pore size distributions of typical filter-aid grades of a single manufacturer determined by a mercury intrusion porosimeter. The optimum filter-aid grade to use for a given system is usually the one having a median pore size most closely matching the median particle size of the turbidity being removed. The particle size of the turbidity can be determined by microscopic methods, or by use of a continuous-stream particle analyzer such as a Coulter Counter or Hi-Ac system.

The amount of filter aid added to the system results in a change in the permeability of the filter cake. For this reason, filtration tests must be run at two or three filter-aid dosages

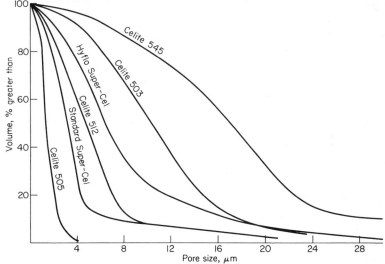

Fig. 3 Pore size distributions of filter-aid grades. *(Johns-Manville)*. Celite, Super-Cel, and Hyflo Super-Cel are trademarks of Johns-Manville Corp. for diatomite filter aids.

in order to determine the optimum filter-aid dosages for a given system. The filter aid adds rigidity to the filter cake, so that we can assume an incompressible filter cake, over a limited range of ΔP, and use Eqs. (3) and (5) in analyzing filtration data. Since all filter cakes are compressible to some degree, there will be some error in the data analysis, but for most engineering work, this approach is good enough.

Serious error has been observed in this approach, however, when the median pore size of the filter aid is considerably larger than the median particle size of the turbidity being filtered. When this occurs, the turbidity particles can migrate within the filter cake pores, giving a nonuniform filter cake permeability. The filtration theory previously developed cannot be used if this is the case. In order to safely use filtration theory, the ratio of turbidity particle size to filter aid pore size should be 1 or higher.

DEPTH FILTRATION

The theory developed for depth filtration has been almost exclusively concerned with the analysis of deep-bed sand filtration. While the same theory could probably be applied to cartridge filtration, it has not been worked out yet. For cartridge filtration, a filtration test is simply performed using a single cartridge, and the results are scaled up to multiple cartridge units to give the required flow rate for the installation.

The basic equation relating the removal of turbidity from a liquid stream in a sand bed is

$$\frac{-\delta C}{\delta L} = \lambda C \tag{6}$$

where C = concentration of particles in suspension expressed as volume/volume
L = distance into the filter from the sand surface
λ = filter coefficient in dimensions of reciprocal length
The value of λ varies not only with time of filtration, but also with depth in the sand bed.
At the start of filtration, assuming a uniform sand bed, we can integrate Eq. (6) to give

$$\frac{C}{C_0} = e^{(-\lambda_0 L)} \tag{7}$$

where C_0 = turbidity concentration at the start
λ_0 = filter coefficient of the clean sand bed
The relationship between the filter coefficient and the amount of turbidity deposited has
been found to be:

$$\frac{\lambda}{\lambda_0} = 1 + R_1\sigma + R_2\sigma^2 + R_3\sigma^3 \tag{8}$$

where R_1, R_2, R_3 = constants dependent upon the nature of the turbidity solids and the
packing of the sand bed, and
σ = volume occupied by the deposited solids per unit of filter volume.
By taking test data on a particular sand bed using the liquid to be filtered, Eqs. (6), (7),
and (8) can be solved in order to calculate the required bed depth and operating time to
give a filtrate containing a specified solids content. There is no known analytical solution
for these equations, but digital computer methods for solving these equations have been
given by Ives.[5]

SURFACE FILTRATION

As was previously mentioned for cartridge filtration, very little theory has been developed
to describe the case of surface filtration. The ways in which turbidity particles interact
with matrix surfaces to plug them are not well understood at present. It has been found
most expedient to simply run filtration tests using the liquid to be filtered. Tests are run
using the matrix having small enough pores to give a suitable filtrate clarity. The volume
of liquid which passes through the matrix before it completely plugs can be scaled up in
direct ratio to the surface area of the test matrix. In this way the required matrix surface for
the designed facility can be estimated.

REFERENCES

1. Carman, P. C., *Trans. Inst. Chem. Eng. (London)*, **16**, 171 (1938).
2. Grace, H. P., *Chem. Eng. Prog.*, **49**, 303 (1953).
3. Tiller, F. M., *Chem. Eng. Prog.*, **49**, 467 (1953).
4. Tiller, F. M., and Cooper, H., *Am. Inst. Chem. Eng. J.*, **8**, 445 (1962).
5. Ives, K. J., *J. Am. Water Works Assoc.*, **52**, 933 (1960).

Filter-Aid Filtration

CLIFFORD W. CAIN, Jr. *Senior Research Engineer, Filtration and Minerals Research, Johns-Manville; Associate Member, American Institute of Chemical Engineers; Member, Filtration Society.*

WHY FILTER-AID FILTRATION?

Over the long run, every segment of an operation must prove itself economically to survive. The use of filter aids arose out of the need to improve the economics of the filtration unit operation. While filtration arose in a rudimentary way out of antiquity, when real engineering hardware began to develop, the need to lengthen the filtration cycle became apparent.

One of the first industries to feel the economic pinch caused by short filtration cycles was sugar refining, where the waxes from the sugar cane stalk plugged filters in short order. After considerable experimentation, it was discovered that crude diatomaceous silica, when mixed with raw sugar liquor, lengthened filtration cycles severalfold while improving the quality of the filtered liquor.

WHAT IS A FILTER AID?

How did the diatomaceous silica help? By providing a great number of microscopic holes for the liquid to flow through. As aptly put by P. C. Carman[1] in 1939, the principal contribution of a filter aid is to give porosity to the filter cake. Porosity ϵ is the ratio of volume of voids to the total filter cake volume, $\epsilon = V_e/V_t$. When particles of dirt (or other unwanted solids) are filtered out, they tend to pack into a dense configuration with but few passages through which liquid can flow. If all of the particles were solid, hard spheres

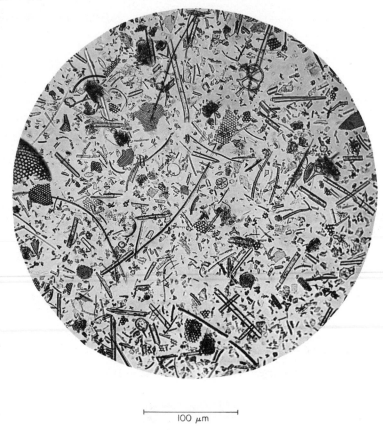

$\vdash\!\!\!\!\dashv$ 100 μm

Fig. 1 Diatomite filter aid.

of the same size, the porosity would be about 0.45. In actual fact, the situation is much worse, with porosities frequently falling in the 0.2 to 0.3 range. When the solids are compressible (which is often the case) porosities may even approach zero at the surface of the filter cake.

Diatomaceous silica is a unique natural material consisting of the skeletal remains of tiny organisms (diatoms). Because of the intricate sizes and shapes of the diatom skeletons, the porosity of this material is in the 0.9 range. Figure 1 shows a photomicrograph of diatomite illustrating the intricate shapes present in the material.

When added in proper amount to a liquid to be filtered, the resultant filter cake assumes the basic structure of the diatomaceous silica with adequate flow channels to keep the cycle going.

AVAILABLE FILTER-AID MATERIALS

While many different materials have been used or proposed as filter aids, only three have reached commercial importance. Of these, diatomaceous silica and perlite are of major importance, while ground wood pulp is used in a number of specialty applications where siliceous materials cannot be used.

Diatomaceous Silica

Diatomaceous silica is found as sedimentary deposits, the most important of which date to the Miocene period. In the United States, the most extensive commercial deposits are

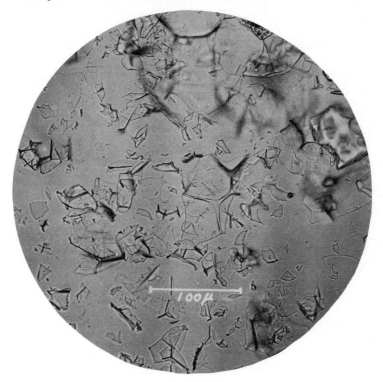

Fig. 2 Perlite filter aid.

found in California, Washington, and Nevada. Smaller deposits have been mined in other parts of the country.

Other commercially important deposits of diatomaceous silica include those of Mexico, Spain, France, Germany, Iceland, Turkey, Kenya, and others.

Processing generally includes some combination of milling, air classification, and calcination to produce the desired product. Natural milled products are buff colored, straight-calcined products are pink to brown, while flux-calcined products are white.

Of special note is the Icelandic deposit which is mined from the bottom of Lake Myvatyn by means of dredging techniques. After beneficiation, dewatering, and drying, the processing is similar to that used for conventionally mined material.

Perlite Filer Aids

Perlite filter aids are made by milling and classifying expanded perlite rock. While this material in no way resembles diatomaceous silica microscopically (see Fig. 2), it does share in having high porosity. As a matter of fact, perlite filter aids can be made having wetted-cake densities as low as 0.128 to 0.16g/cm^3 (8 to 10 lb/ft^3). Most commercial perlite filter aids, however, fall in the low end of the diatomaceous silica range [0.24 to 0.352 g/cm^3 (15 to 22 lb/ft^3)].

Ground Wood Pulp Filter Aids

Ground wood pulp filter aids are generally prepared by ball-milling wood pulp to the desired particle size. Compared with diatomaceous silica and perlite, wood pulp is much more compressible and thus a poorer filter aid when used by itself. As previously stated, its uses are generally specialty uses where siliceous materials cannot be used.

Almost any type of wood pulp could be ground up for filter aids, but generally the material of choice is high-quality chemical pulp.

HOW TO USE FILTER AIDS

Precoat

The most widespread use of filter aids is in what is called precoat filtration, where a thin layer [0.488 to 0.976 kg/m² (0.1 to 0.2 lb/ft²) of filter area] of filter aid is deposited on the filter medium prior to introducing the filter feed to the system. In this use the function of the filter aid is twofold:

1. It protects the filter medium from fouling by the solids removed from the filter feed.

2. It provides a finer matrix to exclude particles from the filtrate by bridging over the larger holes of the filter medium with filter-aid particles.

Body Feed

Body feed (or body aid or admix) is the term used for filter aid which is added to the filter feed to increase the porosity of the filter cake. Body feed may be added batchwise or continuously, depending on the circumstances of the operation. When body feed is used, all of the filter feed must have its portion of filter aid. Even a momentary stoppage of body-feed flow (in the continuous addition mode) can cause the formation of an impermeable layer of solids in the filter cake which effectively ends the cycle. No amount of filter aid added later can undo the damage (in cycle length) caused by the stoppage.

It is important to note that to be effective, body feed must be added in sufficient amount to change the matrix of the filter cake to that of the filter aid rather than of the solids being removed. As a general rule of thumb, the amount of filter-aid admix added must be of the order of twice the amount of solids being removed. A quick calculation will show that this amount of filter aid is reasonable when the material removed is less than 1000 ppm. When the solids to be removed from the liquid exceed a concentration level of about 1000 ppm, the filter-aid cost can become prohibitive unless the product is very expensive and can

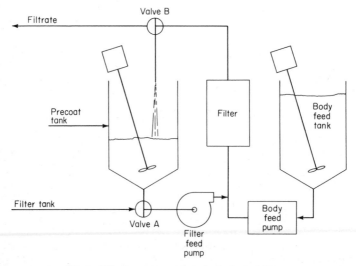

Fig. 3   Filtration system—slurry body feed addition.

bear the increased cost burden. Alternative means of handling such cases will be discussed under Rotary Precoat Filtration.

Filter System

Figure 3 shows an idealized filter system. An alternate dry-feed system is shown in Fig. 4. The precoat tank should have a volume greater than the filter shell such that when liquid is recirculated from the precoat tank to the filter and back, the conical area of the precoat tank remains full.

To start the operation, the precoat filter aid is made up in the body-feed tank. Because of difficulties in handling and pumping, the slurry concentration is usually held to 10% solids [approximately $0.1g/cm^3$ (1 lb/gal)] or less.

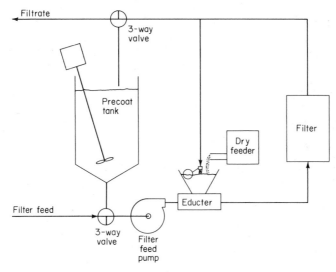

Fig. 4 Filtration system—dry body feed addition.

With valves A and B set for recirculation from the precoat tank to the filter and back, the filter feed pump is started. As soon as the heel of liquid in the precoat tank appears clear (usually 5 to 10 min), body-feed addition is begun and valve A is turned to bring in filter feed. After the precoat tank is filled for the next run, valve B is turned to divert filtrate to the filtrate holding tank.

As a matter of principle, it is important to note that the filter feed should be brought into the filter before all of the precoat has been deposited to avoid forming a layer of low permeability on the surface of the precoat. If the precoating step is carried out until the liquid inside the filter completely clears, the last particles to be deposited tend to form a layer of low permeability that does not "meld" properly with the main body of the filter cake.

Another important principle to be kept in mind in operating a precoat filter system is that once the precoat has been applied, the forward flow of liquid through the system must be continued without interruption to the end of the cycle. As the particles of precoat and cake are deposited, they form an open bridge structure that is stable as long as forward flow continues. If the flow stops even momentarily, the cake either falls from the filter medium or "relaxes" into a more close-packed configuration of lower permeability. In either case, the cycle effectively ends, and the filter must be prepared for a new start.

Rotary Precoat Filtration

Figure 5 shows an idealized sketch of a rotary vacuum precoat filter system. When dealing with liquors containing high concentrations of solid particles (above 1000 ppm), the rotary precoat filter can provide an economic filtration. A filter-aid precoat of 50 to 125 mm (2 to 5 in) thickness is applied to the filter drum. This cake is then dressed by advancing the knife blade to expose a cylindrical surface of clean filter aid. The liquid to be filtered is pumped to the filter bowl and filtered through the precoat layer, causing solid particles to be deposited on the clean face of the filter aid. As the filter drum rotates through the filter feed in the bowl of the filter, more and more solid particles are deposited until a relatively impermeable skin is formed over the face of the filter surface. Finally, the deposited layer of solids emerges from the filter bowl, dries to some extent and is removed by the knife blade. The blade scrapes off the filtered solids plus a very small cut into the clean filter cake. In this way, a clean face of filter cake is again dipped into the filter feed in the bowl as the drum continues to rotate. The continuous immersion of a clean filter cake into the filter feed liquor maintains a relatively high flow rate. When the knife has removed all but 12.7 mm (0.5 in) of precoat, the filter drum must be washed down, and a new precoat applied.

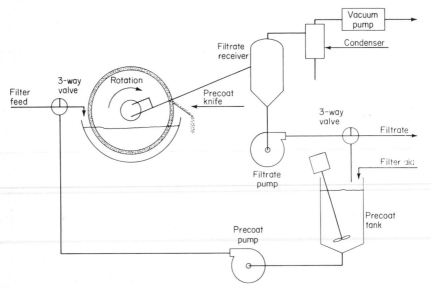

Fig. 5   Rotary precoat filter system.

The major variables to be controlled in this type of filtration are the rate of knife advance and the grade of filter aid to be used. A filter aid having pores which are large relative to the solid particles to be removed will give poor filtration performance because of deep penetration of particles into the precoat. The knife blade will not remove all the turbidity solids, resulting in a continually decreasing filtration rate. The rate of knife advance will be determined primarily by the nature of the turbidity solids being filtered. A rapid enough knife advance must be maintained to keep the filtration rate at an acceptable level, while not cutting too much of the precoat off and causing excessive filter-aid usage. The optimum filter-aid grade and knife advance to be used for a given filtration can be determined experimentally by using a small prototype filter.

Generally speaking, a filter aid can be used to improve the economics of any filtration as long as the liquid is the valuable commodity. When it is necessary to recover a pure solid product, filter-aid usage must generally be ruled out.

REFERENCE

1. Carman, P. C., *Ind. Eng. Chem.*, **31**, 1047 (1939).

Batch Filtration

JAMES F. ZIEVERS, P.E. *Vice-President, Industrial Filter and Pump Mfg. Co.*

INTRODUCTION

Although the terms "batch filtration" and "batch filtration hardware" are fairly commonly used to designate the subject matter of this chapter, they are not correctly used. To explain, an exercise in semantics is in order.

Consider that, when in filtration parlance we use the word "batch," we think of the opposite as "continuous." Almost all continuous-filtration equipment in use is vacuum type. Vacuum filters are typically about 40% submergence[1]—that is, the filter drum is submerged to about 40% of its total surface in the feed pan, as shown in Fig. 1. Using the 40% submergence figure we could then logically say that, at any given time, 60% of the filter is not working. Following this line of reasoning we can then further say that any time the turnaround time is equal to or less than the filter cycle length, the filter surface of a "batch" unit is better utilized than that of a typical (40% submergence) vacuum unit. See Fig. 2.

BATCH FILTRATION HARDWARE

A good way to start a discussion of batch types is to differentiate between some of them. With the exception of nutsches, units with scavenger plates, and plate and frame presses, most batch-type units have a "heel," a volume of unfiltered liquid remaining at the end of a cycle. In the instance of single-unit stations this "heel" must be: (1) cycled to clarity and blown ahead, or (2) blown back to prefilt supply (to be "worked" in next cycle). In either of the two alternatives cited, the consequences involve the use of tankage to ensure continuity of the process stream, and they do not really represent a "batch" situation. Before leaving this consideration of heel, it should be pointed out that in a shutdown situation a continuous unit has a similar heel problem, the liquid remaining in the feed pan. On reflection it would be more accurate to entitle this section *Cyclical Filtration Hardware.*

Fig. 1   Rotary vacuum drum filter. *(Industrial Pump and Filter Mfg. Co.)*

Cyclical types can first be divided into two broad categories:
1. True batch type (no heel)
2. Other cyclical types (heel to be taken care of)
Before we proceed to specific discussion of configuration it must be pointed out that it just is not possible to discuss every variation of every type. It is hoped that the most typical examples have been chosen.

TRUE BATCH TYPES

Nutsche Filter

One of the earliest batch types was the nutsche—German for suckling. This hardware was so named because many configurations are operated by vacuum. The prefilt is placed in the chamber and run through the filter medium by vacuum, gravity, or pressure. The units can be precoated or not. In a sense the simplest form of nutsche is the bag used to strain cooked fruit in jelly making. Even today most nutsches are fabricated by the plants that use them. See Fig. 3.

Closely akin to the nutsche is the horizontal-plate filter. For batch work it is equipped with a scavenger plate used, as the name implies, to scavenge the heel

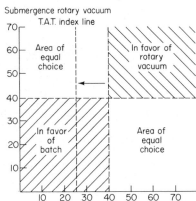

Fig. 2   Areas of equal choice—batch vs. rotary vacuum. *(Industrial Pump and Filter Mfg. Co.)*

left in the filter at the end of the batch or cycle. A great variety of plate designs are available. Usually they are dressed with paper or textile and may be precoated or not as necessity demands. See Fig. 4.

Other Pressure-Tank Batch Types

Two more batch types are shown in Figs. 5 and 6. The first captures the solids on the inside of the tubes. At the end of the batch the last liquid remaining prefilt is blown through the tubes to the "clean side" of the filter. The tubes are commonly lined with paper sheets, paper bags, or textile bags.

The second style features horizontal plates in a horizontal tank and usually a scavenger plate or scavenger tubes at the bottom to handle the heel.

Plate-and-Frame Presses

Probably the most commonly used batch-type unit is the plate-and-frame press. It is truly a versatile tool. See Fig. 7. Successive combinations of plates and frames form, in a sense, a series of bottoms and sides of a nutsche. The plates are dressed with filter cloth which in turn can be precoated or not. There are probably over a hundred different plate-and-frame designs, each for a specific purpose. It is sufficient to say that by porting the plates and frames in different ways for entrance and exit, the press can be set for straight filtration, filtration and blow, filtration and cake washing, etc.

The combination of plates and frames can be locked up to form a pressure vessel by means of a hand-turned screw or by a power-driven unit. See Fig. 8. Each plate can be equipped with an individual outlet and shutoff cock. Starting with any given size unit, plates and frames can be removed to decrease the size of the press to fit it to a given size batch.

Fig. 4   Horizontal plate filter pack ("bundle"). *(Industrial Pump and Filter Mfg. Co.)*

Fig. 3   Nutsche opened and discharging cake. *(Industrial Pump and Filter Mfg. Co.)*

Fig. 5   Solids inside tubular batch unit. *(Industrial Pump and Filter Mfg. Co.)*

Fig. 6   Horizontal leaf batch type. (*Industrial Pump and Filter Mfg. Co.*)

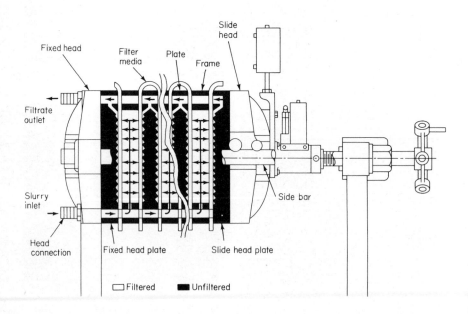

Fig. 7   Plate-and-frame cutaway diagram. (*D. R. Sperry Company.*)

Fig. 8 Plate-and-frame press with plate shifter. *(D. R. Sperry Company.)*

Fig. 9 Bottom-outlet tubular filter. *(Schumachersche Fabrik.)*

More recently, presses have been equipped with automatic power-operated opening, plate shifting, and closing devices which are especially useful when large-size units are contemplated. Figure 8 shows an example.

CYCLICAL-TYPE FILTERS

Plate-and-frame presses, still being truly suited for batch work, make a good transition point in this section. Other filters of cyclical type will be classified by their type of filter elements. There are exceptions, but these general categories are valid:
1. Tube types
2. Leaf types

Tubular-Type Filters

Tubular elements are made of stone or ceramic, carbon and sintered metal, or plastic in one subcategory which we can classify as single layer. The porosity of such elements (and hence clarity of filtrate) can be controlled during element manufacture. Often these elements are used without precoat.

The other subcategory is most commonly used with precoat. These tubular elements typically will have a strength-member substructure in turn covered with a solids retainer which may be textile, elastomer, wire, or wire cloth. Solids are captured on the exterior of the tubes. It is interesting to note that as cake is accumulated on such an element, the surface of the element also increases, partially compensating for the normal pressure drop due to increased cake depth.

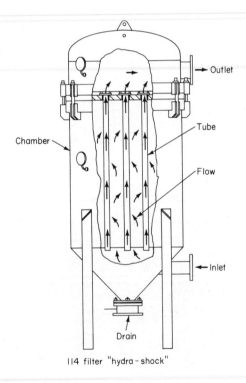

114 filter "hydra-shock"

Fig. 10   Top-outlet tubular filter. (*Industrial Pump and Filter Mfg. Co.*)

Tube outlet can be at the bottom (see Fig. 9) or at the top (see Fig. 10). If the outlet is at the bottom, the tube sheet is often canted to facilitate cake removal.

Where prefilt solids are encountered that may (for example) stick to the tube sheet and cause maintenance problems, the unit can be equipped with a header/lateral tube suspension instead of a tube sheet.

Tubular filters are usually cleaned by reverse flow of filtered liquid or (for example) water. The reverse flow is accomplished by pumping or by sudden release of pressurized gas against a clear liquid reservoir—this is called "shocking." Figure 11 illustrates a typical operational sequence for a tubular filter cleaned by shocking. Note the number of valves is relatively small and the turnaround time is very short. Because of their "clean" hydraulics, tubular units are well suited to high-rate applications. Tube modifications can permit blowdown[3] and cake washing but this is *not* common.

Control valves		
No.	Function	Size
I	Precoat pump suction	
2	Filter inlet from precoat tank	
3	Filter discharge to precoat tank	
4	Vent to precoat tank	
5	Filter inlet from process	
6	Filter discharge to process	
7	Body feed pump suction	
8	Body feed pump discharge	
9	Filter drain (quick opening)	
IO	Precoat tank drain	
II	Body feed tank drain	

Simplified operating procedure chart											
Step	Operation	Control valves x indicates open									Remarks
		I	2	3	4	5	6	7	8	9	
	Off										
	Precoat	x	x	x							A, B, C
	Service					x	x	x	x		D, E
	Prepare to shoc					x					
	Shoc										x	

A. Precoat tank filled with clear liquid and predetermined amount of filter aid thoroughly mixed
B. Precoat pump on. Precoat complete when liquid in precoat tank is clear
C. Open valve 4 to vent air under tube sheet
D. Open valves 5 and 6 before closing valves I, 2 and 3 and turning off precoat pump
E. Body feed pump on

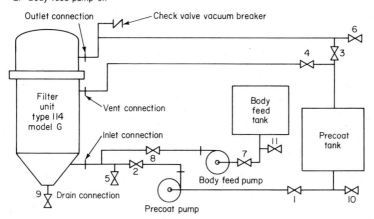

Fig. 11 Valve sequence for shock-cleaning tubular filter. *(Industrial Pump and Filter Co.)*

Leaf-Type Filters

Leaf-type pressure filters are often categorized as vertical or horizontal tank. Vertical tank units are probably efficiently utilized in sizes up to about 55 m² (600 ft²). Usually they are wet-discharge types, but may also be "rigged" for dry discharge. Figures 12 and 13 illustrate both types.

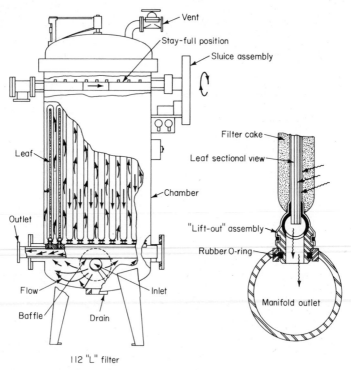

Fig. 12 Vertical "wet discharge" filter. *(Industrial Pump and Filter Mfg. Co.)*

Vertical-tank units occupy less floor space than horizontal-tank units (to be discussed next). They are a little more difficult to work on from a maintenance standpoint, and of necessity elements (leaves) are of different size (see Fig. 12), complicating the spare parts problem somewhat. They can be readily automated.

On the other hand, horizontal-tank vertical-leaf-type units use one size of leaf, require more floor space, and are more readily available to maintenance crews. They can be cleaned "wet" or "dry" and are easily automated. See Fig. 14.

Leaves themselves are built usually of one or more layers of a strength member which is in turn covered with a solids retainer which can be textile, felt, or wire cloth. (See Fig. 15.) The outlet (filtrate exit) can be at the top, side, center, or bottom, but most commonly it is at the center or bottom (see Figs. 16 and 17).

Wet Cake Discharge   If wet discharge is contemplated, the filters are usually sluiced by means of high-pressure liquid. If the leaves are center outlet, usually the leaves are rotated and the sluice header is fixed. If, for example, the leaves are bottom outlet, the leaves stay fixed and the sluice header is moved to spray the liquid across the leaves.

Horizontal-tank vertical-leaf units are also wet-discharged by reslurrying the cake in a filter chamber full of liquid by means of a vibrator attached to the filter leaves.

Dry Cake Discharge   This same vibration method is used to dry-discharge both the horizontal- and vertical-tank-type pressure filters, especially if the leaves are bottom-outlet type.

If the leaves are center-outlet type, vibration may also be used, but more commonly the leaves are rotated against a cutting knife or wire and the dislodged cake is dropped into a trough in the bottom of the filter vessel.

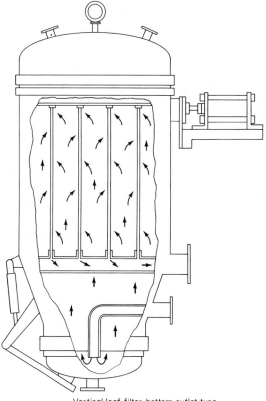

Vertical leaf filter, bottom outlet type

Fig. 13   Vertical "dry discharge" filter. *(Industrial Pump and Filter Mfg. Co.)*

The bottom-outlet-type leaf units usually are opened up by either moving the leaf assembly out (see Fig. 18), or by pulling the chamber away from the leaf assembly. The cake is dislodged onto a belt or into a screw conveyor or cart. Prior to discharge, the cakes can be washed or leached and blown or steamed dry. Leaching to residual of 0.2% is feasible and common. Depending on specific gravities involved, typical cake moisture content on dry discharge is 50%.

In cases where there is a tendency toward gradual buildup of salts or other solids, it has been found extremely helpful to use the same vibration system discussed above to prevent such buildup by incorporating a "preventive maintenance" step in the operating sequence. Figure 19 illustrates a typical dry-discharge operating sequence incorporating a preventive maintenance step.

Multiple Units

To this point the discussion has been of "batch-type" pressure filters as single units. They are commonly arranged in multiple stations or batteries. Figures 20 and 21 illustrate

Simplified operating procedure chart

"O" indicates open or on

Drum control position	Cycle step	Function	Δt	1	2	3	4	5	6	7	8	9	10	11	12	13	14	15	16	17	18	19	Comments
																							Control valves
1	1	Fill and vent	4 M	O		O																	Prec pump on. FS1 terminates step
2	2	Precoat	4 M	O	O																		Prec pump on. timed step
3	3	Run	35 M				O	O															Service pump on CTD on flow control 200 gpm
4 and 5	4	Blow heel	4 M						O	O													FS2 terminates step
6	5	Fill to sweeten off	4 M			O				O	O												Water pump on FS1 terminates step
7 and 8	6	Sweeten off	2 M							O	O	O											Water pump on flow cont 120-150 gpm timed step
9 and 10	7	Blow water							O	O		O											FS-2 terminates step
11	8	Air to dry cake	3 M						O			O											Timed step
12	9	Vent	1 M			O																	Timed step
13	10	Unlock, ring	30 sec			O											O						Pressure safety interlock
14	10	Open ring and extract				O											O	O					LS-1 starts ass'y to open
14	10	Ass'y				O											O	O	O				LS-2 terminates step
15	11	Vibrate	1 M			O									O		O	O	O				Timed step
16	12	Retract ass'y, close				O											O		O				LS-3 starts ring to close
16	12	Ring				O											O	O					LS-4 terminates step
17	12	Lock ring	30 sec			O											O						Timed step
18	13	Sluice	3 M											O		O	O			O	O		Water pump on IO/II work alternately timed step
19-30	14	Off				O								O		O							Selector switch permits cycle to repeat if desired

4-24

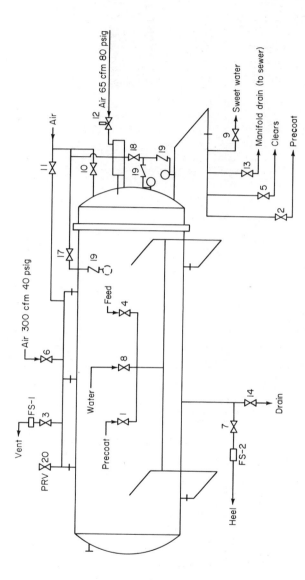

Fig. 14 Valve sequence for dry discharge of horizontal tank filter. (*Industrial Pump and Filter Mfg. Co.*)

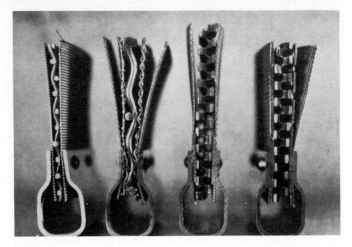

Fig. 15 Frequently used filter leaf constructions. *(Industrial Pump and Filter Mfg Co.)*

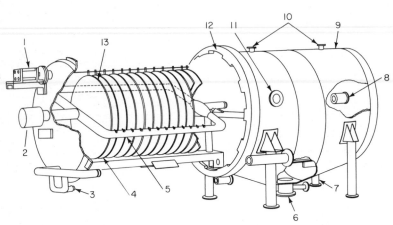

1. Vibrator	8. Sluicing liquor inlet
2. Rotac	9. Chamber
3. Outlet	10. Vent
4. Leaf assembly	11. Sight port
5. Rotating sluice header	12. Rotating ring for quick opening cover
6. Drain	13. Vibrator bar
7. Inlet	

Fig. 16 Cutaway of horizontal tank filter. *(Industrial Pump and Filter Mfg. Co.)*

SIMPLIFIED OPERATIONAL PROCEDURE CHART

Steps	Functions	Δt	\multicolumn{17}{c}{Valve numbers — "O" indicates open or on}																	Motors
			1	2	3	4	5	6	7	8	9	10	11	12	13	14	15	16	17	A	B	C	D
1	Fill for precoat	10 M	O	O	O															O	O	O	
2	Precoat	8–10		O	O	O														O	O	O	
3	Service	–					O	O	O	O	O									O	O		
4	Prepare to blow	15 S					O	O	O	O	O	O								O	O	O	
5	Blow heel	5 M		O				O					O							O	O		
6	Fill to leach cake	8 M		O	O							O		O	O					O	O		
7	Leach	8 M			O							O		O	O					O	O		
8	Vibrate–blow	4 M														O				O	O		
9	Sluice	2 M		O	O												O	O	O		O	O	
10	Off or open				O													O	O				

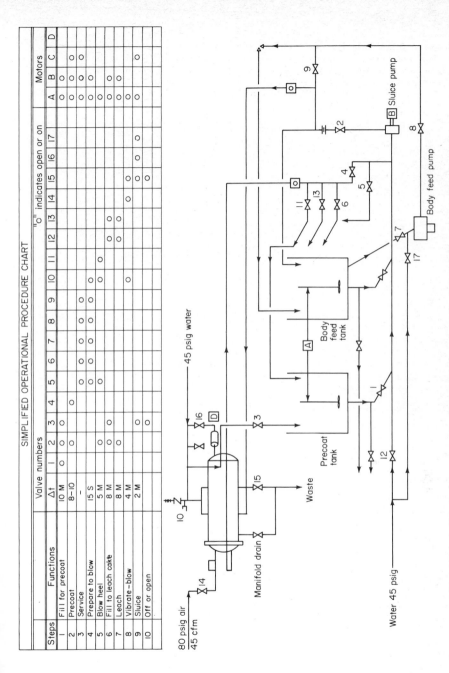

Fig. 17 Typical operating schematic for bottom-outlet sluicing filters. *(Industrial Pump and Filter Mfg Co.)*

respectively a typical battery of tubular filters arranged on a common "service module" of precoat and body-feed tanks and a like system for horizontal-tank vertical-leaf units. Figure 21 should be carefully noted because it shows how heels can be transferred directly from one filter to another in a matter of 4 or 5 min, rather than "dropped" to a tank and repumped to another filter which would probably require (practically) 30 min.

Fig. 18   Horizontal tank filter on test-out. (*Industrial Pump and Filter Mfg. Co.*)

Refer again to Fig. 20. With a station of this type very often it is desired to make a multiple use of (for example) powdered carbon. Depending upon the dictates of the process the (carbon) cake can be dropped to a receiver tank and reapplied to another filter in the battery, or (for example) "first and second contact streams" can be piped to each filter.

DRYING FILTER CAKES

Cakes are commonly dried by blowing with air, gas, or steam. A typical air-consumption figure is 236 cm^3/s (0.5 ft^3/min) at 2.45 to 2.8 kg/cm^2 [35 to 40 lb/in^2 (gage)] for 5 to 10 min. Generally this would produce a crumbly cake. More recently the use of flexibly mounted impervious diaphragms[5] to squeeze dry cake has been reported.[6] The principle of this system is shown in Fig. 22. Especially with compressible cakes, this principle seems to be quite efficient.

WASHING FILTER CAKES

Some fairly accurate data has been reported[7] in regard to cake washing. It is more efficiently accomplished as a dual rate application, the first rate of about 0.61 m^3/(h) (ft^2) [0.25 gal/(min) (ft^2)] being followed by a rate of about 0.244 m^3/(h) (ft^2) [0.1 gal/(min) (ft^2)] after 5 min. As stated earlier, residuals of less than 0.2% in the cake are common. (See Fig. 23.)

Dry cake discharge

Simplified operating procedure chart

Step	Operation	Time	Control valves "O" indicates open															Pumps	Remarks
			1	2	3	4	5	6	7	8	9	10	11	12	13	14	15	P1	
1	N₂ Purge	3 min	O	O	O	O					O			O					
2	Fill and vent	10 min		O	O	O	O	O	O									On	
3	P.M. cycle	1 min			O		O	O	O	O								On	0 psig on gage
4	Recycle for clarity	2–5 min		◑			O	◑	O		O							On	See note 1
5	Filter cycle	30–60 min		◑	O		O	◑	O			O						On	See note 1
6	Filter tank heel	15 min			O	O	O	O			O							On	
7	Push heel forward	5 min										O	O	O					
8	Dry cake	15–20 min				O													See note 2
9	Drop dry cake	30 sec								O									Bundle extracted
	Stop/Or/Repeat																		

Notes: 1. "◑" indicates valves throttled as required. Valve no. 6 approx 1/3 open △

2. On step 8 N₂ or steam can be used (N₂ recommended) N₂ is used on step no. 8 if Ni catalyst is to be reused. Steam (valve 15) is used on step 8 when Ni catalyst is spent.

Fig. 19 Valve operating sequence incorporating "preventive maintenance" steps. (*Industrial Pump and Filter Mfg. Co.*)

NOTE:
Pressure relief devices in accordance with local codes and standard engineering practices.

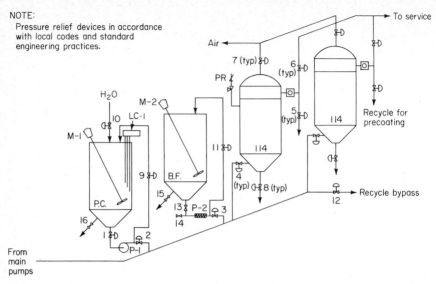

Remarks	
N.C.	Normally closed
N.O.	Normally open
⊗	Open as required
O	Indicates on or open
◐	Indicates open first part of step
◑	Indicates open last part of step
●	Controlled by LC-1
P-1	Pump to be sized 3 minutes plus D.E. injection
LC-1	Liquid level control No. 1
ΔP	Differential pressure
M	Mixer
P	Pump
T	Time

Typical cycle	Step indexed by	Time	Precoat outlet 3" (1)	Precoat to series of filters 3" (2)	Body feed to serie of filters 1/2" (3)	Inlet to filter 4" (4)	Precoat & body feed to filter 4" (5)	Filtered effluent 4" (6)	Air inlet 2" (7)	Drain or discharge 10" (8)	Recycle precoat 3" (9)	Water inlet to precoat tank 2" (10)	Recycle body feed 1/2" (11)	Recycle by-pass 4" (12)	Precoat mixer M-1	Body feed mixer M-2	Precoat injection pump P1	Body feed pump P2	Body feed pump N.O. 1/2" (13)	Waterflush for B/F pump N.C. 1/2" (14)	Auxiliary B/F drain N.C. 2" (15)	Auxiliary P/C drain N.C. 3" (16)
					Automatic valves										Mixers & pumps				Aux. valves			
Step		ΔT	1	2	3	4	5	6	7	8	9	10	11	12	M-1	M-2	P1	P2	13	14	15	16
Fill & vent & add prec.	LC-1	3-5M	◐	◐		O	O			◐	●	O			O	O	O	O	O			
Recycle	T	3-5M		O	O	O									O	O		O	O			
Service	Δport	12-24H		O	O		O								O	O		O	O			
Prepare to shoc	T	15-30S			◑			◑				O	O		O	O		O	O			
Shoc & drain	T	30-60S					O			O		O	O		O	O		O	O			
Off										O					O	O				O	O	O

Fig. 20 Multiple tubular filters on common service module. (*Industrial Pump and Filter Mfg. Co.*)

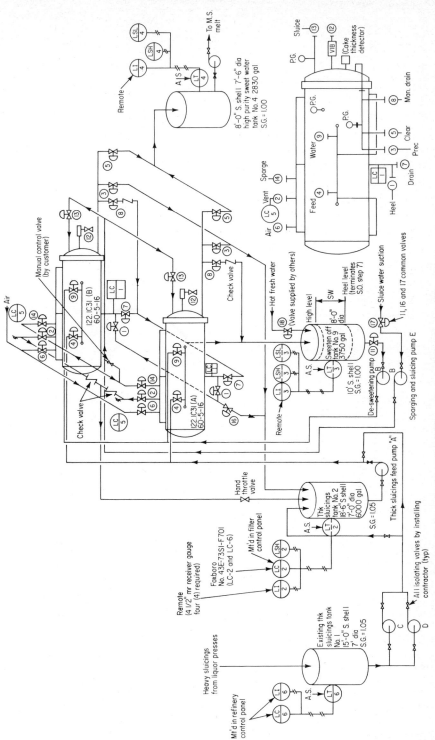

Fig. 21 Multiple horizontal tank filters. (*Industrial Pump and Filter Mfg. Co.*)

SIMPLIFIED OPERATING PROCEDURE CHART

"O" indicates open/on

Drum pos.	Step	Operation	Δt	Control valves																	Pumps					Notes
				1	2	3	4	5	6	7	8	9	11	12	13	14	15	16	17	18	A	B	C	D*	E	
1	1	Fill and vent	5 min	O	O		O													O	O					Note 1 & 14 plus 1.5 min time on opposite filter note 6
2	2	PM. cycle	1 min		O	O	O							O						O	O		O	◐		Timed
3	3	Recirculate	4 min		O	O														O	O		O			Timed
4	4	Service/recycle	0-60 min		O	◐	O	◐												O	O		O	◐		Note 2 & 5
5,6,7	5	Blow heel △5	4 min	O			O		O			O								O	O					Note 4 & 9
8	6	Fill for S.O.	5 min		O						O	O	O					O		O		O				Note 8 & 16
9	7	S.O.	5 min					O				O	O									O				Note 3
10, 11, 12	8	Blow H₂O △5	4 min						O	O	O															LC-1
13	9	Blow dry	6 min						O	O	O															Timed
14	10	Vent	.25							O	O															Timed
15	11	Open filter	1 min		O												O									LS-1A & LS-2 (limit switch on filter)
16	12	Vibrate △5	5.5 min		O									O			O									Timed
17	13	Close assembly	1 min		O												O									LS-3 (limit switch on filter)
17	14	Close ring	.5 min		O										◐	◐										LS-1B (limit switch on filter)
18	15	Sluice and sparge	6 min							O									O						O	Note 15
19–30	16	Standby			O					O																Note 14

NOTE: ∗ Indicates spare parts to be used if main pumps are out of service

Fig. 21 (continued)

Referring once again to our opening comments regarding ratio of cycle length to turnaround time, there are some units that have been reported that, for batch types, are quite close to being continuous. The BMD[8] concept can be applied to either tubular or leaf types and features only 6 to 12 min offstream time each cycle while discharging cake or mud at about toothpaste consistency.

FILTER SELECTION

How does one choose between these excellent processing tools? Of course, there are some obvious physical determinants, such as space available and process stream flux, but a good laboratory test cannot be beaten. An innovative laboratory system was recently

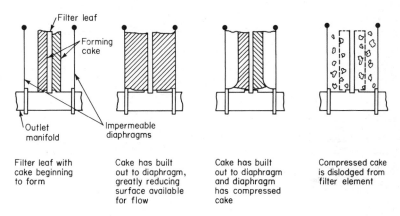

| Filter leaf with cake beginning to form | Cake has built out to diaphragm, greatly reducing surface available for flow | Cake has built out to diaphragm and diaphragm has compressed cake | Compressed cake is dislodged from filter element |

Fig. 22 Mud compaction with diaphragms. *(Industrial Pump and Filter Mfg. Co.)*

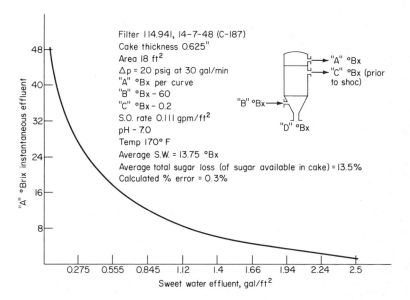

Filter 114.941, 14-7-48 (C-187)
Cake thickness 0.625"
Area 18 ft^2
Δp = 20 psig at 30 gal/min
"A" °Bx per curve
"B" °Bx - 60
"C" °Bx - 0.2
S.O. rate 0.111 gpm/ft^2
pH - 7.0
Temp 170° F
Average S.W. = 13.75 °Bx
Average total sugar loss (of sugar available in cake) = 13.5%
Calculated % error = 0.3%

Fig. 23 Desugaring a filter cake. *(Industrial Pump and Filter Mfg. Co.)*

reported[9] that gives a recorded pressure-drop curve, ability to change flow rate instantaneously (and to see the effect on the curve), freedom from viscosity and specific gravity correction, and means of testing cake washing and drying. The system is simple and appears to be relatively inexpensive.

Figure 24 is a chart that summarizes this chapter in the form of a rough guide to the selection of specific filters for specific jobs.

Usual equipment selection (column key):

1. Nutsche
2. Horizontal plate
3. Solids inside tube
4. Plate and frame
5a. Depth cartridge
5b. High surface cartridge
6a. Solid exterior cartridge
6b. Solid exterior tube and precoat
7a. Leaf exterior tube-no precoat
7b. Leaf vertical tank
7c. Leaf horiz. tank, top or center out / Leaf horiz. tank, bottom out

Problem	1	2	3	4	5a	5b	6a	6b	7a	7b	7c	Notes
1 Cyclical (relatively continuous supply)				x	x	x	x	x	x	x	x	
2 Batch	x	x	x	x								
3 Trap		x	x		x	x						
4 Crystals–Lo resistance	x	x		x			x	x	x	x	x	
5 Washing	x	x		x			x^1	x^1	x^1	x^1	x^1	1. Special care in cycle
6 Recover solids	x	x	x	x					x^2		x	2. Requires spec. tank bottom
7 Recover liquids [3]	x	x	x	x	x	x	x	x	x	x	x	3. Value of "heel" affects choice
8 Hi resistance–compression		x	x	x^4		x^5	x	x	x	x	x	4. Watch leakage
9 Lo solids conc.		x	x		x	x	x	x				5. Only lo solids vol
10 Med solids conc.		x	x	x			x	x	x	x	x	
11 Hi solids conc.			x							x	x	
12 Volatiles [6]		x	x			x	x	x	x	x	x	6. Explosionproof
13 Temp susceptibility [7]		x	x			x	x	x	x	x	x	7. Jacketed
14 Filter aid OK		x	x	x	x	x	x	x	x	x	x	
15 No filter aid OK [8]		x	x	x	x	x	x	x	x	x	x	8. Care in sel. of filter media
16 Slurried disch. OK (2-3% D.S.)						x	x	x	x	x	x	9. Sluice, vibrate or backwash
17 Dry disch. req'd (10-20% D.S.)							x^{10}	x^{10}	x^{10}			10. See U.S.P. 3744633
18 (30-40% D.S.) [11]				x					x^2		x	11. Air or gas blow
19 Full auto-incl. clean							x	x	x	x	x	
20 Dangerous or corrosive							x	x	x	x	x	
21 Lo turnaround time				x			x	x	x	x	x	

(True batch cyclical: columns 1–7c)

Fig. 24   Filter selection chart. (*Industrial Pump and Filter Mfg. Co.*)

REFERENCES

1. "Chemical Engineers Handbook," 3d ed., McGraw Hill, New York, 1950, pp. 776–781.
2. U.S. Patent 3,244,286.
3. U.S. Patent 3,233,739 and U.S. Patent 3,240,347.
4. U.S. Patent 3,212,643 and U.S. Patent 3,224,587.
5. U.S. Patent 3,708,072 and U.S. Patent 3,017,966.
6. Zievers, J. F., Pressure Filtration of Clarifier Underflow II, 69th Meeting American Institute Chemical Engineers, Cincinnati, Ohio, May 17–19, 1971.
7. Boore, Earl A., Research Report 100P12N, Industrial Filter and Pump Mfg. Co., August 1963.
8. Zievers, J. F., and Crain, R. W., Liquid/Solids Separation of Brewery Wastes, Anaheim, California, November 5, 1973.
9. Zievers, J. F., and Schmidt, H. J., Toward Accuracy in Bench Filtration Tests, Fourth Joint Chemical Engineering Conferences, Vancouver, British Columbia, Canada, September 9–12, 1973.

Continuous Filtration

ROBERT C. EMMETT, JR. *Program Manager, Industrial Processing Technology, Eimco BSP Division of Envirotech Corporation; Member, American Institute of Metallurgical Engineers and American Institute of Chemical Engineers*

NOMENCLATURE

C_1 Initial solute concentration in filter cake liquor
C_2 Solute concentration in cake liquor following washing
C_w Solute concentration in wash liquor
G Gas or air flow rate through cake during drying cycle
n Wash ratio, volume of wash per volume of liquor in cake at discharge, or at end of dewatering
   interval during multistage washing
P_a Local atmospheric pressure
P_v Absolute pressure in vacuum system
ΔP Pressure drop across filter cake
R Residual solubles following washing, expressed as a ratio of final and initial concentrations
R_F Cake formation rate, weight dry solids per hour per unit area
S Mass fraction of solids in feed slurry
S_c Mass fraction of solids in undewatered filter cake
W Cake weight per cycle per unit area (dry solids)
V Wash volume per cycle per unit area
θ_d Dry time, min
θ_f Form time, min
θ_t Cycle time, min (or minutes per revolution, min/r)
θ_w Wash time, min
μ Liquor viscosity, cP

INTRODUCTION

In the separation of solids from liquids, the continuous filter is an important tool which is widely used throughout the industrial and municipal fields.

Continuous filters are characterized by having two or more operating zones which perform continuously to carry out these functions: formation of the filter cake, displacement washing of the residual liquor in the filtered solids, dewatering or drying of the filter

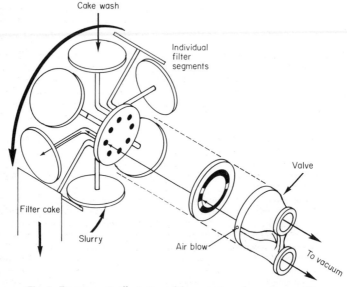

Fig. 1 Diagrammatic illustration of the operation of a continuous filter.

cake, and removal of the solids from the filter. The separated liquid phase is withdrawn continuously through the filter internals during the forming, washing, and dewatering zones.

Continuous filters can have several configurations: drum type, either bottom or top feed; disk filters; or horizontal filters. Various means are provided for filter cake discharge, such as scraper with air-blow, string, continuous media removal, roll discharge, and precoat with doctor knife. Both the choice of filter and the appropriate discharge method are determined largely by the characteristics of the slurry and the objective desired. While

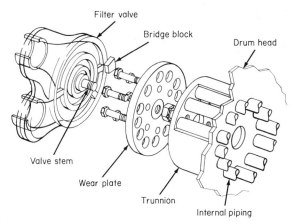

Fig. 2 Component arrangement of a continuous filter valve. *(Eimco BSP Division of Envirotech Corp.)*

a particular slurry might be handled by any number of different types of continuous filters, one single type generally will represent the best choice from an economical and practical standpoint. To make this choice, the design engineer must be familiar not only with the operating characteristics of the various units but also with the properties of the material to be filtered.

The elements of the continuous filter are illustrated schematically in Fig. 1. Operational control of the individual elements is provided through a filter valve, such as shown in Fig. 2. The filter valve is used to regulate the relative duration of each phase of the filter cycle, as well as to isolate the driving force sections of the operation from the discharge and "dead" portions of the cycle. This control is provided through bridge-blocks which can be adjusted to cover or expose the ports connected to the separate sections of the filter. As customarily furnished by the supplier, bridge-blocks will be provided only to isolate the dead and discharge zones. If special operations are required, such as separate "form" and "dry" vacuums, additional bridge-blocks and solution ports can be furnished. When necessary, these modifications can also be made in the field by the user, using bridge-blocks fabricated from metal, wood, Micarta, or plastic, and of such shape as required to provide the necessary function.

The driving force usually employed in filtration is vacuum, supplied by any suitable type of vacuum pump, with pressure differentials ranging from about 50 to 686 mmHg (2 to 27 inHg). A filter can be enclosed in a pressure shell and positive pressure applied as the driving force, or, in some cases, both pressure and vacuum can be utilized. The added cost of this construction as well as the difficulty of maintenance and filter cake removal usually limits pressure filtration to those applications where vacuum may not be practical, such as in high-pressure or high-temperature processing conditions. Gravity may also be employed as a driving force, although generally limited to straining, screening, or thickening operations in which a small amount of solids is separated from a large volume of liquid, and no additional dewatering of the solids is required.

Most continuous filters operate at cycle times in the range of 0.1 to 10 minutes per revolution (min/r). Cycles longer than 10 min/r are rare and often indicate a poor application for a continuous filter, and a batch filter possibly might represent a more economical selection. Variable-speed drives usually are furnished so that the cycle time can be adjusted to suit conditions.

TYPES OF FILTERS

Drum Filters

The rotary drum filter is the most widely used type of continuous unit, and is available in various sizes ranging up to as large as 148 m² (1600 ft²). Bottom-feed drum filters utilize an oscillating rake type of agitator to prevent settling of the solids. The drum is divided into a number of sections, with the typical arc length of each section being about 38 cm (15 in).

The filter medium is held to the drum by rope caulking (or similar means) in division strips, grooves which separate the individual sections. Frequently, wire winding also will be applied circumferentially to the drum in a continuous spiral with spacing between loops of about 50 mm (2 in). Where continuous belt removal is practiced, the seal between sections is provided by the medium itself in contact with a smooth, flat, or slightly raised surface between the sections and at the periphery.

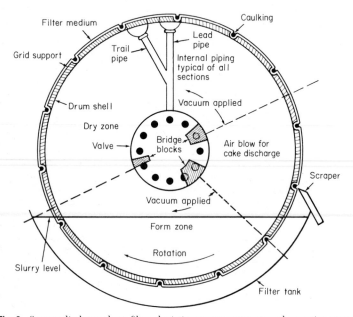

Fig. 3 Scraper discharge drum filter, depicting component parts and operating zones.

The filter medium is supported on each section with sufficient clearance between it and the drum deck to provide an adequate drainage compartment. The filtrate separated from the slurry is removed through pipes spaced at various intervals and connected from each section to an outlet port at the filter valve. While many different types of drainage compartment construction can be used, the preferred system utilizes molded plastic grids which can be inserted easily into each section, providing a minimum of flow restriction and pressure drop, and which also are readily cleaned or replaced.

Drum filters generally are classified according to the feeding arrangement (bottom feed or top feed) and the cake-discharge method used. The characteristics of the slurry and filter cake usually control the cake-discharge method and thus influence the selection of the appropriate type of unit.

Scraper Discharge With the filter medium caulked in place or wire-wound as described, the usual form of cake discharge is by means of a scraper located near the edge of the filter tank (Fig. 3). In most applications, the scraper serves only as a deflector, preventing the cake, which is dislodged by the discharge air blow, from falling back into the slurry. This is particularly true where no wire winding is used, and where rubbing of the medium against the scraper would cause rapid wear. In some applications, where a thin, adherent filter cake is formed, the scraper can be allowed to rest lightly against the wire winding and help in the removal of the filter cake from the medium. In such instances, the scraper edge preferably should be made from a firm, soft material, such as rubber belting. A taut wire may be used in certain applications where physical dislodging of the cake is required, particularly with sticky, cohesive materials.

The factor which largely controls the filtration rate is the minimum cake thickness which can be removed. With filter cakes which might be described as rigid and cohesive, a cake thickness of about 6 mm (¼ in) is considered a minimum for effective removal, although this is related to the filter speed and other conditions. Typical materials meeting this description are mineral concentrates or relatively coarse precipitates, such as gypsum, calcium citrate, etc. Solids which form friable cakes composed of less cohesive materials, such as salts or coal, generally will require greater thicknesses, 13 mm (½ in) or more. Filter cakes composed of fine precipitates such as pigments and magnesium hydroxide,

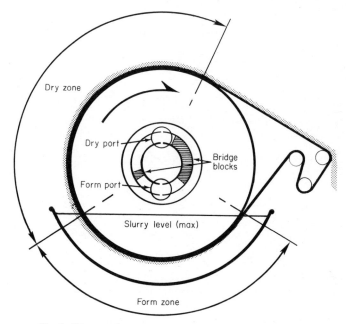

Fig. 4 Diagram of a continuous-belt drum filter operation.

which often produce cracking cakes that tend to adhere to the filter medium, will require cake thicknesses in the order of 10 mm (⅜ in) for effective discharge.

Continuous-Belt Removal System The continuous-belt removal system provides a means for washing the medium continuously during operation in order to prevent blinding. Blinding generally appears in either of two forms: physical lodging of particles or chemical precipitation of insoluble compounds within the medium. Continuous washing of the medium, generally with water, reduces or completely eliminates these conditions.

The continuous-belt drum filter (Fig. 4) generally can maintain a higher filtration rate because of the greater degree of permeability of the medium and more complete discharge of the filter cake. The limiting cake thicknesses which will be discharged from the belt filter are about the same as with scraper filters, although slightly thinner cakes can be removed successfully because there is less adhesion between the filtered solids and the relatively clean medium. If the cake solids can be slurried after filtration, then a sluice discharge can be very effective on the belt filter, allowing the removal of much thinner cakes and thereby making possible a substantial increase in capacity.

The ideal application for the continuous-belt filter is represented by a slurry with the liquid phase saturated with gypsum or similar slightly soluble substance, and which filters relatively slowly, yielding filter cakes 3 to 6 mm (⅛ to ¼ in) thick in cycle times of 3 min or longer. Where a sluice discharge is permissible, maximum rates are possible, since thin cakes can be sluiced off the belt and blinding will be either completely prevented or greatly reduced. A typical example is the residue from leaching of roasted zinc concentrates. In numerous applications in this field, the conversion to belt filters has resulted in an increase in filtration rate ranging from 50 to 100%, compared with standard scraper drum filters.

Provision must be made for handling the waste belt wash water, since this volume can range from 0.75 to 3.75 m³/(h)(m) of filter length [1 to 5 gal/(min)(ft)]. Quite often this waste water can be used for process makeup. Where this is not possible, a closed circuit can be employed with the wash water being discharged to a small settling tank for removal of the

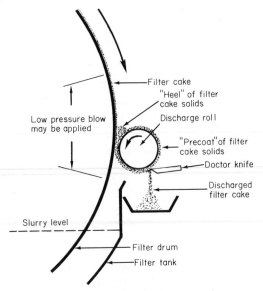

Fig. 5 Operating principle of a roll-discharge mechanism.

solids, and clarified overflow recycled for medium washing. Strainers *must* be included for removal of any solids which could plug the spray nozzles.

Filtrate clarity from a belt filter often will be poorer than from the standard scraper drum filter, principally because of a small amount of leakage at the edges and a higher loss of solids through the clean medium itself. As a general rule, if the filter medium is maintained in good condition, filtrate from a belt filter will contain from 100 to 1000 ppm suspended solids when a relatively tight filter medium is used. By comparision, under the same conditions, a scraper drum filter can produce filtrate solids concentrations of 50 to 300 ppm.

Roll Discharge The roll-discharge drum filter employs a roll mechanism for the removal of tacky filter cakes, such as those composed of clay, magnesium hydroxide, plating wastes, and some fine-size minerals. The discharge mechanism, consisting of a roll which rotates in a direction opposite to the drum rotation and at a surface speed equal to or just slightly faster than that of the drum, utilizes the cohesiveness of the filter cake to cause it to adhere to a surface other than the filter medium. Thus, the cake can be transferred from the drum to the roll, from which it can be removed by a doctor knife or taut wire (Fig. 5).

The roll surface may be coated with up to 25 mm (1 in) of the filter cake solids by retracting the knife or wire from the roll a corresponding distance. This coated surface then is maintained slightly less than one cake thickness away from the drum surface, so that the filter cake is forced to squeeze lightly against the material on the roll and adhere to it. With proper operation, a "heel" of excess solids will form in the nip between the roll and the drum, and will further assist in the removal process.

This heel of material does not build up indefinitely but will maintain an equilibrium condition under normal operation, provided that the filter cake thickness does not change significantly. The amount of material held in this heel can be changed by slight adjustments to the gap between the discharge roll and drum.

If required, a slight air blow can be applied to the section being discharged in order to assist removal of the filter cake. Most materials will be removed merely by prior venting of the discharge section to atmosphere, thus removing any restraining force. Filter cake thicknesses ranging from 1 to 10 mm ($\frac{1}{32}$ to $\frac{3}{8}$ in) are readily discharged with this mechanism, if the properties of the filter cake allow the required adherence. Since this is also a function of the filter cake moisture, the degree of dryness may be significant, and cakes which are either excessively wet or dry may not discharge satisfactorily.

String Discharge   The string-discharge mechanism employs a series of strings, thin chains, or wires spaced approximately 13 mm ($\frac{1}{2}$ in) apart, which are guided from the drum over a discharge roll and back onto the drum with a return roll. These strings will lift the entire filter cake, if it has the required tenacity or "body," from the drum and allow it to be removed at the discharge roll. The minimum cake thicknesses which can be removed will range from 5 to 10 mm ($\frac{3}{16}$ to $\frac{3}{8}$ in), depending upon the properties of the filter cake.

The filter medium rarely is caulked in at every section, but instead is fastened to the drum only at enough points to prevent sagging. Since the medium does not benefit from the cleansing action of the air blow used with the scraper discharge, it is often necessary to use a more open fabric than otherwise would be required in order to reduce blinding. String discharge filters are used principally in the starch and pharmaceutical industries and to some extent in the metallurgical field.

Coilfilter   The coilfilter is similar to the continuous-belt filter in that the filtering medium, either a single or double layer of metal coils approximately 10 mm ($\frac{3}{8}$ in) in diameter, is removed for discharging the cake, followed by washing and return to the drum. The coils, which may contain plastic tubes to reduce air leakage, generally are fabricated of stainless steel, and, since they have a long service life, can be considered a semipermanent medium. Filtrate clarities are poorer than with most other media, and air leakage through the coils requires higher vacuum-pump capacities than with fabric media.

This unit has been applied very successfully in waste treatment, particularly in the filtration of raw sewage and similar materials. These materials often form a cohesive matlike filter cake, and 3 to 5 mm ($\frac{1}{8}$ to $\frac{3}{16}$ in) thicknesses will discharge satisfactorily.

Single-Compartment Filter   This type of filter employs no internal piping but provides suction to the entire drum interior, with perforations through the drum for passage of filtrate. These include "gravity deckers" used in pulp manufacture and the Bird-Young filter.

The latter unit employs a "shoe" located at the discharge zone on the inside of the drum in order to interrupt the vacuum for cake discharge. This internal surface must be machined to provide the close clearance required. Each vacuum section is narrow, only a few inches wide, which facilitates cake discharge. Compressed air introduced through the shoe is used to assist cake discharge. A wash collection trough also may be provided on the inside of the filter to prevent mingling of wash and strong filtrate.

These filters operate at relatively high speed, in the range of 10 to 20 r/min, discharging a thin cake, typically 3 to 6 mm ($\frac{1}{8}$ to $\frac{1}{4}$ in) thick. Their best application is on free-filtering, nonblinding materials such as paper pulp or crystallized salts. The practical maximum size of these filters is limited to an area of approximately 14 m^2 (150 ft^2).

Top-Feed Drum Filters   Free-filtering solids which cannot be suspended by the agitation system used on bottom-feed machines can be dewatered on top-feed drum filters. These units employ a feed compartment formed by a flexible seal in contact with the drum surface at a point 20 to 30° prior to top dead center. Together with side flanges about 152 mm (6 in) high, this construction provides a means for retaining a pool of slurry and forming a relatively thick filter cake. Cake formation necessarily must occur within a few seconds to make this system practical. Dewatering normally occurs from top center for about three-fourths of the drum rotation, at which point the cake is discharged. Metal screen of roughly 30- to 60-mesh opening or similar monofilament synthetic fabric ordinarily is used as the filter medium, since a permeable, nonblinded condition is mandatory for effective operation. Usual applications include crystallized salts, sand, and coarse iron ore.

A unit similar in application is the internal drum filter, in which the medium is fastened to the inside of a drum having a partially opened head through which the feed slurry enters and the filter cake is discharged. While this unit obviates the sealing problems posed by the feed dam on the top-feed machines, cake discharge must take place near the top center, thus reducing the fraction of available area. The use of both top-feed and internal-drum machines largely has been supplanted by horizontal filters.

Top-Feed Filter-Dryer The top-feed filter-dryer combines slurry dewatering and solids drying in one step, and is most commonly used in the production of table salt. The drying section of the filter is surrounded by a hood to which a supply of heated air is introduced. Vacuum draws the hot air through the cake, evaporating the contained moisture. The dry filter cake can be shaved off by fixed blades at various intervals in the direction of rotation in order to prevent overheating of the solids. When operated correctly, this unit often represents the most efficient drying system available, since the exit gas leaving the unit will be at saturated conditions. Furthermore, the final heel of filter cake can be discharged while still slightly moist, mixed with the balance of the hot, dry filter cake, and cooled adiabatically (for example, in a cooling tower) to a temperature of around 49°C (120°F). Steady feed conditions, adequate instrumentation, and good seal and filter medium maintenance are necessary for most efficient operation. From a practical standpoint, the unit is limited to materials which dewater initially to low moistures (10% or less) and form cakes through which a large volume of air can be drawn at low vacuum, e.g., 15 to 30 $m^3/(min)(m^2)$ at 75 to 125 mmHg (50 to 100 cfm/ft² at 3 to 5 inHg).

Continuous Precoat Filter The continuous precoat filter is used primarily as a clarification device to provide consistently perfect filtrate clarity. In addition, it represents a means for discharging extremely thin filter cakes which would not be removed by any of the previously described systems. In construction it is similar to other drum units except that vacuum is maintained over the entire filtering surface, and the actual medium consists of a bed of highly permeable solids, such as diatomaceous earth (Fig. 6). This "precoat" is applied to the filter on a cyclic basis by filtering a dilute suspension of suitable material, depositing a layer 75 to 125 mm (3 to 5 in) thick. Once the medium is applied to the desired thickness, the feed slurry is introduced to the filter and the liquid is drawn through the porous bed while the solids are deposited on the outer surface of the precoat. This filter cake is shaved off by a slowly advancing doctor knife which also trims off a thin layer of precoat material. With proper operation, the doctor knife need advance no more than the amount of penetration of the solids, which will range from 0.05 to 0.2 mm (0.002 to 0.008 in) per revolution of the filter, depending upon the porosity of the precoat material and the characteristics of the solids. On the average, a precoat bed will last from 2 to 4 days, after which the filter must be taken out of service, washed, and re-precoated, an operation requiring approximately 1 to 3 h.

Precoat materials usually consist of permeable substances such as diatomaceous earth or pulverized, expanded perlite. Other materials which can be used when the silica content of the diatomaceous earth or perlite is considered disadvantageous include finely powdered coal, activated charcoal, sawdust, potato starch, and similar substances. With the different grades of perlites and diatomaceous earths, the permeabilities vary greatly, which results in a similar variation in the filtration rate when very dilute suspensions are filtered. If a substantial amount of solids is present in the feed slurry, most of the pressure drop is produced by the filter cake itself, and the rates will not be significantly different, regardless of the choice of precoat material. With the more permeable grades, the penetration of solids will be greater, and this can result in a greater consumption of precoat material per unit quantity of filtrate.

The precoat material is retained by a suitable fabric or screen medium, which is held in place on the drum surface in typical fashion. It is important that blinding of this medium be prevented, since this will cause the precoat to soften and fall off at the blinded zone.

In operation, the most important single factor is to ensure that the filtered solids are removed continuously every revolution by the doctor knife. If incomplete removal should occur, "a breathing" cycle may be initiated in which the precoat bed will shrink away from the knife and result in a reduction in rate, because of the resistance of the filtered solids. Once the knife catches up with the cake and removes this layer of solids, a rapid expansion of the bed will occur and a deep cut of the precoat material results. This cycle generally continues indefinitely until the knife-advance speed is increased to the level required to keep up with the normal penetration of filtered solids. Repeated breathing cycles will cause precoat bed cracking and deterioration of filtrate quality.

A similar effect, although not so prolonged, can occur if other factors cause a change in the pressure differential across the precoat bed, such as an increase in solids concentration in the feed, or variation in the vacuum, resulting from the use of a common vacuum system.

The continuous precoat filter is best suited for handling dilute suspensions containing from 50 to 5000 ppm solids. In these applications, batch-type pressure filters are often uneconomical because of the low filtration rates and the large amount of filter-aid "body

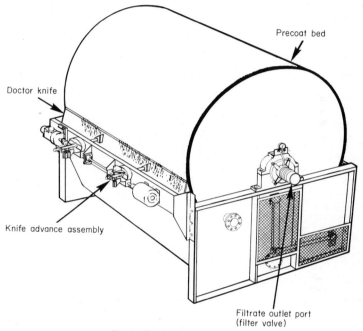

Fig. 6 Continuous precoat filter.

feed" required to maintain a practical filtration rate. Because of the cost of the filter aid, averaging 4 to 10¢/lb, use of the precoat filter is most common in the clarification of relatively valuable solutions, such as fruit juices, algin extract, and pharmaceutical products. It is also used in waste treatment applications where a perfect filtrate clarity is necessary and other alternatives are less practical.

Disk Filters

The disk filter (Fig. 7) consists of a number of vertically mounted disks, each composed of 10 to 30 sectors, supported on a center shaft which also serves as a conduit for the filtrate. Agitation may be provided if the settling characteristics of the material being filtered require it. On the discharge side of the filter, the tank is sectioned into compartments so that the filter cake discharging from each disk can fall vertically to a conveyor belt or other collecting device.

Discharge may be effected by a "snap blow," a sudden pulse of high-pressure air, with the pulse duration ranging from 0.5 to 2 s, and the air pressure at the supply source varying from 1.75 to 3.5 kg/cm² [25 to 50 lb/in² (gage)]. This rapidly inflates the filter bag, causing the cake to be "thrown" loose. Pressure inside the bag should remain low during this operation. Also, a low-pressure, steady blow can be used with heavy filter cakes which release easily. The blow is maintained for the time required for the sector to pass beyond the scraper, which rarely is allowed to contact the filter cake or the medium. With a material such as coal, which is more difficult to discharge, the usual procedure is to set the scraper blades approximately 6 mm (¼ in) from the filter medium, thus allowing the blades to assist in the discharge of the filter cake. Since disk misalignment is a common occurrence on large machines, it is normal to suspend the scrapers from an aligning device which permits them to follow the disk surface and maintain the required spacing. This procedure is limited to noninflatable media, such as stainless steel screen.

Of the continuous units, the disk filter represents the lowest cost per unit area where mild steel, cast iron, or similar materials of construction can be utilized. Where a large amount of area must be provided in a minimum of floor space, the disk filter is the preferred choice, and single units having up to 306 m² (3300 ft²) of area are available.

Disk filters are primarily dewatering devices, and are not used in dissolved value

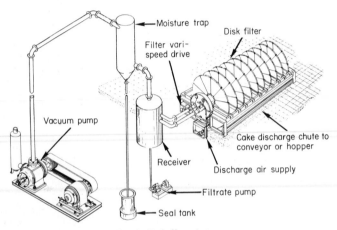

Fig. 7 Disk filter station.

recovery except in rare cases, by filtering, repulping, and refiltering steps. Direct application of wash water generally is difficult and such systems are seldom employed other than on readily washed materials.

Disk filters can be used as thickening devices, such as save-alls in the pulp and paper industry, where the concentrated solids are sent to storage or are mixed with bypassed feed slurry to produce a desired solids concentration for a subsequent operation. Prior to the development of polymer-type flocculants, disk filters operating in a 100% submergence mode sometimes were used as thickening devices, discharging the filtered material directly in the slurry being filtered, thus providing a concentrating action.

One of the principal advantages of this unit is the ease with which damaged or blinded media can be replaced, by simply removing the sectors affected and substituting newly dressed ones. In some instances this can be done without interrupting filtration, although most plants prefer regular shutdown periods for media maintenance.

Disk filter area generally is based on the annular area represented by the diameter of the machine, less the unavailable inner area. If the material being filtered is capable of forming a thick cake, the entire area may be used. With thinner cakes, the material will not bridge across the radial rod and frequently it will not form a dischargeable cake in the inner area, near the sector bell. In some instances, such as in the filtration of iron ore concentrate, the perimeter of both faces of each sector will be sealed off for a width of about 25 mm (1 in) adjacent to the radial rod and sector clamp with a suitable waterproof paint so that filtration cannot take place in this area. This prevents the buildup of a mass of poorly dewatered filter cake which would be apt to discharge only periodically and at excessive moisture content. This could interfere in such operations as pelletizing, where a critical moisture level must be maintained. In these cases, the filter area actually utilized is less than the rated area, and a correction factor ranging from 0.77 to 0.83 should be incorporated with the design calculations to account for this unavailable area.

Horizontal Filters

Continuous horizontal filters generally fall into two broad classes: rotary circular units and belt-type machines. While these filters can be used in most of the applications previously described for drum and disk machines, their most practical service is in handling materials which are difficult to suspend by agitation, or in cases where countercurrent washing is desired on a single machine.

Scroll Discharge Filter The scroll discharge filter is used more frequently for dewatering of relatively fast-filtering, slime-free suspensions. This unit is similar in principle to other vacuum filters except that it utilizes a rapidly rotating scroll to loosen the filter cake

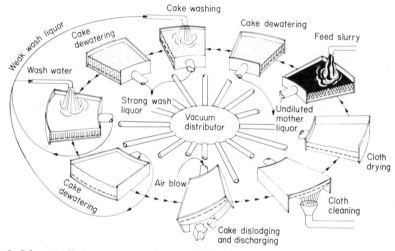

Fig. 8 Schematic illustration of the operation of a tilting-pan filter. (*Bird-Prayon Filter, Bird Machine Company*).

from the medium and elevate it over the rim of the filter. In one version, the rim itself consists of an endless belt which can be removed from contact with the filter at the discharge point in order to facilitate cake removal. A clearance of about 10 mm (3/8 in) between scroll and medium must be maintained in order to prevent damage to the medium. The heel of unremoved solids may be repulped by an air blow beneath the feed slurry in order to prevent accumulation of fine solids, which could reduce the filtration rate. With material that can cause blinding, frequent shutdowns for thorough cleaning may be required. Thus, the unit is best suited for dewatering clean, free-filtering materials, such as crystallized salts, nickel powder from hydrogen reduction, aluminum trihydrate, and similar substances. Unit sizes range from 1 to 8 m (3 to about 25 ft) in diameter, with approximately 80% of the total surface area usable.

Tilting-Pan Filter The tilting-pan filter employs individual horizontal sectors which can rotate on a radial axis for cake discharge (Fig. 8). Thus, complete discharge of the filter cake can be effected, and, if necessary, the filter medium washed while in the inverted position. This filter is well suited to multistage, countercurrent washing of freely filtering substances such as the gypsum precipitate produced in phosphoric acid manufacture. Filter cake thicknesses of 50 to 100 mm (2 to 4 in) are common. Similarly, it is well adapted to dewatering coarse solids which are slightly contaminated with slimes, such as the +30 mesh salt tailings produced in potash manufacture, and where filter cakes up to 152 mm (6 in) thick can be formed.

Approximately 75% of the filter surface area can be considered as usable, with the balance being out of service for cake discharge and medium washing. Of all the continuous vacuum filters, this is probably the most expensive on a unit-area basis, both as to original equipment and installation cost. The multiplicity of moving parts also contributes to a higher maintenance cost.

Horizontal-Belt Filter The horizontal-belt filter utilizes a sectionalized drainage belt driven similarly to a conveyor belt, and supported by a suitable deck (Fig. 9). The deck will include a vacuum section having appropriate connections between the drainage compartments of the belt and the vacuum compartment. The filter medium covers the drainage compartments of the belt, but will travel a separate path on leaving the vacuum section for cake discharge and medium washing. Certain versions of this type of filter omit

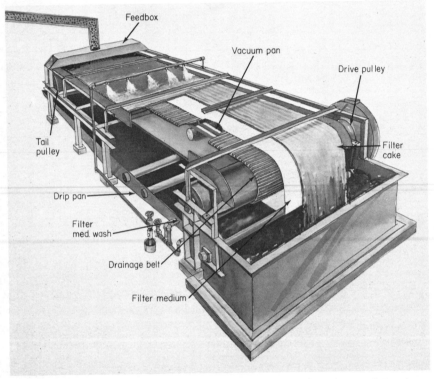

Fig. 9 Horizontal-belt filter. (*Eimco-Extractor, Eimco BSP Division of Envirotech Corp.*)

the drainage belt, employing a porous medium conveyed above a perforated drainage surface as both filtering and transport device.

The major differences between the various units available lie in the construction of the drainage belt, the means employed to retain the filter cake along the length of the belt, and the alignment systems utilized for the filter medium. The filters generally are given an area rating based on the actual available drainage belt area exposed to the vacuum.

As with other horizontal filters, the principal uses are for dewatering and washing of free-filtering, generally coarse substances, such as gypsum from phosphoric acid production, phosphate sands, and sulfite pulp from paper manufacture. Since the wash zones can be separated precisely, which is particularly useful in countercurrent washing, this type of unit also has been applied successfully in difficult washing operations where the filtration rate is relatively low, such as in pigment manufacture.

Relatively high belt speeds, up to 30 m/min (100 ft/min) can be used with cakes up to

152 mm (6 in) thick, and, hence, very high production rates are possible with freely filtering materials.

PRACTICAL APPLICATION OF FILTRATION THEORY

Filtration rate expressions, derived largely from the Hagen-Poiseuille equation, and various empirical correlations can be used reliably in the design of continuous filtration systems, particularly if actual samples are available for testing. The standard testing methods employ a 0.01 m² (0.1 ft²) test leaf to duplicate the cyclic steps of a continuous filter operation. In the use of the data for scale-up, the following procedures can be employed.

Cake Formation Rate

Cake formation rate as a function of filtering time is illustrated in Fig. 10, a logarithmic plot in which the data normally exhibit a slope of −0.5. This is predicted for the ideal case, in which one assumes a negligible resistance from the filter medium. Constant cake thickness lines having a slope of −1.0 are also illustrated.

Deviations from the slope may indicate potential problems in the filtration step. A value between 0 and −0.5 signifies a high medium resistance (such as caused by blinding) or it

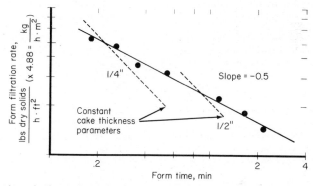

Fig. 10   Filter cake formation rate vs. form time, with parameters of constant cake thickness.

may be a result of hydraulic resistance in the filtrate removal system. A slope of −0.5 to −1.0 suggests an increase in the filter cake specific resistance, possibly because of the formation of a layer of finer solids on the outer cake surface with increasing form times. Or, if the filtrate rate does not show a corresponding decrease, the deviation may be caused by the sloughing off of the solids from the cake, a common occurrence in overflocculated pulps. Corrective action in the last case could consist of a different flocculation procedure to produce a more nearly homogeneous slurry.

In precoat filtration, low slope values (−0.2 to −0.4) are common with very dilute feed-solids concentrations, since the precoat medium resistance forms a large part of the overall flow resistance. Should this be the case, then the bed thickness and the precoat material specific permeability will have an important bearing on the actual rate, and design should consider these factors.

The actual rate must take into account the fraction of form time available. Typically, for a low submergence drum filter, this will be around 30% of the total cycle time. Thus, the actual rate (not including any scale-up factors) will become $0.3R_F$ at a cycle time equal to $\theta_f/0.3$.

Cake Washing

Cake washing involves two considerations: the rate at which wash water passes through the cake, and the effectiveness of removal of original liquor. Figure 11 shows wash time expressed as a function of the correlating factor WV. While a difference in viscosity

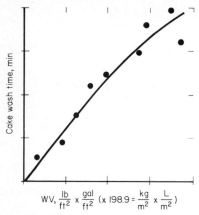

$$WV, \frac{lb}{ft^2} \times \frac{gal}{ft^2} \; (\times \, 198.9 = \frac{kg}{m^2} \times \frac{L}{m^2})$$

Fig. 11 Wash time as a function of wash correlating factor WV for filter leaf test data.

between wash and mother liquor can cause some deviation from a linear relationship, this effect will be minor. For a given cake weight W and a wash volume V, a specific wash time θ_w will be indicated. If the filter under consideration can use 25% of its total area (or cycle) for washing, then $\theta_t = \theta_w/0.25$, and the dry cake rate is equal to $W(60/\theta_t)$, for a rate expressed in hourly units and θ_t in minutes.

The wash effectiveness data may be plotted on linear or semilog graph paper, with the former illustrated in Fig. 12. The wash results can be expressed as a ratio of liquor concentrations, $R = (C_2 - C_w)/(C_1 - C_w)$, as a function of the wash ratio n. The curve shown is typical of most systems in which adsorption of the solute by the solids does not occur.

For a specified final cake solute content, a wash ratio can be selected. From the cake moisture data and a particular value of W, a wash volume then can be calculated, and the rate determined as shown above.

It will be evident that, for a particular filter handling a fixed quantity of solids, a greater degree of washing will be achieved simply by decreasing the cycle time and producing a thinner cake, provided that the thinner cake still can be discharged successfully.

Filter Cake Moisture

Filter cake moisture can be expressed as a function of the approach factor $F_a = \theta_d \, \Delta P \, G/W\mu$ as shown in Fig. 13. At constant vacuum and feed temperature, and at relatively low air rates, the factor can be simplified to θ_d/W. This factor in itself expresses the filtration rate at a specific moisture. For example, on a disk filter, approximately 45% of the filter area is used in dewatering, and, therefore, the available time in each hour for this step is 27 min. The rate then becomes $27W/\theta_d$.

With very fine materials (-10-μm particles), such as precipitates, this relationship is less well defined, particularly if the cake should crack during dewatering.

It will be seen that any one of the above conditions, form rate, wash rate, or dewatering rate may determine the actual full-scale filtration rate, and all functions may have to be checked separately to determine which one is the limiting, or area determining, factor.

Effect of Vacuum Level

The effect of vacuum level on filtration is to increase the cake formation in proportion to the square root of the pressure differential ratio and wash rate in direct proportion (assuming a constant cycle time). Cake moisture usually will be reduced by increasing the vacuum level, but the degree of reduction can be determined only by actual experiment.

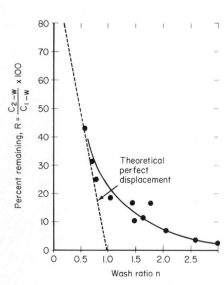

Fig. 12 Characteristic wash effectiveness plot, residual solubles R vs. wash ratio n.

The vacuum air flow at a different pressure differential can be predicted from an expression derived from Darcy's law:

$$G_2 = G_1 \frac{P_{v_1}}{P_{v_2}} \left(\frac{P_{a_2}^2 - P_{v_2}^2}{P_{a_1}^2 - P_{v_1}^2} \right)$$

The equation will not be valid should cake cracking occur.

Feed Solids Concentration

Feed solids concentration has a strong bearing on the cake formation rate, and most systems attempt to maximize this level. Where it is necessary to predict the effect of a change in solids concentration, the following equation can be used:

$$R_{F_2} = R_{F_1} \frac{S_2}{S_1} \left(\frac{1 - S_1/S_c}{1 - S_2/S_c} \right)$$

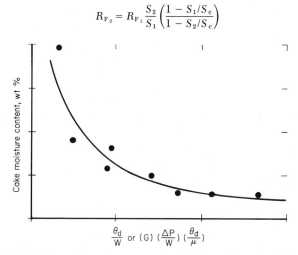

Fig. 13   Filter cake moisture expressed as a function of a correlating factor.

The value of S_c, the mass fraction of solids in the formed but undewatered cake, is best calculated from two sets of known values of R_F and S, rate and feed solids concentration. However, if only one pair of data points is available, S_c can be estimated from cake moisture data. Caution must be used in predicting the rate at much lower solids concentrations, since classification can occur and result in a greatly reduced filtration rate.

VACUUM SYSTEM

Generally, the vacuum system comprises three major items: vacuum pump, filtrate receiver, and filtrate pump. Moisture traps, scrubbers, and condensers may also be included in the vacuum line. Most vacuum pumps will utilize a silencer on the discharge to reduce the noise level and separate any seal liquid from the exit gas.

Vacuum Pumps

The different types of vacuum pumps commonly used in filtration applications include: liquid-piston (or "liquid-ring") pumps, rotary positive-displacement pumps, reciprocating pumps, and centrifugal exhausters. To a much lesser extent, steam ejectors may also be used.

The liquid piston pump can claim these advantages: quiet operation; lower maintenance than other pumps; moisture trap rarely required; condensing capability with hot, saturated inlet gas.

Rotary, positive-displacement pumps include the double-lobe type, twin-screw type, and sliding vane pumps, primarily. These units are somewhat lower in cost than others, deliver more volumetric capacity per horsepower than liquid piston pumps at vacuums below 457 mmHg (18 inHg), and use relatively minor amounts of seal water.

Centrifugal exhausters are best suited to operations at low vacuum, usually less than 254 mmHg (10 inHg), and are particularly useful in installations involving large inlet volumes, 283 m³/min (10,000 ft³/min) and above. Specific advantages include a relatively high efficiency and low maintenance if the exhauster is not exposed to abrasive material in the air stream.

Reciprocating piston-type vacuum pumps are less frequently used at the present date, due to the relatively higher cost per unit of capacity, but do have the advantages of being able to achieve very high vacuums and maintain good volumetric efficiencies at these levels.

Two-stage pumps are commonly used when relatively high vacuums, 610 mmHg (24 inHg) or higher, are required.

Vacuum Piping

In sizing vacuum piping, a design basis of 23 m/s (75 ft/s) will result in a minimum pressure loss over relatively long distances. Ideally, the pressure drop across the vacuum system from filter valve to pump inlet should be below 13 mmHg (0.5 inHg) to provide the most economic use of operating and investment capital.

At high-vacuum operation, less than 127 mmHg (5 inHg) absolute pressure, the effect of temperature on the volume of gas will be very significant, since the gas usually is saturated with water vapor. For example, at a temperature of 38°C (100°F) and an absolute vacuum of 89 mmHg (3.5 inHg), half the inlet gas volume would be water vapor. Cooling this gas stream to 24°C (75°F) by any appropriate means will reduce the volume reaching the vacuum pump inlet by 30%.

With some large pumps, check valves should be installed in the vacuum line to prevent reverse rotation of the pump during shutdown. Safety shutoff switches activated by excess pump temperature or loss of lubricant pressure are useful additions to the system.

Filtrate Receivers

The filtrate receiver provides a means for separating liquid and gas drawn through the filter, as well as a device for separating strong and weak filtrates, regulating vacuum, etc. The receivers generally are sized on vacuum gas handling capacity, and a nominal cross-section velocity of 1.5 m/s (5 ft/s) will allow good separation of liquid and gas phases. With foaming liquids, lower air velocities often are required in order to allow time for the foam layer to dissipate in the receiver.

Vacuum Regulation

If vacuum regulation is to be used, a throttling valve on the air outlet from the receiver represents the most effective approach. A butterfly valve probably is the best choice in this service, since it requires a minimum operating force and can function effectively at nearly closed positions. Proper sizing of the line and the valve is required to ensure best operation. This is particularly true with control on the form receiver, since the total amount of airflow through this receiver (with relatively impermeable filter cakes) is accounted for almost entirely by the amount of air remaining in the filter deck and piping after cake discharge. In some instances, this low volume will dictate the use of a valve and pipeline too small to handle the high airflow rates which occur at start-up, and some difficulty could be experienced in forming an initial filter cake. Since a manual bypass valve usually is desirable when automatic valves are installed, this bypass valve can be made much larger in order to accommodate the greater flow at start-up.

Form zone receivers frequently collect little or no filtrate because of the relatively large holdup of filtrate in each filter section, and the lack of displacing action by air passing through the cake. No liquid may leave the section until the horizontal center line is reached, at which point gravity drainage, assisted by the air passing through the filter cake, removes the liquid.

The receivers should be located as close to the filter valve as practical, and, if possible, at the same level. The lines should slope slightly toward the receiver in order to provide drainage and prevent leakage at the filter valve at shutdown. In some cases the receivers may be located on the floor below the filter, to provide a less obstructed area around the filter. Disadvantages of this arrangement include the greater line pressure drop, a reduction in the available positive head above the filtrate pump, and increased wear on the connecting pipes.

The connecting lines should be equipped with full-port valves for use when the filter must be isolated from the vacuum system. To some extent, these valves can be used for manual vacuum regulation, although control will be poor and wear on the valves high.

A flexible connector is used at the filter valve to join the valve outlet port to the line proceeding to the receiver. Close alignment of these lines is necessary, and separate support of the receiver line must be provided to avoid imposing an uneven load on the filter valve.

Moisture Traps

A moisture trap is recommended with most dry pumps in order to protect the pump from a sudden surge of liquid caused by a failure in the filtrate removal system. Moisture traps generally are sized on the same basis as receivers, and are equipped with a barometric leg for removal of any liquid carried over. A scrubber may be required to prevent carryover of corrosive mists into the vacuum pump.

Filtrate Pumps

In many vacuum filter installations, improper operation of the filter and vacuum system often can be traced to a faulty filtrate pump installation. Any pump selected for this service should have the NPSH capabilities required by the job. It is often preferable to provide at least 3 to 5 m (10 to 15 ft) of pipe between the bottom of the receiver and the pump inlet in order to insure sufficient positive head at this point. Connecting pipes should be large enough to prevent cavitation and flashing. In general, sharp-angle bends and long, horizontal sections of pipe should be avoided. All filtrate pumps should be provided with a check valve on the discharge outlet to prevent air from being pulled back through the pump, particularly at start-up. Similarly, shaft seals must be maintained in good condition to prevent inward air leakage at this point.

Since a centrifugal pump operating in vacuum service can easily become air-locked, a small balance line may be installed between the pump inlet and the top of the vacuum receiver in order to evacuate air which could collect at this point and prevent pump operation.

Receiver-mounted pumps are used frequently in this service, particularly where no head room is available for the installation of a pump beneath the receiver. These pumps will operate on mixtures of gas and liquid without difficulty.

In some instances, it is preferable to avoid pumping the receiver dry, either to prevent aeration of the filtrate or to reduce flow surges. A level control mechanism in the receiver at the desired level can be used to operate a control valve on the filtrate pump discharge.

Barometric Legs

In large installations, or in systems where a significant quantity of abrasive solids is expected in the filtrate, barometric legs are used instead of filtrate pumps. This is particularly true in the beneficiation of iron ore, where filtrates normally contain up to 0.5% of highly abrasive material. Barometric leg size should be based on a liquid velocity of 0.6 to 2 m/s (2 to 6 ft/s). Ideally, the barometric leg should extend vertically from the bottom of the receiver to the seal tank a distance at least equal to the water head equivalent of the local barometric pressure. Where the plant layout absolutely does not allow a vertical drop, the pipe may be sloped as required to reach the level of the seal tank. However, horizontal runs, slopes less than 30°, sharp elbows, and other restrictions should be avoided. When the barometric pipe is used as an air pumping leg, sloped lines must be avoided and if the seal tank is offset, then a horizontal run of pipe should be employed, either at the filter level or at the seal tank level.

MOISTURE REDUCTION SYSTEMS

Mechanical Devices

With some filter cakes, it is possible to lower the filter cake moisture beyond that normally achievable by the use of mechanical attachments which squeeze the cake to compress it, flap it, vibrate it, etc. Compression rolls are most effective on filter cakes which can be squeezed down to a lesser thickness upon the application of pressure. A typical example is sodium bicarbonate, which usually is dewatered on specially constructed, heavy-duty filters utilizing three or four heavily loaded compression rolls to reduce the cake thickness and, correspondingly, the moisture. Vibrating and flapping devices are effective on materials which, upon discharge from a filter, tend to become somewhat liquefied, particularly under the influence of motion, such as on a conveyor belt or in a railroad car. The flapping action rearranges the cake structure and allows moisture to be withdrawn from the filter cake. Typical materials are gypsum precipitate and fine-size concentrates of zinc and copper minerals.

Steam Hoods

By using a closely fitted hood, steam can be applied to a vacuum filter during a portion of the cycle normally used for dewatering with ambient air. If sufficient steam can be pulled into the cake to heat it to an equilibrium temperature, a reduction in moisture of 2 to 4 percentage points usually will be obtained. This effect is attributable principally to better drainage of entrained liquor resulting from the reduction in viscosity of the liquid phase caused by the increased temperature. In addition, after termination of the steam zone, a certain amount of evaporation of entrained moisture will occur because of the sensible heat remaining in the cake. This evaporation can occur either on the filter, during air drying, or following discharge, provided that the hot filter cake is exposed to dry air during the cooling process. Low-pressure steam is suitable for this system, and little benefit is gained by using superheated steam.

On the average, the steam consumption required for maximum reduction will vary between 38 and 75 kg/t (75 and 150 lb/ton) of material processed. The quantity of steam needed can be estimated by calculating the amount which will have to be condensed to raise the filter cake temperature to the average temperature expected in the cake, generally 77 to 93°C (170 to 200°F), and allowing approximately 25% additional steam to account for the excess which will be pulled through the cake.

The principal use of steam drying is in applications where a minor reduction in moisture beyond that which can be achieved by ambient filtration will provide a desired result, such as in the pelletizing of iron ore, where a specific moisture level must be attained for satisfactory pellet formation.

Surfactants

Surfactants also can be used effectively for moisture reduction. Dosages of about 0.25 kg/t (0.5 lb/ton) of solids will reduce moisture 1 to 3%. Cost of these reagents averages around 88¢/kg (40¢/lb).

FILTER CAKE WASHING

Wash Apparatus

Filter cake washing is usually confined to drum filters and horizontal filters. With the latter, the apparatus employed may consist of spray nozzles or simply wash boxes which flood the wash zone. With drum filters, the application of wash requires more care, since, with relatively tight cakes, excess water can run down either side of the drum and interfere with cake discharge or cause dilution in the filter tank. A preferred system uses 5 to 7 wash pipes with flat-spray, low-pressure, flooding nozzles to apply a uniform layer of water on the cake over approximately 25 to 33% of the total area. Line strainers are necessary to keep nozzles from plugging. Should cake cracking tend to occur before washing, this can be prevented by reducing the vacuum in the zone between the slurry level and the initiation of wash, or by using fine, misting-type nozzles in this area.

In certain systems, primarily gold and silver cyanide leach plants, flood washing is used on drum filters, with wash water applied through three or four perforated pipes located near the top center of the filter. The entire area of the filter between its emergence from the slurry to the discharge point is blanketed with water. The cake discharge is, of course, practically equivalent to a sluice discharge, and the cake must be repulped and removed by pumping. In addition, a certain amount of filtrate will blow back through the medium and cause some loss of solution. The system has three advantages: it utilizes about 50% of the filter area for washing, it reduces the vacuum pump requirements, and the wet discharge helps keep the filter medium from blinding.

Countercurrent Washing

Countercurrent washing requires careful consideration of the actual mechanism of wash filtrate advance from one stage to the next. Both the filter and the material being handled must be suited to removal of the filtrate from the system. If the filter employs typical drainage sections, each with separate connecting piping to the filter valve, thin cakes [less than 19 mm (¾ in)] of low air permeability may not produce enough wash filtrate to allow this volume to be separated from the balance of the filtrate. For this reason, horizontal belt filters are well adapted to multistage countercurrent washing, since the filtrate holdup is minimal. Also, a large reservoir of wash filtrate must be avoided between stages since the effect of a plant upset will be greatly magnified by the filtrate retention time across the washing system, due to the delay in returning to equilibrium conditions.

FLOCCULATION

Flocculation can be of considerable benefit in improving continuous filter operation, particularly when very fine size material (less than 2 μm) is present. Inorganic chemicals and polymers (synthetic organic polyelectrolytes as well as natural gums and starches) are used extensively to improve filter performance.

Certain substances, particularly organic wastes such as sewage, are practically unfilterable without a heavy flocculant dosage, which normally is ferric chloride and lime or cationic polymer. Slurries in which most of the solids are coarse (+325 mesh) with a small amount of very fine size material present may have the filtration rate controlled by the latter, particularly if the slurry is somewhat dilute. A small dosage of flocculant, generally 5 to 20 ppm on a solids basis, may be sufficient to flocculate these fines, allowing a uniform filter cake to form at a rate several times greater than otherwise possible.

Flocculant selection is best done on an experimental basis, although certain general rules often apply: nonionic polymers work best in acidic systems, anionics are more effective in neutral, ammoniacal, or slightly alkaline circuits, and cationic or inorganic reagents work best on organic substances. With submicron or colloidal material present, it is often necessary to add an inorganic chemical such as lime, acid, calcium chloride, etc. to the system initially, and after a short retention time, a polymer reagent may be added to provide the complete flocculation required.

Rapid, uniform blending of polymer reagents with slurry is generally necessary, due to the rapid adsorption of the flocculant on fine solids. The rate of mixing is not nearly as critical with inorganic chemicals, as long as uniformity is achieved.

In some systems the effectiveness of the flocculant will be reduced with time, particularly if the slurry is exposed to the mechanical action of the filter agitator. Top-feed units have the advantage of being able to filter a slurry immediately following flocculation, and therefore will obtain the maximum benefit of the treatment.

INSTRUMENTATION

A continuous filter can be instrumented for total automatic control, or merely for monitoring the performance of the unit. The advantage of automatic control is that it avoids the arbitrary variations commonly found in manually regulated filters. Too often, in these cases, the only criterion is that the machine produce filter cake, rather than maximize solute recovery, obtain minimum moisture, or economize on precoat materials.

Full automatic control is recommended on: feed rate, filter tank slurry level, cake wash-water rate, and precoat knife-advance rate. Semiautomatic systems (operator selects control level as required) may be used on filter drive speed, form and dry vacuums, and dosage of flocculant or other conditioning chemicals employed. Monitoring instruments should be used to indicate vacuum levels and feed solids concentration, indicate and record feed and wash rates, and, in multistage washing operations, filtrate flow rates and solute concentrations.

FILTER MEDIA SELECTION

The selection of the best filter medium for a continuous filter is determined largely by experience in a similar application, or, lacking this information, trial-and-error testing.

Since there are hundreds of different fabrics available, a few general rules can provide assistance in narrowing down the choice.

Thread count or mesh is of little use in selecting the filter medium. Air permeability rating, weight or thickness, type of thread (monofilament, multifilament, spun staple, or felt), weave (plain, twill, or sateen), and surface finish are far more important considerations. Similarly, the type of material—nylon, polyester, polyethylene, polypropylene, Saran, cotton, or wool—must be compatible with the system composition and temperature.

In most continuous filter applications, some of the first solids reaching the medium at the onset of filtration will pass through with the filtrate before particles bridge across the openings and form a filter cake, which will then retain all the solids in the slurry. This is a desirable action in most instances, since it produces a more permeable filter cake, permitting a higher rate and better cake discharge than otherwise would be possible. With the exception of the precoat filter, continuous filters cannot achieve perfect clarity except under unusual circumstances, and a certain amount of solids in the filtrate must be expected.

With feed slurries containing a wide particle distribution, it is particularly important that a bleed-through of solids be allowed so that the coarse-size rather than fine-size particles are in immediate contact with the medium. Monofilament media are preferred for such materials. Monofilaments ordinarily will perform better in the presence of agents which cause blinding due to chemical precipitation, although they will not be completely immune to blinding. When both multifilament and monofilament media will work satisfactorily, the former generally will yield a slightly higher filtration rate. With fine-size materials, such as chemical precipitates and clays, a thin, tight, multifilament medium of twill or sateen weave will yield the best results. Plain weaves usually can achieve the best clarities, particularly with spun-staple thread, but they are the most likely to blind. Felt media can produce good clarities and high rates, but have lower resistance to abrasion.

Damaged and blinded filter media preferably should be replaced immediately, since the loss of production usually far outweighs the cost of the media. Should replacement not be convenient at the time, minor repair to holes can be made by applying a patch of the same material, using a waterproof glue such as contact cement. Blinded media can often be restored to their original permeability by removing the causative agent: washing with weak hydrochloric acid (preferably inhibited) will remove most carbonate blinding; gypsum can be leached out with certain proprietary reagents or solutions; trapped insoluble solids often can be removed by brushing, or washing with high-pressure sprays.

Filter medium life depends upon the application as well as upon the preventative maintenance used on the filter. In filtration of iron ore concentrates, average medium life is 2 to 3 weeks; with slow filtering, nonabrasive materials, 6 months' or a year's life is not uncommon. With belt filters, normal media life will vary between 1000 and 2000 h of operation. The most significant factor in the latter case is the operator attention given to belt alignment, medium washing, and general filter operation.

REFERENCES

1. Ruth, B. F., *Ind. Eng. Chem.*, **27**(6), 708–723 (1935).
2. Carman, P. C., *Trans. Inst. Chem. Eng. (London)*, **16**, 168 (1938).
3. Mead, W. J., "Encyclopedia of Chemical Process Equipment," Reinhold, New York, 1964, pp. 417–438.
4. Poole, J. B., and Doyle, D., "Solid-Liquid Separation," Chemical Publishing, New York, 1968.
5. Purchas, D. B., "Industrial Filtration of Liquids" 2d ed., CRC Press, Cleveland, 1971.
6. Tiller, F. M., How to Select Solid-Liquid Separation Equipment, *Chem. Eng.*, Apr. 29, 1974, pp. 116–136.

Section **4.5**

Centrifugation

CHARLES M. AMBLER, P.E. *Registered Professional Engineer; Consultant, Pennwalt Corporation (Sharples-Stokes Division); Fellow, American Institute of Chemical Engineers; Fellow, American Institute of Chemists; Life Member, Association Iron and Steel Engineers.*

NOMENCLATURE

A	Superscript, exponent
B	Superscript, exponent
C	Concentration
D	Diameter (as of rotor)
d	Stokes equivalent diameter of particle
g	Gravitational acceleration
k	Constant of proportionality
Re	Reynolds number
n	Number of spaces between disks in a stack
P	Pressure
Q	Flow rate (volumetric)
r	Radial distance from axis of rotation
r_1	Radius of liquid surface (free interface), or of disk ID
r_2	Outer radius of moving layer, or of disk OD
r_3	Inner radius of bowl wall
r_i	Radius of liquid-liquid interface
r_H	Radius of heavy-liquid discharge
r_L	Radius of light-liquid discharge
RCF	Relative centrifugal force ($\omega^2 r/g$)
S	Volume of liquid/volume of solid
t	Time
V	Volume
v	Velocity
v_g	Velocity in gravitational field
μ	Absolute viscosity
θ	Half occluded angle of disk section
π	Constant (3.1416)
ρ	Density
ρ_H	Density of heavy-liquid phase
ρ_L	Density of light-liquid phase
Σ	Equivalent area of sedimentation centrifuge
ω	Angular velocity (radians/sec)

INTRODUCTION

In the context of this section, centrifugation is defined as the process of resolving multicomponent systems, at least one phase of which is liquid, by the application of centrifugal force. Commercial centrifuges can be divided into two broad types, *sedimentation centrifuges* and *centrifugal filters*, each of which can be further subdivided according to the means provided for advancing and discharging the separated solids and liquid phases (see Tables 1, 2, and 3).

Sedimentation Centrifuges

The sedimentation type of centrifuge simulates and amplifies the force of gravity, by a factor of from 2 to 5 orders of magnitude in the commercial sizes. As in the gravitational field, there must be a difference between the mass of the dispersed particle and that of the liquid phase it displaces to cause this particle to migrate—away from the axis of rotation if the difference is positive, and toward it if the difference is negative. For particles whose Re ($dv\,\rho/\mu$) is <1, this velocity of migration at equilibrium is

$$v = \frac{(\rho_s - \rho_L)\,d^2\omega^2 r}{18\,\mu} \tag{1}$$

in which d is the diameter of a spherical particle or the equivalent diameter of a nonspherical particle. By assuming a pattern of flow through a sedimentation centrifuge, it is possible to relate the centrifuge's geometry and rotational speed to the area of a gravity settling tank capable of performing the same amount of useful separative (clarification) work. If this area function is designated as Σ, for a bottle centrifuge,

$$\Sigma_B = \frac{\omega^2 V}{2g \ln\left(\dfrac{2r_2}{r_2 + r_1}\right)} \tag{2}$$

For a tubular bowl centrifuge,

$$\Sigma_T = \frac{2\pi Z \omega^2}{g} \frac{r_2^2 - r_1^2}{\ln\left(\dfrac{2r_2^2}{r_2^2 + r_1^2}\right)} \tag{3}$$

The simplified form of this, accurate to within 4%, is

$$T = \frac{2\pi Z \omega^2}{g} (\tfrac{3}{4} r_2^2 + \tfrac{1}{4} r_1^2) \tag{4}$$

For a disk bowl centrifuge,

$$D = \frac{2\pi n \omega^2}{3g \tan \theta} (r_2^3 - r_1^3) \tag{5}$$

In a heterogeneous system it is usually convenient to refer to a flow rate at which 50% of the particles of size d are captured and $Q = 2\Sigma v_g$, in which $v_g = [(\rho_S - \rho_L) d^2 g]/18u$. In a system containing particles of uniform size and mass, all particles will be captured when $Q = \Sigma v_g$. Any errors that are introduced by an incorrectly assumed flow pattern are largely self-compensating when centrifuges of similar design operating at a similar level of $\omega^2 r$ are compared. For instance, in the tubular bowl, the solid wall "basket," and the continuous decanter, the flow is like an annular boundary layer defined by r_1, the radius of the free interface, and r_2, which is smaller than the inner radius of the rotor but whose exact value often cannot be accurately determined.

For the disk bowl, the model [Eq. (4)] assumes flow of both liquid and solids in planes through the axis of rotation. This oversimplification results in an indicated efficiency of approximately 50% for this type, but this efficiency factor is constant within the limits expressed in the preceding paragraph.

These models are based on the unhindered settling of a single particle. When the concentration of particles reaches the level where hindered settling occurs, the value of v in Eq. (1) will be reduced. This reduction is a function of the system being centrifuged and will be the same for all similar centrifuges that are being compared with this system. Within these limits,

$$2v_g = \frac{Q_1}{\Sigma_1} = \frac{Q_2}{\Sigma_2} = \cdots = \frac{Q_n}{\Sigma_n} \tag{6}$$

for the same degree of particle removal.

The terminal concentration of the sedimented solids is a function of their average size $\overline{d}$ and effective mass $(\rho_P - \rho_L)$, the applied centrifugal force $\omega^2 r$, and the time t of exposure to this force. On any given system this can be determined only experimentally.

A dispersion of one liquid phase in another can also be separated centrifugally, and the same models, with some modification, are applicable. The separated light and heavy (in terms of density) phases are discharged at different elevations to permit their segregation (see Fig. 4). Hydrostatic balance is maintained by selection of the light- and heavy-phase discharge radii to locate the real or imaginary interface between the phases inside the rotor at the optimum radius. From the centrifugal pressure equation

$$P = \int \rho f(r) \, dr \tag{7}$$

$$\frac{\rho_H}{\rho_L} = \frac{r_i^2 - r_H^2}{r_i^2 - r_L^2} \tag{8}$$

The Σ area operating on the light phase is then calculated between r_1 and r_i, and on the heavy phase between r_i and r_2.

Centrifugal Filters

The centrifugal filter supports the particulate solids phase on a porous septum, usually circular in cross section, through which the liquid phase is free to pass under the action of centrifugal force. The density of the solid phase is important only for calculation of the weight loading in the available volume of the basket. A more important parameter is the permeability of the filter cake under the applied centrifugal force. The cost of centrifugal filters per unit of filter area is relatively high. Except for very special applications, they are generally applied only to the dewatering of relatively free-draining solids. The several batch types (fixed bed) possess considerable flexibility to compensate for process variations, since the several phases of their operating cycle are sequential (see Table 7), and the time for each phase can be adjusted as required to compensate for process variations. The continuous-moving-bed types are less flexible and are generally applied to the separation of relatively free-draining solids. They are limited in the amount of mother liquor and wash liquid that can be separated from the solids.

There are two important considerations in centrifugal filter performance:

1. *Feed slurry concentration.* At 10% solids concentration in the feed, almost 9 lb of liquid must be centrifuged per pound of solids. At 50% concentration in the feed, somewhat less than 1 lb of liquid must be centrifuged per pound of solids in the feed. This difference is directly reflected in the feed time for the fixed-bed type and the feed rate for the moving-bed type that is liquid-limited. This liquid loading also imposes a similar difference in the power required during the feed portion of the cycle. Preconcentration of the feed should always be evaluated to optimize cost per unit of solids being centrifuged.

2. *Particle size.* In the heterogeneous-particle-size systems usually encountered in commercial operation, the smallest 15% of the particles present exert a strong influence on the permeability of the filter cake (and the drain rate through it) and on the final retained liquid content of the cake. Where process conditions permit, performance (in terms of capacity and final dryness) can frequently be substantially improved by classification of the feed in a settling tank or liquid cyclone with recycle of the fines in the overflow back to process, and with a concentrated underflow containing the larger particles for feed to the centrifuge.

Drainage on the centrifugal filter takes place in two regimes: bulk drainage while the cake is in the flooded condition, and film drainage at a much lower rate as the radius of the free liquid interface exceeds the inner radius of the cake. At any given level of centrifugal force, a fixed volume of liquid is retained on the wetted cake particles regardless of time. This level is approximated by

$$S_\infty = k\overline{d}^{-0.5}(\omega^2 r/g)^{-0.5}\rho_L^{-0.25} \tag{9}$$

During film drainage the residual liquid content of the cake is approximated by

$$S = \frac{k^1}{\overline{d}}\left[\frac{\mu}{\rho_L(\omega^2 r/g)}\right]\frac{(r_2 - r_c)^A}{t^B} + S_\infty \tag{10}$$

Coefficient A of the cake thickness $(r_2 - r_c)$ is usually in the range of 0.5 to 1.0 but may be larger for compressible cakes. Coefficient B is usually in the range of 0.3 to 0.5 when t is measured from the time the liquid surface and cake surface coincide.

Rinsing efficiency on centrifugal filters, defined as proportionate displacement of

TABLE 1   Classification of Centrifuges by Size of Dispersed Particles

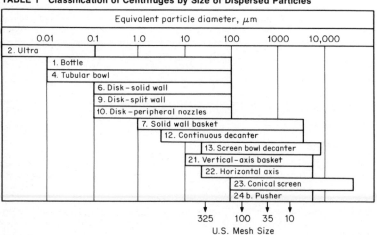

soluble impurities in the mother liquor, is from fair, 50%, to excellent, upward of 95%. The portion of the mother liquor that is retained by porosity of the solid particles can seldom be removed on the centrifugal filter. The time available for rinsing is insufficient to permit the required mass transfer.

Practical Design Approach

Commercial centrifugation covers an extremely broad spectrum of applications ranging from the recovery and concentration of protein molecules such as viruses, with an equivalent diameter of 0.04 μm, to the dewatering of coal in the 25,000-μm (1-in) size range. Table 1 indicates the particle size range to which the centrifuge types of Tables 2 and 3 are generally applicable.

For preliminary screening as to the suitability of the application for centrifugal separation and tentative selection of suitable centrifuge types, the following information is needed:

1. Nature of the liquid phase(s)
   a. Temperature
   b. Viscosity at operating temperature
   c. Density at operating temperature
   d. Vapor pressure at operating temperature
   e. Corrosive characteristics
   f. Fumes are noxious, toxic, inflammable, or none of these
   g. Contact with air is not important, is undesirable, or must be avoided
2. Nature of the solids phase
   a. Particle size and distribution
   b. Particles are amorphous, flocculant, soft, friable, crystalline, or abrasive
   c. Particle size degradation is unimportant, undesirable, or highly critical
   d. Concentration of solids in feed
   e. Density of solids particles
   f. Retained mother liquid content, _____% is tolerable, _____% is desired
   g. Rinsing to further reduce soluble mother liquor impurities is unnecessary or required
3. Quantity of material to be handled per batch or per unit time

TABLE 2   Sedimentation Centrifuges Classified by Flow Patterns

	Sedimentation centrifuges*		
Flow pattern	Type	Centrifugal force†	Capacity‡
Liquid: batch	Analytical and clinical*	3,000	4 L
Solids: batch	Preparatory and ultra	350,000	0.5 L
	Miscellaneous batch	80,000	6 L
Liquid: continuous	Tubular bowl*	16,000	20 gal/min
(interrupted for	Multipass clarifier	6,000	70 gal/min
the removal of	Solid-wall "basket"	1,800	80 gal/min
solids)			
Solids: batch	Disk—solid wall*	9,000	200 gal/min
Liquid: continuous	Disk—valve discharge*	6,000	50 gal/min
Solids: intermittent	Disk—split bowl*	6,000	100 gal/min
Liquid: continuous	Disk—peripheral nozzles*	9,000	400 gal/min
Solids: continuous	Disk—light solids skimmer	6,000	20 gal/min
	Continuous decanter	3,200	
	Sedimentation centrifuges with postdrainage		
Liquid: continuous	Screen bowl decanter	2,200	40 ton/h solids,
Solids: continuous			300 gal/min
			liquid

*Some sedimentation centrifuges are also used for the separation of two liquid phases, usually with the continuous discharge of both.

†Nominal maximum centrifugal force $\omega^2 r/g$ developed, usually less in larger sizes.

‡Nominal maximum capacity of largest size of type, subject to reduction as necessary to meet required performance on a given application.

Centrifuge manufacturers have a fund of nonproprietary application information from which preliminary and frequently final equipment selection can be made based on the above information. When this is not possible, bench-scale screening tests are frequently helpful. Such tests will provide answers to the following questions:

1. Does the system settle under gravity? If so, at what rate?
2. Does it settle in a clinical or bottle centrifuge? If so, in what time at what value of $\omega^2 r/g$ (inch radius × r/min × 0.0000284)?
3. Is the supernatant liquid from items 1 or 2 of satisfactory clarity?
4. What is the nature of the sedimented solids? Are they soft, plastic, or firm?
5. On a Buchner funnel does the system filter rapidly, slowly or not at all?

TABLE 3   Centrifugal Filters Classified by Flow Pattern

Flow pattern	Fixed-bed type	Centrifugal force†	Basket capacity* (under lip ring), ft³
Liquid: continuous (interrupted for discharge of solids) Solids: batch	Vertical axis Manual unload Container unload Knife unload	1200 550 1800	16 20 16
	Horizontal axis Knife unload	1000	20

Flow pattern	Moving-bed type	Centrifugal force†	Solids capacity,‡ lb/h
Liquid: continuous Solids: continuous	Conical screen Wide angle Differential scroll Axial vibration Torsional vibration Oscillating	2400 1800 600 600 600	150,000 300,000
	Cylindrical screen Differential scroll Reciprocating pusher	600 600	80,000 60,000

*Reduce by ⅓ for volume of processed solids ready to be discharged.
†Nominal maximum centrifugal force ($\omega^2 r/g$) developed, usually less in larger sizes.
‡Nominal maximum capacity of largest sizes, subject to reduction as necessary to meet required performance on a given application.

If the system settles or filters, centrifugation should be further investigated. Even when the bench tests are negative, there are certain applications in which centrifugation has little competition. Typically these include the separation and concentration of very small particles such as viruses, where a threshold of separative force must be exceeded to overcome diffusive forces, and the resolution of emulsions that are stabilized by finely divided solids that can only be sedimented away from the interfacial boundaries by centrifugal force. The sedimentation and compaction of fine and flocculent solids can frequently be improved by pH control or the addition of flocculating agents such as polyelectrolytes. The results of these screening tests will frequently lead to a preliminary selection of centrifuge types, and reduce the number that must be tested quantitatively.

Centrifuge manufacturers maintain facilities for testing customers' samples on a scale adequate to size the proper centrifuge for the application and demonstrate the results that can be anticipated and guaranteed. See Table 4. Where the sample is not sufficiently stable to be transported, it is frequently possible to arrange tests in the prospective customer's plant on loan or rental equipment.

The design of centrifuges is an expensive, highly skilled combination of art and science. It is seldom feasible to completely design a centrifuge for a specific application. Rather, the market utilizes existing designs subject to such modifications as are compatible with the process requirements and the estimated volume of the market for the modified design.

Certain applications, e.g., coal dewatering, are of such large volume that some manufacturers may make centrifuges for only a single market.

In designing a system involving a separation stage, such as centrifugation, both the upstream conditions and downstream requirements should be carefully examined. As indicated by Eqs. (1) and (10), both sedimentation and filtration performance can be improved, often substantially, by:

1. Increasing the particle size.
2. Decreasing the suspending liquid viscosity.
3. Increasing the concentration of particulate matter in the feed to the centrifuge when output is measured in rate of solids processed.

Particle size can frequently be increased by control of pH; use of agglomerating or flocculating agents; control of agitation, temperature, and pH during precipitation reactions; or control of crystallization.

Viscosity of the liquid phase can be decreased by elevating the temperature. The viscosity of pure water reduces from 1.005 cP at 20°C to 0.284 cP at 100°C and the proportionate reduction is even greater for solutions and organic liquids.

Concentration can frequently be increased by gravity thickening or hydrocycloning.

For each of these methods of improving centrifuge performance and reducing separation costs, there is an offsetting increment of additional processing cost. The design engineer, being aware of the parameters, must arrive at the optimum compromise.

The downstream disposition of the separated products is of equal importance in the overall design. Complete separation of the phases is seldom practical and usually impossible in a centrifuge.

In the sedimentation type the amount of mother liquor contained in the cake is an inverse function of retention time and applied centrifugal force that approaches an asymptote. This residual liquid contains the same proportion of dissolved impurities as existed in the feed. Further reduction of the soluble impurity concentration is possible only by repulping and reseparation. In the heterogeneous (with respect to particle size) systems usually encountered, total removal of the suspended particles from the liquid phase may be impossible or obtainable only at an economically low throughput rate. In many systems this creates no problem because the effluent is recycled back through the process. In others, it must be considered in the downstream processing.

In the centrifugal filter the retention of mother liquor and dissolved impurities is in accordance with Eqs. (9) and (10). A considerable portion of this can be displaced by washing the cake in the centrifuge. Washing efficiency can approach 100% in some

TABLE 4   Domestic Centrifuge Manufacturers with Test Facilities*

Company name	Principal address
Ametek Inc.	E. Moline, Ill.
Baker Perkins Inc.	Saginaw, Mich.
Leon J. Barrett Co.	Worcester, Mass.
Beckman Instruments Inc., Spinco Div.	Palo Alto, Cal.
Beloit Corp., Jones Div.	Dalton, Mass.
Bird Machine Co.	S. Walpole, Mass.
Centrico Inc.	Northvale, N.J.
Centrifugal & Mechanical Industries Inc.	St. Louis, Mo.
CF&I Engineers Inc.	Denver, Colo.
Chemapec, Inc.	Hoboken, N.J.
Combustion Engineering Inc., Raymond Div.	Chicago, Ill.
The DeLaval Separator Co.	Poughkeepsie, N.Y.
Dorr Oliver Inc.	Stamford, Conn.
Electro-Nuecleonics Inc.	Fairfield, N.J.
Envirotech Corporation, Eimco Machinery Division	Salt Lake City, Utah
Heyl and Patterson Inc.	Pittsburgh, Pa.
IEC Division, Damon Corp.	Needham Heights, Mass.
McNally Pittsburg Inc.	Pittsburg, Kansas
Pennwalt Corporation, Sharples-Stokes Division	Warminster, Pa.
The Pfaudler Co.	Rochester, N.Y.
DuPont Co., Inst. Prod. Div.	Newtown, Conn.
Western States Machine Co.	Hamilton, Ohio

*The author apologizes for any inadvertent omissions.

applications and be as low as 50% in others. The amount of wash liquid that can be used is frequently limited by overall process considerations.

EQUIPMENT—SEDIMENTATION TYPES

Analytical and Clinical Bottle Centrifuge

These apply centrifugal force to the contents of a bottle or other container that is caused to rotate about a vertical axis. They are driven from below by an electric motor or oil turbine through a vertical shaft mounted in appropriate bearings. At the upper end of the shaft a head provides support for and rotates the containers.

Horizontal ("swinging bucket") heads accommodate cups and adapters designed to support a wide variety of containers for specific uses. These cups are mounted in trunnion rings that allow the cups and their contents to swing from a vertical position, while at rest, to a horizontal position when rotating. The center of gravity of the cup and its contents must be below the center line of the trunnions.

Angle heads support the containers at a fixed angle, with respect to the axis of rotation, that does not change during operation. This shortens the path through which the sedimenting particles must travel and reduces the time required for centrifuging.

Many bottle centrifuges also accommodate "basket" heads, perforated for centrifugal filtration and imperforate for continuous flow clarification. Some angle heads are also arranged for continuous liquid flow through the containers so that the quantity of solids accumulated per run is greatly increased.

Except in the smallest sizes, the rotating element is enclosed in a casing for safety. In some, the casing and its contents can be heated to a controlled temperature. In others, refrigeration is provided both to permit subambient-temperature operation and to absorb the heat generated by the windage of the rotor. The thermal effects of windage on ultrahigh speed rotors are minimized by maintaining the casing under vacuum.

Besides providing a convenient way of quickly settling and compacting small solid particles that settle or filter slowly, bottle centrifuges are also used for the separation of refractory emulsions. Their use is specified in a number of ASTM standard petroleum and other tests, a variety of analytical methods, and clinical and public health procedures. When the volume of the separated phase is to be measured accurately, the use of the swinging-bucket type is preferred. It gives an interface at right angles to the axis of the container. The containers are frequently shaped and calibrated to emphasize a small volume of the separated phase; examples are small bore tips on pear-shaped bottles for the determination of small amounts of free moisture and sediment in oil or small bore tops on Babcock test bottles for the determination of butterfat in milk products.

A large variety of containers is available for specific applications ranging in size from 175 μL, or even smaller, to 1.0 L. Localized stressing of glass containers is minimized by supporting them on a plastic cushion and filling the space between them and the cup with a nonvolatile fluid such as glycerin. Plastic bottles of polypropylene, polyallomer, and polycarbonate are less subject to breakage and do not require as careful handling as glass.

Horizontal heads provide from 2 to 16 places in which cups may be located. The number of small containers handled per run can be multiplied by as much as 28 with the use of adapters in each place. Angle heads are available to centrifuge from 4 $\times$ 600 ml blood bags, through 6 $\times$ 250 mL bottles to 60 $\times$ 15 mL bottles. A large variety of drives is available to provide operating speeds of up to 3000 r/min (1980 g at the container tip) in the larger sizes, through 23,400 r/min (38,000 g) on smaller sizes and multispeed head attachments, to 65,000 r/min (429,000 g) on ultrahigh-speed models. In every case the manufacturer's recommendations regarding the relation of maximum operating speed to operating load should be followed.

Accessories include speed control and indication, automatic timers, temperature control and indication, and safety interlocks.

Preparatory Centrifuges and Ultracentrifuge

The term *ultracentrifuge* was originally applied by T. Svedberg to any centrifuge that permitted observation of the contents of the container during the act of centrifuging. It is now more commonly applied to any ultrahigh-force centrifuge, and the Svedberg instrument is known as the analytical (or optical) ultracentrifuge.

Preparatory bottle ultracentrifuges may be of the fixed-angle or swinging-head type as described in the preceding section. They are most commonly applied in biological and biochemical techniques such as the separation and isolation by selective classification of cells and macromolecules.

The analytical ultracentrifuge applies high centrifugal force (up to 372,000 g) to a small sample in a transparent cell. With Schlieren or interference optics, the change in concentration can be measured as a function of time. From this information can be derived molecular weight and shape, sedimentation and diffusion coefficients, and partial specific volume.

Zonal Centrifuge

In analytical and preparatory centrifuges described above, the collected fraction reports to the bottom of the container as a pellet. This pellet is contaminated with larger and heavier particles as well as smaller particles that started from nearer the bottom of the container. In many biochemical procedures, a higher-purity fraction is required that has not been denatured by agglomeration into a pellet. The zonal centrifuge (Fig. 1) is preloaded with a

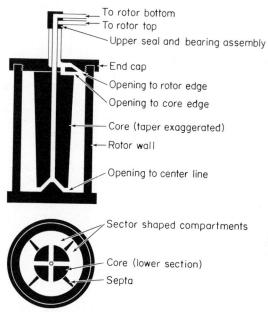

Fig. 1 Zonal centrifuge, type B-16 continuous-flow rotor. (*Spinco Division, Beckman Instruments Inc.*)

compatible solution, such as sucrose, in which a radial density gradient from light near the center to heavy at the periphery is established. The feed is then introduced, either batchwise or continuously, and each component layers itself as an annulus at a zone in the gradient corresponding to its density. The technique is referred to as isopyknic banding and is used for the purification of influenza and other viruses.

Tubular Bowl Centrifuge

The tubular bowl centrifuge, Fig. 2, consists of a hollow cylindrical rotating element, with lower and upper end caps, suspended from and driven by a spindle of designated flexibility. The spindle in turn is supported through radial and thrust bearings on a rigid frame. The drive may be direct, from a steam or air turbine mounted on the upper end of the spindle. Much more often the drive is through a flexible belt from an offset motor and pulley to a pulley that is mounted on the bearings and supports the spindle and rotor through a flexible coupling. A boss on the bottom end cap fits into a sleeve bushing that

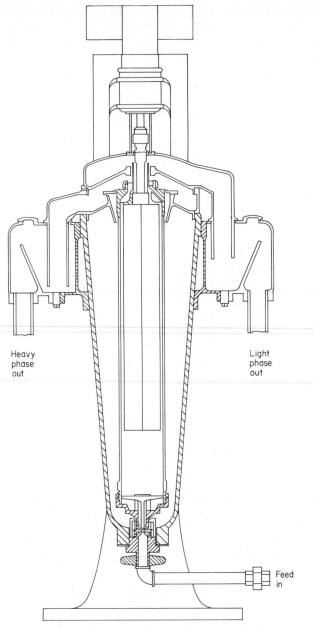

Heavy
phase
out

Light
phase
out

Feed
in

Fig. 2 Tubular bowl centrifuge (separator). *(Sharples Centrifuges, Pennwalt Corporation.)*

provides controlled dampening during acceleration and deceleration and accommodates the excursion of the axis of rotation from the geometrical center to the center of mass as the rotational speed passes above critical. The tubular bowl centrifuge derives its stability above critical speed from its relatively large length-to-diameter (L/D) ratio, in the range of approximately 4.35 to 7.25.

The liquid to be clarified or separated enters the rotor through a centrally located opening in the lower end cap as a free-standing jet from a fixed feed nozzle. It is accelerated to the rotational velocity of the rotor, flows upward as an annulus, and discharges ("overflows") from a larger diameter opening in the upper end cap. The diameter of this opening establishes the free (gas/liquid) interface within the rotor. When used to continuously separate two liquids, an internal baffle, commonly called the L-X disk, provides a separate passage adjacent to the bowl wall to conduct the heavier-phase liquid to a different discharge elevation. The radial position of the liquid/liquid interface within the rotor is controlled by the diameter of the adjustable heavy-phase discharge overflow with respect to the fixed diameter of the light-phase overflow in accordance with the pressure equilibrium equation

$$\frac{\text{Density (light phase)}}{\text{Density (heavy phase)}} = \frac{r_i^2 - r_H^2}{r_i^2 - r_L^2} \qquad (11)$$

The liquid effluent, or separated liquids, is captured in covers mounted on the frame as shown. These carry the liquid(s) to a convenient point discharge and provide some control of windage. An internal septum, frequently in the form of a trifoil, assists in accelerating the feed stream to the rotational velocity of the bowl and provides a stabilizing influence during deceleration. As the annulus of liquid moves upward through the rotor, it is subjected to centrifugal force. Dispersed particles are caused to migrate toward or away from the axis of rotation in accordance with the parameters of Eq. (4). These are captured if they have passed through the moving boundary layer during the time it is within the rotor.

The accumulated solids are recovered manually by detaching the rotor from the drive assembly, removing the bottom end cap and three wingnuts, and washing or scraping the solids out of the cylinder. By virtue of the simplicity of construction, this operation usually requires less than 15 min. Therefore, tubular bowl centrifuges are generally applied to the removal and/or recovery of relatively low concentrations of solids.

Their principal applications are:

The purification of fuel and used lubricating and industrial oils at rates up to 4500 L/h (1200 gal/h). Any separated water is discharged continuously; the separated solids are retained in the rotor for intermittent removal.

The harvesting of bacteria and other microorganisms.

The harvesting of precipitated blood protein fractions.

The clarification of viscous substances such as molten chicle and nitrocellulose solution.

The clarification of essential oils, flavoring extracts, and other food products.

The removal of microcrystalline wax from chilled lubricating oil in naphtha solution with continuous discharge of the separated wax on an aqueous carrier liquid.

The recovery of finely divided metal particles such as silver resulting from the washing of scrap photographic film.

The continuous separation of the soap stock formed by the alkali refining of vegetable and animal fats and oils.

The classification (removal of oversize particles) from pigment slurries.

The sizes of tubular bowl centrifuges are listed in Table 5.

TABLE 5 Size of Tubular Bowl Centrifuges

Diameter, in	Length, in	Max. r/min	Max. RCF	Capacity
1.75	9	50,000	62,400	0.15 to 60 gal/h
4.125	18 to 30	15,000	13,200	to 600 gal/h
5	30	15,000	16,000	to 1200 gal/h

A special model of the 5 in diameter × 30 in long size (the KII zonal centrifuge) develops up to 87,000 g at 3500 r/min.

Multipass Clarifier Centrifuge

The multipass clarifier centrifuge (Fig. 3) functions as a series of interconnected tubular bowls, each of relatively short and constant length and of stepwise increasing diameter, all in the same rotor. The feed enters at the center, flows downward through the smallest tube, then reverses direction to flow upward through the next-larger-diameter tube, reverses again to flow downward through the next-larger diameter, etc. There are up to six such annular passageways, the last being defined by the vertical bowl shell wall. The tube diameters are selected so that each annulus (between adjacent tubes) has the same area so the velocity of flow through it is constant. Since the flow at constant velocity is being subjected to zones of successively greater centrifugal force, the multipass clarifier acts as a classifier with the largest, heaviest particles being deposited in the first pass and the smallest in the last pass. The accumulated solids are removed manually after the bowl has been disassembled. The rotor is mounted on top of a drive assembly similar to that described for solid-wall disk centrifuges.

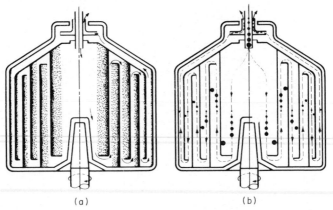

(a)   (b)

Fig. 3 Multipass clarifier centrifuge—"chamber bowl"; (a) flow diagram, (b) stage-by-stage removal of solid matter. (*Centrico Inc.*)

While the liquid discharge can be into open covers, most of the applications for this centrifuge will not tolerate exposure to air. In the hermetic design, the feed enters through a rotary joint on the bottom of the drive spindle. The effluent discharges at the top through a rotary face seal that may require supplementary cooling. In the centripetal pump design the effluent is led to a chamber that is part of the bowl top and rotates with it. Set into this chamber is a stationary closed pump "impeller." This converts the rotational-velocity energy of the contents of the chamber to pressure energy. The back pressure on the central discharge of this impeller controls its depth of submergence in the liquid in the chamber, and this in turn reduces the amount of entrained air in the discharged liquid to almost zero. The principal applications are:

Clarification of fruit juices	to 3000 L/h (800 gal/h)
Clarification of wine	to 5200 L/h (1375 gal/h)
Clarification of beerwort, cold	to 9800 L/h (2600 gal/h
hot	to 4170 L/h (1100 gal/h)
Clarification of beer	to 6800 L/h (1800 gal/h)
Clarification of varnish and lacquer	to 13,265 L/h (3500 gal/h)
Classification of pigment varnish	to 11,370 L/h (3000 gal/h)

The sludge-holding capacity of the largest size is approximately 64 L (17 gal).

Solid-Wall "Basket" Centrifuge

These centrifuges (Fig. 4) have many similarities to the tubular bowl centrifuge, but their *L/D* ratio is much smaller, always less than 0.75 and frequently less than 0.6 effective. The rotor consists of a lip ring at the top, a cylindrical imperforate shell, and a bottom. The bottom may be solid, containing a central nave through which the rotor is driven. More often the nave is surrounded by an open spider through which the accumulated solids can be dropped while the rotor is at low speed or at rest. In this case the remainder of the bowl bottom is solid. The diameter of the opening of the spider must be less than the ID of the lip ring so that the liquid being clarified will be directed to and overflow the lip ring at the top. In the modern configuration this rotor is mounted in the link suspended casing and drive described under *Vertical-Axis Batch Centrifuge*. The feed is directed to the bottom

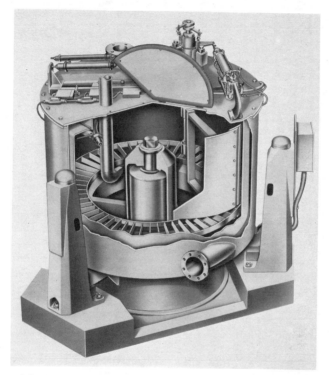

Fig. 4   Solid-wall "basket" centrifuge. *(Sharples Centrifuges, Pennwalt Corporation.)*

of the rotor through an accelerating cone, or its equivalent, mounted on the nave; or through an offset feed pipe that gives it a velocity component in the direction of rotation. With the feed pipe arrangement an accelerator wheel consisting of a set of radially disposed blades is provided to ensure that the feed attains the angular velocity of the rotor. The liquid being centrifuged travels upward as an annular layer and from 60% to over 80% of the volume under the lip ring is available for the accumulation of solids before clarification effectiveness is impaired. A cake detection system can be set to signal when the solids have reached a level at which clarification effectiveness decreases. A radially adjustable skimmer can be used to remove the free supernatant mother liquor from the surface of the cake. This same skimmer is also used to remove soft solids, such as metal hydrates or the biomass from activated sludge, without reducing rotational speed. More-refractory solids are removed manually from the solid bottom rotor while it is at rest, or removed by an unloader plow that is moved into the solids layer while the bowl is rotating at low speed, causing the solids to drop through the open bottom.

The system can be fully automated under timer control with the cake detection system as overriding insurance for variations in feed solids concentration. The resulting cycle is shown in Table 6. The gross flow rate during the feed part of the cycle is determined by the size and rotational speed of the centrifuge and the required degree of capture of the suspended solids. The net capacity of the centrifuge is this gross feed rate times the feed time per cycle divided by the total cycle time.

The cyclic solid-wall centrifuge is primarily used for the recovery and concentration of sludges and precipitates that remain plastic or fluid under centrifugal force of the order of 1500 g. Typical are the metal hydroxide precipitates produced during ECM (electrochemical machining) operation and the biomass of secondary waste activated sludge from domestic and industrial sewage treatment.

The solid bottom design is available in sizes from 300 mm (12 in) diameter by 50 mm (6 in) deep with a holding capacity under lip range of 0.006 m³ (0.2 ft³) to 1200 mm (48 in) diameter by 750 mm (30 in) deep with a holding capacity under the lip ring of 0.45 m³ (16 ft³). The range of centrifugal force is from 1800 g on the smaller sizes to 1300 g on the largest. The bottom-discharge design is available in the larger sizes with the same parameters.

On a typical waste activated sludge, gravity-settled to 1% dry substance concentration, the automated 1200 mm (48 in) × 750 mm (30 in) will recover 90% of the suspended solids

TABLE 6   Solid-Wall "Basket" Centrifuge—Operating Cycle

1. Accelerate
2. Feed (timer or cake detector to terminate)
3. Idle (optional if required to compact cake)
4. Skim supernat
5. Skim soft solids
6. Repeat 2, 3, and 4 (if hard-solids loading permits)
7. Decelerate
8. Unload hard solids (plow or manual), and repeat

at 10% concentration at an instantaneous feed rate of 227 L/min (60 gal/min), and a net throughput rate of 170 L/min (45 gal/min).

Introduction to Disk Centrifuges

The operating principle of the disk centrifuge is to stratify the flow through it into a set of parallel paths and thus decrease the distance a suspended particle must be moved radially by centrifugal force to reach the surface of a disk and be captured. The "disks" are actually truncated cones, flanged at their ID and OD to provide strength and rigidity, and held a fixed distance apart by radially disposed spacer bars. The feed enters the stack at its OD, or through feed holes parallel to the axis of rotation within the stack. From there it flows inward to the ID of the stack from which the clarified liquid, or separated light phase, discharges. Feed holes are almost always used when two immiscible liquids are to be separated. In this case the heavier liquid flows from the holes outward to the OD of the stack. For maximum efficiency the holes should be located so that they coincide with the liquid/liquid interface. Feed holes are used in clarification applications when the solids are difficult to compact and additional retention time to concentrate them is required.

The angle the section through the disk makes with the axis of rotation, the θ terms of Eq. (5), should be the minimum that will still permit the captured solid particles to move freely across the underside of each disk toward its OD and into the holding space in the bowl outside of the stack. It is usually in the range of 35 to 45°, very seldom greater.

The spacing between adjacent disks, measured normal to the disk surface, determines the number of parallel paths in a given disk stack [the n term of Eq. (5)]. It should be as small as possible but must be large enough to permit the countercurrent flow of the liquid phase(s) and the sedimented solids without turbulence and remixing. This puts it in the range of 0.625 mm (0.025 in) for small sizes to 3.125 mm (0.125 in) for the largest sizes for most applications.

In the derivation of Eq. (5), it is assumed that the flow of liquid is radially inward toward the center and the flow of solids is radially outward away from the axis of rotation. Experimentally it has been shown that this assumed flow pattern is incorrect. The observed liquid flow pattern indicates a vortex in each sector between adjacent disk spacers. This reduces the effective area Σ value by a factor of approximately 2.

Solid-Wall Disk Centrifuge

In this configuration (Fig. 5), the sedimented solids are directed to an annular dirt-holding space between the OD of the disk stack and the ID of the bowl shell wall from which they can be removed manually as required, after the rotor has been stopped and disassembled. For separating immiscible liquids, the disk stack is surmounted with a dividing cone, equivalent to the L-X disc of a tubular bowl, that includes a cylindrical extension at its small diameter. This leads the light-phase discharge to a different elevation than that of the heavy phase. The separated phases may be discharged into open covers, led to separate chambers in the rotor from which they can be removed by centripetal pumps, or taken off through rotary seals. The position of the liquid-liquid interface is usually controlled and optimized by adjusting the heavy-phase discharge radius, the r_H dimension of Eq. (6), but in sealed-discharge models may be externally controlled by adjustment of the back pressure on discharge lines.

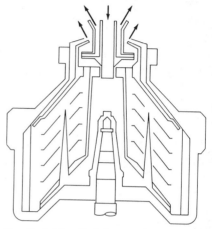

The solid-wall disk rotor is supported on and rotated by a stiff spindle. The spindle is mounted in appropriate thrust and radial bearings, and the assembly is supported and aligned in a rigid frame on which the covers are also mounted. The upper spindle bearing, nearest the rotor, is preloaded radially with springs or elastomeric cushions to provide the required degree of flexibility and dampening. The drive is usually through step-up right-angle gearing from a horizontal-shaft electric motor. This drive train is almost exclusively limited to underdriven centrifuges. It is splash-lubricated and consists of a bearing-grade bronze gear on the motor shaft driving a steel worm shell on the spindle. It is

Fig. 5 Solid-wall disk centrifuge—separator with cetripetal pump light-phase discharge. *(DeLaval Separator Company.)*

limited to the transmission of 25 hp. On larger sizes, V-belt drive from a vertical motor is used.

By far, the largest single application of the solid-wall disk centrifuge is the separation of cream from milk at flow rates up to 22,500 kg/h (50,000 lb/h). They are also used in large numbers for the purification of used lubricating oils, of a variety of used industrial oils, of new and used insulating oils (to raise their dielectric value); for the purification of light and heavy fuel oils, immediately before combustion; and for the continuous separation of soap stock in the process of refining vegetable and animal fats and oils.

The size of the solid-wall disk centrifuge covers a broad spectrum, with disk stacks from a few inches to over 500 mm (20 in) in diameter and with upward of 100 disks in the stack. The corresponding effective capacity for separation or clarification covers the range from a small fraction of a L/min (gal/min) to upward of 760 L/min (200 gal/min). The volume available for holding sedimented solids is defined by the difference between the OD of the disks (or stack) and the ID of the bowl shell times the height of the stack, $(r_3^2 - r_2^2)\pi Z$. Since the rotor must be stopped, disassembled and cleaned manually when this volume has been filled, the use of the solid-wall disk centrifuge is limited to applications in which the content of sedimentable solids is low. Increased dirt-holding capacity can be obtained by:

1. Decreasing the disk stack diameter with a corresponding decrease in the area parameter Σ.

2. Increasing the bowl wall diameter with a corresponding decrease in rotative speed ω, to maintain the same stress in the bowl shell.

3. Increasing the stack and bowl wall height Z, limited by decreasing rotational stability above critical speed as the ellipsoid of inertia approaches a circle.

The small Z/D ratio of the disk centrifuge makes it feasible to slope the bowl walls to direct the sedimented solids to a centrally located zone, from which their discharge can be effected, without greatly increasing the diameter and consequently stress level in the bowl shell at a given rotative speed, Fig. 6. It can be shown that when the height and diameter of the disk stack are controlled by the sloping side walls of the bowl shell, the Σ value of the assembly is optimized when the radius of the ID of the bowl shell

$$r_3 = \frac{4(r_2^3 - r_1^3)}{3r_2^2}$$

Valve-Discharge Disk Centrifuge

These centrifuges permit the accumulation of sedimented solids in the holding space between the disk stack and the bowl shell. At controlled intervals, radially disposed ports

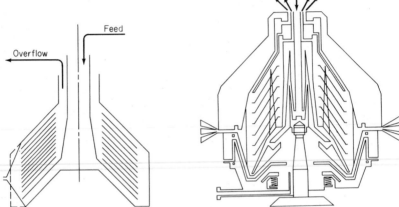

Fig. 6 Disk bowl—straight vs. sloping side wall.   **Fig. 7** Split-bowl desludger centrifuge. (*De-Laval Separator Company.*)

around the periphery of the largest diameter of the shell are caused to open by mechanical or hydraulic force acting on the valve tips that normally seal each port.

This unloading action may be under timed cycle control or triggered by a change in the clarity of the effluent stream. In one version the valve is held closed by an internal flow of supernatant liquid. When this flow is interrupted by the accumulation of sedimented solids, the valve operating liquor drains off through a leak hole and the internal hydrostatic pressure causes the valve to open automatically.

The satisfactory operation of valve-discharge centrifuges is highly dependent on the nature of the sedimented solids. The externally controlled valve bowls are usually applied to operation on relatively soft fluid solids. The automatic valve bowls require the sedimented solids to be plastic and of substantially higher density than the liquid phase. These have the further requirement that the feed flow rate be sufficiently greater than the flow rate through the open ports to restore the valve operating liquor circuit when the solids have been cleared from the bowl.

Valve-discharge disk centrifuges are used for the control of the pulp content of fruit and vegetable juices, the separation of wool grease from wool scouring liquor, the removal of solid fats and waxes from chilled hydrocarbon solutions of vegetable and animal "triglycerides" and mineral oils, and the polishing of rendered animal fats.

To a large extent they have been replaced by the split-bowl desludger centrifuges and the continuous nozzle-discharge centrifuges described in the next two sections.

Split-Bowl "Desludger" Disk Centrifuge

In this design a space is provided between the OD of the disk stack and the sloping walls of the bowl shell for the accumulation of sedimented solids. This space is sealed by an elastomeric ring in the upper part held against the lip ring of a movable piston or sleeve in the lower part. The seal is maintained in different designs by (1) spring pressure as shown in Fig. 7, or (2) the pressure from a continuous flow of an auxiliary operating liquid (usually water) that is continuously being drained at a lower rate through peripheral leak holes, or (3) by a residual pool of operating liquid below the piston that serves the same function as the springs of Fig. 7.

When a suitable load of solids has accumulated, the upper and lower sections of the bowl are caused to separate, exposing an annular slot through which the solids, if of proper plasticity, discharge under centrifugal force assisted by the head of liquid in the bowl. The operating means in design (2) is an interruption in the flow of the operating liquid, and in designs (1) and (3) is the brief application of the operating liquid.

The triggering mechanism may be (1) under time-cycle control, (2) response to a change in effluent clarity, or (3) response to rejection of a small recycle stream of effluent caused by the buildup of solids between the bowl wall and the OD of the equivalent of a dividing cone. Both clarifier and liquid-liquid separator models are available.

The liquid discharge(s) can be into open covers, through rotary seals, or through centripetal pumps, as selected.

The smaller sizes are driven by a horizontal motor through a right-angle gear train; the larger sizes, with hydraulic capacity up to approximately 1000 L/min, are underdriven by belts from an offset vertical shaft motor.

The sludge-holding space between the OD of the disk stack and the bowl shell ranges from a fraction of a liter in the smallest size to approximately 25 L in the largest. The desludging action is quite rapid, requiring no more than 1 or 2 s. The manufacturers usually recommend that the desludging action should occur no more frequently than once a minute. This factor, combined with sludge holding volume, should be considered in sizing such a centrifuge for a given application.

Stainless steels of the AISI 300 series, or their European equivalents, are the usual materials of construction.

Desludging centrifuges are used for the clarification and pulp content control of fruit and vegetable juices, the clarification of extracts such as coffee, the clarification of animal fats, the recovery and concentration of vegetable, milk, and single-cell proteins, and numerous other applications where the solids to be separated are soft, plastic, and not abrasive. They are frequently used on applications in which a high degree of sanitation is needed to avoid bacterial buildup and contamination. At intervals, as required, the feed is interrupted and sanitizing solution passed through the assembly with several desludging cycles. The time for manual disassembly and cleaning is replaced by a few minutes off-the-line time without shutting down the centrifuge.

Peripheral-Nozzle-Discharge Disk Centrifuge

The mechanical complexities of the valve discharge and split bowl desludging centrifuges can be reduced by discharging the sedimented solids continuously from the annular band through ports or nozzles in the outer perimeter of the bowl shell (Fig. 8). The flow rate of this peripheral discharge is fixed by the radial depth of the liquid layer and the rotational speed of the bowl, in accordance with the centrifugal pressure equation, and by the size and number of the nozzles.

For good operation, the minimum allowable nozzle size is twice the diameter of the largest particle to be discharged through it. Larger particles must be removed by pretreatment such as screening. The number of nozzles is controlled by the angle of repose of the sedimenting solids and must be selected so that the accumulation of solids between adjacent nozzles will not build into the disk stack and interfere with its clarification effectiveness.

For once-through operation, the underflow concentration is

$$C_u = C_f Q_f - \frac{C_o Q_o}{Q_u}$$

where u = underflow, f = feed, and o = overflow. The underflow concentration can be increased by recycling a portion of it

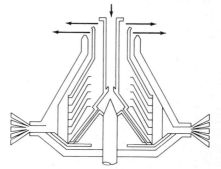

Fig. 8 Peripheral-nozzle-discharge centrifuge. (*DeLaval Separator Company.*)

back through the feed, i.e., increasing the quantity C_f. This method necessitates a reduction in capacity to maintain the same loss of solids to overflow, C_o, or conversely causes an increase in C_o at the same throughput rate. Better performance is obtained by recycling a portion of the nozzle discharge directly back to the nozzles instead of mixing it with the feed. In effect, this preloads the nozzle with underflow that has already passed through it and reduces the net Q_u taken away from the system.

When used as a liquid-liquid-solid separator, the conventional nozzle discharge centrifuge has one characteristic that is different from the other disk centrifuges described above. The volume of the separated heavy-phase liquid must be great enough to satisfy the flow requirement through the fixed size and number of nozzles with sufficient excess to maintain the hydrostatic balance between the liquid phases. One modification permits addition of a supplementary stream of the heavy phase behind the ring dam that controls the liquid-liquid interface position inside the bowl. Any excess beyond that required to satisfy the nozzles and maintain this interface position is automatically rejected.

To conserve the energy required to drive the nozzle discharge centrifuge, the nozzles are directed away from the direction of rotation of the bowl. The flow from them is approximately tangential to the outer perimeter of the bowl shell. In this manner much of the energy required to accelerate the nozzle discharge liquid to the peripheral velocity of the bowl is recovered as in a reaction turbine.

Nozzle-discharge centrifuges are usually constructed of AISI 300 series stainless steels. The discharge nozzles, being subject to erosion from the very high velocity through them (of the order of up to 150 m/s), are constructed of hard abrasion-resistant materials such as tungsten carbide in an appropriate matrix, fused boron carbide, or fused Al_2O_3 (synthetic sapphire), and are easily replaceable. In size they range from 0.5-mm-diameter opening for use on the smaller centrifuges to 3.2 mm (0.125 in) on the larger sizes. The design number of nozzles per centrifuge, depending on its size, ranges from 2 to 24. The velocity of flow through the nozzles is about 90% of the peripheral velocity of their discharge point, from which the volumetric flow through them can be calculated.

The relative separative or clarification capacity of the nozzle-discharge centrifuge can be calculated from the parameters of its disk stack in accordance with Eq. (5). The resultant area function applies only to the discharge overflow in the case of a clarification application. The gross feed rate is the sum of this overflow and the underflow through the nozzles.

Nozzle-discharge centrifuges are available in a range of sizes (referring to the disk stack dimensions) from an r_2 of 7.2 cm with 10 to 25 disks to an r_2 of 30 cm with 125 disks. The smaller sizes (up to 25 hp) are driven through a right-angle gear train, the intermediate sizes (up to 100 hp) are underdriven through V belts or a spur-gear train, and the larger sizes (up to 300 hp and 4000 L/min hydraulic capacity) are suspended and driven from the top through V belts from an offset motor.

Nozzle-discharge centrifuges are used for the concentration of soft and fine solids such as secondary waste activated sewage sludge, vegetable and single-cell proteins, corn gluten, corn and wheat starch, the separation of corn gluten from cornstarch, bakers' and brewers' yeast, classified clay for paper coating, residual solids after liquefaction of coal, clarification of wet-process phosphoric acid, and many more. The liquid-liquid separator models are used for the recovery of citrus oils from peel press liquid, wool grease from wool scouring liquor, fish oil from fish press liquid, and dehydration and purification of heavy fuel oils after water washing and of oil from tar sands, to name only a few applications.

The concentration of solids that can be obtained in the underflow varies widely, from 4 to 5% for activated sludge to upward of 50% for clay.

Disk Centrifuge with Light-Phase Skimmer

In certain separation applications, the separated light phase is too viscous and gummy to flow over the annular weir that forms the r_L dimension of the rotor. The discharge of these can be facilitated on all the above types of disk-centrifuge separators by incorporating a stationary light-phase skimmer nozzle or centripetal pump into the design.

Typical applications are the concentration and recovery of high-butterfat-content cream cheese from hot mix; the separation of crystalline waxes and sterols from chlorinated hydrocarbon solutions; and the separation of unhydrolyzed fat and protein from dextrose (corn syrup). Since these materials tend to plug the disks, it is conventional to establish the phase interface inside the r_1 dimension of a special stack.

Continuous-Scroll Solids Discharge Decanter Centrifuge

This type consists of a solid-wall rotor inside of which a helical screw conveyor is caused to rotate at a slightly lower speed. The rotor may be conical or cylindrical in shape; most commonly it is a combination of the two as shown in Fig. 9. This assembly is rotated about either a horizontal or vertical axis to develop the required centrifugal force. The differential speed between the rotor and the conveyor is obtained through a two-stage planetary gear box (American practice) or a "cyclo" box (European practice). The differential speed is in the range of 0.6 to 2.5% of the rotational speed of the bowl with different gearboxes when the first-stage sun gear is not allowed to rotate. It can be decreased by causing this sun gear to rotate in the direction of rotation of the bowl and increased by driving it in the opposite direction. The gearbox shell containing the ring gears is attached to and rotates with one of the bowl hubs. The other hub, in the horizontal design, includes the driven pulley. The stationary feed tube passes through this pulley and the bore of the hub to deliver the feed slurry into a feed zone inside the conveyor. This rotating assembly is surrounded by a casing to segregate the separated phases and guide them to appropriate discharge areas. A rigid frame serves to support and align the assembly between appropriate bearings. The drive is from an offset motor through V belts. For best performance the horizontal model should be mounted on vibration isolators. The vertical model includes self-contained isolators.

In operation, the slurry to be centrifuged is delivered by gravity head or pump pressure through the stationary feed tube to the rotating feed zone. From here it passes through appropriate openings in the conveyor hub where it is accelerated to the rotating speed of the bowl. As it travels around the spiral of the conveyor toward the liquid discharge or front-end hub, it is subjected to centrifugal force. The sedimented solids are caused to move in the opposite direction by the differential motion of the helical conveyor and transported "up" the sloping beach of the conical end to appropriate discharge openings. The flow of both liquid and solids is continuous.

Generally the feed slurry is introduced at or near to the intersection of the cylindrical and conical sections of the rotor. The effective clarifying length Z of Eq. (4) is the distance from the feed ports to the liquid overflow discharge ports. Even though the liquid flow follows the spiral path of the conveyor, Eq. (4) for a tubular bowl furnishes a good first approximation for comparing the performance of different sizes of this configuration.

It has been shown that the liquid flow is as a surface layer, so that the dimension r_2 is rate sensitive. The total available pond depth from r_1 to the bowl wall must be great enough to accommodate this moving layer and the loading of solids being moved in the opposite direction without interaction at their interface.

For most applications the liquid discharge radius is greater than the solids discharge radius so that the solids being conveyed along the conical section pass over a section of dry beach. Here they are given an opportunity to drain or be washed before reaching the solids discharge ports. When the sedimented solids are difficult to scroll across the free liquid interface and onto the dry beach section, it may be desirable to raise the pond level to or above the solids discharge level. This provides hydraulic pressure to facilitate their discharge. In one modification, this conical section is shrouded so the solids do not pass through a free liquid layer. In certain applications, the time required to compact the solids to equilibrium concentration at a given centrifugal force may be relatively long compared with the time required to remove them from the moving liquid layer. One design introduces the feed next to the hub farthest from the solids discharge ports. The entire bowl length is available for the compaction of the solid fraction and the clarified liquid layer is taken off by a skimmer or modified centripetal pump near the cylindrical/conical intersection. In other modifications, longitudinal pathways inside the conveyor helix provide for annular instead of spiral flow of the liquid layer. Continuous-discharge decanters cover a wide range of sizes from 150 mm (6 in) diameter × 300 mm (12 in) long developing up to 8500 g, through 625 mm (25 in) diameter × 2875 mm (115 in) long developing 3200 g and 900 mm (36 in) diameter × 2400 mm (96 in) long developing 1300 g, to 1350 mm (54 in) diameter × 1800 mm (72 in) long developing 275 × g. Various

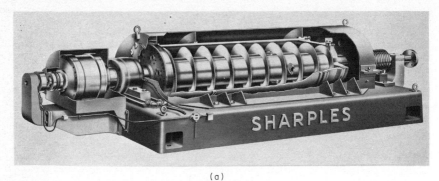

(a)

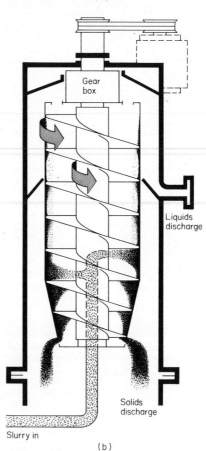

Gear
box

Liquids
discharge

Solids
discharge

Slurry in

(b)

Fig. 9 Continuous scroll discharge decanter. (a) Horizontal *(Sharples Centrifuges, Pennwalt Corporation.)* (b) Vertical *(Sharples Centrifuges, Pennwalt Corporation.)*

designs permit operation at from full vacuum to upward of 11 atm at temperatures from −100 to 300°C. Their hydraulic capacity ranges from 10 gal/min (approximately 40 L/min) on the smallest sizes to over 300 gal/min (approximately 1150 L/min) on the largest. Their solids-handling capacity, which is a function of cake density and may be limited by the torque the gearbox is capable of transmitting to the conveyor, is in the range of 5 to 10 kg/h for the smallest sizes to upward of 40 t/h for the largest. The overriding consideration is always the capacity at which the desired degree of solids capture and residual retained liquid concentration can be obtained.

The continuous decanter was originally conceived for the recovery of nondeformable particulate solids from fluid slurries. Its uses now include the handling of many soft solids that could not be discharged from the earlier models. Typical of the many applications are the dewatering of coal up to 6 mm (¼ in) mesh; the recovery and dewatering with some displacement washing of a variety of crystals and polymeric solids; the classification by particle size of clay, titanium dioxide, and other pigments; the concentration of municipal and industrial wastes; and the clarification (with recovery of proteinaceous solids) of animal and vegetable fats and oils.

The lower limit of size of particles that can be effectively separated by the continuous decanter is in the range of 2 to 20 μm, depending on their relative density. In some applications the capacity and proportion of solids captured as well as their concentration can be greatly improved by flocculation, for example, by the addition of suitable polyelectrolytes. The residual mother liquor content of the discharged solids covers a wide range depending on their particle size (specific surface area) and interstitial retention. It varies from as much as 95% for secondary waste activity sewage sludge to less than 1% for polystyrene beads.

The most common material of construction is stainless steel, followed by carbon steel. Titanium and Hastelloy C construction can be used when the process environment is highly corrosive.

Continuous Screen Bowl Decanter Centrifuge

The screen bowl decanter (Fig. 10) provides a perforated extension, usually cylindrical, on which the separated solids exiting from the beach end of a conventional decanter are given additional time for drainage.

The wet end of the bowl has all the characteristics of a solid bowl decanter of the same dimensions, as described in the previous section. The helical conveyor extends through the inside of the screen section and transports the solids coming up from the beach across the screen. The retention time of the solids on the screen can be up to 10 s in some models, as compared with approximately 2 s on the "dry" section of the beach.

Typically crystalline solids that exit the beach of the solid bowl decanter with 30% retained mother liquor will discharge from the screen bowl decanter at 10% retained mother liquor content. It should be noted that all of this gain is not a result of the

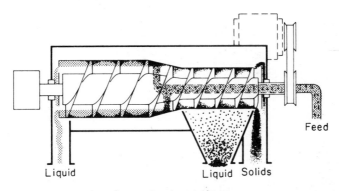

Screen bowl centrifuge

Fig. 10 Continuous-screen bowl decanter centrifuge. (*Sharples-Stokes Division, Pennwalt Corporation.*)

additional retention time. At least some of the solid particles that are smaller than the opening of the slotted bar screen will report to the screen filtrate. The effect is to reduce the specific surface area of the solid particles that are recovered as product.

The concentration of solids in the screen filtrate that have to be recycled or otherwise disposed of may not be inconsiderable. The cost of this secondary treatment vs. the advantage of drier solids must be evaluated in the consideration of the screen bowl decanter for each application.

Screen bowl decanters are used for the separation of paraxylene crystals, the dewatering of polyvinyl chloride, and the dewatering of Trona crystals.

EQUIPMENT—CENTRIFUGAL FILTER TYPES

Centrifugal filters may be divided into two broad categories:
  1. Fixed bed
  2. Moving bed

Fixed-bed filters operate batchwise. The operating cycle may be under entirely manual control, fully automated under timer control, or a combination of the two. The various types differ in the means provided for removing the processed solids.

Moving-bed filters operate essentially continuously. The feed and liquid discharge are continuous, the solids discharge is either continuous (as in the case of the screen bowl decanter described in *Continuous Screen Bowl Decanter Centrifuges* or the wide-angle bowl in *Conical-Screen Continuous Centrifuges*) or intermittent at high frequency, greater than 20 cycles/min. The types differ in the means provided for advancing the solids toward the discharge end of the screen.

Vertical-Axis Batch Centrifuge

These centrifuges are classified as follows:
  Manual unload
  Container unload
  Knife unload

These have in common the necessity of reducing their rotational speed (to zero for the manual or container unload or to less than 100 r/min for the knife unload). The conventional operating cycle is shown in Table 7. Phase 4 may be repeated if required by process considerations. Since the several phases are sequential, reasonably good segregation of the mother liquor and the one or more wash liquors can be accomplished by appropriate valving of the effluent filtrate line.

TABLE 7   Operating Cycle—Batch Basket Centrifuges

1. Accelerate to load speed
2. Load
3. Accelerate to purge or wash speed
4. Wash
5. Accelerate to full speed
6. Dry spin
7. Decelerate to unload speed
8. Unload

Solid-bottom basket centrifuges must be loaded and unloaded at zero speed.

Horizontal-axis basket centrifuges are loaded and unloaded at full speed, so cycle phases 1, 3, 5, and 7 are omitted in their operation.

Manual Unload   This type contains a rotor with a perforated cylindrical shell, lined with an appropriate filter medium, mounted on a bowl bottom, and surmounted by an annular lip ring to retain its contents. The bowl bottom may have a solid floor, in which case the processed solids are removed manually from above through the opening in the lip ring. Or, the bowl bottom can consist of a central nave connected by spokes to an annular solid floor whose ID is less than the ID of the lip ring. In this design, the processed solids are ejected downward through the openings between the spokes.

This rotor is supported and rotated by a spindle that is turned by an appropriate prime mover and the whole is mounted and aligned in a rigid frame that also supports a curb or casing to collect the filtrate and direct it to a convenient outlet. The rotor may be suspended from a frame and driven from above. More often this type is driven from below through belts from an offset motor affixed to the frame or casing.

For light-duty service, with the solid-floor bowl only, the base-bearing drive is used. The spindle housing is supported on a ball-and-socket joint at its lower end. This permits excursion of the upper end of the spindle and the rotor to accommodate some degree of out-of-balance loading. The amount of this excursion is controlled by a dampening arrangement around the upper radial bearing located as close to the bowl bottom as possible. Since the drive motor is fixed to the frame and there is some radial movement of the driven pulley, the amount of torque and power that can be transmitted through this arrangement is limited.

For heavier-duty service, the link-suspended mounting is employed. The main bearing housing forms the center of the casing. The vertical-drive motor is rigidly attached in an offset position to the casing. The center line distance between the motor drive shaft and the driven shaft is fixed, so much more torque and power can be transmitted through multiple V belts. The entire assembly is suspended from the upper end of three vertical posts connected solidly to ground. The suspension may be chains or spring-loaded rods. Out-of-balance forces causing displacement of the casing assembly are largely absorbed in the suspension.

The motor constitutes an overhanging weight that can be compensated for, to a degree, by reducing the distance between the two vertical posts adjacent to it. On the 1200 mm (48 in) diameter × 750 mm (30 in) deep basket size, the limit is reached with the weight of an open-type 60-hp motor.

When greater power input is required, hydraulic motor drive is used. This consists of a relatively lightweight hydraulic motor mounted on the casing and driven through flexible hoses by a variable-volume, constant-speed hydraulic pump that is in turn driven by an electric motor. The output volume of the hydraulic motor, which determines the speed of the centrifuge, is controlled by adjusting the position of its stator. Usually this control is by air pressure so that the speed at each phase of the cycle is infinitely variable.

In one design the drive motor, either electric or hydraulic, is mounted on the curb cover and close-coupled to the rotor. This obviates the need for V belts and a belt tunnel that restricts the area available for the discharging solids.

The earliest form of vertical-axis batch centrifuge is suspended from a drive head supported by appropriate crossbars (Fig. 11). The drive may be through V belts and pulleys from an offset motor or turbine, or direct through a flexible coupling from the prime mover. In either case the weight of the motor is supported entirely by the crossbar. The strength of the crossbar is the only limitation on the size of the motor, and hence the frequency of cycling. In this design, the curb and the unloader knife it supports are fixed to ground. The possibility of relative motion between the basket/drive shaft assembly and the unloader knife shaft assembly is compensated for in the design of the knife blade.

Container Unload   In this type the material to be centrifuged can be preloaded into a separate metal or cloth container. To accept the metal container, which fits snugly into the inside of the bowl shell, the rotor is straight-sided with no lip ring. After the centrifuge cycle is completed, the container and its contents are removed. This can be replaced with a freshly charged container and the cycle repeated with minimum loss of operating time.

In one design, generally used for recovering cutting oil from metal chips and turnings, the wall of the "basket" is not perforated and is tapered to a larger diameter at its upper end. After charging, a loose-fitting lid is fixed to the top and the rotor spun to discharge the oil through the clearance between the shell and its removable top. The rotor itself serves as the container and a number of rotors can be used to segregate the various metals and oils on the machine shop floor.

Knife Unload   For this the open-bottom basket is always used. The unloader knife is mounted on a vertical shaft off-center from the axis of rotation that extends through and is supported and aligned by the curb cover. In older designs the knife blade, with its face parallel to the filter medium, is relatively short with respect to the depth of the basket. At unloading speed, it is rotated into the cake at the top of the basket. The knife acts as an internal boring tool to deliver the knifed cake to the opening in the basket bottom. In more modern designs the knife blade is the same length as the depth of the baskets so that only a simple in-and-out motion is required to unload.

Vertical-axis centrifuge baskets range in size from 125 mm (5 in) to over 2500 mm (100 in) in diameter and from 75 mm (3 in) to 1000 mm (40 in) in depth. The smaller sizes, to

300 mm (12 in) diameter, are frequently supplied as accessories to the analytical and clinical bottle centrifuges. Commercial sizes, accommodating loads of 80 lb/ft³ (1.28/cm³) under the lip ring at maximum operating speed, range from 500 mm (20 in) diameter × 250 mm (10 in) deep for base-bearing solid-bottom types, through 750 mm (30 in) diameter × 375 mm (15 in) deep, 1000 mm × 500 mm (40 in × 20 in), 1200 mm × 750 mm (48 in × 30 in), and 1500 mm × 1000 mm (60 in × 40 in) for link-suspended and overhead-driven bottom-discharge types developing up to 1800 g, to very large diameter solid-bottom baskets for special applications such as textile dewatering after dyeing or cleaning.

The manual-unload types are used for processing relatively small quantities, particularly when it is desired to preserve the integrity of each batch and the value of the product offsets the relatively high attendant manual labor.

The container-unload types are used for the dewatering of washed vegetables, textiles such as skeins of yarn or laundry in separate bags, or where it is necessary to maintain the orientation of articles during centrifuging, for example, in removing excess varnish from fabricated electrical parts.

Of principal interest to the chemical and allied industries are the bottom-discharge types, with full or partial automation. Table 8 shows the range of available sizes. These are used for dewatering and washing (or purging) a variety of crystalline and fibrous solids ranging from fine pharmaceuticals through fiber floc to salt and sugar at capacities from a few hundred kilograms or pounds per hour (1 cycle in an hour or longer) to up to 9000 kg/h (20,000 lb/h) [25 cycles/h of 800 lb/cycle of processed crystals from the 1200 mm × 750 mm (48 in × 30 in) size].

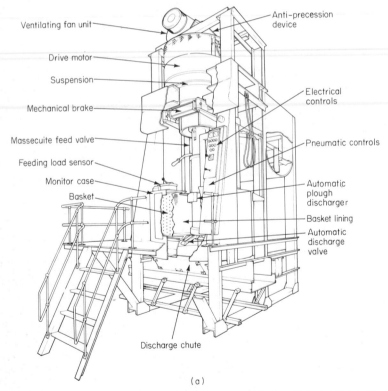

(a)

Fig. 11 Vertical-axis batch centrifuges. (a) Manual unload: (1) basket, (2) basket lining, (3) monitor case, (4) discharge chute, (5) drive motor, (6) ventilating fan unit, (7) suspension, (8) antiprecession device, (9) automatic discharge valve, (10) automatic plough discharge, (11) electrical controls, (12) pneumatic controls, (13) mechanical brake, (14) feeding load sensor, (15) massecuite feed valve. (b) Knife unload—suspended automatic sugar centrifuge. (*Thomas Broadbent and Sons, Ltd.*)

The size of the prime mover controls the nonprocess time of acceleration and deceleration and to a lesser extent unloading time. When the processing time (feeding, rinsing, and final drainage), as determined by the system being centrifuged, is short, the increased price and operating cost of a large prime mover is offset by the gain in productive capacity. As the time required for processing increases, the proportionate advantage of a larger prime mover decreases.

TABLE 8   Typical Knife-Discharge Basket Centrifuge Sizes

Diameter, in	Depth, in	Screen area, ft²	Approximate volume* under lip ring, ft³
20	10	4.4	1
30	15	9.8	3
40	20	14.5	7.4
48	24	25.1	12.8
48	30	31.4	16
60	40	52.4	32

*Lip ring diameters are not standard; consult manufacturers ratings.
Deduct 25 to 33% to establish maximum processed cake volume.

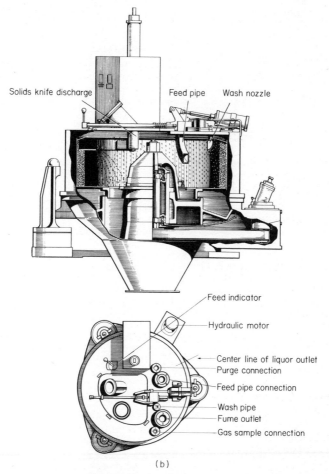

(b)

Fig. 11 (continued)

Rinsing efficiency, defined as the proportionate reduction of soluble mother liquor impurities, ranges from fair (50 to 75%) to excellent (>95%), depending on the application and depth of cake. It is optimized by depositing a uniform depth of cake over the entire filter surface during the feeding phase of the cycle. Various feeding arrangements are available to facilitate this, each with its quota of advantages and disadvantages. Rotative speed control during feeding is likely to be of critical importance to ensuring uniform cake thickness. Too high a speed will deposit the crystals by sedimentation near to where they are introduced into the basket. Too low a speed introduces the risk of mother liquor leaking or splashing into the solids-discharge receiver. In certain critical applications such as sugar purging, this is overcome on the suspended type by a valve or cover that seals the basket floor during loading and is opened mechanically during unloading.

Underdriven basket centrifuges are available to contain noxious and volatile liquids at up to 0.35 kg/cm² (5 lb/in²) (gage).

Horizontal-Axis Batch Centrifuge

These centrifuges (Fig. 12) are rotated on a horizontal axis to permit the introduction of a sloping chute into the opening of the lip ring at the front of the basket. During the unloading phase of the cycle, an unloader knife or peeler is raised or rotated into the processed cake to direct it down the sloping chute and out of the basket and the front casing. The angle of the chute, at least that portion of it that is inside the basket, is controlled by the depth of the basket and the diameter of the opening through the lip ring. A steeper slope to the chute that would facilitate the discharge of the solids is possible only by increasing the lip-ring opening diameter or decreasing the basket depth. Either of these modifications results in a smaller solids loading per cycle.

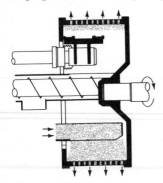

Fig. 12 Horizontal-axis batch centrifuge. *(Sharples Centrifuges, Pennwalt Corporation.)*

Horizontal-axis centrifuges are usually operated at full speed during the processing cycle so that no time is lost or energy consumed for acceleration or deceleration as with the vertical basket centrifuge. The time required for unloading varies from 3 or 4 s to several times this on the various models that are offered. In at least one type, a screw conveyor inside a modified discharge chute facilitates the movement of the solids out of the front of the basket at the expense of considerably increasing the unloading time.

Horizontal basket centrifuges are available in sizes from 10 to 78.75 in (2.0 m) diameter. The Z/D ratio of the basket is usually 0.5, somewhat less in the larger sizes, so the corresponding filter surface area ranges from 0.1 m² (1.1 ft²) to approximately 3.8 m² (41 ft²). The operating cycle per unit charge ranges from 20 s to 15 min or longer, depending on the application. Their cost is approximately twice that of vertical basket centrifuges per square foot of filter area in similar materials of construction, so their widest application is in the area of dewatering solids in the 60- to 325-mesh size range with fair-to-good drainage characteristics. Their high-speed unloading may create undesirable crystal breakage and the formation of an impenetrable glaze on the heel that must be removed intermittently by a screen wash or other means.

Scaling up of batch basket centrifuges, vertical and horizontal, presents a number of interesting problems. At the very least, sufficient datum points must be established to determine the coefficients A and B of Eq. (10) for a given application.

Conical-Screen Continuous Centrifuge

This type of centrifuge is classified into the following categories:
   a. Wide angle
   b. Differential scroll
   c. Axial vibration
   d. Torsional vibration
   e. Oscillating

The vertical- and horizontal-axis fixed-bed centrifugal filters described in the preceding sections process a unit batch through the operations of loading, rinsing, and spinning to dryness. Their operation can be fully automated and their cycles made automatically repetitive. In effect, by the addition of appropriate upstream and downstream surge vessels, complete continuity of operation is obtained. However, the modern requirement in chemical engineering practice is fully continuous operation with minimum inventory of material in process and minimum direct operating labor.

Continuous centrifugal filters fulfill this requirement with some sacrifice in flexibility. This in turn has given rise to a variety of configurations for specific applications. These differ principally in the means provided for the advance of the solid phase toward the discharge end.

The screen bowl decanter can be considered as one of these with a preconcentrator or clarifier section followed by a cylindrical screen section with screw conveyor providing the controlled advance of the solids across the filter medium. In earlier designs the clarifier section was omitted and the slurry fed directly onto the screen. This design has been superseded largely by the pusher centrifuge.

When a slurry is fed onto a conical screen rotating about its axis, centrifugal force provides two vectors, one parallel to the screen surface and the other at right angles to the surface. The former tends to impel the feed toward the large end of the cone, the latter to drive the liquid fraction (and undersize solid particles) through the openings in the screen.

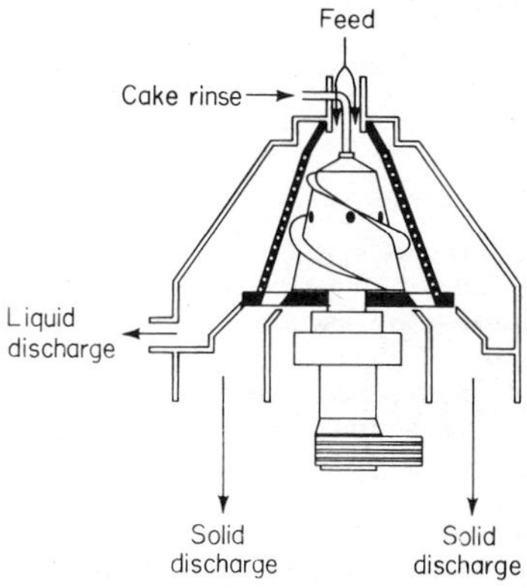

Wide-Angle Screen If the angle between a section through-screen and its axis of rotation is great enough to overcome the coefficient of friction between the solids and the screen, the solids will advance toward the large end of the cone and discharge. If the angle is large, their retention time will be too short for satisfactory drainage to occur through the limited open area of the screen. If the angle is too small, the solids will build up on the screen until their angle of repose is

Fig. 13 Conical-screen continuous centrifuges: wide angle. (*Sharples Centrifuges, Pennwalt Corporation.*)

reached, and drainage through this layer of solids will be greatly restricted. The simplicity of the wide-angle conical screen (Fig. 13) makes it attractive to the chemical engineer. In practice its use has been largely restricted to the recovery of starch (fine particles) from corn fiber, with partial dewatering of the latter; and the separation of molasses from certain grades of sugar that require little or no washing.

One modification that improves its utility incorporates a system of baffles inside of the cone rotating at the same speed. These baffles, which may be adjusted for specific applications, retard the flow rate and increase the time available for drainage.

Differential Scroll In this (Fig. 13b), the angle selected is less than that required to overcome the coefficient of friction between the solids and the screen. The advance of the solids toward the large end is controlled by the differential speed of a conveyor fitted to close clearance inside the conical screen. This conveyor has the form of a modified multipitch helix that retards the slurry flow at the small end and controls the partially dried solids advance rate at the large end. The required differential speed between the screen and the conveyor is usally created by a cyclo gearbox, although in an early design, as used for coal dewatering, two sets of spur gears of slightly different pitch from a common driver provided the differential.

Differential-scroll-discharge centrifuges are used for dewatering a variety of fibrous materials such as fiber from wet corn milling, medium to coarse crystals, and coarser solids such as 28 mesh × 6-mm (¼ in) coal. While they provide relatively short retention time, the high centrifugal force generated [1800 g at the large diameter end on sizes from 250 to 600 mm (10 to 24 in) large-end diameter] provides excellent dryness on solids that reach the equilibrium moisture content of Eq. (9) quickly. Their liquid-handling capacity is limited, so performance is favored by high feed-slurry concentration. Their rinsing efficiency is fair and the added liquid load imposed by a rinse adversely affects capacity.

Solids Advance   The advance of the solids toward the large end of a conical screen can be encouraged by superimposing a vibratory action on the rotating assembly. In effect, this partially fluidizes the bed and reduces the coefficient of friction between it and the filter screen, facilitating the advance of the solids toward the larger-diameter discharge end. In the various designs available, this action is created in three different ways.

Axial Vibration. A system of symmetrical weights imposes a high-frequency vibration along the axis of rotation.

Torsional Vibration. A force is applied at high frequency to alternately slightly accelerate and retard the rotative speed.

Oscillating. The axis of rotation is slightly oscillated at high frequency.

These three types are available in sizes to 1 m (40 in) in diameter. They are generally operated at relatively low speed, to generate less than 500 g. Their liquid-handling capacity is quite limited, so that high feed-slurry concentration is mandatory. Within this limitation, they offer the lowest power consumption per unit of solids processed of any of the other centrifugal filters. Typically, the oscillating advance type is reported to require only 0.2 kWh per ton of solids to dewater 18 mm × 6 mm (¾ in × ¼ in) coal to 6% residual moisture at 135,000 kg/h (150 ton/h) in the largest size.

Most of the conical-screen centrifugal filters are available with either vertical or horizontal axes of rotation and in carbon steel and stainless steel construction.

Cylindrical-Screen Continuous Centrifuge

These consist of a cylindrical filter member supported in a perforated rotor that is cantilever-mounted through an appropriate hub to a horizontal drive shaft. The slurry is introduced near the closed end and the solids while draining and drying are caused to advance along the screen to the open discharge end opposite the hub.

Two methods are used to provide this solid advance motion.

Differential Scroll Advance   In this earlier, and now practically obsolete, design the solids advance is caused by the action of a modified helical conveyor turning inside the filter screen at a different rotative speed. This principle is now applied in the filter section of the screen bowl decanter centrifuge.

Reciprocating Pusher Advance   In this the drive shaft is hollow. In it is contained a second shaft that rotates with it but on which a reciprocating motion can be imparted. This second shaft projects through the main rotor head and carries the pusher mechanism.

In the simplest form, the single-stage pusher centrifuge, the pusher plate is essentially a disk with close clearance to the inside of the cylindrical screen. The feed slurry is introduced continuously through a feed tube inside a distribution cone that is attached to

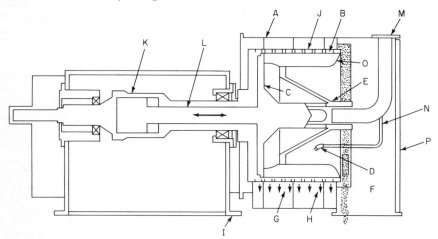

Fig. 14   Cylindrical screen pusher centrifuge: single stage. (*a*) Wet housing, (*b*) screen, (*c*) pusher, (*d*) wash spray, (*e*) rotating inlet funnel, (*f*) dry collector housing, (*g*) opening, (*i*) opening, (*j*) drum, (*k*) hydraulic servomotor, (*l*) piston rod, (*m*) feed pipe, (*n*) wash feed pipe, (*o*) cake, (*p*) access door. (*Baker Perkins Inc.*)

and rotates with the pusher hub. The outer diameter of the cone forms the inside of an annulus through which the solids pass (Fig. 14).

The reciprocating motion of the pusher has a frequency of from 20 to 120 cycles/min and an amplitude of 10.2 to 2.5 cm (4 to 1 in), depending on the diameter of the rotor, which is in the range of 91.5 to 25.4 cm (36 to 10 in) in modern designs, and the application.

In the smaller sizes the reciprocating motion may be created by an eccentric motion at the outboard end of the pusher shaft. More commonly, and always on the larger sizes, this motion occurs from the action of hydraulic pressure on a double-acting hydraulic piston/cylinder combination contained within the drive shaft assembly.

The cake is formed by bulk drainage of the mother liquor from the slurry in the area of the screen swept by the pusher plate. For satisfactory operation, this increment of cake must have sufficient rigidity to transmit the forward thrust of the pusher plate to the increments of cake preceding it and cause the cake to advance with the frequency of the pusher as an annulus along the screen to the discharge end.

The initial thickness of the cake is controlled by the radial clearance between the feed cone and the filter screen. From this point forward the thickness of the cake is largely controlled by the frictional resistance to its advance along the screen. In general, screens are selected that provide the minimum coefficient of friction to the solids in the axial direction. These are usually of the axially slotted or bar screen type. The individual bars are of trapezoidal or modified V shape, with their wide dimension facing the center of rotation to minimize clogging from fine crystals. The slot width between adjacent bars at the inner surface should be the maximum that will support the crystals and prevent an undesirably high proportion of fines from reporting to the filtrate. The permeability of this type of filter medium, with bars of given cross-sectional dimensions, is an exponential function of the slot width, which is usually in the range of 76 to 381 μm (0.003 to 0.015 in).

Single-stage pushers are used to dewater and wash a variety of fibrous solids, such as nitrocellose, and moderately coarse crystals, at least 90% of which should be larger than 100 mesh.

Their washing efficiency is from good to excellent (upward of 98% displacement with less than 0.25 parts wash per unit of cake). The wash is applied through spray nozzles to the surface of the cake. Since the cake advance is intermittent, the wash pattern should cover about 1.5 times the pusher stroke amplitude for maximum efficiency.

The scale-up of pusher centrifuges is subject to a number of variables. A useful simplification, at least for a first approximation, is that the capacity (cake-as-discharged basis) is proportional to $(D_2/D)^{1.4}$.

A critical limitation of the single-stage pusher described above is the limited screen area (that swept by the pusher stroke) available for bulk drainage. Unless the feed rate is adjusted to complete bulk drainage in this area, free liquid will be pushed forward over the previously formed cake causing it to wash out irregularly, unbalancing the centrifuge. The capacity of the conventional single pusher is then highly responsive to feed slurry concentration, and on most applications maximum capacity is realized only when this concentration exceeds 40% wt/wt. In one configuration the pusher plate includes a wide-angle-cone screen. In this a considerable amount of preliminary bulk drainage occurs in advance of the pusher action so that the capacity of this design is somewhat less responsive to feed slurry concentration, and to unavoidable changes in it. Typically in a 750 mm (30 in) diameter size some 285 L/min (75 gal/min) of mother liquor can be discharged through the cone screen.

The length of cylindrical bar screen that can be effectively utilized in the pusher centrifuge is controlled by the nature of the cake that is formed and its frictional resistance to advance across the screen. At some dimension of length, the pusher action will result in increasing the radial thickness of the cake without causing it to advance further toward the discharge end. In the multistage pusher (Fig. 15), the length is divided into a series of sections of increasingly larger diameter toward the discharge end. The incremental radius of each section provides the pusher action on both the forward and return stroke. Rotors of this type are

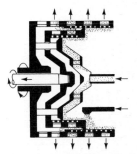

Fig. 15 Cylindrical screen pusher centrifuge, multistage. (Filtration and Separation; *Alfa-Laval Industrietechnik GmbH.*)

available with from 2 to 10 stages so the individual sections can be made as short as necessary to provide transportation of the solid annulus without unduly increasing the radial thickness of the cake. Rapid drainage is also favored by reorientation of the cake and the particles in it as it transits from one stage to the next larger diameter stage.

The applicability of each of the various types of filter centrifuges on particulate solids systems is usually expressed in terms of the average size of the particles. This can lead to problems, since the distribution of size is also important. While somewhat more difficult to evaluate, the actual permeability of the filter cake at the proposed operating centrifugal force and cake thickness is the more important consideration.

For very high permeability cakes, continuous centrifugal filters, particularly of the pusher type, should be considered; for intermediate permeability cakes, horizontal knife-discharge batch centrifuges; for medium-to-low-permeability cakes, vertical-axis batch basket centrifuges. When the cake permeability is very low, centrifugal sedimentation rather than centrifugal filtration should be considered.

Cartridge Filtration

NICHOLAS NICKOLAUS *Vice-President Marketing, Pall Corporation, Glen Cove, N.Y.; Member, American Institute of Chemical Engineers, Filtration Society, Parenteral Drug Association, National Fluid Power Association.*

INTRODUCTION

Many types of separation equipment are available for the removal of solids from fluids. The choice of cartridge filtration is limited to fluids containing no more than 0.01% solids and where cake handling is unnecessary.

As an example, assume a water flow of 75L/h (20 gal/min), solids content of 1% by weight, and a wet cake bulk density of 0.64 g/cm³ (40 lb/ft³). The amount of cake to be handled at the end of 24 h would be 1080 kg (2399 lb), clearly not an application for cartridges. Assume also maximum allowable cake thickness is 3 mm (⅛ in) because of pressure drop considerations. Total required filtration area would be 535 m² (5760 ft²), an obviously uneconomical solution.

Particles to be removed from fluids by cartridge filters range in size from submicron to above 40 micrometers,* the smallest size visible to the unaided eye.

*1 micrometer = 0.000039 in, 0.1 micrometer = 1000 angstrom. Micron is a shortened form for micrometer; its symbol is μ or μm.

Such fine particles often produce an impermeable cake with unusually high resistance (for its thickness). Consequently cartridge filtration is used primarily to clarify low-solids-containing fluids to an optically clear condition, or to a much lower level of contamination, measurable by microscopic or automatic particle-counting techniques, or to a sterile state by quantitatively removing all bacteria.

Typical filter cartridges and housings are shown in Fig. 1. The cartridges are cylindrical in configuration, with either pleated or nonpleated, disposable or cleanable filter media, integrally bonded to plastic or metal hardware. Housings are available in plastic, lined metal, or metal construction for any condition of pressure, temperature, and fluid compatibility and they accept multiple cartridges for large flows. For a description of the many available filter cartridges, see Ref. 1.

Fig. 1 Typical cartridge filters and housings. *(Pall Corporation.)*

CARTRIDGE SELECTION

Virtually all fluids can be clarified by cartridges, which explains the availability of a wide assortment of commercial types. Narrowing the selection requires an examination of the following system parameters:

1. Temperature
2. Pressure
3. Fluid compatibility
4. Effective filter removal rating (efficiency)
5. Fluid viscosity
6. Type of contaminant
7. Quantity of contaminant
8. Maximum allowable pressure drop across the filter assembly
9. Required throughput and flow rate
10. Filter sterilization methods

Some of the parameters, by a process of elimination, will limit the choice to a few types, others will further reduce this choice and the remaining parameters will help to determine required filter area. Finally, filter reliability and economics will invariably eliminate all types but one.

Temperature

Ceramic and metal filters are most common in continuous service for high-temperature applications over 260°C (500°F). Metal filters are used for cryogenic temperatures, although polypropylene[2] cartridges with no elastomeric seals have been used for short periods (several hours) for submicron particulate removal. In the 149°C (300°F) to 260°C (500°F) range, glass fiber or asbestos media (with metal hardware) or polyfluorinated hydrocarbon filters can be used, in addition to metals and ceramics. From above 80°C (180°F) to 149°C (300°F) almost all filter media with metal hardware are functional. At 80°C (180°F) or less the same media but with plastic hardware (for reduced cost or improved fluid compatibility) are functional.

Pressure

The filter housing must be rated for maximum system pressure and temperature. Beyond this, pressure is needed to initiate flow and to maintain flow as the cartridge accumulates contaminant. Pressure must also be available to overcome other resistances in the system from pipe, valves, tanks, etc. High pressures, however, can result in filter blockage, consequently starving downstream components; can cause cartridge structural damage followed by catastrophic system failure or product rejection, or even patient damage, in the case of biomedical filters; can alter filter-medium pore sizes and compromise efficiency; and can deform particles which allow them to work through the medium. Filter assemblies with bypass valves to prevent fluid starvation, with differential pressure alarms to warn against impending filter damage or blockage, or with higher pressure ratings which afford protection against such failures are commercially available. Furthermore, cartridges which do not change in efficiency throughout their life should always be selected. (See Fig. 2.)

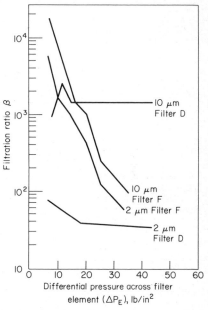

Fig. 2 Cartridge efficiency vs. pressure drop. *(Pall Corporation.)*

A common misconception is that cartridge removal efficiency continues to improve after start-up because the initial cake performs as a supplemental filter medium. This is true for only the most reliable filters. Many will unload or slowly release previously collected solids because the integrity of their medium is degraded by the mechanical stresses imposed at high differential pressures. Data substantiating efficient filter performance throughout cartridge life must be reviewed before selecting a vendor.

Fluid Compatibility

Always verify filter compatibility with the fluid, since disposable cartridges can contain as many as 10 or more materials of construction. Corrosion handbooks and plant experience (as well as that of the manufacturer) should help in this verification. Testing is recommended in critical applications and for proprietary filter materials. A 24-h soak, followed by visual observation and determination of gravimetric weight loss of the filter is the most common test. The bubble point[3] test is also used, and the more quantitative measurement of the forward-flow bubble point[4] test is especially used for sterilizing grade and "ultimate" water (18-M$\Omega \cdot$ cm resistivity) filters. A more pragmatic test is to dynamically pass the fluid through one cartridge at the expected flux rate and sample the fluid and check the filter after approximately 10 min of flow. When necessary, measure changes in fluid pH, surface tension, resistivity, taste, odor, or color. For solvents compatibility, the test fluid may (in addition to the above) be boiled off and the quantity of nonvolatile extractable materials determined. For use with parenteral solutions, food products, or cosmetics, components of the filter should be on approved FDA lists and should have undergone acute oral toxicity, systemic toxicity, implantation, eye irritation, and pyrogen tests as prescribed by the U.S. Pharmacopeia. Long-term stability tests may also be necessary.

Effective Filter Removal Rating (Efficiency)

At this point determine the degree of filtration required, although the supplier will make recommendations. It may be a matter of achieving:

- Optical brilliance—3 to 10 μm absolute
- Sterility—0.2 μm absolute for liquids and 0.2 or 0.45 μm absolute for gases
- Protection of patients from microemboli in transfused blood—40 μm absolute
- Deionized water column protection—10 to 30 μm absolute
- Clarifying of process tap water—30 μm absolute
- Safety or efficacy control the choice.

Concomitant with the need to specify optimum degree of filtration is the difficulty in relating this to actual cartridge performance. The many suppliers use different test methods for rating their filters which often cannot be correlated. The following rating definitions are common:

Nominal filtration rating: an arbitrary micrometer value indicated by filter manufacturers. It is based upon removal of some percentage of particles of a given size or larger, but rarely well defined and consequently not reproducible.

Absolute filtration rating: The diameter of the largest hard spherical particle that will pass through a filter under specified test conditions. This is an indication of the largest opening in the filter element.

Beta ratio[5]: equals total number of particles in influent greater than size a divided by total or number of particles in effluent greater than this same size a.

Filtration ratio[6]: equals number of particles in influent equal to size b divided by number of particles in effluent equal to size b.

Until standard test procedures for evaluating the performance of cartridges and their integrity are available (now being developed by AIChE),[7] the user must have a clear understanding of the removal capability of the cartridge chosen and its relation to the actual service required.

Fluid Viscosity

This parameter will affect filter pressure drop so that type, area, and prefiltration need are influenced by viscosity. For coarse filtration and high viscosities (above 3000 cP), metal-mesh cartridges are selected because of their high permeability and strength. Nonpleated disposable cartridges have been used to viscosities of 20,000 cP but at very low flux rates (< 0.1 gal/min per cartridge) and to only 15 lb/in^2 (gage) (before removal efficiency degrades).

Pressure-drop effect can be reduced by increasing filter area. In continuous service, this is an economical solution since total throughput would increase more than the proportionate cost increase. For fine filtration of viscous oils [3 μm (abs) and less], prefiltration is common to mitigate the usual rapid pressure-drop increase associated with such filtrations. Laboratory tests[8] are the most precise means for determining prefiltration removal rating and ratio of prefilter to final filter area. Ideally both filters should block simultaneously to avoid system shutdown each time only one filter becomes exhausted.

Type of Contaminant

This parameter influences filter type and area. Hard, irregular-shaped particles filter most easily, whereas gelatinous materials are the most difficult! They are best removed by trapping within the interstices of the filter medium, rather than on its surface, where a thin, impermeable cake would form. Such filters are commercially available and should be selected only if they conform to the required degree of filtration. If not, they may be effective as prefilters. For quantitative removal of bacteria from liquids, a 0.2-μm (abs) sterilizing-grade filter must be specified, whereas a 0.2 or 0.45-μm (abs) hydrophobic filter medium is required for sterile filtration of gases. Other contaminants also preclude the selection of all but one or two removal ratings [i.e., yeast removal is reliably achieved with either 0.65 or 1.2 μm (abs) filtration].

Quantity of Contaminant

As contaminant loading increases, area must be increased if throughput to clogging or to maximum available pressure drop is to remain constant. There is no direct correlation between contaminant quantity and increase in area because influent contaminant composition is not constant (except in unique cases) and because flux rate decreases as area increases. Empirical performance data is the best guide for determining if any correlation exists.

As stated in the introduction, cartridge filtration is not optimum or economical for clarifying fluids with a solids content greater than 0.01% by weight.

Maximum Allowable Cartridge Pressure Drop

This parameter, beyond which the filter will fail structurally, is specified by the manufacturer. The user must specify a maximum available filter pressure drop, which should be lower. However, sufficient pressure drop should be allowed for optimum filter dirt capacity to achieve economical cartridge use. The characteristic pressure drop/dirt capacity curve for cartridges is shown in Fig. 3. Note that most of its capacity is consumed before the sharp increase in pressure drop so that available pressure drop should be at least equal to that at the knee of the curve. The most frequently overlooked factors are the pressure losses associated with the housing and internal hardware. These are a function of the square of the flow rate and remain constant for a given flow. Filter-medium pressure drop is variable, however, and increases with contaminant loading. Filter selection

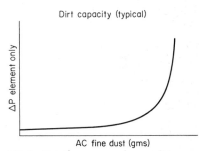

Dirt capacity (typical)

AC fine dust (gms)

Fig. 3   Cartridge pressure drop vs. dirt capacity. *(Pall Corporation.)*

therefore should minimize the ratio of such hardware losses to the total available pressure drop. This optimizes cartridge life by maximizing pressure drop available for passing fluid through the medium and overcoming its growing resistance as contaminant plugs its pores.

Required Throughput and Flow Rate

For batch operation it is good practice to clarify the complete load without interruption. For sterilizing applications it is essential to do so, in order to avoid the hazard of violating the sterile integrity of the system when cartridges are changed in midstream. In such cases required throughput should be part of your specifications. Throughput per cartridge is also best determined by laboratory test, from previous plant filter experience, or from field experience of the cartridge manufacturer. Lab tests are not difficult and are the most reliable, provided influent fluid and contaminant characteristics of test and plant fluid do not differ. Throughput per cartridge is also related to available filter pressure drop. Optimum area therefore must consider initial pressure drop, available and maximum allowable pressure drop, and throughput.

For continuous service a similar review is necessary except that oversizing this assembly will reduce operating costs because of lower flux rates. Oversizing should be balanced against first costs and space limitations. Some typical flux rates are:

Industrial water	2.44 to 9.7 m³/h[1 to 4 gal/(min) (ft²)]
Deionized water	1.22 to 3.66 m³/h[½ to 1½ gal/(min) (ft²)]
Solvents	2.44 to 7.32 m³/h[1 to 3 gal/(min) (ft²)]
Dilute acids and bases	1.22 to 7.32 m³/h[½ to 3 gal/(min) (ft²)]
Vegetable oils	0.61 to 2.44 m³/h[¼ to 1 gal/(min) (ft²)]
Compressed air at 7 kg/cm (gage) [100 lb/in² (gage)]	33 to 148 m³/h(20 to 100 scfm/ft)

Filter Sterilization Methods

To cold sterilize fluids by filtration, the cartridge must be guaranteed to quantitatively remove all bacteria. Furthermore, the filter itself must be capable of being sterilized by either ethylene oxide, steam autoclaving, radiation, or in-situ steam sterilization. The first three are limited to small installations where the filter can be accommodated by the sterilizing equipment and physically handled by the operator. These techniques also suffer from the disadvantage that the filter must be aseptically assembled to the system, a procedure that risks introduction of bacterial contamination. In-situ steam sterilization can be used for all assemblies, regardless of size, so that aseptic installation is not necessary and sterilizing turnaround time is very much reduced. Very few cartridge types will resist this service so the manufacturer should be requested to submit evidence (historical field use) of this capability. Also, in view of the critical nature of sterilizing filtration, all cartridges should be field tested by manufacturer-supplied procedures for verifying the integrity of the assembly before and after use.

A TYPICAL OPERATION

Let us examine the use of the above parameters. Since the sterilization of water is so critical and requires a variety of cartridge types, a sterile distilled water flow sheet is shown (Fig. 4).

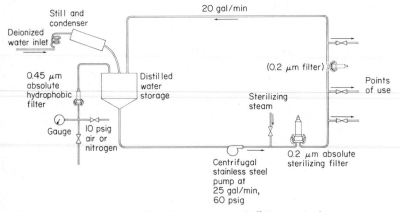

Fig. 4 Distilled sterile water system. *(Pall Corporation.)*

Installation Sanitary stainless steel filter housings (see Fig. 5) and lines with sanitary fittings are recommended. Pressure gauges should be sparingly used downstream of sterilizing filters. Upstream filter drain connections should be fitted with suitable valves and piped to waste or to product recovery. Filter vent connections should be similarly piped and allow connection of a compressed air line for testing filter integrity. Avoid elevated downstream piping and receiving tanks because retained liquids will cause damaging back pressure to the cartridge. Where backflow is unavoidable, install quick-acting, low-pressure-drop check valves at the immediate filter outlet. Pulsating pumps, such as peristaltic, piston, and diaphragm types, should be avoided, since flow surges will cause plastic membrane cracking, paper pleat rupture, or flexing of nonrigid media. Always install surge tanks between such pumps and the filter. When rapid-cycling valves are used (e.g., washing machines) the surge tank should be located between the valves and the filter.

If the raw water has received good primary treatment consisting of flocculation and settling to reduce particulate, chlorine addition to oxidize organics and control bacterial levels, filtration by sand filters to remove unsettled fines, passage through carbon beds to adsorb dissolved organics and residual chlorine, filtration by 10- to 20-μm (abs) cartridges to retain carbon fines and other particulates which will foul the downstream resin beds, and removal of dissolved inorganics by deionization, followed by 0.2 μm (abs) cartridge filtration to remove all particulates 0.2 μm and larger, then the quality of water fed to the still would be exceptionally low in contamination—ideal for cartridge filtration. To produce sterile water the choice of removal rating is mandated at 0.2 μm (abs) for the water system and 0.45 or 0.2 μm (abs) hydrophobic media for the storage tank vent filters (necessary to prevent the ingress of environmental contamination as water level rises and falls in the tank). Temperature is ambient and so is not a factor in filter selection, although more and more sterile water systems are operating at 80°C (176°F) to maintain low bacterial levels. Maximum pump pressure (in this example) is 6.3 kg/cm^2 [60 lb/

in² (gage)] and the specification for the cartridge requires it to remain structurally intact to 60 lb/in² (gage). The sanitary housing is rated by the supplier for 7 kg/cm² [100 lb/in² (gage)] at 120°C (248°F). (Check with your legal department to determine if state law requires ASME code construction and a stamp per BPVC Section VIII for the housing.) Fluid viscosity is relatively unimportant in the case of distilled water and because of extensive pretreatment the quantity of contaminant will be very low, and cartridge life should be relatively high. Assume pressure drop through the recirculation section of the system will be 0.7 kg/cm² [10 lb/in² (gage)] and that at the points of use 1.75 kg/cm² [25 lb/in² (gage)] is necessary because of valving and equipment located there. Consequently 1.75 kg/cm² [25 lb/in² (gage)] is the maximum available for filtration. Assume weekend plant shutdowns and that the system is in continuous use 20 h each day. Cartridges are to be changed no more than once per week. Required

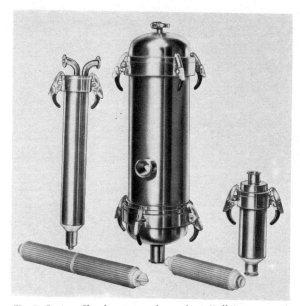

Fig. 5   Sanitary filter housings and cartridges. *(Pall Corporation.)*

weekly throughput is 454,250 L (120,000 gal). Filter performance data (from the supplier) indicates that a 3.7-m² (40-ft²) assembly will have an initial clean pressure drop of 0.35 kg/cm² (5 lb/in²), of which hardware pressure drop will be 0.18 kg/cm² (2.5 lb/in²). From previous experience, the supplier estimates that throughput will be more than 1.136,000 L (300,000 gal) when assembly pressure drop reaches 1.75 kg/cm² (25 lb/in²) (although a test can be completed in one day to verify this). Since the system will be steam sterilized every day during the brief 4 h shutdown, and cartridges are not to be changed more often than weekly, in-situ steam sterilization filter capability as well as filter reuse after multiple steaming cycles should be specified. The specifications for the filter must also include a proven procedure for in-place testing of sterilizing-grade filtration integrity of the assembly because of the critical nature of this filtration.

The maximum withdrawal or fill rate expected for the 3790 L (1000 gal) storage tank is 95 L/h [25 gal/min]. This is equivalent to an air rate of 3.3 scfm, and flow/pressure drop data (from the manufacturer) for the 0.45-μm (abs) hydrophobic filter shows a drop of 0.11 lb/in² (3 inH₂O negative pressure filter sizing is sufficient). This filter should also have in-situ steam sterilization capability. After steam sterilization of the distilled water storage tank, rapid internal cooling and condensation will create an unusually large airflow demand to prevent collapse of this thin-walled vessel. To provide this, a separate oilfree air or nitrogen pressure source should be connected to the sterilizing vent filter (as shown) and flow should be introduced after the steam valve is closed. Recommended pressure is 0.7 kg/cm² [10 lb/in² (gage)]. Vent filter life should be very long, at least 6 months (because of the high area and low contamination levels in controlled environments). Filter integrity can be checked weekly using the manufacturer's recommended forward-flow bubble point (see Fig. 6) procedure.

For long piping runs, additional filtration may be needed close to points of use (dotted figure) to remove particulate and fibers which may show up with time. Sampling of the water at these locations and passing it thru a 0.8-μm analytical membrane, followed by microscopic examination, will determine the need for such a unit(s).[9] This filter should be located in the loop itself, rather than in a stagnant area where bacteria can multiply. Since this will be a 0.2-μm sterilizing filter, downstream of the system sterilizing filter, it can be much smaller in area and sized only for pressure drop.

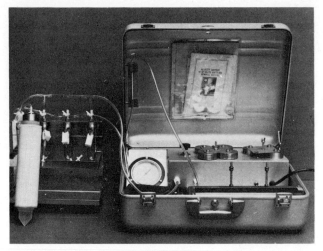

Fig. 6 Forward-flow bubble point test apparatus. *(Pall Corporation.)*

TROUBLESHOOTING

The most common problems will be low particulate count, short filter life, or failure to achieve effluent sterility. Check the following questions:

1. Effluent quality
    a. Was the filter sterilized according to the manufacturer's recommendations?
    b. Was the correct time/temperature relationship established for sterilizing the downstream system?
    c. Were water samples taken aseptically?
    d. Is the flow direction through the installed filter from inlet to outlet? Reverse flow will cause cartridge damage.
    e. Were cartridges checked for correct pore size rating?
    f. Was the assembly checked for integrity before and after filtration using the manufacturer's procedures?[10]
    g. Was assembly inspected to ensure that all cartridges are installed and that cartridge seals are not damaged or seating improperly?
    h. Do cartridges show filter-medium damage or other visible damage at side seal and end caps?
2. Filter life
    a. Was housing vented on start-up? If not, trapped air will prevent complete submersion of the cartridge and full utilization of its area.
    b. Is sufficient pressure drop available?
    c. Was influent analyzed for quantity, identity and type of contaminant to be sure it has not changed from normal conditions? Water characteristics change with the seasons and the condition of treatment equipment.

Cartridge filtration is an ever-changing technology and each year new products are introduced with better fluid compatibility, greater strength, improved efficiencies, higher dirt capacities, and greater reliability and economy. Many critical filtrations would be impossible or extremely costly except for cartridge filters and so the benefits of such improvements in the state of the art must not be overlooked. Conventional filter systems should be periodically reviewed for replacement with the latest in fluid purification technology and expertise.

REFERENCES

1. Nickolaus, N., What, When, and Why of Cartridge Filters, *Filtr. Sep.*, March/April 1975, p. 155.
2. *Filtration News*, Pall Corporation, Summer 1976.
3. Pall, D. B., WADC TR256, May 1956.
4. Pall, D. B., *Bull. Parenter. Drug Assoc.*, **29**(4), July–August 1975.
5. American National Standard Multi-Pass Method for Evaluating the Filtration Performance of A Fine Fluid Power Filter Element, ANSI B93.31-1973.
6. Tsai, C. P., and Farris, J. A., Multi-Pass Constant Flow Filter for Silt Control Hydraulic Filters, Pall Corporation Field Service Report 53, April 1976.
7. Nickolaus, N., Cartridge Filter Evaluation, *Filtr. Eng.*, May/June 1975.
8. Instructions for Use of Disk Filter Test Jigs, Pall Corporation Bulletin PCT-100, March 1975.
9. Quality Assurance Practice and Standards for Bacteria Removal Membrane Filters, bulletin AB800a-1-72, Pall Corporation.
10. Cole, J. C., and Pauli, W. A., Field Experiences in Testing Membrane Filter Integrity by the Forward Flow Test Method, *Bull. Parenter. Drug Assoc.*, **29**(6), (1975).

Section **4.7**

Felt Strainer Bags

ARTHUR C. WROTNOWSKI   *Consultant.*

AREA OF APPLICATION

Felt strainer bags are especially suitable for straining slurries or dispersions. Felt media can effectively remove specific size particles in the 20- to 200-μm size range from practically all known combinations of slurries, dispersions, solutions, or the like.

Felt's industrial suitability is based on two physical situations as follows:

1. Conventional screens, filters, cartridges, or other known methods of particle size classification are not capable of removing a specific size particle in this size range on a practical basis.

2. The felt strainers employed have been designed to maximize solids collection in depth and for a high felt voidage, usually exceeding 85%. These felts are designed to and can intercept and collect in depth a given size particle with practical and usable capacity.

The strainer felts employed can collect particles in a 20- to 200-μm size range because the very fine individual textile fiber itself is used to arrest and hold a specific size particle. From fiber denier, felt density, and the amount of fiber per unit volume, the number of working pores can be calculated. It has been demonstrated that there are multitudinous available and usable pores for particle straining.

Accompanying the large array of specific-size felt pores is a large fraction of voids, i.e., over 85%. It is this physical property, high voidage, of the strainer felt that gives a usable, practical straining capacity. For example, as a given slurry passes through a strainer felt the voids permit the slurry to pass except for the oversize particles.

The filtration categories of felt strainers as described above appear to fit between:

1. Rapid-settling systems
2. Moderate- and slow-settling systems as described by Purchas.[1] In this regard, felt strainer bags are not used for cake filtration or for clarification operations.

From a felt medium–particle size point of view a chemical engineering definition has been given by Grace[2] as follows:

1. *Straining:* removal of particles larger than the pore size.
2. *Filtration:* removal of particles smaller than the pore size.

The latter functions by means of the classical filtration mechanisms: gravitation, interception, diffusion, inertial separation, and electrostatic separation. These definitions are consistent with the observed performance of felt strainers. For example, Wrotnowski[3] by means of a mathematical model predicted and observed consistent particle-size straining performance by a fly ash slurry.

Conventional screens and strainers—including cyclones using related techniques—have a lowest physical practical limit of 50 μm. The felt strainer bag extends the useful range of straining down to 20 μm.

Within the extended straining range, i.e., down to 20 μm, the strainer media maintain good operating characteristics as follows:

1. Because of felt depth straining, high viscosity, high flow rate, and gelatinous particles are handled conveniently.
2. With proper choice of synthetic fibers, broad chemical stability is practical.

Felt straining, although a batch operation, can strain down lower than competitive methods. It therefore can make a batch operation justifiable.

Long runs are usually obtained with felt strainer bags because "oversize" particles are usually not present in large quantities. If too many large particles are present, e.g., over 30%, other methods must be used—not usually including a particle classification method, however.

OPEN SYSTEM DESIGN AND OPERATION

Felt straining action is quite straightforward. Selective screening action requires only that the liquid slurry or dispersion be passed through a given felt.

An elementary bag design is shown in Fig. 1, where enamel is being strained at the 25-μm level. This tie-on arrangement will achieve fine particle screening characteristic of the felt strainer media series (Table 1).

As indicated by the relatively high air permeability felt values in Table 1, the initial particle classification and screening operation occurs at low or no pressure drop ΔP. In the straining cycle, oversize particles are collected in the felt and a filter cake forms which eventually causes a ΔP increase and clogging. In the usual case, the felt bag does not "blind," i.e., become irrelievably clogged. However, it is usually more economical to discard the used felt bag and collected solids rather than to clean it up.

It is good practice to throttle the flow rate, for example, to 2 L/(s)(m^2) [3 gal/(min)(ft^2)]; however, the use of the equilibrium free-flowing rate is common. The tie-on bag design can with reliability operate up to 0.7 kg/cm^2 (10 lb/in^2); however, the bag must be securely fastened to the feed pipe.

A more efficient use of the strainer felt medium is employed by the Snap-Ring® design which will fit and be supported by an adapter head. Note in Fig. 2 that the adapter head has a ledge to support a given size metal ring sewn inside the bag, as shown.

By tilting and flexing the ring can be installed and seated on the support ledge of the adapter. When pulled down by hand, the bag is fully supported and sealed and suitable for full-scale operation up to 0.7 kg/cm^2 (10 lb/in^2) as shown in Fig. 3.

The use of the Snap-Ring bag provides complete usage of the strainer medium, and rapid and easy installation and removal. It is automatically sealed, requiring no hose clamp or other tightening devices, and no other precautionary steps needed.

The Snap-Ring bag design, compared with the tie-on method, is more convenient, reliable, and reproducible and involves less labor and materials expense.

The stainless steel restrainer basket shown in Fig. 4 makes it possible to use the same

TABLE 1   Felt Strainer Bag Media—Air Permeability, Nominal Values

Particle size rating, μ	Air permeability, cfm/ft²/½ inH₂O	Application
	Rayon, viscose (nylon scrim)	
5	50	Solvents, paints, general use
10	100	
15	150	
25	250	
50	300	
100	350	
200	400	
	Polypropylene	
5	50	Acid, alkali, general use
10	100	
25	150	
50	200	
100	300	
	Nylon	
5	50	Heavy-duty alkali and solvents
10	100	
25	150	
50	200	
100	300	
	Nomex	
5	25	Elevated temperature
25	150	
50	250	
100	350	
	Teflon	
30	50	Extreme corrosion and temperature
	Wool-silk	
1	10	Fine-particle retention
	Wool-cotton	
3	20	Fine-particle retention

felt bags up to a pressure drop of 3.5 kg/cm² (50 lb/in²). Standard sizes No. 1 [0.279 m² (3 ft²)] and No. 2 [0.558 m² (6 ft²)] are available.

The strainer basket of Fig. 4 should be installed a bit "high" so that as the bag fills out, the restrainer basket orients downward to effectively restrain the taut strainer bag.

Bag restrainers are used when a given end use benefits by throughout or amount of solids collected from higher ΔP. It has been found that for bench-top testing, the use of felt cones (see Fig. 5) will predict quite well the results obtained in a given production performance. For example, place a given felt cone in a ring stand as shown. Collect and evaluate the effluent obtained by pouring a sample of feed through a felt cone.

These various open system methods are convenient and inexpensive to use.

Fig. 1 Tie-on bag.

CLOSED SYSTEM DESIGN AND OPERATION

The advantages of continuous piping, handling solvents, good housekeeping, and the permanency of hardware require the use of a closed system.

Felt bags employed as high-flow-rate strainers permit the design of compact equipment as pictured in Figs. 6 and 7 (the GAFLO® RB hardware).

The Snap-Ring bag—open system employs a support ledge located on the inner periphery with respect to the bag (see Fig. 2). The same Snap-Ring bag—closed system employed in the RB hardware design has a support ledge located on the outer periphery with respect to the bag, (see Fig. 8).

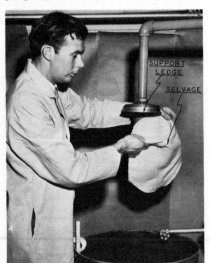

Fig. 2 Snap-Ring® bag and adapter. (*GAF Corp.*)

In both cases the Snap-Ring is a convenience which causes the bag to spread out so that it is available for complete straining performance. However, a significant detail is the location of the felt selvage. In Fig. 2, for the open system, the selvage has been purposefully placed outside. In Fig. 8 the selvage again has been purposefully placed—inside, however. In both cases the sealing mechanism gives more reliable performance when one uniform layer of felt is used. The selvage can be placed either inside or outside a given bag by pushing the felt inside out or the reverse. The felts are symmetrical, being unaffected by direction of flow.

As shown in Fig. 8, items 8 and 9, O rings supply the sealing for the restrainer basket, felt bag, and vessel. Item 7, another O ring, provides the sealing for the feed line.

For ASME specification compliance for hydrostatic testing, 15.75 kg/cm² (225 lb/in²) is applied to the restrainer basket, and 21 kg/cm² (300 lb/in²) is applied independently to the vessel itself. The other RB hardware models are modules of RB-1A and they essentially work on the same principle.

Fig. 3 Four No. 1, Snap-Ring® bags in use—open system. (*GAF Corp.*)

O-Ring Design

An occasional leakage problem can occur because of O-ring leakage caused by uneven tightening of the bolts. According to theory, the mechanism of O-ring sealing is that about two-thirds of the cross-section diameter is contained by snug fit in a rectangular groove. As the opposing flange surface uniformly presses on the O ring, it fills the rectangular groove with enough excess elastomer to perform the primary sealing job.

Fig. 4 Snap-Ring® bag, adapter, and re-strainer—open system. (*GAF Corp.*)

Fig. 5 Felt cone—bench-top testing. (*GAF Corp.*)

The RB hardware restrainer basket, item 3 in Fig. 8, has a double O-ring design to simultaneously seal the pressure vessel shell and the cover plate; a third O ring seals the inlet. The RB hardware counts heavily on O-ring sealing, which in fact gives reliable sealing and which presents a cleanable surface on the restrainer basket flange, a desirable feature at the time of bag changing.

SELECTING THE FELT STRAINER BAG MEDIUM

Without previous felt straining experience, it is best to perform a bench-top test employing a felt cone. The cone can be placed in the support ring of a laboratory stand as indicated in Fig. 5. The applicable control test can then be performed.

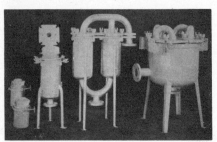

There is, of course, a large variety of particles and liquids in this area of interest. For example, molten wax, dispersed wax, natural gums and varnishes, glue, latex, and paint, during manufacture, often require particles of some nature to be removed to achieve product and quality control. Whether straining delicate flocculent materials, gelatinous particles, or high-density hard solids, oversized particles can usually be strained out by calibrated felt strainer bags.

Fig. 6   RB hardware: RBX, RB-1A, RB-2A, RB-4A (Refer to Table 2). *(GAF Corp.)*

When selecting a felt medium, the investigator should consider the nature of the particle and the process objective. For example, the particle size ratings given in Table 1 were determined by means of a fly ash slurry. Fly ash particles are hard vitreous siliceous spheres which are the slag from burning powdered coal in power plants.

The particle size classification action will probably be somewhat different when some other irregular-shape or gelatinous-type particle is handled. However, it is expected that an optimum strainer felt can be found in the range of sequential-pore-sized felts as given

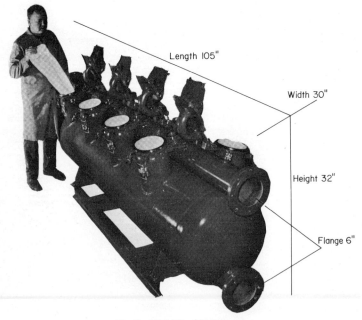

Fig. 7   RB-8A2L. *(GAF Corp.)*

in Table 1. The four felt strainer bag series A, B, C, and D of Table 1 basically perform oversize particle removal at some given level in the 20- to 200-μm range. These four families are based on the synthetic fibers available as needed to meet the general needs of the chemical process industries. The many synthetic fiber diameters are useful. In felt straining, the mechanism variable is principally fiber diameter instead of felt density.

The strainer bag basic construction has a fleece or working staple fiber interlocked onto a strength member, an open scrim. The open scrim permits fine particle passage and use as an unsupported felt strainer bag, and in the case of heavy duty solids removal, it is a

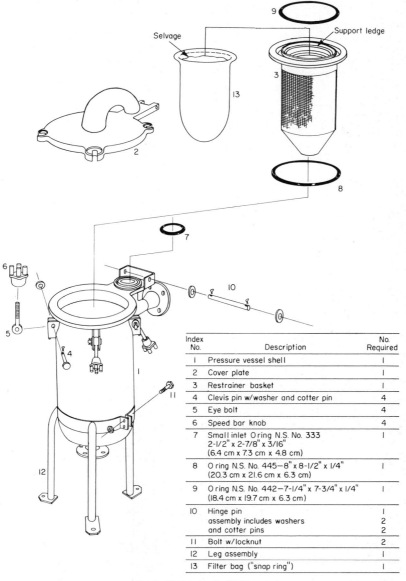

Index No.	Description	No. Required
1	Pressure vessel shell	1
2	Cover plate	1
3	Restrainer basket	1
4	Clevis pin w/washer and cotter pin	4
5	Eye bolt	4
6	Speed bar knob	4
7	Small inlet O ring N.S. No. 333 2-1/2" x 2-7/8" x 3/16" (6.4 cm x 7.3 cm x 4.8 cm)	1
8	O ring N.S. No. 445—8" x 8-1/2" x 1/4" (20.3 cm x 21.6 cm x 6.3 cm)	1
9	O ring N.S. No. 442—7-1/4" x 7-3/4" x 1/4" (18.4 cm x 19.7 cm x 6.3 cm)	1
10	Hinge pin assembly includes washers and cotter pins	1 2 2
11	Bolt w/locknut	2
12	Leg assembly	1
13	Filter bag ("snap ring")	1

Fig. 8   RB-1A detail. *(GAF Corp.)*

tough industrial fabric suitable for general rough use. The use of the scrim does not interfere with particle separation yet provides strength for industrial applications.

Discussion of Table 1

The rayon viscose series is a general-purpose "workhorse" line. It is supported by a strong nylon scrim. Note that both nylon and viscose have similar chemical stability properties. For example, they are stable in caustic and stable in aliphatic and aromatic solvents, and both have histories of being FDA-amenable for use with foodstuff.

The rayon-nylon series is not suitable in acid or biologically active environments.

The polypropylene strainer bags are especially useful in very corrosive acid or alkali chemical exposures except in hot aromatic solvents. Polypropylene is completely stable in biological environments and is excellent for general-purpose use.

The nylon strainer bags are tough and thermally resistant, especially in anhydrous and oxygen-free environments.

Nomex can be used for hot-melt adhesives at 350°F or for other high-temperature uses.

A special strainer bag of Teflon is useful for exotic chemicals; wool-silk and wool-cotton are useful for extra-fine straining principally in solvent or aqueous slurries.

SIZING THE EQUIPMENT

Since the bag system is essentially a batch operation, the size of the equipment determines the frequency of bag changing. However, it is quite possible to get an effective continuous system by using two vessels placed in parallel, each vessel having enough capacity to handle the full flow and solids encountered. While one bag vessel is in use the other can have new bags installed; meanwhile the flow is continuous.

After selecting a felt strainer bag medium, as dictated by chemical compatibility and particle size rating, and determining if a batch or continuous method will be used, the size of the equipment required can be analyzed by:

1. The pressure drop, based on liquid flow rate and viscosity
2. Solids to be removed

In general, and for practical reasons, we are considering large quantities of liquids with relatively small quantities of oversized particles to be removed. By determining the pressure drop under these conditions we essentially size the equipment on the basis of relative power cost.

Closed Filter Bag System

The example given in Table 4 has been calculated for all of the larger hardware given in Table 2. It will be seen that the pressure drop developed by bag and vessel is quite measurable. The pressure drop developed by the solids collection is the main controlling variable for long life.

As textile materials strainer felts can be classified conveniently by air permeability.

TABLE 2   Pressure Vessels—RB Hardware (ASME coded—150 lb/in²)

Item	Identification model no.	Strainer medium area, ft²	Bags employed	Reference figure no.
1	RB-X	1	One RBX	6
2	RB-1A	3	One size No. 1 Snap-Ring*	6
3	RB-2A	6	Two size No. 1 Snap-Rings	6
4	RB-4A	12	Four size No. 1 Snap-Rings	6
5	RB-6A2L	36	Six size No. 2 Snap-Rings	
6	RB-8A2L	48	Eight size No. 2 Snap-Rings	7

*As shown in Figs. 2, 3, 4, and 11.

This value can be converted to various permeabilities as shown in Fig. 9. As a convenience an approximate conversion formula applying to 12 different viscosity units is given in Table 5.

A summary of the procedural steps in Table 4 follows:
1. Convert viscosity units to centipoise.
2. Determine liquid permeability of the felt medium.
3. Determine pressure drop in RB hardware.
4. Calculate the total pressure drop considering total felt area and the individual models.

TABLE 3 Chemical and Thermal Component Compatibilities—10 Examples*

Item	Liquid	Felt strainer medium fiber rating, μ	RB vessel	O ring	Snap-Ring
1	Raw latex	RV-200	Steel	Neoprene	Cadmium-(Cd) plated steel
2	Clay slip	RV-25	Steel	Neoprene	Cd steel
3	Gums and waxes	RV-25	Steel	Neoprene	Cd steel
4	Varnish (hot)	NY-25	Steel	Viton	Cd Steel
5	Magnetic tape	WS-1	316SS	Viton	Cd Steel
6	Metallic paint (aluminum flake)	RV-50	Steel	Neoprene	Cd Steel
7	Hot-melt adhesive	HT-25	Steel	Silicone	Cd Steel
8	Sulfuric acid, 10%	PO-5	PVC-coated steel	Neoprene	316SS
9	Caustic, 10%	PO-5	Steel	Neoprene	Cd steel
10	Enamel	RV-25	Steel	Neoprene	Cd steel

NOTE:

Fiber	Fiber code	Operating temperature, °F	Chemical compatability†		
			Acid	Alkali	Solvent
Rayon viscose	RV	180	Over 5%—poor	Excellent	Excellent
Nylon	NY	220	Poor	Excellent	Excellent
Polypropylene	PO	240	Excellent	Excellent	Aromatics—poor
Nomex	HT	350	Good	Good	Excellent
Wool-silk	WS	180	Poor	Poor	Excellent

*Refer to Fig. 8, RB-1A detail.
†For detail values consult fiber manufacturer.

The total pressure drop results are clean felt and hardware values. From the design flow rate and calculated pressure drop, the energy consumption pumping costs can be calculated for economic comparison for long-time straining with no solids buildup.

From a practical point of view, the average dirt-holding capacity for a No. 1 size bag [0.28 m² (3 ft²)] for ordinary, loam, rust scale, or otherwise granular-type dirt is 2.27 kg (5 lb). The potential dirt removal from a given feed stream is a variable which must be determined experimentally, yet this factor will determine the straining life of the bag and frequency of changing bags, and is the major factor in determining the size of the equipment.

Economic Comparison

In sizing the equipment, the investigator can compare the capital expenditure and the associated rate of depreciation against:
1. The increase in power cost of high equipment-pressure drops.
2. The increase in labor cost resulting from bag changing frequency. (In some cases there may be no labor increase over a fixed labor expense.)

The maximum rate of depreciation for chemical processing equipment is 9 years, based on the Internal Revenue Code Section No. 167 using the Asset Depreciation Range System.

TABLE 4   Procedure for Determining the Clean Felt Pressure Drop in RB Hardware

EXAMPLE

PROBLEM: Determine the clean felt and hardware pressure drop for a water solution thickened with cellulosic sizing.

GIVEN:

1. Desired flow rate, gal/min	100
2. Liquid viscosity, cP	1000
3. Desired level of cleanliness,	
i.e., particle size, μm	50
4. Liquid specific gravity	1.0

PROCEDURE

Step No. 1: Select strainer bag medium

From Table 3, rayon viscose is chemically compatible. From Table 1 the 50-μ rayon viscose medium has an air permeability of 300 cfm/ft²/½ in H₂O.

Step No. 2: Convert felt air permeability to liquid permeability

From Fig. 13 at 1000 cP, the liquid permeability is 5 gpm/ft²/(1.0 lb/ft²) at specific gravity 1.0.

Step No. 3: Hardware Pressure Drop (ΔP_H).

Consider the RB hardware as follows:

Model	Figure No.	Pressure Drop ΔP_H, lb/in², 100 gpm, 1000 cP, specific Gravity 1.0
RB-1A	15	12
RB-2A	16	8
RB-4A	17	5
RB-6A2L	18	2.2
RB-8A2L	19	2.0

Step No. 4: Felt pressure drop ΔP_F

$$\Delta P_F = \frac{R}{LP \times A}$$

where R = desired equipment flow rate, gpm
    LP = clean felt liquid permeability
        (Step No. 2), gpm/ft²/(1.0 lb/ft²)
    A = felt Area (Table 2), ft²

Step No. 5: Total pressure drop ΔP, lb/in²

Clean felt and hardware
    $\Delta P = [\Delta P_H + \Delta P_F]$ Sp. Gr.
    where Sp. Gr. = specific gravity

Model	(ΔP_F)	$+ (\Delta P_H)$	$= (\Delta P)$
RB-1A	$\dfrac{100}{5 \times 3}$	$+ 12.$	$= 18.7$
RB-2A	$\dfrac{100}{5 \times 6}$	$+ 8.$	$= 11.3$
RB-4A	$\dfrac{100}{5 \times 12}$	$+ 5.$	$= 6.7$
RB-6A2L	$\dfrac{100}{5 \times 36}$	$+ 2.2$	$= 2.8$
RB-8A2L	$\dfrac{100}{5 \times 48}$	$+ 2.0$	$= 2.4$

COMMENT ON SOLIDS CAPACITY:

    The average dirt-holding capacity for a given size No. 1 bag (3 ft²) for ordinary loam, rust scale, or other granular-type dirt is 5 lb.

    The clogging or blinding properties of an individual filter cake will dominate "close off." However, the depth straining capacity of a given felt will exceed a flat screen or cloth capacity by 3 to 5 times. Furthermore, use of a controlled-pore-size prestrainer can provide an additional increase of 5 times[5] for the same strainer area.

Open Filter Bag System

The open bag system has a variety of combinations that may be used in emergency or inexpensive arrangements, and quality straining performance can be expected from them. As described above, the main controlling factor is the amount of solids collected in each case, and the design decision for the open system is selection of a functional system of adequate solids capacity. Pressure drop can be calculated in each case from Fig. 9; however, the solids collected determine the capacity.

TABLE 5   Converting Various Viscosity Units to Centipoise (for Subsequent Use on Fig. 9)

1. Viscosity units at 70°F			
Units	Factor	Units	Factor
Demmler No. 1	14.6	Parlin Cup No. 15	98.2
Demmler No. 10	146.	Redwood Admiralty	10.87
Engler	34.5	Redwood Standard	1.095
Ford Cup No. 4	17.4	Saybolt Furol	10
MacMichael	1.92	Saybolt Seconds Universal	1.0
Parlin Cup No. 2	187.0	Stormer	Approx. 13

2. Conversion Formula*

$$\text{Absolute viscosity} = \frac{(\text{viscosity units})\,(\text{factor})\,(\text{Sp. Gr.})}{4.62} \quad \text{cP}$$

*The conversion formula is approximately valid when the numerator is greater than 250.

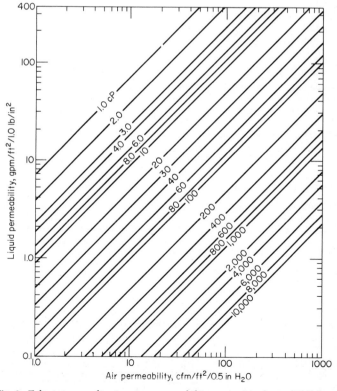

Fig. 9 Felt strainer media viscosity—permeability conversion charts. *(GAF Corp.)*

ALLIED EQUIPMENT AND PIPING ARRANGEMENTS

Both open and closed felt bag straining systems require a minimum amount of external pumps, tanks, and piping. The felt straining process is essentially a low-pressure-drop operation, and therefore low-horsepower motors are employed. It is not unusual to employ. gravity feed. Suction and discharge piping and tanks are required, of course.

The selection of pumps is mainly a function of the liquid being pumped rather than the effect on the felt strainer operation. High-viscosity liquids such as paint and varnish require positive-displacement pumps. Felt straining by means of positive-displacement pumping proceeds normally (see Fig. 1).

Felt strainer bags have been found to be functional with all types of pumps, including those that have oscillatory pressure.

The RB hardware is quite amenable for adaptation as a portable pump and filter rig. The number of applications match the wide number of possible liquids. For example, portable rigs are practical for paint, lube oil, wire drawing oil, lithographic ink, and felt-pen-nib ink formulations. The types of control equipment used to evaluate straining are numerous. Some examples follow:

1. Hegman grind gage—paints
2. Millipore septum—general
3. Andreason pipette—solids settling method
4. Metal screens—general

Open System Piping

For open systems, the piping systems in effect terminate at the bag, where the straining itself occurs, as shown in Figs. 1, 2, 4, and 7. Manifolding as shown in Fig. 3 indicates that many piping variations are possible. Some practical applications include strainer bag discharge into:

1. An electroplating bath
2. A paint or chemical dip
3. Straining of water supply for immediate use
4. Ink straining discharge directly into a fiber drum for shipment

Closed System Piping

The RB hardware is mainly used to transfer strained product to some distant location. Figure 10 has four piping arrangements: recycle, normal run, air evacuation, and evacuation by pumping; these are conventional operations. It is quite practical to employ various lineups such as two vessels in series, the first removing a coarser fraction to promote greater capacity in a second operation. It is also practical to make a continuous system placing two vessels in parallel. While one vessel handles the flow and straining, the second vessel would have a new bag installed to be ready to take the load as needed.

EQUIPMENT AND FELT INSPECTION

The use of strainer bags is deceptively simple. The use of a common felt bag tends to surprise one technically. However, it is now well established that felt straining technology makes a contribution to the particle size classification field and has a place in the chemical process industry. For reproducible and reliable felt straining, a number of details must be under control. With regard to the felt bag there are visual judgment inspections to make, as follows:

1. Check for a quality sewing job, including correct dimensions, uniform seams, etc.
2. Check for uniform felt fabric.
3. Check for proper fiber content, determined by a straining test.
4. In a given felt bag, place the felt selvage opposite the "support ledge" (see Figs. 2 and 8), as discussed under *Closed System Design and Operation*. The uneven selvage will interfere with snap-ring sealing. It is more reliable to have one uniform layer of felt at the point of sealing.

With regard to the equipment inspection, a brief product control checkoff list is given in Table 7. The subjects covered are intended to assure good workmanship and performance reliability.

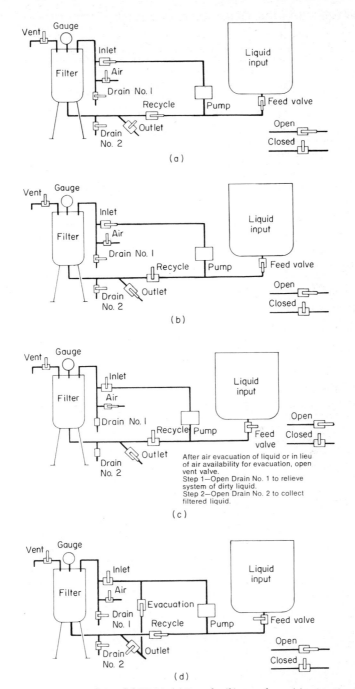

Fig. 10 Piping arrangements for model RB-1A. (*a*) Recycle, (*b*) normal run; (*c*) evacuation (air), (*d*) evacuation (pump). (*GAF Corp.*)

In addition to the information in this section and Table 7 there follows a formal and complete specification itemizing the proper design of felt strainer bag equipment.

TABLE 6   Range of Materials Strained or "Scalped"

Paint, lacquer, varnish, and inks
High viscosity: cellulosic, alginic, carrageenin, and xanthate
Raw latex, dispersions of synthetic resins and polymers
Hydrocarbons and aromatic solvents
Reflective pigments: pearl essence and metallic paint
Dyestuff pastes and pigments
Sandmill "take-off" disperse dyestuff
Abrasives, magnetic tape particles
Biomass
Ceramic clay slip
Monomers, resin, glue
Pectin, gelatin, milk

TABLE 7   Product Control Checkoff List for Model No. RB-1A Vessel*

1. Top flange of vessel, item 1, and mating cover, item 2, must be flat and parallel. Sealing surfaces must be free of scratches, dents, and weld splatter.
2. A straightedge placed on the restrainer basket, item 3, and the inlet must be parallel to the sealing surfaces of items 1 and 2 above.
3. The two weld neck flanges (inlet and feed), item 7, must be square and parallel respectively to the sealing surface of item 1.
4. Vessel opening dimension must be a minimum of 7.156 in.
5. With the restrainer basket, item 3, in place and the cover plate, item 2, closed, the hinge pin, item 10, must not bend.
6. The restrainer basket, item 3, flange thickness must be 0.325 in in thickness.
7. Swing eyebolts must be aligned properly to match slots and counterbore.
8. Angle of cover when open should be between 115° and 125°.

*Refer to Fig. 8.

CHEMICAL AND THERMAL STABILITY

Today's chemical processing industry challenges the engineer using the felt strainer bag system to balance off the choice of materials available against the chemical environment challenging the equipment. Actually there are few hardware components, all of which are readily available in a wide range of materials of construction.

The various strainer bag components with comments follow:

No	Description	Comment on materials
1.	O rings	A large number of elastomers are available. Stability to temperature corrosiveness and solvents is usually possible.
2.	Felt	A large range of textile fibers, mainly stable to chemicals, are available (see Table 1).
3.	Vessel and Snap-Ring	Steel, Type 316 stainless steel, and coatings, e.g., fluorocarbon and PVC.
4.	Adapter head	Delrin, Celcon, polypropylene, and Type 316 stainless steel.

The corrosiveness of CPI liquids fall into a few categories:
   a. *Benign:* water, latex dispersions, natural gums, botanicals.
   b. *Acid, alkali, reducing:* degradation attack and oxidizing chemicals.
   c. *Solvents:* solubility attack.
   d. *Exotic:* extremely corrosive.
All materials of construction have some chemical weakness which makes them susceptible to degradation and breakdown. However, there are many materials available, and chemically and thermally stable materials of construction can usually be chosen.

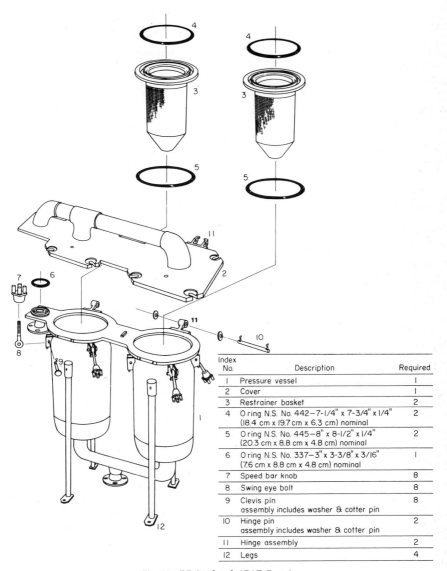

Index No.	Description	Required
1	Pressure vessel	1
2	Cover	1
3	Restrainer basket	2
4	O ring N.S. No. 442—7-1/4" x 7-3/4" x 1/4" (18.4 cm x 19.7 cm x 6.3 cm) nominal	2
5	O ring N.S. No. 445—8" x 8-1/2" x 1/4" (20.3 cm x 8.8 cm x 4.8 cm) nominal	2
6	O ring N.S. No. 337—3" x 3-3/8" x 3/16" (7.6 cm x 8.8 cm x 4.8 cm) nominal	1
7	Speed bar knob	8
8	Swing eye bolt	8
9	Clevis pin assembly includes washer & cotter pin	8
10	Hinge pin assembly includes washer & cotter pin	2
11	Hinge assembly	2
12	Legs	4

Fig. 11   RB-2A detail. *(GAF Corp.)*

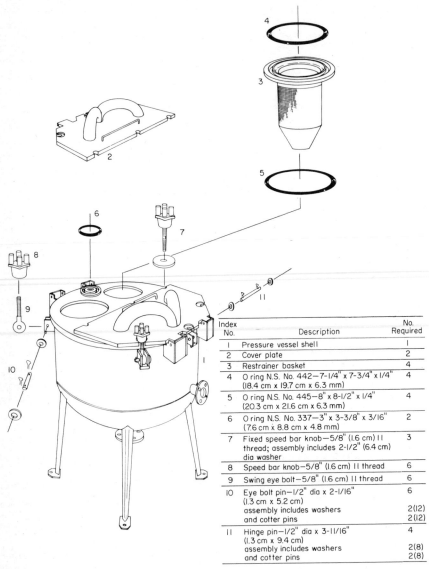

Index No.	Description	No. Required
1	Pressure vessel shell	1
2	Cover plate	2
3	Restrainer basket	4
4	O ring N.S. No. 442—7-1/4" x 7-3/4" x 1/4" (18.4 cm x 19.7 cm x 6.3 mm)	4
5	O ring N.S. No. 445—8" x 8-1/2" x 1/4" (20.3 cm x 21.6 cm x 6.3 mm)	4
6	O ring N.S. No. 337—3" x 3-3/8" x 3/16" (7.6 cm x 8.8 cm x 4.8 mm)	2
7	Fixed speed bar knob—5/8" (1.6 cm) 11 thread; assembly includes 2-1/2" (6.4 cm) dia washer	3
8	Speed bar knob—5/8" (1.6 cm) 11 thread	6
9	Swing eye bolt—5/8" (1.6 cm) 11 thread	6
10	Eye bolt pin—1/2" dia x 2-1/16" (1.3 cm x 5.2 cm) assembly includes washers and cotter pins	6 2(12) 2(12)
11	Hinge pin—1/2" dia x 3-11/16" (1.3 cm x 9.4 cm) assembly includes washers and cotter pins	4 2(8) 2(8)

Fig. 12   RB-4A detail. *(GAF Corp.)*

DESIGN AND PERFORMANCE SPECIFICATION

Introduction

The felt strainer bag system has been developed for liquid straining to a high performance level and has achieved general acceptance as an engineering tool. Important features of the strainer bag system are as follows:

    a. Economic comparison: one bag has the equivalent capacity of several cartridges.

    b. Disposal of filter cake: collected contaminate is contained in the bag.

    c. Filtrate cleanliness: contaminate does not reach the downstream side of the strainer bag.

    d. Labor expense: minimizes downtime.

    e. Time elapsed per cycle: high flow rates.

    f. Pump wear: low pressure drop.

    g. Power cost: less energy consumed in processing.

    h. Warehouse expense: less inventory requirement in units and space.

    i. Minimum operator exposure time possible due to rapid bag change time.

    j. Pressure leaf filters and plate-and-frame filters: roughing filters have been successfully replaced.

    k. 200,000-cP liquids strained on a practical basis.

    l. Above 200 μm woven-mesh bags in the 200 to 800 μm range are economical and practical.

The above features result in less manufacturing, processing, and inventory costs and are associated with higher production efficiency. The strainer bag system is recognized as one of the simplest and most effective straining devices available. The Standards and Design Criteria are hereby established for a complete strainer bag system.

Manufacturers of filter bag systems should meet all of the design requirements. When systems fall short of meeting the basic requirements, full performance of the felt strainer medium cannot be obtained.

The sophistication in the design features that has resulted in the felt strainer bag's simplicity and effectiveness is easily overlooked. There are 10 pertinent design features to consider. These considerations, which should aid in understanding and selecting a strainer bag system, are given below.

Design Criteria for Strainer Bag Pressure Vessels

- Contaminate contained in the bag.
- Postive 360° circumferential sealing of the bag.
- Quick and complete access to the bag.
- No contaminate reaches the downstream side, even during the changing of spent media.
- No residual contaminate build-up.
- Positive vessel sealing under all operating conditions.
- Fixed sealing devices.
- Restrainers should not impair flow or life.
- Easily cleanable.
- Meet ASME and OSHA requirements.

Figures 11 and 12 (pages 4-109 and 4-110) are examples of typical equipment.

REFERENCES

1. Purchas, D. B., "Industrial Filtration of Liquids," Leonard Hill, London, 1967.
2. Grace, H. P., The Art and Science of Liquid Filtration, presented at Jubilee Meeting A.I.C.H.E., Philadelphia, June 25, 1958.
3. Wrotnowski, A. C., *Filt. Sep.*, **5**(5), (1968).
4. Testing Felt, ASTM D461-67.
5. Wrotnowski, A. C., *Filt. Sep.*, **13**(5), 487 (1976).

Section **4.8**

Sedimentation

E. G. KOMINEK *Technical Director, Industrial Water and Waste Operations, Eimco-PMD Division, Envirotech Corporation.*

L. D. LASH *Systems Project Manager, Eimco-PMD Division, Envirotech Corporation.*

DEFINITION OF SEDIMENTATION AND FLOTATION

Sedimentation is the removal of suspended solid particles from a liquid stream by gravitational settling. There are two types of sedimentation: thickening, the purpose of which is to increase the concentration of the feed stream; and clarification, which removes solids from a relatively dilute stream. In thickening and clarification, a feed stream is pumped to a feed well and discharged ideally into a pulp concentration equal to that of the feed. As the solid particles settle, the settling rate decreases as a result of the increase in hindered settling caused by the higher concentration of solid particles.

In sedimentation, the denser particles settle through the less-dense fluid. In air flotation, the solid particles rise to the surface aided by air bubbles. The average density of the sludge is reduced to less than the liquid medium and the air/particle clump rises.

In clarification, the removal of solids from the effluent is of primary concern, i.e., clarity. In thickening, the concentration of underflow sludge is of greater importance. Classification involves splitting a mixture into two streams with one containing particles generally above a specified mesh size and the other containing particles below that size. Flotation by dissolved air (DAF) or induced air (IAF) may result in clarification or thickening, depending upon the operating conditions.

Insoluble solids may be recovered by gravity separation techniques or by air flotation. These also separate immiscible liquids such as oil in water.

In specifying a thickener or clarifier size, the major features to be considered are the cross-sectional area, the depth, and the type of mechanism. In specifying a design basis, three conditions are commonly considered:

1. Solids-handling capacity or unit area, ft^2/tpd dry solids or $lb/(h)(ft^2)$, where tpd represents tons per day
2. Overflow rate, gpm/ft^2 or $gal/(day)(ft^2)$
3. Detention time, h

Additional data required to ascertain mechanical construction and torque are:

1. Specific gravity of solids
2. Size distribution of solids
3. Underflow concentration, wt %
4. Operating temperature
5. Compression zone depth, ft
6. Geographical location

SETTLING CHARACTERISTICS OF SUSPENDED SOLIDS

Suspended solids found in water, sewage, and industrial wastes vary appreciably in size, density, and settling characteristics. Settling rates are also influenced by water temperature and viscosity. Velocities at which particles of sand and silt subside in still water are listed in Table 1. The relationship of viscosity to water temperature is listed in Fig. 1.

The settling characteristics of finer particles ranging from about 0.02 mm to submicron size are virtually impossible to predict. Laboratory tests are necessary to determine settling characteristics. Table 2 classifies sedimentation pulps.[1]

Treatment of particles approaching micron size or less requires an understanding of colloid chemistry. Lyophobic colloids are dispersions of small solid or liquid particles produced by mechanical means. Dispersions of particles less than 1 μm size are classed as sols. Larger particles are considered as suspensions.

Colloid stability depends on the forces of repulsion which exceed the forces of attraction. Properties of colloidal dispersions can be changed by changing particle size distribution by crystal growth, by comminution, or by adding electrolytes or surface active agents.

Particle charge is distributed over two concentric layers of water surrounding a particle: an inner layer of water and ions which is tightly bound to the particle and moves with it through the solution, and an outer layer which is a part of the bulk water phase and moves independently of the particle. Charges of these layers are not directly measurable, but the zeta potential, which is the residual charge at the interface between the layer of bound water and the mobile water phase, can be determined indirectly with commercially available instruments.[2]

To control coagulation by zeta potential, samples of water while being mixed are dosed with different concentrations of coagulant. Zeta potentials are then measured and recorded for floc in each sample. The dosage which produces the desired floc potential value is applied to the treatment plant. The zeta potential value for optimum coagulation must be determined for a given waste water by actual correlation with jar tests or with plant performance. A control point is generally in the range of 0 to 10 mV.

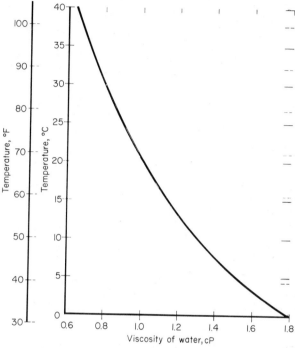

Fig. 1 Change of water viscosity with temperature.

TABLE 1 Velocities at Which Particles of Sand and Silt Subside in Still Water at 10°C and 2.65 g/cm³ Density

Diameter of particle, mm	Classification	Hydraulic subsiding rate, mm/s	Comparable overflow rate, gal/(min)(ft²)
10.0	gravel	1000.0	1475.0
1.0		100.0	148.0
0.6		63.0	93.0
0.4	coarse sand	42.0	62.0
0.2		21.0	31.0
0.1		8.0	11.8
0.06		3.8	5.6
0.04	fine sand	2.1	3.1
0.02		0.62	0.91
0.01		0.154	0.227
0.004	silt	0.0247	0.036*

*Use coagulants.

Visual observation of optimum coagulation indicates the isoelectric point. The pH at this point can usually be the control point for coagulation. It is generally found that the isoelectric point of pH is that at which zeta potential changes from negative to positive.

TABLE 2   Classification of Pulps According to Sedimentation Characteristics

Pulp description	Description of initial sedimentation	Example
Dilute, Class 1; independent particle subsidence	Particles or flocs settle independently. No definite line of subsidence. Settling unhindered. Settling rate mainly dependent upon size of particle or floc.	Turbid water and trade wastes. Silt.
Intermediate, Class 2; phase subsidence	Upper zone of independent particle subsidence. Lower zone of collective subsidence. Line of demarcation not sharp.	Chemical and metallurgical pulps. Raw sewage. Flue dust.
Concentrated, Class 3; collective subsidence or mass subsidence	Definite line of subsidence. Settling rate decreases with increasing concentration of solids. Settling rate retarded by particle or floc interference	Chemical and metallurgical pulps. Activated sludge.
Compact, Class 4; compact subsidence	Flocs or particles in intimate contact subsidence due to compression.	All pulps by sedimentation pass to this.

Flocculation and Coagulation

Coagulation is the driving together of colloidal particles by chemical forces. The process occurs within seconds of the application of the coagulating chemicals to the water or waste. Intense mixing is necessary at the point of chemical application to ensure uniform chemical distribution and exposure of the fine particles to the coagulating agent before the coagulation reaction is completed.

Flocculation is the coalescing of coagulated particles into larger particles. This involves adsorption in metallic hyroxides or bridging which is strengthened by the use of polyelectrolyte. Ideally, compaction will also occur. Flocculation is a much slower reaction than coagulation and is more dependent upon time and amount of agitation.

Practice in design has been to use about 30 s (Ref. 3) to 5 min (Ref. 4) in a rapid mix for coagulation, with relatively high-powered mixing devices. The rapid-mix operation is the initial step in the flocculation process. The flocculating agent or coagulant, commonly a hydrolyzing salt such as aluminum sulfate (alum), is dispersed by turbulent motion through the water being treated. Particle agglomeration and floc erosion are important in the rapid-mix as well as in the slow-mix operation.

Flocculation time should generally be 30 to 60 min. Formation of flocs of appreciable size or high floc volume concentration during the slow-mix operation could be inhibited by intense mixing during coagulant addition.[5]

Optimal rapid-mix design may differ significantly for different types of systems, such as those for water clarification, water softening, industrial waste treatment, and chemical treatment of sanitary sewage.[6]

The average velocity gradient $\overline{G}$ is a measure of shear intensity, with units (ft/s)/ft. When average velocity gradient is multiplied by time t in seconds, the product $\overline{G}t$ is useful as a measure of mixing.[7,8]

For baffled basins[9]:

$$\overline{G} = \sqrt{\frac{62.4\,H}{\mu\,t}}$$

For mechanical agitation:

$$\overline{G} = \sqrt{\frac{550\,P}{V\mu}}$$

where H = head loss due to friction, ft
    μ = viscosity, 16 s/ft^2
    V = volume of basin, ft^3
    P = water horsepower

Flocculating equipment should be capable of imparting the maximum velocity gradient during the process of flocculation to speed agglomeration and compaction of floc without exceeding the cohesion strength of the floc particles. The G value should be greater than 10 to promote flocculation and usually less than 75 to prevent disentegration of the floc by shear.[10] The velocity gradient G varies as the ³⁄₂ power of drive speed. Most mixing units impart much higher G values near their tips than the average in the flocculating chamber.

As a general rule, the traditional maximum tip speed values of 2 ft/s should be used where floc is weak and 4 ft/s where floc is strong. The Ten State Standards recommends peripheral speeds of 0.5 to 2 ft/s for flocculators used for water treatment and 1.5 for flocculation of sanitary sewage.

Solids Contact

Solids contact reactors typically employ variable-speed turbine-type agitators with rapid-mix and mild-agitation zones. The average design velocity gradient $\bar{G}$ for the combined zones is 100 (ft/s)/ft. Agitation in the rapid-mix zone may reach values as high as 300 (ft/s)/ft, while reaction-zone values may vary from 100 (ft/s)/ft near the entrance to very low values at the settling-zone boundary.

In a solids contact unit, effects of a single-mixing compartment subject to short-circuiting are largely overcome by the accelerated reaction resulting from the contact of influent water with slurry during the reaction.

SEDIMENTATION APPLIED THEORY

Pulp Categories

Class I   Dilute pulps settling without an interface. There are two subgroups within this class.

Class II   This is a *two-phase pulp* with a Class I upper zone and a lower zone in mass subsidence. Detention tests determine tank volume for required clarity, and area is found by the Kynch or Coe-Clevenger method.

Class III   *Mass subsidence* with distinct interface. This includes most thickener applications. The interface falls at a constant rate initially, this rate being a function of the concentration. The Kynch or Coe-Clevenger method applies.

Class IV   *Compression*—in this case, floc or particles are in intimate contact and further subsidence is mainly a squeezing function due to the weight of solids and stirring action of the mechanism.

Free-Settling Rates

The terminal velocity of solids settling in a fluid is expressed by

$$V_m = \frac{KD^2(\rho_s - \rho_l)}{\mu_b} \tag{1}$$

where V_m = terminal velocity
    ρ_s = density of solid
    ρ_l = density of fluid
    μ_b = bulk viscosity of fluid
    D = diameter of particles
    K = factor depending on concentration of solids

For free settling (low concentration of solids so they do not hinder each other in settling, Class I pulp, not flocculated) and particles large in relation to molecules of the fluid medium, Stokes' law applies:

$$V_m = \frac{(\rho_s - \rho_l)g D^2}{18\mu_b} \tag{2}$$

where g = gravitational constant.

Thus, the settling rate is proportional to the difference between the density of the particles and the fluid, proportional to the square of the diameter of the particle and inversely proportional to the viscosity of the fluid medium.

The effect of a change in density of the settling particles is proportional to the difference between the density and the surrounding fluid medium.

The effect of temperature on settling rates results from viscosity changes. For water the values of viscosity with respect to temperature are shown in Fig. 1.

When discrete particles are considered, the settling rate is proportional to the square of the diameter of the solid particle. These values can be related to the Reynolds number and friction factor. The shape of the settling particles also has a minor effect.[1]

Clarifiers which are fed dilute concentrations of suspended solids essentially follow the Stokes' law relationship. Without flocculation, the initial settling rate is linear with time. Consequently the velocity of settling in a clarifier, or conversely the allowable rise rate of liquid, may be established by a simple settling test. A plot of interface height vs. time will yield a slope which establishes the clarifier overflow rate.

CLARIFIER SIZING EXAMPLE

For a free-settling pulp (Class I), the settling rate is linear with time with regard to clarification.

Example Calculate clarifier size for a flow rate of 550 gpm.

Tests show that the pulp settled 285 mL linearly in 4 min in a 500-mL graduate 12 in high. Settling rate is $(285/500)$ $(12/4)$ = 6.84 in/4 min = 1.71 in/min. If a 50% scale-up is assumed, settling rate = 0.855 in/min (7.48 gal/ft^3)/12 in/ft = 0.53 gpm/ft^3 = 767 (gal/day)/ft^2.

Use feed well velocity of 1 ft/min, 550 gpm/(7.48 gal/ft^3) = 73.5 ft^3/min. At 1 ft/min, 73.5 ft^2 is required. A 10-ft.-diameter feedwell provides 78.5 ft^2.

$$\text{Clarifier area} = 500 \text{ gpm}/[0.53 \text{ gal}/(\text{ft}^2)(\text{min})] = 1032 \text{ ft}^2.$$

Adding the feed well, 1111 ft^2 are required. A 38-ft-diameter tank provides 1134 ft^2.

If floating material is present, add a skimmer and place a scum baffle 1 ft in from the edge of a 40-ft-diameter tank.

THICKENING THEORY

Coe-Clevenger Method

When solids are concentrated so that hindered settling exists (distinct-interface fall in settling test), every particle or floc in a particular concentration layer ideally settles at a single settling rate. The settling rate is then a function of solids concentration. In a continuous thickener, flow of solids to the underflow is made up of the following two components:

1. CU/A = underflow pumping rate

where C = solids concentration at any particular level in the thickener
U = underflow pumping rate, volume per unit time
A = thickener area

2. VC = flux of solids with respect to the pulp

where V = settling rate of solids (function of concentration)
At any layer, total flux G is as follows:

$$G = \frac{CU}{A} + VC \tag{3}$$

where G = solids mass per unit time per unit area. At steady state,

$$G = \frac{C_u U}{A} \tag{4}$$

where C_u = concentration of underflow.

Combining Eqs. (3) and (4) to eliminate U, we find

$$G = \frac{V}{(1/C) - (1/C_u)} \tag{5}$$

By definition,

$$\text{Unit area} = \frac{1}{G} = \frac{(1/C) - (1/C_u)}{V} \tag{6}$$

This is the Coe-Clevenger expression, normally written as follows:

$$\text{U.A.} = \frac{1.333(F - D)}{R(\text{sp.gr.})_L} = \frac{4(F - D)}{3R(\text{sp.gr.})_L}$$

where U.A. = unit thickening area, ft²/tpd

F = a particular "dilution" at a concentration between feed and underflow in pounds of liquor per pound of solid

D = underflow dilution, pounds of liquid per pound of solid

R = settling rate of interface at dilution F

4/3 = 1.333 = units conversion factor

$(\text{sp.gr.})_L$ = specific gravity of liquid

If a critical layer exists in a thickener, solids passing equals G at steady state. However, steady state is virtually impossible to maintain since no thickener can be exactly at capacity for any time period. Therefore, either the feed rate exceeds the critical G and all concentrations will build upward eventually overflowing feed (thickener area inadequate), or feed = G, at which no critical layer exists. Sufficient area must be supplied so that feed tonnage can pass through all concentrations to the underflow.

The Coe-Clevenger method consists of determing R at a large number of concentrations to determine the limiting case (critical unit area). See Fig. 2.

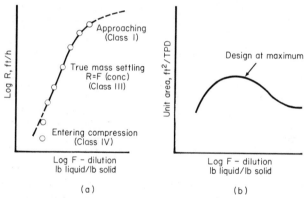

Fig. 2 (*a*) Hindered settling by the Coe-Clevenger method; setting rate vs. dilution. (*b*) Hindered settling by the Coe-Clevenger method; unit area vs. dilution.

Kynch Method

Kynch also assumed settling rate to be a function of concentration only within the hindered or "zone-settling" region. In batch settling, Kynch showed that intermediate concentrations between initial (C_0) and underflow (C_u) propagate upward beginning at the bottom of the graduate. Such concentrations must move upward since solids cannot settle from a layer as fast as they enter. By analyzing the material balance around a layer of constant concentration, Kynch showed that this layer propagates upward at constant velocity until it reaches the liquid-pulp interface. Using this conclusion, Kynch was able to determine the solids concentration at the interface for any point on the settling curve.

Figure 3 illustrates how the Kynch analysis is employed to develop the Coe-Clevenger relationship

$$\text{U.A.} = \frac{1/C - 1/C_u}{V}$$

From this, the following expression is developed:

$$\text{U.A.} = \frac{t_x}{C_0 H_0} \quad \text{ft}^2/\text{tpd}$$

where t_x = time, days
  C_0 = initial concentration, tons of solids per cubic foot of pulp
  H_0 = initial height, ft

The selection of "critical point" to apply the Kynch method is empirical. Experience has shown the proper point is the approximate point of most rapid change in slope.

In thickening of biological sludges, pounds per day per square foot of surface area is conventionally used.

Scale-up factors applied to bench scale Coe-Clevenger or Kynch tests range from 1.2 to 1.5.

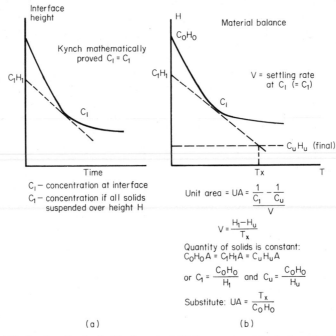

(a)   (b)

Fig. 3 (a) Hindered settling by the Kynch method. (b) Construction of the thickener holdup time.

Thickener Depth Considerations

Compression Class IV pulps are rarely encountered. However, all pulps will eventually pass into compression before leaving a thickener. It is necessary to provide sufficient time (volume) in the lower section of the tank to compress the pulp to the desired concentration. This volume can be determined by graphically integrating the settling curve from the start of compression to the final height.

Thickener Depth Thickeners for Class III pulps are generally quite shallow by comparison with dilute flocculated pulps. The side water depth (SWD) is typically computed as follows:

Feed well immersion	2 ft
Mass subsidence	2 to 6 ft (lower values for fast-settling pulps)
Compression zone	As calculated (generally use 3 ft)

THICKENER SIZING CALCULATION

Problem Thicken a leached ore and calculate thickener size.

The settling test vs. time is performed in a 2-L graduated cylinder. Picket rakes are inserted to maximize the underflow concentration.

A graph showing the relationship between interface height and time is drawn to observe the knee of the curve, as shown in Fig. 4.

The critical point is generally determined by inserting the curve which is the bisector of the angle at the intersection of tangents to the extremities of the curve. However, certain pulps thicken slower, as indicated by a greater curvature to the right of the bisector. Therefore we will draw a tangent to the curve, shown in Fig. 4, at the critical point. This is extended to the desired underflow operating concentration, thus giving the settling time t_x. The unit area is determined from U.A. $= t_x/C_0H_0$. However, a safety factor of 1.35 is used in this case based on judgment and past experience.

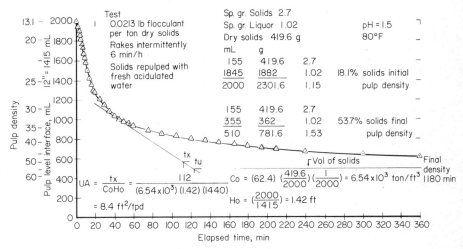

Fig. 4 Experimental calculation of thickener size from sedimentation test.

As seen in Fig. 4, t_x for an underflow concentration of 50% solids is 112 min.

The unit area required is 8.4 ft²/tpd. Utilizing a "highest" tonnage condition of 1000 tpd, the area required would be 8400 ft² by the Kynch method.

A 105-ft-diameter thickener would provide 8660 ft², so this size mechanism will be selected.

The hydraulic loading of a thickener or clarifier may be obtained from the initial linear portion of the settling curve as demonstrated earlier under clarifier size. This rate is not controlling.

The depth of solids that may normally be maintained in a compression zone in a thickener is limited to 3 ft. An additional 5 ft for clarification and transition zones should be adequate. Therefore, an 8-ft SWD was chosen.

Flotation Theory

The flotation process consists of attaching fine gas bubbles to suspended or oily material to reduce specific gravity. Two types of flotation are commonly used: dissolved-air flotation and induced-air flotation.

Dissolved-Air Flotation In dissolved-air flotation, micron-size bubbles are produced by dissolving gas into water at elevated pressures followed by subsequent release to atmospheric pressure. The dissolved gas in excess of saturation is released as extremely fine gas bubbles.

The quantity of gas which will theoretically be released from solution when the pressure is reduced to atmospheric is calculated from the following equation:

$$S = S_g \frac{fP}{14.7} - 1$$

where S = gas released at atmospheric pressure, mg/L
$\quad\quad\ S_g$ = gas saturation at atmospheric conditions, mg/L
$\quad\quad\ P$ = absolute pressure, lb/in² (abs)
$\quad\quad\ f$ = fraction of saturation achieved in the pressurization tank

The amount of air required for effective flotation depends upon the nature of the solids to be removed by flotation. It is generally based upon the air/solids ratio and ranges from 2 to 4% by weight for most applications.

Hydraulic loading rates in flotation applications depend upon the characteristics of the solids to be removed and the effluent quality which is required. (Compliance with EPA guidelines generally requires biological treatment following flotation.) Design rates most commonly used for removal of oily material, organic solids, and fibrous solids would range from 2 to 3.5 gal/(min)(ft²). Removal of flocculant material after coagulation generally requires a separation rate in the range from 1.5 to 2.0 gal/(min)(ft²).

Dissolved air flotation is used for the thickening of waste activated sludge. Flotation thickeners are generally designed for a solids loading rate of 1 to 2 lb/(ft²)(h). The float concentration will range from 3 to 4% dry solids.

Dispersed-Air Flotation   Dispersed-air flotation, or induced-air flotation, has been used for over 50 years for the separation and selective recovery of minerals in the mining industry. It has found wide application for removing oil from oil field brine and ballast water and for other applications where floatable solids can be separated by contact with a large volume of small bubbles.

Dispersed-air flotation induces air into water with motor-driven rotors. Air is applied in the range from 1 to 10 ft³/gal. If suitable polymers, which may be cationic or anionic, are used for coagulation, emulsified oils and colloidal solids can also be removed.

Dispersed-air flotation units have a retention time of about 4 min. This permits installation in locations where other types of equipment could not be used.

Float is removed by means of rotating skimmer paddles which move floating solids over weirs into discharge launders.

Dispersed-air flotation units offer the advantages of minimum space requirements and the production of float which does not contain precipitated metal hydroxides. The effectiveness and the economics of this type of treatment can be demonstrated in either laboratory-model or pilot-plant units.

EQUIPMENT DESIGN FEATURES

Gravity Separation

Horizontal flow units, both rectangular and circular, are most often used for sedimentation applications in the United States.

Flow through a rectangular tank enters at one end, passes a baffle arrangement, and traverses the length of the tank to effluent weirs. In narrow tanks, longitudinal collectors scrape sludge to single or multiple hoppers at one end. In tanks with multiple wide bays, the longitudinal collectors scrape sludge to a cross collector which then moves the sludge to a central hopper.

Circular designs use circular center feed with a scraper sludge removal system (Fig. 5) or hydraulic suction sludge removal system (Fig. 6). Circular-rim-feed clarifiers with rim takeoff or center takeoff and a hydraulic suction sludge removal system are also used.

The basic parameter to which settling tank performance is related is the surface hydraulic loading. Performance is optimized when there is quiescent or nonturbulent

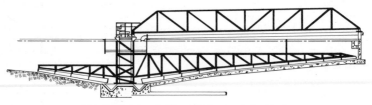

Fig. 5 Circular clarifier with long and short arms.

flow, uniform distribution of velocity over all sections normal to general flow direction, and no resuspension of particles which have settled. Currents induced by inlets, outlets, wind and density differences, or temperature variations can cause short-circuiting and retard settling. Turbulence due to high velocities also retards settling.

If a clarifier is designed for the removal of flocculant solids, the velocity in the influent channel should be kept in the range from 0.5 to 2.0 ft/s. The velocity in the feed well outlet should be in the same range, particularly if fragile floc, such as aluminum hydroxide or activated sludge, is to be removed.

A gravity-type oil-water separator will operate correctly only if the flow is in the laminar range. The flow in a separator 20 ft wide and 6 ft deep would be limited to a velocity of less than 0.08 ft/min to assure a Reynolds number of 2000, considered the upper limit of laminar flow.[1,11]

Oil rises through water even though the flow is turbulent, and when it reaches the surface, it is trapped as part of the oil film unless there is sufficient turbulence to reentrain the oil. The controlling parameter is surface area per unit flow. This concept has been

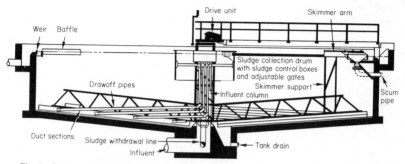

Fig. 6 Suction-type clarifier for biologically degradable sludges. *(Envirotech Corp.)*

demonstrated to correlate well with operating data.[12] A correlation is also made with the corrugated-plate interceptor (CPI) type separator (Heil Process Equipment Corporation) which uses settling plates. (See Fig. 10.) When the plate area is added to the surface area, there is a close correlation with data plotted for American Petroleum Institute (API)-type gravity separators. Complete design of the latter is described in Ref. 13.

Treatment units range from flocculating clarifiers to solids-contact units capable of maintaining up to 5% of precipitated slurry in suspension. The flocculating clarifiers have the advantage of floc formation in the center compartment of a clarifier (Figs. 7 and 8). This avoids floc breakup in transfer and thereby produces a better effluent in most cases.

Sludge blanket units incorporate the features of a flocculating clarifier with a separation zone of increasing area so that a sludge blanket can be maintained in suspension in the lower portion of the outer clarification zone. The sludge blanket is helpful in screening out fine floc particles so that a higher quality effluent will be obtained.

In a solids-contact unit, the raw water or waste is brought into immediate contact with a large volume of circulating and relatively dense previously formed floc or precipitate (Fig. 9). Solids contact induces more rapid coagulation with floc growth. In precipitation reactions, the new precipitate is formed on the existing particles and supersaturation is

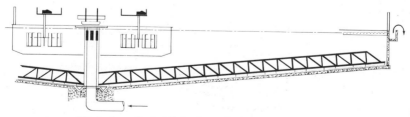

Fig. 7 Large feed well clarifier with flocculating turbines.

minimized. The mixture is sent upward into a reaction cone with 75 to 95% being returned to recirculation with incoming raw water or waste. The remaining 10 to 25% passes under the cone and into the clarification zone.

The slurry level or sludge bed may be carried above or below the bottom of the cone, depending upon which arrangement gives the best results. The recirculation drum draws the heaviest solids from near the tank bottom center, ensuring the recirculation of the densest particles. Mechanical recirculation of a large volume of solids means that change in influent rate does not cause the sludge bed to appreciably drop down, rise up, or blow out.

Design features of solids-contact clarifiers should include[14]:

1. Rapid and complete mixing of chemicals, incoming water, and slurry solids must be provided. This should be comparable to conventional flash-mixing capability and should provide for variable control of $\check{G}t$, usually by adjustment of recirculator speed.

2. Mechanical means for controlled circulation of the solids slurry should be provided with at least a 3:1 range of speeds. The maximum peripheral speed of mixer blades should not exceed 6 ft/s.

3. Means should be provided for measuring and varying the slurry concentration in the contacting zone up to 50% by volume.

4. Sludge discharge systems should allow for easy automation and variation of volumes discharged. Mechanical scraper tip speed should be less than 1 ft/min with speed variation of 3:1.

5. Sludge blanket levels must be kept at a minimum of 5 ft below the water surface.

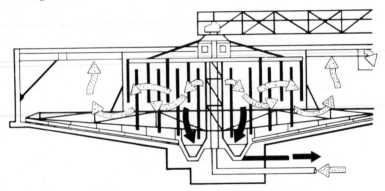

Fig. 8 Flocculating-type reactor clarifier unit. (*Envirotech Corp.*)

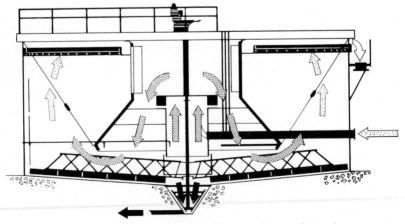

Fig. 9 Solids contact reactor clarifier unit. (*Envirotech Corp.*)

6. Effluent launders should be spaced so as to minimize the horizontal movement of clarified water.

The determination of actual settling rates in any specific design situation will reduce the need for applying a high factor of safety in the design of the sedimentation area. However, as a practical matter, to compensate for the many variables which cannot be predicted or included in a test procedure, a factor of at least 2:1 should be employed.[9]

Solids contact units can be designed with gas diffusers for introducing air or carbon dioxide into the reaction compartment. For cold weather operation, a submerged takeoff can be used so that an ice cap formation on the surface will not interfere with operations. Skimmers can also be installed to remove floating floc, thereby maintaining a higher-quality effluent.[9]

Weir overflow rates will also affect effluent clarity. Weir loading rates range from 10 gal/min per foot of weir length, when light floc particles are being separated, to 20 gal/min per foot, when denser solids, such as calcium carbonate, are being separated.

Unit design to maintain these loading rates requires radial launders on high-rate units and annular launders on others, depending upon the design separation rate.

The overflow weir length required for rectangular clarifiers usually cannot be obtained with a single weir. It is often necessary to provide effluent launders for about one-third of the outlet section of the clarifier.

The bottom of a clarifier is generally designed with a slope of $\frac{1}{12}$ toward the sludge hopper. Rotating scraper trusses are used in circular clarifiers. If the mechanisms are located in square tanks, corner sweep mechanisms attached to the rotating trusses are used. In rectangular tanks, light scrapers connected through chain-and-sprocket drive mechanisms are necessary. If oil or floating solids removal is necessary, scum baffles and skimmer mechanisms are used.

The removal of nitrified biological sludge or biological sludge which tends to bulk favors the use of a hydraulic suction-type solids-removal mechanism (Fig. 6). This is connected to the rotating scraper mechanism with pipe connections into a center compartment. The difference in water level between the compartment and the clarification section is controlled to regulate the hydraulic removal of settled sludge. This design is desirable for separating nitrified biological sludge. Otherwise, with warm water, anoxic bacteria will reduce nitrates to nitrogen, releasing gas bubbles which adhere to floc particles and cause them to rise to the surface.

Scraper Design

As a general rule, scraper design torque in foot-pounds is based upon a multiplier times the tank diameter squared. The multiplier will range from 3, where the removal of light solids such as alum floc or activated sludge is involved, up to 25 for heavy sludges such as in primary clarifiers used in the pulp and paper industry. For heavy-duty service, the scraper drives can be provided with manual, semiautomatic, or automatic lifting mechanisms so that the scraper trusses can be raised if the load of settled solids exceeds the torque capability of the scraper drive.

The use of traction drives, common in Europe, is gaining acceptance in the United States because of the lower design torque requirements. With this design, the scraper truss is driven from the tank wall by a rotating drive unit riding on a track or concrete pad.

Sludge removal from circular clarifiers is most common from a center hopper by means of a sludge pipe under the tank bottom. Some units are designed with an access tunnel below the clarifier. However, OSHA standards require a "see-through tunnel." This adds to the cost of the installation. An alternative is to use center column pumping, with sludge pumps located either in the sludge hopper inside of a caisson, or on the clarifier bridge with suction lines extending down to the sludge hopper.

Clarifier Sizing

Clarifier design rates range from about 250 to 1200 gal/(day)(ft²), depending upon the application and the design water temperature. If a clarifier must handle peak flow rates for several-hour intervals, for example, for the treatment of storm water, the sizing should be based upon peak conditions, taking into consideration the effluent suspended solids concentration which can be tolerated. The EPA guidelines have established a 24-h maximum and 30-day average suspended solids concentrations for most industries. These must be considered in establishing clarifier tank diameters.

Clarifier tank depth depends upon certain factors which are unpredictable. These include density currents, eddy currents and the operation of the sludge removal mechanism. It is recommended that velocity through a sedimentation tank should be between 0.5 and 3 ft/min (Ref. 9).

In the absence of testing, the following typical surface loading rates may be used for sizing tanks:

	Peak Surface Loading,
Chemical	*gal/(day)/(ft²)*
Alum	500–600
Iron	700–800
Lime	1400–1600

Lamella Plate or Tube Separators

Separation rates of solids from liquids may be enhanced by use of inclined planes or tubes. Increases in operating times by 1.4 and increases in solids-flux-handling capacity of 2.4 were obtained in semicontinuous and continuous settlers, respectively, equipped with inclined planes over the same settlers without planes.[15] A layer of clear liquid forms immediately under the inclined plane. The pressure difference, between the suspension (dense) and liquid, forces the clear liquid layer to rise very rapidly, improving separation rates compared with conventional sedimentation devices. The theoretical basis and design calculations are given by Rubin and Zahavi.[15]

Tube and plate settlers are also being used in sedimentation-type units. The presence of inclined tubes increases settling rates by a factor of 1.4 to 2 or higher compared with settling in a conventional vertical vessel under the same conditions. For most new applications, utilization of tube modules may reduce sedimentation volume by one-third or one-half that required by conventional basins.

Installation of tube or plate modules in existing clarifiers will increase the treatment capacity, if they can handle the higher hydraulic load. For new installations, space requirements and the installed cost of the two types of units will be the major factors for evaluating tube- or plate-type settlers against conventional sedimentation units.

Improved feed well design will partially offset the improvements provided by lamella or tube settling devices.

The use of inclined tubes to increase settling rates of solids has been demonstrated on a plant scale.[16] The separation of activated sludge with a density little different from water (about 1.05) was improved.

A corrugated-plate interceptor (CPI) is shown in Fig. 10.

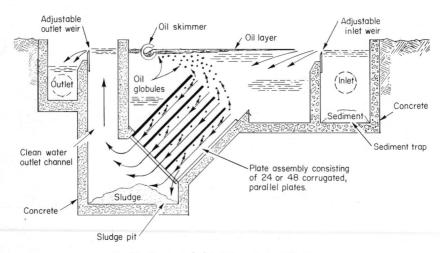

Fig. 10   Corrugated-plate interceptor or CPI oil separator.

Waste water enters a separator bay and flows downward through corrugated plates arranged at an angle of 45° to the horizontal. Oil collects on the underside of the plates and rises to the surface where it is skimmed. Solids settle into a sludge compartment, with the clarified waste discharging into an outlet channel.

Parallel-Plate Oil Separator

The parallel-plate separators are reported to achieve higher oil removal than API separators having the same area. Thus, for the same amount of oil removal, the parallel-plate separator should permit the use of a smaller installation.

The parallel-plate separator reduces the distance that a particle of oil must travel before reaching a collection surface. This consists of a number of parallel plates set at an angle approximately 4 in apart. The oil particles coalesce on the underside of these plates and creep up to the surface of the water. Thus, the maximum distance an oil droplet has to travel before being trapped and coalesced is a few inches.

Dissolved-Air Flotation

In dissolved-air flotation (DAF), air is intimately contacted with an aqueous stream at high pressure which dissolves the air. The pressure on the liquid is reduced through a back-pressure valve, thereby releasing micron-sized bubbles that sweep suspended solids and oil from the polluted stream to the surface of the air flotation unit. Applications include treating effluents from metal finishing, pulp and paper, cold rolling, poultry processing, grease recovery in meat-packing plants, cooking-oil separation from French-fry processing, and refinery API separators. An increasingly important application is the thickening of sludge.

Attachment of gas bubbles to suspended solids or oily materials occurs by several methods, as illustrated in Fig. 11. The suspended solids/gas mixture is carried to the vessel surface after: precipitation of air on the particle; trapping of gas bubbles, as they rise, under a floc particle; and adsorption of the gas by a floc formed or precipitated around the air bubble.

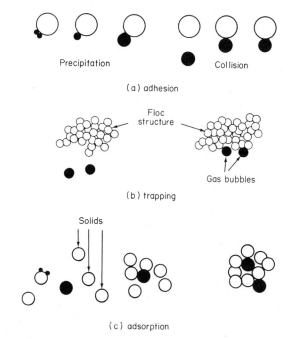

(a) adhesion

(b) trapping

(c) adsorption

Fig. 11 Mechanisms of gas bubble attachment to suspended solids or oily materials.

To dissolve air for flotation, three types of pressurization systems are used. These are illustrated in Fig. 12.

Full-flow or total pressurization is used when the waste water contains large amounts of oily material. The intense mixing occurring in the pressurization system does not affect the treatment results.

Partial-flow pressurization is used where moderate to low concentrations of oily material are present. Again, intense mixing by passage through the pressurization systems does not affect treatment efficiency significantly.

The recycle-flow pressurization system is for treatment of solids or oily materials that would degrade by the intense mixing in the other pressurization systems. This approach is used after chemical treatment of oil emulsions or for clarification and thickening of

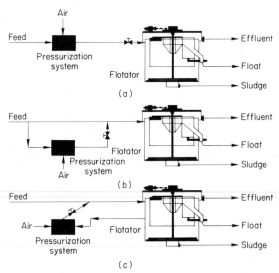

Fig. 12 Pressurization methods for dissolved-air flotation. (*a*) Total; (*b*) partial; (*c*) recycle.

flocculent suspensions. As shown in Fig. 13, the pressurized system is the clarified effluent. Air is dissolved in this stream at an elevated pressure, and mixed with the feed stream at the point of pressure release.

Mixing of streams before they enter the flotation zone results in intimate contact of the pressurized air/water mixture with the suspended solids to effect efficient flotation.

The amount of pressurization flow is based on the air-to-solids weight ratio required for treatment. Ratios ranging from 0.01 to 0.06 have been demonstrated effective. A design air-to-solids ratio of 0.02 is appropriate for many applications.

A simplified drawing of the action of dissolved air flotation is shown in Fig. 13. The solids-laden or oily water influent mixture enters the flotation vessel, and the air-solids mixture rises to the liquid surface. The air-solids mixture has a specific gravity less than that of water. Solids having a specific gravity greater than that of water tend to settle to the bottom and are removed by a rotating scraper arm. Attached to the same shaft is a rotating skimmer blade that removes the floating matter from the surface of the vessel into a skimmings hopper. Clean water passes underneath a skirt and then must leave the vessel through a launder, which is located in the peripheral region.

A portion of the effluent water is recycled for pressurization. Compressed air is introduced into the discharge of the recycle pump, and intimate contact with the water is achieved in the aeration tank. Maximum solubilization efficiency is important at this point. The aerated recycle water is then returned through a back-pressure valve, where the pressurized air is released, and mixed with the influent for flotation.

Flocculants such as synthetic polymers may be used to improve the effectiveness of dissolved-air flotation. Also coagulants such as filter alum may be used to break emulsified oils and to coagulate materials for improved flotation recovery.

A combination of flotation and clarification is achieved in the same vessel by the ClariFlotator clarifier, Fig. 14. Feed, pressurized with air, enters at the bottom and disperses into the top flotation-disk compartment. After pressure release, flotation by small bubbles takes place, with solids and oil carried to the surface. The concentrated float is gathered by skimmers into the scum box. Removal to discharge is by a screw-flight conveyor.

The partially settled effluent from the flotation-disk region passes under the baffle and is then separated in the annular settling zone. Fine particles, after release of the flotation bubbles, settle in this clarification region and are plowed to the center for discharge as a sludge out the bottom. Also, the dense particles that settle initially move to the bottom of the equipment as part of the sludge.

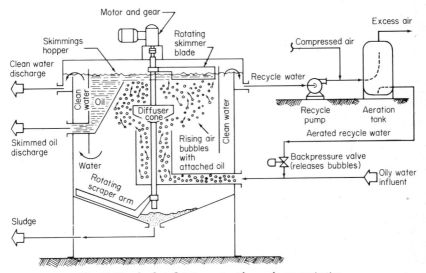

Fig. 13 Dissolved-air flotation unit with recycle pressurization.

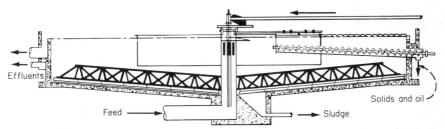

Fig. 14 ClariFlotator clarifier combines flotation and clarification in the same unit. (*Envirotech Corp.*)

Induced-Air Flotation

Flotation by induced air uses devices similar to minerals beneficiation-flotation machines. Air is drawn into the cell by action of the rotor, is mixed with the water, and transformed into minute bubbles. Oil particles and suspended solids attach themselves to the gas bubbles and are borne to the surface of the water. Skimmer paddles push the contaminated froth from the top of the cell into collection launders.

Properly conditioned oily wastes leave nearly 100% oilfree, and in most cases suspended solids are reduced significantly.

A typical design is a four-cell unit, each cell having a retention time of 1 min. Capacity may be determined by running bench tests on the water under consideration. These tests also determine the amount and type of chemical reagent best suited for a particular application.

The operating mechanism is illustrated in Fig. 15. The shaft, driven through a V-belt drive, extends into the tank, at the end of which is mounted a star rotor. A steel standpipe extends down from the bearing assembly plate and has attached to it the disperser and disperser hood. Rotor speed decreases as the size of machine increases, in order to maintain a constant tip speed. Typical rotor speed varies from 437 to 186 r/min.

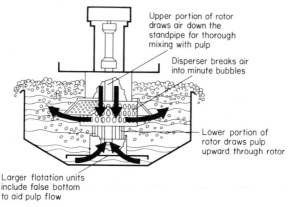

Fig. 15 Induced-air flotation unit. *(Wemco Division, Envirotech Corp.)*

INSTRUMENTATION AND CONTINUOUS PROCESS DESIGN FOR SEDIMENTATION

Instrumentation of Thickeners and Clarifiers

Mechanical torque indicators are supplied on all thickeners and clarifiers. The worm shaft is allowed to float against a calibrated spring; this motion is then utilized to provide torque indication. A continuous torque transmitter and recorder can also be provided.

Motorized lifting devices actuated automatically according to torque can be supplied for both superstructure-supported (type B) and stationary-center-column-supported (type C) thickeners.

Underflow density recorders of the gamma-gauge type can be supplied to control underflow pumping rate and/or feed rate. Mass flow can be recorded by combining the gamma gauge with a flowmeter. Pulp level may be sensed and used as a control, as shown for bubbler tubes.

Thickener Layout

Tanks are generally constructed of concrete, steel, or a combination of these. Earthen bottoms are common in the larger sizes where ground conditions permit. If an impervious slime bed cannot be formed by material settled, the earth basin may be grouted with soil cement or asphalt. Submerged mechanism parts may be steel, rubber- or neoprene-covered, wood, stainless, Hastelloy C, or even titanium, depending on corrosion conditions.

Covered thickeners are common where heat retention is necessary to the process. An inexpensive cover with the drive and mechanism supported from the roof is available.

Arm designs depend on torque required, sand-out problems, presence of gelling muds, scaling tendencies, and corrosion conditions.

Feed wells may be quite simple with thickeners where flows are low and feed tends to plunge and stabilize readily. However, clarifiers require more sophisticated velocity head dispersion to ensure maximum tank effectiveness.

Thickener and clarifier auxiliaries consist primarily of pumping systems. Underflow is commonly removed by the "standard" radial tunnel. Many designs are available to avoid the expense of tunnels, including encased buried pipe, center-column submersible pumps, bridge-mounted pumps with suction lines down through the center column, and underflow piping laid in a radial trench at the bottom.

Where flocculant addition is required, care should be taken to ensure complete dispersion of chemicals, proper addition points, and inclusion of detention for floc growth, if critical.

Process Design Considerations

Personnel attention necessarily varies with the regularity of conditions, the importance of underflow concentrations and overflow clarity, the degree of excess capacity, and the need for flocculants. Maintenance of proper underflow pumping rate to ensure steady state and maximum density requires most of the attention given. Flocculant stock preparation and rate control requires from ½ to 2 h per day. Thickener care is normally included in the duties of a person employed principally in other work. Repairs on other than underflow pumps are negligible, provided that proper torque is supplied with the machine and reasonable safety precautions are maintained—90% of repair costs are a result of carelessness.

Countercurrent decantation (CCD) systems for pregnant liquor recovery with minimum wash dilution are utilized primarily with high-flow systems. From two to six or more stages may be employed. To achieve maximum effectiveness, two predominant features require attention: First, completeness of blending liquors with settled pulp between settling stages determines the "efficiency" of the system. Some pulps are so fluid that simple baffling in radial launders is adequate for mixing. Other pulps require interstage detention tanks with mixers. Improvement in repulp efficiency from 80% to 90% can result in a savings of over $150,000/year. Second, maximizing underflow concentration from each stage is extremely important. Recovery can be increased tremendously with only a 1 or 2% increase in underflow concentration. Adequate thickener size, a suitably designed mechanism, and operating care ensure maximum concentrations.

When process conditions do not permit exposure of liquor to the atmosphere, a simple labyrinth liquid seal may be provided for the drive shaft through the thickener cover.

Surface skimmers are provided on many thickeners and clarifiers when the flow has a tendency to float, or froth, or when oil is present. A simplified arrangement includes skimmer, scum scraper, scum box, and scum baffle.

Temperature effects such as surface freezing can occur where water temperature is near zero, but unless there is a floating problem, the thickener will continue to operate as long as flow is maintained. Freezing at the walls can be a problem in that ice buildup can stall the mechanism. In adverse climates, concrete walls will help minimize wall freeze—or the tanks may be insulated or buried. Drivehead protection in particularly severe climates may be provided simply by proper enclosures. Thermal currents due to convection and wall gradient temperatures can adversely affect clarification but will normally cause no problems with thickener applications.

COUNTERCURRENT DECANTATION CIRCUITS

Countercurrent decantation (CCD) is a method for separation of soluble material from insoluble by successive thickening and washing solids countercurrently.

The technique maintains concentrated liquor, yet efficiently washes the tailings. Also, CCD is effective for washing a chemically precipitated product, generating a minimum of washwater or wastewater.[1]

A common application using sedimentation is a series of gravity thickeners. Alternately, vacuum filters may be used.

The feed slurry is introduced into the first thickener. The solids are thickened to reduce soluble material (in the underflow liquor with the solids) to a minimum. The clarified overflow, containing most of the dissolved solids, is removed. However, in a single stage, the underflow contains a large fraction of the soluble components. Additional stages of thickeners are added to reduce this loss of solubles. Fresh washwater is added only to the last thickener stage. The amount of washwater compared with the feed solids is the wash ratio.

EXAMPLE

An example of CCD calculations is shown in Fig. 16: a uranium mill with 500 tpd solids at 50% solids contains 500 tpd leach liquor.

Set X as the units of soluble uranium (on a U_3O_8 basis) which leaves with the final thickener. Assuming a fresh water wash ratio of 2 to 1, 1000 tpd of fresh water enters the final stage. The example has 55% solids in each thickener underflow, a function of ore characteristics, grind, and other factors.

A liquid material balance around the last stage is as follows, where subscript u is underflow, subscript e is overflow (or effluent), and Q is the mass flow per unit time.

$$Q_{5u} + Q_{7e} = Q_{6e} + Q_{6u}$$
$$\frac{500(0.45)}{0.55} + 1000 = Q_{6e} + \frac{500(0.45)}{0.55}$$
$$Q_{6e} = 1000$$

Around any stage:

$$Q_{(n-1)u} + Q_{(n+1)e} = Q_{ne} + Q_{nu}$$

The dissolved uranium balance is:

$$Q_{5u}C_{5U_3O_8} + Q_7 C_{7U_3O_8} = Q_{6e}C_{6U_3O_8} + Q_{6u}C_{6U_3O_8}$$
$$408(C_{5U_3O_8}) + 0 = 1000\,(C_{6U_3O_8}) + 408\,(C_{6U_3O_8})$$
$$1408C_{6U_3O_8} = 408C_{5U_3O_8}$$

Solving for the soluble concentration in stage 5 in terms of X, we find

$$C_{5U_3O_8} = 3.45X$$

Working through each stage, we find relative amount of soluble U_3O_8 as a multiple of X. Finally, comparing the amount of uranium entering (398 X) with that leaving (x), we determine the soluble loss [(1/398) 100% = 0.25% soluble loss]. The recovery is 1 − soluble loss = 99.75% for six stages.

Calculations are facilitated with a programmable hand calculator. A method for optimizing the number of CCD stages and one for countercurrent leaching[17] has been developed.

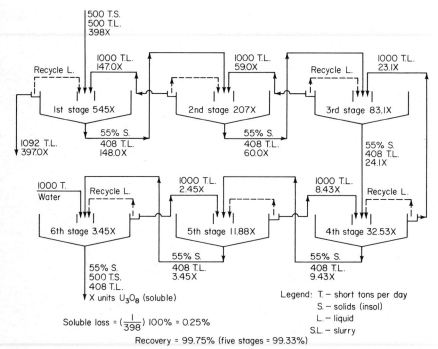

Fig. 16 Countercurrent decantation calculations. Pregnant liquor recovery with six stages CCD and 2 lb water/lb ore; 55% solids in all underflows.

Selection of the number of stages is an economic decision logically based on capital and operation costs plus marginal improvement in recovery per stage added.

The method assumes complete mixing between stages and complete operator effectiveness in controlling conditions. To more accurately represent actual operating conditions an additional stage may be added. (Assume six actual stages produce the results of five theoretical stages.) Mixing between stages can sometimes improve CCD effectiveness. An alternate calculation method introduces a stage efficiency, which may be around 80%.

REFERENCES

1. Perry, R. H., and Chilton, C. H., "Chemical Engineer's Handbook" 5th ed., McGraw-Hill, New York, 1975.
2. "The Laser Zee Meter," Pen Kem Co., Groton-on-Hudson, New York.
3. Hudson, H. E., Jr., and Wolfner, J. P., *J. Am. Water Works Assoc.*, **59**, October 1967, pp. 1257–1267.
4. Letterman, R. D., Quon, J. E., and Gemmell, R. S., *J. Am. Water Works Assoc.*, **62**, pp. 652–658.
5. Camp, T. R., *J. Am. Water Works Assoc.*, **60**, June 1968, pp. 656–673.
6. Vrale, L., and Jordan, R. M., *J. Am. Water Works Assoc.*, **63**, January 1971, pp. 52–58.
7. Camp, T. R., *Trans. ASCE*, **120**(1), 1–16 (1955).
8. Lash, L. D., and Kominek, E. G., *Chem. Eng.*, Environmental Engineering Deskbook.
9. "Water Treatment Plant Design," American Water Works Association, New York, 1969.
10. Shell, G. L., personal communication.
11. Standards of the Hydraulic Institute, 12th ed., Hydraulic Institute, New York, 1969.
12. Thompson, S. J., *Hydrocarbon Process.*, October 1973, pp. 81–83.
13. Manual on Disposal of Refinery Wastes, American Petroleum Institute, Washington, D.C., 1969.
14. Process Design Manual for Suspended Solids Removal, EPA 625/1-75-003a, Environmental Protection Agency, Washington, D.C., 1975.
15. Rubin, E., and Zahave, E., Enhanced Settling Rates of Solid Suspensions in the Presence of Inclined Planes, AIChE Meeting, Houston, March 16–20, 1975.
16. Hansen, S. P., Culp, G. L., and Stukenberg, J. R., *J. Water Poll. Control Fed.*, **41**, August 1969.
17. Chen, N. H., *Chem. Eng.*, **77**, August 24, 1970, pp. 71–74.

Hydrocyclone Separation

ROGER W. DAY *Marketing Director, Solids Control Operations, Geosource, Inc.; formerly Planning Manager, Bird Machine Co.; B.S.Ch. E., Cornell University; member, A.I.Ch.E., S.P.E., and the Filtration Society.*

CHARLES N. GRICHAR *Supervisor of Application Engineering, Solids Control Operations, Geosource Inc.; B.S.Ch.E., University of Houston; member, Filtration Society and A.I.Ch.E.*

INTRODUCTION

The hydrocyclone is a process device which utilizes high centrifugal forces to separate solids from liquid. Usually the hydrocyclone consists of an upper cylindrical section fitted with a tangential feed connection and a lower conical section having a bottom concentric apex opening for solids discharge. The closed top of the upper cylindrical section has a downward protruding vortex finder pipe, ending below the tangential feed location, through which the liquid fraction discharges from the device.

Pressurized slurry is introduced to the hydrocyclone tangentially through the feed connection which creates a downward spiraling action developing high centrifugal forces. As the fluid moves downward, the solids in suspension are driven by the centrifugal force to the inner cyclone wall and progress towards the apex opening at the bottom of the conical section. At the apex, the solids, along with a small portion of liquid, continue on and leave the device by virtue of their inertia. As the apex is approached, the fluid phase reverses axial direction and spirals upward, exiting the hydrocyclone through the vortex finder pipe. Since those solids removed via the apex have been eliminated, the overflow fluid is substantially free of solids or, at a minimum, devoid of the larger or heavier solids fractions. Figure 1 illustrates a typical hydrocyclone in operation.

The intrinsic simplicity of the hydrocyclone suggests its consideration whenever solids/liquid separation problems are encountered. Low initial capital, minimum operating cost, and little maintenance can result in real process economy. In some instances the hydrocyclone by itself can accomplish the necessary separation requirements. Sometimes it may

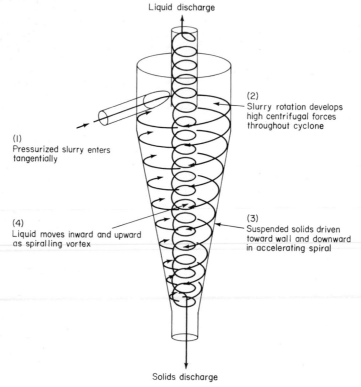

Liquid discharge

(2) Slurry rotation develops high centrifugal forces throughout cyclone

(1) Pressurized slurry enters tangentially

(4) Liquid moves inward and upward as spiralling vortex

(3) Suspended solids driven toward wall and downward in accelerating spiral

Solids discharge

Fig. 1 Operation of the hydrocyclone.

be applied in conjunction with other devices to reduce the overall process cost in accomplishing the separation needs. It should be understood that hydrocyclones are more properly termed classifiers rather than absolute clarifiers; nevertheless, when properly applied and operated, the hydrocyclone can be an appropriate separator of solids from liquids.

SLURRY PROCESS VARIABLES AND MECHANICAL DESIGN CHARACTERISTICS AFFECTING HYDROCYCLONE PERFORMANCE

The ability of any gravity-force machine to effect an adequate solids/liquid separation is governed by Stokes' law. Specifically, the ease of separation is directly proportional to the suspended particle diameter squared times the specific gravity differential between the solid and the liquid phases, and inversely proportional to the viscosity of the continuous liquid phase.[1] These factors can often be found in the literature, but when not available can be developed by simple laboratory techniques—microscopic solids size observation, viscosity measurements, and appropriate gravimetric analysis. Regardless of these known physical properties, it is difficult to apply the information directly because these characteristics may be altered drastically when materials are combined to form a mixture or slurry. The combined phases may exhibit unique properties. For instance, the specific gravity differential may be modified by the degree of solids porosity or the amount of surface-bound liquid held tightly to the accompanying particles. Likewise, the viscosity, when measured in a quiescent state, is quite different from that of the high-shear situation found in hydrocyclones. Furthermore, the presence of the particles themselves creates a pseudo specific gravity effect. Nevertheless, the Stokes' variables represent a means of measuring the applicability of gravity-force equipment to separate a slurry into individual components.

The ability of a hydrocyclone to meet required solids/liquid separation needs is governed by the design variables of the equipment itself. These variables include cone diameter, overall body length, as well as the dimensions of the feed, apex, and vortex openings. The relative importance of each design variable is discussed by a number of authors.[2,3,4] Regardless of the "Stokes" data available and the equipment design formulas that may exist, the suitability of a hydrocyclone to a given process must depend upon existing information known from past experience or upon results developed by laboratory or field testing.

Fortunately, empirical data does exist to help make a preliminary choice as to size and number of cyclones needed to accomplish a specific separation requirement. Figure 2 represents the separation of silica from water with different-size hydrocyclones operating at a 2.8 g/cm² (40 lb/in²) differential pressure. The chart indicates a 150-mm (6-in) hydrocyclone will capture almost all the silica particles of 40-μm size from a water slurry. It will be noted that approximately 50% of the 22-μm size material will be captured and only a few of the 10-μm particles. The nomograph portion of the chart utilizes differential specific gravity and viscosity information to normalize particle size and viscosity of a given feed slurry to an equivalent silica/water situation upon which the chart is based. In other

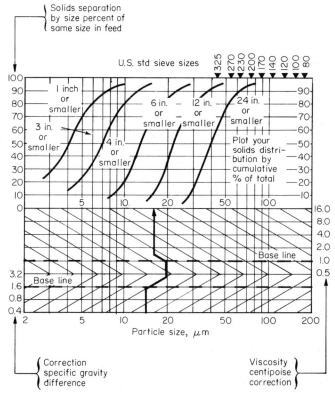

Fig. 2 Separation performance of various sized hydrocyclones.

words, if the Stokes' variables of a particular problem are known, the characteristics of an unknown mixture can be shifted so as to simulate the silica and water slurry, leading to proper hydrocyclone choice.

EXAMPLE ILLUSTRATING THE USE OF FIG. 2

Problem Suppose a solids/liquid slurry has a viscosity of 1.5 cp. The specific gravity of the solids fraction is 4.0 and the particles range in size from 15 to 40μ m. The specific gravity of the liquid phase is 1.0. Find the size hydrocyclone needed to remove essentially all of the solids from the liquid.

Solution Move from the 15 μm entry point to the specific gravity difference base line and correct diagonally upward to 3.0. Then proceed to the viscosity correction base line and correct to a value of 1.5. Extending this final adjustment vertically demonstrates that a 3-in hydrocyclone must be utilized to remove essentially all solids from this slurry.

NOTE: The above chart is for guidance purposes only. Final sizing of cyclones must be determined by tests on the expected slurry or prior experience in similar situations.

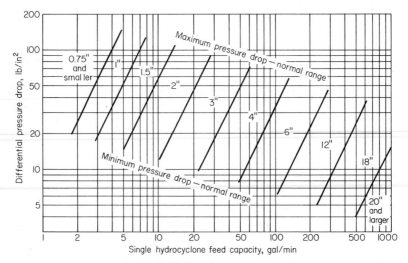

Fig. 3 Hydrocyclone capacity.

TABLE 1   Maximum Capacity of Apex Values

Apex diameter, in	Apex area, in²	Capacity, dry tons per hour
⅝	0.31	2.0 to 4.5
¾	0.44	3.5 to 7.5
1	0.85	6.5 to 12.0
1¼	1.23	9 to 17
1⅝	2.07	15 to 29
2	3.14	22 to 44
2¼	3.98	28 to 56
2½	4.91	35 to 70
2¾	5.94	42 to 84
3	7.07	50 to 100
3½	9.62	100
4	12.52	130
4½	15.90	165

NOTE: Densities of underflow range from 64 to 80% solids and specific gravities of dry solids from 2.6 to 4.2.

If the unknown slurry lends itself to hydrocyclone application and a properly sized device has been selected, the capacity of the unit can be determined as shown in Fig. 3. When a hydrocyclone does not have adequate capacity over the pressure range indicated to handle a given problem, multiple hydrocyclones are manifolded in parallel. The selected hydrocyclone should have adequate solids capacity as determined by the apex dimension. Table 1 relates solids capacity to apex cross-sectional area. Under normal operating conditions, there are limitations of apex size for a given hydrocyclone. Gener-

TABLE 2   Generalized Cyclone Application Categories

Hydrocyclones can be of value in treating virtually any solids-liquid stream. Cited below are a few applications to suggest the versatility of this equipment.

Classifying desired solids	
(a) Degritting and removal of oversized particles	Lime slaking, kaolin, ceramics, pollution wastes
(b) Fractionation of pigments and fillers	Calcium carbonate, mica
(c) Desliming slurries	Potash, coal, minerals
(d) Closed-circuit grinding operations	Calcite, taconite, copper
(e) Classifying tailings for backfill use or recovery	Phosphates, dredging wastes
(f) Processing for proper sizing	Crystallized salts, sand
(g) Product separation	Starch, minerals

Thickening of solids and clarifying fluids	
(a) Desanding and degritting water	Municipal and process waters
(b) Recovery of catalysts	Petroleum cracking, chemical reactions
(c) Thickening of slurries prior to deliquoring	Polymers, coal, urea
(d) Cleaning up fuel for engine consumption	Diesel oil, crude oil
(e) Thickening crystallized slurries	Adipic acid, ammonium sulphate
(f) Desanding fluids to gathering lines in order to reduce system maintenance	Water, crude oil, various slurries
(g) Removing solids from evaporation systems	Organic and inorganic crystals
(h) Cleaning up disposal well and injection fluids to maintain formation permeability	Waterflooding, solution wastes

Treating slurries to benefit other equipment	
(a) Removing oversized particles to reduce plugging	Spraying systems, nozzle centrifuges, coalescers
(b) Partial separation of solids to improve cycle	Basket centrifuge, cartridge/leaf filters
(c) Concentration of slurries to increase capacity and performance	Centrifuges, filters, crystallizers
(d) Removal of abrasive material to reduce maintenance	Centrifuges, pumps, flotation units, other equipment
(e) Cleaning up filtrates to improve performance	Screen centrifuges, filters
(f) Removing solids from process streams to minimize scaling	Piping, heat exchangers
(g) Processing bottoms to improve performance, eliminate cleanout	Storage vessels, settling tanks, lagoons

Cleaning up recirculating fluids	
(a) Cleaning coolant fluids	Machining and grinding operations
(b) Separating oversized matter	Lapping and polishing systems
(c) Maintaining water quality	Cooling towers, air-conditioning systems
(d) Recovering soaps, detergents, solvents, and water	Washing operations
(e) Rejecting sands and silts	Drilling and treatment
(f) Cleaning up fluids used to cool or lubricate	Seals and bearing in pumps and other equipment
(g) Removing solids from oils and fluid baths	Heat treating

ally the apex diameter ranges from approximately ⅓ to ⅔ the vortex finder diameter, which in turn can vary from ¼ to ½ the cone body diameter.

Assuming the calculations and studies lead to a selection of a given hydrocyclone in accordance with the above methods, it is always advisable to conduct large- or intermediate-scale tests to develop the actual performance of the hydrocyclone. Should such tests be impractical, scale-up from physical data or small samples should be left to equipment manufacturers whose background and experience in sizing techniques will lead to more assured and appropriate choice.

APPLICATIONS

Hydrocyclones have been employed in substantial numbers over several decades in pulp and paper, metallurgical processing, coolant fluids in machining operations, drilling mud treatment, and in the process industries generally. Table 2 illustrates the use of the hydrocyclone in a variety of process categories. The details of the process results of these examples are too lengthy to discuss here. Whether a new application or a well-established hydrocyclone application is to be considered, appropriate manufacturers should be contacted, not only to discuss the probable results of the separation, but also to review proper system arrangements including the pumping, piping, and other requirements so necessary for satisfactory hydrocyclone installation and operation.

REFERENCES

1. Tiller, F. M. (ed.), "Theory and Practice of Solid-Liquid Separation" 2d ed., University of Houston Press, Houston, 1975.
2. Bradley, D., "The Hydrocyclone," Pergamon Press, New York, 1965.
3. Perry, R. (ed.), "Chemical Engineers' Handbook," 4th ed., McGraw-Hill, New York, 1969.
4. Rietema, K., and Vewer, C. G., "Cyclones in Industry," Elsevier, New York, 1958.
5. Day, R. W., *Chem. Eng. Prog.*, **69**, (September 1973), pp. 67–72.
6. Herkenhoff, E. C., *Min. Eng. (N.Y.)*, (August 1957), pp. 873–876.

Section **4.10**

Drying of Wet Solids

THEODORE H. WENTZ, M.E., P.E. *Sales Manager, Proctor and Schwartz Division, SCM Corp., Philadelphia, Pa.*

JOHN R. THYGESON, Jr., Ph.D. *Associate Professor of Chemical Engineering, Drexel University.*

PREFACE

In this chapter we describe six basic dryer types which encompass and characterize most of the dying systems presently applied in the food, chemical, pharmaceutical, agricultural, and textile industries. They are the tray dryer, the continuous through-circulation dryer, the rotary dryer, the spray dryer, the flash dryer, and the fluid-bed dryer. The authors show how a particular drying problem can be cast in terms of the characteristic features of one or more of the above dryer types; thus the methodology of dryer design and analysis can, to a considerable extent, be formalized. Accordingly, the objective of this chapter is twofold: (1) to set forth a systematic procedure for selecting the proper type of dryer for a specific application, and (2) to illustrate, by worked examples, how the basic design calculations are made for each of the six dryer types.

In brief, then, the emphasis of this chapter is on the *application* of fundamental drying theory to the practical problem of dryer selection and performance analysis related to cost estimation and to optimal design of the drying system. Consequently, a detailed discussion of dryer hardware, which is adequately covered in the trade literature and in various handbooks, was not considered appropriate for this presentation.

INTRODUCTION

Drying is a physical separation process that has as its objective the removal of a liquid phase from a solid phase by means of thermal energy. The liquid, generally water, is liberated by the process of vaporization rather than by the mechanism of breaking chemical bonds between liquid and solid; i.e., the liquid is not chemically bound to the solid. In most industrial drying applications it is not necessary nor economically feasible to remove every vestige of the liquid from the solid; the commercially dry solid will usually contain a certain amount of residual moisture, the amount of which is determined by a compromise between product quality and economic factors.

Industrial dryers may be classified according to the physical characteristics of the material being dried, the method of transferring the thermal energy to the wet solid, the source of the thermal energy, the method of physical removal of the solvent vapor, and the method of dispersion of the wet solid in the drying operation.

The basic problems that must be resolved by the designer of drying equipment are (1) how to provide an efficient means of supplying energy to the wet solid, (2) how to provide an effective means of removing the solvent vapor produced in the drying operation, and (3) how to recover efficiently the dried solid.

As a consequence of dryer specialization, the selection of the type of dryer appropriate to the specific product to be dried becomes a critical step in the specification and design of the processing plant. The choice of the wrong type of dryer can lead to inefficient operation, reduced product quality, and loss of profit. Accordingly, the buyer of drying equipment would be well advised to discuss particular requirements with an expert in the design and application of industrial dryers.

THEORETICAL CONSIDERATIONS

The fundamental principles underlying the performance and design of industrial drying equipment are, in general, those typical of the science of heat and mass transfer. In all types of dryers herein discussed, the basic characteristics of the equipment may be described by an air-movement system consisting of fans or blowers and appropriate ducting, an energy source, and a system for dispersing and/or conveying the solid phase through the dryer. The purpose of the air-movement system is twofold: to provide for the transfer of energy from the source to the wet solid, and to transport the resulting vapors produced in the drying process away from the dryer. The energy source produces the heat necessary for the drying process. The dispersal or conveyance system transports the solid phase through the drying medium or distributes the solid phase throughout the drying medium. In essence, the theory and design of the dryer are determined by the principles of heat and mass transfer and fluid mechanics that govern the performance of the subsystems described above.

Drying, as a unit operation in chemical engineering, is characterized by the separation (usually partial) of a liquid contained within a solid by the process of vaporization of the liquid into a gas phase. The mechanism of the drying process, as controlled by the principle of heat and mass transfer, consists of the transport of mass from the interior of the solid to the surface, the vaporization of the liquid at or near the surface, and the transport of the vapor into the bulk gas phase. Simultaneously, heat is transferred from the bulk gas phase to the solid phase where all or a portion of it provides for vaporization and the remainder accumulates in the solid as sensible heat. The overall rate by which the above sequence of steps takes place defines the drying rate and is inversely related to the drying time. Since the prediction, or estimation, of the latter is generally mandatory in dryer design, we shall now direct our attention to the analytical description of the drying mechanism. Before doing so, however, it is appropriate to define several terms peculiar to the drying operation.

Basic Definitions

Percent moisture content, bone-dry basis: The mass ratio of water to dry solid multiplied by 100.

Percent moisture content, commercial-dry basis: The mass fraction of water in the commercially dry solid multiplied by 100.

Percent moisture content, wet-weight basis: The mass fraction of water in the wet solid multiplied by 100.

Bound moisture: The amount of water in the solid which exhibits a vapor pressure less than normal for the pure liquid.

Equilibrium moisture: The amount of moisture in the solid that is in thermodynamic equilibrium with its vapor in the gas phase. For given temperature and humidity conditions, the material cannot be dried below its corresponding equilibrium moisture content.

Free moisture content: The moisture content of the material less the equilibrium moisture content for the given temperature and humidity conditions.

Hygroscopic material: A material that may contain bound moisture.

Constant-rate period: The part of the drying process during which the drying rate is constant and is controlled by external rather than internal conditions.

Critical moisture content: The moisture content of the material at the end of the constant-rate period. The critical moisture content is not a unique property of the material but is influenced by its physical shape as well as the conditions of the drying process.

Falling-rate period: The part of the drying during which the drying rate varies in time. Internal factors, i.e., physical and transport properties of the material, control the drying process.

External drying factors: The independent variables associated with the conditions and flow of the gas phase.

Internal drying factors: The properties of the solid that influence the transport of heat and mass within the solid phase.

Rate of drying: The amount of water (usually expressed as pounds) removed per square foot of drying area per hour.

Instantaneous drying rate: The derivative $(dm/d\theta)$ $(1/A_m)$.

Adiabatic drying: The drying process described by a path of constant adiabatic cooling temperature on the psychrometic chart.

Through-circulation drying: The method of drying by which the air passes through a packed bed of solid particles.

Physical Description of the Drying Process

The drying characteristics of wet solids are best described by plotting the average moisture content of the material (usually defined on the bone-dry basis) against elapsed time measured from the beginning of the drying process. Most wet solids exhibit a drying-time curve similar to that of Fig. 1. For the material of interest, the prediction or, more generally, the experimental estimation of this curve must be made before one can begin the design calculations. Moreover, the influence of the internal and external variables of drying on the drying-time curve should be determined in order that an optimal design can be developed. For this reason it is important that a statistically sound and efficient experimental plan be followed in the estimation of the drying curve.

The drying-rate curve can be derived from the drying-time curve by plotting slopes of the latter curve against the corresponding moisture content. The distinctive shape of this plot, shown in Fig. 2, illustrates the constant-rate period, terminating at the critical moisture content, followed by the falling-rate period. The variables that influence the constant-rate period are the so-called external factors consisting of gas mass velocity, thermodymanic state of the gas, transport properties of the gas, and the state of aggregation of the solid phase. Thus, changes in gas temperature, humidity, and flow rate have a profound effect on the drying rate during this period. In contrast, the effect of the external factors on the falling-rate period is much diminished. The controlling factors in this period are the transport properties of the solid and the primary design variable is temperature.

Because the characteristic drying behavior in the two periods markedly differs, the design of the dryer should accordingly take this into account. Materials which exhibit predominantly constant-rate drying, such as thin sheeting and liquid sprays, are subject to different design criteria than materials which exhibit predominantly falling-rate drying. In the context of economics, it is costlier to remove a pound of water in the falling-rate period than it is to remove the same amount of water in the constant-rate period. Accordingly, it is desirable to extend the length of the constant-rate period with respect to that of the falling-rate period as much as practicable. One way to accomplish this is by creating additional drying area through particle size reduction. An analysis of the mechanisms of drying that govern the two periods is essential if the designer is to know in a quantitative sense the effects of the internal and external factors on the overall drying time.

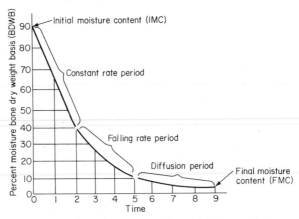

Fig. 1   Classic drying curve. *(Proctor & Schwartz, Inc.)*

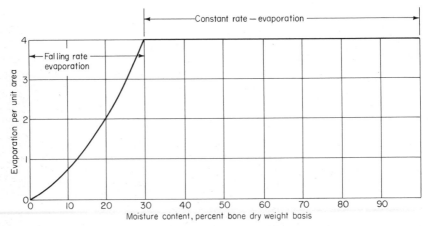

Fig. 2   Drying-rate curve. *(Proctor & Schwartz, Inc.)*

Mechanisms of Drying

The transport processes that characterize the drying of a wet solid in a flowing gas medium are governed by the well-known principles of heat and mass transfer. The translation of these principles into specific mathematical equations is subject to a number of uncertainties because of the complex nature of the solid-phase transport processes. Consequently, it is only for cases of special simplicity that the drying time can be predicted with confidence. In this category fall those materials which exhibit only, or primarily, constant-rate drying behavior. Where the falling-rate period is a significant portion of the total drying time, simplifying assumptions are generally required in order to solve the equations. The governing equations describing the two drying periods may be summarized as follows:

Constant-Rate Period   The rate of drying is

$$\frac{1}{A_m}\frac{dm}{d\theta} = -k_H(H_i - H_b) \tag{1}$$

$$\frac{1}{A_h}\frac{dm}{d\theta} = -\frac{h}{\lambda}(T_b - T_i) \tag{2}$$

Generally, $A_h = A_m$, resulting in a dynamic equilibrium between heat and mass transfer. The expression for the drying time in the constant-rate period is then

$$\theta_c = \int_{m_c}^{m_o} \frac{dm}{A_m k_H(H_i - H_b)} = \int_{m_c}^{m_o} \frac{\lambda dm}{A_h h(T_b - T_i)} \tag{3}$$

For constant-drying conditions, the driving force $(H_i - H_b)$ is constant and integration of Eq. (3) gives

$$\theta_c = \frac{m_o - m_c}{A_m k_H(H_i - H_b)} = \frac{\lambda(m_o - m_c)}{A_h h(T_b - T_i)} \tag{4}$$

For varying drying conditions, H_b (or T_b) must be related to m by the appropriate mass (or heat) balance. Accordingly, $GA(H_b - H_{b_o}) = S(m_o - m)$, and upon the elimination of H_b, Eq. (3) becomes

$$\theta_c = \int_{m_c}^{m_o} \frac{dm}{A_m k_H\left[H_i - \dfrac{S}{GA}(m_o - m) - H_{b_o}\right]}$$

$$= \frac{GA}{SA_m k_H} \ln \left[\frac{B + \dfrac{S}{GA} m_o}{B + \dfrac{S}{GA} m_c} \right] \tag{5}$$

where $B \equiv H_i - H_{b_o} - \dfrac{S}{GA} m_o$.

Falling-Rate Period   The drying rate during this period is a complex function of the transport, physical, and thermodynamic properties of the solid phase as well as of the same properties of the gas phase. Consequently, it is impossible to obtain a general expression for the drying rate that will lead to the prediction of the drying time corresponding to this period. However, in certain circumstances, empirical assumptions result in a simple expression for the rate equation which is amenable to solution. An analysis of the form of the rate equation illustrates the analytical difficulties in predicting the drying time.

Let R = instantaneous drying rate. Then $R = kf(m)$, where k is dependent on temperature, the thermodynamic and transport properties, and the hydrodynamic condition of the gas phase. A graph of the specific form of the above equation generates the characteristic shape of the rate curve of Fig. 2. Given the rate curve, or its analytical expression, the drying time can be calculated by evaluating the integral

$$\int_{m_f}^{m_c} \frac{dm}{kf(m)} = t = \int_{m_f}^{m_c} \frac{dm}{A_m k_H(H_i - H_B)f(m)} \tag{6}$$

For a large class of materials, the rate is linear with respect to moisture content; accordingly, $f(m) = m/m_c$, and the rate expression is analogous to that of a first-order chemical reaction. Extending this analogy, one observes that the constant-rate period is described by a zero-order type of rate expression.

Estimation of the drying time for the falling-rate period, then, depends primarily on experimental data. However, a knowledge of the controlling mechanisms of the drying process during the falling-rate period gives one insight into the dryer performance, and this has practical implications to design.

Theoretical Models of the Drying Process *Constant-Rate Period.* Drying during this period is controlled by the principles of simultaneous heat and mass transfer applied to a liquid-gas interface in dynamic equilibrium with a bulk gas phase. Correlations of heat-transfer coefficients, summarized below, permit the designer to estimate the drying time during this period.

Parallel flow:

$$h_c = \frac{0.01 G^{0.8}}{D_p^{0.2}}$$

Impingement flow (from slots, nozzles, or perforated plates):

$$h_c = \alpha\, G^{0.78}$$

where α varies from about 0.3 to 0.02, depending upon plate open area and geometry. For most-efficient operation, plate open area is about 2.5% of the total heat-transfer area and the plate is located at a distance of 4 to 6 hole diameters from the heat transfer area. In this case, $\alpha = 0.2$ (approximately).

Through-flow (packed bed):

$$h_c = 0.11 \frac{G^{0.59}}{D_p^{0.41}} \quad \text{for} \quad \frac{D_p G}{\mu} > 350$$

$$h_c = 0.15 \frac{G^{0.49}}{D_p^{0.51}} \quad \text{for} \quad \frac{D_p G}{\mu} < 350$$

The heat-transfer coefficient calculated from the appropriate correlation above is then substituted into Eq. (3) to obtain the estimated drying time for the constant rate period.

Falling-Rate Period. Three models for the movement of moisture in porous solids, which characterize the falling-rate drying period, may be summarized as follows:

1. Transport by concentration-driven liquid diffusion
2. Transport by capillary forces
3. Transport by concentration-driven vapor diffusion

On the basis of the liquid diffusion model, the drying rate, expressed as lb/(h)(lb dry solid), is (for long drying times)

$$\frac{dm}{d\theta} = -\frac{\pi^2 D_L}{4d^2}(m)$$

which upon integration gives

$$m = \frac{8m_c}{\pi^2} \exp\left[-D_L\theta(\pi/2d)^2\right] \tag{7}$$

Thus, Eq. (7) indicates that the drying rate, when internal liquid diffusion controls, is directly proportional to the moisture content, which is consistent with the empirical linear falling-rate model described above.

The *vapor diffusion model* postulates that moisture migration occurs entirely by gaseous diffusion along the pores of the solid. Harmathy[1] presents a theoretical analysis of drying during the pendular state that is based on this model.

The *capillary model* assumes that the moisture transport in the solid is controlled by capillary action. Based on this model the instantaneous drying rate is consistent with

$$\frac{dm}{d\theta} = -Km$$

where $K = \dfrac{(dm/d\theta)}{m_c}$ constant rate

The estimated drying times derived from the above are as follows.

Liquid diffusion model:
$$\theta_f = \frac{4D_p^2}{D_L\pi^2} \ln \frac{m_c}{m} \tag{8}$$

Capillary model:
$$\theta_f = \frac{\rho_s D_p \lambda m_c}{h(T - T_i)} \ln \frac{m_c}{m} \tag{9}$$

Materials obeying Eq. (8) include soap, gelatin, glue, wood, starch, textiles, paper, and clay. Materials obeying Eq. (9) include sand, paint, pigments, minerals, and other coarse granular solids.

Psychrometric Considerations

Adiabatic drying occurs when the heat lost by the hot gas is entirely used in vaporizing liquid from the wet solid. Accumulation of heat in the solid or loss of heat to the surroundings is, by definition, negligible. The adiabatic process can be graphically described on the humidity chart by following a line of constant adiabatic saturation temperature. The equation of the line is

$$H_s - H = \frac{C_s}{\lambda}(T - T_s) \tag{10}$$

where T_s is the temperature of the solid and is equal to the adiabatic saturation temperature. Most industrial drying systems deviate more or less from adiabatic operation; however, in certain practical cases adiabatic behavior is assumed in order that preliminary calculations can be carried out with the aid of the humidity chart.

For air-water vapor systems, the line of constant adiabatic saturation temperature is, under normal conditions of temperature and humidity, virtually identical to the line of constant wet-bulb temperature. Accordingly, under adiabatic drying conditions, the wet-bulb temperature of the air, as it cools and humidifies, remains constant and is given by the equation

$$H_w - H = \frac{h}{\lambda k_H}(T - T_w) \tag{11}$$

By comparing Eqs. (10) and (11), one observes that the equality between the wet-bulb temperature and the adiabatic saturation temperature can be true only if

$$h/k_H = C_s$$

which is approximately the case for the system air-water vapor. For other systems, e.g., organic vapors in air, the magnitude of the psychrometric ratio, $h/k_H C_s$, deviates considerably from 1. In the case of wet cylinders in turbulent gas streams, Wasan and Wilke[2] propose the following correlation:

$$\frac{h}{k_H Cs} = \frac{[1 + 0.7(N_{sc}^{0.77} - 1)]}{[1 + 0.7(N_{pr}^{0.77} - 1)]} \tag{12}$$

Some specific values of the psychrometric ratio, for the specified organic vapor in air, are carbon tetrachloride, 0.51; benzene, 0.54; toluene, 0.47; o-xylene, 0.49. In these cases the adiabatic drying process will follow the line of constant adiabetic saturation temperature rather than the line of constant wet-bulb temperature.

The thermodynamic limitations imposed on the drying process are clearly illustrated by means of the humidity chart. During the adiabatic drying process, the thermodynamic state of the gas-vapor mixture changes in accordance with its movement along a line of constant adiabatic saturation temperature. However, the limit of this movement is fixed by the interception of this line by the 100% saturation curve. When this point is reached in the drying process, all drying ceases because the gas-vapor mixture is in thermodynamic equilibrium with the solid phase. According to the criterion of efficiency of energy utilization, the above situation would conform to optimal design conditions, since the exhaust gas would be incapable of any additional drying.

However, as the thermodynamic state of the gas-vapor mixture approaches equilibrium, the rate of drying proportionally decreases, and very near the equilibrium state the drying rate is exceedingly low. Thus, opposing objectives of maximization of energy utilization and minimization of capital cost, i.e., minimization of drying time, require an optimal design. Specifically, the designer must determine at what point along the path of constant adiabatic saturation temperature to terminate the process. Moreover, the particular adiabatic saturation line on which to operate must be selected by the designer, and this choice is dictated by the optimal recycle ratio. Thus, the lower the ratio of exhaust to recycle gas, the higher the humidity maintained in the dryer. Once again the optimal design is a compromise between the objectives of minimum capital cost and maximum energy utilization.

As the recycle ratio approaches 1, the drying medium approaches the condition of pure vapor. In the limit of 100% recycle, a condition of virtually closed-cycle operation is manifested and maximum efficiency of energy utilization is achieved. For this case, the temperature at which vaporization of water occurs approaches 212°F at atmospheric pressure.

In certain industrial dryers, e.g., batch tray dryers, the thermodynamic state of the gas phase remains substantially constant over the drying cycle. This is a consequence of the exceedingly small quantity of water vaporized into a relatively large mass of the gas phase. In this situation, there exists a critical boundary separating two thermodynamic regions of operation: below 212°F, the gas phase has limited solubility for water; at 212°F and above, the gas phase exhibits infinite solubility for water. When the liquid is an organic substance, the critical temperature is the boiling point of the liquid at 1 atm. Thus, below the critical temperature the effect of humidity on dryer performance is more pronounced than its effect at temperatures above the critical. Accordingly, adequate dryer exhaust capacity is an important design factor in the case of low-temperature drying.

CLASSIFICATION AND SELECTION OF INDUSTRIAL DRYERS

Classification

Industrial dryers may be classified according to (1) mode of operation, (2) physical properties of the material to be dried, (3) method of conveyance or dispersal of the material through the dryer, and (4) method of supplying energy to the material undergoing drying.

Mode of Operation   This category refers to the nature of the production schedule. For large-scale production the appropriate dryer is of the *continuous type* with continuous flow of the material into and out of the dryer. Conversely, for small production requirements, *batch-type* operation is generally desirable.

Physical Properties of the Material   The physical state of the feed is probably the most important factor in the selection of the dryer type. The wet feed may vary from a liquid solution, a paste, or filter cake to free-flowing powders, granulations, and fibrous and nonfibrous solids. The design of the dryer is greatly influenced by the properties of the feed; thus dryers handling similar feeds have many design characteristics in common. A comprehensive classification of commercial dryers based on materials handled is given in Perry's handbook.[3]

Method of Conveyance   In many cases, the physical state of the feed dictates the method of conveyance of the material through the dryer; however, when the feed is capable of being preformed, the handling characteristics of the feed may be modified so that the method of conveyance can be selected with greater flexibility. Generally, the mode of conveyance correlates with the physical properties of the feed.

Method of Energy Supply   Where the energy is supplied to the material by convective heat transfer from a hot gas flowing past the material, the dryer is classified as a convection type. Conduction-type dryers are those in which the heat is transferred to the material by the direct contact of the latter with a hot metal surface.

Basic Dryer Types

Most industrial dryers can be categorized in terms of six basic dryer types: rotary, tray, continuous through-circulation, spray, flash, and fluid bed. The physical characteristics of these dryers and their classification with respect to the aforementioned categories are described below.

Rotary Dryer   This basic dryer is most suitable for free-flowing granular solids. Typical products dried in a rotary dryer include ammonium sulfate, nitrate, and phosphate fertilizer salts, sand, fluorspar, and vinyl resins. Its advantages include low capital cost, fairly close temperature control, drying and calcining in the same unit, high thermal efficiency, low labor requirement, and moderate drying times. Some disadvantages are difficulty of sealing, high structural load, tendency to create dust, which requires subsequent dust collection equipment, product buildup on interior walls, and nonuniform residence time.

The method of conveyance of the product is by the combined effects of gravity and the rotation of the dryer shell, which is slightly inclined to the horizontal. Heat is supplied by hot air flowing in cocurrent or countercurrent direction to the movement of the product. The normal mode of operation is continuous.

Tray Dryer   Typical materials dried in the tray dryer include filter cake, dyestuffs, pharmaceuticals, and almost any material in small quantities. Tray dryers are specifically appropriate for drying batches of a large number of products, usually of considerable value. Advantages of this type of dryer are capability of handling fragile products, no loss of product during drying, low space requirement, ease of cleaning, and close control of drying conditions. Some disadvantages include high labor requirements and long drying time.

The product is contained in solid-bottom or perforated trays, over or through which hot air flows. Because of the relatively high mass flow of air and low drying rate, the drying conditions tend to be uniform throughout the drying space.

Through-Circulation, Continuous Dryers   This type of dryer is suitable for materials that are capable of forming a pervious packed bed, either through preforming or by virtue of their original shape and stability. Some typical products dried in a through-circulation continuous dryer include catalyst pellets; chemicals such as pigments, carbonates, stearates, uranium precipitate, and urea resin; minerals such as kaolin clays, coal pellets, and nickel ore briquettes; synthetic and natural fibers (staple and tow) such as nylon, rayon, acrylic, polyester, and wool; ceramics such as slate aggregates and glass pellets; and food products including nuts, fruits, vegetables, cereals, and animal foods.

The essential features of this dryer are the feeding device, which generally consists of a preformer such as an extruder and conveyor to feed and uniformly distribute the product on the main perforated-plate or wire-mesh conveyor, and an air circulation system which forces hot air through the packed bed of wet particles. Because of the large surface-to-volume ratio of the bed and the intimate contact between air and particles in the bed, high drying rates are achieved.

Advantages of the system are high production rates per unit plant floor space; good control of product quality; zone control of temperature, humidity, and air flow; high thermal efficiency; and integral product-cooling capability. Disadvantages include high capital cost, poor control of fines, and frequent maintenance.

Products which have inherent granular shape or can be extruded, granulated, briquetted, or otherwise preformed to produce a particle that will maintain its integrity in a packed bed are generally amenable to drying in a through-circulation system.

Spray Dryers   Spray dryers are adapted to handle liquids, slurries, or, in general, any product that is pumpable. Typical products that are spray dried include dairy products, coffee, tea, eggs, and chemicals such as organic and inorganic salts, pigments, detergents, latex, and cellulose.

The material to be dried is atomized by a nozzle or disk-type atomizer and dispersed as a fine spray into a vertical cylindrical chamber with a conical bottom through which hot gas flows. The droplets quickly vaporize, leaving a dry solid residue which is discharged from the main chamber. Particles entrained in the exiting gas stream are removed in a cyclone separator or bag collector.

The major advantages of spray dryers are short drying times (2 to 20 s), adaptability to heat-sensitive products, and control of particle size and density which is desirable from the standpoint of product flow properties and rapid dehydration. Disadvantages are low solids content, critical maintenance of atomizing equipment, and tendency to product buildup on interior walls.

Flash Dryers In some respects, flash drying is similar to spray drying in that in each process the wet product is dispersed as fine particles into a flowing hot stream. The high-velocity gas stream transports the solid particles through the drying chamber (generally a long vertical tube) to a cyclone separator or a bag collector where the dried particles are removed.

As in the case of the spray dryer, the advantages of the flash dryer are extremely short drying times and a low product discharge temperature. In the flash dryer the atomizing device is eliminated, but often a high-speed mill must be substituted to disperse the wet solids. There are limitations on the particle size that can be handled, since only surface water will be removed in the short holding time. Other disadvantages include erosion of internal surfaces, and the danger in the drying of flammable materials at high temperature required for good thermal efficiency.

Typical products dried in flash dryers include plastic resins, distiller's spent grain, starch, clays, calcium carbonate, coal, and ammonium sulfate.

Fluid-Bed Dryer When a gas stream flows up through a bed of particulate solids there exists a critical gas velocity at which the bed abruptly expands and the individual particles become suspended in the gas phase. In this state the bed is said to be fluidized. If the fluidizing agent is a hot gas and the particles are wet, the conditions are such that rapid drying will occur. Since each particle is surrounded by the hot gas and the relative velocity between particles and gas is high, high rates of heat transfer are achieved and, consequently, drying times are very short.

Products falling within the size range of 1- to 30-mm particle diameter are generally suitable for fluidized drying. These include granular and crystalline materials in the pharmaceutical, fine-chemical, plastics, and food processing industries. Some typical products are sodium and potassium chloride, urea, polymer chips, coal, NPK fertilizers, sand, clay granules, asphalt, and iron ore. Fluid-bed dryers can be designed for either batch or continuous operation.

Some advantages of fluid-bed dryers are rapid and uniform heat transfer, relatively short drying times, and small floor-space requirements. Disadvantages include high power cost, nonuniform residence time, and poor control of fines.

Selection of Dryer Type

The choice of the best type of dryer to use for a particular application is generally dictated by the following factors: (1) the nature of the product, both physical and chemical, (2) the value of the product, (3) the scale of production, (4) available heating media, (5) product quality considerations, and (6) space requirements. For application of each of these factors in the selection process a systematic procedure is recommended. One such procedure involves the following five steps:

1. Formulate the drying problem as completely as possible.
2. Collect all available data related to the problem.
3. Define the critical factors, constraints, and limitations associated with the particular product and with available resources.
4. Make a preliminary identification of the appropriate drying systems.
5. On the basis of the foregoing, select the optimal drying system and determine its cost effectiveness.

In step 1 the specific requirements and variables are explicitly identified. Thus the problem might be stated as follows: It is desired to dry product X (98% pure) at an initial moisture content of 100% (bone-dry basis) to a final moisture content of 1.0% (bone-dry basis). The initial moisture content may vary between 90 and 110% (bone-dry basis) and the final between 0.50 and 2.0% (bone-dry basis). The desired production rate is 10,000 lb of dry solids per 24 h on the basis of a 24-h working day.

The important information derived from step 1 may be summarized as follows:
   a. Identification of the product and its purity.
   b. Specification of initial and final moisture content.
   c. Specification of range of variation of initial and final moisture content.
   d. Specification and basis of production rate.

Having stated the problem, in step 2 one investigates previous experience related to the drying of the particular product of interest or of a similar material. This provides insight into the drying behavior of the product and identifies one or more feasible approaches to the solution of the problem. In addition, the physical and chemical properties of both the wet feed and dry product are established. These include: physical state of the feed (filter cake, granulations, crystals, extrusions, briquettes, slurry, paste, powder, etc.) including size, shape, and flow characteristics; chemical state of the feed (pH, water of crystallization, chemical structure, degree of toxicity of vapor or solid, corrosive properties, inflammability of vapor or solid, explosive limits of vapor); physical properties of dry product (dusting characteristics, friability, flow characteristics, and bulk density). Finally, available drying data in the form of prior laboratory results, pilot-plant performance data, or full-scale plant data on the drying of similar materials should be obtained.

In step 3 any particular hazards related to the handling of the product (wet or dry) should be specifically and quantitatively identified. For example, if a toxic vapor is produced in the drying operation above a certain temperature, this may dictate the operating gas temperature. If the vapor produced in the drying operation forms an explosive mixture with air, the dryer exhaust system must be designed to maintain the concentration of the vapor below a fraction of its explosive limit. In addition, any characteristics of the product that present potential problems should be recognized. Thus, if the product has a low melting point, it may have a tendency to stick to the conveyer or to form an impervious conglomeration during drying. Or if the product or its vapor is corrosive, special materials of construction may be required. Frequently, the quality of the dry product is adversely affected if local wet spots exist, even though the average moisture content falls within the specified limits. Thus, the degree of uniformity of drying will be, in this case, an important consideration in the selection process. Another critical factor is the material-handling characteristics of the feed. Relevant questions to ask are these: Can the feed be extruded, granulated, briquetted, pelletized, or preformed in some other fashion in order to form a packed bed? Can the feed be fluidized or atomized or dispersed and pneumatically conveyed? The handling characteristics of the partially dry product can also be critical if dusting or physical degradation of the product is excessive, especially if the dust is toxic or the product is extremely valuable, in which case significant loss of product by leakage would be intolerable.

Other critical factors to consider are the value of the product with regard to maintaining product quality, the details of the production schedule, preliminary and subsequent operations in the production line, space limitations, and temperature limitation of the preferred heating medium. Of extreme importance is the safety of the system. Does it depend on high temperature to maintain an acceptable level of thermal efficiency, and what will occur in an upset condition?

At this point a preliminary selection of dryer type can normally be made. Usually this amounts to the identification of several dryer types that would appear to be appropriate. This can be accomplished by simply comparing the properties and critical factors identified in steps 2 and 3 with the characteristic features of the industrial dryers classified according to the categories defined previously. This constitutes step 4.

In the final step 5, the optimal dryer type is identified and the appropriate experimental program and design calculations initiated. The completion of step 5 usually requires some laboratory-scale experimentation and a preliminary cost analysis. Thus, the ultimate choice is usually that which is dictated by minimum total cost. However, it should be noted that a detailed economic analysis might lead to a selection based on maximum profit rather than minimum cost.

Laboratory Testing Program

Once the appropriate dryer types have been identified, one must make a preliminary estimate of cost of each feasible dryer. This requires some degree of laboratory experimentation to determine the drying time. It is essential in the laboratory test that a representative sample of the wet feed be used and that the test conditions simulate as closely as possible the conditions characteristic of the commercial-size dryer. In addition, the experimental method for measuring product moisture content should be clearly defined and should be consistent with that used by the product buyer.

Scale-up of laboratory data to that applicable to plant-scale operation is a critical step and requires considerable experience. Accordingly, the dryer vendor is probably the one most competent to make this decision. The process of scaling up the laboratory test drying time to the commercial drying time is subject to considerable uncertainties and subjective factors, the effects of which are not quantitatively predictable. Thus, the magnitude of the scale-up factor is based primarily on experience and is a function of the specific dryer type being simulated and of the drying and physical characteristics of the material being dried. The value of the scale-up factor can vary from about 1.2 to 2.0 or greater, depending on variables such as the degree of uniformity of the feed and the rapidity of drying. The designer should consult the vendor for an appropriate scale-up factor for typical drying systems.

Once the most appropriate dryer type has been determined, an efficient laboratory testing program should be designed to establish the optimal operating conditions. An effective experimental plan will elicit the critical data required by the designer and will ensure that the ultimate commercial system is the one most suited to meet the client's requirements.

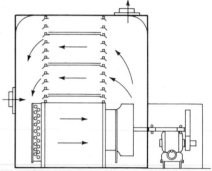

Fig. 3   Typical cross-circulation tray dryer. (*Proctor & Schwartz, Inc.*)

DRYER DESIGN AND PERFORMANCE

Tray Dryer

Tray dryers are adaptable to the drying of almost any material that can be contained in a tray. A variation of this is the rack-type dryer. Generally, batch operation is used; however, there are many continuous tray or rack dryers in industry today. Figure 3 illustrates the air circulation system and tray arrangement in a typical batch atmospheric tray dryer. Heat is usually supplied by steam coils or gas burners located inside the dryer, and the heated air is recirculated over the surface of the stacked trays. Drying times may vary from hours to days. Because of the labor required to load and unload the trays, materials that exhibit short drying times are not normally dried in a tray dryer except where production is very limited or of an experimental nature. Figure 4 shows a typical truck-tray dryer arrangement.

Design Equations   In the typical design problem it is desired to determine the amount of tray area required for a specified production rate. The necessary data (usually obtained from the laboratory tests) are the drying time for the given initial and final moisture contents and the desired tray loading. The design equation is

$$A = \frac{P(t + t_d)}{L}$$

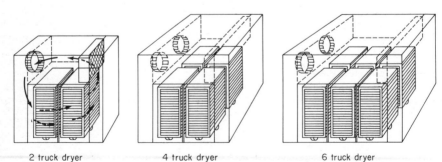

2 truck dryer   4 truck dryer   6 truck dryer

Fig. 4   Typical truck dryer arrangement. (*Proctor & Schwartz, Inc.*)

where t is the commercial drying time, t_d is the downtime for loading and unloading the dryer, L is the tray loading in pounds of dry solid per square foot of tray, and P is the production rate in pounds of dry solid per hour. The number of trays is then readily determined for the specified tray size. The latter is usually standardized; however, where the trays are manually loaded in the dryer, the maximum weight of the loaded tray may represent a constraint on tray size, or alternatively, on tray loading.

Performance Characteristics   Typical performance data are shown in Table 1.

Effect of Air Velocity. Air mass velocity over the tray surface has its maximum influence during the constant-rate drying period, which normally represents a small fraction of the total drying period. During the period of maximum effect, the drying rate varies as the 0.8 power of the air velocity. Maintaining air velocity throughout the rest of the drying cycle enhances uniformity of drying over the entire tray area.

Effect of Tray Loading. For a homogeneous slab, the effect of thickness on the drying time will depend on the controlling mechanism of drying. If liquid diffusion controls, the drying time will vary approximately as the square of the slab thickness, whereas, if capillary forces control, the drying time will vary directly as the thickness.

In general the effect of cake thickness is more complex than that described above because of the difference in mechanisms controlling the constant-rate and falling-rate periods. The rate during the former period is independent of cake thickness; consequently, as the thickness decreases, the influence of cake thickness decreases since the fraction of the total period governed by constant-rate drying increases. A qualitative analysis of the effect of cake thickness on the production rate (rather than the drying time) shows that an optimal cake thickness exists, and, in general, the optimum value can best be predicted experimentally rather than analytically. Cake thickness of 2.5 to 5.0 cm are generally considered to be optimal.

Effects of Temperature and Humidity. These effects are analytically predictable for the constant-rate period according to the general theory described under Theoretical Considerations. As falling-rate drying takes over, however, the role of temperature and humidity is complex and is of decreasing importance to the drying rate. Temperature and humidity, however, can have a significant role in eliminating or reducing "case hardening." Generally temperatures are maintained at a relatively low level to prevent product degradation during the long drying cycles.

TABLE 1   Comparative Performance Data on Basic Dryer Types

	Basic dryer type					
	Tray*	Conveyor†	Rotary‡	Spray†	Flash‡	Fluid bed‡
Product	Filter cake	Clay	Sand	TiO₂	Spent grain	Coal
Drying time, min	1320	9.5	12	<1.0	<1.0	2.0
Inlet gas temperature, °F	300	420	1650	490	1200	1000
Initial moisture, % dry basis	233	25	6	100	150	16
Final moisture, % dry basis	1	5.3	0.045	0	14	7.5
Product loading, lb dry/ft²	3.25	16.60	N.A.	N.A.	N.A.	21 in deep
Gas velocity, ft/min	500	295	700	50	2000	1000
Product dispersion in gas	Slab	Packed bed	Gravity flow	Spray	Dispersed	Fluid bed
Characteristic product shape	Thin slab	Extrusion	Granules	Spherical drops	Grains	½-in particles
Capacity, lb evap./(h)(dryer area)	0.34	20.63	1.35§	0.27§	10§	285
Energy consumed, Btu/lb evap.	3000	1700	2500	1300	1900	2000
Fan hp/(lb evap./h)	0.042	0.0049	0.0071	0.019	0.017	0.105

*Perry, "Chemical Engineers' Handbook," 4th ed., p. 20-7.
†Proctor and Schwartz, Division of SCM.
‡Williams-Gardner, "Industrial Drying," 1971, pp. 75, 149, 168, 193.
§lb evap./(h)(dryer, volume).

TiO_2

Operating and Maintenance Problems. Some operating problems adversely affecting dryer performance are "case hardening," the formation of a highly resistant skin on the surface of the cake which drastically decreases the drying rate, and cake shrinkage, which can significantly reduce the heat- and mass-transfer area, and thereby increase the drying time.

Tray dryers require little maintenance because of the relatively few mechanically moving parts.

TYPICAL DESIGN PROBLEM

Calcium carbonate filter cake is to be dried in a tray dryer from an initial moisture content of 50% bone-dry basis (BDB) to a final moisture content of 4.0% BDB. The desired production rate is 200 lb of dried product per 8-h day. The laboratory testing program has established the optimal drying conditions tabulated below:

Dry-bulb temperature	250°F
Wet-bulb temperature	110°F
Air velocity	500 ft/min over tray surface
Initial moisture content	50% BDB
Final moisture content	5.0% BDB
Loading	0.50 lb bone dry solid per square foot of tray area
Tray spacing	2 in
Test drying time	1.75 h

Design calculations

1. *Determination of batch time:* Based on experience, a scale-up factor of 2.0 is used to obtain the commercial drying time. Thus, commercial drying time = $2 \times 1.75 = 3.5$ h. Assuming 30 min downtime for loading and unloading the dryer, one obtains a batch time of $3.5 + 0.5 = 4.0$ h, thus providing two batches per 8-h day.

2. *Determination of total tray area required:* Pounds of bone-dry solid charged per batch/loading $= (100/1.05)/0.50 = 190.50$ ft² of tray area.

3. *Number of trays required:* A standard tray, 2 by 2.5 ft, is selected. Thus,

$$\text{Number of trays} = 190.50/2.0 \times 2.5 = 38.10 \text{ or } 40 \text{ trays}$$

Thus, one 40-tray dryer is required.

4. *Airflow rate:* For a drying compartment 20 trays high spaced 2 in apart and two trays deep, the area for airflow is $(2/12)$ (2 trays deep) (2-ft-long tray) (20 trays) = 13.33 ft². The volumetric airflow rate is $(13.33)(500) = 6665$ ft³/min. The estimated pressure drop around the flow circuit is 2 inH₂O at 70°F. The performance requirement of the circulation fan is, then, 7000 ft³/min at 2 inH₂O static pressure.

5. *Calculation of steam consumption:* The maximum steam consumption rate will occur at the beginning of the batch cycle; however, by staggering the loading of the trays, a uniform steam consumption rate corresponding to the average drying rate over the cycle time can be achieved.

Assuming a maximum drying rate of 0.25 lb water per square foot of tray area per hour (obtained by measuring the slope of the test drying-time curve), we calculate the heat balance as follows.

Heat balance

Evaporation load:	
$0.25 (5 \times 40)(970)$	= 48,500 Btu/h
Vapor load:	
$0.25 (5 \times 40)(250 - 110)(0.5)$	= 3,500 Btu/h
Sensible heat of product and trays:	
$150 (0.50)(110 - 60)$	= 3,750 Btu/h
Radiation loss (10% of above)	= 5,575 Btu/h
Total	61,325 Btu/h

Exhaust air rate

The exhaust air rate required to maintain a wet-bulb temperature of 110°F during the maximum-drying-rate part of the cycle is calculated as follows:

$$\text{Temperature drop of air} = \frac{48,500 + 3500 + 3750}{(7000)(60)(0.056)}$$
$$= 2.37 \quad (3°\text{F})$$
$$\text{Exhaust temperature} = 250 - 3 = 247°\text{F}$$

The exhaust air conditions are, then, 247°F dry-bulb and 110°F wet-bulb. The exhaust air rate is now obtained by writing a mass balance on water for the maximum-rate part of the cycle. Thus,

$$G(H_{\text{exh}} - H_{\text{makeup}}) = R_{\text{max}} = \text{maximum drying rate}$$

where G = mass rate of bone-dry air in the exhaust and H = humidity, pounds water vapor per pound dry air.

$$G(0.025 - 0.010) = 0.25 \, (2 \times 2.5) \, 40 = 50$$
$$G = 3333 \text{ lb dry air per hour or 739 ft}^3/\text{min at 70°F}$$

the total energy requirement is

$$61325 + 3333 \, (0.24)(247 - 70) = 202{,}910 \text{ Btu/h}$$

For saturated steam at 100 lb/in² (gage) as the source of energy, the steam consumption is

$$\frac{202{,}910}{889} = 228 \text{ lb steam per hour}$$

which represents the maximum steam rate. The ratio of steam to water evaporated is

$$\frac{228}{50} = 4.56$$

The low thermal efficiency is a consequence of the low evaporation load and the high makeup air load.

Continuous, Through-Circulation Dryer

Products in the form of extrusions, granulations, pellets, briquettes, or any other physical form suitable for packed-bed operation can be dried in a through-circulation dryer. Commercial drying times generally vary from several minutes to an hour or more.

Figures 5, 6, and 7 show the air circulation and conveyor systems in a through-circulation dryer. Bed loading and superficial air velocities are constrained by the pressure drop through the bed, the maximum value of which is about 2.0 inH₂O at 70°F. Normal superficial velocities through the bed range from 50 to 300 ft/min. Bed depths usually range from 1 to 6 in.

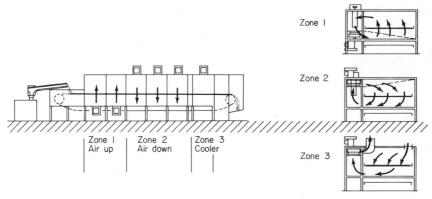

Fig. 5 Basic conveyor dryer configuration. (*Proctor & Schwartz, Inc.*)

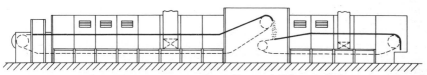

Fig. 6 Two-stage conveyor dryer. (*Proctor & Schwartz, Inc.*)

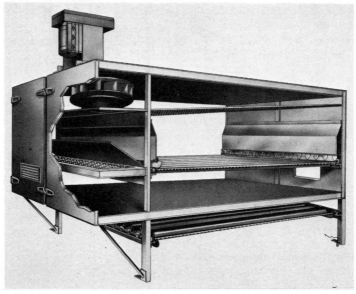

Fig. 7   Cutaway view of conveyor dryer. *(Proctor & Schwartz, Inc.)*

Design Equations   The required dryer holding capacity C is given by $C = Pt$ where P is the production rate in pounds of dry solid per hour and t is the commercial drying time. The necessary conveyor area A is then $A = Pt/L$, where L is the bed loading in pounds of dry solid per square foot of conveyor area. If W is the effective width of the conveyor and B the effective dryer length, then $B = A/W$. Standard conveyor widths range from 2 to 12 ft.

Performance Characteristics   Performance data for a through-circulation dryer are shown in Table 1.

Effect of Air Velocity: The influence of air velocity on the drying time can be predicted with confidence for shallow beds undergoing constant-rate drying. The equation is $t = C/G^{0.59}$, where C is an experimentally determined constant. In the falling-rate period the velocity effect is a complex function of the properties of the material being dried, the moisture content of the material, and the geometry of the bed. However, a general rule is that the magnitude of the effect decreases with the moisture content of the material and is always less than its value for the constant-rate period. In the case of deep beds, as defined by a temperature drop of the air flowing through the bed of 20 F° or more, the effect of air velocity is complicated by the coupling effect of temperature difference and air mass flow rate. This can be quantitatively stated for the constant-rate period as follows:

$$t = \frac{C_1}{G} (1 - e^{-\beta})^{-1}$$

where C_1 and C_2 are experimentally obtained constants and $\beta \equiv C_2 G^{-0.2}$. When the bed is partially in the constant-rate period and partially in the falling-rate period, the influence of air velocity is difficult to predict. Since the power consumption of the circulation fan goes up as the cube of the mass velocity, there is a significant price to be paid for increased production by increased air velocity.

Effect of Bed Loading: For shallow beds the drying time is only slightly influenced by bed loading; however, the production rate varies directly with this variable. In the case of deep beds, a significant variation in drying conditions occurs throughout the bed, resulting in a decrease in the average drying rate per unit volume of bed as the bed depth increases. If the bed is strictly in the constant-rate period, then the effect of bed loading on average drying flux is

$$R_{\text{avg}} = \frac{C_1'}{L}(1 - e^{-C_2'L})$$

where C_1' and C_2' are experimentally determined constants. Although the average drying flux decreases, an increase in the bed loading may result in a net increase in the production rate because of the increase in dryer holding capacity. An analysis of the interacting effects of gas velocity and bed loading on dryer performance is presented in Ref. 4. When both constant-rate and falling-rate periods appertain, the effect of bed loading on the overall drying rate is a complex function of those factors which control the mechanisms of the two periods. As a consequence, it is difficult to predict quantitatively the influence of bed loading without resorting to some degree of experimentation.

Effect of Particle Size: Particle size is primarily dictated by the requirements of stability and permeability of the bed. When an acceptable bed can be formed for a range of particle sizes, that size which results in maximum overall drying rate with minimum air pressure drop should be selected. Generally, the optimum particle size must be determined experimentally. Qualitatively, as particle size decreases, drying rate and gas pressure drop increases for a given bed loading. Quantitatively, when the bed is strictly in the constant-rate period, the drying flux varies as $D_p^{-0.41}$ for the particle Reynolds number $D_pG/\mu > 350$ and as $D_p^{-0.51}$ for $D_pG/\mu < 350$.

Effect of Gas Temperature and Humidity: For a shallow bed, the influence of gas temperature and humidity is the same as for the case of tray drying under constant gas conditions. For deeper beds undergoing constant-rate drying, the gas temperature decreases and the gas humidity increases according to the condition of constant adiabatic cooling temperature. Accordingly, the local drying flux declines as the gas proceeds through the bed. In this situation (constant-rate period throughout the bed) the gas temperature profile is

$$T_g = T_s + (T_{go} - T_s)\exp(-h'Z/GC_g)$$

and the average drying flux is

$$R_{\text{avg}} = \frac{GC_g}{\rho_B\lambda L}(T_{go} - T_s)[1 - \exp(-h'L/GC_g)]$$

where h' is the volumetric heat transfer coefficient
G is the gas mass velocity
L is the bed depth
λ is the latent heat of vaporization
C_g is the gas specific heat
T_s is the adiabatic cooling temperature
ρ_B is the bulk density of the bed
T_{go} is the gas inlet temperature

As drying proceeds, the top of the bed (for downward flow) eventually reaches its critical moisture content and falling-rate drying is initiated. A falling-rate drying zone is thus established at the top of the bed while the remainder of the bed is still undergoing constant-rate drying. The analysis of this period of the drying cycle between strictly constant-rate and strictly falling-rate drying is a complicated one and is best treated experimentally. Fortunately, it is only in a few cases of practical interest where this intermediate drying period is a significant fraction of the total drying cycle; usually the complete cycle can be treated as the sum of a strictly constant-rate period and a strictly falling-rate period. In the falling-rate period, the influence of gas temperature and humidity is best determined experimentally, since the physical and transport properties of the material being dried are involved in a complex way. In general, it can be stated that both gas temperature and gas humidity are of lesser significance to the falling-rate period than to the constant-rate period. By zoned construction of the dryer, this behavior can be optimally treated for each part of the drying cycle in the design and operation of the commercial system.

Several attempts to formulate analytical models of the through-circulation system are described in the literature. The success of these models in predicting the drying time is greatly limited by the simplifications and assumptions on the basis of which the models were derived. Optimization studies of the through-circulation dryer suffer from similar inadequacies and have been applied, up to the present, to constant-rate systems only.

Troubleshooting   When dryer performance fails to meet design specifications, likely sources of the problem are a malfunctioning feeding system, a change in the characteristics of the feed, a decrease in drying temperature, or inadequate exhaust capacity. It is essential that periodic maintenance be carried out to ensure that operating conditions conform to design specifications. Accordingly, frequent cleaning of the conveyor to prevent clogging of air passages and of the air filter, if used, is advisable if maximum production is to be achieved. To implement this, dryer accessibility is an important design factor.

Auxiliary Equipment   The feeding arrangement is an essential ancillary feature of the through-circulation dryer since optimum performance of the system depends on the capability to form a stable, permeable bed subject to minimal energy consumption through pressure drop. Moreover, the feeding system should provide a uniform distribution of the wet product over the conveyor.

Where preforming devices are employed, optimal performance depends upon the physical characteristics of the feed. Methods of preforming some typical materials are tabulated in Ref. 3. A well-designed laboratory experimental program will determine the preforming characteristics of the wet product and will identify the most appropriate preformer and its performance characteristics. Some typical preforming devices and their operating features are described in Ref. 5. In general, products that can be granulated, extruded, briquetted, pelletized, sliced, diced, or reduced in size by a hammer mill are susceptible to drying in a through-circulation system. Extrusible materials include those dewatered by a filter press, rotary vacuum filter, or a centrifuge. Some materials that can be compacted by pelletizing or briquetting are mineral bag-house dust and carbonaceous materials such as coal and charcoal. In other cases, materials such as gelatin and soap slurries can be solidified on a chill roll or jacketed extruder prior to drying. Precooking extruders are used for grain and cereal products and dewatering screws discharge synthetic rubbers in crumb form. Oscillating spreaders of the belt or vibrating type are usually used to distribute the wet product uniformly onto the dryer conveyor.

Although the proper preformer is an essential component of the through-circulation drying system, the capital cost and energy consumption of the preformer are significant factors in the economic evaluation of this type of drying system.

Mechanical Features   The main dryer conveyor is usually made up of sectional hinged perforated plates or wire-mesh screens. Conveyor widths may vary from 2 to 12 ft. Conveyor loadings of from 2 to 100 lb/ft² are typical of this system. The circulation fans are of the centrifugal type and are capable of working against a pressure of up to 6 inH_2O. The length of the dryer is divided into independently controlled zones to provide an optimal schedule of temperature, humidity, and gas velocity for each part of the drying cycle. Alternate up and down direction of airflow can be maintained in order to achieve uniform drying throughout the bed. Internal fines recovery systems minimize product loss. The unitized construction of the dryer permits preassembled shipment and allows for expansion of production by the addition of prefabricated units to an existing installation.

TYPICAL DESIGN PROBLEM

It is desired to dry kaolin clay in a continuous through-circulation dryer. Laboratory tests on the preforming characteristics of the wet product discharged from a rotary vacuum filter indicate that a rolling extruder type of feed is most appropriate. The following experimental data (optimum) are to be used as the basis for the preliminary design of the system:

Air inlet temperature	420°F
Air wet-bulb temperature	150°F
Air velocity through the bed	300 ft/min
Initial moisture content	25.0% (bone-dry basis)
Final moisture content	5.26% (bone-dry basis)
Equilibrium moisture content	0
Test drying time	5.5 min
Air pressure drop through bed	0.85 inH_2O
Average particle diameter	3/8 in
Average particle length	1.0 in
Bulk density of bed (dry)	44 lb/ft³
Void fraction of bed	0.35
Bed loading (dry)	18.0 lb/ft²

The desired production rate is 50,000 lb of dry product (at 5.26%) per hour.

1. *Calculation of dryer size.* On the basis of a scale-up factor of 1.73, the commercial drying time is 9.5 min. The required effective drying area is

$$\frac{50,000}{1.0526} \times \frac{9.5}{60} \times \frac{1}{18.0} = 418 \text{ ft}^2$$

For a conveyor width of 10 ft, the length of dryer is 418/10 = 41.8 ft, or three dryer compartments each 14 ft long.

2. *Selection of circulation fans:* The volumetric flow rate per fan with three fans for each compartment is 300 (14) (10)/3 = 14,000 ft³/min. An estimation of air pressure drop around the airflow circuit gives approximately 4 inH_2O as the pressure against which each fan has to work. For this performance a 30-in-diameter centrifugal-type fan with a 10-hp drive is adequate.

3. *Heat load and exhaust calculations:* The average evaporative load is (50,000/1.0526)(0.25 − 0.0526) = 9377 lb water evaporated per hour. However, this is not the maximum drying flux, which occurs in the first compartment. The recommended procedure is to write separate heat balances around each of the independently controlled dryer compartments and to calculate the exhaust and steam consumption rates for each zone accordingly.

Thus in this example it is decided to divide the dryer into three individually controlled zones. The evaporation load in each zone is obtained by dividing the experimental drying time curve into three segments of equal time duration, and by picking off from the curve the initial and final moisture contents for each segment. Accordingly, these values are found to be

Zone 1: 25 to 16.4%
Zone 2: 16.4 to 9.8%
Zone 3: 9.8 to 5.26%

and the respective evaporation loads are 4085, 3135, and 2157 lb water per hour for a total of 9377. The energy and exhaust requirements are then calculated for each compartment as follows:

Heat balance—compartment 1

Evaporation:	4085(1008)	= 4,118,000 Btu/h
Sensible heat of vapor:	4085(250 − 150)0.5	= 204,000
Sensible heat of liquid:	4085(150 − 60)1.0	= 368,000
Sensible heat solid:	47,500(100 − 60)0.25	= 687,000
Residual liquid:	5292(100 − 60)1.0	
Total (exclusive of radiation losses and exhaust):		= 5,377,000 Btu/h

The humidity of the air exiting the bed, a portion of which is exhausted, is

$$\text{Humidity of air entering bed} + \text{pickup} = 0.130 + \frac{4085}{(42000)(60)(0.0376)}$$
$$= 0.173 \text{ lb water/lb dry air}$$

The exhaust air rate is then determined by writing a mass balance on water. Thus,

$$G(H_{exh} - H_{makeup}) = E \quad \text{(evaporation rate)}$$
$$G(0.173 - 0.0193) = 4085$$
$$G = 27,000 \text{ lb/h dry air, exhaust}$$

The recycle ratio for the first compartment is:

$$\frac{(42,000)(60)(0.0376) - 27,000}{(42,000)(60)(0.0376)} = 0.72$$

In order to determine the heat required for the makeup air, one first calculates the temperature drop through the bed. Thus

$$\Delta T = \frac{5,377,000}{(42,000)(60)(0.0376)(0.24)} = 236°F$$

The exhaust air temperature is then

$$420 - 236 = 184°F$$

Thus, the exhaust air condition is 184°F, 40% relative humidity. The estimated heat loss due to radiation from external surfaces to the surroundings is 80,000 Btu/h. Thus, the total energy requirement for compartment 1 is 5,377,000 + 80,000 + 27000(0.25)(184 − 70) = 6,226,000 Btu/h.

Heat balance—compartment 2

Evaporation:	3135(1008)	= 3,160,000 Btu/h
Sensible heat of vapor:	3135(250 − 150)0.5	= 157,000
Sensible heat of liquid:	7790(150 − 100)1.0	= 389,000
Sensible heat of solid:	47,500(150 − 100)0.25	= 594,000
Total (exclusive of radiation losses and exhaust):		= 4,300,000 Btu/h

The exhaust air humidity is

$$0.130 + \frac{3135}{42000(60)(0.0376)} = 0.163 \text{ lb water/lb dry air}$$

and the exhaust air rate is

$$\frac{3135}{0.163 - 0.0193} = 21,800 \text{ lb/h dry air}$$

Recycle ratio is

$$\frac{42000(60)(0.0376) - 21,800}{42000(60)(0.0376)} = 0.77$$

Temperature drop through bed is

$$\frac{4,300,000}{42000(60)(0.0376)(0.24)} = 189°F$$

Exhaust temperature: 420 − 189 = 231°F
Exhaust air condition: 231°F, 14% relative humidity
Estimated loss by radiation: 60,000 Btu/h
Makeup-air heating load = 21,800(231 − 70)(0.24)
$\qquad\qquad$ = 842,000 Btu/h
Total heat load = 4,300,000 + 60,000 + 842,000
$\qquad\qquad$ = 5,200,000 Btu/h

Heat balance—compartment 3

Evaporation:	2157(1008)	= 2,174,000 Btu/h
Sensible heat of vapor:	2157(250 − 150)(0.5)	= 108,000
Sensible heat of solid:	50,000(200 − 150)0.25	= 625,000
Total (exclusive of radiation losses and exhaust):		= 2,907,000 Btu/h

Humidity of exhaust: $0.130 + \dfrac{2157}{42,000(60)(0.0376)} = 0.153$ lb water/lb dry air

Exhaust air rate: $\dfrac{2157}{0.153 - 0.019} = 16,100$ lb/h dry air

Recycle ratio: $\dfrac{42,000(60)(0.0376) - 16,100}{42,000(60)(0.0376)} = 0.83$

Temperature drop through bed:

$$\frac{2,907,000}{42,000(60)(0.0376)(0.24)} = 128°F$$

Exhaust temperature: 420 − 128 = 292°F
Exhaust air condition: 292°F, 5% relative humidity
Estimated loss by radiation: 80,000 Btu/h
Makeup-air heating load = 16,100(292 − 70)(0.24)
$\qquad\qquad$ = 858,000 Btu/h
Total heating load = 2,907,000 + 80,000 + 858,000
$\qquad\qquad$ = 3,845,000 Btu/h

Summary of Calculations

Compartment	1	2	3	Grand totals
Total heat load per compartment, Btu/h	6,226,000	5,200,000	3,845,000	15,271,000
Exhaust air rate, lb/h dry air	27,000	21,800	16,100	64,900
Exhaust, ft³/min at cond.	9360	7993	6359	23,800
Recycle ratio	0.72	0.77	0.83	0.70
Btu/lb water evaporated				1629

Rotary Dryer

This type of dryer is suitable for free-flowing, nonsticking materials of relatively small particle size. Rotary dryers are of two general types—direct and indirect. A direct-heat continuous rotary dryer consists of a rotating cylindrical shell, slightly inclined to the horizontal, through which hot gas flows in cocurrent or in countercurrent direction to the flow of the product. The rotating shell is typically up to 10 ft in diameter and between 4 and 15 diameters in length, and may be equipped with internal, longitudinal flights spaced about the circumference of the cylinder. Various flight designs are used, depending upon the handling characteristics of the product. For example, the flights may extend continuously the entire length of the dryer, or they may be offset every 2 to 6 ft. Since the purpose of the flight is to lift and shower the wet material through the hot gas, the shape of the flights is an important design consideration. For free-flowing granular materials, a radial flight with a 90° lip is effective, whereas, for sticky materials, a straight radial flight is satisfactory. In operation, the shell rotates at 4 to 5 r/min and the retention time of the product varies from 5 min to 2 h. The gas velocity throughout the cylinder varies from 4.9 to 9.8 ft/s. Figures 8 and 9 illustrate the mechanical features of a typical direct rotary dryer. The feed system may consist of an inclined chute fed from a rotary valve, a vibrating feeder, or a belt conveyor—or, for very wet material, a screw conveyor may be used. The dried product is normally discharged through a counterbalanced flap or rotary valve into the collecting device. The drying air is heated in a heat exchanger; in moderate- and low-

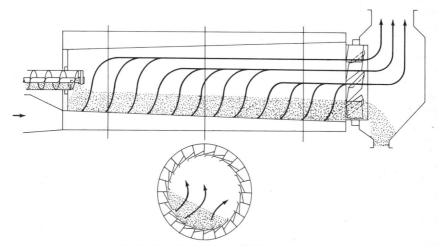

Fig. 8   Typical rotary dryer. *(FMC Corporation.)*

temperature operation, this may consist of a steam-heated tempering coil, while for high-temperature drying, a direct-fired (oil or gas) combustion furnace directly connected to the dryer supplies hot combustion gas to the dryer. Where product contamination by the combustion gas is a possibility, an indirect-fired heating unit is used. The hot gas is forced through the drying cylinder by an induced-draft centrifugal-type fan which is appropriately arranged to provide either cocurrent or countercurrent flow, that is, in the direction of product flow or in a direction counter to product flow. Typically, a dust collector (cyclone type, or filters and scrubbers) recovers the fines from the exhaust gas.

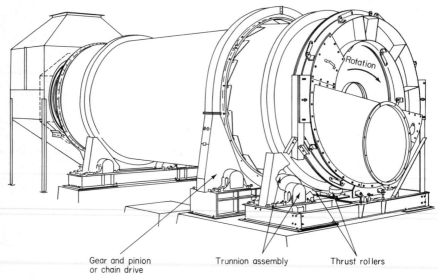

Gear and pinion   Trunnion assembly   Thrust rollers
or chain drive

Fig. 9 Rotary dryer. *(FMC Corporation.)*

In the case of indirect rotary dryers, the material is dried by direct contact with a hot metal surface, such as steam- or hot-water-heated tubes, over which the product flows in a thin layer. A sufficient flow of air is maintained through the dryer to remove the water vapor. Indirect rotary dryers are applicable in the situation where the material undergoing drying cannot be exposed to combustion gases, or where excessive dust carry-over in the combustion gas will occur.

Design Equations: Retention Time An estimation of product residence time is difficult to obtain because of the complex interaction of the following factors:

Percentage loading
Number of flights
Design of flights
Slope of the dryer from the horizontal
Speed of rotation of the dryer shell
Length of dryer (effective)
Diameter of dryer
Physical properties of the material
Air velocity within the dryer

The effect of each of the above factors on the residence time is discussed qualitatively by Williams-Gardner.[5] Empirical expressions, based on experimental work, for the residence time are:

$$t = \frac{KL}{nDS} + Yv \tag{13}$$

where t = retention time, min
  L = effective length of dryer, ft

$n = $ r/min
$D = $ diameter of shell, ft
$S = $ slope of shell, in/ft
$v = $ air velocity, ft/min
K and $Y = $ constants

The constants K and Y depend on certain design characteristics such as number of flights and flight design, size and density of particle, and method of operating the dryer.

For single-shell direct dryers,

$$t = \frac{6KL}{nDS} \qquad (14)$$

where 10 to 15% flight holdup is assumed. K has a value of 0.52 to 2.0 for counterflow dryers and 0.2 to 0.17 for parallel-flow dryers.

Friedman and Marshall[6] suggest the following relationship for the residence time t in minutes:

$$t = \frac{0.23L}{SN^{0.9}D} \pm 0.6 \frac{BLG}{F} \qquad (15)$$

where $B = 5\,D_p^{-0.5}$ is a constant whose value depends upon the material being dried
$D_p = $ weighted average particle size of material in micrometers
$F = $ feed rate in pounds dry material per hour per square foot of dryer cross section
$S = $ slope, ft/ft
$N = $ speed, r/min
$L = $ dryer length, ft
$G = $ air mass velocity, lb/(h)(ft²)
$D = $ dryer diameter, ft

For countercurrent flow, the sign in the expression above is positive, and for cocurrent flow it is negative. Alternatively, Saeman and Mitchell[7] recommend

$$t = \frac{L}{aND(S + bv_m)} \qquad \text{min} \qquad (16)$$

where a and b are constants and v_m is negative for countercurrent flow. The product bv_m is the equivalent slope due to the displacement of the falling material by the airstream. This should not exceed half the actual slope in countercurrent dryers. Peck and Wasan[8] theoretically derived an equation for the resistance time. Their expression is

$$t = \frac{L}{CDN[(Kv^n/\cos \alpha) - \tan \alpha]} \qquad (17)$$

where the constant C depends upon flight design, K is a drag coefficient (particle to air), and α is the angle of inclination of the dryer.

The means by which the rotary dryer performs two basic functions determine the approach to be used in the design analysis: (1) transporting the material through the drying zone, and (2) exchanging heat and mass between air and product. Because of the complex relation between product and airflow, the prediction of heat- and mass-transfer coefficients must, of necessity, be based on empiricism. Knowing the volumetric heat-transfer coefficient U_a, one can calculate the dryer volume V from the expression

$$V = \frac{q}{U_a \,\Delta T_m}$$

where $q = $ the total heat transferred, Btu/h, and $\Delta T_m = $ true mean temperature difference between the hot gas and the product. The general empirical form of the coefficient is

$$U_a = KG^n/D$$

where $K = $ a constant
$G = $ gas mass velocity in pounds per hour per dryer cross-sectional area
$D = $ shell diameter, ft
$n = $ a constant

The value of $n = 0.67$ is probably the most representative of commercial equipment. The *Chemical Engineer's Handbook* (5th ed.),[3] recommends the following expression:

$$q_t = 0.4LDG^{0.67} \, \Delta T_m$$

where D = dryer diameter, ft
  L = dryer length, ft
  ΔT_m = log mean of the drying gas wet-bulb depression at the inlet end and exit end of the dryer shell

Design Procedure

1. *Calculation of Dryer Diameter:* The maximum permissible gas mass velocity is usually that value above which intolerable dust carry-over occurs. On the basis of specified inlet and outlet gas temperatures, the volumetric gas flow rate can be obtained by writing heat and mass balances. Then, if it is assumed that the cross-sectional area available for airflow is 85% of the total crosssection, the dryer diameter can be readily calculated.

2. *Heat Balance:* This can be written as follows:

$$E[\lambda + 0.45(T_{out} - T_{in})] + SC_s(t_{in} - t_{out}) + SX_{in}(1.0)(T_{in} - t_{in})$$
$$+ SX_{out}(1.0)(t_{out} - t_{in}) + q_{rad} = q_T$$

where E = evaporation rate, lb/h
  λ = latent heat of water at T_w, °F
  T_{out} = exit gas temperature, °F
  T_w = gas wet-bulb temperature, °F
  t_{in} = inlet temperature of product, °F
  t_{out} = exit temperature of product, °F
  S = product rate, pounds bone-dry solid per hour
  X_{in} = moisture content of wet product, lb water/lb dry solid
  X_{out} = moisture content of dry product, lb water/lb dry solid
  C_s = specific heat of dry product, Btu/(lb)(°F)
  q_{rad} = heat lost from dryer by radiation from external surfaces, Btu/h
  q_T = total heat transferred, Btu/h

3. *Calculation of Outlet Gas Temperature:* Practical experience indicates that for direct rotary dryers the number of heat-transfer units, N_T, should be between 1.5 and 2 for efficient operation. The number of heat-transfer units is defined as follows:

$$N_T = \frac{T_{in} - T_{out}}{\Delta T_m}$$

where T_{in} = inlet gas temperature, T_{out} = outlet gas temperature, and ΔT_m = the overall mean temperature difference between gas and material. In the case of high product-moisture content, ΔT_m can be defined as the log mean average of the wet-bulb depression of the inlet and exit gas. Thus,

$$N_T = \ln \frac{T_{in} - T_w}{T_{out} - T_w}$$

and

$$T_{out} = T_w + (T_{in} - T_w)e^{-N_T}, \text{ where } N_T \text{ is between 1.5 and 2.0.}$$

4. *Calculation of Gas Mass Rate:* If it is assumed that radiation and convection heat losses are about 7.5% of the evaporative and material heat load as calculated above, the total heat load q_T is obtained. Then, the gas mass flow rate is given by

$$G = \frac{q_T}{C_g(T_{in} - T_{out})}$$

where G = mass flow rate of the gas, lb/h, and C_g is the specific heat of the gas, Btu/(lb)(°F).

5. *Calculation of Exit-Gas Humidity:* A check should be made to ensure that the exit humidity is such that saturation is not exceeded. Accordingly, the exit gas humidity X_{out} is given by

$$X_{out} = X_{in} + \frac{E}{G}$$

where E/G is the moisture pickup by the gas, lb water/lb dry air, and E = the evaporation rate, lb/h of water. Thus, having calculated the temperature and humidity of the exit gas, we have completely determined its thermodynamic state.

6. *Calculation of Dryer Length:* The length can be calculated theoretically from the equation

$$L = \frac{2.5 q_T}{DG^{0.67} \Delta T_m}$$

The ratio L/D, based on the foregoing calculations, should be within the range 4 to 10. If it is not, another value of N_T is chosen in order to obtain a value of the ratio that is within the desired range.

7. *Calculation of Retention Time:* The retention time θ is defined as H/S, where H is the material holdup in the dryer and S is the production rate in pounds of dry solid per hour. For good design, the holdup should be about 7 to 8% of the volume of the dryer. Thus,

$$\theta = \frac{0.075 V \rho_s}{S}$$

where ρ_s is the bulk density of the dry product.

8. *Alternative Calculation of Dryer Volume:* In general it is desirable to obtain the retention time by experiment rather than by the theoretical approach described above. Accordingly, on the basis of pilot-plant data, the experimental retention time required is available to the designer. The necessary volume of dryer is then given by

$$V = \frac{\theta S}{0.075 \rho_s}$$

A reliable experimental estimate of the retention time is almost mandatory in cases of extremely low final moisture content and high final temperature of the product. The scale-up of pilot-plant data to provide design information applicable to the industrial-size dryer is a process that requires considerable experience and good judgment on the part of the designer.

9. *Cocurrent vs. Countercurrent Operation:* The decision to design for cocurrent operation depends upon the following factors:

1. Heat sensitivity of the product
2. Drag effect contribution to solids flow rate
3. Low moisture content of entrained dust

Since the gas and product leave at the same end of the dryer, the temperature difference (gas to product) at this point is minimum and represents a "pinch point." Thus, the final section of the dryer is inefficient in terms of drying potential. Conversely, countercurrent operation ensures a more uniform distribution of the temperature difference; consequently, the efficiency of drying is maximized. Other factors that are controlling in the case of countercurrent flow are

1. Compatibility of dry product with high temperature
2. Drag effect inhibiting product flow rate
3. Poor control of final product temperature
4. Entrainment of wet dust particles

Performance Characteristics of Direct-Type Dryers Typical performance data are shown in Table 1. Thermal efficiencies vary from 30 to 55% for steam-heated dryers and from 50 to 75% for fuel-fired dryers. The evaporation rate of steam-heated dryers varies from 0.2 to 2.0 lb water/ft³. Fuel-fired direct dryers will evaporate from 2.0 to 7.0 lb water/ft³ of dryer volume.

Effect of Gas Velocity. Gas velocity affects dryer performance in several ways, both direct and indirect. With respect to the former, gas velocity has a significant effect on the heat-transfer coefficient; with respect to the latter, gas velocity influences product retention time and degree of product entrainment.

Effect of Rate of Rotation. The retention time is inversely proportional to the rate of rotation. The rotational speed (r/min) times dryer diameter (ft) usually lies between 25 and 35.

Effect of Slope of Shell. For a given rotation speed, as the dryer slope increases, the retention time decreases. For the range of slopes and rotation speeds commonly used ($\frac{1}{4}$ to $\frac{3}{4}$ in/ft and 2 to 7 r/min) the retention time is inversely proportional to the slope.

Effect of Dryer Loading. The percentage of dryer loading, i.e., the ratio of holdup in the dryer to the dryer volume, influences the retention time. Optimum dryer loading lies between 8 to 12% of the dryer volume. The retention time and loading are related by the following:

$$HL = S\theta$$

where H = holdup per unit length, lb/ft
L = dryer length, ft
S = product mass rate, lb/h
θ = retention time, h

Power Requirements. The power required to rotate the dryer shell is given by

$$P = 4.5(10^{-4})W_T v_r + 1.2(10^{-4})BDfN$$

where P = power, kW
W_T = total weight of rotating parts of dryer, kg
v_r = peripheral speed of carrying rollers, m/s
B = mass holdup, kg
D = diameter of shell, m
f = average number of flights per revolution of dryer shell
N = speed of rotation of shell, r/min

TYPICAL DESIGN PROBLEM

A direct rotary dryer is to be designed to dry extruded catalyst pellets ($\frac{3}{8}$ in diameter by $\frac{3}{8}$ in long) at the rate of 1000 lb of dry material per hour from an initial moisture content of 0.60 lb water/lb dry solid to a final moisture content of 0.05 lb/lb. From performance studies of a pilot-plant-size rotary dryer, the following data are available:

Inlet air temperature	320°F
Inlet wet-bulb temperature	110°F
Final air temperature	160°F
Average air mass velocity	36.0 lb/(ft²)(min)
Retention time of product	20 min
Initial moisture content	0.60 lb/lb
Final moisture content	0.05 lb/lb
Final product temperature	110°F
Final air wet-bulb temperature	108°F
Direction of airflow	Cocurrent
Physical properties of product	
Bulk density	34 lb/ft³
Specific heat	0.24 Btu/(lb)(°F)
Product is nonsticking	

A preliminary design of the dryer is desired.

Solution

Cocurrent operation is to be used because of the temperature sensitivity of the dry product.

Dryer Heat Load:

$$E = \frac{1,000}{1.05}(0.60 - 0.05) = 523.8 \text{ lb/h}$$

Heat for evaporation: 523.8(1030)		= 540,000 Btu/h
Heat for vapor:	523.8(160 − 110)0.45 =	12,000
Heat for liquid:	523.8(110 − 70)1.0 =	21,000
Heat for product		
Solid:	953(0.24)(110 − 70) =	9,000
Water:	48(1.0)(110 − 70) =	2,000
		584,000 Btu/h

Estimating heat loss by radiation to be 10% of the above figure we obtain for the total heat load q_T = 640,000 Btu/h.

Calculation of Air-Mass Flow Rate:

On the basis of the pilot-plant data the exit air temperature is assumed to be 160°F. The required air mass rate G is then

$$G = \frac{64000}{0.24(320 - 160)} = 16,700 \text{ lb/h dry air}$$

Calculation of Dryer Diameter:
Tests indicate that minimum entrainment of product occurs when an air mass velocity of 36.0 lb/(ft²)(min) is used. Assuming that 85% of the dryer cross section is free area, we obtain the required shell diameter D. Thus

$$D = \sqrt{\frac{16700(4)}{0.85(36)(60)\pi}} = 3.4 \text{ ft}$$

Calculation of Exit Air Humidity:

$$[X_{in}(= 0.01)] + \frac{523.8}{16,700} = X_{out} = 0.041 \text{ lb/lb}$$

Percent relative humidity = 20 (from psychrometric chart)

Calculation of Number of Transfer Units, N_T:

$$N_T = \ln \frac{T_{in} - T_w}{T_{out} - T_w} = \ln \frac{320 - 110}{160 - 110} = 1.16$$

Calculation of Dryer Length, L:
Assuming a dryer holdup of 7.5%, we obtain the volume V. Thus

$$V = \frac{\theta S}{0.075\rho_s} = \frac{20/60(1000)}{0.075(34)} = 131 \text{ ft}^3$$

Then

$$L = \frac{131(4)}{\pi(3.4)^2} = 14.4 \text{ ft}$$

and

$$\frac{L}{D} = 4.24$$

Steam Consumption:
The energy load for the heat exchanger is determined by the amount of heat necessary to raise the temperature of outside air from 60 to 320°F. Accordingly,

$$q_{\text{heat exchanger}} = 16,700(320 - 60)0.24 = 1,040,000 \text{ Btu/h}$$

Assuming steam pressure of 175 lb/in² (gage) is available, we find that the steam consumption is 1,040,000/845 = 1230 lb/h. The thermal efficiency of the dryer is

$$\frac{q_T - q_r}{q_{\text{heat exchanger}}} = \frac{585,000}{1,040,000} = 0.56 \text{ or } 56\%$$

Horsepower Requirements:
Total power required includes that of the dryer fan and the shell drive. The power input to the fan is a function of the volumetric airflow rate and the pressure drop over the entire system including the gas cleaning equipment. In the example above, a 10-hp centrifugal fan and 3-hp dryer drive would be adequate for estimation purposes. As a general approximation, the total horsepower is estimated to fall within the range of $0.5D^2$ to $1.0D^2$, where D is the diameter of the dryer in feet. The speed of rotation of the shell can be approximated by the rule that $25 \leqslant ND \leqslant 35$, where N is in revolutions/minute and D is the diameter in feet. Thus, in this example, N is 30/3.4 ≈ 9 r/min.

Spray Dryer

The spray dryer provides a fourfold function: dispersal of the product in the form of fine atomized droplets, pneumatic transport of the product through the drying zone, supplying energy to the product in order to accomplish drying, and separation of dried product from the gas stream. Accordingly, a complete design and performance analysis of the spray-drying system should consider the aspects of atomization of the liquid feed, residence time distribution in the drying chamber, mechanism of heat and mass transfer to the product droplets and the surrounding gas, and mechanisms of gas-solid separation.

Two basic spray-drying systems are illustrated diagrammatically in Figs. 10 and 11. The feed is pumped from the product feed tank to the atomization unit which consists of either a number of small orifice nozzles or a spinning disk. The purpose of this device is to create a uniform spray of small droplets (usually in the 10 to 20 μm range). Air or gas is forced through a direct-fired combustion chamber, or over steam-heated coils, by a centrifugal blower, and is delivered to the drying chamber inlet at a temperature ranging between 200 and 1400°F. The contact between the finely dispersed droplets and the hot gas results in rapid drying (drying time can be on the order of a fraction of a second). The dry particles settle through the drying chamber and are pneumatically conveyed from the product discharge, often at the bottom of the drying chamber, to a cyclone or bag collector where the product is separated from the carrier air.

The relative flow direction of air and product can be varied to provide several different airflow configurations. Figure 10 depicts cocurrent operation. In countercurrent flow, the hot air enters at the opposite end of the chamber from the product atomization point. Although this design maximizes the value of the mean temperature driving force, it has the disadvantage, in the case of temperature-sensitive materials, that a lower inlet-gas temperature must be used than that in the cocurrent dryer. In the mixed-flow dryer, the hot air enters tangentially, spirals down the side of the cone, reverses direction, and flows

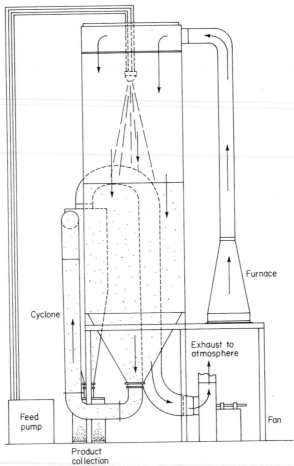

Furnace

Cyclone

Exhaust to
atmosphere

Feed
pump

Fan

Product
collection

Fig. 10 Nozzle-atomization spray dryer. *(Proctor & Schwartz, Inc.)*

upward through the center of the chamber to leave at the top. The product is sprayed into the rising air stream. This type of flow pattern results in high evaporative capacity and relatively low entrainment of dust.

Atomization System. Three types of atomizers are in general use. Single-fluid pressure nozzles atomize the liquid by virtue of the high pressure of the feed delivered by the pump. Pressures may range from 400 to 10,000 lb/in². The higher the pressure, the finer is the atomization and the higher is the capacity of the nozzle. Nozzle orifice diameters may vary from 0.013 to 0.15 in. Two-fluid nozzles, appropriate for low-capacity operations, require a second fluid (air or steam) at 60 to 100 lb/in² (gage) to atomize the feed. Disk atomizers use a whirling disk to atomize the feed. Disk diameters range from 2 to 14 in and rotation speeds from 3000 to 50,000 r/min. Capacities up to 60,000 lb/h of feed can be obtained from a single atomizer. Disk atomization is generally applicable to suspensions and pastes that would erode and plug nozzles.

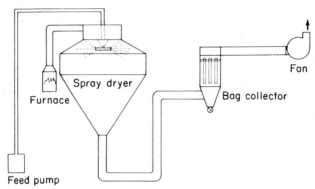

Fig. 11   Disk-atomization spray dryer. (*Proctor & Schwartz, Inc.*)

Separation System. Two systems are in general use to separate the dry product from the air. In the first, the dry material is allowed to settle in the coned-shaped bottom of the drying chamber where it is discharged by a rotary valve. The exhaust air containing some entrained product flows to a cyclone collector, bag filter, wet scrubber, or an elecrostatic precipitator, or combinations of the foregoing. In the second system, the exhaust air containing all of the product is discharged from the drying chamber and sent to the separation equipment where total recovery of the product occurs.

Product Characteristics   An important advantage of spray drying is related to the physical properties of the dried product. The formation of a spherical particle, solid or hollow, is a desirable and unique feature of the spray-drying process. This shape is conducive to a free-flowing product of controlled bulk density. Moreover, a hollow, spherical product correlates with a low bulk density and high solubility. The properties of particle size distribution, bulk density, and agglomeration are usually of greatest interest and can be controlled within limits by appropriate manipulation of the design variables.

Performance Characteristics   Performance data for a typical spray drying system are presented in Table 1. Volumetric evaporative capacities generally lie in the range 1 to 5 kg/(h)(m³) for drying temperatures between 150°C and 200°C.

Estimation of Drying Time. Particle residence times may range up to 30 s. Actual drying times can be crudely estimated from the theoretical expression for the drying time of a pure liquid drop of fixed diameter D_p:

$$\theta = \frac{\lambda W \rho_s D_p^2}{12 k_f (T_a - T_s)}$$

where θ = drying time, h
   λ = latent heat of evaporation, Btu/lb
   W = moisture content of the drop, lb/lb of dry solid
   ρ_s = density of dry particle, lb/ft³
$T_a - T_s$ = temperature difference between drop and gas, °F

k_f = thermal conductivity of the gas film, Btu/(h)(ft²)(°F/ft). On the basis of the above equation, a drop 10 μm in diameter would theoretically require a drying time of 0.0042 s at a ΔT of 275°F.

The estimation of drying rates in spray dryers can be obtained if an expression for the heat-transfer coefficient (gas to drop) is available. One such correlation in wide use is that suggested by Ranz and Marshall:[9]

$$\frac{h_c d_p}{k_f} = 2.0 + 0.6 \left(\frac{c_p \mu_g}{k_f}\right)^{0.33} \left(\frac{d_p U \rho_g}{\mu_g}\right)^{0.5}$$

where h_c = heat-transfer coefficient, k_f = thermal conductivity of gas film, c_p = heat capacity of gas, ρ_g = density of gas, d_p = drop diameter, μ_g = viscosity of gas, and U is drop velocity relative to that of the air.

The evaporation rate for a drop is then given by

$$\frac{dW}{dt} = \frac{h_c A \ \Delta T}{\lambda} = \frac{h_c \pi D_p^2 \ \Delta T}{\lambda} \tag{18}$$

where W is the mass of the drop,

$$W = \frac{\pi D_p^3 \rho_d}{6}$$

and ρ_d = density of the drop, assumed to be pure liquid.

The integration of Eq. (18) between an initial diameter D_0 and a final diameter D_1 yields, for the drying time

$$\theta = \frac{\lambda \rho_d (D_0^2 - D_1^2)}{8 k_f \ \Delta T_m}$$

where ΔT_m = log mean temperature difference, and where $h_c = 2k_f/D_p$ based on negligible relative velocity.

When the relative velocity between gas and droplet is taken into account, the following equation for the drying time of a pure liquid droplet is obtained:

$$\theta = \frac{\lambda \rho_d}{2 k_f \ \Delta T_m} \int_{D_1}^{D_0} \frac{D \ d(D)}{2 + 0.6 \mathrm{Re}^{0.5} P_r^{0.33}}$$

This equation was integrated by Duffie and Marshall[10] for droplets falling at terminal velocity within the Reynolds number range of 0.2 to 500. The procedure of predicting drying time for sprays of pure liquid drops is outlined by Masters.[11]

When dissolved solids are present in the droplet, a falling-rate period may be exhibited during the evaporation process. This case is considered by Masters, as well as the case in which the droplets contain insoluble solids.

Estimation of Residence Time. The volume of the drying chamber divided by the volumetric gas flow rate (calculated at the outlet gas temperature) gives the minimum residence of the product. The actual particle residence time of most of the droplets is much higher than this because of recirculating gas flow patterns and nonuniform gas velocity.

Theoretical Prediction of Droplet Velocity Vector. Because of the complex interaction of a number of design factors such as nozzle configuration and type, chamber size and geometry, size distribution of the droplets, and gas flow pattern, it is impossible to predict precisely the droplet movement in the drying chamber. At best one can make approximations of limiting velocity profiles based on idealized flow conditions. Droplet transport in the drying chamber may be analyzed in terms of droplet trajectory from the atomizer, deceleration, and terminal velocity under gas flow conditions. Masters considers the three flow regimes and summarizes the important hydrodynamic equations.

Spray trajectory pattern is a function of the type of atomizer. For disk-type atomizers, Frazer et al.[12] give the radial distance in feet at which 99% of the spray falls 3 ft below the atomizer as

$$R_{\max} = 7.2 d^{0.21} \frac{M^{0.2}}{N^{0.16}}$$

where d = wheel/disk diameter, M = feed rate, and N = atomizer speed of rotation. In the case of nozzle atomizers, the droplets are ejected with a high velocity component in a conical pattern generally in the direction of the nozzle axis. Radial velocity component is usually low and spray trajectory to the wall is limited.

The spray release velocity for different types of atomizers is summarized by Masters.[11] Rapid deceleration of the droplets occurs in the neighborhood of the atomizer. An equation for the deceleration time in a turbulent flow region is given by Masters. The terminal velocity V_f is

$$V_f = \left[\frac{4}{3}\frac{(\rho_w - \rho_a)gD}{C_D\rho_a}\right]^{1/2} \quad \text{for a spherical drop}$$

where C_D = drag coefficient
  ρ_w = density of droplet
  ρ_a = density of air
  g = gravitational acceleration
  D = particle diameter

In spray dryers, terminal velocities can well be within the semiturbulent regime, in which case

$$V_f = \frac{0.153D^{1.14}g^{0.71}(\rho_w - \rho_a)}{(\rho_a)^{0.29}(\mu_a)^{0.4}}$$

Distance-time relationships for single spherical droplets are given by Masters. The trajectory time to the drying chamber wall is given (for a disk atomizer) by

$$\theta = \frac{(R_c - r_d/2)^2}{2.4V_0(b'/r_d)}\frac{1}{2}$$

where R_c = chamber radius, ft
  V_0 = the droplet release velocity from the atomizer
  r_d = disk atomizer wheel/disk radius, ft
  b' = width of annular-slot air orifice, ft = $(M_L/\rho_a\pi DV)$
  V = induced airstream velocity, ft/h;
  D = drop diameter, ft
  M_L = liquid mass feed rate

For pressure nozzles, the trajectory time is given by

$$\theta = \left(\frac{64}{\pi}\right)^{2/3}\frac{V^{2/3}}{6.4V_{vo}d_o'}$$

where V = drying chamber volume, ft^3
  V_{vo} = axial velocity of liquid at nozzle orifice, ft/h
and $d_o' = d_o(\rho_w/\rho_a)^{1/2}$, ft, d_o = diameter of nozzle orifice

The above results for the trajectory time apply to chamber diameters under 5 ft.

Atomization Characteristics and Performance. The characteristics of the spray, such as droplet size distribution, release velocity, and dispersal area have profound influence on the rate and uniformity of drying and on the physical properties of the dry product. Moreover, the type of atomizer determines to a great extent the geometry of the drying chamber. Thus, a disk atomizer, which produces an umbrella-shaped spray, requires a chamber of large diameter, whereas a pressure nozzle results in a smaller diameter chamber but with greater depth.

The *disk or centrifugal atomizer* uses the energy developed by centrifugal force to create a spray. The liquid film is ejected from the rim of the disk at a velocity of 200 to 600 ft/s, whereupon it breaks up into droplets. Droplet size prediction for smooth flat vaneless disks indicates that the maximum drop diameter is proportional to N^{-1}, $\gamma^{1/2}$, ρ^{-1}, and D^{-1}, where N = speed of rotation of the disc, γ = surface tension, ρ = liquid density, and D = diameter of disk. For vaned atomizer wheels, the droplet diameter varies with $N^{-0.6}$, $D^{-0.2}$, $Q^{0.2}$, $b^{-0.1}$, $n^{-0.1}$, $\mu^{-0.1}$, $\rho^{-0.5}$, and $\gamma^{0.1}$, where Q = feet rate, b = vane height, n = number of vanes, μ = feed viscosity. Available correlations for the prediction of droplet size are summarized by Masters. Size distribution of sprays is best predicted by the method of Herring and Marshall[13] according to which the group

$$\frac{D_p (N\,D)^{0.83}(nh)^{0.12}}{M_L^{0.24}} \times 10^{-4}$$

is plotted on square-root probability paper against the cumulative volume percent less than D_p, the given droplet size, where D = wheel diameter, and M_L = liquid mass feed rate. Data, however, are largely lacking for liquids other than water. For scale-up purposes at a constant size distribution, the correlation of Friedman, Gluckert, and Marhsall[14] is recommended. Accordingly

$$D_{vs} = K' r \left(\frac{M_p}{\rho N r^2}\right)^{0.6} \left(\frac{\mu}{M_p}\right)^{0.2} \left(\frac{\gamma \rho n h}{M_p^2}\right)^{0.1} \tag{19}$$

where K' = constant ($\sim$0.37)
$\quad D_{vs}$ = Sauter mean diameter, ft
$\quad r$ = radius of wheel, ft
$\quad M_p$ = mass feed rate per unit of wet periphery, lb/(min)(ft)
$\quad \gamma$ = surface tension, lb/min^{-2}
$\quad \mu$ = vicosity, lb/(ft)(min)
$\quad \rho$ = density, lb/ft^3
$\quad D_{max} = 3.0\,D_{vs}$

Thus, for a change in feed rate, the vane height h or number of vanes n is adjusted to maintain the same conditions of M_p. When this is done, the predicted mean and maximum droplet sizes remain unaffected. The following example, from Masters, illustrates the method of scale-up.

EXAMPLE

An atomizer with a 9-in wheel spins at 18,000 r/min. The wheel contains 240 vanes ($\frac{3}{8}$-in height). The feed rate is 2200 lb/h. It is desired to scale up the data to a feed rate of 3300 lb/h, at the same time maintaining the original droplet size distribution in the spray. Predict the original mean droplet diameter and specify the design conditions for the scaled-up atomizer. The feed properties are $\rho = 70$ lb/ft^3, $\gamma = 74$ dyn/cm, $\mu = 1$ cP.

Solution

$$\text{Total wetted periphery} = \frac{240(0.375)}{12} = 7.5 \text{ ft}$$

$$M_p = \frac{2200}{60(7.5)} = 4.9 \text{ lb/(min)(ft)}$$

Substituting into Eq. (19), we obtain for the Sauter mean droplet size $D_{vs} = 75.5 \ \mu$m. In order to maintain the feed rate per wet periphery constant, the vane height is increased to $(3300/2200) \times 0.375 = 0.563$ in, and because of the last term of Eq. (14) the mean droplet diameter is increased by only 4%.

For a given size atomizer, the change in evaporative capacity requires a corresponding adjustment in wheel speed of rotation. The required wheel speed to produce the same dried particle size for a specified feed condition is given by

$$N_2^{0.6} = N_1^{0.6} \left(\frac{E_2}{E_1}\right)^{0.2}$$

where N_2 is the adjusted wheel speed to produce a capacity corresponding to the dryer evaporative capacity E_2.

Pressure nozzles produce hollow conical spray patterns; however, solid cone spray patterns can be produced by balancing the rotational motion with an axial liquid velocity component. The mass flow rate M_L is a function of the orifice area and the pressure of the feed. Thus, $M_L = C_D A \ (2gP\rho)^{1/2}$, where C_D is the orifice coefficient, P = pressure drop, A = orifice area, and ρ = density of the liquid feed.

The effect of operating variables on droplet size is such that the latter decreases with an increase of pressure to about the 0.3 power of the pressure, increases with the 0.20 power of the viscosity, increases with the square of the orifice diameter, and is only slightly affected by feed viscosity. The range of droplet size obtained with this type of atomizer is extensive, and correlations to predict nozzle performance are available for given nozzle designs. Many sprays exhibit a square root normal distribution of drop sizes. The following empirical correlation applies to centrifugal pressure nozzles:

$$D_{vs} = 286(d_o + 0.17) \exp\left[\frac{13}{U_v} - 0.0094U_T\right]$$

where d_o = diameter of nozzle orifice, in
    U_v = axial velocity, ft/s
    U_T = tangential velocity, ft/s, where:

$$U_v = \frac{Q}{\pi(r_o^2 - r_c^2)} \quad \text{and} \quad U_T = \frac{Q}{A_{sw}}$$

in which Q = volumetric feed rate
    r_o = orifice radius
    r_c = air core radius
    A_{sw} = area of flow into swirl chamber

Masters summarizes other correlations for the prediction of mean droplet size.

In the case of two-fluid nozzle atomizers, small droplet sizes are produced over a wide range of feed rates. Droplet size can be varied significantly by adjustment of the feed-air ratio at the nozzle head. Maximum spray angle is about 70 to 80° at maximum feed rate and air pressure. The Sauter mean droplet size in micrometers is given by:

$$D_{vs} = \frac{1410}{V_a}\left(\frac{\gamma}{\rho}\right)^{0.5} + 191\left[\frac{\mu}{(\gamma\rho)^{0.5}}\right]^{0.45} \times \left(\frac{Q_l}{Q_a}10^3\right)^{1.5}$$

where V_a = velocity of air, ft/s, Q_l/Q_a = ratio of volumetric rates of liquid to air; γ is in dyn/cm, and μ is in cP.

Design Considerations.

1. Heat balance: For specified inlet and outlet gas temperatures, the heat balance gives the required gas mass flow rate through the dryer. The quantities included in the heat balance are:

$$\begin{aligned}
\text{Heat for evaporation} &= E\lambda = F(m_0 - m_1)\lambda &&= q_1 \\
\text{Heat for vapor} &= E(0.45)(T_1 - T_s) &&= q_2 \\
\text{Heat for liquid} &= F(M_0 - M_1)(C_{p_l})(T_s - t_0) &&= q_3 \\
\text{Heat for dry product} &= F(1 + M_1)(c_{p_s})(t_1 - t_0) &&= q_4 \\
\text{Heat losses} &= AU_a(T_{avg} - T_{amb}) &&= q_5
\end{aligned}$$

and $q_D = q_1 + q_2 + q_3 + q_4 + q_5$ = dryer heat load.

The gas rate G is obtained as follows:

$$G = \frac{q_T}{C_g(T_0 - T_1)}$$

The humidity H_1 of the outlet gas is then given by

$$\frac{E}{G} + H_0 = H_1$$

where H_0 is the humidity of the inlet gas. The total heat load Q_T of the system is that required to heat the air from ambient temperature to the inlet temperature. Thus

$$Q_T = GC_g(T_0 - T_{amb})$$

Defining the overall thermal efficiency $\eta_{overall}$ as the fraction of the total heat supplied to the dryer that is used in the evaporation process, we have

$$\eta_{overall} = \frac{T_0 - T_2}{T_0 - T_{amb}}$$

where T_2 = the outlet gas temperature if the drying process were purely adiabatic. Alternatively,

$$\eta_{overall} = \frac{q_D - q_5}{q_T}$$

2. Residence time in drying chamber: The minimum residence time of the product is given by V/Gv where V = volume of the drying chamber, G = mass flow rate of dry gas, and v = specific volume of the gas, defined as cubic feet of humid gas at outlet conditions per pound of *dry* gas. Thus, for a conical-based drying chamber of cylindrical height h' and a cone angle of 60°, the minimum residence time t_r is

$$t_r = \frac{0.7854 D_{CH}^2(h' + 0.2886 D_{CH})}{Gv}$$

where D_{CH} is the chamber diameter.

Size of Drying Chamber. The design of the drying chamber is fundamentally dictated by (1) the required time to dry the product to the desired final moisture content, (2) the radial component of the velocity of the spray droplets leaving the atomizer, and (3) the time required to dry the product to a moisture content just below its sticking point. Gas flow pattern, droplet size and velocity distributions, and desired final product temperature and moisture content are the design parameters that have a direct bearing on the critical factors stated above. If the droplet trajectory characteristics are such that the larger droplets have experienced insufficient drying time when contact with the chamber wall takes place, then buildup of semiwet product on the wall is apt to occur. To limit this, a judicious choice of chamber diameter as influenced by gas flow pattern and production rate and size distribution of the spray must be made. This decision is based primarily on experience combined with operating data obtained on a similar unit. Unfortunately, theory in the form of mathematical formulas and correlations is of little help. As a rough approximation, the estimated radial velocity of an average droplet, is given by

$$U_r = \frac{V_t^2(\rho_w - \rho_a)D^2}{18\mu_a r}$$

for a vaned atomizer wheel, where

$$V_t = V_{t(wall)}\left(\frac{r}{R_c}\right)^{0.5} = \text{air velocity in the tangential direction}$$

A rough estimate of the transit time of a droplet before it reaches the chamber wall can be obtained by dividing the chamber radius by the integrated average radial velocity. This gives the following expression for the transit time of a droplet in traveling from r_1 to r_2:

$$t = \frac{18\mu_a \ln (r_2/r_1)}{D^2\omega^2(\rho_w - \rho_a)}$$

where ω = average angular velocity. If the radial transit time is greater than, or equal to, the drying time for the droplet, the sticking of those particles striking the chamber wall will not occur.

Notwithstanding the foregoing, it must be emphasized that the optimal design of the drying chamber, i.e., its shape and size, can be determined only from reliable experimental data. The critical design information, which can be obtained realistically only by experiment, is: (1) percent feed solids that can be handled, (2) inlet and outlet temperatures to the dryer, (3) airflow rate, and (4) required exposure time.

EXAMPLE: PERFORMANCE ANALYSIS OF SPRAY DRYER

The following operating data were obtained from a spray dryer:

Inlet air: 100,000 ft³/h, 600°F, 760 mmHg, 40°F dew point
Inlet solution: 300 lb/h, 15 wt % solids, 70°F
Outlet air: 200°F, 760 mmHg, 100°F dew point
Outlet solid: 130°F

Calculate the thermal efficiency of the dryer assuming no recirculation of the exit air and an air supply temperature of 70°F.

Solution

Procedure: A mass balance on water around the dryer gives the moisture content of the product. The energy requirements for the drying operation can then be established, and the thermal efficiency can be calculated by writing an overall heat balance.

Calculations:
Mass balance on water:

$$H_2O \text{ in inlet air} + H_2O \text{ in feed} = H_2O \text{ in outlet air} + H_2O \text{ in product}$$

From the psychrometric chart we obtain the inlet and outlet air humidities of 0.0050 and 0.0428 lb H_2O/lb dry air, respectively. Also, since the humid volume of the inlet air is, from the psychrometric chart, 26.90 ft³ of moist air per pound of dry air, the mass flow rate of air through the dryer is 100,000/26.90 = 3718 lb/h dry air.

Then the mass balance is

$$3718(0.0050) + 300(0.85) = 3718(0.0428) + 300(0.15)X_p$$

Thus $X_p = 2.543$ lb water/lb dry product.

Energy required for drying:

Evaporation	140.56(1054)	= 148,161 Btu/h
Vapor	140.56(200 − 70)0.5	= 9,137
Solid	300(0.15)(0.2)(130 − 70) =	540
Water, liquid	300(0.15)(2.543)(1.0)	= 114
Total heat load =		158,000 Btu/h

From psychrometric chart:

Enthalpy of air entering dryer = 134.2 Btu/lb dry air
Enthalpy of air leaving dryer = 80.5

Heat balance around dryer proper:
Total enthalpy (inlet air) − Total enthalpy (outlet air)
= energy lost to surrounding + heat content of product = 3718 (134.2 − 80.5) = 200,000 Btu/h
Heat lost to surroundings by radiation, etc.
= 200000 − 540 − 114 = 199,350 Btu/h
Total heat supplied = 3718 (0.24)(600 − 70)
= 473,000 Btu/h

Thermal efficiency:
$$\frac{158,000}{473,000} (100) = 33.40\%$$

Flash Dryer

In its essential features the flash dryer is similar to the spray dryer with the exception that the feed to the former is a particulate solid rather than an atomized liquid. The feed—usually a free-flowing powder or granulations—is injected into a hot gas stream at the bottom of a vertical pipe which serves as the drying chamber. The wet feed, pneumatically conveyed up the drying leg, flashes its liquid moisture into the gas stream, causing a significant increase of the velocity as the gas flows along the drying chamber. The dry solid is separated from the gas stream in a cyclone, while the gas and vapor are exhausted from the system by means of a centrifugal vent fan. Bag collectors or scrubbers are often used for final collection.

The basic system consists of an air heater, a feeding device, vertical drying leg, cyclone or other collector, and exhaust fan. See Fig. 12. Size reduction of the feed, such as milling, may be necessary if the wet product is too large to be suspended directly in the flowing gas. The drying gas may be indirectly heated air or the combustion gas from a direct-fired furnace. The feeding mechanism may be a screw conveyor or vibratory conveyor feeding through a rotary air lock or a venturi section. Some recycling of the dry material to mix with the wet feed may be necessary in order to obtain an optimum initial moisture content. Drying times typical of flash dryers are less than 10 s and may be as low as 1 s or less. These values correspond to particle residence times of the same order of magnitude. Normal gas velocities in the drying leg are in the range of 1000 to 6000 ft/min. Devices are often employed to increase hold time for heavier particles that have a longer drying time.

Typical Design and Operating Data   For an evaporative capacity range of 1000 to 20,000 lb of water per hour, the fuel requirement range is 1,750,000 to 35,000,000 Btu/h. For these conditions, the power requirement will vary from 20 to 285 kW. The corresponding approximate space requirements are a floor area of 15 × 20 ft × 25 ft high at the lower end of the range and 48 × 50 × 100 ft at the upper end. Inlet gas temperatures are generally high, similar to inlet temperatures used in spray drying. The use of high inlet temperatures leads to higher thermal efficiencies.

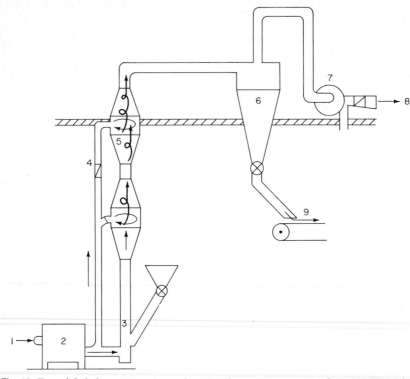

Fig. 12 Typical flash dryer. (1) Fresh air inlet; (2) air heater; (3) mainstream drying air; (4) auxiliary drying air; (5) dryer; (6) separator; (7) main circulating fan; (8) exhaust air outlet; (9) product delivery. (*Proctor & Schwartz, Inc.*)

DESIGN EXAMPLE

It is desired to dry a particular product in a flash dryer. The production rate is 2500 lb bone-dry solid per hour. Design conditions established from laboratory studies on a test unit are tabulated below.

Initial moisture content, dry basis	54%
Final moisture content, dry basis	19%
Inlet air temperature, °F	600
Outlet air temperature, °F	200
Average residence time, s	1.2
Product temperature, °F	150
Maximum allowable product temperature, °F	175
Inlet air humidity, lb water/lb dry air	0.01

Calculate:
  1. Volumetric flow rate of air
  2. Energy consumption
  3. Overall thermal efficiency
  4. Size of drying chamber

Solution

  1. Assuming a 5% heat loss (heat lost to surroundings by radiation and sensible heat leaving in the dry product), one first calculates the enthalpies of the inlet and outlet air streams. Thus,
    Inlet air enthalpy (Based on a reference state of liquid water at 32°F and dry air at 0°F):

$$H_{in} = 0.24T + Y_{in}[1075.5 + 0.45(T - 32)]$$

where H = enthalpy, Btu/lb dry air and Y = humidity, lb water vapor/lb dry air.

$$H_{in} = 0.24(600) + 0.01[1075.5 + 0.45(600 - 32)]$$
$$= 157.31 \text{ Btu/lb dry air}$$

Out air enthalpy:

$$H_{out} = 0.24(200) + Y_{out}[1075.5 + 0.45(200 - 32)]$$
$$= 48 + 1151Y_{out}$$

But

$$H_{out} = H_{in} - 0.05H_{in} = 0.95(157.31) = 149.4$$

Solving for Y_{out}, one obtains $Y_{out} = 0.0881$. Now, by a *mass balance* on water, one obtains

$$G(Y_{out} - Y_{in}) = E = \text{evaporation rate}$$

where G = mass rate of dry air through the drying chamber. Thus,

$$G = \frac{E}{Y_{out} - Y_{in}} = \frac{2500(0.54 - 0.01)}{0.0881 - 0.01}$$
$$= 17,000 \text{ lb dry air per hour}$$

The humid volume of the air at inlet conditions is 27.1 ft³ moist air per pound of dry air. Thus, the volumetric flow rate is

$$17,000 \times 27.1 \times \frac{1}{60} = 7700 \text{ ft}^3/\text{min}$$

2. The heat load for the dryer is

Evaporation:	$2500(0.54 - 0.01)(1075.5)$	=	1,425,000 Btu/h
Vapor:	$1325(0.45)(230 - 32)$	=	118,000
Product:	$2500(0.2)(150 - 32)$	=	59,000
Total heat load			1,600,000 Btu/h

3. Total energy input to system (for supply air at 50°F) is $17,000(157.31 - 22.84) = 2,300,000$ Btu/h. where the enthalpy of the supply air is 22.84. Overall thermal efficiency is, therefore,

$$\frac{1,600,000}{2,300,000} \times 100 = 70\%$$

4. Assume an average velocity of 2000 ft/min in the drying chamber.

Diameter of vertical lag:
$$7700 \times \frac{1}{2000} \times \frac{(4)(144)}{\pi} = D^2 = 706 \text{ in}^2$$
$$D = 27 \text{ in}$$

Length of vertical leg:
$$2000 \times \frac{1.2}{60} = 40 \text{ ft}$$

Fluid-Bed Dryer

In the fluid-bed dryer a granular or, in general, particulate solid phase is partially dispersed in a vertically flowing gas stream as a consequence of the buoyancy effect of the gas. Because of the agitation in the fluidized bed, good mixing of the solid phase is usually achieved. Moreover, the turbulent activity in the bed produces high rates of heat transfer between gas and solid phases, and results in a uniformity of the solid-phase temperature throughout the dryer. Although the agitation of the product can be violent, the individual particles experience gentle treatment with little attrition.

The basic fluid-bed drying system consists of a feeding device such as a screw conveyor or star rotary valve, a fluidizing chamber which may be round, square, or rectangular in cross section and which contains a perforated distribution plate at the bottom, a centrifugal blower inducing heated air through the plenum chamber underneath the distribution plate, a heating source such as steam coils or a direct-fired furnace, a cyclone dust collector, and a bag house or scrubber. Vibrating conveyors are often used to move the product through the drying chamber. Feed inlet and product drawoff are usually on opposite sides of the drying chamber. The fluidizing and drying medium, generally hot air, is fed to the plenum chamber and is uniformly distributed over the cross section of the drying chamber by the perforated plate or other porous medium. The pressure drop of the air across the distribution plate and the bed is on the order of 20 to 40 inH₂O. Figure 13 shows basic fluid-bed drying systems. Figure 14 shows modes of fluidization.

Fluidizing Velocity   The normal operating range of the superficial gas velocity is about 25 to 300 ft/min. A consideration of the relation between the pressure drop across the bed and the superficial gas velocity through the bed is essential to the understanding of the operation of the fluid-bed dryer. As the gas velocity is increased from zero, the bed undergoes a transformation from a fixed bed to a partially or completely fluidized bed. The incipient fluidizing velocity corresponds to that velocity for which the pressure drop is sufficient to support the weight of the bed. A plot of the log of pressure drop against the

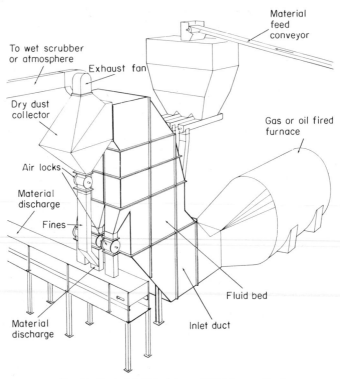

Fig. 13   Typical fluid-bed dryer. *(FMC Corporation.)*

log of the superficial velocity yields a straight line of positive slope over that range of operation corresponding to fixed-bed drying. Beyond the incipient fluidizing velocity, the bed markedly expands at constant pressure drop as the velocity is increased. The bed is now said to be in a state of aggregative fluidization. The flow regimes of the aggregative fluidized system are characterized by "bubbling" followed by "slugging" at gas velocities much greater than the incipient fluidizing velocity. In the bubbling regime, bubbles of the gas rise through the fluidized bed and burst at the top, thereby creating a circulation flow of particles in the bed. As the gas velocity increases further, the bubbles grow in size and in frequency of appearance, causing greater expansion of the bed. When the diameter of the bubble is equal to the diameter of the containing column, the behavior of the system is referred to as "slugging." A further increase of velocity causes particles to be carried out of the container by the gas stream. At this point the void fraction of the bed may be as high as 0.98 and the behavior is termed dilute-phase fluidization. For best results, the fluid-bed dryer is operated with a superficial gas velocity that is about twice the incipient fluidizing velocity. This places the system in the "bubbling" regime corresponding to dense-phase fluidization.

In lieu of an experimental determination, the following expression gives the theoretical value of the incipient velocity:

$$\mathrm{Re}_0 = 25.7[(1 + 5.53 \times 10^{-5}G_a)^{0.5} - 1]$$

where

$$\mathrm{Re}_0 = \frac{V_0\, d_p}{\mu}$$

is the Reynolds number at incipient fluidization point, and

$$G_a = \rho_f(\rho_s - \rho_f)g\, d_p^3/\mu^2 \equiv \text{Galileo number.}$$

where d_p is the equivalent spherical diameter of the particle. However, it is recommended that the optimum operating gas velocity for a specific case be determined experimentally.

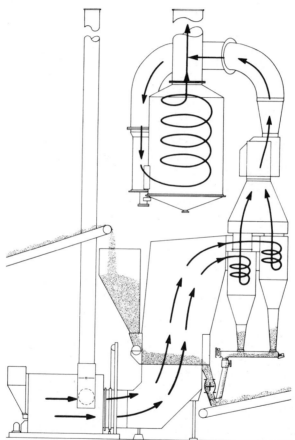

Fig. 14 Flow pattern of fluid-bed dryer. *(FMC Corporation.)*

Design Procedure   The minimum bed area is calculated on the basis of the specified inlet air temperature, the total evaporation rate, and the operating gas velocity. The temperature and humidity of the exit gas will correspond to the equilibrium gas conditions for the product moisture content. Thus, by writing material and heat balances, one obtains the volumetric flow rate of the gas, and this number divided by the minimum fluidizing velocity gives the minimum cross-sectional area of the drying chamber. The design area of the chamber will be in excess of this value and is evaluated from the experimental conditions.

The holdup or required holding volume of the drying chamber can be obtained if the drying time and the bulk density of the fluidized bed are known from laboratory data or from operating data on similar full-scale dryers.

Calculation of Minimum Bed Area. From a mass balance on water:

$$Y_{out} = \frac{W}{G}(X_{in} - X_{out}) + Y_{in} \tag{20}$$

where Y = gas humidity, lb water/lb dry gas
  X = product moisture content, lb water/lb dry solid
  G = dry gas rate, lb dry gas/hour
  W = product rate, lb dry solid/hour

A heat balance gives

$$C_m G(T_{in} - T_{out}) = \lambda_{T_{out}} W(X_{in} - X_{out}) + W(t_{out} - t_{in})[C_p + C_l X_{in}] \tag{21}$$

where T = gas temperature
  t = product temperature
  C_m = average heat capacity of gas
  C_p = average heat capacity of solid
  C_l = average heat capacity of liquid
  $\lambda_{T_{out}}$ = heat of vaporation at T_{out}

Finally, the equilibrium condition between the exit gas and product is

$$T_{out} = t_{out}$$
$$Y_{out} = f(X_{out}) \tag{22}$$

where Y_{out} is the equilibrium humidity for the equilibrium moisture content X_{out}. If the product equilibrium moisture content is essentially zero for any gas humidity, then $Y_{out} = Y_{sat}$ = saturation humidity. In the previous three equations, the unknowns are T_{out}, t_{out}, Y_{out}, and G. In the case where $Y_{out} = Y_{sat}$, the psychrometric chart provides the missing fourth equation. G is then calculated by a trial-and-error solution of the above equations. Finally, the bed area A is obtained from

$$A = G v_0/V_0$$

where v_0 = humid volume of the inlet gas.

Calculation of Required Dryer Holdup:

If τ is the required average retention time in the bed, w the production rate, and M the holdup in the dryer, then

$$M = w\tau$$

Given the void fraction ϵ_f of the fluidized bed, the bed volume is

$$V = M/\rho_s(1 - \epsilon_f)$$

where ρ_s is the density of the individual particle. The height of clear space above the fluidized bed should be about 1.5 times the fluidized bed height.

DESIGN EXAMPLE

It is desired to dry a granulated fertilizer from 15% (dry basis) to 2% (dry basis) using a continuous fluid-bed dryer. The inlet air temperature is 400°F, the operating fluidizing velocity is 250 ft/min and the retention time is 7 min. The operating gas velocity used is approximately twice that required for incipient fluidization. The retention time is obtained from pilot-plant studies.

The production rate is 8000 lb/h of the wet feed.

Calculation of minimum bed diameter

On the basis of pilot-plant studies, the bed temperature is estimated to be 170°F for an inlet temperature of 400°F. At this temperature the equilibrium relative humidity of the exit air is 48%, which is obtained from the equilibrium moisture curve for the product at the selected bed temperature.

A heat balance gives

$$0.24G(400 - 170) = 996(8000)\frac{(0.15)}{1.15}(0.15 - 0.02) + 8000\frac{(1.0)}{(1.15)}(170 - 60)[0.20 + 1.0(0.15)]$$

from which $G = 7300$ lb/h dry air, and feed temperature is assumed to be 60°F. A mass balance on water is then written to obtain the humidity of the exit air. Accordingly,

$$Y_{out} = \frac{8000}{7300}(0.15 - 0.02)\frac{1.0}{1.15} + 0.01$$

where $Y_{in} = 0.01$ lb water/lb dry air. Thus, $Y_{out} = 0.1339$ lb water/lb dry gas. The corresponding relative humidity, from the psychrometric chart, is approximately 48%. Had the calculated outlet humidity been different from the value corresponding to the equilibrium value, the inlet air temperature would have to be adjusted until agreement is reached. It should also be noted that the drying process in this example is nonadiabatic, since the bed temperature is greater than the wet-bulb temperature of the inlet air. This is a consequence of the presence of bound moisture rather than surface moisture in the product.

The minimum cross-sectional area of the bed is obtained thus:

$$D = \left[\frac{4G}{60\rho_f \pi \, V_f}\right]^{0.5} = \left[\frac{4(7300)}{60(0.0455)(\pi)(250)}\right]^{0.5}$$
$$= 3.69 \text{ ft}$$

Calculation of fluid-bed height

From the pilot-plant studies it is determined that the bed expands by 50% in reaching the fluidized state. The bulk density of the settled bed is 40 lb wet product per cubic foot. Thus, the bed volume is

$$M/\rho_b = \frac{w\tau}{\rho_b} = \frac{(8000)(7)}{(40)(60)} = 23.3 \text{ ft}^3$$

in the settled state. The fluid bed height is then

$$(23.3)(1.5)/\pi D^2 = (23.3)(1.5)/\pi(3.69)^2/4.$$

Fluid bed height = 3.27 ft.

GENERAL CONSIDERATIONS

A number of considerations are common to most types of dryers and are therefore discussed generally.

Alternative Fuels and Heat Sources

Alternative fuels and heat sources are of increasing concern to the dryer designer and operator. Direct firing of natural gas involving the full entrainment of the products of combustion gave a cheap, reliable, clean, safe source of heat for many dryer applications. Extended dryer temperatures can be obtained and combustion equipment is efficient and easily maintained. All of the thermal units released from the fuel go directly into the dryer. Other than the use of waste heat from other processes, natural gas was the least expensive source of heat for drying applications and became widely used in the 1960s and early '70s. The dwindling supply of natural gas and the setting of national priorities has forced regulating agencies to reduce the amount of gas available to present processors and to deny many applications for additional gas for new process lines. In many areas alternate sources of heat must be selected for new applications and as conversion or standby sources for existing installations to avoid production curtailment or midwinter shutdown.

Clean waste heat in either liquid or gaseous form should be used wherever possible. Corrosives in effluents from other processes must be avoided. Particulate matter can be removed from off-gases from rotary kilns by reducing the temperature with an atmospheric bleed to a safe level to employ a bag filter before introducing the gases into the dryer.

Liquid propane is a good standby fuel for direct natural-gas-fired systems. Propane can be purchased at a time when production exceeds demand to get a more favorable price. This is usually in the summer, so it must be stored for use in the winter when natural gas curtailments are more likely to occur. In use it is converted from liquid to gaseous form and generally diluted with air to bring the heat content down to natural gas levels to permit combustion without altering the gas-firing equipment. Consideration of the use of propane as a prime fuel source is usually not practical because of the higher initial cost, erratic supply-demand basis and high storage costs.

Steam is a viable alternate source of heat but is temperature-limited and less fuel efficient than direct firing of a dryer. Boiler efficiency and transmission losses must enter into the cost calculation in addition to the basic combustion efficiency. The use of cheaper, more dependable fuel supplies can help to offset the higher-steam-generating costs. Heat exchangers in the form of tubes or coils are employed in the dryer, and where air is to be heated, a 50°F or better approach to steam temperature is easily obtained.

Higher temperatures approaching 600°F can be obtained with hot oil boilers utilizing heat-transfer fluids exchanging with the air through internal dryer coils. These heat-transfer fluids usually operate in the liquid phase at relatively low pressure. A wide choice of fuels can be used to fire the boiler.

Direct oil firing is gaining favor because of easy availability of light No. 2 fuel oil that is relatively clean burning. As with direct gas firing, the products of combustion mix directly with the circulating drying media. Combustion gas-oil burners are available to permit the use of light fuel oils on a standby basis. Although fuel oil is currently more expensive than natural gas on an available-heat basis, national policy will bring it into parity with gas. The decision to direct-oil-fire must be made cautiously, taking a number of factors into consideration including the availability of reliable sources of clean oil, the entry into the dryer of unburned hydrocarbons, the exposure of the product being dried to possible contamination, and the higher maintenance levels required for oil combustion equipment.

Reliable pulverized-coal-firing equipment should start to appear and should be suitable for limited drying applications.

Electric heat is usually ruled out by economics where high evaporative water loads are required of a dryer. The relatively low fossil-fuel conversion rates of electric power to usable thermal units in the dryer makes electricity as a heat source prohibitively expensive in most applications. Radiant and convection heaters can be used in limited applications. Conversion of electric power to radio frequency waves or microwaves for the purpose of drying further reduces the efficiency and is prohibitive for all but the most unusual applications.

Indirect firing of oil to eliminate possible product contamination is possible. The gas-to-gas heat exchangers required for convection drying applications are inefficient, requiring large surface areas resulting in expensive bulky exchangers with alloy tubes to limit burnout. Relatively high stack temperatures reduce the dryer fuel efficiency. This method of heating a dryer is generally economically unattractive.

Proper insulation of dryers is of increasing concern, with the escalating costs of fossil fuels generally used as a heat source for the evaporation of water. Conduction of heat from the interior of the dryer to the room or atmosphere must be minimized to maintain fuel-efficient operations. Industrial hygiene and safety also dictate effective insulation, and the increasing use of air conditioning in processing spaces makes it mandatory.

Glass or mineral wool bats or blankets are available in a number of grades, thicknesses, and densities to give the required insulating effect and prevent settling. They are useful in all but the highest-temperature drying applications or where specialty high-temperature insulations should be selected. Insulating foams generally do not have sufficient temperature range to be useful.

Pollution Controls

Atmospheric pollution from a dryer can be of four major types—sound, heat, odor, and particulate. Sound is generally not a problem although all dryers must meet government codes relating to limiting decibel levels. Proper selection of fans, combustion equipment, and mechanical auxiliaries will usually keep sound levels within acceptable limits, provided the equipment is located in an area where there are not highly sound-reflective walls nearby or other equipment operating at a high sound level where the addition of the drying machine will add to the crescendo effect.

Most commonly used dryers have a gaseous effluent stream vented to the atmosphere. The amount of vent gas is dictated by the amount of water vapor removed from the product, the volume of products of combustion if a direct-firing system is used, and in the case of solvent removal or high dust loading in the drying media, the maintenance of concentrations below acceptable levels as a percent of the lower explosive limit. A further constraint is the temperature of the exhaust gas as it relates to its ability to hold water vapor in acceptable amounts to maintain reasonable levels of drying and fuel efficiency.

Heat pollution from a dryer exhaust has not generally been a problem which conflicts with governmental regulations, provided the exhaust is vented at high enough levels above the surrounding community to avoid local complaints. This can also be true of high vapor loading if it is carried off at a high enough level that condensation is not a problem. Special care must be exercised if the condensibles are corrosive. If this is the case, scrubbers may have to be used to condense the corrosive vapors.

Odor pollution problems too will vary with the plant location relative to the local community. Odors are not easily removed from a gaseous exhaust. They can vary from organic compounds to volatilized oils and solvent vapors. Scrubbers are not generally effective except where the odors are condensible. Catalytic units may remove or alter the odor. The only positive way to remove odors is incineration involving elevating the exhaust gas temperature in the range of 1500 to 1800°F for a sufficient time to ensure oxidation. Even with efficient heat-recovery equipment this method is heat intensive and becomes very expensive. Scrubbing too is expensive. For this reason, governmental agencies are backing off from strict odor enforcement regulations where the odors are of the nuisance type. Where noxious compounds are involved great care must be exercised in controlling the effluent.

Particulate pollution can be controlled in a number of ways including cyclones, scrubbers, and filters, all of which are dealt with in other chapters of this handbook.

Other Types of Dryers

The spectrum of available dryer types is a broad one, most of them being specialty types with limited application. They fall into a number of categories:

Conduction dryers transfer heat through shells, jackets, tubes, and screws in varying ways. Steam, hot water, hot oil, or combustion gases provide the required heat. These dryers can be very efficient, for the water vapors are often drawn off with little or no heated air required for mass transfer. Care must be exercised to be sure there is no product degradation from contact with the hot surfaces or product buildup on the surfaces which will scale off or reduce heat transfer. With conduction drum dryers, a thin film of slurry or paste is doctored onto the drum continuously and scraped off when dry. A jacketed trough and hollow screw to advance the product through the trough is another type. Others have jacketed shells with scrapers advancing the product within the shell.

Radiant dryers convey the products on a belt where they are exposed to radiant energy from direct-gas-fired burners heating parabolic refractories to infrared temperatures. Others use infrared electric bulbs or electric strip heaters mounted in parabolic reflectors to radiate the heat to the product to be dried. Electrically heated quartz tubes are also used. Radiant dryers depend on heat in the infrared range—around 1600°F—radiating from the heat source to the product which absorbs the heat for the evaporation of moisture. They are suitable for sheet materials or thin beds since there is very limited penetration of the heat. Absorbtivity depends on surface color, texture, and reflectivity. Overheating and nonuniform drying can be a problem.

Vacuum dryers are usually batch type and are used for heat-sensitive materials or products where final moisture requirements cannot be met with atmospheric means. Capital and operating costs are high so that they are not used where conventional dryers will suffice.

Freeze dryers are a form of vacuum dryer used to remove moisture from heat-sensitive products that are introduced into the dryer in a frozen state. Food and other products dried in this way may have better flavor retention and more rapid rehydration characteristics.

Microwave and radio-frequency dryers work in conjunction with an electronic generator producing electrical waves of a controlled length. The product to be dried is continuously passed through an enclosure into which the waves are introduced in such a way that they pass through the product, generating heat within the product from the vibration caused by the wave passage. This heat increases the vapor pressure of the water to be removed, forcing it to the surface where convection drying takes place. Wavelengths are controlled to transmit maximum energy to the water within the product rather than the product itself. This type of drying has been applied to products with deeply diffused water that is difficult to reach with ordinary drying methods or where space limitations prevail. Care must be exercised to assure uniform transmittal of energy in the form of electrical waves.

Continuous shelf dryers utilize rotating circular trays or scrapers to cascade products from the upper to the lower shelves. They can be of the convection or conduction type.

REFERENCES

1. Harmathy, T. Z., *Ind. Eng. Chem. Fundam.*, **8**, 92 (1969).
2. Wilke, C. R., and Wasan, D. T., AIChE-I.Ch.E. Symp. Series **6**, 21 (1965).
3. Perry, J. H. (ed.), "Chemical Engineers' Handbook" (5th ed.), McGraw-Hill, New York, 1975.
4. Thygeson, J. R., and Grossmann, E. D., *AIChE J.*, **16**, 749 (1970).
5. Williams-Gardner, A., "Industrial Drying," Leonard Hill, London, 1971.
6. Friedman, S. J., and Marshall, W. R., Jr., *Chem. Eng. Prog.*, **45**, 482, 573 (1949).
7. Saeman, W. C., and Mitchell, T. R., *Chem. Eng. Prog.*, **50**, 467 (1954).
8. Peck, R. E., and Wasan, D. T., *Adv. Chem. Eng.*, Vol. 9 (1974).
9. Ranz, W. E., and Marshall, W. R., Jr., *Chem. Eng. Prog.*, **48**, 141 (1952).
10. Duffie, J. A., and Marshall, W. R., Jr., *Chem. Eng. Prog.*, **49**, 417, 480 (1953).
11. Masters, K., "Spray Drying," CRC Press, Cleveland, 1972.
12. Frazer, R. P., Eisenklam, E. D., and Dombrowski, N., *Br. Chem. Eng.*, **2**, 196 (1957).
13. Herring, W. M., and Marshall, W. R., Jr., *AIChE J.* **1**, 200 (1955).
14. Friedman, S. J., Gluckert, F. A., and Marshall, W. R., Jr., *Chem. Eng. Prog.*, **48**, 181 (1952).

BIBLIOGRAPHY

For general information on the basic dryer types, the following secondary sources are recommended.
SPRAY DRYERS
Masters, K., "Spray Drying," CRC Press, Cleveland, 1972.
ROTARY DRYERS
Drew, T. B., and Hoopes, J. W. (eds.), "Advances in Chemical Engineering," vol. 9, chapter by R. E. Peck and D. T. Wasan.
FLUID-BED DRYERS
Davidson, J. F., and Harrison, D. (eds.), "Fluidization," Academic Press, New York 1971.
FLASH DRYERS
Nonhebel, G., and Moss, A. A. H., "Drying of Solids in the Chemical Industry," CRC Press, Cleveland, 1971.
THROUGH-CIRCULATION AND TRAY DRYERS
Keey, R. B., "Drying—Principles and Practice," Pergamon Press, New York, 1972.

Part **5**

Solid Mixtures

Leaching

RAJARAM K. PRABHUDESAI, M.Sc.(Tech.),
M.Chem.E., Ph.D. *Senior Process Engineer, Stauffer
Chemical Company, Dobbs Ferry, N.Y.; Member,
A.I.Ch.E.; Registered Professional Engineer*

INTRODUCTION

Leaching is a separation technique that is often employed to remove a solute from a solid mixture with the help of a liquid solvent. It originally referred to removal of a soluble component through a fixed bed, but now it is applied to solid-liquid extractions generally. The separation operation involving washing of a solute adhering to the surface of a solid is also considered solid-liquid extraction. Because of its application in several industries, leaching is known by such other names as decoction (use of solvent at its boiling point), lixiviation, percolation, infusion, elutriation, decantation, and settling. The presence of a solid component distinguishes all these techniques from liquid-liquid extraction.

DESIGN CONSIDERATIONS

Two steps are always involved in solid-liquid extraction:

1. Contact of liquid solvent with the solid to effect transfer of solute from the solid to the solvent
2. Separation of resulting solution from the residual solid

Two other auxiliary operations are involved; namely: (a) preparation of the solid for extraction and (b) recovery of solute from the solvent, usually by evaporation or distillation of solvent from the solute. To meet these objectives, the designer has to select the most practical and economical contacting and separation equipment, batch or continuous. Choice of equipment and method of operation for a particular extraction depend upon various factors such as (1) physical characteristics of the solids to be leached and (2) quantity and size of the solid in relation to the amount of solute to be recovered.

Physical Characteristics of Solids

A knowledge of the physical characteristics of the carrier solid is very important to determine whether it needs prior treatment to make the solute more accessible to the solvent. Prior treatment may involve crushing, grinding, cutting into pieces, or re-forming into special shapes such as flakes.

Solute particles may exist in the inert solid in a variety of ways. It may exist on the surface of the solid, may be surrounded by a matrix of inert material, may be chemically combined, or may exist inside cells as in the case of many vegetable and animal bodies.

Solute adhering to the solid surface is readily removable by the solvent. When the solute exists in pores surrounded by a matrix of inert material, the solvent has to diffuse to the interior of the solid to capture solute and then diffuse out before a separation can result. In such cases, subdivision of the solid by crushing, grinding, or cutting increases the surface exposed to the solvent. However, reduction of solids to finer particle size has its limitations. In some instances, the amount of solute to be recovered is small in relation to the amount of material to be treated, in which case grinding becomes uneconomical. Too-fine division may result in packing of solids during extraction, preventing free flow of the solvent through the solid bed. In such a case, the extraction is much more difficult, especially when finely divided solids are treated in an unagitated state. Dispersion of the particles in liquid solvent by agitation permits thorough contacting of the solid with the solvent. Agitation, while giving good extraction, may cause suspension of fine particles in outflowing solution, which may subsequently require a difficult filtration or clarification step.

In the case of materials with cellular structure, if the cell walls remain intact, the leaching action involves osmotic passage of the solute through cell walls. It is, however, impractical and undesirable to grind materials to rupture cell walls, since this may result in extraction of some other undesirable material in addition to the desired solute, creating a purification problem. Therefore, instead of resorting to excessive subdivision, many solids of porous structure are cut into wedge-shaped slices called *cossettes,* as in the case of sugar beets, or are crushed and reshaped into flakes, as in the case of vegetable seeds, to obtain increased surface, which permits free flow of solvent through the solid and allows a more selective extraction.

The mechanism of leaching may involve simple physical solution, or solution may be made possible by a chemical reaction. Therefore, the rates of diffusion of solvent into the mass of solid to be leached, of solute into the solvent or of extract solution out of inert solid, or a combination of these rates, may be significant in overall rate of extraction. The chemical reaction rate or membranous resistance may also be factors. Frequently, it is possible to establish some overall equilibrium or pseudoequilibrium relationship leading to an overall stage efficiency of the extraction systems without evaluating the individual rates and equilibrium relationships.

Solvent Selection

Properties of a solvent such as its boiling point, density, and viscosity affect its suitability and selectivity to effect more complete and economical extraction of the desired solute. The extent and rate of solution of the desired solute and other impurities in the solid by a solvent influence considerably the size of an extractor, overall operating costs, and type and cost of separation, as well as solvent and solute recovery equipment and the quality of products. The solvent should be selective, relatively cheap, nontoxic, and readily available. Very often, a compromise in selection of the solvent is required.

Temperature of Leaching

Higher temperatures result in higher solubility of the solute in solvent and therefore are desirable to realize higher solute concentrations in the extract. Increased rates of leaching are obtained because of lower viscosity of the liquid and larger diffusivities of the solute and the solvent. Higher temperatures, in the case of some natural materials, such as coffee, tea, and sugar beets, may, however, result in excessive extraction of undesirable material. Higher-temperature operation may not be permissible because of unacceptable solvent losses and for safety considerations. Again, a compromise is necessary in selection of temperature.

METHODS OF SOLID-LIQUID CONTACTING AND SOLID-SOLUTION SEPARATION

Two types of solid-liquid contacting methods may be distinguished. In fixed-bed contacting, the solid particles are stationary, while dispersed contact involves motion of solid particles relative to each other and also relative to the liquid. Solvent may be contacted with a fixed bed of solids in three different ways (Fig. 1): (1) spray percolation, (2) full immersion, and (3) intermittent drainage. In the spray-percolation method, the solvent is sprayed on the solid and allowed to drain through it continuously. In the full-immersion method, the solid remains submerged in liquid and may be treated with a batch of solvent or continuous flow of solvent. The third method involves intermittent drainage of the solvent from the solids. A dispersed contact is usually effected by suitable agitation.

The simplest fixed-bed contacting equipment consists of an open tank with a perforated bottom in which solvent is allowed to percolate by gravity through the bed of undissolved solids. An example of dispersed contact equipment is a leaching tank in which the solids are dispersed in the solvent by agitation and then allowed to separate from extract solution by settling. Separation of extract solution from the solids may be effected in the same tank or a separate settling unit, or by the use of various types of filtration equipment. When volatile solvents are used or when percolation under gravity is too slow, a closed percolation tank with a solvent circulation pump may be used.

Separation of solvent and solute may involve operations such as clarification, evaporation, and distillation.

METHODS OF OPERATION OF EXTRACTION SYSTEMS

The three principal types of operating methods used in leaching systems are indicated in Fig. 2. The single stage (Fig. 2a) represents the complete operation of contacting the

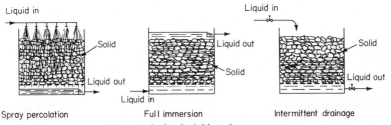

Fig. 1 Methods of solid-liquid contacting.

solids feed and fresh solvent and subsequent mechanical separation. This is rarely encountered in industrial practice because of the low recovery of solute obtained and the relatively dilute solution produced. Efficiency of extraction is somewhat improved by dividing the solvent into a number of smaller portions and then effecting successive multiple contacts of the solvent portions instead of only one contact of the whole amount of solvent with the solids.[1]

In the multistage cocurrent (parallel) system shown in Fig. 2b, fresh solvent and solids feed are contacted in the first stage. Underflow from the first stage is sent to the second stage, where it is contacted with more fresh solvent. This scheme is repeated in all succeeding stages. In the continuous countercurrent multistage system shown in Fig. 2c, the underflow and overflow streams flow countercurrent to each other. This system allows high recovery of solute with a highly concentrated product solution because the concentrated solution leaves the system after contact with fresh solids. Multistage countercurrent contact between the solids and solvent may be obtained by actual movement of the solids by some means, countercurrent to the direction of solvent flow from stage to stage, or it may be simulated with a number of stages in which solids remain stationary. Thus, the

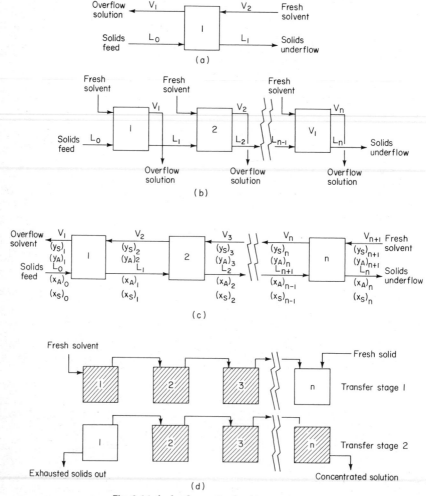

Fig. 2 Methods of operating leaching equipment.

batch countercurrent multiple-contact system of Fig. 2*d* consists of a number of batch contact units arranged in a circle or in a line, called the extraction or diffusion battery. The main feature of this system is that solids remain stationary in each tank or diffusion unit but are subjected to multiple contacts with extracts of diminishing concentration. The final contact of most nearly exhausted solids is with fresh solvent, while concentrated solution leaving the system is in contact with fresh solids in another tank. Countercurrent motion in the battery is obtained by advancing receiving and discharge tanks one at a time, when solids are charged and removed. When the extraction process has gone through several cycles of operation, the concentrations of solution and in the solid in each tank nearly resemble the values of concentrations obtained in a true multistage counter-current extraction system. If the extraction units are arranged in a straight line instead of a circle, it becomes convenient to add or eliminate a unit to or from the battery at a later time if needed.

EXTRACTION EQUIPMENT

Percolation Equipment

Coarse solids are leached by percolation in fixed-bed or moving-bed equipment. Both open and closed tanks with false bottoms are used. Solids to be leached are dumped into the vessel to a uniform depth and then treated with the solvent by percolation, immersion, or intermittent drainage methods. Tanks should be filled with solid of as uniform particle size as possible. This will allow maximum void space, resulting in low pressure drop for the flow of solvent and less channeling through the solid bed.

Closed percolation vessels are required when pressure drop is too high for gravity flow of solvent, when evaporation losses of solvent are to be avoided, or when above-boiling-point temperatures are desired. The solvent may be circulated through the tank by pumping or the leaching may be carried out in the closed vessel without circulation of the solvent. Such closed vessels are called *diffusers*.

In multibatch countercurrent extraction, a number of batch extraction vessels, termed an *extraction or diffusion battery*, are used. Figure 3 illustrates the principle of a

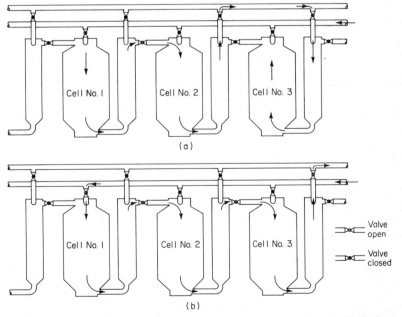

Fig. 3 Principle of diffusion battery for extraction of sugar from beets. (*a*) Filling period; (*b*) drawing period. (*From Badger and Banchero, "Introduction to Chemical Engineering," McGraw-Hill, New York, 1951.*)

diffusion battery.[2] The vessels may be of the type described above or of special design for battery operation. Each is charged with solids. The solvent is transferred from vessel to vessel in a given order to treat progressively the less exhausted solids, and leaves the system as extract solution from the vessel, which is charged with the fresh solids. The most extracted solid after contact with fresh solvent is discharged from the vessel, which is then refilled with fresh solid. Extraction of tannin from wood or bark is carried out in this manner. In the sugar industry, a battery of closed vessels is used for water extraction of sugar from sugar beets.[3]

Sometimes the extraction battery may consist of a number of fixed-bed packed columns which have greater height relative to diameter. For example, extraction of coffee solubles is carried out with hot water in a system of 5 to 10 column percolators (Fig. 4). The

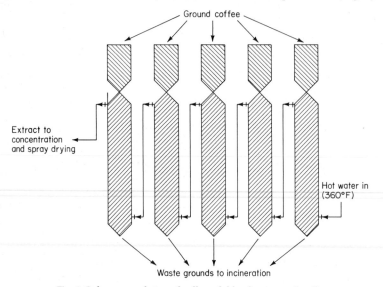

Fig. 4 Column percolation of coffee solubles from ground coffee.

percolators may be operated batchwise or as semicontinuous units with countercurrent flow of hot water, either top to bottom or the reverse. The feedwater temperatures range from 310 to 360°F. Sometimes the columns are hot-water jacketed. As usual, the most exhausted percolator is emptied and removed from the system and an empty one is charged with fresh coffee, thus advancing the battery by one percolator at a time to simulate countercurrent operation.

Moving-Bed Continuous Percolation Systems

Percolation leaching of solids is also carried out in moving-bed equipment. Cottonseeds, soybeans, peanuts, rice bran, castor beans, and many other vegetable seeds are treated with organic solvents for extraction of vegetable oils which they contain. Usually, the seeds are required to be formed into flakes or roles for most advantageous leaching. This generally involves very costly pretreatment which consists of dehulling, precooking, and adjustment of water content. The miscella or oil solvent solution containing a small amount of finely divided solids is subjected to separation operations, to remove solids and to remove solvent from oil. The Bollman extractor[4] (Fig. 5) is perforated-basket-type equipment. Solids are charged to perforated baskets attached to a chain conveyor. As the baskets descend (right side in figure), the solids are leached in cocurrent flow by solvent-oil solution called "half miscella," pumped from the left-half compartment at the bottom of the vessel and sprayed over the baskets at the top. The liquid percolates through the solids from basket to basket, and collects in the right-half compartment of the vessel bottom from where it is removed as a strong solution usually termed "full miscella." When baskets ascend in the left half of the vessel, they move countercurrently to liquid flow and the solids are extracted in this section by fresh solvent to obtain half miscella, or weak oil-solvent solution. Because the solids are not agitated and because final miscella moves cocurrently, the Bollman extractor permits the use of thin flakes while producing an extract solution which is relatively free of suspended fine solids. It is, however, a countercurrent device only partially, and because of channeling has low stage efficiency.

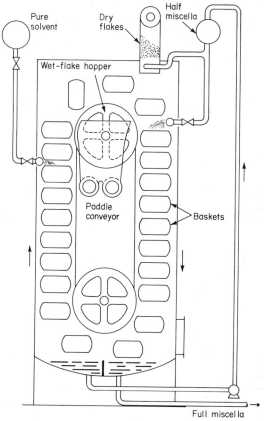

Fig. 5 Bollman extractor. [*From* Chem. Eng., **58**(*1*), *127 (1951)*.]

Some drainage time is provided at the top before the baskets are dumped. A horizontal version of the Bollman extractor is also in use.[5] In another variation, the baskets remain stationary while the filling, leaching, and dumping equipment revolves. The basket extractors are comparatively large, expensive, and complicated devices. Accurate control of flake size, thickness, and bulk density is required. In spite of these disadvantages, the basket extractors are widely accepted because they produce, without auxiliary equipment, clear miscella and well-drained extracted flake, and allow large quantities of solids to be handled continuously.

Rotocel[6-8] is a percolation extractor (Fig. 6a and b) which obtains countercurrent extraction of solids like a batch countercurrent multiple-contact battery. It consists of a rotor divided into wedge-shaped cells or compartments. The rotor turns slowly into a very tight tank, and permits continuous introduction and discharge of solids. As the rotor revolves, each cell or compartment passes in turn under a special device for feeding the seeds and then under a series of sprays of leaching solvent. After leaching, the solids are automatically dumped into one of the stationary compartments below, from which they are continuously conveyed away. The solvent from each spray percolates downward through the solid and the supporting screen into the corresponding compartment of the

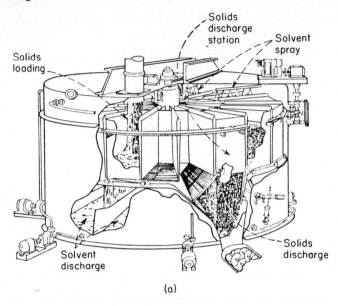

(a)

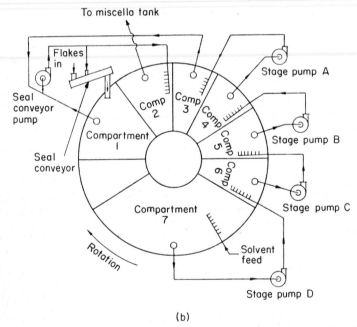

(b)

Fig. 6 (*a*) Rotocel extractor. (*b*) Principle of Rotocel extractor. [*From* Chem. Eng. **58**(*1*), *127 (1951).*]

lower tank, from which it is dumped to the next spray. The strongest solution is taken from fresh solids as product extract. The entire machine is enclosed in a vaportight housing to prevent escape of solvent vapors.

The Kennedy extractor,[4] shown in Fig. 7, consisting of a series of tubs, subjects the solids to impeller-induced mechanical movement from tub to tub, but operates substantially as a percolator. Drainage occurs when the solids are lifted by the impeller blade above the liquid level in a tub and dumped into the next one. The extraction is countercurrent as the solids are moved from tub to tub in the direction opposite to the movement of solvent.

Tilting pan filters and horizontal filters[9] are also commonly used for leaching. They operate on a principle very similar to that of a Rotocel extractor.

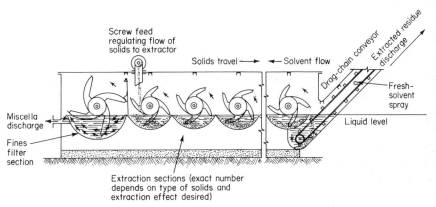

Fig. 7 Kennedy extractor. [*From* Chem. Eng., 58*(1)*, 127 *(1951)*.]

A typical solvent extraction process system for soybeans[10] is shown in Fig. 8. The process may be considered to consist of three major operations: (1) preparation of beans, (2) extraction of the oil from the beans, and (3) purification of oil by recovery of solvent. A related operation is to recover the solvent from the extracted meal. The beans are cracked, dehulled, and then conditioned by means of a rotary steam-tube dryer equipped with water sprays so that water may be added or removed. The conditioned beans (moisture content of 9.5 to 10%) are passed between rolls to form flakes. The flakes are carried by a paddle conveyor and fed to the baskets of the Bollman extractor housed in vaportight steel tank about 45 ft high through a vaportight double hopper. Solvent used is a mixture of hexanes and has a boiling range of 146 to 156°F. The solvent recovery system consists of atmospheric pressure evaporators followed by a falling film evaporator. The 90% oil solution produced by evaporation is then subjected to vacuum steam distillation in a series of two vacuum columns to yield up to 99.95% oil, which is pumped out from the bottom of the column to a storage tank. The economics of such an extraction system is dominated by the investment in and operating costs of the auxiliary equipment required for preparation of the beans for extraction and for postextraction purification of oil and recovery of solvent.

Gravity settling tanks, operated as thickeners, can serve as continuous contact and separation equipment for leaching fine solids. Thickeners are equipment to continuously increase the ratio of solid to liquid in a dilute suspension of finely divided solids by settling and decantation, producing a clear liquid and a thickened sludge. A Dorr-Oliver thickener, consisting of a single compartment, is shown in Fig. 9. The thin slurry of liquid and suspended solids are fed to the tank through a feed port at the top center without allowing mixing of slurry with the clear liquid at the top of the tank. The solids settle from the slurry to the bottom of the tank. The settled sludge is gently directed toward the conical discharge bottom by four sets of plow blades or rakes. The sludge is pumped out from the conical bottom by means of diaphragm pump. The supernatant clear liquid overflows into a launder built around the upper periphery of the tank. Such thickeners are built in various sizes and materials of construction.

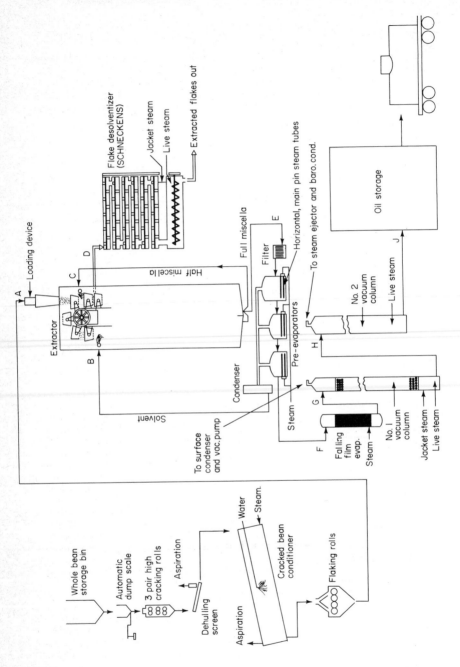

Fig. 8 A typical extraction system—extraction of oil from soybeans. [*From* Ind. Eng. Chem., **2**, 186, (1948). *Copyright by the American Chemical Society.*]

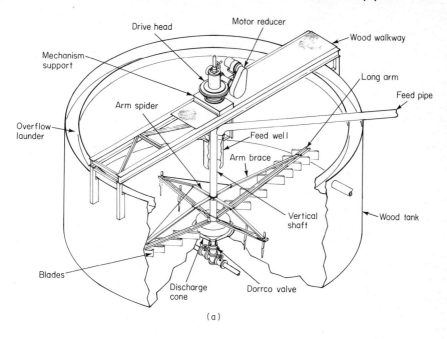

(a)

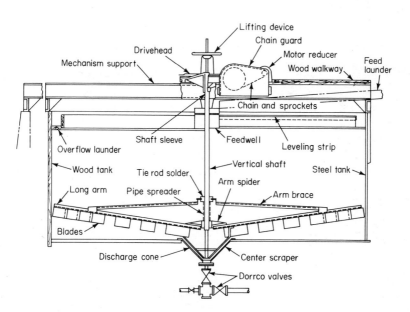

Fig. 9 Dorr-Oliver thickener. (*a*) Top view; (*b*) side view. (*Dorr-Oliver Inc.*)

Continuous dispersed solids leaching is also carried out in other equipment of special design. The Bonotto extractor[11-13] (Fig. 10), a vertical plate-extractor, consists of a column divided into cylindrical compartments by equispaced horizontal plates, and employs scraper arms to move solids over the plates. Alternate plates have radial openings staggered 180° from each other. Solids are fed to the top of the extraction column and are caused to fall to each lower plate in succession. The solvent enters at the bottom of the column and flows upward, and extract solution leaves the column at the top.

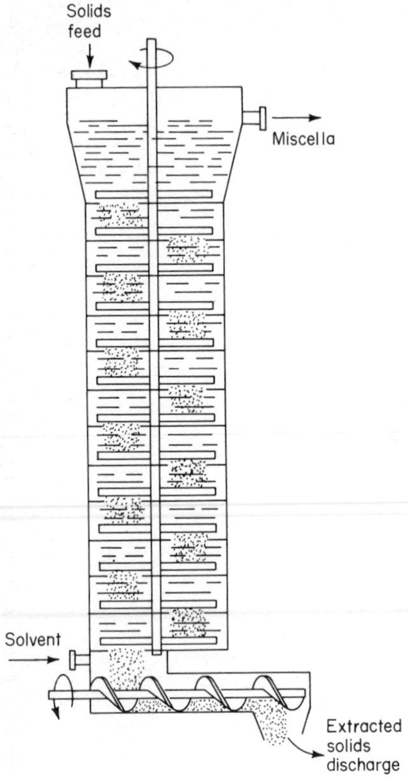

Agitated Extraction Vessels

The main drawbacks of percolation leaching in fixed beds are channeling of solvent and slow incomplete leaching. For coarse solids, many types of stirred vessels[14] are used (Fig. 11). Closed cylindrical vessels are arranged vertically (Fig. 11a) and are fitted with power-driven solids agitators and false bottoms to drain the extract solution. In some cases (such as recovery of tallows and greases), the extraction vessel is horizontal, with the agitator on a horizontal shaft (Fig. 11b). Horizontal drums rotated on rollers are also frequently used. Most of these devices are operated batchwise but are also arranged in batteries for countercurrent batch leaching.

Leaching by Solids Dispersion

Finely divided solids may be dispersed in solvents by agitation. For batch operation, a variety of agitated vessels are used. The Pachuca tank[15] (Fig. 12), extensively used in metallurgical industry, is a tall cylindrical vessel provided with an air lift to obtain agitation by vertical circulation of solids. These tanks may be constructed of wood, metal, or concrete, or may be steel tanks lined with inert material. After leaching is completed, airflow and consequently

Fig. 10 Bonotto extractor. (*From* Chem. Eng., *Mar. 15, 1965, p. 157.*)

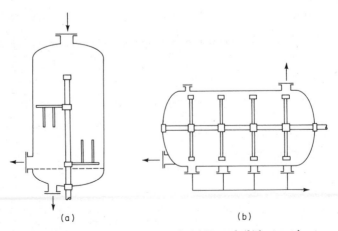

Fig. 11 Agitated extraction vessels. (*a*) Vertical; (*b*) horizontal.

agitation are stopped, solids are allowed to settle, and the supernatant extract solution is decanted off by siphoning from the top.

It is also possible to use standard turbine-type agitators to suspend finely divided solids. After leaching has been accomplished, the solids are allowed to settle in the same tank or a separate vessel and the extract solution is withdrawn by siphoning at the top or through discharge pipes located in the side of the tank. If the finely divided solids settle to a compressible sludge, the solution retention by the sludge will be considerable. Multiple agitation and settling with several batches of solvent may then be required to recover the solute, which may be done in countercurrent fashion. Alternatively, the solids may be filtered and washed in a filter. Impeller-agitated tanks can be operated as continuous leaching tanks, singly or in series.

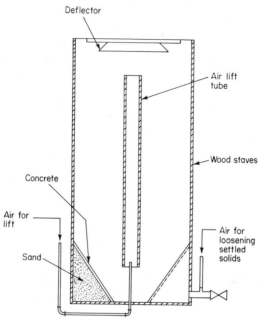

Fig. 12 Pachuca tank.

Screw-Conveyor Extractors

This type of continuous equipment is often classed with percolation equipment although there may be considerable agitation of solids during their conveyance by the screw. The Hildebrandt[13,16] total-immersion extractor (Fig. 13a) uses three separate screw conveyors to move solids in three parts of a U-shaped extraction vessel. The helix surface is perforated so that solvent can pass through the unit countercurrent to the movement of solids. The screw conveyors rotate at different speeds in such a manner that solids are compacted on their way toward the discharge port. However, the possibility of solvent loss and feed overflow restricts the screw-conveyor's successful use to light, permeable solids. Another simpler version[13,17] of this type equipment (Fig. 13b) uses a horizontal screw section for leaching and a second screw in an inclined section for washing, draining, and discharging the extracted solids.

Continuous Multistage Countercurrent Leaching Systems

By far the most important method of leaching is the continuous countercurrent method using multiple stages. Even in extraction batteries, where the solids are not physically moved from stage to stage, the solids in one extraction vessel are treated by a succession of liquids of constantly diminishing concentrations as if they are moved from stage to stage in a countercurrent system.

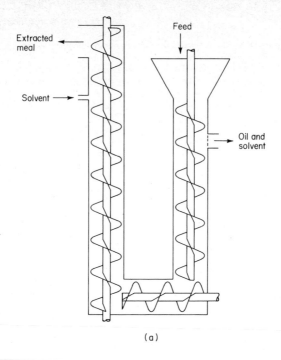

(a)

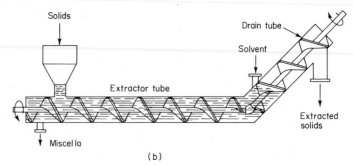

(b)

Fig. 13 (*a*) Hildebrandt extractor. (*b*) Two-screw extractor. (*From* Chem. Eng., *March 15, 1965, p. 157.*)

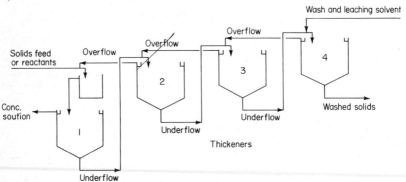

Fig. 14 Continuous countercurrent decantation system.

Continuous countercurrent multistage extraction can be advantageously used when it is possible to suspend the solids to produce pumpable slurries. The separation of solution from solids may be accomplished either by gravity settling and decantation when filtration is difficult or by use of filtration equipment such as continuous-drum filters, centrifuges, or perforated-basket centrifugals.

Continuous Countercurrent Decantation Systems

A number of Dorr thickeners described above may be used in a series cascade for continuous countercurrent washing of the finely divided solids free of adhering solute. A similar cascade may also be used to wash the solids formed during chemical reactions, as in the manufacture of phosphoric acid, by treatment of phosphate rock with sulfuric acid, or in the manufacture of caustic soda by causticizing soda ash. A simple arrangement is shown in Fig. 14. Capacity of each thickener in the unit is so related to the quantity of material handled that any one particle of solid remains in the thickener long enough to be completely leached or to allow the reaction to proceed to completion. The mixture of solid and liquid from the last thickener goes to the first thickener. The overflow from this thickener is the product concentrated solution. Underflow from this thickener is sent to a second thickener, that from the second to a third, and so on to the last thickener, where fresh solvent is added. In operating such a system, a careful control of rates of flow of both underflow and overflow are necessary so as not to disturb prevailing steady-state conditions.

For small decantation plants, where ground area is a limiting factor, it is possible to obtain a countercurrent cascade of thickeners built in superimposed fashion in a single shell as in case of the Dorr-balanced tray thickener.[5]

Interstage filtration allows use of a much smaller amount of solvent and accomplishes theoretically the same degree of solute removal as the countercurrent decantation system without filters. It also allows a better separation of solid from the solution, resulting in a higher recovery of the solute. This filtration system has some disadvantages, however. Because of the variation in cake thickness in improperly operated filtration systems, leaching may not be uniform. Filters take much less space than equivalent thickeners, but a filter is a more expensive device than a thickener.

The Sherwin-Williams system[4] (Fig. 15) for the leaching of vegetable seeds uses

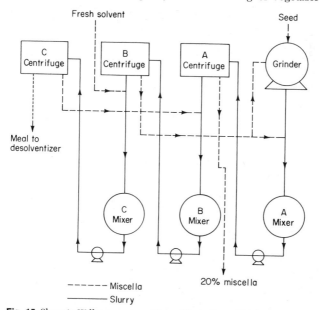

Fig. 15 Sherwin-Williams system. [*From* Chem. Eng., **58**(*1*), *127*, (*1951*).]

centrifuges for the separation of extract solution from the solids. The system essentially consists of a number of stages, each of which consists of an agitation tank and associated centrifuge and a pump.

For many vegetable materials, such as tea and coffee, the inert solids absorb a weight of solvent many times their own weight and therefore require use of larger amounts of solvent so that the slurry is still pumpable even at the stage where fresh feed is added. This results in dilute product solution and therefore subsequent higher costs for evaporation. In some cases, such as that of tea, this difficulty can be overcome by a split-feed arrangement. One such arrangement[18] for extraction of soluble solids from tea is shown in Fig. 16. In this system, to obtain concentrated solutions and still have pumpable slurry at the stage where fresh tea is fed to the system, the entire feed is not added to one agitation tank but is divided and added to a number of tanks, so that the amount of fresh solids coming in contact with the concentrated solution at a time is small. This allows retention

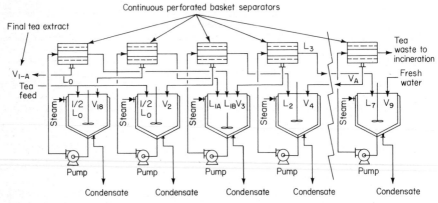

Fig. 16 Continuous countercurrent multistage system for extraction of tea solubles (split-feed arrangement). (Coca-Cola Co.—Atlanta, Ga.)

of pumpability of the mixture, at the same time allowing a concentrated overflow to be produced. It increases, however, the number of agitation tanks and associated equipment required. Usually more than a two-way split would not be economical.

CONTROL AND INSTRUMENTION IN LEACHING SYSTEMS

In most leaching systems, the most critical factors are maintenance of constant fluid flows, constant pressures, and temperatures. The objective is to maintain steady conditions, and to avoid extraction of undesirable material and loss of solvent for economic and safety reasons. Fortunately, these factors allow precise and automatic regulation with ordinary controlling instruments. Recording instruments, although not essential, may serve a useful purpose in maintaining a record of operating conditions which would be helpful in studying plant performance.

DESIGN CALCULATION METHODS

Ideal Stage and Equilibrium in Leaching

As in other mass-transfer operations, leaching calculations are based on the concept of an "equilibrium" or "ideal" stage which is defined as a stage from which the resultant solution is of the same composition as the solution adhering to the solids leaving the stage. This definition of an ideal stage, however, may not be interpreted as an indication that the equilibrium of a saturated condition between the solution and the solute in or on the carrier solids exists. In leaching, true equilibrium between solution and solute is rarely involved because there is not usually sufficient time of contact for complete dissolution of the solute. Also, it is impractical to make separation of solid and solution perfect, and therefore solids leaving the stage will always retain some liquid and its associated dissolved solute. In case the solute is adsorbed by the solids, imperfect settling or draining will result in lowering of stage efficiency, although equilibrium between the solution and solid phases is established. One, therefore, needs a practical overall stage efficiency factor in addition to the number of ideal stages to obtain the actual number of stages for a given leaching operation. It is simplest, however, to use practical equilibrium data which take the stage efficiencies into account directly.

Calculation Procedures for Purposes of Design

A given solid-liquid extraction system is assumed to consist of the following three components:

1. Inert, insoluble solids
2. A single solute which may be a single solid or liquid or a mixture of soluble components
3. A solvent which dissolves the solute but has little or no effect on the inert solid

The solubility of inerts and adsorption of solute by the inert solid can be accounted for in the calculations, if data are available on solubility of inerts in the solvent and on adsorption of solute by inerts as a function of solution concentration. The computation of the number of ideal stages required is then based on material balances, knowledge of solution quantity retained by the inert solids as a function of solute concentration in solution and the definition of an ideal stage. Either stage-to-stage algebraic methods, graphical methods, or in some special situations, analytical methods, can be employed. Because of its relative importance, only the application of these methods to the design calculations of a continuous countercurrent multistage system is discussed in what follows. For information concerning the application of these methods to other extraction systems such as single-stage leaching and crosscurrent leaching, reference is made to additional literature.

Overall Material Balance and Equation of Operating Line

In Fig. 2 are also shown the material balance and the nomenclature commonly used to describe the various streams and concentrations in a continuous countercurrent multistage solid-liquid extraction system. When the system is operating under steady-state conditions, the equations for material balances over the entire system are as follows:

For total solution:

$$V_{n+1} + L_0 = V_1 + L_n \tag{1}$$

For solute:

$$V_{n+1}(y_A)_{n+1} + L_0(x_A)_0 = L_n(x_A)_n + V_1(y_A)_1 \tag{2}$$

where L_n = total mass of underflow leaving a stage
V_n = total mass of overflow leaving a stage
x_A = mass fraction of solute in underflow
y_A = mass fraction of solvent in overflow
x_S = mass fraction of solvent in underflow
y_S = mass fraction of solvent in overflow
X = solute-to-solution ratio = solute/(solute + solvent)
Y = inerts-to-solution ratio = inerts/(inerts + solution)
n = subscript denoting the number of the stage where the stream originates

Eliminating V_{n+1} and solving for $(y_A)_{n+1}$ gives the following equation for the operating line:

$$(y_A)_{n+1} = \frac{L_n}{1 + (V_n - L_0)}(x_A)_{n+1} + \frac{V_0(y_A)_0 - L_0(x_A)_0}{L_0 + V_0 - L_0} \tag{3}$$

The operating line as represented by Eq. (3) is a straight line if L_n, the underflow from each stage, is a constant quantity. If L_n is not constant, the slope of the operating line varies from stage to stage. Algebraic, graphical, and analytical methods are employed in conjunction with the operating line equation to calculate the number of equilibrium stages required for a given extraction problem.

Algebraic Method

This method starts with calculation of the quantities and compositions of all the terminal streams, using a convenient quantity of one of the terminal streams as the basis of calculations. Material balance and stream compositions are then computed for a terminal ideal stage at either end of the extraction cascade, using equilibrium and solution retention data. Calculations are then repeated for each successive ideal stage from one end of the system to the other until an ideal stage is obtained which corresponds to the desired terminal conditions. The number of actual stages may then be obtained by dividing the number of ideal stages by the overall stage efficiency. Quantity of solution retained by inerts depends upon the properties of the solution, especially the viscosity, and is determined experimentally under conditions similar to those which are considered for commercial operation. Any solid-liquid extraction problem can be solved by the algebraic method, but it is tedious, especially when a large number of stages are involved. The method, as applied to leaching, is illustrated below by an example.

EXAMPLE 1

Oil is to be extracted from meal by means of benzene in a continuous countercurrent extractor. The unit is to treat 1000 lb of meal based on completely exhausted solids per hour. The untreated meal contains 400 lb oil and no benzene. The final product solution obtained from the operation is to contain 60% oil, and 90% of oil in the underflow feed is to be recovered. Assume no carry-over of meal (inert basis) into the overflow solution. Test data (Fig. 17 and Table 1) indicate that constant underflow cannot be assumed. Calculate the number of ideal stages required.

Solution: For 1000 lb meal per hour (inert basis)
    1. Calculation of terminal conditions:
        Solute in underflow feed, L_0 = 400 lb (oil)
        Solute in underflow discharge, L_n = 0.1 × 400 = 40 lb
        Benzene in underflow feed, L_0 = 0 lb
        By overall material balance, oil in overflow solution = 400 − 40 = 360 lb
Therefore, amount of overflow solution, V_1 = 360/0.60 = 600 lb/h
    2. Material balance on stage 1. From experimental data, by graphical interpolation (Fig. 17): Solution in stream L_1 (leaving stage 1) corresponding to a solute (oil) concentration of 0.6 in overflow solution V_1 is 0.595 lb/lb inert meal
Therefore, solution in L_1 = 0.595(1000) = 595 lb/h
Since stream L_1 is in equilibrium with V_1, oil in L_1 = 0.6 × 595 = 357 lb/h
By solution balance, V_2 = 600 + 595 − 425 = 710 lb/h
By solute balance, oil in V_2 = 360 + 357 − 400 = 317 lb/h.
    3. Material balances for ideal stages 2 to n: the same procedure used for the first stage is followed

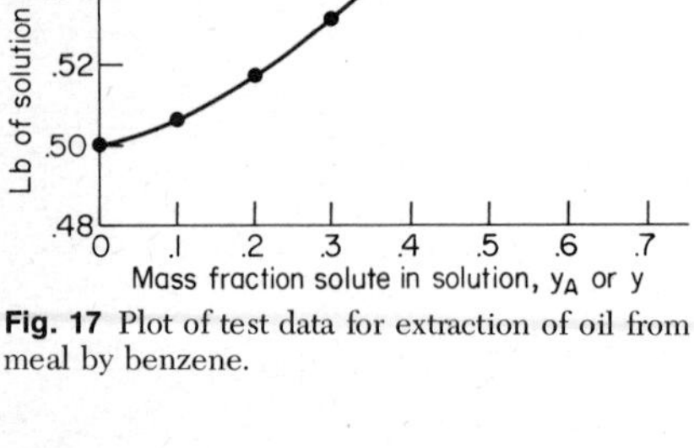

Fig. 17 Plot of test data for extraction of oil from meal by benzene.

in making calculations for ideal stages 2, 3, . . . , n, in the cascade until the desired terminal conditions are obtained at the nth stage, where n is the total number of stages required. The calculations are summarized in Table 2.

An examination of Table 2 shows that five stages reduce the solute content in the underflow to better than the desired degree, but an exact specification of the terminal condition of solvent benzene feed is not obtained at the fifth stage. Usually trial-and-error calculations are necessary to obtain an integral number of stages if practical equilibrium data are employed, and for this, a readjustment of terminal conditions is required. In practice, an integral number of stages has to be used, and therefore, the number of stages is chosen such that better than desired terminal conditions are indicated by calculation. In the present case, five stages and about 695 lb of benzene solvent will give operating conditions slightly better than desired.

Graphical Method

The graphical method of calculation is simply a graphical representation of the material balances and equilibrium data, and it is theoretically equivalent to the algebraic method. Its principal advantages are that it simplifies calculations and permits visualization of the process variables and their effect on the operation. The three-component system consisting of solute, inert solid, and solvent at constant temperature may be represented on various types of diagrams such as equilateral triangular, right triangular, modified Ponchon-Savarit, and McCabe-Thiele plots which take into account the interrelationship among the three variables.[2,16,19–22] Computations and graphical representation can be made on equilateral triangular coordinates for any ternary system,[21] but it is preferable to use either right-triangular or modified Ponchon-Savarit diagrams because they offer a choice of scales for both the ordinate and abscissa and avoid crowding of construction into one corner as in the case of the equilateral-triangular diagram. The right triangular

TABLE 1   Test Data and Calculation of Underflow Compositions

Test data							Calculated compositions				
		Conc. lb solute (oil)			lb/lb inerts			Mass fraction in underflow			
$F_n = \dfrac{\text{lb solution}}{\text{lb oilfree meal}}$		lb sol. (or X)	$1/F_n$ (or Y)	Solute	Solvent	Underflow	Solute	Solvent	Inerts		
0.500		0.0	2.00	0.000	0.500	1.500	0	0.333	0.670		
0.505		0.1	1.98	0.0505	0.4545	1.505	0.0336	0.302	0.664		
0.515		0.2	1.94	0.1030	0.4120	1.515	0.0682	0.272	0.660		
0.530		0.3	1.89	0.159	0.371	1.530	0.1039	0.242	0.6541		
0.550		0.4	1.82	0.220	0.330	1.550	0.1419	0.213	0.6451		
0.571		0.5	1.75	0.2855	0.2855	1.571	0.1817	0.1817	0.6366		
0.595		0.6	1.68	0.357	0.238	1.595	0.224	0.1492	0.6268		
0.620		0.7	1.613	0.434	0.186	1.620	0.268	0.1148	0.6172		

diagram is preferred by many because the concentrations of various streams are somewhat easier to visualize than in other diagrams.

Typical Equilibrium Diagrams

Right-triangular and Ponchon-Savarit diagrams are shown in Fig. 18. Figure 18a is a right-triangular plot in which the mass fraction of solvent (x_s or y_s) is plotted against the mass fraction of solute (x_A or y_A). Figure 18b represents a plot of inerts-to-solution ratio Y against solute-to-solution ratio X and can be considered a modification of the Ponchon-Savarit diagram as used in distillation and adsorption calculations.

In Fig. 18a and b, EF represents the locus of overflow solutions for the case where the overflow stream contains no inerts. $E'F'$ represents the overflow streams containing some inert solids, either by entrainment or by partial solubility in the overflow solution. Lines GF, GL, and GM represent loci of stream compositions for the three different conditions, viz., constant underflow, variable underflow, and constant solvent-to-inerts in underflow. In Fig. 18a, the constant underflow line GM is parallel to EF, the hypotenuse of the triangle, whereas GF passes through the right-hand vertex representing 100% solute. In Fig. 18b, the constant-underflow line GM is parallel to the abscissa, and GF passes through the point on the abscissa representing the composition of the clear solution adhering to the inert solids.

If all the solute is in solution and if the solution adhering to the solids has the same composition as that of the overflow stream, then lines such as AC and BD of Fig. 18a and b represent the ideal equilibrium tie lines. In the case of the right-triangular diagram, the equilibrium tie lines pass through the origin O (representing 100% inerts). In Fig. 18b, the equilibrium tie lines are vertical. For nonequilibrium conditions or for equilibrium conditions with selective adsorption, the tie lines would be displaced from their equilibrium position such as AC' and BD'. Point C' is to the right of C, if the solute concentration in the overflow solution is less than that in the underflow solution adhering to the

TABLE 2 Summary of Stage-to-Stage Calculations for Example 1

| | Underflow leaving stage n, L_n | | | | | | | Solution Entering Stage $n - V_{n+1}$ | | | | |
| | Quantities, lb/h | | | | Composition in mass fractions | | | Quantities, lb/h | | | Composition in mass fractions | |
n	Total L_n	Solution V_n	Solute $(Lx_A)_n$	Benzene $(Lx_S)_n$	Solute $(Lx_A)_n$ (oil)	Benzene $(x_S)_n$	Inert meal $(x_I)_n$	Total V_{n+1}	Solute $(Vy_A)_{n+1}$ (oil)	Benzene $(Vy_S)_{n+1}$	Solute $(y_A)_{n+1}$	Benzene $(y_S)_{n+1}$
0	1400	400	400	0.0	0.2857	0.00	0.7143	600	360	240	0.6000	0.400
1	1595	595	357	238	0.2238	0.1492	0.6270	770	317	453	0.4120	0.588
2	1552	552	227	325	0.1463	0.2094	0.6443	727	187	540	0.2572	0.7428
3	1523	523	135	388	0.0886	0.2548	0.6566	698	95	603	0.1361	0.8639
4	1508	508	69	439	0.0458	0.2911	0.6631	683	29	654	0.0425	0.9575
5	1501	501	21	480	0.0140	0.320	0.6660	676	Negligible			

solids. Unequal concentrations in the two solutions indicate insufficient contact time, and/ or preferential adsorption of one of the components on the inert solids. Tie lines such as AC' may be considered "practical tie lines," if data on underflow and overflow composition are obtained experimentally under conditions simulating actual operation, particularly with respect to contact time, agitation, and particle size of solids.

In certain cases of leaching, the solute is of limited solubility and the overflow solution reaches saturation. This situation can also be treated by the above methods. In this case, the solvent input to the nth stage should be such that a saturated solution is obtained as effluent from stage 1 (i.e., from the system). If saturation is obtained in more than one stage, all such stages, except one, are unnecessary since no more solute is obtained in these stages because of the saturation condition of the solution.

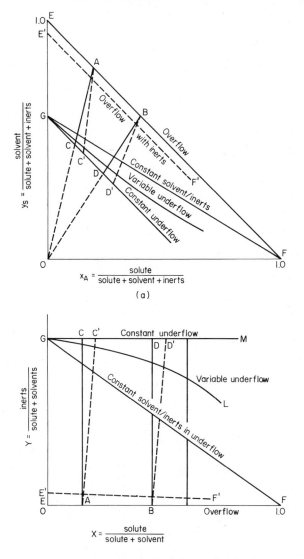

Fig. 18 Typical equilibrium diagrams. (*a*) Right-triangular diagram; (*b*) Ponchon-Savarit diagram.

Figure 19 shows the methods of graphical calculation of stages using right-triangular and Ponchon-Savarit diagrams for a countercurrent system. Terminal concentrations are established, thus fixing points A, B, C, and D of Fig. 19 for the respective diagrams. Stepping of stages is done in the usual manner by following difference point (Δ point) lines and tie lines until the desired terminal concentration is reached. In Fig. 19, four stages are indicated. Use of practical tie lines in the construction will give actual stages required. Similarly, ideal equilibrium tie lines will yield ideal stages, in which case an overall efficiency factor must be obtained to calculate the actual number of stages

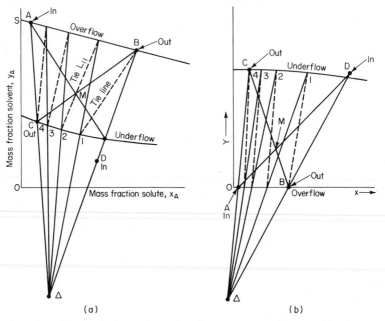

Fig. 19 Procedure of graphical solution of a number of stages. Stepping of stages on (*a*) right-triangular diagrams and (*b*) Ponchon-Savarit diagram.

required. If the number of stages is known and one of the terminal concentrations is the unknown quantity, a trial-and-error method must be used to get the unknown concentration.

In Figure 19a, CB intersects AD at point M, which represents the addition point and the weighted average compositions of all terminal streams combined. In the right-triangular diagram (Fig. 19a) the lengths of the line segments are proportional to the stream weights as follows:

$$\frac{MD}{AM} = \frac{\text{mass of underflow}}{\text{mass of overflow feed}}$$

$$\frac{CM}{MB} = \frac{\text{mass of underflow discharge}}{\text{mass of overflow effluent}}$$

whereas in the Ponchon-Savarit diagram (Fig. 19b) the same relationships will be as follows:

$$\frac{AM}{AD} = \frac{\text{mass of underflow feed}}{\text{mass of overflow feed}}$$

$$\frac{CM}{MB} = \frac{\text{mass of overflow effluent}}{\text{mass of underflow discharge}}$$

These relationships may be used to locate the addition point and thereby one of the terminal points, if the coordinates of the other three terminal points are known. Example 2 illustrates the application of the graphical method.

EXAMPLE 2

Assume same problem data as in Example 1.

Solution 1: Right-triangular diagram   Underflow compositions are calculated from the test data given in Table 1 and are given in the last two columns of the table. They are plotted as underflow line GL in Fig. 20a. Composition of the underflow discharge is determined as follows:

Basis: 1400 lb/h underflow feed
Solute in underflow feed = 400 lb/h
Oilfree meal (inerts) in underflow feed = 1000 lb/h
Underflow discharge contains 40 lb of oil/1000 lb of inerts or 0.04 lb/lb oilfree meal.

From Table 1, by interpolation in column 5,

$$\frac{\text{lb benzene}}{\text{lb oilfree meal}} = 0.5 - \frac{0.04}{0.0505} \times (0.5 - 0.4545) = 0.464$$

Therefore, the underflow discharge composition is:

	lb/lb inerts	*Mass fraction*
Solute (oil)	0.40	0.0266
Solvent (benzene)	0.464	0.3085
Oilfree meal	1.0	0.6649
	1.504	

Alternatively, one can establish the underflow composition graphically. This involves marking the composition of underflow on a solvent-free basis, on the abscissa of Fig. 20a, then drawing a straight line from this point to point S, representing pure solvent. The

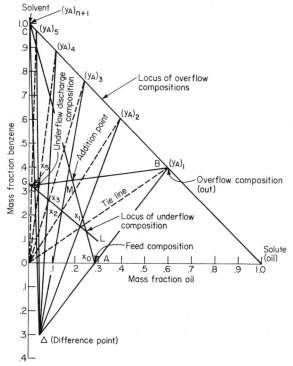

Fig. 20a Solution of Example 2 on right-triangular diagram.

point of intersection of this line with the underflow curve GL represents the composition of the underflow discharge.

After terminal composition points $A, B, C,$ and D have been established on the diagram, as explained above, they are connected by lines BA and CD which are extended to intersect, as shown in the diagram. The point of intersection of BA and CD is the difference point. Following the difference point lines and tie lines, in the conventional manner, gives the number of equilibrium stages required. Since practical equilibrium data have been used, the stage efficiency is already accounted for and hence there must be an integral number of stages. The unknown quantity in this example is number of stages required, and therefore a trial-and-error adjustment of the effluent concentrations or amount of solvent will be necessary to obtain an integral number of stages.

The quantity of oilfree benzene required may be obtained graphically. A straight line joining (x_A) and $(y_A)_{n+1}$ and a straight line joining $(x_A)_n$ and $(y_A)_n$ are constructed. The intersection point of these lines, M, is the addition point. The ratio of the mass of benzene required to the mass of oilfree meal is obtained by determining the ratio $(x_A)_0 M/M(y_A)_{n+1}$ on the diagram.

By a material balance, the amount of solvent required is 704 lb/h (Example 1, algebraic solution) for the terminal conditions of the problem (the graphical method gives solvent quantity of 700 lb/h). It is seen from Fig. 20a that four equilibrium stages are not sufficient and five equilibrium stages give a slightly higher recovery than desired.

Solution 2: Ponchon-Savarit Diagram Data in Table 1 provide underflows also. Values of X are given in column 2 of the table, and corresponding values of Y are given in column 3. In Fig. 20b, Y values are

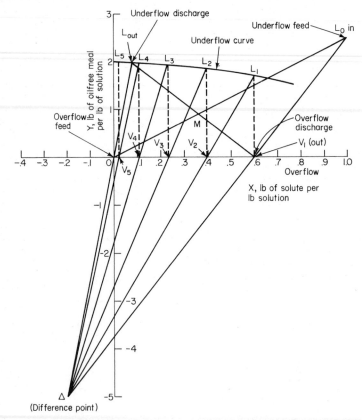

Fig. 20b Solution of Example 2 on Ponchot-Savarit diagram.

plotted against corresponding X values as curve GL. Since overflows from each stage are assumed to be inertfree, for overflows, $Y = 0$ for all X. Hence, the Y vs. X line for overflows coincides with the abscissa. Terminal conditions are then established on the diagram in the following manner:

L_0, fresh feed

$$X = \frac{\text{oil}}{\text{oil} + \text{benzene}} = \frac{400}{400 + 0} = 1$$

$$Y = \frac{\text{oilfree meal}}{\text{oil} + \text{benzene}} = \frac{1000}{400 + 0} = 2.5$$

V_{n+1}, overflow benzene feed:

$$X = 0 \quad \text{(oilfree solvent)}$$
$$Y = 0 \quad \text{(inertfree solution)}$$

V_1, overflow effluent:

$$X = 0.6$$
$$Y = 0$$

For underflow discharge, the ratio Y/X is the ratio oilfree meal/oil in the stream and is $1000/40 = 25$. This is the slope of a line through O which cuts underflow curve GL in L_n. M, the difference point, is obtained by extending AB and CD to intersect at point M. Tie lines are vertical since $(x_A)_n = (y_A)_n$. The number of stages are then obtained by following difference point lines and tie lines as shown in Fig. 20b. In this case, also, a trial-and-error solution will be required to obtain an integral number of stages.

Other graphical procedures[22-26] to take into account special situations have been developed. A combined graphical and analytical method has been devised by Ruth[27] to take full advantage of both methods. Slow rates of solution must be accounted for with experimentally determined underflow curves.

Analytical Methods

Analytical methods to obtain the number of equilibrium stages require certain simplifying assumptions and, therefore, are not general in application. When applicable, they can be advantageously used to study the effect of different variables quickly and comparatively with less effort. The analytical methods that were followed by Baker[28] and by McCabe and Smith[16] were developed for multistage countercurrent leaching, using the ideal stage concept. They assume that (1) the solution adhering to the solids in the underflow has the same composition as the overflow from that stage, and (2) the underflow rate between stages or the solvents-to-inerts ratio is constant between stages.

It is not possible to apply these equations to an entire extraction cascade if L_0, the underflow feed to the first stage, is not the same as L_n, the underflow from stage to stage within the system. Equations[24] have been derived for this situation but it is simpler to calculate, by material balances, the performance of the first stage and then apply these equations to obtain the number of remaining stages in the cascade. Baker's method[28] is particularly useful for calculating terminal concentrations if the number of stages is known. The McCabe-Smith[16] method facilitates calculation of number of stages directly by the use of known solution concentrations.

For constant underflow or constant solution-to-inerts ratio, the McCabe-Smith equation becomes

$$n - 1 = \frac{\log \dfrac{y_{n+1} - x_n}{y_2 - x_1}}{\log \dfrac{y_{n+1} - y_2}{x_n - x_1}} \tag{4}$$

For constant solvent-to-inerts ratio, Eq. (4) becomes

$$n - 1 = \frac{\log \dfrac{y'_{n+1} - x'_n}{y'_2 - y'_2}}{\log \dfrac{y'_{n+1} - y'_2}{x'_n - x'_1}} \tag{5}$$

where the symbols not defined before are:

> x = mass fraction of solute in solution adhering to inerts or on an inertfree basis
> y = mass fraction of solute in overflow solution, inertfree basis
> x' = mass ratio of solute to solvent in underflow
> y' = mass ratio of solute to solvent in overflow

Subscripts, as before, refer to the stage where the stream originates.

EXAMPLE 3: McCABE-SMITH METHOD

One hundred tons of underflow feed containing 20 tons of solute, 2 tons of water, and 78 tons of inert material are to be leached with water to give an overflow-effluent concentration of 15 wt% solute and a 95% recovery of solute. The underflow from each stage carries 0.5 lb solution/lb inerts. Calculate the number of ideal stages required.

Solution Basis, 100 tons of underflow feed. Since total underflow is constant, Eq. (4) applies. Overall solute material balance gives the following:

Solute in overflow discharge = 0.95(20) = 19 tons.
Concentration of overflow discharge = y_1 = 0.15.
Therefore, weight of overflow discharge = 19/0.15 = 126.67 tons.
Weight of solution adhering to inert solids in underflow leaving each stage = 0.5(78) = 39 tons.
Therefore, water to stage n = 126.67 + 39 − 22 = 143.67 tons.
Concentration of the overflow leaving stage 1 is equal to that of solution adhering to the solids leaving this stage.
Therefore, weight of solute in underflow leaving stage 1 = (0.15)(39) = 5.85 tons.
Solute balance around stage 1 gives solute entering this stage = 19 + 5.85 − 20 = 4.85 tons.
Concentrations are therefore as follows:
Solute in underflow from stage 1: x_1 = 5.85/39 = 0.15
Solute in overflow to stage 1: y_2 = 4.85/143.67 = 0.0388
Solute in underflow discharge: x_n = 1/39 = 0.0256
Solute in solvent feed: y_{n+1} = 0
Substitution in Eq. (4) then gives

$$n - 1 = \frac{\log \dfrac{0 - 0.0256}{0.0338 - 0.15}}{\log \dfrac{0 - 0.0338}{0.0256 - 0.15}} = 1.6$$

Therefore, n = 2.6 ideal stages.

In the above example, the performance of the first stage is calculated separately, and then the Eq. (4) is applied to the remaining $n − 1$ stages. The same procedure is to be followed in application of Eq. (5) to the case of constant solvent-to-inerts ratio.

Chen's method[29] makes the same assumptions as the methods of Baker and McCabe-Smith but yields a single formula which is valid for both constant underflow and constant solvent-to-inerts ratio if appropriate units are used:

$$n = \frac{\left[\log 1 + (r - 1) \dfrac{x_1 - y_n + 1}{x_n - y_n + 1} \right]}{\log r} \tag{6}$$

where $n = V_n + 1/IF$
> F = pounds solution retained per pound solid (constant)
> I = flow rate of inert solid
> V_{n+1} = fresh solvent flow rate to the nth stage

Chen[29] has also developed a simplified equation for the case of constant underflow and equilibrium relationship of the type $y_n = mx_{n+B}$.

Analytical Methods for Variable Underflow For the case of variable underflow, a modification of Baker's method has been developed by Grosberg.[24] In some cases, the underflow is variable, but the solution retention by the inerts is such that the reciprocal of the retained solution is linear with solution concentration. If the data show a slight curvature, they can be treated as several straight lines. If the straight-line relationship is represented by $1/F_n = A + BX_n$, Chen's[29] equation with the present notation becomes

$$\frac{x_1 + b + \beta_2}{x_1 + b + \beta_1} = \frac{y_{n+1} + b + \beta_2}{y_{n+1} + b + \beta_1} \left(\frac{\beta_2}{\beta_1} \right)^n \tag{7}$$

where $a = -\dfrac{A + By_{n+1}r_1}{B(r_1 - 1)}$

$b = \dfrac{Ar_1 + Bx_n}{B(r_1 - 1)}$

$c = \dfrac{A(x_1 - r_1 y_{n+1})}{B(r_1 - 1)}$

$r_1 = \dfrac{V_{n+1}}{F_1 I}$

The form of solution of Eq. (4) depends upon the values of β_1 and β_2, which are the roots of the equation

$$\beta = \frac{(a - b) \pm \sqrt{(a + b)^2 - 4c}}{2}$$

EXAMPLE 4: CHEN'S METHOD OF VARIABLE UNDERFLOW

Assume the same conditions as in Example 1. Selecting two points $F_n = 0.5$ at $x = 0$ and $F_n = 0.595$ at $x = 0.6$ and substituting these values in the equation $1/F_n = A + Bx_n$, observe that $A = 2$ and $B = -0.533$. Next, assume $x_n = 0.08$. From test data, by interpolation, solution retained per pound of oilfree meal = 0.504 lb. Therefore, solution retained by oilfree meal = $0.504 \times 1000 = 504$ lb. Mass fraction of oil in solution = $40/504 = .0794$, which checks closely with the assumed value.

Now,

$$\frac{1}{F_n} = A + Bx_n$$
$$= 2 - 0.533(0.0794) = 1.96$$

Therefore, $F_n = 0.51$.

Composition of overflow discharge solution: $y_1 = 0.6$

Composition of fresh solvent $y_{n+1} = 0$

$$r_1 = V_{n+1}/IF_n = \frac{700}{1000(.51)} = 1.373$$

$$a = \frac{2 + 0.533 \times 0 \times 1.373}{-0.533(1.373 - 1)} = 10.06$$

$$b = \frac{2(1.373) - 0.533(0.0794)}{-0.533(1.373 - 1)} = -13.6$$

$$c = \frac{2(0.0794 - 1.373 \times 0)}{-0.533(1.373 - 1)} = -0.799$$

Therefore,

$$\beta = \frac{10.06 - (-13.6) \pm (10.06 - 13.6)^2 - 4(-0.799)}{2}$$

Solving for β gives $\beta_2 = 13.815$ and $\beta_1 = 9.845$.

Substitution of these values in Eq. (5) gives

$$\frac{0.6 - 13.6 + 13.815}{0.6 - 13.6 + 9.845} = \frac{0 - 13.6 + 13.845}{0 - 13.6 + 9.845} \left(\frac{13.815}{9.845} \right)^n$$

from which $n = 4.1$ stages. Both algebraic and graphical methods gave slightly higher values for n.

Other Analytical Procedures In the case of constant underflow and a straight equilibrium line of slope m, it is possible to use the absorption factor method.[5] Sao and Tsujimo[30] have developed methods and prepared charts for calculation of multistage leaching systems in which adsorption of solute on the solids may not be neglected and where equilibrium in each stage is not reached. Brunische-Olson[31] has derived mathematical expressions for several cases of stepwise countercurrent extraction.

Calculation of Batch Countercurrent Multiple-Contact System

Hawley[1,29] has developed formulas which show the relationship between number of cells and number of treatments in terms of various typical extraction operations, both for separate and simultaneous pumping of the solvent in the cells. When the time required for each operation is known, it is possible to use these formulas to determine the number of separate cells or extractors required to furnish a certain number of treatments for each charge. This subject has been discussed in detail by Lerman.[17]

For countercurrent multibatch system, there are two distinct types of operation. In type 1 operation, the liquid inflow to the system and liquid withdrawal from the system are carried out during different liquid-transfer periods. Type 2 operation involves liquid inflow to and discharge from the extraction system during the same liquid-transfer period. The period of liquid feed and discharge to and from the system alternates with the period of liquid-transfer between cells without the extracting liquid entering or leaving the system as a whole. Variations of these two types are possible, and can be used to increase or decrease the treatments in special cases. Some of these variations differ from the countercurrent flow.

For type 1 operation, both Hawley[32] and Ravenscroft[25] have shown that N_B, the number of batch units in a batch countercurrent multiple system, is related to N_C, the number of equilibrium contact units in an equivalent continuous countercurrent system, by the equation

$$N_B = \frac{N_C + 1}{2} \tag{8}$$

while for type 2 operation, Lerman[17] has shown that N_B is given by the equation

$$N_B = N_C/2 \tag{9}$$

Where N_B and N_C are defined as in Eq. (8). Therefore, the number of units required in a batch countercurrent multiple-contact system can be obtained by first calculating the number of equilibrium stages required for an equivalent continuous countercurrent system by methods described earlier and then applying Eq. (5) or (6) for the type of operation chosen.

Calculation of Moving-Bed Extraction Systems

Stagewise calculation methods can be applied equally to screw- and basket-type extractors, where solid material is moved countercurrent to the flow of solvent by defining an "equilibrium length." An equilibrium length is defined as the length of a section of an extraction unit such that the extract solution leaving the section has the same concentration as the solution entrained by the solid leaving that section. Analytical or graphical solution yields the number of equilibrium lengths required instead of the number of stages. The actual value of equilibrium length must be determined experimentally in a given case by methods similar to those employed to determine HETP (height equivalent to a theoretical plate) for distillation or absorption in packed columns.

REFERENCES

1. Hawley, L. F., *Ind. Eng. Chem.*, **9**, 866 (1917).
2. Badger, W. L., and Banchero, J. T., "Introduction to Chemical Engineering," McGraw-Hill, New York, 1936.
3. McGennis, R. A., "Beet Sugar Technology," Reinhold, New York, 1951.
4. Cofield, E. P. T., *Chem. Eng.* **58**(1), 127 (1951).
5. Treybal, R. E., "Mass Transfer Operations" (2d ed.), McGraw-Hill, 1968, p. 646.
6. Anderson, E. T., and McCubbins, K., *J. Am. Oil Chem. Soc.*, **31**, 475 (1954).
7. Karnofsky, G., *Chem. Eng.*, **8**, 108 (1950).
8. McCubbins, K., and Rite, G. J., *Chem. Ind. (London)*, **66**, 354 (1950).
9. Gastrock, E. A., et al., *Ind. Eng. Chem.*, **49**, 921, 930 (1957).
10. Kenyon, R. L., et al., *Ind. Eng. Chem.*, **2**, 186 (1948).
11. Bible, C. W., *Mech. Eng.*, **63**, 357 (1941).

12. Markley, K. S., and Longtin, B., "Soybean Chemistry and Technology," Chemical Publishing, 1944, pp. 175–182.
13. Rickles, R. N., *Chem. Eng.*, March 15, 1965, p. 157.
14. Rushton, J. H., and Maloney, L. H., *J. Met.*, **6**; 1199 (1954).
15. Lamont, A. G. W., *Can. J. Chem. Eng.*, **36**, 153 (1958).
16. Mc Cabe, N. L., and Smith, V., "Unit Operations of Chemical Engineering," McGraw-Hill, New York, 1967, p. 769.
17. Lerman, F., "Encylopedia of Chemical Technology," **6**, 91–122 (1951).
18. Prabhudesai, R. K., research report, Coca-Cola Co., Atlanta, 1967.
19. Armstrong, R. T., and Kammermayer, K., *Ind. Eng. Chem.* **34**, 1288 (1942).
20. Brown, G. G., et al., "Unit Operations," Wiley, New York, 1956, pp. 277–293.
21. Elgin, J. C., Trans. AIChE, *32*, 457 (1936).
22. Kinney, G. F., *Ind. Eng. Chem.*, **34**, 1102 (1942).
23. Fitch, B., *Ind. Eng. Chem.*, **58**(10), 18 (1966).
24. Grosberg, J. A., *Ind. Eng. Chem.*, **42**, 155 (1950).
25. Ravenscroft, E. A., *Ind. Eng. Chem.*, **28**, 851 (1936).
26. Scheibel, E. G., *Chem. Eng. Prog.*, **49**, 356 (1953). *AIME Trans.*, **200**, 1199 (1954).
27. Ruth, B. F., *Chem. Eng. Prog.*, **44**, 72 (1948).
28. Baker, E. M., *Chem. Metall. Eng.*, **42**, 669 (1935).
29. Chen, N. H., *Chem. Eng.*, **71**(24), 125 (1964).
30. Sao, E. and Tsujimo, A., *Chem. Eng. (Jpn)* **14**, 264 (1950); **15**, 101, 343 (1951).
31. Brunische-Olson, "Solid-Liquid Extraction," NYT, Nordisk Forlag Arnold Busch, Copenhagen, 1962.
32. Hawley, L. F., *Ind. Eng. Chem.*, **12**, 493 (1920).
33. Donald, M. B., *Trans. Inst. Chem. Eng. (London)*, **15**, 77 (1937).
34. George, W. J., *Chem. Eng.*, **66**(2), 111 (1959).
35. Goss, W. H., *J. Am. Oil. Chem. Soc.*, **23**, 348 (1946).
36. Maloney, J. O., and Schubert, A. E., *Trans. AIChE*, **36**, 741 (1946).

Section **5.2**

Flotation

FRANK F. APLAN *Mineral Processing Section,*
Department of Mineral Engineering, The Pennsylvania
State University

FLOTATION

Froth flotation is a major means of achieving solid-solid separations, though mutations of the process are also used to affect solid-liquid separations (dissolved air flotation) and liquid-liquid separations (foam fractionation).

Description of the Process

Flotation finds its greatest application in ore separations. In a typical such process, the ore is finely ground in water, usually to a nominal -250 μm, and introduced into a large stirred tank called a *flotation cell,* where the water-ore slurry is conditioned with a frothing agent to create a copious supply of bubbles and a collecting agent designed to coat the desired minerals with a hydrophobic coating so that they can be attached to air bubbles. The ore pulp, at approximately 25% solids, is then sparged with air, and the air bubbles created, together with their attached mineral particles, rise to the surface of the pulp where they are removed from the system as the concentrate. The undesired or "gangue" mineral particles are removed from the flotation cell as a tailing product.

Magnitude

Flotation is an important separation process in the mining industry, and the United States Bureau of Mines reported in 1975 that a total of 252 flotation plants processed 423 million tons of ore. The installed capacity of these plants was about 1.6 million tons per day or nearly one-half billion tons per year. The growth of the industry has been spectacular, for in 1960 only 198 million tons were treated by the process. Some idea of the magnitude of this industry can be obtained from Table 1, which shows the quantities of the principal ores treated by the froth flotation process in 1975. The U.S. Bureau of Mines has also reported that 1.77 million pounds of reagents valued at $87.7 million were used in the year 1975. Water usage is large, amounting to 527 billion gallons or 1270 gallons per ton treated. Most of this water is, of necessity, recycled. Energy requirements amounted to

TABLE 1   Principal Ores Treated by the Froth Flotation Process in the United States in 1975*

Ore treated	Millions of tons
Copper, copper-molybdenum, and molybdenum ores	260†
Lead-zinc (some also containing copper)	19.2
Iron ores	28.6
Phosphate	75.1
Potash	13.9
Glass sand	7.3
Feldspar-mica-quartz	2.6
Coal	13.1

*SOURCE: U.S. Bureau of Mines, "Minerals Yearbook," 1975.
†Approximate.

15.8 kWh per ton of ore treated, though most of this is used in grinding the ore to flotation size.

Sources of Information

A recent flotation symposium volume edited by M. C. Fuerstenau[1] provides a wealth of data on the theory and practice of the process. In addition there are older, but still very useful books, by D. W. Fuerstenau;[2] Gaudin;[3] Glembotskii, Klassen, and Plaksin;[4] Klassen and Mokrousov;[5] Lemlich;[6] Sutherland and Wark;[7] and Taggart;[8] and review articles by Aplan[9] and Bloecher.[10] Industrial bulletins such as those by the American Cyanamid Co.,[11] Denver Equipment Co.,[12] and Dow Chemical Co.[13] are also particularly helpful. These sources provide an in-depth coverage of all phases of the subject of froth flotation.

PHASES IN FLOTATION

Because of the wide variation in industrial practice, it is first necessary to understand some of the principles involved in the flotation process. There are three phases involved in froth flotation: gas, liquid, and solid. For a bubble which has become attached to a hydrophobic surface, these phases are related by the Young equation:

$$\gamma_{SG} = \gamma_{SL} + \gamma_{LG} \cos \theta$$

where γ_{SG}, γ_{SL}, and γ_{LG} are the surface free energies at the solid-gas, solid-liquid, and liquid-gas phases, respectively, and θ is the angle of contact at the solid-gas contact measured through the liquid phase. The contact angle θ thus indicates the degree of bubble attachment to a surface and provides a simple laboratory means of predicting flotation behavior under a wide variety of conditions.

Gas Phase

The gas used is invariably air. While, theoretically, any gas can be used, it has been found that oxygen often plays a special role in controlled surface oxidation to allow for ease of attachment of the collectors.

Liquid Phase

The liquid usually used in flotation is water, though seawater and saturated brines are used in a few instances. However, it is by manipulating the composition of the liquid phase that much of the selectivity is achieved in flotation. For this reason flotation engineers have largely focused their attention on the nature of reagents dissolved in the liquid as a means of achieving selective attachment of flotation reagents onto mineral surfaces.

Solid Phase

As might be expected, the nature of the solid phase will dictate the choice of the flotation collector to be used. From Table 2 it may be seen that thio collectors are used for sulfide minerals and fatty acids are used for the slightly soluble minerals. Insoluble oxide

TABLE 2   Common Collectors Used in Mineral Flotation

Mineral class	Mineral samples	Collector type	Typical collector quantity used, lb/ton
Sulfide minerals	$CuFeS_2$, PbS	Xanthate Dithiophosphate Thiocarbamate	0.05 to 0.1
Slightly soluble minerals	$BaSO_4$, CaF_2	Fatty acids	0.5 to 1
Insoluble oxide minerals	Fe_2O_3, SiO_2	Amines Sulfonates Fatty acids	0.5 to 1
Naturally floatable minerals	Coal, sulfur	Fuel oil*	0.1 to 1

*Often, only a frother is necessary.

minerals such as hematite, Fe_2O_3, and silica, SiO_2, are typically floated with collectors of the colloidal electrolyte class, amines, sulfonates, and fatty acids.

FLOTATION REAGENTS

These reagents may be classified as frothing, collecting, and modifying agents.

Frothers

As the name implies, a frother is used to create a generous supply of bubbles. These bubbles must be of the appropriate size to contact and lift the minerals to be collected to the top of the ore pulp. The froth generated must be tenacious enough to suspend the collected minerals until they are removed from the cell but not so tenacious that the froth cannot later be destroyed for ease of pumping and transport to the next operation. As with all flotation reagents, they must be as inexpensive as possible. These criteria are met by a series of natural and synthetic compounds of the alcohol class. Pine oil and cresylic acid have long been popular, though MIBC (methyl isobutyl carbinol; also called methyl amyl alcohol or, more properly, 4-methyl-2-pentanol) and polypropylene glycol methyl ethers have gained great popularity.

Collectors

Collecting agents, also called *promoters,* are the basis of the flotation process. They are heteropolar organic compounds with an ionic or neutral functional group which can attach to the mineral, and a hydrocarbon end which attaches to the air bubble. The collector gives the mineral to be collected a hydrophobic coating and thus serves as a bridge between the mineral and the air bubble. Collectors may be divided into two classes: (1) those that attach by chemical reaction or chemisorption with the surface, and (2) those that are held to the mineral surface by electrostatic attraction.

Attachment by Chemisorption   The collectors that function by chemisorption are by far the most commonly used industrially, and they are used to float large tonnages of sulfide ores, phosphate, barite, fluorite, etc. For the flotation of sulfide ores, thio collectors are used. These are generally compounds of the xanthate, dithiophosphate, or thiocarbamate class (see Table 2), though a wide variety of similar thio compounds are potentially useful. They generally have a short hydrocarbon chain, C_2 to C_6 being typical of industrial reagents. In a very simplified way, they may be thought of as chemisorbing onto the mineral surface, and metal xanthate and similar compounds show K_{SP} values of $\sim 10^{-18}$. Their attachment to the mineral results in a surface which is strongly hydrophobic, and it is thus easily attached to an air bubble. Only small amounts of these thio collectors are needed; generally 0.05 to 0.1 lb per ton of ore treated. On the other hand, since the K_{SP} of zinc xanthate is $\sim 10^{-9}$, sphalerite, ZnS, cannot be directly floated with the usual thio collectors unless excessive reagent quantities are used.

Phosphate ores, barite ($BaSO_4$), fluorite (CaF_2), and other slightly soluble minerals ($K_{SP} \sim 10^{-10}$) are floated by the chemisorption of a fatty acid onto the surface of the mineral. Normally the hydrocarbon chain length of these collectors is C_{12} to C_{18}, and, industrially, oleic and similar fatty acids are used for reasons of economy. Approximately 1.0 lb of this collector is used per ton of ore treated.

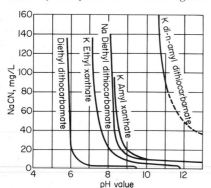

Fig. 1 Contact curves for chalcopyrite with standard equivalent concentrations of various thio collectors. *(From Principles of Flotation, Australiasian Institute of Mining and Metallurgy.)*

While the nature of these chemisorbed collectors may be generalized, it is not to be assumed that all thio collectors react the same with all sulfide minerals just because the metal thio compounds have similar K_{SP} values. Obviously, the nature of both the ionic and the polar groups of the collector exerts an influence as does the nature of the solid.

Figure 1 shows that the mineral chalcopyrite, $CuFeS_2$, responds in different ways to various thio collectors. For these curves, which were constructed from contact angle experiments, flotation is possible only to the left of the curve. A specific knowledge of the system is thus necessary in order to achieve selectivity between mineral species, since there are many gross and subtle differences among thio collectors, sulfide minerals, and their interactions. The same holds true for the chemisorption of fatty acids onto slightly soluble minerals.

Attachment by Electrostatic Attraction   Another method of collector attachment is by electrostatic attraction. In this instance a positive surface will attract an anionic collector while a negative surface will attract a cationic collector. The procedure can be applied to the flotation of any charged solid, but it is generally used for the flotation of insoluble solids, such as SiO_2, where direct chemisorption of the collector onto the surface is not readily possible. The tonnage of minerals floated by this method is small though the potential is great.

All minerals in solution are charged, and hence they attract ions of the opposite charge in order to establish electrical neutrality. A certain class of ions, called potential-determining ions, can change the sign of the charge. For slightly soluble minerals these potential-determining ions are generally the lattice ions. Thus, for example, barite, $BaSO_4$, may be charged positively in a Ba^{2+} solution and negatively in a SO_4^{2-} solution. For highly insoluble minerals such as Fe_2O_3, SiO_2, and TiO_2, the potential determining ions are H^+ and OH^-, and thus the pH of the solution controls the charge on these minerals in solution. The pH at which these minerals change their sign of charge is known as the point of zero charge, PZC. The PZC of several common minerals is given in Table 3. Note that the PZC for different minerals varies over a very broad range from pH 2 to 11. Selectivity between species is often possible because of this circumstance.

The collectors used are colloidal electrolytes, generally the anionic sodium salt of fatty acids, sulfates, or sulfonates or the cationic hydrochloride or acetate salt of an amine. A fairly long hydrocarbon chain is used, generally C_{12} to C_{18}, or an aryl-alkyl compound. The quantity of collector used is about 0.5 to 1.0 lb per ton of ore treated, and this is about 10 times greater than that used for the flotation of sulfide minerals by thio collectors.

The model was first delineated by Fuerstenau and Modi, and they showed that the flotation of alumina can be controlled by changing either the sign of the charge on the mineral or the ionic nature of the collector. Figure 2 clearly shows the relationship between surface charge and the ionic nature of the collector used to achieve flotation. Note that below the PZC, where corundum is positively charged, flotation is best achieved with the anionic carboxylic acid, whereas above the PZC the corundum, now charged negatively, is floated with the cationic amine.

It must be cautioned that practical application of the electrostatic theory has many complications due to the nature and concentration of the collector. Then, too, the presence of very fine particles (such as clays) of opposite charge to that of the mineral to be floated can lead to their attachment to the mineral surface which, in turn, will inhibit collector attachment. For this reason it is usual practice to deslime the ore pulp prior to the introduction of flotation collectors which function in this manner.

TABLE 3   Point of Zero Charge of Several Common Minerals

Mineral	pH for point of zero charge
Quartz	~2.0
Kaolinite clay	3.4
Hematite	6.7
Corundum	9.4
Asbestos (Chrysotile)	10.5

SOURCE: Aplan and Fuerstenau in "Froth Flotation—50th Anniversary Volume," D. W. Fuerstenau (ed.), AIME, New York, 1962; Smith and Trivedi, *Trans. AIME*, **255**, 69 (1974).

Modifiers

This class of reagent is generally used to achieve selectivity between species. For convenience they are subdivided into activators, depressants, pH regulators, and flocculating and dispersing agents.

Activators   These reagents are used to modify the surface of a mineral to enhance collector attachment. The classic example is the use of copper sulfate to activate sphalerite when a xanthate or other thio collector is used. As mentioned previously, a short chain xanthate will not attach to the surface of sphalerite at the usual concentration of collector used. However, by adding Cu^{2+}, which will adsorb onto and diffuse into the sphalerite surface, the thio collector may be readily attached to the surface.

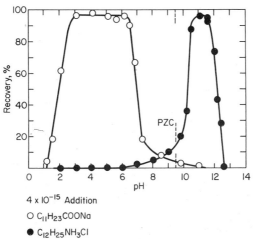

4 x 10^{-15} Addition

○ $C_{11}H_{23}COONa$

● $C_{12}H_{25}NH_3Cl$

Fig. 2   Effect of pH and collector type on the recovery of corundum. (*From* Trans. AIME, **217**, *381, 1960.*)

Sphalerite is a special case, and other metal ions used for mineral activation function in a rather different manner. Near the pH of precipitation of a heavy metal ion in dilute solution, the predominant first hydroxy complex adsorbs onto the surface. The hydroxide coating then "activates" the surface so a collector such as a sulfonate or a fatty acid soap may be attached.

Depressants   These reagents are used to inhibit collector attachment. For sulfide mineral flotation, OH^-, CN^-, and HS^- ions exert the greatest influence, and the former two are used extensively in industry. Each mineral is influenced somewhat differently. The influence of pH and CN^- on the collector-mineral system may be seen in Fig. 1.

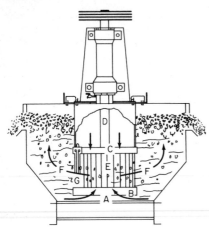

Fig. 3   Fagergren flotation machine.

Since flotation occurs to the left of the curves, cyanide is seen to have a very substantial effect on the flotation of chalcopyrite.

Many other reagents may also be used to modify the surface in such a way as to prevent collector attachment. For example, oxidizing agents such as steam or sodium hypochlorite are used, as is strong acid, to destroy collector coatings on the minerals to be depressed. In addition, reagents such as sodium silicate, which coat the surface of, especially, silicate minerals, are also used as depressants.

pH Regulating Agents   These exert a great influence on the flotation of various mineral species, and pH control is the single most important method used to achieve selectivity between various mineral species. Its primary function is to modify the surface of minerals, but it also controls the ionization of some collectors.

Flocculating and Dispersing Agents   Dispersing agents are often used to prevent flocculation and thus assure that each mineral acts as a discrete entity until separated. Flocculating agents may be used to remove fine particulates or to selectively flocculate one class of fine minerals for ease of subsequent flotation.

INDUSTRIAL TECHNOLOGY

Flotation Machines

The standard flotation machine is a large stirred vat. A typical example is shown in Fig. 3. Industrially, the most commonly used cells today are the Agitair, Denver D-R, and Fagergren cells manufactured by the Galigher Company, Denver Equipment Company, and Envirotech Corporation, respectively. Cell volumes vary considerably with the tonnage to be treated, but they generally range from 40 to 1000 ft³ with 200 to 500 ft³ cells being common in the large base metal concentrators. Large cells have gained wide acceptance in recent years. Units of 12 to 20 cells in series are common, and the number of cells is selected in order to achieve the appropriate residence time and to ensure that all of the valuable mineral has been floated.

Industrial Practice

Copper-Molybdenum Concentration   A schematic flow sheet for a mill of this type is given in Fig. 4. This is typical of large plants which may process 20,000 to 120,000 tons/day each. Typical reagent practice for a separation of this type is given in Table 4. The reagents used are lime, a frother, and a thio collector, such as a xanthate, dithiophosphate, or thiocarbamate. The lime serves to ensure selectivity between species and to precipitate small amounts of Cu^{2+} or other ions which may either accidentally activate undesired minerals or consume the thio collector by precipitation in solution.

After the ore has been crushed and ground to a nominal −65 mesh, it is classified in a hydraulic classifier, such as a hydrocyclone, to remove coarser particles for regrinding. Reagents are added and the ground ore is then sent to a bank of rougher flotation cells. Here, the bulk of the valuable sulfide minerals is floated while most of the gangue minerals and the pyrite are depressed. To ensure that essentially all of the copper and molybdenum sulfide minerals are floated, the rougher tailing may be sent to a scavenger circuit where more intense operating conditions, such as additional collector, are employed to ensure nearly complete recovery. This scavenger concentrate is then returned to the rougher circuit, and the nonfloated material is the final mill tailing.

The rougher concentrate is then recleaned several times to reject unwanted gangue particles which have accidentally floated with the desired minerals. The cleaner tailings are typically sent back to the preceding unit for reprocessing. The final product contains largely copper sulfide minerals together with some molybdenite, MoS_2. Molybdenite frequently occurs in trace amounts in the large porphyry copper deposits of the western

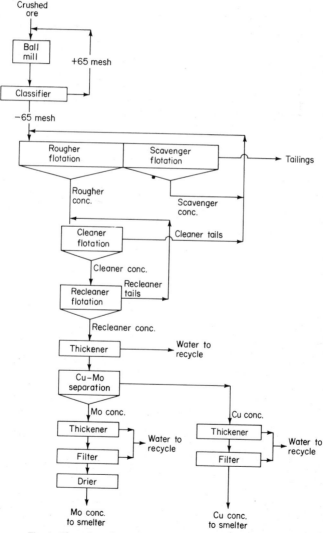

Fig. 4 Flow sheet for typical Cu-Mo sulfide concentrator.

United States. The valuable copper and molybdenum sulfides are then separated by using an incipient roast, steam, or oxidizing agents to destroy the thio collector coating on the surface. Fuel oil and a frother are then used to selectively float the molybdenite from the copper minerals. The two separate concentrates are then thickened and filtered. The copper concentrate is typically not dried because of the proximity of a smelter, though the molybdenite concentrate is often dried prior to shipment to a smelter.

TABLE 4   Typical Reagents and Specific Metallurgical Data for a Copper-Molybdenum Concentrator

Reagents	Point of addition	Approximate quantity added, lb/ton	Function
Potassium amyl xanthate	Ball mills and flotation cells	0.05	Collector
MIBC	Ball mills and flotation cells	0.05	Frother
Lime	Ball mills	to pH 9	Pyrite depressant
Steam	Conditioner ahead of Cu-Mo separation		Remove collector coating
Fuel Oil	Cu-Mo separation flotation cells	1.0	MoS$_2$ collector

		Metallurgical data (actual plant practice)*		
Product	Tons/day	% Cu Total	% MoS$_2$	% Cu distribution
Cu concentrate	863	27.30	0.164	84.4†
Mo concentrate	5.7	1.82	42.16	
Tailings	54,075	0.08†		15.6
Mill feed	54,944	0.504	0.014	100.00

*SOURCE: T. Ramsey, "The Cyprus Pima Concentrator," Chap. 38 in "Flotation—A. M. Gaudin Memorial Volume, M. C. Fuerstenau (ed.), AIME, New York, 1976.
†Much of the Cu in the tailing is present in oxide minerals not recoverable by this process. Sulfide Cu recovery exceeds 90%.

Any pyrite or silicious gangue minerals which still remain in the final cleaner concentrate normally report to the copper concentrate and serve as a diluent. The copper concentrate produced normally contains about 25% copper. Specific metallurgical data for one such plant is given in Table 4.

Circuitry in a given plant may be much more complex than that shown and there are an infinite number of mutations to the basic circuitry to tailor-make a process for a specific ore. The most common modification is to add a regrind circuit prior to the cleaning circuits. This is done in those instances when the grind is not sufficiently fine to liberate the valuable minerals from their attached gangue particles.

Typical Flotation Procedures   Each ore deposit requires a separate procedure, although, as previously mentioned, various classes of minerals are treated in the same general manner. These separation procedures must take cognizance not only of the valuable ore minerals but also of the gangue minerals. Table 5 shows typical procedures used to float various types of ore (lead-zinc, fluorspar, and sylvite) and coal. The literature is voluminous but the references give details for the froth flotation of various specific minerals.

Nonore Applications   While froth flotation of ores and coal represents the bulk of the tonnage treated, the use of the flotation process in other areas is potentially great and the increased use of this process may be expected. The removal of bitumins from tar sands, removal of ink from paper, and the separation of impurities from grains are rather obvious

TABLE 5 Typical Flotation Procedures

Lead–zinc

Ore: Galena, PbS, and sphalerite, ZnS, with pyrite, FeS$_2$ in siliceous or calcareous gangue.

Procedure: Grind to −65 mesh. Float PbS while depressing ZnS and FeS$_2$ with cyanide, zinc sulfate, and lime. Following PbS flotation, ZnS activated with copper sulfate and FeS$_2$ depressed by lime.

Reagents	Point of addition	Approximate quantity added, lb/ton	Function
Sodium cyanide	Ball mill and flotation cells	0.05 to 0.2	ZnS, FeS$_2$ depressant
Zinc sulfate	Ball mill and flotation cells	0.1 to 2	ZnS depressant
Lime	Ball mill and flotation cells	to pH 9	ZnS, FeS$_2$ depressant
Xanthate	Ball mill and flotation cells	0.1	PbS collector
MIBC	Flotation cells	0.05 to 0.1	Frother
Copper sulfate	Zinc conditioner	0.2 to 1.0	ZnS activator
Lime	Zinc flotation cells	to pH 9 to 12	FeS$_2$ depressant
MIBC	Zinc flotation cells	0.03	Frother

Fluorspar

Ore: Fluorite, CaF$_2$, in calcareous or siliceous gangue. Sulfide minerals may also be present.

Procedure: Grind to −65 mesh. CaF$_2$ floated from CaCO$_3$ with oleic acid as collector, and quebracho as CaCO$_3$ depressant. Sulfides, if present, are floated previously with xanthate.

Reagent	Point of addition	Approximate reagent quantity added, lb/ton	Function
Quebracho	First conditioner ahead of flotation	0.2 to 0.5	CaCO$_3$ depressant
Soda ash	First conditioner ahead of flotation	to pH 8.0 to 9.5	pH regulator
Sodium silicate	First conditioner ahead of flotation	0.2 to 1.0	Silicate mineral depressant
Oleic acid	Second conditioner and flotation cells	1.0	CaF$_2$ collector, frother

Coal

Ore: Coal

Procedure: −28 mesh coal fines floated from ash constituents with frother and/or fuel oil.

Reagents	Point of addition	Approximate quantity added, lb/ton	Function
MIBC and fuel oil	Flotation cells	0.1 to 0.5	Frother
Fuel oil	Flotation cells	0.1 to 3.0	Collector

Soluble salts, KCl–NaCl

Ore: Sylvite, KCl, and halite, NaCl, also containing some clay.

Procedure: Grind to −30 mesh. Float KCl from NaCl with amine in saturated brine. Clay "blocked" with guar gum prior to addition of amine.

Reagent	Point of addition	Approximate quantity added lb/ton	Function
Guar gum	Flotation conditioner	0.1	Prevent adsorption of amine on clay
Tallow amine acetate	Second conditioner	0.15 to 0.4	KCl collector
MIBC	Flotation cells	0.05 to 0.1	Frother

extensions. Ion and precipitate flotation is described in detail by Lemlich,[6] and Somasundaran[14] has an excellent review of foam separation methods for the separation of cations, anions, surfactants, algae, bacteria, proteins, colloidal particles, etc. Foam fractionation may be used to separate organic compounds from water, such as naphthalene from coke plant discharges, cutting oils from machine-shop aqueous discharge, etc. Dissolved air flotation[15] is finding great application in the removal of particulate solids, oils, and surfactants from plant waste water discharges, and it has found particularly wide acceptance in the removal of sewage solids and paper mill effluents. In this process, the plant effluent water is first pressurized with air. When the pressure is released, the air, now supersaturated in the water, precipitates onto the surface of the particles, and the attached air bubble carries the particle to the surface for removal. The process is particularly effective for the removal of fine particulates from dilute suspensions. Because of environmental considerations, this process may be expected to grow substantially.

All of these flotation processes function in the same manner; provision must be made to create bubbles and to attach the ion, molecule, or particle to the bubble created. The general principles elucidated above for froth flotation apply to the process mutations as well.

REFERENCES

1. Fuerstenau, M. C. (ed.), "Flotation—A. M. Gaudin Memorial Volume," 2 vols., AIME, New York, 1976.
2. Fuerstenau, D. W. (ed.), "Froth Flotation—50th Anniversary Volume," AIME, New York, 1962.
3. Gaudin, A. M., "Flotation," 2d ed., McGraw-Hill, New York, 1957.
4. Glembotskiii, V. A., Klassen, V. I., and Palksin, in *"Flotation,"* translated by R. E. Hammond, Primary Sources, New York, 1963.
5. Klassen, V. I., and Mokrousov, V. A., "An Introduction to the Theory of Flotation," translated by J. Leja and G. W. Poling, Butterworths, London, 1963.
6. Lemlich, R. (ed.), "Adsorptive Bubble Separation Techniques," Academic Press, New York, 1972.
7. Sutherland, K. L., and Wark, I. W., "Principles of Flotation," Australasian Institute of Mining and Metallurgy, Melbourne, 1955.
8. Taggart, A. F., "Handbook of Mineral Dressing," Wiley, New York, 1945.
9. Aplan, F. F., "Flotation," in Kirk-Othmer, "Encyclopedia of Chemical Technology," New York, Vol. 9, Wiley, 1966, pp. 380–398.
10. Bloecher, F. W., "Flotation," in "Chemical Engineer's Handbook," J. H. Perry (ed.), McGraw-Hill, New York, 1963, pp. 21–70.
11. American Cyanamid Co., "Mining Chemicals Handbook," Wayne, N.J., 1976.
12. Denver Equipment Co., "Mineral Processing Flowsheets," 2d ed., Denver, 1965.
13. Dow Chemical Co., "Flotation Fundamentals and Mining Chemicals," Midland, Mich., 1976.
14. Somasundaran, P., *Sep. Purif. Methods*, 1(1), 117 (1972).
15. Eckenfelder, W. W., "Flotation," in "Industrial Water Pollution Control," McGraw-Hill, New York, 1960, pp. 52–61.

Part **6**

Gas-Solid Mixtures

Section **6.1**

Gas-Solid Separations

JOSEPH H. MAAS, Ch.E., P.E. *Independent Chemical Engineer and Consultant*

INTRODUCTION TO GAS-SOLID MIXTURES

The separation of solid dispersoids of various particle sizes and shapes that frequently are found with gases in the form of dusts, catalyst fines, fly ash, and other similarly mixed combinations, is the subject of this section of the handbook. Such mixtures are defined as being in a condition of pneumatic transport, the solids and gases being conveyed forward at some uniform rate of flow before separation.

Gas-solid systems occur in solids-rich combinations as fluidized solids from dense solid fluidization, as in a boiling bed, or from lean-phase fluidization, as in pneumatic transport of solids with the flowing gas. The solids-rich condition falls within the physical boundary of the fluidized bed. The lean-phase systems exist under less definite conditions since the dispersed particles, unless of a small size that permits Brownian motion, will be undergoing gravitational settling continually. Insofar as a process feed for separation is concerned, the plausibility of a uniform rate of flow should be examined when a design is considered, so that any presettling of solids is taken into account before the design.

The mechanisms of separation have been described,[1] but an adequate functional description[3] of commercial-type separation methods exploit the following possibilities:

Gravitation, as settling chambers
Inertial, as baffled chambers, or centrifugal apparatus
Filtration, as bag filters
Sprays, as scrubbing towers
Electrical, as electrostatic precipitators

The usual properties of gas-solid mixtures required for design of a separation process include the properties and quantities of the gas and solids entering as the feed. Aside from any unusual characteristics that must be taken into account, such as agglomeration of particles, poisonous properties, or any other tendencies common in plant design, the size analysis of the solid particles is the most significant.

The most probable method to apply in making a size analysis of the solid particles is described by and shown in Table 1. The methods used will naturally depend upon the particles that require separation, according to the design requirements. Information on the density of the particles as well as the particle sizes and distribution will be required.

TABLE 1   Examples of Size-Analysis Methods and Equipment*

Particle size, μ	General method	Examples† of specific instruments
37 and larger	Dry-sieve analysis	Tyler Ro-Tap, Alpine Jet sieve
10 and larger	Wet-sieve analysis	Buckbee-Mears sieves
1–100	Optical microscope	Zeiss, Bausch & Lomb, Nikon microscopes
	Microscope with scanner and counter	Millipore IIMC system
	Dry gravity sedimentation	Roller analyzer, Sharples micromerograph
	Wet gravity sedimentation	Andreasen pipet
	Electrolyte resistivity change	Coulter counter
0.2–20	Light scattering	Royco
	Cascade impactor	Brink, Anderson, Casella, Lundgren impactors
	Wet centrifugal sedimentation	M.S.A.-Whitby analyzer
0.01–10	Ultracentrifuge	Goetz aerosol spectrometer
	Transmission electron microscope	Philips, RCA, Hitachi, Zeiss, Metropolitan-Vickers, Siemens microscopes
	Scanning electron microscope	Reist & Burgess‡ system

*Krockta and Lucas, *Air Pollution Control Assoc. J.*, **22**, 461 (1972).

†This table gives examples of specific equipment. It is not intended to be a complete listing, nor is it intended as an endorsement of any instrument.

‡Reist and Burgess, "Development of an Automatic Particle Assaying Instrument Utilizing a Scanning Microscope," Paper No. 69–124, Air Pollution Control Association Meeting, New York, 1969.

SOURCE: Perry and Chilton, Chemical Engineer's Handbook," 5th ed., McGraw-Hill, New York, 1973.

It is common to divide the solid dispersoids into the categories of "dusts" and "fumes," where the former are above 1 μm (one-millionth of a meter) and the latter below 1 μm in diameter, as an approximation. Fume particles also are described as not being redispersed by an air blast, while dusts may be. The aspects of particle size determination, with sampling procedures for process dispersoids in gas and atmosphere-polluting dusts, and the general subject of interpretation of existing information on these systems is extended in the references.[1]

SEPARATION BY GRAVITATIONAL METHODS

The following conditions generally apply:
Particle size limitation, 10 μm, 0.0003937 in diameter
Spherical particles assumed, sphericity = 1.0
The Reynolds number, dimensionless and based on particle diameter, justifies the use of one of the three equations below for calculation of terminal falling velocity of a specified particle size:

Stokes' law of Settling	Re = 2.0 or less
Stokes-Cunningham law	Re = 2.0 to 500
Newton (free-falling)	Re = 500 to 200,000

Stokes' law is the most conservative and may often be used without resorting to a correlation of Reynolds numbers vs. drag coefficients.

Reynolds number here is a modified number, based on particle size and terminal velocity of fall:

$$\text{Re} = \frac{D_{ps}u_t\rho}{\mu} \times 1488$$

where D_{ps} = particle diameter (Stokes' law equivalent) at terminal velocity u_t and viscosity μ in cP. (The symbol μ also frequently represents microns or millionths of meter.)

Stokes' law of settling, which applies when Re = 2.0 or less, is

$$u_t = \frac{18.5D_{ps}^2(\rho_s - \rho)}{\mu} \qquad \text{ft/s}$$

For this particular equation it will be found convenient to use the following system of dimensions:

u_t = terminal velocity of fall, ft/s
D_{ps} = spherical particle diameter, in
ρ_s = true particle density, lb/ft³
ρ = density of gas through which particle falls, lb/ft³
μ = viscosity of gas at existing conditions of fall, cP

A more general method for the calculation of terminal velocities of particles is frequently used by employing a plot (see Fig. 1) giving a relation between drag coefficient and Reynolds number, covering a wide range of conditions and including particle shapes of spheres, disks, and cylinders as well. The solutions by this method include trial-and-error computation, but this may be simplified by methods of solution offered.[1] Portions of this correlating curve would apply to the individual laws in settling—Stokes, Stokes-Cunningham, and Newton. Alternatively, if portions of the correlating curve are approximated as straight lines, the simplified expressions for the settling laws mentioned are possible, eliminating trial-and-error-type computations. See (Ref. 1) Table 5-26.

The drag coefficients used with the Reynolds number, both as dimensionless variables, are interrelated in the curve; see Fig. 1. An example of a consistent set of dimensional units for the variables is the following:

$$C = \frac{F_d/A_p}{\rho u^2/2g_c} \qquad \text{dimensionless}$$

where C = drag coefficient
F_d = resistance to motion of particle in gas in this application, and the motion is in the direction of settling, lb force

A_p = area of particle projected on plane normal to direction of settling, ft²
ρ = gas density at settling conditions, lb/ft³
u = settling velocity, same as u_t of Stokes, terminal ft/s
g_c = dimensional constant, 32.17 (lb· ft)/(lb force)(s²)
Re = $(D_p u \rho)/(\mu)$, Reynolds number, dimensionless
D_p = diameter of particle, ft
μ = gas viscosity at settling conditions, lb/(ft)(s) or viscosity, (centipoise)/1488

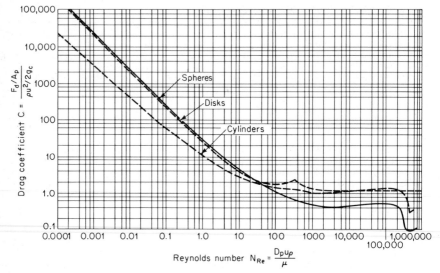

Fig. 1 Drag coefficients for spheres, disks, and cylinders. [*From Lapple and Shepherd*, Ind. Eng. Chem., **32**, 605 (1940).]

A_p = area of particle projected on plane normal to direction of motion, ft²
C = overall drag coefficient, dimensionless
D_p = diameter of particle, ft
F_d = drag of resistance to motion of body in fluid, lb force
g_c = dimensional constant, 32.17 (lb)(ft)/(lb force)(s²)
N_{Re} = Reynolds number, dimensionless
u = relative velocity between particle and main body of fluid, ft/s
μ = fluid viscosity, lb/(ft)(s)
p = fluid density, lb/ft³

Approximate equations for the drag coefficients of spheres, using the above nomenclature and starting with the Stokes' law range, where Re = 2.0 or less,

$$C = \frac{24}{\text{Re}}$$

and

$$F_D = \frac{3\pi u_t D_p}{g_c}$$

For the range Re = 2.0 to 500, the Stokes-Cunningham equation

$$C = \frac{18.5}{(\text{Re})^{0.6}}$$

applies and

$$F_D = 2.31\pi(u_t D_p)^{1.4}\mu^{0.6}\rho^{0.4} \times \left(\frac{1}{g_c}\right)$$

For the range Re = 500 to 200,000, Newton's law applies

$$C = 0.44$$

and $$F_D = 0.055(u_t D_p)^2 \times (1/g_c)$$

See Ref. 4, p. 152, for a more detailed treatment.

DESIGN OF A DUST COLLECTION CHAMBER

A gravitational type of dust collector of rectangular shape is to be designed to remove dust particles from a directly fired dryer that is currently discharging these particles from a flue duct into a stack, and then to the atmosphere. It is observed that there are no solids deposited in the duct or stack as long as the gas velocity exceeds 10 ft/s, giving support to the criterion that settled dust may be redispersed only at gas velocities above this flow rate.

This observation is used to set a maximum superficial velocity through the system for design, but the final dimensions are not limited by this condition, as will be seen. No baffling is used, but longitudinal baffles could be used in such instances to reduce the time of fall and thus reduce the size of the chambers if suitable dust removal apparatus is considered.[1]

The properties of the solids, assumed uniformly dispersed at the feed inlet, range from particle sizes of 0.0005 to 0.004 in diameter, spherical shape assumed. A true density is 125 lb/ft³; the bulk density is 40 lb/ft³.

The gas properties are flue gas 30.0 mol wt, 300°F and 14.8 lb/in²(abs) at flowing conditions, The density is calculated to be 0.00550 lb/ft³ and the viscosity 0.0238 cP at these conditions.

The gas flow rate is 10,000 acfm, or 695.6 scfm, and contains 1.0 grains/ft³, 1.428 lb/min (7000 grains = 1 lb). This will require a maximum removal of 2057 lb/day or 51.4 ft³ (bulk volume) removal per day.

A calculation of the Reynolds numbers for the smallest and largest particles is

$D_p = 0.0005$ in
Re = $(0.0005/12) \times 0.0537 \times 0.0243 \times (1488/0.0238)$
   = 0.0034   (which is below 2.0 and in Stokes' law range)
$D_p = 0.004$ in
Re = $(0.004/12) \times 0.0537 \times 1.554 \times (1488/0.0238)$
   = 1.74   (which is below 2.0 and in Stokes' law range)

The settling velocities above from Stokes' Law are

$D_p =$ 0.0005 in
$u_t =$ $18.5(0.0005)^2 \times (125.0 - 0.0537)/0.0238$
   =   0.0243 ft/s
$D_p =$ 0.004 in
$u_t =$ $18.5(0.004)^2 \times (125.0 - 0.0537)/0.0238$
   =   1.554 ft/s

If the chamber height is set at 6 ft to facilitate cleaning, and the superficial velocity is 10 ft/s through the chamber to the stack, the time of fall for the 0.004-in-diameter particle is 6/1.554, or 3.861 s. In this time a particle at the roof of the chamber would travel 3.861 × 10, or 38.61 ft, before falling the full height. The width would be 2.78 ft for the 10-ft maximum velocity permitted, but for practical reasons a chamber less than 38.6 ft, or say half as long, could be provided if the width was doubled for a 5 ft/s velocity to give the particle the same falling time in a chamber only 19.3 ft long.

Performance of Gravity Settling Chambers

With the usual simplifying assumptions valid, the performance of gravity settling chambers, designated by η, is

$$\eta = \frac{u_t L_s}{H_s V_s} = \frac{u_t B_s L_s}{q}$$

where L_s = length of settler, ft
    H_s = height of settler, ft
    V_s = superficial gas velocity through settler, ft/s
    B_s = width of settler, ft
    q = gas flow rate, acfm

The calculation of performance based on the largest particle size, 0.004 in, and the 38.6 ft chamber length is

$$\eta = \frac{1.554 \times 38.6}{6.0 \times 10.0}$$
$$= 1.0, \text{ or } 100\%$$

Adjustment to shorten the length by half and double the width does not change the performance here. It was originally intended to design for 100% settling of the largest particles.

Similarly, for the 0.0005-in-diameter particles,

$$\eta = \frac{0.0243 \times 38.6}{6.0 \times 10.0}$$
$$= 0.016, \text{ or } 1.6\% \text{ recovery (small as expected)}$$

The formula above for the performance value η, if separated into two functions, may be regarded as (L_s/V_s) divided by (H_s/u_t), or the time alloted to the fall (L_s/V_s) divided by the time required (H_s/u_t) for the particle (of designated size) to fall the height of the chamber.

The performance η for gravity settling chambers may also be interpreted as the percentage ratio of solids that settle out at given chamber lengths; thus at constant gas velocity when 100% is settled at 38.6 ft as above, then 50% would be settled at 19.3 ft, provided the same gas velocity is preserved.

Safety factors and miscellaneous conditions in design may make it advisable to test the particles encountered for both size and settling rate in still air (or gas) for design safety. It is sometimes recommended that because of eddy currents possibly present in the system, the settling velocity for design purposes be taken as one-half the calculated value.

In a design case the dust particles may be described as occurring in various percentages over ranges of size distribution, say in some material balance form such as the following (which will find later use):

Particle dia., μm	
0 to 5	26%
5 to 10	10%
10 to 20	21%
20 +	43%

It would then be possible to select the separation point desired by choosing the smallest particle size to be removed and the percentage of removal (recovery) desired, and, with the calculation procedures described, proceed to size the equipment according to the method being considered.

The tacit assumption made in the design when the vessel was sized was that a maximum height of any particle fall would be the 6 ft height set by practical considerations of cleanout convenience and perhaps other desires. If the initial inlet position of the dust-laden mixture is taken as 3 ft or half the height of the chamber, and the solid particles are taken to be evenly distributed at what may be called a *point source* and to enter the chamber at evenly distributed angles, fanning out in different trajectory angles, it may be seen that the design assumptions are very approximate indeed. It would not be surprising in performance if the dropout of smaller particles reflected a fall from an initial position of 3 ft (or less) for approximately half of the small particle content present. Similarly, a smaller portion of the larger particles may reflect a delayed condition of falling when they have initial trajectories that propel them upward, resulting in the particles reaching the top and then starting their fall some distance downstream from the inlet. Practical allowances can be made, however, and these may be dependent upon the individual performance needs for the case at hand.

SEPARATION BY INERTIAL MEANS—IMPINGEMENT ON BAFFLES

Impingement separation depends upon the collection of solids on the surface of an intercepting body, while the gases are deflected or pass by through the free area necessarily provided between the intercepting devices, to maintain a continuous flow through the system.

The prediction and measurement of the separation is described in terms of target efficiency of the installed baffles or intercepting devices. In complex shapes, baffling performance would be determined experimentally.

It is to be noted that a large number of small targets will produce more separation than a smaller number of large targets of the same total area. This may be accounted for by noting that a large target must deflect the gas over a longer path after impingement of the solids takes place. For this reason the baffling dimension D_b, called the representative diameter, is characteristic of baffle size, and is taken as independent of other settler dimensions.

At the design level, the objective is to economically obtain the degree of target efficiency desired under pressure-drop conditions that are tolerable. Simplifying assumptions are that the gas flow is not significantly distorted by the location of nearby collecting members, and that in using the particle settling velocity (here as a correlating variable), the particle shape and simplifying assumptions in the settling equations are valid.

Particle size limitation is around 10 μm (0.00039 in) as a minimum. Particles of large size may produce erosive wear of equipment even if they are easily separated. System pressure drop in commercial operations may range from 0.1 to 1.5 inH$_2$O. Pressure drop is calculated best by estimating the velocity head lost from the probable equivalent reversals in the course of the gas passage through the system.

The target efficiency is a function of the dimensionless group

$$\text{Separation number} = \frac{u_t V_0}{g_L D_b}$$

where u_t = terminal settling velocity of particle under Stokes' law, ft/s
  V_0 = average velocity of gas, ft/s
  g_L = local acceleration due to gravity, ft/s^2
  D_b = representative diameter of baffle or body impinged upon, ft [i.e., a line of baffles of 2-in width each would have a representative diameter of (2/12) ft provided the baffles are spaced on about 2.5-in. centers or farther]

The curve in Fig. 2 is used to obtain the target efficiency η_t.

DESIGN OF A BAFFLED DUST COLLECTOR-CHAMBER

Properties of a gas-solids dispersoid used in the gravity settler design previously are used here. The example is one where Stokes' law holds, as previously shown. The representative diameter of the baffles is 2 in. Baffles are made of a simple standard material such as steel tubing or strips. (The diameter and material were chosen in order to free the design from the more complicated shapes such

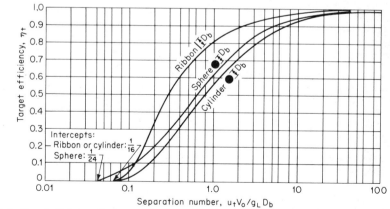

Fig. 2 Target efficiency of spheres, cylinders, and ribbons. [*Langmuir and Blodgett, U.S. Army Air Forces Tech. Rept. 5418, Feb. 19, 1946 (U.S. Department of Commerce, Office of Technical Services PB 27565.)*]

as may be found in commercial equipment, where a high degree of specialization exists. The presentation here illustrates the principles and a calculation method.)

The separator is assumed to be made of 2-in-diameter (or width) baffles set on 2.5-in centers, with vertical height of 6 ft; the total duct width is 11.1 ft, calculated to give a superficial gas velocity of 2.5 ft/s. The velocity at the constrictions between the baffles or tubes is 10.0 ft/s, and the average gas velocity is taken as 6.25 ft/s. The inertial separation number for use with the Langmuir and Blodgett correlation is

$$
\begin{aligned}
N_{si} &= \frac{u_t V_0}{g_L D_b} \\
&= \frac{1.55 \times 6.25}{32.2 \times 0.1667} \\
&= 1.81
\end{aligned}
$$

which corresponds to $\eta_t = 0.75$ or 75% target efficiency (see Fig. 2).

If four sets of baffles are installed in series, then the total recovery in four successive passes through a baffle section will result in a recovery of 0.996, or 99.6%, for the particle size of 0.004-in diameter and with the terminal velocity u_t of 1.55 ft/s.

The pressure drop through the system may be calculated by the use of velocity head estimation on examining the flow path. The velocity head is a measure of the static head (feet of gas at flowing conditions) equivalent to the kinetic energy of the flowing stream. The maximum gas velocity at the constricted point between baffles, 10.0 ft/s, is used. The velocity head requirement for each baffle section is taken to be equivalent to the pressure drop created by the halt of the flow on impingement of most of the gas and the resumption of flow, or say 2 velocity heads per baffle, amounting to 8 velocity heads total for the system.

One velocity head is calculated as feet of gas at conditions of flow from the general expression

$$
\text{Head} = \frac{V^2}{2g}
$$

or
$$
\frac{(10.0)^2}{64.4} = 1.553 \text{ ft of gas.}
$$

The velocity head may be calculated in inches of water directly from the formula:

$$
\text{Head} = 0.00298 \, \rho V^2 \quad \text{in} H_2O
$$

where ρ = gas density at flowing conditions, lb/ft^3
V = gas velocity, ft/s

Thus

$$
\begin{aligned}
\text{Head} &= 0.00298(0.0537)(10.0)^2 \\
&= 0.0160 \text{ or} \\
8 \text{ velocity heads} &= 0.128 \text{ in} H_2O
\end{aligned}
$$

Performance and Mechanical Characteristics of Impingement Separators

The performance of baffled chamber separators may be expressed through the use of the η expression for the settling chamber performance evaluation, from the ratio of the time allotted (in this case the time of passage through the baffled chamber) divided by the seconds required for a particle of a given size to fall the height of the baffled chamber. This comparison should be made on the same gas velocity or mass velocity and same volumetric flow for each type compared.

With such a ratio, it is then possible to express the baffled chamber in terms of equivalent length of an unbaffled one, or to express the baffle performance similarly as an equivalent length per baffle.

SEPARATION BY INERTIAL MEANS—CYCLONE SEPARATORS

Centrifugal separating devices consist of cyclone separators, rotational-flow dust precipitators, and mechanical centrifugal separators, the cyclones being most widely used. The smallest particle size removable is about 5 μm, although smaller sizes as low as 0.1 μm have been separated in cases where particle agglomeration takes place.

The cyclone separator is uniquely conceived so that the solid particles are separated from gases under centrifugal-flow path conditions to produce much greater separation forces than obtained by gravitational methods. The flow path is patterned so that gases and solids are discharged separately and with no appreciable reentrainment.

The general mechanical arrangement of the parts is illustrated in Fig. 4. Details of various types of cyclones are available.[1,3] Pertinent flow properties and velocity profiles associated with centrifugal flow, as described in the field of fluid mechanics are available.[4]

Probably the most conspicuous property of centrifugal flow is the velocity distribution as examined radially across the flow. If the wall boundary, where the velocity is zero (from Prandl) is omitted, the effect of viscous shear in the fluid will tend to give higher velocities *near the inside* of a centrifugal path (see Fig. 4). In fluid mechanics this diagram is for an irrotational condition of flow and is a two-dimensional condition. It may also be described as a quasirepresentation of the flow.

This radial-flow distribution may be represented as an expression that states the tangential component of velocity will vary in inverse proportion to the radial distance, or $V_{ct} \propto r^{-n}$, where $n = 1$ under ideal flow, but may be 0.52 as measured in the cylindrical section of the typical cyclone shown in Fig. 3.

In a qualitative description, the cyclone may be said to produce favorable separation forces, if its particle paths are of relatively short distances, and the inner walls of its cylindrical and conical sections also provide impingement area.

Here, as when many small-diameter baffles are used in preference to fewer larger units, the use of more cyclones in a parallel arrangement at the same pressure drop will give better efficiency of particle removal for a given flow. This is reasonable, since an increase in the number of cyclones results in separation at smaller radial distances in the centrifugal flows produced, and shorter particle travel distances, before impingement on the inner walls of the cyclones.

From the proportional areas and dimensions of the equipment, see Fig. 4; the

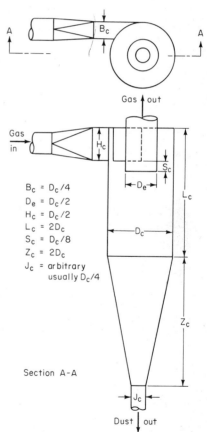

$$B_c = D_c/4$$
$$D_e = D_c/2$$
$$H_c = D_c/2$$
$$L_c = 2D_c$$
$$S_c = D_c/8$$
$$Z_c = 2D_c$$
$$J_c = \text{arbitrary}$$
$$\text{usually } D_c/4$$

Fig. 4 Cyclone separator proportions. (*From Perry and Chilton, "Chemical Engineers' Handbook," 5th ed., McGraw-Hill, New York, 1973.*)

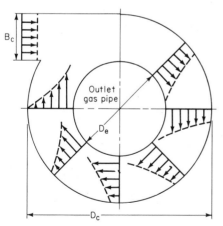

This flow is representative of flow in top section, see fig. 4.

Note: Velocity increases at inner radius except on wall. Neglect wall effect.

Fig. 3 Velocity distribution in centrifugal flow within cyclone separators.

entering gas at the rectangular inlet first enters an annular space, tangentially to the inner wall of the cylindrical section at the top, and is also bounded inside the centrifugal path by the outside of the vertical gas outlet pipe, which is concentric with the cylinder.

Although the inlet gas flow after entrance is only bounded on the sides and top, and is free to pass downward (which it does, following a spiral path on the outer diameter similar to the upward-directed center vortex), it is of some value to note possible gross velocity changes of flow according to the approximate cross-sectional areas presented. It should be noted that the inner and outer spiral paths of flow have the same sense of rotation and that the inner flow of the clean gas enters the outlet gas pipe below the lowest level of the rectangular inlet.

The rectangular inlet provides the lowest cross-sectional area of flow. In the downward direction, the annular space is pi (3.14) times this area, and the gas outlet, presumably even at some lower static pressure, has only half the area for upward flow. The exact boundary conditions between the slow-rotating outer vortex and fast-rotating inner vortex, moving in their downward and upward directions respectively, do not tend to contaminate the outgoing gases, since the inner vortex rotation is reported to pass, to some extent, the residual small particles back into the outer vortex, resulting in final entrainment of some very small solids.

Mechanical Description of Cyclone Separators

Proportional dimensions are shown in Fig. 4. The following statements summarize the essential mechanical characteristics of cyclones:

Vertically cylindrical top section in a typical case, but variations with top and middle conical sections are possible.[1] Slanted installation is also possible.[1]

Lower conical section below cylindrical section tapering downward into the solids discharge outlet in a typical case.

Tangential inlet for dust-laden gas of rectangular cross-sectional shape.

Gas outlet pipe vertical and concentric with top cylindrical section starting below the rectangular inlet where dust-laden gas enters, partially separating the two vortex flows of gases downward and upward.

The cone supplies volume to hold a solids bed that may be continuously or intermittently removed.[1]

The application of cyclones to fluid-bed systems is the subject of Ref. 1, p. 20–70.

The use of purge gas installations, when solids are removed under hazardous conditions, should be considered.[1]

The pressure drop is reported to decrease if the inner side of the inlet duct is extended into the cylindrical section of the cyclone. Other observers[1] note that the separation efficiency may drop when this is done.

Precautions are necessary in the prevention of leakage of air into the cyclone and leakage of gas plus dust out of the cyclone; either will reduce collection efficiency. For intermittent withdrawal of dust, an airtight receiver is used; for continuous operation, a star valve or special types of locking gate valves are used.

Fans are usually required with cyclones and may be placed to handle the dust-laden gas or the cleaned gas from either upstream or downstream installations. The calculation of pressure drop caused by cyclone flow resistance will be shown later, but a less obvious calculation is the determination of the fan loading requirements to bring the gases either into the fan itself, or into the cyclone inlet pipe if the fan is exhausting gas from the device. If a fan and cyclone installation is to be placed in a location that can accommodate the gas out at some very low velocity, such as a stack, then there are some extra velocity heads required for the fan to take in the slow-moving gas, as a suction head requirement (at the existing pressure), to be added to the cyclone pressure drop in the fan specification.

Design of Cyclone Separators

The scope of this subject is extensive, and the careful treatment of the subject, as collected in the various editions of the *Chemical Engineers' Handbook*,[1-3] is used in preliminary estimations of engineering needs. A typical design example is given based on data and material taken from Ref. 1, pp. 20–82 to 20–85, but it should be noted that the separations reported by Anderson in table 20–36 is from a centrifuge of different proportions than the example here. The material balance in the data of Anderson was used in this example so that later comparisons are possible.

This design example is concerned with the following subjects:
Pressure drop through cyclone
Minimum size of particles separated
Cut size of particles for calculation of separation efficiency
Calculation of separation efficiency
Calculation of overall collection efficiency

Pressure drop as calculated is defined as the head that a fan must develop when taking in air or gas (at specified suction pressure), and discharging it to the entrance of the cyclone in the amount, and at a pressure condition, that is adequate to overcome frictional and static resistance characteristic of the path. Cyclone pressure drop is chiefly the result of frictional resistance and is calculated empirically on the principal that cyclones built with proportional dimensions have equal pressure drops for the same inlet conditions and inlet velocity.

Probably because of the frictional character of the cyclone pressure drop, the symbol F_{cv} is carried in the pressure drop calculation. It is applicable only to cyclones with proportional dimensions, as in Fig. 4. The friction loss or pressure drop in the cyclone, as the minimum static head requirement for a fan operating at design flow rate, is

$$F_{cv} = \frac{KB_cH_c}{D_e^2} \qquad \text{(number of cyclone inlet heads)}$$

where K is a constant of value 16.0. The formula is used when cyclone separator dimensions fall within the ranges $B_c/D_c = \frac{1}{12}$ to $\frac{1}{4}$, $H_c/D_c = \frac{1}{4}$ to $\frac{1}{2}$, and $D_e/D_c = \frac{1}{4}$ to $\frac{1}{2}$.
Note: B_c, H_c, and D_e indicate cyclone separator proportions, as in Fig. 3.

Since F_{cv} is dimensionless and represents velocity *heads* at the cyclone inlet, it is meant to apply over velocity ranges, and may be expressed as inches of water (pressure drop) by

$$h_{vi} = 0.00298\rho V_c^2 \qquad \text{inH}_2\text{O for one velocity head}$$

The calculation of the minimun size of particle separated will give a preliminary view of the choice of cyclones for the separation at hand. The Rosin, Rammler, and Intelmann equation is used and is based on Stokes' law range, spherical particles, and a gas stream path of a fixed number of turns at constant spiral velocity with no mixing action.

For nomenclature defined below, the minimum particle size in the English system of units is

$$D_p, \text{min} = \left[\frac{9\mu B_c}{\pi N_{tc} V_c(\rho_s - \rho)} \right]^{0.5} \qquad \text{feet}$$

where V_c = cyclone inlet velocity, average based on inlet area ($B_c \times H_c$), ft^2
μ = viscosity, lb/(ft)(s) of gas, or cP/1488.
N_{tc} = number of turns of 360° made spirally downward in cyclone; $N_{tc} = 5.0$ approximately, as in Fig. 4
ρ_s = density of solid particles, lb/ft^3
ρ = density of gas, lb/ft^3

EXAMPLE

Dust-laden air at 60°F enters a cyclone at approximately 4 inH$_2$O pressure (above atmospheric). The density of air is taken as 0.0755 lb/ft^3 neglecting moisture in air or dust content and its velocity is 44 ft/s. Then

$$F_{cv} = KB_cH_c/D_e^2$$

Since $K = 16.0$,

$$F_{cv} = \frac{16 \times 1 \times 2}{2^2} = 8.0 \text{ velocity heads}$$

or $\qquad P_i = 8.0 \times 0.00298 \times 0.0755 \times (44)^2 = 3.48 \text{ inH}_w\text{O}$

which is approximately the pressure (4 in) assumed at the cyclone inlet, and typical for a fan (centrifugal compressor) application.

The cyclone is to be designed to handle 10,000 scfm of gas at 60°F, with 2.0 grains/ft³, normally with two cyclones operating and a third provided for a spare. Under some circumstances, all three can be operated. The basic design is for two in use at $V_c = 44.0$ ft/s, to conform with the data by Anderson (Ref. 1, p. 20–85), although 50 ft/s is commonly used. For the 5000 scfm rate on one cyclone and cross-sectional entrance area $B_c H_c$ or $2B_c^2 = 1.89$ ft², or $B_c = 0.97$ ft (which would be probably standardized to an entrance 1 ft wide and 2 ft long), the entrance velocity is 42 ft/s as calculated, and pressure drop is 3.2 inH₂O vs. 3.5 inH₂O at 44 ft/s.

With the dimension of B_c established at 1.0 ft; the gas viscosity for 60°F air equal to 0.0179 cP, or $(0.0179/1488)$ lb/(ft) × (s) = 0.120×10^{-4}; density of solids (true) = $3 \times 62.4 = 187$ lb/ft³; and gas density = 0.0755 lb/ft³, the minimum particle size that should be completely separated is

$$D_p, \min = \left[\frac{9 \times 1.20 \times 10^{-5} \times 1.00}{\pi \times 5 \times 42. \times (187.21 - 0.0755)} \right]^{0.5}$$
$$= 2.96 \times 10^{-5} \text{ ft}$$
$$= 9.01 \times 10^{-4} \text{ cm, or } 9.0 \text{ } \mu m$$

The inlet dust particle size distribution is given by Ref. 1, table 20–36 as follows:

Particle diameter, μm	Cumulative % larger than size
5	74
10	64
20	43

The cyclone diameter for this design would be 4 ft at the 5000 scfm rate and inlet dimensions of $B_c = 1.0$ ft, $H_c = 2.0$ ft.

The cut size D_{pc}, for the same nomenclature as above, but with N_{tc} the same as N_e and equal to 5.0 (approximately), is

$$D_{pc} = \left[\frac{9\mu B_c}{2\pi N_c V_c (\rho_s - \rho)} \right]^{0.5}$$

or

$$D_{pc} = \left[\frac{9 \times 1.20 \times 10^{-5} \times 1.0}{2 \times 3.14 \times 5 \times 42(187.21 - 0.0755)} \right]^{0.5}$$
$$D_{pc} = 2.09 \times 10^{-5} \text{ ft}$$
$$= 6.37 \times 10^{-4} \text{ cm, or } 6.37 \text{ } \mu m$$

By definition, this particle size should correspond to a fractional efficiency of 50% (see Fig. 5). This plot in Fig. 5 of fractional weight collection efficiency η vs. particle size ratio D_p/D_{pc} may be used to calculate the collection efficiency over the whole particle size range, taken in steps to obtain an overall collection efficiency from separate particle sizes and amounts of each present if complete analysis is available on the dust entering the cyclone. Lacking complete particle size analysis, an approximation is possible (see Fig. 6 and Table 2) from the cumulative percent of material present above, given particle sizes as shown for this previous example.

It is assumed that an even distribution of particles of a given size exists as the curve Fig. 6 represents, where particle size is plotted vs. cumulative values, using mid-percent points, in this case every 5%. The particle size is made the variable and the steps of division (at 5%) are constant. From the micron size of the particle, and size at each step as an average, the values of $D_p D_{pc}$ are obtained, using the calculated cut-point diameter (of 6.37 μm) above.

From this division of cumulative percent data on the incoming dust to the cyclone, and the collection efficiency at each 5% step, taken as a mid-percent point, the overall collection efficiency η is obtained, either from the direct calculation, or from a plot as recommended; see Fig. 7.

In order to carry out the example it was necessary to extend the efficiency curve to smaller

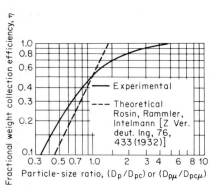

Fig. 5 Separation efficiency of cyclones (*From Perry and Chilton, "Chemical Engineers' Handbook, 5th ed.," McGraw-Hill, New York, 1973.*)

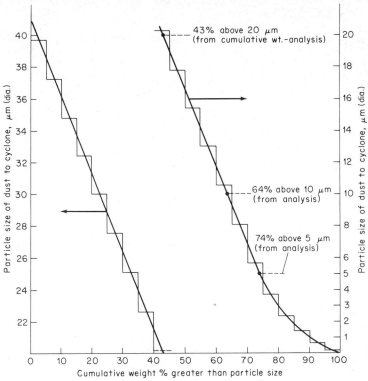

Fig. 6 Estimation of particle size distribution from mid-percent steps.

TABLE 2 Calculation of Overall Collection Efficiency from Size Distribution of Dust Entering Cyclone

Fraction* above size	Particle size, μm	D_p/D_{cp}† ($D_{cp} = 6.37$ μm)	Collection/ Efficiency η	Fractional recovery per step
1.00	0.0	0.000	0.000	0.000
0.95	0.20	0.031	0.000	0.000
0.90	0.60	0.094	0.009	0.000
0.85	1.40	0.220	0.045	0.002
0.80	2.3	0.361	0.110	0.004
0.75	3.7	0.580	0.250	0.013
0.70	5.7	0.894	0.490	0.025
0.65	8.1	1.27	0.610	0.030
0.60	10.6	1.66	0.730	0.037
0.55	13.0	2.04	0.810	0.040
0.50	15.4	2.42	0.900	0.045
0.45	17.8	2.79	0.910	0.046
0.40	20.2	3.17	0.920	0.046
0.35	22.6	3.55	0.950	0.048
0.30	25.1	3.94	0.960	0.048
0.25	27.6	4.33	0.970	0.049
0.20	30.0	4.71	0.980	0.049
0.15	32.4	5.08	0.990	0.049
0.10	34.8	5.45	0.099	0.049
0.05	37.2	5.84	0.995	0.050
0.00	39.7	6.23	0.995	0.050

*Cumulative. $D_{cp} = 6.37$ μm, for 50% recovery.
†60.0% recovery.

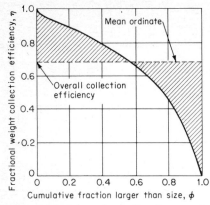

Fig. 7 Calculation of overall collection efficiency. *(From Perry and Chilton, "Chemical Engineers' Handbook," 5th ed., McGraw-Hill, New York, 1973.)*

diameter ratios than covered by the correlation, but this was done in a location where the curve is relatively straight. The overall efficiency is 68.0%, from the summation of each recovery at the particle sizes calculated at each step. A plot similar to that in the recommended method[1] shows some irregularity from the large amount of very small particles which flatten out the lower portion, and would extend this condition if still more were present.

The efficiency of 68.0% for the 4-ft-diameter cyclone could be improved by putting in multiple units, as in the "multiclone collector" which apparently gives recoveries of 70% for 2-ft-diameter cyclones, 83% for 9-in-diameter, and 90% for the 6-in-diameter units.

Commercial Types of Cyclones

The immense variation possible in cyclones and in mechanical arrangements to meet commercial needs can only be indicated generally here. Several typical ones using the true cyclone flow pattern, as distinct from the rotational-flow dust precipitator or the mechanical types where a rotating part provides the centrifugal action, may be summarized as follows.

Uniflow Cyclone   The uniflow cyclone has a gas entrance at top and vertical arrangement with swirl vanes in top cylindrical section. Cleaned gases leave at the base through a centrally located pipe, while dust is trapped in the annular section between the gas outlet pipe and the vertical cylindrical body of the vessel. Solids are removed from annular base section by exhausting 5 to 20% of gas entering. This purge gas is necessarily an additional load as recycle.[4]

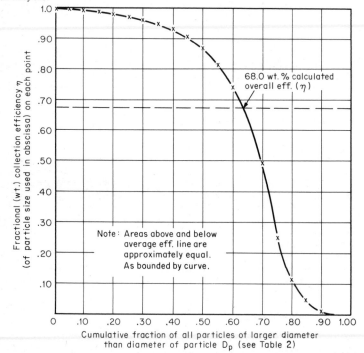

Fig. 8 Determination of collection efficiency from data on dust entering cyclone.

Duclone Cyclone   Contains two vertical cone-shaped sections with a vortex shield between.[1]

Sirrocco Type D Collector   This contains three vertical cone-shaped sections and a helical inlet which provides a relatively large clean gas outlet at the top. Solids removed from base through a blast gate.[1]

Van Tongeren Cyclone   This device utilizes an eddy current in the inlet section to trap finer particles, passing these downward though the so-called shave-off dust channel, where the finer particles then reenter the main flow at a lower level, in the cyclone.

Multiclone Collector   This is a group of vertically arranged small-diameter cyclones with a common dust hopper. A multiple group of small-diameter cyclones provides small-diameter centrifugal paths, and smaller travel distances for the particles to move before impingement, so both the minimum diameter of separated particles and the cut diameter are smaller, resulting in smaller particle separation and improved separation efficiency.

Dustex Miniature Collector Assembly   This is a group of small-diameter cyclones set at an angle from the vertical and discharging into a common hopper as in the multiclone collector above. Tube sheets divide the inlet gas and receive the dust discharged into the hopper from each cyclone unit. The cleaned gas leaves the individual cyclones from a side opening and enters the space between the inlet tube sheet and the hopper tube sheet, the space between being an outlet duct for the cleaned gas. This arrangement has many mechanical advantages, but the nonvertical installation may cause some difficulties with solids movement when a degree of agglomeration is possible.

Rotational-Flow Dust Precipitator   This device is a centrifugal separator that is unique in that the maximum tangential velocity is at the outside radius of the cyclone, this condition being established by the injection of secondary gases in a tangential and downward direction around the outer periphery of the centrifugal gas flow. The injection is made by a number of nozzles, placed at various levels and around the circumference of the cylindrical precipitator vessel. The dust-laden gas enters the crude gas pipe at the bottom, goes through a spin-vane element for an initial rotation, then passes upward where it meets the injected gases to get additional rotational force on the outer rim of the gas flow. In this way the tangential component of flow will increase as the velocity distribution goes to the outer radii of the separator. Mathematically stated, the velocity component $V_{ct} \propto r^n$, when $n = +1$ here, while in vortex flow $n = -1$ theoretically. There is an annular space between the cylinder wall and the gas inlet pipe formed by the pipe extending inward, and this space is used as a hopper to trap out the solids collected. The solids are removed by skimming off the air used for the secondary air injection by means of a blower. As in any condition of recycle this will add extra equipment volume, but may result in separation advantages as well, which is indicated by possible cut diameters as low as 0.5 μm.[1]

Commercial Types of Mechanical Centrifugal Separators

Two types are common.

Type D Rotoclone   This is a combined dust separator and exhaust fan, with blades shaped to direct separated dust to an annular slot at the outside circumference of the vanes and then to the dust hopper. Some gas flow through the dust hopper passes back to the suction end of the fan for facilitating the flow of dust into the hopper. The fan blades also perform as an exhauster as well as a dust interceptor and mover. Air leaves from the outer rim of the fan housing which is shaped as a centrifugal scroll.

Sirrocco Cinder Fan   This device collects the dust from the dust-laden gas as it enters, using the concave side of radially placed blades around the shaft. The shaft is also used for the fan, which takes suction on the cleaned air and passes it out of the system. Dust is passed out from the periphery of the blades, through a skimmer slot, and provision for the recirculation of some gas is made to pass solids with the recirculated gas on to a secondary collector where there is a means of solids removal.

SEPARATION BY FILTRATION

Mechanical filters exhibit separation through the mechanism termed *flow-line interception*, but effects of inertial impingement and gravity settling are also evident. These latter effects account for an initial deposition of even very small dust particles within the pores of a clean filter at the start of its use, when the clean pore openings are much larger than most of the particles being deposited. This initial precoating effect is particularly important in the use of woven filter cloth where the passage route through the filter is shorter and less devious than in filters of felt construction, and possibly has even larger pores to start with. The precoating period may require only a few seconds when there is a fair amount of dust present in the gas. The temporary passage of some small particles through the filter during this period may be acceptable, but if not, some temporary preconditioning is possible. After this, the filtration very quickly assumes pressure-drop conditions characteristic of the dust cake as it builds up, rather than of the clean filter cloth.

Filters used in gas-solids separations may be woven or felted fabrics of natural or synthetic fibers. Filtrations may also be made with granular solids in the form of stationary or moving beds. Many other types of materials that are porous, or are capable of providing a screening effect after weaving or fabrication, may be found suitable for certain filtration applications.

It is customary to list the properties of the fibers used in the making of common filter cloths, so that from the fiber properties one can determine temperature limits in operation, corrosiveness, flammability, and perhaps comparative strength. For the dry service indicated, some fiber properties for a number of materials are available in Ref. 1, p. 20–94, table 20–39.

Flow-resistance factors and the conventional air permeability values for woven and felted filter fabrics are given in Ref. 1, p. 20–90, table 20–37 and Ref. 1, p. 20–96, table 20–40 (woven and felted respectively).

The cake resistance observed in the filtration of some solids is also available.[1] This data has been converted to fit into a similar correlation as in the reference. For this combination of data see Fig. 9. The data mentioned above will be used in the example of a design for a filtering operation.

A mechanical filter is usually required when final separation of dust must exceed a 99% efficiency but it may also be used as a cleanup operation following other separation techniques. For large-scale operations, mechanical filters may be found available in standard sizes of about 1000 or 2000 ft^2 of filter surface each. They may be installed in parallel arrangement to any desired capacity.

In some cases it is possible to estimate the size of a filtration unit from literature data at hand as in the example given. A survey of overall filtration operation shows charge rates may fall in the range of 1 to 8 ft^3/min per square foot of filter cloth area, while 3 ft^3/min is regarded as a maximum rate for dusts of small particle size or for high dust loads (above 8 grains/ft^3).

The physical size of the installation may be roughly estimated after the filter surface and frequency of the shaking period is estimated. The size is determined by the limitation of the pressure drop that the filtration can be carried out at (6 inH$_2$O is a possible limit). Assume that a bag filter with 15-ft-long bags of 6 in diameter are spaced on center distances of approximately 1 ft and that several feet of elevation are required to fit in a storage hopper. This should permit a size to be estimated for an installation and build an image of the physical size of the installation.

The mechanical filter is usually chosen from a standard type of commercial unit, and the vendors of such units have most of the specific data from their experience. In this effort to follow separation techniques, it is required to make comparisons between methods, rather than between the types of mechanical filters available, although in making any comparison between methods, the best available designs should be used.

General Characteristics of Mechanical Filters

Standard commercial types of filters use a bag or an envelope, the latter being basically a flat rectangular-shaped filter covering a frame, which is also called a retainer for the filter cloth. The term *cylinder* is also applied to filters of the bag type since these elongated cylinders or bags may be open at both ends to make cleaning by air jets blowing into the top a possibility. Alternately, bags may be cylindrical or oval in cross-sectional shape.

Envelope filters, with a frame for support and with large flat surfaces exposed, do not have the cleaning possibilities of the bag filters, which can be given more shaking, or can be collapsed with no particular harm. If air blowing in reverse direction to filtration flow is used, cleaning can be made more effective. Therefore the envelope filter is best used on dusts that are easily shaken or removed from the cloth surface. The dust is removed in the

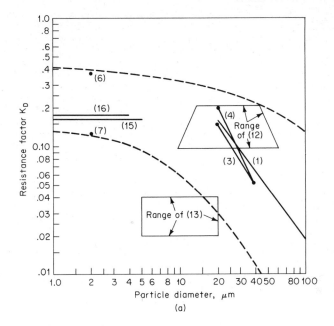

(a)

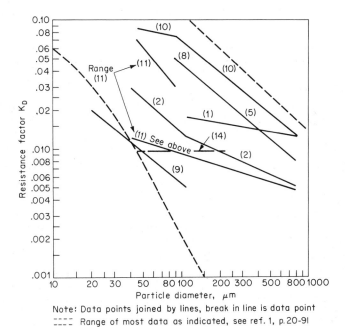

Note: Data points joined by lines, break in line is data point
- - - - Range of most data as indicated, see ref. 1, p. 20-91

(b)

Fig. 9 Resistance factors for dust layers. (1) Granite dust, (2) foundry dust, (3) gypsum dust, (4) feldspar dust, (5) stone dust, (6) lampblack dust, (7) zinc oxide dust, (8) wood dust, (9) resin, (10) oats, (11) corn, (12) minus 200 mesh coal dust, (13) cellulose acetate dust flocculate, (14) pipeline dust, (15) zinc ore roaster fines, (16) talc dust. (Data *from Williams, Hatch, and Greenberg. See Ref. 1, pp. 20–91.*)

envelope type by beating the screen that supports the filter and by shaking or rocking the frame with a mechanism.

Bag filters have a variety of methods for dust cake removal, and some are used in combinations. The shaking out of dust must be carried out when the flow of gas is shut off in some types of mechanical filters, therefore in these cases a spare filter must be provided. If normal operation is carried out between two units, one-half sparing rather than a full sparing is used.

The various shaking arrangements for bag filters consist of the simple case of supporting the bags by holding the bags in a vertical position with the top end closed off. The gas and dust enter the inside and the dust deposits in the form of a cake inside the bag. The gas is shut off for cleaning and the bags are shaken from the top support frame so the solids drop down into the lower bin. There is enough play in the bags to permit shaking without tearing. In some cases, clean air may be used in addition to shaking to assist in the cleanup. This may be done as a reverse-flow operation, and may be arranged with automatically timed opening and closing of dampers or valves.

Another arrangement collapses the bags by a reverse flow of gas or air, and high-frequency sound waves combined with gas flow dislodge the cake. Another method uses a reverse flow of air to collapse the bags and then sends into the bags jets of compressed air from the top, which move the dust downward into the bin in slugs. The air is described here as forming bubbles and the action does not noticeably cause any reentrainment of dust particles from the cake.

The use of a blow ring in some bag- or cylinder-type mechanical filters permits continuous operation while cleaning. Dust is collected on the inside, and the blow ring travels up and down along the outer surface of the bag. The blow ring has an inside slot that is used to blow gas or air against the wall of the bag. The ring is tight enough to partially collapse the bag in order to break the dust cake and provide a close seal so the gas blown in is fully delivered through the filter in blowing back.

Dust may be collected on the outside of filter tubes (bags) if a support is provided inside to prevent collapse of the filter. In some cases it is possible to remove the cake from the outer wall by periodically using a jet of compressed air from the inside of the filter to produce a shock that breaks the cake from the outer wall whence it can settle into the bin.

The envelope- or frame-type of filter collects the dust on the outer wall of the filter, as might be expected, since the outer wall is easier to get to in this arrangement.

Design of Mechanical Filters

The following example indicates the type of information and calculation methods used to select and size a mechanical filter.

EXAMPLE

The following design conditions are typical:

Dust stream to be filtered: 10,000 scfm of air, carrying 8 grains/scfm of gypsum dust (gypsum, $CaSO_4 \cdot 2H_2O$), operation at 2000°F, atmospheric discharge, 6 in H_2O inlet pressure from fan, design for maximum of 6 in delta pressure.

Density of gypsum is 145 lb/ft³ (true density), bulk density is 72.5 lb/ft³, and voids are represented from this, since $\epsilon_v = 0.5$ for dusts at 20 or 40 μm(D_p) sizes, which are under consideration.

Resistance factors for the filtration of two operations are given, these described as being typical of dusts of 20 and 40 μm, respectively, and flows with pressure-drop values are calculated for each size as a uniform particle mixture.

The resistance factors were obtained at a filtering temperature of 70°F, where gas viscosity is 0.0181 cP; at 200°F the viscosity is 0.0216 cP. The resistance factors are 0.15 and 0.05 for the 20- and 40-μm operations, respectively.

The Carman equation[1] is used with Fig. 10 to assist calculation; the plausibility of voidage and resistance factors will be examined. Since two separate operations are given this will be an advantage.

The Carman equation is

$$K_d = \left[\frac{160.0}{\phi_s^2 D_p \mu^2 \rho_s} \right] \left[\frac{1 - \epsilon_v}{(\epsilon_v)^3} \right] \quad \text{in} H_2O/(cP)(gr/ft^2)(ft/min)$$

Substitution for the 20 μm filtration yields

$$0.15 = \frac{160.0}{(0.806)^2 (20)(0.0181)^2 (145 \times 7000)} \cdot \frac{1 - \epsilon_v}{(\epsilon_v)^3}$$

so that

$$\frac{1 - \epsilon_v}{(\epsilon_v)^3} = 4.05$$

From Fig. 10 this corresponds very closely to 50% voids, or $\epsilon_v = 0.50$, as indicated by the bulk density given in the analysis. Similarly for 40 μm

$$0.05 = \frac{160.0}{(0.806)^2(40)(0.0181)^2(145 \times 7000)} \cdot \frac{1 - \epsilon_v}{(\epsilon_v)^3}$$

or

$$\frac{1 - \epsilon_v}{(\epsilon_v)^3} = 2.70$$

which corresponds to 0.55 voids.

The above indications seem in good enough agreement to assume that the data are consistent for preliminary design. Figure 9 shows this to fall in the general area of all data.

The Carman equation requires a correction when particle size is approximately 5 μm or less,

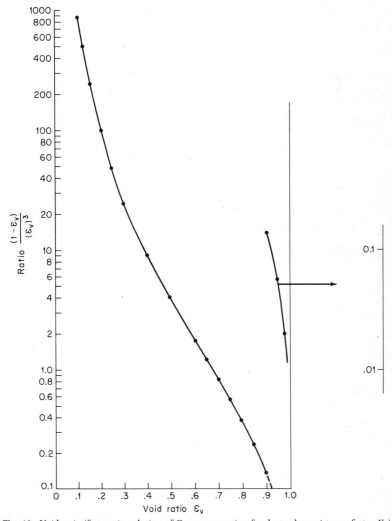

Fig. 10 Void ratio (for use in solution of Carman equation for dust cake resistance factor K_D).

approaching the mean free path size, which is near 0.021 μm (0.021 $\times$ 10^{-4} cm). For this correlation the mean free path, based on bimolecular gases, may be approximated by this expression:

$$\lambda_m = 0.021 \times [(460 + T_F)/492] \times [(14.7/P_A)]$$

or by correcting the mean free path distance by the ratios of absolute temperature and pressure for the operating conditions used. In the expression, T_F is °F and P_A is lb/in²(abs).

This is called the slip-flow correction and it tends to reduce the resistance factor for the dust layer filtration. The correction factor obtained from the following expression must be used by dividing it into the resistance factor K_d, for a corrected value:

$$\text{Factor for division} = \left[1 + K_s \left(\frac{1 - \epsilon_v}{\epsilon_v} \right) \left(\frac{\lambda_m}{\phi_s D_p} \right) \right]$$

where K_s is a constant for the correction with the value of 15. The other terms used are covered in the nomenclature defined previously. Using the voids ratio ϵ from the 20-μm case above, where $(1 - \epsilon_v)/\epsilon_v$ = 4.05, the correction, which is expected to be small, is then

$$\text{Correction} = 1 + (4.05) \left[\frac{(0.021)(530/492)(14.7/14.9)}{0.806 \times 20} \right]$$
$$= 1.0056 \quad \text{for 20-}\mu\text{m case}$$

$$\text{Correction} = 1 + (2.70) \left[\frac{(0.021)(530/492)(14.7/14.9)}{0.806 \times 10} \right]$$
$$= 1.00186 \quad \text{for 40-}\mu\text{m case}$$

Either correction is probably not significant, in that fluctuations in practice exceed this effect in most cases.

With filtration at filter inlet pressure and temperature, the 10,000 scfm is calculated to be 14,400 acfm and the dust content is 5.54 gr/acfm. The K_d values are then recalculated and a 50% void ($\epsilon = 0.5$) is used. Filtration here is at 200°F with gas viscosity = 0.0216 DF.

For the 20-μm case:

$$K_d = \left[\frac{160.0}{(0.806)^2(20)(0.0216)^2(145 \times 7000)} \right] \times (4.05)$$
$$= 0.1053$$

Corrected according to the correction factor above,

$$K_d = 0.1053/1.084 = 0.972$$

which will be used for design of the filtration of 20-μm dust. Note that a drop in the resistance implies a more favorable separation condition, as the lower value of the resistance factor for the 40-μm case also indicates.

For the 40-μm case:

$$K_d = \frac{160.0}{(0.806)^2(40)(0.0216)^2(145 \times 7000)} \times 2.70$$
$$= 0.00059$$

The corrected K_d is (0.0351/1.002), or

$$K_d = 0.0350 \text{ for the 40-}\mu\text{m case}$$

The pressure drop through the cake deposit on the filter is expressed in terms of the grains of dust per square foot of filter surface at the time, since the bulk density is not constant through the depth of the layer. The equation of flow is

$$\Delta p_i = K_d \mu_c m_s V_f$$

where Δp_i = pressure drop, inH$_2$O
μ_c = viscosity, cP at flowing temperature of gas
m_s = grains of dust per square foot of filter surface
V_f = actual and superficial gas velocity through filter cloth, ft/min (or ft³/min per square foot of filter surface)

The pressure drop for the two cases at 5 ft/min after ½ h of continuous filtration is

$$\Delta p_i = 0.0972 \times 0.0216 \times 798 \times 5.0 = 8.37 \text{ inH}_2\text{O}$$

for 20-μm particle filtration.

$$\Delta p_i = 0.0525 \times 0.0216 \times 798 \times 5.0 = 4.52 \text{ inH}_2\text{O}$$

for 40-μm particle filtration.

These selected conditions related to the design in that practically 15,000 acfm of gas is handled (14,400 cfm), and the 5.0 ft/min velocity of gas would be handled in three filters of 1000 ft² each. The velocity of 5.0 ft/min also fits into some moderate operation with shaking intervals of 20 to 40 min. The 798 grains per square foot of filter surface is from the flow of dust at 5.54 grains per cubic foot of actual gas flow, after ½ h accumulation.

A time ratio will show that the 20-μm case will have a 6 inH$_2$O pressure drop in 21 min and the 40-μm size will filter 40 min. The pressure drop from the cloth screen is negligible. For this case it is calculated as

$$\Delta p = K_c \mu_c V_f$$
$$= 0.84 \times 0.0216 \times 5.0 = 0.090 \text{ inH}_2\text{O}$$

The pressure drop through cloth and filter cake varies in a linear way with the gas velocity, this being referred to as a streamline condition of flow. It is based on a criterion that the value of $\rho V_f / \mu_c$ should not exceed 100, and for this case it is equal to 36 at 5 ft/min gas velocity.

It appears that a conservative design in this case would be three filters at 1000 ft^2 each, plus one spare.

SEPARATION OF GASES AND SOLIDS BY SCRUBBING

Commercial separations that involve a contacting action between dust-laden gases and liquids to bring the solids into a slurry form for removal must utilize methods of low energy requirements. Energy consumption is reflected in pressure drop required to operate the various types of scrubbers; those of the lowest pressure drop, such as spray towers, are limited to particle size separation above 10 μm, while others at progressively higher pressure-drop characteristics will give particle separation down to as low as 0.5 μm (see Ref. 1, p. 20-98, table 20-41). Favorable gas contacting methods are therefore hoped to be found for a design among the following types of scrubbing devices, alone or in some combination:

Spray Towers

Spray towers with spray nozzles are designed for effective droplet size distribution.

Tray Towers

Tray towers have elutriated sieve trays with baffles to produce aeration and spray droplets by utilizing pressure drop for the passage of gases through the tray perforations into the liquid passing over the tray. The baffles, each located above a perforation, provide impingement surface to provide the turbulence required. Other aspects of such a tower design may include gas humidification sprays at the gas inlet, and entrainment separators for the liquid droplets present in the effluent gases. Most of the dust removal here actually takes place in the region of the spraying droplets above the tray, after the gases have been brought to a high local velocity induced by the pressure drop established as the dust-laden gases are forced through the tray perforations.

Cyclone Spray Scrubbers

Cyclone spray scrubbers bring inlet dust-laden gases to velocities of approximately 50 to 75 ft/s so that water sprayed from some central location within will supply droplets that are swirled around within the cyclone. The resulting impacts of dust particles and droplets of liquid result in the formation of a slurry which is separated from the gas stream through the centrifugal forces produced. Energy input is chiefly from the relatively high entrance gas velocity, supplied by fans or blowers, and ranges from 2 to 10 inH$_2$O in water pressure drop for most applications.

Wet-Approach-Type Scrubbers

Wet-approach-type scrubbers also require relatively high gas inlet velocities, but the liquid is introduced as a fluid at low pressure and the gas provides the motive power to break up the liquid into drops. There is the simple application of using a pipe, or using a venturi shape to carry the gas and produce the droplet dispersion effective for the operation. This type of scrubbing may be effectively combined with a cyclonic separator to remove liquid droplets from the effluent gas, as an entrainer. Such scrubbers may require provision for pressure drop ranging from 5 to 100 inH$_2$O.

Among the wet-approach type of scrubbers there is the so-called venturi type (see Ref.

1, p. 20–102, fig. 20–120), where the initial mixing of gas and liquid is done in a venturi-shaped housing, wherein the liquid enters tangentially to meet the gas which is directed downward towards the throat of the venturi-shaped housing (installed in a vertical position). Slurry, gas, and the liquid load will leave the venturi mixer to enter a cyclone for gross separation. Solids can be removed as a slurry or decanted, if reuse of liquid material or water is considered.

Water-Jet Scrubbers

Water-jet scrubbers, in contrast to spray towers, are adaptations of the evactor principle used in steam ejectors to pull vacuum on vessels or towers. Here a high-velocity stream of water leaves the nozzle of the jet, entraining the entering dust-laden gas and discharging the mixture in a turbulent state into a sump for possible decantation, and then the liquid is recirculated. The cleaned gas is passed out. This operation could be combined with a final entrainment device for the effluent gas, but one preferably of low pressure drop characteristics.

Orifice Scrubbers

Orifice scrubbers are characterized by gas flow contacting the surface of a pool of liquid to produce turbulence and dispersion of the liquid into droplets, so the dust-laden gas must be either blown into the equipment or the clean gases must be drawn out by an exhaust fan to produce the required turbulence. This classification includes a wide variety of equipment, but in all cases the gas is never forced deeply into the liquid pool, but merely impinges with the dispersed droplets to provide the separation of dust particles from the gases.

Mechanical Scrubbers

Mechanical scrubbers include a variety of arrangements, often arranged so that liquid enters the inlet of a fan and is dispersed while the gas is being accelerated to a required velocity. Some types employ rotating screens to pass through and pick up the liquid from a pool at the base of the rotator (see Ref. 1, p. 20–102, figs. 20–122 and 20–123).

Fibrous-Bed Scrubbers

Fibrous-bed scrubbers are intended for separation of submicron-size particles using a thin felted fiber filter, or in some applications a Dacron polyester (0.195 in thick) which is elutriated during the passage of gas and particles through the fiber. Unlike typical filtration, the particles are washed from the downstream side of felted cloth, impaction being produced without clogging the fibrous cloth.

Packed-Bed Scrubbers

Packed-bed scrubbers of various types may be regarded as versions of elutriated packed towers, but for dust-laden gases these must be arranged to carry off the solids deposited. This may be done by adding sprays to the conventional wash given to the packed bed, for example, by placing sprays beneath the packing support plate. The packing beds may sometimes be kept in a partially flooded state. One innovation uses a floating bed of low-density spheres to facilitate the washing out of collected solids, assisted by the motion of the packing when the gas passes through it.

Of the nine general types of scrubbing devices outlined above only a limited number can be used in the realy large operations, and some operations may exceed 500,000 ft³/min, as in a 500 ton/day fertilizer plant. As mentioned above, this involves pressure-drop requirements, and 0.1 inH_2O differential will require 0.157 theoretical horsepower per 1000 ft⁶/min, so scrubbing devices using several inches of water as their requirement would become large power users in the bigger installations.

SEPARATION BY INERTIAL IMPINGEMENT OF SOLID PARTICLES AND LIQUID DROPLETS AS IN SCRUBBING

Although flow-line interception and Brownian diffusion are possible, the predominant mechanism of scrubbing is the inertial impaction of a relatively high-density particle into a liquid droplet. This leads to the visualization of a projectile-target action, but since the relative velocity between the particle and droplet is important, scrubbing may be dependent upon the acceleration of either the particle or the droplet (or both) toward the point of impact. In other words, the target may be moved to the projectile, as in the case of large droplets being driven towards the solid particles in the gases.

It is customary to describe target efficiency as the fraction of particles collected by a receiving body, from that part of the gas stream that is necessarily swept by the body, but more exactly stated, it is the fraction of particles that are collected from a gas volume determined by the projected area of the receptor body. The target efficiency η is correlated with a separation number, as in Fig. 2 where η was first introduced as applicable to the collection of particles on baffles. Here also a characteristic diameter (if spherical particles and receiving droplets are assumed) is used in the equation.

It is not immediately evident that target efficiency is decreased as the diameter of the droplet D_b in the separation number is increased (see Fig. 2 and Ref. 1, p. 20–97, fig. 20–111). Here again, it should be noted the gas must be passed around the particle after solids deposition, and this associated gas that must sweep the receiving body is greater in amount for the larger particles for the same efficiency, and thus there is proportionately more gas to disrupt the neighboring particle reception, or even to dislodge other entrained particles in the droplet. It may also be reasoned independently that a particle reaching a smaller-diameter droplet (a smaller target, so to speak) must be representative of a higher target efficiency. Drop size is important, but other factors such as the difference between the rates of fall of the solid particle and the droplet will increase efficiency, with drop size tending to give higher target efficiency because of increased impact velocity, when impact velocity is the result of droplet fall.

The overall correlation of scrubber performance with power usage in the processes has been made on the assumption that scrubbers will perform approximately according to this factor power usage, regardless of the numerous variations possible. This assumption is valuable in design comparisons.[6]

The Stefan flow effect is demonstrated by the preaddition of steam (two to three times the amount required to saturate the incoming gases). This preaddition induces condensation of water on the particles in the gas when contacting with cool water sprays, and thus causes a size increase that favors separation.

It may be noted that most variables in scrubbing process design are difficult to even approximate, for example, such variables as water droplet distributions in sprays and above the perforations in sieve trays and in centrifugal apparatus. The assumption of spherical shapes of particles and droplets is also highly approximate. Even so, the characteristics of suitable apparatus may be visualized well enough to be of value. For this reason some tables have been prepared to assist in evaluations, particularly those requiring target efficiency values, with the assumption made that makers of spray devices will supply the needed apparatus, within limits.[1]

DESIGN OF A SPRAY TOWER FOR SCRUBBING

The choice of a wet separation is often based on dust properties that make it desirable to handle the removed solids as a wet slurry. This choice may be the result of the dust having toxic properties, being of an acid or caustic nature, or even having some unusual property such as a color, and therefore being capable of staining objects in the environment.

Water flow requirement for a spray tower is estimated from the application of target efficiency methods to the dust removal (from the gas) by the water in some dispersed form, by spraying or providing trays. If the dust load is very concentrated, extra water may be required for the slurry removal. See Fig. 11.

The handling of slurry and the reuse of water after decantation might involve liquid-solid separators as used in sewage and refinery sludge.

Performance expected in a spray tower will have its limitation at particle sizes around 10 μm, as shown below. To remove these particles at around 98% efficiency will require about four perfect contacts and considerable water circulation. Some particles below 10-μm size will be removed through the other mechanisms mentioned above. This limitation is a practical one, established in order to keep tower diameters within economic bounds, with 1 to 3 ft/s superficial gas velocity through the tower. The gas velocity will directly affect the entrainment of smaller particles in the effluent gases and permit carry over of smaller droplets as well, if too high.

Entrainment devices should be considered for installation in the top section of spray towers and evaluated in terms of their pressure-drop requirements in the process. These should reduce tower height qualitatively speaking, since the tower could conceivably be designed for a higher velocity of gas flow with entrainment removal. The devices for entrainment may be baffles, or some means of inducing centrifugal flow of gases before their exit, or a metal mesh pad made of stainless steel or other material. Pressure-drop characteristics of the mesh are usually obtained from vendors' experience, although approximation is possible from basic impingement considerations.

The tower, or both tower and spray, for an operation should have several levels of spray

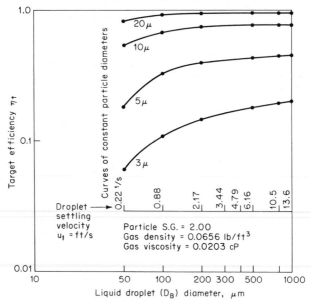

Fig. 11 Target efficiencies in spray towers.

injection because of possible plugging of some sprays during runs, as well as for the staging effect. If water is reused after simple decantation this point is increasingly important. If low system pressure conditions exist, say less than several inches of water above atmospheric pressure, it may be feasible to consider a large sump at the base of the tower to receive the solids and effluent water through a sealed pit rather than use a conventional tower with a dished bottom. If approximately 3 ft is allowed between spray levels with sprays at 10 levels maximum, and several feet are allotted to the gas inlet pipe entrance at the base and top disengagement space or entrainment separator installation, a tower of approximately 50 ft in height could develop. The use of pumps capable of handling slurry must also be considered here.

The sprays used should have the capability of delivering droplets in the 500- to 1000-μm size range, and in commercial spray devices some portion of the droplets will be around 100-μm size; see Ref. 1, p. 18–63, and p. 20–97, fig. 20–112. If the target efficiency in a spray tower is considered to be dependent upon impact velocity of falling droplets and dry particles the target efficiency is greatest between the 1000- and 500-μm range of drop size. The larger drops with greater settling velocity become in some cases as efficient as drops of smaller diameter and resultingly lower settling velocity.

While this correlation, Fig. 2, describes the impact principle, the effects of the jet spray before the droplets fall, and their subsequent breakup into smaller particles, cannot be taken into account here, nor can any ultimate saturation of droplets by solid particles be estimated. For high dust concentrations, this point of ultimate saturation of a droplet with solids is possibly redundant if the amount of water required to wash down and carry the slurry from the tower is considerable.

Target efficiency values that may be calculated from Fig. 2, and Table 3 is helpful in the

TABLE 3 Particle and Droplet Properties for Scrubbing

$T = 150°F$
$\rho_{gas} = 0.0656$ lb/ft³
$\mu_{gas} = 0.0203$ cP

Sphere diameters, μm	Sphere volume, (μm)³	Sphere volume, ft³	Projected area/sphere, (μm)²	Projected area/sphere, ft²	Weight/sphere (S.G. = 2.00), Gr	Weight/sphere (S.G. = 1.00), lb	No. spheres per cu. ft., 1 gr/ft³ of gas (S.G. = 2.00)	No. spheres per sq. ft. of area projected	Particle velocity u_t (S.G. = 2.00), ft/s	Droplet velocity u_t (S.G. = 1.00), ft/s	Re modified, for particle fall, u_t	Re modified, for droplet fall, u_t	Sphere diameter, in
1	.524	5.64×10^{-16}	0.7854	8.45×10^{-12}	4.92×10^{-10}	3.52×10^{-14}	2.03×10^{9}	1.18×10^{11}	1.76×10^{-4}	8.80×10^{-5}	2.78×10^{-6}	1.39×10^{-6}	3.94×10^{-5}
3	$.141 \times 10^{2}$	1.52×10^{-14}	$.707 \times 10^{1}$	7.61×10^{-11}	1.32×10^{-8}	9.50×10^{-13}	$.752 \times 10^{8}$	1.31×10^{10}	1.58×10^{-3}	7.92×10^{-4}	7.47×10^{-5}	3.75×10^{-5}	1.18×10^{-4}
5	$.655 \times 10^{2}$	7.05×10^{-14}	$.196 \times 10^{2}$	2.11×10^{-10}	6.16×10^{-8}	4.40×10^{-12}	1.62×10^{7}	4.73×10^{9}	4.41×10^{-3}	2.20×10^{-3}	3.48×10^{-4}	1.74×10^{-4}	1.97×10^{-4}
10	$.524 \times 10^{3}$	5.64×10^{-13}	$.785 \times 10^{2}$	8.45×10^{-10}	4.92×10^{-7}	3.52×10^{-11}	2.03×10^{6}	1.18×10^{9}	1.76×10^{-2}	8.80×10^{-3}	2.77×10^{-3}	1.39×10^{-3}	3.94×10^{-4}
20	$.419 \times 10^{4}$	4.51×10^{-12}	$.314 \times 10^{3}$	3.38×10^{-9}	3.93×10^{-6}	2.81×10^{-10}	2.54×10^{5}	2.96×10^{8}	7.05×10^{-2}	3.52×10^{-2}	2.22×10^{-2}	1.11×10^{-2}	7.87×10^{-4}
50	$.655 \times 10^{5}$	$.705 \times 10^{-10}$	$.196 \times 10^{4}$	2.11×10^{-8}	6.16×10^{-5}	4.40×10^{-9}	1.62×10^{4}	4.73×10^{7}	4.40×10^{-1}	2.20×10^{-1}	3.47×10^{-1}	1.74×10^{-1}	1.97×10^{-3}
100	$.524 \times 10^{6}$	5.64×10^{-10}	$.785 \times 10^{4}$	8.45×10^{-8}	4.92×10^{-4}	3.52×10^{-8}	2.03×10^{3}	1.18×10^{7}	1.76	8.80×10^{-1}	2.78	1.39	3.94×10^{-3}
200	$.419 \times 10^{7}$	4.51×10^{-9}	$.314 \times 10^{5}$	$.338 \times 10^{-6}$	3.93×10^{-3}	2.81×10^{-7}	2.54×10^{2}	2.96×10^{6}	(int.*)3.54	(int.*) 2.17	11.2	6.84	7.87×10^{-3} Int.*
300	$.141 \times 10^{8}$	1.52×10^{-8}	$.707 \times 10^{5}$	$.761 \times 10^{-6}$	1.33×10^{-2}	9.50×10^{-7}	$.752 \times 10^{2}$	1.31×10^{6}	5.62	3.44	26.6	16.3	1.18×10^{-2}
400	$.335 \times 10^{8}$	3.61×10^{-8}	$.126 \times 10^{6}$	1.35×10^{-6}	3.15×10^{-2}	2.25×10^{-6}	3.17×10^{1}	7.39×10^{5}	7.84	4.79	49.6	30.3	1.58×10^{-2}
500	$.655 \times 10^{8}$	7.05×10^{-8}	$.196 \times 10^{6}$	2.11×10^{-6}	6.15×10^{-2}	4.40×10^{-6}	1.62×10^{1}	4.73×10^{5}	10.1	6.16	79.7	48.6	1.97×10^{-2}
800	$.268 \times 10^{9}$	2.89×10^{-7}	$.503 \times 10^{6}$	5.41×10^{-6}	.2522	1.80×10^{-5}	3.97	1.85×10^{5}	17.2	10.5	217.	133.	3.15×10^{-2}
1000	$.524 \times 10^{9}$	5.64×10^{-7}	$.785 \times 10^{6}$	8.45×10^{-6}	.4924	3.52×10^{-5}	2.03	1.18×10^{5}	22.2	13.6	351.	215.	3.94×10^{-2}

*Use intermediate settling formula when Re is in 3.0 to 500 range.

preliminary computations. In the design example, particles of up to 50 μm fall into the range of the target efficiency correlation; see Fig. 2. The particles above 50 μm are easily predictable; otherwise the reference mentioned in Fig. 2 can be examined.

The amount of water used in a spray tower, for some required separation on a given dust with particle distribution known, may be approximated to some extent by considering the amount of water needed to provide a target surface formed as a projected area of droplets of some given size, say 1000 μm. These falling droplets may be viewed as providing target interception for the impact of dust particles moving upward with the gases in the tower. Table 3 gives the number of spheres per square foot of projected area and the weight in pounds of 1000-μm particles, from which it can be determined that 4.16 lb of water is required to form a droplet barrier over 1 ft^2 of tower cross-sectional area. Similarly, a droplet of 100-μm diameter would require only 0.416 lb of water.

It then becomes necessary to determine the number of such barriers or contacts that must be set up, if the particle removal requires several stages of contact because the target efficiency value is less than desired. There is also the question of the number of feet of flowing gas that may be served by a stage. Gases of unusually high dust content may require an excess of water to carry off the heavy slurry that could form.

In the example below, if 1000-μm droplets cover the tower cross section, every passing foot of dust-laden gas is given this wash, and 10 washes (stages) are included, then a tower using 65 lb/in^2 (gage) discharge pressure could conceivably be 50 ft high (for the 10 stages) and require about 6.8 hp (theoretical) per 1000 ft^3 (actual) of gas handled. On the basis of the power principle, this is still a moderate requirement compared with some scrubbing devices, although this spray tower is limited to about 10-μm particle separation if significant recovery is expected.

In the example shown it is considered sufficient to outline the conditions without giving particle distribution of the dust since this has been examined in the design of cyclones as in Fig. 6, but an impingement case is examined to illustrate the order of calculations that must be performed to obtain target efficiency values. When these are determined for the particle sizes of interest and the effect of stages is included, at least an approximate recovery may be predicted.

EXAMPLE

An explicit example of a spray tower operation to be considered is the following. A dust-laden gas mixture considered as dry air contains 15 grains of dust per standard cubic foot, of specific gravity = 2.00 (true density). Flow rate is 15,000 scfm, giving an actual gas flow of 17,400 cfm, based on 14.9 lb/in^2(abs) and 150°F supply. The gas density is taken as 0.0656 lb/ft^3, and the viscosity is 0.0203 cP. A tower is provided for a gas velocity of 3.00 ft/s and is 11.5 ft in diameter, to the nearest half-foot. The dust content is then 13 grains per cubic foot of flowing gas. A complete barrier of 1000-μm droplets as a projected surface across the tower cross section of 103.9 ft^2 will require 51.9 gal. of water. If every foot of gas passing through the tower is sprayed under this condition, the water used would be 3114 gal/min.

Starting with the conditions of gas flow above, the separation of a 10-μm particle is considered. Stokes' law terminal velocity u_t is calculated, using the formula. Then the Reynolds number is calculated to see if it falls in the Stokes' region for later use with Fig. 2, to obtain target efficiency values. Some references give a Reynolds number limit here as below 0.3, and others set it at 3.0. Stokes' law must hold here for particles, but not necessarily for the droplet. When Stokes' law does not apply to the droplet, then the Stokes-Cunningham formula is used for settling rate where droplets are over 200 μm. The Stokes-Cunningham formula giving velocity directly is from Ref. 4, p. 155, which is used for the droplet settling rate, and not confined to the Stokes' law range (see Fig. 2). The intermediate formula of Stokes-Cunningham is:

$$u_t = \frac{0.153 a_e^{0.71} D_p^{1.14}(\rho_p - \rho)^{0.71}}{\rho^{0.29}\mu^{0.43}}$$

where $a_e = g = 32.2$.

From the data in Table 3, the case of a 10-μm spherical particle and a 1000-μm micron droplet is examined, starting with Stokes' law and formula

$$u_t = 18.5 \, D_{ps}^2(\rho_s - \rho)/\mu$$

where D_{ps} is in inches and μ is in centipoise. For the 10-μm particle,

$$u_t = 18.5 \times 3.94 \times 10^{-4}(124.8 - 0.0656)/0.0203$$
$$= 1.76 \times 10^{-2} \text{ ft/s}$$

from which the Reynolds number is calculated as

$$\text{Re} = D_p \times \rho_g \times u_t \times (1488/\mu)$$

for D_p in feet. Then

$$Re = (3.94/12) \times 10^{-4} \times 0.0656 \times (1.76 \times 10^{-2})/(1488/0.0203)$$
$$= 2.77 \times 10^{-3}$$

For the droplet at 1000-μm diameter, the above intermediate formula of Stokes-Cunningham is used, since it will be shown that this is justified by the final calculation of the Reynolds number. With number substituted, the formula becomes

$$u_t = \frac{0.153(32.2)^{0.71}(3.94 \times 10^{-2}/_{12})^{1.14} \times (62.4 - 0.0656)^{0.71}}{(0.0656)^{0.29} \times (0.0203/1488)^{0.43}}$$
$$= 13.6 \text{ ft/s}$$

The Reynolds number is

$$Re = (3.94/12) \times 10^{-2} \times 0.0656 \times 13.6 \times (1488/0.0203)$$
$$= 215$$

which is above 3 and below 500, justifying the use of the Stokes-Cunningham formula.

The separation number is then obtained for use with Fig. 2, to obtain the target efficiency of—in this case—the 10-μm particle impinging upon the 1000-μm droplet. The relative velocity is obtained from the difference of the settling rates above;

$$V_0 = 13.6 - (1.76 \times 10^{-2})$$

which is essentially the velocity of fall of the droplet at 13.6 ft/s.

From Fig. 2 the separation number is

$$N = \frac{(u_t)_{\text{particle}} \times V_0}{32.2 \times D_b}$$

or

$$N = \frac{(1.76 \times 10^{-2})(13.6)}{32.2 \times (3.94/12) \times 10^{-2}}$$
$$= 2.26$$

From Fig. 2, this separation number corresponds to a target efficiency η of 0.78, or 78% recovered.

DESIGN OF A PERFORATED TRAY TOWER FOR SCRUBBING

This application of perforated trays, also described as elutriated sieve trays, depends in principle upon inducing a gas velocity of approximately 75 ft/s so that a pressure drop

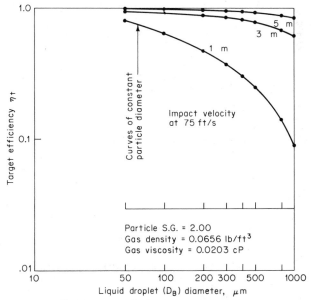

Fig. 12 Target efficiencies from perforated trays.

through the tray and perforations passes the gas upward. Water flowing across the tray in the conventional fashion is forced into foam and droplets of water to provide the liquid droplet impingement surface. Baffles are used to aid this action and have special shapes and dimensions to give favorable droplet size.

Target efficiencies are calculated for the impingement of particles at this jet velocity of 75 ft/s, see Fig. 12, in much the same procedure as for spray towers.

The fact that the particle velocity is high does not disqualify it as being in the Stokes' law range, which is determined by its settling velocity, so Fig. 2 can be used for the target efficiency of the particle. The droplet size distribution must be that of a tray type whose design involves some experience not covered in this section (see Ref. 8).

REFERENCES

1. Perry, R. H., and Chilton, C. H., "Chemical Engineers' Handbook," 5th ed., McGraw-Hill, New York, 1975.
2. Perry, R. H., Chilton, C. H., and Kirkpatrick, S. D., "Chemical Engineers' Handbook," 4th ed., McGraw-Hill, New York, 1963.
3. Perry, "Chemical Engineers' Handbook," 2d ed., McGraw-Hill, New York, 1941.
4. McCabe, W. L., and Smith, J. C., "Unit Operations of Chemical Engineering," 3d ed., McGraw-Hill, New York, 1975.
5. Swanson, W. M., "Fluid Mechanics," Holt, Rinehart and Winston, New York, 1970.
6. Semrau, *Ind. Eng. Chem.*, **50**, 1615 (1958); *Air Pollut. Cont. Assoc. J.*, **13**, 587 (1963).
7. Lapple and Kamack, *Chem. Eng. Prog.*, **51**, 110 (1955).
8. Nukiyama-Tanasawa, *Trans. Soc. Mech. Eng. Japan*, **5**, 63 (1939); also Ref. 1, p. 20–100.

Section **6.2**

Electrostatic Precipitators

G. G. SCHNEIDER and T. I. HORZELLA *Enviro*
Energy Corporation, Burbank, Calif.

P. J. STRIEGL *Amex Corporation, Los Angeles, Calif.*

INDUSTRY STANDARDS

The separation of particulate matter from industrial gases is a relatively old art. While it has been relatively common knowledge among engineers that charged particles will migrate to oppositely charged bodies, it was not until the early 1900s that Frederick Gardner Cottrell developed a high-voltage transformer-rectifier system capable of really effective particle separation.

Originally contemplated as a device to be utilized by smelter operators and cement plant operators, the electrostatic precipitator today has become the mainstay of most industrial processes involving hot, dusty gases. The application of the machinery has changed rather drastically considering the original design efficiencies of 90% dust removal. Today we see large gas volumes in excess of 4 million cubic feet of gas per minute from which the particulate must be removed to the extent of 99.5% or better.

It is important to note that the electrostatic precipitator is a volume-sensitive machine. It is also a piece of apparatus in which many of the static internal components are interrelated. That is, the flat collecting surfaces are related to the discharge electrode system, which is in turn related to the high-voltage energizer capability. Since the machine is volume-sensitive, the engineer must recognize that the specifications which he or she draws together apply to a specific custom-designed piece of apparatus whose performance will change as volume to it changes.

Engineers will recognize that a boiler, for example, can be bought for a rated steam flow and it will produce a reasonably predictable gas volume. Ordinarily a precipitator efficiency is specified at the rated gas flow. Should, for some reason, the plant depart from expected practice and produce more gas at the stack than has been anticipated, the precipitator efficiency will deteriorate as the volume increases over design volumes.

In many cases, we see some degree of "overkill" built into specifications to account for the intuitive feeling that plant operators will exceed the rating of the equipment.[1] It is not surprising, therefore, to see some degree of overkill put into plant designs utilizing electrostatic precipitator apparatus. To have the precipitator be the limiting factor on the ability of the plant to produce would be a most costly and wasteful exercise. Generally, therefore, on existing and new projects, one is required to approach the problem with a great deal of care, and with the desire to anticipate any future problems that would cause the design base to deviate from that which was used at the inception of the project.

PRACTICAL DESIGN APPROACH

Theoretical Basis

It is possible to develop theoretical designs based on the physics of electrical precipitation which still would not function well as a commercial piece of hardware. Our purpose, therefore, is to concentrate on basic design information only as it affects the industrial application of electrostatic precipitators to the more common gas cleaning problems. The bulk of air-pollution-control apparatus is designed to remove particulate matter rather than mist or fume materials. Particulate matter will occur in concentrations ranging from 0.5 to 15 grains/acf.

In most industrial applications, we are interested in the particle size only insofar as it affects the capability of the air-pollution-control equipment. Most air-pollution-control apparatus utilizes externally applied forces, such as centrifugal force (in cyclones) and water atomizing forces in wet scrubbers.

In electrostatic precipitation, the force applied to the particles is electrical, and this provides the mechanism for removing the particulate matter from the gas stream. The electrostatic precipitator is a volume-sensitive piece of apparatus in that the larger the volume to be handled, the larger will the commercial apparatus be for the same efficiency. In the same manner, should an electrostatic precipitator device handle less gas than the unit was originally designed for, the efficiency will rise.[2]

Precipitators are selected according to the Deutsch equation or some of its modifications, and to numerous experience factors:

$$\eta = 1 - e^{-w(A/V)}$$

where η = collection efficiency (decimal)
   w = migration velocity, cm/s (ft/s)
   A = area of collecting electrodes, m² (ft²)
   V = gas flow, m³/s (ft³/s)
   e = natural logarithm base

The units of the above equation are used by most U.S. manufacturers of precipitators to express the migration velocity in whole numbers. In solving the equation, we generally are aware of the efficiency required to produce the residual outlet dust concentration necessary to keep the plant in compliance. We are also aware of the gas flow, and from our experience file, the migration velocity is selected. This allows for the calculation of A, the plate area. It is with this basic information that a precipitator design is put together.

Practical Basis

The realization of theoretical plate area requirements will not, by itself, give the designer a precipitator "size." The other variables that must be considered would involve the selection of a gas velocity through the precipitator which would have an effect on the number of gas passages. A decision would be made on the number of fields to be supplied and a decision would be made on the plate curtain height to be utilized.

There would be some input on the method of electrical energization, and finally the decision of single or multiple chambers would be made.

For any particular set of design conditions, there are many alternate equipment arrangements that would satisfy the conditions. The final decision from a number of acceptable alternatives would probably be the decision that provides the best cost benefit. The basic size, however, derives from the precipitation expression.[3]

This apparently straightforward equation solution must be tempered by a more clear understanding of the nature of w, the drift velocity. This term is meant to express the average speed with which a particle will approach the oppositely charged (grounded) collecting plate. While it can be calculated from a theoretical basis, there are so many variables involved in the successful precipitation of dust particles that it is most difficult for the designer to have security in the value calculated. Theoretically, the drift velocity can be expressed as

$$w = \frac{aE_oE_p}{6\pi\,\eta}$$

in which, for a conducting particle,

w = migration velocity, cm/s
a = radius of dust particle, cm
E_o = electrical field strength at the point where the particle received its equilibrium charge, statvolts/cm
E_p = field strength at point of precipitation, statvolts/cm
η = gas viscosity, P

Terminology

Because of the complexity of electrostatic precipitator equipment, the industry, in the recent past, agreed to a standard terminology for the equipment. A review of Fig. 1 indicates the basic arrangement. Definitions of parts are as follows. (Refer to Fig. 2.)[4]

Chamber   This is the longitudinal gastight subdivision of the precipitator. A precipitator without an internal dividing wall is a single-chamber unit. A precipitator with a single dividing wall is a two-chamber unit, etc.

Bus Section   Any portion of a precipitator that can be independently energized. This would be by subdivision of the high-voltage system and arrangement of the high-voltage support insulators and frames.

Cells (in width)   Bus sections arranged perpendicular to gas flow. A four-cell precipitator would contain four parallel electrically separated high-voltage systems in each field.

Fields (in depth)   Bus sections arranged in the direction of gas flow. A three-field precipitator would contain three electrically separated high-voltage systems in each cell. Note from the above that the nomenclature develops from the high-voltage discharge system. It is possible, for example, to use a collecting surface of, say, 15 ft in length and subdivide it electrically through the use of two independent high-voltage systems, one of which is 9 ft long and the other 6 ft long. This would make it a two-field installation even though it used only one collecting surface.

In large modern installations, the designer usually calls for a maximum number of bus sections and as many as eight fields in depth. The purpose is to minimize the effect on efficiency (and outlet dust loading) when a bus section grounds out because of a failure in the high-voltage electrode system. High-voltage electrodes fail either through burning and subsequent grounding or through dust buildup in the hopper accumulating high enough to touch the high-voltage frame and short-circuit. Large utility boiler precipitators with as many as 120 bus sections would show negligible loss of efficiency if one or two short-circuits developed. See Fig. 3.

Fig. 1 Precipitator with insulation and cladding transformer-rectifiers mounted on top. (*Environmental Elements Corp., subsidiary of Koppers Co.*)

Energization

The energization of the bus sections is conventionally accomplished through the use of high-voltage transformer-rectifier (T-R) sets using silicon diode rectifiers. The sets are rated at approximately 60 kV output with the capacity to deliver 500 to 2000 mA. They charge the precipitator from the negative pole and are connected to deliver full-wave or half-wave power to the bus sections. Pipe conductors are used most frequently. Oil-filled high-voltage cables are used occasionally. All sets are equipped with transfer switches to allow for cutout of individual bus sections should a fault develop in that section. In addition, the half-wave sets are equipped for optional full-wave operation. All sets are equipped with solid-state automatic controls which will sense spark rate or intensity and adjust the T-R sets so that they will operate at the highest optimum conditions. Saturable reactor controls are relatively common, but newer designs generally utilize SCR controls, which react faster and consume less power. (See Fig. 4.)

In determining the capacity of the individual set, many designers call for less current capacity on the inlet fields than on the outlet fields. This is because the heavier dust load in the inlet limits the voltage and current that can be applied to the gas. As the gas becomes cleaner and the potential for arcing less, the power input can be increased and better performance on fine particulate realized.

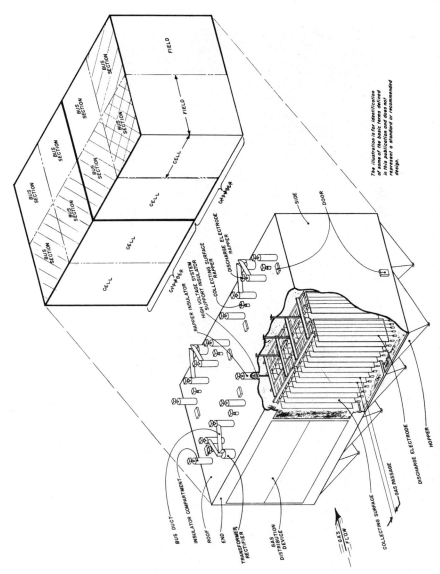

Fig. 2 Cutaway diagram illustrating precipitator terminology. (Reprinted with permission of Industrial Gas Cleaning Institute from Publication E-P1, Revised Edition, October 1967.)

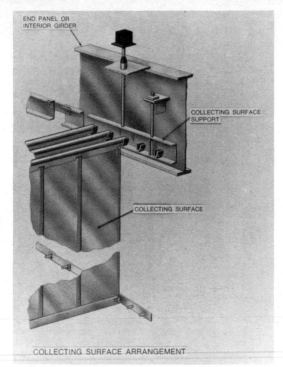

Fig. 3 Cutaway view showing how collecting curtains are placed within the precipitator. *(Environmental Elements Corp., subsidiary of Koppers Co.)*

Fig. 4 Transformer-rectifier double-half-wave connected. *(NWL Transformers.)*

Ratings are generally based upon the nameplate milliamp rating per 1000 ft² of collecting surface served (mA/1000 ft²). Therefore, it is not uncommon for engineers to select sets on inlet fields rated at, say, 55 mA/1000 ft². On the outlet fields, it is not uncommon to provide 90 to 100 mA/1000 ft². Many older units have been supplied with smaller sets than they needed, and plant operators have found that efficiency can be improved by adding T-R capacity to the outlet fields.

As a practical matter, w is a composite of so many physical conditions as to make the confidence level of a calculated figure very low.[5] For example, w is affected by the efficiency level (derated for very high efficiencies), dust load, particle size, dust resistivity, chemical composition, equipment geometry, and many other factors. The prudent designer, therefore, does not rely completely on a calculated w in working up the

Fig. 5   Discharge electrode top framer. (*Environmental Elements Corp., subsidiary of Koppers Co.*)

precipitator design. There are two avenues of approach, which are relatively successful in designing a precipitator, and both methods reinforce the thought that design is still more art than science.

One approach that has been shown to be good is the establishment of a pilot unit of sufficient size to give reasonable results when scaled up to commercial size. A good pilot unit should utilize collecting plates that approach the same length that a commercial full-size unit would use. This means something in excess of 20 ft high. Additionally, the rapping schedule (number of curtains rapped by a single rapper) should be close to commercial practice, say, five curtains per rapper. Three fields, independently powered, will give good correlation to full-size practice. Insulation should be good to be sure thermal conditions in the test unit are the same as would be finally experienced. Even with these precautions, the data obtained should be carefully examined since there is some tendency for a closely controlled pilot unit to operate better than the full-scale plant. For many new processes, the use of a pilot-plant provides the only guidance to the designer, and a well-run program can reduce the chances of grief with the commercial unit.

The other option open to designers is to utilize information developed on similar processes and equipment and take advantage of the background data developed. Similar processes and efficiency data for processes such as found in cement, paper, chemical, and steel mill practice frequently can provide a suitable design base. In some power generation plants, it has been possible to ship coal from a newly opened mine to a similar boiler installation and observe the ash precipitation characteristics in that boiler's collector.

The whole purpose of the above procedures is to work up a precipitator size based upon a rationale that will result in a successful commercial installation operating at design efficiency. Sometimes designers will prepare specifications calling for a minimum plate area and then add a safety factor to it of 10% or so. Tests for proof of design efficiency are then run with the additional 10% of plate area deenergized.

Gas Flow

Once having selected the approximate plate area, the designer will then look to the gas flow. There has been much written about the correct handling of industrial gases to an electrostatic precipitator. This is a result of the nature of the process involved, which is frequently variable and which is always complicated by the load of dust entrained in the flow and the maldistribution of dust and gas caused by complex flue shapes. The basic principle on which the precipitator operates involves the moving of small particles from the area of the discharge electrode to the collecting plate. If the gas velocity in any one of the multiple parallel passages formed by the collecting plates exceeds the design gas velocity, some particles will not have adequate time to reach the collecting electrode. Once having deposited on the collecting plate, the particle is held there by electrical and other forces and gradually a layer of dust builds up sufficiently to impair electrical operation if it is not removed. Normal practice is to vibrate or rap the plate and cause the deposited dust to slide into hoppers located beneath the equipment. Good gas-flow patterns are critical to avoid reentrainment of the particles into the gas stream.

It is for this reason that much of the literature describes even gas flow or well-distributed gas flow as critical to precipitator operation. Excessive gas velocity may cause reentrainment from the collecting plates and hoppers or may cause unbalanced flow through the apparatus. The hoppers located beneath the apparatus are frequently well baffled to prevent "sneak-by." Velocities in this area have been measured at 2 or 3 times design gas velocity, and the reentrainment of collected dust from this area is one of the major reasons for poor performance of precipitator equipment. In the design, therefore, the engineer generally selects gas velocities in the range of 2.5 to 6.5 ft/s, depending upon the application and the efficiency requirements. Finely divided fume will have lower velocities, and higher efficiency units will have lower velocities also.

The other design feature which would influence equipment size is the length of electrical field when compared with the height of the plate curtain. This L/H ratio, frequently called aspect ratio, should be in excess of 1. That is, a 30-ft-high curtain should be used in conjunction with a 30-ft-long series of electrical fields. In modern precipitator design technology, aspect ratios as high as 2 are used when very high efficiency requirements are to be met or when there is some degree of redundancy in the apparatus as required by the purchaser for reasons of "security."

ELECTROSTATIC PRECIPITATORS—GENERAL DESIGN

The electrostatic precipitator apparatus generally consists of a plate-type unit, although there are some models (usually referred to as "wet" or "mist" precipitation devices) utilizing round tubes. By far, the larger quantity of apparatus is the flat-plate-type unit handling gases in the hot, dry state. The hot precipitator itself consists of a casing fabricated from mild steel and suitably stiffened to withstand internal negative or positive pressure. Under the casing are the hoppers which are utilized to collect and remove the precipitated material from the gas stream. Dust conveyors are generally connected to hoppers.

Within the casing are located flat plates with varying heights. That is, depending on the gas flow, the plate height is varied from 24 to 48 ft. The plates are generally of modular construction—6, 9, 12, or 15 ft long—and are arranged to provide a specific number of gas passages. The plates are generally on 9- to 12-in centers. The entire system is carefully hung so that it is plumb and the plates are generally equipped with fins to provide shielding of the dust when rapped.

Hung between the plates is a discharge electrode system, often consisting of vertical steel wires (Fig. 6). The discharge electrode system is electrically insulated from the collecting plate system, and, in the United States, the practice has been to utilize a wire approximately ⅒ in in diameter suspended from an angle frame at the top and held taut by a 15- to 25-lb cast-iron weight at the bottom. The discharge electrode frame is stabilized to reduce motion in the gas stream. Stabilization consists of ceramic insulators or internal truss arrangements. European practice for discharge electrode systems is somewhat different from U.S. practice in that Europeans interpose a harp-type rigid frame in the center of the gas passage. This generally calls for somewhat wider spacing of collecting plates and, on heavy sparking applications, reputedly gives longer electrode life. Heavy sparking will cause electrodes to oscillate and possibly break.

The high-voltage electrode system is energized by a transformer-rectifier system utilizing silicon diode circuitry. Controls for the electrical apparatus are generally of the "peak-seeking" type in which the apparatus will attempt to raise the voltage of the machines to the highest point at which it will operate stably without sparking.

Fig. 6 Cutaway of wire-weight type of electrostatic precipitator using finplates and with "penthouse" roof. (*Research-Cottrell Inc.*)

APPLICATIONS

Metallurgical Refining

The production of copper and associated metals is one of the oldest industries to successfully utilize electrostatic precipitators. Emanation of great quantities of dust and fumes from metallurgical practices has resulted in electrostatic precipitator service since the early 1900s. This application has been enhanced somewhat by the fact that dust collected has a value, and, customarily, electrostatic precipitators can pay for themselves in a very short time period.

The particulate collection problems associated with metallurgical refining are extreme, in that temperatures are high, operations are cyclical, corrosion is common, and sulfate deposition or sublimation can provide extreme operating difficulties due to their stickiness. The application of precipitators in the metallurgical business, therefore, is dependent upon a great deal of knowledge on how stringent technical problems can be overcome.

Metallurgical processing has traditionally been one of the most difficult for dust and fume control apparatus. Varying dust loads, temperatures, pressures, and the inherently cyclical operation all contribute to this very demanding engineering and technical applications problem. Particular attention to electrical supply and to the type of electrode rapping gear is important. Even though this is a relatively old art, there still appears to be ample room for imaginative solutions to successful operation of gas cleaning machinery in this industry.

Pulp and Paper Industry

There are many sources in most kraft mills requiring particulate collection. The largest of the sources, however, arises from the black liquor recovery boiler which loses salt cake (Na_2SO_4) in fairly substantial quantities. The application of electrostatic precipitators to this source is not new, but there have been a number of changes in the art requiring substantial refinements in precipitator design.

Sulfuric Acid

In the production of sulfuric acid from concentrates or pyrites roasting, the flue gas is cleaned of most of its particulate through a series of cyclones and then is scrubbed down to a temperature of about 110°F saturated. In the process, a small amount of the SO_2 is converted to sulfuric acid mist. Because the contact acid plant which follows is susceptible to corrosive attack and poisoning of the catalyst, it is important that the acid mist be virtually completely removed from the gas stream.

While there have been a number of techniques developed for removal of the acid mist particles, their fineness and the fact that they are free flowing when precipitated has largely resulted in electrostatic precipitators for the service. In the case of acid precipitation, dust loadings must be low to ensure that encrustations do not build up on the collecting elements. Dust loadings of 0.3 grain/scf or less are appropriate.

Power

The cost of power generation has been affected significantly by the necessity to provide advanced control methods to stack emissions from coal-burning plants. The current campaign against air pollution classes six controllable compounds. These are:

1. Particulate matter (fly ash)
2. Sulfur oxides (SO_2, SO_3)
3. Nitrogen oxides (NO_x)
4. Oxidants, such as ozone
5. Carbon monoxide and hydrocarbons

The use of electrostatic precipitators to collect fly ash from coal-burning power plants has been relatively standard for 50 years. In fact, many curves have been developed over the years which relate efficiency requirements to purchase price. The use of electrostatic precipitators to collect 90 to 97% of the fly ash would result in a relatively flat curve. However, as the efficiency requirement reaches 99%, the curves become asymptotic. The requirements for a precipitator to collect 99% result in a unit being twice as big (and twice as costly) as one for 90% efficiency, given the same fly ash and flue gas composition at a given temperature, humidity, and flow rate.

Today there are few electrostatic precipitators in the power generation field that are specified with less than 99% removal, and they are required to produce less than 0.02 grain/scf at the stack.

ACCESSORIES

There are several accessories to the electrostatic precipitator which directly affect its proper functioning. It is difficult to give an order of importance to the effect of such accessories, and each one must be evaluated in the specific design.

Flues and Ductwork

Both the inlet and outlet gas ductwork will have a direct influence in the gas flow distribution as it moves through the precipitator.[6]

Because of space limitations, in most real-life applications, the compromised design of flues and precipitator inlet transition may result in gas turbulence and dust settling before the gas enters the precipitator, and uneven distribution of the gas flow across the face of the precipitator.

Dust settlement within the duct and inlet nozzle to the precipitator will lead to trouble

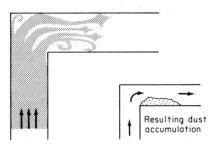

Fig. 7 Dust accumulation in flue without corrective turning vanes. *(Reprinted with permission of Air Pollution Control Association, "Ductwork Arrangement Criteria for Electrostatic Precipitator without Model Study," R. Zarfoss, 1969.)*

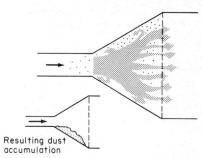

Fig. 8 Flow pattern diagram. *(Reprinted with permission of Air Pollution Control Association, "Ductwork Arrangement Criteria for Electrostatic Precipitator without Model Study," R. Zarfoss, 1969.)*

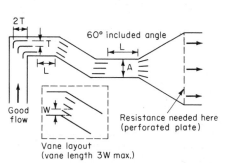

Fig. 9 Flow pattern diagram. *(Reprinted with permission of Air Pollution Control Association, "Ductwork Arrangement Criteria for Electrostatic Precipitator without Model Study," R. Zarfoss, 1969.)*

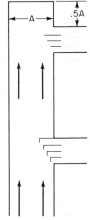

Fig. 10 Technique for branching off large header. *(Reprinted with permission of Air Pollution Control Association, "Ductwork Arrangement Criteria for Electrostatic Precipitator without Model Study," R. Zarfoss, 1969.)*

caused by excessive load to the duct and duct support structure and by gas distribution distortions as the dust deposits accumulate.

Figure 7 shows how deposits of dust are caused by a 90° gas flow change. Such a duct configuration is a perfect dust settling chamber.

Figure 8 points out that the expansion of the flue into a gas inlet nozzle will most likely cause dust deposits if no proper gas distribution devices are incorporated into the flue and nozzle.

The electrostatic precipitator, because of its principle of operation, is most sensitive to uneven gas velocities across its inlet plenum.

In order to maintain equal gas velocity across the duct or precipitator inlet nozzle, vanes, which divide the duct into smaller ducts, have been very successfully used. Figure 9 shows a typical use of curved, flat, and diverging vanes in a properly designed duct.

Flat vanes should be used in turns between 30° and 45°, but curved vanes should be exclusively used when the turn is larger than 45°.

In every turn, with or without vanes, a stratification and turbulence of the gas flow is imparted. A straight run of duct after the turn is required to return the gas flow pattern to a homogeneous one. As a rule of thumb, the length of the straight run L is equal to (refer to Fig. 10):

Without turning vanes	3 to 5 times A (duct height)
With flat vanes	1 to 2 times A
With curved vanes	½ to 1 times A

In addition, care must be taken to select the length of the vanes so that they will completely cover the projected duct facing the vanes. The length of the trailing edge of the vane should be about twice the distance T between individual vanes. However, such length must be limited to 2 ft to obtain best results. The number and length of the vanes in any given turn can thus be calculated to obtain consistent results. When a large header must be branched off, it again is important to provide suitable vanes as shown in Fig. 10. The main reason for discussing gas-flow-modifying vanes in regard to electrostatic precipitators is because of gas flow effects on the precipitator performance.

The next flow-improvement device is the diffuser plate(s) installed directly at the precipitator inlet. It corrects the turbulence caused by the gas velocity reduction from about 60 ft/s in the flues to about 4 to 10 ft/s through the precipitator. The diffuser is a vertical perforated plate which usually has 2-in-diameter holes. The diffuser plate creates a pressure drop which, while very low, is sufficient to ensure that each opening in the plate will see a relatively equal amount of gas flow. In addition, the diffuser plate breaks up large turbulences and creates, as the gas goes through the perforations, a multitude of minuscule turbulences which rapidly fade. This action allows a very smooth flow of gas parallel to the dust-collecting plates.

In many instances, large precipitator installations are provided with diffuser plates which can be adjusted to correct flow imbalance. This can be accomplished by the use of sliding plates or bars that close off or restrict flow in some areas of the inlet nozzles. These adjustments are made prior to start-up with the fan pulling air and the precipitator deenergized. Flow measurement or velocity measurement traverses are utilized and when flow balancing is accomplished, the sliding plates are welded in place.[6] Some typical duct expansion arrangements and perforated plates are shown in Fig. 11.

Dust Evacuation

Many well-designed electrostatic precipitators have failed to perform or have been a source of high maintenance because of poor coordination between precipitator and dust evacuation equipment designers. Theoretically, the dust collected by a precipitator is simple to calculate and amounts to X lb/min. However, each precipitator hopper will receive various amounts of dust. Depending on its location in the direction of gas flow, the first row of hoppers usually receives about 80 to 90% of the total dust collected by the precipitator. The amount of dust received by the next row of hoppers decreases exponentially, the last row receiving a minimal amount of dust.

Theoretically, the dust evacuation system could be designed to handle the calculated amounts of dust. However, any malfunction in the first or second fields may reduce the collected dust substantially, causing a higher load to the next hopper. If the dust evacuation system for the next hopper is of inadequate capacity, the hopper will fill up and the level of collected dust may build up to a point where it reaches the discharge and collection electrodes. This will cause short-circuits and may damage both discharge and collecting electrodes. In addition, it is not uncommon that the dust accumulated in the hopper will have a tendency to agglomerate, forming pieces large enough to block off the hopper discharge openings, or the agglomerate, if hard enough, may cause plugging of screw conveyors and other material handling equipment. At other times, dust left in the hopper for extended periods of time may (and often does) cement and bridge in the hopper, requiring shutdown of the system to enter the hopper and remove the bridged material. Dust evacuation systems used in electrostatic precipitators may be continuous or intermittent.

Intermittent evacuation systems are used where the dust load is relatively small and can be kept in the precipitator hoppers without agglomerating. Periodically, daily, every shift or more often, each hopper is unloaded into a truck and disposed of. It may also be unloaded by belt or screw conveyors which operate intermittently.

During the dust unloading operation, it is important to prevent in-leakage of air through the hopper discharge opening. A rotary airlock or a gravity-operated tipping valve may be required.

In continuous dust-discharging systems, screw conveyors, belt conveyors, pneumatic conveyors, and drag conveyors (kraft recovery boilers) are most common. In all cases, a rotary air lock, gravity- or mechanically operated tipping valve, or other automatically operated feeder, is required to prevent in-leakage of air.

As discussed above, air locks and conveyors must be conservatively designed to ensure

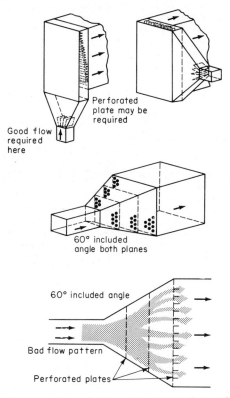

Fig. 11 Techniques for duct expansion and diffuser plates. (*a*) Technique for expanding gas while making a 90° turn; (*b*) technique for divergence in two planes; (*c*) divergence in one plane (design has significant flow correction ability). (*Reprinted with permission of Air Pollution Control Association, "Ductwork Arrangement Criteria for Electrostatic Precipitator without Model Study," R. Zarfoss, 1969.*)

peak load capacity and they must be of highest quality to minimize breakdowns which may shut off the complete operation.

As with all types of moving and material handling equipment, a strictly kept maintenance schedule will probably be the best insurance for continuous and reliable service.

Expansion Joints

The electrostatic precipitator, because of its large size and the large range of its temperature fluctuations, is normally subjected to extreme expansion forces from flues and from its own structure and supports. In addition, seismic conditions require allowance for movements of the structures with minimal possible damage to the equipment, flues, stacks, etc. A precipitator operating in a power plant may be subject to temperature variations from 0 to 700°F. Considering a gas flow in the order of 2,000,000 acfm, the precipitator may consist of two tiers, one above the other. Each tier may be about 60 ft long, 30 ft high, and 80 ft wide. The total height of the structure will be about 130 ft. The expected expansion in such a structure could amount to 4 in in the width of the equipment and about 3 in in the height. This will impose very substantial unit stress changes. Without provisions for expansion joints, permanent damage will occur.

Expansion joints used industrially range from simple slip joints to more elaborate metal bellows or asbestos fabric joints. Figure 12 shows typical camera corner metallic bellows which are used as single units to absorb longitudinal expansion or as tandem arrangements to allow multidirectional movements. Figure 13 shows a round bellows in tandem arrangement as used for round ducts. The asbestos fabric joint (Fig. 14) has increased in popularity as materials of construction are developed that will handle higher temperatures and corrosive atmospheres. Its main advantage is that it will absorb large movements in a single joint. Figure 15 shows several arrangements of the fabric. Note that in all of them, as well as in the metallic joints, a metallic liner inside the ductwork is provided to minimize dust accumulations in the bellows. For fabric joints, the liner eliminates the fluttering effect caused by the gas flow inside the duct. The material of construction of the fabric depends, of course, on the temperature and chemical composition of the gas. The fabric is arranged in layers.

Fig. 12   Metal expansion joint. (*Tuboflex Inc.*)

Dampers

Dampers are used in connection with electrostatic precipitators to either regulate the distribution of the gas to different chambers of the equipment or shut off a chamber for inspection or maintenance.

The typical damper used in blowers to adjust the gas flow or to reduce the start-up horsepower is not adequate for most electrostatic precipitator installations.

Fig. 13   Circular high-pressure expansion joints with toggle piece. (*Tuboflex Inc.*)

For flow control purposes, a multileaf damper is adequate provided the following requirements are met:

1. Leaks through leaf shaft bearings must not occur. The shaft seals must be capable of handling the levels of dust, corrosion, and temperature for the application. At the same time, they must not bind and make the damper inoperative.

2. The leaves of the damper must be capable of handling dusty gas and its erosive action.

3. The damper operators, either manual or motorized, must be of a heavy-duty type capable of handling the relatively large forces required for adjustment.

4. The frame of the damper must be sturdy enough to allow good flanging with the ductwork to prevent gas leaks.

Fig. 14   Asbestos cloth expansion joints for power plant service. (*Tuboflex Inc.*)

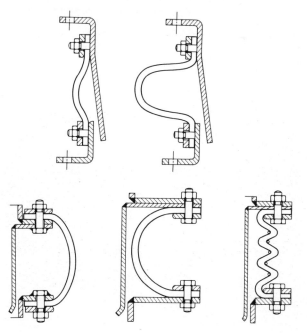

Fig. 15   Various arrangements of cloth joint manufacture. (*Tuboflex Inc.*)

For gas flow shutoff purposes, it is imperative that gas leakage will not occur since any leakage will make safe servicing of a chamber of the precipitator impossible. Temperature and toxicity of the gas are the main offenders. In addition, if the damper allows leakage of ambient air to the system, it will just cause unnecessary gas flow into the system, creating

Fig. 16 Slide gate guillotine-type motor-operated damper. *(Koppers Co.)*

additional operating costs and often a source of corrosion due to gas cooling and moisture condensation.

Slide gates, guillotine dampers, goggle valves, and specially designed leaf dampers are successfully used for shutoff purposes. The criteria to make a shutoff damper acceptable are:

1. There must be no gas leakage. Only a handful of damper manufacturers have acquired the skill to design and fabricate equipment to meet this requirement.

2. The damper must be capable of remaining in operation despite the presence of dust which can block guiding frames, jacking screw guides, etc.

3. The expansion due to gas temperature changes must be included in the design without affecting its tight sealing capability.

Figure 16 depicts a damper with a tension screw operator which is provided with a thrust bearing that allows the screw to be in tension both opening and closing. Thus, the screw cannot bend, since all compression forces are eliminated. The nut arrangement contains a scraper which automatically cleans the ball screw threads. Figure 17 shows a goggle valve and its highly engineered fabrication to ensure tightness at high gas pressures. Figure 18 depicts a double-louver damper with pressurized chamber for full isolation required for maintenance of individual precipitator chambers without complete plant shutdown.

Insulation

Insulation is required in an electrostatic precipitator to:

   1. Minimize the condensation of moisture or acid inside the precipitator. Condensation is the source of metal corrosion and accumulation of dust.

   2. Maintain the temperature of the collected dust.

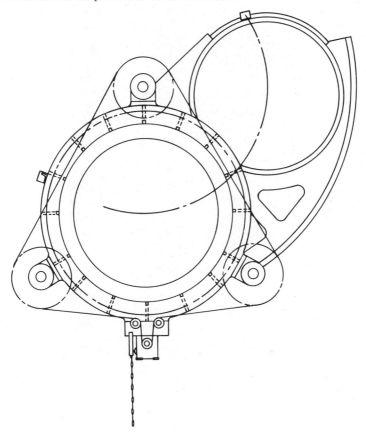

Fig. 17   Goggle valve. *(Koppers Co.)*

Typical is the case of dust hoppers of precipitators in coal-fired boilers. Coal fly ash when hot is free-flowing and easily moves out of the hoppers. If cool, however, fly ash becomes tacky and agglomerates, causing bridging in the hoppers. During start-up operations, the hopper metal surfaces are cold and often the collected dust will adhere to cold metal surfaces. To correct this problem, insulation alone is often insufficient and hopper heating is required. Such heating is accomplished by attaching electric heating pads or steam tracing under the insulation. While this is expensive, it is mandatory for many installations.

Vibrators

Dust discharge from hoppers may be substantitally improved by using hopper vibrators. Vibrators cause a steady flow of the dust down the hopper walls and corners, thus minimizing the chances of the dust building up and permanently adhering to the walls.

This is always a good base for hopper bridging. In most cases, the precipitator manufacturer will have experience with the specific dust and should be able to recommend whether vibrators are required. If it is desired to avoid the expense of hopper vibrators, their installation may be postponed until after their need becomes apparent. However, provision for their mounting and baseplate should be made with the original equipment to avoid costly field alteration if vibrators are needed.

Fig. 18   Double-frame multiple-louver damper. *(Tuboflex, Inc.)*

Conditioning Agents

It has been known for quite some time that traces of various substances in the gas or in the dust itself have a marked influence on the performance of electrostatic precipitators. The gas moisture also has a tremendous effect in some specific processes.[7]

Early in commercial precipitator development, the effect of conditioning was discovered at a Garfield, Utah, copper smelter. It was observed that the precipitator performed well when the SO_3 content of the gas was high, but when the SO_3 content of the gas was low, the performance decreased continuously and eventually the precipitator ceased to collect dust. At the same plant, it was also observed that water sprayed into the gas stream when SO_3 content was low provided an improvement of precipitator performance.

Later it was discovered that the addition of traces of NH_3 to the gas stream from catalytic petroleum cracking had a similar improvement in dust collection.

Today, gas conditioning is known to have a direct effect on the dust resistivity, and commercial processes have been developed and equipment is marketed specifically for gas conditioning.

The dust resistivity is affected not only by the presence or absence of various substances in the dust or in the gas stream, but also by gas temperature. Often it is possible to select the proper gas temperature as an answer to dust resistivity problems. The best known application using this approach is that of coal-fired boilers using low-sulfur-containing coal.

When precipitators were first applied to such operation using the know-how gained from high-sulfur-fired boilers, it was discovered that the precipitators were too small to attain the desired dust collection efficiency. Basically, the "migration velocity" of the dust particles was much lower than that of fly ash from high-sulfur coal.

Research and full-size plant experience have demonstrated that there are three basic approaches to the fly ash resistivity problem[8]:

1. Install the precipitator where the gas temperature is high as upstream of the air heater (about 600° to 700°F).

2. Install the precipitator downstream of the air heater (about 300°F), but design it for a low migration velocity—4 to 5 cm/s (0.132 to 0.165 ft/s) vs. 8 to 10 cm/s (0.264 to 1.65 ft/s) for a high-sulfur coal.

3. Install the precipitator downstream of the air heater, using high migration velocities as for high-sulfur coal, but add a conditioning agent to the gas.

The selection of a hot precipitator (case 1) vs. a cold precipitator (case 2) must be made on a case-to-case basis. The evaluation must include:

(a) Cost of flues (a hot precipitator requires more flues because gases must return to the air heater).

(b) Cost of insulation (both flues and precipitator need more insulation for the hot precipitator casing).

(c) Cost of precipitator (the cold precipitator is larger than the hot one, despite the fact that the hot precipitator handles about 40% more gas flow because of its temperature).

(d) Structural supports.

(e) Layout of equipment.

(f) Other considerations, such as the problems sometimes encountered with colder fly ash which is tacky and more difficult to rap off the collecting plates and is more likely to cause hopper bridging and similar problems.

The use of conditioning, either in the form of traces of various chemicals and/or water added to the gas stream, represents, of course, additional operating costs and conditioning equipment investment and maintenance.

In the case of SO_3 conditioning, SO_3 gas in trace amounts is added to the gas stream upstream of the precipitator. The gaseous SO_3 can be supplied by a liquid SO_3 source and evaporated prior to injection into the gas stream. Some installations have been made where liquid sulfuric acid was evaporated. Finally, possibly the best approach is combustion of sulfur into SO_2 and catalytic oxidation into SO_3. Package plants are commercially available for the latter approach.

Conditioning of gas by the addition of water is practical and very effective for cement kiln operations where the process is being converted from gas- or oil-fired to coal-fired. The amount of water vapor produced by gas or oil combustion is not produced by coal combustion and precipitator performance is reduced.

In water conditioning, it is important that the water be completely evaporated and that the water spray system does not produce large water droplets which form agglomerates with the dust. This may result in the formation of mud in the ducts, spray chamber, and precipitator. A well-designed water-conditioning tower will produce a gas containing about 10 to 15% moisture without water droplets that have not evaporated by the time the gas enters the precipitator. Proper selection of a water-conditioning tower includes proper selection of the tower size, spray-nozzle type and arrangement, and suitable automatic controls.

Other methods of conditioning may include additions of conditioning agents in solid form to the fuel.[9]

FINANCIAL CONSIDERATIONS

Air-pollution-control standards in the 1970s have required the design and installation of precipitators which recover over 99% of the particulate matter in the gas stream. In prior years, efficiencies in the range of 90 to 98% were common and, consequently, the financial decisions on precipitator configuration, design, and operation were based on experience from those precipitators. In general, because of the lower efficiency requirements, variations in design configuration were construed to have a minimal impact on the cost of purchasing, installing, and operating a precipitator. As an illustration, Fig. 19 relates the precipitator size to efficiency on an arithmetic scale to depict the relatively minor increase in size for efficiencies up to 98% and the dramatic increase in size as efficiencies over 98% are specified.

A slight change in efficiency requirements in the 99+% range thus has a significant impact on costs, and, therefore, consideration should be given to optimizing the design consistent with cost and technical requirements.

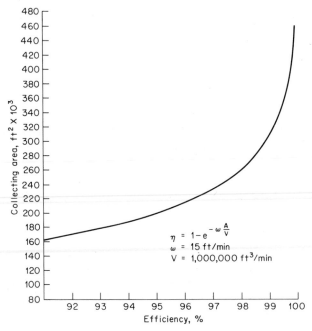

$$\eta = 1 - e^{-\omega \frac{A}{V}}$$
$$\omega = 15 \text{ ft/min}$$
$$V = 1{,}000{,}000 \text{ ft}^3/\text{min}$$

Fig. 19 Precipitator collecting area vs. efficiency.

Configuration

The Deutsch-Anderson equation is generally utilized to determine the collecting surface area to achieve the required efficiency for a given gas volume. The packaging of the required collecting surface into a given precipitator can result in various costs. Table 1 illustrates the variations in cost for three different configurations containing the same

TABLE 1   Variations in Precipitator Costs with Configuration Changes

		Configuration cost change	
Cost item	24' H × 21' W × 27' L	24' H × 24' W × 24' L	21' H × 27' W × 24' L
Casing	Base	+ 1.9%	+ 6.2%
Hoppers	Base	+ 8.2%	+23.8%
Alternate	Base	+ 5.7%	+15.4%
Collecting surface	Base	+ 1.8%	+ 6.2%
Electrodes	Base	+ 1.6%	+14.3%
Inlet/outlet nozzle	Base	+ 5.9%	+11.8%
Gas distribution	Base	+14.8%	+12.6%
Composite	Base	+ 3.5%	+ 9.4%

There are other costs such as electrical, insulation, ladder, and platform costs, which also have an impact on the configuration cost.

collecting surface area. The base precipitator selected is 24 ft high × 21 ft wide × 27 ft long. Varying the width, length, or height by 3 ft can impact the selected cost items by as much as 9.4%.

The possible configurations, which could enclose the same collecting surface area, are significant in that one precipitator manufacturer can package the same collecting surface area, similar to the one evaluated here, into 35 different "standard" combinations. While

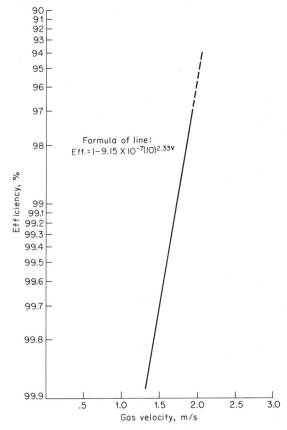

Fig. 20 Efficiency vs. gas velocity. Fly ash precipitator tests.

common-sense criteria would preclude using some of the combinations, the optimization of cost and technical considerations requires the testing of a number of alternative packages.

Configuration Changes vs. Efficiency

In addition to the cost impact of configuration changes, there are variations in efficiency which are not recognized in the Deutsch-Anderson equation. Two of these variations are: (1) gas velocity, which is a function of gas volume per unit time, and the precipitator entry area presented to the gas stream; and (2) contact or gas retention time, which is a function of precipitator length and gas velocity.

Collection efficiency for fly ash is related to the gas velocity by the following formula, as shown in Fig. 20,

$$\text{Efficiency} = 1 - 9.15 \times 10^{-7}(10)^{2.33v}$$

where v = gas velocity, m/s

This formula demonstrates that, for a given gas volume, the width and height of a precipitator affect efficiency in that a larger cross-sectional area will result in a lower gas velocity and thus improve efficiency. However, to maintain the same collecting surface as determined by the Deutsch-Anderson equation, the precipitator length is shortened with the lower gas velocity. Thus, a shortening has a deleterious effect on collection efficiency since the contact time may be reduced (Fig. 21).

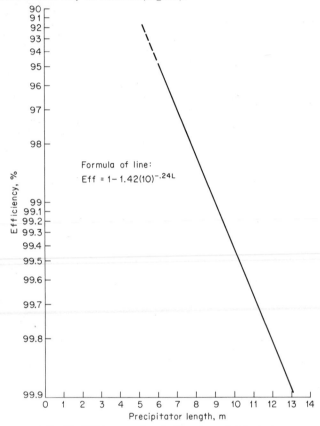

Fig. 21 Efficiency vs. length. Fly ash precipitator tests.

The length of a fly ash precipitator is related to collection efficiency by the following formula:

$$\text{Efficiency} = 1 - 1.42(10)^{-.24L}$$

where L = precipitator length, m

Thus, as the precipitator length increases, the efficiency improves, but again, the width or height must correspondingly be reduced to maintain the same collection efficiency.

Retention Time

The retention time of the gas stream is a function of gas velocity and precipitator length as follows:

$$\text{Retention time} = \frac{\text{length}}{\text{gas velocity}}$$

For fly ash precipitators, the gas retention time is related to loss by an empirically derived formula of:

$$\text{Retention time} = 4.8(\text{loss})^{-0.15}$$

Refer to Fig. 22. If the retention time is known for a particular efficiency, then the ratio of precipitator length to gas velocity is specified for the technical requirements. The optimization of the physical parameters of length, width, and height, in conjunction with the technical requirements, can then be performed.

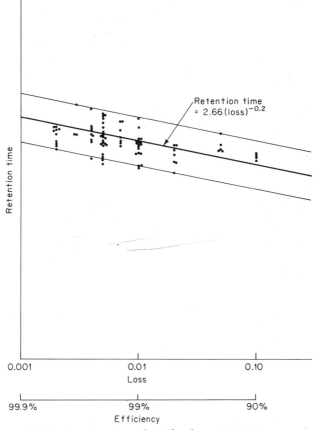

Fig. 22 Retention time vs. loss. Fly ash precipitator tests.

It is also of interest to note that the Deutsch-Anderson equations can be transposed into terms of precipitator length and gas velocity to yield the following relationship for precipitators with 9-in collecting surface spacing:

$$\text{Efficiency} = 1 - e^{-0.044wL/v_g}$$

where L = precipitator length, ft
$\quad v_g$ = gas velocity, ft/s
$\quad w$ = migration velocity, ft/s

Sectionalization and Costs

The collection effeciency of a precipitator is also related to the ability to charge the incoming particulate matter, and one of the significant variables is the power input. The efficiency has been shown to be related to power by the formula

$$\text{Efficiency} = 1 - e^{-0.59Pc/V}$$

where P_c = corona power, W
$\quad V$ = gas volume, ft³/s

Increasing the corona power results in an improved efficiency, but the limiting factor on corona power is the sparking between the electrodes and the collecting plates. The operating mode is to increase the corona power on the electrodes until sparking occurs at a preselected rate. Precipitators have automoatic controls to maintain a spark rate that maximizes the corona power.

Precipitators are assembled with a normal tolerance of ±¼ in between the electrodes and plates.

In a precipitator energized by a single T-R set, the corona power is thus limited by those

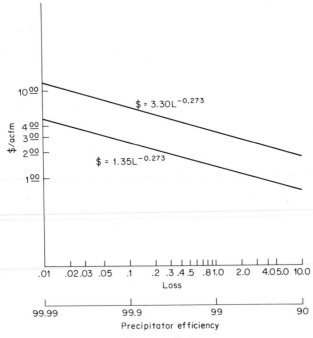

Fig. 23   Installed cost.

few electrodes which are closest to the plates. As additional energizing fields are added, the probability increases that a given field will have less of a variation in electrode spacing and therefore be able to carry a higher average corona power. The limiting factor is the relatively high cost of added sectionalization vs. the efficiency of alternate means of achieving the same efficiency. Equating the two formulas for efficiency as follows shows a possible trade-off of corona power with additional collecting area:

$$\text{Efficiency} = 1 - e^{\omega A/V_1} = 1 - e^{0.59P/V_2}$$

where ω = 15.35 ft/min
V_1 = 122,880 cfm
V_2 = 2048 ft³/s
A = collecting area, ft²
P_c = corona power

Solving the above equation shows that 1 W of corona power is equivalent to 0.43 ft² of collecting area and a cost trade-off is possible by comparing the costs of achieving additional corona power vs. increasing the collecting surface area.

Total Installed Cost

The total installed cost of a precipitator is a function of the efficiency configuration, dust loading, location, etc., and therefore a considerable range of values for a given volume of gas exists. In Fig. 23 a range of values derived from large fly ash precipitators is presented.

The curve is based on 1974–75 data for 1,000,000 cfm precipitators. The actual range of values may fall on either side of this range as the sample size from which these costs are derived is increased.

This discussion is of interest since the commercial sizing of electrostatic precipitation equipment is not customarily accomplished in a technically rigorous manner. For example, when a boiler, kiln, or fan manufacturer responds to a set of engineering specifications, the engineer can expect proposals from various qualified manufacturers for approximately the same size equipment. Selection of sizing parameters for electrostatic precipitators is largely an art, and frequently the consulting engineer will receive bids from suppliers in which there is a 2:1 variation in size of equipment for the same duty. There exists an opportunity to develop significantly more background data. The data will then allow better statistical approaches to basic size selections. This, in turn, will lead to optimization of financially sensitive equipment alternatives and reduction in the variation in size and price from manufacturers without sacrificing reliability.

REFERENCES

1. Cooper, J., and Schneider, G. G., *Chem. Eng.,* September 30, 1974.
2. White, W. J., "Industrial Electrostatic Precipitation," Addison-Wesley, Reading, Mass. 1963.
3. Sproull, W. T., "Air Pollution and Its Control," Exposition Press, Jericho, N.Y., 1970.
4. "Terminology for Electrostatic Precipitators," Industrial Gas Cleaning Institute, 1967.
5. Schneider, G. G., "Hot Side Precipitators for Coal Fired Boilers," Third Annual Thermal Power Conference, Pullman, Washington, 1972.
6. Zarfoss, J. R., "Ductwork Arrangement Criteria for Electrostatic Precipitator without Model Study," Air Pollution Control Association, 1969.
7. Archer, W. E., *Power Eng.,* December, 1972.
8. Schneider, G. G., Horzella, T. I., Cooper, J., and Striegl, P. J., *Chem. Eng.,* May 26, 1975.
9. Strauss, W., "Industrial Gas Cleaning," Pergamon Press, New York, 1966.

ADDITIONAL REFERENCES

Baxter, W. A., "Electrostatic Precipitator Design for Western Coals," Air Pollution Control Conference, Knoxville, Tennessee, 1975.

Stenby, E. W., York, J. L., and Campbell, K. S., "Particulate Removal When Burning Western Coal", 78th National Meeting of AIChE, Salt Lake City, Utah, 1974.

Index

Index

3

DATE DUE

DEC 17	NOV 7 '88		
JAN 80	NOV 23 '83		
MAR 17 '80	MAY 0 7 1984		
DEC 8			
2 FEB '81	DEC 13		
20 APR '17	MAY 29 '86		
28 MAY '81	NOV 24 '87		
26 MAY 2	FEB 20 '88		
SEP 14 '81	JUN 17 '88		
MAN 22 '82			
MAY 17			
OCT 11 '82			
NOV '3 '8			
NOV 14 '83'			
APR 18			
13 SEP '8			

DATE DUE

FEB 9 '90			
OCT 9 '90			
DEC 7 '91			
FEB 19 '92			
MAY 5 '93			
NOV 12 '93			
FEB 11 '94			
APR 8 '94			
MAY 5 '94			
OCT 10 '94			
261-2500		Printed in USA